Your Guide to Success in Math

Complete Step 0 as soon as you begin your math course.

STEP 0: PLAN YOUR SEMESTER

- ☐ Register for the online part of the course (if there is one) as soon a
- ☐ Fill in your Course and Contact information on this pull-out card.
- ☐ Write important dates from your syllabus on the Semester Organizer on this pull-out card.

Follow Steps 1–4 during your course. Your instructor will tell you which resources to use—and when—in the textbook or eText, *MyMathGuide* workbook, videos, and MyMathLab. Use these resources for extra help and practice.

STEP 1: LEARN THE SKILLS AND CONCEPTS

- ☐ Read the **textbook** or **eText,** listen to your instructor's lecture, and/or watch the **videos.** You can work in ***MyMathGuide*** as you do this. As you are learning:
  - ☐ Take notes, write down your questions, and save all your work (including homework solutions, quizzes, and tests) to review throughout the course.
  - ☐ Work the *Skill to Review* exercises at the beginning of each section.
  - ☐ Stop and do the *Margin* and *Guided Solution Exercises* as directed.
  - ☐ Watch the videos. Answer the *Interactive Your Turn* questions in the videos and in *MyMathGuide.*

STEP 2: CHECK YOUR UNDERSTANDING

- ☐ Answer the *Reading Checks* in the section exercise sets or in MyMathLab.
- ☐ Explore the concepts using the *Active Learning Figures* in MyMathLab.

STEP 3: DO YOUR HOMEWORK

- ☐ Plan to spend 2 hours studying and doing homework for every hour of class.
- ☐ Complete your assigned homework from the textbook and/or in MyMathLab.
  - ☐ When doing homework from the textbook, use the answer section to check your work.
  - ☐ When doing homework in MyMathLab, use the Learning Aids, such as Help Me Solve This and View an Example, as needed, working toward being able to complete exercises without the aids.

STEP 4: REVIEW AND TEST YOUR UNDERSTANDING

- ☐ Work the exercises in the *Mid-Chapter Review.*
- ☐ Make your own chapter study sheet by doing the *Chapter Summary and Review.*
- ☐ Take the *Chapter Test* as a practice exam. To watch an instructor solve each problem, go to the Chapter Test Prep Videos in MyMathLab or on YouTube (search "BittingerBasic" and click on "Channels").

Use the ***Studying for Success*** tips in the text and the MyMathLab **Study Skills modules** (with videos, tips, and activities) to help you develop effective time-management, note-taking, test-prep, and other skills.

Student Organizer

Course Information

Course Number: ______________ Name: ______________________________

Location: ______________________________ Days/Time: ______________________

Contact Information

Contact	Name	Email	Phone	Office Hours	Location
Instructor					
Tutor					
Math Lab					
Classmate					
Classmate					

Semester Organizer

Week	Homework	Quizzes and Tests	Other

Basic College Mathematics with Early Integers

THIRD EDITION

MARVIN L. BITTINGER

Indiana University Purdue University Indianapolis

JUDITH A. PENNA

PEARSON

Boston Columbus Indianapolis New York San Francisco Upper Saddle River
Amsterdam Cape Town Dubai London Madrid Milan Munich Paris Montréal Toronto
Delhi Mexico City São Paulo Sydney Hong Kong Seoul Singapore Taipei Tokyo

Editorial Director	Christine Hoag
Editor in Chief	Maureen O'Connor
Executive Editor	Cathy Cantin
Senior Managing Editor	Karen Wernholm
Senior Production Supervisor	Ron Hampton
Composition	PreMediaGlobal
Production Services	Jane Hoover/Lifland et al., Bookmakers
Editorial Services	Martha K. Morong/Quadrata, Inc.
Art Editor and Photo Researcher	The Davis Group, Inc.
Manager, Multimedia Production	Christine Stavrou
Associate Producer	Jonathan Wooding
Executive Content Manager	Rebecca Williams (MathXL)
Senior Content Developer	John Flanagan (TestGen)
Marketing Manager	Rachel Ross
Marketing Assistant	Kelly Cross
Senior Manufacturing Buyer	Debbie Rossi
Text Designer	The Davis Group, Inc.
Associate Design Director	Andrea Nix
Senior Designer/Cover Design	Barbara Atkinson
Cover Photograph	© Rich Carey/Shutterstock

Photo Credits
Photo credits appear on page vi.

Library of Congress Cataloging-in-Publication Data
Bittinger, Marvin L.
Basic college mathematics with early integers / Marvin L. Bittinger, Indiana University Purdue University Indianapolis, Judith A. Penna.—Third edition.
pages cm
Includes index.
ISBN: 978-0-321-92234-2
1. Algebra—Textbooks. 2. Numbers, Natural—Textbooks. 3. Numbers, Natural—Problems, exercises, etc. I. Penna, Judith A. II. Title.
QA152.3.B528 2014
510—dc23 2013033494

1 2 3 4 5 6 7 8 9 10—CRK—16 15 14 13

www.pearsonhighered.com

ISBN-13: 978-0-321-92234-2
ISBN-10: 0-321-92234-4

Contents

4 Fraction Notation and Mixed Numerals 187

5 Decimal Notation 255

Ratio and Proportion 329

Percent Notation 381

Data, Graphs, and Statistics 453

Measurement 499

Geometry 553

Algebra: Solving Equations and Problems 621

PHOTO CREDITS

Cover Rich Carey/Shutterstock **p. 3** Tina Manley/Alamy **p. 8** (left) Anton Balazh/Shutterstock, (right) NASA **p. 55** Courtesy of Barbara Johnson **p. 66** (left) Yevgenia Gorbulsky/Fotolia, (right) Courtesy of Geri Davis **p. 82** Sebastian Duda/Fotolia **p. 87** Ed Metz/Shutterstock **p. 90** Dave King/Dorling Kindersley, Ltd. **p. 100** Mellowbox/Fotolia **p. 111** Ivan Alvarado/Reuters **p. 122** Brian Snyder/Reuters **p. 146** Cynoclub/Fotolia **p. 151** Petr84/Shutterstock **p. 152** Larry Roberg/Fotolia **p. 168** (left) UW Images/Fotolia, (right) Image 100/Corbis/Glow Images **p. 172** Phaitoon Sutunyawatchai/Shutterstock **p. 177** (left) Simon Greig/Fotolia, (right) Rigucci/Shutterstock **p. 182** Michael Jung/Fotolia **p. 186** Marce Clemens/Shutterstock **p. 200** (left) Christophe Fouquin/Fotolia, (right) Jim West/Glow Images **p. 201** Ursule/Fotolia **p. 221** Interfoto/Alamy **p. 222** Ruud Morijn/Fotolia **p. 230** Ruud Morijn/Fotolia **p. 234** Joshua Lott/Reuters **p. 235** Mark Duncan/AP Images **p. 236** Essam Al-Sudani/AFP/Getty Images **p. 243** Tom Sears **p. 249** Image Source/Glow Images **p. 253** I. Pilon/Shutterstock **p. 256** EPA European Pressphoto Agency/Alamy **p. 258** Daily Mail/Rex/Alamy **p. 263** (left) MSPhotographic/Fotolia, (right) Zai Aragon/Fotolia **p. 277** (left) Stockshooter/Alamy, (right) Estima/Fotolia **p. 281** (left) Michaklootwijk/Fotolia, (right) Santi Visalli/Glow Images **p. 301** (top left) Rtimages/Shutterstock, (bottom left) Scott Kane/Icon SMI, CAX/Newscom, (bottom right) ZUMA Press, Inc./Alamy **p. 310** TheFinalMiracle/Fotolia **p. 311** Iain Masterton/AGE Fotostock/SuperStock **p. 312** Comstock/Getty Images **p. 316** (left) PCN Photography/Alamy, (right) Christopher Sadowski/Splash News/Newscom **p. 322** Edward Rozzo/Corbis/Glow Images **p. 323** Jochen Tack/Glow Images **p. 332** NBAE/Getty Images **p. 335** (left) Sam72/Shutterstock, (right) Alfred Pasieka/Science Source **p. 337** Carsten Reisinger/Fotolia **p. 338** ZUMA Press, Inc./Alamy **p. 339** Howard Shooter/Dorling Kindersley, Ltd. **p. 341** (top left) Gerard Sioen/Gamma-Rapho/Getty Images, (top right) Rudie/Fotolia, (bottom left) Eye of Science/Science Source/Photo Researchers, Inc., (bottom right) Jose Garcia/Fotolia **p. 350** (left) RIRF Stock/Shutterstock, (right) Geri Lynn Smith/Shutterstock **p. 353** Ioannis Ioannou/Shutterstock **p. 354** M. Timothy O'Keefe/Alamy **p. 357** Antoni Murcia/Shutterstock **p. 359** (left) Pictorial Press, Ltd./Alamy, (right) Kim D. French/Fotolia **p. 360** (left) Monkey Business/Fotolia, (right) Fancy Collection/SuperStock **p. 362** (left) Stephen Meese/Fotolia, (right) mattjeppson/Fotolia **p. 363** (left) EPA European Pressphoto Agency/Alamy, (right) ZUMA Press, Inc./Alamy **p. 366** (bottom left) Courtesy of Geri Davis, (right) Courtesy of Elaine Somers **p. 369** (left and right) Courtesy of Geri Davis **p. 374** Imagebroker/Alamy **p. 375** (top right) Martin Shields/Alamy, (bottom right) Mark Bonham/Shutterstock **p. 382** John Dorton/Shutterstock **p. 383** Tina Jeans/Shutterstock **p. 386** (left) Karin Hildebrand Lau/Shutterstock, (right) Susan The/Fotolia **p. 387** (left) iStockphoto/Thinkstock, (right) iStockphoto/Thinkstock **p. 390** Bbbar/Fotolia **p. 398** Nata-Lia/Shutterstock **p. 399** (top right) Rehan Qureshi/Shutterstock, (bottom right) Studio D/Fotolia **p. 403** Gina Sanders/Fotolia **p. 413** (left to right) Uabels/Shutterstock; Winfried Wisniewski/AGE Fotostock; William Mullins/Alamy; Seitre/Nature Picture Library; Corbis/AGE Fotostock; Kevin Schafer/Alamy; Andy Rouse/Nature Picture Library **p. 415** Visions of America, LLC/Alamy **p. 416** Francis Vachon/Alamy **p. 417** David Taylor/Alamy **p. 419** Daniel Borzynski/Alamy **p. 420** (left) ZUMA Press, Inc./Alamy, (right) Image Source Plus/Alamy **p. 421** iStockphoto/Thinkstock **p. 422** Ashley Cooper Pics/Alamy **p. 423** (left) Rob Wilson/Shutterstock, (right) Wollertz/Shutterstock **p. 428** Flonline Digitale Bildagentur GmbH/Alamy **p. 429** Gmcgill/Fotolia **p. 431** (left) Pavel L. Photo and Video/Shutterstock, (right) Brian Jackson/Fotolia **p. 432** (left) Auremar/Fotolia, (right) Christina Richards/Shutterstock **p. 437** Valua Vitaly/Shutterstock **p. 440** (left) Dmitry Vereshchagin/Fotolia, (right) Andres Rodriguez/Fotolia **p. 451** Robert F. Balazik/Shutterstock **p. 452** AP Images **p. 455** Yuri Arcurs/Shutterstock **p. 457** Mike Wulf/Cal Sport Media/Newscom **p. 460** (left) Narumol Pug/Fotolia, (right) Courtesy of Barbara Johnson **p. 483** All Canada Photos/Alamy **p. 491** Robert Daly/Caia Images/Glow Images **p. 497** Kristoffer Tripplaar/Alamy **p. 503** Mark Schwettmann/Shutterstock **p. 508** Golddc/Shutterstock **p. 513** (top) Klaus Rademaker/Shutterstock, (right) The Natural History Museum/Alamy **p. 514** Gaspar Janos/Shutterstock **p. 515** (left) Orhan Çam/Fotolia, (right) Hsieh Chang-Che/Fotolia **p. 516** (left) Courtesy of Marvin Bittinger, (right) Ian Halperin/UPI/Newscom **p. 522** Tyler Olson/Shutterstock **p. 523** (left) Viktoriya Field/Shutterstock, (right) Urbanlight/Shutterstock **p. 525** Jake Lyell/Alamy **p. 526** Greg C. Grace/Alamy **p. 532** (bottom left) Blend Images/SuperStock, (top right) Kiri/Fotolia **p. 543** (left) Hemis/Alamy, (right) Jerry Ballard/Alamy **p. 548** Gpointstudio/Shutterstock **p. 551** Laborant/Shutterstock **p. 552** Ikonoklast_hh/Fotolia **p. 563** Thierry Roge/Reuters **p. 575** (left) Maisna/Fotolia, (right) Wisconsin DNR **p. 588** (left) Luisa Fernanda Gonzalez/Shutterstock, (right) AP Images **p. 589** (left) Imagebroker/Alamy, (top right) Lev1977/Fotolia, (bottom right) Graham Prentice/Shutterstock **p. 619** Clem Murray/MCT/Newscom **p. 656** Simon Kwong/Reuters **p. 660** Magic Mountain/Associated Press **p. 663** (left) Rodney Todt/Alamy, (right) Corbis/Superstock **p. 664** (left) Courtesy of the 500 Festival Mini Marathon, (right) Lars Lindblad/Shutterstock **p. 666** (top) Elena Yakusheva/Shutterstock, (bottom left) Courtesy of Barbara Johnson, (bottom right) Studio 8/Pearson Education, Inc. **p. 672** ZUMA Press, Inc./Alamy **p. 680** Imagebroker/Alamy

Index of Applications

Consumer

Domestic

Economics

Education

Engineering

Environment

Finance

Food

Geometry

Government

Health/Medicine

Labor

Miscellaneous

Physics

Social Sciences

Sports/Entertainment

Statistics/Demographics

Technology

Transportation

Preface

The Bittinger Program

Math hasn't changed, but students—and the way they learn it—have.

Basic College Mathematics with Early Integers, 3rd Edition, continues the Bittinger tradition of objective-based, guided learning, while integrating timely updates to the proven pedagogy. In this edition, there is a greater emphasis on guided learning and helping students get the most out of all of the course resources available with the Bittinger program, including new opportunities for mobile learning.

The program has expanded to include these comprehensive new teaching and learning resources: ***MyMathGuide*** **workbook**, **To-the-Point Objective Videos**, and enhanced, media-rich **MyMathLab** courses. Feedback from instructors and students motivated these and several other significant improvements: a new design to support guided learning, new figures and photos to help students visualize both concepts and applications, and many new and updated real-data applications to bring the math to life.

With so many resources available in so many formats, the trusted guidance of the Bittinger team on *what to do* and *when* will help today's math students stay on task. Students are encouraged to use **Your Guide to Success in Math**, a four-step learning path and checklist available on the handy reference card in the front of this text and in MyMathLab. The guide will help students identify the resources in the textbook, supplements, and MyMathLab that support *their* learning style, as they develop and retain the skills and conceptual understanding they need to succeed in this and future courses.

In this preface, a look at the key new *and* hallmark resources and features of the *Basic College Mathematics with Early Integers* program—including the textbook/eText, video program, *MyMathGuide* workbook, and MyMathLab—is organized around **Your Guide to Success in Math**. This will help instructors direct students to the tools and resources that will help them most in a traditional lecture, hybrid, lab-based, or online environment.

NEW AND HALLMARK FEATURES IN RELATION TO Your Guide to Success in Math

STEP 1 Learn the Skills and Concepts

Students have several options for learning, reviewing, and practicing the math concepts and skills.

Textbook/eText

- ☐ **Skill to Review.** At the beginning of nearly every text section, *Skill to Review* offers a just-in-time review of a previously presented skill that relates to the new material in the section. Section and objective references are included for the student's convenience, and two practice exercises are provided for review and reinforcement.
- ☐ **Margin Exercises.** For each objective, problems labeled "Do Exercise . . ." give students frequent opportunities to solve exercises while they learn.

- ☐ ***New!*** **Guided Solutions.** Nearly every section has *Guided Solution* margin exercises with fill-in blanks at key steps in the problem-solving process.
- ☐ ***Enhanced!*** **MyMathLab.** MyMathLab now includes *Active Learning Figures* for directed exploration of concepts; more problem types, including *Reading Checks* and *Guided Solutions*; and new, objective-based videos. (See pp. xvi–xix for a detailed description of the features of MyMathLab.)
  - ☐ ***New!*** **Skills Checks.** In the Learning Path for Ready-to-Go MyMathLab, each chapter begins with a brief assessment of students' mastery of the prerequisite skills needed to learn the new material in the chapter. Based on the results of this pre-test, a personalized homework set is designed to help each student prepare for the chapter.
- ☐ ***New!*** **To-the-Point Objective Videos.** This is a comprehensive new program of objective-based, interactive videos that are incorporated into the Learning Path in MyMathLab and can be used hand-in-hand with the *MyMathGuide* workbook.
  - ☐ ***New!*** **Interactive Your Turn Exercises.** For each objective in the videos, students solve exercises and receive instant feedback on their work.
- ☐ ***New!*** ***MyMathGuide: Notes, Practice, and Video Path.*** This is an objective-based workbook (available printed and in MyMathLab) for guided, hands-on learning. It offers vocabulary, skill, and concept review—along with problem-solving practice—with space to show work and write notes. Incorporated in the Learning Path in MyMathLab, it can be used together with the To-the-Point Objective Video program, instructor lectures, and the textbook.

STEP 2 Check Your Understanding

Throughout the program, students have frequent opportunities to check their work and confirm that they understand each skill and concept before moving on to the next topic.

- ☐ ***New!*** **Reading Checks.** At the beginning of each set of section exercises in the text, students demonstrate their grasp of the skills and concepts.
- ☐ ***New!*** **Active Learning Figures.** In MyMathLab, Active Learning Figures guide students in exploring math concepts and reinforcing their understanding.
- ☐ **Translating for Success.** In the text and in MyMathLab, these activities offer students extra practice with the important first step of the process for solving applied problems.

STEP 3 Do Your Homework

Basic College Mathematics with Early Integers, 3rd Edition, has a wealth of proven and updated exercises. Pre-built assignments are available for instructors in MyMathLab, and they are pre-assigned and incorporated into the Learning Path in the Ready-to-Go course.

- ☐ **Skill Maintenance.** In each section, these exercises offer a thorough review of the math in the preceding text.
- ☐ **Synthesis Exercises.** To help build critical-thinking skills, these section exercises require students to use what they know and combine learning objectives from the current section with those from previous sections.

STEP 4 Review and Test Your Understanding

Students have a variety of resources to check their skills and understanding along the way and to help them prepare for tests.

- ☐ **Mid-Chapter Review.** Mid-way through each chapter, students work a set of exercises (*Concept Reinforcement, Guided Solutions, Mixed Review,* and *Understanding*

Through Discussion and Writing) to confirm that they have grasped the skills and concepts covered in the first half before moving on to new material.

- ☐ **Summary and Review.** This resource provides an in-text opportunity for active learning and review for each chapter. *Vocabulary Reinforcement, Concept Reinforcement,* objective-based *Study Guide* (examples paired with similar exercises), *Review Exercises* (including *Synthesis* problems), and *Understanding Through Discussion and Writing* are included in these comprehensive chapter reviews.
- ☐ **Chapter Test.** Chapter Tests offer students the opportunity for comprehensive review and reinforcement prior to taking their instructor's exam. **Chapter Test-Prep Videos** (in MyMathLab and on YouTube) show step-by-step solutions to the Chapter Tests.
- ☐ **Cumulative Review.** Following every chapter beginning with Chapter 3, a Cumulative Review revisits skills and concepts from all preceding chapters to help students retain previously learned material.

Study Skills

Developing solid time-management, note-taking, test-taking, and other study skills is key to student success in math courses (as well as professionally and personally). Instructors can direct students to related study skills resources as needed.

- ☐ *New!* **Student Study Reference.** This pull-out card at the front of the text is perforated, three-hole-punched, and binder-ready for convenient reference. It includes **Your Guide to Success in Math** course checklist, **Student Organizer**, and **At a Glance**, a list of key information and expressions for quick reference as students work exercises and review for tests.
- ☐ *New!* **Studying for Success.** Checklists of study skills—designed to ensure that students develop the skills they need to succeed in math, school, and life—are integrated throughout the text at the beginning of selected sections.
- ☐ *New!* **Study Skills Modules.** In MyMathLab, interactive modules address common areas of weakness, including time-management, test-taking, and note-taking skills. Additional modules support career-readiness.

Learning Math in Context

- ☐ *New!* **Applications.** Throughout the text in examples and exercises, real-data applications encourage students to see and interpret the mathematics that appears every day in the world around them. Applications that use real data are drawn from business and economics, life and physical sciences, medicine, technology, and areas of general interest such as sports and daily life. New applications include "Fastest-Growing Occupations" (p. 68), "Training Regimens" (p. 230), "Media Usage" (p. 331), and "*The Hobbit: An Unexpected Journey*" (p. 359). For a complete list of applications, please refer to the Index of Applications (p. vii).

BREAKTHROUGH
To improving results

MyMathLab

Ties the Complete Learning Program Together

MyMathLab® Online Course (access code required)
MyMathLab from Pearson is the world's leading online resource in mathematics, integrating interactive homework, assessment, and media in a flexible, easy to use format. MyMathLab delivers **proven results** in helping individual students succeed. It provides **engaging experiences** that personalize, stimulate, and measure learning for each student. And it comes from an **experienced partner** with educational expertise and an eye on the future.

MyMathLab for Developmental Mathematics

Prepared to go wherever you want to take your students.

Personalized Support for Students

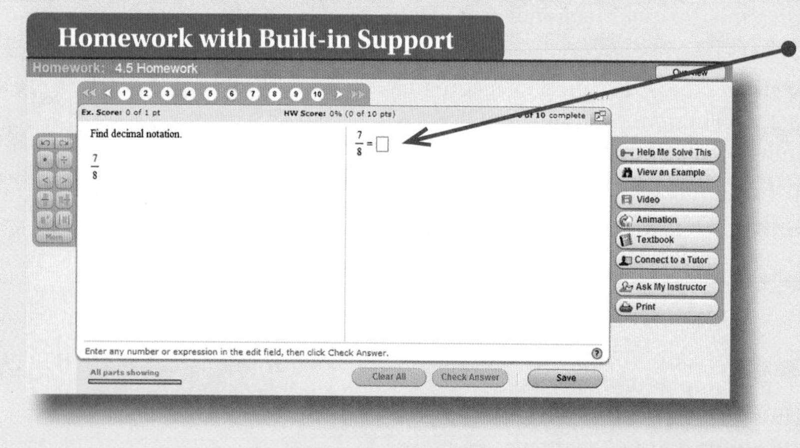

Exercises: The homework and practice exercises in MyMathLab are correlated to the exercises in the textbook, and they regenerate algorithmically to give students unlimited opportunities for practice and mastery. The software offers immediate, helpful feedback when students enter incorrect answers.

Multimedia Learning Aids: Exercises include guided solutions, sample problems, animations, videos, and eText access for extra help at point of use.

Expert Tutoring: Although many students describe the whole of MyMathLab as "like having your own personal tutor," students using MyMathLab do have access to live tutoring from qualified math instructors.

Personalized Homework

To help students achieve mastery, MyMathLab can generate **personalized homework** based on individual performance on tests or quizzes. Personalized homework allows students to focus on topics they have not yet mastered.

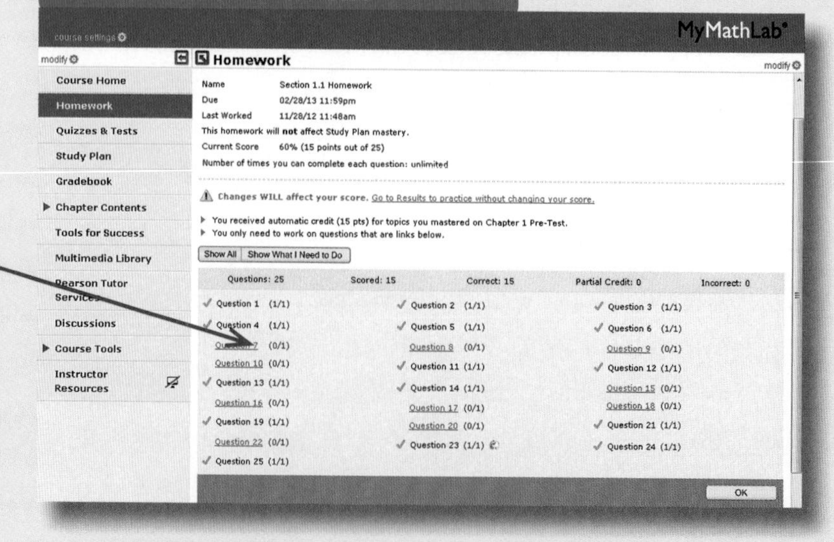

The **Adaptive Study Plan** makes studying more efficient and effective for every student. Performance and activity are assessed continually in real time. The data and analytics are used to provide personalized content—reinforcing concepts that target each student's strengths and weaknesses.

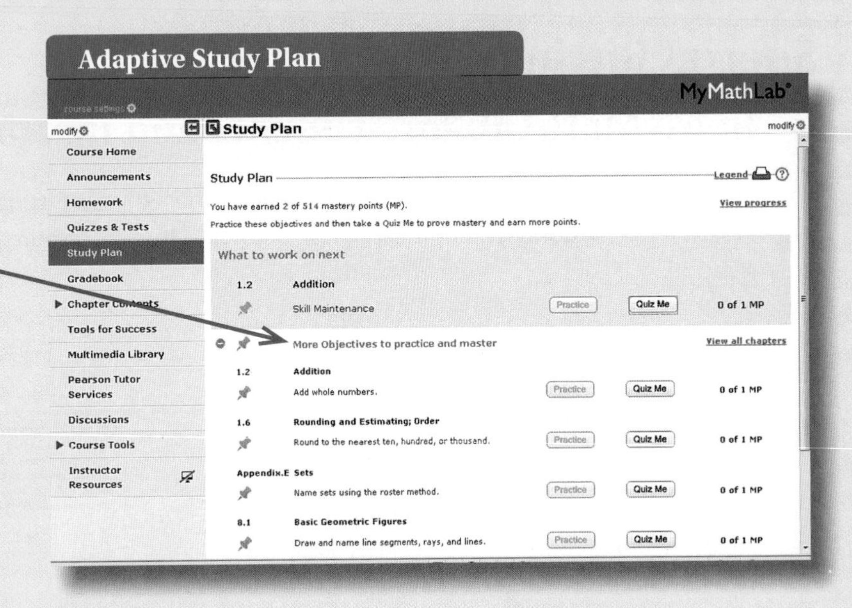

Flexible Design, Easy Start-Up, and Results for Instructors

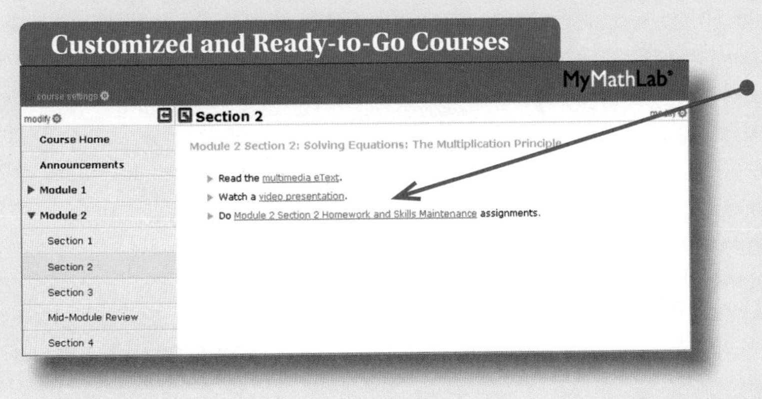

Instructors can modify the site navigation and insert their own directions on course-level landing pages; also, a **custom MyMathLab** course can be built that reorganizes and structures the course material by chapters, modules, units—whatever the need may be.

Ready-to-Go courses include pre-assigned homework, quizzes, and tests to make it even easier to get started. The Bittinger Ready-to-Go courses include new *Mid-Chapter Reviews* and *Reading Check Assignments*, plus a four-step Learning Path on each section-level landing page to help instructors direct students where to go and what resources to use.

The **comprehensive online gradebook** automatically tracks students' results on tests, quizzes, and homework and in the study plan. Instructors can use the gradebook to quickly intervene if students have trouble, or to provide positive feedback on a job well done. The data within MyMathLab is easily exported to a variety of spreadsheet programs, such as Microsoft Excel. Instructors can determine which points of data to export and then analyze the results to determine success.

New features, such as **Search/Email by criteria**, make the gradebook a powerful tool for instructors. With this feature, instructors can easily communicate with both at-risk and successful students. They can search by score on specific assignments, non-completion of assignments within a given time frame, last login date, or overall score.

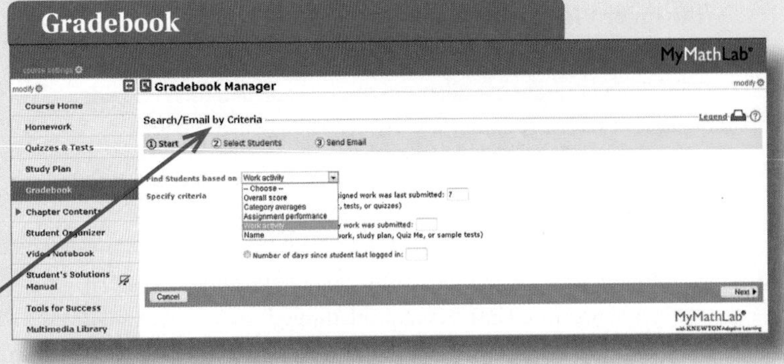

Special Bittinger Resources
in MyMathLab for Students and Instructors

In addition to robust course delivery, MyMathLab offers the full Bittinger eText, additional Bittinger Program features, and the entire set of instructor and student resources in one easy-to-access online location.

New! Active Learning Figures

In MyMathLab, Active Learning Figures guide students in exploring math concepts and reinforcing their understanding. Instructors can use Active Learning Figures in class or as media assignments in MyMathLab to guide students to explore math concepts and reinforce their understanding.

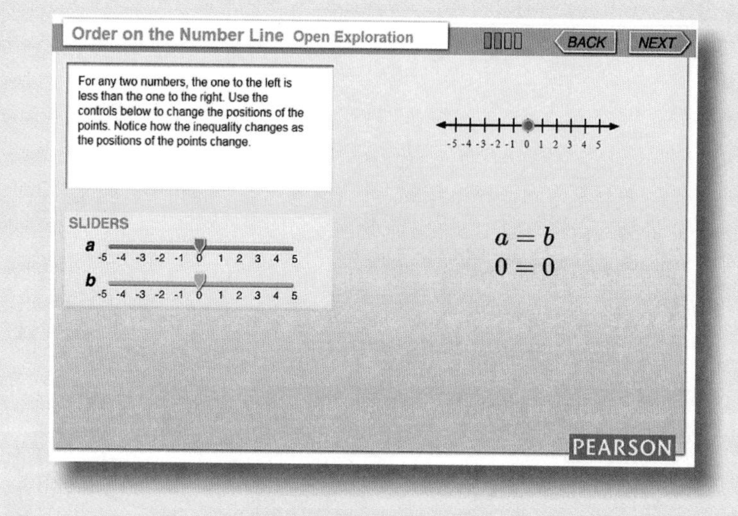

New! Four-Step Learning Path

Each of the section-level landing pages in the Ready-to-Go MyMathLab course includes a Learning Path that aligns with *Your Guide to Success in Math* to link students directly to the resources they should use when they need them. This also allows instructors to point students to the best resources to use at particular times.

New! Integrated Bittinger Video Program and *MyMathGuide* workbook

Bittinger Video Program*

(DVD ISBN: 978-0-321-92280-9)

The Video Program is available in MyMathLab and on DVD and includes closed captioning and the following video types:

New! To-the-Point Objective Videos. These objective-based, interactive videos are incorporated into the Learning Path in MyMathLab and can be used along with the *MyMathGuide* workbook.

Chapter Test Prep Videos. The Chapter Test Prep Videos let students watch instructors work through step-by-step solutions to all the Chapter Test exercises from the textbook. Chapter Test Prep Videos are also available on YouTube™ (search using author name and book title).

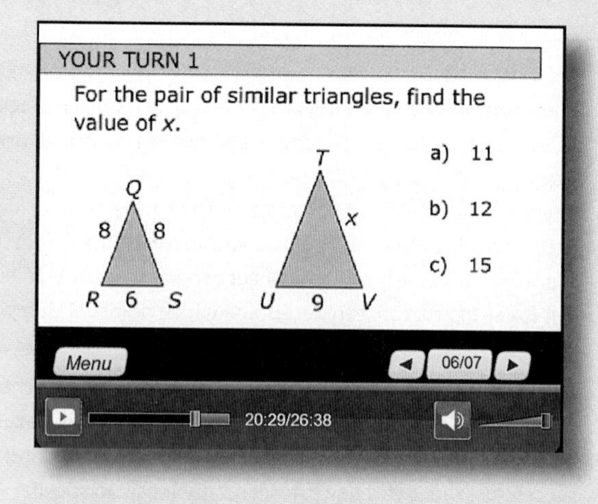

New! *MyMathGuide: Notes, Practice, and Video Path* workbook*

(Printed Workbook ISBN: 978-0-321-86863-3)

This objective-based workbook for guided, hands-on learning offers vocabulary, skill, and concept review—along with problem-solving practice—with space to show work and write notes. Incorporated in the Learning Path in MyMathLab, *MyMathGuide* can be used together with the To-the-Point Objective Video program, instructor lectures, and the textbook. Instructors can assign To-the-Point Objective Videos in MyMathLab in conjunction with the *MyMathGuide* workbook.

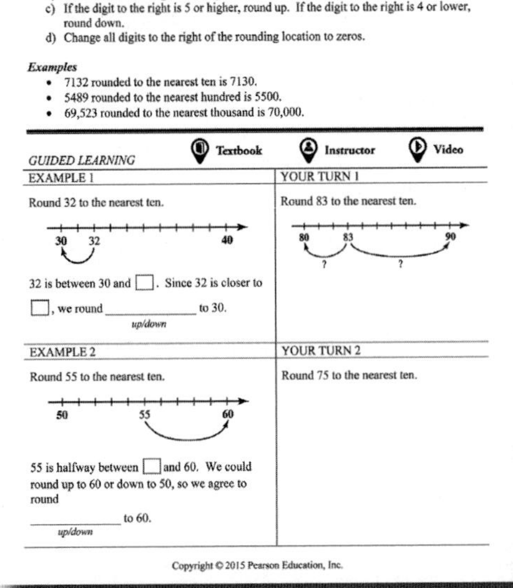

Section 1.6 | *Rounding and Estimating; Order* 1

Rounding

ESSENTIALS

Rounding can be done by looking at the number line or using the following rule.

To round to a certain place:
a) Locate the digit in that place.
b) Consider the next digit to the right.
c) If the digit to the right is 5 or higher, round up. If the digit to the right is 4 or lower, round down.
d) Change all digits to the right of the rounding location to zeros.

Examples
- 7132 rounded to the nearest ten is 7130.
- 5489 rounded to the nearest hundred is 5500.
- 69,523 rounded to the nearest thousand is 70,000.

GUIDED LEARNING Textbook Instructor Video

EXAMPLE 1	YOUR TURN 1
Round 32 to the nearest ten. 30 32 40 32 is between 30 and ☐. Since 32 is closer to ☐, we round ______ (up/down) to 30.	Round 83 to the nearest ten. 80 83 90 ? ?
EXAMPLE 2	**YOUR TURN 2**
Round 55 to the nearest ten. 50 55 60 55 is halfway between ☐ and 60. We could round up to 60 or down to 50, so we agree to round ______ (up/down) to 60.	Round 75 to the nearest ten.

Copyright © 2015 Pearson Education, Inc.

MyMathLab

course settings

modify

Study Skills: Time Management; Coping with Stress and Anxiety; Concentrating when you Read and Study; Concept Maps and Flashcards; Effective Note Taking; Preparing for and Taking Exams

Math-Reading Connections

Time Management

Review Materials
- View an instructional video about Time Management.
- View notes and tips on Getting Organized.
- View notes and tips on Using a Planner.
- View notes and tips on Procrastination

Activities
- Print and complete the Time Management Self Assessment. You can hand in your completed assessment to your instructor.
- Print and complete the activity on Getting Organized.
- View the interactive activity on Using a Calendar.
- Print and complete the scenario on Overcoming Procrastination.

Post-Test
- Go to the Quizzes & Tests page.
- From the **Sample Tests** section of the page, click **Time Management Post-Test**.

Study Skills Modules

In MyMathLab, interactive modules address common areas of weakness, including time-management, test-taking, and note-taking skills. Additional modules support career readiness. Instructors can assign module material with a post-quiz.

Additional Resources in MyMathLab

For Students

Student's Solutions Manual*
(ISBN: 978-0-321-92294-6)
By Judith A. Penna

This manual contains completely worked-out annotated solutions for all the odd-numbered exercises in the section-level exercise sets in the text. It also includes fully worked-out annotated solutions for all the exercises (odd- and even-numbered) in the Mid-Chapter Reviews, the Summary and Reviews, the Chapter Tests, and the Cumulative Reviews.

For Instructors

Annotated Instructor's Edition**
(ISBN: 978-0-321-92281-6)

This version of the text includes answers to all exercises presented in the book.

Instructor's Resource Manual with Tests and Mini Lectures**
(download only)
By Laurie Hurley

This manual includes resources designed to help both new and experienced instructors with course preparation and classroom management. This includes chapter-by-chapter teaching tips and support for media supplements. It contains two multiple-choice tests per chapter, six free-response tests per chapter, and eight final exams.

Instructor's Solutions Manual**
(download only)
By Judith A. Penna

This manual contains brief solutions to the even-numbered exercises in the section-level exercise sets. It also includes fully worked-out annotated solutions for all the exercises (odd- and even-numbered) in the Mid-Chapter Reviews, the Chapter Tests, and the Cumulative Reviews.

PowerPoint® Lecture Slides**
(download only)

These slides present key concepts and definitions from the text.

To learn more about how MyMathLab combines proven learning applications with powerful assessment, visit www.mymathlab.com or contact your Pearson representative.

*Printed supplements or DVDs are also available for separate purchase through MyMathLab, MyPearsonStore.com, or other retail outlets. They can also be value-packed with a textbook or MyMathLab code at a discount.

**Also available in print or for download from the Instructor Resource Center (IRC) on www.pearsonhighered.com.

Acknowledgments

Our deepest appreciation to all of you who helped to shape this edition by reviewing and spending time with us on your campuses. In particular, we would like to thank the following reviewers:

Simone Aeshliman, *St. Cloud Technical and Community College*
Afsheen Akbar, *Bergen Community College*
Kory Ambrosich, *Phoenix College*
Morgan Arnold, *Central Georgia Technical College*
Connie Buller, *Metropolitan Community College*
Erin Cooke, *Gwinnett Technical College*
Kay Davis, *Del Mar College*
Edward Dillon, *Century Community and Technical College*
Beverlee Drucker, *Northern Virginia Community College*
Sabine Eggleston, *Edison State College*
Dylan Faullin, *Dodge City Community College*
Anne Fischer, *Tulsa Community College, Metro Campus*
Rebecca Gubitti, *Edison State College*
Exie Hall, *Del Mar College*
Stephanie Houdek, *St. Cloud Technical Institute*
Linda Kass, *Bergen Community College*
Chauncey Keaton, *Central Georgia Technical College*
Dorothy Marshall, *Edison State College*
Kimberley McHale, *Heartland Community College*
Arda Melkonian, *Victor Valley College*
Christian Miller, *Glendale Community College*
Christine Mirbaha, *Community College of Baltimore County-Dundalk*
Joan Monaghan, *County College of Morris*
Louise Olshan, *County College of Morris*
Deborah Poetsch, *County College of Morris*
Thomas Pulver, *Waubonsee Community College*
Nimisha Raval, *Central Georgia Technical College*
Nicole Saporito, *Luzerne County Community College*
Jane Serbousek, *Northern Virginia Community College*
Alexis Thurman, *County College of Morris*
Melanie Walker, *Bergen Community College*

Endless hours of hard work by Martha Morong, Jane Hoover, and Geri Davis have led to products of which we are immensely proud. Strong support has also come from Laurie Hurley for the *Instructor's Resource Manual* and for accuracy checking, along with checkers Holly Martinez, Joanne Koratich, and Patty LaGree, and from proofreader Monroe Street. Michelle Lanosga assisted with applications research. We also wish to recognize Nelson Carter, who wrote video scripts.

In addition, a number of people at Pearson have contributed in special ways to the development and production of this textbook, including the Developmental Math team: Senior Production Supervisor Ron Hampton, Senior Designer Barbara Atkinson, and Associate Media Producer Jonathan Wooding. Executive Editor Cathy Cantin and Marketing Manager Rachel Ross encouraged our vision and provided marketing insight.

Whole Numbers

STUDYING FOR SUCCESS *Getting Off to a Good Start*

- ☐ Your syllabus for this course is extremely important. Read it carefully, noting required texts and materials.
- ☐ If there is an online component for your course, register for it as soon as possible.
- ☐ At the front of the text, you will find a Student Organizer card. This pullout card will help you keep track of important dates and useful contact information.

1.1 Standard Notation

OBJECTIVES

- **a** Give the meaning of digits in standard notation.
- **b** Convert from standard notation to expanded notation.
- **c** Convert between standard notation and word names.

We study mathematics in order to be able to solve problems. In this section, we study how numbers are named. We begin with the concept of place value.

a PLACE VALUE

The numbers of jobs available in 2010 for several professions are shown in the following table.

PROFESSION	NUMBER OF JOBS, 2010
Registered nurses	2,737,400
Radiologic technologists	219,900
Radiation therapists	16,900

SOURCE: U.S. Department of Labor, Bureau of Labor Statistics

A **digit** is a number 0, 1, 2, 3, 4, 5, 6, 7, 8, or 9 that names a place-value location. For large numbers, digits are separated by commas into groups of three, called **periods**. Each period has a name: *ones, thousands, millions, billions, trillions,* and so on. To understand the number of jobs for registered nurses in the table above, we can use a **place-value chart**, as shown below.

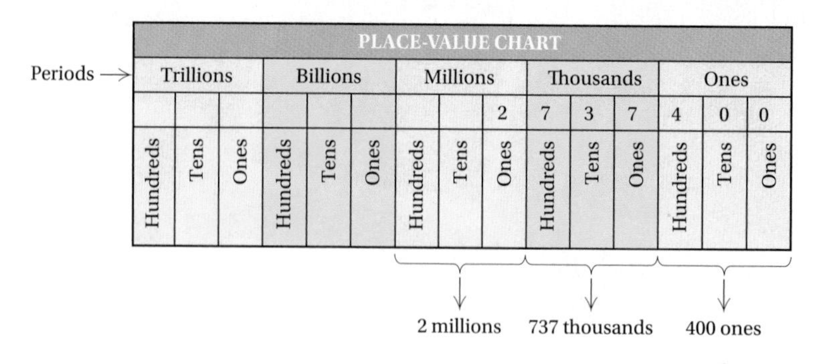

PLACE-VALUE CHART

Periods →	Trillions			Billions			Millions			Thousands			Ones		
									2	7	3	7	4	0	0
	Hundreds	Tens	Ones	Hundreds	Tens	Ones	Hundreds	Tens	Ones	Hundreds	Tens	Ones	Hundreds	Tens	Ones

EXAMPLES In each of the following numbers, what does the digit 8 mean?

1. 278,342 — 8 thousands
2. 872,342 — 8 hundred thousands
3. 28,343,399,223 — 8 billions
4. 98,413,099 — 8 millions
5. 6328 — 8 ones

Do Margin Exercises 1–6 (in the margin at right).

What does the digit 2 mean in each number?

1. 526,555
2. 265,789
3. 42,789,654
4. 24,789,654
5. 8924
6. 5,643,201

EXAMPLE 6 *Charitable Organizations.* Since its founding in 1881 by Clara Barton, the American Red Cross has been the nation's best-known emergency response organization. As part of a worldwide organization, the American Red Cross also aids victims of devastating natural disasters. For the fiscal year ending June 2011, the total revenue of the American Red Cross was $3,452,960,387. What digit names the number of ten millions?
Source: charitynavigator.org

Ten millions → 3,452,960,387

The digit 5 is in the ten millions place, so 5 names the number of ten millions.

Do Exercise 7.

7. *Government Payroll.* In March 2011, the total payroll for all state employees in the United States was $19,971,861,990. What digit names the number of ten billions?
Source: *2011 Annual Survey of Public Employment and Payroll*

b CONVERTING FROM STANDARD NOTATION TO EXPANDED NOTATION

Heifer International is a charitable organization whose mission is to work with communities to end hunger and poverty and care for the earth by providing farm animals to impoverished families around the world. Consider the data in the following table.

GEOGRAPHICAL AREAS OF NEED	NUMBER OF FAMILIES ASSISTED DIRECTLY AND INDIRECTLY BY HEIFER INTERNATIONAL IN 2011
Africa	220,275
Americas	934,871
Asia, South Pacific	407,640
Central and Eastern Europe	344,945

SOURCE: *Heifer International 2011 Annual Report*

Answers

1. 2 ten thousands 2. 2 hundred thousands 3. 2 millions 4. 2 ten millions 5. 2 tens 6. 2 hundreds 7. 1

Write expanded notation.

8. 2718 mi, the length of the Congo River in Africa

GS

2718 = 2 ____ + 7 ____
+ ____ ten + ____ ones

9. 344,945, the number of families in Central and Eastern Europe assisted by Heifer International in 2011

10. 1670 ft, the height of the Taipei 101 Tower in Taiwan

11. 104,094 square miles, the area of Colorado

The number of families assisted in the Americas was 934,871. This number is expressed in **standard notation**. We write **expanded notation** for 934,871 as follows:

934,871 = 9 hundred thousands + 3 ten thousands
+ 4 thousands + 8 hundreds
+ 7 tens + 1 one.

EXAMPLE 7 Write expanded notation for 1815 ft, the height of the CN Tower in Toronto, Canada.

1815 = 1 thousand + 8 hundreds + 1 ten + 5 ones

EXAMPLE 8 Write expanded notation for 407,640, the number of families in Asia and the South Pacific assisted by Heifer International in 2011.

407,640 = 4 hundred thousands + 0 ten thousands
+ 7 thousands + 6 hundreds + 4 tens + 0 ones

or

4 hundred thousands + 7 thousands + 6 hundreds + 4 tens

◀ **Do Exercises 8–11.**

C CONVERTING BETWEEN STANDARD NOTATION AND WORD NAMES

We often use **word names** for numbers. When we pronounce a number, we are speaking its word name. Russia won 82 medals in the 2012 Summer Olympics in London, Great Britain. A word name for 82 is "eighty-two." Word names for some two-digit numbers like 36, 51, and 72 use hyphens. Others like that for 17 use only one word, "seventeen."

2012 Summer Olympics Medal Count

COUNTRY	GOLD	SILVER	BRONZE	TOTAL
United States of America	46	29	29	104
People's Republic of China	38	27	23	88
Russia	24	26	32	82
Great Britain	29	17	19	65
Germany	11	19	14	44

SOURCE: espn.go.com

Answers

8. 2 thousands + 7 hundreds + 1 ten + 8 ones
9. 3 hundred thousands + 4 ten thousands + 4 thousands + 9 hundreds + 4 tens + 5 ones
10. 1 thousand + 6 hundreds + 7 tens + 0 ones, or 1 thousand + 6 hundreds + 7 tens
11. 1 hundred thousand + 0 ten thousands + 4 thousands + 0 hundreds + 9 tens + 4 ones, or 1 hundred thousand + 4 thousands + 9 tens + 4 ones

Guided Solution:
8. thousands, hundreds, 1, 8

EXAMPLES Write a word name.

9. 46, the number of gold medals won by the United States

Forty-six

10. 19, the number of silver medals won by Germany

Nineteen

11. 104, the total number of medals won by the United States

One hundred four

Do Exercises 12–14. ▶

For word names for larger numbers, we begin at the left with the largest period. The number named in the period is followed by the name of the period; then a comma is written and the next number and period are named. Note that the name of the ones period is not included in the word name for a whole number.

EXAMPLE 12 Write a word name for 46,605,314,732.

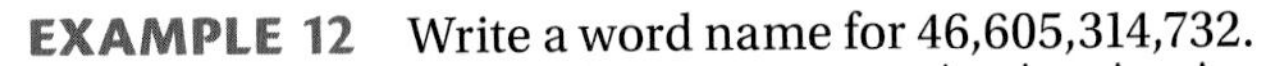

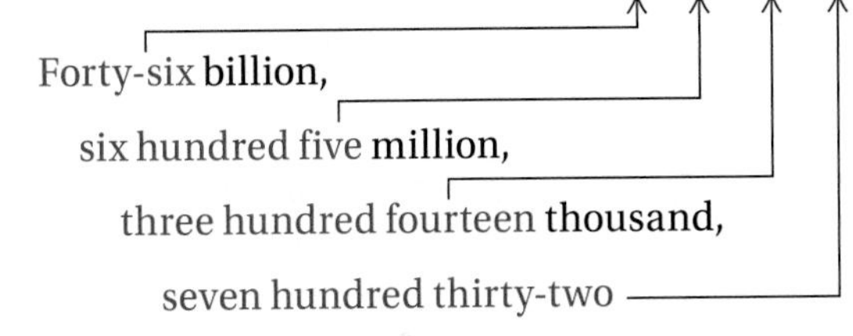

The word "and" *should not* appear in word names for whole numbers. Although we commonly hear such expressions as "two hundred *and* one," the use of "and" is not, strictly speaking, correct in word names for whole numbers. For decimal notation, it is appropriate to use "and" for the decimal point. For example, 317.4 is read as "three hundred seventeen *and* four tenths."

Do Exercises 15–18. ▶

EXAMPLE 13 Write standard notation.

Five hundred six million,
three hundred forty-five thousand,
two hundred twelve

Standard notation is 506,345,212.

Do Exercise 19. ▶

Write a word name. (Refer to the chart on the previous page.)

12. 65, the total number of medals won by Great Britain

13. 14, the number of bronze medals won by Germany

14. 38, the number of gold medals won by the People's Republic of China

Write a word name.

15. 204

16. $44,640, the average annual wage for athletic trainers in the United States in 2012
Source: U.S. Bureau of Labor Statistics

GS **17.** 1,879,204
One ____, eight hundred ____ thousand, two hundred ____

18. 7,052,428,785, the world population in 2012
Source: U.S. Census Bureau

19. Write standard notation.
Two hundred thirteen million, one hundred five thousand, three hundred twenty-nine

Answers

12. Sixty-five **13.** Fourteen **14.** Thirty-eight **15.** Two hundred four **16.** Forty-four thousand, six hundred forty **17.** One million, eight hundred seventy-nine thousand, two hundred four **18.** Seven billion, fifty-two million, four hundred twenty-eight thousand, seven hundred eighty-five **19.** 213,105,329

Guided Solution:
17. Million, seventy-nine, four

1.1 Exercise Set

For Extra Help MyMathLab® MathXL® PRACTICE WATCH READ REVIEW

Reading Check

Complete each statement with the correct word from the following list.

digit expanded period standard

RC1. In 983, the ________________ 9 represents 9 hundreds.

RC2. In 615,702, the number 615 is in the thousands ________________.

RC3. The phrase "3 hundreds + 2 tens + 9 ones" is ________________ notation for 329.

RC4. The number 721 is written in ________________ notation.

a What does the digit 5 mean in each number?

1. 235,888

2. 253,777

3. 1,488,526

4. 500,736

Movie Receipts. The final movie of the Harry Potter series, *Harry Potter and the Deathly Hallows: Part II*, grossed $1,328,111,219 worldwide.

Source: Nash Information Services, LLC

What digit names the number of:

5. thousands?

6. millions?

7. ten millions?

8. hundred thousands?

b Write expanded notation.

Stair-Climbing Races. The figure below shows the number of stairs in four buildings in which stair-climbing races are held. In Exercises 9–12, write expanded notation for the number of stairs in each race.

Stair-Climbing Races

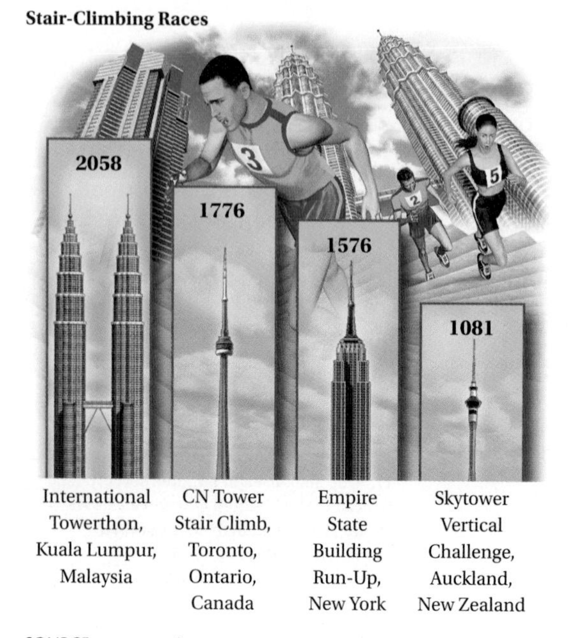

SOURCE: towerrunning.com

9. 2058 steps in the International Towerthon, Kuala Lumpur, Malaysia

10. 1776 steps in the CN Tower Stair Climb, Toronto, Ontario, Canada

11. 1576 steps in the Empire State Building Run-Up, New York City, New York

12. 1081 steps in the Skytower Vertical Challenge, Auckland, New Zealand

13. 5702 **14.** 3097 **15.** 93,986 **16.** 38,453

Population. The table below shows the populations of four countries in 2012. In Exercises 17–20, write expanded notation for the population of the given country.

Four Most Populous Countries in the World

COUNTRY	POPULATION, 2012
China	1,343,239,923
India	1,205,073,612
United States	313,847,465
Indonesia	248,645,008

SOURCE: *CIA World Factbook*

17. 1,343,239,923 for China

18. 1,205,073,612 for India

19. 248,645,008 for Indonesia

20. 313,847,465 for the United States

C Write a word name.

21. 85 **22.** 48 **23.** 88,000 **24.** 45,987

25. 123,765 **26.** 111,013 **27.** 7,754,211,577 **28.** 43,550,651,808

29. *English Language Learners.* In the 2007–2008 academic year, there were 701,799 English language learners in Texas schools. Write a word name for 701,799.

Source: U.S. Department of Education

30. *College Football.* The 2012 Rose Bowl game was attended by 91,245 fans. Write a word name for 91,245.

Source: bizjournals.com

31. *Auto Racing.* Dario Franchitti, winner of the 2012 Indianapolis 500 auto race, won a prize of $2,474,280. Write a word name for 2,474,280.

Source: sbnation.com

32. *Busiest Airport.* In 2010, the world's busiest airport, Hartsfield-Jackson Atlanta International Airport, hosted 89,331,622 passengers. Write a word name for 89,331,622.

Source: Airports Council International

Write each number in standard notation.

33. Six hundred thirty-two thousand, eight hundred ninety-six

34. Three hundred fifty-four thousand, seven hundred two

35. Fifty thousand, three hundred twenty-four

36. Seventeen thousand, one hundred twelve

37. Two million, two hundred thirty-three thousand, eight hundred twelve

38. Nineteen million, six hundred ten thousand, four hundred thirty-nine

39. Eight billion

40. Seven hundred million

41. Forty million

42. Twenty-six billion

43. Thirty million, one hundred three

44. Two hundred thousand, seventeen

Write standard notation for the number in each sentence.

45. *Pacific Ocean.* The area of the Pacific Ocean is sixty-four million, one hundred eighty-six thousand square miles.

46. The average distance from the sun to Neptune is two billion, seven hundred ninety-three million miles.

Synthesis

To the student and the instructor: The Synthesis exercises found at the end of every exercise set challenge students to combine concepts or skills studied in the section or in preceding parts of the text. Exercises marked with a calculator symbol are meant to be solved using a calculator.

47. How many whole numbers between 100 and 400 contain the digit 2 in their standard notation?

48. (calculator) What is the largest number that you can name on your calculator? How many digits does that number have? How many periods?

Addition

a ADDITION OF WHOLE NUMBERS

To answer questions such as “How many?”, “How much?”, and “How tall?”, we often use whole numbers. The set, or collection, of **whole numbers** is

0, 1, 2, 3, 4, 5, 6, 7, 8, 9, 10, 11, 12, ...

The set goes on indefinitely. There is no largest whole number, and the smallest whole number is 0. Each whole number can be named using various notations. The set 1, 2, 3, 4, 5, . . . , without 0, is called the set of **natural numbers**.

Addition of whole numbers corresponds to combining things together.

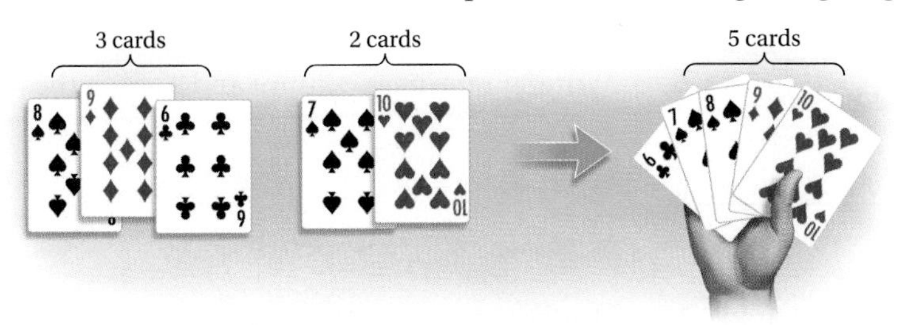

We say that the **sum** of 3 and 2 is 5. The numbers added are called **addends**. The addition that corresponds to the figure above is

$$\underset{\text{Addend}}{3} + \underset{\text{Addend}}{2} = \underset{\text{Sum}}{5}.$$

To add whole numbers, we add the ones digits first, then the tens, then the hundreds, then the thousands, and so on.

EXAMPLE 1 Add: 878 + 995.

Place values are lined up in columns.

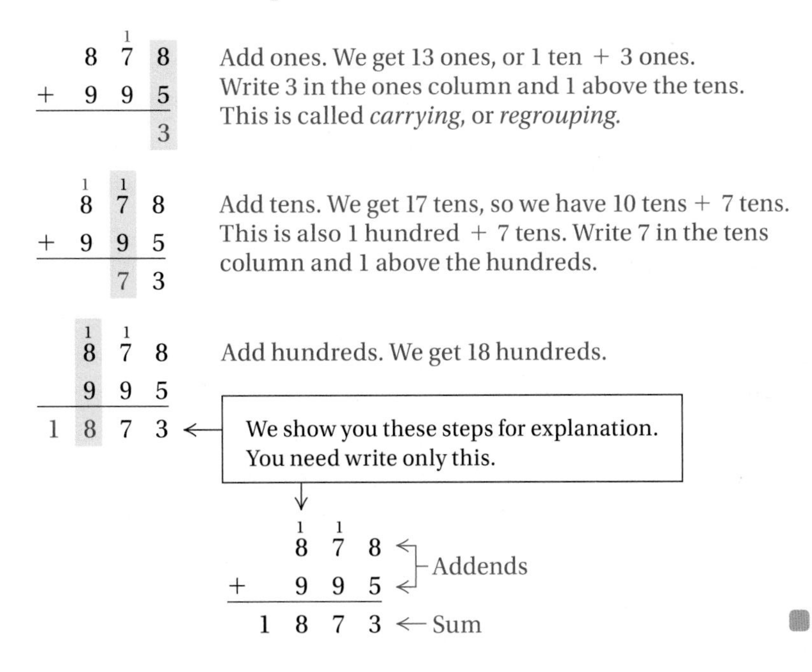

OBJECTIVES

- **a** Add whole numbers.
- **b** Use addition in finding perimeter.

SKILL TO REVIEW

Objective 1.1a: Give the meaning of digits in standard notation.

In each of the following numbers, what does the digit 4 mean?

1. 8342

2. 14,976

Answers

Skill to Review:

1. 4 tens
2. 4 thousands

How do we perform an addition of three numbers, like $2 + 3 + 6$? We could do it by adding 3 and 6, and then 2. We can show this with parentheses:

$$2 + (3 + 6) = 2 + 9 = 11.$$ Parentheses tell what to do first.

We could also add 2 and 3, and then 6:

$$(2 + 3) + 6 = 5 + 6 = 11.$$

Either way the result is 11. It does not matter how we group the numbers. This illustrates the **associative law of addition**, $a + (b + c) = (a + b) + c$. We can also add whole numbers in any order. That is, $2 + 3 = 3 + 2$. This illustrates the **commutative law of addition**, $a + b = b + a$. Together, the commutative and associative laws tell us that to add more than two numbers, we can use any order and grouping we wish. Adding 0 to a number does not change the number: $a + 0 = 0 + a = a$. That is, $6 + 0 = 0 + 6 = 6$, or $198 + 0 = 0 + 198 = 198$. We say that 0 is the **additive identity**.

EXAMPLE 2 Add: $391 + 1276 + 789 + 5498$.

```
      2
    3 9 1
  1 2 7 6
    7 8 9
+ 5 4 9 8
---------
        4
```

Add ones. We get 24, so we have 2 tens + 4 ones. Write 4 in the ones column and 2 above the tens.

```
    3 2
    3 9 1
  1 2 7 6
    7 8 9
+ 5 4 9 8
---------
      5 4
```

Add tens. We get 35 tens, so we have 30 tens + 5 tens. This is also 3 hundreds + 5 tens. Write 5 in the tens column and 3 above the hundreds.

```
  1 3 2
    3 9 1
  1 2 7 6
    7 8 9
+ 5 4 9 8
---------
    9 5 4
```

Add hundreds. We get 19 hundreds, so we have 10 hundreds + 9 hundreds. This is also 1 thousand + 9 hundreds. Write 9 in the hundreds column and 1 above the thousands.

```
  1 3 2
    3 9 1
  1 2 7 6
    7 8 9
+ 5 4 9 8
---------
  7 9 5 4
```

Add thousands. We get 7 thousands.

◀ Do Exercises 1–4.

Add.

1. $6203 + 3542$

2. GS
```
  7968
+ 5497
```

```
  1 1 ▢
  7 9 6 8
+ 5 4 9 7
---------
▢ ▢ ,4 ▢ 5
```

3.
```
  9804
+ 6378
```

4.
```
  1932
  6723
  9878
+ 8941
```

Answers

1. 9745 2. 13,465 3. 16,182
4. 27,474

Guided Solution:

2.
```
  1 1 1
  7 9 6 8
+ 5 4 9 7
---------
 13,4 6 5
```

CALCULATOR CORNER

Adding Whole Numbers This is the first of a series of *optional* discussions on using a calculator. A calculator is *not* a requirement for this textbook. Check with your instructor about whether you are allowed to use a calculator in the course.

There are many kinds of calculators and different instructions for their usage. Be sure to consult your users manual.

To add whole numbers on a calculator, we use the [+] and [=] keys. After we press [=], the sum appears on the display.

EXERCISES Use a calculator to find each sum.

1. $73 + 48$
2. $925 + 677$
3. $826 + 415 + 691$
4. $253 + 490 + 121$

b FINDING PERIMETER

Addition can be used when finding perimeter.

PERIMETER

The distance around an object is its **perimeter.**

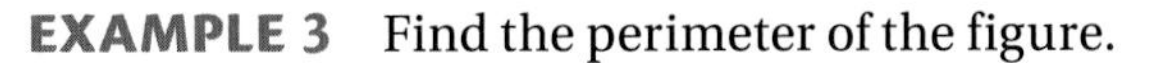

EXAMPLE 3 Find the perimeter of the figure.

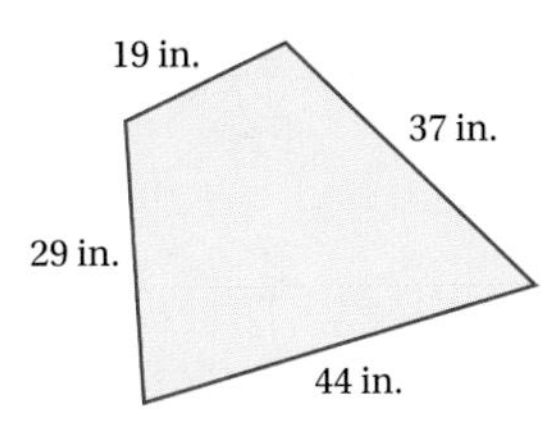

We add the lengths of the sides:

$$\text{Perimeter} = 29 \text{ in.} + 19 \text{ in.} + 37 \text{ in.} + 44 \text{ in.}$$
$$= 129 \text{ in.}$$

The perimeter of the figure is 129 in. (inches).

Do Exercises 5 and 6. ▶

EXAMPLE 4 Lucas Oil Stadium in Indianapolis has a unique retractable roof. When the roof is opened (retracted) in good weather to create an open-air stadium, the opening approximates a rectangle 588 ft long and 300 ft wide. Find the perimeter of the opening.

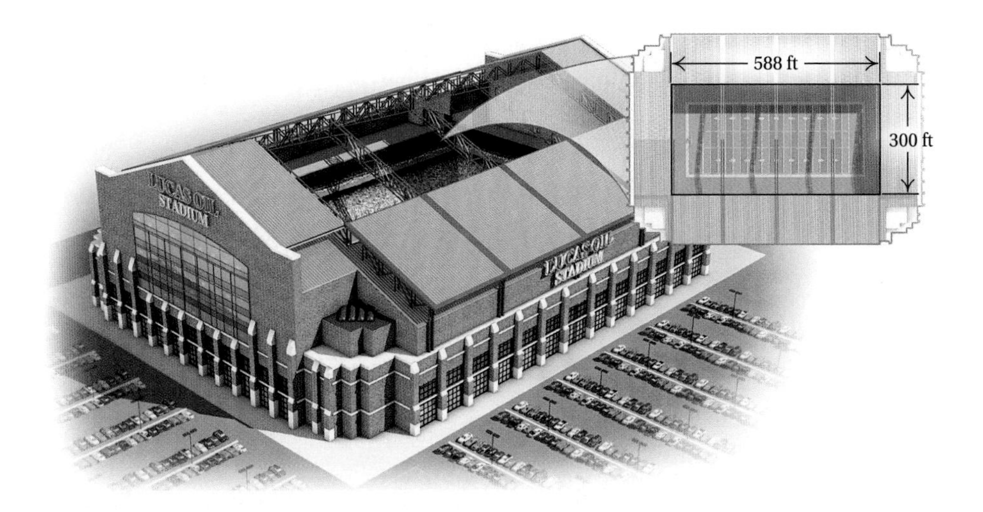

Opposite sides of a rectangle have equal lengths, so this rectangle has two sides of length 588 ft and two sides of length 300 ft.

$$\text{Perimeter} = 588 \text{ ft} + 300 \text{ ft} + 588 \text{ ft} + 300 \text{ ft}$$
$$= 1776 \text{ ft}$$

The perimeter of the opening is 1776 ft.

Do Exercise 7. ▶

Find the perimeter of each figure.

GS **5.**

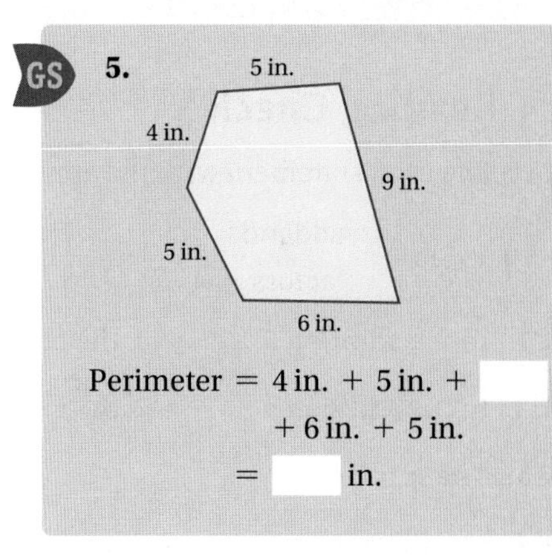

Perimeter = 4 in. + 5 in. + ▭ + 6 in. + 5 in.
= ▭ in.

6.

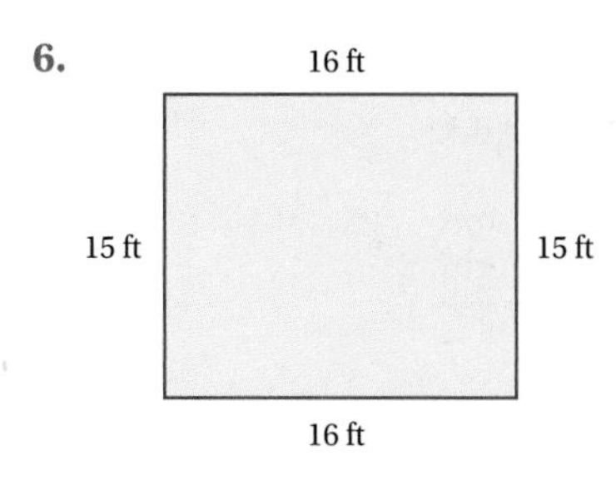

7. ***Index Cards.*** Two standard sizes for index cards are 3 in. by 5 in. and 5 in. by 8 in. Find the perimeter of each type of card.

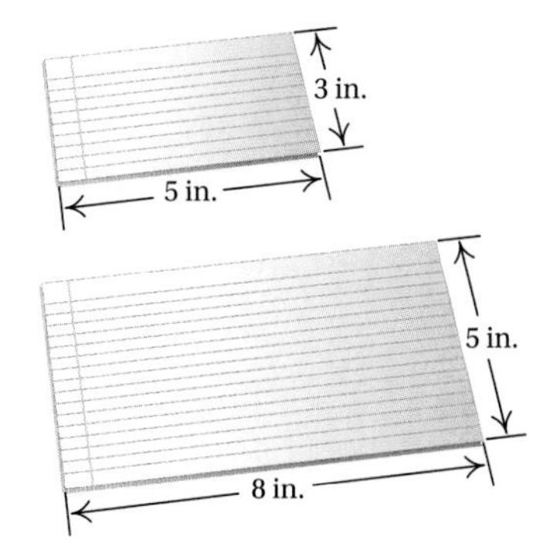

Answers

5. 29 in. **6.** 62 ft **7.** 16 in.; 26 in.

Guided Solution:
5. 9 in., 29

1.2 Exercise Set

For Extra Help
MyMathLab®
MathXL® PRACTICE WATCH READ REVIEW

Reading Check

Complete each statement with the appropriate word or number from the following list. Not every choice will be used.

0	addends	law	product
1	factors	perimeter	sum

RC1. In the addition $5 + 2 = 7$, the numbers 5 and 2 are ________________.

RC2. In the addition $5 + 2 = 7$, the number 7 is the ________________.

RC3. The sum of ________________ and any number a is a.

RC4. The distance around an object is its ________________.

a Add.

1. $364 + 23$

2. $1521 + 348$

3. $86 + 78$

4. $73 + 69$

5. $1716 + 3482$

6. $7503 + 2683$

7. $99 + 1$

8. $999 + 11$

9. $8113 + 390$

10. $271 + 3338$

11. $356 + 4910$

12. $280 + 34{,}902$

13. $3870 + 92 + 7 + 497$

14. $10{,}120 + 12{,}989 + 5738$

15. $4825 + 1783$

16. $3654 + 2700$

17. $23{,}443 + 10{,}989$

18. $45{,}879 + 21{,}786$

19. $77{,}543 + 23{,}767$

20. $99{,}999 + 112$

21. $45 + 25 + 36 + 44 + 80$

22. $38 + 27 + 32 + 14 + 76$

23. $12{,}070 + 2{,}954 + 3{,}400$

24. $42{,}487 + 83{,}141 + 36{,}712$

25. $4835 + 729 + 9204 + 8986 + 7931$

26. $989 + 566 + 834 + 920 + 703$

b Find the perimeter of each figure.

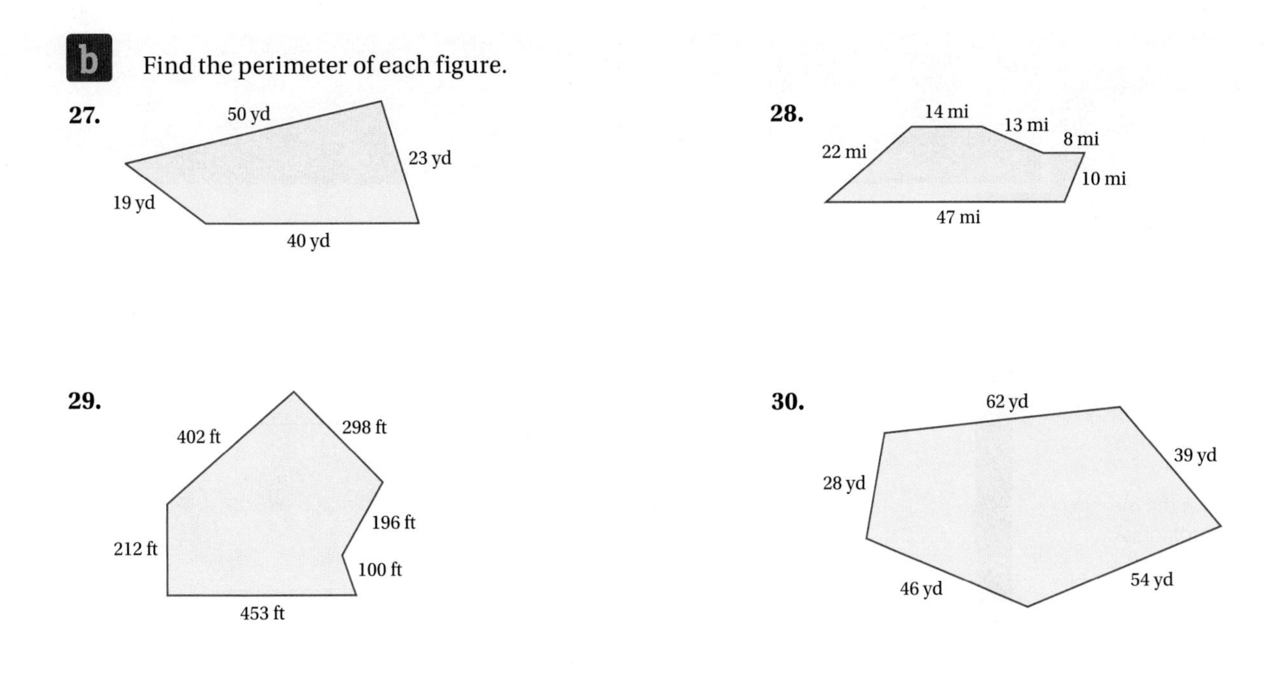

31. Find the perimeter of a standard hockey rink.

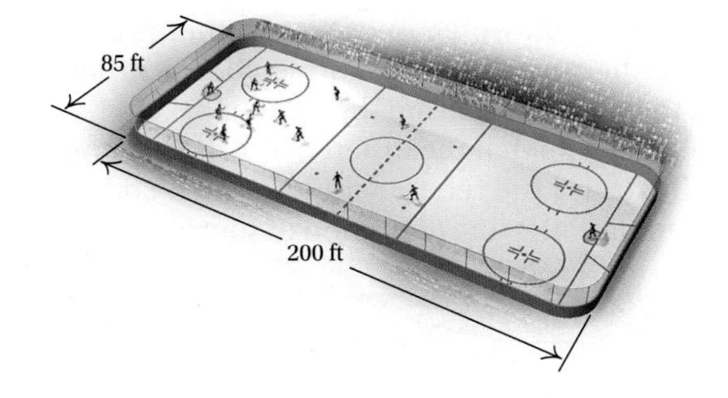

32. In Major League Baseball, how far does a batter travel when circling the bases after hitting a home run?

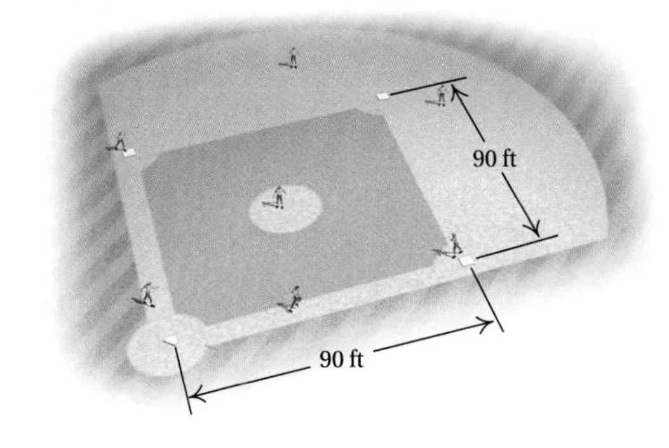

Skill Maintenance

The exercises that follow begin an important feature called *Skill Maintenance exercises.* These exercises provide an ongoing review of topics previously covered in the book. You will see them in virtually every exercise set. It has been found that this kind of continuing review can significantly improve your performance on a final examination.

33. What does the digit 8 mean in 486,205? [1.1a]

34. The population of the world is projected to be 9,346,399,468 in 2050. Write a word name for 9,346,399,468. [1.1c]

Source: U.S. Census Bureau

Synthesis

35. A fast way to add all the numbers from 1 to 10 inclusive is to pair 1 with 9, 2 with 8, and so on. Use a similar approach to add all numbers from 1 to 100 inclusive.

1.3 Subtraction

OBJECTIVE

a Subtract whole numbers.

SKILL TO REVIEW

Objective 1.1a: Give the meaning of digits in standard notation.

Consider the number 328,974.

1. What digit names the number of hundreds?
2. What digit names the number of ones?

a SUBTRACTION OF WHOLE NUMBERS

Subtraction is finding the difference of two numbers. Suppose you purchase 6 tickets for a concert and give 2 to a friend.

The subtraction that represents this situation is

$$\underset{\text{Minuend}}{6} - \underset{\text{Subtrahend}}{2} = \underset{\text{Difference}}{4}.$$

The **minuend** is the number from which another number is being subtracted. The **subtrahend** is the number being subtracted. The **difference** is the result of subtracting the subtrahend from the minuend.

In the subtraction above, note that the difference, 4, is the number we add to 2 to get 6. This illustrates the relationship between addition and subtraction and leads us to the following definition of subtraction.

SUBTRACTION

The difference $a - b$ is that unique whole number c for which $a = c + b$.

We see that $6 - 2 = 4$ because $4 + 2 = 6$.

To subtract whole numbers, we subtract the ones digits first, then the tens digits, then the hundreds, then the thousands, and so on.

EXAMPLE 1 Subtract: $9768 - 4320$.

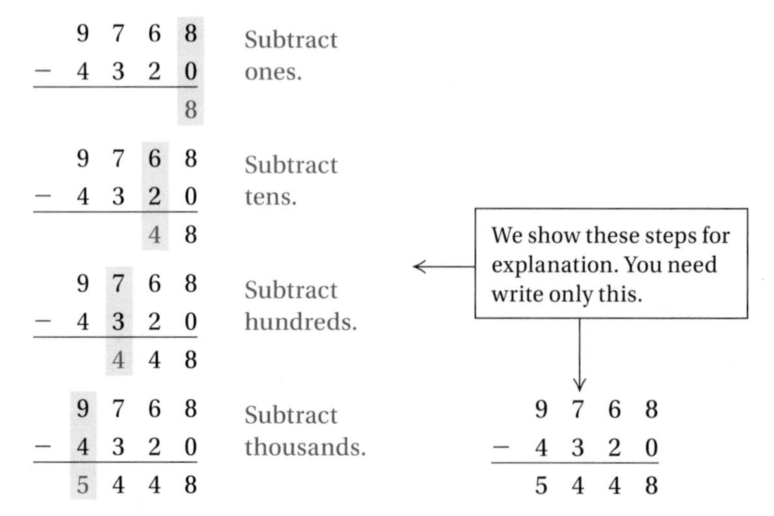

Answers

Skill to Review:
1. 9 **2.** 4

Because subtraction is defined in terms of addition, we can use addition to *check* subtraction.

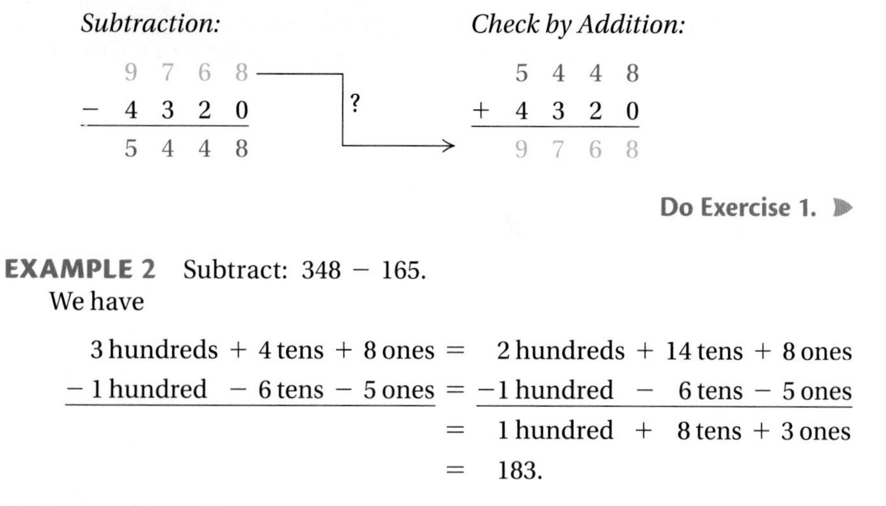

Do Exercise 1. ▶

GS **1.** Subtract. Check by adding.

$$\begin{array}{r} 7893 \\ -\ 4092 \\ \hline \end{array}$$

$$\begin{array}{r} 7\ 8\ 9\ 3 \\ -\ 4\ 0\ 9\ 2 \\ \hline \square\ 8\ \square\ 1 \end{array}$$

Check:

$$\begin{array}{r} \square \\ +\ 4092 \\ \hline \square \end{array}$$

EXAMPLE 2 Subtract: 348 − 165.

We have

$$\begin{aligned} &3 \text{ hundreds} + 4 \text{ tens} + 8 \text{ ones} = 2 \text{ hundreds} + 14 \text{ tens} + 8 \text{ ones} \\ &-1 \text{ hundred} - 6 \text{ tens} - 5 \text{ ones} = -1 \text{ hundred} - 6 \text{ tens} - 5 \text{ ones} \\ &= 1 \text{ hundred} + 8 \text{ tens} + 3 \text{ ones} \\ &= 183. \end{aligned}$$

First, we subtract the ones.

$$\begin{array}{r} 3\ 4\ 8 \\ -\ 1\ 6\ 5 \\ \hline 3 \end{array}$$ Subtract ones.

We cannot subtract the tens because there is no whole number that when added to 6 gives 4. To complete the subtraction, we must *borrow* 1 hundred from 3 hundreds and regroup it with the 4 tens. Then we can do the subtraction 14 tens − 6 tens = 8 tens.

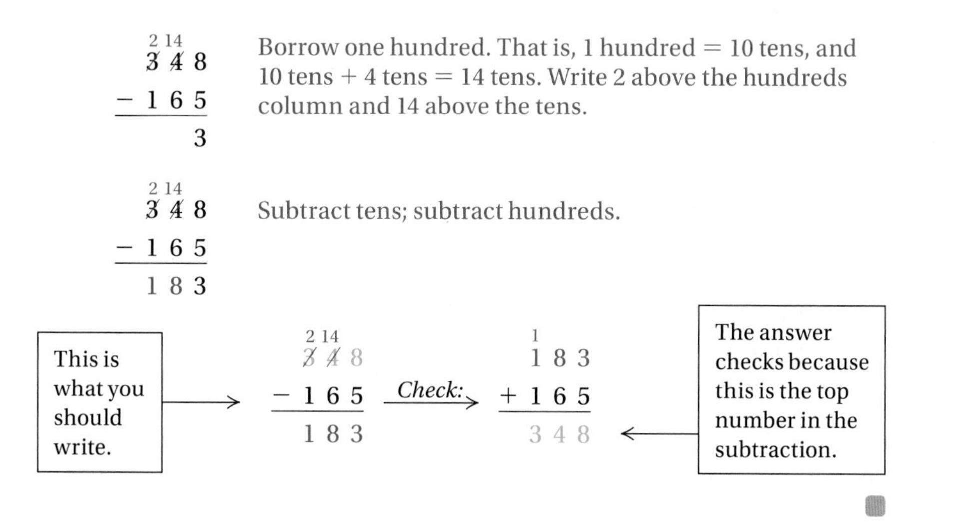

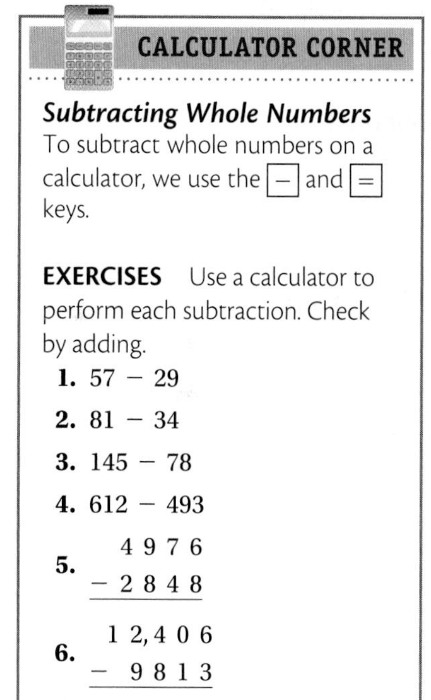

Subtracting Whole Numbers To subtract whole numbers on a calculator, we use the [−] and [=] keys.

EXERCISES Use a calculator to perform each subtraction. Check by adding.

1. 57 − 29

2. 81 − 34

3. 145 − 78

4. 612 − 493

5. $\begin{array}{r} 4976 \\ -\ 2848 \\ \hline \end{array}$

6. $\begin{array}{r} 12{,}406 \\ -\ 9813 \\ \hline \end{array}$

Answer

1. 3801

Guided Solution:

1. 3, 0; 3801, 7893

25. $\begin{array}{r} 12,647 \\ -\ \ 4,899 \\ \hline \end{array}$

26. $\begin{array}{r} 16,222 \\ -\ \ 5,888 \\ \hline \end{array}$

27. $\begin{array}{r} 51,342 \\ -47,198 \\ \hline \end{array}$

28. $\begin{array}{r} 32,194 \\ -29,236 \\ \hline \end{array}$

29. $\begin{array}{r} 80 \\ -24 \\ \hline \end{array}$

30. $\begin{array}{r} 90 \\ -78 \\ \hline \end{array}$

31. $\begin{array}{r} 690 \\ -236 \\ \hline \end{array}$

32. $\begin{array}{r} 803 \\ -418 \\ \hline \end{array}$

33. $\begin{array}{r} 6808 \\ -3059 \\ \hline \end{array}$

34. $\begin{array}{r} 9405 \\ -\ \ 258 \\ \hline \end{array}$

35. $\begin{array}{r} 2300 \\ -\ \ 109 \\ \hline \end{array}$

36. $\begin{array}{r} 7500 \\ -3604 \\ \hline \end{array}$

37. 90,237 − 47,209

38. 84,703 − 298

39. 101,734 − 5760

40. 15,017 − 7809

41. $\begin{array}{r} 6007 \\ -1589 \\ \hline \end{array}$

42. $\begin{array}{r} 8003 \\ -\ \ 599 \\ \hline \end{array}$

43. $\begin{array}{r} 39,000 \\ -37,695 \\ \hline \end{array}$

44. $\begin{array}{r} 17,000 \\ -11,598 \\ \hline \end{array}$

45. 10,008 − 19

46. 40,006 − 147

47. 50,001 − 1984

48. 30,004 − 6749

Skill Maintenance

Add. [1.2a]

49. 567 + 778

50. 901 + 23

51. 12,885 + 9807

52. 9909 + 1011

53. Write a word name for 6,375,602. [1.1c]

54. Write expanded notation for 9103. [1.1b]

Synthesis

55. Fill in the missing digits to make the subtraction true:

9,□48,621 − 2,097,□81 = 7,251,140.

56. 🖩 Subtract: 3,928,124 − 1,098,947.

Multiplication

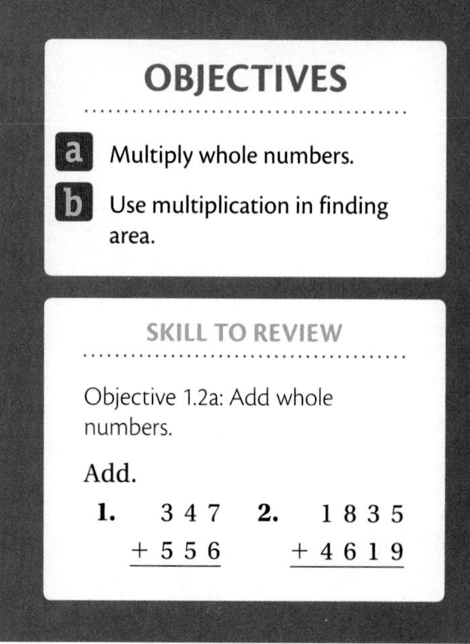

a MULTIPLICATION OF WHOLE NUMBERS

Repeated Addition

The multiplication 3×5 corresponds to this repeated addition.

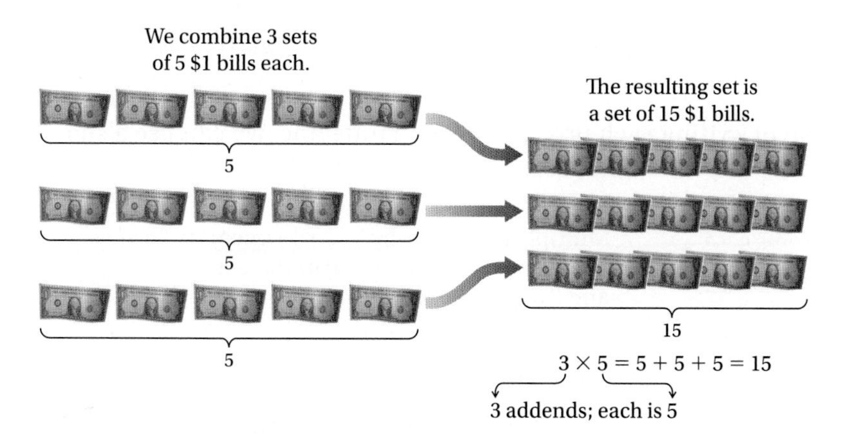

The numbers that we multiply are called **factors**. The result of the multiplication is called a **product**.

$$3 \times 5 = 15$$

3 → Factor, 5 → Factor, 15 → Product

Rectangular Arrays

Multiplications can also be thought of as rectangular arrays. Each of the following corresponds to the multiplication 3×5.

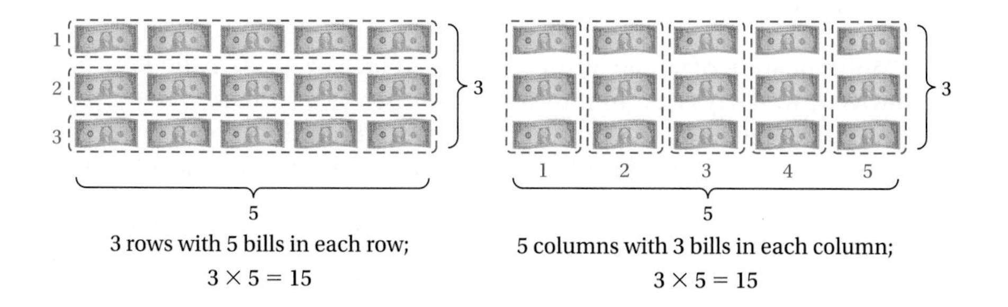

When you write a multiplication corresponding to a real-world situation, you should think of either a rectangular array or repeated addition. In some cases, it may help to think both ways.

We have used an "×" to denote multiplication. A dot " · " is also commonly used. (Use of the dot is attributed to the German mathematician Gottfried Wilhelm von Leibniz over three centuries ago.) Parentheses are also used to denote multiplication. For example,

$$3 \times 5 = 3 \cdot 5 = (3)(5) = 3(5) = 15.$$

Answers

Skill to Review:
1. 903 **2.** 6454

The product of 0 and any whole number is 0: $0 \cdot a = a \cdot 0 = 0$. For example, $0 \cdot 3 = 3 \cdot 0 = 0$. Multiplying a number by 1 does not change the number: $1 \cdot a = a \cdot 1 = a$. For example, $1 \cdot 3 = 3 \cdot 1 = 3$. We say that 1 is the **multiplicative identity**.

EXAMPLE 1 Multiply: 5×734.

We have

$$\begin{array}{rl} 7\,3\,4 & \\ \times \quad 5 & \\ \hline 2\,0 & \leftarrow \text{Multiply the 4 ones by 5: } 5 \times 4 = 20. \\ 1\,5\,0 & \leftarrow \text{Multiply the 3 tens by 5: } 5 \times 30 = 150. \\ 3\,5\,0\,0 & \leftarrow \text{Multiply the 7 hundreds by 5: } 5 \times 700 = 3500. \\ \hline 3\,6\,7\,0 & \leftarrow \text{Add.} \end{array}$$

Instead of writing each product on a separate line, we can use a shorter form.

$$\begin{array}{r} {}^{2} \\ 7\,3\,4 \\ \times \quad 5 \\ \hline 0 \end{array}$$

Multiply the 4 ones by 5: $5 \cdot (4 \text{ ones}) = 20$ ones $= 2$ tens $+ 0$ ones. Write 0 in the ones column and 2 above the tens.

$$\begin{array}{r} {}^{1\,2} \\ 7\,3\,4 \\ \times \quad 5 \\ \hline 7\,0 \end{array}$$

Multiply the 3 tens by 5 and add 2 tens: $5 \cdot (3 \text{ tens}) = 15$ tens, 15 tens $+ 2$ tens $= 17$ tens $= 1$ hundred $+ 7$ tens. Write 7 in the tens column and 1 above the hundreds.

$$\begin{array}{r} {}^{1\,2} \\ 7\,3\,4 \\ \times \quad 5 \\ \hline 3\,6\,7\,0 \end{array}$$

Multiply the 7 hundreds by 5 and add 1 hundred: $5 \cdot (7 \text{ hundreds}) = 35$ hundreds, 35 hundreds $+ 1$ hundred $= 36$ hundreds.

$$\begin{array}{r} {}^{1\,2} \\ 7\,3\,4 \\ \times \quad 5 \\ \hline 3\,6\,7\,0 \end{array}$$

You should write only this.

Do Exercises 1–4.

Multiply.

1. $\begin{array}{r} 58 \\ \times\ 2 \\ \hline \end{array}$

2. $\begin{array}{r} 37 \\ \times\ 4 \\ \hline \end{array}$

3. $\begin{array}{r} 823 \\ \times\ 6 \\ \hline \end{array}$

4. $\begin{array}{r} 1348 \\ \times\ 5 \\ \hline \end{array}$

GS

$$\begin{array}{r} \square\,{}^{2}\,\square \\ 1\,3\,4\,8 \\ \times \quad 5 \\ \hline \square\,7\,\square\,0 \end{array}$$

Multiplication of whole numbers is based on a property called the **distributive law**. It says that to multiply a number by a sum, $a \cdot (b + c)$, we can multiply each addend by a and then add like this: $(a \cdot b) + (a \cdot c)$. Thus, $a \cdot (b + c) = (a \cdot b) + (a \cdot c)$. For example, consider the following.

$4 \cdot (2 + 3) = 4 \cdot 5 = 20$ — Adding first; then multiplying

$4 \cdot (2 + 3) = (4 \cdot 2) + (4 \cdot 3) = 8 + 12 = 20$ — Multiplying first; then adding

The results are the same, so $4 \cdot (2 + 3) = (4 \cdot 2) + (4 \cdot 3)$.

Let's find the product 51×32. Since $32 = 2 + 30$, we can think of this product as

$$51 \times 32 = 51 \times (2 + 30) = (51 \times 2) + (51 \times 30).$$

That is, we multiply 51 by 2, then we multiply 51 by 30, and finally we add. We can write our work in columns.

Answers

1. 116 2. 148 3. 4938 4. 6740

Guided Solution:

4. $\begin{array}{r} {}^{1\,2\,4} \\ 1\,3\,4\,8 \\ \times \quad 5 \\ \hline 6\,7\,4\,0 \end{array}$

```
   5 1
 × 3 2
 1 0 2     Multiplying by 2
1 5 3 0    Multiplying by 30. (We write a 0 and then multiply 51 by 3.)
```

You may have learned that such a 0 need not be written. You may omit it if you wish. If you do omit it, remember, when multiplying by tens, to start writing the answer in the tens place.

We add to obtain the product.

```
   5 1
 × 3 2
 1 0 2
1 5 3 0
1 6 3 2    Adding to obtain the product
```

EXAMPLE 2 Multiply: 457×683.

```
    5 2
    6 8 3
  × 4 5 7
  4 7 8 1        Multiplying 683 by 7

    4 1
    5 2
    6 8 3
  × 4 5 7
  4 7 8 1
3 4 1 5 0        Multiplying 683 by 50

      3 1
      4 1
      5 2
      6 8 3
    × 4 5 7
    4 7 8 1
  3 4 1 5 0
2 7 3 2 0 0      Multiplying 683 by 400. (We write 00 and then multiply 683 by 4.)
3 1 2,1 3 1      Adding
```

Do Exercises 5–8. ▶

EXAMPLE 3 Multiply: 306×274.

Note that $306 = 3$ hundreds $+ 6$ ones.

```
   2 7 4
 × 3 0 6
 1 6 4 4     Multiplying by 6
8 2 2 0 0    Multiplying by 3 hundreds. (We write 00 and then multiply 274 by 3.)
8 3,8 4 4    Adding
```

Do Exercises 9–12. ▶

CALCULATOR CORNER

Multiplying Whole Numbers To multiply whole numbers on a calculator, we use the [×] and [=] keys.

EXERCISES Use a calculator to find each product.

1. 56×8

2. 845×26

3. $5 \cdot 1276$

4. $126(314)$

5. 3760×48

6. 5218×453

Multiply.

5. 45×23

6. 48×63

7. 746×62

8. 245×837

Multiply.

9. 472×306

10. 408×704

11. 2344×6005

12. 1006×703

Answers

5. 1035 **6.** 3024 **7.** 46,252 **8.** 205,065 **9.** 144,432 **10.** 287,232 **11.** 14,075,720 **12.** 707,218

a Multiply.

1. $\begin{array}{r} 65 \\ \times \quad 8 \\ \hline \end{array}$

2. $\begin{array}{r} 87 \\ \times \quad 4 \\ \hline \end{array}$

3. $\begin{array}{r} 94 \\ \times \quad 6 \\ \hline \end{array}$

4. $\begin{array}{r} 76 \\ \times \quad 9 \\ \hline \end{array}$

5. $3 \cdot 509$

6. $7 \cdot 806$

7. $7(9229)$

8. $4(7867)$

9. $90(53)$

10. $60(78)$

11. $(47)(85)$

12. $(34)(87)$

13. $\begin{array}{r} 87 \\ \times 10 \\ \hline \end{array}$

14. $\begin{array}{r} 2340 \\ \times 1000 \\ \hline \end{array}$

15. $\begin{array}{r} 96 \\ \times 20 \\ \hline \end{array}$

16. $\begin{array}{r} 800 \\ \times 700 \\ \hline \end{array}$

17. $\begin{array}{r} 643 \\ \times \quad 72 \\ \hline \end{array}$

18. $\begin{array}{r} 777 \\ \times \quad 77 \\ \hline \end{array}$

19. $\begin{array}{r} 444 \\ \times \quad 33 \\ \hline \end{array}$

20. $\begin{array}{r} 549 \\ \times \quad 88 \\ \hline \end{array}$

21. $\begin{array}{r} 564 \\ \times 458 \\ \hline \end{array}$

22. $\begin{array}{r} 432 \\ \times 375 \\ \hline \end{array}$

23. $\begin{array}{r} 853 \\ \times 936 \\ \hline \end{array}$

24. $\begin{array}{r} 346 \\ \times 659 \\ \hline \end{array}$

25. $\begin{array}{r} 6428 \\ \times 3224 \\ \hline \end{array}$

26. $\begin{array}{r} 8928 \\ \times 3172 \\ \hline \end{array}$

27. $\begin{array}{r} 3482 \\ \times \quad 104 \\ \hline \end{array}$

28. $\begin{array}{r} 6408 \\ \times 6064 \\ \hline \end{array}$

29. $\begin{array}{r} 876 \\ \times 345 \\ \hline \end{array}$

30. $\begin{array}{r} 355 \\ \times 299 \\ \hline \end{array}$

31. $\begin{array}{r} 7889 \\ \times 6224 \\ \hline \end{array}$

32. $\begin{array}{r} 6521 \\ \times 3449 \\ \hline \end{array}$

33. $\begin{array}{r} 5608 \\ \times\ 4500 \\ \hline \end{array}$

34. $\begin{array}{r} 4506 \\ \times\ 7800 \\ \hline \end{array}$

35. $\begin{array}{r} 5006 \\ \times\ 4008 \\ \hline \end{array}$

36. $\begin{array}{r} 6009 \\ \times\ 2003 \\ \hline \end{array}$

b Find the area of each region.

37.

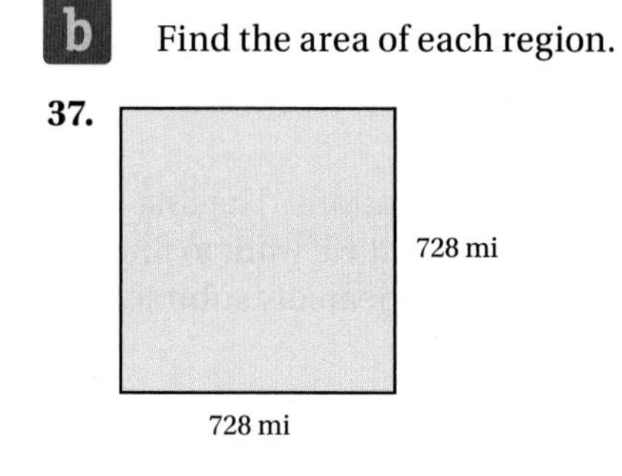

38.

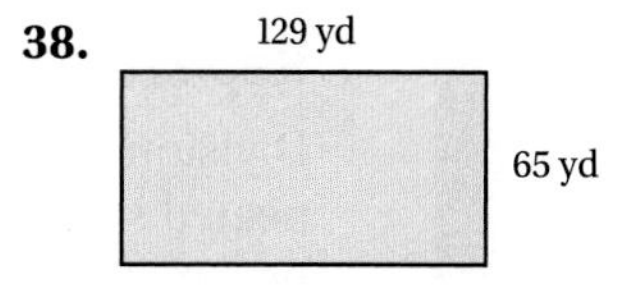

39. Find the area of the region formed by the base lines on a Major League Baseball diamond.

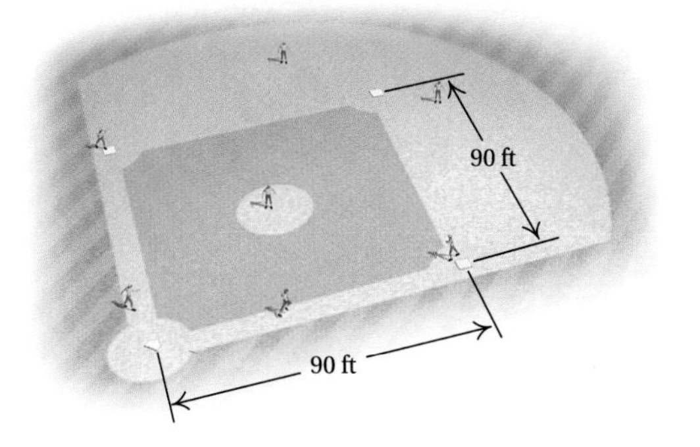

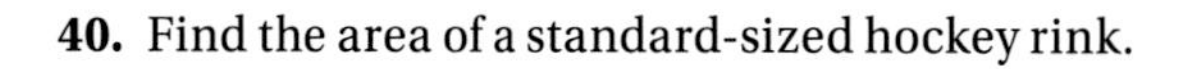

40. Find the area of a standard-sized hockey rink.

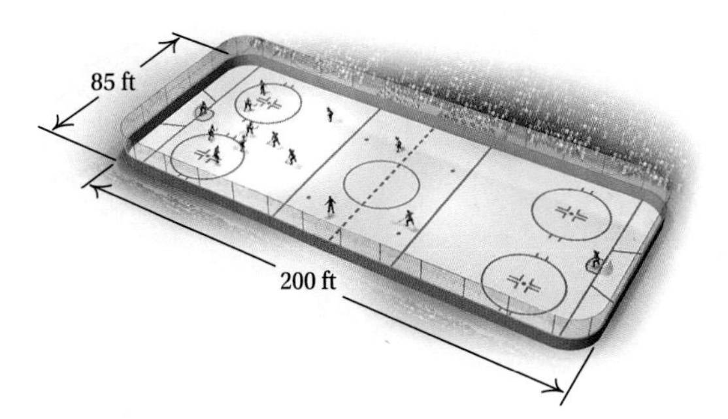

Skill Maintenance

Add. [1.2a]

41. $\begin{array}{r} 4908 \\ 5667 \\ +\ 2110 \\ \hline \end{array}$

42. $\begin{array}{r} 9876 \\ 876 \\ 76 \\ +\ \ 6 \\ \hline \end{array}$

Subtract. [1.3a]

43. $\begin{array}{r} 9876 \\ -\ \ 987 \\ \hline \end{array}$

44. $\begin{array}{r} 340{,}798 \\ -\ \ 86{,}679 \\ \hline \end{array}$

45. What does the digit 4 mean in 9,482,157? [1.1a]

46. What digit in 38,026 names the number of hundreds? [1.1a]

47. Write expanded notation for 12,847. [1.1b]

48. Write a word name for 7,432,000. [1.1c]

Synthesis

49. 🖩 An 18-story office building is box-shaped. Each floor measures 172 ft by 84 ft with a 20-ft by 35-ft rectangular area lost to an elevator and a stairwell. How much area is available as office space?

1.5 Division

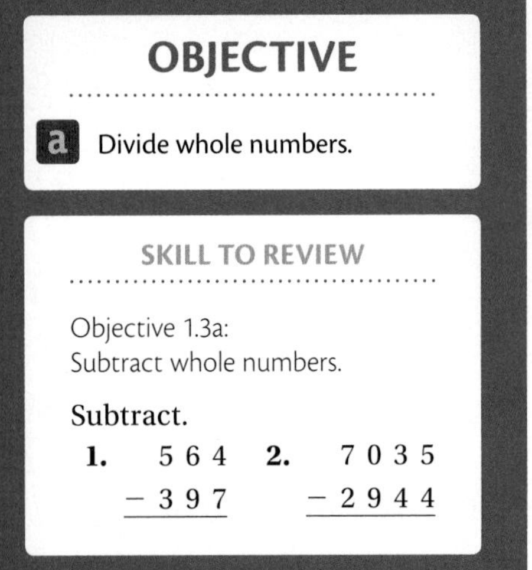

OBJECTIVE

a Divide whole numbers.

SKILL TO REVIEW

Objective 1.3a:
Subtract whole numbers.

Subtract.

1. $\begin{array}{r} 564 \\ -\ 397 \\ \hline \end{array}$ **2.** $\begin{array}{r} 7035 \\ -\ 2944 \\ \hline \end{array}$

a DIVISION OF WHOLE NUMBERS

Repeated Subtraction

Division of whole numbers applies to two kinds of situations. The first is repeated subtraction. Suppose we have 20 doughnuts, and we want to find out how many sets of 5 there are. One way to do this is to repeatedly subtract sets of 5 as follows.

Since there are 4 sets of 5 doughnuts each, we have

$$\underset{\text{Dividend}}{20} \div \underset{\text{Divisor}}{5} = \underset{\text{Quotient}}{4}.$$

The division $20 \div 5$ is read "20 divided by 5." The **dividend** is 20, the **divisor** is 5, and the **quotient** is 4. We divide the *dividend* by the *divisor* to get the *quotient.* We can also express the division $20 \div 5 = 4$ as

$$\frac{20}{5} = 4 \quad \text{or} \quad 5\overline{)20}^{\,4}.$$

Rectangular Arrays

We can also think of division in terms of rectangular arrays. Consider again the 20 doughnuts and division by 5. We can arrange the doughnuts in a rectangular array with 5 rows and ask, "How many are in each row?"

We can also consider a rectangular array with 5 doughnuts in each column and ask, "How many columns are there?" The answer is still 4.

In each case, we are asking, "What do we multiply 5 by in order to get 20?"

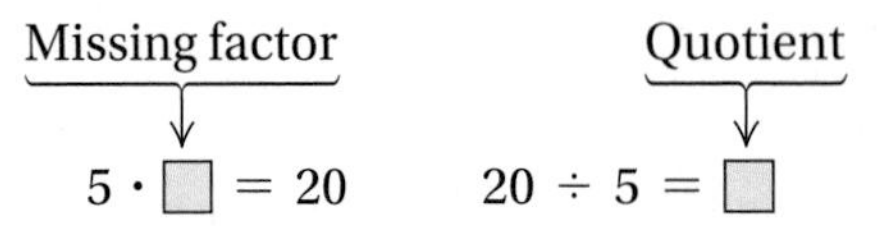

This leads us to the following definition of division.

Answers

Skill to Review:
1. 167 **2.** 4091

DIVISION

The quotient $a \div b$, where b is not 0, is that unique number c for which $a = b \cdot c$.

This definition shows the relation between division and multiplication. We see, for instance, that

$$20 \div 5 = 4 \text{ because } 20 = 5 \cdot 4.$$

This relation allows us to use multiplication to check division.

EXAMPLE 1 Divide. Check by multiplying.

a) $16 \div 8$ **b)** $\frac{36}{4}$ **c)** $7\overline{)56}$

We do so as follows.

a) $16 \div 8 = 2$ *Check*: $8 \cdot 2 = 16$.

b) $\frac{36}{4} = 9$ *Check*: $4 \cdot 9 = 36$.

c) $7\overline{)56}$ with quotient 8 *Check*: $7 \cdot 8 = 56$.

Do Exercises 1–3.

Let's consider some basic properties of division.

DIVIDING BY 1

Any number divided by 1 is that same number: $a \div 1 = \frac{a}{1} = a$.

For example, $6 \div 1 = 6$ and $\frac{15}{1} = 15$.

DIVIDING A NUMBER BY ITSELF

Any nonzero number divided by itself is 1: $a \div a = \frac{a}{a} = 1, \quad a \neq 0$.

For example, $7 \div 7 = 1$ and $\frac{22}{22} = 1$.

DIVIDENDS OF 0

Zero divided by any nonzero number is 0: $0 \div a = \frac{0}{a} = 0, \quad a \neq 0$.

For example, $0 \div 14 = 0$ and $\frac{0}{3} = 0$.

Do Exercises 4–7.

Divide. Check by multiplying.

1. $9\overline{)45}$

2. $27 \div 3$

3. $\frac{48}{6}$

Divide.

4. $\frac{9}{9}$ **5.** $\frac{8}{1}$

6. $\frac{0}{12}$ **7.** $\frac{28}{28}$

Answers

1. 5 **2.** 9 **3.** 8 **4.** 1 **5.** 8 **6.** 0 **7.** 1

EXCLUDING DIVISION BY 0

Division by 0 is not defined: $a \div 0$, or $\frac{a}{0}$, is **not defined**.

For example, $16 \div 0$, or $\frac{16}{0}$, is not defined.

8. Divide if possible: $0 \div 2$. If not possible, write "not defined."

$0 \div 2$ means ___ divided by ___.

Since zero divided by any nonzero number is 0, $0 \div 2 =$ ___.

9. Divide if possible: $7 \div 0$. If not possible, write "not defined."

$7 \div 0$ means ___ divided by ___.

Since division by 0 is not defined, $7 \div 0$ is ___.

Why can't we divide by 0? Suppose the number 4 could be divided by 0. Then if $\square$ were the answer, we would have

$$4 \div 0 = \square,$$

and since 0 times any number is 0, we would have

$$4 = \square \cdot 0 = 0. \quad \text{False!}$$

Thus, the only possible number that could be divided by 0 would be 0 itself. But such a division would give us any number we wish. For instance,

$$\left.\begin{array}{lll} 0 \div 0 = 8 & \text{because} & 0 = 8 \cdot 0; \\ 0 \div 0 = 3 & \text{because} & 0 = 3 \cdot 0; \\ 0 \div 0 = 7 & \text{because} & 0 = 7 \cdot 0. \end{array}\right\} \quad \text{All true!}$$

We avoid the preceding difficulties by agreeing to exclude division by 0.

Do Exercises 8–9.

Division with a Remainder

Suppose everyone in a group of 22 people wants to ride a roller coaster. If each car on the ride holds 6 people, the group will fill 3 cars and there will be 4 people left over.

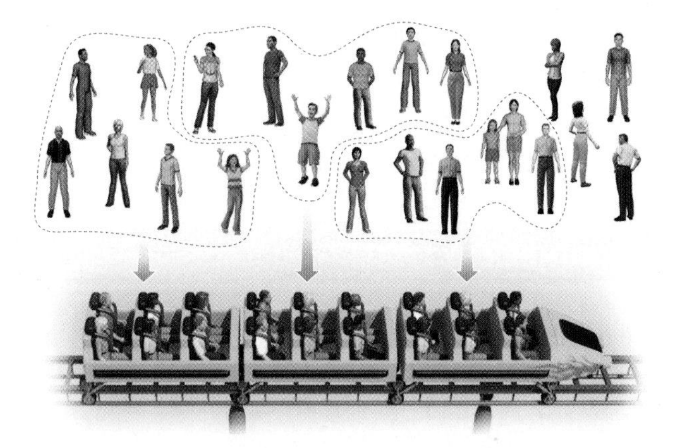

We can think of this situation as the following division. The people left over are the **remainder**.

$$\begin{array}{r} 3 \leftarrow \text{Quotient} \\ 6\overline{)22} \\ \underline{18} \\ 4 \leftarrow \text{Remainder} \end{array}$$

Answers

8. 0 **9.** Not defined

Guided Solutions:

8. 0, 2, 0

9. 7, 0, not defined

We express the result as

$$\underset{\text{Dividend}}{22} \div \underset{\text{Divisor}}{6} = \underset{\text{Quotient}}{3}\,\text{R}\,\underset{\text{Remainder}}{4}.$$

Note that

Quotient · Divisor + Remainder = Dividend.

Thus we have

$3 \cdot 6 = 18$ Quotient · Divisor

and $18 + 4 = 22.$ Adding the remainder. The result is the dividend.

We now show a procedure for dividing whole numbers.

EXAMPLE 2 Divide and check: $4\overline{)\,3\,4\,5\,7}$.

First, we try to divide the first digit of the dividend, 3, by the divisor, 4. Since $3 \div 4$ is not a whole number, we consider the first *two* digits of the dividend.

$$\begin{array}{r} 8 \\ 4\overline{)\,3457} \\ \underline{32} \\ 2 \end{array}$$

Since $4 \cdot 8 = 32$ and 32 is smaller than 34, we write an 8 in the quotient above the 4. We also write 32 below 34 and subtract.

What if we had chosen a number other than 8 for the first digit of the quotient? Suppose we had used 7 instead of 8 and subtracted $4 \cdot 7$, or 28, from 34. The result would have been $34 - 28$, or 6. Because 6 is larger than the divisor, 4, we know that there is at least one more factor of 4 in 34, and thus 7 is too small. If we had used 9 instead of 8, then we would have tried to subtract $4 \cdot 9$, or 36, from 34. That difference is not a whole number, so we know 9 is too large. When we subtract, the difference must be smaller than the divisor.

Let's continue dividing.

$$\begin{array}{r} 86 \\ 4\overline{)\,3457} \\ \underline{32} \\ 25 \\ \underline{24} \\ 1 \end{array}$$

Now we bring down the 5 in the dividend and consider $25 \div 4$. Since $4 \cdot 6 = 24$ and 24 is smaller than 25, we write 6 in the quotient above the 5. We also write 24 below 25 and subtract. The difference, 1, is smaller than the divisor, so we know that 6 is the correct choice.

$$\begin{array}{r} 864 \\ 4\overline{)\,3457} \\ \underline{32} \\ 25 \\ \underline{24} \\ 17 \\ \underline{16} \\ 1 \end{array}$$

We bring down the 7 and consider $17 \div 4$. Since $4 \cdot 4 = 16$ and 16 is smaller than 17, we write 4 in the quotient above the 7. We also write 16 below 17 and subtract.

1 ← The remainder is 1.

Check: $864 \cdot 4 = 3456$ and $3456 + 1 = 3457$.

The answer is 864 R 1.

Do Exercises 10–12.

Divide and check.

10. $3\overline{)\,2\,3\,9}$

11. $5\overline{)\,5\,8\,6\,4}$

12. $6\overline{)\,3\,8\,5\,5}$

Answers

10. 79 R 2 **11.** 1172 R 4
12. 642 R 3

EXAMPLE 3 Divide: $8904 \div 42$.

Because 42 is close to 40, we think of the divisor as 40 when we make our choices of digits in the quotient.

```
         2
42 ) 8 9 0 4   ← Think: 89 ÷ 40. We try 2. Multiply 42 · 2 and
     8 4 ↓       subtract. Then bring down the 0.
       5 0
```

```
         2 1
42 ) 8 9 0 4
     8 4
       5 0     ← Think: 50 ÷ 40. We try 1. Multiply 42 · 1 and
       4 2 ↓     subtract. Then bring down the 4.
         8 4
```

```
         2 1 2
42 ) 8 9 0 4
     8 4
       5 0
       4 2
         8 4   ← Think: 84 ÷ 40. We try 2. Multiply 2 · 42 and subtract.
         8 4
           0
```

The remainder is 0, so the answer is 212.

Divide.

13. $45\overline{)6030}$

14. $52\overline{)3288}$

Do Exercises 13 and 14.

Caution!

Be careful to keep the digits lined up correctly when you divide.

CALCULATOR CORNER

Dividing Whole Numbers To divide whole numbers on a calculator, we use the [÷] and [=] keys.

When we enter $453 \div 15$, the display reads [30.2]. Note that the result is not a whole number. This tells us that there is a remainder. The number 30.2 is expressed in decimal notation. The symbol "." is called a decimal point. Although it is possible to use the number to the right of the decimal point to find the remainder, we will not do so here.

EXERCISES Use a calculator to perform each division.

1. $19\overline{)532}$
2. $7\overline{)861}$
3. $9367 \div 29$
4. $12{,}276 \div 341$

Answers

13. 134 **14.** 63 R 12

Zeros in Quotients

EXAMPLE 4 Divide: 6341 ÷ 7.

```
      9
 7 )6 3 4 1  ← Think: 63 ÷ 7 = 9. The first digit in the quotient is
    6 3 ↓       9. We do not write the 0 when we find 63 − 63. Bring
        4       down the 4.
```

```
      9 0
 7 )6 3 4 1
    6 3   ↓
        4 1  ← Think: 4 ÷ 7. If we subtract a group of 7's, such
                as 7, 14, 21, etc., from 4, we do not get a whole
                number, so the next digit in the quotient is 0.
                Bring down the 1.
```

```
      9 0 5
 7 )6 3 4 1
    6 3
        4 1  ← Think: 41 ÷ 7. We try 5. Multiply 7 · 5
        3 5     and subtract.
          6  ← The remainder is 6.
```

The answer is 905 R 6.

Do Exercises 15 and 16. ▶

EXAMPLE 5 Divide: 8169 ÷ 34.

Because 34 is close to 30, we think of the divisor as 30 when we make our choices of digits in the quotient.

```
         2
 3 4 )8 1 6 9  ← Think: 81 ÷ 30. We try 2. Multiply 34 · 2 and
      6 8 ↓       subtract. Then bring down the 6.
      1 3 6
```

```
         2 4
 3 4 )8 1 6 9
      6 8
      1 3 6    ← Think: 136 ÷ 30. We try 4. Multiply 34 · 4 and
      1 3 6       subtract. The difference is 0, so we do not write it.
            9     Bring down the 9.
```

```
         2 4 0
 3 4 )8 1 6 9
      6 8
      1 3 6
      1 3 6
            9  ← Think: 9 ÷ 34. If we subtract a group of 34's,
            0     such as 34 or 68, from 9, we do not get a whole
            9     number, so the last digit in the quotient is 0.
               ← The remainder is 9.
```

The answer is 240 R 9.

Do Exercises 17 and 18. ▶

Divide.

15. $6\overline{)4846}$

16. $7\overline{)7616}$

Divide.

17. $27\overline{)9724}$

18. $56\overline{)44,847}$

Answers

15. 807 R 4 16. 1088
17. 360 R 4 18. 800 R 47

1.5 Exercise Set

For Extra Help
MyMathLab® MathXL® PRACTICE WATCH READ REVIEW

Reading Check

Match each word from the following list with the indicated part of the division.

dividend divisor quotient remainder

RC1. A ______________

RC2. B ______________

RC3. C ______________

RC4. D ______________

$$\begin{array}{r} 29 \leftarrow (A) \\ (D) \rightarrow 8\overline{)235} \leftarrow (B) \\ \underline{16} \\ 75 \\ \underline{72} \\ 3 \leftarrow (C) \end{array}$$

a Divide, if possible. If not possible, write "not defined."

1. $72 \div 6$

2. $54 \div 9$

3. $\frac{23}{23}$

4. $\frac{37}{37}$

5. $22 \div 1$

6. $\frac{56}{1}$

7. $\frac{0}{7}$

8. $\frac{0}{32}$

9. $\frac{16}{0}$

10. $74 \div 0$

11. $\frac{48}{8}$

12. $\frac{20}{4}$

Divide.

13. $277 \div 5$

14. $699 \div 3$

15. $864 \div 8$

16. $869 \div 8$

17. $4\overline{)1228}$

18. $3\overline{)2124}$

19. $6\overline{)4521}$

20. $9\overline{)9110}$

21. $297 \div 4$

22. $389 \div 2$

23. $738 \div 8$

24. $881 \div 6$

25. $5\overline{)8515}$

26. $3\overline{)6027}$

27. $9\overline{)8888}$

28. $8\overline{)4139}$

29. $127{,}000 \div 10$

30. $127{,}000 \div 100$

31. $127{,}000 \div 1000$

32. $4260 \div 10$

33. $70\overline{)3692}$

34. $20\overline{)5798}$

35. $30\overline{)875}$

36. $40\overline{)987}$

37. $852 \div 21$

38. $942 \div 23$

39. $85\overline{)7672}$

40. $54\overline{)2729}$

41. $111\overline{)3219}$

42. $102\overline{)5612}$

43. $8\overline{)843}$

44. $7\overline{)749}$

45. $5\overline{)8047}$

46. $9\overline{)7273}$

47. $5\overline{)5036}$

48. $7\overline{)7074}$

49. $1058 \div 46$

50. $7242 \div 24$

51. $3425 \div 32$

52. $48\overline{)4899}$

53. $24\overline{)8880}$

54. $36\overline{)7563}$

55. $28\overline{)17{,}067}$

56. $36\overline{)28{,}929}$

57. $80\overline{)24{,}320}$

58. $90\overline{)88{,}560}$

59. $285\overline{)999{,}999}$

60. $306\overline{)888{,}888}$

61. $456\overline{)3{,}679{,}920}$

62. $803\overline{)5{,}622{,}606}$

Skill Maintenance

Subtract. [1.3a]

63. $\begin{array}{r} 4908 \\ -\ 3667 \\ \hline \end{array}$

64. $\begin{array}{r} 88{,}777 \\ -\ 22{,}333 \\ \hline \end{array}$

Multiply. [1.4a]

65. $\begin{array}{r} 198 \\ \times\ 100 \\ \hline \end{array}$

66. $\begin{array}{r} 268 \\ \times\ 35 \\ \hline \end{array}$

Use the following figure for Exercises 67 and 68.

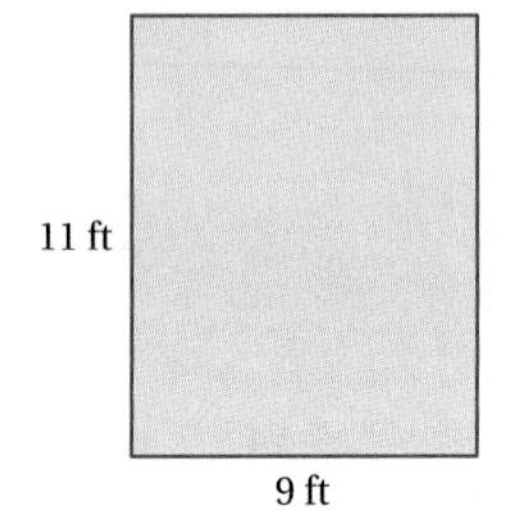

67. Find the perimeter of the figure. [1.2b]

68. Find the area of the figure. [1.4b]

Synthesis

69. Complete the following table.

a	b	$a \cdot b$	$a + b$
	68	3672	
84			117
		32	12

70. Find a pair of factors whose product is 36 and:

a) whose sum is 13.
b) whose difference is 0.
c) whose sum is 20.
d) whose difference is 9.

71. A group of 1231 college students is going to use buses to take a field trip. Each bus can hold 42 students. How many buses are needed?

72. 🖩 Fill in the missing digits to make the equation true:

$$34{,}584{,}132 \div 76\square = 4\square{,}386.$$

Mid-Chapter Review

Concept Reinforcement

Determine whether each statement is true or false.

______ **1.** If $a - b = c$, then $b = a + c$. [1.3a]

______ **2.** We can think of the multiplication 4×3 as a rectangular array containing 4 rows with 3 items in each row. [1.4a]

______ **3.** We can think of the multiplication 4×3 as a rectangular array containing 3 columns with 4 items in each column. [1.4a]

______ **4.** The product of two whole numbers is always greater than either of the factors. [1.4a]

______ **5.** Zero divided by any nonzero number is 0. [1.5a]

______ **6.** Any number divided by 1 is the number 1. [1.5a]

Guided Solutions

GS Fill in each blank with the number that creates a correct statement or solution.

7. Write a word name for 95,406,237. [1.1c]

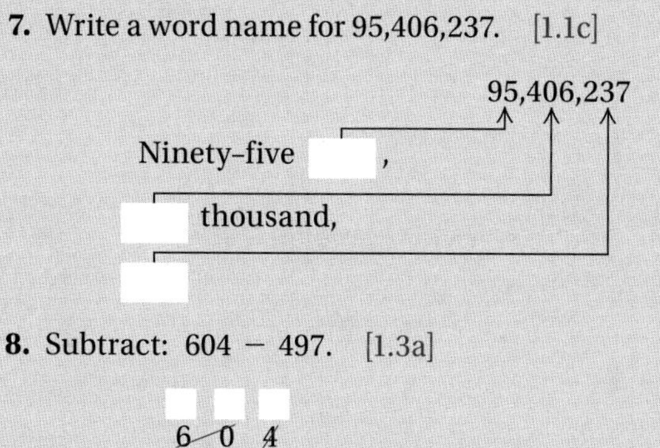

8. Subtract: $604 - 497$. [1.3a]

$$\begin{array}{r} 6\ 0\ 4 \\ -\ 4\ 9\ 7 \\ \hline \end{array}$$

Mixed Review

In each of the following numbers what does the digit 6 mean? [1.1a]

9. 2698

10. 61,204

11. 146,237

12. 586

Consider the number 306,458,129. What digit names the number of: [1.1a]

13. tens?

14. millions?

15. ten thousands?

16. hundreds?

Write expanded notation. [1.1b]

17. 5602

18. 69,345

Write a word name. [1.1c]

19. 136

20. 64,325

Write standard notation. [1.1c]

21. Three hundred eight thousand, seven hundred sixteen

22. Four million, five hundred sixty-seven thousand, two hundred sixteen

Add. [1.2a]

23. $\begin{array}{r} 316 \\ +\ 482 \\ \hline \end{array}$

24. $\begin{array}{r} 593 \\ +\ 437 \\ \hline \end{array}$

25. $\begin{array}{r} 2638 \\ +\ 5284 \\ \hline \end{array}$

26. $\begin{array}{r} 4617 \\ 2436 \\ +\ \ 481 \\ \hline \end{array}$

Subtract. [1.3a]

27. $\begin{array}{r} 786 \\ -\ 321 \\ \hline \end{array}$

28. $\begin{array}{r} 624 \\ -\ 285 \\ \hline \end{array}$

29. $\begin{array}{r} 3602 \\ -\ 1748 \\ \hline \end{array}$

30. $\begin{array}{r} 5004 \\ -\ \ 676 \\ \hline \end{array}$

Multiply. [1.4a]

31. $\begin{array}{r} 36 \\ \times\ \ 6 \\ \hline \end{array}$

32. $\begin{array}{r} 567 \\ \times\ \ 28 \\ \hline \end{array}$

33. $\begin{array}{r} 407 \\ \times\ 325 \\ \hline \end{array}$

34. $\begin{array}{r} 9435 \\ \times\ \ 602 \\ \hline \end{array}$

Divide. [1.5a]

35. $4\overline{)1012}$

36. $38\overline{)4261}$

37. $60\overline{)1399}$

38. $56\overline{)8095}$

39. Find the perimeter of the figure (m stands for "meters"). [1.2b]

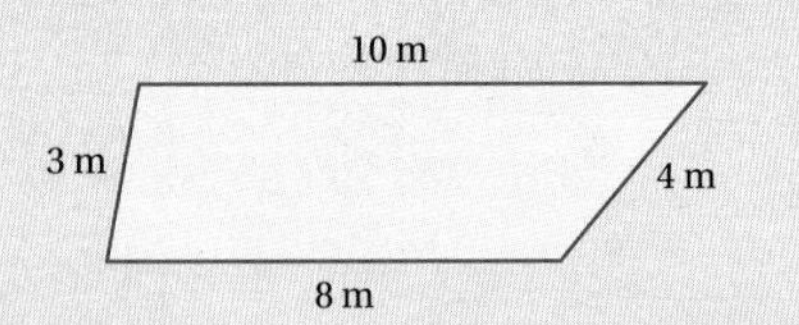

40. Find the area of the region. [1.4b]

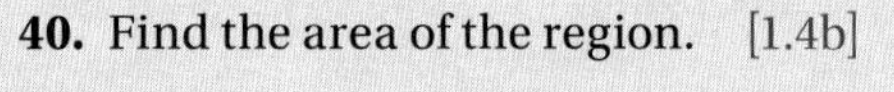

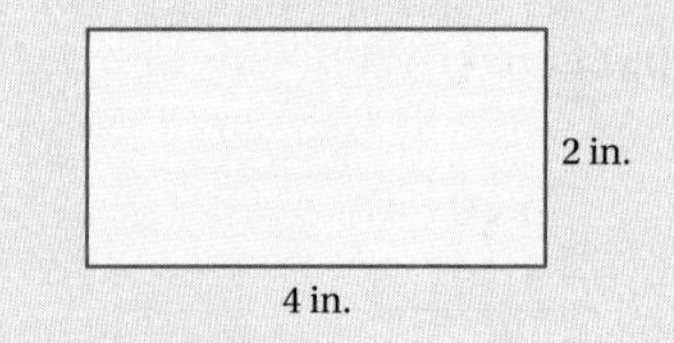

Understanding Through Discussion and Writing

To the student and the instructor: The Discussion and Writing exercises are meant to be answered with one or more sentences. They can be discussed and answered collaboratively by the entire class or by small groups.

41. Explain in your own words what the associative law of addition means. [1.2a]

42. Is subtraction commutative? That is, is there a commutative law of subtraction? Why or why not? [1.3a]

43. Describe a situation that corresponds to each multiplication: $4 \cdot \$150$; $\$4 \cdot 150$. [1.4a]

44. Suppose a student asserts that "$0 \div 0 = 0$ because nothing divided by nothing is nothing." Devise an explanation to persuade the student that the assertion is false. [1.5a]

STUDYING FOR SUCCESS *Using the Textbook*

- ☐ Study the step-by-step solutions to the examples, paying attention to the extra explanations shown in red.
- ☐ Stop and do the margin exercises when directed to do so. Answers are at the bottom of the margin, so you can check your work right away.
- ☐ The objective symbols **a**, **b**, **c**, and so on, allow you to refer to the appropriate place in the text whenever you need to review a topic.

Rounding and Estimating; Order 1.6

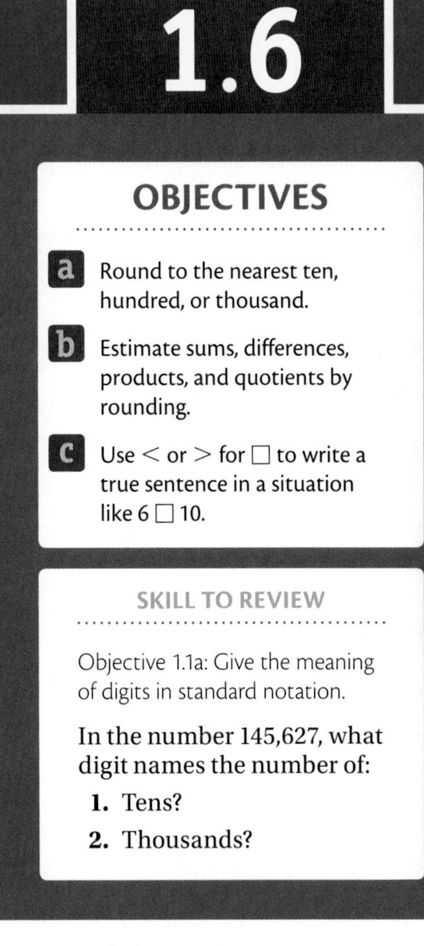

OBJECTIVES

a Round to the nearest ten, hundred, or thousand.

b Estimate sums, differences, products, and quotients by rounding.

c Use < or > for □ to write a true sentence in a situation like 6 □ 10.

SKILL TO REVIEW

Objective 1.1a: Give the meaning of digits in standard notation.

In the number 145,627, what digit names the number of:

1. Tens?
2. Thousands?

a ROUNDING

We round numbers in various situations when we do not need an exact answer. For example, we might round to see if we are being charged the correct amount in a store. We might also round to check if an answer to a problem is reasonable or to check a calculation done by hand or on a calculator.

To understand how to round, we first look at some examples using the number line. The number line displays numbers at equally spaced intervals.

EXAMPLE 1 Round 47 to the nearest ten.

47 is between 40 and 50. Since 47 is closer to 50, we round up to 50.

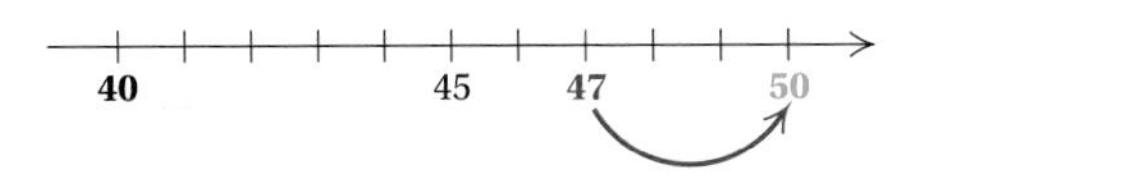

EXAMPLE 2 Round 42 to the nearest ten.

42 is between 40 and 50. Since 42 is closer to 40, we round down to 40.

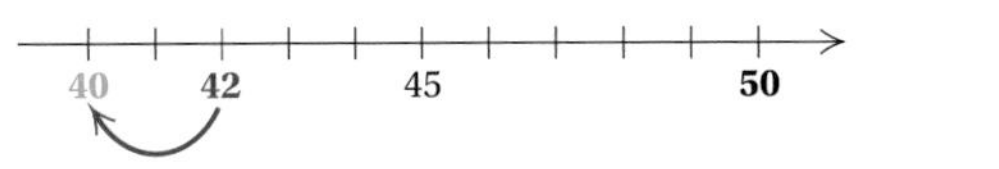

EXAMPLE 3 Round 45 to the nearest ten.

45 is halfway between 40 and 50. We could round 45 down to 40 or up to 50. We agree to round up to 50.

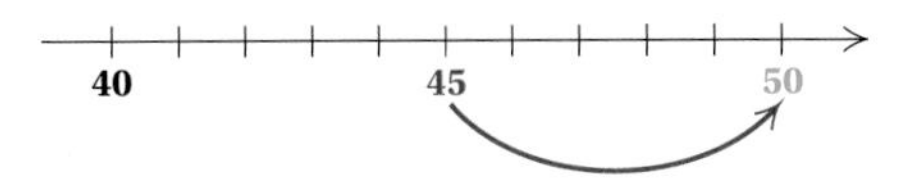

When a number is halfway between rounding numbers, round up.

Do Margin Exercises 1–7.

Based on these examples, we can state a rule for rounding whole numbers.

Round to the nearest ten.

1. 37

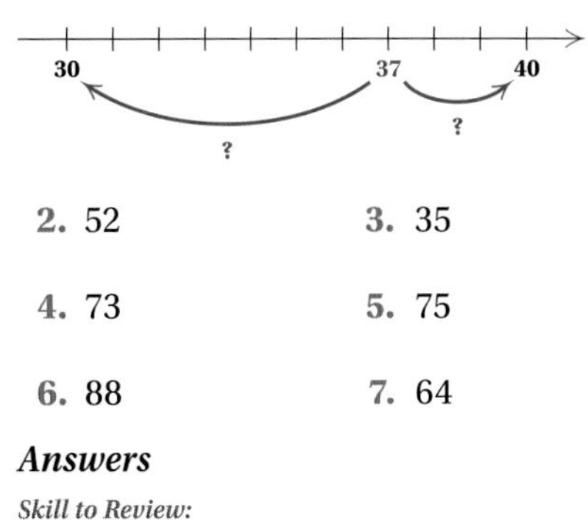

2. 52
3. 35
4. 73
5. 75
6. 88
7. 64

Answers

Skill to Review:
1. 2 2. 5

Margin Exercises:
1. 40 2. 50 3. 40 4. 70 5. 80 6. 90 7. 60

ROUNDING WHOLE NUMBERS

To round to a certain place:

a) Locate the digit in that place.

b) Consider the next digit to the right.

c) If the digit to the right is 5 or higher, round up. If the digit to the right is 4 or lower, round down.

d) Change all digits to the right of the rounding location to zeros.

EXAMPLE 4 Round 6485 to the nearest ten.

a) Locate the digit in the tens place, 8.

6 4 8 5
    ↑

b) Consider the next digit to the right, 5.

6 4 8 5

c) Since that digit, 5, is 5 or higher, round 8 tens up to 9 tens.

d) Change all digits to the right of the tens digit to zeros.

6 4 9 0 ← This is the answer.

Do Exercises 8–11.

Round to the nearest ten.

8. 137 **9.** 473

10. 235 **11.** 285

EXAMPLE 5 Round 6485 to the nearest hundred.

a) Locate the digit in the hundreds place, 4.

6 4 8 5
  ↑

b) Consider the next digit to the right, 8.

6 4 8 5

c) Since that digit, 8, is 5 or higher, round 4 hundreds up to 5 hundreds.

d) Change all digits to the right of hundreds to zeros.

6 5 0 0 ← This is the answer.

Do Exercises 12–15.

Round to the nearest hundred.

12. 641 **13.** 759

14. 1871 **15.** 9325

Caution!

It is incorrect in Example 6 to round from the ones digit over, as follows:

6485 → 6490 → 6500 → 7000.

Note that 6485 is closer to 6000 than it is to 7000.

EXAMPLE 6 Round 6485 to the nearest thousand.

a) Locate the digit in the thousands place, 6.

6 4 8 5
↑

b) Consider the next digit to the right, 4.

6 4 8 5

c) Since that digit, 4, is 4 or lower, round down, meaning that 6 thousands stays as 6 thousands.

d) Change all digits to the right of thousands to zeros.

6 0 0 0 ← This is the answer.

Do Exercises 16–19.

Round to the nearest thousand.

16. 7896 **17.** 8459

18. 19,343 **19.** 68,500

Answers

8. 140 **9.** 470 **10.** 240 **11.** 290
12. 600 **13.** 800 **14.** 1900 **15.** 9300
16. 8000 **17.** 8000 **18.** 19,000 **19.** 69,000

Sometimes rounding involves changing more than one digit in a number.

EXAMPLE 7 Round 78,595 to the nearest ten.

a) Locate the digit in the tens place, 9.

7 8 , 5 9 5
↑

b) Consider the next digit to the right, 5.

7 8 , 5 9 5

c) Since that digit, 5, is 5 or higher, round 9 tens to 10 tens. We think of 10 tens as 1 hundred + 0 tens and increase the hundreds digit by 1, to get 6 hundreds + 0 tens. We then write 6 in the hundreds place and 0 in the tens place.

d) Change the digit to the right of the tens digit to zero.

7 8 , 6 0 0 ← This is the answer.

Note that if we round this number to the nearest hundred, we get the same answer.

Do Exercises 20 and 21. ▶

b ESTIMATING

Estimating can be done in many ways. In general, an estimate rounded to the nearest ten is more accurate than one rounded to the nearest hundred, and an estimate rounded to the nearest hundred is more accurate than one rounded to the nearest thousand, and so on.

EXAMPLE 8 Estimate this sum by first rounding to the nearest ten:

78 + 49 + 31 + 85.

We round each number to the nearest ten. Then we add.

$$\begin{array}{r} 78 \\ 49 \\ 31 \\ +\ 85 \\ \hline \end{array} \qquad \begin{array}{r} 80 \\ 50 \\ 30 \\ +\ 90 \\ \hline 250 \end{array} \leftarrow \text{Estimated answer}$$

Do Exercises 22 and 23. ▶

EXAMPLE 9 Estimate the difference by first rounding to the nearest thousand: 9324 − 2849.

We have

$$\begin{array}{r} 9324 \\ -\ 2849 \\ \hline \end{array} \qquad \begin{array}{r} 9000 \\ -\ 3000 \\ \hline 6000 \end{array} \leftarrow \text{Estimated answer}$$

Do Exercises 24 and 25. ▶

20. Round 48,968 to the nearest ten, hundred, and thousand.

21. Round 269,582 to the nearest ten, hundred, and thousand.

22. Estimate the sum by first rounding to the nearest ten. Show your work.

$$\begin{array}{r} 74 \\ 23 \\ 35 \\ +\ 66 \\ \hline \end{array}$$

23. Estimate the sum by first rounding to the nearest hundred. Show your work.

$$\begin{array}{r} 650 \\ 685 \\ 238 \\ +\ 168 \\ \hline \end{array}$$

24. Estimate the difference by first rounding to the nearest hundred. Show your work.

$$\begin{array}{r} 9285 \\ -\ 6739 \\ \hline \end{array}$$

25. Estimate the difference by first rounding to the nearest thousand. Show your work.

$$\begin{array}{r} 23{,}278 \\ -\ 11{,}698 \\ \hline \end{array}$$

Answers

20. 48,970; 49,000; 49,000
21. 269,580; 269,600; 270,000
22. 70 + 20 + 40 + 70 = 200
23. 700 + 700 + 200 + 200 = 1800
24. 9300 − 6700 = 2600
25. 23,000 − 12,000 = 11,000

In the sentence $7 - 5 = 2$, the equals sign indicates that $7 - 5$ is the *same* as 2. When we round to make an estimate, the outcome is rarely the same as the exact result. Thus we cannot use an equals sign when we round. Instead, we use the symbol $\approx$. This symbol means "**is approximately equal to**." In Example 9, for instance, we could have written

$$9324 - 2849 \approx 6000.$$

EXAMPLE 10 Estimate the following product by first rounding to the nearest ten and then to the nearest hundred: 683×457.

Nearest ten:

$$\begin{array}{r} 680 \\ \times\ 460 \\ \hline 40800 \\ 272000 \\ \hline 312{,}800 \end{array} \qquad \begin{array}{l} 683 \approx 680 \\ 457 \approx 460 \end{array}$$

Nearest hundred:

$$\begin{array}{r} 700 \\ \times\ 500 \\ \hline 350{,}000 \end{array} \qquad \begin{array}{l} 683 \approx 700 \\ 457 \approx 500 \end{array}$$

Exact:

$$\begin{array}{r} 683 \\ \times\ 457 \\ \hline 4781 \\ 34150 \\ 273200 \\ \hline 312{,}131 \end{array}$$

We see that rounding to the nearest ten gives a better estimate than rounding to the nearest hundred.

◀ Do Exercise 26.

26. Estimate the product by first rounding to the nearest ten and then to the nearest hundred. Show your work.

GS

$$\begin{array}{r} 837 \\ \times\ 245 \\ \hline \end{array}$$

Nearest ten:

$$\begin{array}{r} 840 \\ \times\ \square \\ \hline 42000 \\ \square\square\square000 \\ \hline \square\square\square{,}000 \end{array}$$

Nearest hundred:

$$\begin{array}{r} 800 \\ \times\ \square \\ \hline \square\square0{,}000 \end{array}$$

EXAMPLE 11 Estimate the following quotient by first rounding to the nearest ten and then to the nearest hundred: $12{,}238 \div 175$.

Nearest ten:

$$\begin{array}{r} 68 \\ 180\overline{)12{,}240} \\ 1080 \\ \hline 1440 \\ 1440 \\ \hline 0 \end{array}$$

Nearest hundred:

$$\begin{array}{r} 61 \\ 200\overline{)12{,}200} \\ 1200 \\ \hline 200 \\ 200 \\ \hline 0 \end{array}$$

The exact answer is 69 R 163. Again we see that rounding to the nearest ten gives a better estimate than rounding to the nearest hundred.

◀ Do Exercise 27.

27. Estimate the quotient by first rounding to the nearest hundred. Show your work.

$$64{,}534 \div 349$$

Answers

26. $840 \times 250 = 210{,}000$;
$800 \times 200 = 160{,}000$
27. $64{,}500 \div 300 = 215$

Guided Solution:
26. Nearest ten: 250, 1, 6, 8, 2, 1, 0;
Nearest hundred: 200, 1, 6

The next two examples show how estimating can be used in making financial decisions.

EXAMPLE 12 *Tuition.* Ellen plans to take 12 credit hours of classes next semester. If she takes the courses on campus, the cost per credit hour is \$248. Estimate, by rounding to the nearest ten, the total cost of tuition.

We have

$$\begin{array}{r} 250 \\ \times \quad 10 \\ \hline 2500. \end{array}$$

The tuition will cost about \$2500.

Do Exercise 28.

28. *Tuition.* If Ellen takes courses online, the cost per credit hour is \$198. Estimate, by rounding to the nearest ten, the total cost of 12 credit hours of classes.

EXAMPLE 13 *Purchasing a New Car.* Jon and Joanna are shopping for a new car. They are considering buying a Ford Focus S sedan. The base price of the car is \$16,200. A 6-speed automatic transmission package can be added to this, as well as several other options, as shown in the chart below. Jon and Joanna want to stay within a budget of \$18,000.

Estimate, by rounding to the nearest hundred, the cost of the Focus with the automatic transmission package and all other options and determine whether this will fit within their budget.

FORD FOCUS S SEDAN	PRICE
Basic price	\$16,200
6-speed automatic transmission	\$1,095
Remote start system (requires purchase of automatic transmission)	\$445
SYNC basic (includes Bluetooth and USB input jacks)	\$295
Cargo management	\$115
Graphics package	\$375
Exterior protection package	\$245

SOURCE: motortrend.com

First, we list the base price of the car and then the cost of each of the options. We then round each number to the nearest hundred and add.

$$\begin{array}{rr} 16,200 & 16,200 \\ 1,095 & 1,100 \\ 445 & 400 \\ 295 & 300 \\ 115 & 100 \\ 375 & 400 \\ +\quad 245 & +\quad 200 \\ \hline & 18,700 \end{array} \leftarrow \text{Estimated cost}$$

The estimated cost is \$18,700. This exceeds Jon and Joanna's budget of \$18,000, so they will have to forgo at least one option.

Do Exercises 29 and 30.

Refer to the chart in Example 13 to do Margin Exercises 29 and 30.

29. By eliminating at least one option, determine how Jon and Joanna can buy a Focus S sedan and stay within their budget. Keep in mind that purchasing the remote-start system requires purchasing the automatic transmission.

30. Elizabeth and C.J. are also considering buying a Focus S sedan. Estimate, by rounding to the nearest hundred, the cost of this car with automatic transmission, SYNC basic, and exterior protection package.

Answers

28. \$2000 **29.** Eliminate either the automatic transmission and the remote-start system or the remote-start system and the graphics package. There are other correct answers as well. **30.** \$17,800

C ORDER

A sentence like $8 + 5 = 13$ is called an **equation**. It is a *true* equation. The equation $4 + 8 = 11$ is a *false* equation.

A sentence like $7 < 11$ is called an **inequality**. The sentence $7 < 11$ is a *true* inequality. The sentence $23 > 69$ is a *false* inequality.

Some common **inequality symbols** follow.

INEQUALITY SYMBOLS

$<$ means "is less than"

$>$ means "is greater than"

$\neq$ means "is not equal to"

We know that 2 is not the same as 5. We express this by the sentence $2 \neq 5$. We also know that 2 is less than 5. We symbolize this by the expression $2 < 5$. We can see this order on the number line: 2 is to the left of 5. The number 0 is the smallest whole number.

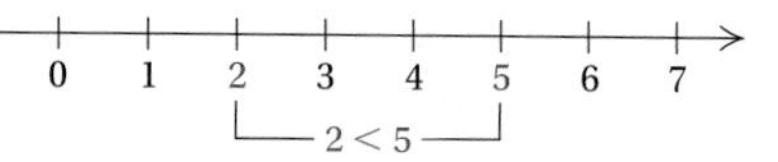

ORDER OF WHOLE NUMBERS

For any whole numbers a and b:

1. $a < b$ (read "a is less than b") is true when a is to the left of b on the number line.
2. $a > b$ (read "a is greater than b") is true when a is to the right of b on the number line.

EXAMPLE 14 Use $<$ or $>$ for $\square$ to write a true sentence: $7 \square 11$.

6 7 8 9 10 11 12 13

Since 7 is to the left of 11 on the number line, $7 < 11$.

EXAMPLE 15 Use $<$ or $>$ for $\square$ to write a true sentence: $92 \square 87$.

86 87 88 89 90 91 92 93

Since 92 is to the right of 87 on the number line, $92 > 87$.

Do Exercises 31–36.

Use $<$ or $>$ for $\square$ to write a true sentence. Draw the number line if necessary.

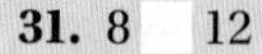

31. 8 ___ 12

Since 8 is to the ___ of 12 on the number line, 8 ___ 12.

GS

32. $12 \square 8$

33. $76 \square 64$

34. $64 \square 76$

35. $217 \square 345$

36. $345 \square 217$

Answers

31. $<$ **32.** $>$ **33.** $>$ **34.** $<$
35. $<$ **36.** $>$

Guided Solution:
31. left, $<$

1.6 Exercise Set

For Extra Help MyMathLab® MathXL® PRACTICE WATCH READ REVIEW

Reading Check

Determine whether each statement is true or false.

________ **RC1.** When rounding to the nearest hundred, if the digit in the tens place is 5 or higher, we round up.

________ **RC2.** When rounding 3500 to the nearest thousand, we should round down.

________ **RC3.** An estimate made by rounding to the nearest thousand is more accurate than an estimate made by rounding to the nearest ten.

________ **RC4.** Since 78 rounded to the nearest ten is 80, we can write $78 \approx 80$.

a Round to the nearest ten.

1. 48 **2.** 532 **3.** 463 **4.** 8945

5. 731 **6.** 54 **7.** 895 **8.** 798

Round to the nearest hundred.

9. 146 **10.** 874 **11.** 957 **12.** 650

13. 9079 **14.** 4645 **15.** 32,839 **16.** 198,402

Round to the nearest thousand.

17. 5876 **18.** 4500 **19.** 7500 **20.** 2001

21. 45,340 **22.** 735,562 **23.** 373,405 **24.** 6,713,255

b Estimate each sum or difference by first rounding to the nearest ten. Show your work.

25. $\begin{array}{r} 78 \\ +\ 92 \\ \hline \end{array}$

26. $\begin{array}{r} 62 \\ 97 \\ 46 \\ +\ 81 \\ \hline \end{array}$

27. $\begin{array}{r} 8074 \\ -\ 2347 \\ \hline \end{array}$

28. $\begin{array}{r} 673 \\ -\ 28 \\ \hline \end{array}$

Estimate each sum by first rounding to the nearest ten. State if the given sum seems to be incorrect when compared to the estimate.

29. $\begin{array}{r} 45 \\ 77 \\ 25 \\ +\ 56 \\ \hline 343 \end{array}$

30. $\begin{array}{r} 41 \\ 21 \\ 55 \\ +\ 60 \\ \hline 177 \end{array}$

31. $\begin{array}{r} 622 \\ 78 \\ 81 \\ +\ 111 \\ \hline 932 \end{array}$

32. $\begin{array}{r} 836 \\ 374 \\ 794 \\ +\ 938 \\ \hline 3947 \end{array}$

Estimate each sum or difference by first rounding to the nearest hundred. Show your work.

33. $\begin{array}{r} 7348 \\ +9247 \\ \hline \end{array}$

34. $\begin{array}{r} 568 \\ 472 \\ 938 \\ +402 \\ \hline \end{array}$

35. $\begin{array}{r} 6852 \\ -1748 \\ \hline \end{array}$

36. $\begin{array}{r} 9438 \\ -2787 \\ \hline \end{array}$

Estimate each sum by first rounding to the nearest hundred. State if the given sum seems to be incorrect when compared to the estimate.

37. $\begin{array}{r} 216 \\ 84 \\ 745 \\ +595 \\ \hline 1640 \end{array}$

38. $\begin{array}{r} 481 \\ 702 \\ 623 \\ +1043 \\ \hline 1849 \end{array}$

39. $\begin{array}{r} 750 \\ 428 \\ 63 \\ +205 \\ \hline 1446 \end{array}$

40. $\begin{array}{r} 326 \\ 275 \\ 758 \\ +943 \\ \hline 2302 \end{array}$

Estimate each sum or difference by first rounding to the nearest thousand. Show your work.

41. $\begin{array}{r} 9643 \\ 4821 \\ 8943 \\ +7004 \\ \hline \end{array}$

42. $\begin{array}{r} 7648 \\ 9348 \\ 7842 \\ +2222 \\ \hline \end{array}$

43. $\begin{array}{r} 92{,}149 \\ -22{,}555 \\ \hline \end{array}$

44. $\begin{array}{r} 84{,}890 \\ -11{,}110 \\ \hline \end{array}$

Estimate each product by first rounding to the nearest ten. Show your work.

45. $\begin{array}{r} 45 \\ \times 67 \\ \hline \end{array}$

46. $\begin{array}{r} 51 \\ \times 78 \\ \hline \end{array}$

47. $\begin{array}{r} 34 \\ \times 29 \\ \hline \end{array}$

48. $\begin{array}{r} 63 \\ \times 54 \\ \hline \end{array}$

Estimate each product by first rounding to the nearest hundred. Show your work.

49. $\begin{array}{r} 876 \\ \times 345 \\ \hline \end{array}$

50. $\begin{array}{r} 355 \\ \times 299 \\ \hline \end{array}$

51. $\begin{array}{r} 432 \\ \times 199 \\ \hline \end{array}$

52. $\begin{array}{r} 789 \\ \times 434 \\ \hline \end{array}$

Estimate each quotient by first rounding to the nearest ten. Show your work.

53. $347 \div 73$

54. $454 \div 87$

55. $8452 \div 46$

56. $1263 \div 29$

Estimate each quotient by first rounding to the nearest hundred. Show your work.

57. 1165 ÷ 236

58. 3641 ÷ 571

59. 8358 ÷ 295

60. 32,854 ÷ 748

Planning a Vacation. Most cruise ships offer a choice of rooms at varying prices, as well as additional packages and shore excursions at each port. The table below lists room prices for a seven-day Mediterranean cruise, as well as prices for several additional packages and excursions.

ROOMS	PRICE
Suite	$856
Balcony	$686
Ocean View	$586
Interior	$536
ADDITIONAL PACKAGES	**PRICE**
Spa	$115
Specialty Dining	$129
Beverage	$79
EXCURSIONS	**PRICE**
Athens, Greece; Private Tour	$289
Venice, Italy	$95
Istanbul, Turkey	$130
Pisa, Italy; Biking Tour	$199

61. Estimate the total price of a cruise with an ocean view room, a spa package, and an Istanbul excursion. Round each price to the nearest hundred dollars.

62. Estimate the total price of a cruise with a balcony room, no additional packages, a Venice excursion, and a biking tour in Pisa. Round each price to the nearest hundred dollars.

63. Antonio has a budget of $1000 for a Mediterranean cruise. He would like a balcony room, a specialty dining package, a private tour of Athens, and a tour of Venice. Estimate the total price of this cruise by rounding each price to the nearest hundred dollars. Can he afford his choices?

64. Alyssa has a budget of $1400 for a Mediterranean cruise. She is planning to book an interior room and would like to go on all the excursions listed. Estimate the total price of this cruise by rounding each price to the nearest hundred dollars. Does her budget cover her choices?

65. If you were going on a Mediterranean cruise and had a budget of $1500, what options would you choose? Decide on the options you would like and estimate the total price by rounding each price to the nearest hundred dollars. Could you afford all your chosen options?

66. If you were going on a Mediterranean cruise and had a budget of $1200, what options would you choose? Decide on the options you would like and estimate the total price by rounding each price to the nearest hundred dollars. Could you afford all your chosen options?

67. *Mortgage Payments.* To pay for their new home, Tim and Meribeth will make 360 payments of \$751.55 each. In addition, they must add an escrow amount of \$112.67 to each payment for insurance and taxes.

a) Estimate the total amount they will pay by rounding the number of payments, the amount of each payment, and the escrow amount to the nearest ten.

b) Estimate the total amount they will pay by rounding the number of payments, the amount of each payment, and the escrow amount to the nearest hundred.

68. *Conference Expenses.* The cost to attend a three-day teachers' conference is \$245, and a hotel room costs \$169 a night. One year, 489 teachers attended the conference, and 315 rooms were rented for two nights each.

a) Estimate the total amount spent by the teachers by rounding the cost of attending the conference, the cost of a hotel room, the number of teachers, and the number of rooms to the nearest ten.

b) Estimate the total amount spent by the teachers by rounding the cost of attending the conference, the cost of a hotel room, the number of teachers, and the number of rooms to the nearest hundred.

69. *Banquet Attendance.* Tickets to the annual awards banquet for the Riviera Swim Club cost \$28 each. Ticket sales for the banquet totaled \$2716. Estimate the number of people who attended the banquet by rounding the cost of a ticket to the nearest ten and the total sales to the nearest hundred.

70. *School Fundraiser.* For a school fundraiser, Charlotte sells trash bags at a price of \$11 per roll. If her sales totaled \$2211, estimate the number of rolls she sold by rounding the price per roll to the nearest ten and the total sales to the nearest hundred.

C Use $<$ or $>$ for □ to write a true sentence. Draw the number line if necessary.

71. 0 □ 17

72. 32 □ 0

73. 34 □ 12

74. 28 □ 18

75. 1000 □ 1001

76. 77 □ 117

77. 133 □ 132

78. 999 □ 997

79. 460 □ 17

80. 345 □ 456

81. 37 □ 11

82. 12 □ 32

New Book Titles. The number of new book titles published in the United States in each of three recent years is shown in the table below. Use this table to do Exercises 83 and 84.

YEAR	NEW BOOK TITLES
2009	1,335,475
2010	4,134,519
2011	1,532,623*

SOURCE: R. R. Bowker
*Projected

83. Write an inequality to compare the numbers of new titles published in 2009 and in 2010.

84. Write an inequality to compare the numbers of new titles published in 2010 and in 2011.

85. *Public Schools.* The number of public schools in the United States increased from 97,382 in 2006 to 98,817 in 2010. Write an inequality to compare these numbers of schools.

Public Schools

Number of schools: 99,000; 98,500; 98,000; 97,500; 97,000
Year: 2006, 2007, 2008, 2009, 2010
97,382; 98,817

SOURCE: National Center for Education Statistics

86. *Life Expectancy.* The life expectancy of a female in the United States in 2020 is predicted to be about 82 years and that of a male about 77 years. Write an inequality to compare these life expectancies.

Life Expectancy in the United States

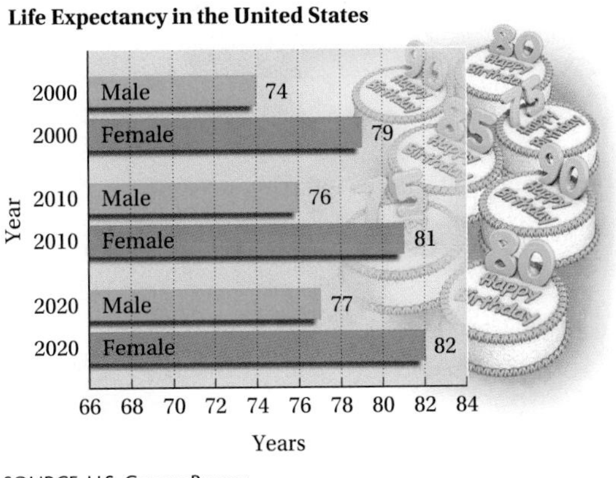

SOURCE: U.S. Census Bureau

Skill Maintenance

Add. [1.2a]

87. $\begin{array}{r} 67{,}789 \\ +\ 18{,}965 \\ \hline \end{array}$

88. $\begin{array}{r} 9002 \\ +\ 4587 \\ \hline \end{array}$

Subtract. [1.3a]

89. $\begin{array}{r} 67{,}789 \\ -\ 18{,}965 \\ \hline \end{array}$

90. $\begin{array}{r} 9002 \\ -\ 4587 \\ \hline \end{array}$

Multiply. [1.4a]

91. $\begin{array}{r} 46 \\ \times\ 37 \\ \hline \end{array}$

92. $\begin{array}{r} 306 \\ \times\ 58 \\ \hline \end{array}$

Divide. [1.5a]

93. $328 \div 6$

94. $4784 \div 23$

Synthesis

95.–98. ▦ Use a calculator to find the sums and the differences in each of Exercises 41–44. Then compare your answers with those found using estimation. Even when using a calculator it is possible to make an error if you press the wrong buttons, so it is a good idea to check by estimating.

Solving Equations

OBJECTIVES

a Solve simple equations by trial.

b Solve equations like $t + 28 = 54$, $28 \cdot x = 168$, and $98 \cdot 2 = y$.

SKILL TO REVIEW

Objective 1.5a: Divide whole numbers.

Divide.

1. $1008 \div 36$
2. $675 \div 15$

a SOLUTIONS BY TRIAL

Let's find a number that we can put in the blank to make this sentence true:

$$9 = 3 + \square.$$

We are asking "9 is 3 plus what number?" The answer is 6.

$$9 = 3 + 6$$

◀ **Do Margin Exercises 1 and 2.**

A sentence with $=$ is called an **equation**. A **solution** of an equation is a number that makes the sentence true. Thus, 6 is a solution of

$9 = 3 + \square$ because $9 = 3 + 6$ is true.

However, 7 is not a solution of

$9 = 3 + \square$ because $9 = 3 + 7$ is false.

◀ **Do Exercises 3 and 4.**

We can use a letter in an equation instead of a blank:

$$9 = 3 + n.$$

We call n a **variable** because it can represent any number. If a replacement for a variable makes an equation true, the replacement is a solution of the equation.

> **SOLUTIONS OF AN EQUATION**
>
> A **solution of an equation** is a replacement for the variable that makes the equation true. When we find all the solutions, we say that we have **solved** the equation.

EXAMPLE 1 Solve $y + 12 = 27$ by trial.

We replace y with several numbers.

If we replace y with 13, we get a false equation: $13 + 12 = 27$.
If we replace y with 14, we get a false equation: $14 + 12 = 27$.
If we replace y with 15, we get a true equation: $15 + 12 = 27$.

No other replacement makes the equation true, so the solution is 15. ■

EXAMPLES Solve.

2. $7 + n = 22$
(7 plus what number is 22?)
The solution is 15.

3. $63 = 3 \cdot x$
(63 is 3 times what number?)
The solution is 21.

◀ **Do Exercises 5–8.**

Find a number that makes each sentence true.

1. $8 = 1 + \square$
2. $\square + 2 = 7$
3. Determine whether 7 is a solution of $\square + 5 = 9$.
4. Determine whether 4 is a solution of $\square + 5 = 9$.

Solve by trial.

5. $n + 3 = 8$
6. $x - 2 = 8$
7. $45 \div 9 = y$
8. $10 + t = 32$

Answers

Skill to Review:
1. 28 2. 45

Margin Exercises:
1. 7 2. 5 3. No 4. Yes 5. 5 6. 10 7. 5 8. 22

b SOLVING EQUATIONS

We now begin to develop more efficient ways to solve certain equations. When an equation has a variable alone on one side and a calculation on the other side, we can find the solution by carrying out the calculation.

EXAMPLE 4 Solve: $x = 245 \times 34$.

To solve the equation, we carry out the calculation.

$$\begin{array}{r} 245 \\ \times \ 34 \\ \hline 980 \\ 7350 \\ \hline 8330 \end{array} \qquad \begin{aligned} x &= 245 \times 34 \\ x &= 8330 \end{aligned}$$

The solution is 8330.

Do Exercises 9–12. ▶

Solve.

9. $346 \times 65 = y$

10. $x = 2347 + 6675$

11. $4560 \div 8 = t$

12. $x = 6007 - 2346$

Look at the equation

$$x + 12 = 27.$$

We can get x alone by subtracting 12 *on both sides*. Thus,

$$\begin{aligned} x + 12 - 12 &= 27 - 12 && \text{Subtracting 12 on both sides} \\ x + 0 &= 15 && \text{Carrying out the subtraction} \\ x &= 15. \end{aligned}$$

SOLVING $x + a = b$

To solve $x + a = b$, subtract a on both sides.

If we can get an equation in a form with the variable alone on one side, we can "see" the solution.

EXAMPLE 5 Solve: $t + 28 = 54$.

We have

$$\begin{aligned} t + 28 &= 54 \\ t + 28 - 28 &= 54 - 28 && \text{Subtracting 28 on both sides} \\ t + 0 &= 26 \\ t &= 26. \end{aligned}$$

To check the answer, we substitute 26 for t in the original equation.

$$\text{Check:} \quad \begin{array}{c|l} t + 28 & = 54 \\ \hline 26 + 28 & ?\ 54 \\ 54 & \end{array} \qquad \text{TRUE}$$

Since 54 = 54 is true, 26 checks.

The solution is 26.

Do Exercises 13 and 14. ▶

Solve. Be sure to check.

GS 13. $x + 9 = 17$

$x + 9 - \square = 17 - 9$

$x = \square$

Check: $\begin{array}{c|l} x + 9 & = 17 \\ \hline \square + 9 & ?\ 17 \\ \square & \end{array}$

Since 17 = 17 is $\square$, the answer checks. The solution is $\square$.

14. $77 = m + 32$

Answers

9. 22,490 **10.** 9022 **11.** 570 **12.** 3661 **13.** 8 **14.** 45

Guided Solution:

13. 9, 8, 8, 17; true; 8

EXAMPLE 6 Solve: $182 = 65 + n$.

We have

$$182 = 65 + n$$
$$182 - 65 = 65 + n - 65 \quad \text{Subtracting 65 on both sides}$$
$$117 = 0 + n \quad \text{65 plus } n \text{ minus 65 is } 0 + n.$$
$$117 = n.$$

Check: $182 = 65 + n$

$182 \ ? \ 65 + 117$

$\quad\quad 182$ TRUE

The solution is 117.

15. Solve: $155 = t + 78$. Be sure to check.

◀ **Do Exercise 15.**

EXAMPLE 7 Solve: $7381 + x = 8067$.

We have

$$7381 + x = 8067$$
$$7381 + x - 7381 = 8067 - 7381 \quad \text{Subtracting 7381 on both sides}$$
$$x = 686.$$

Check: $7381 + x = 8067$

$7381 + 686 \ ? \ 8067$

8067 TRUE

The solution is 686.

Solve. Be sure to check.

16. $4566 + x = 7877$

17. $8172 = h + 2058$

◀ **Do Exercises 16 and 17.**

We now learn to solve equations like $8 \cdot n = 96$. Look at

$8 \cdot n = 96$.

We can get n alone by dividing by 8 *on both sides*. Thus,

$$\frac{8 \cdot n}{8} = \frac{96}{8} \quad \text{Dividing by 8 on both sides}$$
$$n = 12. \quad \text{8 times } n \text{ divided by 8 is } n.$$

To check the answer, we substitute 12 for n in the original equation.

Check: $8 \cdot n = 96$

$8 \cdot 12 \ ? \ 96$

96 TRUE

Since $96 = 96$ is a true equation, 12 is the solution of the equation.

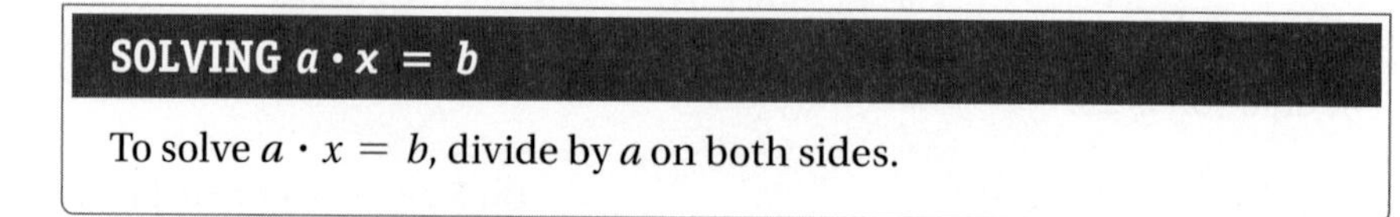

SOLVING $a \cdot x = b$

To solve $a \cdot x = b$, divide by a on both sides.

Answers

15. 77 16. 3311 17. 6114

EXAMPLE 8 Solve: $10 \cdot x = 240$.

We have

$$10 \cdot x = 240$$

$$\frac{10 \cdot x}{10} = \frac{240}{10} \quad \text{Dividing by 10 on both sides}$$

$$x = 24.$$

Check: $10 \cdot x = 240$

$10 \cdot 24 \ ? \ 240$

240 | TRUE

The solution is 24.

EXAMPLE 9 Solve: $5202 = 9 \cdot t$.

We have

$$5202 = 9 \cdot t$$

$$\frac{5202}{9} = \frac{9 \cdot t}{9} \quad \text{Dividing by 9 on both sides}$$

$$578 = t.$$

Check: $5202 = 9 \cdot t$

$5202 \ ? \ 9 \cdot 578$

| 5202 TRUE

The solution is 578.

Do Exercises 18–20. ▶

EXAMPLE 10 Solve: $14 \cdot y = 1092$.

We have

$$14 \cdot y = 1092$$

$$\frac{14 \cdot y}{14} = \frac{1092}{14} \quad \text{Dividing by 14 on both sides}$$

$$y = 78.$$

The check is left to the student. The solution is 78.

EXAMPLE 11 Solve: $n \cdot 56 = 4648$.

We have

$$n \cdot 56 = 4648$$

$$\frac{n \cdot 56}{56} = \frac{4648}{56} \quad \text{Dividing by 56 on both sides}$$

$$n = 83.$$

The check is left to the student. The solution is 83.

Do Exercises 21 and 22. ▶

Solve. Be sure to check.

18. $8 \cdot x = 64$

GS **19.** $144 = 9 \cdot n$

$$\frac{144}{9} = \frac{9 \cdot n}{\square}$$

$$\square = n$$

Check: $144 = 9 \cdot n$

$144 \ ? \ 9 \cdot \square$

| $\square$

Since $144 = 144$ is ____,
the answer checks.
The solution is ____.

20. $5152 = 8 \cdot t$

Solve. Be sure to check.

21. $18 \cdot y = 1728$

22. $n \cdot 48 = 4512$

Answers

18. 8 **19.** 16 **20.** 644 **21.** 96 **22.** 94

Guided Solution:

19. 9, 16, 16, 144; true; 16

1.7 Exercise Set

For Extra Help
MyMathLab® MathXL® PRACTICE WATCH READ REVIEW

Reading Check

Match each word with its definition from the list on the right.

RC1. Equation ________

RC2. Solution ________

RC3. Solved ________

RC4. Variable ________

a) A replacement for the variable that makes an equation true

b) A letter that can represent any number

c) A sentence containing $=$

d) An equation for which we have found all solutions

a Solve by trial.

1. $x + 0 = 14$

2. $x - 7 = 18$

3. $y \cdot 17 = 0$

4. $56 \div m = 7$

b Solve. Be sure to check.

5. $x = 12{,}345 + 78{,}555$

6. $t = 5678 + 9034$

7. $908 - 458 = p$

8. $9007 - 5667 = m$

9. $16 \cdot 22 = y$

10. $34 \cdot 15 = z$

11. $t = 125 \div 5$

12. $w = 256 \div 16$

13. $13 + x = 42$

14. $15 + t = 22$

15. $12 = 12 + m$

16. $16 = t + 16$

17. $10 + x = 89$

18. $20 + x = 57$

19. $61 = 16 + y$

20. $53 = 17 + w$

21. $3 \cdot x = 24$

22. $6 \cdot x = 42$

23. $112 = n \cdot 8$

24. $162 = 9 \cdot m$

25. $3 \cdot m = 96$

26. $4 \cdot y = 96$

27. $715 = 5 \cdot z$

28. $741 = 3 \cdot t$

29. $8322 + 9281 = x$

30. $9281 - 8322 = y$

31. $47 + n = 84$

32. $56 + p = 92$

33. $45 \cdot 23 = x$

34. $23 \cdot 78 = y$

35. $x + 78 = 144$

36. $z + 67 = 133$

37. $6 \cdot p = 1944$

38. $4 \cdot w = 3404$

39. $567 + x = 902$

40. $438 + x = 807$

41. $234 \cdot 78 = y$

42. $10{,}534 \div 458 = q$

43. $18 \cdot x = 1872$

44. $19 \cdot x = 6080$

45. $40 \cdot x = 1800$

46. $20 \cdot x = 1500$

47. $2344 + y = 6400$

48. $9281 = 8322 + t$

49. $m = 7006 - 4159$

50. $n = 3004 - 1745$

51. $165 = 11 \cdot n$

52. $660 = 12 \cdot n$

53. $58 \cdot m = 11{,}890$

54. $233 \cdot x = 22{,}135$

55. $491 - 34 = y$

56. $512 - 63 = z$

Skill Maintenance

Divide. [1.5a]

57. $1283 \div 9$

58. $1278 \div 9$

59. $17\overline{)5678}$

60. $17\overline{)5689}$

Use $>$ or $<$ for $\square$ to write a true sentence. [1.6c]

61. $123 \square 789$

62. $342 \square 339$

63. $688 \square 0$

64. $0 \square 11$

65. Round 6,375,602 to the nearest thousand. [1.6a]

66. Round 6,375,602 to the nearest ten. [1.6a]

Synthesis

Solve.

67. 🖩 $23{,}465 \cdot x = 8{,}142{,}355$

68. 🖩 $48{,}916 \cdot x = 14{,}332{,}388$

Applications and Problem Solving

OBJECTIVE

a Solve applied problems involving addition, subtraction, multiplication, or division of whole numbers.

SKILL TO REVIEW

Objective 1.6b: Estimate sums, differences, products, and quotients by rounding.

Estimate each sum or difference by first rounding to the nearest thousand. Show your work.

1. $\begin{array}{r} 367{,}982 \\ +\ 43{,}495 \\ \hline \end{array}$

2. $\begin{array}{r} 9287 \\ -\ 3502 \\ \hline \end{array}$

a A PROBLEM-SOLVING STRATEGY

One of the most important ways in which we use mathematics is as a tool in solving problems. To solve a problem, we use the following five-step strategy.

FIVE STEPS FOR PROBLEM SOLVING

1. **Familiarize** yourself with the problem situation.
2. **Translate** the problem to an equation using a variable.
3. **Solve** the equation.
4. **Check** to see whether your possible solution actually fits the problem situation and is thus really a solution of the problem.
5. **State** the answer clearly using a complete sentence and appropriate units.

The first of these five steps, becoming familiar with the problem, is probably the most important.

THE FAMILIARIZE STEP

- If the problem is presented in words, read and reread it carefully until you understand what you are being asked to find.
- Make a drawing, if it makes sense to do so.
- Write a list of the known facts and a list of what you wish to find out.
- Assign a letter, or *variable*, to the unknown.
- Organize the information in a chart or a table.
- Find further information. Look up a formula, consult a reference book or an expert in the field, or do research on the Internet.
- Guess or estimate the answer and check your guess or estimate.

Answers

Skill to Review:

1. $368{,}000 + 43{,}000 = 411{,}000$
2. $9000 - 4000 = 5000$

EXAMPLE 1 *Video Game Platforms.* The following table shows the total number of units of eight popular video game platforms sold worldwide, as of December 8, 2012. Three of these are made by PlayStation®. Find the total number of PlayStation units sold.

PLATFORM	GLOBAL SALES
Nintendo 3DS	682,396
Xbox 360	614,353
Nintendo Wii U	610,384
PlayStation® 3	576,565
Nintendo Wii	252,484
Nintendo DS	178,209
PlayStation® Vita	152,049
PlayStation® Portable	54,677

Source: www.vgchartz.com

1. **Familiarize.** First, we assign a letter, or variable, to the number we wish to find. We let $p =$ the total number of video game units sold by PlayStation. Since we are combining numbers, we will add.
2. **Translate.** We translate to an equation:

Number of PlayStation 3 units	plus	Number of PlayStation Vita units	plus	Number of PlayStation Portable units	is	Total number of PlayStation units
↓	↓	↓	↓	↓	↓	↓
576,565	$+$	152,049	$+$	54,677	$=$	p

3. **Solve.** We solve the equation by carrying out the addition.

$$576{,}565 + 152{,}049 + 54{,}677 = p$$
$$783{,}291 = p$$

$$\begin{array}{r} 576{,}565 \\ 152{,}049 \\ +\ \ 54{,}677 \\ \hline 783{,}291 \end{array}$$

4. **Check.** We check our result by rereading the original problem and seeing if 783,291 answers the question. Since we are looking for a total, we could repeat the addition calculation. We could also check whether the answer is reasonable. In this case, since the total is greater than any of the three separate sales numbers, the result seems reasonable. Another way to check is to estimate the expected result and compare the estimate with the calculated result. If we round each PlayStation sales number to the nearest ten thousand and add, we have

$$580{,}000 + 150{,}000 + 50{,}000 = 780{,}000.$$

Since $780{,}000 \approx 783{,}291$, our result again seems reasonable.

5. **State.** The total number of the three PlayStation units sold worldwide as of December 8, 2012, is 783,291.

Do Exercises 1–3. ▶

Refer to the table in Example 1 to do Margin Exercises 1–3.

1. Find the total number of Nintendo units sold.
2. Find the total number of units sold for the four most popular game platforms listed in the table.
3. Find the total number of units sold for all the game platforms listed in the table.

Answers

1. 1,723,473 units 2. 2,483,698 units
3. 3,121,117 units

EXAMPLE 2 *Travel Distance.* Abigail is driving from Indianapolis to Salt Lake City to attend a family reunion. The distance from Indianapolis to Salt Lake City is 1634 mi. In the first two days, she travels 1154 mi to Denver. How much farther must she travel?

1. **Familiarize.** We first make a drawing or at least visualize the situation. We let d = the remaining distance to Salt Lake City.

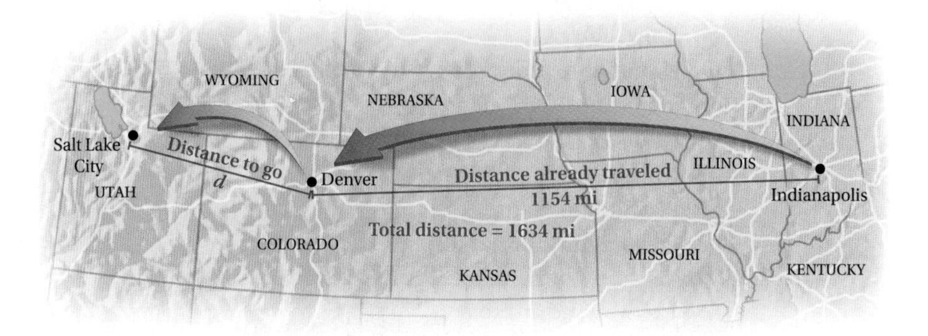

2. **Translate.** We want to determine how many more miles Abigail must travel. We translate to an equation:

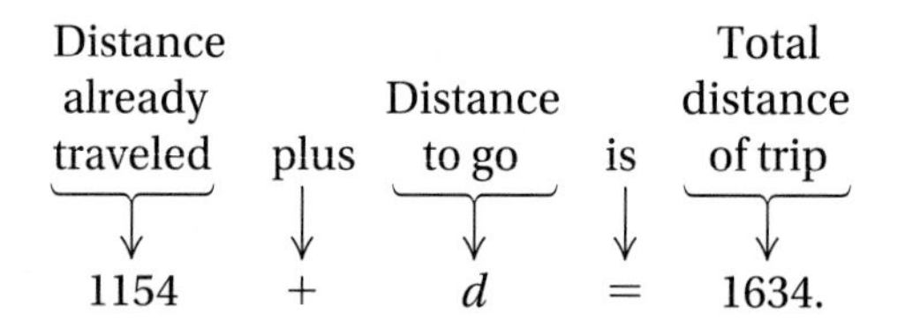

3. **Solve.** To solve the equation, we subtract 1154 on both sides.

$$1154 + d = 1634$$
$$1154 + d - 1154 = 1634 - 1154$$
$$d = 480$$

$$\begin{array}{r} \overset{5\,13}{1\,\not{6}\,\not{3}\,4} \\ -\,1\,1\,5\,4 \\ \hline 4\,8\,0 \end{array}$$

4. **Check.** We check our answer of 480 mi in the original problem. This number should be less than the total distance, 1634 mi, and it is. We can add the distance traveled, 1154, and the distance left to go, 480: $1154 + 480 = 1634$. We can also estimate:

$$1634 - 1154 \approx 1600 - 1200$$
$$= 400 \approx 480.$$

The answer, 480 mi, checks.

5. **State.** Abigail must travel 480 mi farther to Salt Lake City.

Do Exercise 4.

---

4. Reading Assignment. William has been assigned 234 pages of reading for his history class. He has read 86 pages. How many more pages does he have to read?

GS

1. **Familiarize.** Let p = the number of pages William still has to read.
2. **Translate.**

Pages already read	plus	Number of pages to read	is	Total number of pages
86	+	☐	=	☐

3. **Solve.**

$$86 + p = 234$$
$$86 + p - \square = 234 - 86$$
$$p = \square$$

4. **Check.** If William reads 148 more pages, he will have read a total of $86 + 148$ pages, or ☐ pages.
5. **State.** William has ☐ more pages to read.

Answer

4. 148 pages

Guided Solution:

4. p, 234; 86, 148; 234; 148

EXAMPLE 3 *Total Cost of Chairs.* What is the total cost of 6 Adirondack chairs if each one costs $169?

1. **Familiarize.** We make a drawing and let $C =$ the cost of 6 chairs.

2. **Translate.** We translate to an equation:

Number of chairs	times	Cost of each chair	is	Total cost
↓	↓	↓	↓	↓
6	×	169	=	C.

3. **Solve.** This sentence tells us what to do. We multiply.

$$6 \times 169 = C$$
$$1014 = C$$

$$\begin{array}{r} 1\,6\,9 \\ \times \quad 6 \\ \hline 1\,0\,1\,4 \end{array}$$

4. **Check.** We have an answer, 1014, that is greater than the cost of any one chair, which is reasonable. We can also check by estimating:

$$6 \times 169 \approx 6 \times 170 = 1020 \approx 1014.$$

The answer checks.

5. **State.** The total cost of 6 chairs is $1014.

Do Exercise 5. ▶

5. *Total Cost of Gas Grills.* What is the total cost of 14 gas grills, each with 520 sq in. of total cooking surface, if each one costs $398?

EXAMPLE 4 *Area of an Oriental Rug.* The dimensions of the oriental rug in the Fosters' front hallway are 42 in. by 66 in. What is the area of the rug?

1. **Familiarize.** We let $A =$ the area of the rug and use the formula for the area of a rectangle, $A = \text{length} \cdot \text{width} = l \cdot w$. Since we usually consider length to be larger than width, we will let $l = 66$ in. and $w = 42$ in.

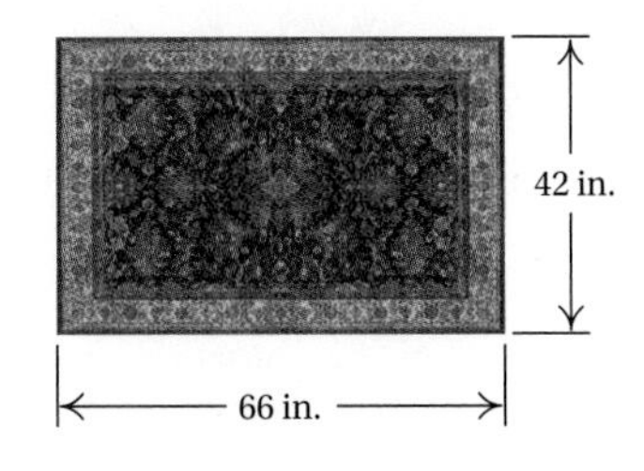

2. **Translate.** We substitute in the formula:

$$A = l \cdot w = 66 \cdot 42.$$

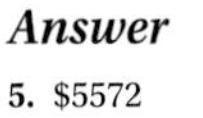

Answer

5. $5572

3. **Solve.** We carry out the multiplication.

$$A = 66 \cdot 42 \qquad \begin{array}{r} 6\,6 \\ \times\ 4\,2 \\ \hline 1\,3\,2 \\ 2\,6\,4\,0 \\ \hline 2\,7\,7\,2 \end{array}$$
$$A = 2772$$

4. **Check.** We can repeat the calculation. We can also round and estimate:

$$66 \times 42 \approx 70 \times 40 = 2800 \approx 2772.$$

The answer checks.

5. **State.** The area of the rug is 2772 sq in.

6. *Bed Sheets.* The dimensions of a flat sheet for a queen-size bed are 90 in. by 102 in. What is the area of the sheet?

◀ **Do Exercise 6.**

EXAMPLE 5 *Packages of Gum.* A candy company produces 3304 sticks of gum. How many 12-stick packages can be filled? How many sticks will be left over?

1. **Familiarize.** We make a drawing to visualize the situation and let n = the number of 12-stick packages that can be filled. The problem can be considered as repeated subtraction, taking successive sets of 12 sticks and putting them into n packages.

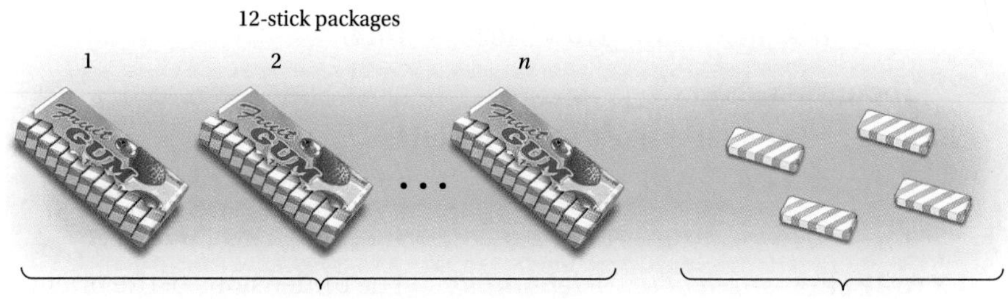

2. **Translate.** We translate to an equation:

Number of sticks	divided by	Number in each package	is	Number of packages
↓	↓	↓	↓	↓
3304	÷	12	=	n.

3. **Solve.** We solve the equation by carrying out the division.

$$3304 \div 12 = n$$
$$275 \text{ R } 4 = n$$

$$\begin{array}{r} 2\,7\,5 \\ 12\,\overline{)\,3\,3\,0\,4} \\ 2\,4 \\ \hline 9\,0 \\ 8\,4 \\ \hline 6\,4 \\ 6\,0 \\ \hline 4 \end{array}$$

Answer

6. 9180 sq in.

4. **Check.** We can check by multiplying the number of packages by 12 and adding the remainder, 4:

$$12 \cdot 275 = 3300,$$
$$3300 + 4 = 3304.$$

5. **State.** Thus, 275 twelve-stick packages of gum can be filled. There will be 4 sticks left over.

Do Exercise 7. ▶

7. *Packages of Gum.* The candy company in Example 5 also produces 6-stick packages. How many 6-stick packages can be filled with 2269 sticks of gum? How many sticks will be left over?

EXAMPLE 6 *Automobile Mileage.* A 2013 Toyota Matrix gets 26 miles per gallon (mpg) in city driving. How many gallons will it use in 4758 mi of city driving?

Source: Toyota

1. **Familiarize.** We make a drawing and let $g =$ the number of gallons of gasoline used in 4758 mi of city driving.

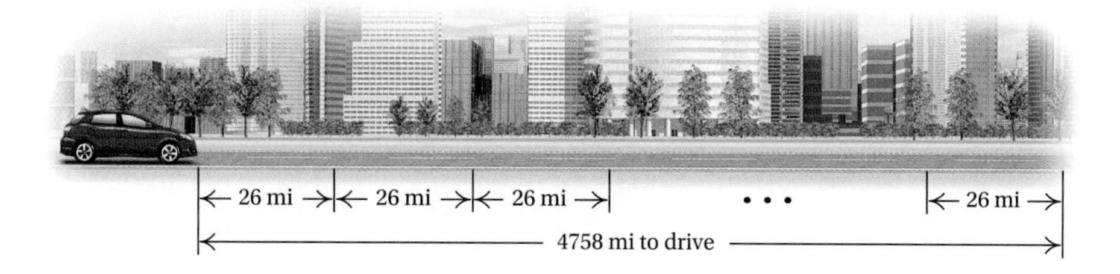

2. **Translate.** Repeated addition or multiplication applies here.

$$\underbrace{\text{Number of miles per gallon}}_{26} \underbrace{\text{times}}_{\cdot} \underbrace{\text{Number of gallons used}}_{g} \underbrace{\text{is}}_{=} \underbrace{\text{Number of miles driven}}_{4758}$$

3. **Solve.** To solve the equation, we divide by 26 on both sides.

$$26 \cdot g = 4758$$
$$\frac{26 \cdot g}{26} = \frac{4758}{26}$$
$$g = 183$$

$$\begin{array}{r} 183 \\ 26\overline{)4758} \\ \underline{26} \\ 215 \\ \underline{208} \\ 78 \\ \underline{78} \\ 0 \end{array}$$

4. **Check.** To check, we multiply 183 by 26.

$$\begin{array}{r} 183 \\ \times 26 \\ \hline 1098 \\ 3660 \\ \hline 4758 \end{array}$$

The answer checks.

5. **State.** The Toyota Matrix will use 183 gal of gasoline.

Do Exercise 8. ▶

8. *Automobile Mileage.* A 2013 Toyota Matrix gets 32 miles per gallon (mpg) in highway driving. How many gallons will it use in 2528 mi of highway driving?

Source: Toyota

Answers

7. 378 packages with 1 stick left over
8. 79 gal

Multistep Problems

Sometimes we must use more than one operation to solve a problem, as in the following example.

EXAMPLE 7 *Weight Loss.* To lose one pound, you must burn about 3500 calories in excess of what you already burn doing your regular daily activities. The following chart shows how long a person must engage in several types of exercise in order to burn 100 calories. For how long would a person have to run at a brisk pace in order to lose one pound?

1. **Familiarize.** This is a multistep problem. We will first find how many hundreds are in 3500. This will tell us how many times a person must run for 8 min in order to lose one pound. Then we will find the total number of minutes required for the weight loss.
   We let $x =$ the number of hundreds in 3500 and $t =$ the time it takes to lose one pound.
2. **Translate.** We translate to two equations.

100 calories	times	How many hundreds	is	3500
↓	↓	↓	↓	↓
100	$\cdot$	x	$=$	3500

Number of hundreds	times	8 minutes	is	Time to lose one pound
↓	↓	↓	↓	↓
x	$\cdot$	8	$=$	t

3. Solve. We divide by 100 on both sides of the first equation to find x.

$$100 \cdot x = 3500$$
$$\frac{100 \cdot x}{100} = \frac{3500}{100}$$
$$x = 35$$

$$\begin{array}{r} 35 \\ 100\,\overline{)\,3500} \\ \underline{300} \\ 500 \\ \underline{500} \\ 0 \end{array}$$

Then we use the fact that $x = 35$ to find t.

$$x \cdot 8 = t$$
$$35 \cdot 8 = t$$
$$280 = t$$

$$\begin{array}{r} 35 \\ \underline{\times \quad 8} \\ 280 \end{array}$$

4. Check. Suppose you run for 280 min. For every 8 min of running, you burn 100 calories. Since $280 \div 8 = 35$, there are 35 groups of 8 min in 280 min, so you will burn $35 \times 100 = 3500$ calories.

5. State. You must run for 280 min, or 4 hr 40 min, at a brisk pace in order to lose one pound.

Do Exercise 9. ▶

The key words, phrases, and concepts in the following table are useful when translating the problems to equations.

Key Words, Phrases, and Concepts

ADDITION (+)	SUBTRACTION (−)	MULTIPLICATION (×)	DIVISION (÷)
add	subtract	multiply	divide
added to	subtracted from	multiplied by	divided by
sum	difference	product	quotient
total	minus	times	repeated subtraction
plus	less than	of	missing factor
more than	decreased by	repeated addition	finding equal quantities
increased by	take away	rectangular arrays	
	how much more		

The following tips are also helpful in problem solving.

PROBLEM-SOLVING TIPS

1. Look for patterns when solving problems.
2. When translating in mathematics, consider the dimensions of the variables and constants in the equation. The variables that represent length should all be in the same unit, those that represent money should all be in dollars or all in cents, and so on.
3. Make sure that units appear in the answer whenever appropriate and that you completely answer the original problem.

GS **9. Weight Loss.** Use the information in Example 7 to determine how long an individual must swim at a brisk pace in order to lose one pound.

1. **Familiarize.** Let $x =$ the number of hundreds in 3500. Let $t =$ the time it takes to lose one pound.
2. **Translate.**

   $100 \cdot x = \_\_\_$

   $x \cdot \_\_\_ = t$

3. **Solve.** From Example 7, we know that $x = \_\_\_$.

   $x \cdot 2 = t$

   $\_\_\_ \cdot 2 = t$

   $\_\_\_ = t$

4. **Check.** Since $70 \div 2 = 35$, there are ___ groups of 2 min in 70 min. Thus, you will burn $35 \times 100 =$ ___ calories.
5. **State.** You must swim for ___ min, or 1 hr ___ min, in order to lose one pound.

Answer

9. 70 min, or 1 hr 10 min

Guided Solution:

9. 3500, 2; 35, 35, 70, 35, 3500, 70, 10

Translating for Success

1. *Brick-Mason Expense.* A commercial contractor is building 30 two-unit condominiums in a retirement community. The brick-mason expense for each building is \$10,860. What is the total cost of bricking the buildings?

2. *Heights.* Dean's sons are on the high school basketball team. Their heights are 73 in., 69 in., and 76 in. How much taller is the tallest son than the shortest son?

3. *Account Balance.* James has \$423 in his checking account. Then he deposits \$73 and uses his debit card for purchases of \$76 and \$69. How much is left in the account?

4. *Purchasing a Computer.* A computer is on sale for \$423. Jenny has only \$69. How much more does she need to buy the computer?

5. *Purchasing Coffee Makers.* Sara purchases 8 coffee makers for the newly remodeled bed-and-breakfast that she manages. If she pays \$52 for each coffee maker, what is the total cost of her purchase?

The goal of these matching questions is to practice step (2), Translate, of the five-step problem-solving process. Translate each word problem to an equation and select a correct translation from equations A–O.

A. $8 \cdot 52 = n$

B. $69 \cdot n = 76$

C. $73 - 76 - 69 = n$

D. $423 + 73 - 76 - 69 = n$

E. $30 \cdot 10{,}860 = n$

F. $15 \cdot n = 195$

G. $69 + n = 423$

H. $n = 10{,}860 - 300$

I. $n = 423 \div 69$

J. $30 \cdot n = 10{,}860$

K. $15 \cdot 195 = n$

L. $n = 52 - 8$

M. $69 + n = 76$

N. $15 \div 195 = n$

O. $52 + n = 60$

Answers on page A-2

6. *Hourly Rate.* Miller Auto Repair charges \$52 per hour for labor. Jackson Auto Care charges \$60 per hour. How much more does Jackson charge than Miller?

7. *College Band.* A college band with 195 members marches in a 15-row formation in the home-coming halftime performance. How many members are in each row?

8. *Shoe Purchase.* A college football team purchases 15 pairs of shoes at \$195 a pair. What is the total cost of this purchase?

9. *Loan Payment.* Kendra's uncle loans her \$10,860, interest free, to buy a car. The loan is to be paid off in 30 payments. How much is each payment?

10. *College Enrollment.* At the beginning of the fall term, the total enrollment in Lakeview Community College was 10,860. By the end of the first two weeks, 300 students had withdrawn. How many students were then enrolled?

1.8 Exercise Set

For Extra Help MyMathLab® MathXL® PRACTICE WATCH READ REVIEW

Reading Check

List the steps of the problem-solving strategy in order, using the choices given below. The last step is already listed.

Check Familiarize Solve Translate

RC1. 1. ______________.

RC2. 2. ______________.

RC3. 3. ______________.

RC4. 4. ______________.

5. State.

Towers Never Built. The buildings shown in the figure below were designed but never completed. Use the information to do Exercises 1–4.

Towers Never Built

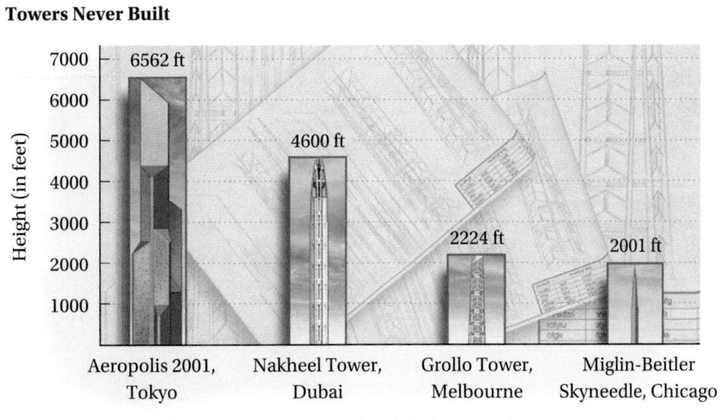

SOURCE: http://en.wikipedia.org/wiki/Proposed_tall_buildings_and_structures

1. How much taller would the Aeropolis 2001 have been than the Nakheel Tower?

2. How much taller would the Grollo Tower have been than the Miglin-Beitler Skyneedle?

3. The Willis Tower (formerly the Sears Tower) is the tallest building in Chicago. If the Miglin-Beitler Skyneedle had been built, it would have been 551 ft higher than the Willis Tower. What is the height of the Willis Tower?

4. The Burj Khalifa is the tallest building in Dubai. If the Nakheel Tower had been built, it would have been 1883 ft higher than the Burj Khalifa. What is the height of the Burj Khalifa?

5. *Caffeine Content.* An 8-oz serving of Red Bull energy drink contains 76 milligrams of caffeine. An 8-oz serving of brewed coffee contains 19 more milligrams of caffeine than the energy drink. How many milligrams of caffeine does the 8-oz serving of coffee contain?

Source: The Mayo Clinic

6. *Caffeine Content.* Hershey's 6-oz milk chocolate almond bar contains 25 milligrams of caffeine. A 20-oz bottle of Coca-Cola has 32 more milligrams of caffeine than the Hershey bar. How many milligrams of caffeine does the 20-oz bottle of Coca-Cola have?

Source: *National Geographic*, "Caffeine," by T. R. Reid, January 2005

7. A carpenter drills 216 holes in a rectangular array to construct a pegboard. There are 12 holes in each row. How many rows are there?

8. Lou arranges 504 entries on a spreadsheet in a rectangular array that has 36 rows. How many entries are in each row?

9. *Olympics.* There were 302 events in the 2012 Summer Olympics in London, England. This was 259 more events than there were in the first modern Olympic games in Athens, Greece, in 1896. How many events were there in 1896?

Sources: *USA Today* research; infoplease.com

10. *Drilling Activity.* In 2011, there were 984 rotary rigs drilling for crude oil in the United States. This was 687 more rigs than were active in 2007. Find the number of active rotary oil rigs in 2007.

Source: Energy Information Administration

11. *Boundaries between Countries.* The boundary between mainland United States and Canada including the Great Lakes is 3987 mi long. The length of the boundary between the United States and Mexico is 1933 mi. How much longer is the Canadian border?

Source: U.S. Geological Survey

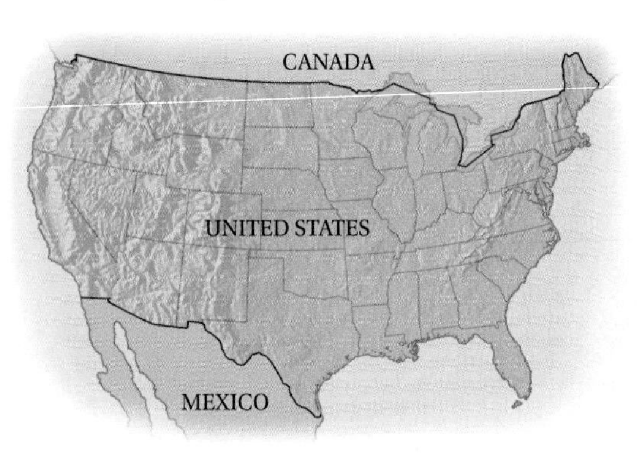

12. *Longest Rivers.* The longest river in the world is the Nile in Africa at about 4135 mi. The longest river in the United States is the Missouri–Mississippi at about 3860 mi. How much longer is the Nile?

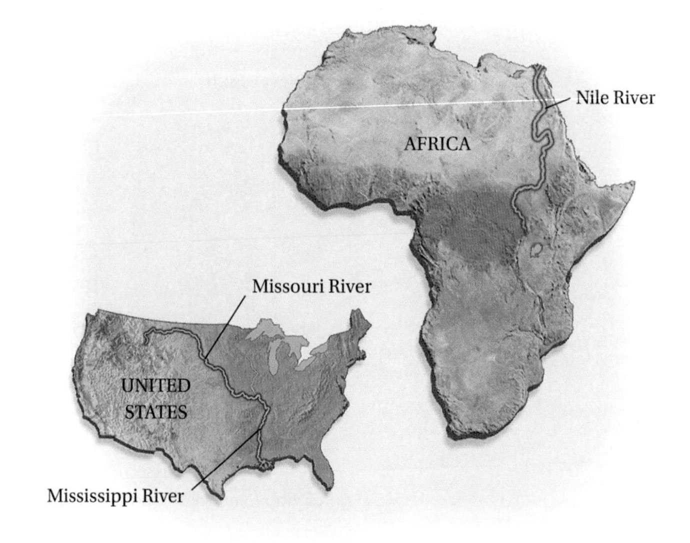

13. *Pixels.* A high-definition television (HDTV) screen consists of small rectangular dots called *pixels*. How many pixels are there on a screen that has 1080 rows with 1920 pixels in each row?

14. *Crossword Puzzle.* The *USA Today* crossword puzzle is a rectangle containing 15 rows with 15 squares in each row. How many squares does the puzzle have altogether?

15. There are 24 hr in a day and 7 days in a week. How many hours are there in a week?

16. There are 60 min in an hour and 24 hr in a day. How many minutes are there in a day?

Housing Costs. The graph below shows the average monthly rent for a one-bedroom apartment in several cities in November 2012. Use this graph to do Exercises 17–22.

Average Rent for a One-Bedroom Apartment

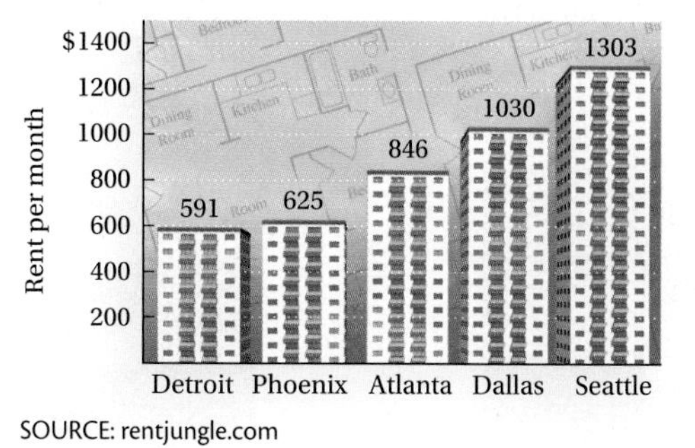

SOURCE: rentjungle.com

17. How much higher is the average monthly rent in Seattle than in Dallas?

18. How much lower is the average monthly rent in Phoenix than in Atlanta?

19. Phil, Scott, and Julio plan to rent a one-bedroom apartment in Detroit immediately after graduation, sharing the rent equally. What average monthly rent can each of them expect to pay?

20. Maria and her sister Theresa plan to share a one-bedroom apartment in Dallas, dividing the monthly rent equally between them. About how much can each of them expect to pay?

21. On average, how much rent would a tenant pay for a one-bedroom apartment in Detroit during a 12-month period?

22. On average, how much rent would a tenant pay for a one-bedroom apartment in Seattle during a 6-month period?

23. *Colonial Population.* Before the establishment of the U.S. Census in 1790, it was estimated that the colonial population in 1780 was 2,780,400. This was an increase of 2,628,900 from the population in 1680. What was the colonial population in 1680?

Source: *Time Almanac*

24. *Interstate Speed Limits.* The speed limit for passenger cars on interstate highways in rural areas in Montana is 75 mph. This is 10 mph faster than the speed limit for trucks on the same roads. What is the speed limit for trucks?

25. *Yard-Sale Profit.* Ruth made $312 at her yard sale and divided the money equally among her four grandchildren. How much did each child receive?

26. *Paper Measures.* A quire of paper consists of 25 sheets, and a ream of paper consists of 500 sheets. How many quires are in a ream?

27. *Parking Rates.* The most expensive parking in the United States is found in midtown New York City, where the average rate is $585 per month. This is $545 per month more than in the city with the least expensive rate, Bakersfield, California. What is the average monthly parking rate in Bakersfield?

Source: Colliers International

28. *Trade Balance.* In 2011, international visitors spent $153,000,000,000 traveling in the United States, while Americans spent $110,200,000,000 traveling abroad. How much more was spent by visitors to the United States than by Americans traveling abroad?

Source: U.S. Office of Travel and Tourism Industries

29. *Refrigerator Purchase.* Gourmet Deli has a chain of 24 restaurants. It buys a commercial refrigerator for each store at a cost of $1019 each. Determine the total cost of the purchase.

30. *Microwave Purchase.* Each room in the new dorm at Bridgeway College has a small kitchen. To furnish the kitchens, the college buys 96 microwave ovens at $88 each. Determine the total cost of the purchase.

31. *"Seinfeld."* A local television station plans to air the 177 episodes of the long-running comedy series "Seinfeld." If the station airs 5 episodes per week, how many full weeks will pass before it must begin re-airing previously shown episodes? How many unaired episodes will be shown the following week before the previously aired episodes are rerun?

32. *"Everybody Loves Raymond."* The popular television comedy series "Everybody Loves Raymond" had 208 scripted episodes and 2 additional episodes consisting of clips from previous shows. A local television station plans to air the 208 scripted episodes, showing 5 episodes per week. How many full weeks will pass before it must begin re-airing episodes? How many unaired episodes will be shown the following week before the previously aired episodes are rerun?

33. *Crossword Puzzle.* The *Los Angeles Times* crossword puzzle is a rectangle containing 441 squares arranged in 21 rows. How many columns does the puzzle have?

34. *Mailing Labels.* A box of mailing labels contains 750 labels on 25 sheets. How many labels are on each sheet?

35. *Automobile Mileage.* The 2013 Hyundai Tucson GLS gets 30 miles per gallon (mpg) in highway driving. How many gallons will it use in 7080 mi of highway driving?

Source: Hyundai

36. *Automobile Mileage.* The 2013 Volkswagen Jetta (5 cylinder) gets 24 miles per gallon (mpg) in city driving. How many gallons will it use in 3960 mi of city driving?

Source: Volkswagen of America, Inc.

37. *High School Court.* The standard basketball court used by high school players has dimensions of 50 ft by 84 ft.

a) What is its area?

b) What is its perimeter?

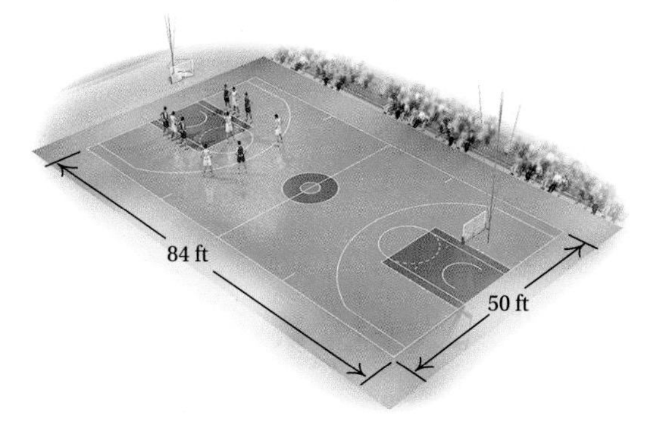

38. *College Court.* The standard basketball court used by college players has dimensions of 50 ft by 94 ft.

a) What is its area?

b) What is its perimeter?

c) How much greater is the area of a college court than a high school court? (See Exercise 37.)

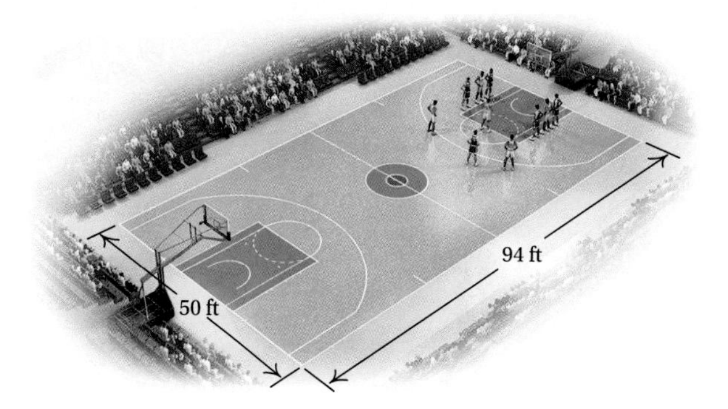

39. Copies of this book are usually shipped from the warehouse in cartons containing 24 books each. How many cartons are needed to ship 1344 books?

40. The H. J. Heinz Company ships 16-oz bottles of ketchup in cartons containing 12 bottles each. How many cartons are needed to ship 528 bottles of ketchup?

41. *Map Drawing.* A map has a scale of 215 mi to the inch. How far apart *in reality* are two cities that are 3 in. apart on the map? How far apart *on the map* are two cities that, in reality, are 1075 mi apart?

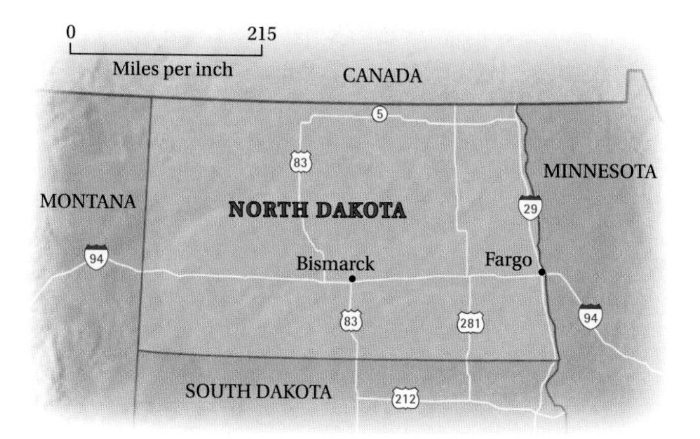

42. *Map Drawing.* A map has a scale of 288 mi to the inch. How far apart *on the map* are two cities that, in reality, are 2016 mi apart? How far apart *in reality* are two cities that are 8 in. apart on the map?

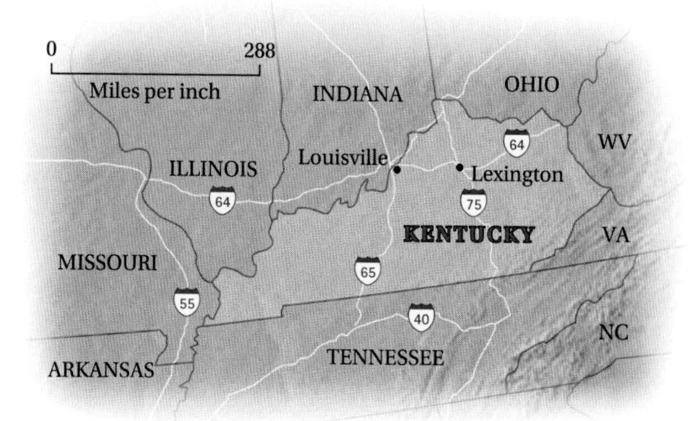

43. *Loan Payments.* Dana borrows $5928 for a used car. The loan is to be paid off in 24 equal monthly payments. How much is each payment (excluding interest)?

44. *Home Improvement Loan.* The Van Reken family borrows $7824 to build a detached garage next to their home. The loan is to be paid off in equal monthly payments of $163 (excluding interest). How many months will it take to pay off the loan?

Refer to the information in Example 7 to do Exercises 45 and 46.

45. For how long must you do aerobic exercises in order to lose one pound?

46. For how long must you bicycle at 9 mph in order to lose one pound?

New Jobs. Many of the fastest-growing occupations in the United States require education beyond a high school diploma. The following table lists some of these and gives the projected numbers of new jobs expected to be created between 2010 and 2020. Use the information in the table for Exercises 47 and 48.

New Jobs Created, 2010–2020

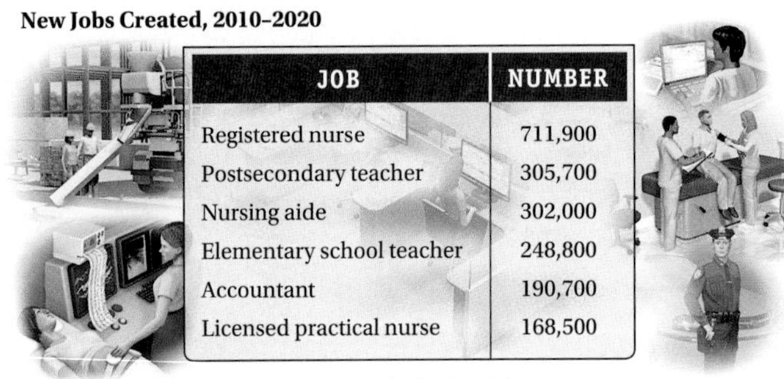

JOB	NUMBER
Registered nurse	711,900
Postsecondary teacher	305,700
Nursing aide	302,000
Elementary school teacher	248,800
Accountant	190,700
Licensed practical nurse	168,500

SOURCE: U.S. Bureau of Labor Statistics

47. The U.S. Bureau of Labor Statistics predicts that between 2010 and 2020, there will be 1,014,100 more new jobs created for registered nurses, nursing aides, and licensed practical nurses than there will be for physicians. How many new jobs will be created for physicians between 2010 and 2020?

48. The U.S. Bureau of Labor Statistics predicts that between 2010 and 2020, there will be 484,600 more new jobs created for postsecondary teachers and elementary teachers than there will be for secondary teachers. How many new jobs will be created for secondary teachers between 2010 and 2020?

49. *Seating Configuration.* The seats in the Boeing 737-500 airplanes in United Airlines' North American fleet are configured with 2 rows of 4 seats across in first class and 16 rows of 6 seats across in economy class. Determine the total seating capacity of one of these planes.

Source: United Airlines

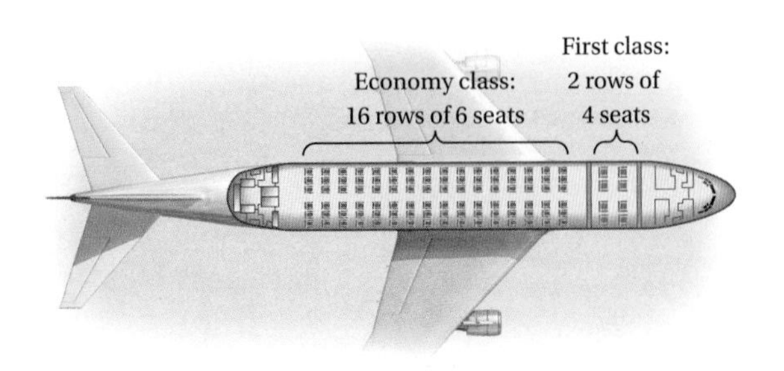

50. *Seating Configuration.* The seats in the Airbus 320 airplanes in United Airlines' North American fleet are configured with 3 rows of 4 seats across in first class and 21 rows of 6 seats across in economy class. Determine the total seating capacity of one of these planes.

Source: United Airlines

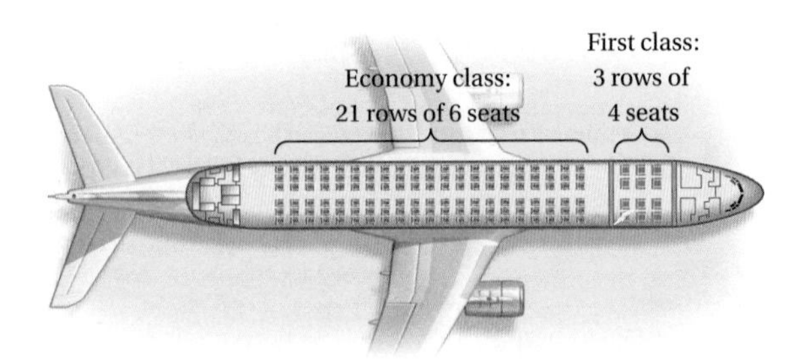

51. Elena buys 5 video games at \$64 each and pays for them with \$10 bills. How many \$10 bills does it take?

52. Pedro buys 5 video games at \$64 each and pays for them with \$20 bills. How many \$20 bills does it take?

53. The balance in Meg's bank account is \$568. She uses her debit card for purchases of \$46, \$87, and \$129. Then she deposits \$94 in the account after returning a textbook. How much is left in her account?

54. The balance in Dylan's bank account is \$749. He uses his debit card for purchases of \$34 and \$65. Then he makes a deposit of \$123 from his paycheck. What is the new balance?

55. *Bones in the Hands and Feet.* There are 27 bones in each human hand and 26 bones in each human foot. How many bones are there in all in the hands and feet?

56. An office for adjunct instructors at a community college has 6 bookshelves, each of which is 3 ft wide. The office is moved to a new location that has dimensions of 16 ft by 21 ft. Is it possible for the bookshelves to be put side by side on the 16-ft wall?

Skill Maintenance

57. Add: [1.2a]

$$\begin{array}{r} 6\,2\,5\,4 \\ 1\,5\,3\,7 \\ +\ \ 4\,8\,2 \\ \hline \end{array}$$

58. Subtract: [1.3a]

$$\begin{array}{r} 9\,6\,0\,2 \\ -\ 1\,8\,4\,3 \\ \hline \end{array}$$

59. Multiply: [1.4a]

$$\begin{array}{r} 3\,4\,0\,5 \\ \times\ 2\,3\,7 \\ \hline \end{array}$$

60. Divide: [1.5a]

$$32\,\overline{)\,4\,7\,0\,8}$$

61. Find the perimeter of the figure. [1.2b]

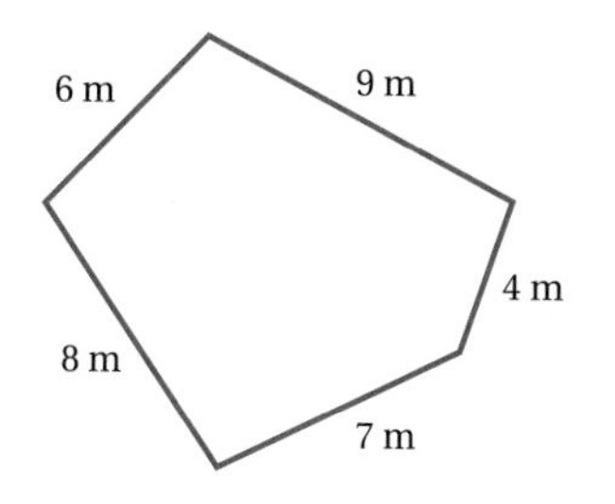

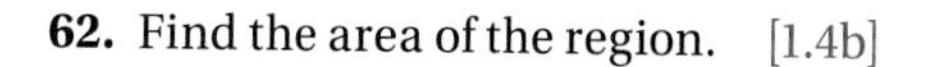

62. Find the area of the region. [1.4b]

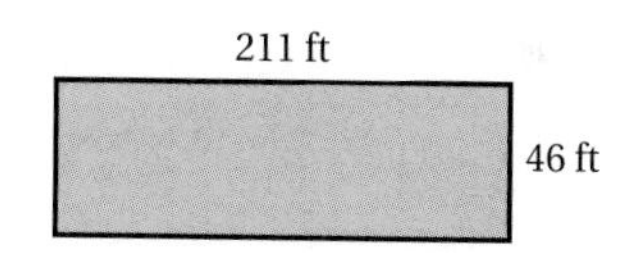

63. Estimate 238×596 by rounding to the nearest hundred. [1.6b]

64. Solve: $x + 15 = 81$. [1.7b]

Synthesis

65. *Speed of Light.* Light travels about 186,000 miles per second (mi/sec) in a vacuum such as in outer space. In ice it travels about 142,000 mi/sec, and in glass it travels about 109,000 mi/sec. In 18 sec, how many more miles will light travel in a vacuum than in ice? than in glass?

66. Carney Community College has 1200 students. Each instructor teaches 4 classes, and each student takes 5 classes. There are 30 students and 1 instructor in each classroom. How many instructors are there at Carney Community College?

1.9 Exponential Notation and Order of Operations

OBJECTIVES

a Write exponential notation for products such as $4 \cdot 4 \cdot 4$.

b Evaluate exponential notation.

c Simplify expressions using the rules for order of operations.

d Remove parentheses within parentheses.

SKILL TO REVIEW

Objective 1.4a: Multiply whole numbers.

Multiply.

1. $5 \times 5 \times 5$

2. $2 \times 2 \times 2 \times 2 \times 2$

a WRITING EXPONENTIAL NOTATION

Consider the product $3 \cdot 3 \cdot 3 \cdot 3$. Such products occur often enough that mathematicians have found it convenient to create a shorter notation, called **exponential notation**, for them. For example,

$$\underbrace{3 \cdot 3 \cdot 3 \cdot 3}_{4 \text{ factors}} \text{ is shortened to } 3^4. \quad \leftarrow \text{exponent; } 3 \text{ is the base}$$

We read exponential notation as follows.

NOTATION	WORD DESCRIPTION
3^4	"three to the fourth power," or "the fourth power of three"
5^3	"five-cubed," or "the cube of five," or "five to the third power," or "the third power of five"
7^2	"seven squared," or "the square of seven," or "seven to the second power," or "the second power of seven"

The wording "seven squared" for 7^2 is derived from the fact that a square with side s has area A given by $A = s^2$.

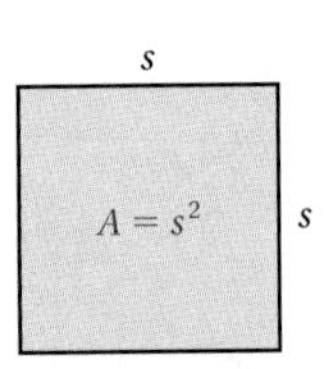

An expression like $3 \cdot 5^2$ is read "three times five squared," or "three times the square of five."

EXAMPLE 1 Write exponential notation for $10 \cdot 10 \cdot 10 \cdot 10 \cdot 10$.

Exponential notation is 10^5. 5 is the *exponent*. 10 is the *base*.

EXAMPLE 2 Write exponential notation for $2 \cdot 2 \cdot 2$.

Exponential notation is 2^3.

Do Margin Exercises 1–4.

Write exponential notation.

1. $5 \cdot 5 \cdot 5 \cdot 5$

2. $5 \cdot 5 \cdot 5 \cdot 5 \cdot 5 \cdot 5$

3. $10 \cdot 10$

4. $10 \cdot 10 \cdot 10 \cdot 10$

Answers

Skill to Review:
1. 125 **2.** 32

Margin Exercises:
1. 5^4 **2.** 5^6 **3.** 10^2 **4.** 10^4

b EVALUATING EXPONENTIAL NOTATION

We evaluate exponential notation by rewriting it as a product and then computing the product.

EXAMPLE 3 Evaluate: 10^3.

$10^3 = 10 \cdot 10 \cdot 10 = 1000$

Caution!

10^3 does not mean $10 \cdot 3$. ■

EXAMPLE 4 Evaluate: 5^4.

$5^4 = 5 \cdot 5 \cdot 5 \cdot 5 = 625$

Do Exercises 5–8. ▶

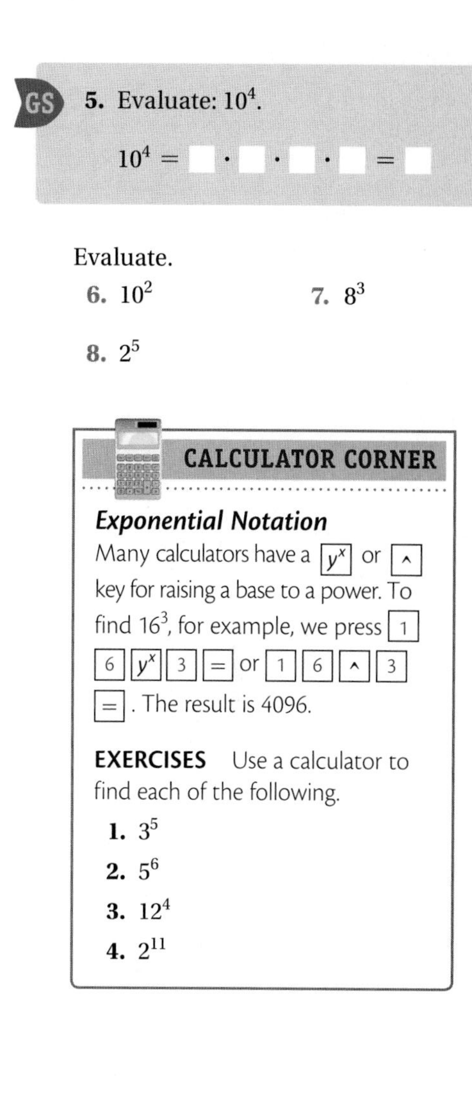

GS **5.** Evaluate: 10^4.

$10^4 = \square \cdot \square \cdot \square \cdot \square = \square$

Evaluate.

6. 10^2

7. 8^3

8. 2^5

CALCULATOR CORNER

Exponential Notation

Many calculators have a $\boxed{y^x}$ or $\boxed{\wedge}$ key for raising a base to a power. To find 16^3, for example, we press $\boxed{1}\ \boxed{6}\ \boxed{y^x}\ \boxed{3}\ \boxed{=}$ or $\boxed{1}\ \boxed{6}\ \boxed{\wedge}\ \boxed{3}\ \boxed{=}$. The result is 4096.

EXERCISES Use a calculator to find each of the following.

1. 3^5
2. 5^6
3. 12^4
4. 2^{11}

c SIMPLIFYING EXPRESSIONS

Suppose we have a calculation like the following:

$3 + 4 \cdot 8.$

How do we find the answer? Do we add 3 to 4 and then multiply by 8, or do we multiply 4 by 8 and then add 3? In the first case, the answer is 56. In the second, the answer is 35. We agree to compute as in the second case:

$3 + 4 \cdot 8 = 3 + 32 = 35.$

The following rules are an agreement regarding the order in which we perform operations. These are the rules that computers and most scientific calculators use to do computations.

RULES FOR ORDER OF OPERATIONS

1. Do all calculations within parentheses ( ), brackets [ ], or braces { } before operations outside.
2. Evaluate all exponential expressions.
3. Do all multiplications and divisions in order from left to right.
4. Do all additions and subtractions in order from left to right.

EXAMPLE 5 Simplify: $16 \div 8 \cdot 2$.

There are no parentheses or exponents, so we begin with the third step.

$$\left.\begin{aligned} 16 \div 8 \cdot 2 &= 2 \cdot 2 \\ &= 4 \end{aligned}\right\}$$ Doing all multiplications and divisions in order from left to right ■

EXAMPLE 6 Simplify: $7 \cdot 14 - (12 + 18)$.

$7 \cdot 14 - (12 + 18) = 7 \cdot 14 - 30$ Carrying out operations inside parentheses

$= 98 - 30$ Doing all multiplications and divisions

$= 68$ Doing all additions and subtractions

Do Exercises 9–12. ▶

Simplify.

9. $93 - 14 \cdot 3$

10. $104 \div 4 + 4$

11. $25 \cdot 26 - (56 + 10)$

12. $75 \div 5 + (83 - 14)$

Answers

5. 10,000 **6.** 100 **7.** 512 **8.** 32 **9.** 51 **10.** 30 **11.** 584 **12.** 84

Guided Solution:

5. 10, 10, 10, 10, 10,000

EXAMPLE 7 Simplify and compare: $23 - (10 - 9)$ and $(23 - 10) - 9$.

We have

$$23 - (10 - 9) = 23 - 1 = 22;$$
$$(23 - 10) - 9 = 13 - 9 = 4.$$

We can see that $23 - (10 - 9)$ and $(23 - 10) - 9$ represent different numbers. Thus subtraction is not associative.

◀ **Do Exercises 13 and 14.**

Simplify and compare.

13. $64 \div (32 \div 2)$ and $(64 \div 32) \div 2$

14. $(28 + 13) + 11$ and $28 + (13 + 11)$

EXAMPLE 8 Simplify: $7 \cdot 2 - (12 + 0) \div 3 - (5 - 2)$.

$7 \cdot 2 - (12 + 0) \div 3 - (5 - 2)$

$= 7 \cdot 2 - 12 \div 3 - 3$ — Carrying out operations inside parentheses

$= 14 - 4 - 3$ — Doing all multiplications and divisions in order from left to right

$= 10 - 3$
$= 7$ — Doing all additions and subtractions in order from left to right

◀ **Do Exercise 15.**

15. Simplify: **GS**

$9 \times 4 - (20 + 4) \div 8 - (6 - 2)$.

$9 \times 4 - (20 + 4) \div 8 - (6 - 2)$

$= 9 \times 4 - \square \div 8 - \square$

$= \square - 24 \div 8 - 4$

$= 36 - \square - 4$

$= \square - 4$

$= \square$

EXAMPLE 9 Simplify: $15 \div 3 \cdot 2 \div (10 - 8)$.

$15 \div 3 \cdot 2 \div (10 - 8)$

$= 15 \div 3 \cdot 2 \div 2$ — Carrying out operations inside parentheses

$= 5 \cdot 2 \div 2$
$= 10 \div 2$
$= 5$ — Doing all multiplications and divisions in order from left to right

◀ **Do Exercises 16–18.**

Simplify.

16. $5 \cdot 5 \cdot 5 + 26 \cdot 71 - (16 + 25 \cdot 3)$

17. $30 \div 5 \cdot 2 + 10 \cdot 20 + 8 \cdot 8 - 23$

18. $95 - 2 \cdot 2 \cdot 2 \cdot 5 \div (24 - 4)$

EXAMPLE 10 Simplify: $4^2 \div (10 - 9 + 1)^3 \cdot 3 - 5$.

$4^2 \div (10 - 9 + 1)^3 \cdot 3 - 5$

$= 4^2 \div (1 + 1)^3 \cdot 3 - 5$ — Subtracting inside parentheses

$= 4^2 \div 2^3 \cdot 3 - 5$ — Adding inside parentheses

$= 16 \div 8 \cdot 3 - 5$ — Evaluating exponential expressions

$= 2 \cdot 3 - 5$
$= 6 - 5$ — Doing all multiplications and divisions in order from left to right

$= 1$ — Subtracting

◀ **Do Exercises 19–21.**

Simplify.

19. $5^3 + 26 \cdot 71 - (16 + 25 \cdot 3)$

20. $(1 + 3)^3 + 10 \cdot 20 + 8^2 - 23$

21. $81 - 3^2 \cdot 2 \div (12 - 9)$

Answers

13. 4; 1 **14.** 52; 52 **15.** 29 **16.** 1880 **17.** 253 **18.** 93 **19.** 1880 **20.** 305 **21.** 75

Guided Solution:

15. 24, 4, 36, 3, 33, 29

EXAMPLE 11 Simplify: $2^9 \div 2^6 \cdot 2^3$.

$2^9 \div 2^6 \cdot 2^3 = 512 \div 64 \cdot 8$ — Since there are no parentheses, we evaluate the exponential expressions.

$= 8 \cdot 8$
$= 64$ — Doing all multiplications and divisions in order from left to right

Do Exercise 22. ▶

22. Simplify: $2^3 \cdot 2^8 \div 2^9$.

Averages

In order to find the average of a set of numbers, we use addition and then division. For example, the average of 2, 3, 6, and 9 is found as follows.

$$\text{Average} = \frac{2 + 3 + 6 + 9}{4} = \frac{20}{4} = 5$$

The number of addends is 4.

Divide by 4.

The fraction bar acts as a grouping symbol, so

$\dfrac{2 + 3 + 6 + 9}{4}$ is equivalent to $(2 + 3 + 6 + 9) \div 4$.

Thus we are using order of operations when we compute an average.

AVERAGE

The **average** of a set of numbers is the sum of the numbers divided by the number of addends.

EXAMPLE 12 *National Parks.* Since 1995, four national parks have been established in the United States. The sizes of these parks are shown in the figure below. Determine the average size of the four parks.

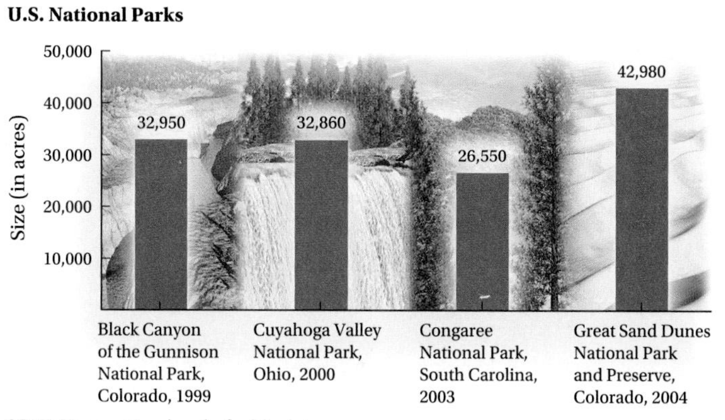

SOURCE: us-national-parks.findthedata.org

CALCULATOR CORNER

Order of Operations We can test whether a calculator is programmed to follow the rules for order of operations by entering a calculation such as $3 + 4 \cdot 2$. Using the rules for order of operations, the value of this expression is 11. If a calculator returns the result 14 instead of 11, that calculator performs operations as they are entered rather than following the rules for order of operations. When using such a calculator, we have to enter the operations in the order in which we want them performed.

EXERCISES Simplify.

1. $84 - 5 \cdot 7$
2. $80 + 50 \div 10$
3. $3^2 + 9^2 \div 3$
4. $4^4 \div 64 - 4$
5. $15 \cdot 7 - (23 + 9)$
6. $(4 + 3)^2$

Answer

22. 4

23. *Average Number of Career Hits.* The numbers of career hits of five Hall of Fame baseball players are given in the graph below. Find the average number of career hits of all five.

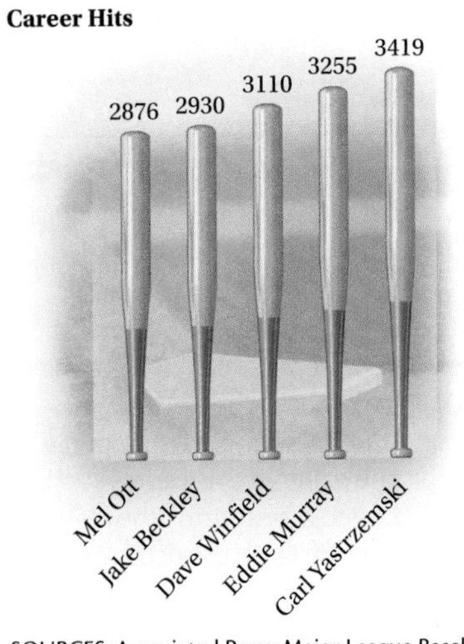

SOURCES: Associated Press; Major League Baseball

The average is given by

$$\frac{32{,}950 + 32{,}860 + 26{,}550 + 42{,}980}{4} = \frac{135{,}340}{4} = 33{,}835.$$

The average size of the four national parks is 33,835 acres.

◀ **Do Exercise 23.**

d REMOVING PARENTHESES WITHIN PARENTHESES

When parentheses occur within parentheses, we can make them different shapes, such as [ ] (called "brackets") and { } (called "braces"). All of these have the same meaning. When parentheses occur within parentheses, computations in the innermost ones are to be done first.

EXAMPLE 13 Simplify: $[25 - (4 + 3) \cdot 3] \div (11 - 7)$.

$[25 - (4 + 3) \cdot 3] \div (11 - 7)$

$= [25 - 7 \cdot 3] \div (11 - 7)$	Doing the calculations in the innermost parentheses first
$= [25 - 21] \div (11 - 7)$	Doing the multiplication in the brackets
$= 4 \div 4$	Subtracting
$= 1$	Dividing

EXAMPLE 14 Simplify: $16 \div 2 + \{40 - [13 - (4 + 2)]\}$.

$16 \div 2 + \{40 - [13 - (4 + 2)]\}$

$= 16 \div 2 + \{40 - [13 - 6]\}$	Doing the calculations in the innermost parentheses first
$= 16 \div 2 + \{40 - 7\}$	Again, doing the calculations in the innermost brackets
$= 16 \div 2 + 33$	Subtracting inside the braces
$= 8 + 33$	Dividing
$= 41$	Adding

◀ **Do Exercises 24 and 25.**

Simplify.

24. $9 \times 5 + \{6 \div [14 - (5 + 3)]\}$

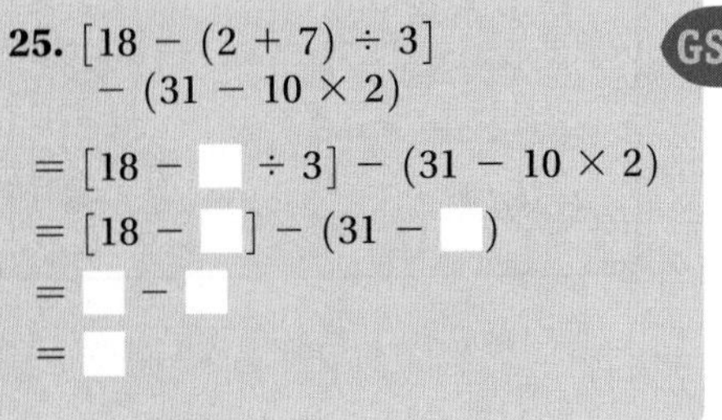

25. $[18 - (2 + 7) \div 3] - (31 - 10 \times 2)$ GS

$= [18 - \square \div 3] - (31 - 10 \times 2)$

$= [18 - \square] - (31 - \square)$

$= \square - \square$

$= \square$

Answers

23. 3118 hits **24.** 46 **25.** 4

Guided Solution:

25. 9, 3, 20, 15, 11, 4

1.9 Exercise Set

For Extra Help
MyMathLab® MathXL® PRACTICE WATCH READ REVIEW

Reading Check

Complete each statement by choosing the correct word or number from below the blank.

RC1. In the expression 5^3, the number 3 is the ________________.
base/exponent

RC2. The expression 9^2 can be read "nine ________________."
cubed/squared

RC3. To calculate $10 - 4 \cdot 2$, we perform the ________________ first.
multiplication/subtraction

RC4. To find the average of 7, 8, and 9, we add the numbers and divide the sum by ________________.
2/3

a Write exponential notation.

1. $3 \cdot 3 \cdot 3 \cdot 3$

2. $2 \cdot 2 \cdot 2 \cdot 2 \cdot 2$

3. $5 \cdot 5$

4. $13 \cdot 13 \cdot 13$

5. $7 \cdot 7 \cdot 7 \cdot 7 \cdot 7$

6. $9 \cdot 9$

7. $10 \cdot 10 \cdot 10$

8. $1 \cdot 1 \cdot 1 \cdot 1$

b Evaluate.

9. 7^2

10. 5^3

11. 9^3

12. 8^2

13. 12^4

14. 10^5

15. 3^5

16. 2^6

c Simplify.

17. $12 + (6 + 4)$

18. $(12 + 6) + 18$

19. $52 - (40 - 8)$

20. $(52 - 40) - 8$

21. $1000 \div (100 \div 10)$

22. $(1000 \div 100) \div 10$

23. $(256 \div 64) \div 4$

24. $256 \div (64 \div 4)$

25. $(2 + 5)^2$

26. $2^2 + 5^2$

27. $(11 - 8)^2 - (18 - 16)^2$

28. $(32 - 27)^3 + (19 + 1)^3$

29. $16 \cdot 24 + 50$

30. $23 + 18 \cdot 20$

31. $83 - 7 \cdot 6$

32. $10 \cdot 7 - 4$

33. $10 \cdot 10 - 3 \cdot 4$

34. $90 - 5 \cdot 5 \cdot 2$

35. $4^3 \div 8 - 4$

36. $8^2 - 8 \cdot 2$

37. $17 \cdot 20 - (17 + 20)$

38. $1000 \div 25 - (15 + 5)$

39. $6 \cdot 10 - 4 \cdot 10$

40. $3 \cdot 8 + 5 \cdot 8$

41. $300 \div 5 + 10$

42. $144 \div 4 - 2$

43. $3 \cdot (2 + 8)^2 - 5 \cdot (4 - 3)^2$

44. $7 \cdot (10 - 3)^2 - 2 \cdot (3 + 1)^2$

45. $4^2 + 8^2 \div 2^2$

46. $6^2 - 3^4 \div 3^3$

47. $10^3 - 10 \cdot 6 - (4 + 5 \cdot 6)$

48. $7^2 + 20 \cdot 4 - (28 + 9 \cdot 2)$

49. $6 \cdot 11 - (7 + 3) \div 5 - (6 - 4)$

50. $8 \times 9 - (12 - 8) \div 4 - (10 - 7)$

51. $120 - 3^3 \cdot 4 \div (5 \cdot 6 - 6 \cdot 4)$

52. $80 - 2^4 \cdot 15 \div (7 \cdot 5 - 45 \div 3)$

53. $2^3 \cdot 2^8 \div 2^6$

54. $2^7 \div 2^5 \cdot 2^4 \div 2^2$

55. Find the average of \$64, \$97, and \$121.

56. Find the average of four test grades of 86, 92, 80, and 78.

57. Find the average of 320, 128, 276, and 880.

58. Find the average of \$1025, \$775, \$2062, \$942, and \$3721.

d Simplify.

59. $8 \times 13 + \{42 \div [18 - (6 + 5)]\}$

60. $72 \div 6 - \{2 \times [9 - (4 \times 2)]\}$

61. $[14 - (3 + 5) \div 2] - [18 \div (8 - 2)]$

62. $[92 \times (6 - 4) \div 8] + [7 \times (8 - 3)]$

63. $(82 - 14) \times [(10 + 45 \div 5) - (6 \cdot 6 - 5 \cdot 5)]$

64. $(18 \div 2) \cdot \{[(9 \cdot 9 - 1) \div 2] - [5 \cdot 20 - (7 \cdot 9 - 2)]\}$

65. $4 \times \{(200 - 50 \div 5) - [(35 \div 7) \cdot (35 \div 7) - 4 \times 3]\}$

66. $15(23 - 4 \cdot 2)^3 \div (3 \cdot 25)$

67. $\{[18 - 2 \cdot 6] - [40 \div (17 - 9)]\} + \{48 - 13 \times 3 + [(50 - 7 \cdot 5) + 2]\}$

68. $(19 - 2^4)^5 - (141 \div 47)^2$

Skill Maintenance

Solve. [1.7b]

69. $x + 341 = 793$

70. $4197 + x = 5032$

71. $7 \cdot x = 91$

72. $1554 = 42 \cdot y$

73. $6000 = 1102 + t$

74. $10{,}000 = 100 \cdot t$

Solve. [1.8a]

75. *Colorado.* The state of Colorado is roughly the shape of a rectangle that is 273 mi by 382 mi. What is its area?

76. On a long four-day trip, a family bought the following amounts of gasoline for their motor home: 23 gal, 24 gal, 26 gal, and 25 gal. How much gasoline did they buy in all?

Synthesis

Each of the answers in Exercises 77–79 is incorrect. First find the correct answer. Then place as many parentheses as needed in the expression in order to make the incorrect answer correct.

77. $1 + 5 \cdot 4 + 3 = 36$

78. $12 \div 4 + 2 \cdot 3 - 2 = 2$

79. $12 \div 4 + 2 \cdot 3 - 2 = 4$

80. Use one occurrence each of 1, 2, 3, 4, 5, 6, 7, 8, and 9, in order, and any of the symbols $+, -, \cdot, \div$, and ( ) to represent 100.

CHAPTER

1 Summary and Review

Vocabulary Reinforcement

In each of Exercises 1–8, fill in the blank with the correct term from the given list. Some of the choices may not be used and some may be used more than once.

1. The distance around an object is its ______________. [1.2b]
2. The ______________ is the number from which another number is being subtracted. [1.3a]
3. For large numbers, ______________ are separated by commas into groups of three, called ______________. [1.1a]
4. In the sentence $28 \div 7 = 4$, the ______________ is 28. [1.5a]
5. In the sentence $10 \times 1000 = 10{,}000$, 10 and 1000 are called ______________ and 10,000 is called the ______________. [1.4a]
6. The number 0 is called the ______________ identity. [1.2a]
7. The sentence $3 \times (6 \times 2) = (3 \times 6) \times 2$ illustrates the ______________ law of multiplication. [1.4a]
8. We can use the following statement to check division: quotient · ______________ + ______________ = ______________. [1.5a]

associative
commutative
addends
factors
area
perimeter
minuend
subtrahend
product
digits
periods
additive
multiplicative
dividend
quotient
remainder
divisor

Concept Reinforcement

Determine whether each statement is true or false.

______________ 1. $a > b$ is true when a is to the right of b on the number line. [1.6c]

______________ 2. Any nonzero number divided by itself is 1. [1.5a]

______________ 3. For any whole number a, $a \div 0 = 0$. [1.5a]

______________ 4. Every equation is true. [1.7a]

______________ 5. The rules for order of operations tell us to multiply and divide before adding and subtracting. [1.9c]

______________ 6. The average of three numbers is the middle number. [1.9c]

Study Guide

Objective 1.1a Give the meaning of digits in standard notation.

Example What does the digit 7 mean in 2,379,465?

2,3 7 9,4 6 5

7 means 7 ten thousands.

Practice Exercise

1. What does the digit 2 mean in 432,079?

Objective 1.2a Add whole numbers.

Example Add: 7368 + 3547.

$$\begin{array}{r} {}^{1\ 1} \\ 7368 \\ +\ 3547 \\ \hline 10{,}915 \end{array}$$

Practice Exercise

2. Add: 36,047 + 29,255.

Objective 1.3a Subtract whole numbers.

Example Subtract: 8045 − 2897.

$$\begin{array}{r} {}^{7\ 9\ \overset{13}{\cancel{3}}\ 15} \\ \cancel{8}\cancel{0}\cancel{4}\cancel{5} \\ -\ 2897 \\ \hline 5148 \end{array}$$

Practice Exercise

3. Subtract: 4805 − 1568.

Objective 1.4a Multiply whole numbers.

Example Multiply: 57 × 315.

$$\begin{array}{rl} {}^{2}_{1\ 3} & \\ 315 & \\ \times\ 57 & \\ \hline 2205 & \leftarrow 315 \times 7 \\ 15750 & \leftarrow 315 \times 50 \\ \hline 17{,}955 & \end{array}$$

Practice Exercise

4. Multiply: 329 × 684.

Objective 1.5a Divide whole numbers.

Example Divide: 6463 ÷ 26.

$$\begin{array}{r} 248 \\ 26\overline{)6463} \\ \underline{52} \\ 126 \\ \underline{104} \\ 223 \\ \underline{208} \\ 15 \end{array}$$

The answer is 248 R 15.

Practice Exercise

5. Divide: 8519 ÷ 27.

Objective 1.6a Round to the nearest ten, hundred, or thousand.

Example Round to the nearest thousand.

6 4 7 1
↑

The digit 6 is in the thousands place. We consider the next digit to the right. Since the digit, 4, is 4 or lower, we round down, meaning that 6 thousands stays as 6 thousands. Change all digits to the right of the thousands digit to zeros. The answer is 6000.

Practice Exercise

6. Round 36,468 to the nearest thousand.

Objective 1.6c Use $<$ or $>$ for □ to write a true sentence in a situation like 6 □ 10.

Example Use $<$ or $>$ for □ to write a true sentence:

34 □ 29.

Since 34 is to the right of 29 on the number line,

$34 > 29$.

Practice Exercise

7. Use $<$ or $>$ for □ to write a true sentence:

78 □ 81.

Objective 1.7b Solve equations like $t + 28 = 54$, $28 \cdot x = 168$, and $98 \cdot 2 = y$.

Example Solve: $y + 12 = 27$.

$$y + 12 = 27$$
$$y + 12 - 12 = 27 - 12$$
$$y + 0 = 15$$
$$y = 15$$

The solution is 15.

Practice Exercise

8. Solve: $24 \cdot x = 864$.

Objective 1.9b Evaluate exponential notation.

Example Evaluate: 5^4.

$5^4 = 5 \cdot 5 \cdot 5 \cdot 5 = 625$

Practice Exercise

9. Evaluate: 6^3.

Review Exercises

The review exercises that follow are for practice. Answers are given at the back of the book. If you miss an exercise, restudy the objective indicated in red next to the exercise or on the direction line that precedes it.

1. What does the digit 8 mean in 4,678,952? [1.1a]

2. In 13,768,940, what digit tells the number of millions? [1.1a]

Write expanded notation. [1.1b]

3. 2793

4. 56,078

5. 4,007,101

Write a word name. [1.1c]

6. 67,819

7. 2,781,427

Write standard notation. [1.1c]

8. Four hundred seventy-six thousand, five hundred eighty-eight

9. *Subway Ridership.* Ridership on the New York City Subway system totaled one billion, six hundred forty million in 2011.

Source: Metropolitan Transit Authority

Add. [1.2a]

10. $7304 + 6968$

11. $27{,}609 + 38{,}415$

12. $2703 + 4125 + 6004 + 8956$

13. $\begin{array}{r} 91{,}426 \\ +\ \ 7{,}495 \\ \hline \end{array}$

Subtract. [1.3a]

14. $8045 - 2897$

15. $9001 - 7312$

16. $6003 - 3729$

17. $\begin{array}{r} 37{,}405 \\ -\ 19{,}648 \\ \hline \end{array}$

Multiply. [1.4a]

18. $17{,}000 \cdot 300$

19. $7846 \cdot 800$

20. $726 \cdot 698$

21. $587 \cdot 47$

22. $\begin{array}{r} 8305 \\ \times\ \ 642 \\ \hline \end{array}$

Divide. [1.5a]

23. $63 \div 5$

24. $80 \div 16$

25. $7\overline{)6394}$

26. $3073 \div 8$

27. $60\overline{)286}$

28. $4266 \div 79$

29. $38\overline{)17{,}176}$

30. $14\overline{)70{,}112}$

31. $52{,}668 \div 12$

Round 345,759 to the nearest: [1.6a]

32. Hundred.

33. Ten.

34. Thousand.

35. Hundred thousand.

Use < or > for $\square$ to write a true sentence. [1.6c]

36. $67 \ \square\ 56$

37. $1 \ \square\ 23$

Estimate each sum, difference, or product by first rounding to the nearest hundred. Show your work. [1.6b]

38. $41{,}348 + 19{,}749$

39. $38{,}652 - 24{,}549$

40. $396 \cdot 748$

Solve. [1.7b]

41. $46 \cdot n = 368$

42. $47 + x = 92$

43. $1 \cdot y = 58$

44. $24 = x + 24$

45. Write exponential notation: $4 \cdot 4 \cdot 4$. [1.9a]

Evaluate. [1.9b]

46. 10^4

47. 6^2

Simplify. [1.9c, d]

48. $8 \cdot 6 + 17$

49. $10 \cdot 24 - (18 + 2) \div 4 - (9 - 7)$

50. $(80 \div 16) \times [(20 - 56 \div 8) + (8 \cdot 8 - 5 \cdot 5)]$

51. Find the average of 157, 170, and 168. [1.9c]

Solve. [1.8a]

52. *Computer Purchase.* Natasha has $196 and wants to buy a computer for $698. How much more does she need?

53. Toni has $406 in her checking account. She is paid $78 for a part-time job and deposits that in her checking account. How much is then in her account?

54. *Lincoln-Head Pennies.* In 1909, the first Lincoln-head pennies were minted. Seventy-three years later, these pennies were first minted with a decreased copper content. In what year was the copper content reduced?

55. A beverage company packed 228 cans of soda into 12-can cartons. How many cartons were filled?

56. An apartment builder bought 13 gas stoves at $425 each and 13 refrigerators at $620 each. What was the total cost?

57. An apple farmer keeps bees in her orchard to help pollinate the apple blossoms. The bees from an average beehive can pollinate 30 surrounding trees during one growing season. A farmer has 420 trees. How many beehives does she need to pollinate all of them?

Source: Jordan Orchards, Westminster, PA

58. *Olympic Trampoline.* Shown below is an Olympic trampoline. Determine the area and the perimeter of the trampoline. [1.2b], [1.4b]

Source: International Trampoline Industry Association, Inc.

59. A chemist has 2753 mL of alcohol. How many 20-mL beakers can be filled? How much will be left over?

60. A family budgeted $7825 a year for food and clothing and $2860 for entertainment. The yearly income of the family was $38,283. How much of this income remained after these two allotments?

61. Simplify: $7 + (4 + 3)^2$. [1.9c]

A. 32 B. 56
C. 151 D. 196

62. Simplify: $7 + 4^2 + 3^2$. [1.9c]

A. 32 B. 56
C. 130 D. 196

63. $[46 - (4 - 2) \cdot 5] \div 2 + 4$ [1.9d]

A. 6 B. 20
C. 114 D. 22

Synthesis

64. Determine the missing digit d. [1.4a]

$$\begin{array}{r} 9\,d \\ \times\ \ d\,2 \\ \hline 8\,0\,3\,6 \end{array}$$

65. Determine the missing digits a and b. [1.5a]

$$\begin{array}{r} 9\ a\ 1 \\ 2\ b\ 1\,\overline{)\,2\ 3\ 6{,}4\ 2\ 1} \end{array}$$

66. A mining company estimates that a crew must tunnel 2000 ft into a mountain to reach a deposit of copper ore. Each day, the crew tunnels about 500 ft. Each night, about 200 ft of loose rocks roll back into the tunnel. How many days will it take the mining company to reach the copper deposit? [1.8a]

Understanding Through Discussion and Writing

1. Is subtraction associative? Why or why not? [1.3a]

2. Explain how estimating and rounding can be useful when shopping for groceries. [1.6b]

3. Write a problem for a classmate to solve. Design the problem so that the solution is "The driver still has 329 mi to travel." [1.8a]

4. Consider the expressions $9 - (4 \cdot 2)$ and $(3 \cdot 4)^2$. Are the parentheses necessary in each case? Explain. [1.9c]

CHAPTER 1 Test

For Extra Help For step-by-step test solutions, access the Chapter Test Prep Videos in MyMathLab® or on YouTube (search "BittingerBasicEl" and click on "Channels").

1. In the number 546,789, which digit tells the number of hundred thousands?

2. Write expanded notation: 8843.

3. Write a word name: 38,403,277.

Add.

4. $\begin{array}{r} 6811 \\ +\ 3178 \\ \hline \end{array}$

5. $\begin{array}{r} 45{,}889 \\ +\ 17{,}902 \\ \hline \end{array}$

6. $\begin{array}{r} 1239 \\ 843 \\ 301 \\ +\ 782 \\ \hline \end{array}$

7. $\begin{array}{r} 6203 \\ +\ 4312 \\ \hline \end{array}$

Subtract.

8. $\begin{array}{r} 7983 \\ -\ 4353 \\ \hline \end{array}$

9. $\begin{array}{r} 2974 \\ -\ 1935 \\ \hline \end{array}$

10. $\begin{array}{r} 8907 \\ -\ 2059 \\ \hline \end{array}$

11. $\begin{array}{r} 23{,}067 \\ -\ 17{,}892 \\ \hline \end{array}$

Multiply.

12. $\begin{array}{r} 4568 \\ \times\ 9 \\ \hline \end{array}$

13. $\begin{array}{r} 8876 \\ \times\ 600 \\ \hline \end{array}$

14. $\begin{array}{r} 65 \\ \times\ 37 \\ \hline \end{array}$

15. $\begin{array}{r} 678 \\ \times\ 788 \\ \hline \end{array}$

Divide.

16. $15 \div 4$

17. $420 \div 6$

18. $89\overline{)8633}$

19. $44\overline{)35{,}428}$

Round 34,528 to the nearest:

20. Thousand.

21. Ten.

22. Hundred.

Estimate each sum, difference, or product by first rounding to the nearest hundred. Show your work.

23. $\begin{array}{r} 23{,}649 \\ +\ 54{,}746 \\ \hline \end{array}$

24. $\begin{array}{r} 54{,}751 \\ -\ 23{,}649 \\ \hline \end{array}$

25. $\begin{array}{r} 824 \\ \times\ 489 \\ \hline \end{array}$

Use $<$ or $>$ for $\square$ to write a true sentence.

26. $34 \square 17$

27. $117 \square 157$

Solve.

28. $28 + x = 74$

29. $169 \div 13 = n$

30. $38 \cdot y = 532$

31. $381 = 0 + a$

Solve.

32. *Calorie Content.* An 8-oz serving of whole milk contains 146 calories. This is 63 calories more than the number of calories in an 8-oz serving of skim milk. How many calories are in an 8-oz serving of skim milk?

Source: *American Journal of Clinical Nutrition*

33. A box contains 5000 staples. How many staplers can be filled from the box if each stapler holds 250 staples?

34. *Largest States.* The following table lists the five largest states in terms of their land area. Find the total land area of these states.

STATE	AREA (in square miles)
Alaska	571,951
Texas	261,797
California	155,959
Montana	145,552
New Mexico	121,356

Sources: U.S. Department of Commerce; U.S. Census Bureau

35. *Pool Tables.* The Bradford™ pool table made by Brunswick Billiards comes in three sizes of playing area, 50 in. by 100 in., 44 in. by 88 in., and 38 in. by 76 in.

Source: Brunswick Billiards

a) Determine the perimeter and the playing area of each table.
b) By how much does the area of the largest table exceed the area of the smallest table?

36. *Hostess Ding Dongs®.* Hostess packages its Ding Dong snack cakes in 12-packs. How many 12-packs can it fill with 22,231 cakes? How many will be left over?

37. *Office Supplies.* Morgan manages the office of a small graphics firm. He buys 3 black inkjet cartridges at $15 each and 2 photo inkjet cartridges at $25 each. How much does the purchase cost?

38. Write exponential notation: $12 \cdot 12 \cdot 12 \cdot 12$.

Evaluate.

39. 7^3

40. 10^5

Simplify.

41. $35 - 1 \cdot 28 \div 4 + 3$

42. $10^2 - 2^2 \div 2$

43. $(25 - 15) \div 5$

44. $2^4 + 24 \div 12$

45. $8 \times \{(20 - 11) \cdot [(12 + 48) \div 6 - (9 - 2)]\}$

46. Find the average of 97, 99, 87, and 89.

A. 93 **B.** 124 **C.** 186 **D.** 372

Synthesis

47. An open cardboard container is 8 in. wide, 12 in. long, and 6 in. high. How many square inches of cardboard are used?

48. Use trials to find the single-digit number a for which

$$359 - 46 + a \div 3 \times 25 - 7^2 = 339.$$

49. Cara spends $229 a month to repay her student loan. If she has already paid $9160 on the 10-year loan, how many payments remain?

Integers

STUDYING FOR SUCCESS *Make the Most of Your Homework*

- ☐ Double-check that you have copied an exercise correctly.
- ☐ Check that your answer is reasonable.
- ☐ Include correct units with answers.

2.1 The Integers

OBJECTIVES

a State the integer that corresponds to a real-world situation.

b Graph integers on the number line.

c Determine which of two integers is greater and indicate which, using $<$ or $>$.

d Find the absolute value of an integer.

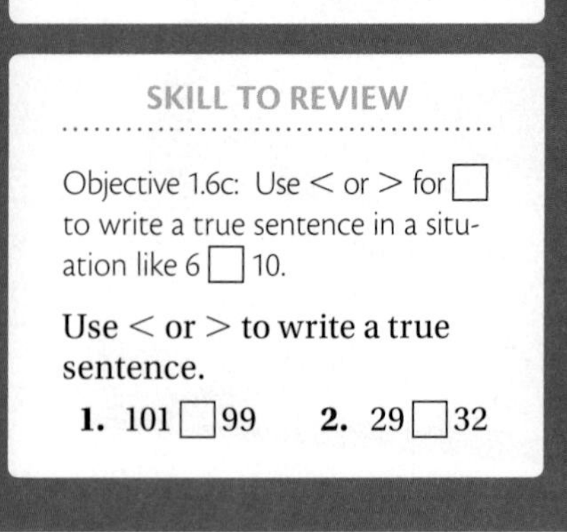

SKILL TO REVIEW

Objective 1.6c: Use $<$ or $>$ for $\square$ to write a true sentence in a situation like $6 \square 10$.

Use $<$ or $>$ to write a true sentence.

1. $101 \square 99$ **2.** $29 \square 32$

In this section, we introduce the *integers*. To describe integers, we start with the whole numbers, 0, 1, 2, 3, and so on. For each number 1, 2, 3, and so on, we obtain a new number to the left of zero on the number line:

> For the number 1, there will be an *opposite* number -1 (negative 1).
>
> For the number 2, there will be an *opposite* number -2 (negative 2).
>
> For the number 3, there will be an *opposite* number -3 (negative 3), and so on.

The **integers** consist of the whole numbers and these new numbers. We picture them on the number line as follows.

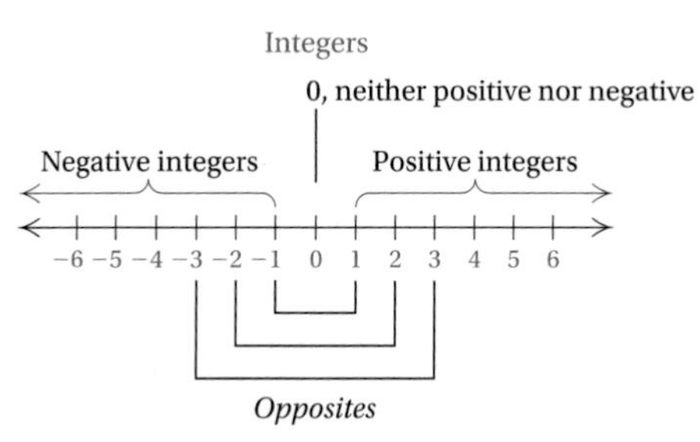

We call the integers to the left of zero **negative integers**. The natural numbers are called **positive integers**. Zero is neither positive nor negative. We call -1 and 1 **opposites** of each other. Similarly, -2 and 2 are opposites, -3 and 3 are opposites, -100 and 100 are opposites, and 0 is its own opposite. Opposite pairs of numbers like -3 and 3 are the same distance from 0. The integers extend infinitely on the number line to the left and right of zero.

INTEGERS

The **integers:** $\ldots, -5, -4, -3, -2, -1, 0, 1, 2, 3, 4, 5, \ldots$

a INTEGERS AND THE REAL WORLD

Integers correspond to many real-world problems and situations. The following examples will help you get ready to translate problem situations that involve integers to mathematical language.

Answers

Skill to Review:
1. $>$ **2.** $<$

EXAMPLE 1 Tell which integer corresponds to this situation: The temperature is 4 degrees below zero.

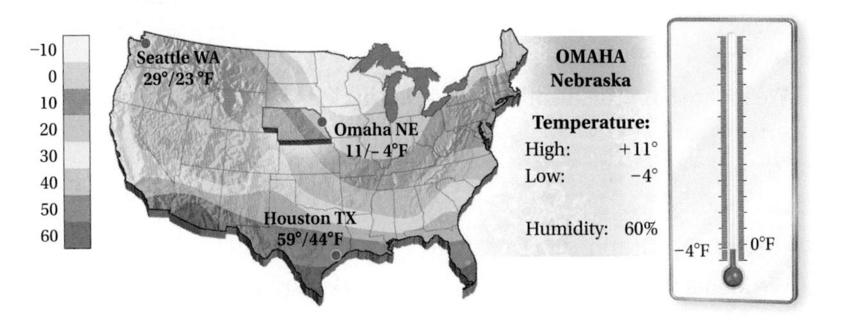

The integer -4 corresponds to the situation. The temperature is $-4°$. ■

EXAMPLE 2 *Water Level.* Tell which integer corresponds to this situation: As the water level of the Mississippi River fell during the drought of 2012, barge traffic was restricted, causing a severe decline in shipping volumes. On August 24, the river level at Greenville, Mississippi, was 10 ft below normal.

Source: Rick Jervis, *USA TODAY*, August 24, 2012

The integer -10 corresponds to the drop in water level. ■

EXAMPLE 3 *Stock Price Change.* Tell which integers correspond to this situation: Hal owns a stock whose price decreased from \$27 per share to \$11 per share over a recent time period. He owns another stock whose price increased from \$20 per share to \$22 per share over the same time period.

The integer -16 corresponds to the decrease in the value of the first stock. The integer 2 represents the increase in the value of the second stock.

Do Exercises 1–5. ▶

Tell which integers correspond to each situation.

1. *High and Low Temperatures.* As of 2010, the highest recorded temperature in Illinois was 117°F on July 14, 1954, in East St. Louis. The lowest recorded temperature in Illinois was 36°F below zero on January 5, 1999, in Congerville.

   Source: National Climate Data Center, NESDIS, NOAA, U.S. Dept. of Commerce

2. *Stock Price Decrease.* The price of a stock decreased from \$41 per share to \$38 per share over a recent period.

3. At 10 sec before liftoff, ignition occurs. At 148 sec after liftoff, the first stage is detached from the rocket.

4. The halfback gained 8 yd on first down. The quarterback was sacked for a 5-yd loss on second down.

5. A submarine dove 120 ft, rose 50 ft, and then dove 80 ft.

Answers

1. 117; -36 **2.** -3 **3.** -10; 148
4. 8; -5 **5.** -120; 50; -80

b GRAPH OF INTEGERS

To **graph** a number means to find and mark its point on the number line.

EXAMPLE 4 Graph: 5.

We locate 5 on the number line and mark its point with a dot.

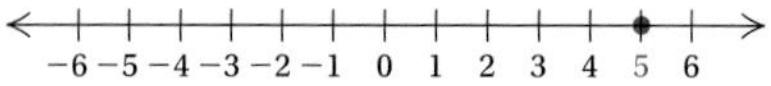

EXAMPLE 5 Graph: 0.

We locate 0 on the number line and mark its point with a dot.

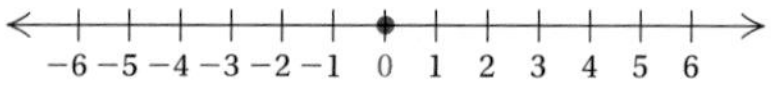

EXAMPLE 6 Graph: −3.

We locate −3 on the number line and mark its point with a dot.

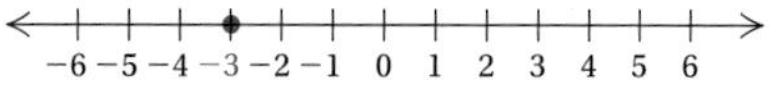

◀ **Do Exercises 6–8.**

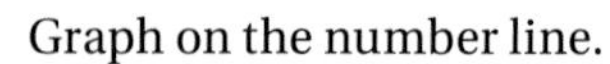

Graph on the number line.

6. −4

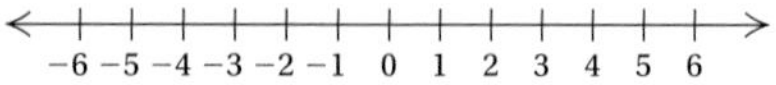

7. 1

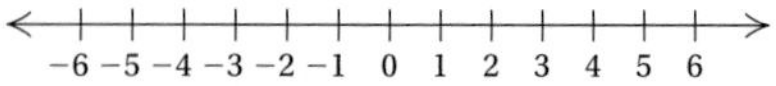

8. 4

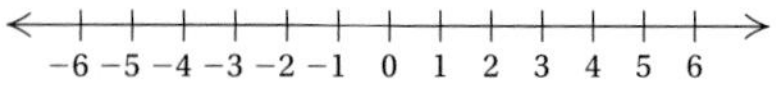

c THE INTEGERS AND ORDER

The integers are named in order on the number line. (See Section 1.6.) For any two numbers on the line, the one to the left is less than the one to the right.

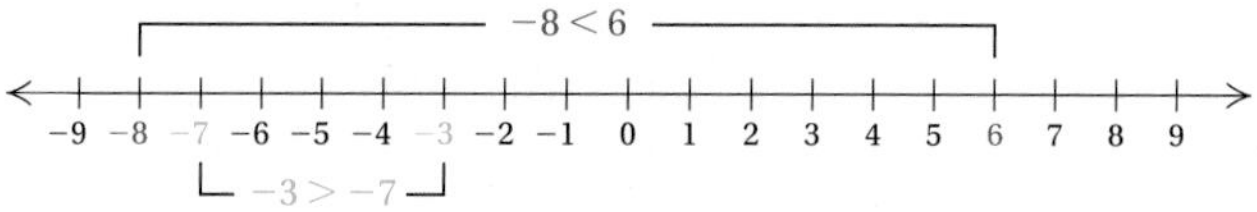

We use the symbol **<** to mean "**is less than**." The sentence $-8 < 6$ means "−8 is less than 6." The symbol **>** means "**is greater than**." The sentence $-3 > -7$ means "−3 is greater than −7."

EXAMPLES Use either < or > for □ to write a true sentence.

7. −7 □ 3 — Since −7 is to the left of 3, we have $-7 < 3$.
8. 6 □ −12 — Since 6 is to the right of −12, we have $6 > -12$.
9. −18 □ −5 — Since −18 is to the left of −5, we have $-18 < -5$.
10. 25 □ 23 — Since 25 is to the right of 23, we have $25 > 23$.
11. −8 □ 0 — Since −8 is to the left of 0, we have $-8 < 0$.
12. −11 □ −13 — Since −11 is to the right of −13, we have $-11 > -13$.

◀ **Do Exercises 9–14.**

Use either < or > for □ to write a true sentence.

9. −3 □ 7
10. −8 □ −5
11. 7 □ −10
12. 0 □ −15
13. 79 □ 63
14. −9 □ −8

Answers

6. 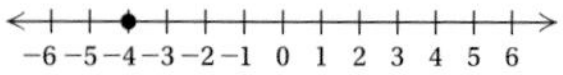

7. 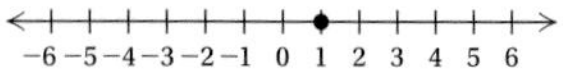

8. 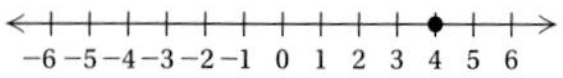

9. < 10. < 11. > 12. >
13. > 14. <

d ABSOLUTE VALUE

From the number line, we see that numbers like 4 and -4 are the same distance from zero. Distance is always a nonnegative number. We call the distance of a number from zero the **absolute value** of the number.

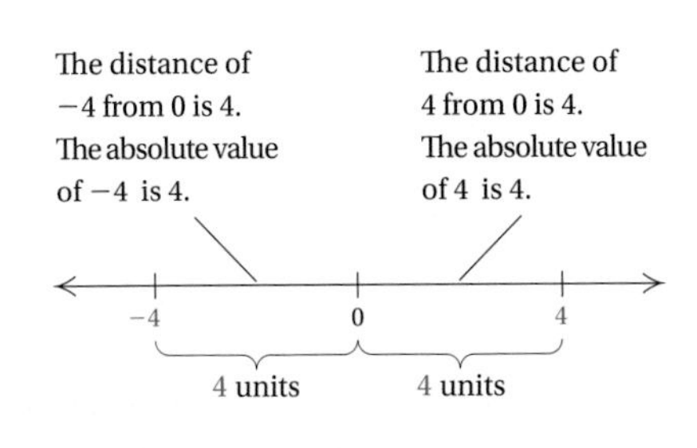

ABSOLUTE VALUE

The **absolute value** of a number is its distance from zero on the number line. We use the symbol $|x|$ to represent the absolute value of a number x.

EXAMPLES Find the absolute value.

13. $|-7|$ The distance of -7 from 0 is 7, so $|-7| = 7$.

14. $|12|$ The distance of 12 from 0 is 12, so $|12| = 12$.

15. $|0|$ The distance of 0 from 0 is 0, so $|0| = 0$.

16. $|45| = 45$

17. $|-101| = 101$

Do Exercises 15–20. ▶

Find the absolute value.

15. $|8|$

16. $|0|$

17. $|-9|$

18. $|-84|$

19. $|157|$

20. $|-225|$

To find the absolute value of a number:

a) If a number is negative, its absolute value is its opposite.

b) If a number is positive or zero, its absolute value is the same as the number.

Caution!

Absolute value is a distance, so the absolute value of a number is always positive or zero. Absolute value cannot be negative.

Answers

15. 8 **16.** 0 **17.** 9 **18.** 84
19. 157 **20.** 225

2.1 Exercise Set

For Extra Help

MyMathLab® MathXL® PRACTICE WATCH READ REVIEW

Reading Check

Use the number line below, on which the letters name numbers, for Exercises RC1–RC8.

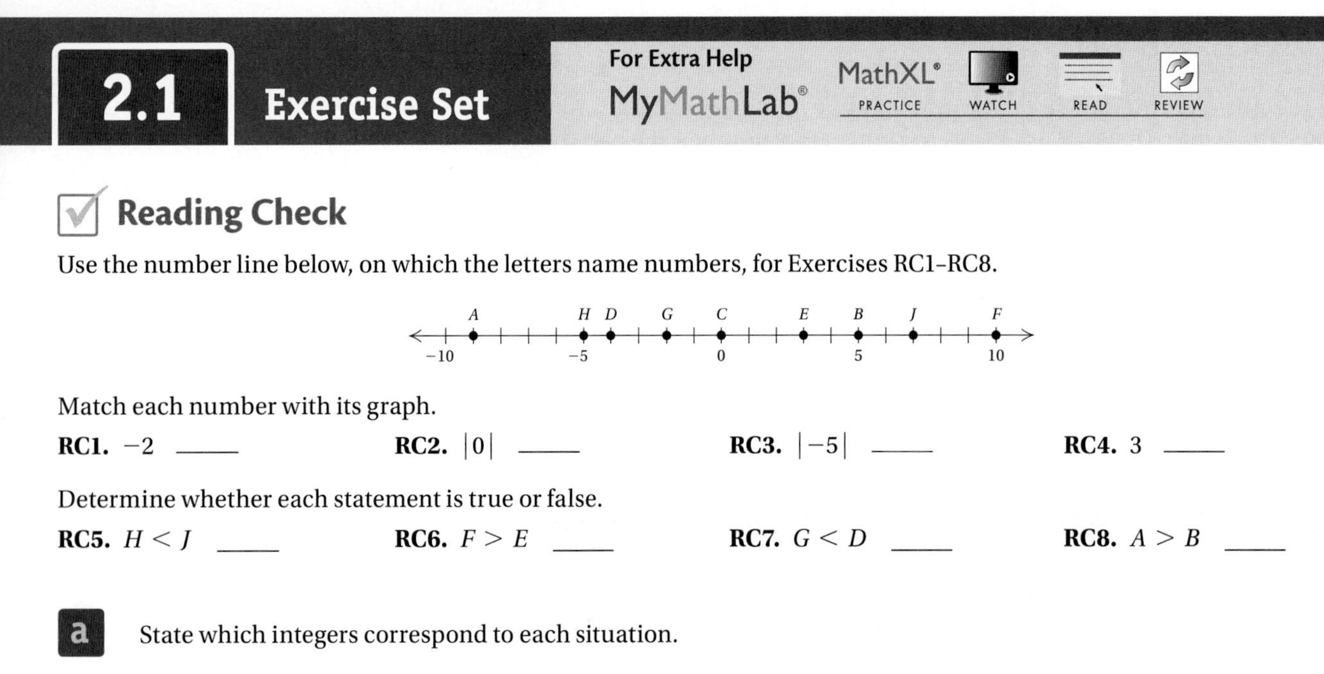

Match each number with its graph.

RC1. -2 ____ **RC2.** $|0|$ ____ **RC3.** $|-5|$ ____ **RC4.** 3 ____

Determine whether each statement is true or false.

RC5. $H < J$ ____ **RC6.** $F > E$ ____ **RC7.** $G < D$ ____ **RC8.** $A > B$ ____

a State which integers correspond to each situation.

1. On Wednesday, the temperature was 24° above zero. On Thursday, it was 2° below zero.

2. A student deposited her tax refund of $750 in a savings account. Two weeks later, she withdrew $125 to pay technology fees.

3. *Temperature Extremes.* The highest temperature ever created on Earth was 950,000,000°F. The lowest temperature ever created was approximately 460°F below zero.

Source: *The Guinness Book of Records*

4. *Extreme Climate.* Verkhoyansk, a river port in northeast Siberia, has the most extreme climate on the planet. Its average monthly winter temperature is 58.5°F below zero, and its average monthly summer temperature is 56.5°F.

Source: *The Guinness Book of Records*

5. *Empire State Building.* The Empire State Building has a total height, including the lightning rod at the top, of 1454 ft. The foundation depth is 55 ft below ground level.

Source: www.empirestatebuildingfacts.com

6. *Shipwreck.* There are numerous shipwrecks to explore near Bermuda. One of the most frequently visited wrecks is *L'Herminie*, a French warship that sank in 1837. This wreck is 35 ft below the surface.

Source: www./10best.com/interests/adventure/scuba-diving-in-pirate-territory/

b Graph each number on the number line.

7. -5

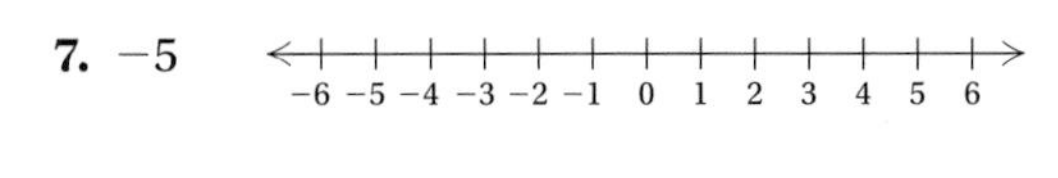

8. 3

9. 2

−6 −5 −4 −3 −2 −1 0 1 2 3 4 5 6

10. -1

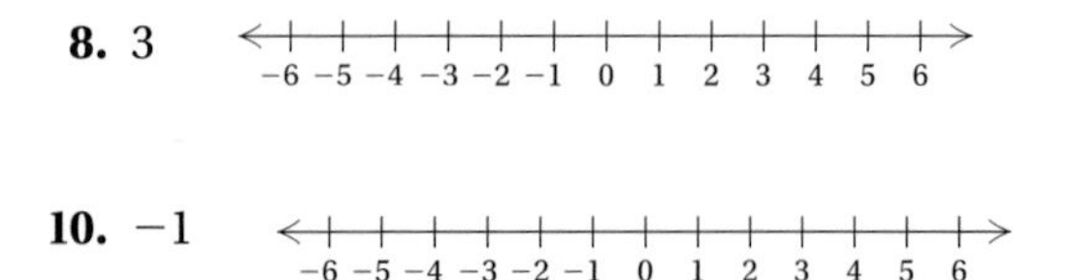

c Use either $<$ or $>$ for $\square$ to write a true sentence.

11. $8 \square 0$

12. $3 \square 0$

13. $-8 \square 3$

14. $6 \square -6$

15. $8 \square -8$

16. $0 \square -9$

17. $-10 \square -5$

18. $-4 \square -3$

19. $-5 \square -11$

20. $-3 \square -4$

21. $-6 \square -5$

22. $-10 \square -14$

23. $-7 \square 0$

24. $-3 \square -2$

25. $1 \square -15$

26. $2 \square -12$

d Find the absolute value.

27. $|-3|$

28. $|-6|$

29. $|18|$

30. $|0|$

31. $|325|$

32. $|-4|$

33. $|-29|$

34. $|217|$

35. $|-300|$

36. $|-47|$

37. $|53|$

38. $|-76|$

Skill Maintenance

Add. [1.2a]

39. $\begin{array}{r} 9182 \\ +4367 \\ \hline \end{array}$

40. $\begin{array}{r} 278 \\ +829 \\ \hline \end{array}$

41. $\begin{array}{r} 32,047 \\ +18,562 \\ \hline \end{array}$

Subtract. [1.3a]

42. $\begin{array}{r} 651 \\ -432 \\ \hline \end{array}$

43. $\begin{array}{r} 1207 \\ -\ 948 \\ \hline \end{array}$

44. $\begin{array}{r} 43,004 \\ -34,226 \\ \hline \end{array}$

Synthesis

Use either $<$, $>$, or $=$ for $\square$ to write a true sentence.

45. $|-5| \square |-2|$

46. $|4| \square |-7|$

47. $|-8| \square |8|$

48. List in order from the least to the greatest: $-6, 7, -10, |-6|, 2^2, |3|, 1^6, -5, 0$.

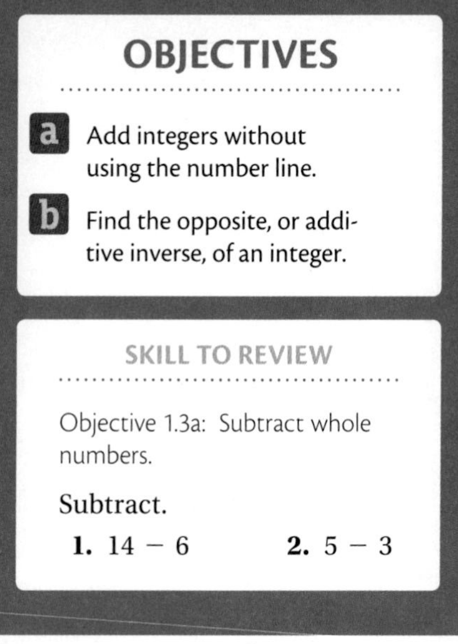

2.2 Addition of Integers

OBJECTIVES

a Add integers without using the number line.

b Find the opposite, or additive inverse, of an integer.

SKILL TO REVIEW

Objective 1.3a: Subtract whole numbers.

Subtract.

1. $14 - 6$ **2.** $5 - 3$

We now consider addition of integers. First, to gain an understanding, we add using the number line. Then we consider rules for addition.

Addition on the Number Line

To find $a + b$, we start at 0, move to a, and then move according to b.

a) If b is positive, we move from a to the right.

b) If b is negative, we move from a to the left.

c) If b is 0, we stay at a.

EXAMPLE 1 Add: $3 + (-5)$.

We start at 0 and move to 3. Then we move 5 units left since -5 is negative.

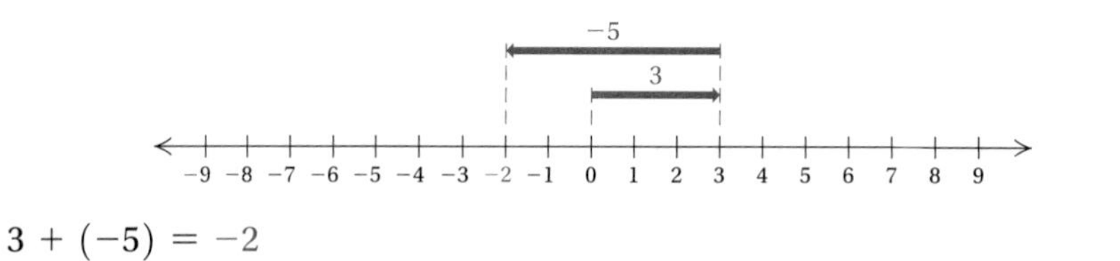

$3 + (-5) = -2$

EXAMPLE 2 Add: $-4 + (-3)$.

We start at 0 and move to -4. Then we move 3 units left since -3 is negative.

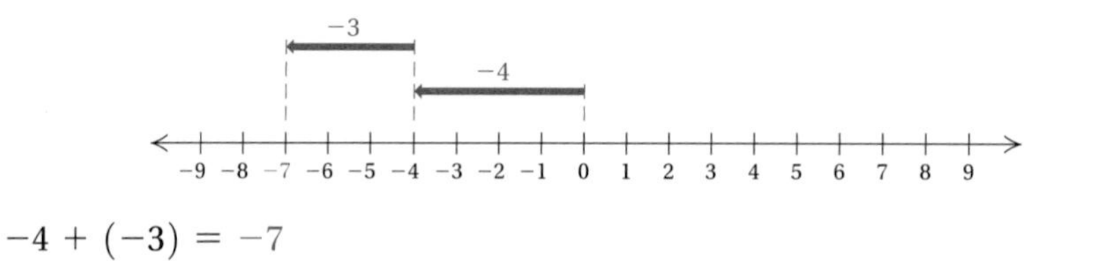

$-4 + (-3) = -7$

EXAMPLE 3 Add: $-4 + 9$.

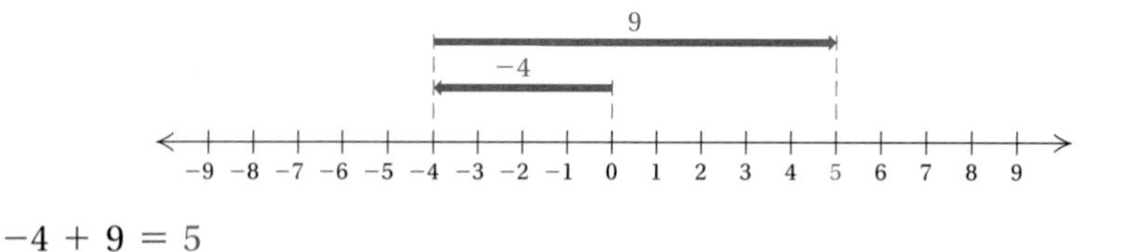

$-4 + 9 = 5$

EXAMPLE 4 Add: $-6 + 0$.

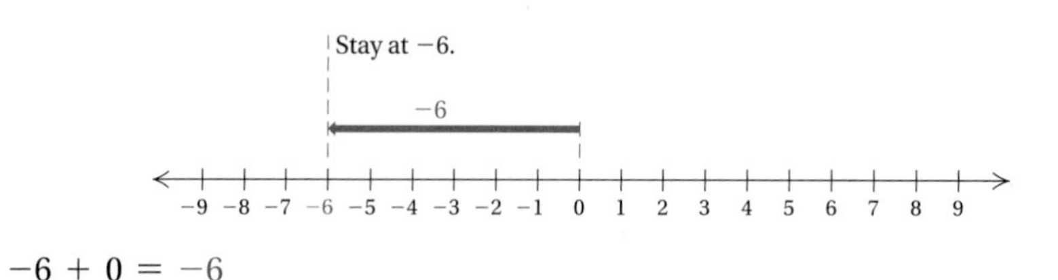

$-6 + 0 = -6$

Do Margin Exercises 1–6.

Add using the number line.

1. $0 + (-3)$

−6 −5 −4 −3 −2 −1 0 1 2 3 4 5 6

2. $1 + (-4)$

−6 −5 −4 −3 −2 −1 0 1 2 3 4 5 6

3. $-3 + (-2)$

−6 −5 −4 −3 −2 −1 0 1 2 3 4 5 6

4. $-3 + 7$

−6 −5 −4 −3 −2 −1 0 1 2 3 4 5 6

5. $-2 + 2$

−6 −5 −4 −3 −2 −1 0 1 2 3 4 5 6

6. $-1 + (-3)$

−6 −5 −4 −3 −2 −1 0 1 2 3 4 5 6

Answers

Skill to Review:
1. 8 **2.** 2

Margin Exercises:
1. -3 **2.** -3 **3.** -5 **4.** 4 **5.** 0 **6.** -4

a ADDITION WITHOUT THE NUMBER LINE

You may have noticed some patterns in the preceding examples. These lead us to rules for adding without using the number line that are more efficient for adding larger or more complicated numbers.

RULES FOR ADDITION OF INTEGERS

1. *Positive numbers*: Add the same way you do whole numbers. The answer is positive.
2. *Negative numbers*: Add absolute values. The answer is negative.
3. *A positive number and a negative number*:
   - If the numbers have the same absolute value, the answer is 0.
   - If the numbers have different absolute values, subtract the smaller absolute value from the larger. Then:
     a) If the positive number has the greater absolute value, the answer is positive.
     b) If the negative number has the greater absolute value, the answer is negative.
4. *One number is zero*: The sum is the other number.

Rule 4 is known as the **identity property of 0**. It says that for any number a, $a + 0 = a$.

EXAMPLES Add without using the number line.

5. $-12 + (-7) = -19$ Add the absolute values, 12 and 7, getting 19. Make the answer *negative*, -19.

6. $-1 + 8 = 7$ The absolute values are 1 and 8. The difference is 7. The positive number has the larger absolute value, so the answer is *positive*, 7.

7. $-36 + 21 = -15$ The absolute values are 36 and 21. The difference is 15. The negative number has the larger absolute value, so the answer is *negative*, -15.

8. $5 + (-5) = 0$ The numbers have the same absolute value. The sum is 0.

9. $-7 + 0 = -7$ One number is zero. The sum is the other number, -7.

10. $-9 + 3 = -6$

11. $15 + (-4) = 11$

12. $-13 + (-27) = -40$

Do Exercises 7–20.

Suppose we wish to add several numbers, some positive and some negative, such as $15 + (-2) + 7 + 14 + (-5) + (-12)$. How can we proceed?

Add without using the number line.

7. $-5 + (-6)$

8. $-9 + (-3)$

9. $-4 + 6$

10. $-7 + 3$

11. $5 + (-7)$

12. $-20 + 20$

13. $-11 + (-11)$

14. $10 + (-7)$

15. $0 + (-26)$

16. $-6 + 8$

17. $-4 + (-5)$

18. $-8 + 2$

19. $9 + (-7)$

20. $-19 + (-11)$

Answers

7. -11 8. -12 9. 2 10. -4 11. -2 12. 0 13. -22 14. 3 15. -26 16. 2 17. -9 18. -6 19. 2 20. -30

The commutative and associative laws hold for integers. Thus we can change grouping and order as we please when adding. For instance, we can group the positive numbers together and the negative numbers together and add them separately. Then we add the two results.

EXAMPLE 13 Add: $15 + (-2) + 7 + 14 + (-5) + (-12)$.

a) $15 + 7 + 14 = 36$ Adding the positive numbers
b) $-2 + (-5) + (-12) = -19$ Adding the negative numbers
c) $36 + (-19) = 17$ Adding the results of (a) and (b)

We can also add the numbers in any other order we wish, say, from left to right as follows:

$$\begin{aligned} 15 + (-2) + 7 + 14 + (-5) + (-12) &= 13 + 7 + 14 + (-5) + (-12) \\ &= 20 + 14 + (-5) + (-12) \\ &= 34 + (-5) + (-12) \\ &= 29 + (-12) \\ &= 17 \end{aligned}$$

Do Exercises 21–24.

Add.

21. $(-15) + (-37) + 25 + 42 + (-59) + (-14)$

22. $42 + (-81) + (-28) + 24 + 18 + (-31)$

23. $-2 + (-10) + 6 + (-7)$

24. $-35 + 17 + 14 + (-27) + 31 + (-12)$

b OPPOSITES, OR ADDITIVE INVERSES

Suppose we add two numbers that are **opposites**, such as 6 and -6. The result is 0. When opposites are added, the result is always 0. Opposites are also called **additive inverses**. Every integer has an opposite, or additive inverse.

OPPOSITES, OR ADDITIVE INVERSES

Two numbers whose sum is 0 are called **opposites**, or **additive inverses**, of each other.

EXAMPLES Find the opposite, or additive inverse, of each number.

14. 34 The opposite of 34 is -34 because $34 + (-34) = 0$.
15. -8 The opposite of -8 is 8 because $-8 + 8 = 0$.
16. 0 The opposite of 0 is 0 because $0 + 0 = 0$.

Do Exercises 25–30.

Find the opposite, or additive inverse.

25. -4
26. 8
27. -7
28. -34
29. 0
30. 12

To name an opposite, we use the symbol $-$, as follows.

SYMBOLIZING OPPOSITES

The opposite, or additive inverse, of a number a can be named $-a$ (read "the opposite of a," or "the additive inverse of a").

Note that if we take a number, say 8, and find its opposite, -8, and then find the opposite of the result, we will have the original number, 8, again.

Answers

21. -58 22. -56 23. -13 24. -12
25. 4 26. -8 27. 7 28. 34 29. 0
30. -12

THE OPPOSITE OF THE OPPOSITE

The opposite of the opposite of a number is the number itself. (The additive inverse of the additive inverse of a number is the number itself.) That is, for any number a,

$$-(-a) = a.$$

EXAMPLE 17 Evaluate $-x$ and $-(-x)$ when $x = 16$.

We replace x in each case with 16.

a) If $x = 16$, then $-x = -16 = -16$. The opposite of 16 is -16.

b) If $x = 16$, then $-(-x) = -(-16) = 16$. The opposite of the opposite of 16 is 16.

EXAMPLE 18 Evaluate $-x$ and $-(-x)$ when $x = -3$.

We replace x in each case with -3.

a) If $x = -3$, then $-x = -(-3) = 3$.

b) If $x = -3$, then $-(-x) = -(-(-3)) = -(3) = -3$.

Note that in Example 18 we used an extra set of parentheses to show that we are substituting the negative number -3 for x. Symbolism like $- -x$ is not considered meaningful.

Do Exercises 31–34.

Evaluate $-x$ and $-(-x)$ when:

31. $x = 14$.

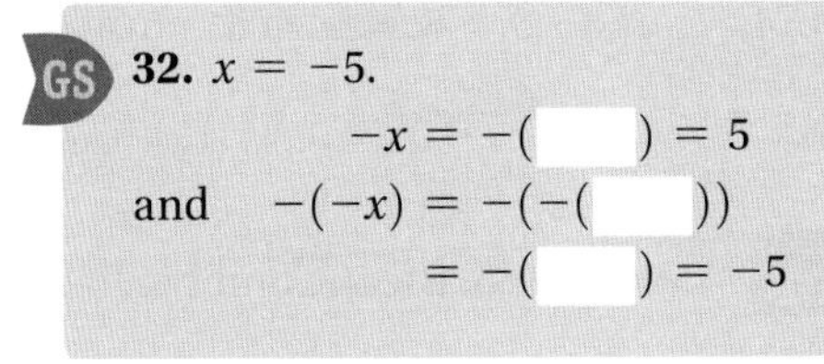

32. $x = -5$.

$-x = -(\quad) = 5$

and $-(-x) = -(-(\quad))$

$= -(\quad) = -5$

33. $x = -9$. **34.** $x = 2$.

A symbol such as -8 is usually read "negative 8." It could be read "the additive inverse of 8," because the additive inverse of 8 is negative 8. It could also be read "the opposite of 8," because the opposite of 8 is -8. Thus a symbol like -8 can be read in more than one way. A symbol like $-x$, which has a variable, should be read "the opposite of x" or "the additive inverse of x" and *not* "negative x," because we do not know whether x represents a positive number, a negative number, or 0.

We can use the symbolism $-a$ to restate the definition of opposite, or additive inverse.

OPPOSITES, OR ADDITIVE INVERSES

For any integer a, the opposite, or additive inverse, of a, denoted $-a$, is such that

$$a + (-a) = (-a) + a = 0.$$

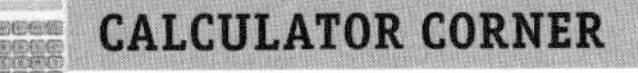

CALCULATOR CORNER

Negative Numbers On many calculators, we can enter negative numbers using the [+/−] key. This allows us to perform calculations with integers. On some calculators, this key is labeled [(−)]. To enter -8, for example, we press [8] [+/−]. To find the sum $-14 + (-9)$, we press [1] [4] [+/−] [+] [9] [+/−] [=]. The result is -23. Note that it is not necessary to use parentheses when entering this expression.

EXERCISES Add.

1. $-4 + 17$
2. $3 + (-11)$
3. $-2 + (-10)$

Signs of Numbers

A negative number is sometimes said to have a "negative sign." A positive number is said to have a "positive sign." When we replace a number with its opposite, we can say that we have "changed its sign."

EXAMPLES Change the sign. (Find the opposite.)

19. -3 $\quad -(-3) = 3$

20. -13 $\quad -(-13) = 13$

21. 0 $\quad -(0) = 0$

22. 14 $\quad -(14) = -14$

Do Exercises 35–38.

Find the opposite. (Change the sign.)

35. -4 **36.** -21

37. 0 **38.** 46

Answers

31. $-14; 14$ **32.** $5; -5$ **33.** $9; -9$ **34.** $-2; 2$ **35.** 4 **36.** 21 **37.** 0 **38.** -46

Guided Solution:

32. $-5; -5, 5$

2.2 Exercise Set

For Extra Help
MyMathLab® MathXL® PRACTICE WATCH READ REVIEW

Reading Check

Fill in each blank with either "left" or "right" so that the statement describes the steps for adding the given numbers on the number line.

RC1. To add $7 + 2$, start at 0, move ______________ to 7, and then move 2 units ______________. The sum is 9.

RC2. To add $-3 + (-5)$, start at 0, move ______________ to -3, and then move 5 units ______________. The sum is -8.

RC3. To add $4 + (-6)$, start at 0, move ______________ to 4, and then move 6 units ______________. The sum is -2.

RC4. To add $-8 + 3$, start at 0, move ______________ to -8, and then move 3 units ______________. The sum is -5.

a Add. Do not use the number line except as a check.

1. $-9 + 2$

2. $-5 + 2$

3. $-10 + 6$

4. $4 + (-3)$

5. $-8 + 8$

6. $4 + (-4)$

7. $-3 + (-5)$

8. $-6 + (-8)$

9. $-7 + 0$

10. $-10 + 0$

11. $0 + (-27)$

12. $0 + (-36)$

13. $17 + (-17)$

14. $-20 + 20$

15. $-17 + (-25)$

16. $-23 + (-14)$

17. $18 + (-18)$

18. $-13 + 13$

19. $-18 + 18$

20. $11 + (-11)$

21. $8 + (-5)$

22. $-7 + 8$

23. $-4 + (-5)$

24. $10 + (-12)$

25. $13 + (-6)$

26. $-3 + 14$

27. $-25 + 25$

28. $40 + (-40)$

29. $63 + (-18)$

30. $85 + (-65)$

31. $-6 + 4$

32. $-3 + 2$

33. $-2 + (-5)$

34. $-7 + (-6)$

35. $-22 + 3$

36. $35 + (-19)$

37. $-5 + (-7) + 6$

38. $-10 + (-8) + 3$

39. $-4 + 7 + (-4)$

40. $-1 + 20 + (-1)$

41. $75 + (-14) + (-17) + (-5)$

42. $28 + (-44) + 17 + 31 + (-94)$

43. $-44 + (-3) + 95 + (-5)$

44. $24 + 3 + (-44) + (-8) + 63$

45. $98 + (-54) + 113 + (-998) + 44 + (-612) + (-18) + 334$

46. $-455 + (-123) + 1026 + (-919) + 213 + 111 + (-874)$

b Find the opposite, or additive inverse.

47. 24

48. −84

49. −26

50. 36

Find $-x$ when:

51. $x = 9$.

52. $x = -26$.

53. $x = -14$.

54. $x = 52$.

Find $-(-x)$ when:

55. $x = -65$.

56. $x = 31$.

57. $x = 5$.

58. $x = -18$.

Change the sign. (Find the opposite.)

59. −14

60. 33

61. 10

62. −17

Skill Maintenance

Multiply. [1.4a]

63. $\begin{array}{r} 192 \\ \times\ 18 \\ \hline \end{array}$

64. $\begin{array}{r} 264 \\ \times\ 519 \\ \hline \end{array}$

65. $\begin{array}{r} 6403 \\ \times\ 708 \\ \hline \end{array}$

Divide. [1.5a]

66. $6\overline{)2451}$

67. $54\overline{)6904}$

68. $404\overline{)89{,}615}$

Round 641,539 to the nearest: [1.6a]

69. Ten.

70. Hundred.

71. Thousand.

Synthesis

72. For what integers x is $-x$ positive?

73. For what integers x is $-x$ negative?

Tell whether each sum is positive, negative, or zero.

74. If $n = m$ and n is negative, then $-n + (-m)$ is ______________.

75. If n is positive and m is negative, then $-n + m$ is ______________.

2.3 Subtraction of Integers

OBJECTIVES

a Subtract integers and simplify combinations of additions and subtractions.

b Solve applied problems involving addition and subtraction of integers.

a SUBTRACTION

We now consider subtraction of integers.

SUBTRACTION

The difference $a - b$ is the number c for which $a = b + c$.

Consider, for example, $45 - 17$. *Think*: What number can we add to 17 to get 45? Since $45 = 17 + 28$, we know that $45 - 17 = 28$. Let's consider an example whose answer is a negative number.

EXAMPLE 1 Subtract: $3 - 7$.

Think: What number can we add to 7 to get 3? The number must be negative. Since $7 + (-4) = 3$, we know the number is -4: $3 - 7 = -4$. That is, $3 - 7 = -4$ because $7 + (-4) = 3$.

Do Exercises 1–3.

Subtract.

1. $-6 - 4$

   Think: What number can be added to 4 to get -6:

   $\square + 4 = -6?$

2. $-7 - (-10)$

   Think: What number can be added to -10 to get -7:

   $\square + (-10) = -7?$

3. $-7 - (-2)$

   Think: What number can be added to -2 to get -7:

   $\square + (-2) = -7?$

The definition above does not provide the most efficient way to do subtraction. We can develop a faster way to subtract. As a rationale for the faster way, let's compare $3 + 7$ and $3 - 7$ on the number line.

To find $3 + 7$ on the number line, we start at 0, move to 3, and then move 7 units to the *right* since 7 is positive.

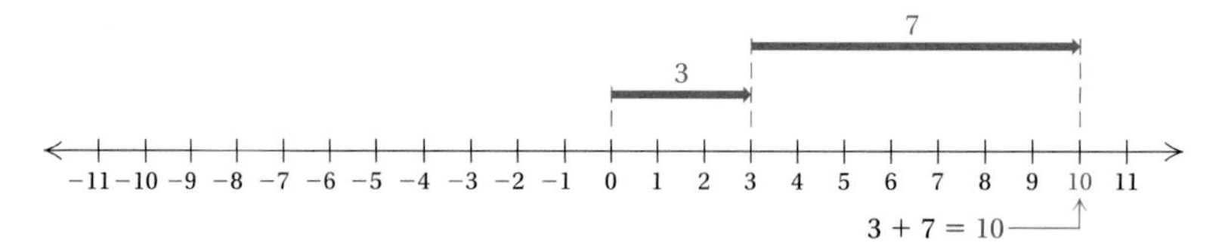

To find $3 - 7$ we do the "opposite" of adding 7: From 3, we move 7 units to the *left* to subtract 7. This is the same as *adding* the opposite of 7, -7, to 3.

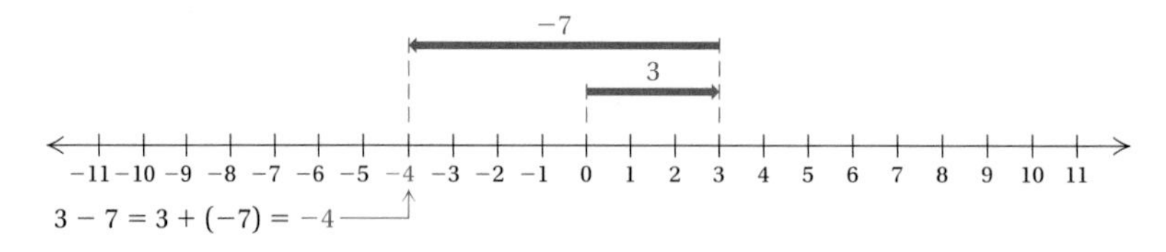

Do Exercises 4–6.

Subtract. Use the number line, doing the "opposite" of addition.

4. $-4 - (-3)$

   −6 −5 −4 −3 −2 −1 0 1 2 3 4 5 6

5. $-4 - (-6)$

   −6 −5 −4 −3 −2 −1 0 1 2 3 4 5 6

6. $5 - 9$

   −6 −5 −4 −3 −2 −1 0 1 2 3 4 5 6

Look for a pattern in the following examples.

SUBTRACTING	ADDING AN OPPOSITE
$5 - 8 = -3$	$5 + (-8) = -3$
$-6 - 4 = -10$	$-6 + (-4) = -10$
$-7 - (-10) = 3$	$-7 + 10 = 3$
$-7 - (-2) = -5$	$-7 + 2 = -5$

Answers

1. −10 2. 3 3. −5 4. −1
5. 2 6. −4

Do Exercises 7–10. ▶

Perhaps you have noticed that we can subtract by adding the opposite of the number being subtracted. This can always be done.

SUBTRACTING BY ADDING THE OPPOSITE

For any numbers a and b,

$$a - b = a + (-b).$$

(To subtract, add the opposite, or additive inverse, of the number being subtracted.)

This is the method generally used for quick subtraction of integers.

EXAMPLES Subtract.

2. $2 - 6 = 2 + (-6) = -4$ — The opposite of 6 is -6. We change the subtraction to addition and add the opposite. *Check*: $-4 + 6 = 2$.

3. $4 - (-9) = 4 + 9 = 13$ — The opposite of -9 is 9. We change the subtraction to addition and add the opposite. *Check*: $13 + (-9) = 4$.

4. $-4 - (-3) = -4 + 3 = -1$ — Adding the opposite. *Check*: $-1 + (-3) = -4$.

5. $-10 - (-13) = -10 + 13 = 3$ — Adding the opposite. *Check*: $3 + (-13) = -10$.

Do Exercises 11–16. ▶

EXAMPLES Subtract by adding the opposite of the number being subtracted.

6. $3 - 5$ *Think*: "Three minus five is three plus the opposite of five."

$3 - 5 = 3 + (-5) = -2$

7. $-16 - 7$ *Think*: "Negative sixteen minus seven is negative sixteen plus the opposite of seven."

$-16 - 7 = -16 + (-7) = -23$

8. $8 - (-9)$ *Think*: "Eight minus negative nine is eight plus the opposite of negative nine."

$8 - (-9) = 8 + 9 = 17$

9. $-11 - (-6)$ *Think*: "Negative eleven minus negative six is negative eleven plus the opposite of negative six."

$-11 - (-6) = -11 + 6 = -5$

Do Exercises 17–21. ▶

Complete the addition and compare with the subtraction.

7. $4 - 6 = -2$;
$4 + (-6) =$ ________

8. $-3 - 8 = -11$;
$-3 + (-8) =$ ________

9. $-5 - (-9) = 4$;
$-5 + 9 =$ ________

10. $-5 - (-3) = -2$;
$-5 + 3 =$ ________

Subtract.

GS **11.** $2 - 8$
$= 2 + (\quad) = \quad$

12. $-6 - 10$

13. $2 - 8$

14. $-8 - (-11)$

15. $-8 - (-8)$

16. $2 - (-5)$

Subtract by adding the opposite of the number being subtracted.

17. $3 - 11$

18. $12 - 5$

GS **19.** $-12 - (-9)$
$= -12 + \quad = \quad$

20. $-7 - 10$

21. $-4 - (-4)$

Answers

7. -2 **8.** -11 **9.** 4 **10.** -2 **11.** -6
12. -16 **13.** -6 **14.** 3 **15.** 0 **16.** 7
17. -8 **18.** 7 **19.** -3 **20.** -17
21. 0

Guided Solutions:
11. $-8, -6$ **19.** $9, -3$

When several additions and subtractions occur together, we can make them all additions.

EXAMPLES Simplify.

10. $8 - (-4) - 2 - (-4) + 2 = 8 + 4 + (-2) + 4 + 2$ Adding the opposites

$= 16$

11. $3 - (-6) + 2 - (-4) = 3 + 6 + 2 + 4 = 15$

12. $9 - (-1) - 7 - 11 = 9 + 1 + (-7) + (-11) = -8$

◀ **Do Exercises 22–24.**

Simplify.

22. $-6 - (-2) - (-4) - 12 + 3$

23. $20 - 16 - (-22) + 21 - 45$

24. $-10 + 7 - (-4) - (-11)$

b APPLICATIONS AND PROBLEM SOLVING

Let's now see how we can use addition and subtraction of integers to solve applied problems.

EXAMPLE 13 *Surface Temperatures on Mars.* Surface temperatures on Mars vary from $-128°C$ during the polar night to $27°C$ at the equator at midday when Mars is at the point in its orbit closest to the sun. Find the difference between the highest value and the lowest value in this temperature range.

Source: Mars Institute

We let $D =$ the difference in the temperatures. Then the problem translates to the following subtraction:

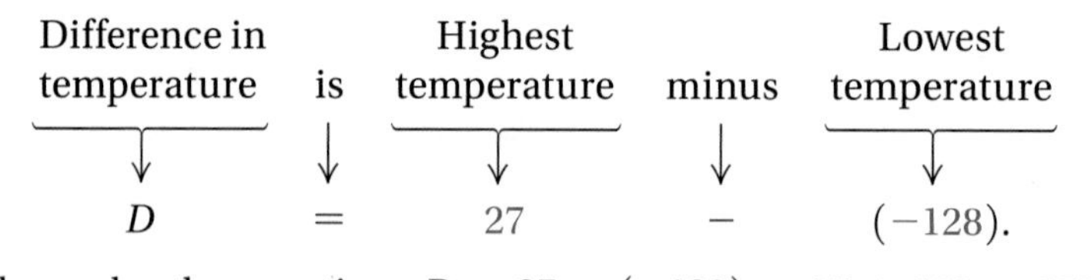

We then solve the equation: $D = 27 - (-128) = 27 + 128 = 155.$

The difference in the temperatures is 155°C.

◀ **Do Exercise 25.**

25. *Temperature Extremes.* The highest temperature ever recorded in the United States was 134°F in Greenland Ranch, California, on July 10, 1913. The lowest temperature ever recorded was $-80°F$ in Prospect Creek, Alaska, on January 23, 1971. How much higher was the temperature in Greenland Ranch than the temperature in Prospect Creek?

Source: National Oceanographic and Atmospheric Administration

Answers

22. -9 **23.** 2 **24.** 12 **25.** 214°F

2.3 Exercise Set

For Extra Help
MyMathLab®
MathXL® PRACTICE | WATCH | READ | REVIEW

Reading Check

Match each expression with the expression from the list at the right that names the same number.

RC1. $18 - 6$ _____ **a)** $18 + 6$

RC2. $-18 - (-6)$ _____ **b)** $-18 + 6$

RC3. $-18 - 6$ _____ **c)** $18 + (-6)$

RC4. $18 - (-6)$ _____ **d)** $-18 + (-6)$

a Subtract.

1. $3 - 7$

2. $5 - 10$

3. $0 - 7$

4. $0 - 8$

5. $-8 - (-2)$

6. $-6 - (-8)$

7. $-10 - (-10)$

8. $-8 - (-8)$

9. $12 - 16$

10. $14 - 19$

11. $20 - 27$

12. $26 - 7$

13. $-9 - (-3)$

14. $-6 - (-9)$

15. $-11 - (-11)$

16. $-14 - (-14)$

17. $8 - (-3)$

18. $-7 - 4$

19. $-6 - 8$

20. $6 - (-10)$

21. $-4 - (-9)$

22. $-14 - 2$

23. $2 - 9$

24. $2 - 8$

25. $0 - 5$

26. $0 - 10$

27. $-5 - (-2)$

28. $-3 - (-1)$

29. $2 - 25$

30. $18 - 63$

31. $-42 - 26$

32. $-18 - 63$

33. $-71 - 2$

34. $-49 - 3$

35. $24 - (-92)$

36. $48 - (-73)$

37. $-2 - 0$

38. $-6 - 1$

39. $3 - 8$

40. $5 - 9$

41. $2 - 3$

42. $4 - 12$

43. $-3 - 2$

44. $-8 - 5$

45. $13 - (-9)$

46. $15 - (-7)$

47. $6 - (-13)$

48. $2 - (-4)$

49. $-14 - 6$

50. $-7 - 9$

51. $1 - 9$

52. $1 - 7$

53. $11 - 21$

54. $30 - 51$

55. $7 - 10$

56. $8 - (-9)$

57. $16 - 23$

58. $-38 - (-12)$

59. $-47 - (-17)$

60. $-12 + 12$

61. $-9 - (-9)$

62. $8 - (-16)$

63. $122 - 123$

64. $20 - (-21)$

Simplify.

65. $18 - (-15) - 3 - (-5) + 2$

66. $22 - (-18) + 7 + (-42) - 27$

67. $-31 + (-28) - (-14) - 17$

68. $-43 - (-19) - (-21) + 25$

69. $-93 - (-84) - 41 - (-56)$

70. $84 + (-99) + 44 - (-18) - 43$

71. $-5 - (-30) + 30 + 40 - (-12)$

72. $14 - (-50) + 20 - (-32)$

73. $132 - (-21) + 45 - (-21)$

74. $81 - (-20) - 14 - (-50) + 53$

b Solve.

75. *"Flipping" Houses.* Buying run-down houses, fixing them up, and reselling them is referred to as "flipping." Charlie and Sophia flipped four houses in a recent year. The profits and losses are shown in the following bar graph. Find the sum of the profits and losses.

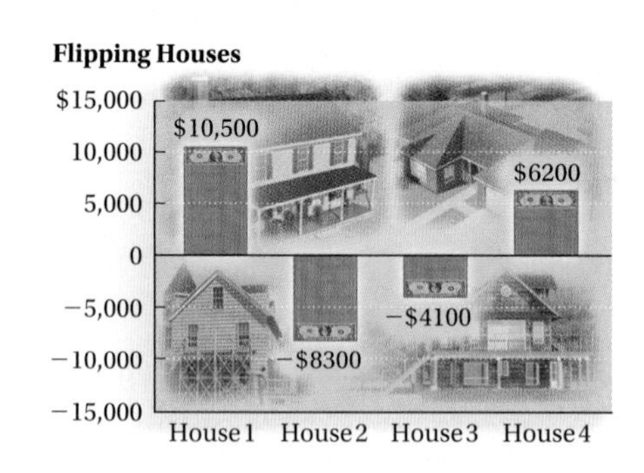

76. *Elevations in Asia.* The elevation of the highest point in Asia, Mt. Everest, on the border between Nepal and Tibet, is 29,035 ft. The lowest elevation, at the Dead Sea, between Israel and Jordan, is −1348 ft. What is the difference in the elevations of the two locations?

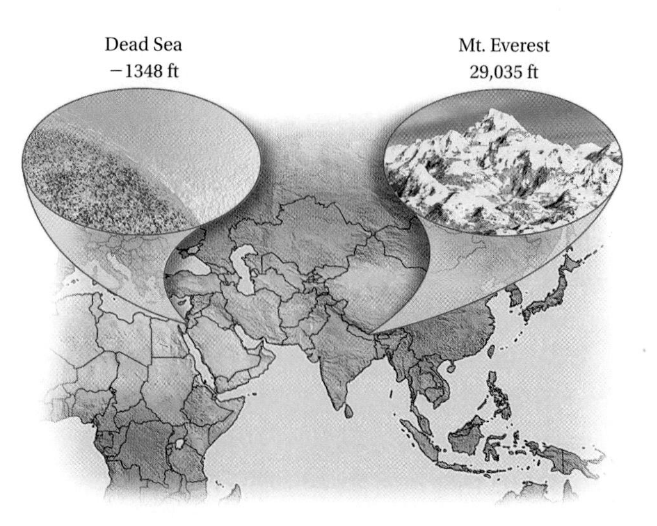

77. *Difference in Elevation.* The highest elevation in Japan is 3776 m above sea level at Fujiyama. The lowest elevation in Japan is 4 m below sea level at Hachirogata. Find the difference in the elevations.

Source: *The CIA World Factbook 2012*

78. *Copy-Center Account.* Rachel's copy-center bill for July was $327. She made a payment of $200 and then made $48 worth of copies in August. How much did she then owe on her account?

79. *Temperature Changes.* One day the temperature in Lawrence, Kansas, is 32° at 6:00 A.M. It rises 15° by noon, but falls 50° by midnight when a cold front moves in. What is the final temperature?

80. *Stock Price Changes.* On a recent day, the price of a stock opened at a value of $61. It rose $5, dropped $7, and then rose $3. Find the value of the stock at the end of the day.

81. Francisca has charged a total of $476 on her credit card, but she then returns a sweater that cost $128. How much does she now owe on her credit card?

82. Jacob has $825 in a checking account. What is the balance in his account after he writes a check for $920 to pay for a laptop?

83. *Tallest Mountain.* The tallest mountain in the world, when measured from base to peak, is Mauna Kea (White Mountain) in Hawaii. From its base 19,684 ft below sea level in the Hawaiian Trough, it rises 33,480 ft. What is the elevation of the peak?

Source: *The Guinness Book of Records*

84. *Low Points on Continents.* The lowest point in Africa is Lake Assal, which is 512 ft below sea level. The lowest point in South America is the Valdes Peninsula, which is 131 ft below sea level. How much lower is Lake Assal than the Valdes Peninsula?

Source: National Geographic Society

Skill Maintenance

Evaluate. [1.9b]

85. 4^3

86. 5^3

Simplify. [1.9c]

87. $5 \cdot 4 + 9$

88. $3 \cdot 16 - (7 - 1) \div 6 - (10 - 4)$

89. $2 + (5 + 3)^2$

90. $27 - 2^3 \cdot 3$

Solve. [1.8a]

91. How many 12-oz cans of soda can be filled with 96 oz of soda?

92. A case of soda contains 24 bottles. If each bottle contains 12 oz, how many ounces of soda are in the case?

Synthesis

Tell whether each statement is true or false for all integers m and n. If false, give an example to show why.

93. $-n = 0 - n$

94. $n - 0 = 0 - n$

95. If $m \neq n$, then $m - n \neq 0$.

96. If $m = -n$, then $m + n = 0$.

97. If $m + n = 0$, then m and n are opposites.

98. If $m - n = 0$, then $m = -n$.

99. $m = -n$ if m and n are opposites.

100. If $m = -m$, then $m = 0$.

Mid-Chapter Review

Concept Reinforcement

Determine whether each statement is true or false.

______________ **1.** All of the natural numbers are integers. [2.1a]

______________ **2.** If $a > b$, then a lies to the left of b on the number line. [2.1c]

______________ **3.** The absolute value of a number is always nonnegative. [2.1d]

Guided Solutions

GS Fill in each blank with the number that creates a correct statement or solution.

4. Evaluate $-x$ and $-(-x)$ when $x = -4$. [2.2b]

$-x = -(\square) = \square$;

$-(-x) = -(-(\square)) = -(\square) = \square$

Subtract. [2.3a]

5. $5 - 13 = 5 + (\square) = \square$

6. $-6 - (-7) = -6 + \square = \square$

Mixed Review

State the integers that correspond to the situation. [2.1a]

7. Jerilyn deposited $450 in her checking account. Later that week she wrote a check for $79.

8. The temperature in Coatsville rose 20° in the morning. It fell 23° overnight.

Graph the number on the number line. [2.1b]

9. -3

−6 −5 −4 −3 −2 −1 0 1 2 3 4 5 6

10. 0

−6 −5 −4 −3 −2 −1 0 1 2 3 4 5 6

Use either $<$ or $>$ for $\square$ to write a true sentence. [2.1c]

11. $-6 \square 6$

12. $-5 \square -3$

13. $-9 \square -10$

14. $5 \square 0$

Find the absolute value. [2.1d]

15. $|15|$

16. $|-18|$

17. $|0|$

18. $|-12|$

Find the opposite, or additive inverse, of the number. [2.2b]

19. -5

20. 7

21. 0

22. -49

23. Evaluate $-x$ when x is -19. [2.2b]

24. Evaluate $-(-x)$ when x is 2. [2.2b]

Compute and simplify. [2.2a], [2.3a]

25. $7 + (-9)$

26. $-3 + 1$

27. $3 + (-3)$

28. $-8 + (-9)$

29. $2 + (-12)$

30. $-4 + (-3)$

31. $-14 + 5$

32. $19 + (-21)$

33. $-4 - 6$

34. $5 - (-11)$

35. $-1 - (-3)$

36. $12 - 24$

37. $-8 - (-4)$

38. $-1 - 5$

39. $12 - 14$

40. $6 - (-7)$

41. $16 - (-9) - 20 - (-4)$

42. $-4 + (-10) - (-3) - 12$

43. $17 - (-25) + 15 - (-18)$

44. $-9 + (-3) + 16 - (-10)$

Solve. [2.3b]

45. *Temperature Change.* In a chemistry lab, Ben works with a substance whose initial temperature is 25°C. During an experiment, the temperature falls to −8°C. Find the difference between the two temperatures.

46. *Stock Price Change.* The price of a stock opened at $56. During the day it dropped $3, then rose $1, and dropped $6. Find the price of the stock at the end of the day.

Understanding Through Discussion and Writing

47. Find an example of a real-world situation that can be represented by a negative integer. [2.1a]

48. Explain why the absolute value of a number cannot be negative. [2.1d]

49. Explain in your own words why the sum of two negative numbers is always negative. [2.2a]

50. If a negative number is subtracted from a positive number, will the result always be positive? Why or why not? [2.3a]

STUDYING FOR SUCCESS *Your Textbook as a Resource*

- ☐ Study any drawings. Observe details in any sketches or graphs that accompany the explanations.
- ☐ Note the careful use of color to indicate substitutions and highlight steps in a multi-step solution.

2.4 Multiplication of Integers

OBJECTIVE

a Multiply integers.

a MULTIPLICATION

Multiplication of integers is very much like multiplication of whole numbers. The only difference is that we must determine whether the answer is positive or negative.

Multiplication of a Positive Number and a Negative Number

This number decreases by 1 each time.

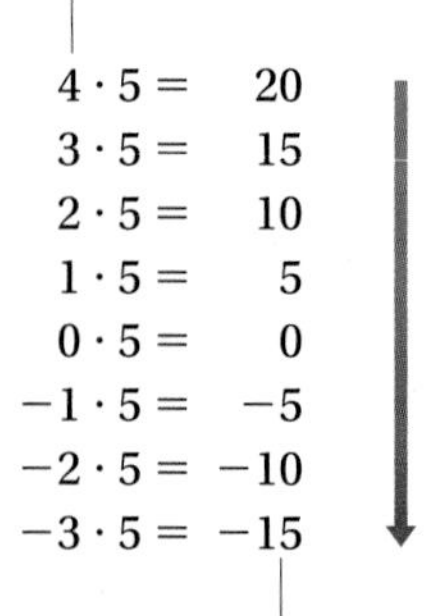

This number decreases by 5 each time.

To see how to multiply a positive number and a negative number, consider the pattern shown at the left.

According to this pattern, it looks as though the product of a negative number and a positive number is negative. That is the case, and we have the first part of the rule for multiplying integers.

> **THE PRODUCT OF A POSITIVE NUMBER AND A NEGATIVE NUMBER**
>
> To multiply a positive number and a negative number, multiply their absolute values. The answer is negative.

EXAMPLES Multiply.

1. $8(-5) = -40$ **2.** $-3 \cdot 7 = -21$ **3.** $(-8)6 = -48$

◀ **Do Exercises 1–6.**

Multiply.

1. $-3 \cdot 6$ **2.** $20 \cdot (-5)$

3. $4 \cdot (-20)$ **4.** $-7 \cdot 8$

5. $4(-7)$ **6.** $8(-4)$

Multiplication of Two Negative Numbers

How do we multiply two negative numbers? Again we look for a pattern.

This number decreases by 1 each time. This number increases by 5 each time.

$4 \cdot (-5) = -20$
$3 \cdot (-5) = -15$
$2 \cdot (-5) = -10$
$1 \cdot (-5) = -5$
$0 \cdot (-5) = 0$
$-1 \cdot (-5) = 5$
$-2 \cdot (-5) = 10$
$-3 \cdot (-5) = 15$

Answers

1. -18 **2.** -100 **3.** -80
4. -56 **5.** -28 **6.** -32

According to the pattern, it looks as though the product of two negative numbers is positive. That is actually so, and we have the second part of the rule for multiplying real numbers.

THE PRODUCT OF TWO NEGATIVE NUMBERS

To multiply two negative numbers, multiply their absolute values. The answer is positive.

Do Exercises 7–12. ▶

To multiply two nonzero numbers:

a) Multiply the absolute values.

b) If the signs are the same, the answer is positive.

c) If the signs are different, the answer is negative.

EXAMPLES Multiply.

4. $(-3)(-4) = 12$ **5.** $-6(2) = -12$ **6.** $-5(-9) = 45$

Do Exercises 13–16. ▶

Multiplying More Than Two Numbers

When multiplying more than two integers, we can choose order and grouping as we please, using the commutative and associative laws.

EXAMPLES Multiply.

7. $-8 \cdot 2(-3) = -16(-3)$ Multiplying the first two numbers
$= 48$ Multiplying the results

8. $-8 \cdot 2(-3) = 24 \cdot 2$ Multiplying the negative numbers. Every pair of negative numbers gives a positive product.
$= 48$

9. $-3(-2)(-5)(4) = 6(-5)(4)$ Multiplying the first two numbers
$= (-30)4 = -120$

10. $-2 \cdot 2(-3)(-4) = -4 \cdot 12$ Multiplying the first two numbers and the last two numbers
$= -48$

11. $-5 \cdot (-2) \cdot (-3) \cdot (-6) = 10 \cdot 18 = 180$

12. $(-3)(-5)(-2)(-3)(-6) = (-30)(18) = -540$ ■

Considering that the product of a pair of negative numbers is positive, we can see the following pattern in the results of Examples 11 and 12.

The product of an even number of negative numbers is positive.
The product of an odd number of negative numbers is negative.

Do Exercises 17–22. ▶

Multiply.

7. $-3 \cdot (-4)$

8. $-16 \cdot (-2)$

9. $-7 \cdot (-5)$

10. $-7(-9)$

11. $-3(-12)$

12. $-6(-4)$

Multiply.

13. $5(-6)$

14. $(-5)(-6)$

15. $(-3) \cdot 10$

16. $(-4)(15)$

Multiply.

17. $5 \cdot (-3) \cdot 2$

18. $-3 \times (-4) \times (-2)$

19. $-2 \cdot (-10) \cdot (-5)$

GS **20.** $-2 \cdot (-5) \cdot (-4) \cdot (-3)$
$= \square \cdot (-4) \cdot (-3)$
$= \square \cdot (-3)$
$= \square$

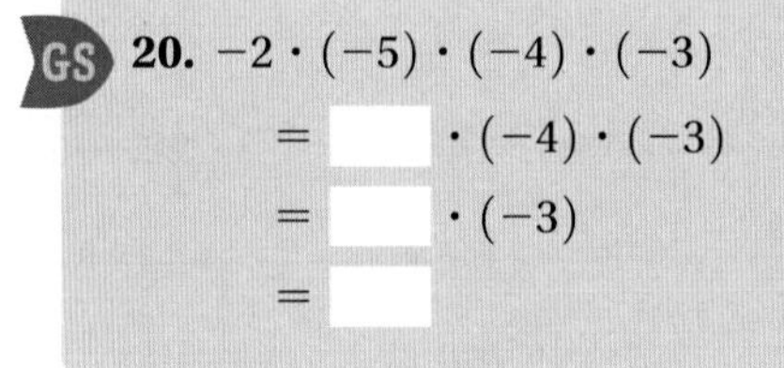

21. $(-4)(-5)(-2)(-3)(-1)$

22. $(-1)(-1)(-2)(-3)(-1)(-1)$

Answers

7. 12 **8.** 32 **9.** 35 **10.** 63 **11.** 36 **12.** 24 **13.** −30 **14.** 30 **15.** −30 **16.** −60 **17.** −30 **18.** −24 **19.** −100 **20.** 120 **21.** −120 **22.** 6

Guided Solution:
20. 10, −40, 120

2.4 Exercise Set

For Extra Help
MyMathLab®
MathXL® PRACTICE · WATCH · READ · REVIEW

Reading Check

Fill in each blank with either "positive" or "negative."

RC1. To multiply a positive number and a negative number, we multiply their absolute values. The answer is ______________.

RC2. To multiply two negative numbers, we multiply their absolute values. The answer is ______________.

RC3. The product of an even number of negative numbers is ______________.

RC4. The product of an odd number of negative numbers is ______________.

a Multiply.

1. $-8 \cdot 2$

2. $-3 \cdot 5$

3. $8 \cdot (-3)$

4. $5 \cdot (-2)$

5. $-9 \cdot 8$

6. $-20 \cdot 3$

7. $-8 \cdot (-2)$

8. $-4 \cdot (-5)$

9. $-7 \cdot (-6)$

10. $-9 \cdot (-2)$

11. $15 \cdot (-8)$

12. $-11 \cdot (-10)$

13. $-14 \cdot 17$

14. $-13 \cdot (-15)$

15. $-25 \cdot (-48)$

16. $39 \cdot (-43)$

17. $-3 \cdot (-28)$

18. $97 \cdot (-2)$

19. $4 \cdot (-3)$

20. $3 \cdot (-22)$

21. $-6 \cdot (-4)$

22. $-5 \cdot (-6)$

23. $-7 \cdot (-3)$

24. $-4 \cdot (-32)$

25. $23 \cdot (-3)$

26. $7 \cdot (-9)$

27. $-3 \cdot (-9)$

28. $-25 \cdot (-8)$

29. -6×3

30. -8×6

31. $-9 \cdot 5$

32. $-8 \cdot 9$

33. $7 \cdot (-4) \cdot (-3) \cdot 5$

34. $9 \cdot (-2) \cdot (-6) \cdot 7$

35. $-3 \cdot 2 \cdot (-6)$

36. $-8 \cdot (-4) \cdot (-3)$

37. $-3 \cdot (-4) \cdot (-5)$

38. $-2 \cdot (-5) \cdot (-7)$

39. $-2 \cdot (-5) \cdot (-3) \cdot (-5)$

40. $-3 \cdot (-5) \cdot (-2) \cdot (-1)$

41. $5(-18)$

42. $-35(-2)$

43. $-7 \cdot (-21) \cdot 13$

44. $-14 \cdot 34 \cdot 12$

45. $-4 \cdot (-2) \cdot 7$

46. $-8 \cdot (-3) \cdot (-5)$

47. $-3(-2)(5)$

48. $-7(-5)(-2)$

49. $4 \cdot (-4) \cdot (-5) \cdot (-12)$

50. $-2 \cdot (-3) \cdot (-4) \cdot (-5)$

51. $7 \cdot (-7) \cdot 6 \cdot (-6)$

52. $8 \cdot (-8) \cdot (-9) \cdot (-9)$

53. $(-5)(8)(-3)(-2)$

54. $(4)(-2)(-3)(2)$

55. $(-6)(-7)(-8)(-9)(-10)$

56. $7 \cdot (-6) \cdot 5 \cdot (-4) \cdot 3 \cdot (-2) \cdot 1 \cdot (-1)$

Skill Maintenance

Round 237,456 to the nearest: [1.6a]

57. Ten.

58. Hundred.

59. Thousand.

Solve. [1.7b]

60. $16 + x = 29$

61. $14 \cdot y = 42$

62. Simplify: $3 + 6[18 - (12 + 3)]$. [1.9d]

Synthesis

63. What must be true of a and b if $-ab$ is to be **(a)** positive? **(b)** zero? **(c)** negative?

2.5 Division of Integers and Order of Operations

OBJECTIVES

a Divide integers.

b Solve applied problems involving multiplication and division of integers.

c Simplify expressions using rules for order of operations.

SKILL TO REVIEW

Objective 1.9c: Simplify expressions using the rules for order of operations.

Simplify.

1. $100 \div 4 - 6$
2. $3 \cdot 8 - 2 \cdot 9$

We now consider division of integers. The definition of division results in rules for division that are the same as those for multiplication.

a DIVISION OF INTEGERS

DIVISION

The quotient $a \div b$, or $\frac{a}{b}$, where $b \neq 0$, is that unique number c for which $a = b \cdot c$.

EXAMPLES Divide, if possible. Check each answer.

1. $14 \div (-7) = -2$ — *Think*: What number multiplied by -7 gives 14? That number is -2. *Check*: $(-2)(-7) = 14$.
2. $-32 \div (-4) = 8$ — *Think*: What number multiplied by -4 gives -32? That number is 8. *Check*: $8(-4) = -32$.
3. $-20 \div 5 = -4$ — *Think*: What number multiplied by 5 gives -20? That number is -4. *Check*: $-4 \cdot 5 = -20$.
4. $\frac{-17}{0}$ is **not defined.** — *Think*: What number multiplied by 0 gives -17? There is no such number because the product of 0 and *any* number is 0.

◀ **Do Margin Exercises 1–3.**

Divide.

1. $6 \div (-3)$ *Think*: What number multiplied by -3 gives 6?
2. $\frac{-15}{-3}$ *Think*: What number multiplied by -3 gives -15?
3. $-24 \div 8$ *Think*: What number multiplied by 8 gives -24?

The rules for division are the same as those for multiplication.

To multiply or divide two numbers (where the divisor is nonzero):

a) Multiply or divide the absolute values.

b) If the signs are the same, the answer is positive.

c) If the signs are different, the answer is negative.

◀ **Do Exercises 4–6.**

Divide.

4. $\frac{-72}{-8}$ 5. $\frac{30}{-5}$ 6. $\frac{56}{-7}$

Excluding Division by 0

Example 4 shows why we cannot divide -17 by 0. We can use the same argument to show why we cannot divide any nonzero number b by 0. Consider $b \div 0$. We look for a number that when multiplied by 0 gives b. There is no such number because the product of 0 and any number is 0. Thus we cannot divide a nonzero number b by 0.

On the other hand, if we divide 0 by 0, we look for a number c such that $0 \cdot c = 0$. But $0 \cdot c = 0$ for any number c. Thus it appears that $0 \div 0$ could be any number we choose. Getting any answer we want when we divide 0 by 0 would be very confusing. Thus we agree that division by 0 is not defined.

Answers

Skill to Review:
1. 19 2. 6

Margin Exercises:
1. −2 2. 5 3. −3 4. 9 5. −6 6. −8

EXCLUDING DIVISION BY 0

Division by 0 is not defined:

$a \div 0$ is not defined for all integers a.

Dividing 0 by Other Numbers

Note that $0 \div 8 = 0$ because $0 = 0 \cdot 8$.

DIVIDENDS OF 0

Zero divided by any nonzero number is 0:

$0 \div a = 0, \quad a \neq 0.$

EXAMPLES Divide.

5. $0 \div (-6) = 0$ **6.** $0 \div 12 = 0$ **7.** $-3 \div 0$ is not defined.

Do Exercises 7 and 8.

Divide, if possible.

7. $-5 \div 0$

8. $0 \div (-3)$

b APPLICATIONS AND PROBLEM SOLVING

We can use multiplication and division of integers to solve applied problems.

EXAMPLE 8 *Mine Rescue.* The San Jose copper and gold mine near Copiapó, Chile, caved in on August 5, 2010, trapping 33 miners. Each miner was safely brought out in a specially designed capsule that could be lowered into the mine at -137 feet per minute. It took approximately 15 min to lower the capsule to the miners' location. Determine how far below the surface of the earth the miners were trapped.

Source: Reuters News

Since the capsule moved -137 feet per minute and it took 15 min to reach the miners, we have the depth d given by

$$d = 15 \cdot (-137) = -2055.$$

Thus the miners were trapped at -2055 ft.

EXAMPLE 9 *Chemical Reaction.* During a chemical reaction, the temperature in a beaker decreased every minute by the same number of degrees. The temperature was 56°F at 10:10 A.M. By 10:42 A.M., the temperature had dropped to -8°F. By how many degrees did it change each minute?

We first determine by how many degrees d the temperature changed altogether. We subtract -8 from 56:

$$d = 56 - (-8) = 56 + 8 = 64.$$

Answers

7. Not defined 8. 0

9. *Chemical Reaction.* During a chemical reaction, the temperature in a beaker increased by 3°C every minute. If the temperature was −17°C at 1:10 P.M., when the reaction began, what was the temperature at 1:34 P.M.?

10. *Chemical Reaction.* During a chemical reaction, the temperature in a beaker decreased every minute by the same number of degrees. The temperature was 71°F at 2:12 P.M. By 2:29 P.M., the temperature had changed to −14°F. By how many degrees did it change each minute?

The temperature changed a total of 64°. We can express this as −64° since the temperature dropped.

The amount of time t that passed was $42 - 10$, or 32 min. Thus the number of degrees T that the temperature dropped each minute is given by

$$T = d \div t = -64 \div 32 = -2.$$

The change was −2°F per minute.

Do Exercises 9 and 10.

C ORDER OF OPERATIONS

When several operations are to be done in a calculation or a problem involving integers, we apply the same rules that we did in Section 1.9. We repeat them here for review. If you did not study that section before, you should do so before continuing.

RULES FOR ORDER OF OPERATIONS

1. Do all operations within grouping symbols before operations outside.
2. Evaluate all exponential expressions.
3. Do all multiplications and divisions in order from left to right.
4. Do all additions and subtractions in order from left to right.

EXAMPLE 10 Simplify: $-34 \cdot 56 - 17$.

There are no parentheses or exponents so we start with the third step.

$-34 \cdot 56 - 17 = -1904 - 17$ Carrying out all multiplications and divisions in order from left to right

$= -1921$ Carrying out all additions and subtractions in order from left to right

EXAMPLE 11 Simplify: $256 \div 2^4 \cdot (-5)$.

There are no calculations within grouping symbols so we start with the second step.

$256 \div 2^4 \cdot (-5) = 256 \div 16 \cdot (-5)$ Evaluating the exponential expression

$= 16 \cdot (-5)$ Doing all multiplications and divisions in order from left to right

$= -80$

Answers

9. 55°C 10. −5°F per minute

EXAMPLE 12 Simplify: $2^4 + 51 \cdot 4 - (37 + 23 \cdot 2)$.

$2^4 + 51 \cdot 4 - (37 + 23 \cdot 2)$

$= 2^4 + 51 \cdot 4 - (37 + 46)$	Carrying out all operations inside parentheses, first multiplying 23 by 2, following the rules for order of operations within the parentheses
$= 2^4 + 51 \cdot 4 - 83$	Completing the addition inside parentheses
$= 16 + 51 \cdot 4 - 83$	Evaluating exponential expressions
$= 16 + 204 - 83$	Doing all multiplications
$= 220 - 83$	Doing all additions and subtractions in order from left to right
$= 137$	

EXAMPLE 13 Simplify: $[-64 \div (-16) \div (-2)] \div [2^3 - 3^2]$.

$$[-64 \div (-16) \div (-2)] \div [2^3 - 3^2] = [4 \div (-2)] \div [8 - 9]$$
$$= -2 \div (-1)$$
$$= 2$$

Do Exercises 11–14. ▶

Simplify.

11. $23 - 42 \cdot 30$

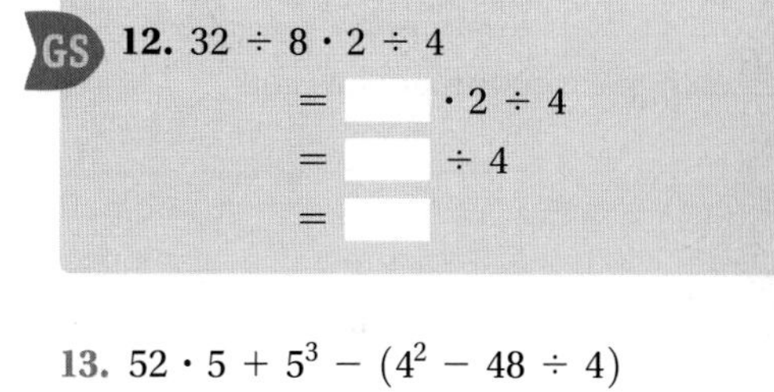

GS 12. $32 \div 8 \cdot 2 \div 4$

$= \square \cdot 2 \div 4$

$= \square \div 4$

$= \square$

13. $52 \cdot 5 + 5^3 - (4^2 - 48 \div 4)$

14. $[5 - 10 - 5 \cdot 23] \div (2^3 + 3^2 - 7)$

Answers

11. -1237 **12.** 2 **13.** 381 **14.** -12

Guided Solution:
12. 4, 8, 2

Translating for Success

1. *Gas Mileage.* After driving his compact car for 759 mi on the first day of a trip, Nathan noticed that it took 23 gal of gasoline to fill the tank. How many miles per gallon did the compact car get?

2. *Apartment Rent.* Cecilia needs $4500 for tuition. This is two times as much as she needs for apartment rent. How much does she need for the rent?

3. *Purchase Price.* The price of an office copier is $4500 and the sales tax is $150. Find the total purchase price.

4. *Change in Elevation.* The lowest elevation in Australia, Lake Eyre, is 52 ft below sea level. The highest elevation in Australia, Mt. Kosciusko, is 7310 ft. Find the difference in elevation between the highest point and the lowest point.

5. *Cell-Phone Bill.* Jeff's cell-phone bill for September was $73. He made a payment of $52. How much did he then owe on his cell-phone bill?

The goal of these matching questions is to practice step (2), Translate, of the five-step problem-solving process. Translate each problem to an equation and select a correct translation from equations A–O.

A. $7 \cdot x = 588$

B. $52 + x = 73$

C. $2 \cdot x = 4500$

D. $150 \cdot x = 4500$

E. $23(759) = x$

F. $4500 = 150 + x$

G. $4500 + 150 = x$

H. $x = 150 \cdot 4500$

I. $23 \cdot x = 759$

J. $7 \div 588 = x$

K. $x = 23 + 759$

L. $x = 7310 - (-52)$

M. $x = 588 \cdot 7$

N. $2(4500) = x$

O. $52 \cdot x = 7310$

Answers on page A-3

6. *Camp Sponsorships.* Donations of $52 per camper are needed for camp sponsorships at Lazy Day Summer Camp. Two weeks prior to camp, $7310 had been collected for sponsorships. How many campers can be enrolled?

7. *Drain Pipe.* A construction engineer has 588 ft of flexible drain pipe. How many 7-ft lengths can be cut from the total amount of pipe available?

8. *Savings-Account Balance.* After $150 interest has been added to Kelly's savings account, the new balance is $4500. Find the previous balance.

9. *Loan Payments.* Matt's student loan totals $4500. If he repays $150 each month, how many months will it take him to repay the loan?

10. *Drain Pipe.* A subcontractor needs 588 pieces of pipe for a large irrigation project. If each piece must be 7 ft long, how many feet of pipe have to be purchased?

2.5 Exercise Set

For Extra Help
MyMathLab® MathXL® PRACTICE WATCH READ REVIEW

Reading Check

Determine whether each statement is true or false.

_______________ **RC1.** The quotient $a \div b$, where $b \neq 0$, is that unique number c for which $a = b \cdot c$.

_______________ **RC2.** When a positive number is divided by a negative number, the answer is negative.

_______________ **RC3.** $0 \div (-5)$ is not defined.

_______________ **RC4.** $0 \div 0 = 1$.

a Divide, if possible. Check each answer.

1. $36 \div (-6)$

2. $42 \div (-7)$

3. $26 \div (-2)$

4. $24 \div (-12)$

5. $-16 \div 8$

6. $-18 \div (-2)$

7. $-48 \div (-12)$

8. $-72 \div (-9)$

9. $-72 \div 9$

10. $-50 \div 25$

11. $-100 \div (-50)$

12. $-200 \div 8$

13. $-108 \div 9$

14. $-64 \div (-8)$

15. $200 \div (-25)$

16. $-390 \div (-13)$

17. $75 \div 0$

18. $0 \div (-5)$

19. $81 \div (-9)$

20. $-145 \div (-5)$

b Solve.

21. *Drop in Temperature.* The temperature in Osgood was 62°F at 2:00 P.M. It dropped 6°F per hour for the next 4 hr. What was the temperature at the end of the 4-hr period?

22. *Juice Consumption.* Oliver bought a 64-oz container of cranberry juice and drank 8 oz per day for a week. How much juice was left in the container at the end of the week?

23. *Stock Price.* The price of a stock began the day at $32 per share and dropped $2 per hour for 3 hr. What was the price of the stock after 3 hr?

24. *Population Decrease.* The population of a rural town was 12,500. It decreased 380 each year for 4 years. What was the population of the town after 4 years?

25. *Diver's Position.* After diving 95 m below sea level, a diver rises at a rate of 7 meters per minute for 9 min. At that point, where is the diver in relation to the surface?

26. *Bank Account Balance.* Karen had \$234 in her bank account. After she used her debit card to make seven purchases at \$39 each, what was the balance in her bank account?

C Simplify.

27. $8 - 2 \cdot 3 - 9$

28. $8 - (2 \cdot 3 - 9)$

29. $(8 - 2 \cdot 3) - 9$

30. $(8 - 2)(3 - 9)$

31. $16 \cdot (-24) + 50$

32. $10 \cdot 20 - 15 \cdot 24$

33. $2^4 + 2^3 - 10$

34. $40 - 3^2 - 2^3$

35. $5^3 + 26 \cdot 71 - (16 + 25 \cdot 3)$

36. $4^3 + 10 \cdot 20 + 8^2 - 23$

37. $4 \cdot 5 - 2 \cdot 6 + 4$

38. $4 \cdot (6 + 8)(4 + 3)$

39. $4^3 \div 8$

40. $5^3 - 7^2$

41. $8(-7) + 6(-5)$

42. $10(-5) + 1(-1)$

43. $19 - 5(-3) + 3$

44. $14 - 2(-6) + 7$

45. $9 \div (-3) + 16 \div 8$

46. $-32 - 8 \div 4 - (-2)$

47. $-4^2 + 6$

48. $-5^2 + 7$

49. $-8^2 - 3$

50. $-9^2 - 11$

51. $12 - 20^3$

52. $20 + 4^3 \div (-8)$

53. $2 \times 10^3 - 5000$

54. $-7(3^4) + 18$

55. $6[9 - (3 - 4)]$

56. $8[(6 - 13) - 11]$

57. $-1000 \div (-100) \div 10$

58. $256 \div (-32) \div (-4)$

59. $8 - (7 - 9)$

60. $(8 - 7) - 9$

61. $(10 - 6^2) \div (3^2 + 2^2)$

62. $(-3 - 5^3 - 4^3) \div (6^2 - 10^2)$

63. $[20(8 - 3) - 4(10 - 3)] \div [10(2 - 6) + 2(7 + 4)]$

64. $[(3 - 5)^2 - 4(5 - 13)] \div [(12 - 9)^2 + (11 - 14)^2]$

Skill Maintenance

What does the digit 8 mean in each number? [1.1a]

65. 4,678,952

66. 8,473,901

67. 7148

68. 23,803

Subtract. [1.3a]

69. $9001 - 6798$

70. $2037 - 1189$

71. $67{,}113 - 29{,}874$

72. $12{,}327 - 476$

Solve. [1.8a]

73. A playing field is 78 ft long and 64 ft wide. What is its area? its perimeter?

74. A painter buys three 5-gal pails of exterior paint at $112 each and eight 1-gal cans of interior paint at $28 each. How much is spent altogether for the paint?

Synthesis

Simplify.

75. $[195 + (-15)^3] \div [195 - 7 \cdot 5^2]$

76. $[19 - 17^2] \div [13^2 - 34]$

Determine the sign of each expression if m is negative and n is positive.

77. $-n \div (-m)$

78. $-n \div m$

79. $-[n \div (-m)]$

80. $-(-n \div m)$

81. $-[-n \div (-m)]$

CHAPTER

2 Summary and Review

Vocabulary Reinforcement

In Exercises 1–6, fill in each blank with the correct term from the given list. Some of the choices may not be used.

1. The ______________ are . . . , −3, −2, −1, 0, 1, 2, 3, ... [2.1a]
2. The ______________ value of a number is its distance from zero on the number line. [2.1d]
3. Numbers such as −3 and 3 are called ______________, or ______________ inverses. [2.2b]
4. The ______________ $a - b$ is the number c for which $a = b + c$. [2.3a]
5. The product of two negative numbers is ______________. [2.4a]
6. The quotient of a positive number and a negative number is ______________. [2.5a]

sum
difference
product
quotient
integers
whole numbers
additive
multiplicative
opposites
absolute
positive
negative

Concept Reinforcement

Determine whether each statement is true or false.

______________ 1. The sum of two negative numbers is positive. [2.2a]

______________ 2. The sum of a number and its opposite is 0. [2.2b]

______________ 3. The product of an even number of negative numbers is negative. [2.4a]

______________ 4. The opposite of the opposite of a number is the number itself. [2.2b]

Study Guide

Objective 2.1b Graph integers on the number line.

Example Graph −3 on the number line.

We locate −3 on the number line and mark its point with a dot.

−6 −5 −4 −3 −2 −1 0 1 2 3 4 5 6

Practice Exercise

1. Graph 4 on the number line.

−6 −5 −4 −3 −2 −1 0 1 2 3 4 5 6

Objective 2.1c Determine which of two real numbers is greater and indicate which, using $<$ or $>$.

Example Use either $<$ or $<$ for $\square$ to write a true sentence.

$$-5 \square -8$$

Since -5 is to the right of -8 on the number line, we have $-5 > -8$.

Practice Exercise

2. Use either $<$ or $>$ for $\square$ to write a true sentence.

$$-7 \square 1$$

Objective 2.1d Find the absolute value of an integer.

Example Find. **a)** $|-9|$ **b)** $|35|$

a) The number is negative, so we make it positive.

$$|-9| = 9$$

b) The number is positive, so the absolute value is the same as the number.

$$|35| = 35$$

Practice Exercise

3. Find.

a) $|-17|$ **b)** $|14|$

Objective 2.2a Add integers without using the number line.

Example Add without using the number line: $-15 + 9$.

We have a negative number and a positive number. The absolute values are 15 and 9. Their difference is 6. The negative number has the larger absolute value, so the answer is negative.

$$-15 + 9 = -6$$

Example Add without using the number line: $-8 + (-9)$.

We have two negative numbers. We add the absolute values, 8 and 9. The answer is negative.

$$-8 + (-9) = -17$$

Practice Exercises

Add without using the number line.

4. $6 + (-9)$

5. $-5 + (-3)$

Objective 2.3a Subtract integers.

Example Subtract: $8 - 12$.

We add the opposite of the number being subtracted.

$$8 - 12 = 8 + (-12)$$
$$= -4$$

Practice Exercise

6. Subtract: $6 - (-8)$.

Objective 2.4a Multiply integers.

Example Multiply: $-6(-4)$.

The signs are the same, so the answer is positive.

$$-6(-4) = 24$$

Example Multiply: $-5(7)$.

The signs are different, so the answer is negative.

$$-5(7) = -35$$

Practice Exercises

Multiply.

7. $-9(-8)$

8. $6(-15)$

Objective 2.5a Divide integers.

Example Divide: $-36 \div (-4)$.

The signs are the same, so the answer is positive.

$-36 \div (-4) = 9$

Example Divide: $-21 \div 7$.

The signs are different, so the answer is negative.

$-21 \div 7 = -3$

Practice Exercises

Divide.

9. $-32 \div (-8)$

10. $48 \div (-12)$

Objective 2.5c Simplify expressions using rules for order of operations.

Example Simplify: $3^2 - 24 \div 2 - (4 + 2 \cdot 8)$.

$$\begin{aligned} &3^2 - 24 \div 2 - (4 + 2 \cdot 8) \\ &= 3^2 - 24 \div 2 - (4 + 16) \\ &= 3^2 - 24 \div 2 - 20 \\ &= 9 - 24 \div 2 - 20 \\ &= 9 - 12 - 20 = -3 - 20 = -23 \end{aligned}$$

Practice Exercise

11. Simplify: $4 - 8^2 \div (10 - 6)$.

Review Exercises

1. State the integers that correspond to this situation: Josh earned $620 for one week's work. While driving to work one day, he received a speeding ticket for $125. [2.1a]

Simplify. [2.1d]

2. $|-38|$

3. $|7|$

4. $|0|$

5. $-|-2|$

Use either $<$ or $>$ for $\square$ to write a true sentence. [2.1c]

6. $-3 \square 10$

7. $-1 \square -6$

8. $11 \square -12$

9. $-2 \square -1$

Graph the number on the number line. [2.1b]

10. 6

−6 −5 −4 −3 −2 −1 0 1 2 3 4 5 6

11. −2

−6 −5 −4 −3 −2 −1 0 1 2 3 4 5 6

Find the opposite, or additive inverse, of each number. [2.2b]

12. 8

13. −14

14. 0

15. −23

16. Evaluate $-x$ when x is -34. [2.2b]

17. Evaluate $-(-x)$ when x is 5. [2.2b]

Compute and simplify.

18. $4 + (-7)$ [2.2a]

19. $-8 + 1$ [2.2a]

20. $6 + (-9) + (-8) + 7$ [2.2a]

21. $-4 + 5 + (-12) + (-4) + 10$ [2.2a]

22. $-3 - (-7)$ [2.3a]

23. $-9 - 5$ [2.3a]

24. $-4 - 4$ [2.3a]

25. $-9 \cdot (-6)$ [2.4a]

26. $-3(13)$ [2.4a]

27. $7 \cdot (-8)$ [2.4a]

28. $3 \cdot (-7) \cdot (-2) \cdot (-5)$ [2.4a]

29. $35 \div (-5)$ [2.5a]

30. $-51 \div 17$ [2.5a]

31. $-42 \div (-7)$ [2.5a]

Simplify. [2.5c]

32. $-12(-3) - 2^3 - (-9)(-10)$

33. $2(-3 - 12) - 8(-7)$

34. $625 \div (-25) \div 5$

35. $-16 \div 4 - 30 \div (-5)$

36. $9[(7 - 14) - 13]$

Solve.

37. Chang's total assets are $2140. Then he borrows $2500. What are his total assets now? [2.3b]

38. *Stock Price.* The price of a stock opened at \$78 per share and dropped by \$2 per hour for 8 hr. What was the price of the stock after 8 hr? [2.5b]

39. On the first, second, and third downs, a football team had these gains and losses: 5-yd gain, 12-yd loss, and 15-yd gain, respectively. Find the total gain or loss. [2.3b]

40. *Bank Account Balance.* Yuri had \$68 in his bank account. After using his debit card to buy seven equally priced tee shirts, the balance in his account was $-\$65$. What was the price of each shirt? [2.5b]

41. Compute and simplify: $8 - (-5) - 7 - (-9)$. [2.3a]

A. -13 **B.** -3
C. 15 **D.** 29

42. Simplify: $-3 \cdot 4 - 12 \div 4$. [2.5c]

A. -16 **B.** -15
C. 0 **D.** 6

Synthesis

43. The following are examples of consecutive integers: $4, 5, 6, 7, 8$; $-13, -12, -11, -10$. [2.3b], [2.5b]

a) Express the number 8 as the sum of 16 consecutive integers.
b) Find the product of the 16 consecutive integers in part (a).

44. Insert one pair of parentheses to make the following a true statement: [2.5c]

$$9 - 3 - 4 + 5 = 15.$$

Simplify.

45. $-|8 - (-4 \div 2) - 3 \cdot 5|$ [2.1d], [2.5c]

46. $(|-6 - 3| + 3^2 - |-3|) \div (-3)$ [2.1d], [2.5c]

Understanding Through Discussion and Writing

1. What rule have we developed that would tell you the sign of $(-7)^8$ and $(-7)^{11}$ without doing the computations? Explain. [2.4a]

2. Under what circumstances will the sum of a positive integer and a negative integer be negative? [2.2a]

3. Jake enters $18 \div 2 \cdot 3$ on his calculator and expects the result to be 3. What mistake is he making? [2.5c]

4. Write a problem for a classmate to solve. Design the problem so that the solution is "The temperature dropped to $-9°$F." [2.3b], [2.5b]

Test

For Extra Help For step-by-step test solutions, access the Chapter Test Prep Videos in MyMathLab® or on YouTube (search "BittingerBasicEl" and click on "Channels").

Use either $<$ or $>$ for $\square$ to write a true sentence.

1. $-4 \square 0$

2. $-3 \square -8$

3. $-7 \square -8$

4. $-1 \square 1$

Simplify.

5. $|-7|$

6. $|94|$

7. $-|-27|$

Find the opposite, or additive inverse.

8. 23

9. -14

10. Evaluate $-x$ when x is -8.

11. Graph -6 on the number line.

$-6\ -5\ -4\ -3\ -2\ -1\ 0\ 1\ 2\ 3\ 4\ 5\ 6$

Compute and simplify.

12. $31 - (-47)$

13. $-8 + 4 + (-7) + 3$

14. $-13 + 15$

15. $2 - (-8)$

16. $32 - 57$

17. $18 + (-3)$

18. $4 \cdot (-12)$

19. $-8 \cdot (-3)$

20. $-45 \div 5$

21. $-63 \div (-7)$

22. $64 \div (-16)$

23. $-2(16) - [2(-8) - 5^3]$

24. *Difference in Elevation.* The lowest elevation in Australia, Lake Eyre, is 15 m below sea level. The highest elevation in Australia, Mount Kosciuszko, is 2229 m. Find the difference in elevation between the highest point and the lowest point.

Source: *The CIA World Factbook 2012*

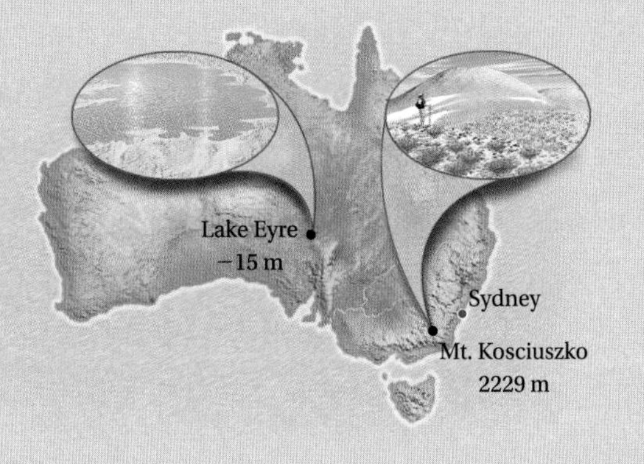

25. Isabella kept track of the changes in the stock market over a period of 5 weeks. By how many points had the market risen or fallen over this time?

WEEK 1	WEEK 2	WEEK 3	WEEK 4	WEEK 5
Down 13 pts	Down 16 pts	Up 36 pts	Down 11 pts	Up 19 pts

26. *Population Decrease.* The population of Stone City was 18,600. It decreased by 420 each year for 6 years. What was the population of the city after 6 years?

27. *Chemical Reaction.* During a chemical reaction, the temperature in a beaker decreased every minute by the same number of degrees. The temperature was 17°C at 11:08 A.M. By 11:25 A.M., the temperature had dropped to −17°C. By how many degrees did it change each minute?

28. Evaluate $-(-x)$ when x is 14.

A. −14 **B.** 196
C. 0 **D.** 14

Synthesis

29. Simplify: $|-27 - 3(4)| - |-36| + |-12|$.

30. The deepest point in the Pacific Ocean is the Marianas Trench with a depth of 11,033 m. The deepest point in the Atlantic Ocean is the Puerto Rico Trench with a depth of 8648 m. How much higher is the Puerto Rico Trench than the Marianas Trench?

Source: Defense Mapping Agency, Hydrographic/Topographic Center

31. Find the next three numbers in each sequence.

a) 6, 5, 3, 0, ____, ____, ____

b) 14, 10, 6, 2, ____, ____, ____

c) −4, −6, −9, −13, ____, ____, ____

d) 64, −32, 16, −8, ____, ____, ____

CHAPTER 3

Fraction Notation: Multiplication and Division

STUDYING FOR SUCCESS *Learning Resources on Campus*

- ☐ There may be a learning lab or a tutoring center for drop-in tutoring.
- ☐ There may be group tutoring sessions for this specific course.
- ☐ The math department may have a bulletin board or network that provides contact information for private tutors.

3.1 Factorizations

OBJECTIVES

a Determine whether one number is a factor of another, and find the factors of a number.

b Find some multiples of a number, and determine whether a number is divisible by another.

c Given a number from 1 to 100, tell whether it is prime, composite, or neither.

d Find the prime factorization of a composite number.

SKILL TO REVIEW

Objective 1.5a: Divide whole numbers.

Divide.

1. $329 \div 8$

2. $23\overline{)1081}$

In this chapter, we begin our work with fractions and fraction notation. *Factoring* is an important skill in working with fractions. For example, in order to simplify $\frac{12}{32}$, it is important that we be able to factor 12 and 32, as follows:

$$\frac{12}{32} = \frac{4 \cdot 3}{4 \cdot 8}.$$

Then we "remove" a factor of 1:

$$\frac{4 \cdot 3}{4 \cdot 8} = \frac{4}{4} \cdot \frac{3}{8} = 1 \cdot \frac{3}{8} = \frac{3}{8}.$$

a FACTORS AND FACTORIZATIONS

Here we consider only the **natural numbers** 1, 2, 3, and so on.

Let's look at the product $3 \cdot 4 = 12$. We say that 3 and 4 are **factors** of 12. When we divide 12 by 3, we get a remainder of 0. We say that the divisor 3 is a **factor** of the dividend 12.

FACTOR

- In the product $a \cdot b$, a and b are called **factors**.
- If we divide Q by d and get a remainder of 0, then the divisor d is a **factor** of the dividend Q.

EXAMPLE 1 Determine by long division **(a)** whether 6 is a factor of 198 and **(b)** whether 15 is a factor of 198.

a)
$$\begin{array}{r} 33 \\ 6\overline{)198} \\ \underline{18} \\ 18 \\ \underline{18} \\ 0 \end{array}$$
0 ← Remainder is 0.

The remainder is 0, so 6 is a factor of 198.

b)
$$\begin{array}{r} 13 \\ 15\overline{)198} \\ \underline{15} \\ 48 \\ \underline{45} \\ 3 \end{array}$$
3 ← Not 0

The remainder is not 0, so 15 is not a factor of 198.

◀ **Do Margin Exercises 1 and 2.**

Determine whether the second number is a factor of the first.

1. 72; 8 **2.** 2384; 28

Answers

Skill to Review:
1. 41 R 1 **2.** 47

Margin Exercises:
1. Yes **2.** No

Consider $12 = 3 \cdot 4$. We say that $3 \cdot 4$ is a **factorization** of 12. Similarly, $6 \cdot 2$, $12 \cdot 1$, $2 \cdot 2 \cdot 3$, and $1 \cdot 3 \cdot 4$ are also factorizations of 12. Since $a = a \cdot 1$, every number has a factorization, and every number has factors. For some numbers, the factors consist of only the number itself and 1. For example, the only factorization of 17 is $17 \cdot 1$, so the only factors of 17 are 17 and 1.

List all the factors of each number.

3. 10

EXAMPLE 2 List all the factors of 70.

We list as many "two-factor" factorizations as we can. We check sequentially the numbers 1, 2, 3, and so on, to see if we can form any factorizations:

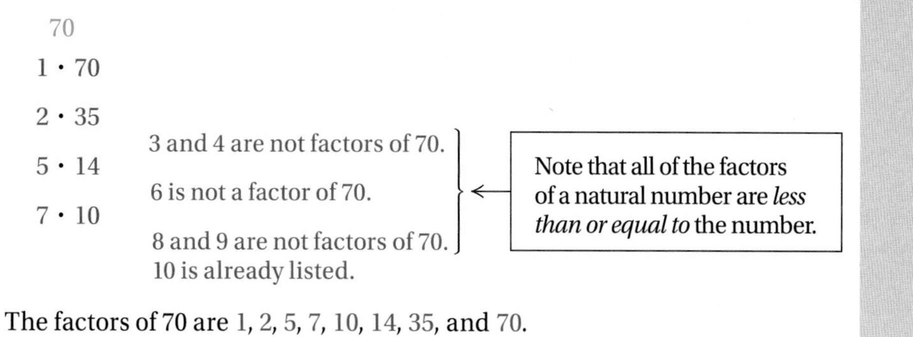

The factors of 70 are 1, 2, 5, 7, 10, 14, 35, and 70.

Do Exercises 3–6.

GS **4.** 45

$1 \cdot$ ___

2 is not a factor of 45.

$3 \cdot$ ___

4 is not a factor of 45.

$5 \cdot$ ___

6, 7, and 8 are not factors of 45.

9 is already listed.

The factors of 45 are 1, ___ , 5, ___ ,15, and ___ .

5. 62

6. 24

b MULTIPLES AND DIVISIBILITY

A **multiple** of a natural number is a product of that number and some natural number. For example, some multiples of 2 are

2 (because $2 = 1 \cdot 2$);
4 (because $4 = 2 \cdot 2$);
6 (because $6 = 3 \cdot 2$);
8 (because $8 = 4 \cdot 2$).

Note that all of the multiples of a number are *greater than or equal to* the number.

We find multiples of 2 by counting by twos: 2, 4, 6, 8, and so on. We can find multiples of 3 by counting by threes: 3, 6, 9, 12, and so on.

EXAMPLE 3 Show that each of the numbers 8, 12, 20, and 36 is a multiple of 4.

$8 = 2 \cdot 4 \qquad 12 = 3 \cdot 4 \qquad 20 = 5 \cdot 4 \qquad 36 = 9 \cdot 4$

Do Exercises 7 and 8.

7. Show that each of the numbers 5, 45, and 100 is a multiple of 5.

8. Show that each of the numbers 10, 60, and 110 is a multiple of 10.

EXAMPLE 4 Multiply by 1, 2, 3, and so on, to find ten multiples of 7.

$1 \cdot 7 = 7 \qquad 3 \cdot 7 = 21 \qquad 5 \cdot 7 = 35 \qquad 7 \cdot 7 = 49 \qquad 9 \cdot 7 = 63$

$2 \cdot 7 = 14 \qquad 4 \cdot 7 = 28 \qquad 6 \cdot 7 = 42 \qquad 8 \cdot 7 = 56 \qquad 10 \cdot 7 = 70$

Do Exercise 9.

9. Multiply by 1, 2, 3, and so on, to find ten multiples of 5.

Answers

3. 1, 2, 5, 10 **4.** 1, 3, 5, 9, 15, 45
5. 1, 2, 31, 62 **6.** 1, 2, 3, 4, 6, 8, 12, 24
7. $5 = 1 \cdot 5$; $45 = 9 \cdot 5$; $100 = 20 \cdot 5$
8. $10 = 1 \cdot 10$; $60 = 6 \cdot 10$; $110 = 11 \cdot 10$
9. 5, 10, 15, 20, 25, 30, 35, 40, 45, 50

Guided Solution:
4. 45, 15, 9; 3, 9, 45

DIVISIBILITY

The number a is **divisible** by another number b if there exists a number c such that $a = b \cdot c$. The statements "a is **divisible** by b," "a is a **multiple** of b," and "b is a **factor** of a" all have the same meaning.

Thus, 27 is *divisible* by 3 because $27 = 3 \cdot 9$. We can also say that 27 is a *multiple* of 3, and 3 is a *factor* of 27.

EXAMPLE 5 Determine **(a)** whether 45 is divisible by 9 and **(b)** whether 45 is divisible by 4.

a) We divide 45 by 9.

$$\begin{array}{r} 5 \\ 9\overline{)45} \\ \underline{45} \\ 0 \end{array} \leftarrow \text{Remainder is 0.}$$

Because the remainder is 0, 45 is divisible by 9.

b) We divide 45 by 4.

$$\begin{array}{r} 11 \\ 4\overline{)45} \\ \underline{4\ } \\ 5 \\ \underline{4} \\ 1 \end{array} \leftarrow \text{Not 0}$$

Since the remainder is not 0, 45 is not divisible by 4.

◀ Do Exercises 10–12.

10. Determine whether 16 is divisible by 2.

$$\begin{array}{r} 8 \\ \square\overline{)16} \\ \underline{16} \\ \square \end{array}$$

Since the remainder is ☐, 16 ☐ divisible by 2. (is/is not)

11. Determine whether 125 is divisible by 5.

12. Determine whether 125 is divisible by 6.

C PRIME AND COMPOSITE NUMBERS

PRIME AND COMPOSITE NUMBERS

- A natural number that has exactly two *different* factors, only itself and 1, is called a **prime number**.
- The number 1 is *not* prime.
- A natural number, other than 1, that is not prime is **composite**.

EXAMPLE 6 Determine whether the numbers 1, 2, 7, 8, 9, 11, 18, 39, and 59 are prime, composite, or neither.

The number 1 is not prime. It does not have two different factors.

The number 2 is prime. It has only the factors 2 and 1.

The numbers 7, 11, and 59 are prime. Each has only two factors, itself and 1.

The number 8 is not prime. It has the factors 1, 2, 4, and 8 and is composite.

The numbers 9, 18, and 39 are composite. Each has more than two factors.

Thus we have

Prime: 2, 7, 11, 59;
Composite: 8, 9, 18, 39;
Neither: 1.

◀ Do Exercise 13.

13. Tell whether each number is prime, composite, or neither.

1, 2, 6, 12, 13, 19, 41, 65, 73, 99

Answers

10. Yes **11.** Yes **12.** No **13.** 2, 13, 19, 41, and 73 are prime; 6, 12, 65, and 99 are composite; 1 is neither.

Guided Solution:
10. 2, 0; 0, is

The number 2 is the *only* even prime number. It is also the smallest prime number. The number 0 is neither prime nor composite, but 0 is *not* a natural number and thus is not considered here. We are considering only natural numbers. The number 1 is the only natural number that is neither prime nor composite.

The table at right lists the prime numbers from 2 to 97. These prime numbers will be the most helpful to you in this text.

A TABLE OF PRIME NUMBERS FROM 2 TO 97

2, 3, 5, 7, 11, 13, 17, 19, 23, 29, 31, 37, 41, 43, 47, 53, 59, 61, 67, 71, 73, 79, 83, 89, 97

d PRIME FACTORIZATIONS

When we factor a composite number into a product of primes, we find the **prime factorization** of the number. To do this, we consider the primes

2, 3, 5, 7, 11, 13, 17, 19, 23, and so on,

and determine whether a given number is divisible by the primes.

EXAMPLE 7 Find the prime factorization of 39.

a) We divide by the first prime, 2.

$$\begin{array}{r} 19 \\ 2\overline{)39} \\ \underline{2} \\ 19 \\ \underline{18} \\ 1 \end{array}$$

Because the remainder is not 0, 2 is not a factor of 39, and 39 is not divisible by 2.

b) We divide by the next prime, 3.

$$\begin{array}{r} 13 \quad R = 0 \\ 3\overline{)39} \end{array}$$

The remainder is 0, so we know that $39 = 3 \cdot 13$. Because 13 is a prime, we are finished. The prime factorization is

$$39 = 3 \cdot 13.$$

EXAMPLE 8 Find the prime factorization of 220.

a) We divide by the first prime, 2.

$$\begin{array}{r} 110 \quad R = 0 \\ 2\overline{)220} \end{array} \qquad 220 = 2 \cdot 110$$

b) Because 110 is composite, we continue to divide, starting with 2 again.

$$\begin{array}{r} 55 \quad R = 0 \\ 2\overline{)110} \end{array} \qquad 220 = 2 \cdot 2 \cdot 55$$

c) Since 55 is composite and is not divisible by 2 or 3, we divide by the next prime, 5.

$$\begin{array}{r} 11 \quad R = 0 \\ 5\overline{)55} \end{array} \qquad 220 = 2 \cdot 2 \cdot 5 \cdot 11$$

Because 11 is prime, we are finished. The prime factorization is

$$220 = 2 \cdot 2 \cdot 5 \cdot 11.$$

CALCULATOR CORNER

Divisibility and Factors We can use a calculator to determine whether one number is divisible by another number or whether one number is a factor of another number. For example, to determine whether 387 is divisible by 9, we press [3] [8] [7] [÷] [9] [=]. The display is [43]. Since 43 is a natural number, we know that 387 is a multiple of 9; that is, $387 = 43 \cdot 9$. Thus, 387 is divisible by 9, and 9 is a factor of 387.

EXERCISES For each pair of numbers, determine whether the second number is a factor of the first number.

1. 502; 8

2. 651; 21

3. 3875; 25

4. 1047; 14

We abbreviate our procedure as follows.

$$\begin{array}{r} 11 \\ 5\overline{)55} \\ 2\overline{)110} \\ 2\overline{)220} \end{array}$$

$220 = 2 \cdot 2 \cdot 5 \cdot 11$

Because multiplication is commutative, a factorization such as $2 \cdot 2 \cdot 5 \cdot 11$ could also be expressed as $5 \cdot 2 \cdot 2 \cdot 11$ or $2 \cdot 5 \cdot 11 \cdot 2$ (or, in exponential notation, as $2^2 \cdot 5 \cdot 11$ or $11 \cdot 2^2 \cdot 5$), but the prime factors are the same in each case. For this reason, we agree that any of these is "the" prime factorization of 220.

> Every number has just one (unique) prime factorization.

Caution!

Keep in mind the difference between finding all the factors of a number and finding the prime factorization. In Example 9, the prime factorization of 72 is $2 \cdot 2 \cdot 2 \cdot 3 \cdot 3$. The factors of 72 are 1, 2, 3, 4, 6, 8, 9, 12, 18, 24, 36, and 72.

EXAMPLE 9 Find the prime factorization of 72.

We can do divisions "up" as follows:

$$\begin{array}{rl} 3 & \leftarrow \text{Prime quotient} \\ 3\overline{)\ 9} & \\ 2\overline{)18} & \\ 2\overline{)36} & \\ 2\overline{)72} & \leftarrow \text{Begin here} \end{array}$$

$72 = 2 \cdot 2 \cdot 2 \cdot 3 \cdot 3$

Another way to find the prime factorization of 72 uses a **factor tree** as follows. Begin by determining any factorization you can, and then continue factoring. Any one of the following factor trees gives the prime factorization of 72:

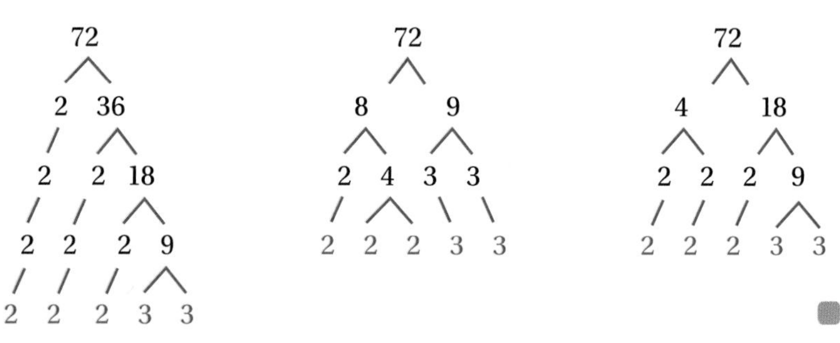

EXAMPLE 10 Find the prime factorization of 189.

We can use a string of successive divisions or a factor tree. Since 189 is not divisible by 2, we begin with 3.

$$\begin{array}{r} 7 \\ 3\overline{)21} \\ 3\overline{)63} \\ 3\overline{)189} \end{array}$$

$189 = 3 \cdot 3 \cdot 3 \cdot 7$

189
3 63
3 7 9
3 7 3 3

◀ Do Exercises 14–21.

Find the prime factorization of each number.

14. 6 **15.** 12

16. 45 **17.** 98

18. 126 **19.** 144

20. 1960 **21.** 1925

Answers

14. $2 \cdot 3$ **15.** $2 \cdot 2 \cdot 3$ **16.** $3 \cdot 3 \cdot 5$
17. $2 \cdot 7 \cdot 7$ **18.** $2 \cdot 3 \cdot 3 \cdot 7$
19. $2 \cdot 2 \cdot 2 \cdot 2 \cdot 3 \cdot 3$ **20.** $2 \cdot 2 \cdot 2 \cdot 5 \cdot 7 \cdot 7$
21. $5 \cdot 5 \cdot 7 \cdot 11$

3.1 Exercise Set

Reading Check

Determine whether each statement is true or false.

_______________ **RC1.** One factorization of 20 is $4 \cdot 5$.

_______________ **RC2.** One multiple of 15 is 30.

_______________ **RC3.** The number 18 is divisible by 5.

_______________ **RC4.** A prime number has exactly two different factors.

_______________ **RC5.** The smallest prime number is 1.

_______________ **RC6.** The prime factorization of 30 is $3 \cdot 10$.

a Determine whether the second number is a factor of the first.

1. 52; 14

2. 52; 13

3. 625; 25

4. 680; 16

List all the factors of each number.

5. 18

6. 16

7. 54

8. 48

9. 4

10. 9

11. 1

12. 13

13. 98

14. 100

15. 255

16. 120

b Multiply by 1, 2, 3, and so on, to find ten multiples of each number.

17. 4

18. 11

19. 20

20. 50

21. 3

22. 5

23. 12

24. 13

25. 10

26. 6

27. 9

28. 14

29. Determine whether 26 is divisible by 6.

30. Determine whether 48 is divisible by 8.

31. Determine whether 1880 is divisible by 8.

32. Determine whether 4227 is divisible by 3.

33. Determine whether 256 is divisible by 16.

34. Determine whether 102 is divisible by 4.

35. Determine whether 4227 is divisible by 9.

36. Determine whether 4143 is divisible by 7.

c Determine whether each number is prime, composite, or neither.

37. 1 **38.** 2 **39.** 9 **40.** 19

41. 11 **42.** 27 **43.** 29 **44.** 49

d Find the prime factorization of each number.

45. 8 **46.** 16 **47.** 14 **48.** 15 **49.** 42

50. 32 **51.** 25 **52.** 40 **53.** 50 **54.** 62

55. 169 **56.** 140 **57.** 100 **58.** 110 **59.** 35

60. 70 **61.** 72 **62.** 86 **63.** 77 **64.** 99

65. 2884 **66.** 484 **67.** 51 **68.** 91 **69.** 1200

70. 1800 **71.** 273 **72.** 675 **73.** 1122 **74.** 6435

Skill Maintenance

What does the digit 4 mean in each number? [1.1a]

75. 134,895 **76.** 4,682,013 **77.** 29,745 **78.** 13,248,957

Round 34,562 to the nearest: [1.6a]

79. Ten. **80.** Hundred.

Round 2,428,497 to the nearest: [1.6a]

81. Thousand. **82.** Ten.

Synthesis

83. *Factors and Sums.* The top number in each column of the table below can be factored as a product of two numbers whose sum is the bottom number in the column. For example, in the first column, 56 has been factored as $7 \cdot 8$, and $7 + 8 = 15$. Fill in the blank spaces in the table.

PRODUCT	56	63	36	72	140	96		168	110			
FACTOR	7									9	24	3
FACTOR	8						8	8		10	18	
SUM	15	16	20	38	24	20	14		21			24

Divisibility

3.2

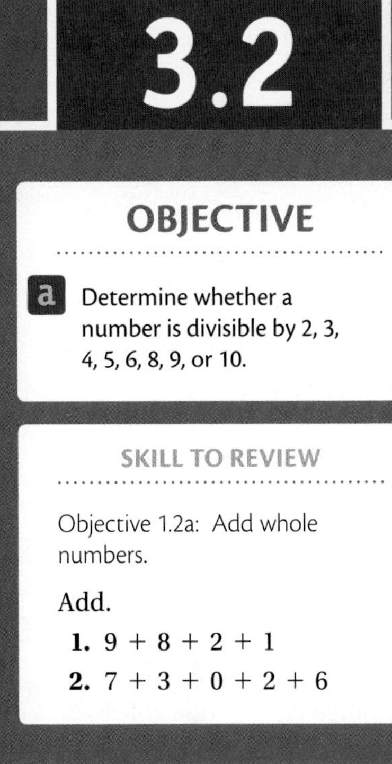

Suppose you are asked to find the simplest fraction notation for

$$\frac{117}{225}.$$

Since the numbers are quite large, you might feel that the task is difficult. However, both the numerator and the denominator are divisible by 9. If you knew this, you could factor and simplify as follows:

$$\frac{117}{225} = \frac{9 \cdot 13}{9 \cdot 25} = \frac{9}{9} \cdot \frac{13}{25} = 1 \cdot \frac{13}{25} = \frac{13}{25}.$$

How did we know that both numbers have 9 as a factor? There is a simple test for determining this.

In this section, you learn quick ways to determine whether a number is divisible by 2, 3, 4, 5, 6, 8, 9, or 10. This will make simplifying fraction notation much easier.

a RULES FOR DIVISIBILITY

Divisibility by 2

You may already know the test for divisibility by 2.

BY 2

A number is **divisible by 2** (is *even*) if it has a ones digit of 0, 2, 4, 6, or 8 (that is, it has an even ones digit).

Let's see why this test works. Consider 354, which is

3 hundreds + 5 tens + 4.

Hundreds and tens are both multiples of 2. If the last digit is a multiple of 2, then the entire number is a multiple of 2.

EXAMPLES Determine whether each number is divisible by 2.

1. 355 is *not* divisible by 2; 5 is *not* even.
2. 4786 is divisible by 2; 6 is even.
3. 8990 is divisible by 2; 0 is even.
4. 4261 is *not* divisible by 2; 1 is *not* even.

Do Margin Exercises 1–4.

Determine whether each number is divisible by 2.

1. 84
2. 59
3. 998
4. 2225

Answers

Skill to Review:
1. 20 2. 18

Margin Exercises:
1. Yes 2. No 3. Yes 4. No

Divisibility by 3

BY 3

A number is **divisible by 3** if the sum of its digits is divisible by 3.

Let's illustrate why the test for divisibility by 3 works. Consider 852; since $852 = 3 \cdot 284$, 852 is divisible by 3.

$$\begin{aligned} 852 &= 8 \cdot 100 + 5 \cdot 10 + 2 \cdot 1 \\ &= 8(99 + 1) + 5(9 + 1) + 2(1) \\ &= 8 \cdot 99 + 8 \cdot 1 + 5 \cdot 9 + 5 \cdot 1 + 2 \cdot 1 \end{aligned}$$

Using the distributive law: $a(b + c) = a \cdot b + a \cdot c$

Since 99 and 9 are each a multiple of 3, we see that $8 \cdot 99$ and $5 \cdot 9$ are multiples of 3. This leaves $8 \cdot 1 + 5 \cdot 1 + 2 \cdot 1$, or $8 + 5 + 2$. If $8 + 5 + 2$, the sum of the digits, is divisible by 3, then 852 is divisible by 3.

EXAMPLES Determine whether each number is divisible by 3.

5. 18 $1 + 8 = 9$
6. 93 $9 + 3 = 12$
7. 201 $2 + 0 + 1 = 3$

Each is divisible by 3 because the sum of its digits is divisible by 3.

8. 256 $2 + 5 + 6 = 13$

The sum of the digits, 13, is *not* divisible by 3, so 256 is *not* divisible by 3.

Do Exercises 5–8.

Determine whether each number is divisible by 3.

5. 111 **6.** 1111

7. 309

GS **8.** 17,216

Add the digits:

$1 + 7 + \_\_ + 1 + 6 = \_\_$

Since 17 __ (is/is not) divisible by 3, the number 17,216 __ (is/is not) divisible by 3.

Divisibility by 6

A number divisible by 6 is a multiple of 6. But $6 = 2 \cdot 3$, so the number is also a multiple of 2 and 3. Since 2 and 3 have no factors in common, we have the following.

BY 6

A number is **divisible by 6** if its ones digit is 0, 2, 4, 6, or 8 (is even) and the sum of its digits is divisible by 3.

EXAMPLES Determine whether each number is divisible by 6.

9. 720

Because 720 is even, it is divisible by 2. Also, $7 + 2 + 0 = 9$, so 720 is divisible by 3. Thus, 720 is divisible by 6.

720 (Even) $7 + 2 + 0 = 9$ (Divisible by 3)

10. 73

73 is *not* divisible by 6 because it is *not* even.

11. 256

Although 256 is even, it is *not* divisible by 6 because the sum of its digits, $2 + 5 + 6$, or 13, is *not* divisible by 3.

Do Exercises 9–12.

Determine whether each number is divisible by 6.

9. 420 **10.** 106

11. 321 **12.** 444

Answers

5. Yes **6.** No **7.** Yes **8.** No **9.** Yes **10.** No **11.** No **12.** Yes

Guided Solution:
8. 2, 17, is not, is not

Divisibility by 9

The test for divisibility by 9 is similar to the test for divisibility by 3.

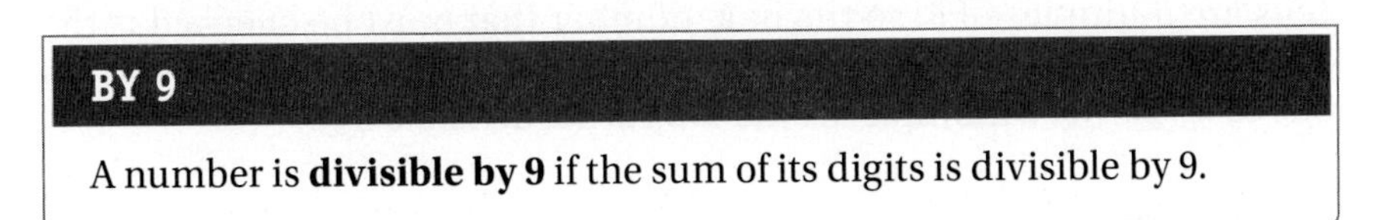

BY 9

A number is **divisible by 9** if the sum of its digits is divisible by 9.

EXAMPLES Determine whether each number is divisible by 9.

12. 6984

Because $6 + 9 + 8 + 4 = 27$ and 27 is divisible by 9, 6984 is divisible by 9.

13. 322

Because $3 + 2 + 2 = 7$ and 7 is *not* divisible by 9, 322 is *not* divisible by 9.

Do Exercises 13–16. ▶

Determine whether each number is divisible by 9.

13. 16 **14.** 117

15. 930 **16.** 29,223

Divisibility by 10

BY 10

A number is **divisible by 10** if its ones digit is 0.

We know that this test works because the product of 10 and *any* number has a ones digit of 0.

EXAMPLES Determine whether each number is divisible by 10.

14. 3440 is divisible by 10 because the ones digit is 0.

15. 3447 is *not* divisible by 10 because the ones digit is not 0.

Do Exercises 17–20. ▶

Determine whether each number is divisible by 10.

17. 305 **18.** 847

19. 300 **20.** 8760

Divisibility by 5

BY 5

A number is **divisible by 5** if its ones digit is 0 or 5.

EXAMPLES Determine whether each number is divisible by 5.

16. 220 is divisible by 5 because the ones digit is 0.

17. 475 is divisible by 5 because the ones digit is 5.

18. 6514 is *not* divisible by 5 because the ones digit is neither 0 nor 5.

Do Exercises 21–24. ▶

Determine whether each number is divisible by 5.

21. 5780 **22.** 3427

23. 34,678 **24.** 7775

Let's see why the test for 5 works. Consider 7830:

$$7830 = 10 \cdot 783 = 5 \cdot 2 \cdot 783.$$

Since 7830 is divisible by 10 and 5 is a factor of 10, 7830 is divisible by 5.

Answers

13. No **14.** Yes **15.** No **16.** Yes
17. No **18.** No **19.** Yes **20.** Yes
21. Yes **22.** No **23.** No **24.** Yes

Consider 6734:

6734 = 673 tens + 4.

Tens are multiples of 5, so the only number that must be checked is the ones digit. If the last digit is a multiple of 5, then the entire number is. In this case, 4 is *not* a multiple of 5, so 6734 is *not* divisible by 5.

Divisibility by 4

The test for divisibility by 4 is similar to the test for divisibility by 2.

BY 4

A number is **divisible by 4** if the number named by its last *two* digits is divisible by 4.

EXAMPLES Determine whether the number is divisible by 4.

19. 8212 is divisible by 4 because 12 is divisible by 4.
20. 5216 is divisible by 4 because 16 is divisible by 4.
21. 8211 is *not* divisible by 4 because 11 is *not* divisible by 4.
22. 7538 is *not* divisible by 4 because 38 is *not* divisible by 4.

Do Exercises 25–28.

Determine whether each number is divisible by 4.

25. 216 **26.** 217

27. 5862

28. 23,524
The number named by the last two digits is ____.
Since 24 ____ (is/is not) divisible by 4, the number 23,524 ____ (is/is not) divisible by 4.

To see why the test for divisibility by 4 works, consider 516:

516 = 5 hundreds + 16.

Hundreds are multiples of 4. If the number named by the last two digits is a multiple of 4, then the entire number is a multiple of 4.

Divisibility by 8

The test for divisibility by 8 is an extension of the tests for divisibility by 2 and 4.

BY 8

A number is **divisible by 8** if the number named by its last *three* digits is divisible by 8.

EXAMPLES Determine whether the number is divisible by 8.

23. 5648 is divisible by 8 because 648 is divisible by 8.
24. 96,088 is divisible by 8 because 88 is divisible by 8.
25. 7324 is *not* divisible by 8 because 324 is *not* divisible by 8.
26. 13,420 is *not* divisible by 8 because 420 is *not* divisible by 8.

Do Exercises 29–32.

Determine whether each number is divisible by 8.

29. 7564 **30.** 7864

31. 17,560 **32.** 25,716

A Note about Divisibility by 7

There are several tests for divisibility by 7, but all of them are more complicated than simply dividing by 7. If you want to test for divisibility by 7, simply divide by 7, either by hand or using a calculator.

Answers

25. Yes **26.** No **27.** No **28.** Yes
29. No **30.** Yes **31.** Yes **32.** No

Guided Solution:
28. 24; is, is

3.2 Exercise Set

For Extra Help
MyMathLab®
MathXL® PRACTICE | WATCH | READ | REVIEW

Reading Check

Match the beginning of each divisibility rule with the appropriate ending from the list at the right.

RC1. A number is divisible by 2 if ______________.
RC2. A number is divisible by 3 if ______________.
RC3. A number is divisible by 4 if ______________.
RC4. A number is divisible by 5 if ______________.
RC5. A number is divisible by 6 if ______________.
RC6. A number is divisible by 8 if ______________.
RC7. A number is divisible by 9 if ______________.
RC8. A number is divisible by 10 if ______________.

a) the sum of its digits is divisible by 3
b) the sum of its digits is divisible by 9
c) it has an even ones digit
d) its ones digit is 0 or 5
e) its ones digit is 0
f) it has an even ones digit and the sum of its digits is divisible by 3
g) the number named by its last two digits is divisible by 4
h) the number named by its last three digits is divisible by 8

a For Exercises 1–16, answer yes or no and give a reason based on the tests for divisibility.

1. Determine whether 84 is divisible by 3.

2. Determine whether 467 is divisible by 9.

3. Determine whether 5553 is divisible by 5.

4. Determine whether 2004 is divisible by 6.

5. Determine whether 671,500 is divisible by 10.

6. Determine whether 6120 is divisible by 5.

7. Determine whether 1773 is divisible by 9.

8. Determine whether 3286 is divisible by 3.

9. Determine whether 21,687 is divisible by 2.

10. Determine whether 64,091 is divisible by 10.

11. Determine whether 32,109 is divisible by 6.

12. Determine whether 9840 is divisible by 2.

13. Determine whether 126,930 is divisible by 4.

14. Determine whether 546,106 is divisible by 8.

15. Determine whether 796,840 is divisible by 8.

16. Determine whether 298,736 is divisible by 4.

For Exercises 17–24, test each number for divisibility by 2, 3, 4, 5, 6, 8, 9, and 10.

17. 6825

18. 12,600

19. 119,117

20. 2916

21. 127,575

22. 25,088

23. 9360

24. 143,507

To answer Exercises 25–32, consider the following numbers.

56	200	75	35
324	42	812	402
784	501	2345	111,111
55,555	3009	2001	1005

25. Which of the above are divisible by 3?

26. Which of the above are divisible by 2?

27. Which of the above are divisible by 5?

28. Which of the above are divisible by 4?

29. Which of the above are divisible by 8?

30. Which of the above are divisible by 6?

31. Which of the above are divisible by 10?

32. Which of the above are divisible by 9?

To answer Exercises 33–40, consider the following numbers.

305	313,332	876	64,000
1101	7624	1110	9990
13,205	111,126	5128	126,111

33. Which of the above are divisible by 2?

34. Which of the above are divisible by 3?

35. Which of the above are divisible by 6?

36. Which of the above are divisible by 5?

37. Which of the above are divisible by 9?

38. Which of the above are divisible by 8?

39. Which of the above are divisible by 10?

40. Which of the above are divisible by 4?

Skill Maintenance

Solve. [1.7b]

41. $56 + x = 194$

42. $y + 124 = 263$

43. $3008 = x + 2134$

44. $18 \cdot t = 1008$

45. $24 \cdot m = 624$

46. $338 = a \cdot 26$

Solve. [1.8a]

47. Marty's car has a five-speed transmission and gets 33 mpg in city driving. How many gallons of gas will it use to travel 1485 mi of city driving?

48. There are 60 min in 1 hr. How many minutes are there in 72 hr?

Synthesis

Find the prime factorization of each number. Use divisibility tests where applicable.

49. 7800

50. 2520

51. 2772

52. 1998

53. Fill in the missing digits of the number

95,☐☐8

so that it is divisible by 99.

54. A passenger in a taxicab asks for the driver's company number. The driver says abruptly, "Sure—you can have my number. Work it out: If you divide it by 2, 3, 4, 5, or 6, you will get a remainder of 1. If you divide it by 11, the remainder will be 0, and no driver has a company number that meets these requirements and is smaller than this one." Determine the number.

Fractions and Fraction Notation

3.3

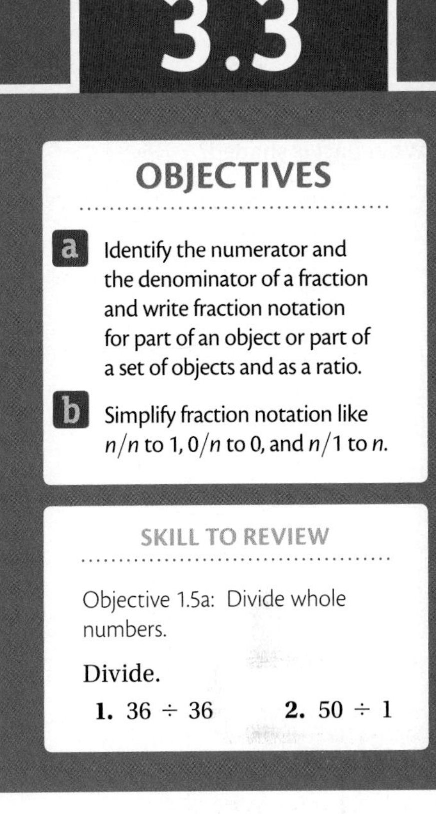

OBJECTIVES

a Identify the numerator and the denominator of a fraction and write fraction notation for part of an object or part of a set of objects and as a ratio.

b Simplify fraction notation like n/n to 1, $0/n$ to 0, and $n/1$ to n.

SKILL TO REVIEW

Objective 1.5a: Divide whole numbers.

Divide.

1. $36 \div 36$ **2.** $50 \div 1$

The study of arithmetic begins with the set of whole numbers. But we also need to be able to use fractional parts of numbers such as halves, thirds, fourths, and so on. Here is an example.

Households in auto-dependent locations spend about $\frac{1}{4}$ of their income on transportation costs, while location-efficient households (those with easy access to public transportation) can hold transportation costs to $\frac{1}{10}$ of their income.

Auto-Dependent Households

$\frac{1}{4}$ Transportation $\frac{3}{4}$ Remaining

Location-Efficient Households

$\frac{1}{10}$ Transportation $\frac{9}{10}$ Remaining

SOURCE: Based on data from U.S. Department of Transportation, Federal Highway Administration

a FRACTIONS AND THE REAL WORLD

Numbers like those above are written in **fraction notation**. The top number is called the **numerator** and the bottom number is called the **denominator**.

EXAMPLE 1 Identify the numerator and the denominator.

$$\frac{7}{8} \quad \begin{array}{l} \leftarrow \text{Numerator} \\ \leftarrow \text{Denominator} \end{array}$$

Do Margin Exercises 1–3. ▶

Let's look at various situations that involve fractions.

Fractions as a Partition of an Object Divided into Equal Parts

Consider a candy bar divided into 5 equal sections. If you eat 2 sections, you have eaten $\frac{2}{5}$ of the candy bar. The denominator 5 tells us the unit, $\frac{1}{5}$. The numerator 2 tells us the number of equal parts we are considering, 2.

For each fraction, identify the numerator and the denominator.

1. About $\frac{1}{5}$ of people age 5 and older in the United States speak a language other than English at home.

   Source: 2010 American Community Survey

2. About $\frac{4}{5}$ of the parts on a Toyota Camry were produced in the United States.

   Source: "Made in America: Which Car Creates the Most Jobs?" by David Muir and Sharyn Alfonsi, on abcnews.go.com

3. It is projected that $\frac{19}{100}$ of the U.S. population in 2050 will be foreign-born.

   Source: Pew Research Center

Answers

Skill to Review:
1. 1 **2.** 50

Margin Exercises:
1. Numerator: 1; denominator: 5
2. Numerator: 4; denominator: 5
3. Numerator: 19; denominator: 100

EXAMPLE 2 What part is shaded?

There are 8 equal parts. This tells us the unit, $\frac{1}{8}$. The *denominator* is 8. We have 5 of the units shaded. This tells us the *numerator*, 5. Thus,

$$\frac{5}{8} \begin{array}{l} \leftarrow \text{5 units are shaded.} \\ \leftarrow \text{The unit is } \frac{1}{8}. \end{array}$$

is shaded.

EXAMPLE 3 What part is shaded?

There are 18 equal parts. Thus the unit is $\frac{1}{18}$. The denominator is 18. We have 7 units shaded. This tells us the numerator, 7. Thus, $\frac{7}{18}$ is shaded.

Do Exercises 4–7.

The markings on a ruler use fractions.

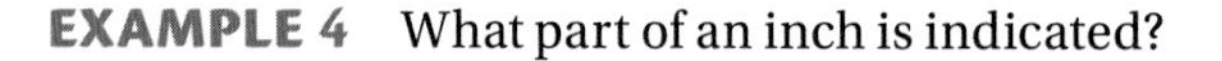

EXAMPLE 4 What part of an inch is indicated?

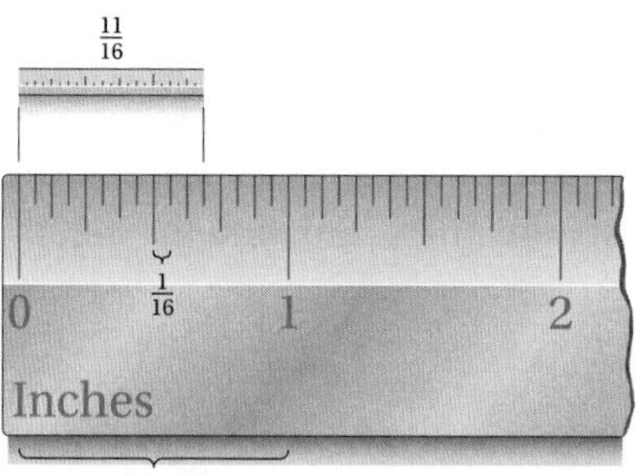

Each inch on the ruler shown above is divided into 16 equal parts. The marked section extends to the 11th mark. Thus, $\frac{11}{16}$ of an inch is indicated.

Do Exercise 8.

What part is shaded?

4.

5. 1 mile

6. 1 gallon

7.

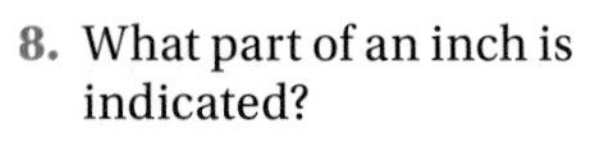

8. What part of an inch is indicated?

Answers

4. $\frac{5}{6}$ **5.** $\frac{1}{3}$ **6.** $\frac{3}{4}$ **7.** $\frac{8}{15}$ **8.** $\frac{15}{16}$

Fractions greater than or equal to 1, such as $\frac{24}{24}$, $\frac{10}{3}$, and $\frac{5}{4}$, correspond to situations like the following.

EXAMPLE 5 What part is shaded?

a)

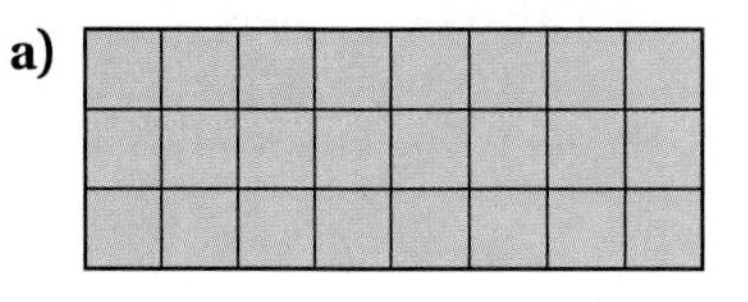

b)

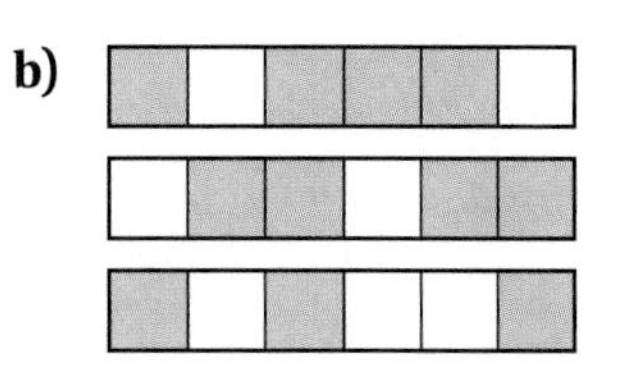

a) The rectangle is divided into 24 equal parts. Thus the unit is $\frac{1}{24}$. The denominator is 24. All 24 equal parts are shaded. This tells us that the numerator is 24. Thus, $\frac{24}{24}$ is shaded.

b) Each rectangle is divided into 6 parts. Thus the unit is $\frac{1}{6}$. The denominator is 6. We see that 11 of the equal units are shaded. This tells us that the numerator is 11. Thus, $\frac{11}{6}$ is shaded.

EXAMPLE 6 *Ice-Cream Roll-up Cake.* What part of an ice-cream roll-up cake is shaded?

Each cake is divided into 6 equal slices. The unit is $\frac{1}{6}$. The *denominator* is 6. We see that 13 of the slices are shaded. This tells us that the *numerator* is 13. Thus, $\frac{13}{6}$ are shaded.

Do Exercises 9–11. ▶

What part is shaded?

9.

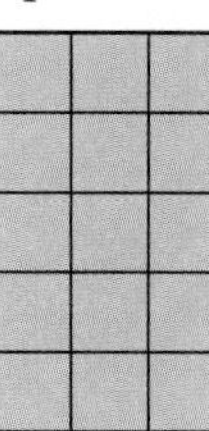

10.

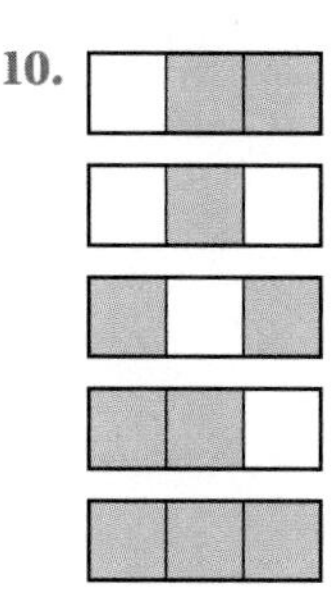

GS 11.

1 gallon

2 gallons

Each gallon is divided into ☐ equal parts.

The unit is $\frac{1}{\square}$.

There are ☐ equal units shaded.

The part that is shaded is $\frac{\square}{\square}$.

Negative fractions can also be used to represent real-world situations. If the water level in a reservoir drops five-eighths inch, for example, the fraction $-\frac{5}{8}$ (read "negative five-eighths") represents this situation.

Do Exercise 12. ▶

12. Write a fraction that represents the situation:

The blacktop on the playground sank one-fourth inch.

Fractions larger than or equal to 1, such as $\frac{13}{6}$ or $\frac{9}{9}$, are sometimes referred to as "improper" fractions. We will not use this terminology because notation such as $\frac{27}{8}$, $\frac{11}{3}$, and $\frac{4}{4}$ is quite "proper" and very common in algebra.

Fractions like $\frac{2}{3}$ are not integers. A number system called the **rational numbers** contains integers and fractions. The rational numbers consist of quotients of integers with nonzero divisors.

Answers

9. $\frac{15}{15}$ 10. $\frac{10}{3}$ 11. $\frac{7}{4}$ 12. $-\frac{1}{4}$

Guided Solution:

11. 4, 4, 7, $\frac{7}{4}$

The following are rational numbers:

$$\frac{2}{3},\quad -\frac{2}{3},\quad \frac{7}{1},\quad 4,\quad -3,\quad 0,\quad \frac{23}{-8}.$$

The number $-\frac{2}{3}$ (read "negative two-thirds") can also be named $\frac{-2}{3}$ or $\frac{2}{-3}$. That is,

$$-\frac{a}{b} = \frac{-a}{b} = \frac{a}{-b}.$$

RATIONAL NUMBERS

The **rational numbers** consist of all numbers that can be named in the form $\frac{a}{b}$, where a and b are integers and b is not 0.

Fractions as Ratios

A **ratio** is a quotient of two quantities. We can express a ratio with fraction notation. (We will consider ratios in more detail in Chapter 6.)

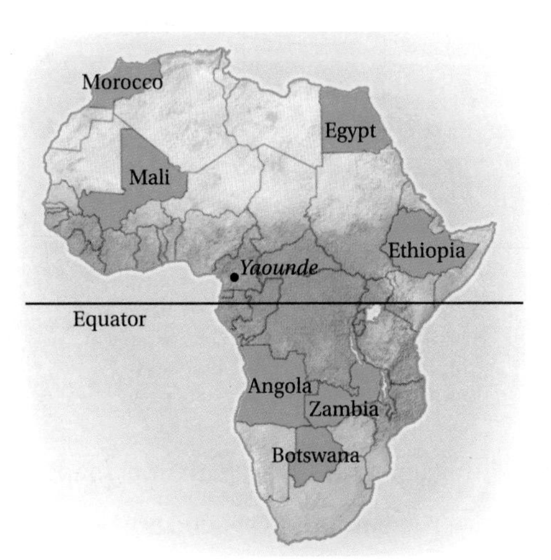

EXAMPLE 7 *Countries of Africa.* What part of this group of countries is north of the equator? south of the equator?

Angola	Mali
Bostwana	Morocco
Egypt	Zambia
Ethiopia	

There are 7 countries in the set, and 4 of them, Egypt, Ethiopia, Mali, and Morocco, are north of the equator. Thus, 4 of 7, or $\frac{4}{7}$, are north of the equator. The 3 remaining countries are south of the equator. Thus, $\frac{3}{7}$ are south of the equator.

Do Exercise 13.

13. What part of the set of countries in Example 7 is west of Yaounde?

b SOME FRACTION NOTATION FOR INTEGERS

Fraction Notation for 1

The number 1 corresponds to situations like those shown here. If we divide an object into n parts and take n of them, we get all of the object (1 whole object).

1

$\frac{2}{2}$

$\frac{4}{4}$

THE NUMBER 1 IN FRACTION NOTATION

$\frac{n}{n} = 1,\quad$ for any integer n that is not 0.

Answer

13. $\frac{2}{7}$

EXAMPLES Simplify.

8. $\frac{5}{5} = 1$

9. $\frac{-9}{-9} = 1$

10. $\frac{23}{23} = 1$

Do Exercises 14–19. ▶

Fraction Notation for 0

Consider the fraction $\frac{0}{4}$. This corresponds to dividing an object into 4 parts and taking none of them. We get 0.

> **THE NUMBER 0 IN FRACTION NOTATION**
>
> $\frac{0}{n} = 0,$ for any integer n that is not 0.

EXAMPLES Simplify.

11. $\frac{0}{1} = 0$

12. $\frac{0}{-9} = 0$

13. $\frac{0}{23} = 0$

Fraction notation with a denominator of 0, such as $n/0$, is meaningless because we cannot speak of an object being divided into *zero* parts. This corresponds to division by 0 being not defined.

> **A DENOMINATOR OF 0**
>
> $\frac{n}{0}$ is not defined for any integer n.

Do Exercises 20–25. ▶

Other Whole Numbers

Consider the fraction $\frac{4}{1}$. This corresponds to taking 4 objects and dividing each into 1 part. (In other words, we do not divide them.) We have 4 objects.

> **ANY INTEGER IN FRACTION NOTATION**
>
> Any integer divided by 1 is the integer itself. That is,
>
> $\frac{n}{1} = n,$ for any integer n.

EXAMPLES Simplify.

14. $\frac{2}{1} = 2$

15. $\frac{9}{1} = 9$

16. $\frac{-34}{1} = -34$

Do Exercises 26–29. ▶

Simplify.

14. $\frac{1}{1}$

15. $\frac{-4}{-4}$

16. $\frac{-34}{-34}$

17. $\frac{100}{100}$

18. $\frac{2347}{2347}$

19. $\frac{-103}{-103}$

Simplify, if possible.

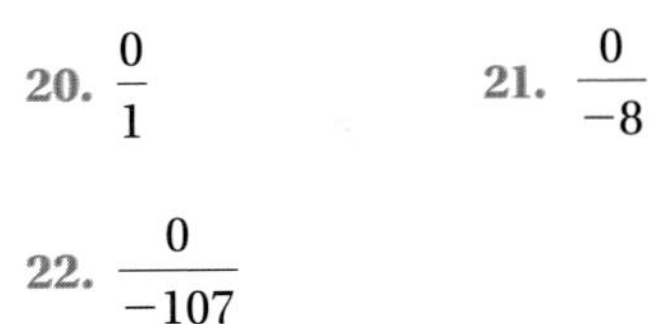

20. $\frac{0}{1}$

21. $\frac{0}{-8}$

22. $\frac{0}{-107}$

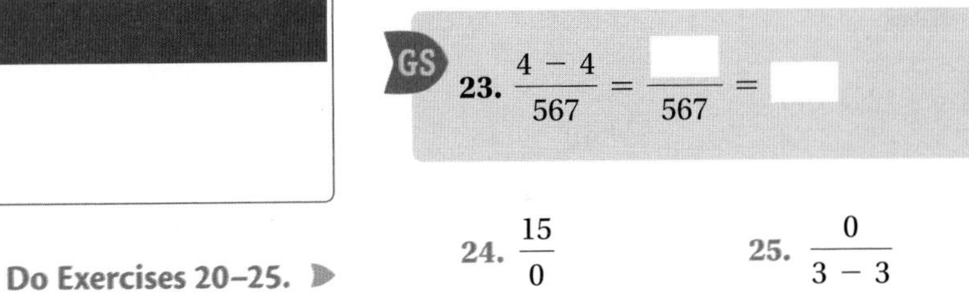

GS **23.** $\frac{4-4}{567} = \frac{\square}{567} = \square$

24. $\frac{15}{0}$

25. $\frac{0}{3-3}$

Simplify.

26. $\frac{8}{1}$

27. $\frac{-10}{1}$

28. $\frac{-346}{1}$

29. $\frac{24-1}{23-22}$

Answers

14. 1 **15.** 1 **16.** 1 **17.** 1 **18.** 1 **19.** 1 **20.** 0 **21.** 0 **22.** 0 **23.** 0 **24.** Not defined **25.** Not defined **26.** 8 **27.** −10 **28.** −346 **29.** 23

Guided Solution:

23. 0, 0

3.3 Exercise Set

For Extra Help MyMathLab® MathXL® PRACTICE WATCH READ REVIEW

Reading Check

Match each expression with the appropriate description or value from the list at the right.

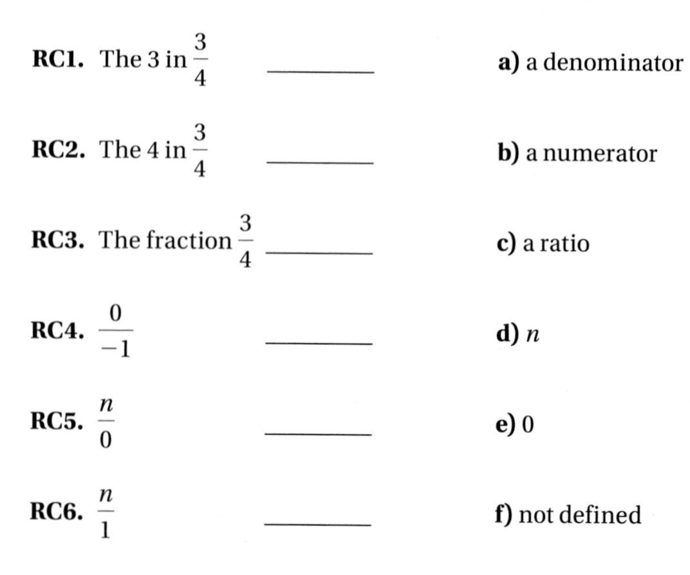

RC1. The 3 in $\frac{3}{4}$ ________ **a)** a denominator

RC2. The 4 in $\frac{3}{4}$ ________ **b)** a numerator

RC3. The fraction $\frac{3}{4}$ ________ **c)** a ratio

RC4. $\frac{0}{-1}$ ________ **d)** n

RC5. $\frac{n}{0}$ ________ **e)** 0

RC6. $\frac{n}{1}$ ________ **f)** not defined

a Identify the numerator and the denominator.

1. $\frac{3}{4}$ **2.** $\frac{9}{10}$ **3.** $\frac{11}{2}$ **4.** $\frac{18}{5}$ **5.** $\frac{0}{7}$ **6.** $\frac{1}{13}$

What part of each object or set of objects is shaded?

7.

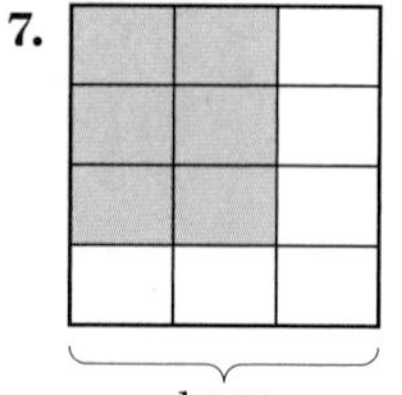

8.

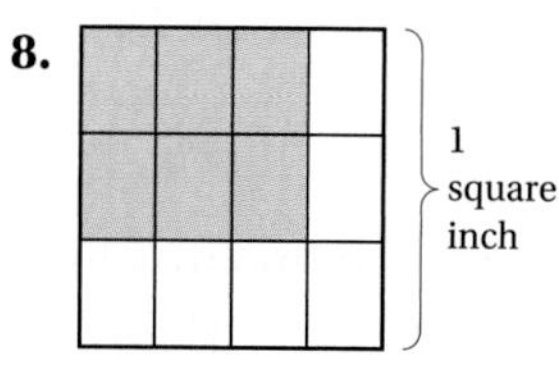

9.

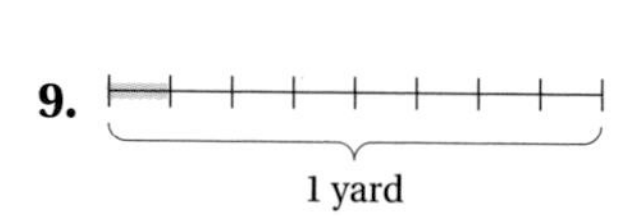

10. 1 mile

11.

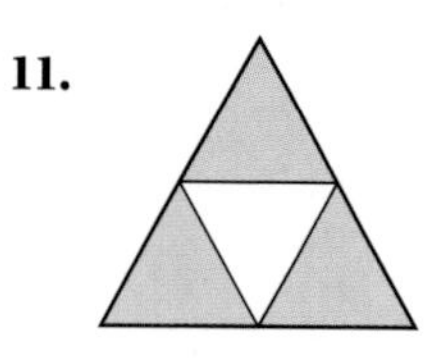

12.

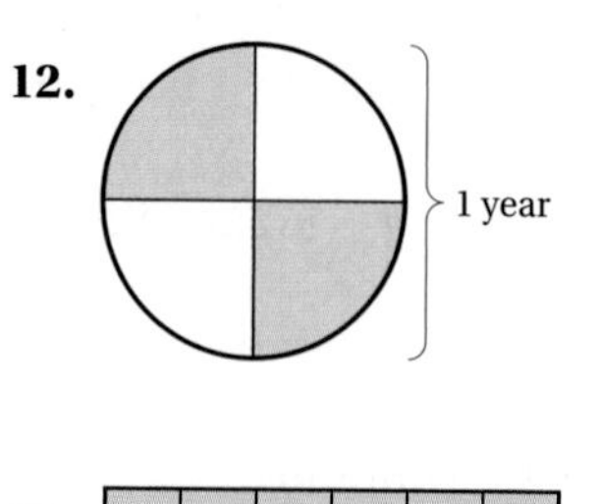

13.

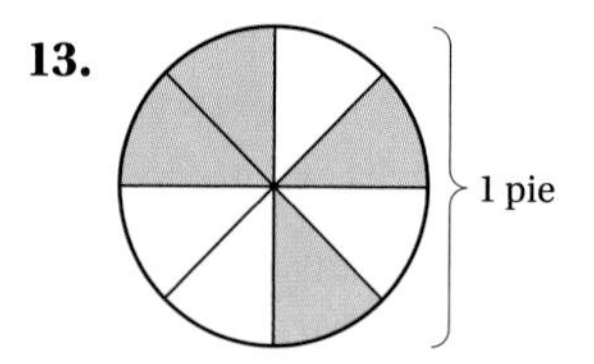

14.

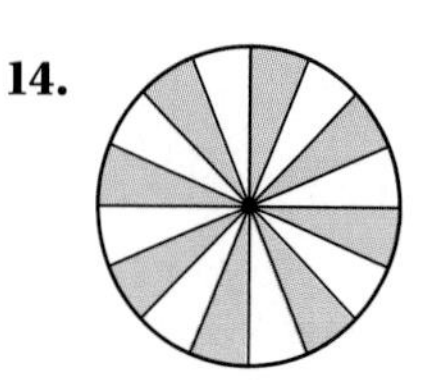

15.

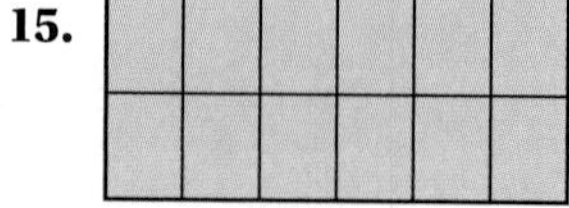

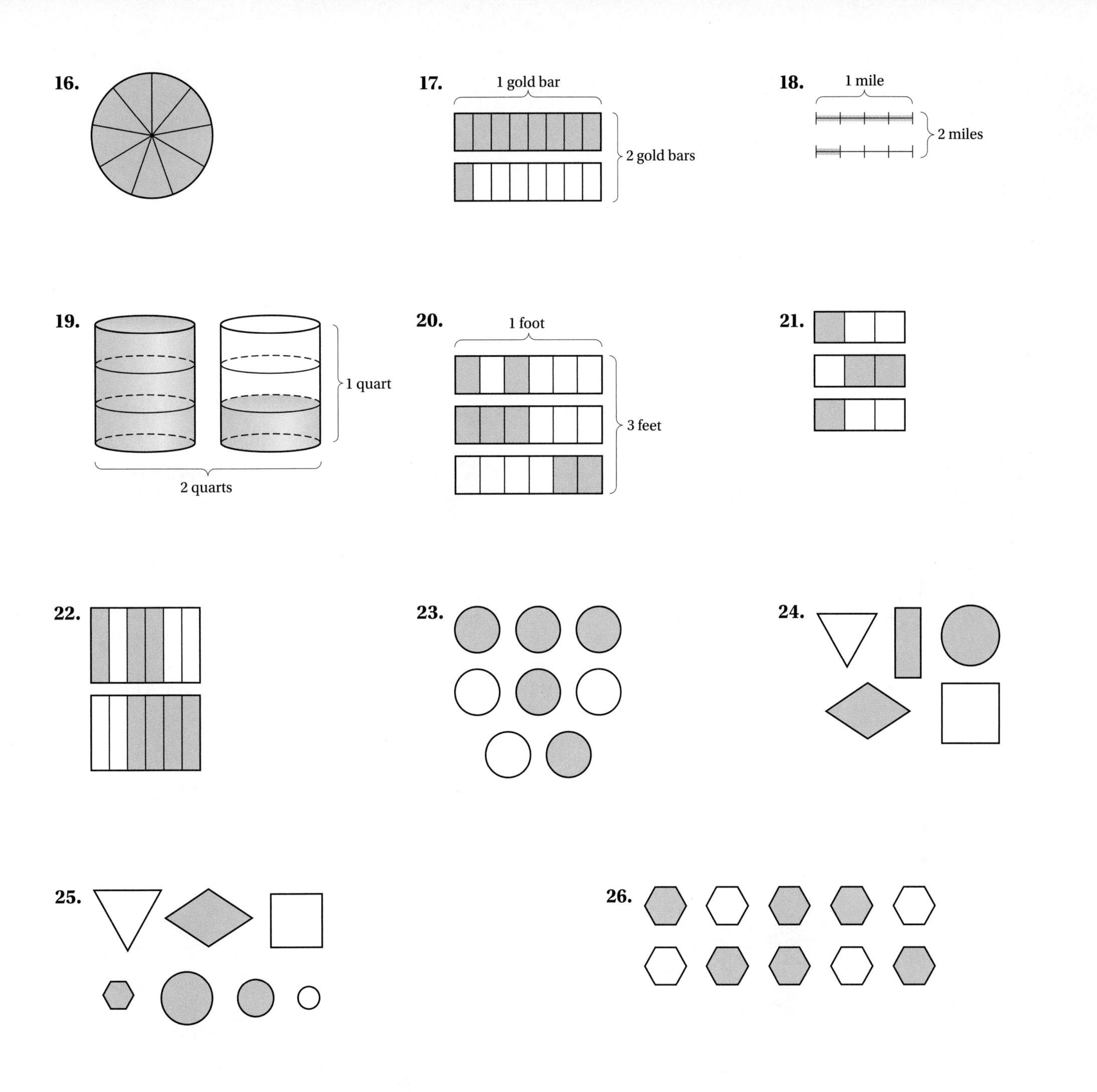

Write a fraction that represents the situation.

27. The water level in the well dropped one-third foot.

28. There is a depression in the road's surface that is five-sixteenths inch deep.

What part of an inch is indicated?

29.

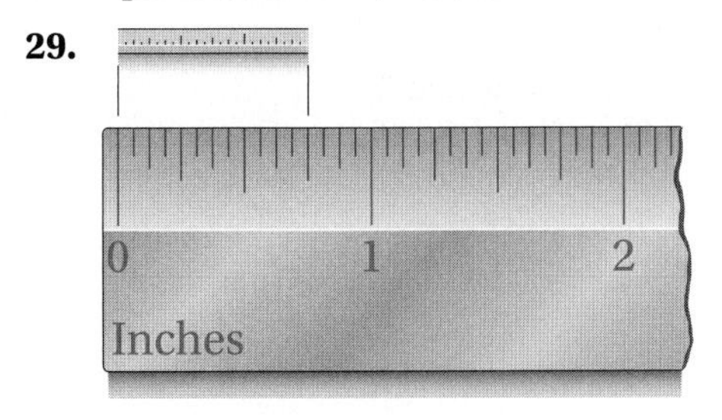

30.

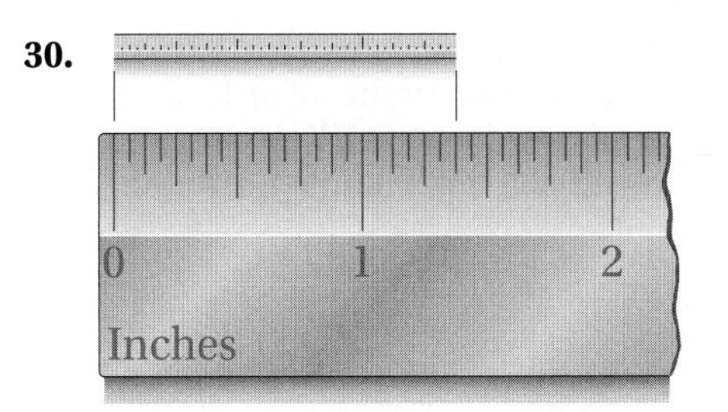

31. **32.**

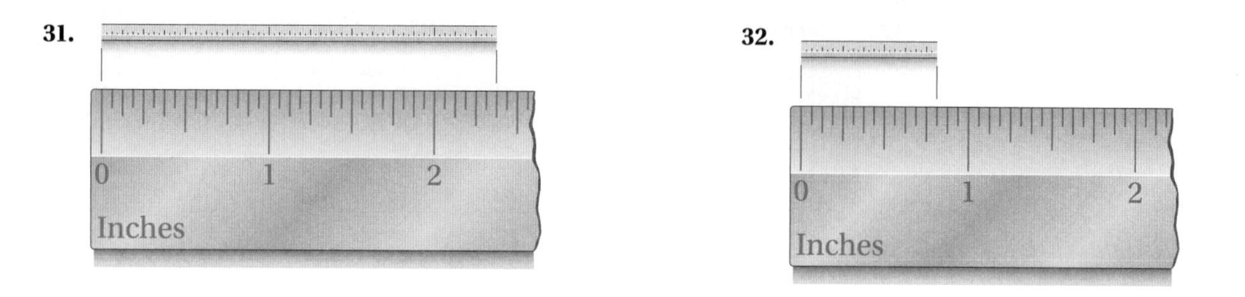

For each of Exercises 33–36, give fraction notation for the amount of gas **(a)** in the tank and **(b)** used from a full tank.

33.

34.

35.

36.

37. For the following set of animals, what is the ratio of:

a) puppies to the total number of animals?
b) puppies to kittens?
c) kittens to the total number of animals?
d) kittens to puppies?

38. For the following set of sports equipment, what is the ratio of:

a) basketballs to footballs?
b) footballs to basketballs?
c) basketballs to the total number of balls?
d) total number of balls to basketballs?

39. Bryce delivers car parts to auto service centers. On Thursday he had 15 deliveries scheduled. By noon he had delivered only 4 orders. What is the ratio of:

a) orders delivered to total number of orders?
b) orders delivered to orders not delivered?
c) orders not delivered to total number of orders?

40. *Gas Mileage.* A Volkswagen Passat TDI® SE will travel 473 mi on 11 gal of gasoline in highway driving. What is the ratio of:

a) miles driven to gasoline used?
b) gasoline used to miles driven?

Source: vw.com

For Exercises 41 and 42, use the following bar graph, which shows the number of registered nurses per 100,000 residents in each of twelve states or districts.

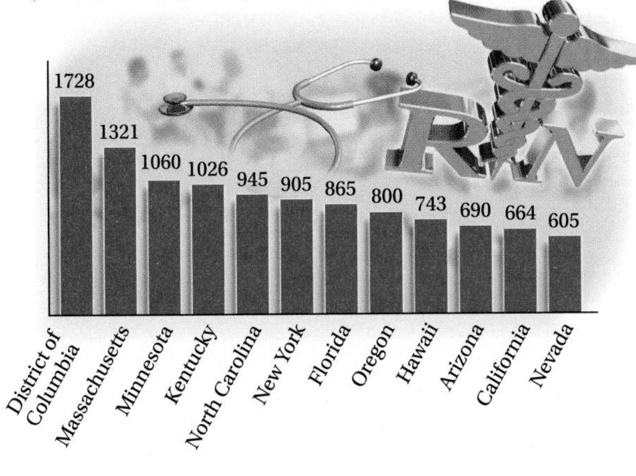

SOURCES: www.statehealthfacts.org; Kaiser Family Foundation, 2011

41. What is the ratio of registered nurses to 100,000 residents in the given state or district?

a) Minnesota
b) Hawaii
c) Florida
d) Kentucky
e) New York
f) District of Columbia

42. What is the ratio of registered nurses to 100,000 residents in the given state?

a) North Carolina
b) California
c) Massachusetts
d) Nevada
e) Arizona
f) Oregon

For Exercises 43 and 44, use the following set of states, as illustrated in the map.

Alabama	Nebraska	West Virginia
Arkansas	South Dakota	Wisconsin
Illinois		

43. What part of this group of states is east of the Mississippi River?

44. What part of this group of states is north of Nashville, Tennessee?

b Simplify.

45. $\frac{0}{8}$

46. $\frac{-8}{-8}$

47. $\frac{8-1}{9-8}$

48. $\frac{16}{1}$

49. $\frac{-20}{-20}$

50. $\frac{-20}{1}$

51. $\frac{45}{45}$

52. $\frac{11-1}{10-9}$

53. $\frac{0}{238}$

54. $\frac{238}{1}$

55. $\frac{238}{238}$

56. $\frac{0}{-16}$

57. $\frac{3}{3}$

58. $\frac{56}{56}$

59. $\frac{87}{87}$

60. $\frac{98}{98}$

61. $\frac{18}{18}$

62. $\frac{0}{-18}$

63. $\frac{-18}{1}$

64. $\frac{8-8}{1247}$

65. $\frac{729}{0}$

66. $\frac{1317}{0}$

67. $\frac{5}{6-6}$

68. $\frac{13}{10-10}$

Skill Maintenance

Add. [1.2a]

69. $\begin{array}{r} 57{,}877 \\ +\ 32{,}406 \\ \hline \end{array}$

70. $\begin{array}{r} 8004 \\ 6789 \\ 7720 \\ +\ 6851 \\ \hline \end{array}$

Subtract. [1.3a]

71. $\begin{array}{r} 9060 \\ -4387 \\ \hline \end{array}$

72. $\begin{array}{r} 7800 \\ -2462 \\ \hline \end{array}$

Multiply. [1.4a]

73. $\begin{array}{r} 217 \\ \times\ 30 \\ \hline \end{array}$

74. $\begin{array}{r} 538 \\ \times\ 27 \\ \hline \end{array}$

Add. [2.2a]

75. $7 + (-14)$

76. $-6 + (-9)$

Synthesis

What part of each object is shaded?

77.

78.

79.

80.

Shade or mark each figure to show $\frac{3}{5}$.

81.

82.

83.

84.

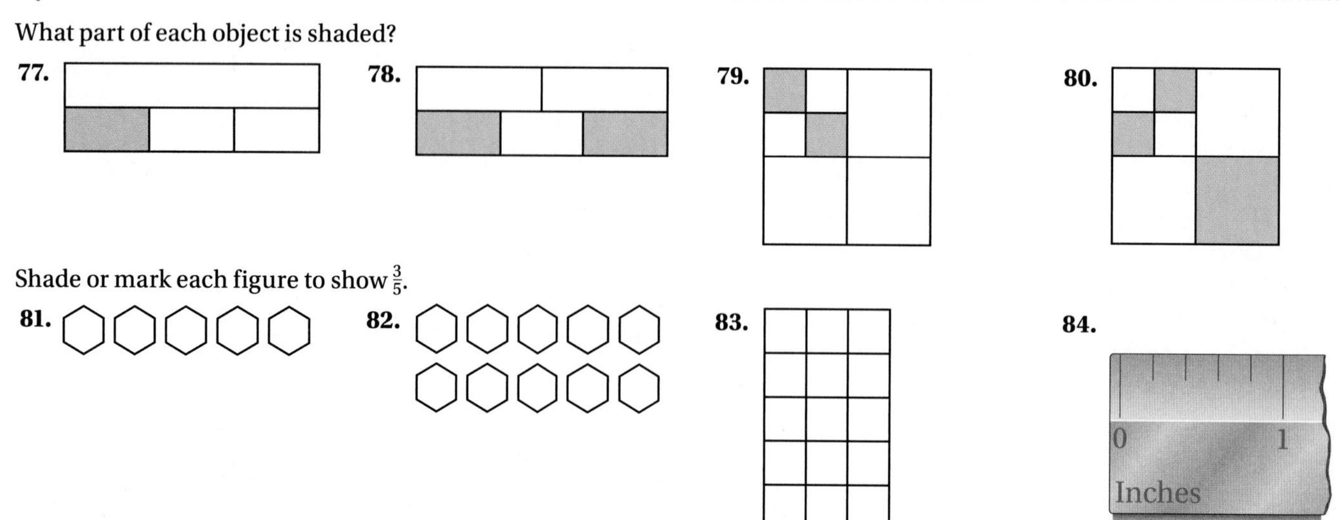

Multiplication and Applications

3.4

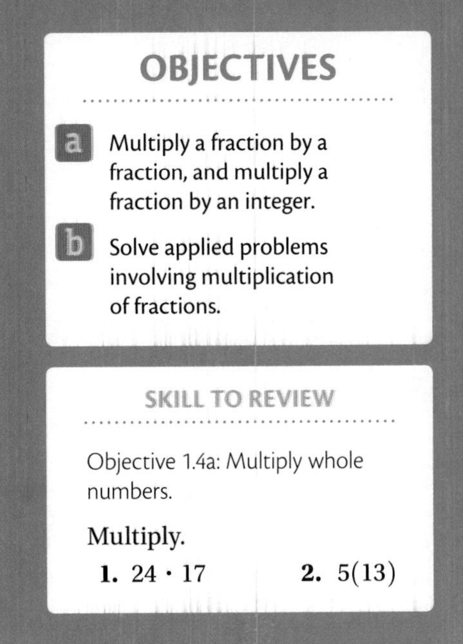

OBJECTIVES

a Multiply a fraction by a fraction, and multiply a fraction by an integer.

b Solve applied problems involving multiplication of fractions.

SKILL TO REVIEW

Objective 1.4a: Multiply whole numbers.

Multiply.

1. $24 \cdot 17$ 2. $5(13)$

 MULTIPLICATION USING FRACTION NOTATION

Let's visualize the product of two fractions. We consider the multiplication

$$\frac{3}{5} \cdot \frac{3}{4}.$$

This is equivalent to finding $\frac{3}{5}$ of $\frac{3}{4}$. We first consider an object and take $\frac{3}{4}$ of it. We divide the object into 4 equal parts using vertical lines and take 3 of them. That is shown by the shading below.

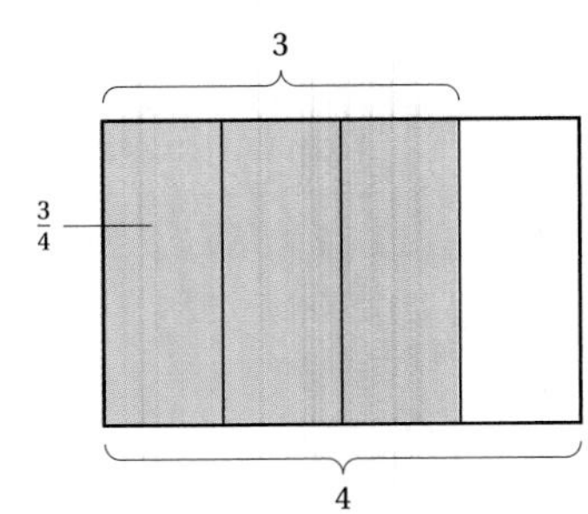

Next, we take $\frac{3}{5}$ of the shaded area. We divide it into 5 equal parts using horizontal lines and take 3 of them. That is shown by the darker shading below.

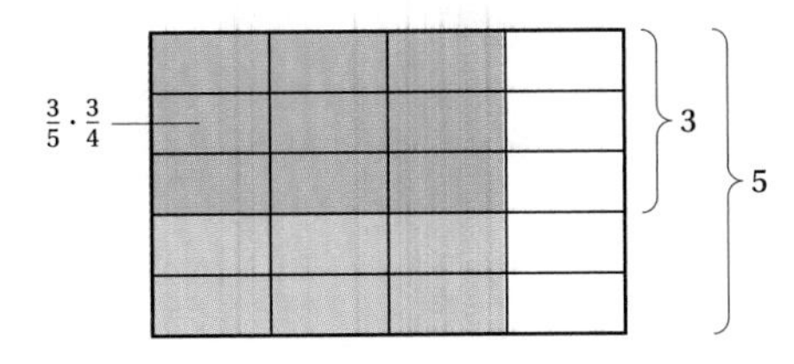

The entire object has now been divided into 20 parts, and we have shaded 9 of them twice. Thus we see that $\frac{3}{5}$ of $\frac{3}{4}$ is $\frac{9}{20}$, or

$$\frac{3}{5} \cdot \frac{3}{4} = \frac{9}{20}.$$

The figure above shows a rectangular array inside a rectangular array. The number of pieces in the entire array is $5 \cdot 4$ (the product of the denominators). The number of pieces shaded a second time is $3 \cdot 3$ (the product of the numerators). The product is represented by 9 pieces out of a set of 20, or $\frac{9}{20}$, which is the product of the numerators over the product of the denominators. This leads us to a statement of the procedure for multiplying a fraction by a fraction.

Do Margin Exercise 1.

1. Draw a diagram like the one at left to show the multiplication $\frac{2}{3} \cdot \frac{4}{5}$.

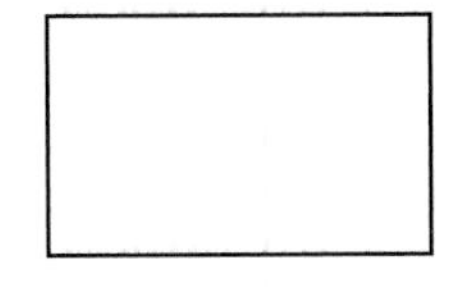

Answers

Skill to Review:
1. 408 2. 65

Margin Exercise:
1.

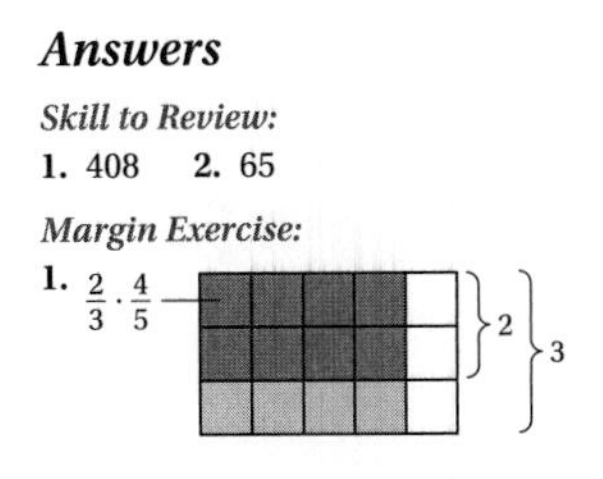

We find a product such as $\frac{9}{7} \cdot \frac{3}{4}$ as follows.

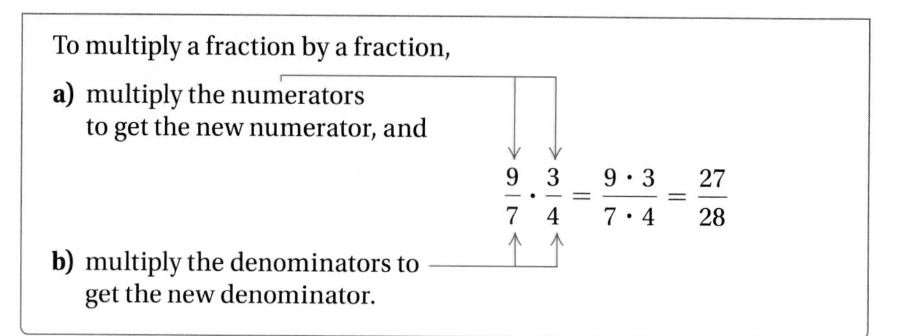

EXAMPLES Multiply.

1. $\frac{5}{6} \times \frac{7}{4} = \frac{5 \times 7}{6 \times 4} = \frac{35}{24}$

Skip writing this step whenever you can.

2. $\frac{3}{5} \cdot \frac{7}{8} = \frac{3 \cdot 7}{5 \cdot 8} = \frac{21}{40}$

3. $\frac{3}{5} \cdot \left(-\frac{4}{7}\right) = -\frac{3 \cdot 4}{5 \cdot 7} = -\frac{12}{35}$

4. $\frac{1}{4} \cdot \frac{1}{3} = \frac{1}{12}$

◀ **Do Exercises 2–5.**

Multiply.

2. $\frac{3}{8} \cdot \frac{5}{7} = \frac{3 \cdot 5}{8 \cdot \square} = \frac{\square}{\square}$ GS

3. $\frac{4}{3} \times \frac{8}{5}$

4. $\frac{3}{10} \cdot \frac{1}{10}$

5. $\frac{5}{2} \cdot \left(-\frac{9}{4}\right)$

Multiplication by an Integer

When multiplying a fraction by an integer, we first express the integer in fraction notation. We find a product such as $6 \cdot \frac{4}{5}$ as follows:

$$6 \cdot \frac{4}{5} = \frac{6}{1} \cdot \frac{4}{5} \qquad 6 = \frac{6}{1}$$

$$= \frac{6 \cdot 4}{1 \cdot 5} \qquad \text{Multiplying}$$

$$= \frac{24}{5}.$$

EXAMPLES Multiply.

5. $5 \times \frac{3}{8} = \frac{5}{1} \times \frac{3}{8} = \frac{5 \times 3}{1 \times 8} = \frac{15}{8}$

6. $\frac{2}{7} \cdot 13 = \frac{2}{7} \cdot \frac{13}{1} = \frac{2 \cdot 13}{7 \cdot 1} = \frac{26}{7}$

7. $10 \cdot \left(-\frac{1}{3}\right) = \frac{10}{1} \cdot \left(-\frac{1}{3}\right) = -\frac{10 \cdot 1}{1 \cdot 3} = -\frac{10}{3}$

◀ **Do Exercises 6–8.**

Multiply.

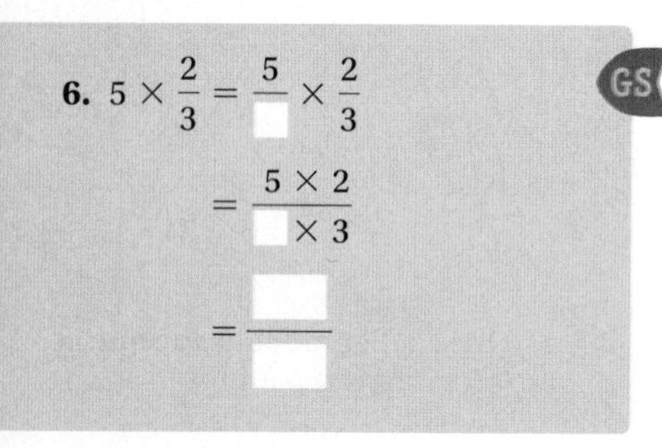

7. $-\frac{3}{8} \cdot 11$

8. $23 \cdot \left(-\frac{2}{5}\right)$

Answers

2. $\frac{15}{56}$ 3. $\frac{32}{15}$ 4. $\frac{3}{100}$
5. $-\frac{45}{8}$ 6. $\frac{10}{3}$ 7. $-\frac{33}{8}$ 8. $-\frac{46}{5}$

Guided Solutions:
2. 7, $\frac{15}{56}$ 6. 1, 1, $\frac{10}{3}$

b APPLICATIONS AND PROBLEM SOLVING

Many problems that can be solved by multiplying fractions can be thought of in terms of rectangular arrays.

EXAMPLE 8 A real estate developer owns a plot of land and plans to use $\frac{4}{5}$ of the plot for a small strip mall and parking lot. Of this, $\frac{2}{3}$ will be needed for the parking lot. What part of the plot will be used for parking?

1. **Familiarize.** We first make a drawing to help familiarize ourselves with the problem. The land may not be rectangular, but we can think of it as a rectangle. The strip mall, including the parking lot, uses $\frac{4}{5}$ of the plot. We shade $\frac{4}{5}$ as shown on the left below. The parking lot alone uses $\frac{2}{3}$ of the part we just shaded. We shade that as shown on the right below.

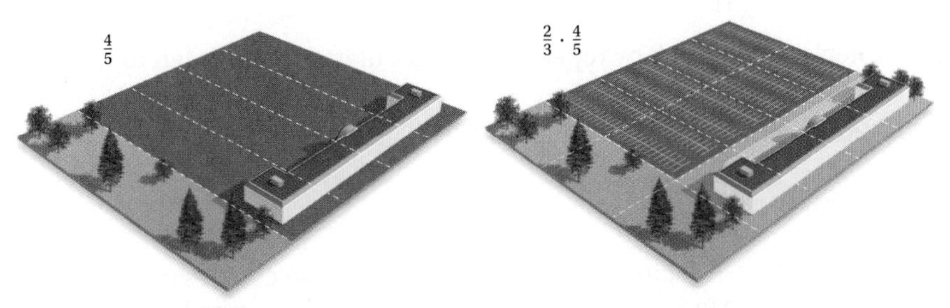

2. **Translate.** We let $n =$ the part of the plot that is used for parking. We are taking "two-thirds of four-fifths." The word "of" corresponds to multiplication. Thus the following multiplication sentence corresponds to the situation:

$$\frac{2}{3} \cdot \frac{4}{5} = n.$$

3. **Solve.** The number sentence tells us what to do. We multiply:

$$\frac{2}{3} \cdot \frac{4}{5} = \frac{2 \cdot 4}{3 \cdot 5} = \frac{8}{15}.$$

Thus, $\frac{8}{15} = n.$

4. **Check.** We can do a partial check by noting that the answer is a fraction less than 1, which we expect since the developer is using only part of the original plot of land. Thus, $\frac{8}{15}$ is a reasonable answer. We can also check this in the figure above, where we see that 8 of 15 parts represent the parking lot.
5. **State.** The parking lot takes up $\frac{8}{15}$ of the plot of land.

Do Exercise 9. ▶

9. A developer plans to set aside $\frac{3}{4}$ of the land in a housing development as open (undeveloped) space. Of this, $\frac{1}{2}$ will be green (natural) space. What part of the land will be green space?

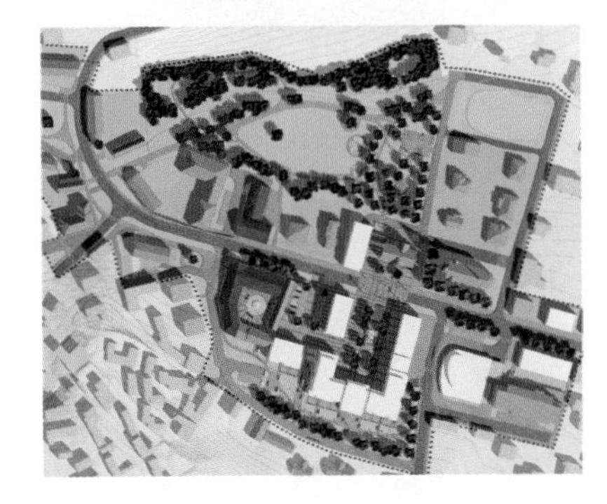

Answer

9. $\frac{3}{8}$

EXAMPLE 9 *Area of a Cranberry Bog.* The length of a rectangular cranberry bog is $\frac{9}{16}$ mi. The width is $\frac{3}{8}$ mi. What is the area of the bog?

1. **Familiarize.** Recall that area is length times width. We let A = the area of the cranberry bog.
2. **Translate.** Next, we translate:

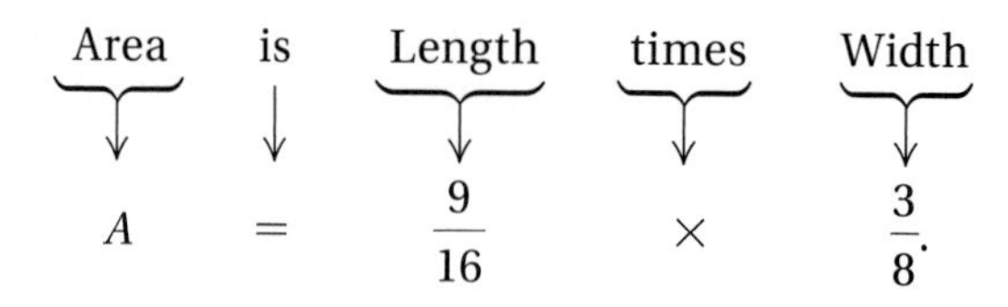

$$A = \frac{9}{16} \times \frac{3}{8}.$$

3. **Solve.** The sentence tells us what to do. We multiply:

$$A = \frac{9}{16} \cdot \frac{3}{8} = \frac{9 \cdot 3}{16 \cdot 8} = \frac{27}{128}.$$

4. **Check.** We check by repeating the calculation. This is left to the student.
5. **State.** The area is $\frac{27}{128}$ square mile (mi^2).

10. *Area of a Ceramic Tile.* The length of a rectangular ceramic tile inlaid on a countertop is $\frac{4}{9}$ ft. The width is $\frac{2}{9}$ ft. What is the area of the tile?

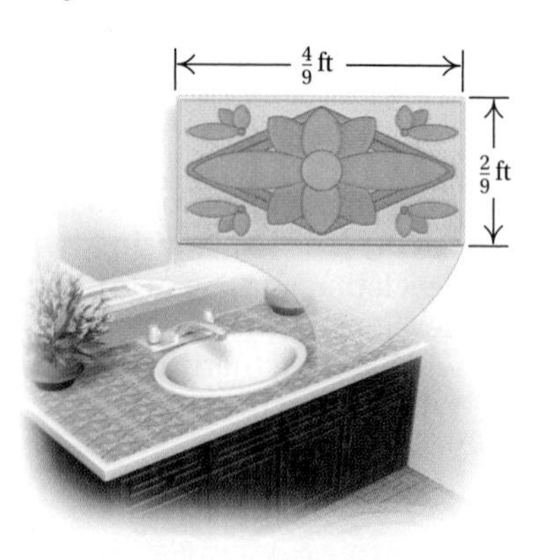

◀ **Do Exercise 10.**

EXAMPLE 10 A recipe for oatmeal chocolate chip cookies calls for $\frac{3}{4}$ cup of brown sugar. Monica is making $\frac{1}{2}$ of the recipe. How much brown sugar should she use?

1. **Familiarize.** We first make a drawing or at least visualize the situation. We let n = the amount of brown sugar that Monica should use.

$\frac{3}{4}$ cup in the recipe $\quad$ $\frac{1}{2} \cdot \frac{3}{4}$ cup in $\frac{1}{2}$ the recipe

2. **Translate.** We are finding $\frac{1}{2}$ of $\frac{3}{4}$, so the multiplication sentence $\frac{1}{2} \cdot \frac{3}{4} = n$ corresponds to the situation.
3. **Solve.** We carry out the multiplication:

$$\frac{1}{2} \cdot \frac{3}{4} = \frac{1 \cdot 3}{2 \cdot 4} = \frac{3}{8}.$$

Thus, $\frac{3}{8} = n$.

4. **Check.** We check by repeating the calculation. This is left to the student.
5. **State.** Monica should use $\frac{3}{8}$ cup of brown sugar.

11. Of the students at Overton Junior College, $\frac{1}{8}$ participate in sports and $\frac{3}{5}$ of these play football. What fractional part of the students play football?

◀ **Do Exercise 11.**

Answers

10. $\frac{8}{81}\text{ ft}^2$ 11. $\frac{3}{40}$

3.4 Exercise Set

For Extra Help

MyMathLab®

MathXL® PRACTICE

WATCH READ REVIEW

Reading Check

Determine whether each statement is true or false.

_______________ **RC1.** Multiplying $\frac{1}{2} \cdot \frac{3}{5}$ is the same as finding $\frac{1}{2}$ of $\frac{3}{5}$.

_______________ **RC2.** When we multiply two fractions, the new numerator is the product of the numerators in the two fractions.

_______________ **RC3.** The whole number 6 can be written $\frac{6}{1}$.

_______________ **RC4.** The product of two fractions can be smaller than either of the two fractions.

a Multiply.

1. $\frac{2}{5} \cdot \frac{2}{3}$

2. $\frac{3}{4} \cdot \frac{3}{5}$

3. $10 \cdot \frac{7}{9}$

4. $9 \cdot \frac{5}{8}$

5. $\frac{7}{8} \cdot \frac{7}{8}$

6. $\frac{4}{5} \cdot \frac{4}{5}$

7. $-\frac{2}{3} \times \frac{1}{5}$

8. $-\frac{3}{5} \times \frac{1}{5}$

9. $\frac{8}{7} \cdot \frac{5}{3}$

10. $\frac{11}{2} \cdot \frac{9}{8}$

11. $\frac{2}{5} \cdot (-3)$

12. $\frac{3}{5} \cdot (-4)$

13. $\frac{1}{2} \cdot \frac{1}{3}$

14. $\frac{1}{6} \cdot \frac{1}{4}$

15. $17 \times \frac{5}{6}$

16. $\frac{3}{7} \cdot 40$

17. $\frac{1}{10} \cdot \left(-\frac{7}{10}\right)$

18. $\frac{3}{10} \cdot \left(-\frac{7}{100}\right)$

19. $-\frac{2}{5} \cdot (-1)$

20. $-2 \cdot \left(-\frac{1}{3}\right)$

21. $-\frac{2}{3} \cdot \frac{7}{13}$

22. $-\frac{3}{11} \cdot \frac{4}{5}$

23. $5 \times \frac{1}{8}$

24. $4 \times \frac{1}{5}$

25. $\frac{1}{4} \times \frac{1}{10}$

26. $\frac{21}{4} \cdot \frac{7}{5}$

27. $\frac{8}{3} \cdot \frac{20}{9}$

28. $\frac{1}{3} \times \frac{1}{10}$

29. $-\frac{14}{15} \cdot \left(-\frac{13}{19}\right)$

30. $-\frac{12}{13} \cdot \left(-\frac{12}{13}\right)$

31. $\frac{3}{4} \cdot \frac{3}{4}$

32. $\frac{3}{7} \cdot \frac{4}{5}$

33. $\frac{2}{11} \cdot (-4)$

34. $\frac{2}{5} \cdot (-3)$

b Solve.

35. *Hair Bows.* It takes $\frac{5}{3}$ yd of ribbon to make a hair bow. How much ribbon is needed to make 8 bows?

36. A gasoline can holds $\frac{5}{2}$ gal. How much will the can hold when it is $\frac{1}{2}$ full?

37. *Basketball: High School to Pro.* One of 35 high school basketball players plays college basketball. One of 75 college players plays professional basketball. What fractional part of high school basketball players play professional basketball?
Source: National Basketball Association

38. *Football: High School to Pro.* One of 42 high school football players plays college football. One of 85 college players plays professional football. What fractional part of high school football players play professional football?
Source: National Football League

39. *Slices of Pizza.* One slice of a pizza is $\frac{1}{8}$ of the pizza. How much of the pizza is $\frac{1}{2}$ slice?

40. *Tossed Salad.* The recipe for a tossed salad calls for $\frac{3}{4}$ cup of sliced almonds. How much is needed to make $\frac{1}{2}$ of the recipe?

41. *Floor Tiling.* The floor of a room is being covered with tile. An area $\frac{3}{5}$ of the length and $\frac{3}{4}$ of the width is covered. What fraction of the floor has been tiled?

42. A rectangular table top measures $\frac{4}{5}$ m long by $\frac{3}{5}$ m wide. What is its area?

Skill Maintenance

Solve. [1.8a]

43. *Morel Mushrooms.* During the spring, Kate's Country Market sold 43 pounds of fresh morel mushrooms. The mushrooms sold for $22 a pound. Find the total amount Kate took in from the sale of the mushrooms.

44. Sandy can type 62 words per minute. How long will it take her to type 12,462 words?

45. Write exponential notation: $4 \cdot 4 \cdot 4 \cdot 4 \cdot 4$. [1.9a]

46. Evaluate: 2^4. [1.9b]

Simplify. [1.9c]

47. $8 \cdot 12 - (63 \div 9 + 13 \cdot 3)$

48. $(10 - 3)^4 + 10^3 \cdot 4 - 10 \div 5$

Synthesis

Multiply. Write the answer using fraction notation.

49. $\frac{341}{517} \cdot \frac{209}{349}$

50. $\left(\frac{57}{61}\right)^3$

51. $\left(\frac{2}{5}\right)^3\left(-\frac{7}{9}\right)$

52. $\left(-\frac{1}{2}\right)^5\left(\frac{3}{5}\right)$

Simplifying

3.5

a MULTIPLYING BY 1

Recall the following:

$$1 = \frac{1}{1} = \frac{2}{2} = \frac{3}{3} = \frac{4}{4} = \frac{10}{10} = \frac{45}{45} = \frac{100}{100} = \frac{n}{n}.$$

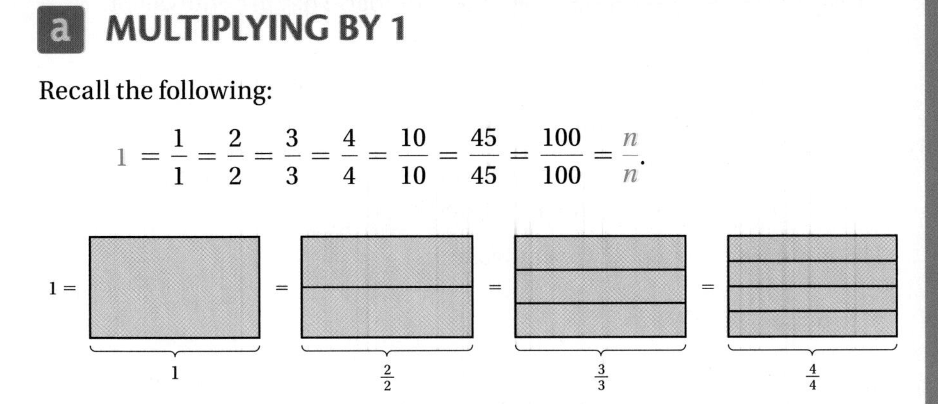

The multiplicative identity states that for any number a, $1 \cdot a = a \cdot 1 = a$. Since any nonzero number divided by itself is 1, we can state the multiplicative identity using fraction notation.

MULTIPLICATIVE IDENTITY FOR FRACTIONS

When we multiply a number by 1, we get the same number:

$$a = a \cdot 1 = a \cdot \frac{n}{n} = a.$$

For example, $\frac{3}{5} = \frac{3}{5} \cdot 1 = \frac{3}{5} \cdot \frac{4}{4} = \frac{12}{20}$. Since $\frac{3}{5} = \frac{12}{20}$, we say that $\frac{3}{5}$ and $\frac{12}{20}$ are **equivalent fractions**.

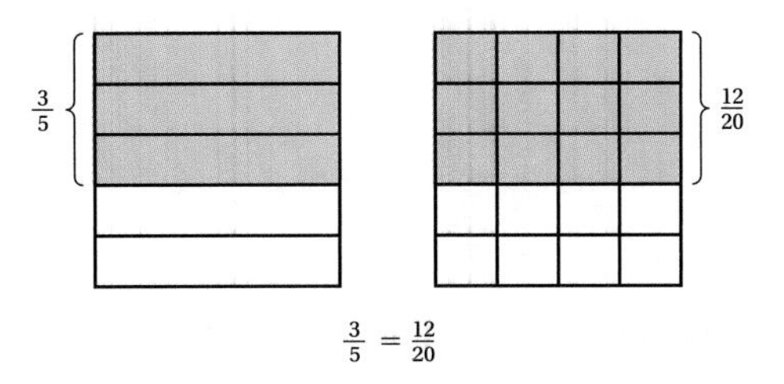

$\frac{3}{5} = \frac{12}{20}$

Do Margin Exercises 1–4.

Suppose we want to find another name for $\frac{2}{3}$, one that has a denominator of 9. We can multiply by 1 to find equivalent fractions. Since $9 = 3 \cdot 3$, we choose $\frac{3}{3}$ for 1 in order to get a denominator of 9:

$$\frac{2}{3} = \frac{2}{3} \cdot 1 = \frac{2}{3} \cdot \frac{3}{3} = \frac{2 \cdot 3}{3 \cdot 3} = \frac{6}{9}.$$

$\frac{2}{3}$ $\frac{6}{9}$

$\frac{2}{3} = \frac{6}{9}$

OBJECTIVES

a Multiply a number by 1 to find fraction notation with a specified denominator.

b Simplify fraction notation.

c Use the test for equality to determine whether two fractions name the same number.

SKILL TO REVIEW

Objective 3.1d: Find the prime factorization of a composite number.

Find the prime factorization.

1. 84 **2.** 2250

Multiply.

1. $\frac{1}{2} \cdot \frac{8}{8}$ **2.** $\frac{3}{5} \cdot \frac{10}{10}$

3. $-\frac{13}{25} \cdot \frac{4}{4}$ **4.** $\frac{8}{3} \cdot \frac{25}{25}$

Answers

Skill to Review:
1. $2 \cdot 2 \cdot 3 \cdot 7$ **2.** $2 \cdot 3 \cdot 3 \cdot 5 \cdot 5 \cdot 5$

Margin Exercises:
1. $\frac{8}{16}$ **2.** $\frac{30}{50}$ **3.** $-\frac{52}{100}$ **4.** $\frac{200}{75}$

Find another name for each number, but with the denominator indicated. Use multiplying by 1.

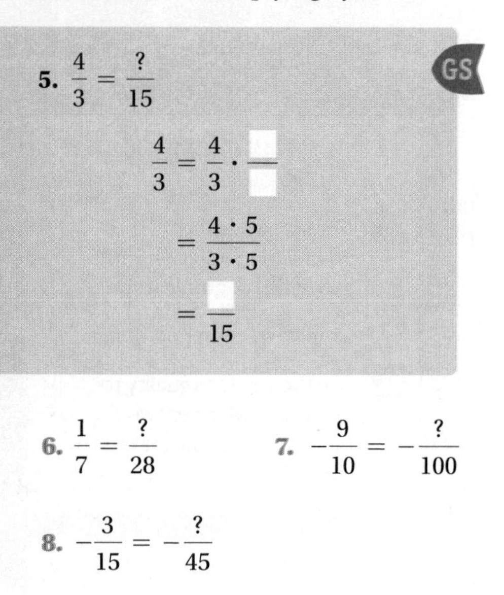

GS

5. $\frac{4}{3} = \frac{?}{15}$

$\frac{4}{3} = \frac{4}{3} \cdot \frac{\square}{\square}$

$= \frac{4 \cdot 5}{3 \cdot 5}$

$= \frac{\square}{15}$

6. $\frac{1}{7} = \frac{?}{28}$

7. $-\frac{9}{10} = -\frac{?}{100}$

8. $-\frac{3}{15} = -\frac{?}{45}$

EXAMPLE 1 Find a name for $\frac{2}{5}$ with a denominator of 35.

Since $5 \cdot 7 = 35$, we multiply by $\frac{7}{7}$:

$$\frac{2}{5} = \frac{2}{5} \cdot \frac{7}{7} = \frac{2 \cdot 7}{5 \cdot 7} = \frac{14}{35}.$$

We say that $\frac{2}{5}$ and $\frac{14}{35}$ represent the same number. They are equivalent. ■

EXAMPLE 2 Find a name for $-\frac{2}{5}$ with a denominator of 35.

Since $5 \cdot 7 = 35$, we multiply by $\frac{7}{7}$:

$$-\frac{2}{5} = -\frac{2}{5} \cdot \frac{7}{7} = -\frac{2 \cdot 7}{5 \cdot 7} = -\frac{14}{35}.$$

The numbers $-\frac{2}{5}$ and $-\frac{14}{35}$ are equivalent.

◀ Do Exercises 5–8.

b SIMPLIFYING FRACTION NOTATION

All of the following are names for three-fourths:

$$\frac{3}{4}, \frac{6}{8}, \frac{9}{12}, \frac{12}{16}, \frac{15}{20}.$$

We say that $\frac{3}{4}$ is **simplest** because it has the smallest numerator and the smallest denominator. That is, the numerator and the denominator have no common factor other than 1.

To simplify, we reverse the process of multiplying by 1:

$$\frac{12}{18} = \frac{2 \cdot 6}{3 \cdot 6} \quad \begin{array}{l} \leftarrow \text{Factoring the numerator} \\ \leftarrow \text{Factoring the denominator} \end{array}$$

$$= \frac{2}{3} \cdot \frac{6}{6} \qquad \text{Factoring the fraction}$$

$$= \frac{2}{3} \cdot 1 \qquad \frac{6}{6} = 1$$

$$= \frac{2}{3}. \qquad \text{Removing a factor of 1: } \frac{2}{3} \cdot 1 = \frac{2}{3}$$

EXAMPLES Simplify.

3. $\frac{8}{20} = \frac{2 \cdot 4}{5 \cdot 4} = \frac{2}{5} \cdot \frac{4}{4} = \frac{2}{5}$

4. $\frac{2}{6} = \frac{1 \cdot 2}{3 \cdot 2} = \frac{1}{3} \cdot \frac{2}{2} = \frac{1}{3}$

The number 1 allows for pairing of factors in the numerator and the denominator.

◀ Do Exercises 9–12.

Simplify.

9. $\frac{2}{8}$

10. $-\frac{24}{18}$

11. $\frac{10}{12}$

12. $\frac{15}{80}$

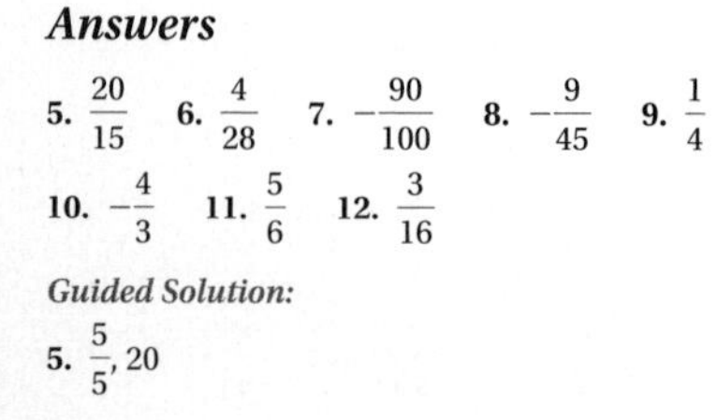

Answers

5. $\frac{20}{15}$ 6. $\frac{4}{28}$ 7. $-\frac{90}{100}$ 8. $-\frac{9}{45}$ 9. $\frac{1}{4}$
10. $-\frac{4}{3}$ 11. $\frac{5}{6}$ 12. $\frac{3}{16}$

Guided Solution:

5. $\frac{5}{5}$, 20

The use of prime factorizations can be helpful for simplifying when numerators and/or denominators are large numbers.

EXAMPLE 5 Simplify: $\frac{90}{84}$.

$$\frac{90}{84} = \frac{2 \cdot 3 \cdot 3 \cdot 5}{2 \cdot 2 \cdot 3 \cdot 7}$$ Factoring the numerator and the denominator into primes

$$= \frac{2 \cdot 3 \cdot 3 \cdot 5}{2 \cdot 3 \cdot 2 \cdot 7}$$ Changing the order so that like primes are above and below each other

$$= \frac{2}{2} \cdot \frac{3}{3} \cdot \frac{3 \cdot 5}{2 \cdot 7}$$ Factoring the fraction

$$= 1 \cdot 1 \cdot \frac{3 \cdot 5}{2 \cdot 7}$$

$$= \frac{3 \cdot 5}{2 \cdot 7}$$ Removing factors of 1

$$= \frac{15}{14}$$

The tests for divisibility are very helpful in simplifying fraction notation. We could have shortened the preceding example had we noted that 6 is a factor of both the numerator and the denominator. Then we would have

$$\frac{90}{84} = \frac{6 \cdot 15}{6 \cdot 14} = \frac{6}{6} \cdot \frac{15}{14} = \frac{15}{14}.$$

EXAMPLE 6 Simplify: $\frac{603}{207}$.

At first glance this looks difficult. But, using the test for divisibility by 9 (sum of digits is divisible by 9), we find that both the numerator and the denominator are divisible by 9. Thus we write both numbers with a factor of 9:

$$\frac{603}{207} = \frac{9 \cdot 67}{9 \cdot 23} = \frac{9}{9} \cdot \frac{67}{23} = \frac{67}{23}.$$

EXAMPLE 7 Simplify: $-\frac{660}{1140}$.

Using the tests for divisibility, we have

$$-\frac{660}{1140} = -\frac{66 \cdot 10}{114 \cdot 10} = -\frac{66}{114} \cdot \frac{10}{10} = -\frac{66}{114}$$ Both 660 and 1140 are divisible by 10.

$$= -\frac{11 \cdot 6}{19 \cdot 6} = -\frac{11}{19} \cdot \frac{6}{6} = -\frac{11}{19}.$$ Both 66 and 114 are divisible by 6.

Do Exercises 13–19.

Simplify.

13. $\frac{35}{40}$

14. $-\frac{24}{21}$

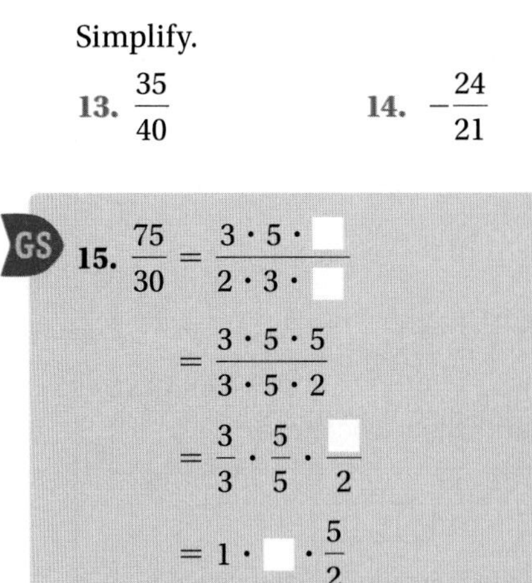

GS 15. $\frac{75}{30} = \frac{3 \cdot 5 \cdot \square}{2 \cdot 3 \cdot \square}$

$= \frac{3 \cdot 5 \cdot 5}{3 \cdot 5 \cdot 2}$

$= \frac{3}{3} \cdot \frac{5}{5} \cdot \frac{\square}{2}$

$= 1 \cdot \square \cdot \frac{5}{2}$

$= \frac{\square}{\square}$

16. $-\frac{75}{300}$

17. $\frac{280}{960}$

18. $\frac{1332}{2880}$

19. Simplify each fraction in this circle graph.

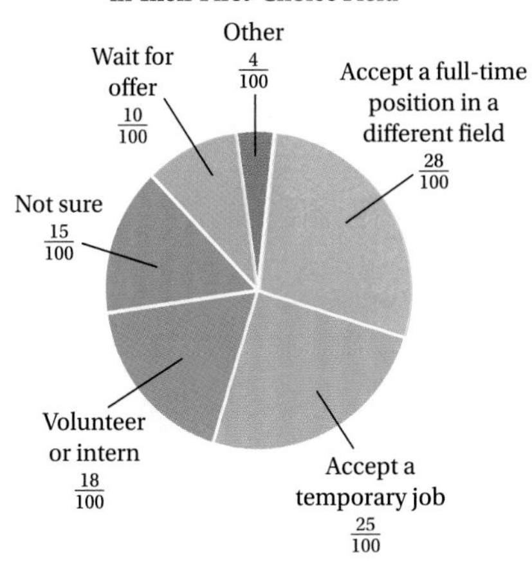

SOURCE: Yahoo Hotjobs survey

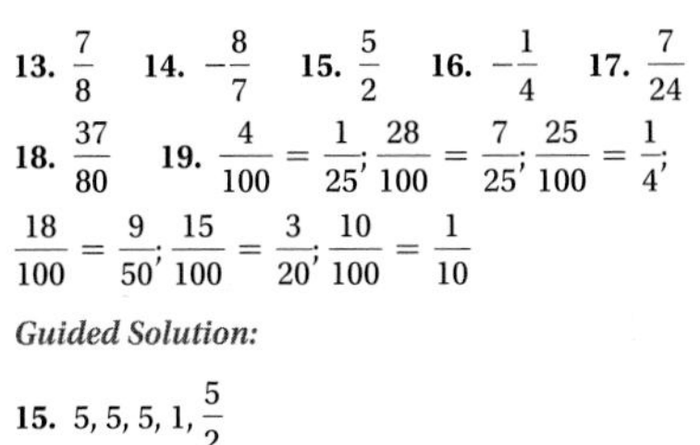

Answers

13. $\frac{7}{8}$ 14. $-\frac{8}{7}$ 15. $\frac{5}{2}$ 16. $-\frac{1}{4}$ 17. $\frac{7}{24}$
18. $\frac{37}{80}$ 19. $\frac{4}{100} = \frac{1}{25}$; $\frac{28}{100} = \frac{7}{25}$; $\frac{25}{100} = \frac{1}{4}$; $\frac{18}{100} = \frac{9}{50}$; $\frac{15}{100} = \frac{3}{20}$; $\frac{10}{100} = \frac{1}{10}$

Guided Solution:

15. 5, 5, 5, 1, $\frac{5}{2}$

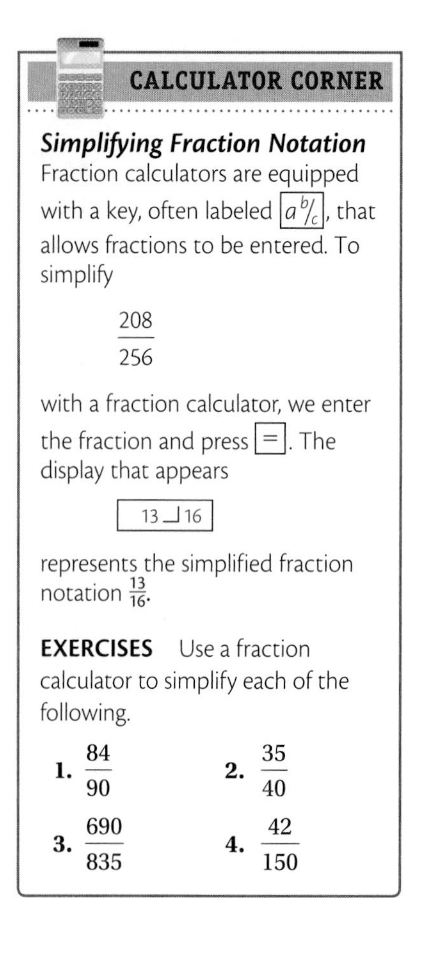

CALCULATOR CORNER

Simplifying Fraction Notation Fraction calculators are equipped with a key, often labeled $\boxed{a\,^b\!/_c}$, that allows fractions to be entered. To simplify

$$\frac{208}{256}$$

with a fraction calculator, we enter the fraction and press $\boxed{=}$. The display that appears

$$\boxed{13 \lrcorner 16}$$

represents the simplified fraction notation $\frac{13}{16}$.

EXERCISES Use a fraction calculator to simplify each of the following.

1. $\frac{84}{90}$ **2.** $\frac{35}{40}$

3. $\frac{690}{835}$ **4.** $\frac{42}{150}$

Canceling

Canceling is a shortcut that you may have used for removing a factor of 1 when working with fraction notation. With *great* concern, we mention it as a possibility for speeding up your work. Canceling may be done only when removing common factors in numerators and denominators. Each such pair allows us to remove a factor of 1 in a fraction.

Our concern is that canceling be done with care and understanding. In effect, slashes are used to indicate factors of 1 that have been removed. For instance, Example 5 might have been done faster as follows:

$$\frac{90}{84} = \frac{2 \cdot 3 \cdot 3 \cdot 5}{2 \cdot 2 \cdot 3 \cdot 7}$$ Factoring the numerator and the denominator

$$= \frac{\cancel{2} \cdot \cancel{3} \cdot 3 \cdot 5}{2 \cdot \cancel{2} \cdot \cancel{3} \cdot 7}$$ When a factor of 1 is noted, it is canceled as shown: $\frac{2}{2} \cdot \frac{3}{3} = 1$.

$$= \frac{3 \cdot 5}{2 \cdot 7} = \frac{15}{14}.$$

Caution!

The difficulty with canceling is that it is often applied incorrectly in situations like the following:

$$\frac{\cancel{2} + 3}{\cancel{2}} = 3; \qquad \frac{\cancel{4} + 1}{\cancel{4} + 2} = \frac{1}{2}; \qquad \frac{1\cancel{5}}{\cancel{5}4} = \frac{1}{4}.$$

Wrong! Wrong! Wrong!

The correct answers are

$$\frac{2 + 3}{2} = \frac{5}{2}; \qquad \frac{4 + 1}{4 + 2} = \frac{5}{6}; \qquad \frac{15}{54} = \frac{3 \cdot 5}{3 \cdot 18} = \frac{3}{3} \cdot \frac{5}{18} = \frac{5}{18}.$$

In each situation, the number canceled was not a factor of 1. Factors are parts of products. For example, in $2 \cdot 3$, 2 and 3 are factors, but in $2 + 3$, 2 and 3 are *not* factors. Canceling may not be done when sums or differences are in numerators or denominators, as shown here. **If you cannot factor, you cannot cancel! If in doubt, do not cancel!**

C A TEST FOR EQUALITY

When denominators are the same, we say that fractions have a **common denominator**. When fractions have a common denominator, we can compare them by comparing numerators. Suppose we want to compare $\frac{3}{6}$ and $\frac{2}{4}$. First, we find a common denominator. To do this, we multiply each fraction by 1, using the denominator of the other fraction to form the symbol for 1. We multiply $\frac{3}{6}$ by $\frac{4}{4}$ and $\frac{2}{4}$ by $\frac{6}{6}$:

$$\frac{3}{6} = \frac{3}{6} \cdot \frac{4}{4} = \frac{3 \cdot 4}{6 \cdot 4} = \frac{12}{24};$$ Multiplying by $\frac{4}{4}$

$$\frac{2}{4} = \frac{2}{4} \cdot \frac{6}{6} = \frac{2 \cdot 6}{4 \cdot 6} = \frac{12}{24}.$$ Multiplying by $\frac{6}{6}$

Once we have a common denominator, 24, we compare the numerators. And since these numerators are both 12, the fractions are equal:

$$\frac{3}{6} = \frac{2}{4}.$$

Note in the preceding that if $\frac{3}{6} = \frac{2}{4}$, then $3 \cdot 4 = 6 \cdot 2$. This tells us that we need to check only the products $3 \cdot 4$ and $6 \cdot 2$ to compare the fractions.

A TEST FOR EQUALITY

Two fractions are equal if their cross products are equal.

We multiply these two numbers: $3 \cdot 4$.

We multiply these two numbers: $6 \cdot 2$.

$$\frac{3}{6} \square \frac{2}{4}$$

We call $3 \cdot 4$ and $6 \cdot 2$ **cross products**. Since the cross products are the same—that is, $3 \cdot 4 = 6 \cdot 2$—we know that

$$\frac{3}{6} = \frac{2}{4}.$$

If a sentence $a = b$ is true, it means that a and b name the same number. If a sentence $a \neq b$ (read "a is not equal to b") is true, it means that a and b do *not* name the same number.

EXAMPLE 8 Use $=$ or $\neq$ for $\square$ to write a true sentence:

$$\frac{6}{7} \square \frac{7}{8}.$$

We multiply these two numbers: $6 \cdot 8 = 48$.

We multiply these two numbers: $7 \cdot 7 = 49$.

$$\frac{6}{7} \square \frac{7}{8}$$

Because $48 \neq 49$, $\frac{6}{7}$ and $\frac{7}{8}$ do not name the same number. Thus,

$$\frac{6}{7} \neq \frac{7}{8}.$$

EXAMPLE 9 Use $=$ or $\neq$ for $\square$ to write a true sentence:

$$-\frac{6}{10} \square \frac{-3}{5}.$$

We rewrite $-\frac{6}{10}$ as $\frac{-6}{10}$ and then check cross products.

We multiply these two numbers: $-6 \cdot 5 = -30$.

We multiply these two numbers: $10 \cdot (-3) = -30$.

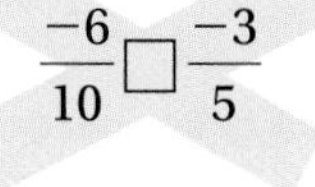

Because the cross products are the same, we have

$$\frac{-6}{10} = \frac{-3}{5}, \text{ or } -\frac{6}{10} = \frac{-3}{5}.$$

Do Exercises 20 and 21. ▶

Use $=$ or $\neq$ for $\square$ to write a true sentence.

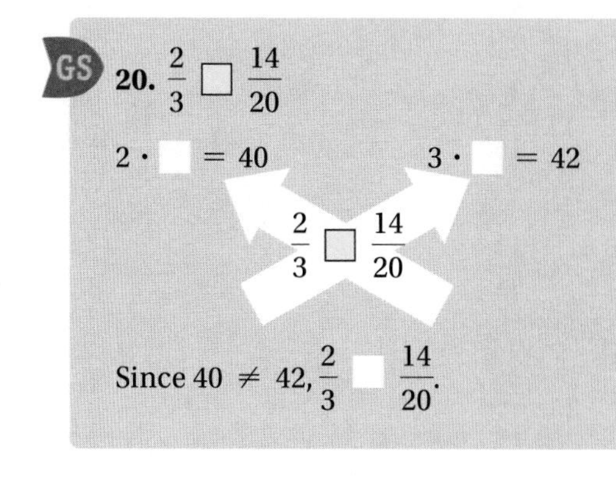

21. $\frac{-2}{6} \square -\frac{3}{9}$

Answers

20. $\neq$ **21.** $=$

Guided Solution:

20. 20, 14; $\neq$

3.5 Exercise Set

For Extra Help
MyMathLab® MathXL® PRACTICE WATCH READ REVIEW

Reading Check

Complete each statement with the appropriate word from the following list.

common cross equivalent simplify

RC1. ________________ fractions name the same number.

RC2. To ________________ a fraction, we find a fraction that names the same number and that has a numerator and a denominator with no common factor.

RC3. The fractions $\frac{2}{7}$ and $\frac{4}{7}$ have a ________________ denominator.

RC4. Two fractions are equal if their ________________ products are equal.

a Find another name for the given number, but with the denominator indicated. Use multiplying by 1.

1. $\frac{1}{2} = \frac{?}{10}$

2. $\frac{1}{6} = \frac{?}{18}$

3. $\frac{5}{8} = \frac{?}{32}$

4. $\frac{2}{9} = \frac{?}{18}$

5. $-\frac{9}{10} = -\frac{?}{30}$

6. $-\frac{5}{6} = -\frac{?}{48}$

7. $\frac{5}{12} = \frac{?}{48}$

8. $\frac{5}{3} = \frac{?}{45}$

9. $\frac{17}{18} = \frac{?}{-54}$

10. $\frac{11}{16} = \frac{?}{-256}$

11. $\frac{7}{22} = \frac{?}{132}$

12. $\frac{10}{21} = \frac{?}{126}$

b Simplify.

13. $\frac{2}{4}$

14. $\frac{4}{8}$

15. $-\frac{6}{8}$

16. $-\frac{8}{12}$

17. $\frac{3}{15}$

18. $\frac{8}{10}$

19. $\frac{-24}{8}$

20. $\frac{-36}{9}$

21. $\frac{18}{24}$

22. $\frac{42}{48}$

23. $\frac{14}{16}$

24. $\frac{15}{25}$

25. $\frac{12}{10}$

26. $\frac{16}{14}$

27. $\frac{16}{48}$

28. $\frac{100}{-20}$

29. $\frac{150}{-25}$

30. $\frac{19}{76}$

31. $-\frac{17}{51}$

32. $-\frac{425}{525}$

33. $\frac{540}{810}$

34. $\frac{1000}{1080}$

35. $-\frac{210}{2700}$

36. $-\frac{300}{2250}$

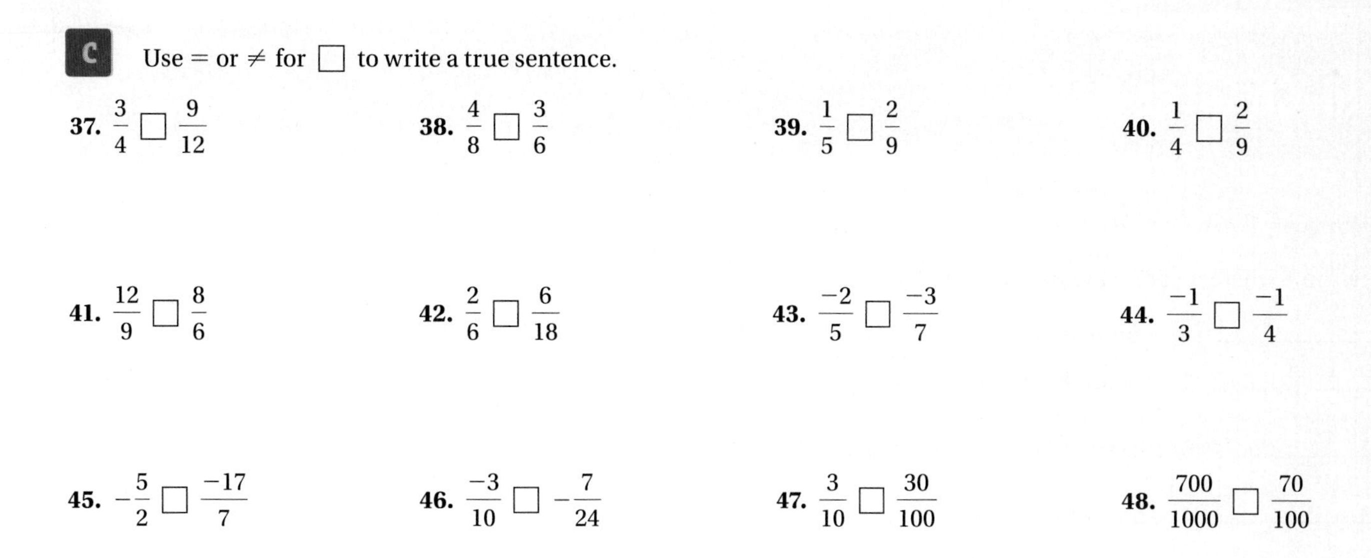

C Use = or ≠ for □ to write a true sentence.

37. $\frac{3}{4} \square \frac{9}{12}$

38. $\frac{4}{8} \square \frac{3}{6}$

39. $\frac{1}{5} \square \frac{2}{9}$

40. $\frac{1}{4} \square \frac{2}{9}$

41. $\frac{12}{9} \square \frac{8}{6}$

42. $\frac{2}{6} \square \frac{6}{18}$

43. $\frac{-2}{5} \square \frac{-3}{7}$

44. $\frac{-1}{3} \square \frac{-1}{4}$

45. $-\frac{5}{2} \square \frac{-17}{7}$

46. $\frac{-3}{10} \square -\frac{7}{24}$

47. $\frac{3}{10} \square \frac{30}{100}$

48. $\frac{700}{1000} \square \frac{70}{100}$

The following circle graph shows the fractional part of each day that the average person spends in various activities. For Exercises 49–52, simplify the fraction associated with the activity.

Time Use

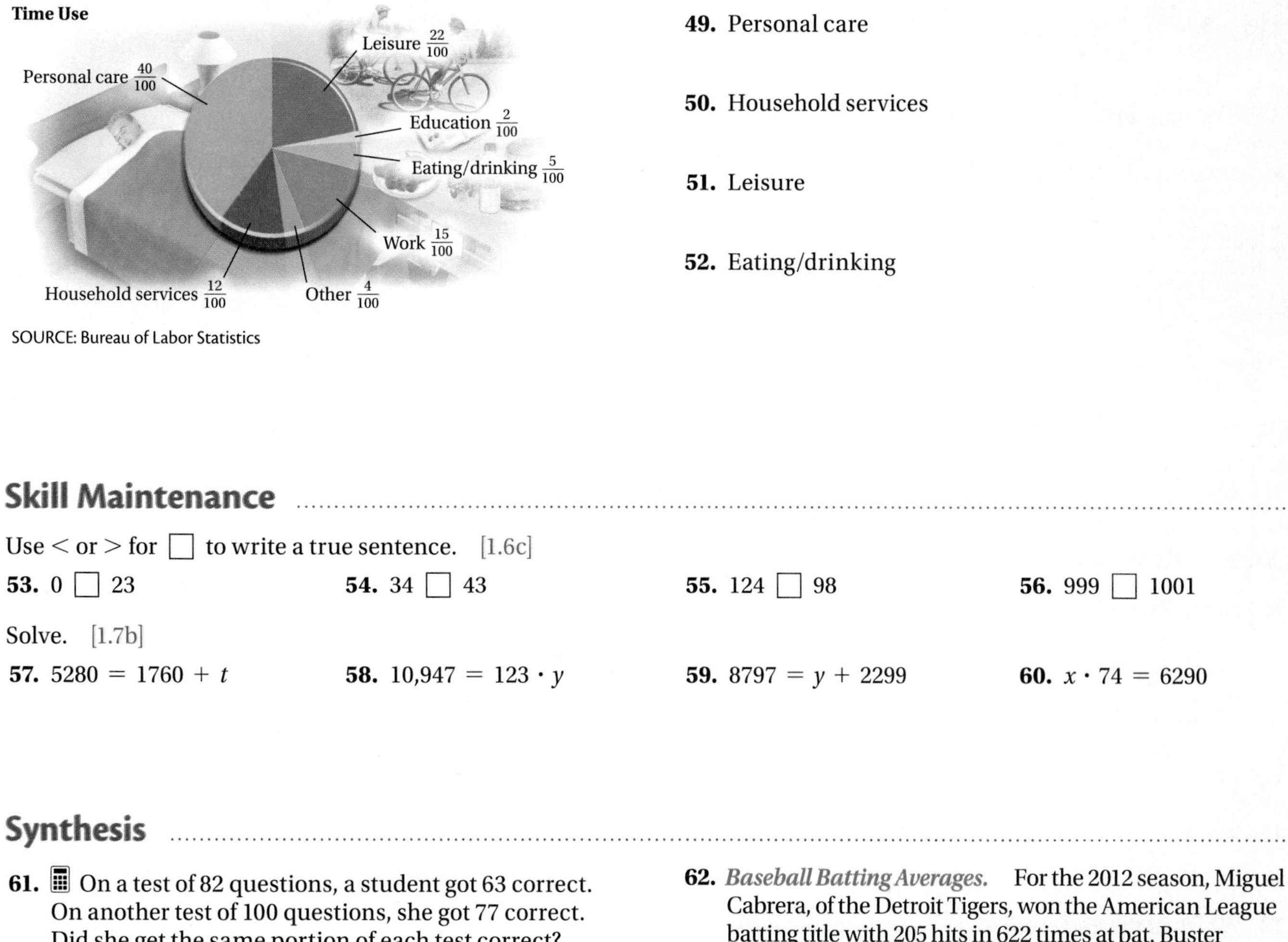

SOURCE: Bureau of Labor Statistics

49. Personal care

50. Household services

51. Leisure

52. Eating/drinking

Skill Maintenance

Use < or > for □ to write a true sentence. [1.6c]

53. 0 □ 23

54. 34 □ 43

55. 124 □ 98

56. 999 □ 1001

Solve. [1.7b]

57. $5280 = 1760 + t$

58. $10{,}947 = 123 \cdot y$

59. $8797 = y + 2299$

60. $x \cdot 74 = 6290$

Synthesis

61. 🖩 On a test of 82 questions, a student got 63 correct. On another test of 100 questions, she got 77 correct. Did she get the same portion of each test correct? Why or why not?

62. *Baseball Batting Averages.* For the 2012 season, Miguel Cabrera, of the Detroit Tigers, won the American League batting title with 205 hits in 622 times at bat. Buster Posey, of the San Francisco Giants, won the National League title with 178 hits in 530 times at bat. Did they have the same fraction of hits per times at bat (batting average)? Why or why not?

Source: Major League Baseball

Mid-Chapter Review

Concept Reinforcement

Determine whether each statement is true or false.

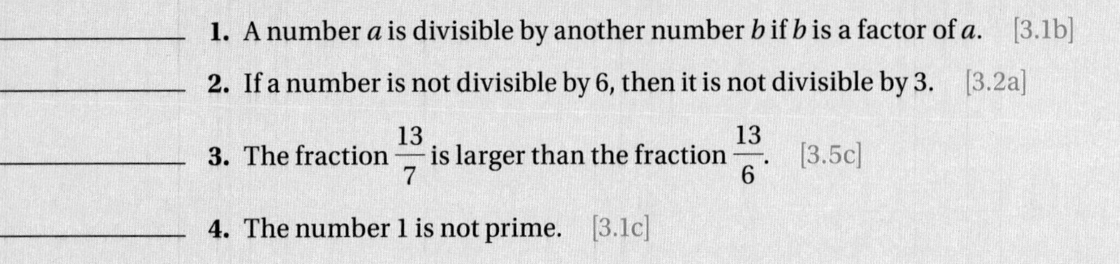

_______________ **1.** A number a is divisible by another number b if b is a factor of a. [3.1b]

_______________ **2.** If a number is not divisible by 6, then it is not divisible by 3. [3.2a]

_______________ **3.** The fraction $\frac{13}{7}$ is larger than the fraction $\frac{13}{6}$. [3.5c]

_______________ **4.** The number 1 is not prime. [3.1c]

Guided Solutions

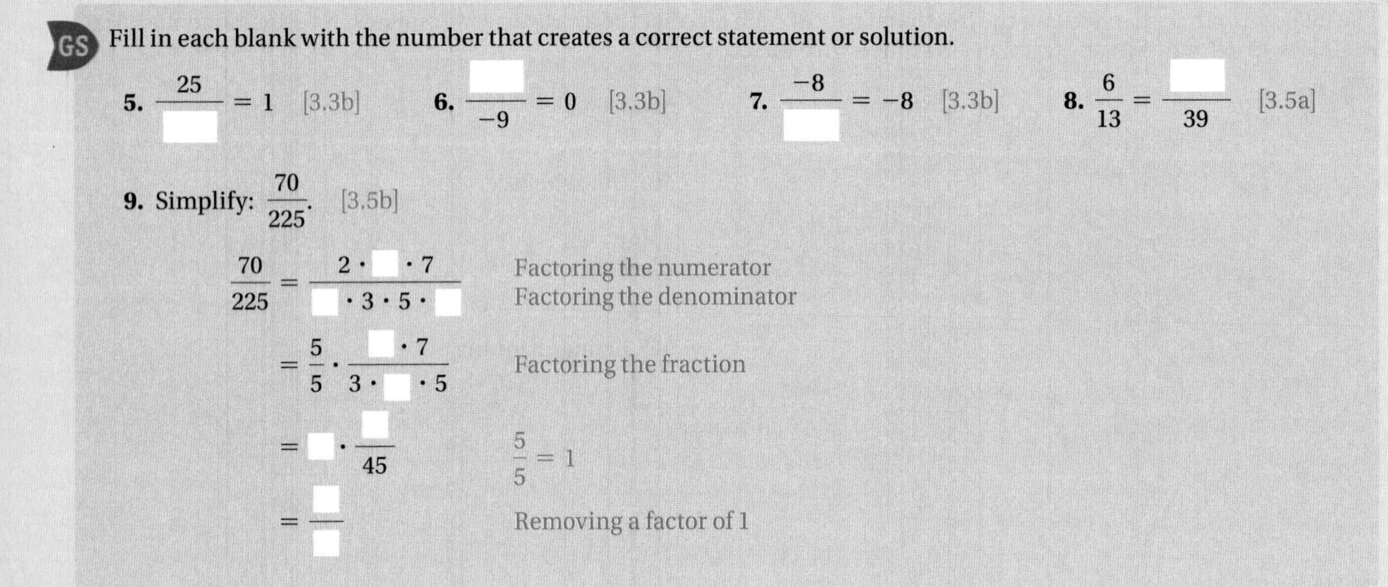

GS Fill in each blank with the number that creates a correct statement or solution.

5. $\frac{25}{\square} = 1$ [3.3b]

6. $\frac{\square}{-9} = 0$ [3.3b]

7. $\frac{-8}{\square} = -8$ [3.3b]

8. $\frac{6}{13} = \frac{\square}{39}$ [3.5a]

9. Simplify: $\frac{70}{225}$. [3.5b]

$$\frac{70}{225} = \frac{2 \cdot \square \cdot 7}{\square \cdot 3 \cdot 5 \cdot \square}$$ Factoring the numerator; Factoring the denominator

$$= \frac{5}{5} \cdot \frac{\square \cdot 7}{3 \cdot \square \cdot 5}$$ Factoring the fraction

$$= \square \cdot \frac{\square}{45}$$ $\frac{5}{5} = 1$

$$= \frac{\square}{\square}$$ Removing a factor of 1

Mixed Review

To answer Exercises 10–14, consider the following numbers. [3.2a]

84	132	594	350
300	500	120	14,850
17,576	180	1125	504
224	351	495	1632

10. Which of the above are divisible by 2 but not by 10?

11. Which of the above are divisible by 4 but not by 8?

12. Which of the above are divisible by 4 but not by 6?

13. Which of the above are divisible by 3 but not by 9?

14. Which of the above are divisible by 4, 5, and 6?

Determine whether each number is prime, composite, or neither. [3.1c]

15. 61 **16.** 2 **17.** 91 **18.** 1

Find all the factors of each composite number. Then find the prime factorization of the number. [3.1a], [3.1d]

19. 160 **20.** 222 **21.** 98 **22.** 315

What part of each object or set of objects is shaded? [3.3a]

23.

24.

Multiply. [3.4a]

25. $7 \cdot \frac{1}{9}$ **26.** $\frac{4}{15} \cdot \frac{2}{3}$ **27.** $\frac{5}{11} \cdot (-8)$ **28.** $-\frac{3}{4} \cdot \left(-\frac{7}{8}\right)$

Simplify. [3.3b], [3.5b]

29. $\frac{24}{60}$ **30.** $\frac{220}{60}$ **31.** $\frac{-17}{-17}$

32. $\frac{0}{23}$ **33.** $\frac{54}{-186}$ **34.** $\frac{36}{20}$

35. $\frac{75}{630}$ **36.** $\frac{315}{435}$ **37.** $\frac{-14}{0}$

Use = or ≠ for □ to write a true sentence. [3.5c]

38. $\frac{3}{7} \; \square \; \frac{48}{112}$ **39.** $\frac{19}{3} \; \square \; \frac{95}{18}$

40. *Job Applications.* Of every 200 online job applications started, only 25 are reviewed by a hiring manager. What is the ratio of applications reviewed to applications started? [3.3a]

Source: Talent Function Group LLC, in "Your Résumé vs. Oblivion," wsj.com, 1/24/12

41. *Area of an Ice-Skating Rink.* The length of a rectangular ice-skating rink in the atrium of a shopping mall is $\frac{7}{100}$ mi. The width is $\frac{3}{100}$ mi. What is the area of the rink? [3.4b]

Understanding Through Discussion and Writing

42. Explain a method for finding a composite number that contains exactly two factors other than itself and 1. [3.1c]

43. Which of the years from 2000 to 2020, if any, also happen to be prime numbers? Explain at least two ways in which you might go about solving this problem. [3.2a]

44. Explain in your own words when it *is* possible to cancel and when it *is not* possible to cancel. [3.5b]

45. Can fraction notation be simplified if the numerator and the denominator are two different prime numbers? Why or why not? [3.5b]

STUDYING FOR SUCCESS *A Valuable Resource—Your Instructor*

- ☐ Don't be afraid to ask questions in class. Other students probably have the same questions you do.
- ☐ Visit your instructor during office hours if you need additional help.
- ☐ Many instructors welcome e-mails from students who have questions.

3.6 Multiplying, Simplifying, and Applications

OBJECTIVES

a Multiply and simplify using fraction notation.

b Solve applied problems involving multiplication of fractions.

SKILL TO REVIEW

Objective 3.2a: Determine whether a number is divisible by 2, 3, 4, 5, 6, 8, 9, or 10.

Determine whether each number is divisible by 9.

1. 486 **2.** 129

a MULTIPLYING AND SIMPLIFYING USING FRACTION NOTATION

It is often possible to simplify after we multiply. To make such simplifying easier, it is usually best not to carry out the products in the numerator and the denominator immediately, but to factor and simplify first. Consider the product

$$\frac{3}{8} \cdot \frac{4}{9}.$$

We proceed as follows:

$$\frac{3}{8} \cdot \frac{4}{9} = \frac{3 \cdot 4}{8 \cdot 9}$$ We write the products in the numerator and the denominator, but we do not carry them out.

$$= \frac{3 \cdot 2 \cdot 2}{2 \cdot 2 \cdot 2 \cdot 3 \cdot 3}$$ Factoring the numerator and the denominator

$$= \frac{3 \cdot 2 \cdot 2 \cdot 1}{2 \cdot 2 \cdot 2 \cdot 3 \cdot 3}$$ Using the identity property of 1 to insert the number 1 as a factor

$$= \frac{3 \cdot 2 \cdot 2}{3 \cdot 2 \cdot 2} \cdot \frac{1}{2 \cdot 3}$$ Factoring the fraction

$$= 1 \cdot \frac{1}{2 \cdot 3}$$

$$= \frac{1}{2 \cdot 3}$$ Removing a factor of 1

$$= \frac{1}{6}.$$

The procedure could have been shortened had we noticed that 4 is a factor of the 8 in the denominator:

$$\frac{3}{8} \cdot \frac{4}{9} = \frac{3 \cdot 4}{8 \cdot 9} = \frac{3 \cdot 4}{4 \cdot 2 \cdot 3 \cdot 3} = \frac{3 \cdot 4}{3 \cdot 4} \cdot \frac{1}{2 \cdot 3} = 1 \cdot \frac{1}{2 \cdot 3} = \frac{1}{2 \cdot 3} = \frac{1}{6}.$$

> To multiply and simplify:
>
> **a)** Write the products in the numerator and the denominator, but do not carry them out.
>
> **b)** Factor the numerator and the denominator.
>
> **c)** Factor the fraction to remove a factor of 1, if possible.
>
> **d)** Carry out the remaining products.

Answers

Skill to Review:
1. Yes **2.** No

EXAMPLES Multiply and simplify.

$$\textbf{1. } \frac{2}{3}\cdot\frac{9}{4}=\frac{2\cdot 9}{3\cdot 4}=\frac{2\cdot 3\cdot 3}{3\cdot 2\cdot 2}=\frac{2\cdot 3}{2\cdot 3}\cdot\frac{3}{2}=1\cdot\frac{3}{2}=\frac{3}{2}$$

$$\textbf{2. } 40\cdot\left(-\frac{7}{8}\right)=-\frac{40\cdot 7}{8}=-\frac{5\cdot 8\cdot 7}{8\cdot 1}=-\frac{5\cdot 7}{1}\cdot\frac{8}{8}=-\frac{5\cdot 7}{1}\cdot 1$$

$$=-\frac{5\cdot 7}{1}=-35$$

Caution!

Canceling can be used as follows for these examples.

$$\textbf{1. } \frac{2}{3}\cdot\frac{9}{4}=\frac{2\cdot 9}{3\cdot 4}=\frac{\cancel{2}\cdot\cancel{3}\cdot 3}{\cancel{3}\cdot\cancel{2}\cdot 2}=\frac{3}{2}$$

Removing a factor of 1: $\frac{2\cdot 3}{2\cdot 3}=1$

$$\textbf{2. } 40\cdot\left(-\frac{7}{8}\right)=-\frac{40\cdot 7}{8}=-\frac{5\cdot\cancel{8}\cdot 7}{\cancel{8}\cdot 1}=-\frac{5\cdot 7}{1}$$

$$=-35$$

Removing a factor of 1: $\frac{8}{8}=1$

Remember: If you can't factor, you can't cancel!

Do Exercises 1–4.

Multiply and simplify.

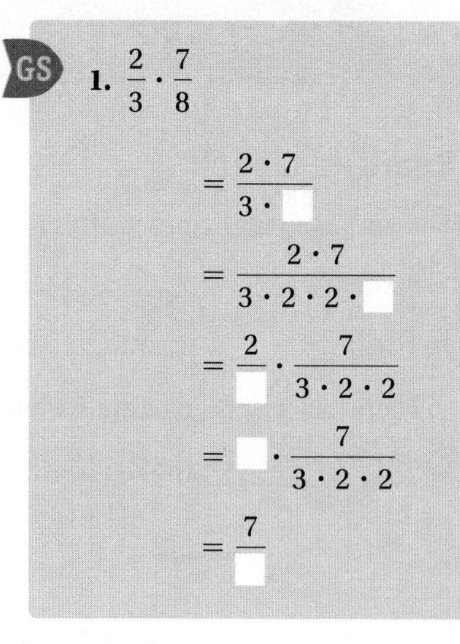

GS **1.** $\frac{2}{3}\cdot\frac{7}{8}$

$$=\frac{2\cdot 7}{3\cdot \square}$$

$$=\frac{2\cdot 7}{3\cdot 2\cdot 2\cdot \square}$$

$$=\frac{2}{\square}\cdot\frac{7}{3\cdot 2\cdot 2}$$

$$=\square\cdot\frac{7}{3\cdot 2\cdot 2}$$

$$=\frac{7}{\square}$$

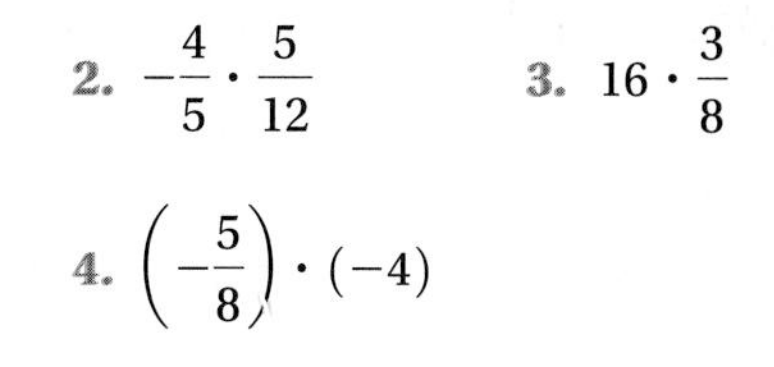

2. $-\frac{4}{5}\cdot\frac{5}{12}$

3. $16\cdot\frac{3}{8}$

4. $\left(-\frac{5}{8}\right)\cdot(-4)$

b APPLICATIONS AND PROBLEM SOLVING

EXAMPLE 3 *Landscaping.* Celina's Landscaping uses $\frac{2}{3}$ lb of peat moss when planting a rosebush. How much will be needed to plant 21 rosebushes?

1. **Familiarize.** We let $n =$ the number of pounds of peat moss needed. Each rosebush requires $\frac{2}{3}$ lb of peat moss, so repeated addition, or multiplication, applies.
2. **Translate.** The problem translates to the following equation:

$$n = 21\cdot\frac{2}{3}.$$

3. **Solve.** To solve the equation, we carry out the multiplication:

$$n = 21\cdot\frac{2}{3}=\frac{21}{1}\cdot\frac{2}{3}=\frac{21\cdot 2}{1\cdot 3}\qquad\text{Multiplying}$$

$$=\frac{3\cdot 7\cdot 2}{1\cdot 3}=\frac{3}{3}\cdot\frac{7\cdot 2}{1}=14.$$

4. **Check.** We check by repeating the calculation. (This is left to the student.) We can also ask if the answer seems reasonable. We are putting less than a pound of peat moss on each bush, so the answer should be less than 21. Since 14 is less than 21, we have a partial check. The number 14 checks.
5. **State.** Celina's Landscaping will need 14 lb of peat moss to plant 21 rosebushes.

Do Exercise 5.

5. *Candy.* Chocolate Delight sells $\frac{4}{5}$-lb boxes of truffles. How many pounds of truffles will be needed to fill 85 boxes?

85 boxes

$\frac{4}{5}$ pound of truffles in each box

Answers

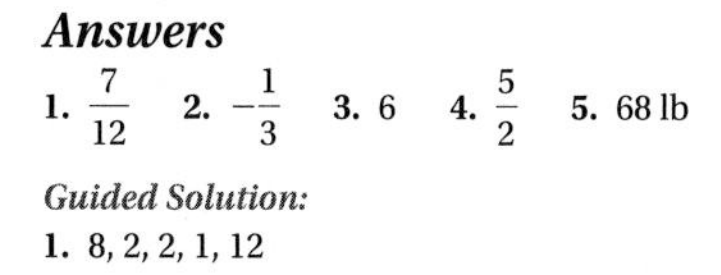

1. $\frac{7}{12}$ **2.** $-\frac{1}{3}$ **3.** 6 **4.** $\frac{5}{2}$ **5.** 68 lb

Guided Solution:
1. 8, 2, 2, 1, 12

3.6 Exercise Set

For Extra Help
MyMathLab® MathXL® PRACTICE WATCH READ REVIEW

Reading Check

Complete each step in the process for multiplying and simplifying using fraction notation.

RC1. a) Write the ______________ in the numerator and the denominator, but do not carry them out.

RC2. b) ______________ the numerator and the denominator.

RC3. c) Factor the fraction to remove a factor of ______________, if possible.

RC4. d) ______________ the remaining products.

a Multiply and simplify. Don't forget to simplify!

1. $\frac{2}{3} \cdot \frac{1}{2}$

2. $\frac{3}{8} \cdot \frac{1}{3}$

3. $\frac{7}{8} \cdot \frac{1}{7}$

4. $\frac{4}{9} \cdot \frac{1}{4}$

5. $-\frac{1}{8} \cdot \frac{4}{5}$

6. $-\frac{2}{5} \cdot \frac{1}{6}$

7. $\frac{1}{4} \cdot \frac{2}{3}$

8. $\frac{4}{6} \cdot \frac{1}{6}$

9. $\frac{12}{5} \cdot \frac{9}{8}$

10. $\frac{16}{15} \cdot \frac{5}{4}$

11. $\frac{10}{9} \cdot \left(-\frac{7}{5}\right)$

12. $\frac{25}{12} \cdot \left(-\frac{4}{3}\right)$

13. $9 \cdot \frac{1}{9}$

14. $4 \cdot \frac{1}{4}$

15. $\frac{1}{3} \cdot 3$

16. $\frac{1}{6} \cdot 6$

17. $\left(-\frac{7}{10}\right) \cdot \left(-\frac{10}{7}\right)$

18. $\left(-\frac{8}{9}\right) \cdot \left(-\frac{9}{8}\right)$

19. $\frac{7}{5} \cdot \frac{5}{7}$

20. $\frac{2}{11} \cdot \frac{11}{2}$

21. $\frac{1}{4} \cdot 8$

22. $\frac{1}{3} \cdot 18$

23. $-24 \cdot \frac{1}{6}$

24. $-16 \cdot \frac{1}{2}$

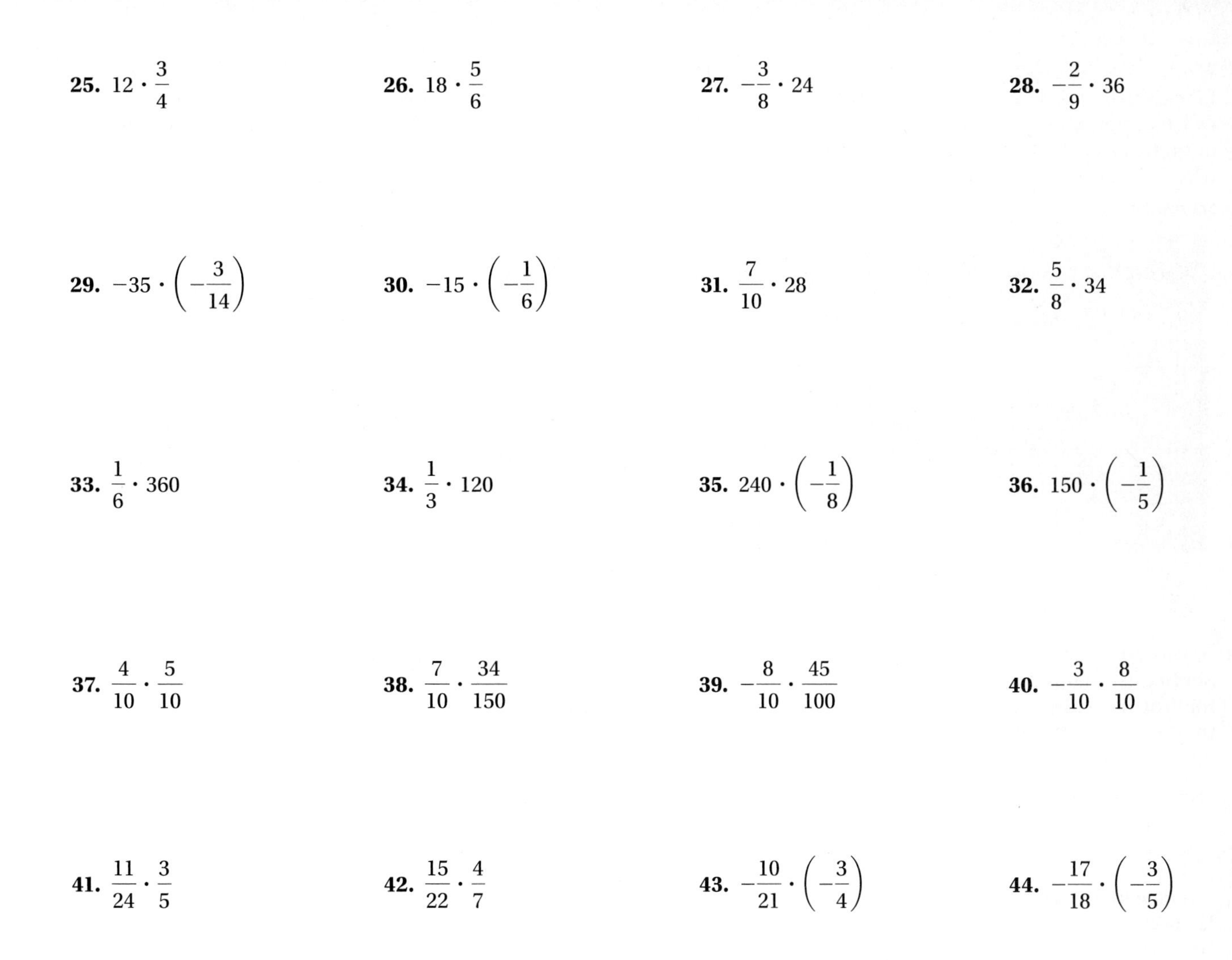

25. $12 \cdot \frac{3}{4}$

26. $18 \cdot \frac{5}{6}$

27. $-\frac{3}{8} \cdot 24$

28. $-\frac{2}{9} \cdot 36$

29. $-35 \cdot \left(-\frac{3}{14}\right)$

30. $-15 \cdot \left(-\frac{1}{6}\right)$

31. $\frac{7}{10} \cdot 28$

32. $\frac{5}{8} \cdot 34$

33. $\frac{1}{6} \cdot 360$

34. $\frac{1}{3} \cdot 120$

35. $240 \cdot \left(-\frac{1}{8}\right)$

36. $150 \cdot \left(-\frac{1}{5}\right)$

37. $\frac{4}{10} \cdot \frac{5}{10}$

38. $\frac{7}{10} \cdot \frac{34}{150}$

39. $-\frac{8}{10} \cdot \frac{45}{100}$

40. $-\frac{3}{10} \cdot \frac{8}{10}$

41. $\frac{11}{24} \cdot \frac{3}{5}$

42. $\frac{15}{22} \cdot \frac{4}{7}$

43. $-\frac{10}{21} \cdot \left(-\frac{3}{4}\right)$

44. $-\frac{17}{18} \cdot \left(-\frac{3}{5}\right)$

b Solve.

Construction. The *pitch* of a screw is the distance between its threads. With each complete rotation, the screw goes in or out a distance equal to its pitch. Use this information to do Exercises 45 and 46.

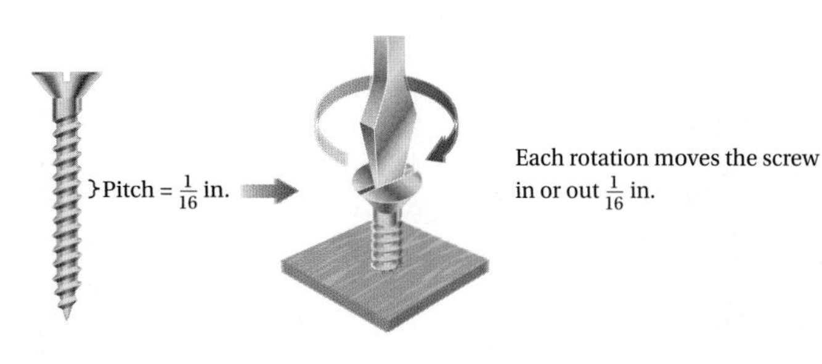

45. The pitch of a screw is $\frac{1}{16}$ in. How far will it go into a piece of oak when it is turned 10 complete rotations clockwise?

46. The pitch of a screw is $\frac{3}{32}$ in. How far will it come out of a piece of plywood when it is turned 10 complete rotations counterclockwise?

47. *World Silver Supply.* The total world supply of silver for new uses in 2011 was about 1040 million ounces. Of this, approximately $\frac{1}{4}$ was scrap silver being reused or repurposed. How many ounces of silver for new uses was supplied as scrap silver?

Source: Based on information from the *World Silver Survey 2012*

48. *Substitute Teaching.* After Jack completes 60 hr of teacher training in college, he can earn \$75 for working a full day as a substitute teacher. How much will he receive for working $\frac{3}{5}$ of a day?

49. *Mailing-List Changes.* The United States Postal Service estimates that $\frac{4}{25}$ of the addresses on a mailing list will change in one year. A business has a mailing list of 3000 people. After one year, how many addresses on that list will be incorrect?

Source: Based on information from usps.com

50. *Shy People.* Sociologists have determined that $\frac{2}{5}$ of the people in the world are shy. A sales manager is considering 650 people for an aggressive sales position. How many of these people might be shy?

51. A recipe for piecrust calls for $\frac{2}{3}$ cup of flour. A baker is making $\frac{1}{2}$ of the recipe. How much flour should the baker use?

52. Of the students in the freshman class, $\frac{4}{5}$ have digital cameras; $\frac{1}{4}$ of these students also join the college photography club. What fraction of the students in the freshman class join the photography club?

53. A house worth \$154,000 is assessed for $\frac{3}{4}$ of its value. What is the assessed value of the house?

54. Roxanne's tuition was \$4600. A loan was obtained for $\frac{3}{4}$ of the tuition. How much was the loan?

55. *Map Scaling.* On a map, 1 in. represents 240 mi. What distance does $\frac{2}{3}$ in. represent?

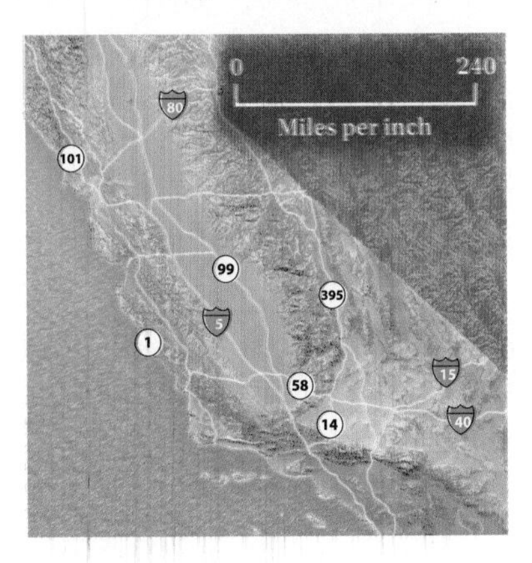

56. *Map Scaling.* On a map, 1 in. represents 120 mi. What distance does $\frac{3}{4}$ in. represent?

57. *Household Budgets.* A family has an annual income of \$42,000. Of this, $\frac{1}{5}$ is spent for food, $\frac{1}{4}$ for housing, $\frac{1}{10}$ for clothing, $\frac{1}{14}$ for savings, $\frac{1}{5}$ for taxes, and the rest for other expenses. How much is spent for each?

58. *Household Budgets.* A family has an annual income of \$28,140. Of this, $\frac{1}{5}$ is spent for food, $\frac{1}{4}$ for housing, $\frac{1}{10}$ for clothing, $\frac{1}{14}$ for savings, $\frac{1}{5}$ for taxes, and the rest for other expenses. How much is spent for each?

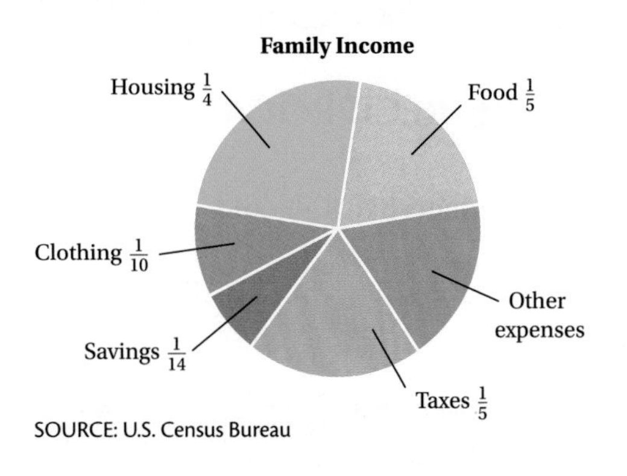

Skill Maintenance

Add. [1.2a], [2.2a]

59. $7246 + 1341$

60. $24 + 2683$

61. $-14 + 5$

62. $-6 + (-7)$

Subtract. [1.3a], [2.3a]

63. $9001 - 6798$

64. $2037 - 1189$

65. $8 - 12$

66. $-4 - (-4)$

Multiply. [1.4a], [2.4a]

67. $2 \cdot 13$

68. $8 \cdot 32$

69. $17 \cdot (-25)$

70. $-5 \cdot (-6)$

Divide. [1.5a], [2.5a]

71. $0 \div 22$

72. $22 \div 1$

73. $7140 \div (-35)$

74. $-56 \div (-7)$

Synthesis

Multiply and simplify. Use a list of prime numbers or a fraction calculator.

75. 🖩 $\frac{201}{535} \cdot \frac{4601}{6499}$

76. 🖩 $-\frac{5767}{3763} \cdot \frac{159}{395}$

77. *College Profile.* Of students entering a college, $\frac{7}{8}$ have completed high school and $\frac{2}{3}$ are older than 20. If $\frac{1}{7}$ of all students are left-handed, what fraction of students entering the college are left-handed high school graduates over the age of 20?

78. *College Profile.* Refer to the information in Exercise 77. If 480 students are entering the college, how many of them are left-handed high school graduates 20 yr old or younger?

79. *College Profile.* Refer to Exercise 77. What fraction of students entering the college did not graduate from high school, are 20 yr old or younger, and are left-handed?

3.7 Division and Applications

OBJECTIVES

a Find the reciprocal of a number.

b Divide and simplify using fraction notation.

c Solve equations of the type $a \cdot x = b$ and $x \cdot a = b$, where a and b may be fractions.

d Solve applied problems involving division of fractions.

SKILL TO REVIEW

Objective 1.7b: Solve equations like $t + 28 = 54$, $28 \cdot x = 168$, and $98 \cdot 2 = y$.

Solve.

1. $88 = 8 \cdot y$
2. $16 \cdot x = 1152$

Find the reciprocal.

1. $\frac{2}{5}$
2. $\frac{10}{7}$
3. 9
4. $-\frac{1}{5}$

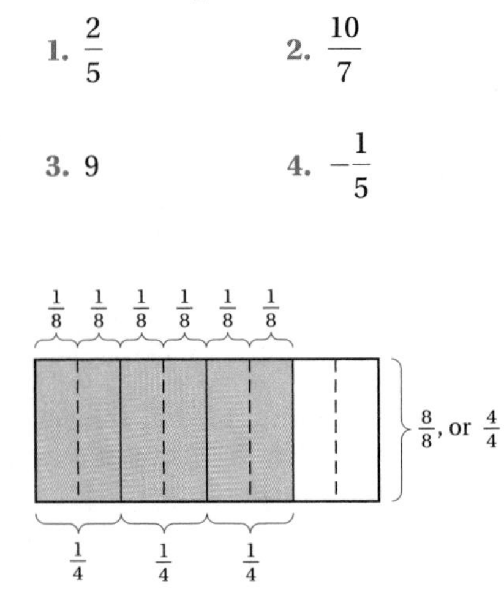

Answers

Skill to Review:
1. 11 2. 72

Margin Exercises:
1. $\frac{5}{2}$ 2. $\frac{7}{10}$ 3. $\frac{1}{9}$ 4. -5

a RECIPROCALS

Products like $8 \cdot \frac{1}{8}$ and $\frac{2}{3} \cdot \frac{3}{2}$ simplify to 1:

$$8 \cdot \frac{1}{8} = \frac{8}{1} \cdot \frac{1}{8} = \frac{8 \cdot 1}{1 \cdot 8} = \frac{8}{8} = 1; \qquad \frac{2}{3} \cdot \frac{3}{2} = \frac{2 \cdot 3}{3 \cdot 2} = \frac{6}{6} = 1.$$

RECIPROCALS

If the product of two numbers is 1, we say that they are **reciprocals** of each other. To find the reciprocal of a fraction, interchange the numerator and the denominator.

$$\text{Number: } \frac{3}{4} \longrightarrow \text{Reciprocal: } \frac{4}{3}$$

EXAMPLES Find the reciprocal.

1. The reciprocal of $\frac{4}{5}$ is $\frac{5}{4}$. $\quad \frac{4}{5} \cdot \frac{5}{4} = \frac{20}{20} = 1$

2. The reciprocal of 24 is $\frac{1}{24}$. $\quad$ Think of 24 as $\frac{24}{1}$: $\frac{24}{1} \cdot \frac{1}{24} = \frac{24}{24} = 1.$

3. The reciprocal of $-\frac{1}{3}$ is -3. $\quad -\frac{1}{3} \cdot (-3) = -\frac{1}{3} \cdot \left(-\frac{3}{1}\right) = \frac{3}{3} = 1$

◀ Do Margin Exercises 1–4.

Does 0 have a reciprocal? If it did, it would have to be a number x such that $0 \cdot x = 1$. But 0 times any number is 0. Thus we have the following.

0 HAS NO RECIPROCAL

The number 0, or $\frac{0}{n}$, has no reciprocal. $\left(\text{Recall that } \frac{n}{0} \text{ is not defined.}\right)$

b DIVISION

Consider the division $\frac{3}{4} \div \frac{1}{8}$. We are asking how many $\frac{1}{8}$'s are in $\frac{3}{4}$. From the figure at left, we see that there are six $\frac{1}{8}$'s in $\frac{3}{4}$. Thus,

$$\frac{3}{4} \div \frac{1}{8} = 6.$$

We can check this by multiplying:

$$6 \cdot \frac{1}{8} = \frac{6}{1} \cdot \frac{1}{8} = \frac{6}{8} = \frac{2 \cdot 3}{2 \cdot 4} = \frac{2}{2} \cdot \frac{3}{4} = \frac{3}{4}.$$

Here is a faster way to do this division:

$$\frac{3}{4} \div \frac{1}{8} = \frac{3}{4} \cdot \frac{8}{1} = \frac{3 \cdot 8}{4 \cdot 1} = \frac{24}{4} = 6.$$ Multiplying by the reciprocal of the divisor

To divide fractions, multiply the dividend by the reciprocal of the divisor:

$$\frac{2}{5} \div \frac{3}{4} = \frac{2}{5} \cdot \frac{4}{3} = \frac{2 \cdot 4}{5 \cdot 3} = \frac{8}{15}.$$

EXAMPLES Divide and simplify.

4. $\frac{5}{6} \div \frac{2}{3} = \frac{5}{6} \cdot \frac{3}{2} = \frac{5 \cdot 3}{6 \cdot 2} = \frac{5 \cdot 3}{3 \cdot 2 \cdot 2} = \frac{3}{3} \cdot \frac{5}{2 \cdot 2} = \frac{5}{2 \cdot 2} = \frac{5}{4}$

5. $-\frac{7}{8} \div \frac{1}{16} = -\frac{7}{8} \cdot 16 = -\frac{7 \cdot 16}{8} = -\frac{7 \cdot 2 \cdot 8}{8 \cdot 1} = -\frac{7 \cdot 2}{1} \cdot \frac{8}{8}$

$= -\frac{7 \cdot 2}{1} = -14$

6. $\frac{2}{5} \div 6 = \frac{2}{5} \cdot \frac{1}{6} = \frac{2 \cdot 1}{5 \cdot 6} = \frac{2 \cdot 1}{5 \cdot 2 \cdot 3} = \frac{2}{2} \cdot \frac{1}{5 \cdot 3} = \frac{1}{5 \cdot 3} = \frac{1}{15}$

Caution!

Canceling can be used as follows for Examples 4–6.

4. $\frac{5}{6} \div \frac{2}{3} = \frac{5}{6} \cdot \frac{3}{2} = \frac{5 \cdot 3}{6 \cdot 2} = \frac{5 \cdot \cancel{3}}{\cancel{3} \cdot 2 \cdot 2} = \frac{5}{2 \cdot 2} = \frac{5}{4}$ Removing a factor of 1: $\frac{3}{3} = 1$

5. $-\frac{7}{8} \div \frac{1}{16} = -\frac{7}{8} \cdot 16 = -\frac{7 \cdot 16}{8} = -\frac{7 \cdot \cancel{8} \cdot 2}{\cancel{8} \cdot 1}$ Removing a factor of 1: $\frac{8}{8} = 1$

$= -\frac{7 \cdot 2}{1} = -14$

6. $\frac{2}{5} \div 6 = \frac{2}{5} \cdot \frac{1}{6} = \frac{2 \cdot 1}{5 \cdot 6} = \frac{\cancel{2} \cdot 1}{5 \cdot \cancel{2} \cdot 3} = \frac{1}{5 \cdot 3} = \frac{1}{15}$ Removing a factor of 1: $\frac{2}{2} = 1$

Remember: if you can't factor, you can't cancel!

Do Exercises 5–8.

What is the explanation for multiplying by a reciprocal when dividing? Let's consider $\frac{2}{3} \div \frac{7}{5}$. We multiply by 1. The name for 1 that we will use is $(5/7)/(5/7)$; it comes from the reciprocal of $\frac{7}{5}$.

$$\frac{2}{3} \div \frac{7}{5} = \frac{\frac{2}{3}}{\frac{7}{5}} = \frac{\frac{2}{3}}{\frac{7}{5}} \cdot 1 = \frac{\frac{2}{3}}{\frac{7}{5}} \cdot \frac{\frac{5}{7}}{\frac{5}{7}} = \frac{\frac{2}{3} \cdot \frac{5}{7}}{\frac{7}{5} \cdot \frac{5}{7}} = \frac{\frac{2}{3} \cdot \frac{5}{7}}{1} = \frac{2}{3} \cdot \frac{5}{7} = \frac{10}{21}$$

Thus, $\frac{2}{3} \div \frac{7}{5} = \frac{2}{3} \cdot \frac{5}{7} = \frac{10}{21}$.

Do Exercise 9.

Divide and simplify.

GS **5.** $\frac{6}{7} \div \frac{3}{4} = \frac{6}{7} \cdot \frac{\square}{\square}$

$= \frac{6 \cdot 4}{7 \cdot 3}$

$= \frac{2 \cdot 3 \cdot 2 \cdot \square}{7 \cdot 3}$

$= \frac{3}{\square} \cdot \frac{2 \cdot 2 \cdot 2}{7}$

$= \frac{2 \cdot 2 \cdot 2}{7}$

$= \frac{\square}{7}$

6. $-\frac{2}{3} \div \frac{1}{4}$

7. $\frac{4}{5} \div 8$

8. $-60 \div \left(-\frac{3}{5}\right)$

9. Divide by multiplying by 1:

$$\frac{\frac{4}{5}}{\frac{6}{7}}$$

Answers

5. $\frac{8}{7}$ **6.** $-\frac{8}{3}$ **7.** $\frac{1}{10}$ **8.** 100 **9.** $\frac{14}{15}$

Guided Solution:

5. $\frac{4}{3}$, 2, 3, 8

c SOLVING EQUATIONS

Now let's solve the equations $a \cdot x = b$ and $x \cdot a = b$, where a and b may be fractions. We proceed as we did with equations involving whole numbers. We divide by a on both sides.

EXAMPLE 7 Solve: $\frac{4}{3} \cdot x = \frac{6}{7}$.

We have

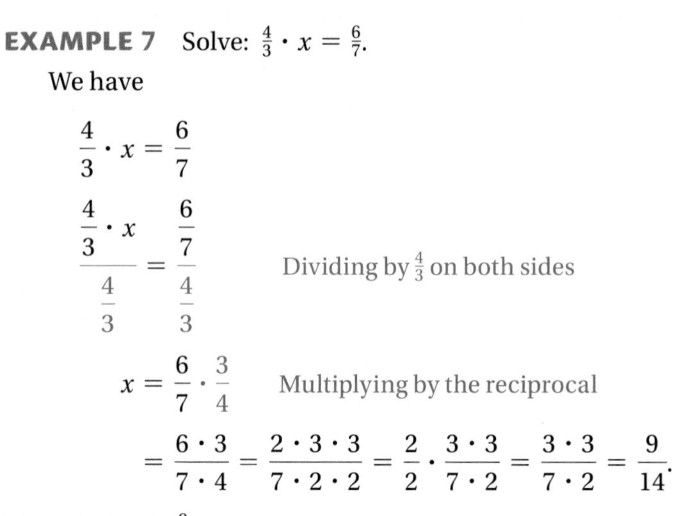

$$\frac{4}{3} \cdot x = \frac{6}{7}$$

$$\frac{\frac{4}{3} \cdot x}{\frac{4}{3}} = \frac{\frac{6}{7}}{\frac{4}{3}} \qquad \text{Dividing by } \tfrac{4}{3} \text{ on both sides}$$

$$x = \frac{6}{7} \cdot \frac{3}{4} \qquad \text{Multiplying by the reciprocal}$$

$$= \frac{6 \cdot 3}{7 \cdot 4} = \frac{2 \cdot 3 \cdot 3}{7 \cdot 2 \cdot 2} = \frac{2}{2} \cdot \frac{3 \cdot 3}{7 \cdot 2} = \frac{3 \cdot 3}{7 \cdot 2} = \frac{9}{14}.$$

The solution is $\frac{9}{14}$. ■

Solve.

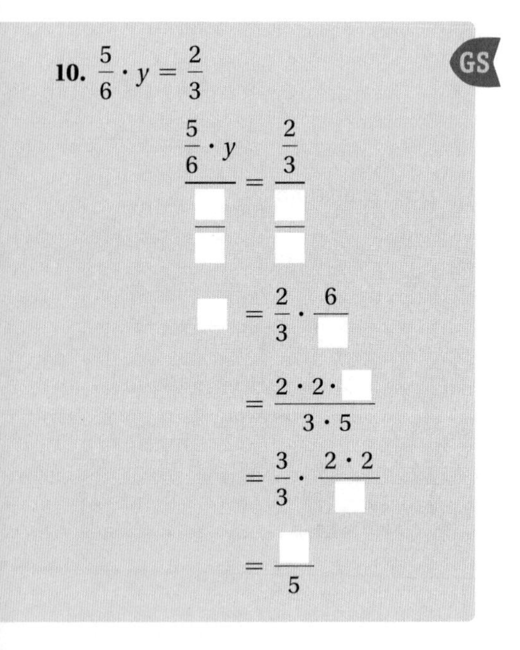

10. $\frac{5}{6} \cdot y = \frac{2}{3}$ **GS**

$$\frac{\frac{5}{6} \cdot y}{\frac{\square}{\square}} = \frac{\frac{2}{3}}{\frac{\square}{\square}}$$

$$\square = \frac{2}{3} \cdot \frac{6}{\square}$$

$$= \frac{2 \cdot 2 \cdot \square}{3 \cdot 5}$$

$$= \frac{3}{3} \cdot \frac{2 \cdot 2}{\square}$$

$$= \frac{\square}{5}$$

11. $n \cdot \frac{3}{4} = -24$

EXAMPLE 8 Solve: $t \cdot \frac{4}{5} = -80$.

Dividing by $\frac{4}{5}$ on both sides, we get

$$t = -80 \div \frac{4}{5} = -80 \cdot \frac{5}{4} = \frac{-80 \cdot 5}{4} = \frac{4 \cdot (-20) \cdot 5}{4 \cdot 1}$$

$$= \frac{4}{4} \cdot \frac{-20 \cdot 5}{1} = \frac{-20 \cdot 5}{1} = -100.$$

The solution is -100.

◀ **Do Exercises 10 and 11.**

d APPLICATIONS AND PROBLEM SOLVING

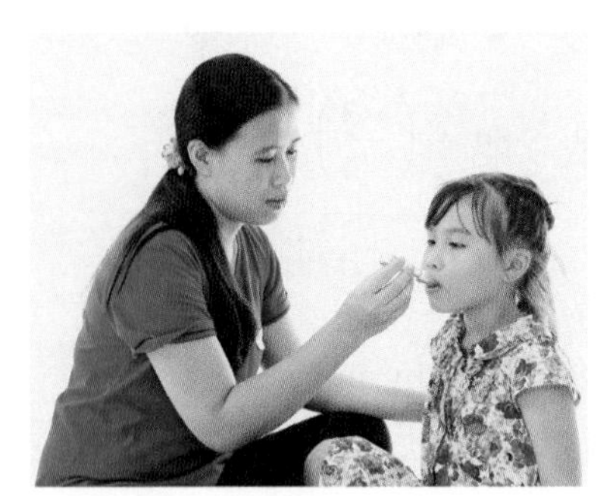

EXAMPLE 9 *Doses of an Antibiotic.* How many doses, each containing $\frac{15}{4}$ milliliters (mL), can be obtained from a bottle of a children's antibiotic that contains 60 mL?

1. **Familiarize.** We are asking the question "How many $\frac{15}{4}$'s are in 60?" Repeated addition will apply here. We make a drawing. We let n = the number of doses in all.

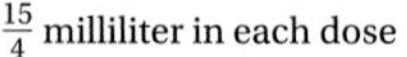

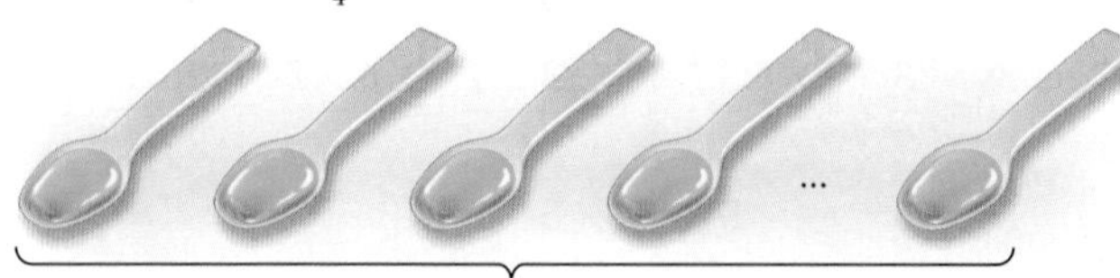

2. **Translate.** The equation that corresponds to the situation is

$$n = 60 \div \frac{15}{4}.$$

Answers

10. $\frac{4}{5}$ **11.** -32

Guided Solution:

10. $\frac{5}{6}, \frac{5}{6}; y, 5, 3, 5, 4$

3. **Solve.** We solve the equation by carrying out the division:

$$n = 60 \div \frac{15}{4} = 60 \cdot \frac{4}{15} = \frac{60}{1} \cdot \frac{4}{15}$$

$$= \frac{60 \cdot 4}{1 \cdot 15} = \frac{4 \cdot 15 \cdot 4}{1 \cdot 15} = \frac{15}{15} \cdot \frac{4 \cdot 4}{1} = 1 \cdot 16 = 16.$$

4. **Check.** We check by multiplying the number of doses by the size of the dose: $16 \cdot \frac{15}{4} = 60$. The answer checks.

5. **State.** There are 16 doses in a 60-mL bottle of the antibiotic.

Do Exercise 12. ▶

12. Each loop in a spring uses $\frac{21}{8}$ in. of wire. How many loops can be made from 210 in. of wire?

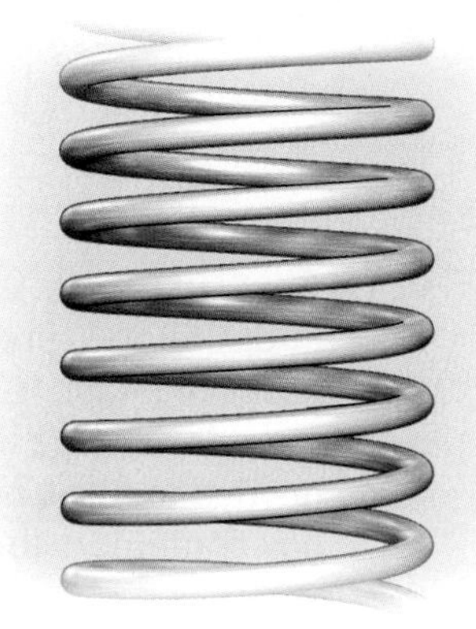

EXAMPLE 10 *Bicycle Paths.* The city of Indianapolis has adopted the *Indianapolis Bicycle Master Plan* as a strategy for creating an environment where bicycling is a safe, practical, and enjoyable transportation choice. After the city finished constructing 60 mi of bike paths and on-road bike lanes, the master plan was $\frac{3}{10}$ complete. What is the total number of miles of bicycling surface that the city of Indianapolis plans to construct?

Source: *Indianapolis Bicycle Master Plan*, June 2012

1. **Familiarize.** We ask: "60 mi is $\frac{3}{10}$ of what length?" We make a drawing or at least visualize the problem. We let b = the total number of miles of bicycling surface in the master plan.

2. **Translate.** We translate to an equation:

Fraction completed	of	Total miles planned	is	Amount completed
↓	↓	↓	↓	↓
$\frac{3}{10}$	$\cdot$	b	$=$	60.

3. **Solve.** We divide by $\frac{3}{10}$ on both sides and carry out the division:

$$b = 60 \div \frac{3}{10} = \frac{60}{1} \cdot \frac{10}{3} = \frac{60 \cdot 10}{1 \cdot 3} = \frac{3 \cdot 20 \cdot 10}{1 \cdot 3} = \frac{3}{3} \cdot \frac{20 \cdot 10}{1} = 200.$$

4. **Check.** We determine whether $\frac{3}{10}$ of 200 is 60: $\frac{3}{10} \cdot 200 = 60$. The answer, 200, checks.

5. **State.** The *Indianapolis Bicycle Master Plan* calls for 200 mi of bicycling surface.

Do Exercise 13. ▶

13. *Sales Trip.* Ed Jacobs sells soybean seeds to seed companies. After he had driven 210 mi, $\frac{5}{6}$ of his sales trip was completed. How long was the total trip?

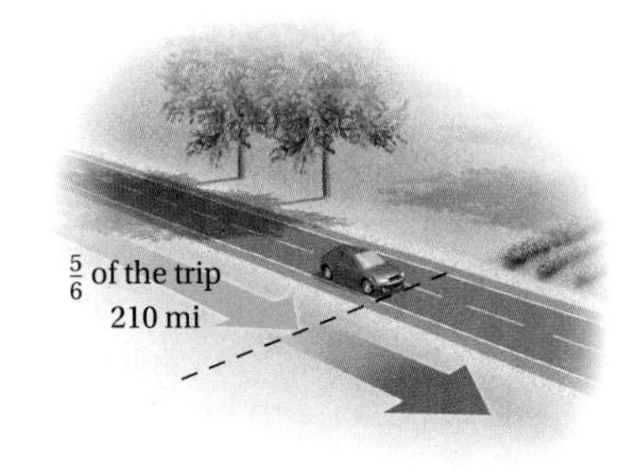

Answers

12. 80 loops **13.** 252 mi

Translating for Success

The goal of these matching questions is to practice step (2), Translate, of the five-step problem-solving process. Translate each problem to an equation and select a correct translation from equations A–O.

1. *Boxes of Candy.* Jane's Fudge Shop is preparing gift boxes of fudge. How many pounds of fudge will be needed to fill 80 boxes if each box contains $\frac{5}{16}$ lb?

2. *Gallons of Gasoline.* On the third day of a business trip, a sales representative used $\frac{4}{5}$ of a tank of gasoline. If the tank holds 20 gal of gasoline, how many gallons were used on the third day?

3. *Purchasing a Shirt.* Tom received $36 for his birthday. If he spends $\frac{3}{4}$ of the gift on a new shirt, what is the cost of the shirt?

4. *Checkbook Balance.* The balance in Sam's checking account is $1456. He writes a check for $28 and makes a deposit of $52. What is the new balance?

5. *Boxes of Candy.* Jane's Fudge Shop prepared 80 lb of fudge for gift boxes. If each box contains $\frac{5}{16}$ lb, how many boxes can be filled?

A. $x = \frac{3}{4} \cdot 36$

B. $28 \cdot x = 52$

C. $x = 80 \cdot \frac{5}{16}$

D. $x = 1456 \div 28$

E. $x = 20 - \frac{4}{5}$

F. $20 = \frac{4}{5} \cdot x$

G. $x = 12 \cdot 28$

H. $x = \frac{4}{5} \cdot 20$

I. $\frac{3}{4} \cdot x = 36$

J. $x = 1456 - 52 - 28$

K. $x \div 28 = 1456$

L. $x = 52 - 28$

M. $x = 52 \cdot 28$

N. $x = 1456 - 28 + 52$

O. $\frac{5}{16} \cdot x = 80$

Answers on page A-5

6. *Gasoline Tank.* A gasoline tank contains 20 gal when it is $\frac{4}{5}$ full. How many gallons can it hold when full?

7. *Knitting a Scarf.* It takes Rachel 36 hr to knit a scarf. She can knit only $\frac{3}{4}$ hr per day because she is taking 16 hr of college classes. How many days will it take her to knit the scarf?

8. *Bicycle Trip.* On a recent 52-mi bicycle trip, David stopped to make a cell-phone call after completing 28 mi. How many more miles did he bicycle after the call?

9. *Crème de Menthe Thins.* Andes Candies L.P. makes Crème de Menthe Thins. How many 28-piece packages can be filled with 1456 pieces?

10. *Cereal Donations.* The Williams family donates 28 boxes of cereal weekly to the local Family in Crisis Center. How many boxes does this family donate in one year?

3.7 Exercise Set

For Extra Help
MyMathLab® MathXL® PRACTICE WATCH READ REVIEW

Reading Check

Determine whether each statement is true or false.

________ **RC1.** The numbers $-\frac{1}{7}$ and -7 are reciprocals.

________ **RC2.** The number 1 has no reciprocal.

________ **RC3.** To divide fractions, we multiply the dividend by the reciprocal of the divisor.

________ **RC4.** To solve $\frac{2}{5} \cdot x = \frac{3}{8}$, we divide by $\frac{3}{8}$ on both sides.

a Find the reciprocal of each number.

1. $\frac{5}{6}$

2. $\frac{7}{8}$

3. 6

4. 4

5. $\frac{1}{6}$

6. $\frac{1}{4}$

7. $-\frac{10}{3}$

8. $-\frac{17}{4}$

b Divide and simplify.

Don't forget to simplify!

9. $\frac{3}{5} \div \frac{3}{4}$

10. $\frac{2}{3} \div \frac{3}{4}$

11. $-\frac{3}{5} \div \frac{9}{4}$

12. $-\frac{6}{7} \div \frac{3}{5}$

13. $\frac{4}{3} \div \frac{1}{3}$

14. $\frac{10}{9} \div \frac{1}{3}$

15. $-\frac{1}{3} \div \left(-\frac{1}{6}\right)$

16. $-\frac{1}{4} \div \left(-\frac{1}{5}\right)$

17. $\frac{3}{8} \div 3$

18. $\frac{5}{6} \div 5$

19. $\frac{12}{7} \div 4$

20. $\frac{18}{5} \div 2$

21. $12 \div \left(-\frac{3}{2}\right)$

22. $24 \div \left(-\frac{3}{8}\right)$

23. $28 \div \frac{4}{5}$

24. $40 \div \frac{2}{3}$

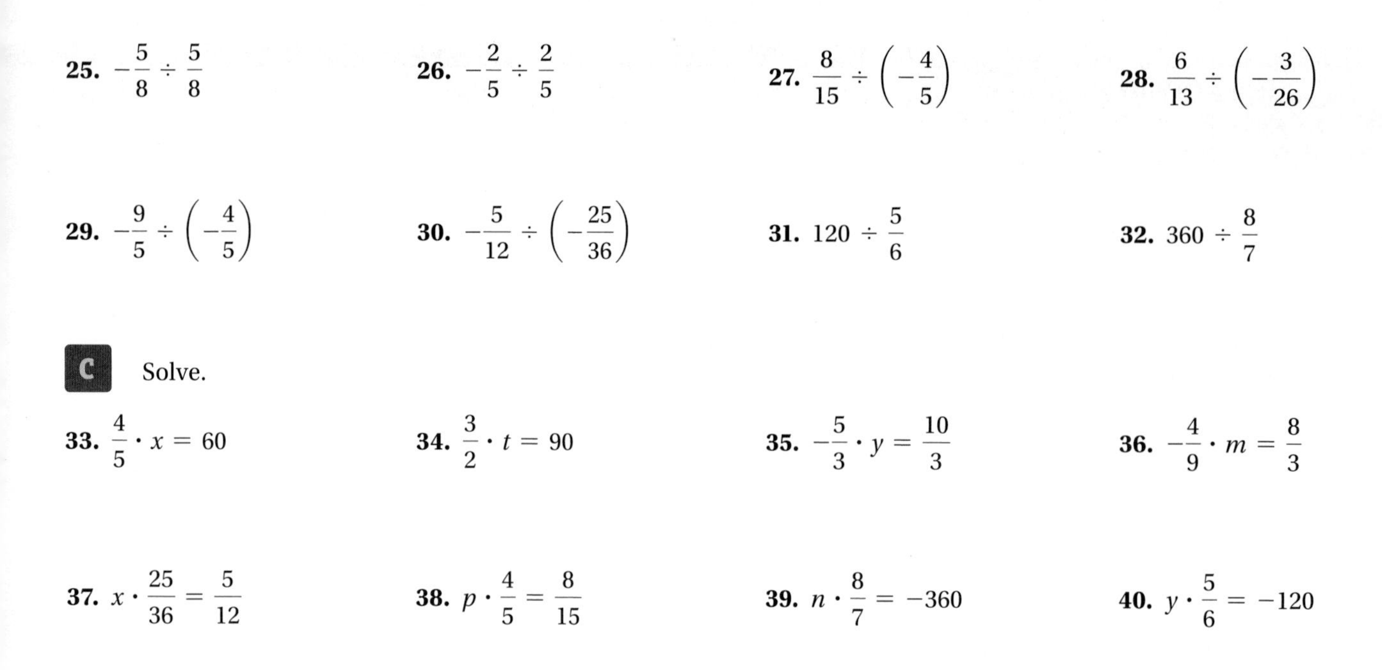

25. $-\frac{5}{8} \div \frac{5}{8}$

26. $-\frac{2}{5} \div \frac{2}{5}$

27. $\frac{8}{15} \div \left(-\frac{4}{5}\right)$

28. $\frac{6}{13} \div \left(-\frac{3}{26}\right)$

29. $-\frac{9}{5} \div \left(-\frac{4}{5}\right)$

30. $-\frac{5}{12} \div \left(-\frac{25}{36}\right)$

31. $120 \div \frac{5}{6}$

32. $360 \div \frac{8}{7}$

c Solve.

33. $\frac{4}{5} \cdot x = 60$

34. $\frac{3}{2} \cdot t = 90$

35. $-\frac{5}{3} \cdot y = \frac{10}{3}$

36. $-\frac{4}{9} \cdot m = \frac{8}{3}$

37. $x \cdot \frac{25}{36} = \frac{5}{12}$

38. $p \cdot \frac{4}{5} = \frac{8}{15}$

39. $n \cdot \frac{8}{7} = -360$

40. $y \cdot \frac{5}{6} = -120$

d Solve.

41. *Extension Cords.* An electrical supplier sells rolls of SJO 14-3 cable to a company that makes extension cords. It takes $\frac{7}{3}$ ft of cable to make each cord. How many extension cords can be made with a roll of cable containing 2240 ft of cable?

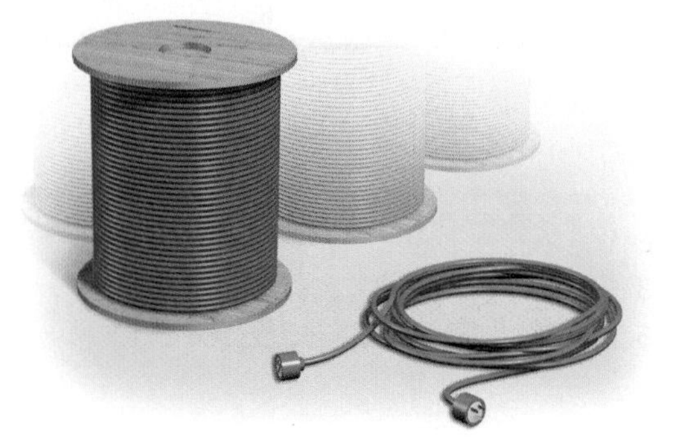

42. Benny uses $\frac{2}{5}$ gram (g) of toothpaste each time he brushes his teeth. If Benny buys a 30-g tube, how many times will he be able to brush his teeth?

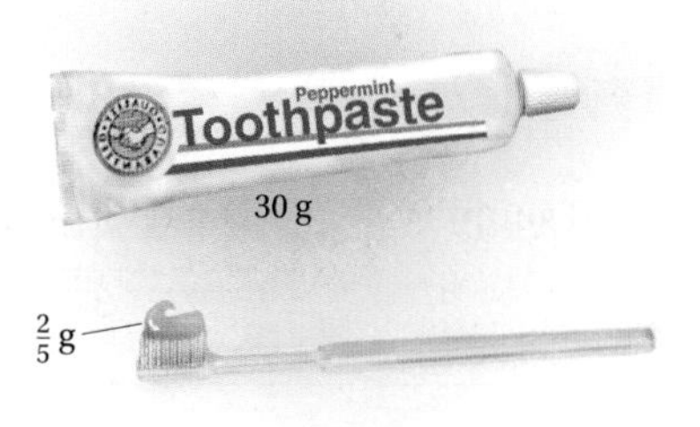

43. A pair of basketball shorts requires $\frac{3}{4}$ yd of nylon. How many pairs of shorts can be made from 24 yd of nylon?

44. A child's baseball shirt requires $\frac{5}{6}$ yd of fabric. How many shirts can be made from 25 yd of fabric?

45. How many $\frac{2}{3}$-cup sugar bowls can be filled from 16 cups of sugar?

46. How many $\frac{2}{3}$-cup cereal bowls can be filled from 10 cups of cornflakes?

47. A bucket had 12 L of water in it when it was $\frac{3}{4}$ full. How much could it hold when full?

48. A tank had 20 L of gasoline in it when it was $\frac{4}{5}$ full. How much could it hold when full?

49. Yoshi Teramoto sells hardware tools. After driving 180 kilometers (km), he has completed $\frac{5}{8}$ of a sales trip. How long is the total trip? How many kilometers are left to drive?

50. A piece of coaxial cable $\frac{4}{5}$ meter (m) long is to be cut into 8 pieces of the same length. What is the length of each piece?

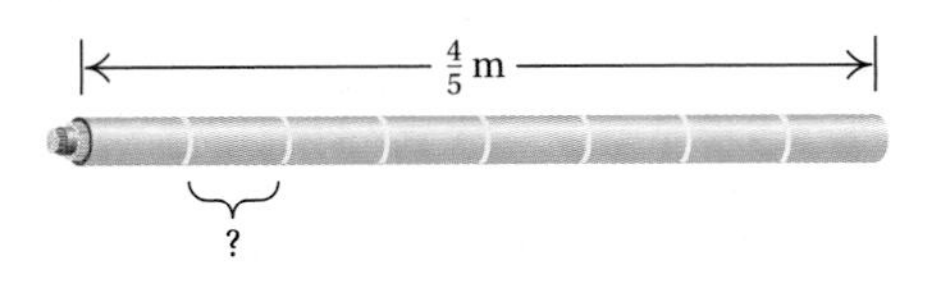

Pitch of a Screw. The pitch of a screw is the distance between its threads. With each complete rotation, the screw goes in or out a distance equal to its pitch. Use this information to do Exercises 51 and 52.

51. After a screw has been turned 8 complete rotations, it is extended $\frac{1}{2}$ in. into a piece of wallboard. What is the pitch of the screw?

52. The pitch of a screw is $\frac{3}{32}$ in. How many complete rotations are necessary to drive the screw $\frac{3}{4}$ in. into a piece of pine wood?

Skill Maintenance

Solve. [1.8a]

53. *New York Road Runners.* The New York Road Runners, a group that runs the New York City Marathon, began in 1958 with 40 members. The membership had increased to 60,000 in 2012. How many more members were there in 2012 than in 1958?

Source: New York Road Runners

54. *Gas Mileage.* The 2013 Chevrolet Corvette coupe gets 26 mpg in highway driving. How many gallons will it use in 1846 mi of highway driving?

Source: Chevrolet

55. *Associate's Degrees.* About 849,452 associate's degrees were earned in the United States in 2010. Of these, 640,113 degrees were conferred by public institutions. How many associate's degrees were conferred by private institutions?

Source: nces.ed.gov

56. *Bachelor's Degrees and Master's Degrees.* About 693,025 master's degrees were earned in the United States in 2010. This is 956,989 fewer than the number of bachelor's degrees earned the same year. How many bachelor's degrees were earned in 2010?

57. *Tiananmen Square.* Tiananmen Square in Beijing, China, is the largest public square in the world. The length of the rectangular region is 963 yd and the width is 547 yd. What is its area? its perimeter?

58. A landscaper buys 13 small maple trees and 17 small oak trees for a project. A maple costs \$23 and an oak costs \$37. How much is spent altogether for the trees?

Synthesis

Simplify. Use a list of prime numbers.

59. 🖩 $\frac{711}{1957} \div \frac{10,033}{13,081}$

60. 🖩 $\frac{8633}{7387} \div \left(-\frac{485}{581}\right)$

61. $\left(\frac{9}{10} \div \frac{2}{5} \div \frac{3}{8}\right)^2$

62. $\dfrac{\left(\frac{3}{7}\right)^2 \div \frac{12}{5}}{\left(\frac{2}{9}\right)\left(\frac{9}{2}\right)}$

63. If $\frac{1}{3}$ of a number is $\frac{1}{4}$, what is $\frac{1}{2}$ of the number?

CHAPTER

3 Summary and Review

Vocabulary Reinforcement

Fill in each blank with the correct term from the list at the right. Some of the choices may not be used.

1. For any number a, $a \cdot 1 = a$. The number 1 is the ______________ identity. [3.5a]
2. In the product $10 \cdot \frac{3}{4}$, 10 and $\frac{3}{4}$ are called ______________. [3.1a], [3.4a]
3. A natural number that has exactly two different factors, only itself and 1, is called a(n) ______________ number. [3.1c]
4. In the fraction $\frac{4}{17}$, we call 17 the ______________. [3.3a]
5. Since $\frac{2}{5}$ and $\frac{6}{15}$ are two names for the same number, we say that $\frac{2}{5}$ and $\frac{6}{15}$ are ______________ fractions. [3.5a]
6. The product of 6 and $\frac{1}{6}$ is 1. We say that 6 and $\frac{1}{6}$ are ______________. [3.7a]
7. Since $20 = 4 \cdot 5$, we say that $4 \cdot 5$ is a ______________ of 20. [3.1a]
8. Since $20 = 4 \cdot 5$, we say that 20 is a ______________ of 5. [3.1b]

equivalent
additive
multiplicative
reciprocals
factors
prime
composite
numerator
denominator
factorization
variables
multiple

Concept Reinforcement

Determine whether each statement is true or false.

______________ 1. For any natural number n, $\frac{n}{n} > \frac{0}{n}$. [3.3b]

______________ 2. A number is divisible by 10 if its ones digit is 0 or 5. [3.2a]

______________ 3. If a number is divisible by 9, then it is also divisible by 3. [3.2a]

______________ 4. The fraction $\frac{13}{6}$ is larger than the fraction $\frac{11}{6}$. [3.5c]

Study Guide

Objective 3.1a Find the factors of a number.

Example Find the factors of 84.

We find as many "two-factor" factorizations as we can.

$1 \cdot 84$ $4 \cdot 21$

$2 \cdot 42$ $6 \cdot 14$

$3 \cdot 28$ $7 \cdot 12$ ← Since 8, 9, 10, and 11 are not factors, we are finished.

The factors are 1, 2, 3, 4, 6, 7, 12, 14, 21, 28, 42, and 84.

Practice Exercise

1. Find the factors of 104.

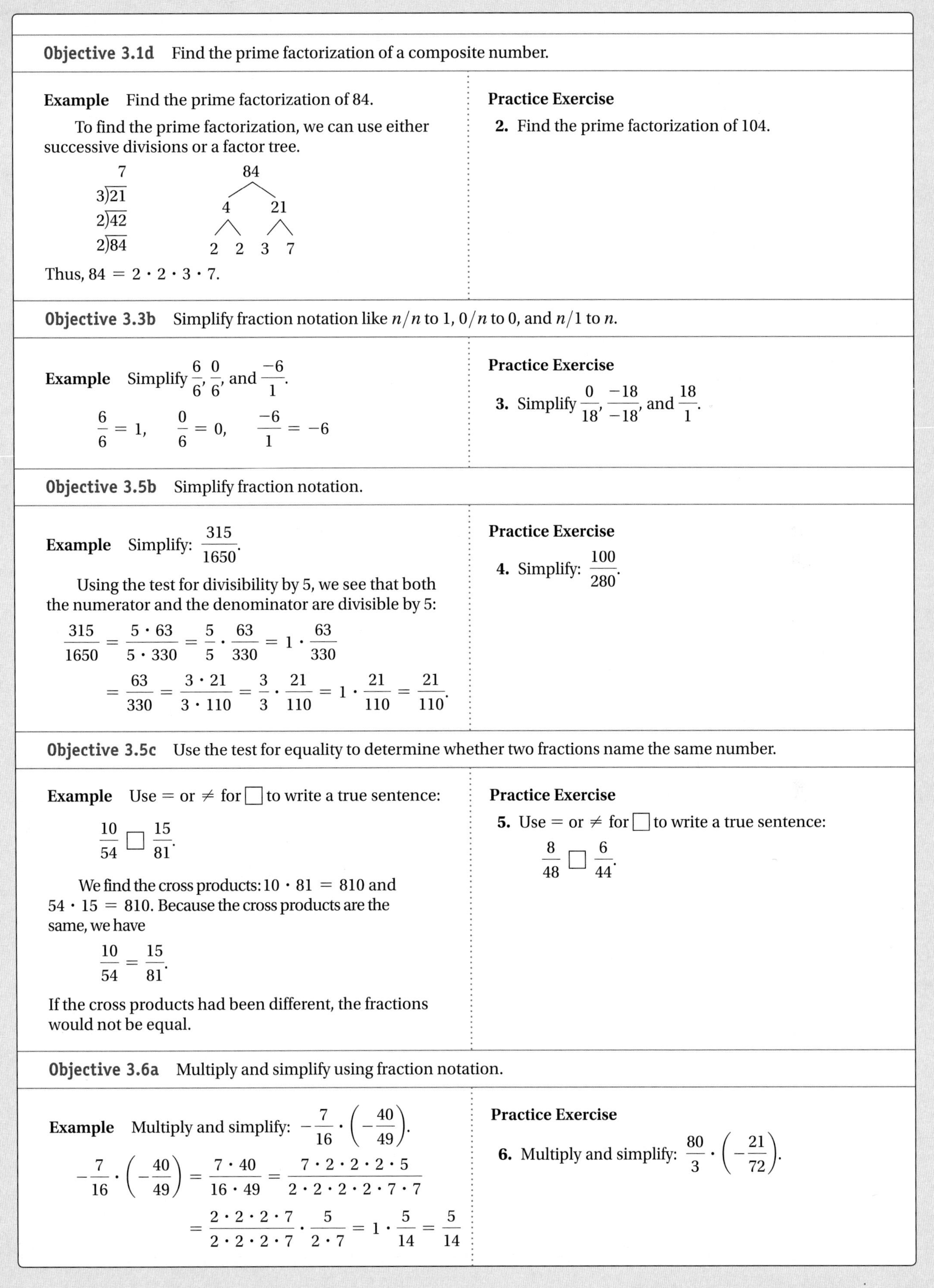

Objective 3.1d Find the prime factorization of a composite number.

Example Find the prime factorization of 84.

To find the prime factorization, we can use either successive divisions or a factor tree.

$$\begin{array}{r} 7 \\ 3\overline{)21} \\ 2\overline{)42} \\ 2\overline{)84} \end{array} \qquad \begin{array}{c} 84 \\ 4 \quad 21 \\ 2 \;\; 2 \;\; 3 \;\; 7 \end{array}$$

Thus, $84 = 2 \cdot 2 \cdot 3 \cdot 7$.

Practice Exercise

2. Find the prime factorization of 104.

Objective 3.3b Simplify fraction notation like n/n to 1, $0/n$ to 0, and $n/1$ to n.

Example Simplify $\frac{6}{6}, \frac{0}{6}$, and $\frac{-6}{1}$.

$$\frac{6}{6} = 1, \qquad \frac{0}{6} = 0, \qquad \frac{-6}{1} = -6$$

Practice Exercise

3. Simplify $\frac{0}{18}, \frac{-18}{-18}$, and $\frac{18}{1}$.

Objective 3.5b Simplify fraction notation.

Example Simplify: $\frac{315}{1650}$.

Using the test for divisibility by 5, we see that both the numerator and the denominator are divisible by 5:

$$\frac{315}{1650} = \frac{5 \cdot 63}{5 \cdot 330} = \frac{5}{5} \cdot \frac{63}{330} = 1 \cdot \frac{63}{330}$$

$$= \frac{63}{330} = \frac{3 \cdot 21}{3 \cdot 110} = \frac{3}{3} \cdot \frac{21}{110} = 1 \cdot \frac{21}{110} = \frac{21}{110}.$$

Practice Exercise

4. Simplify: $\frac{100}{280}$.

Objective 3.5c Use the test for equality to determine whether two fractions name the same number.

Example Use $=$ or $\neq$ for $\square$ to write a true sentence:

$$\frac{10}{54} \square \frac{15}{81}.$$

We find the cross products: $10 \cdot 81 = 810$ and $54 \cdot 15 = 810$. Because the cross products are the same, we have

$$\frac{10}{54} = \frac{15}{81}.$$

If the cross products had been different, the fractions would not be equal.

Practice Exercise

5. Use $=$ or $\neq$ for $\square$ to write a true sentence:

$$\frac{8}{48} \square \frac{6}{44}.$$

Objective 3.6a Multiply and simplify using fraction notation.

Example Multiply and simplify: $-\frac{7}{16} \cdot \left(-\frac{40}{49}\right)$.

$$-\frac{7}{16} \cdot \left(-\frac{40}{49}\right) = \frac{7 \cdot 40}{16 \cdot 49} = \frac{7 \cdot 2 \cdot 2 \cdot 2 \cdot 5}{2 \cdot 2 \cdot 2 \cdot 2 \cdot 7 \cdot 7}$$

$$= \frac{2 \cdot 2 \cdot 2 \cdot 7}{2 \cdot 2 \cdot 2 \cdot 7} \cdot \frac{5}{2 \cdot 7} = 1 \cdot \frac{5}{14} = \frac{5}{14}$$

Practice Exercise

6. Multiply and simplify: $\frac{80}{3} \cdot \left(-\frac{21}{72}\right)$.

Objective 3.7b Divide and simplify using fraction notation.

Example Divide and simplify: $\frac{9}{20} \div \frac{18}{25}$.

$$\frac{9}{20} \div \frac{18}{25} = \frac{9}{20} \cdot \frac{25}{18} = \frac{9 \cdot 25}{20 \cdot 18} = \frac{3 \cdot 3 \cdot 5 \cdot 5}{2 \cdot 2 \cdot 5 \cdot 2 \cdot 3 \cdot 3}$$

$$= \frac{3 \cdot 3 \cdot 5}{3 \cdot 3 \cdot 5} \cdot \frac{5}{2 \cdot 2 \cdot 2} = 1 \cdot \frac{5}{8} = \frac{5}{8}$$

Practice Exercise

7. Divide and simplify: $\frac{9}{4} \div \frac{45}{14}$.

Objective 3.7d Solve applied problems involving division of fractions.

Example A rental car had 18 gal of gasoline when its gas tank was $\frac{6}{7}$ full. How much could the tank hold when full?

The equation that corresponds to the situation is

$$\frac{6}{7} \cdot g = 18.$$

We divide by $\frac{6}{7}$ on both sides and carry out the division:

$$g = 18 \div \frac{6}{7} = \frac{18}{1} \cdot \frac{7}{6} = \frac{18 \cdot 7}{1 \cdot 6}$$

$$= \frac{3 \cdot 6 \cdot 7}{1 \cdot 6} = \frac{6}{6} \cdot \frac{3 \cdot 7}{1} = 1 \cdot \frac{21}{1} = 21.$$

The rental car can hold 21 gal of gasoline.

Practice Exercise

8. A flower vase has $\frac{7}{4}$ cups of water in it when it is $\frac{3}{4}$ full. How much can it hold when full?

Review Exercises

Find all the factors of each number. [3.1a]

1. 60 **2.** 176

3. Multiply by 1, 2, 3, and so on, to find ten multiples of 8. [3.1b]

4. Determine whether 924 is divisible by 11. [3.1b]

5. Determine whether 1800 is divisible by 16. [3.1b]

Determine whether each number is prime, composite, or neither. [3.1c]

6. 37 **7.** 1 **8.** 91

Find the prime factorization of each number. [3.1d]

9. 70 **10.** 30

11. 45 **12.** 150

13. 648 **14.** 5250

To do Exercises 15–22, consider the following numbers:

140	716	93	2802
95	2432	330	711
182	4344	255,555	
475	600	780	

Which of the above are divisible by the given number? [3.2a]

15. 3 **16.** 2

17. 4 **18.** 8

19. 5 **20.** 6

21. 9 **22.** 10

23. Identify the numerator and the denominator of $\frac{2}{7}$. [3.3a]

What part of each object is shaded? [3.3a]

24.

25.

26. What part of the set of objects is shaded? [3.3a]

27. For a committee in the United States Senate that consists of 3 Democrats and 5 Republicans, what is the ratio of: [3.3a]

a) Democrats to Republicans?
b) Republicans to Democrats?
c) Democrats to the total number of committee members?

Simplify. [3.3b], [3.5b]

28. $\frac{12}{30}$

29. $\frac{7}{28}$

30. $\frac{-23}{-23}$

31. $\frac{0}{-25}$

32. $\frac{1170}{1200}$

33. $\frac{18}{1}$

34. $\frac{9}{-27}$

35. $-\frac{88}{184}$

36. $\frac{18}{0}$

37. $\frac{48}{8}$

38. $\frac{140}{490}$

39. $-\frac{288}{2025}$

40. Simplify the fractions on this circle graph, if possible. [3.5b]

Museums in the United States

Science $\frac{15}{100}$

History $\frac{38}{100}$

Art $\frac{24}{100}$

Other $\frac{23}{100}$

Use $=$ or $\neq$ for $\square$ to write a true sentence. [3.5c]

41. $\frac{3}{5} \square \frac{4}{6}$

42. $\frac{-4}{7} \square -\frac{8}{14}$

43. $-\frac{4}{5} \square -\frac{5}{6}$

44. $\frac{4}{3} \square \frac{28}{21}$

Multiply and simplify. [3.4a], [3.6a]

45. $4 \cdot \frac{3}{8}$

46. $\frac{7}{3} \cdot 24$

47. $-9 \cdot \frac{5}{18}$

48. $\frac{6}{5} \cdot (-20)$

49. $\frac{3}{4} \cdot \frac{8}{9}$

50. $\frac{5}{7} \cdot \frac{1}{10}$

51. $-\frac{3}{7} \cdot \frac{14}{9}$

52. $\frac{1}{4} \cdot \frac{2}{11}$

53. $\frac{4}{25} \cdot \frac{15}{16}$

54. $-\frac{11}{3} \cdot \left(-\frac{30}{77}\right)$

Find the reciprocal. [3.7a]

55. $\frac{4}{5}$

56. -3

57. $\frac{1}{9}$

58. $-\frac{47}{36}$

Divide and simplify. [3.7b]

59. $6 \div \frac{4}{3}$

60. $-\frac{5}{9} \div \frac{5}{18}$

61. $\frac{1}{6} \div \frac{1}{11}$

62. $\frac{3}{14} \div \left(-\frac{6}{7}\right)$

63. $-\frac{1}{4} \div \left(-\frac{1}{9}\right)$

64. $180 \div \frac{3}{5}$

65. $\frac{23}{25} \div \frac{23}{25}$

66. $-\frac{2}{3} \div \left(-\frac{3}{2}\right)$

Solve. [3.7c]

67. $\frac{5}{4} \cdot t = \frac{3}{8}$

68. $x \cdot \frac{2}{3} = -160$

Solve. [3.6b], [3.7d]

69. A road crew repaves $\frac{1}{12}$ mi of road each day. How long will it take the crew to repave a $\frac{3}{4}$-mi stretch of road?

70. *Level of Education and Median Income.* The median yearly income of someone with an associate's degree is approximately $\frac{3}{4}$ of the median income of someone with a bachelor's degree. If the median income for those with bachelor's degrees is \$42,780, what is the median income of those with associate's degrees?

Source: U.S. Census Bureau

71. After driving 600 km, the Youssi family has completed $\frac{3}{5}$ of their vacation. How long is the total trip?

72. Molly is making a pepper steak recipe that calls for $\frac{2}{3}$ cup of green bell peppers. How much would be needed to make $\frac{1}{2}$ recipe? 3 recipes?

73. Bernardo earns \$105 for working a full day. How much does he receive for working $\frac{1}{7}$ of a day?

74. A book bag requires $\frac{4}{5}$ yd of fabric. How many bags can be made from 48 yd?

75. Solve: $\frac{2}{13} \cdot x = -\frac{1}{2}$. [3.7c]

A. $-\frac{1}{13}$ B. -13

C. $-\frac{4}{13}$ D. $-\frac{13}{4}$

76. Multiply and simplify: $\frac{15}{26} \cdot \frac{13}{90}$. [3.6a]

A. $\frac{195}{234}$ B. $\frac{1}{12}$

C. $\frac{3}{36}$ D. $\frac{13}{156}$

Synthesis

77. In the division below, find a and b. [3.7b]

$$\frac{19}{24} \div \frac{a}{b} = \frac{187{,}853}{268{,}224}$$

78. A prime number that remains a prime number when its digits are reversed is called a **palindrome prime**. For example, 17 is a palindrome prime because both 17 and 71 are primes. Which of the following numbers are palindrome primes? [3.1c]

13, 91, 16, 11, 15, 24, 29, 101, 201, 37

Understanding Through Discussion and Writing

1. A student incorrectly insists that $\frac{2}{5} \div \frac{3}{4}$ is $\frac{15}{8}$. What mistake is he probably making? [3.7b]

2. Use the number 9432 to explain why the test for divisibility by 9 works. [3.2a]

3. A student claims that "taking $\frac{1}{2}$ of a number is the same as dividing by $\frac{1}{2}$." Explain the error in this reasoning. [3.7b]

4. On p. 149, we explained, using words and pictures, why $\frac{3}{5} \cdot \frac{3}{4}$ equals $\frac{9}{20}$. Present a similar explanation of why $\frac{2}{3} \cdot \frac{4}{7}$ equals $\frac{8}{21}$. [3.4a]

5. Without performing the division, explain why $5 \div \frac{1}{7}$ is a greater number than $5 \div \frac{2}{3}$. [3.5c], [3.7b]

6. If a fraction's numerator and denominator have no factors (other than 1) in common, can the fraction be simplified? Why or why not? [3.5b]

CHAPTER 3 Test

For Extra Help For step-by-step test solutions, access the Chapter Test Prep Videos in MyMathLab® or on YouTube (search "BittingerBasicEl" and click on "Channels").

1. Find all the factors of 300.

Determine whether each number is prime, composite, or neither.

2. 41

3. 14

Find the prime factorization of the number.

4. 18

5. 60

6. Determine whether 1784 is divisible by 8.

7. Determine whether 784 is divisible by 9.

8. Determine whether 5552 is divisible by 5.

9. Determine whether 2322 is divisible by 6.

10. Identify the numerator and the denominator of $\frac{4}{5}$.

11. What part is shaded?

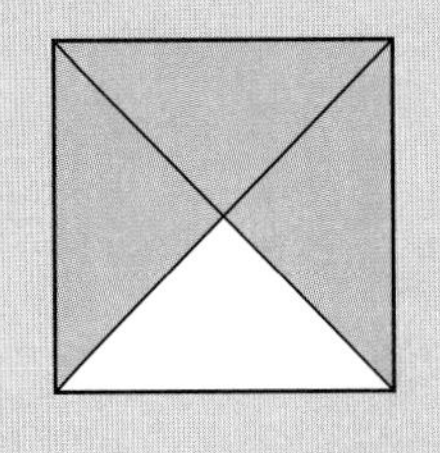

12. What part of the set is shaded?

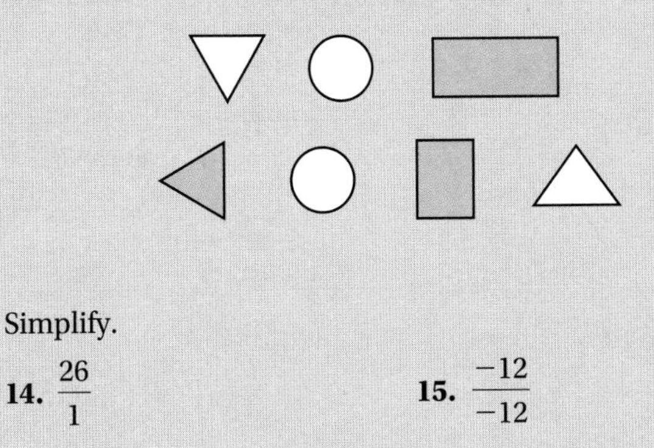

13. *Business Days.* There are approximately 259 business days in a 365-day year.

a) What is the ratio of business days to days in the year?

b) What is the ratio of nonbusiness days to days in the year?

Simplify.

14. $\frac{26}{1}$

15. $\frac{-12}{-12}$

16. $\frac{0}{16}$

17. $-\frac{12}{24}$

18. $\frac{42}{7}$

19. $\frac{9}{0}$

20. $\frac{7}{2-2}$

21. $-\frac{72}{108}$

Use $=$ or $\neq$ for $\square$ to write a true sentence.

22. $\frac{3}{4} \square \frac{6}{8}$

23. $-\frac{5}{4} \square -\frac{9}{7}$

Multiply and simplify.

24. $\frac{4}{3} \cdot 24$

25. $-5 \cdot \frac{3}{10}$

26. $\frac{2}{3} \cdot \frac{15}{4}$

27. $-\frac{22}{15} \cdot \left(-\frac{5}{33}\right)$

Find the reciprocal.

28. $\frac{5}{8}$

29. $-\frac{1}{4}$

30. 18

Divide and simplify.

31. $\frac{1}{5} \div \frac{1}{8}$

32. $12 \div \left(-\frac{2}{3}\right)$

33. $\frac{24}{5} \div \frac{28}{15}$

Solve.

34. $\frac{7}{8} \cdot x = -56$

35. $t \cdot \frac{2}{5} = \frac{7}{10}$

36. There are 7000 students at La Poloma College, and $\frac{5}{8}$ of them live in dorms. How many live in dorms?

37. A strip of taffy $\frac{9}{10}$ m long is cut into 12 equal pieces. What is the length of each piece?

38. A thermos of iced tea held 3 qt of tea when it was $\frac{3}{5}$ full. How much tea could it hold when full?

39. The pitch of a screw is $\frac{1}{8}$ in. How far will it go into a piece of walnut when it is turned 6 complete rotations?

40. In which figure does the shaded part represent $\frac{7}{6}$ of the figure?

A.

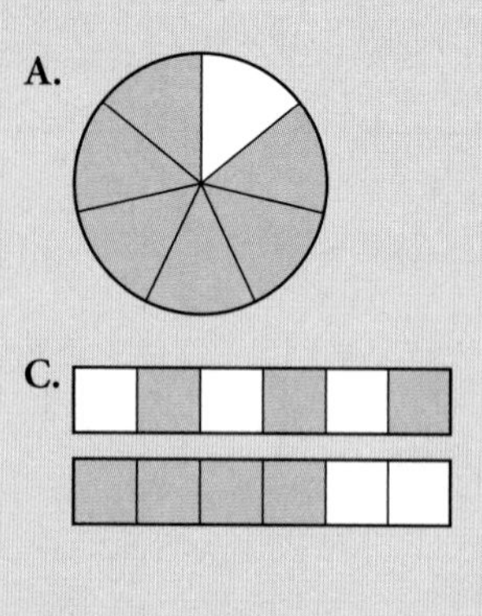

B.

C.

D.

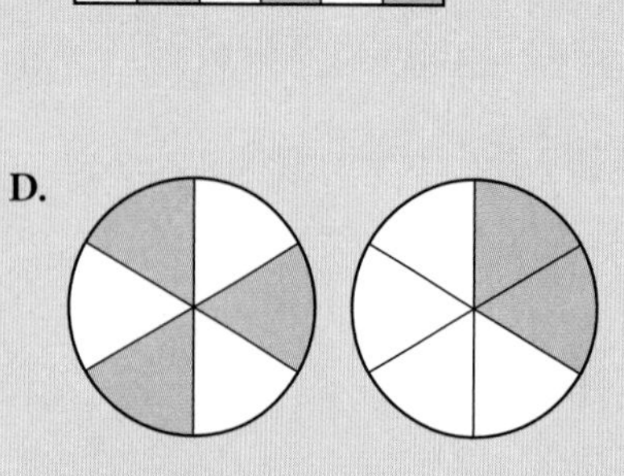

Synthesis

41. Grandma Hammons left $\frac{2}{3}$ of her $\frac{7}{8}$-acre apple farm to Karl. Karl gave $\frac{1}{4}$ of his share to his oldest daughter, Eileen. How much land did Eileen receive?

42. Simplify: $\left(\frac{3}{8}\right)^2 \div \frac{6}{7} \cdot \frac{2}{9} \div (-5)$.

CHAPTERS

1–3 Cumulative Review

1. Write a word name: 7,453,062.

2. What does the digit 4 mean in 23,746,591?

Simplify.

3. $\frac{5}{1}$

4. $\frac{-17}{-17}$

5. $-\frac{56}{42}$

6. $\frac{32}{0}$

Find the reciprocal.

7. 8

8. $-\frac{3}{2}$

Calculate and simplify.

9. $4658 + 729$

10. $-3 + (-9)$

11. $5042 - 3658$

12. $-9 - 7$

13. $457 \cdot 36$

14. $-8 \cdot 12$

15. $-\frac{5}{6} \cdot 15$

16. $\frac{3}{7} \cdot \frac{14}{9}$

17. $10{,}846 \div 24$

18. $-56 \div (-7)$

19. $-\frac{4}{7} \div 28$

20. $\frac{3}{4} \div \frac{9}{16}$

21. $8^2 \div 8 \cdot 2 - (2 + 2 \cdot 7)$

22. $108 \div 9 - [3(18 - 5 \cdot 3)]$

23. $-20 - 10 \div 5 + 2^3$

24. $(8 - 10^2) \div (5^2 - 2)$

25. Find the average of four test scores of 85, 91, 80, and 88.

26. Round 165,739 to the nearest thousand.

Evaluate.

27. 9^2

28. 5^3

29. 2^4

Use either $<$ or $>$ for $\square$ to write a true sentence.

30. $-26 \; \square \; 2$

31. $19 \; \square \; 17$

Find the absolute value.

32. $|33|$ **33.** $|-86|$ **34.** $|0|$

Find the opposite, or additive inverse.

35. 29 **36.** -144

37. Evaluate $-(-x)$ when x is -7.

38. Graph 8 on the number line.

$-2\ -1\ 0\ 1\ 2\ 3\ 4\ 5\ 6\ 7\ 8\ 9\ 10$

Determine whether each number is prime, composite, or neither.

39. 29 **40.** 1 **41.** 24

Find the prime factorization of each number.

42. 36 **43.** 135 **44.** 1350

45. Determine whether 6330 is divisible by 2; by 3; by 4; by 5; by 6; by 8; by 9; by 10.

Use = or ≠ for □ to write a true sentence.

46. $\frac{5}{8} \square \frac{7}{11}$ **47.** $-\frac{5}{4} \square -\frac{20}{16}$

Solve.

48. $17 + x = 61$ **49.** $\frac{3}{5} \cdot y = -\frac{9}{25}$

50. *Halley's Comet.* Halley's Comet passes by Earth every 76 yr. It last appeared to people on Earth in 1986. When will it appear again?

Source: *The Cambridge Factfinder*

51. *Fundraiser.* To support a school fundraiser, Mary Ann bought 4 rolls of wrapping paper at \$5 each and 2 candles at \$8 each. Find the total cost of Mary Ann's order.

52. *Chemical Reaction.* The temperature of a chemical compound was 5°C at 9:00 A.M. During a reaction, it dropped 2°C per minute until 9:11 A.M. What was the temperature at 9:11 A.M.?

53. *Working Students.* An instructor determines that $\frac{2}{5}$ of the 35 students in her accounting class at Beckwith Community College have full-time jobs. How many students have full-time jobs?

Synthesis

54. Find a pair of factors whose product is 100 and:

a) whose sum is 25.
b) whose sum is -29.
c) whose difference is 0.

Fraction Notation and Mixed Numerals

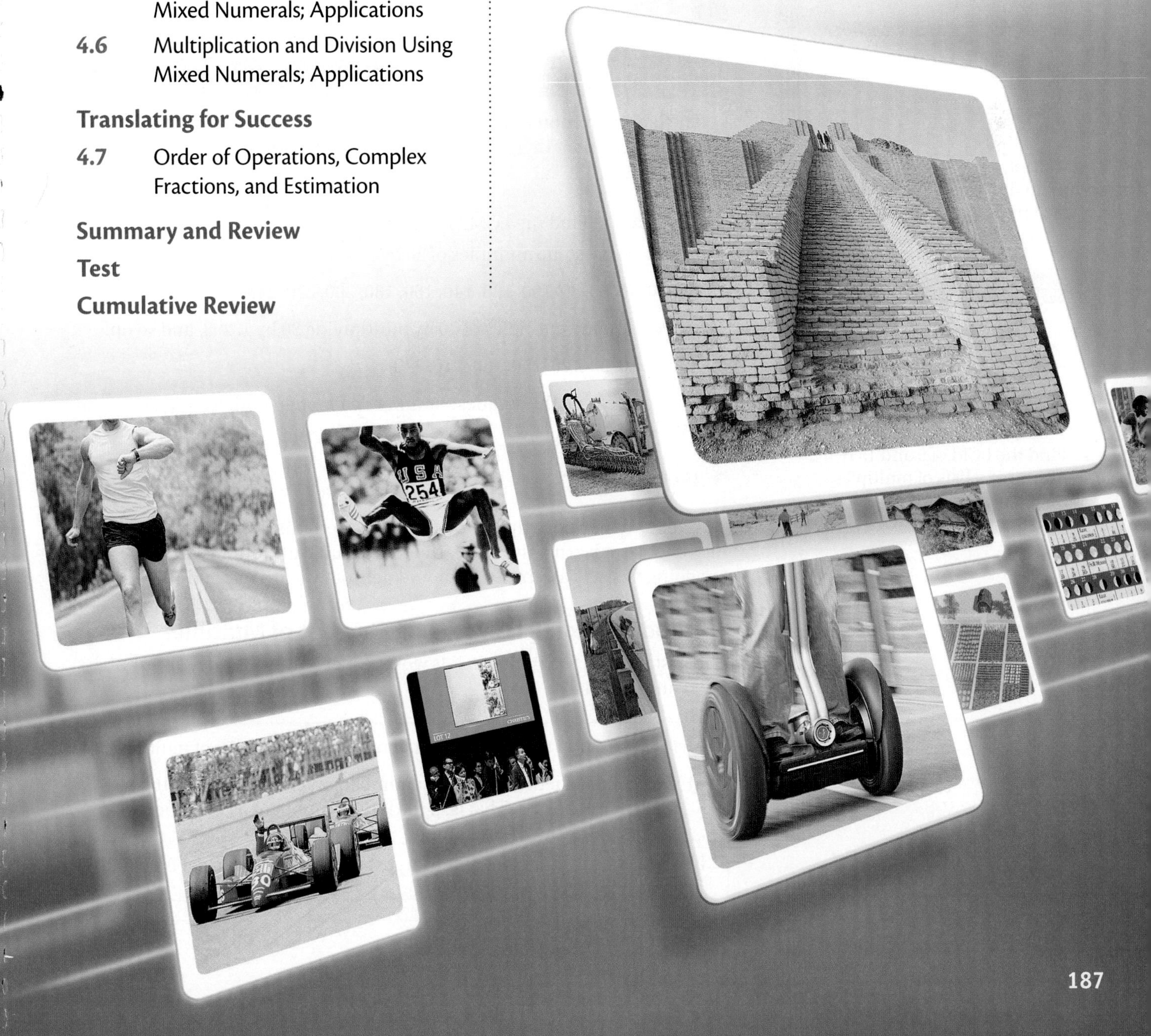

STUDYING FOR SUCCESS *Preparing for a Test*

- ☐ Make up your own test questions as you study.
- ☐ Do an overall review of the chapter focusing on the objectives and the examples.
- ☐ Do the exercises in the Mid-Chapter Review and in the end-of-chapter Review.
- ☐ Take the chapter test at the end of the chapter.

4.1 Least Common Multiples

OBJECTIVE

a Find the least common multiple, or LCM, of two or more numbers.

SKILL TO REVIEW

Objective 3.1b: Find some multiples of a number.

Multiply by 1, 2, 3, and so on, to find six multiples of each number.

1. 8 **2.** 25

When we want to add or subtract fractions, the fractions must share a common denominator. If necessary, we find equivalent fractions using the **least common denominator (LCD)**, or **least common multiple (LCM)** of the denominators.

a FINDING LEAST COMMON MULTIPLES

LEAST COMMON MULTIPLE, LCM

The **least common multiple**, or LCM, of two natural numbers is the smallest number that is a multiple of both numbers.

EXAMPLE 1 Find the LCM of 20 and 30.

First, we list some multiples of 20 by multiplying 20 by 1, 2, 3, and so on:

20, 40, 60, 80, 100, 120, 140, 160, 180, 200, 220, 240, ...

Then we list some multiples of 30 by multiplying 30 by 1, 2, 3, and so on:

30, 60, 90, 120, 150, 180, 210, 240, ...

Now we determine the smallest number *common* to both lists. The LCM of 20 and 30 is 60.

1. Find the LCM of 9 and 15 by examining lists of multiples.

◀ **Do Margin Exercise 1.**

Next, we develop three more efficient methods for finding LCMs. You may choose to learn only one method. (Consult with your instructor.) If you are going to study algebra, you should definitely learn method 2.

Method 1: Finding LCMs Using One List of Multiples

The first method for finding LCMs works especially well when the numbers are relatively small.

Method 1. To find the LCM of a set of numbers using a list of multiples:

a) Determine whether the largest number is a multiple of the others. If it is, it is the LCM. That is, if the largest number has the others as factors, the LCM is that number.

b) If not, check multiples of the largest number until you get one that is a multiple of each of the others.

Answers

Skill to Review:
1. 8, 16, 24, 32, 40, 48
2. 25, 50, 75, 100, 125, 150

Margin Exercise:
1. 45

EXAMPLE 2 Find the LCM of 12 and 15.

a) 15 is the larger number, but it is not a multiple of 12.

b) Check multiples of 15:

$2 \cdot 15 = 30,$ Not a multiple of 12

$3 \cdot 15 = 45,$ Not a multiple of 12

$4 \cdot 15 = 60.$ A multiple of 12: $5 \cdot 12 = 60$

The LCM $= 60$.

EXAMPLE 3 Find the LCM of 8 and 32.

32 is a multiple of 8 $(4 \cdot 8 = 32)$, so the LCM $= 32$.

EXAMPLE 4 Find the LCM of 10, 20, and 50.

a) 50 is a multiple of 10 but not a multiple of 20.

b) Check multiples of 50:

$2 \cdot 50 = 100.$ A multiple of 10 and of 20: $10 \cdot 10 = 100$ and $5 \cdot 20 = 100$

The LCM $= 100$.

Do Exercises 2–5.

Find the LCM.

2. 10, 15

3. 6, 8

4. 5, 10

5. 20, 45, 80

Method 2: Finding LCMs Using Prime Factorizations

A second method for finding LCMs uses prime factorizations. Consider again 20 and 30. Their prime factorizations are $20 = 2 \cdot 2 \cdot 5$ and $30 = 2 \cdot 3 \cdot 5$. Any multiple of 20 will have to have *two* 2's as factors and *one* 5 as a factor. Any multiple of 30 will need to have *one* 2, *one* 3, and *one* 5 as factors. The smallest number satisfying these conditions is

Two 2's, one 5; $2 \cdot 2 \cdot 3 \cdot 5$ is a multiple of 20.

$2 \cdot 2 \cdot 3 \cdot 5.$

One 2, one 3, one 5; $2 \cdot 2 \cdot 3 \cdot 5$ is a multiple of 30.

Thus the LCM of 20 and 30 is $2 \cdot 2 \cdot 3 \cdot 5$, or 60. It has all the factors of 20 and all the factors of 30, but the factors are not repeated when they are common to both numbers.

Note that each prime factor is used the greatest number of times that it occurs in either of the individual factorizations.

> *Method 2.* To find the LCM of a set of numbers using prime factorizations:
>
> **a)** Write the prime factorization of each number.
>
> **b)** Create a product, using each prime factor the greatest number of times that it occurs in any one factorization.

EXAMPLE 5 Find the LCM of 6 and 8.

a) Write the prime factorization of each number.

$6 = 2 \cdot 3, \quad 8 = 2 \cdot 2 \cdot 2$ The prime factors are 2's and 3's.

Answers

2. 30 **3.** 24 **4.** 10 **5.** 720

Use prime factorizations to find each LCM.

6. 8, 10

7. 18, 40

GS

a) $18 = 2 \cdot 3 \cdot$ ☐

$40 = 2 \cdot 2 \cdot 2 \cdot$ ☐

b) Consider the factor 2. The greatest number of times that 2 occurs in any one factorization is ☐ times. Write 2 as a factor ☐ times.

$2 \cdot 2 \cdot 2 \cdot ?$

Consider the factor 3. The greatest number of times that 3 occurs in any one factorization is ☐ times. Write 3 as a factor ☐ times.

$2 \cdot 2 \cdot 2 \cdot 3 \cdot 3 \cdot ?$

Consider the factor 5. The greatest number of times that 5 occurs in any one factorization is ☐ time. Write 5 as a factor ☐ time.

$2 \cdot 2 \cdot 2 \cdot 3 \cdot 3 \cdot 5$

LCM = ☐

8. 32, 54

b) Create a product with the prime factors 2 and 3, using each the greatest number of times that it occurs in any one factorization.

Consider the factor 2. The greatest number of times that 2 occurs in any one factorization is three times. We write 2 as a factor three times.

$2 \cdot 2 \cdot 2 \cdot ?$

Consider the factor 3. The greatest number of times that 3 occurs in any one factorization is one time. We write 3 as a factor one time.

$2 \cdot 2 \cdot 2 \cdot 3$

Since there are no other prime factors in either factorization, the

LCM is $2 \cdot 2 \cdot 2 \cdot 3$, or 24. ■

EXAMPLE 6 Find the LCM of 24 and 36.

a) Write the prime factorization of each number.

$24 = 2 \cdot 2 \cdot 2 \cdot 3, \quad 36 = 2 \cdot 2 \cdot 3 \cdot 3$ The prime factors are 2's and 3's.

b) Create a product with the prime factors 2 and 3, using each the greatest number of times that it occurs in any one factorization.

Consider the factor 2. The greatest number of times that 2 occurs in any one factorization is three times. We write 2 as a factor three times.

$2 \cdot 2 \cdot 2 \cdot ?$

Consider the factor 3. The greatest number of times that 3 occurs in any one factorization is two times. We write 3 as a factor two times.

$2 \cdot 2 \cdot 2 \cdot 3 \cdot 3$

Since there are no other prime factors in either factorization, the

LCM is $2 \cdot 2 \cdot 2 \cdot 3 \cdot 3$, or 72.

◀ **Do Exercises 6–8.**

EXAMPLE 7 Find the LCM of 81, 90, and 84.

a) Write the prime factorization of each number.

$81 = 3 \cdot 3 \cdot 3 \cdot 3,$

$90 = 2 \cdot 3 \cdot 3 \cdot 5,$

$84 = 2 \cdot 2 \cdot 3 \cdot 7$

The prime factors are 2's, 3's, 5's, and 7's.

b) Create a product with the prime factors 2, 3, 5, and 7, using each the greatest number of times that it occurs in any one factorization.

Consider the factor 2. The greatest number of times that 2 occurs in any one factorization is two times. We write 2 as a factor two times.

$2 \cdot 2 \cdot ?$

Consider the factor 3. The greatest number of times that 3 occurs in any one factorization is four times. We write 3 as a factor four times.

$2 \cdot 2 \cdot 3 \cdot 3 \cdot 3 \cdot 3 \cdot ?$

Answers

6. 40 **7.** 360 **8.** 864

Guided Solution:

7. **(a)** 3, 5; **(b)** three, three; two, two; one, one; 360

Consider the factor 5. The greatest number of times that 5 occurs in any one factorization is one time. We write 5 as a factor one time.

$$2 \cdot 2 \cdot 3 \cdot 3 \cdot 3 \cdot 3 \cdot 5 \cdot ?$$

Consider the factor 7. The greatest number of times that 7 occurs in any one factorization is one time. We write 7 as a factor one time.

$$2 \cdot 2 \cdot 3 \cdot 3 \cdot 3 \cdot 3 \cdot 5 \cdot 7$$

Since there are no other prime factors in any of the factorizations, the

LCM is $2 \cdot 2 \cdot 3 \cdot 3 \cdot 3 \cdot 3 \cdot 5 \cdot 7$, or 11,340.

Do Exercise 9.

9. Find the LCM of 24, 35, and 45.

EXAMPLE 8 Find the LCM of 8 and 9.

We write the prime factorization of each number.

$$8 = 2 \cdot 2 \cdot 2, \qquad 9 = 3 \cdot 3$$

Note that the two numbers, 8 and 9, have no common prime factor. When this is the case, the LCM is just the product of the two numbers. Thus the LCM is $2 \cdot 2 \cdot 2 \cdot 3 \cdot 3$, or $8 \cdot 9$, or 72.

Do Exercises 10 and 11.

Find the LCM.

10. 4, 9

11. 6, 25

EXAMPLE 9 Find the LCM of 7 and 21.

We write the prime factorization of each number. Because 7 is prime, it has no prime factorization.

$$7 = 7, \qquad 21 = 3 \cdot 7$$

Note that 7 is a factor of 21. We stated earlier that if one number is a factor of another, the LCM is the larger of the numbers. Thus the LCM is $7 \cdot 3$, or 21.

Do Exercises 12 and 13.

Find the LCM.

12. 3, 18

13. 12, 24

Let's reconsider Example 7 using exponents. We write the prime factorizations of 81, 90, and 84 using exponential notation. The largest exponents indicate the greatest number of times that 2, 3, 5, and 7 occur as factors.

$81 = 3 \cdot 3 \cdot 3 \cdot 3 = 3^4$; 4 is the largest exponent of 3 in any of the factorizations.

$90 = 2 \cdot 3 \cdot 3 \cdot 5 = 2^1 \cdot 3^2 \cdot 5^1$; 1 is the largest exponent of 5 in any of the factorizations.

$84 = 2 \cdot 2 \cdot 3 \cdot 7 = 2^2 \cdot 3^1 \cdot 7^1$. 2 is the largest exponent of 2 and 1 is the largest exponent of 7 in any of the factorizations.

Thus the LCM is $2^2 \cdot 3^4 \cdot 5^1 \cdot 7^1$, or 11,340.

EXAMPLE 10 Find the LCM of 25, 40, and 45 using exponential notation.

a) Write the prime factorization of each number.

$$25 = 5 \cdot 5 = 5^2,$$
$$40 = 2 \cdot 2 \cdot 2 \cdot 5 = 2^3 \cdot 5,$$
$$45 = 3 \cdot 3 \cdot 5 = 3^2 \cdot 5$$

Answers

9. 2520 **10.** 36 **11.** 150 **12.** 18 **13.** 24

b) Create a product with the prime factors 2, 3, and 5. Lining up all the powers of 2, all the powers of 3, and all the powers of 5 can help us construct the LCM.

$$\begin{aligned} 25 &= \qquad\qquad 5^2 \\ 40 &= 2^3 \cdot \qquad 5 \\ 45 &= \qquad 3^2 \cdot 5 \end{aligned}$$

The LCM is formed by choosing the greatest power of each factor, so the

$$\text{LCM} = 2^3 \cdot 3^2 \cdot 5^2 = 1800.$$

14. Use exponents to find the LCM of 24, 35, and 45.

15. Redo Margin Exercises 6–8 using exponents.

◀ **Do Exercises 14 and 15.**

Method 3: Finding LCMs Using Division by Primes

The third method is especially useful for finding the LCM of three or more numbers.

> *Method 3.* To find the LCM using division by primes:
>
> **a)** First look for any prime that divides at least two of the numbers with no remainder. Then divide, bringing down any numbers not divisible by the prime. If you cannot find a prime that divides at least two of the numbers, then the LCM is the product of the numbers.
>
> **b)** Repeat the process until you can divide no more—that is, until there are no two numbers divisible by the same prime.

EXAMPLE 11 Find the LCM of 48, 72, and 80.

We first look for any prime that divides any two of the numbers with no remainder. Then we divide. We repeat the process, bringing down any numbers not divisible by the prime, until we can divide no more—that is, until there are no two numbers divisible by the same prime:

```
2 | 48 72 80
3 | 24 36 40
2 |  8 12 40     40 is not divisible by 3.
2 |  4  6 20
2 |  2  3 10
     1  3  5     3 is not divisible by 2.
```

No two of 1, 3, and 5 are divisible by the same prime. We stop here. The LCM is

$$2 \cdot 3 \cdot 2 \cdot 2 \cdot 2 \cdot 1 \cdot 3 \cdot 5, \quad \text{or} \quad 720.$$

Find the LCM using division by primes.

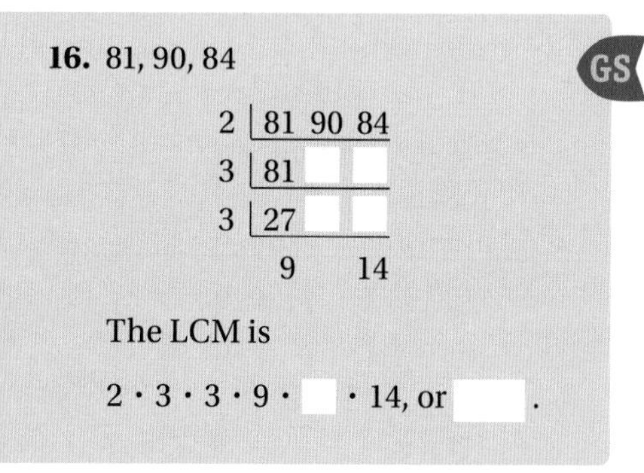

17. 24, 35, 45

18. 12, 75, 80, 120

EXAMPLE 12 Find the LCM of 12, 18, 21, and 40.

```
3 | 12 18 21 40
2 |  4  6  7 40
2 |  2  3  7 20
     1  3  7 10
```

No two of 1, 3, 7, and 10 are divisible by the same prime. We stop here. The LCM is $3 \cdot 2 \cdot 2 \cdot 1 \cdot 3 \cdot 7 \cdot 10$, or 2520.

◀ **Do Exercises 16–18.**

Answers

14. $2^3 \cdot 3^2 \cdot 5 \cdot 7$, or 2520 15. $2^3 \cdot 5$, or 40; $2^3 \cdot 3^2 \cdot 5$, or 360; $2^5 \cdot 3^3$, or 864 16. 11,340 17. 2520 18. 1200

Guided Solution:
16. 45, 42, 15, 14, 5; 5, 11,340

4.1 Exercise Set

For Extra Help MyMathLab® MathXL® PRACTICE WATCH READ REVIEW

Reading Check

Determine whether each statement is true or false.

_______________ **RC1.** The least common denominator of two fractions is the LCM of their denominators.

_______________ **RC2.** If one number is a multiple of a second number, the larger number is the LCM of the two numbers.

_______________ **RC3.** If two numbers have no common prime factor, then the LCM of the numbers is their product.

_______________ **RC4.** LCMs cannot be found using prime factorizations.

a Find the LCM of each set of numbers.

1. 2, 4

2. 3, 15

3. 10, 25

4. 10, 15

5. 20, 40

6. 8, 12

7. 18, 27

8. 9, 11

9. 30, 50

10. 24, 36

11. 30, 40

12. 21, 27

13. 18, 24

14. 12, 18

15. 60, 70

16. 35, 45

17. 16, 36

18. 18, 20

19. 32, 36

20. 36, 48

21. 2, 3, 5

22. 3, 5, 7

23. 5, 18, 3

24. 6, 12, 18

25. 24, 36, 12

26. 8, 16, 22

27. 5, 12, 15

28. 12, 18, 40

29. 9, 12, 6

30. 8, 16, 12

31. 180, 100, 450, 60

32. 18, 30, 50, 48

33. 8, 48

34. 16, 32

35. 5, 50

36. 12, 72 **37.** 11, 13 **38.** 13, 14 **39.** 12, 35 **40.** 23, 25

41. 54, 63 **42.** 56, 72 **43.** 81, 90 **44.** 75, 100 **45.** 36, 54, 80

46. 22, 42, 51 **47.** 39, 91, 108, 26 **48.** 625, 75, 500, 25 **49.** 2000, 3000 **50.** 300, 4000

Applications of LCMs: Planet Orbits. Jupiter, Saturn, and Uranus all revolve around the sun. Jupiter takes 12 years, Saturn 30 years, and Uranus 84 years to make a complete revolution. On a certain night, you look at Jupiter, Saturn, and Uranus and wonder how many years it will take before they have the same position again. (*Hint*: To find out, you find the LCM of 12, 30, and 84. It will be that number of years.)

Source: *The Handy Science Answer Book*

51. How often will Jupiter and Saturn appear in the same direction in the night sky as seen from the earth?

52. How often will Jupiter and Uranus appear in the same direction in the night sky as seen from the earth?

53. How often will Saturn and Uranus appear in the same direction in the night sky as seen from the earth?

54. How often will Jupiter, Saturn, and Uranus appear in the same direction in the night sky as seen from the earth?

Skill Maintenance

55. Multiply and simplify: $\frac{6}{5} \cdot (-15)$. [3.6a]

56. Divide: $7865 \div 132$. [1.5a]

57. Divide and simplify: $\frac{4}{5} \div \frac{7}{10}$. [3.7b]

58. Add: 23,456 + 5677 + 4002. [1.2a]

59. Multiply: 2118×3001. [1.4a]

60. Subtract: 80,004 − 2305. [1.3a]

Synthesis

61. A pencil company uses two sizes of boxes, 5 in. by 6 in. and 5 in. by 8 in. These boxes are packed in bigger cartons for shipping. Find the width and the length of the smallest carton that will accommodate boxes of either size without any room left over. (Each carton can contain only one type of box and all boxes must point in the same direction.)

62. Consider 8 and 12. Determine whether each of the following is the LCM of 8 and 12. Tell why or why not.

a) $2 \cdot 2 \cdot 3 \cdot 3$
b) $2 \cdot 2 \cdot 3$
c) $2 \cdot 3 \cdot 3$
d) $2 \cdot 2 \cdot 2 \cdot 3$

Addition and Applications

4.2

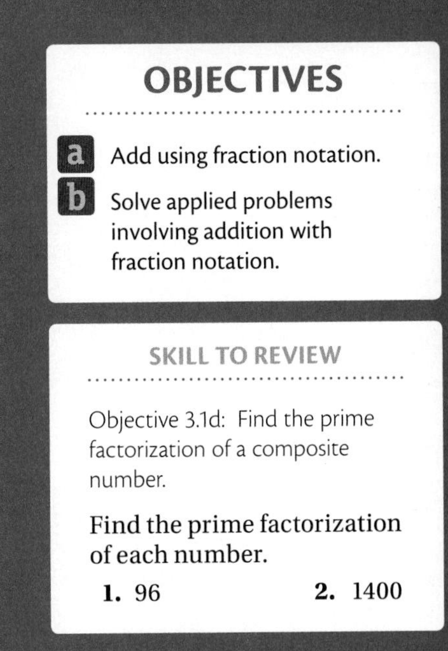

OBJECTIVES

a Add using fraction notation.

b Solve applied problems involving addition with fraction notation.

SKILL TO REVIEW

Objective 3.1d: Find the prime factorization of a composite number.

Find the prime factorization of each number.

1. 96 **2.** 1400

a ADDITION USING FRACTION NOTATION

Like Denominators

Addition using fraction notation corresponds to combining or putting like things together, just as addition with whole numbers does. For example,

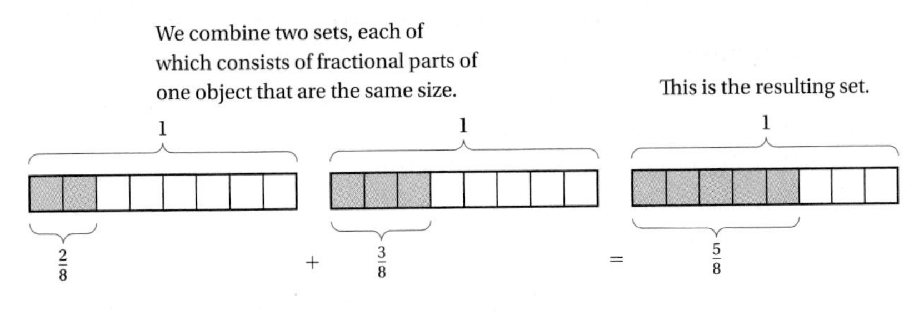

$$2 \text{ eighths} + 3 \text{ eighths} = 5 \text{ eighths},$$

or $\quad 2 \cdot \frac{1}{8} + 3 \cdot \frac{1}{8} = 5 \cdot \frac{1}{8}, \quad$ or $\quad \frac{2}{8} + \frac{3}{8} = \frac{5}{8}.$

We see that to add when denominators are the same, we add the numerators and keep the denominator.

Do Margin Exercise 1. ▶

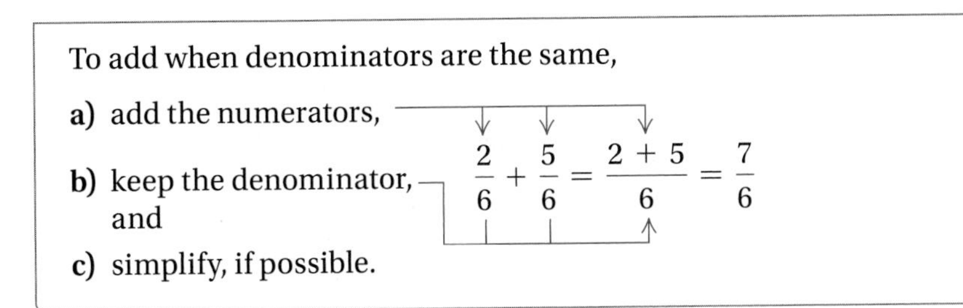

To add when denominators are the same,

a) add the numerators,

b) keep the denominator, and

c) simplify, if possible.

$$\frac{2}{6} + \frac{5}{6} = \frac{2+5}{6} = \frac{7}{6}$$

EXAMPLES Add and simplify.

1. $\frac{2}{4} + \frac{1}{4} = \frac{2+1}{4} = \frac{3}{4}$ No simplifying is possible.

2. $\frac{11}{6} + \frac{3}{6} = \frac{11+3}{6} = \frac{14}{6} = \frac{2 \cdot 7}{2 \cdot 3} = \frac{2}{2} \cdot \frac{7}{3} = 1 \cdot \frac{7}{3} = \frac{7}{3}$ Here we simplified.

3. $\frac{3}{12} + \frac{-5}{12} = \frac{3 + (-5)}{12} = \frac{-2}{12} = \frac{2 \cdot (-1)}{2 \cdot 6} = \frac{2}{2} \cdot \frac{-1}{6}$

$= 1 \cdot \frac{-1}{6} = \frac{-1}{6}, \text{ or } -\frac{1}{6}$ Recall that $\frac{-a}{b} = -\frac{a}{b}$.

Do Exercises 2–5. ▶

Different Denominators

What do we do when denominators are different? We can find a common denominator by multiplying by 1. Consider adding $\frac{1}{6}$ and $\frac{3}{4}$. There are many common denominators that can be obtained. Let's look at two possibilities.

1. Find $\frac{1}{5} + \frac{3}{5}$.

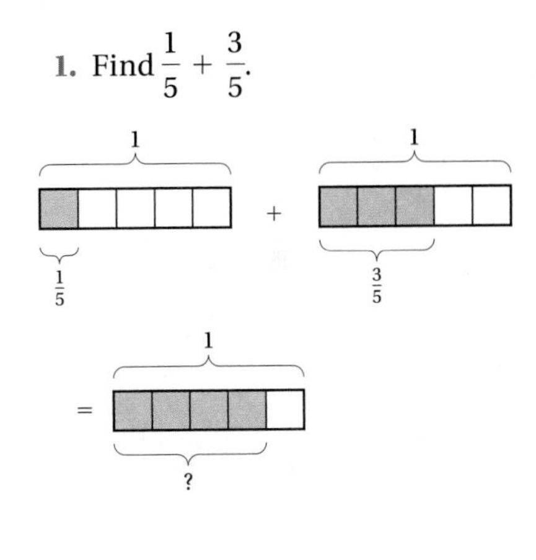

Add and simplify.

2. $\frac{5}{13} + \frac{9}{13}$

3. $\frac{1}{3} + \frac{2}{3}$

4. $\frac{5}{12} + \frac{1}{12}$

5. $\frac{9}{16} + \frac{-3}{16}$

Answers

Skill to Review:

1. $2 \cdot 2 \cdot 2 \cdot 2 \cdot 2 \cdot 3$ **2.** $2 \cdot 2 \cdot 2 \cdot 5 \cdot 5 \cdot 7$

Margin Exercises:

1. $\frac{4}{5}$ **2.** $\frac{14}{13}$ **3.** 1 **4.** $\frac{1}{2}$ **5.** $\frac{3}{8}$

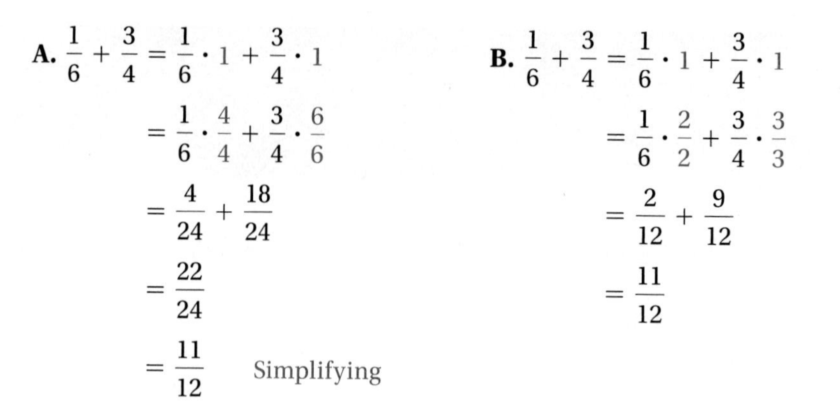

A. $\frac{1}{6} + \frac{3}{4} = \frac{1}{6} \cdot 1 + \frac{3}{4} \cdot 1$

$= \frac{1}{6} \cdot \frac{4}{4} + \frac{3}{4} \cdot \frac{6}{6}$

$= \frac{4}{24} + \frac{18}{24}$

$= \frac{22}{24}$

$= \frac{11}{12}$ Simplifying

B. $\frac{1}{6} + \frac{3}{4} = \frac{1}{6} \cdot 1 + \frac{3}{4} \cdot 1$

$= \frac{1}{6} \cdot \frac{2}{2} + \frac{3}{4} \cdot \frac{3}{3}$

$= \frac{2}{12} + \frac{9}{12}$

$= \frac{11}{12}$

We had to simplify at the end in (A) but not in (B). In (B), we used the least common multiple of the denominators, 12, as the common denominator. That number is called the **least common denominator**, or **LCD**. We may still need to simplify when using the LCD, but it is usually easier than when we use a larger common denominator.

To add when denominators are different:

a) Find the least common multiple of the denominators. That number is the least common denominator, LCD.

b) Multiply by 1, using an appropriate notation, n/n, to express each number in terms of the LCD.

c) Add the numerators, keeping the same denominator.

d) Simplify, if possible.

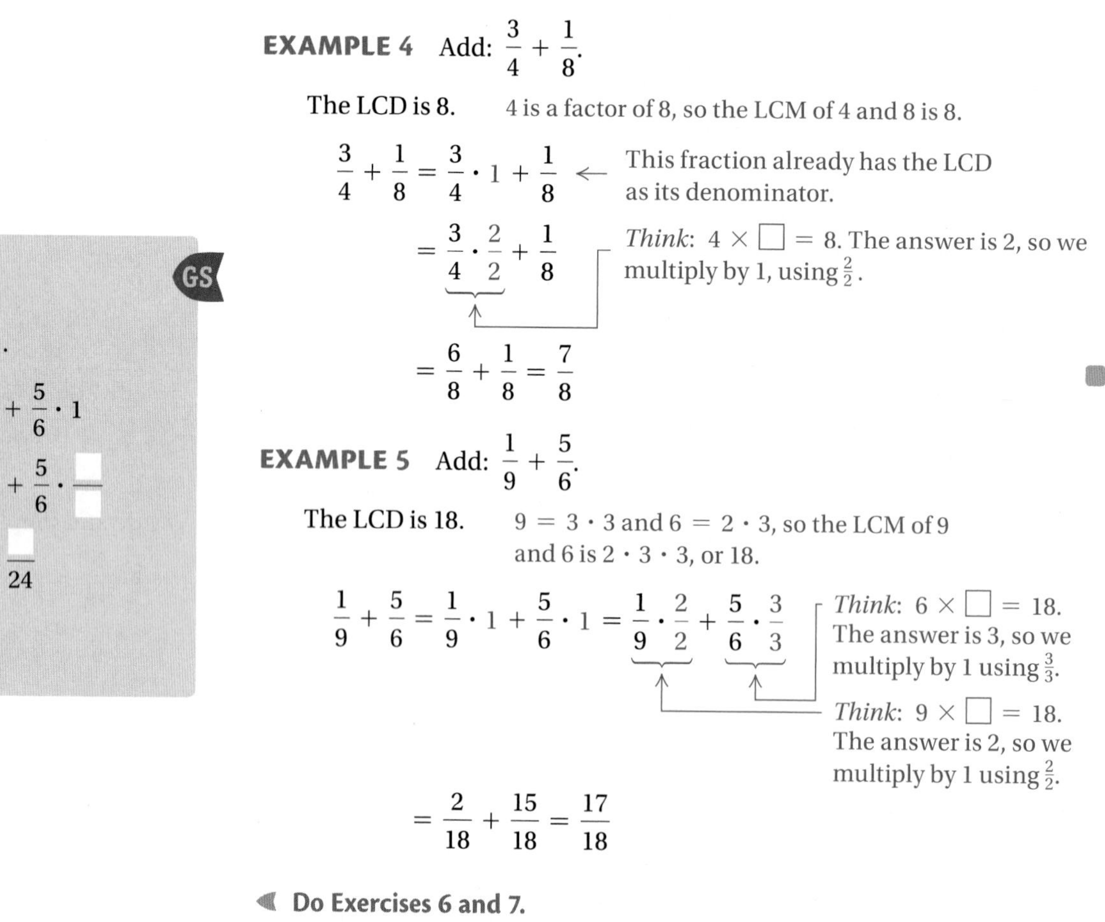

EXAMPLE 4 Add: $\frac{3}{4} + \frac{1}{8}$.

The LCD is 8. 4 is a factor of 8, so the LCM of 4 and 8 is 8.

$\frac{3}{4} + \frac{1}{8} = \frac{3}{4} \cdot 1 + \frac{1}{8}$ ← This fraction already has the LCD as its denominator.

$= \frac{3}{4} \cdot \frac{2}{2} + \frac{1}{8}$ *Think*: $4 \times \square = 8$. The answer is 2, so we multiply by 1, using $\frac{2}{2}$.

$= \frac{6}{8} + \frac{1}{8} = \frac{7}{8}$

EXAMPLE 5 Add: $\frac{1}{9} + \frac{5}{6}$.

The LCD is 18. $9 = 3 \cdot 3$ and $6 = 2 \cdot 3$, so the LCM of 9 and 6 is $2 \cdot 3 \cdot 3$, or 18.

$\frac{1}{9} + \frac{5}{6} = \frac{1}{9} \cdot 1 + \frac{5}{6} \cdot 1 = \frac{1}{9} \cdot \frac{2}{2} + \frac{5}{6} \cdot \frac{3}{3}$

Think: $6 \times \square = 18$. The answer is 3, so we multiply by 1 using $\frac{3}{3}$.

Think: $9 \times \square = 18$. The answer is 2, so we multiply by 1 using $\frac{2}{2}$.

$= \frac{2}{18} + \frac{15}{18} = \frac{17}{18}$

Do Exercises 6 and 7.

Add.

6. $\frac{2}{3} + \frac{1}{6}$

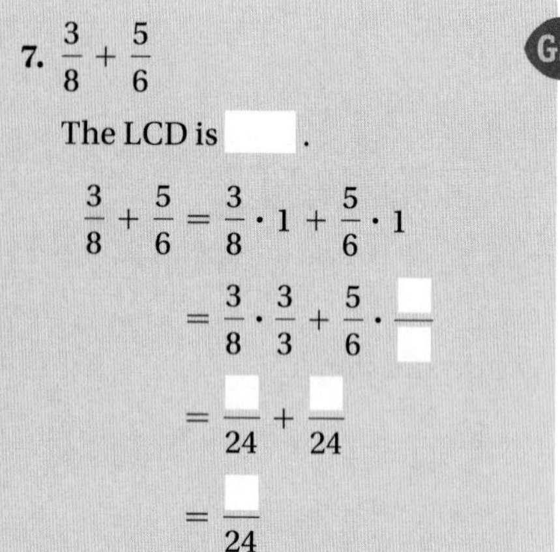

7. $\frac{3}{8} + \frac{5}{6}$

The LCD is ____.

$\frac{3}{8} + \frac{5}{6} = \frac{3}{8} \cdot 1 + \frac{5}{6} \cdot 1$

$= \frac{3}{8} \cdot \frac{3}{3} + \frac{5}{6} \cdot \frac{\square}{\square}$

$= \frac{\square}{24} + \frac{\square}{24}$

$= \frac{\square}{24}$

Answers

6. $\frac{5}{6}$ 7. $\frac{29}{24}$

Guided Solution:

7. 24; $\frac{4}{4}$, 9, 20, 29

EXAMPLE 6 Add: $\frac{5}{9} + \frac{11}{18}$.

The LCD is 18. 9 is a factor of 18, so the LCM is 18.

$$\frac{5}{9} + \frac{11}{18} = \frac{5}{9} \cdot \frac{2}{2} + \frac{11}{18} = \frac{10}{18} + \frac{11}{18}$$

$$= \frac{21}{18} = \frac{3 \cdot 7}{3 \cdot 6} = \frac{3}{3} \cdot \frac{7}{6} = \frac{7}{6} \quad \text{Simplifying}$$

Do Exercise 8.

8. Add: $\frac{1}{6} + \frac{7}{18}$.

EXAMPLE 7 Add: $\frac{1}{10} + \frac{3}{100} + \frac{-7}{1000}$.

Since 10 and 100 are factors of 1000, the LCD is 1000. Then

$$\frac{1}{10} + \frac{3}{100} + \frac{-7}{1000} = \frac{1}{10} \cdot \frac{100}{100} + \frac{3}{100} \cdot \frac{10}{10} + \frac{-7}{1000}$$

$$= \frac{100}{1000} + \frac{30}{1000} + \frac{-7}{1000} = \frac{100 + 30 + (-7)}{1000} = \frac{123}{1000}.$$

Do Exercise 9.

9. Add: $\frac{4}{10} + \frac{-1}{100} + \frac{3}{1000}$.

EXAMPLE 8 Add: $\frac{13}{70} + \frac{11}{21} + \frac{8}{15}$.

We use prime factorizations to determine the LCM of 70, 21, and 15:

$$70 = 2 \cdot 5 \cdot 7, \quad 21 = 3 \cdot 7, \quad 15 = 3 \cdot 5.$$

The LCD is $2 \cdot 3 \cdot 5 \cdot 7$, or 210.

$$\frac{13}{70} + \frac{11}{21} + \frac{8}{15} = \frac{13}{2 \cdot 5 \cdot 7} \cdot \frac{3}{3} + \frac{11}{3 \cdot 7} \cdot \frac{2 \cdot 5}{2 \cdot 5} + \frac{8}{3 \cdot 5} \cdot \frac{7 \cdot 2}{7 \cdot 2}$$

The LCM of 70, 21, and 15 is $2 \cdot 3 \cdot 5 \cdot 7$. In each case, think of which factors are needed to get the LCD. Then multiply by 1 to obtain the LCD in each denominator.

$$= \frac{13 \cdot 3}{2 \cdot 5 \cdot 7 \cdot 3} + \frac{11 \cdot 2 \cdot 5}{3 \cdot 7 \cdot 2 \cdot 5} + \frac{8 \cdot 7 \cdot 2}{3 \cdot 5 \cdot 7 \cdot 2}$$

$$= \frac{39}{2 \cdot 3 \cdot 5 \cdot 7} + \frac{110}{2 \cdot 3 \cdot 5 \cdot 7} + \frac{112}{2 \cdot 3 \cdot 5 \cdot 7}$$

$$= \frac{261}{2 \cdot 3 \cdot 5 \cdot 7} = \frac{3 \cdot 3 \cdot 29}{2 \cdot 3 \cdot 5 \cdot 7} \quad \text{Factoring the numerator}$$

$$= \frac{3}{3} \cdot \frac{3 \cdot 29}{2 \cdot 5 \cdot 7} = \frac{3 \cdot 29}{2 \cdot 5 \cdot 7} = \frac{87}{70}$$

Do Exercises 10 and 11.

Add.

10. $\frac{7}{10} + \frac{2}{21} + \frac{6}{7}$

11. $\frac{5}{18} + \frac{7}{24} + \frac{-11}{36}$

Answers

8. $\frac{5}{9}$ 9. $\frac{393}{1000}$ 10. $\frac{347}{210}$ 11. $\frac{19}{72}$

b APPLICATIONS AND PROBLEM SOLVING

EXAMPLE 9 *Construction.* A contractor requires his subcontractors to use two layers of subflooring under a ceramic tile floor. First, the subcontractors install a $\frac{3}{4}$-in. layer of oriented strand board (OSB). Then a $\frac{1}{2}$-in. sheet of cement board is mortared to the OSB. The mortar is $\frac{1}{8}$-in. thick. What is the total thickness of the two installed subfloors?

1. Familiarize. We first make a drawing. We let T = the total thickness of the subfloors.

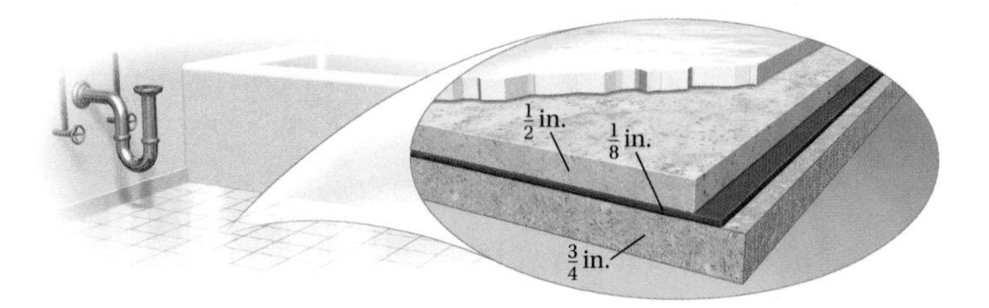

2. Translate. The problem can be translated to an equation as follows.

OSB	plus	Mortar	plus	Cement board	is	Total thickness
$\frac{3}{4}$	$+$	$\frac{1}{8}$	$+$	$\frac{1}{2}$	$=$	T

3. Solve. To solve the equation, we carry out the addition. The LCM of the denominators is 8 because 2 and 4 are factors of 8. We multiply by 1 in order to obtain the LCD:

$$\frac{3}{4} + \frac{1}{8} + \frac{1}{2} = T$$

$$\frac{3}{4} \cdot \frac{2}{2} + \frac{1}{8} + \frac{1}{2} \cdot \frac{4}{4} = T$$

$$\frac{6}{8} + \frac{1}{8} + \frac{4}{8} = T$$

$$\frac{11}{8} = T.$$

4. Check. We check by repeating the calculation. We also note that the sum should be larger than any of the individual measurements, which it is. This tells us that the answer is reasonable.

5. State. The total thickness of the installed subfloors is $\frac{11}{8}$ in.

◀ Do Exercise 12.

---

12. Catering. A caterer prepares a mixed berry salad with $\frac{7}{8}$ qt of strawberries, $\frac{3}{4}$ qt of raspberries, and $\frac{5}{16}$ qt of blueberries. What is the total amount of berries in the salad? GS

1. Familiarize. Let T = the total amount of berries in the salad.

2. Translate. To find the total amount, we add.

$$\frac{7}{\square} + \frac{3}{\square} + \frac{5}{\square} = T$$

3. Solve. The LCD is ____.

$$\frac{7}{8} \cdot \frac{2}{2} + \frac{3}{4} \cdot \frac{\square}{\square} + \frac{5}{16} = T$$

$$\frac{\square}{16} + \frac{\square}{16} + \frac{5}{16} = T$$

$$\frac{\square}{16} = T$$

4. Check. The answer is reasonable because it is larger than any of the individual amounts.

5. State. The salad contains a total of ____ qt of berries.

Answer

12. $\frac{31}{16}$ qt

Guided Solution:

12. 8, 4, 16; 16; $\frac{4}{4}$, 14, 12, 31; $\frac{31}{16}$

4.2 Exercise Set

For Extra Help MyMathLab® MathXL® PRACTICE WATCH READ REVIEW

Reading Check

Determine whether each statement is true or false.

________ **RC1.** Before we can add two fractions, they must have the same denominator.

________ **RC2.** To add fractions, we add numerators and add denominators.

________ **RC3.** If we use the LCD to add fractions, we never need to simplify the result.

________ **RC4.** Adding fractions with different denominators involves multiplying at least one fraction by 1.

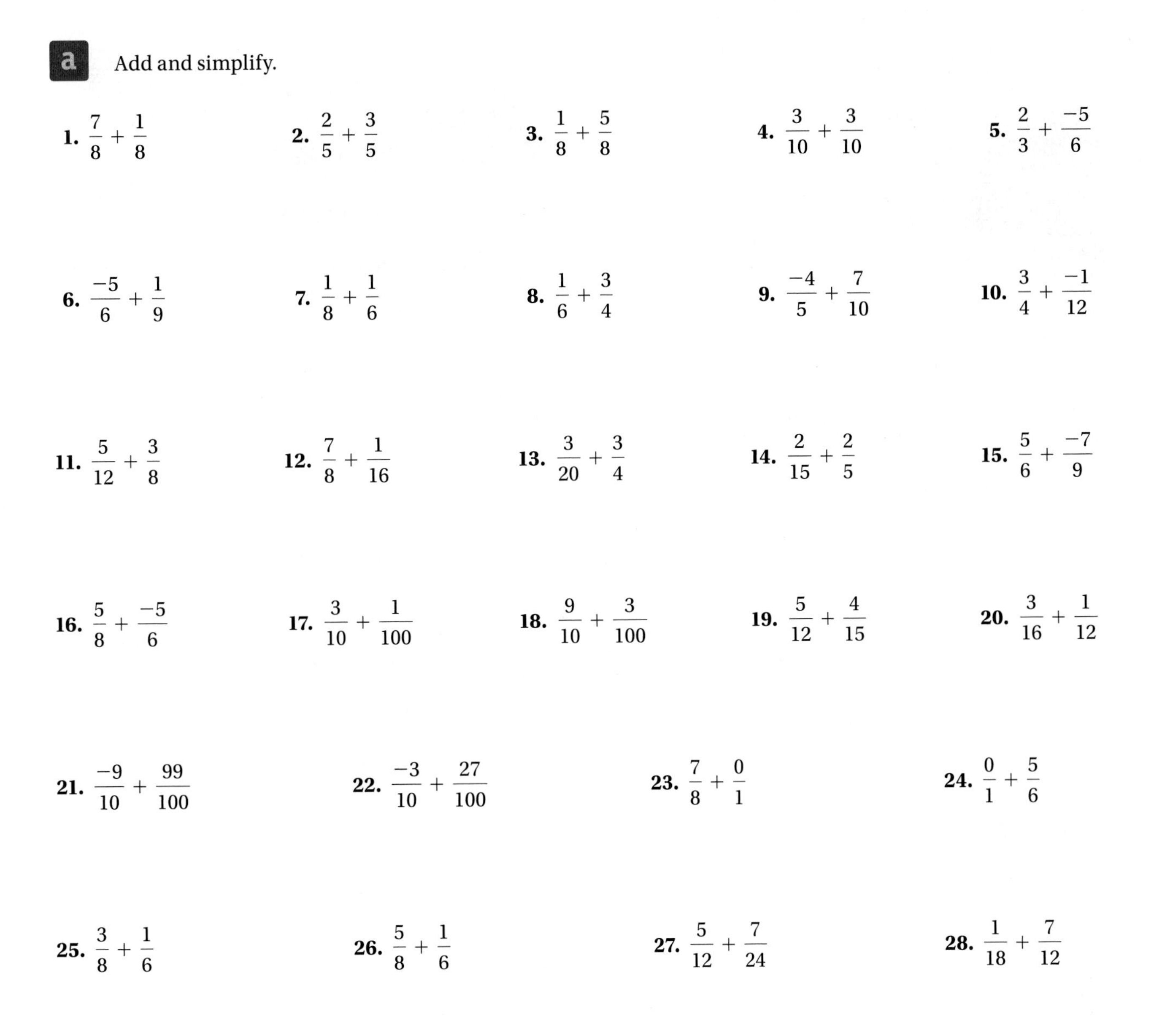

a Add and simplify.

1. $\frac{7}{8}+\frac{1}{8}$

2. $\frac{2}{5}+\frac{3}{5}$

3. $\frac{1}{8}+\frac{5}{8}$

4. $\frac{3}{10}+\frac{3}{10}$

5. $\frac{2}{3}+\frac{-5}{6}$

6. $\frac{-5}{6}+\frac{1}{9}$

7. $\frac{1}{8}+\frac{1}{6}$

8. $\frac{1}{6}+\frac{3}{4}$

9. $\frac{-4}{5}+\frac{7}{10}$

10. $\frac{3}{4}+\frac{-1}{12}$

11. $\frac{5}{12}+\frac{3}{8}$

12. $\frac{7}{8}+\frac{1}{16}$

13. $\frac{3}{20}+\frac{3}{4}$

14. $\frac{2}{15}+\frac{2}{5}$

15. $\frac{5}{6}+\frac{-7}{9}$

16. $\frac{5}{8}+\frac{-5}{6}$

17. $\frac{3}{10}+\frac{1}{100}$

18. $\frac{9}{10}+\frac{3}{100}$

19. $\frac{5}{12}+\frac{4}{15}$

20. $\frac{3}{16}+\frac{1}{12}$

21. $\frac{-9}{10}+\frac{99}{100}$

22. $\frac{-3}{10}+\frac{27}{100}$

23. $\frac{7}{8}+\frac{0}{1}$

24. $\frac{0}{1}+\frac{5}{6}$

25. $\frac{3}{8}+\frac{1}{6}$

26. $\frac{5}{8}+\frac{1}{6}$

27. $\frac{5}{12}+\frac{7}{24}$

28. $\frac{1}{18}+\frac{7}{12}$

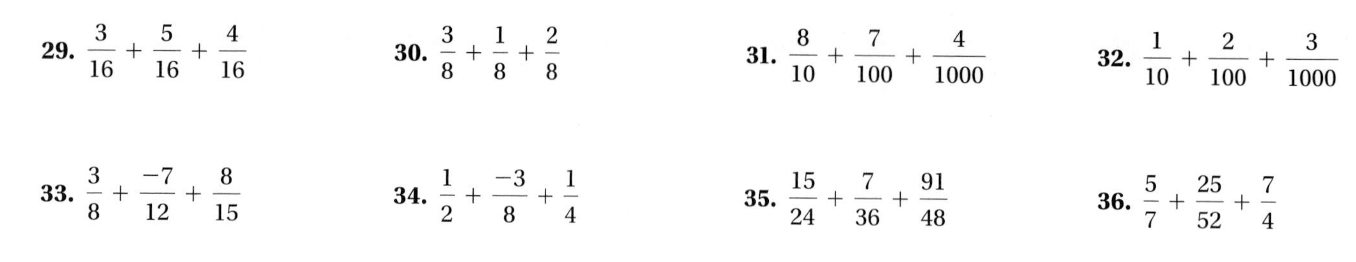

29. $\frac{3}{16}+\frac{5}{16}+\frac{4}{16}$

30. $\frac{3}{8}+\frac{1}{8}+\frac{2}{8}$

31. $\frac{8}{10}+\frac{7}{100}+\frac{4}{1000}$

32. $\frac{1}{10}+\frac{2}{100}+\frac{3}{1000}$

33. $\frac{3}{8}+\frac{-7}{12}+\frac{8}{15}$

34. $\frac{1}{2}+\frac{-3}{8}+\frac{1}{4}$

35. $\frac{15}{24}+\frac{7}{36}+\frac{91}{48}$

36. $\frac{5}{7}+\frac{25}{52}+\frac{7}{4}$

b Solve.

37. *Riding a Segway®.* Tate rode a Segway® Personal Transporter $\frac{5}{6}$ mi to the library, then $\frac{3}{4}$ mi to class, and then $\frac{3}{2}$ mi to his part-time job. How far did he ride his Segway®?

38. *Volunteering.* For a community project, an earth science class volunteered one hour per day for three days to join the state highway beautification project. The students collected trash along a $\frac{4}{5}$-mi stretch of highway the first day, a $\frac{5}{8}$-mi stretch the second day, and a $\frac{1}{2}$-mi stretch the third day. How many miles along the highway did they clean?

39. *Caffeine.* To cut back on their caffeine intake, Michelle and Gerry mix caffeinated and decaffeinated coffee beans before grinding for a customized mix. They mix $\frac{3}{16}$ lb of decaffeinated beans with $\frac{5}{8}$ lb of caffeinated beans. What is the total amount of coffee beans in the mixture?

40. *Purchasing Tea.* Alyse bought $\frac{1}{3}$ lb of orange pekoe tea and $\frac{1}{2}$ lb of English cinnamon tea. How many pounds of tea did she buy?

41. *Culinary Arts.* The campus culinary arts department is preparing brownies for the international student reception. Students in the catering program iced the $\frac{11}{16}$-in. $\left(\frac{11}{16}''\right)$ brownies with a $\frac{5}{32}$-in. $\left(\frac{5}{32}''\right)$ layer of butterscotch icing. What is the thickness of the iced brownie?

42. *Carpentry.* A carpenter glues two kinds of plywood together. He glues a $\frac{1}{4}$-in. $\left(\frac{1}{4}''\right)$ piece of cherry plywood to a $\frac{3}{8}$-in. $\left(\frac{3}{8}''\right)$ piece of less expensive plywood. What is the total thickness of these pieces?

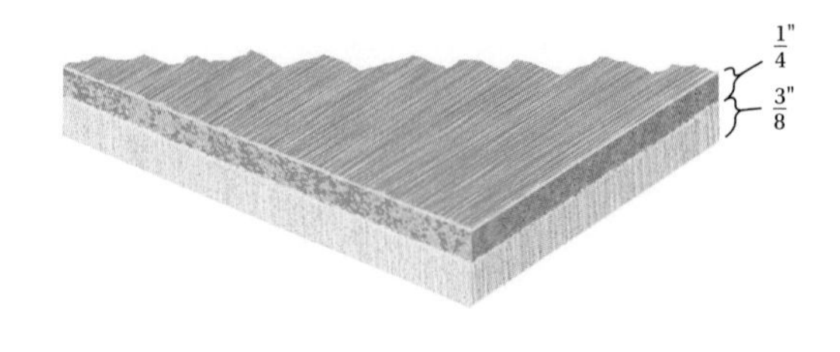

43. A tile $\frac{5}{8}$ in. thick is glued to a board $\frac{7}{8}$ in. thick. The glue is $\frac{3}{32}$ in. thick. How thick is the result?

44. A baker used $\frac{1}{2}$ lb of flour for rolls, $\frac{1}{4}$ lb for donuts, and $\frac{1}{3}$ lb for cookies. How much flour was used?

45. *Meteorology.* On Monday, April 15, it rained $\frac{1}{2}$ in. in the morning and $\frac{3}{8}$ in. in the afternoon. How much did it rain altogether?

46. *Medicine.* Janine took $\frac{1}{5}$ g of ibuprofen before lunch and $\frac{1}{2}$ g after lunch. How much did she take altogether?

47. *Hiking.* A park naturalist hiked $\frac{3}{5}$ mi to a lookout, another $\frac{3}{10}$ mi to an osprey's nest, and finally $\frac{3}{4}$ mi to a campsite. How far did the naturalist hike?

48. *Triathlon.* A triathlete runs $\frac{7}{8}$ mi, canoes $\frac{1}{3}$ mi, and swims $\frac{1}{6}$ mi. How many miles does the triathlete cover?

49. *Culinary Arts.* A recipe for strawberry punch calls for $\frac{1}{5}$ qt of ginger ale and $\frac{3}{5}$ qt of strawberry soda. How much liquid is needed? If the recipe is doubled, how much liquid is needed? If the recipe is halved, how much liquid is needed?

50. *Construction.* A cubic meter of concrete mix contains 420 kg of cement, 150 kg of stone, and 120 kg of sand. What is the total weight of a cubic meter of the mix? What part is cement? stone? sand? Add these fractional amounts. What is the result?

Skill Maintenance

51. *Serving of Cheesecake.* At the Cheesecake Factory, a piece of cheesecake is $\frac{1}{12}$ of cheesecake. How much of the cheesecake is $\frac{1}{2}$ piece? [3.6b]

Source: The Cheesecake Factory

52. *Milk Production.* Holstein's Dairy produced 4578 oz of milk one morning. How many 16-oz cartons could be filled? How much milk would be left over? [1.8a]

53. *Honey Production.* In 2011, 176,462,000 lb of honey was produced in the United States. The two states with the greatest honey production were North Dakota and California. North Dakota produced 46,410,000 lb of honey, and California produced 27,470,000 lb. How many more pounds of honey were produced in North Dakota than in California? [1.8a]

Source: U.S. Department of Agriculture

54. *Community Garden.* The Bingham community garden is to be split into 16 equally sized plots. If the garden occupies $\frac{3}{4}$ acre of land, how large will each plot be? [3.7d]

Synthesis

55. A guitarist's band is booked for Friday and Saturday nights at a local club. The guitarist is part of a trio on Friday and part of a quintet on Saturday. Thus the guitarist is paid one-third of one-half the weekend's pay for Friday and one-fifth of one-half the weekend's pay for Saturday. What fractional part of the band's pay did the guitarist receive for the weekend's work? If the band was paid $1200, how much did the guitarist receive?

4.3 Subtraction, Order, and Applications

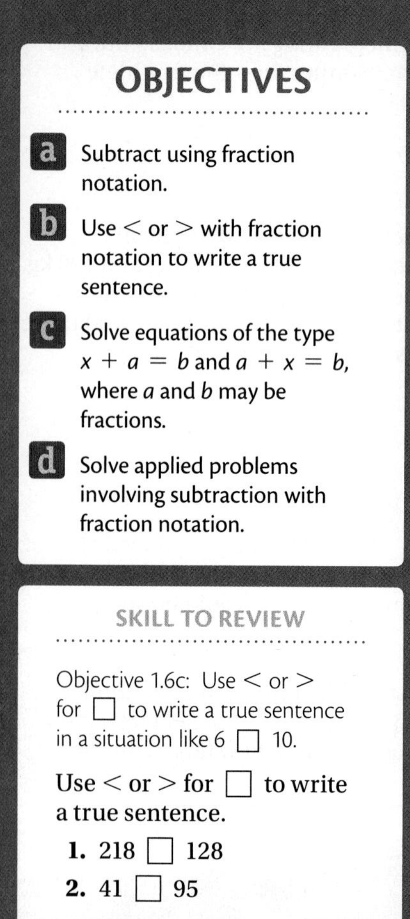

OBJECTIVES

a Subtract using fraction notation.

b Use $<$ or $>$ with fraction notation to write a true sentence.

c Solve equations of the type $x + a = b$ and $a + x = b$, where a and b may be fractions.

d Solve applied problems involving subtraction with fraction notation.

SKILL TO REVIEW

Objective 1.6c: Use $<$ or $>$ for □ to write a true sentence in a situation like 6 □ 10.

Use $<$ or $>$ for □ to write a true sentence.

1. 218 □ 128

2. 41 □ 95

a SUBTRACTION USING FRACTION NOTATION

Like Denominators

Let's consider the difference $\frac{4}{8} - \frac{3}{8}$.

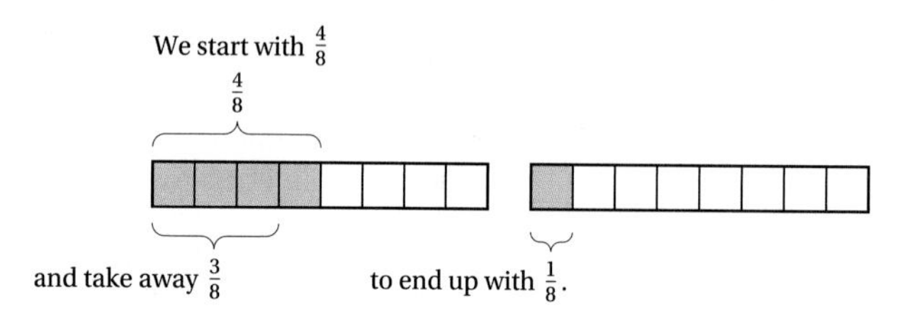

We start with 4 eighths and take away 3 eighths:

4 eighths $-$ 3 eighths $=$ 1 eighth,

or $\quad 4 \cdot \frac{1}{8} - 3 \cdot \frac{1}{8} = \frac{1}{8}, \quad$ or $\quad \frac{4}{8} - \frac{3}{8} = \frac{1}{8}.$

> To subtract when denominators are the same,
>
> **a)** subtract the numerators,
>
> **b)** keep the denominator, and
>
> **c)** simplify, if possible.
>
> $$\frac{7}{10} - \frac{4}{10} = \frac{7-4}{10} = \frac{3}{10}$$

EXAMPLES Subtract and simplify.

1. $\frac{32}{12} - \frac{25}{12} = \frac{32 - 25}{12} = \frac{7}{12}$

2. $\frac{7}{10} - \frac{3}{10} = \frac{7 - 3}{10} = \frac{4}{10} = \frac{2 \cdot 2}{5 \cdot 2} = \frac{2}{5} \cdot \frac{2}{2} = \frac{2}{5} \cdot 1 = \frac{2}{5}$

3. $\frac{25}{12} - \frac{32}{12} = \frac{25 - 32}{12} = \frac{-7}{12}$, or $-\frac{7}{12}$

◀ **Do Margin Exercises 1–3.**

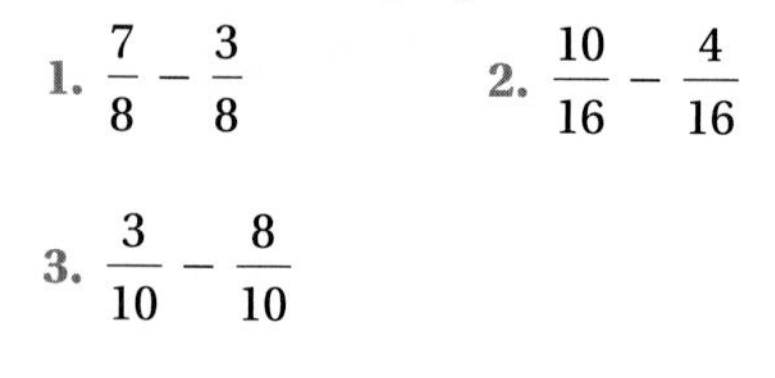

Subtract and simplify.

1. $\frac{7}{8} - \frac{3}{8}$

2. $\frac{10}{16} - \frac{4}{16}$

3. $\frac{3}{10} - \frac{8}{10}$

Different Denominators

> To subtract when denominators are different:
>
> **a)** Find the least common multiple of the denominators. That number is the least common denominator, LCD.
>
> **b)** Multiply by 1, using an appropriate notation, n/n, to express each number in terms of the LCD.
>
> **c)** Subtract the numerators, keeping the same denominator.
>
> **d)** Simplify, if possible.

Answers

Skill to Review:
1. $>$ 2. $<$

Margin Exercises:
1. $\frac{1}{2}$ 2. $\frac{3}{8}$ 3. $-\frac{1}{2}$

EXAMPLE 4 Subtract: $\frac{2}{5} - \frac{3}{8}$.

The LCM of 5 and 8 is 40, so the LCD is 40.

$$\frac{2}{5} - \frac{3}{8} = \frac{2}{5} \cdot \frac{8}{8} - \frac{3}{8} \cdot \frac{5}{5}$$

Think: $8 \times \square = 40$. The answer is 5, so we multiply by 1, using $\frac{5}{5}$.

Think: $5 \times \square = 40$. The answer is 8, so we multiply by 1, using $\frac{8}{8}$.

$$= \frac{16}{40} - \frac{15}{40} = \frac{16 - 15}{40} = \frac{1}{40}$$

Do Exercise 4.

4. Subtract: $\frac{3}{4} - \frac{2}{3}$.

EXAMPLE 5 Subtract: $\frac{7}{12} - \frac{5}{6}$.

Since 12 is a multiple of 6, the LCM of 6 and 12 is 12. The LCD is 12.

$$\frac{7}{12} - \frac{5}{6} = \frac{7}{12} - \frac{5}{6} \cdot \frac{2}{2}$$

$$= \frac{7}{12} - \frac{10}{12} = \frac{7 - 10}{12} = \frac{-3}{12}$$

$$= \frac{3 \cdot (-1)}{3 \cdot 4} = \frac{3}{3} \cdot \frac{-1}{4} = \frac{-1}{4}, \text{ or } -\frac{1}{4}$$

Do Exercises 5 and 6.

Subtract.

GS 5. $\frac{5}{6} - \frac{1}{9}$

The LCD is $\square$.

$$\frac{5}{6} - \frac{1}{9} = \frac{5}{6} \cdot \frac{3}{3} - \frac{1}{9} \cdot \frac{\square}{\square}$$

$$= \frac{\square}{18} - \frac{\square}{18}$$

$$= \frac{\square}{18}$$

6. $\frac{3}{10} - \frac{4}{5}$

EXAMPLE 6 Subtract: $\frac{17}{24} - \frac{4}{15}$.

We use prime factorizations to determine the LCM of 24 and 15:

$24 = 2 \cdot 2 \cdot 2 \cdot 3,$

$15 = 3 \cdot 5.$

The LCD is $2 \cdot 2 \cdot 2 \cdot 3 \cdot 5$, or 120.

$$\frac{17}{24} - \frac{4}{15} = \frac{17}{2 \cdot 2 \cdot 2 \cdot 3} \cdot \frac{5}{5} - \frac{4}{3 \cdot 5} \cdot \frac{2 \cdot 2 \cdot 2}{2 \cdot 2 \cdot 2}$$

The LCM of 24 and 15 is $2 \cdot 2 \cdot 2 \cdot 3 \cdot 5$. In each case, we multiply by 1 to obtain the LCD.

$$= \frac{17 \cdot 5}{2 \cdot 2 \cdot 2 \cdot 3 \cdot 5} - \frac{4 \cdot 2 \cdot 2 \cdot 2}{3 \cdot 5 \cdot 2 \cdot 2 \cdot 2}$$

$$= \frac{85}{120} - \frac{32}{120} = \frac{53}{120}$$

Do Exercise 7.

7. Subtract: $\frac{11}{28} - \frac{5}{16}$.

b ORDER

When two fractions share a common denominator, the fractions can be compared by comparing numerators.

Since $4 > 3$, $\frac{4}{5} > \frac{3}{5}$.

Since $3 < 4$, $\frac{3}{5} < \frac{4}{5}$.

Answers

4. $\frac{1}{12}$ 5. $\frac{13}{18}$ 6. $-\frac{1}{2}$ 7. $\frac{9}{112}$

Guided Solution:

5. 18; $\frac{2}{2}$, 15, 2, 13

8. Use $<$ or $>$ for $\square$ to write a true sentence:

$$\frac{3}{8} \square \frac{5}{8}.$$

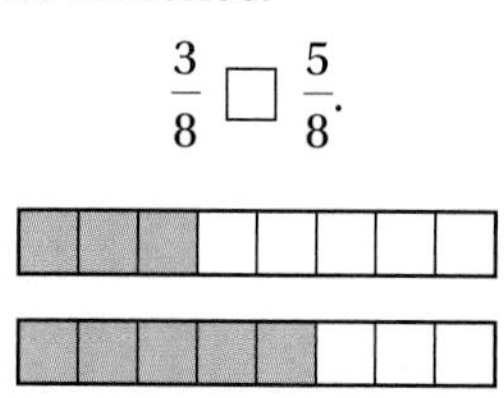

9. Use $<$ or $>$ for $\square$ to write a true sentence:

$$\frac{7}{10} \square \frac{6}{10}.$$

Use $<$ or $>$ for $\square$ to write a true sentence.

10. $\frac{2}{3} \square \frac{5}{8}$

11. $\frac{-3}{4} \square \frac{-8}{12}$

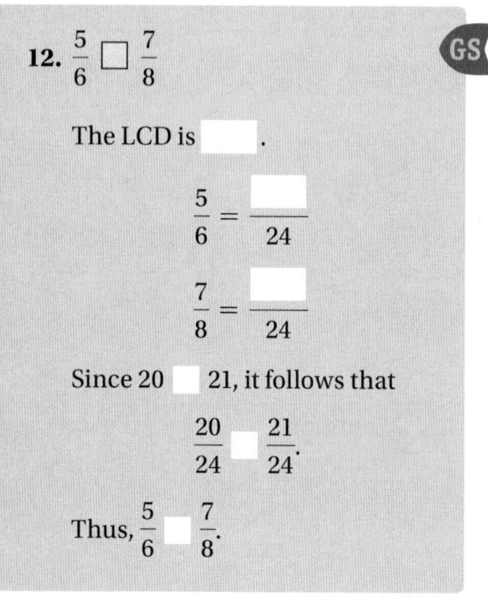

12. $\frac{5}{6} \square \frac{7}{8}$ GS

The LCD is ____.

$\frac{5}{6} = \frac{\square}{24}$

$\frac{7}{8} = \frac{\square}{24}$

Since 20 ☐ 21, it follows that

$\frac{20}{24} \square \frac{21}{24}.$

Thus, $\frac{5}{6} \square \frac{7}{8}.$

Answers

8. $<$ 9. $>$ 10. $>$ 11. $<$ 12. $<$

Guided Solution:

12. 24; 20, 21; $<$, $<$; $<$

To compare two fractions that have a common denominator, compare the numerators:

$$\frac{6}{7} > \frac{2}{7}; \quad \frac{2}{7} < \frac{6}{7}.$$

◀ **Do Exercises 8 and 9.**

When denominators are different, we cannot compare numerators until we multiply by 1 to make the denominators the same.

EXAMPLE 7 Use $<$ or $>$ for $\square$ to write a true sentence:

$$\frac{2}{5} \square \frac{3}{4}.$$

The LCD is 20. We have

$$\frac{2}{5} \cdot \frac{4}{4} = \frac{8}{20}; \quad \text{We multiply by 1 using } \frac{4}{4} \text{ to get the LCD.}$$

$$\frac{3}{4} \cdot \frac{5}{5} = \frac{15}{20}. \quad \text{We multiply by 1 using } \frac{5}{5} \text{ to get the LCD.}$$

Now that the denominators are the same, 20, we can compare the numerators. Since $8 < 15$, it follows that $\frac{8}{20} < \frac{15}{20}$, so

$$\frac{2}{5} < \frac{3}{4}.$$

EXAMPLE 8 Use $<$ or $>$ for $\square$ to write a true sentence:

$$\frac{-89}{100} \square \frac{-9}{10}.$$

The LCD is 100. We write $\frac{-9}{10}$ with a denominator of 100 to make the denominators the same.

$$\frac{-9}{10} \cdot \frac{10}{10} = \frac{-90}{100} \quad \text{We multiply by } \frac{10}{10} \text{ to get the LCD.}$$

Since $-89 > -90$, it follows that $\frac{-89}{100} > \frac{-90}{100}$, so

$$\frac{-89}{100} > \frac{-9}{10}.$$

◀ **Do Exercises 10–12.**

C SOLVING EQUATIONS

Now let's solve equations of the form $x + a = b$ or $a + x = b$, where a and b may be fractions. Proceeding as we have before, we subtract a on both sides of the equation.

EXAMPLE 9 Solve: $x + \frac{1}{4} = -\frac{3}{5}$.

$$x + \frac{1}{4} - \frac{1}{4} = -\frac{3}{5} - \frac{1}{4}$$ Subtracting $\frac{1}{4}$ on both sides

$$x + 0 = -\frac{3}{5} \cdot \frac{4}{4} - \frac{1}{4} \cdot \frac{5}{5}$$ The LCD is 20. We multiply by 1 to get the LCD.

$$x = -\frac{12}{20} - \frac{5}{20}$$

$$x = -\frac{7}{20}$$

Do Exercises 13 and 14. ▶

Solve.

13. $x + \frac{2}{3} = \frac{5}{6}$ 14. $\frac{3}{5} + t = -\frac{7}{8}$

d APPLICATIONS AND PROBLEM SOLVING

EXAMPLE 10 *Phases of the Moon.* The moon rotates in such a way that the same side always faces the earth. Throughout a lunar cycle, the portion of the moon that appears illuminated increases from nearly none (new moon) to nearly all (full moon), then decreases back to nearly none. These *phases* of the moon can be described by fractions between 0 and 1, indicating the portion of the moon illuminated. The partial calendar from August 2013 shows the fraction of the moon illuminated at midnight, Eastern Standard Time, for each day.

How much more of the moon appeared illuminated on August 18, 2013, than on August 15, 2013?

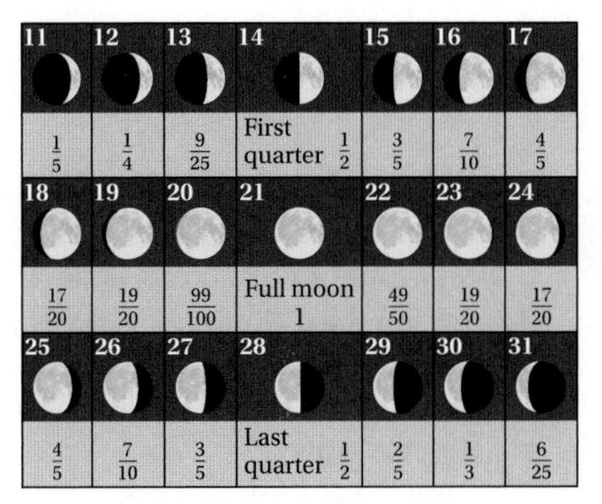

SOURCE: Astronomical Applications Department, U.S. Naval Observatory, Washington, DC 20392-5420

1. **Familiarize.** From the calendar, we see that $\frac{3}{5}$ of the moon was illuminated on August 15 and that $\frac{17}{20}$ of the moon was illuminated on August 18. We let m = the additional part of the moon that appeared illuminated on August 18.

2. **Translate.** We translate to an equation.

Amount illuminated on August 15	plus	Additional amount illuminated	is	Amount illuminated on August 18
↓	↓	↓	↓	↓
$\frac{3}{5}$	$+$	m	$=$	$\frac{17}{20}$

3. **Solve.** To solve the equation, we subtract $\frac{3}{5}$ on both sides:

$$\frac{3}{5} + m = \frac{17}{20}$$

$$\frac{3}{5} + m - \frac{3}{5} = \frac{17}{20} - \frac{3}{5}$$ Subtracting $\frac{3}{5}$ on both sides

$$m + 0 = \frac{17}{20} - \frac{3}{5} \cdot \frac{4}{4}$$ The LCD is 20. We multiply by 1 to obtain the LCD.

$$m = \frac{17}{20} - \frac{12}{20} = \frac{5}{20} = \frac{5 \cdot 1}{5 \cdot 4} = \frac{5}{5} \cdot \frac{1}{4} = \frac{1}{4}.$$

4. **Check.** To check, we add:

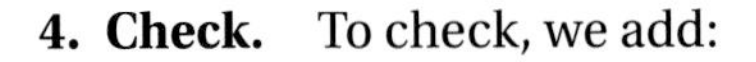

$$\frac{3}{5} + \frac{1}{4} = \frac{3}{5} \cdot \frac{4}{4} + \frac{1}{4} \cdot \frac{5}{5} = \frac{12}{20} + \frac{5}{20} = \frac{17}{20}.$$

5. **State.** On August 18, $\frac{1}{4}$ more of the moon was illuminated than on August 15.

Do Exercise 15. ▶

15. *Phases of the Moon.* Use the calendar in Example 10 to find how much less of the moon appeared illuminated on August 31, 2013, than on August 23, 2013.

Answers

13. $\frac{1}{6}$ 14. $-\frac{59}{40}$ 15. $\frac{71}{100}$

Translating for Success

The goal of these matching questions is to practice step (2), Translate, of the five-step problem-solving process. Translate each word problem to an equation and select a correct translation from equations A–O.

A. $\frac{3}{4} \cdot 64 = x$

B. $\frac{1}{3} \cdot 12{,}000 = x$

C. $\frac{1}{3} + \frac{2}{5} = x$

D. $\frac{2}{5} + x = \frac{7}{9}$

E. $\frac{2}{5} \cdot \frac{7}{9} = x$

F. $\frac{3}{4} \cdot x = 64$

G. $\frac{4}{7} = x + \frac{11}{12}$

H. $\frac{2}{5} = x + \frac{7}{9}$

I. $\frac{4}{7} \cdot \frac{11}{12} = x$

J. $\frac{3}{4} \cdot x = 128$

K. $\frac{1}{3} \cdot x = 12{,}000$

L. $\frac{1}{3} + \frac{2}{5} + x = 1$

M. $\frac{4}{3} \cdot 64 = x$

N. $\frac{4}{7} + x = \frac{11}{12}$

O. $\frac{1}{3} + x = \frac{2}{5}$

Answers on page A-6

1. *Packaging.* One-Stop Postal Center orders bubble wrap in 64-yd rolls. On average, $\frac{3}{4}$ yd is used per small package. How many small packages can be prepared with 2 rolls of bubble wrap?

2. *Distance from College.* The post office is $\frac{7}{9}$ mi from the community college. The medical clinic is $\frac{2}{5}$ as far from the college as the post office is. How far is the clinic from the college?

3. *Swimming.* Andrew swims $\frac{7}{9}$ mi every day. One day he swims $\frac{2}{5}$ mi by 11:00 A.M. How much farther must Andrew swim to reach his daily goal?

4. *Tuition.* The average tuition at Waterside University is \$12,000. If a loan is obtained for $\frac{1}{3}$ of the tuition, how much is the loan?

5. *Thermos Bottle Capacity.* A thermos bottle holds $\frac{11}{12}$ gal. How much is in the bottle when it is $\frac{4}{7}$ full?

6. *Cutting Rope.* A piece of rope $\frac{11}{12}$ yd long is cut into two pieces. One piece is $\frac{4}{7}$ yd long. How long is the other piece?

7. *Planting Corn.* Each year, Prairie State Farm plants 64 acres of corn. With good weather, $\frac{3}{4}$ of the planting can be completed by April 20. How many acres can be planted by April 20 with good weather?

8. *Painting Trim.* A painter used $\frac{1}{3}$ gal of white paint for the trim in the library and $\frac{2}{5}$ gal in the family room. How much paint was used for the trim in the two rooms?

9. *Lottery Winnings.* Sally won \$12,000 in a state lottery and decided to give the net amount after taxes to three charities. One received $\frac{1}{3}$ of the net amount, and a second received $\frac{2}{5}$. What fractional part of the net amount did the third charity receive?

10. *Reading Assignment.* When Lowell had read 64 pages of his political science assignment, he had completed $\frac{3}{4}$ of his required reading. How many total pages were assigned?

4.3 Exercise Set

For Extra Help MyMathLab® MathXL® PRACTICE WATCH READ REVIEW

Reading Check

Complete each statement with the appropriate word or words from the following list. A word may be used more than once or not at all.

denominator numerator
denominators numerators

RC1. To subtract fractions with like denominators, we subtract the ____________ and keep the ____________.

RC2. To subtract fractions when denominators are different, we find the LCM of the ____________.

RC3. To subtract fractions when denominators are different, we multiply one or both fractions by 1 to make the ____________ the same.

RC4. To compare two fractions with like denominators, we compare their ____________.

a Subtract and simplify.

1. $\frac{5}{6} - \frac{1}{6}$

2. $\frac{5}{8} - \frac{3}{8}$

3. $\frac{11}{12} - \frac{2}{12}$

4. $\frac{17}{18} - \frac{11}{18}$

5. $\frac{1}{8} - \frac{3}{4}$

6. $\frac{1}{9} - \frac{2}{3}$

7. $\frac{1}{8} - \frac{1}{12}$

8. $\frac{1}{6} - \frac{1}{8}$

9. $\frac{5}{6} - \frac{4}{3}$

10. $\frac{1}{16} - \frac{7}{8}$

11. $\frac{3}{4} - \frac{3}{28}$

12. $\frac{2}{5} - \frac{2}{15}$

13. $\frac{3}{4} - \frac{3}{20}$

14. $\frac{5}{6} - \frac{1}{2}$

15. $\frac{1}{20} - \frac{3}{4}$

16. $\frac{4}{16} - \frac{3}{4}$

17. $\frac{5}{12} - \frac{2}{15}$

18. $\frac{9}{10} - \frac{11}{16}$

19. $\frac{6}{10} - \frac{7}{100}$

20. $\frac{9}{10} - \frac{3}{100}$

21. $\frac{7}{15} - \frac{3}{25}$

22. $\frac{18}{25} - \frac{4}{35}$

23. $\frac{99}{100} - \frac{9}{10}$

24. $\frac{78}{100} - \frac{11}{20}$

25. $-\frac{2}{3} - \frac{1}{8}$

26. $-\frac{3}{4} - \frac{1}{2}$

27. $\frac{3}{5} - \frac{1}{2}$

28. $\frac{5}{6} - \frac{2}{3}$

29. $\frac{3}{8} - \frac{5}{12}$

30. $\frac{2}{9} - \frac{7}{12}$

31. $\frac{7}{8} - \frac{1}{16}$

32. $\frac{5}{12} - \frac{5}{16}$

33. $\frac{4}{15} - \frac{17}{25}$

34. $\frac{7}{24} - \frac{11}{18}$

35. $\frac{23}{25} - \frac{112}{150}$

36. $\frac{89}{90} - \frac{53}{120}$

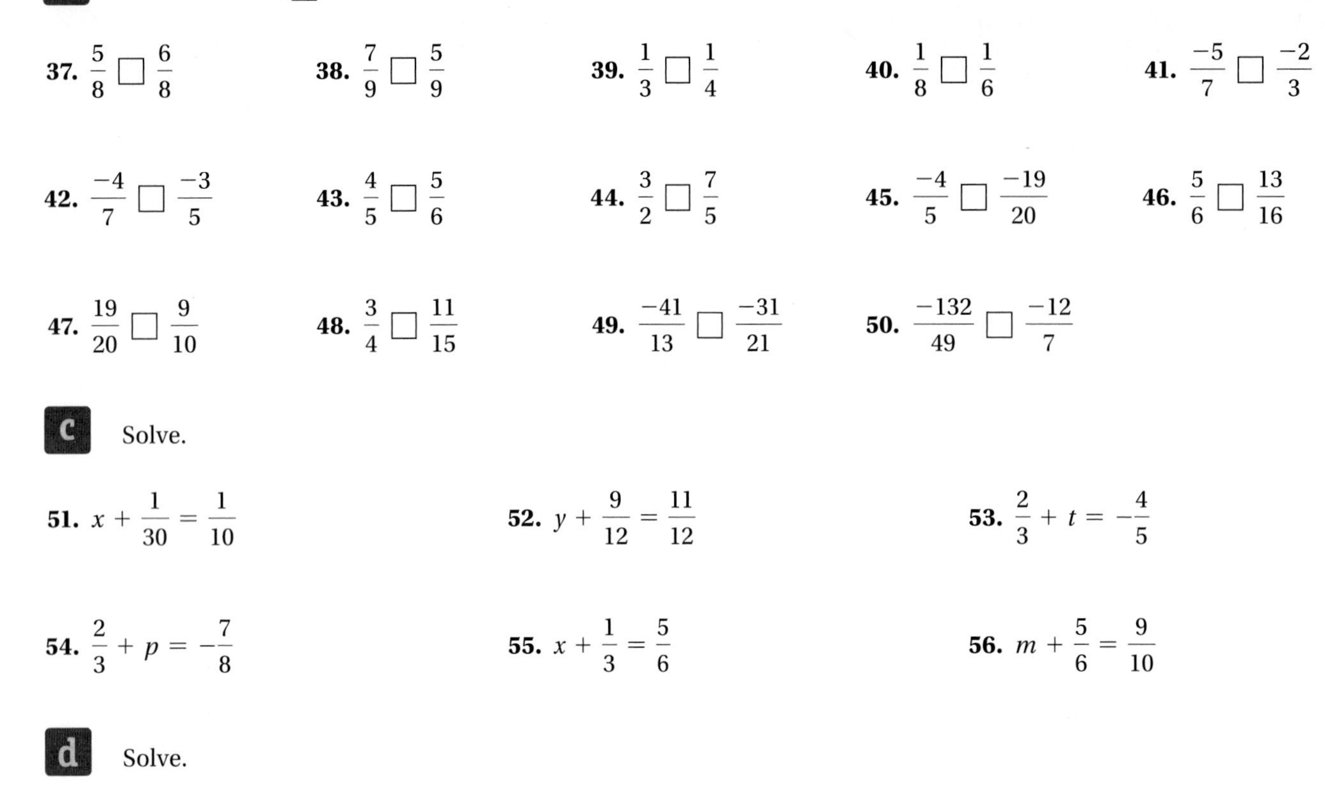

b Use < or > for □ to write a true sentence.

37. $\frac{5}{8} \square \frac{6}{8}$

38. $\frac{7}{9} \square \frac{5}{9}$

39. $\frac{1}{3} \square \frac{1}{4}$

40. $\frac{1}{8} \square \frac{1}{6}$

41. $\frac{-5}{7} \square \frac{-2}{3}$

42. $\frac{-4}{7} \square \frac{-3}{5}$

43. $\frac{4}{5} \square \frac{5}{6}$

44. $\frac{3}{2} \square \frac{7}{5}$

45. $\frac{-4}{5} \square \frac{-19}{20}$

46. $\frac{5}{6} \square \frac{13}{16}$

47. $\frac{19}{20} \square \frac{9}{10}$

48. $\frac{3}{4} \square \frac{11}{15}$

49. $\frac{-41}{13} \square \frac{-31}{21}$

50. $\frac{-132}{49} \square \frac{-12}{7}$

c Solve.

51. $x + \frac{1}{30} = \frac{1}{10}$

52. $y + \frac{9}{12} = \frac{11}{12}$

53. $\frac{2}{3} + t = -\frac{4}{5}$

54. $\frac{2}{3} + p = -\frac{7}{8}$

55. $x + \frac{1}{3} = \frac{5}{6}$

56. $m + \frac{5}{6} = \frac{9}{10}$

d Solve.

57. For a research paper, Kaitlyn spent $\frac{3}{4}$ hr searching the Internet on google.com and $\frac{1}{3}$ hr on chacha.com. How many more hours did she spend on google.com than on chacha.com?

58. As part of a fitness program, Deb swims $\frac{1}{2}$ mi every day. One day she had already swum $\frac{1}{5}$ mi. How much farther should Deb swim?

59. The tread depth of an IRL Indy Car Series tire is $\frac{3}{32}$ in. Tires for a normal car have a tread depth of $\frac{5}{16}$ in. when new and are considered bald at $\frac{1}{16}$ in. How much deeper is the tread depth of an Indy Car tire than that of a bald tire for a normal car?

Sources: Indy500.com; *Consumer Reports*

60. Gerry uses $\frac{1}{3}$ lb of fresh mozzarella cheese and $\frac{1}{4}$ lb of grated Parmesan cheese on a homemade margherita pizza. How much more mozzarella cheese does he use than Parmesan cheese?

61. From a $\frac{4}{5}$-lb wheel of cheese, a $\frac{1}{4}$-lb piece was served. How much cheese remained on the wheel?

62. A baker has a dispenser containing $\frac{15}{16}$ cup of icing and puts $\frac{1}{12}$ cup on a cinnamon roll. How much icing remains in the dispenser?

63. At a party, three friends, Ashley, Cole, and Lauren, shared a big tub of popcorn. Within 30 min, the tub was empty. Ashley ate $\frac{7}{12}$ of the tub while Lauren ate only $\frac{1}{6}$ of the tub. How much did Cole eat?

64. A small community garden was divided among four local residents. Based on the time they could spend on their garden sections and their individual crop plans, each resident received a different-size plot to tend. One received $\frac{1}{4}$ of the garden, the second $\frac{1}{16}$, and the third $\frac{3}{8}$ of the garden. How much did the fourth gardener receive?

Skill Maintenance

Divide, if possible. If not possible, write "not defined." [1.5a], [3.3b]

65. $\frac{-38}{-38}$

66. $\frac{38}{0}$

67. $\frac{124}{0}$

68. $\frac{124}{31}$

Divide and simplify. [3.7b]

69. $\frac{3}{7} \div \left(-\frac{9}{4}\right)$

70. $-\frac{9}{10} \div \left(-\frac{3}{5}\right)$

71. $7 \div \frac{1}{3}$

72. $\frac{1}{4} \div 8$

Multiply and simplify. [3.6a]

73. $18 \cdot \frac{2}{3}$

74. $\frac{5}{12} \cdot (-6)$

75. $\frac{7}{10} \cdot \frac{5}{14}$

76. $\left(-\frac{1}{10}\right) \cdot \left(-\frac{20}{5}\right)$

Synthesis

Solve.

77. 🖩 $x + \frac{16}{323} = \frac{10}{187}$

78. 🖩 $x + \frac{7}{253} = \frac{12}{299}$

79. As part of a rehabilitation program, an athlete must swim and then walk a total of $\frac{9}{10}$ km each day. If one lap in the swimming pool is $\frac{3}{80}$ km, how far must the athlete walk after swimming 10 laps?

80. *Mountain Climbing.* A mountain climber, beginning at sea level, climbs $\frac{3}{5}$ km, descends $\frac{1}{4}$ km, climbs $\frac{1}{3}$ km, and then descends $\frac{1}{7}$ km. At what elevation does the climber finish?

Simplify. Use the rules for order of operations.

81. $\frac{7}{8} - \frac{1}{10} \times \frac{5}{6}$

82. $\frac{2}{5} + \frac{1}{6} \div 3$

83. $\left(\frac{2}{3}\right)^2 - \left(\frac{3}{4}\right)^2$

84. $-5 \times \frac{3}{7} - \frac{1}{7} \times \frac{4}{5}$

Use $<$, $>$, or $=$ for $\square$ to write a true sentence.

85. 🖩 $\frac{37}{157} + \frac{19}{107} \ \square \ \frac{6941}{16{,}799}$

86. 🖩 $\frac{12}{97} + \frac{67}{139} \ \square \ \frac{8167}{13{,}289}$

87. *Microsoft Interview.* The following is a question taken from an employment interview with Microsoft. Try to answer it. "Given a gold bar that can be cut exactly twice and a contractor who must be paid one-seventh of a gold bar every day for seven days, how should the bar be cut?"

Source: *Fortune Magazine*, January 8, 2001

Mixed Numerals

OBJECTIVES

a Convert between mixed numerals and fraction notation.

b Divide whole numbers, writing the quotient as a mixed numeral.

SKILL TO REVIEW

Objective 1.5a: Divide whole numbers.

Divide.

1. $735 \div 16$
2. $23\overline{)6023}$

Convert to a mixed numeral.

1. $1 + \frac{2}{3} = \square\frac{\square}{\square}$

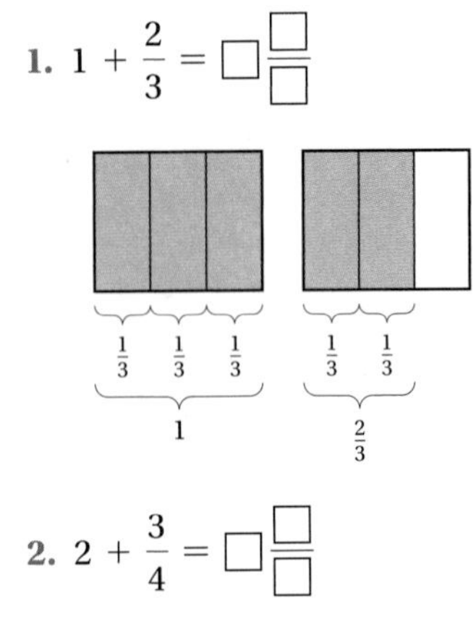

2. $2 + \frac{3}{4} = \square\frac{\square}{\square}$

3. $12 + \frac{2}{7}$

Answers

Skill to Review:
1. 45 R 15 2. 261 R 20

Margin Exercises:
1. $1\frac{2}{3}$ 2. $2\frac{3}{4}$ 3. $12\frac{2}{7}$

a MIXED NUMERALS

The following figure illustrates the use of a **mixed numeral**. The bolt shown is $2\frac{3}{8}$ in. long. The length is given as a whole-number part, 2, and a fractional part less than 1, $\frac{3}{8}$. We can also represent the measurement of the bolt with fraction notation as $\frac{19}{8}$, but the meaning or interpretation of such a symbol is less understandable or less easy to visualize than that of mixed numeral notation.

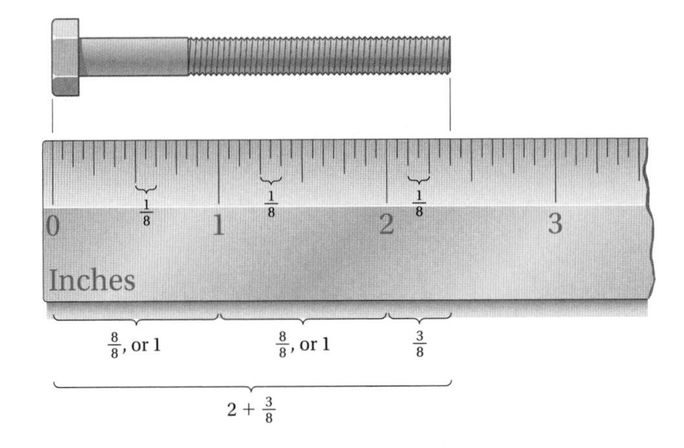

A mixed numeral like $2\frac{3}{8}$ represents a sum:

$$2\frac{3}{8} \quad \text{means} \quad 2 + \frac{3}{8}.$$

This is a whole number. This is a fraction less than 1.

EXAMPLES Convert to a mixed numeral.

1. $7 + \frac{2}{5} = 7\frac{2}{5}$

2. $4 + \frac{3}{10} = 4\frac{3}{10}$

◀ Do Margin Exercises 1–3.

The notation $2\frac{3}{4}$ has a plus sign left out. To aid in understanding, we sometimes write the missing plus sign. This is especially helpful when we convert a mixed numeral to fraction notation.

EXAMPLES Convert to fraction notation.

3. $2\frac{3}{4} = 2 + \frac{3}{4}$ Inserting the missing plus sign

$= \frac{2}{1} + \frac{3}{4}$ $2 = \frac{2}{1}$

$= \frac{2}{1} \cdot \frac{4}{4} + \frac{3}{4}$ Finding a common denominator

$= \frac{8}{4} + \frac{3}{4} = \frac{11}{4}$

4. $4\frac{3}{10} = 4 + \frac{3}{10} = \frac{4}{1} + \frac{3}{10} = \frac{4}{1} \cdot \frac{10}{10} + \frac{3}{10} = \frac{40}{10} + \frac{3}{10} = \frac{43}{10}$

Do Exercises 4 and 5. ▶

Convert to fraction notation.

4. $4\frac{2}{5}$

5. $6\frac{1}{10}$

Let's now consider a faster method for converting a mixed numeral to fraction notation.

> To convert from a positive mixed numeral to fraction notation:
>
> ⓐ Multiply the whole number by the denominator: $4 \cdot 10 = 40$.
>
> ⓑ Add the result to the numerator: $40 + 3 = 43$.
>
> ⓒ Keep the denominator.
>
> $4\frac{3}{10} = \frac{43}{10}$ (ⓐ, ⓑ, ⓒ)

Convert to fraction notation. Use the faster method.

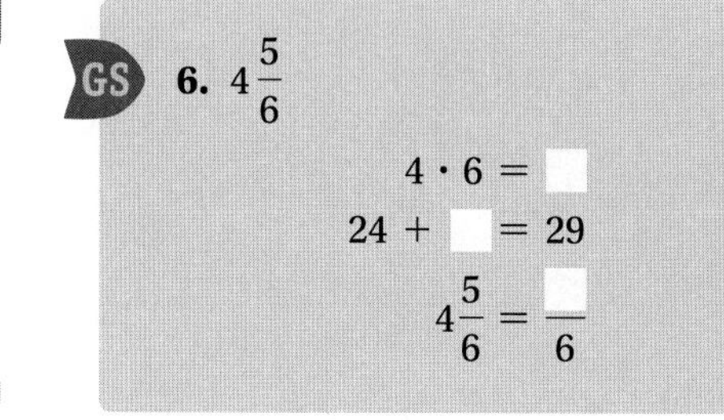

EXAMPLES Convert to fraction notation.

5. $6\frac{2}{3} = \frac{20}{3}$ $\quad 6 \cdot 3 = 18, 18 + 2 = 20$

6. $8\frac{2}{9} = \frac{74}{9}$ $\quad 8 \cdot 9 = 72, 72 + 2 = 74$

To convert a negative mixed numeral to fraction notation, we proceed as follows.

EXAMPLE 7 Convert to fraction notation: $-10\frac{7}{8}$.

$$-10\frac{7}{8} = -10 + \left(-\frac{7}{8}\right) = -\frac{10}{1} \cdot \frac{8}{8} - \frac{7}{8} = -\frac{80}{8} - \frac{7}{8} = -\frac{87}{8}$$

Do Exercises 6–9. ▶

7. $9\frac{1}{4}$

8. $-20\frac{2}{3}$

9. $1\frac{9}{13}$

Writing Mixed Numerals

We can find a mixed numeral for $\frac{5}{3}$ as follows:

$$\frac{5}{3} = \frac{3}{3} + \frac{2}{3} = 1 + \frac{2}{3} = 1\frac{2}{3}.$$

In terms of objects, we can think of $\frac{5}{3}$ as $\frac{3}{3}$, or 1, plus $\frac{2}{3}$, as shown below.

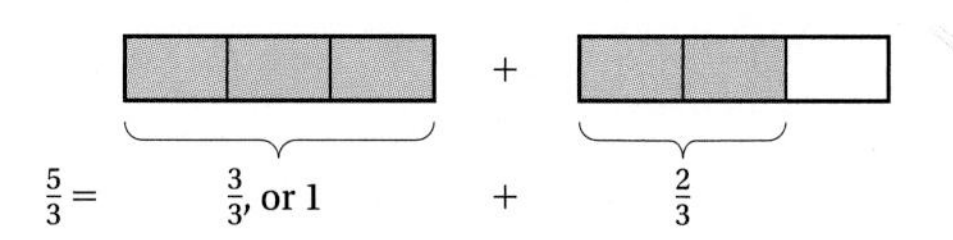

Fraction symbols like $\frac{5}{3}$ also indicate division; $\frac{5}{3}$ means $5 \div 3$. Let's divide the numerator by the denominator.

$$\begin{array}{r} 1 \\ 3\overline{)5} \\ \underline{3} \\ 2 \end{array} \leftarrow 2 \div 3 = \frac{2}{3}$$

Thus, $\frac{5}{3} = 1\frac{2}{3}$.

Answers

4. $\frac{22}{5}$ 5. $\frac{61}{10}$ 6. $\frac{29}{6}$ 7. $\frac{37}{4}$ 8. $-\frac{62}{3}$ 9. $\frac{22}{13}$

Guided Solution:
6. 24, 5, 29

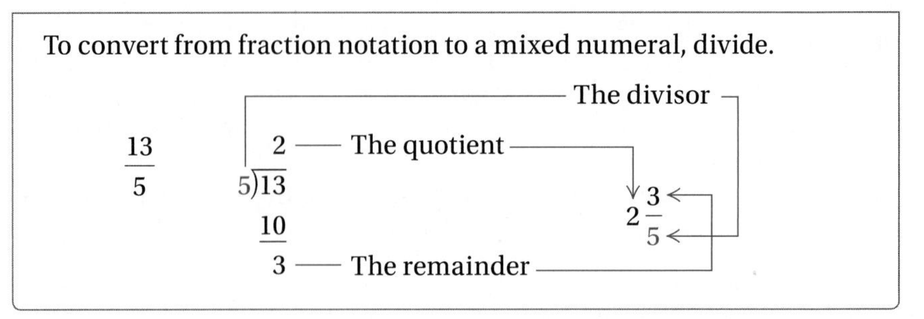

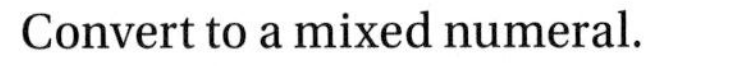
Convert to a mixed numeral.

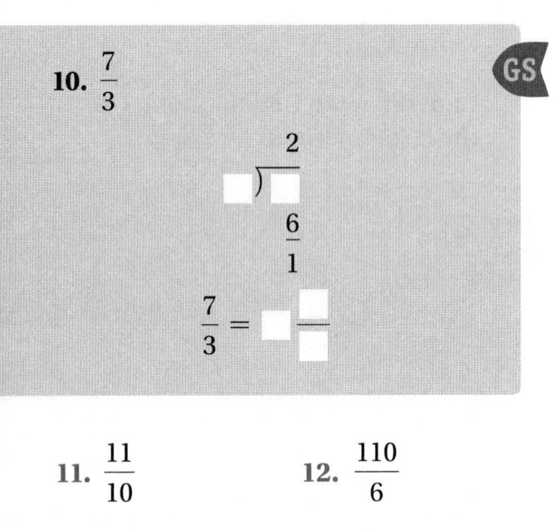

11. $\frac{11}{10}$

12. $\frac{110}{6}$

13. $\frac{229}{18}$

EXAMPLES Convert to a mixed numeral.

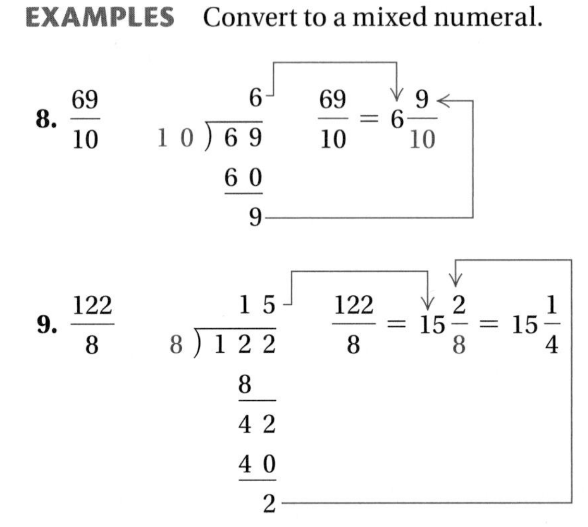

◀ **Do Exercises 10–13.**

The same type of procedure also works with negative numbers. Of course, the result will be a negative mixed numeral.

EXAMPLE 10 Convert $-\frac{17}{5}$ to a mixed numeral.

Since

$$5\overline{)17}\quad \text{quotient } 3,\ 15,\ \text{remainder } 2$$

we have

$$\frac{17}{5} = 3\frac{2}{5}.$$

Then

$$-\frac{17}{5} = -3\frac{2}{5}.$$

◀ **Do Exercises 14 and 15.**

A fraction larger than 1, such as $\frac{27}{8}$, is sometimes referred to as an "improper" fraction. We will not use this terminology because notation such as $\frac{27}{8}$, $\frac{11}{9}$, and $\frac{89}{10}$ is quite "proper" and very common in algebra.

Convert to a mixed numeral.

14. $-\frac{25}{4}$

15. $-\frac{33}{9}$

Answers

10. $2\frac{1}{3}$ 11. $1\frac{1}{10}$ 12. $18\frac{1}{3}$ 13. $12\frac{13}{18}$
14. $-6\frac{1}{4}$ 15. $-3\frac{2}{3}$

Guided Solution:

10. 3, 7, $2\frac{1}{3}$

b WRITING MIXED NUMERALS FOR QUOTIENTS

It is quite common when dividing whole numbers to write the quotient using a mixed numeral. The remainder is the numerator of the fraction part of the mixed numeral.

EXAMPLE 11 Divide. Write a mixed numeral for the answer.

$$42\overline{)8915}$$

We first divide as usual.

$$\begin{array}{r} 212 \\ 42\overline{)8915} \\ \underline{84} \\ 51 \\ \underline{42} \\ 95 \\ \underline{84} \\ 11 \end{array} \qquad \frac{8915}{42} = 212\frac{11}{42}$$

The answer is $212\frac{11}{42}$.

Do Exercises 16 and 17. ▶

Divide. Write a mixed numeral for the answer.

16. $6\overline{)4846}$

17. $45\overline{)6053}$

Answers

16. $807\frac{2}{3}$ 17. $134\frac{23}{45}$

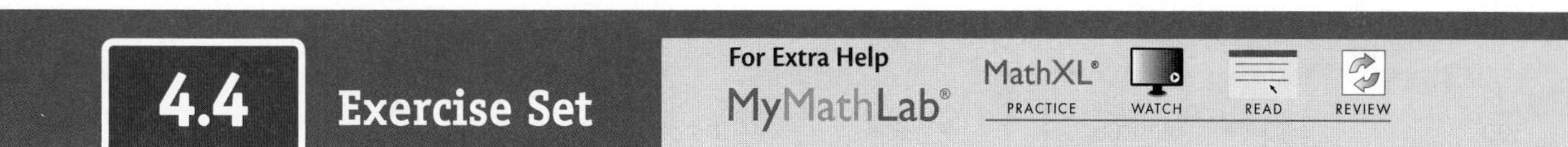

Reading Check

Determine whether each statement is true or false.

_______________ **RC1.** A mixed numeral consists of a whole-number part and a fraction less than 1.

_______________ **RC2.** The mixed numeral $5\frac{1}{4}$ represents $5 + \frac{1}{4}$.

_______________ **RC3.** It is never appropriate to use fraction notation such as $\frac{33}{25}$.

_______________ **RC4.** When a quotient is written as a mixed numeral, the divisor is the denominator of the fraction part (assuming that the fraction has not been simplified).

a Solve.

1. *Stone Walkways.* In order to construct a stone paver walkway, Cheryl ordered $2\frac{1}{2}$ tons of light-colored fieldstone pavers for the main portion of the walkway, $1\frac{7}{8}$ tons of dark-colored stone pavers for edging, and $1\frac{3}{4}$ tons of slab-type stones for steps. Convert $2\frac{1}{2}$, $1\frac{7}{8}$, and $1\frac{3}{4}$ to fraction notation.

2. *Greenhouse Dimensions.* A community college horticulture department builds a greenhouse that measures $32\frac{1}{2}$ ft × $20\frac{5}{6}$ ft × $11\frac{3}{4}$ ft. Convert $32\frac{1}{2}$, $20\frac{5}{6}$, and $11\frac{3}{4}$ to fraction notation.

3. *Quilt Design.* A quilt design requires three different fabrics. The quilter determines that she needs $\frac{17}{4}$ yd of a dominant fabric, $\frac{10}{3}$ yd of a contrasting fabric, and $\frac{9}{8}$ yd of a border fabric. Convert $\frac{17}{4}$, $\frac{10}{3}$, and $\frac{9}{8}$ to mixed numerals.

4. *Bake Sale.* The Valley Township Fire Department organized a bake sale as a fund-raiser for the local library. Each pie was cut into 8 pieces and each cake into 12 pieces. Sales totaled 73 pieces of pie, or $\frac{73}{8}$ pies, and 55 pieces of cake, or $\frac{55}{12}$ cakes. Convert $\frac{73}{8}$ and $\frac{55}{12}$ to mixed numerals.

Convert to fraction notation.

5. $5\frac{2}{3}$ **6.** $3\frac{4}{5}$ **7.** $3\frac{1}{4}$ **8.** $6\frac{1}{2}$

9. $-10\frac{1}{8}$ **10.** $-20\frac{1}{5}$ **11.** $5\frac{1}{10}$ **12.** $9\frac{1}{10}$

13. $20\frac{3}{5}$ **14.** $30\frac{4}{5}$ **15.** $-9\frac{5}{6}$ **16.** $-8\frac{7}{8}$

17. $7\frac{3}{10}$ **18.** $6\frac{9}{10}$ **19.** $1\frac{5}{8}$ **20.** $1\frac{3}{5}$

21. $-12\frac{3}{4}$ **22.** $-15\frac{2}{3}$ **23.** $-4\frac{3}{10}$ **24.** $-5\frac{7}{10}$

25. $2\frac{3}{100}$ **26.** $5\frac{7}{100}$ **27.** $66\frac{2}{3}$ **28.** $33\frac{1}{3}$

29. $-5\frac{29}{50}$ **30.** $-84\frac{3}{8}$ **31.** $101\frac{5}{16}$ **32.** $205\frac{3}{14}$

Convert to a mixed numeral.

33. $\frac{18}{5}$ **34.** $\frac{17}{4}$ **35.** $\frac{14}{3}$ **36.** $\frac{39}{8}$ **37.** $-\frac{27}{6}$

38. $-\frac{30}{9}$ **39.** $\frac{57}{10}$ **40.** $\frac{89}{10}$ **41.** $-\frac{53}{7}$ **42.** $-\frac{59}{8}$

43. $\frac{45}{6}$ **44.** $\frac{50}{8}$ **45.** $\frac{46}{4}$ **46.** $\frac{39}{9}$ **47.** $-\frac{12}{8}$

48. $-\frac{28}{6}$ **49.** $\frac{757}{100}$ **50.** $\frac{467}{100}$ **51.** $-\frac{345}{8}$ **52.** $-\frac{223}{4}$

b Divide. Write a mixed numeral for the answer.

53. $8\overline{)869}$

54. $3\overline{)2126}$

55. $5\overline{)3091}$

56. $9\overline{)9110}$

57. $21\overline{)852}$

58. $85\overline{)7670}$

59. $102\overline{)5612}$

60. $46\overline{)1081}$

61. $35\overline{)80{,}243}$

62. $152\overline{)26{,}107}$

Skill Maintenance

63. Round to the nearest hundred: 45,765. [1.6a]

64. Round to the nearest ten: 45,765. [1.6a]

Simplify. [3.5b]

65. $\frac{200}{375}$

66. $\frac{63}{75}$

67. $-\frac{160}{270}$

68. $-\frac{6996}{8028}$

Use $=$ or $\neq$ for $\square$ to write a true sentence. [3.5c]

69. $\frac{3}{8} \square \frac{6}{16}$

70. $\frac{-7}{10} \square \frac{-2}{3}$

Find the reciprocal of each number. [3.7a]

71. $\frac{9}{7}$

72. $-\frac{1}{8}$

Solve. [1.7b]

73. $48 \cdot t = 1680$

74. $10{,}000 = m + 3593$

Synthesis

Write a mixed numeral.

75. 🖩 $\frac{128{,}236}{541}$

76. 🖩 $\frac{103{,}676}{349}$

77. $\frac{56}{7} + \frac{2}{3}$

78. $-\frac{72}{12} - \frac{5}{6}$

79. There are $\frac{366}{7}$ weeks in a leap year.

80. There are $\frac{365}{7}$ weeks in a year.

Mid-Chapter Review

Concept Reinforcement

Determine whether each statement is true or false.

__________ **1.** If $\frac{a}{b} > \frac{c}{b}$, $a, b, c > 0$, then $a > c$. [4.3b]

__________ **2.** All positive mixed numerals represent numbers larger than 1. [4.4a]

__________ **3.** The least common multiple of two natural numbers is the smallest number that is a factor of both. [4.1a]

__________ **4.** To add fractions when denominators are the same, we keep the numerator and add the denominators. [4.2a]

Guided Solutions

GS Fill in each blank with the number that creates a correct solution.

5. Subtract: $\frac{11}{42} - \frac{3}{35}$. [4.3a]

$$\frac{11}{42} - \frac{3}{35} = \frac{11}{2 \cdot \square \cdot 7} - \frac{3}{\square \cdot 7}$$ Factoring the denominators

$$= \frac{11}{2 \cdot 3 \cdot 7} \cdot \left(\frac{\square}{\square}\right) - \frac{3}{5 \cdot 7} \cdot \left(\frac{2 \cdot 3}{2 \cdot 3}\right)$$ Multiplying by 1 to get the LCD

$$= \frac{11 \cdot \square}{2 \cdot 3 \cdot 7 \cdot \square} - \frac{3 \cdot 2 \cdot 3}{5 \cdot 7 \cdot 2 \cdot 3}$$ Multiplying

$$= \frac{\square}{2 \cdot 3 \cdot 5 \cdot 7} - \frac{\square}{2 \cdot 3 \cdot 5 \cdot 7}$$ Simplifying

$$= \frac{\square - \square}{2 \cdot 3 \cdot 5 \cdot 7} = \frac{\square}{\square}$$ Subtracting and simplifying

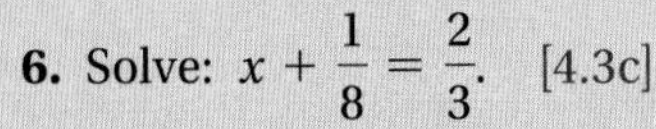

6. Solve: $x + \frac{1}{8} = \frac{2}{3}$. [4.3c]

$$x + \frac{1}{8} = \frac{2}{3}$$

$$x + \frac{1}{8} - \square = \frac{2}{3} - \square$$ Subtracting on both sides

$$x + 0 = \frac{2}{3} \cdot \frac{\square}{\square} - \frac{1}{8} \cdot \frac{\square}{\square}$$ Multiplying by 1 to get the LCD

$$x = \frac{\square}{24} - \frac{\square}{24}$$ Simplifying and multiplying

$$x = \frac{\square}{\square}$$ Subtracting

Mixed Review

7. Match each set of numbers in the first column with its least common multiple in the second column by drawing connecting lines. [4.1a]

45 and 50	120
50 and 80	720
30 and 24	400
18, 24, and 80	450
30, 45, and 50	

Calculate and simplify. [4.2a], [4.3a]

8. $\frac{1}{5} + \frac{7}{45}$

9. $\frac{5}{6} + \frac{2}{3} + \frac{7}{12}$

10. $\frac{1}{6} - \frac{2}{9}$

11. $\frac{-5}{18} + \frac{1}{15}$

12. $\frac{19}{48} - \frac{11}{30}$

13. $\frac{3}{7} + \frac{15}{17}$

14. $\frac{229}{720} - \frac{5}{24}$

15. $\frac{2}{35} - \frac{8}{65}$

Solve.

16. Miguel jogs for $\frac{4}{5}$ mi, rests, and then jogs for another $\frac{2}{3}$ mi. How far does he jog in all? [4.2b]

17. One weekend, Kirby spent $\frac{39}{5}$ hr playing two iPod games—Brain Challenge: Cerebral Burn and Scrabble: Go for a Triple Word Score. She spent $\frac{11}{4}$ hr playing Scrabble. How many hours did she spend playing Brain Challenge? [4.3d]

18. Arrange in order from smallest to largest: $\frac{4}{9}, \frac{3}{10}, \frac{2}{7}$, and $\frac{1}{5}$. [4.3b]

19. Solve: $\frac{2}{5} + x = \frac{9}{16}$. [4.3c]

20. Divide: $15\overline{)263}$. Write a mixed numeral for the answer. [4.4b]

21. Which of the following is fraction notation for $9\frac{3}{8}$? [4.4a]

A. $\frac{27}{8}$ **B.** $\frac{93}{8}$ **C.** $\frac{75}{8}$ **D.** $\frac{80}{3}$

22. Which of the following is mixed numeral notation for $\frac{39}{4}$? [4.4a]

A. $35\frac{1}{4}$ **B.** $\frac{4}{39}$ **C.** $9\frac{3}{4}$ **D.** $36\frac{3}{4}$

Understanding Through Discussion and Writing

23. Is the LCM of two numbers always larger than either number? Why or why not? [4.1a]

24. Explain the role of multiplication when adding using fraction notation with different denominators. [4.2a]

25. A student made the following error:

$$\frac{8}{5} - \frac{8}{2} = \frac{8}{3}.$$

Find at least two ways to convince him of the mistake. [4.3a]

26. Are the numbers $2\frac{1}{3}$ and $2 \cdot \frac{1}{3}$ equal? Why or why not? [4.4a]

STUDYING FOR SUCCESS *Taking a Test*

- ☐ Read each question carefully. Know what the question is before you answer it.
- ☐ Try to do all the questions the first time through, marking those to recheck if you have time at the end.
- ☐ Pace yourself.
- ☐ Write your answers in a neat and orderly manner.

4.5 Addition and Subtraction Using Mixed Numerals; Applications

OBJECTIVES

a Add using mixed numerals.

b Subtract using mixed numerals.

c Solve applied problems involving addition and subtraction with mixed numerals.

SKILL TO REVIEW

Objective 3.5b: Simplify fraction notation.

Simplify.

1. $\dfrac{18}{32}$ 2. $\dfrac{78}{117}$

a ADDITION USING MIXED NUMERALS

To add mixed numerals, we first add the fractions. Then we add the whole numbers.

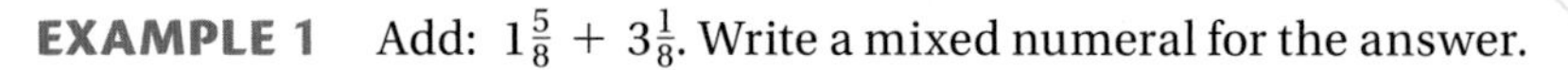

EXAMPLE 1 Add: $1\frac{5}{8} + 3\frac{1}{8}$. Write a mixed numeral for the answer.

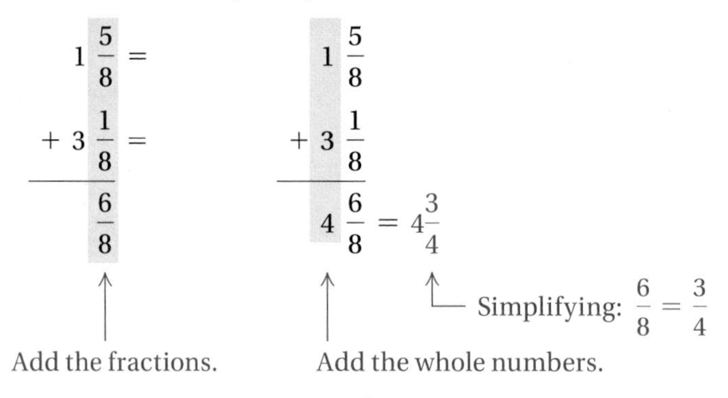

$$\begin{array}{rcl} 1\frac{5}{8} & = & 1\frac{5}{8} \\ +\,3\frac{1}{8} & = & +\,3\frac{1}{8} \\ \hline \frac{6}{8} & & 4\frac{6}{8} = 4\frac{3}{4} \end{array}$$

Add the fractions. Add the whole numbers. Simplifying: $\frac{6}{8} = \frac{3}{4}$

Sometimes we must write the fractional parts with a common denominator before we can add.

EXAMPLE 2 Add: $5\frac{2}{3} + 3\frac{5}{6}$. Write a mixed numeral for the answer.

The LCD is 6.

$$\begin{array}{rcl} 5\frac{2}{3}\cdot\frac{2}{2} & = & 5\frac{4}{6} \\ +\,3\frac{5}{6} & = & +\,3\frac{5}{6} \\ \hline & & 8\frac{9}{6} = 8 + \frac{9}{6} \end{array}$$

$= 8 + 1\frac{1}{2}$ Writing $\frac{9}{6}$ as a mixed numeral; $\frac{9}{6} = 1\frac{3}{6} = 1\frac{1}{2}$

$= 9\frac{1}{2}$

◀ **Do Margin Exercises 1 and 2.**

Add.

1. $2\dfrac{3}{10} + 5\dfrac{1}{10}$

2. $8\dfrac{2}{5} + 3\dfrac{7}{10}$

Answers

Skill to Review:
1. $\frac{9}{16}$ 2. $\frac{2}{3}$

Margin Exercises:
1. $7\frac{2}{5}$ 2. $12\frac{1}{10}$

EXAMPLE 3 Add: $10\frac{5}{6} + 7\frac{3}{8}$.

The LCD is 24.

$$\begin{aligned} 10\frac{5}{6}\cdot\frac{4}{4} &= 10\frac{20}{24} \\ +\ 7\frac{3}{8}\cdot\frac{3}{3} &= +\ 7\frac{9}{24} \\ \hline 17\frac{29}{24} &= 17 + \frac{29}{24} \\ &= 17 + 1\frac{5}{24} \quad \text{Writing } \tfrac{29}{24} \text{ as a mixed numeral, } 1\tfrac{5}{24} \\ &= 18\frac{5}{24} \end{aligned}$$

Do Exercise 3.

3. Add.

$$\begin{array}{r} 9\frac{3}{4} \\ +\ 3\frac{5}{6} \\ \hline \end{array}$$

b SUBTRACTION USING MIXED NUMERALS

Subtraction of mixed numerals is a lot like addition; we subtract the fractions and then the whole numbers.

EXAMPLE 4 Subtract: $7\frac{3}{4} - 2\frac{1}{4}$.

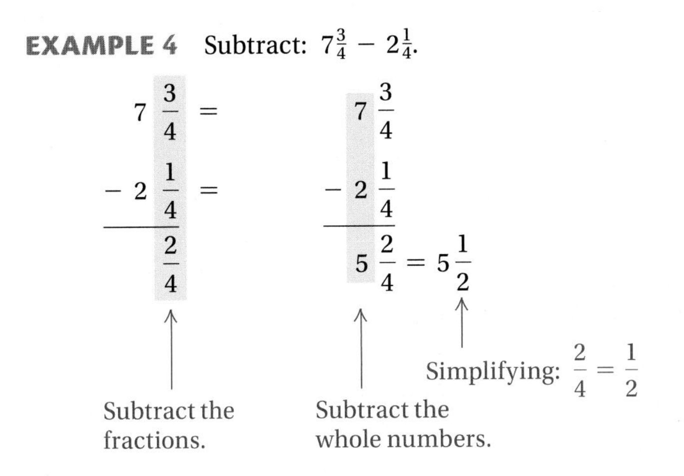

EXAMPLE 5 Subtract: $9\frac{4}{5} - 3\frac{1}{2}$.

The LCD is 10.

$$\begin{aligned} 9\frac{4}{5}\cdot\frac{2}{2} &= 9\frac{8}{10} \\ -\ 3\frac{1}{2}\cdot\frac{5}{5} &= -\ 3\frac{5}{10} \\ \hline &\quad\ 6\frac{3}{10} \end{aligned}$$

Do Exercises 4 and 5.

Subtract.

4.

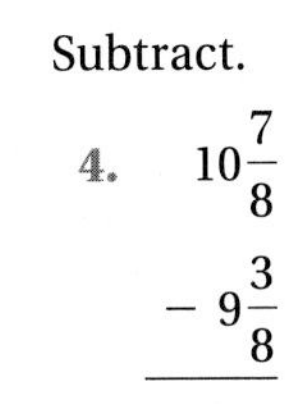

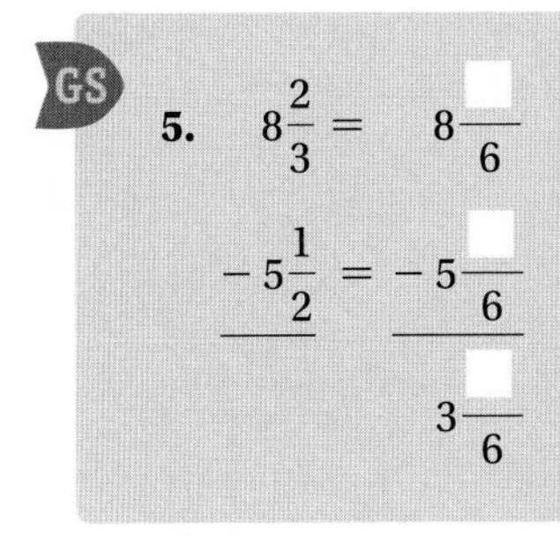

Answers

3. $13\frac{7}{12}$ 4. $1\frac{1}{2}$ 5. $3\frac{1}{6}$

Guided Solution:

5. 4, 3, 1

EXAMPLE 6 Subtract: $7\frac{1}{6} - 2\frac{1}{4}$.

The LCD is 12.

$$\begin{array}{rcr} 7\frac{1}{6}\cdot\frac{2}{2} & = & 7\frac{2}{12} \\ -\,2\frac{1}{4}\cdot\frac{3}{3} & = & -\,2\frac{3}{12} \\ \hline \end{array}$$

We cannot subtract $\frac{3}{12}$ from $\frac{2}{12}$.
← We borrow 1, or $\frac{12}{12}$, from 7:
$7\frac{2}{12} = 6 + 1 + \frac{2}{12} = 6 + \frac{12}{12} + \frac{2}{12} = 6\frac{14}{12}$.

We can write this as

$$\begin{array}{rcr} 7\frac{2}{12} & = & 6\frac{14}{12} \\ -\,2\frac{3}{12} & = & -\,2\frac{3}{12} \\ \hline & & 4\frac{11}{12}. \end{array}$$

6. Subtract.

$$\begin{array}{r} 8\frac{1}{9} \\ -\,4\frac{5}{6} \\ \hline \end{array}$$

◀ **Do Exercise 6.**

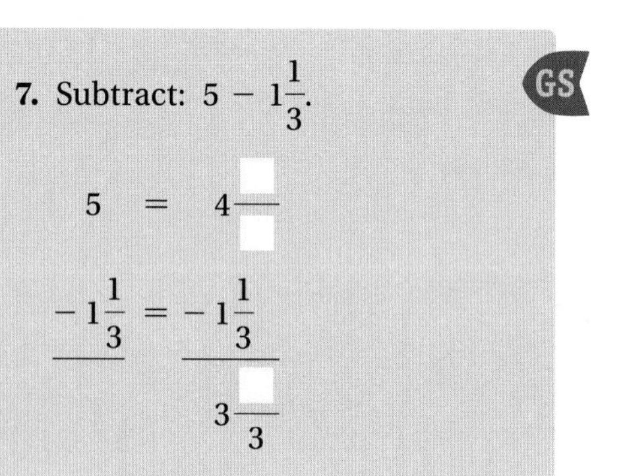
7. Subtract: $5 - 1\frac{1}{3}$.

$$\begin{array}{rcr} 5 & = & 4\frac{\square}{\square} \\ -\,1\frac{1}{3} & = & -\,1\frac{1}{3} \\ \hline & & 3\frac{\square}{3} \end{array}$$

EXAMPLE 7 Subtract: $12 - 9\frac{3}{8}$.

$$\begin{array}{rcr} 12 & = & 11\frac{8}{8} \\ -\,9\frac{3}{8} & = & -\,9\frac{3}{8} \\ \hline & & 2\frac{5}{8} \end{array}$$

$12 = 11 + 1 = 11 + \frac{8}{8} = 11\frac{8}{8}$

◀ **Do Exercise 7.**

C APPLICATIONS AND PROBLEM SOLVING

EXAMPLE 8 *Widening a Driveway.* Sherry and Woody are widening their existing $17\frac{1}{4}$-ft driveway by adding $5\frac{9}{10}$ ft on one side. What is the new width of the driveway?

1. **Familiarize.** We let $w =$ the new width of the driveway.
2. **Translate.** We translate as follows:

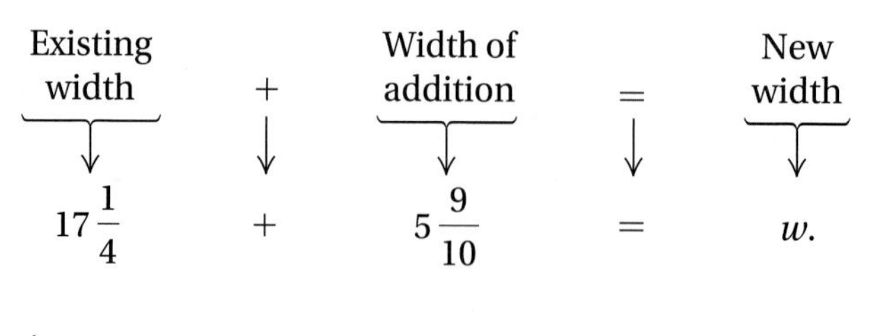

Answers

6. $3\frac{5}{18}$ 7. $3\frac{2}{3}$

Guided Solution:

7. $\frac{3}{3}$, 2

3. **Solve.** The translation tells us what to do. We add. The LCD is 20.

$$\begin{array}{rcrcr} 17\frac{1}{4} & = & 17\frac{1}{4}\cdot\frac{5}{5} & = & 17\frac{5}{20} \\ +\ 5\frac{9}{10} & = & +\ 5\frac{9}{10}\cdot\frac{2}{2} & = & +\ 5\frac{18}{20} \\ \hline & & & & 22\frac{23}{20} = 23\frac{3}{20} \end{array}$$

Thus, $w = 23\frac{3}{20}$.

4. **Check.** We check by repeating the calculation. We also note that the answer is larger than either of the given widths, which means that the answer is reasonable.

5. **State.** The new width of the driveway is $23\frac{3}{20}$ ft.

Do Exercise 8. ▶

8. *Travel Distance.* On a two-day business trip, Paul drove $213\frac{7}{10}$ mi the first day and $107\frac{5}{8}$ mi the second day. What was the total distance that Paul drove?

EXAMPLE 9 *Men's Long-Jump World Records.* On October 18, 1968, Bob Beamon set a world record of $29\frac{3}{16}$ ft for the long jump, a record that was not broken for nearly 23 years. This record-setting jump was significantly longer than the previous one of $27\frac{19}{48}$ ft, accomplished on May 29, 1965, by Ralph Boston. How much longer was Beamon's jump than Boston's?

Source: "Long jump world record progression," en.wikipedia.org

1. **Familiarize.** The phrase "how much longer" indicates subtraction. We let $w =$ the difference in the world records.

2. **Translate.** We translate as follows:

$$\underbrace{\text{Beamon's jump}}_{29\frac{3}{16}} \ \ - \ \ \underbrace{\text{Boston's jump}}_{27\frac{19}{48}} \ \ = \ \ \underbrace{\text{Difference in length}}_{w.}$$

3. **Solve.** To solve the equation, we carry out the subtraction. The LCD is 48.

$$\begin{array}{rcrcrcr} 29\frac{3}{16} & = & 29\frac{3}{16}\cdot\frac{3}{3} & = & 29\frac{9}{48} & = & 28\frac{57}{48} \\ -\ 27\frac{19}{48} & = & -\ 27\frac{19}{48} & = & -\ 27\frac{19}{48} & = & -\ 27\frac{19}{48} \\ \hline & & & & & & 1\frac{38}{48} = 1\frac{19}{24} \end{array}$$

Thus, $w = 1\frac{19}{24}$.

4. **Check.** To check, we add the difference, $1\frac{19}{24}$, to Boston's jump:

$$27\frac{19}{48} + 1\frac{19}{24} = 27\frac{19}{48} + 1\frac{38}{48}$$
$$= 28\frac{57}{48} = 29\frac{9}{48} = 29\frac{3}{16}.$$

This checks.

5. **State.** Beamon's jump was $1\frac{19}{24}$ ft longer than Boston's.

Do Exercise 9. ▶

9. *Nail Length.* A 30d nail is $4\frac{1}{2}$ in. long. A 5d nail is $1\frac{3}{4}$ in. long. How much longer is the 30d nail than the 5d nail? (The "d" stands for "penny," which was used years ago in England to specify the number of pennies needed to buy 100 nails. Today, "penny" is used only to indicate the length of the nail.)

Source: *Pocket Ref,* 2nd ed., by Thomas J. Glover, p. 280, Sequoia Publishing, Inc., Littleton, CO

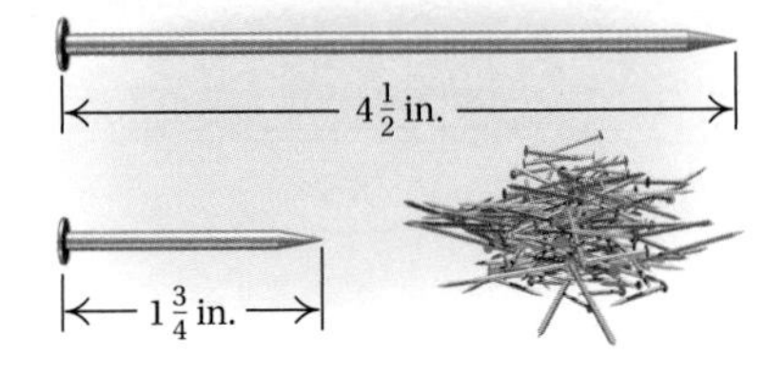

Answers

8. $321\frac{13}{40}$ mi 9. $2\frac{3}{4}$ in.

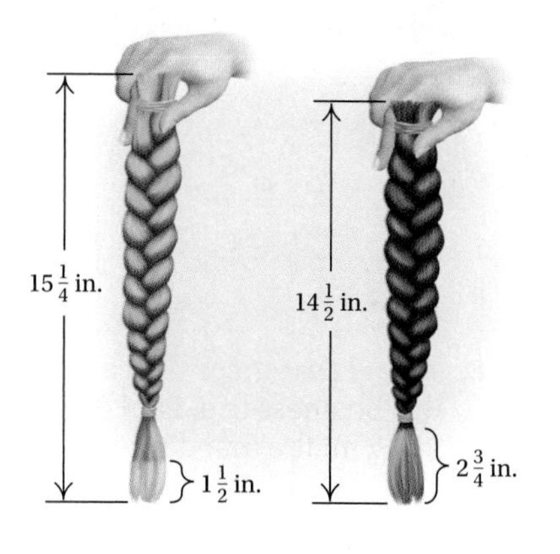

EXAMPLE 10 *Locks of Love.* Locks of Love, Inc., is a non-profit organization that provides hairpieces for children who have lost their hair as a result of medical conditions. The minimum length for hair that is donated is 10 in. Karissa and Cayla allowed their hair to grow in order to donate it. The length cut from Karissa's hair was $15\frac{1}{4}$ in., and the length cut from Cayla's hair was $14\frac{1}{2}$ in. After the hair was cut, it was discovered that the ends of each lock had highlighting that needed to be trimmed. Because of the highlighting, $1\frac{1}{2}$ in. was cut from Karissa's lock of hair and $2\frac{3}{4}$ in. was cut from Cayla's. In all, what was the total usable length of hair that Karissa and Cayla donated?

1. **Familiarize.** We let l = the total usable length of hair that Karissa and Cayla donated.
2. **Translate.** The length l is the sum of the lengths that were cut, minus the sum of the lengths that were trimmed from the locks. Thus we have

$$l = (15\tfrac{1}{4} + 14\tfrac{1}{2}) - (1\tfrac{1}{2} + 2\tfrac{3}{4}).$$

3. **Solve.** This is a three-step problem.

a) We first add the two lengths $15\frac{1}{4}$ and $14\frac{1}{2}$.

$$\begin{array}{rcr} 15\frac{1}{4} & = & 15\frac{1}{4} \\ +\ 14\frac{1}{2} & = & +\ 14\frac{2}{4} \\ \hline & & 29\frac{3}{4} \end{array}$$

b) Next, we add the two lengths $1\frac{1}{2}$ and $2\frac{3}{4}$.

$$\begin{array}{rcr} 1\frac{1}{2} & = & 1\frac{2}{4} \\ +\ 2\frac{3}{4} & = & +\ 2\frac{3}{4} \\ \hline & & 3\frac{5}{4} = 4\frac{1}{4} \end{array}$$

c) Finally, we subtract $4\frac{1}{4}$ from $29\frac{3}{4}$.

$$\begin{array}{r} 29\frac{3}{4} \\ -\ \ 4\frac{1}{4} \\ \hline 25\frac{2}{4} = 25\frac{1}{2} \end{array}$$

Thus, $l = 25\frac{1}{2}$.

4. **Check.** We can check by doing the problem a different way. We can subtract the trimmed length from each lock, then add the adjusted lengths together.

Karissa's lock	Cayla's lock	Sum of lengths
$15\frac{1}{4} = 14\frac{5}{4}$	$14\frac{1}{2} = 13\frac{6}{4}$	$13\frac{3}{4}$
$-\ 1\frac{1}{2} = -\ 1\frac{2}{4}$	$-\ 2\frac{3}{4} = -\ 2\frac{3}{4}$	$+\ 11\frac{3}{4}$
$13\frac{3}{4}$	$11\frac{3}{4}$	$24\frac{6}{4} = 25\frac{1}{2}$

We obtained the same answer, so our answer checks.

5. **State.** The sum of the usable lengths of hair donated was $25\frac{1}{2}$ in.

◀ **Do Exercise 10.**

10. *Liquid Fertilizer.* There is $283\frac{5}{8}$ gal of liquid fertilizer in a fertilizer application tank. After applying $178\frac{2}{3}$ gal to a soybean field, the farmer requests that Braden's Farm Supply deliver an additional 250 gal to the tank. How many gallons of fertilizer are in the tank after the delivery?

Answer

10. $354\frac{23}{24}$ gal

Exercise Set

For Extra Help MyMathLab® MathXL® PRACTICE WATCH READ REVIEW

Reading Check

Match each addition or subtraction with the correct first step from the following list.

a) Add the fractions.

b) Write the fractional parts with a common denominator.

c) Rename 5 as $4\frac{9}{9}$.

d) Borrow 1 from 5 and add it to $\frac{1}{9}$.

____ **RC1.** $5\frac{1}{9} - 3\frac{4}{9}$

____ **RC2.** $5\frac{4}{9} + 3\frac{1}{9}$

____ **RC3.** $5\frac{4}{9} + 3\frac{1}{18}$

____ **RC4.** $5 - 3\frac{1}{9}$

a Add. Write a mixed numeral for the answer.

1. $20 + 8\frac{3}{4}$

2. $37 + 18\frac{2}{3}$

3. $129\frac{7}{8} + 56$

4. $2003\frac{4}{11} + 59$

5. $2\frac{7}{8} + 3\frac{5}{8}$

6. $4\frac{5}{6} + 3\frac{5}{6}$

7. $1\frac{1}{4} + 1\frac{2}{3}$

8. $4\frac{1}{3} + 5\frac{2}{9}$

9. $8\frac{3}{4} + 5\frac{5}{6}$

10. $4\frac{3}{8} + 6\frac{5}{12}$

11. $3\frac{2}{5} + 8\frac{7}{10}$

12. $5\frac{1}{2} + 3\frac{7}{10}$

13. $5\frac{3}{8} + 10\frac{5}{6}$

14. $\frac{5}{8} + 1\frac{5}{6}$

15. $12\frac{4}{5} + 8\frac{7}{10}$

16. $15\frac{5}{8} + 11\frac{3}{4}$

17. $14\frac{5}{8} + 13\frac{1}{4}$

18. $16\frac{1}{4} + 15\frac{7}{8}$

19. $7\frac{1}{8} + 9\frac{2}{3} + 10\frac{3}{4}$

20. $45\frac{2}{3} + 31\frac{3}{5} + 12\frac{1}{4}$

b Subtract. Write a mixed numeral for the answer.

21. $4\frac{1}{5} - 2\frac{3}{5}$

22. $5\frac{1}{8} - 2\frac{3}{8}$

23. $6\frac{3}{5} - 2\frac{1}{2}$

24. $7\frac{2}{3} - 6\frac{1}{2}$

25. $34\frac{1}{3} - 12\frac{5}{8}$

26. $23\frac{5}{16} - 16\frac{3}{4}$

27. $21 - 8\frac{3}{4}$

28. $42 - 3\frac{7}{8}$

29. $34 - 18\frac{5}{8}$

30. $23 - 19\frac{3}{4}$

31. $21\frac{1}{6} - 13\frac{3}{4}$

32. $42\frac{1}{10} - 23\frac{7}{12}$

33. $14\frac{1}{8} - \frac{3}{4}$

34. $28\frac{1}{6} - 5$

35. $25\frac{1}{9} - 13\frac{5}{6}$

36. $23\frac{5}{16} - 14\frac{7}{12}$

c Solve.

37. *Planting Flowers.* A landscaper planted $4\frac{1}{2}$ flats of impatiens, $6\frac{2}{3}$ flats of snapdragons, and $3\frac{3}{8}$ flats of phlox. How many flats did she plant altogether?

38. A plumber uses two pipes, each of length $51\frac{5}{16}$ in., and one pipe of length $34\frac{3}{4}$ in. when installing a shower. How much pipe was used in all?

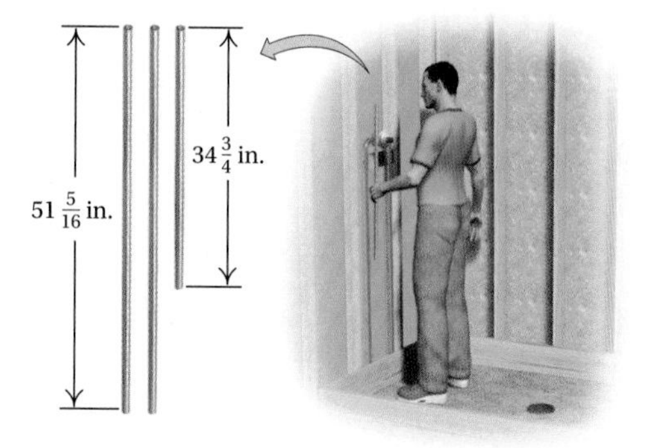

39. For a family party, Dana bought packages of cheese weighing $1\frac{2}{3}$ lb and $5\frac{3}{4}$ lb. What was the total weight of the cheese?

40. Marsha's Butcher Shop sold packages of sliced turkey breast weighing $1\frac{1}{3}$ lb and $4\frac{3}{5}$ lb. What was the total weight of the meat?

41. Casey's beagle is $14\frac{1}{4}$ in. from shoulder to floor, and her basset hound is $13\frac{5}{16}$ in. from shoulder to floor. How much shorter is her basset hound?

42. *Winterizing a Swimming Pool.* To winterize their swimming pool, the Jablonskis are draining the water into a nearby field. The distance to the field is $103\frac{1}{2}$ ft. Because their only hose measures $62\frac{3}{4}$ ft, they need to buy an additional hose. How long must the new hose be?

43. *Upholstery Fabric.* Executive Car Care sells 45-in. upholstery fabric for car restoration. Art bought $9\frac{1}{4}$ yd and $10\frac{5}{6}$ yd for two car projects. How many yards did Art buy?

44. *Painting.* A painter used $1\frac{3}{4}$ gal of paint for the Garcias' living room and $1\frac{1}{3}$ gal for their family room. How much paint was used in all?

45. *Sewing from a Pattern.* Using 45-in. fabric, Regan needs $1\frac{3}{8}$ yd for a dress, $\frac{5}{8}$ yd of contrasting fabric for the band at the bottom, and $3\frac{3}{8}$ yd for a coordinating jacket. How many yards of 45-in. fabric are needed in all?

46. *Sewing from a Pattern.* Using 45-in. fabric, Sarah needs $2\frac{3}{4}$ yd for a dress and $3\frac{1}{2}$ yd for a coordinating jacket. How many yards of 45-in. fabric are needed in all?

47. Kim Park is a computer technician. One day, she drove $180\frac{7}{10}$ mi away from Los Angeles for a service call. The next day, she drove $85\frac{1}{2}$ mi back toward Los Angeles for another service call. How far was she then from Los Angeles?

48. Jose is $4\frac{1}{2}$ in. taller than his daughter, Teresa. Teresa is $66\frac{2}{3}$ in. tall. How tall is Jose?

49. Creative Glass sells a framed beveled mirror as shown below. Its dimensions are $30\frac{1}{2}$ in. wide by $36\frac{5}{8}$ in. high. What is the perimeter of (total distance around) the framed mirror?

50. *Book Size.* One standard book size is $8\frac{1}{2}$ in. by $9\frac{3}{4}$ in. What is the perimeter of (total distance around) the front cover of such a book?

51. Rene is $5\frac{1}{4}$ in. taller than his son, who is $72\frac{5}{6}$ in. tall. How tall is Rene?

52. A Boeing 767 flew 640 mi on a nonstop flight. On the return flight, it landed after having flown $320\frac{3}{10}$ mi. How far was the plane from its original point of departure?

Find the perimeter of (distance around) each figure.

53.

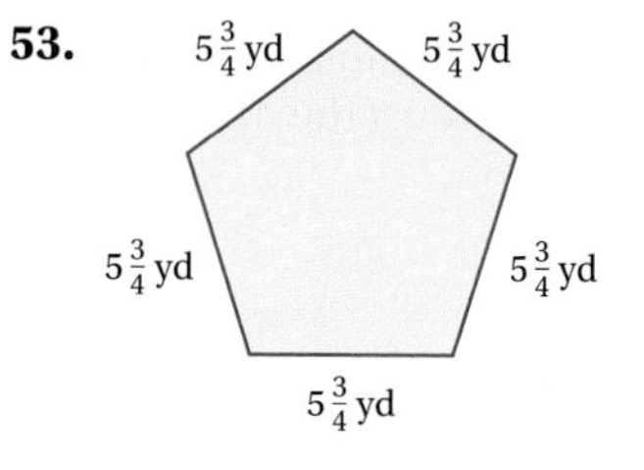

54.

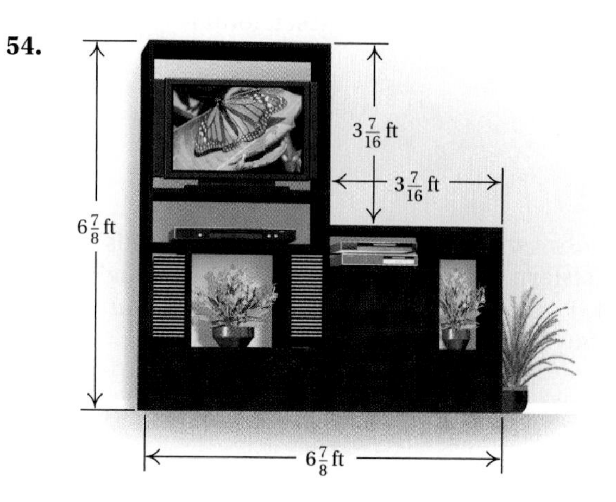

Find the length d in each figure.

55.

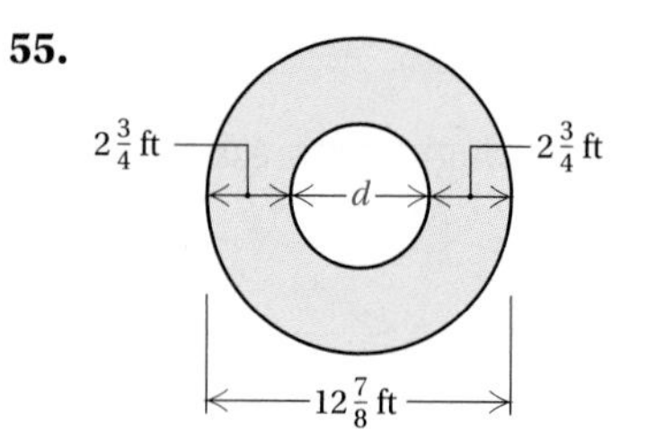

56.

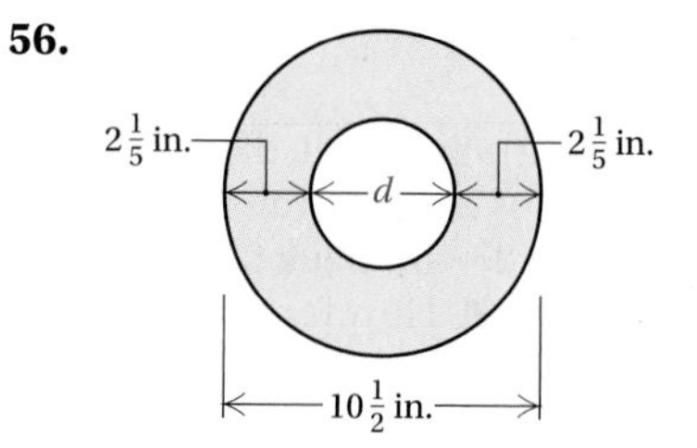

57. *Stone Bench.* Baytown Village Stone Creations is making a custom stone bench as shown below. The recommended height for the bench is 18 in. The depth of the stone bench is $3\frac{3}{8}$ in. Each of the two supporting legs is made up of three stacked stones. Two of the stones measure $3\frac{1}{2}$ in. and $5\frac{1}{4}$ in. How much must the third stone measure?

58. *Window Dimensions.* The Sanchez family is replacing a window in their home. The original window measures $4\frac{5}{6}$ ft $\times$ $8\frac{1}{4}$ ft. The new window is $2\frac{1}{3}$ ft wider. What are the dimensions of the new window?

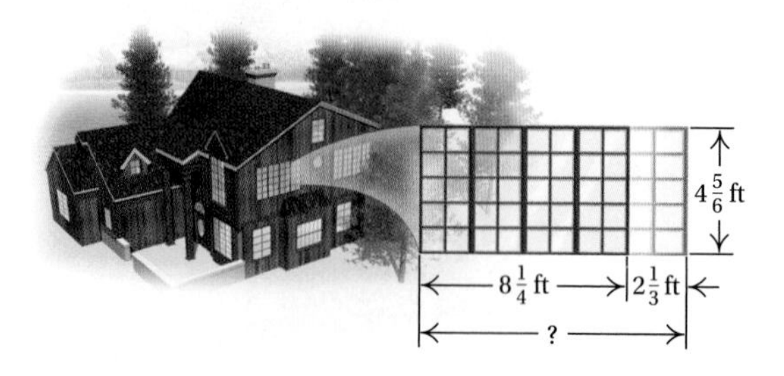

59. *Carpentry.* When cutting wood with a saw, a carpenter must take into account the thickness of the saw blade. Suppose that from a piece of wood 36 in. long, a carpenter cuts a $15\frac{3}{4}$-in. length with a saw blade that is $\frac{1}{8}$ in. thick. How long is the piece that remains?

60. *Cutco Cutlery.* The Essentials 5-piece set sold by Cutco contains three knives: $7\frac{5}{8}''$ Petite Chef, $6\frac{3}{4}''$ Petite Carver, and $2\frac{3}{4}''$ Paring Knife. How much larger is the blade of the Petite Chef than that of the Petite Carver? than that of the Paring Knife?

Source: Cutco Cutlery Corporation

61. *Interior Design.* Eric worked $10\frac{1}{2}$ hr over a three-day period on an interior design project. If he worked $2\frac{1}{2}$ hr on the first day and $4\frac{1}{5}$ hr on the second, how many hours did Eric work on the third day?

62. *Painting.* Geri had $3\frac{1}{2}$ gal of paint. It took $2\frac{3}{4}$ gal to paint the family room. She estimated that it would take $2\frac{1}{4}$ gal to paint the living room. How much more paint did Geri need?

63. *Fly Fishing.* Bryn is putting together a fly fishing line and uses $58\frac{5}{8}$ ft of slow-sinking fly line and $8\frac{3}{4}$ ft of leader line. She uses $\frac{3}{8}$ ft of the slow-sinking fly line to connect the two lines. The knot used to connect the fly to the leader line uses $\frac{1}{6}$ ft of the leader line. How long is the finished fly fishing line?

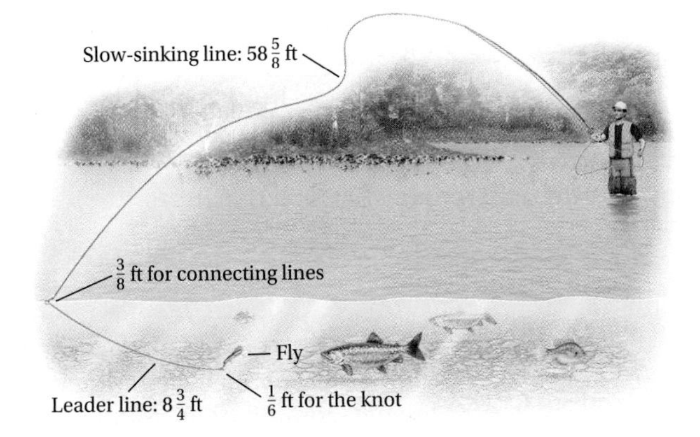

64. Find the smallest length of a bolt that will pass through a piece of tubing with an outside diameter of $\frac{1}{2}$ in., a washer $\frac{1}{16}$ in. thick, a piece of tubing with a $\frac{3}{4}$-in. outside diameter, another washer, and a nut $\frac{3}{16}$ in. thick.

Skill Maintenance

Determine whether the first number is divisible by the second. [3.2a]

65. 9993 by 3

66. 9993 by 9

67. 2345 by 9

68. 2345 by 5

69. 2335 by 10

70. 7764 by 6

71. 18,888 by 8

72. 18,888 by 4

73. Write expanded notation for 38,125. [1.1b]

74. Write a word name for 2,005,689. [1.1c]

75. Write exponential notation for $9 \cdot 9 \cdot 9 \cdot 9$. [1.9a]

76. Evaluate: 3^4. [1.9b]

Synthesis

Perform the indicated operation. Write a mixed numeral for the answer.

77. $-5\frac{1}{4} + \left(-3\frac{3}{8}\right)$

78. $-8\frac{1}{3} + \left(-2\frac{5}{6}\right)$

79. $-4\frac{5}{12} - 6\frac{5}{8}$

80. $-9\frac{1}{2} - 7\frac{3}{5}$

4.6 Multiplication and Division Using Mixed Numerals; Applications

OBJECTIVES

a Multiply using mixed numerals.

b Divide using mixed numerals.

c Solve applied problems involving multiplication and division with mixed numerals.

SKILL TO REVIEW

Objective 3.7b: Divide and simplify using fraction notation.

Divide and simplify.

1. $85 \div \frac{17}{5}$

2. $\frac{7}{65} \div \left(-\frac{21}{25}\right)$

Multiply.

1. $6 \cdot 3\frac{1}{3}$

2. $2\frac{1}{2} \cdot \frac{3}{4}$

Multiply.

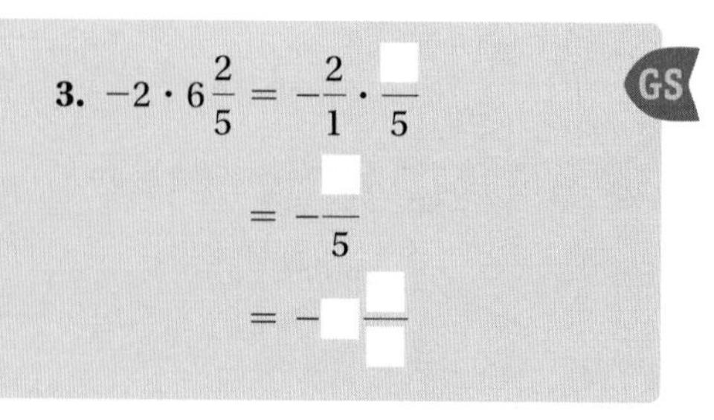

4. $3\frac{1}{3} \cdot 2\frac{1}{2}$

Answers

Skill to Review:

1. 25 2. $-\frac{5}{39}$

Margin Exercises:

1. 20 2. $1\frac{7}{8}$ 3. $-12\frac{4}{5}$ 4. $8\frac{1}{3}$

Guided Solution:

3. 32, 64, $12\frac{4}{5}$

a MULTIPLICATION USING MIXED NUMERALS

Carrying out addition and subtraction with mixed numerals is usually easier if the numbers are left as mixed numerals. With multiplication and division, however, it is easier to convert the numbers to fraction notation first.

MULTIPLICATION USING MIXED NUMERALS

To multiply using mixed numerals, first convert to fraction notation and multiply. Then convert the answer to a mixed numeral, if appropriate.

EXAMPLE 1 Multiply: $6 \cdot 2\frac{1}{2}$.

$$6 \cdot 2\frac{1}{2} = \frac{6}{1} \cdot \frac{5}{2} = \frac{6 \cdot 5}{1 \cdot 2} = \frac{2 \cdot 3 \cdot 5}{2 \cdot 1} = \frac{2}{2} \cdot \frac{3 \cdot 5}{1} = 1 \cdot \frac{3 \cdot 5}{1} = 15$$

Convert the numbers to fraction notation first.

EXAMPLE 2 Multiply: $3\frac{1}{2} \cdot \frac{3}{4}$.

$$3\frac{1}{2} \cdot \frac{3}{4} = \frac{7}{2} \cdot \frac{3}{4} = \frac{21}{8} = 2\frac{5}{8}$$

Recall that common denominators are *not* required when multiplying fractions.

◀ **Do Margin Exercises 1 and 2.**

EXAMPLE 3 Multiply: $-8 \cdot 4\frac{2}{3}$.

$$-8 \cdot 4\frac{2}{3} = -\frac{8}{1} \cdot \frac{14}{3} = -\frac{112}{3} = -37\frac{1}{3}$$

EXAMPLE 4 Multiply: $2\frac{1}{4} \cdot 5\frac{2}{3}$.

$$2\frac{1}{4} \cdot 5\frac{2}{3} = \frac{9}{4} \cdot \frac{17}{3} = \frac{3 \cdot 3 \cdot 17}{2 \cdot 2 \cdot 3} = \frac{3}{3} \cdot \frac{3 \cdot 17}{2 \cdot 2} = \frac{51}{4} = 12\frac{3}{4}$$

◀ **Do Exercises 3 and 4.**

Caution!

Note that $2\frac{1}{4} \cdot 5\frac{2}{3} \neq 10\frac{2}{12}$. A common error is to multiply the whole numbers and then the fractions. This does not give the correct answer, $12\frac{3}{4}$, which is found by converting to fraction notation first.

b DIVISION USING MIXED NUMERALS

The division $1\frac{1}{2} \div \frac{1}{6}$ is shown here. This division means "How many $\frac{1}{6}$'s are in $1\frac{1}{2}$?" We see that the answer is 9.

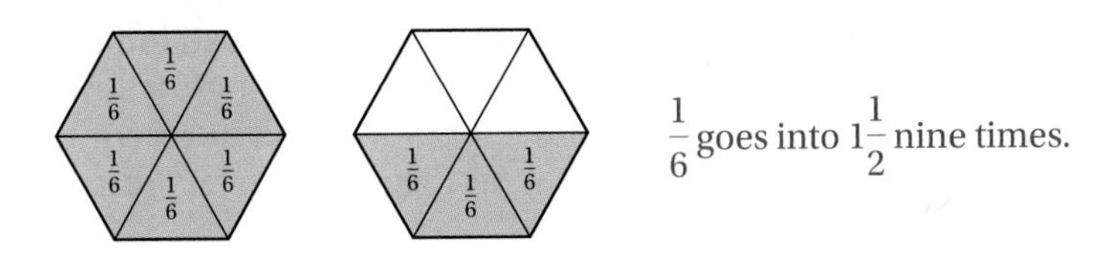

$\frac{1}{6}$ goes into $1\frac{1}{2}$ nine times.

When we divide using mixed numerals, we convert to fraction notation first. Recall that to divide by a fraction, we multiply by its reciprocal.

$$1\frac{1}{2} \div \frac{1}{6} = \frac{3}{2} \div \frac{1}{6} = \frac{3}{2} \cdot \frac{6}{1}$$
$$= \frac{3 \cdot 6}{2 \cdot 1} = \frac{3 \cdot 3 \cdot 2}{2 \cdot 1} = \frac{3 \cdot 3}{1} \cdot \frac{2}{2} = \frac{3 \cdot 3}{1} \cdot 1 = 9$$

DIVISION USING MIXED NUMERALS

To divide using mixed numerals, first write fraction notation and divide. Then convert the answer to a mixed numeral, if appropriate.

EXAMPLE 5 Divide: $32 \div 3\frac{1}{5}$.

$$32 \div 3\frac{1}{5} = \frac{32}{1} \div \frac{16}{5} \quad \text{Writing fraction notation}$$
$$= \frac{32}{1} \cdot \frac{5}{16} = \frac{32 \cdot 5}{1 \cdot 16} = \frac{2 \cdot 16 \cdot 5}{1 \cdot 16} = \frac{16}{16} \cdot \frac{2 \cdot 5}{1} = 1 \cdot \frac{2 \cdot 5}{1} = 10$$

Remember to multiply by the reciprocal.

Do Exercise 5.

EXAMPLE 6 Divide: $35 \div 4\frac{1}{3}$.

$$35 \div 4\frac{1}{3} = \frac{35}{1} \div \frac{13}{3} = \frac{35}{1} \cdot \frac{3}{13} = \frac{105}{13} = 8\frac{1}{13}$$

Do Exercise 6.

EXAMPLE 7 Divide: $2\frac{1}{3} \div 1\frac{3}{4}$.

$$2\frac{1}{3} \div 1\frac{3}{4} = \frac{7}{3} \div \frac{7}{4} = \frac{7}{3} \cdot \frac{4}{7} = \frac{7 \cdot 4}{3 \cdot 7} = \frac{7}{7} \cdot \frac{4}{3} = 1 \cdot \frac{4}{3} = \frac{4}{3} = 1\frac{1}{3}$$

EXAMPLE 8 Divide: $-1\frac{3}{5} \div \left(-3\frac{1}{3}\right)$

$$-1\frac{3}{5} \div \left(-3\frac{1}{3}\right) = -\frac{8}{5} \div \left(-\frac{10}{3}\right) = -\frac{8}{5} \cdot \left(-\frac{3}{10}\right)$$
$$= \frac{8 \cdot 3}{5 \cdot 10} = \frac{2 \cdot 4 \cdot 3}{5 \cdot 2 \cdot 5} = \frac{2}{2} \cdot \frac{4 \cdot 3}{5 \cdot 5} = \frac{12}{25}$$

Do Exercises 7 and 8.

5. Divide: $84 \div 5\frac{1}{4}$.

6. Divide: $26 \div 3\frac{1}{2}$.

Divide.

GS 7. $2\frac{1}{4} \div 1\frac{1}{5}$

$= \frac{\square}{4} \div \frac{\square}{5}$

$= \frac{9}{4} \cdot \frac{5}{\square}$

$= \frac{3 \cdot 3 \cdot 5}{2 \cdot 2 \cdot 2 \cdot \square}$

$= \frac{\square}{\square} \cdot \frac{3 \cdot 5}{2 \cdot 2 \cdot 2}$

$= \frac{15}{\square}$

$= \square\frac{\square}{\square}$

8. $1\frac{3}{4} \div \left(-2\frac{1}{2}\right)$

Answers

5. 16 6. $7\frac{3}{7}$ 7. $1\frac{7}{8}$ 8. $-\frac{7}{10}$

Guided Solution:

7. 9, 6, 6, 3, $\frac{3}{3}$, 8, $1\frac{7}{8}$

C APPLICATIONS AND PROBLEM SOLVING

EXAMPLE 9 *Training Regimens.* Fitness trainers suggest training regimens for athletes who are preparing to run marathons and mini-marathons. One suggested twelve-week regimen combines days of short, easy running with other days of cross-training, rest, and long-distance running. During week nine, this regimen calls for a long-distance run of 10 mi, which is $2\frac{1}{2}$ times the length of the long-distance run recommended for week one. What is the length of the long-distance run recommended for week one?

Source: shape.com

1. **Familiarize.** We ask the question "10 is $2\frac{1}{2}$ times what number?" We let $r =$ the length of the long-distance run recommended for week one.
2. **Translate.** The problem can be translated to an equation.

$$\underbrace{\text{Length of run for week nine}}_{10} \quad \underset{=}{\text{is}} \quad \underset{2\frac{1}{2}}{2\frac{1}{2}} \quad \underset{\cdot}{\text{times}} \quad \underbrace{\text{Length of run for week one}}_{r}$$

3. **Solve.** To solve the equation, we divide on both sides.

$$10 = \frac{5}{2} \cdot r \qquad \text{Converting } 2\frac{1}{2} \text{ to fraction notation}$$

$$10 \div \frac{5}{2} = r \qquad \text{Dividing by } \frac{5}{2} \text{ on both sides}$$

$$10 \cdot \frac{2}{5} = r \qquad \text{Multiplying by the reciprocal of } \frac{5}{2}$$

$$4 = r \qquad \text{Simplifying: } 10 \cdot \frac{2}{5} = \frac{20}{5} = 4$$

4. **Check.** If the length of the long-distance run recommended for week one is 4 mi, we find the length of the run recommended for week nine by multiplying 4 by $2\frac{1}{2}$.

$$2\frac{1}{2} \cdot 4 = \frac{5}{2} \cdot 4 = \frac{20}{2} = 10$$

The answer checks.

5. **State.** The regimen recommends a long-distance run of 4 mi for week one.

Do Exercises 9 and 10.

Solve.

9. Kyle's pickup truck travels on an interstate highway at 65 mph for $3\frac{1}{2}$ hr. How far does it travel?

10. Holly's minivan traveled 302 mi on $15\frac{1}{10}$ gal of gas. How many miles per gallon did it get?

EXAMPLE 10 *Flooring.* Ann and Tony plan to redo the floor of their living room. Since part of the room will be covered by a rug, they want to lay a hardwood floor only on the part of the room not under the rug. If the room is $22\frac{1}{2}$ ft by $15\frac{1}{2}$ ft and the rug is 9 ft by 12 ft, how much hardwood flooring do they need? How much hardwood flooring would it take to cover the entire floor of the room?

Answers

9. $227\frac{1}{2}$ mi 10. 20 mpg

1. **Familiarize.** We draw a diagram and let $B =$ the area of the room, $R =$ the area of the rug, and $H =$ the area to be covered by hardwood flooring.
2. **Translate.** This is a multistep problem. We first find the area of the room, B, and the area of the rug, R. Then $H = B - R$. We find each area using the formula for the area of a rectangle: $A = l \times w$.
3. **Solve.** We carry out the calculations.

$$B = \text{length} \times \text{width} \qquad R = \text{length} \times \text{width}$$
$$= 22\frac{1}{2} \cdot 15\frac{1}{2} \qquad = 12 \cdot 9$$
$$\qquad \qquad = 108 \text{ sq ft}$$
$$= \frac{45}{2} \cdot \frac{31}{2}$$
$$= \frac{1395}{4} = 348\frac{3}{4} \text{ sq ft}$$

Then $H = B - R$
$$= 348\frac{3}{4} \text{ sq ft} - 108 \text{ sq ft}$$
$$= 240\frac{3}{4} \text{ sq ft}$$

4. **Check.** We can perform a check by repeating the calculations.
5. **State.** Ann and Tony will need $240\frac{3}{4}$ sq ft of hardwood flooring. It would take $348\frac{3}{4}$ sq ft of hardwood flooring to cover the entire floor of the room.

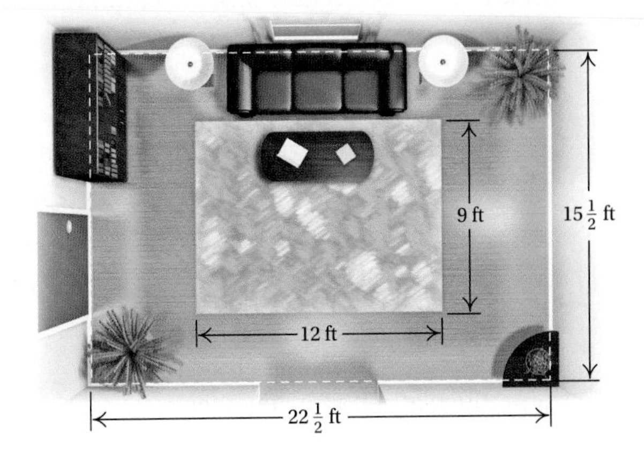

Do Exercise 11.

11. *Koi Pond.* Colleen designed a koi fish pond for her backyard. Using the dimensions shown in the diagram below, determine the area of Colleen's backyard remaining after the pond was completed.
Sources: en.wikipedia.org; pondliner.com

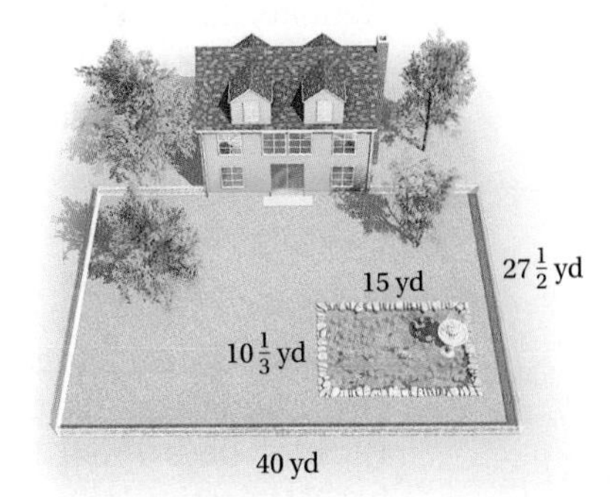

CALCULATOR CORNER

Operations on Fractions and Mixed Numerals Fraction calculators can add, subtract, multiply, and divide fractions and mixed numerals. The [a b/c] key is used to enter fractions and mixed numerals. The fraction $\frac{3}{4}$ is entered by pressing [3] [a b/c] [4], and it appears on the display as [3 ⌋ 4]. The mixed numeral $1\frac{5}{16}$ is entered by pressing [1] [a b/c] [5] [a b/c] [1] [6], and it is displayed as [1 ⌋ 5 ⌋ 16]. Fraction results that are greater than 1 are always displayed as mixed numerals. To express the result for $1\frac{5}{16}$ as a fraction, we press [SHIFT] [d/c]. We get [21 ⌋ 16], or $\frac{21}{16}$. Some calculators display fractions and mixed numerals in the way in which we write them.

EXERCISES Perform each calculation. Give the answer in fraction notation.

1. $\frac{1}{3} + \frac{1}{4}$
2. $\frac{7}{5} - \frac{3}{10}$
3. $\frac{15}{4} \cdot \frac{7}{12}$
4. $\frac{4}{5} \div \frac{8}{3}$

Perform each calculation. Give the answer as a mixed numeral.

5. $4\frac{1}{3} + 5\frac{4}{5}$
6. $9\frac{2}{7} - 8\frac{1}{4}$
7. $2\frac{1}{3} \cdot 4\frac{3}{5}$
8. $10\frac{7}{10} \div 3\frac{5}{6}$

Answer

11. 945 sq yd

Translating for Success

1. *Raffle Tickets.* At the Happy Hollow Camp Fall Festival, Rico and Becca, together, spent \$270 on raffle tickets that sell for $\$\frac{9}{20}$ each. How many tickets did they buy?

2. *Irrigation Pipe.* Jed uses two pipes, one of which measures $5\frac{1}{3}$ ft, to repair the irrigation system in the Aguilars' lawn. The total length of the two pipes is $8\frac{7}{12}$ ft. How long is the other pipe?

3. *Vacation Days.* Together, Oscar and Claire have 36 vacation days a year. Oscar has 22 vacation days per year. How many does Claire have?

4. *Enrollment in Japanese Classes.* Last year at the Lakeside Community College, 225 students enrolled in basic mathematics. This number is $4\frac{1}{2}$ times as many as the number who enrolled in Japanese. How many enrolled in Japanese?

5. *Bicycling.* Cole rode his bicycle $5\frac{1}{3}$ mi on Saturday and $8\frac{7}{12}$ mi on Sunday. How far did he ride on the weekend?

The goal of these matching questions is to practice step (2), Translate, of the five-step problem-solving process. Translate each word problem to an equation and select a correct translation from equations A–O.

A. $13\frac{11}{12} = x + 5\frac{1}{3}$

B. $\frac{3}{4} \cdot x = 1\frac{2}{3}$

C. $270 - \frac{20}{9} = x$

D. $225 = 4\frac{1}{2} \cdot x$

E. $98 \div 2\frac{1}{3} = x$

F. $22 + x = 36$

G. $x = 4\frac{1}{2} \cdot 225$

H. $x = 5\frac{1}{3} + 8\frac{7}{12}$

I. $22 \cdot x = 36$

J. $x = \frac{3}{4} \cdot 1\frac{2}{3}$

K. $5\frac{1}{3} + x = 8\frac{7}{12}$

L. $\frac{9}{20} \cdot 270 = x$

M. $1\frac{2}{3} + \frac{3}{4} = x$

N. $98 - 2\frac{1}{3} = x$

O. $\frac{9}{20} \cdot x = 270$

Answers on page A-7

6. *Deli Order.* For a promotional open house for contractors last year, the Bayside Builders Association ordered 225 turkey sandwiches. Due to increased registrations this year, $4\frac{1}{2}$ times as many sandwiches are needed. How many sandwiches should be ordered?

7. *Dog Ownership.* In Sam's community, $\frac{9}{20}$ of the households own at least one dog. There are 270 households. How many own dogs?

8. *Magic Tricks.* Samantha has 98 ft of rope and needs to cut it into $2\frac{1}{3}$-ft pieces to be used in a magic trick. How many pieces can be cut from the rope?

9. *Painting.* Laura needs $1\frac{2}{3}$ gal of paint to paint the ceiling of the exercise room and $\frac{3}{4}$ gal of the same paint for the bathroom. How much paint does Laura need?

10. *Chocolate Fudge Bars.* A recipe for chocolate fudge bars that serves 16 includes $1\frac{2}{3}$ cups of sugar. How much sugar is needed for $\frac{3}{4}$ of this recipe?

4.6 Exercise Set

For Extra Help
MyMathLab® MathXL® PRACTICE WATCH READ REVIEW

Reading Check

Determine whether each statement is true or false.

__________ **RC1.** To multiply using mixed numerals, we first convert to fraction notation.

__________ **RC2.** To divide using mixed numerals, we first convert to fraction notation.

__________ **RC3.** The product of mixed numerals is generally written as a mixed numeral, unless it is a whole number or less than 1.

__________ **RC4.** To divide fractions, we multiply by the reciprocal of the divisor.

a Multiply. Write a mixed numeral for the answer.

1. $8 \cdot 2\frac{5}{6}$

2. $5 \cdot 3\frac{3}{4}$

3. $3\frac{5}{8} \cdot \left(-\frac{2}{3}\right)$

4. $6\frac{2}{3} \cdot \left(-\frac{1}{4}\right)$

5. $3\frac{1}{2} \cdot 2\frac{1}{3}$

6. $4\frac{1}{5} \cdot 5\frac{1}{4}$

7. $3\frac{2}{5} \cdot 2\frac{7}{8}$

8. $2\frac{3}{10} \cdot 4\frac{2}{5}$

9. $-4\frac{7}{10} \cdot 5\frac{3}{10}$

10. $-6\frac{3}{10} \cdot 5\frac{7}{10}$

11. $-20\frac{1}{2} \cdot 10\frac{1}{5} \cdot \left(-4\frac{2}{3}\right)$

12. $-21\frac{1}{3} \cdot 11\frac{1}{3} \cdot \left(-3\frac{5}{8}\right)$

b Divide. Write a mixed numeral for the answer.

13. $20 \div 3\frac{1}{5}$

14. $18 \div 2\frac{1}{4}$

15. $8\frac{2}{5} \div 7$

16. $3\frac{3}{8} \div 3$

17. $-4\frac{3}{4} \div 1\frac{1}{3}$

18. $-5\frac{4}{5} \div 2\frac{1}{2}$

19. $1\frac{7}{8} \div 1\frac{2}{3}$

20. $4\frac{3}{8} \div 2\frac{5}{6}$

21. $5\frac{1}{10} \div 4\frac{3}{10}$

22. $4\frac{1}{10} \div 2\frac{1}{10}$

23. $-20\frac{1}{4} \div (-90)$

24. $-12\frac{1}{2} \div (-50)$

 Solve.

25. *Art Prices.* A 1966 Andy Warhol portrait of Marlon Brando on his motorcycle sold for \$5 million at Christie's in 2003. On November 14, 2012, the work was sold again by Christie's for about $4\frac{4}{5}$ times the amount paid in 2003. How much was paid for the portrait in 2012?

Source: Businessweek.com

26. *Spreading Grass Seed.* Emily seeds lawns for Sam's Superior Lawn Care. When she walks at a rapid pace, the wheel on the broadcast spreader completes $150\frac{2}{3}$ revolutions per minute. How many revolutions does the wheel complete in 15 min?

27. *Population.* The population of Alabama is $6\frac{2}{3}$ times that of Alaska. The population of Alaska is approximately 720,000. What is the population of Alabama?

Source: U.S. Census Bureau

28. *Population.* The population of New York is $32\frac{1}{2}$ times the population of Wyoming. The population of Wyoming is approximately 600,000. What is the population of New York?

Source: U.S. Census Bureau

29. *Mural.* A student artist painted a mural on the wall under a bridge. The dimensions of the mural are $6\frac{2}{3}$ ft by $9\frac{3}{8}$ ft. What is the area of the mural?

30. *Sodium Consumption.* The average American woman consumes $1\frac{2}{5}$ tsp of sodium each day. How much sodium do 10 average American women consume in one day?

Source: Based on information from mayoclinic.com

31. *Sidewalk.* A sidewalk alongside a garden at the conservatory is to be $14\frac{2}{5}$ yd long. Rectangular stone tiles that are each $1\frac{1}{8}$ yd long are used to form the sidewalk. How many tiles are used?

32. *Aeronautics.* Most space shuttles orbit the earth once every $1\frac{1}{2}$ hr. How many orbits are made every 24 hr?

33. *Weight of Water.* The weight of water is $62\frac{1}{2}$ lb per cubic foot. What is the weight of $5\frac{1}{2}$ cubic feet of water?

34. *Weight of Water.* The weight of water is $62\frac{1}{2}$ lb per cubic foot. What is the weight of $2\frac{1}{4}$ cubic feet of water?

35. *Temperature.* Fahrenheit temperature can be obtained from Celsius (Centigrade) temperature by multiplying by $1\frac{4}{5}$ and adding 32°. What Fahrenheit temperature corresponds to a Celsius temperature of 20°?

36. *Temperature.* Fahrenheit temperature can be obtained from Celsius (Centigrade) temperature by multiplying by $1\frac{4}{5}$ and adding 32°. What Fahrenheit temperature corresponds to the Celsius temperature of boiling water, 100°?

37. *Apple Net Income.* Apple, Inc., reported net income of about \$8,000,000,000 in fiscal year 2009. In fiscal year 2012, Apple's net income was about $5\frac{1}{4}$ times that amount. What was Apple's net income for fiscal year 2012?

38. *Median Income.* Median household income in the United States was about \$12,000 in 1975. By 2011, median household income was $4\frac{1}{6}$ times that amount. What was the median household income in 2011?

Source: U.S. Census Bureau

39. *Average Speed in Indianapolis 500.* Arie Luyendyk won the Indianapolis 500 in 1990 with a record average speed of about 186 mph. This record held through 2012 and is about $2\frac{12}{25}$ times the average speed of the first winner, Ray Harroun, in 1911. What was the average speed in the first Indianapolis 500?

Source: Indianapolis Motor Speedway

40. *Population.* The population of Cleveland is about $1\frac{1}{3}$ times the population of Cincinnati. In 2013, the population of Cincinnati was approximately 294,750. What was the population of Cleveland in 2013?

Source: U.S. Census Bureau

41. *Doubling a Recipe.* The chef of a five-star hotel is doubling a recipe for chocolate cake. The original recipe requires $2\frac{3}{4}$ cups of flour and $1\frac{1}{3}$ cups of sugar. How much flour and sugar will she need?

42. *Half of a Recipe.* A caterer is following a salad dressing recipe that calls for $1\frac{7}{8}$ cups of mayonnaise and $1\frac{1}{6}$ cups of sugar. How much mayonnaise and sugar will he need if he prepares $\frac{1}{2}$ of the amount of salad dressing?

43. *Mileage.* A car traveled 213 mi on $14\frac{2}{10}$ gal of gas. How many miles per gallon did it get?

44. *Mileage.* A car traveled 385 mi on $15\frac{4}{10}$ gal of gas. How many miles per gallon did it get?

45. *Weight of Water.* The weight of water is $62\frac{1}{2}$ lb per cubic foot. How many cubic feet would be occupied by 25,000 lb of water?

46. *Weight of Water.* The weight of water is $8\frac{1}{3}$ lb per gallon. Harry rolls his lawn with an 800-lb capacity roller. Express the water capacity of the roller in gallons.

47. *Servings of Salmon.* A serving of filleted fish is generally considered to be about $\frac{1}{3}$ lb. How many servings can be prepared from $5\frac{1}{2}$ lb of salmon fillet?

48. *Servings of Tuna.* A serving of fish steak (cross section) is generally $\frac{1}{2}$ lb. How many servings can be prepared from a cleaned $18\frac{3}{4}$-lb tuna?

Find the area of each shaded region.

49.

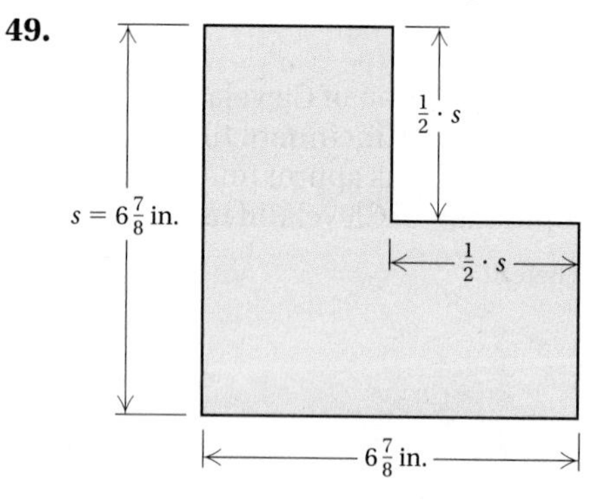

50.

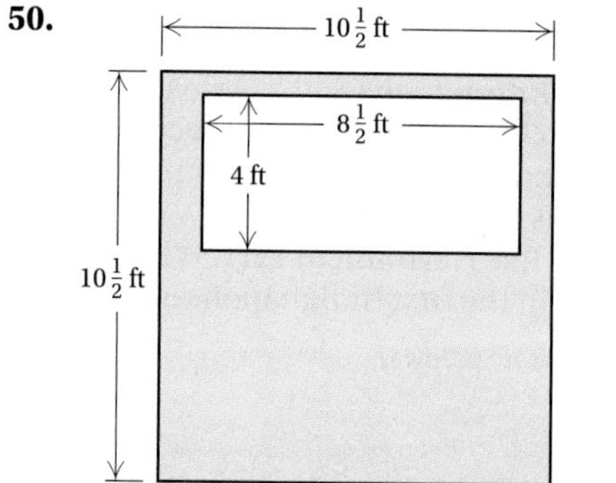

51. *Building a Ziggurat.* The dimensions of all of the square bricks that King Nebuchadnezzar used over 2500 years ago to build ziggurats were $13\frac{1}{4}$ in. $\times$ $13\frac{1}{4}$ in. $\times$ $3\frac{1}{4}$ in. What is the perimeter and the area of the $13\frac{1}{4}$ in. $\times$ $13\frac{1}{4}$ in. side? of the $13\frac{1}{4}$ in. $\times$ $3\frac{1}{4}$ in. side?

Source: www.eartharchitecture.org

52. *Word Processing.* For David's design report, he needs to create a table containing two columns, each $1\frac{1}{2}$ in. wide, and five columns, each $\frac{3}{4}$ in. wide. Will this table fit on a piece of standard paper that is $8\frac{1}{2}$ in. wide? If so, how wide will each side margin be if the margins on each side are to be of equal width?

Skill Maintenance

Solve.

53. *Checking Account Balance.* The balance in Laura's checking account is \$457. She uses her debit card to buy a digital picture frame that costs \$49. Find the new balance in her checking account. [1.8a]

54. Anita buys 12 gift cards at \$15 each and pays for them with \$20 bills. How many \$20 bills does it take? [1.8a]

55. About $\frac{9}{25}$ of all pizzas that Americans order have pepperoni as a topping. If Americans eat 350 slices of pizza every second, how many of those slices are topped with pepperoni? [3.6b]

Source: inventors.about.com

56. A batch of fudge requires $\frac{3}{4}$ cup of sugar. How much sugar is needed to make 12 batches? [3.6b]

57. After her company was restructured, Meghan's pay was $\frac{9}{10}$ of what it had been. If she is now making \$32,850 a year, what was she making before the reorganization? [3.7d]

58. Rick's Market sells Swiss cheese in $\frac{3}{4}$-lb packages. How many packages can be made from a 12-lb slab of cheese? [3.7d]

59. Find the perimeter of the figure. [1.2b]

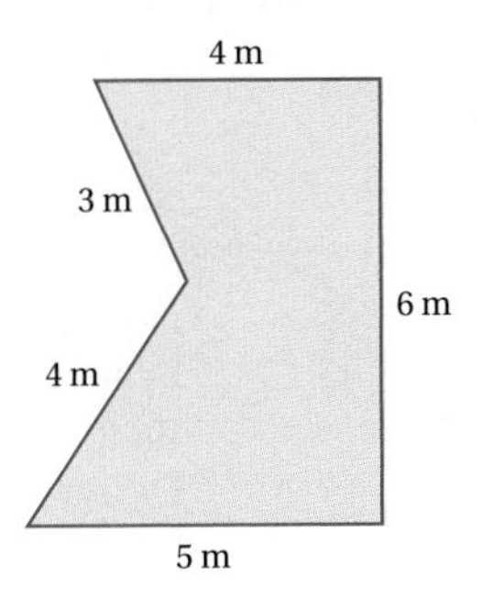

60. Find the area of the region. [1.4b]

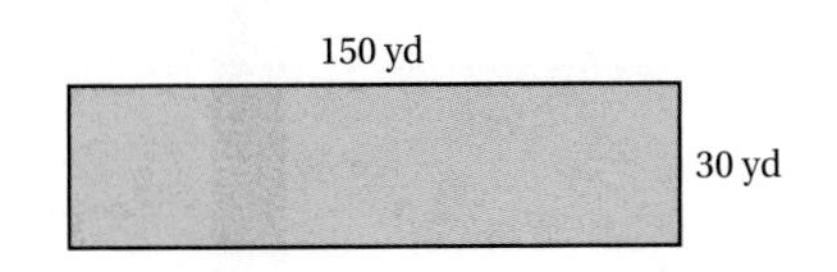

Synthesis

Multiply. Write the answer as a mixed numeral whenever possible.

61. 🖩 $15\frac{2}{11} \cdot 23\frac{31}{43}$

62. 🖩 $17\frac{23}{31} \cdot 19\frac{13}{15}$

Simplify.

63. $8 \div \frac{1}{2} + \frac{3}{4} - \left(5 - \frac{5}{8}\right)^2$

64. $\left(\frac{5}{9} - \frac{1}{4}\right) \times 12 + \left(4 - \frac{3}{4}\right)^2$

65. $\frac{1}{3} \div \left(\frac{1}{2} - \frac{1}{5}\right) \times \frac{1}{4} + \frac{1}{6}$

66. $\frac{7}{8} - 1\frac{1}{8} \times \frac{2}{3} - \frac{9}{10} \div \frac{3}{5}$

67. $4\frac{1}{2} \div 2\frac{1}{2} + 8 - 4 \div \frac{1}{2}$

68. $6 - 2\frac{1}{3} \times \frac{3}{4} + \frac{5}{8} \div \frac{2}{3}$

Order of Operations, Complex Fractions, and Estimation

OBJECTIVES

a Simplify expressions using the rules for order of operations.

b Simplify complex fractions.

c Estimate with fraction notation and mixed numerals.

SKILL TO REVIEW

Objective 1.9c: Simplify expressions using the rules for order of operations.

Simplify.

1. $22 - 3 \cdot 4$

2. $6(4 + 1)^2 - 3^4 \div 3$

Simplify.

1. $\frac{2}{5} \cdot \frac{5}{8} + \frac{1}{4}$

2. $\frac{1}{3} \cdot \frac{3}{4} \div \frac{5}{8} - \frac{1}{10}$ GS

$= \frac{\square}{12} \div \frac{5}{8} - \frac{1}{10} = \frac{3}{12} \cdot \frac{\square}{5} - \frac{1}{10}$

$= \frac{3 \cdot 2 \cdot 2 \cdot \square}{3 \cdot 2 \cdot 2 \cdot 5} - \frac{1}{10} = \frac{\square}{5} - \frac{1}{10}$

$= \frac{\square}{10} - \frac{1}{10} = \frac{\square}{10}$

3. Simplify: $\frac{3}{4} \cdot 16 + 8\frac{2}{3}$.

Answers

Skill to Review:

1. 10 2. 123

Margin Exercises:

1. $\frac{1}{2}$ 2. $\frac{3}{10}$ 3. $20\frac{2}{3}$, or $\frac{62}{3}$

Guided Solution:

2. 3, 8, 2, 2, 4, 3

a ORDER OF OPERATIONS; FRACTION NOTATION AND MIXED NUMERALS

The rules for order of operations that we use with whole numbers apply when we are simplifying expressions involving fraction notation and mixed numerals. For review, these rules are listed below.

RULES FOR ORDER OF OPERATIONS

1. Do all calculations within parentheses before operations outside.
2. Evaluate all exponential expressions.
3. Do all multiplications and divisions in order from left to right.
4. Do all additions and subtractions in order from left to right.

EXAMPLE 1 Simplify: $\frac{1}{6} + \frac{2}{3} \div \frac{1}{2} \cdot \frac{5}{8}$.

$\frac{1}{6} + \frac{2}{3} \div \frac{1}{2} \cdot \frac{5}{8} = \frac{1}{6} + \frac{2}{3} \cdot \frac{2}{1} \cdot \frac{5}{8}$ Doing the division first by multiplying by the reciprocal of $\frac{1}{2}$

$= \frac{1}{6} + \frac{2 \cdot 2 \cdot 5}{3 \cdot 1 \cdot 8}$ Doing the multiplications in order from left to right

$= \frac{1}{6} + \frac{\cancel{2} \cdot \cancel{2} \cdot 5}{3 \cdot 1 \cdot \cancel{2} \cdot \cancel{2} \cdot 2}$ Factoring in order to simplify

$= \frac{1}{6} + \frac{5}{6}$ Removing a factor of 1: $\frac{2 \cdot 2}{2 \cdot 2} = 1$; simplifying

$= \frac{6}{6}$, or 1 Doing the addition

◀ Do Margin Exercises 1 and 2.

EXAMPLE 2 Simplify: $\frac{2}{3} \cdot 24 - 11\frac{1}{2}$.

$\frac{2}{3} \cdot 24 - 11\frac{1}{2} = \frac{2 \cdot 24}{3 \cdot 1} - 11\frac{1}{2}$ Doing the multiplication first

$= \frac{2 \cdot \cancel{3} \cdot 8}{\cancel{3} \cdot 1} - 11\frac{1}{2}$ Factoring the fraction

$= 2 \cdot 8 - 11\frac{1}{2}$ Removing a factor of 1: $\frac{3}{3} = 1$

$= 16 - 11\frac{1}{2}$ Completing the multiplication

$= 4\frac{1}{2}$, or $\frac{9}{2}$ Doing the subtraction

◀ Do Exercise 3.

EXAMPLE 3 Simplify: $\left(\frac{7}{8}-\frac{1}{3}\right)\times 48+\left(13+\frac{4}{5}\right)^2$.

$$\left(\frac{7}{8}-\frac{1}{3}\right)\times 48+\left(13+\frac{4}{5}\right)^2$$

$$=\left(\frac{7}{8}\cdot\frac{3}{3}-\frac{1}{3}\cdot\frac{8}{8}\right)\times 48+\left(13\cdot\frac{5}{5}+\frac{4}{5}\right)^2$$ Carrying out operations inside parentheses first. To do so, we first multiply by 1 to obtain each LCD.

$$=\left(\frac{21}{24}-\frac{8}{24}\right)\times 48+\left(\frac{65}{5}+\frac{4}{5}\right)^2$$

$$=\frac{13}{24}\times 48+\left(\frac{69}{5}\right)^2$$ Completing the operations within parentheses

$$=\frac{13}{24}\times 48+\frac{4761}{25}$$ Evaluating the exponential expression next

$$=26+\frac{4761}{25}$$ Doing the multiplication

$$=26+190\frac{11}{25}$$ Converting to a mixed numeral

$$=216\frac{11}{25},\ \text{or}\ \frac{5411}{25}$$ Adding

Answers can be given using either fraction notation or mixed numerals.

Do Exercise 4. ▶

4. Simplify:
$\left(\frac{2}{3}+\frac{3}{4}\right)\div 2\frac{1}{3}-\left(\frac{1}{2}\right)^3$.

b COMPLEX FRACTIONS

A **complex fraction** is a fraction in which the numerator and/or the denominator contains one or more fractions. The following are some examples of complex fractions.

$$\frac{\frac{7}{3}}{2}$$ ← The numerator contains a fraction.

$$\frac{\frac{1}{5}}{\frac{9}{10}}$$ ← The numerator contains a fraction. ← The denominator contains a fraction.

Since a fraction bar represents division, complex fractions can be rewritten using the division symbol ÷.

EXAMPLE 4 Simplify: $\frac{\frac{7}{3}}{2}$

$$\frac{\frac{7}{3}}{2}=\frac{7}{3}\div 2$$ Rewriting using a division symbol

$$=\frac{7}{3}\cdot\frac{1}{2}$$ Multiplying by the reciprocal of the divisor

$$=\frac{7\cdot 1}{3\cdot 2}$$ Multiplying numerators and multiplying denominators

$$=\frac{7}{6}$$ This expression cannot be simplified. ■

Answer

4. $\frac{27}{56}$

EXAMPLE 5 Simplify: $\dfrac{-\frac{1}{5}}{\frac{9}{10}}$.

$$\frac{-\frac{1}{5}}{\frac{9}{10}} = -\frac{1}{5} \div \frac{9}{10} \qquad \text{Rewriting using a division symbol}$$

$$= -\frac{1}{5} \cdot \frac{10}{9} \qquad \text{Multiplying by the reciprocal of the divisor}$$

$$= -\frac{1 \cdot 2 \cdot \cancel{5}}{\cancel{5} \cdot 3 \cdot 3} \qquad \text{Multiplying numerators and multiplying denominators; factoring}$$

$$= -\frac{2}{9} \qquad \text{Removing a factor of 1 and simplifying}$$

◀ **Do Exercises 5 and 6.**

Simplify.

5. $\dfrac{10}{\frac{5}{8}}$

6. $\dfrac{-\frac{7}{5}}{\frac{10}{7}}$

When the numerator or denominator of a complex fraction consists of more than one term, first simplify that numerator or denominator separately.

EXAMPLE 6 Simplify: $\dfrac{\frac{2}{3} - \frac{1}{2}}{1\frac{7}{8}}$.

$$\frac{\frac{2}{3} - \frac{1}{2}}{1\frac{7}{8}} = \frac{\frac{4}{6} - \frac{3}{6}}{\frac{15}{8}} \qquad \text{Writing the fractions in the numerator with a common denominator; Writing the mixed numeral in the denominator as a fraction}$$

$$= \frac{\frac{1}{6}}{\frac{15}{8}} \qquad \text{Subtracting in the numerator of the complex fraction}$$

$$= \frac{1}{6} \div \frac{15}{8} \qquad \text{Rewriting using a division symbol}$$

$$= \frac{1}{6} \cdot \frac{8}{15} \qquad \text{Multiplying by the reciprocal of the divisor}$$

$$= \frac{1 \cdot \cancel{2} \cdot 2 \cdot 2}{\cancel{2} \cdot 3 \cdot 3 \cdot 5} \qquad \text{Multiplying numerators and multiplying denominators; factoring}$$

$$= \frac{4}{45} \qquad \text{Removing a factor of 1: } \frac{2}{2} = 1$$

◀ **Do Exercises 7 and 8.**

Simplify.

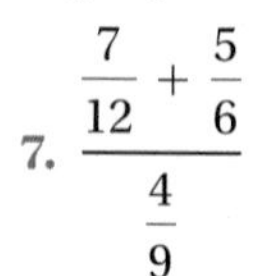

7. $\dfrac{\frac{7}{12} + \frac{5}{6}}{\frac{4}{9}}$

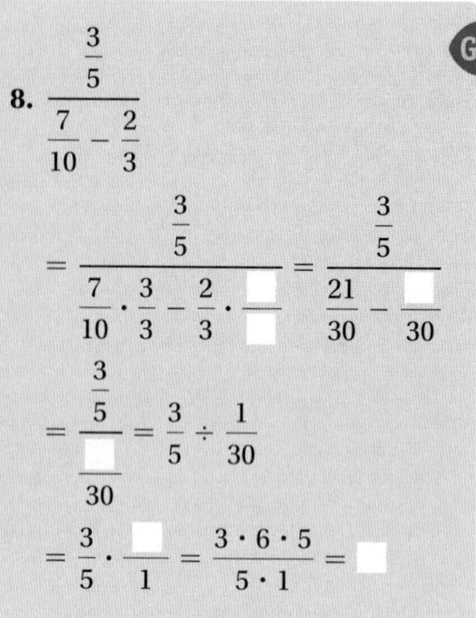

GS

8. $\dfrac{\frac{3}{5}}{\frac{7}{10} - \frac{2}{3}}$

$= \dfrac{\frac{3}{5}}{\frac{7}{10} \cdot \frac{3}{3} - \frac{2}{3} \cdot \frac{\square}{\square}} = \dfrac{\frac{3}{5}}{\frac{21}{30} - \frac{\square}{30}}$

$= \dfrac{\frac{3}{5}}{\frac{\square}{30}} = \dfrac{3}{5} \div \dfrac{1}{30}$

$= \dfrac{3}{5} \cdot \dfrac{\square}{1} = \dfrac{3 \cdot 6 \cdot 5}{5 \cdot 1} = \square$

EXAMPLE 7 *Harvesting Walnut Trees.* A woodland owner decided to harvest five walnut trees in order to improve the growing conditions of the remaining trees. The logs she sold measured $7\frac{5}{8}$ ft, $8\frac{1}{4}$ ft, $8\frac{3}{4}$ ft, $9\frac{1}{8}$ ft, and $10\frac{1}{2}$ ft. What is the average length of the logs?

Answers

5. 16 6. $-\frac{49}{50}$ 7. $\frac{51}{16}$, or $3\frac{3}{16}$ 8. 18

Guided Solution:

8. $\frac{10}{10}$, 20, 1, 30, 18

Recall that to compute an average, we add the numbers and then divide the sum by the number of addends. We have

$$\frac{7\frac{5}{8} + 8\frac{1}{4} + 8\frac{3}{4} + 9\frac{1}{8} + 10\frac{1}{2}}{5} = \frac{7\frac{5}{8} + 8\frac{2}{8} + 8\frac{6}{8} + 9\frac{1}{8} + 10\frac{4}{8}}{5}$$

$$= \frac{42\frac{18}{8}}{5} = \frac{42\frac{9}{4}}{5} = \frac{\frac{177}{4}}{5} \qquad \text{Adding, simplifying, and converting to fraction notation}$$

$$= \frac{177}{4} \cdot \frac{1}{5} \qquad \text{Multiplying by the reciprocal}$$

$$= \frac{177}{20} = 8\frac{17}{20}. \qquad \text{Converting to a mixed numeral}$$

The average length of the logs is $8\frac{17}{20}$ ft.

Do Exercises 9–11. ▶

9. Rachel has triplets. Their birth weights are $3\frac{1}{2}$ lb, $2\frac{3}{4}$ lb, and $3\frac{1}{8}$ lb. What is the average weight of her babies?

10. Find the average of $\frac{1}{2}$, $\frac{1}{3}$, and $\frac{5}{6}$.

11. Find the average of $\frac{3}{4}$ and $\frac{4}{5}$.

C ESTIMATION WITH FRACTION NOTATION AND MIXED NUMERALS

EXAMPLES Estimate each of the following as 0, $\frac{1}{2}$, or 1.

8. $\frac{2}{17}$

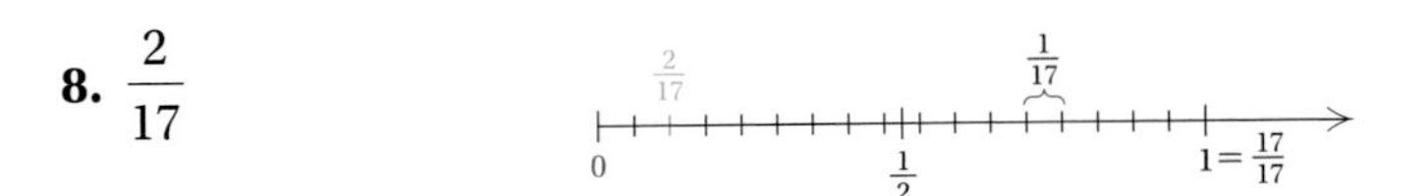

A fraction is close to 0 when the numerator is small in comparison to the denominator. Since 2 is small in comparison to 17, we have $\frac{2}{17} \approx 0$.

9. $\frac{11}{23}$

A fraction is close to $\frac{1}{2}$ when the denominator is about twice the numerator. Twice the numerator of this fraction is 22, and 22 is close to 23. Thus, $\frac{11}{23} \approx \frac{1}{2}$.

10. $\frac{43}{41}$

A fraction is close to 1 when the numerator is nearly equal to the denominator. Since 43 is nearly equal to 41, we have $\frac{43}{41} \approx 1$.

Do Exercises 12–15. ▶

Estimate each of the following as 0, $\frac{1}{2}$, or 1.

12. $\frac{3}{59}$

13. $\frac{61}{59}$

14. $\frac{29}{59}$

15. $\frac{57}{59}$

EXAMPLE 11 Estimate $16\frac{8}{9} + 11\frac{2}{13} - 4\frac{22}{43}$ by estimating each mixed numeral as a whole number or as a mixed numeral where the fractional part is $\frac{1}{2}$.

We estimate each mixed numeral. Then we calculate:

$$16\frac{8}{9} + 11\frac{2}{13} - 4\frac{22}{43} \approx 17 + 11 - 4\frac{1}{2}$$

$$= 28 - 4\frac{1}{2} = 23\frac{1}{2}.$$

Do Exercises 16–18. ▶

Estimate each of the following by estimating each mixed numeral as a whole number or as a mixed numeral where the fractional part is $\frac{1}{2}$ and estimating each fraction as 0, $\frac{1}{2}$, or 1.

16. $5\frac{9}{10} + 26\frac{1}{2} - 10\frac{3}{29}$

17. $10\frac{7}{8} \cdot \left(25\frac{11}{13} - 14\frac{1}{9}\right)$

18. $\left(10\frac{4}{5} + 7\frac{5}{9}\right) \div \frac{17}{30}$

Answers

9. $3\frac{1}{8}$ lb 10. $\frac{5}{9}$ 11. $\frac{31}{40}$ 12. 0 13. 1
14. $\frac{1}{2}$ 15. 1 16. $22\frac{1}{2}$ 17. 132 18. 37

4.7 Exercise Set

For Extra Help MyMathLab® MathXL® PRACTICE WATCH READ REVIEW

Reading Check

Match the beginning of each statement with the correct ending from the list at the right so that the rules for order of operations are listed in the correct order.

RC1. Do all ________.

RC2. Evaluate all ________.

RC3. Do all ________.

RC4. Do all ________.

a) multiplications and divisions in order from left to right

b) additions and subtractions in order from left to right

c) calculations within parentheses

d) exponential expressions

a Simplify.

1. $\frac{5}{8} \div \frac{1}{4} - \frac{2}{3} \cdot \frac{4}{5}$

2. $\frac{4}{7} \cdot \frac{7}{15} + \frac{2}{3} \div 8$

3. $\frac{3}{4} - \frac{2}{3} \cdot \left(\frac{1}{2} + \frac{2}{5}\right)$

4. $\frac{3}{4} \div \frac{1}{2} \cdot \left(\frac{8}{9} - \frac{2}{3}\right)$

5. $28\frac{1}{8} - 5\frac{1}{4} + 3\frac{1}{2}$

6. $10\frac{3}{5} - 4\frac{1}{10} - 1\frac{1}{2}$

7. $\frac{7}{8} \div \frac{1}{2} \cdot \frac{1}{4}$

8. $\frac{7}{10} \cdot \frac{4}{5} \div \frac{2}{3}$

9. $\left(\frac{2}{3}\right)^2 - \frac{1}{3} \cdot 1\frac{1}{4}$

10. $\left(\frac{3}{4}\right)^2 + 3\frac{1}{2} \div 1\frac{1}{4}$

11. $\frac{1}{2} - \left(\frac{1}{2}\right)^2 + \left(\frac{1}{2}\right)^3$

12. $-1 + \frac{1}{4} + \left(\frac{1}{4}\right)^2 - \left(\frac{1}{4}\right)^3$

13. $\left(\frac{2}{3} + \frac{3}{4}\right) \div \left(\frac{5}{6} - \frac{1}{3}\right)$

14. $\left(\frac{3}{5} - \frac{1}{2}\right) \div \left(\frac{3}{4} - \frac{3}{10}\right)$

15. $\left(\frac{1}{2} + \frac{1}{3}\right)^2 \cdot 144 - \frac{5}{8} \div 10\frac{1}{2}$

16. $\left(3\frac{1}{2} - 2\frac{1}{3}\right)^2 + 6 \cdot 2\frac{1}{2} \div 32$

b Simplify.

17. $\dfrac{\frac{3}{8}}{\frac{11}{8}}$

18. $\dfrac{\frac{1}{8}}{\frac{3}{4}}$

19. $\dfrac{-4}{\frac{6}{7}}$

20. $\dfrac{-\frac{3}{8}}{12}$

21. $\dfrac{\frac{1}{40}}{\frac{1}{50}}$

22. $\dfrac{\frac{7}{9}}{\frac{3}{9}}$ **23.** $\dfrac{\frac{1}{10}}{10}$ **24.** $\dfrac{28}{\frac{7}{4}}$ **25.** $\dfrac{-\frac{5}{18}}{1\frac{2}{3}}$ **26.** $\dfrac{2\frac{1}{5}}{-\frac{7}{10}}$

27. $\dfrac{\frac{5}{9}-\frac{1}{6}}{\frac{2}{3}}$ **28.** $\dfrac{\frac{7}{12}}{\frac{5}{8}-\frac{1}{4}}$ **29.** $\dfrac{\frac{3}{8}-\frac{1}{4}}{\frac{7}{8}-\frac{1}{2}}$ **30.** $\dfrac{\frac{3}{5}-\frac{1}{2}}{\frac{1}{2}-\frac{2}{5}}$

31. Find the average of $\frac{2}{3}$ and $\frac{7}{8}$.

32. Find the average of $\frac{1}{4}$ and $\frac{1}{5}$.

33. Find the average of $\frac{1}{6}$, $\frac{1}{8}$, and $\frac{3}{4}$.

34. Find the average of $\frac{4}{5}$, $\frac{1}{2}$, and $\frac{1}{10}$.

35. Find the average of $3\frac{1}{2}$ and $9\frac{3}{8}$.

36. Find the average of $10\frac{2}{3}$ and $24\frac{5}{6}$.

37. *Hiking the Appalachian Trail.* Ellen camped and hiked for three consecutive days along a section of the Appalachian Trail. The distances she hiked on the three days were $15\frac{5}{32}$ mi, $20\frac{3}{16}$ mi, and $12\frac{7}{8}$ mi. Find the average of these distances.

38. *Vertical Leaps.* Eight-year-old Zachary registered vertical leaps of $12\frac{3}{4}$ in., $13\frac{3}{4}$ in., $13\frac{1}{2}$ in., and 14 in. Find his average vertical leap.

39. *Black Bear Cubs.* Black bears typically have two cubs. In January 2007 in northern New Hampshire, a black bear sow gave birth to a litter of 5 cubs. This is so rare that Tom Sears, a wildlife photographer, spent 28 hr per week for six weeks watching for the perfect opportunity to photograph this family of six. At the time of this photo, an observer estimated that the cubs weighed $7\frac{1}{2}$ lb, 8 lb, $9\frac{1}{2}$ lb, $10\frac{5}{8}$ lb, and $11\frac{3}{4}$ lb. What was the average weight of the cubs?

Source: Andrew Timmins, New Hampshire Fish and Game Department, *Northcountry News*, Warren, NH; Tom Sears, photographer

40. *Acceleration.* The results of a *Road & Track* road acceleration test for five cars are given in the graph below. The test measures the time in seconds required to go from 0 mph to 60 mph. What was the average time?

Acceleration: 0 mph to 60 mph

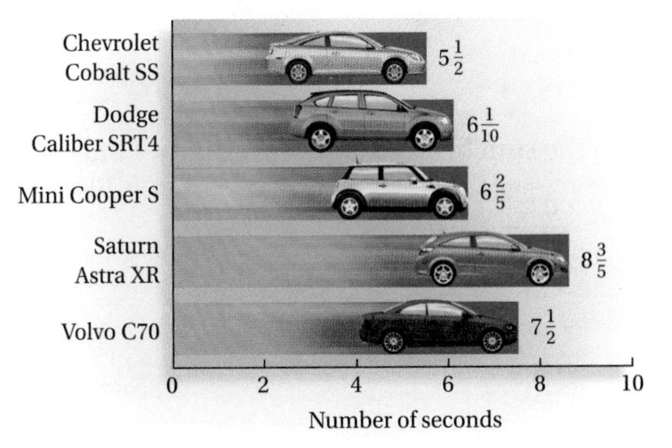

SOURCE: *Road & Track*, October 2008, pp.156–157

C Estimate each of the following as 0, $\frac{1}{2}$, or 1.

41. $\frac{2}{47}$ **42.** $\frac{5}{12}$ **43.** $\frac{7}{8}$ **44.** $\frac{1}{13}$ **45.** $\frac{6}{11}$ **46.** $\frac{11}{13}$

47. $\frac{7}{15}$ **48.** $\frac{1}{16}$ **49.** $\frac{7}{100}$ **50.** $\frac{5}{9}$ **51.** $\frac{19}{20}$ **52.** $\frac{4}{5}$

Estimate each of the following by estimating each mixed numeral as a whole number or as a mixed numeral where the fractional part is $\frac{1}{2}$ and by estimating each fraction as 0, $\frac{1}{2}$, or 1.

53. $2\frac{7}{8}$

54. $12\frac{5}{6}$

55. $\frac{4}{5} + \frac{7}{8}$

56. $\frac{1}{12} \cdot \frac{7}{15}$

57. $24 \div 7\frac{8}{9}$

58. $43\frac{16}{17} \div 11\frac{2}{13}$

59. $7\frac{29}{60} + 10\frac{12}{13} \cdot 24\frac{2}{17}$

60. $1\frac{5}{8} + 1\frac{27}{28} \cdot 6\frac{35}{74}$

61. $16\frac{1}{5} \div 2\frac{1}{11} + 25\frac{9}{10} - 4\frac{11}{23}$

62. $96\frac{2}{13} \div 5\frac{19}{20} + 3\frac{1}{7} \cdot 5\frac{18}{21}$

Skill Maintenance

Simplify.

63. $12 + 30 \div 3 - 2$ [1.9c]

64. $5 \cdot 2^2 \div 10$ [1.9c]

65. $10^2 - [3 \cdot 2^4 \div (10 - 2) + 5 \cdot 2]$ [1.9d]

66. $(10 + 3 \cdot 4 \div 6)^2 - 11 \cdot 2^2$ [1.9c]

67. List all the factors of 42. [3.1a]

68. Determine whether 114 is divisible by 7. [3.1b]

69. Classify the given numbers as prime, composite, or neither. [3.1c]

1, 5, 7, 9, 14, 23, 43

70. Find the prime factorization of 150. [3.1d]

Synthesis

71. 🖩 In the sum below, a and b are digits. Find a and b.

$$\frac{a}{17} + \frac{1b}{23} = \frac{35a}{391}$$

72. 🖩 Consider only the numbers 3, 4, 5, and 6. Assume each can be placed in a blank in the following.

$$\square + \frac{\square}{\square} \cdot \square = ?$$

What placement of the numbers in the blanks yields the largest sum?

73. 🖩 Consider only the numbers 2, 3, 4, and 5. Assume each is placed in a blank in the following.

$$\frac{\square}{\square} + \frac{\square}{\square} = ?$$

What placement of the numbers in the blanks yields the largest sum?

74. 🖩 Use a calculator to arrange the following in order from smallest to largest.

$$\frac{3}{4}, \frac{17}{21}, \frac{13}{15}, \frac{7}{9}, \frac{15}{17}, \frac{13}{12}, \frac{19}{22}$$

Vocabulary Reinforcement

Complete each statement with the correct term from the given list. Some of the choices may not be used and some may be used more than once.

1. The ____________ of two numbers is the smallest number that is a multiple of both numbers. [4.1a]
2. A ____________ represents a sum of a whole number and a fraction less than 1. [4.4a]
3. To multiply using mixed numerals, we first convert to ____________ notation. [4.6a]
4. A ____________ contains a fraction in its numerator and/or denominator. [4.7b]
5. To add fractions, the ____________ of the fractions must be the same. [4.2a]
6. The least common denominator of two fractions is the ____________ of the denominators of the fractions. [4.2a]
7. When finding the LCM of a set of numbers using prime factorizations, we use each prime number the ____________ number of times that it appears in any one factorization. [4.1a]
8. To compare two fractions with a common denominator, we compare their ____________. [4.3b]

greatest
least
numerators
denominators
fraction
decimal
mixed numeral
complex fraction
least common multiple
greatest common factor

Concept Reinforcement

Determine whether each statement is true or false.

____________ 1. The mixed numeral $5\frac{2}{3}$ can be represented by the sum $5 \cdot \frac{3}{3} + \frac{2}{3}$. [4.4a]

____________ 2. The least common multiple of two numbers is always larger than or equal to the larger number. [4.1a]

____________ 3. The sum of any two mixed numerals is a mixed numeral. [4.5a]

____________ 4. The product of any two mixed numerals is greater than 1. [4.6a]

Study Guide

Objective 4.1a Find the least common multiple, or LCM, of two or more numbers.

Example Find the LCM of 105 and 90.

$105 = 3 \cdot 5 \cdot 7,$

$90 = 2 \cdot 3 \cdot 3 \cdot 5;$

$\text{LCM} = 2 \cdot 3 \cdot 3 \cdot 5 \cdot 7 = 630$

Practice Exercise

1. Find the LCM of 52 and 78.

Objective 4.2a Add using fraction notation.

Example Add: $\frac{5}{24} + \frac{7}{45}$.

$$\frac{5}{24} + \frac{7}{45} = \frac{5}{2 \cdot 2 \cdot 2 \cdot 3} \cdot \frac{3 \cdot 5}{3 \cdot 5} + \frac{7}{3 \cdot 3 \cdot 5} \cdot \frac{2 \cdot 2 \cdot 2}{2 \cdot 2 \cdot 2}$$
$$= \frac{75}{360} + \frac{56}{360} = \frac{131}{360}$$

Practice Exercise

2. Add: $\frac{19}{60} + \frac{11}{36}$.

Objective 4.3a Subtract using fraction notation.

Example Subtract: $\frac{7}{12} - \frac{11}{60}$.

$$\frac{7}{12} - \frac{11}{60} = \frac{7}{12} \cdot \frac{5}{5} - \frac{11}{60} = \frac{35}{60} - \frac{11}{60}$$
$$= \frac{35 - 11}{60} = \frac{24}{60} = \frac{2 \cdot \cancel{12}}{5 \cdot \cancel{12}} = \frac{2}{5}$$

Practice Exercise

3. Subtract: $\frac{29}{35} - \frac{5}{7}$.

Objective 4.3b Use $<$ or $>$ with fraction notation to write a true sentence.

Example Use $<$ or $>$ for $\square$ to write a true sentence:

$$\frac{5}{12} \square \frac{9}{16}.$$

Writing both fractions with the LCD, 48, we have

$$\frac{20}{48} \square \frac{27}{48}.$$

Since $20 < 27$, $\frac{20}{48} < \frac{27}{48}$ and thus $\frac{5}{12} < \frac{9}{16}$.

Practice Exercise

4. Use $<$ or $>$ for $\square$ to write a true sentence:

$$\frac{3}{13} \square \frac{5}{12}.$$

Objective 4.3c Solve equations of the type $x + a = b$ and $a + x = b$, where a and b may be fractions.

Example Solve: $x - \frac{1}{6} = -\frac{5}{8}$.

$$x - \frac{1}{6} = -\frac{5}{8}$$
$$x - \frac{1}{6} + \frac{1}{6} = -\frac{5}{8} + \frac{1}{6}$$
$$x = -\frac{5}{8} \cdot \frac{3}{3} + \frac{1}{6} \cdot \frac{4}{4}$$
$$x = -\frac{15}{24} + \frac{4}{24} = -\frac{11}{24}$$

Practice Exercise

5. Solve: $\frac{2}{9} + x = -\frac{9}{11}$.

Objective 4.4a Convert between mixed numerals and fraction notation.

Example Convert $2\frac{5}{13}$ to fraction notation: $2\frac{5}{13} = \frac{31}{13}$.

Example Convert $-\frac{40}{9}$ to a mixed numeral:

$$-\frac{40}{9} = -4\frac{4}{9}.$$

Practice Exercises

6. Convert $8\frac{2}{3}$ to fraction notation.

7. Convert $-\frac{47}{6}$ to a mixed numeral.

Objective 4.5b Subtract using mixed numerals.

Example Subtract: $3\frac{3}{8} - 1\frac{4}{5}$.

$$\begin{array}{rcrcr} 3\frac{3}{8} & = & 3\frac{15}{40} & = & 2\frac{55}{40} \\ -1\frac{4}{5} & = & -1\frac{32}{40} & = & -1\frac{32}{40} \\ \hline & & & & 1\frac{23}{40} \end{array}$$

Practice Exercise

8. Subtract: $10\frac{5}{7} - 2\frac{3}{4}$.

Objective 4.6a Multiply using mixed numerals.

Example Multiply: $7\frac{1}{4} \cdot \left(-5\frac{3}{10}\right)$. Write a mixed numeral for the answer.

$$7\frac{1}{4} \cdot \left(-5\frac{3}{10}\right) = \frac{29}{4} \cdot \left(-\frac{53}{10}\right)$$

$$= -\frac{1537}{40} = -38\frac{17}{40}$$

Practice Exercise

9. Multiply: $-4\frac{1}{5} \cdot 3\frac{7}{15}$.

Objective 4.6c Solve applied problems involving multiplication and division with mixed numerals.

Example The population of New York is $3\frac{1}{3}$ times that of Missouri. The population of New York is approximately 19,000,000. What is the population of Missouri?

Translate:

$$19{,}000{,}000 = 3\frac{1}{3} \cdot x.$$

Solve:

$$19{,}000{,}000 = \frac{10}{3} \cdot x$$

$$\frac{19{,}000{,}000}{\frac{10}{3}} = \frac{\frac{10}{3} \cdot x}{\frac{10}{3}}$$

$$19{,}000{,}000 \cdot \frac{3}{10} = x$$

$$5{,}700{,}000 = x.$$

The population of Missouri is about 5,700,000.

Practice Exercise

10. The population of Louisiana is $2\frac{1}{2}$ times the population of West Virginia. The population of West Virginia is approximately 1,800,000. What is the population of Louisiana?

Objective 4.7a Simplify expressions using the rules for order of operations.

Example Simplify: $\left(\frac{4}{5}\right)^2 - \frac{1}{5} \cdot 2\frac{1}{8}$.

$$\left(\frac{4}{5}\right)^2 - \frac{1}{5} \cdot 2\frac{1}{8} = \frac{16}{25} - \frac{1}{5} \cdot \frac{17}{8}$$

$$= \frac{16}{25} - \frac{17}{40} = \frac{16}{25} \cdot \frac{8}{8} - \frac{17}{40} \cdot \frac{5}{5}$$

$$= \frac{128}{200} - \frac{85}{200} = \frac{43}{200}$$

Practice Exercise

11. Simplify: $\frac{3}{2} \cdot 1\frac{1}{3} \div \left(\frac{2}{3}\right)^2$.

Objective 4.7c Estimate with fraction notation and mixed numerals.

Example Estimate

$$3\tfrac{2}{5} + \tfrac{7}{10} + 6\tfrac{5}{9}$$

by estimating each mixed numeral as a whole number or as a mixed numeral where the fractional part is $\frac{1}{2}$ and by estimating each fraction as 0, $\frac{1}{2}$, or 1.

$$3\frac{2}{5} + \frac{7}{10} + 6\frac{5}{9} \approx 3\frac{1}{2} + \frac{1}{2} + 6\frac{1}{2} = 10\frac{1}{2}$$

Practice Exercise

12. Estimate

$$1\frac{19}{20} + 3\frac{1}{8} - \frac{8}{17}$$

by estimating each mixed numeral as a whole number or as a mixed numeral where the fractional part is $\frac{1}{2}$ and by estimating each fraction as 0, $\frac{1}{2}$, or 1.

Review Exercises

Find the LCM. [4.1a]

1. 12 and 18

2. 18 and 45

3. 3, 6, and 30

4. 26, 36, and 54

Add and simplify. [4.2a]

5. $\frac{6}{5} + \frac{3}{8}$

6. $\frac{-5}{16} + \frac{1}{12}$

7. $\frac{6}{5} + \frac{11}{15} + \frac{3}{20}$

8. $\frac{1}{1000} + \frac{19}{100} + \frac{7}{10}$

Subtract and simplify. [4.3a]

9. $\frac{5}{9} - \frac{2}{9}$

10. $\frac{7}{8} - \frac{3}{4}$

11. $\frac{2}{9} - \frac{11}{27}$

12. $-\frac{5}{6} - \frac{2}{9}$

Use $<$ or $>$ for $\square$ to write a true sentence. [4.3b]

13. $\frac{4}{7} \square \frac{5}{9}$

14. $\frac{-11}{13} \square \frac{-8}{9}$

Solve. [4.3c]

15. $x + \frac{2}{5} = \frac{7}{8}$

16. $\frac{1}{2} + y = -\frac{9}{10}$

Convert to fraction notation. [4.4a]

17. $7\frac{1}{2}$

18. $8\frac{3}{8}$

19. $4\frac{1}{3}$

20. $-10\frac{5}{7}$

Convert to a mixed numeral. [4.4a]

21. $\frac{7}{3}$

22. $\frac{27}{4}$

23. $-\frac{63}{5}$

24. $\frac{7}{2}$

Divide. Write a mixed numeral for the answer. [4.4b]

25. $9\overline{)7896}$

26. $23\overline{)10{,}493}$

Add. Write a mixed numeral for the answer where appropriate. [4.5a]

27. $5\frac{3}{5} + 4\frac{4}{5}$

28. $8\frac{1}{3} + 3\frac{2}{5}$

29. $5\frac{5}{6} + 4\frac{5}{6}$

30. $2\frac{3}{4} + 5\frac{1}{2}$

Subtract. Write a mixed numeral for the answer where appropriate. [4.5b]

31. $12 - 4\frac{2}{9}$

32. $9\frac{3}{5} - 4\frac{13}{15}$

33. $10\frac{1}{4} - 6\frac{1}{10}$

34. $24 - 10\frac{5}{8}$

Multiply. Write a mixed numeral for the answer where appropriate. [4.6a]

35. $6 \cdot 2\frac{2}{3}$

36. $5\frac{1}{4} \cdot \left(-\frac{2}{3}\right)$

37. $2\frac{1}{5} \cdot 1\frac{1}{10}$

38. $-2\frac{2}{5} \cdot 2\frac{1}{2}$

Divide. Write a mixed numeral for the answer where appropriate. [4.6b]

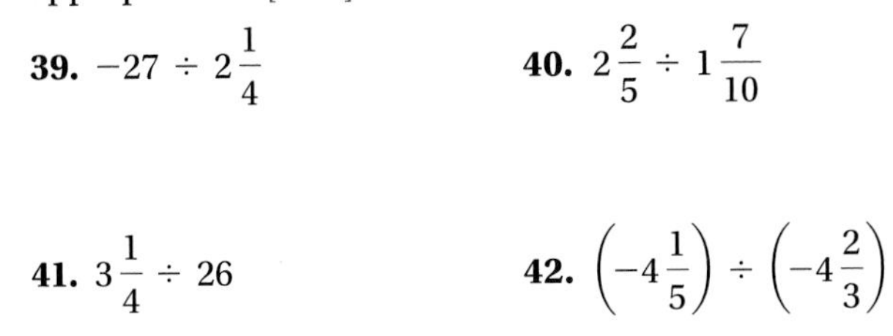

39. $-27 \div 2\frac{1}{4}$

40. $2\frac{2}{5} \div 1\frac{7}{10}$

41. $3\frac{1}{4} \div 26$

42. $\left(-4\frac{1}{5}\right) \div \left(-4\frac{2}{3}\right)$

Solve. [4.2b], [4.3d], [4.5c], [4.6c]

43. *Sewing.* Gloria wants to make a dress and a jacket. She needs $1\frac{5}{8}$ yd of 60-in. fabric for the dress and $2\frac{5}{8}$ yd for the jacket. How many yards in all does Gloria need to make the outfit?

44. What is the sum of the areas in the figure below?

$9\frac{1}{2}$ in.

12 in.

A

B

$7\frac{1}{2}$ in.

$8\frac{1}{2}$ in.

45. In the figure above, how much larger is the area of rectangle A than the area of rectangle B?

46. *Carpentry.* A board $\frac{9}{10}$ in. thick is glued to a board $\frac{4}{5}$ in. thick. The glue is $\frac{3}{100}$ in. thick. How thick is the result?

47. *Turkey Servings.* Turkey contains $1\frac{1}{3}$ servings per pound. How many pounds are needed for 32 servings?

48. *Cake Recipe.* A wedding-cake recipe requires 12 cups of shortening. Being calorie-conscious, the wedding couple decides to reduce the shortening by $3\frac{5}{8}$ cups and replace it with prune purée. How many cups of shortening are used in their new recipe?

49. *Painting a Border.* Katie hired an artist to paint a decorative border around the top of her son's bedroom. The artist charges $20 per foot. The room measures $11\frac{3}{4}$ ft × $9\frac{1}{2}$ ft. What is Katie's cost for the project?

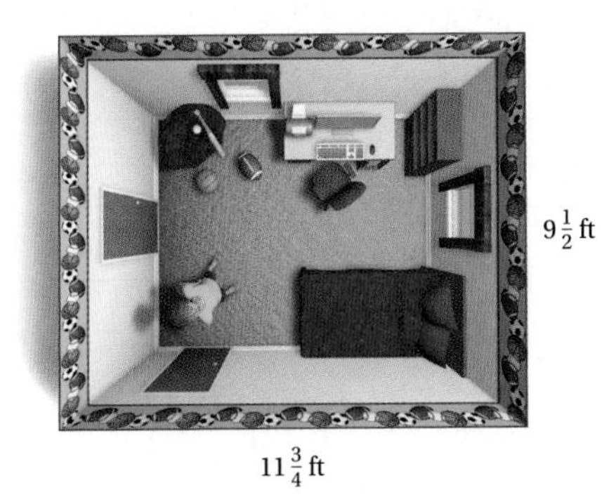

50. *Running.* Janelle has mapped a $1\frac{1}{2}$-mi running route in her neighborhood. One Saturday, she ran this route $2\frac{1}{2}$ times. How many miles did she run?

51. *Humane Society Pie Sale.* Green River's Humane Society recently hosted its annual pie sale. Each of the 83 pies donated was cut into 6 pieces. At the end of the evening, 382 pieces of pie had been sold. How many pies were sold? How many were left over? Express your answers in mixed numerals.

Simplify each expression using the rules for order of operations. [4.7a]

52. $\frac{1}{8} \div \frac{1}{4} + \frac{1}{2}$

53. $\frac{4}{5} - \frac{1}{2} \cdot \left(1 + \frac{1}{4}\right)$

54. $20\frac{3}{4} - 1\frac{1}{2} \times 12 + \left(\frac{1}{2}\right)^2$

Simplify. [4.7b]

55. $\dfrac{-\frac{2}{3}}{\frac{5}{6}}$

56. $\dfrac{10}{\frac{1}{2} - \frac{1}{6}}$

57. Find the average of $\frac{1}{2}, \frac{1}{4}, \frac{1}{3}$, and $\frac{1}{5}$. [4.7b]

Estimate each of the following as 0, $\frac{1}{2}$, or 1. [4.7c]

58. $\frac{29}{59}$

59. $\frac{2}{59}$

60. $\frac{61}{59}$

Estimate by estimating each mixed numeral as a whole number or as a mixed numeral where the fractional part is $\frac{1}{2}$ and by estimating each fraction as 0, $\frac{1}{2}$, or 1. [4.7c]

61. $6\frac{7}{8}$

62. $10\frac{2}{17}$

63. $\frac{11}{12} \cdot 5\frac{6}{13}$

64. $\frac{1}{15} \cdot \frac{2}{3}$

65. $\frac{6}{11} + \frac{5}{6} + \frac{31}{29}$

66. $32\frac{14}{15} + 27\frac{6}{7} - 4\frac{25}{28} \cdot 6\frac{37}{76}$

67. Simplify: $\frac{1}{4} + \frac{2}{5} \div 5^2$. [4.7a]

A. $\frac{133}{500}$
B. $\frac{3}{500}$
C. $\frac{117}{500}$
D. $\frac{5}{2}$

68. Solve: $x + \frac{2}{3} = -5$. [4.3c]

A. $\frac{17}{3}$
B. $\frac{13}{3}$
C. $-\frac{13}{3}$
D. $-\frac{17}{3}$

Synthesis

69. *Orangutan Circus Act.* Yuri and Olga are orangutans that perform in a circus by riding bicycles around a circular track. It takes Yuri 6 min and Olga 4 min to make one trip around the track. Suppose they start at the same point and then complete their act when they again reach the same point. How long is their act? [4.1a]

70. Place the numbers 3, 4, 5, and 6 in the boxes in order to make a true equation: [4.5a]

$$\frac{\square}{\square} + \frac{\square}{\square} = 3\frac{1}{4}.$$

Understanding Through Discussion and Writing

1. Is the sum of two mixed numerals always a mixed numeral? Why or why not? [4.5a]
2. Write a problem for a classmate to solve. Design the problem so that its solution is found by performing the multiplication $4\frac{1}{2} \cdot 33\frac{1}{3}$. [4.6c]
3. A student insists that $3\frac{2}{5} \cdot 1\frac{3}{7} = 3\frac{6}{35}$. What mistake is he making and how should he have proceeded? [4.6a]
4. Discuss the role of least common multiples in adding and subtracting with fraction notation. [4.2a], [4.3a]
5. Find a real-world situation that fits this equation:
$$2 \cdot 15\frac{3}{4} + 2 \cdot 28\frac{5}{8} = 88\frac{3}{4}.$$ [4.5c], [4.6c]
6. A student insists that $5 \cdot 3\frac{2}{7} = (5 \cdot 3) \cdot \left(5 \cdot \frac{2}{7}\right)$. What mistake is she making and how should she have proceeded? [4.6a], [4.7a]

CHAPTER 4 Test

For Extra Help For step-by-step test solutions, access the Chapter Test Prep Videos in MyMathLab® or on YouTube (search "BittingerBasicEl" and click on "Channels").

Find the LCM.

1. 16 and 12

2. 15, 40, and 50

Add and simplify.

3. $\frac{1}{2} + \frac{5}{2}$

4. $\frac{-7}{8} + \frac{2}{3}$

5. $\frac{7}{10} + \frac{19}{100} + \frac{31}{1000}$

Subtract and simplify.

6. $\frac{5}{6} - \frac{3}{6}$

7. $\frac{5}{6} - \frac{3}{4}$

8. $-\frac{17}{24} - \frac{1}{15}$

Solve.

9. $\frac{1}{4} + y = 4$

10. $x + \frac{2}{3} = \frac{11}{12}$

11. Use < or > for □ to write a true sentence:

$\frac{6}{7} \square \frac{21}{25}$.

Convert to fraction notation.

12. $3\frac{1}{2}$

13. $-9\frac{7}{8}$

Convert to a mixed numeral.

14. $\frac{9}{2}$

15. $-\frac{74}{9}$

Divide. Write a mixed numeral for the answer.

16. $11\overline{)1789}$

Add. Write a mixed numeral for the answer.

17. $6\frac{2}{5} + 7\frac{4}{5}$

18. $9\frac{1}{4} + 5\frac{1}{6}$

Subtract. Write a mixed numeral for the answer.

19. $10\frac{1}{6} - 5\frac{7}{8}$

20. $14 - 7\frac{5}{6}$

Multiply. Write a mixed numeral for the answer, if appropriate.

21. $9 \cdot 4\frac{1}{3}$

22. $-6\frac{3}{4} \cdot \frac{2}{3}$

Divide. Write a mixed numeral for the answer, if appropriate.

23. $2\frac{1}{3} \div 1\frac{1}{6}$

24. $2\frac{1}{12} \div (-75)$

25. ***Weightlifting.*** In 2002, Hossein Rezazadeh of Iran did a clean and jerk of 263 kg. This amount was about $2\frac{1}{2}$ times his body weight. How much did Rezazadeh weigh?

Source: *The Guinness Book of Records*, 2005

26. ***Book Order.*** An order of books for a math course weighs 220 lb. Each book weighs $2\frac{3}{4}$ lb. How many books are in the order?

27. *Carpentry.* The following diagram shows a middle drawer support guide for a cabinet drawer. Find each of the following.

a) The short length a across the top

b) The length b across the bottom

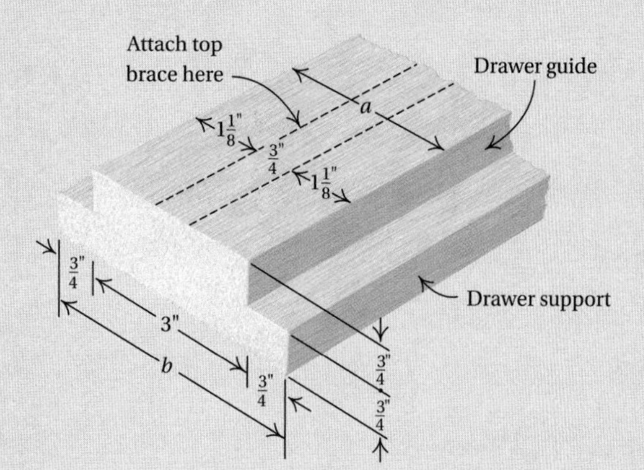

28. *Carpentry.* In carpentry, some pieces of plywood that are called "$\frac{3}{4}$-inch" plywood are actually $\frac{11}{16}$ in. thick. How much thinner is such a piece than its name indicates?

29. *Women's Dunks.* The first three women in the history of college basketball to dunk a basketball are listed below, along with their heights and universities:

Michelle Snow, $6\frac{5}{12}$ ft, Tennessee;
Charlotte Smith, $5\frac{11}{12}$ ft, North Carolina;
Georgeann Wells, $6\frac{7}{12}$ ft, West Virginia.

Find the average height of these women.

Source: *USA Today,* 11/30/00, p. 3C

Simplify.

30. $\frac{2}{3} + 1\frac{1}{3} \cdot 2\frac{1}{8}$

31. $1\frac{1}{2} - \frac{1}{2}\left(\frac{1}{2} \div \frac{1}{4}\right) + \left(\frac{1}{2}\right)^2$

32. $\dfrac{\frac{2}{3} + \frac{1}{6}}{5}$

Estimate each of the following as 0, $\frac{1}{2}$, or 1.

33. $\frac{3}{82}$

34. $\frac{93}{91}$

35. Estimate the following by estimating the mixed numeral as a whole number or as a mixed numeral where the fractional part is $\frac{1}{2}$.

$$256 \div 15\frac{19}{21}$$

36. Find the LCM of 12, 36, and 60.

A. 6 **B.** 12
C. 60 **D.** 180

Synthesis

37. The students in a math class can be organized into study groups of 8 each so that no students are left out. The same class of students can also be organized into groups of 6 so that no students are left out.

a) Find some class sizes for which this will work.

b) Find the smallest such class size.

38. Rebecca walks 17 laps at her health club. Trent walks 17 laps at his health club. If the track at Rebecca's health club is $\frac{1}{7}$ mi long, and the track at Trent's is $\frac{1}{8}$ mi long, who walks farther? How much farther?

CHAPTERS

1–4 Cumulative Review

Solve.

1. *Cross-Country Skiing.* During a three-day holiday weekend trip, David and Sally Jean cross-country skied $3\frac{2}{3}$ mi on Friday, $6\frac{1}{8}$ mi on Saturday, and $4\frac{3}{4}$ mi on Sunday.
   a) Find the total number of miles they skied.
   b) Find the average number of miles they skied per day. Express your answer as a mixed numeral.

2. How many people can receive equal $16 shares from a total of $496?

3. A recipe calls for $\frac{4}{5}$ tsp of salt. How much salt should be used for $\frac{1}{2}$ recipe? for 5 recipes?

4. How many pieces, each $2\frac{3}{8}$ ft long, can be cut from a piece of wire 38 ft long?

5. An emergency food pantry fund contains $423. From this fund, $148 and $167 are withdrawn for expenses. How much is left in the fund?

6. In a walkathon, Jermaine walked $\frac{9}{10}$ mi and Oleta walked $\frac{3}{4}$ mi. What was the total distance they walked?

What part is shaded?

7.

8.

Calculate and simplify.

9. $\begin{array}{r} 3704 \\ +\,5278 \\ \hline \end{array}$

10. $\begin{array}{r} 7605 \\ -\,3087 \\ \hline \end{array}$

11. $-27 + 12$

12. $\frac{3}{8} + \frac{1}{24}$

13. $-20 - (-6)$

14. $-\frac{3}{4} - \frac{1}{3}$

15. $\begin{array}{r} 2\frac{3}{4} \\ +\,5\frac{1}{2} \\ \hline \end{array}$

16. $\begin{array}{r} 2\frac{1}{3} \\ -\,1\frac{1}{6} \\ \hline \end{array}$

17. $15 \cdot (-5)$

18. $\frac{9}{10} \cdot \frac{5}{3}$

19. $-18 \cdot \left(-\frac{5}{6}\right)$

20. $2\frac{1}{5} \div \frac{3}{10}$

Divide. Write the answer in the form 34 R 7.

21. $6\overline{)4290}$

22. $45\overline{)2531}$

23. Write a mixed numeral for the answer in Exercise 22.

24. In the number 2753, what digit names tens?

25. *Room Carpeting.* The Chandlers are carpeting an L-shaped family room consisting of one rectangle that is $8\frac{1}{2}$ ft by 11 ft and another rectangle that is $6\frac{1}{2}$ ft by $7\frac{1}{2}$ ft.

a) Find the area of the carpet.

b) Find the perimeter of the carpet.

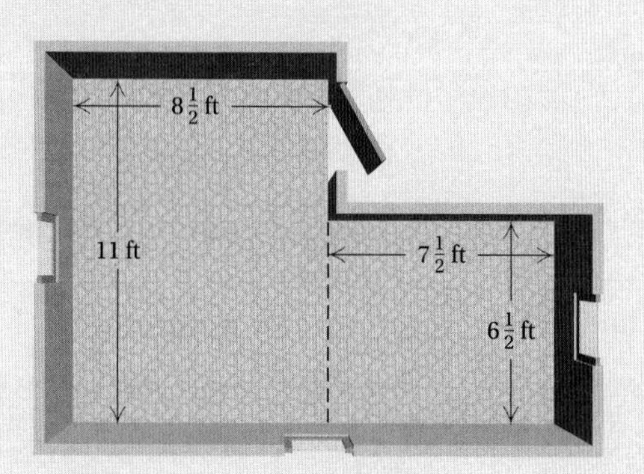

26. Round 38,478 to the nearest hundred.

27. Find the LCM of 18 and 24.

28. Simplify:

$$\left(\frac{1}{2} + \frac{2}{5}\right)^2 \div 3 + 6 \times \left(2 + \frac{1}{4}\right).$$

Use $<$, $>$, or $=$ for □ to write a true sentence.

29. $\frac{4}{5}$ □ $\frac{4}{6}$

30. $\frac{3}{13}$ □ $\frac{9}{39}$

31. $\frac{-3}{7}$ □ $\frac{-5}{12}$

Estimate each of the following as 0, $\frac{1}{2}$, or 1.

32. $\frac{29}{30}$

33. $\frac{15}{29}$

34. $\frac{2}{43}$

Simplify.

35. $\frac{36}{45}$

36. $\frac{0}{-27}$

37. $\frac{320}{10}$

38. Convert to fraction notation: $4\frac{5}{8}$.

39. Convert to a mixed numeral: $\frac{17}{3}$.

Solve.

40. $x + 24 = 117$

41. $x + \frac{7}{9} = \frac{4}{3}$

42. $\frac{7}{9} \cdot t = -\frac{4}{3}$

43. $y = 32{,}580 \div 36$

44. *Matching.* Match each item in the first column with the appropriate item in the second column by drawing connecting lines. There can be more than one correct correspondence for an item.

Factors of 68	12, 54, 72, 300
Factorization of 68	2, 3, 17, 19, 23, 31, 47, 101
Prime factorization of 68	$2 \cdot 2 \cdot 17$
Numbers divisible by 6	$2 \cdot 34$
Numbers divisible by 8	8, 16, 24, 32, 40, 48, 64, 864
Numbers divisible by 5	1, 2, 4, 17, 34, 68
Prime numbers	70, 95, 215

Synthesis

45. Find the smallest prime number that is larger than 2000.

CHAPTER

5

Decimal Notation

STUDYING FOR SUCCESS *Time Management*

- ☐ As a rule, budget 2–3 hours for doing homework and studying for every hour you spend in class.
- ☐ Balance work and study carefully. Be honest with yourself about how much time you have available to work, attend class, and study.
- ☐ Make an hour-by-hour schedule of your week, planning time for leisure as well as work and study.
- ☐ Use your syllabus to help you plan your time. Transfer project deadlines and test dates to your calendar.

5.1 Decimal Notation, Order, and Rounding

OBJECTIVES

a Given decimal notation, write a word name.

b Convert between decimal notation and fraction notation.

c Given a pair of numbers in decimal notation, tell which is larger.

d Round decimal notation to the nearest thousandth, hundredth, tenth, one, ten, hundred, or thousand.

SKILL TO REVIEW

Objective 1.6a: Round to the nearest ten, hundred, or thousand.

Round 4735 to the nearest:

1. Ten.
2. Hundred.

Recall that the set of **rational numbers** consists of the integers, . . . , -3, -2, -1, 0, 1, 2, 3, . . . , and fractions like $\frac{1}{2}$, $-\frac{5}{3}$, $\frac{9}{8}$, $-\frac{17}{10}$, and so on. (See Section 3.3.) In this chapter, we use **decimal notation** to represent the set of rational numbers. For example, $\frac{1}{2}$ can be written as 0.5 in decimal notation and $3\frac{1}{10}$ can be written as 3.1.

The word *decimal* comes from the Latin word *decima*, meaning a tenth part. Since our counting system is based on tens, decimal notation is a natural extension of a system with which we are already familiar.

a DECIMAL NOTATION AND WORD NAMES

One model of the Magellan GPS navigation system sells for \$249.98. The dot in \$249.98 is called a **decimal point**. Since \$0.98, or 98¢, is $\frac{98}{100}$ of a dollar, it follows that

$$\$249.98 = \$249 + \$0.98.$$

Also, since \$0.98, or 98¢, has the same value as 9 dimes + 8 cents and 1 dime is $\frac{1}{10}$ of a dollar and 1 cent is $\frac{1}{100}$ of a dollar, we can write

$$249.98 = 2 \cdot 100 + 4 \cdot 10 + 9 \cdot 1 + 9 \cdot \frac{1}{10} + 8 \cdot \frac{1}{100}.$$

This is an extension of the expanded notation for whole numbers that we used in Chapter 1. The place values are 100, 10, 1, $\frac{1}{10}$, $\frac{1}{100}$, and so on. We can see this on a **place-value chart**. The value of each place is $\frac{1}{10}$ as large as that of the one to its left.

Let's consider decimal notation using a place-value chart to represent 2.0677 min, the women's 200-meter backstroke world record held by Missy Franklin from the United States.

PLACE-VALUE CHART							
Hundreds	Tens	Ones	Ten*ths*	Hundred*ths*	Thousand*ths*	Ten-Thousand*ths*	Hundred-Thousand*ths*
100	10	1	$\frac{1}{10}$	$\frac{1}{100}$	$\frac{1}{1000}$	$\frac{1}{10,000}$	$\frac{1}{100,000}$
		2 .	0	6	7	7	

Answers

Skill to Review:
1. 4740 2. 4700

The decimal notation 2.0677 means

$$2 + \frac{0}{10} + \frac{6}{100} + \frac{7}{1000} + \frac{7}{10,000}, \quad \text{or} \quad 2\frac{677}{10,000}.$$

We read both 2.0677 and $2\frac{677}{10,000}$ as

"Two *and* six hundred seventy-seven ten-thousandths."

We can also read 2.0677 as

"Two *point* zero six seven seven."

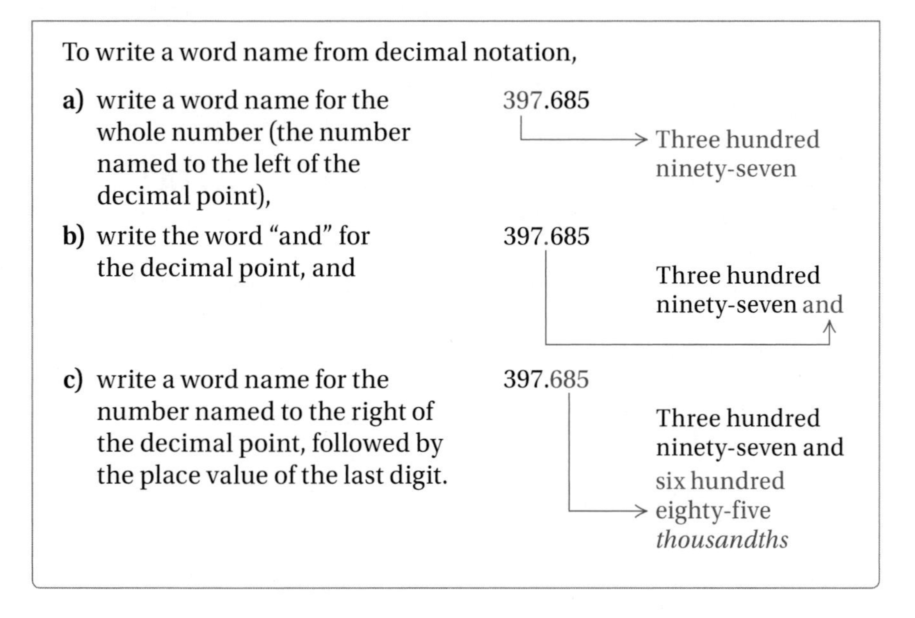

To write a word name from decimal notation,

a) write a word name for the whole number (the number named to the left of the decimal point),

397.685 → Three hundred ninety-seven

b) write the word "and" for the decimal point, and

397.685 → Three hundred ninety-seven and

c) write a word name for the number named to the right of the decimal point, followed by the place value of the last digit.

397.685 → Three hundred ninety-seven and six hundred eighty-five *thousandths*

EXAMPLE 1 *Median Age.* The median age of residents in Maine is 42.7 years. The median age of residents in Utah is 29.2 years. Write word names for 42.7 and 29.2.

Source: U.S. Census Bureau

Forty-two and seven tenths

Twenty-nine and two tenths

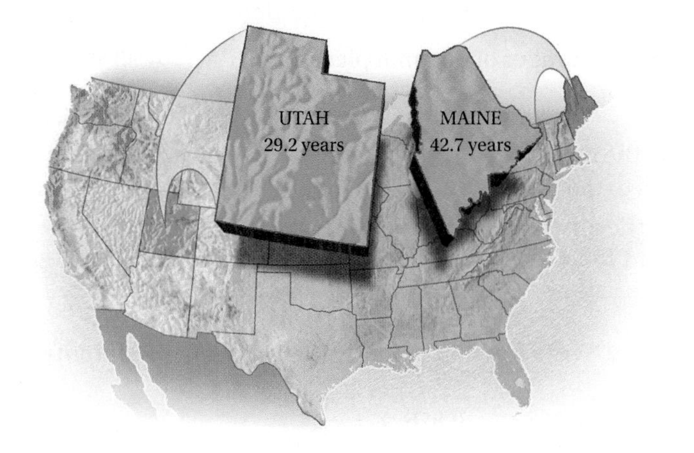

EXAMPLE 2 Write a word name for 410.87.

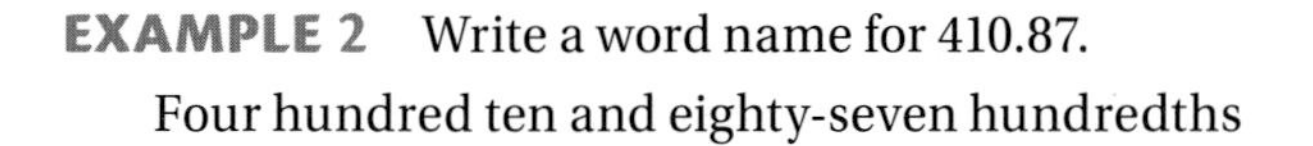

Four hundred ten and eighty-seven hundredths

EXAMPLE 3 *Record Price.* Edvard Munch's painting *The Scream* sold for \$119.9225 million at Sotheby's in New York City on May 2, 2012. This price set a world record as the highest paid for a work of art sold at auction. Write a word name for 119.9225.

Source: Sotheby's

One hundred nineteen and nine thousand, two hundred twenty-five ten-thousandths

EXAMPLE 4 Write a word name for −1788.045.

Negative one thousand, seven hundred eighty-eight and forty-five thousandths

◀ **Do Exercises 1–6.**

b CONVERTING BETWEEN DECIMAL NOTATION AND FRACTION NOTATION

Given decimal notation, we can convert to fraction notation as follows:

$$9.875 = 9 + \frac{8}{10} + \frac{7}{100} + \frac{5}{1000}$$

$$= 9 \cdot \frac{1000}{1000} + \frac{8}{10} \cdot \frac{100}{100} + \frac{7}{100} \cdot \frac{10}{10} + \frac{5}{1000}$$

$$= \frac{9000}{1000} + \frac{800}{1000} + \frac{70}{1000} + \frac{5}{1000} = \frac{9875}{1000}.$$

Decimal notation → 9.875 ← 3 decimal places

Fraction notation → $\frac{9875}{1000}$ ← 3 zeros

> To convert from decimal to fraction notation,
>
> a) count the number of decimal places, — 4.98 — 2 places
>
> b) move the decimal point that many places to the right, and — 4.98. — Move 2 places.
>
> c) write the answer over a denominator with a 1 followed by that number of zeros. — $\frac{498}{100}$ — 2 zeros

EXAMPLE 5 Write fraction notation for 0.876. Do not simplify.

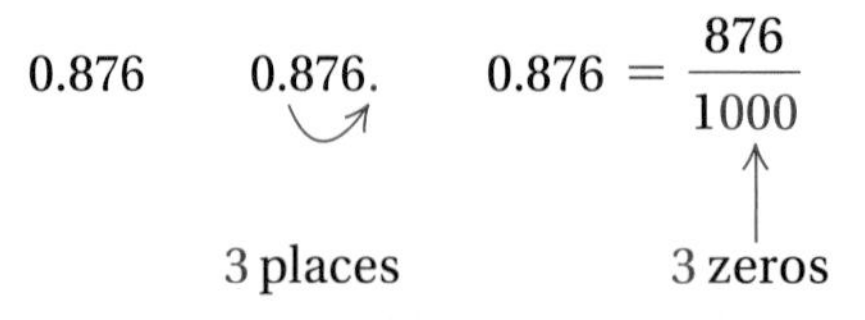

For a number like 0.876, we generally write a 0 before the decimal point to draw attention to the presence of the decimal point.

1. *Life Expectancy.* The life expectancy at birth in Kenya in 2011 was 61.3 years for males and 64.2 years for females. Write word names for 61.3 and 64.2.

   Source: World Health Organization

2. *10,000-Meter Record.* Wang Junxia of China holds the women's world record for the 10,000-meter run: 29.5297 min. Write a word name for 29.5297.

Write a word name for each number.

3. 245.89

4. −34.0064

5. 31,079.756

6. *Temperature Extremes.* The lowest average annual temperature in the United States is −12.6°C. It occurs in Barrow, Alaska.

   Source: *Time Almanac* 2005

Answers

1. Sixty-one and three tenths; sixty-four and two tenths
2. Twenty-nine and five thousand, two hundred ninety-seven ten-thousandths
3. Two hundred forty-five and eighty-nine hundredths
4. Negative thirty-four and sixty-four ten-thousandths
5. Thirty-one thousand, seventy-nine and seven hundred fifty-six thousandths
6. Negative twelve and six tenths

EXAMPLE 6 Write fraction notation for 56.23. Do not simplify.

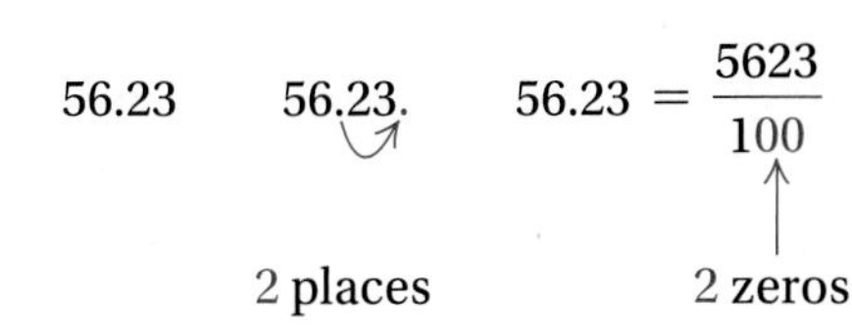

56.23   56.23. (2 places) $56.23 = \frac{5623}{100}$ (2 zeros)

EXAMPLE 7 Write fraction notation for −1.5018. Do not simplify.

−1.5018   −1.5018. (4 places) $-1.5018 = -\frac{15{,}018}{10{,}000}$ (4 zeros)

Do Exercises 7–10. ▶

Write fraction notation.

GS **7.** 0.896

0.896. ☐ places

$0.896 = \frac{896}{1\square}$

8. 23.78

9. −5.6789

10. 1.9

If fraction notation has a denominator that is a power of ten, such as 10, 100, 1000, and so on, we convert to decimal notation by reversing the procedure that we used before.

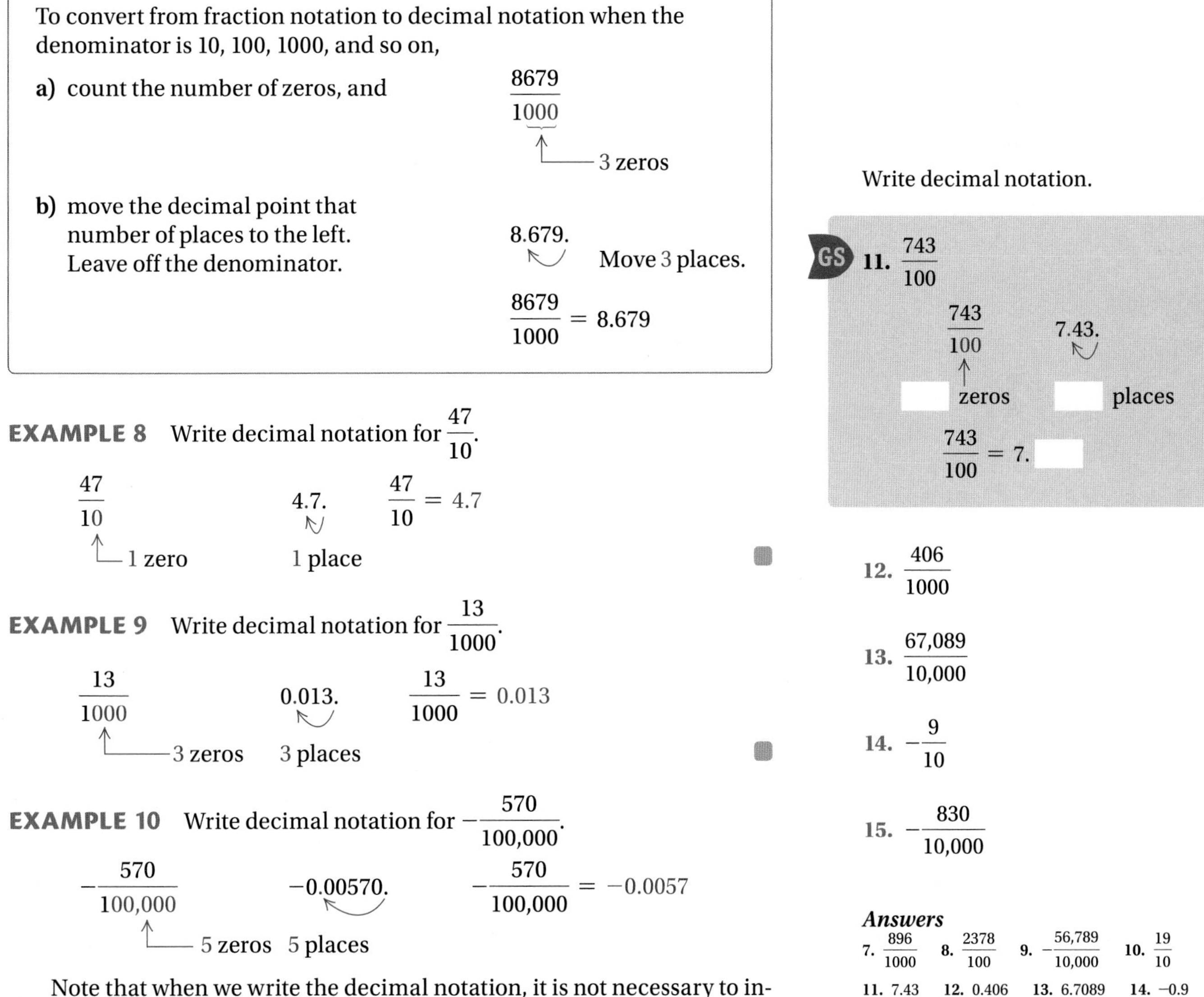

To convert from fraction notation to decimal notation when the denominator is 10, 100, 1000, and so on,

a) count the number of zeros, and $\frac{8679}{1000}$ — 3 zeros

b) move the decimal point that number of places to the left. Leave off the denominator.   8.679. Move 3 places. $\frac{8679}{1000} = 8.679$

EXAMPLE 8 Write decimal notation for $\frac{47}{10}$.

$\frac{47}{10}$ (1 zero)   4.7. (1 place) $\frac{47}{10} = 4.7$

EXAMPLE 9 Write decimal notation for $\frac{13}{1000}$.

$\frac{13}{1000}$ (3 zeros)   0.013. (3 places) $\frac{13}{1000} = 0.013$

EXAMPLE 10 Write decimal notation for $-\frac{570}{100{,}000}$.

$-\frac{570}{100{,}000}$ (5 zeros)   −0.00570. (5 places) $-\frac{570}{100{,}000} = -0.0057$

Note that when we write the decimal notation, it is not necessary to include the 0 that follows the 7.

Do Exercises 11–15. ▶

Write decimal notation.

GS **11.** $\frac{743}{100}$

$\frac{743}{100}$ ☐ zeros   7.43. ☐ places

$\frac{743}{100} = 7.\square$

12. $\frac{406}{1000}$

13. $\frac{67{,}089}{10{,}000}$

14. $-\frac{9}{10}$

15. $-\frac{830}{10{,}000}$

Answers

7. $\frac{896}{1000}$ **8.** $\frac{2378}{100}$ **9.** $-\frac{56{,}789}{10{,}000}$ **10.** $\frac{19}{10}$
11. 7.43 **12.** 0.406 **13.** 6.7089 **14.** −0.9
15. −0.083

Guided Solutions:
7. 3, 000 **11.** 2, 2, 43

When denominators are numbers other than 10, 100, and so on, we will use another method for conversion. It will be considered in Section 5.5.

If a mixed numeral has a fraction part with a denominator that is a power of ten, such as 10, 100, or 1000, and so on, we first write the mixed numeral as a sum of a whole number and a fraction. Then we convert to decimal notation.

EXAMPLE 11 Write decimal notation for $23\frac{59}{100}$.

$$23\frac{59}{100} = 23 + \frac{59}{100} = 23 \text{ and } \frac{59}{100} = 23.59$$

Write decimal notation.

16. $4\frac{3}{10}$

17. $283\frac{71}{100}$

18. $-456\frac{13}{1000}$

EXAMPLE 12 Write decimal notation for $-772\frac{129}{10{,}000}$.

First we consider $772\frac{129}{10{,}000}$.

$$772\frac{129}{10{,}000} = 772 + \frac{129}{10{,}000} = 772 \text{ and } \frac{129}{10{,}000} = 772.0129$$

Since $772\frac{129}{10{,}000} = 772.0129$, we have $-772\frac{129}{10{,}000} = -772.0129$.

◀ **Do Exercises 16–18.**

C ORDER

To understand how to compare numbers in decimal notation, consider 0.85 and 0.9. First note that $0.9 = 0.90$ because $\frac{9}{10} = \frac{90}{100}$. Then $0.85 = \frac{85}{100}$ and $0.90 = \frac{90}{100}$. Since $\frac{85}{100} < \frac{90}{100}$, it follows that $0.85 < 0.90$. This leads us to a quick way to compare two numbers in decimal notation.

COMPARING POSITIVE DECIMALS

To compare two positive numbers in decimal notation, start at the left and compare corresponding digits moving from left to right. If two digits differ, the number with the larger digit is the larger of the two numbers. To ease the comparison, extra zeros can be written to the right of the last decimal place.

EXAMPLE 13 Which of 2.109 and 2.1 is larger?

Think.

2.109 ⟶ 2.109

2.1 ⟶ 2.100

Same — Different; $9 > 0$

Thus, 2.109 is larger than 2.1. That is, $2.109 > 2.1$.

Answers

16. 4.3 **17.** 283.71 **18.** −456.013

EXAMPLE 14 Which of 0.09 and 0.108 is larger?

Think.

0.09 ⟶ 0.090

0.108 ⟶ 0.108

Same └ Different; $1 > 0$

Thus, 0.108 is larger than 0.09. That is, $0.108 > 0.09$.

Do Exercises 19–24. ▶

Which number is larger?

19. 2.04, 2.039

20. 0.06, 0.008

21. 0.5, 0.58

22. 1, 0.9999

23. 0.8989, 0.09898

24. 21.006, 21.05

How do we compare negative decimals? Using the number line, we see that $-1 > -2$. Similarly, $-1.4 > -1.5$.

This leads to the following procedure.

COMPARING NEGATIVE DECIMALS

To compare two negative numbers in decimal notation, start at the left and compare corresponding digits. When two digits differ, the number with the smaller digit is the larger of the two numbers.

EXAMPLE 15 Which of -5.6 and -5.63 is larger?

Think.

-5.6 ⟶ -5.60

-5.63 ⟶ -5.63

Same └ Different; $0 < 3$

Thus, -5.6 is larger than -5.63. That is, $-5.6. > -5.63$.

Do Exercises 25–28. ▶

Which number is larger?

25. $-7.56, -6.18$

26. $-21.006, -21.05$

27. $-0.44, -0.46$

28. $-13.94, -13.79$

d ROUNDING

Rounding of numbers in decimal notation is done as for whole numbers. To understand, we first consider an example using the number line. It might help to review Section 1.6.

EXAMPLE 16 Round 0.37 to the nearest tenth.

Here is part of a number line.

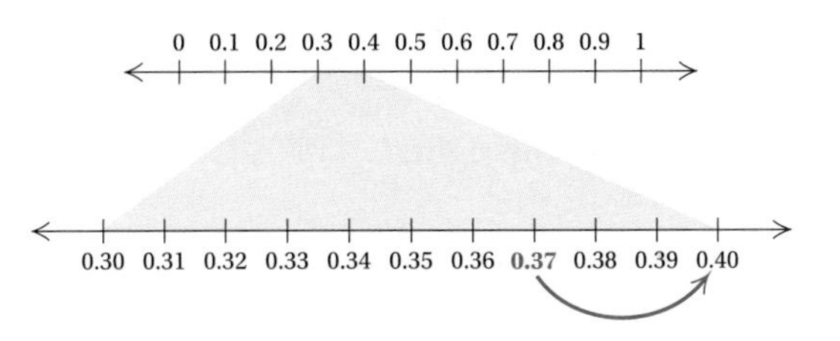

We see that 0.37 is closer to 0.40 than to 0.30. Thus, 0.37 rounded to the nearest tenth is 0.4. ■

Answers

19. 2.04 **20.** 0.06 **21.** 0.58 **22.** 1 **23.** 0.8989 **24.** 21.05 **25.** -6.18 **26.** -21.006 **27.** -0.44 **28.** -13.79

ROUNDING DECIMAL NOTATION

To round to a certain place:

a) Locate the digit in that place.
b) Consider the next digit to the right.
c) If the digit to the right is 5 or higher, round up. If the digit to the right is 4 or lower, round down.

EXAMPLE 17 Round 3872.2459 to the nearest tenth.

a) Locate the digit in the tenths place, 2.

3 8 7 2 . 2 4 5 9
↑

b) Consider the next digit to the right, 4.

3 8 7 2 . 2 4 5 9
↑

c) Since that digit, 4, is 4 or lower, round down.

3 8 7 2 . 2 ⟵ This is the answer.

Caution!

3872.3 is not a correct answer to Example 17. It is *incorrect* to round from the ten-thousandths digit over to the tenths digit, as follows:

3872.246 → 3872.25 → 3872.3.

EXAMPLE 18 Round 3872.2459 to the nearest thousandth, hundredth, tenth, one, ten, hundred, and thousand.

Thousandth:	3872.246	One:	3872	Thousand:	4000
Hundredth:	3872.25	Ten:	3870		
Tenth:	3872.2	Hundred:	3900		

EXAMPLE 19 Round −14.8973 to the nearest hundredth.

a) Locate the digit in the hundredths place, 9.

−1 4 . 8 9 7 3
↑

b) Consider the next digit to the right, 7.

−1 4 . 8 9 7 3
↑

c) Since that digit, 7, is 5 or higher, round the digit 9 up. When we make the hundredths digit a 10, we carry 1 to the tenths place.

The answer is −14.90. The 0 in −14.90 indicates that the answer is correct to the nearest hundredth. Note that when we rounded the digit 9 up, we actually rounded −14.8973 down. That is, $-14.90 < -14.8973$.

EXAMPLE 20 Round 0.008 to the nearest tenth.

a) Locate the digit in the tenths place, 0.

0 . 0 0 8
↑

b) Consider the next digit to the right, 0.

0 . 0 0 8
↑

c) Since that digit, 0, is less than 5, round down.

The answer is 0.0.

Do Exercises 29–47.

Round to the nearest tenth.

29. 2.76 **30.** 13.85

31. 234.448 **32.** 7.009

Round to the nearest hundredth.

33. 0.636 **34.** 7.834

35. 34.675 **36.** 0.025

Round to the nearest thousandth.

37. 0.9434 **38.** 8.0038

39. 43.1119 **40.** 37.4005

Round 7459.3548 to the nearest:

41. Thousandth.

42. Hundredth.

43. Tenth.

44. One.

45. Ten. (*Caution:* "Tens" are not "tenths.")

46. Hundred.

47. Thousand.

Answers

29. 2.8 **30.** 13.9 **31.** 234.4 **32.** 7.0
33. 0.64 **34.** 7.83 **35.** 34.68 **36.** 0.03
37. 0.943 **38.** 8.004 **39.** 43.112
40. 37.401 **41.** 7459.355 **42.** 7459.35
43. 7459.4 **44.** 7459 **45.** 7460
46. 7500 **47.** 7000

5.1 Exercise Set

For Extra Help
MyMathLab® | MathXL® PRACTICE | WATCH | READ | REVIEW

Reading Check

Name the digit that represents each place value in the number 436.81205.

RC1. Hundred-thousandths ________

RC2. Thousandths ________

RC3. Tens ________

RC4. Ten-thousandths ________

RC5. Tenths ________

RC6. Hundreds ________

RC7. Hundredths ________

RC8. Ones ________

a

1. *Currency Conversion.* One Japanese yen was worth about $0.0119 in U.S. currency recently. Write a word name for 0.0119.

   Source: finance.yahoo.com

2. *Currency Conversion.* One U.S. dollar was worth 0.9949 Canadian dollars recently. Write a word name for 0.9949.

   Source: finance.yahoo.com

3. *Birth Rate.* There were 137.6 triplet births per 100,000 live births in a recent year in the United States. Write a word name for 137.6.

   Source: U.S. Centers for Disease Control and Prevention

4. *Coffee Consumption.* Annual per capita coffee consumption in Finland is 26.7 lb. In Spain, annual per capita coffee consumption is only 9.4 lb. Write word names for 26.7 and 9.4.

   Source: International Coffee Organization

5. *Stock Price.* Apple, Inc.'s stock price was recently $519.22 per share. Write a word name for 519.22.

   Source: NASDAQ

6. *Indianapolis 500.* Dario Franchitti won the 2012 Indianapolis 500 race with a time of 21 hr 58 min 51.2532 sec. In second place, Scott Dixon finished 0.0295 sec later. Write a word name for 0.0295.

   Source: Indianapolis Motor Speedway

Write a word name for the number in each sentence.

7. One gallon of paint is equal to 3.785 L of paint.

8. *Water Weight.* One gallon of water weighs 8.35 lb.

Write a word name.

9. -34.891

10. -27.1245

b Write fraction notation. Do not simplify.

11. 8.3

12. 203.6

13. 3.56

14. 0.17

15. 20.003

16. 1.509

17. -1.0008

18. -2.0114

19. 37.2

20. 4567.2

21. -0.00013

22. -6.14057

Write decimal notation.

23. $\frac{8}{10}$

24. $\frac{51}{10}$

25. $\frac{3798}{1000}$

26. $\frac{780}{1000}$

27. $-\frac{889}{100}$

28. $-\frac{92}{100}$

29. $\frac{19}{100,000}$

30. $\frac{56,788}{100,000}$

31. $\frac{78}{10,000}$

32. $\frac{13,463}{10,000}$

33. $-\frac{376,193}{1,000,000}$

34. $-\frac{8,953,074}{1,000,000}$

35. $2\frac{9}{10}$

36. $9243\frac{1}{10}$

37. $3\frac{98}{1000}$

38. $4\frac{67}{1000}$

39. $99\frac{44}{100}$

40. $67\frac{83}{100}$

41. $-2\frac{1739}{10,000}$

42. $-2256\frac{3059}{10,000}$

c Which number is larger?

43. 0.06, 0.58

44. 0.008, 0.8

45. 0.905, 0.91

46. 42.06, 42.1

47. $-0.0009, -0.001$

48. $-7.067, -7.054$

49. 234.07, 235.07

50. 0.99999, 1

51. $0.004, \frac{4}{100}$

52. $\frac{73}{10}, 0.73$

53. $-0.432, -0.4325$

54. $-0.8437, -0.84384$

d Round to the nearest tenth.

55. 0.11

56. 0.85

57. -0.49

58. -0.5394

59. 2.7449

60. 4.78

61. -123.65

62. -36.049

Round to the nearest hundredth.

63. 0.893 **64.** 0.675 **65.** 0.6666 **66.** 6.524

67. −0.995 **68.** −207.9976 **69.** −0.094 **70.** −11.4246

Round to the nearest thousandth.

71. 0.3246 **72.** 0.6666 **73.** −17.0014 **74.** −123.4562

75. 10.1011 **76.** 0.1161 **77.** −9.9989 **78.** −67.100602

Round 809.5732 to the nearest:

79. Hundred. **80.** Tenth. **81.** Thousandth.

82. Hundredth. **83.** One. **84.** Ten.

Round 34.54389 to the nearest:

85. Ten-thousandth. **86.** Thousandth. **87.** Hundredth.

88. Tenth. **89.** One. **90.** Ten.

Skill Maintenance

Round 6172 to the nearest: [1.6a]

91. Ten. **92.** Hundred. **93.** Thousand.

Find the prime factorization. [3.1d]

94. 2000 **95.** 1530

Simplify. [3.5b]

96. $\frac{30}{126}$ **97.** $\frac{120}{100}$

Synthesis

98. Arrange the following numbers in order from smallest to largest.

−0.99, −0.099, −1, −0.9999, −0.89999, −1.00009, −0.909, −0.9889

99. Arrange the following numbers in order from smallest to largest.

−2.1, −2.109, −2.108, −2.018, −2.0119, −2.0302, −2.000001

Truncating. There are other methods of rounding decimal notation. To round using **truncating**, we drop off all decimal places past the rounding place, which is the same as changing all digits to the right to zeros. For example, rounding 6.78093456285102 to the ninth decimal place, using truncating, gives us 6.780934562. Use truncating to round each of the following to the fifth decimal place—that is, the hundred-thousandth place.

100. 6.78346623 **101.** 6.783465902 **102.** 99.999999999 **103.** 0.030303030303

5.2 Addition and Subtraction

OBJECTIVES

a Add using decimal notation for positive numbers.

b Subtract using decimal notation for positive numbers.

c Add and subtract negative decimals.

d Solve equations of the type $x + a = b$ and $a + x = b$, where a and b may be in decimal notation.

SKILLS TO REVIEW

Objective 1.2a: Add whole numbers.
Objective 1.3a: Subtract whole numbers.

Add or subtract.

1. $\begin{array}{r} 387 \\ +\ 254 \\ \hline \end{array}$

2. $\begin{array}{r} 4023 \\ -\ 1667 \\ \hline \end{array}$

Add.

1. $\begin{array}{r} 0.847 \\ +\ 10.07 \\ \hline \end{array}$

2. $\begin{array}{r} 2.1 \\ 0.73 \\ +\ 31.368 \\ \hline \end{array}$

Add.

3. 0.02 + 4.3 + 0.649

4. 0.12 + 3.006 + 0.4357

5. 0.4591 + 0.2374 + 8.70894

Answers

Skills to Review:
1. 641 **2.** 2356

Margin Exercises:
1. 10.917 **2.** 34.198 **3.** 4.969
4. 3.5617 **5.** 9.40544

a ADDITION

Adding positive decimal numbers is similar to adding whole numbers. First, we line up the decimal points so that we can add corresponding place-value digits. Then we add digits from the right. For example, we add the thousandths, then the hundredths, and so on, carrying if necessary. If desired, we can write extra zeros to the right of the decimal point so that the number of places is the same in all of the addends.

EXAMPLE 1 Add: 56.314 + 17.78.

$\begin{array}{r} 56.314 \\ +\ 17.780 \\ \hline \end{array}$ Lining up the decimal points in order to add
Writing an extra zero to the right of the decimal point

$\begin{array}{r} 56.314 \\ +\ 17.780 \\ \hline 4 \end{array}$ Adding thousandths

$\begin{array}{r} 56.314 \\ +\ 17.780 \\ \hline 94 \end{array}$ Adding hundredths

$\begin{array}{r} \scriptstyle 1 \\ 56.314 \\ +\ 17.780 \\ \hline .094 \end{array}$ Adding tenths
We get 10 tenths = 1 one + 0 tenths, so we carry the 1 to the ones column.
Writing a decimal point in the answer

$\begin{array}{r} \scriptstyle 1\ 1 \\ 56.314 \\ +\ 17.780 \\ \hline 4.094 \end{array}$ Adding ones
We get 14 ones = 1 ten + 4 ones, so we carry the 1 to the tens column.

$\begin{array}{r} \scriptstyle 1\ 1 \\ 56.314 \\ +\ 17.780 \\ \hline 74.094 \end{array}$ Adding tens

◀ Do Margin Exercises 1–2.

EXAMPLE 2 Add: 3.42 + 0.237 + 14.1.

$\begin{array}{r} 3.420 \\ 0.237 \\ +\ 14.100 \\ \hline 17.757 \end{array}$ Lining up the decimal points and writing extra zeros
Adding

◀ Do Exercises 3–5.

Now we consider the addition 3456 + 19.347. Keep in mind that any whole number has an "unwritten" decimal point at the right that can be followed by zeros. For example, 3456 can also be written 3456.000. When adding, we can always write in the decimal point and extra zeros if desired.

EXAMPLE 3 Add: 3456 + 19.347.

$$\begin{array}{r} \scriptstyle 1 \quad\quad\quad\quad\quad \\ 3456.000 \\ +\quad 19.347 \\ \hline 3475.347 \end{array}$$

Writing in the decimal point and extra zeros
Lining up the decimal points
Adding

Do Exercises 6 and 7. ▶

Add.

6. 789 + 123.67

GS 7. 45.78 + 2467 + 1.993

$$\begin{array}{r} \square\,1\,\square\;\;1\;\;\, \\ 45.780 \\ 2467.000 \\ +\quad 1.993 \\ \hline 2\,\square\,14.\square\,73 \end{array}$$

b SUBTRACTING POSITIVE DECIMALS

Subtracting positive decimals is similar to subtracting whole numbers. First, we line up the decimal points so that we can subtract corresponding place-value digits. Then we subtract digits from the right. For example, we subtract the thousandths, then the hundredths, the tenths, and so on, borrowing if necessary.

EXAMPLE 4 Subtract: 56.314 − 17.78.

$$\begin{array}{r} 56.314 \\ -\,17.780 \\ \hline \end{array}$$

Lining up the decimal points in order to subtract
Writing an extra 0

$$\begin{array}{r} 56.314 \\ -\,17.780 \\ \hline 4 \end{array}$$

Subtracting thousandths

$$\begin{array}{r} \scriptstyle 2\;11\;\; \\ 56.\not{3}\not{1}4 \\ -\,17.780 \\ \hline 34 \end{array}$$

Borrowing tenths to subtract hundredths
Subtracting hundredths

$$\begin{array}{r} \scriptstyle 12\;\;\;\;\; \\ \scriptstyle 5\;\;\not{2}\;11\;\; \\ 5\not{6}.\not{3}\not{1}4 \\ -\,17.780 \\ \hline .534 \end{array}$$

Borrowing ones to subtract tenths
Subtracting tenths; writing a decimal point

$$\begin{array}{r} \scriptstyle 15\;12\;\;\;\;\; \\ \scriptstyle 4\;\not{5}\;\;\not{2}\;11\;\; \\ \not{5}\not{6}.\not{3}\not{1}4 \\ -\,17.780 \\ \hline 8.534 \end{array}$$

Borrowing tens to subtract ones
Subtracting ones

$$\begin{array}{r} \scriptstyle 15\;12\;\;\;\;\; \\ \scriptstyle 4\;\not{5}\;\;\not{2}\;11\;\; \\ \not{5}\not{6}.\not{3}\not{1}4 \\ -\,17.780 \\ \hline 38.534 \end{array}$$

Subtracting tens

Check by adding:

$$\begin{array}{r} \scriptstyle 1\;1\;\;1\;\;\;\;\;\;\; \\ 38.534 \\ +\,17.780 \\ \hline 56.314 \end{array}$$

The answer checks because this is the top number in the subtraction.

Do Exercises 8 and 9. ▶

Subtract.

GS 8. 37.428 − 26.674

$$\begin{array}{r} \square\;\;\;\;\;\;\; \\ \scriptstyle 6\;\;\;\not{3}\;\square \\ 37.\not{4}\,\not{2}\,8 \\ -\,26.674 \\ \hline 1\,\square.7\,\square\,4 \end{array}$$

9.
$$\begin{array}{r} 0.347 \\ -\,0.008 \\ \hline \end{array}$$

Answers

6. 912.67 7. 2514.773 8. 10.754
9. 0.339

Guided Solutions:
7. 1, 1, 5, 7 8. 13, 12, 0, 5

EXAMPLE 5 Subtract: 13.07 − 9.205.

```
12
 2  10 6 10
1 3.0 7 0     Writing an extra zero
−  9.2 0 5
   3.8 6 5    Subtracting
```

EXAMPLE 6 Subtract: 23.08 − 5.0053.

```
 1 13    7 9 10
 2 3.0 8 0 0    Writing two extra zeros
−   5.0 0 5 3
 1 8.0 7 4 7    Subtracting
```

Check by adding:

```
 1      1 1
 1 8.0 7 4 7
+  5.0 0 5 3
 2 3.0 8 0 0
```

Subtract.

10. 1.2345 − 0.7

11. 0.9564 − 0.4392

12. 7.37 − 0.00008

Do Exercises 10–12.

When subtraction involves a whole number, again keep in mind that there is an "unwritten" decimal point that can be written in if desired. Extra zeros can also be written in to the right of the decimal point.

EXAMPLE 7 Subtract: 456 − 2.467.

```
     5  9 9 10
 4 5 6.0 0 0    Writing in the decimal point and extra zeros
−      2.4 6 7
 4 5 3.5 3 3    Subtracting
```

Subtract.

13. 1277 − 82.78

14. 5 − 0.0089

GS

```
           10
 5.0 0 0 0
−0.0 0 8 9
 4.   9 1
```

Do Exercises 13 and 14.

C ADDING AND SUBTRACTING NEGATIVE DECIMALS

Negative decimals are added or subtracted just like negative integers.

ADDITION INVOLVING NEGATIVE NUMBERS

1. *Negative numbers*: Add absolute values. The answer is negative.
2. *A positive and a negative number*:
   - If the numbers have the same absolute value, the answer is 0.
   - If the numbers have different absolute values, subtract the smaller absolute value from the larger. Then:
     a) If the positive number has the greater absolute value, the answer is positive.
     b) If the negative number has the greater absolute value, the answer is negative.
3. *One number is zero*: The sum is the other number.

Answers

10. 0.5345 11. 0.5172 12. 7.36992
13. 1194.22 14. 4.9911

Guided Solution:
14. 4999, 9, 1

EXAMPLE 8 Add: $-5.103 + (-6.275)$.

$$\begin{array}{r} 5.103 \\ +\ 6.275 \\ \hline 11.378 \end{array}$$

The absolute values are 5.103 and 6.275.

Adding the absolute values

The answer is negative: $-5.103 + (-6.275) = -11.378$.

EXAMPLE 9 Add: $-9.48 + 2.37$.

$$\begin{array}{r} 9.48 \\ -\ 2.37 \\ \hline 7.11 \end{array}$$

Finding the difference of the absolute values

The negative number has the greater absolute value, so the answer is negative: $-9.48 + 2.37 = -7.11$.

Do Exercises 15–18.

Add.

15. $12.064 + (-7.591)$

16. $-2.81 + (-8.432)$

17. $7.42 + (-9.38)$

18. $-4.201 + 7.36$

To subtract, we add the opposite of the number being subtracted.

EXAMPLE 10 Subtract: $-3.1 - 4.8$.

$-3.1 - 4.8 = -3.1 + (-4.8)$ Adding the opposite of 4.8

$= -7.9$ The sum of two negative numbers is negative.

EXAMPLE 11 Subtract: $-7.9 - (-8.5)$.

$-7.9 - (-8.5) = -7.9 + 8.5$ Adding the opposite of -8.5

$= 0.6$ Finding the difference of the absolute values. The answer is positive since 8.5 has the greater absolute value.

Do Exercises 19–22.

Subtract.

19. $9.25 - 13.41$

20. $-5.72 - 4.19$

21. $9.8 - (-2.6)$

22. $-5.9 - (-3.2)$

d SOLVING EQUATIONS

Now let's solve equations $x + a = b$ and $a + x = b$, where a and b may be in decimal notation. Proceeding as we have before, we subtract a on both sides.

EXAMPLE 12 Solve: $x + 28.89 = 74.567$.

We have

$x + 28.89 - 28.89 = 74.567 - 28.89$ Subtracting 28.89 on both sides

$x = 45.677.$

$$\begin{array}{r} 74.567 \\ 28.890 \\ \hline 45.677 \end{array}$$

The solution is 45.677.

Answers

15. 4.473 **16.** -11.242 **17.** -1.96 **18.** 3.159 **19.** -4.16 **20.** -9.91 **21.** 12.4 **22.** -2.7

EXAMPLE 13 Solve: $0.8879 + y = 9.0026$.

We have

$$0.8879 + y - 0.8879 = 9.0026 - 0.8879 \quad \text{Subtracting 0.8879 on both sides}$$
$$y = 8.1147.$$

$$\begin{array}{r} \overset{8}{\cancel{9}}.\overset{9}{\cancel{0}}\overset{9}{\cancel{0}}\overset{\overset{11}{\cancel{1}}}{2}\overset{16}{\cancel{6}} \\ -0.8879 \\ \hline 8.1147 \end{array}$$

The solution is 8.1147.

◀ **Do Exercises 23 and 24.**

Solve.

23. $x + 17.78 = 56.314$

24. $8.906 + t = 23.07$

EXAMPLE 14 Solve: $x + 7.4 = 5.6$.

We have

$$x + 7.4 - 7.4 = 5.6 - 7.4 \quad \text{Subtracting 7.4 on both sides}$$
$$x = -1.8. \quad 5.6 - 7.4 = 5.6 + (-7.4) = -1.8$$

The solution is -1.8.

◀ **Do Exercise 25.**

25. Solve: $10.6 + y = 2.8$.

Answers

23. 38.534 **24.** 14.164 **25.** -7.8

5.2 Exercise Set

For Extra Help MyMathLab® MathXL® PRACTICE WATCH READ REVIEW

Reading Check

Complete each subtraction and its check by selecting a number from the list at right.

RC1. $\begin{array}{r} 23.7 \\ -\ 1.876 \\ \hline \square \end{array}$ Check: $\begin{array}{r} 21.824 \\ +\ 1.876 \\ \hline \square \end{array}$

RC2. $\begin{array}{r} 187.623 \\ -\ 40.9 \\ \hline \square \end{array}$ Check: $\begin{array}{r} 146.723 \\ +\ \square \\ \hline 187.623 \end{array}$

23.7
187.623
21.824
40.9
1.876
146.723

a Add.

1. $\begin{array}{r} 316.25 \\ +\ 18.12 \\ \hline \end{array}$

2. $\begin{array}{r} 641.803 \\ +\ 14.935 \\ \hline \end{array}$

3. $\begin{array}{r} 659.403 \\ +\ 916.812 \\ \hline \end{array}$

4. $\begin{array}{r} 4203.28 \\ +\ 3.39 \\ \hline \end{array}$

5. $\begin{array}{r} 9.104 \\ +\ 123.456 \\ \hline \end{array}$

6. $\begin{array}{r} 81.008 \\ +\ 3.409 \\ \hline \end{array}$

7. $20.0124 + 30.0124$

8. $0.687 + 0.9$

9. $39 + 1.007$

10. $2.3 + 0.729 + 23$

11. $\begin{array}{r} 47.8 \\ 219.852 \\ 43.59 \\ +\ 666.713 \\ \hline \end{array}$

12. $\begin{array}{r} 13.72 \\ 9.112 \\ 6542.7908 \\ +\ 23.901 \\ \hline \end{array}$

13. $0.34 + 3.5 + 0.127 + 768$

14. $17 + 3.24 + 0.256 + 0.3689$

15. $99.6001 + 7285.18 + 500.042 + 870$

16. $65.987 + 9.4703 + 6744.02 + 1.0003 + 200.895$

b Subtract.

17. $\begin{array}{r} 51.31 \\ -\ 2.29 \\ \hline \end{array}$

18. $\begin{array}{r} 44.345 \\ -\ 3.105 \\ \hline \end{array}$

19. $\begin{array}{r} 92.341 \\ -\ 6.42 \\ \hline \end{array}$

20. $\begin{array}{r} 97.01 \\ -\ 3.15 \\ \hline \end{array}$

21. $\begin{array}{l} \ \ 2.5 \\ -\ 0.0025 \\ \hline \end{array}$

22. $\begin{array}{r} 39.0 \\ -\ 0.28 \\ \hline \end{array}$

23. $\begin{array}{l} \ \ 3.4 \\ -\ 0.003 \\ \hline \end{array}$

24. $\begin{array}{r} 2.8 \\ -\ 2.08 \\ \hline \end{array}$

25. $28.2 - 19.35$

26. $100.16 - 0.118$

27. $34.07 - 30.7$

28. $36.2 - 16.28$

29. $8.45 - 7.405$

30. $3.801 - 2.81$

31. $6.003 - 2.3$

32. $9.087 - 8.807$

33. $1 - 0.0098$

34. $2 - 1.0908$

35. $100 - 0.34$

36. $624 - 18.79$

37. $7.48 - 2.6$

38. $18.4 - 5.92$

39. $3 - 2.006$

40. $263.7 - 102.08$

41. $19 - 1.198$

42. $2548.98 - 2.007$

43. $65 - 13.87$

44. $45 - 0.999$

45. $$\begin{array}{r} 32.7978 \\ -\ 0.0592 \\ \hline \end{array}$$

46. $$\begin{array}{r} 0.49634 \\ -\ 0.12678 \\ \hline \end{array}$$

47. $$\begin{array}{l} \ \ 6.07 \\ -\ 2.0078 \\ \hline \end{array}$$

48. $$\begin{array}{l} \ \ 1.0 \\ -\ 0.9999 \\ \hline \end{array}$$

C Add or subtract, as indicated.

49. $-5.02 + 1.73$

50. $-4.31 + 7.66$

51. $12.9 - 15.4$

52. $27.2 - 31.9$

53. $-2.9 + (-4.3)$

54. $-7.49 - 1.82$

55. $-4.301 + 7.68$

56. $-5.952 + 7.98$

57. $-12.9 - 3.7$

58. $-8.7 - 12.4$

59. $-2.1 - (-4.6)$

60. $-4.3 - (-2.5)$

61. $14.301 + (-17.82)$

62. $13.45 + (-18.701)$

63. $7.201 - (-2.4)$

64. $2.901 - (-5.7)$

65. $96.9 + (-21.4)$

66. $43.2 + (-10.9)$

67. $-8.9 - (-12.7)$

68. $-4.5 - (-7.3)$

69. $-4.9 - 5.392$

70. $89.3 - 92.1$

71. $14.7 - 23.5$

72. $-7.201 - 1.9$

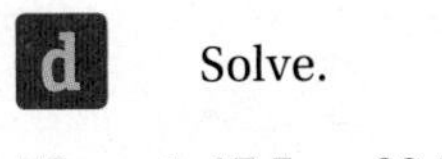

Solve.

73. $x + 17.5 = 29.15$

74. $t + 50.7 = 54.07$

75. $17.95 + p = 402.63$

76. $w + 1.3004 = 47.8$

77. $x + 2349 = -17{,}684.3$

78. $1830.4 + t = -23{,}067$

79. $13{,}083.3 = x + 12{,}500.33$

80. $100.23 = 67.8 + z$

81. $643.2 + w = 109.6$

82. $7375.4 + z = 984.56$

83. $y + 2.39 = -4.6$

84. $x + 16.47 = -15.49$

Skill Maintenance

85. Round 34,567 to the nearest thousand. [1.6a]

86. Round 34,496 to the nearest thousand. [1.6a]

Subtract.

87. $\frac{13}{24} - \frac{3}{8}$ [4.3a]

88. $\frac{8}{9} - \frac{2}{15}$ [4.3a]

89. $8805 - 2639$ [1.3a]

90. $8005 - 2639$ [1.3a]

Solve.

91. A serving of filleted fish is generally considered to be about $\frac{1}{3}$ lb. How many servings can be prepared from $5\frac{1}{2}$ lb of flounder fillet? [4.6c]

92. A photocopier technician drove $125\frac{7}{10}$ mi away from Scottsdale for a repair call. The next day he drove $65\frac{1}{2}$ mi back toward Scottsdale for another service call. How far was the technician from Scottsdale? [4.5c]

Synthesis

93. A student presses the wrong button when using a calculator and adds 235.7 instead of subtracting it. The incorrect answer is 817.2. What is the correct answer?

5.3 Multiplication

OBJECTIVES

a Multiply using decimal notation.

b Convert from notation like 45.7 million to standard notation, and convert between dollars and cents.

SKILL TO REVIEW

Objective 1.4a: Multiply whole numbers.

Multiply.

1. 42×63 **2.** 716×58

a MULTIPLICATION

Let's find the product

$$2.3 \times 1.12.$$

To understand how we find such a product, we first convert each factor to fraction notation. Next, we multiply the numerators and then divide by the product of the denominators.

$$2.3 \times 1.12 = \frac{23}{10} \times \frac{112}{100} = \frac{23 \times 112}{10 \times 100} = \frac{2576}{1000} = 2.576$$

Note the number of decimal places.

$$\begin{array}{rl} 1.12 & \text{(2 decimal places)} \\ \times\ 2.3 & \text{(1 decimal place)} \\ \hline 2.576 & \text{(3 decimal places)} \end{array}$$

Now consider

$$0.011 \times 15.0002 = \frac{11}{1000} \times \frac{150{,}002}{10{,}000} = \frac{1{,}650{,}022}{10{,}000{,}000} = 0.1650022.$$

Note the number of decimal places.

$$\begin{array}{rl} 15.0002 & \text{(4 decimal places)} \\ \times\ 0.011 & \text{(3 decimal places)} \\ \hline 0.1650022 & \text{(7 decimal places)} \end{array}$$

To multiply using decimals:

a) Ignore the decimal points and multiply as though both factors were integers.

b) Then place the decimal point in the result. The number of decimal places in the product is the sum of the numbers of places in the factors. (Count places from the right.)

0.8×0.43

$$\begin{array}{r} \scriptstyle 2 \\ 0.43 \\ \times\ 0.8 \\ \hline 344 \end{array}$$ Ignore the decimal points for now.

$$\begin{array}{rl} 0.43 & \text{(2 decimal places)} \\ \times\ 0.8 & \text{(1 decimal place)} \\ \hline 0.344 & \text{(3 decimal places)} \end{array}$$

EXAMPLE 1 Multiply: 8.3×74.6.

a) We ignore the decimal points and multiply as though factors were integers.

$$\begin{array}{r} 74.6 \\ \times\ 8.3 \\ \hline 2238 \\ 59680 \\ \hline 61918 \end{array}$$

Answers

Skill to Review:
1. 2646 **2.** 41,528

b) We place the decimal point in the result. The number of decimal places in the product is the sum of the numbers of places in the factors, 1 + 1, or 2.

```
      7 4.6     (1 decimal place)
  ×     8.3     (1 decimal place)
    2 2 3 8
  5 9 6 8 0
  6 1 9.1 8     (2 decimal places)
```

Do Exercise 1. ▶

1. Multiply.

```
    8 5.4
  ×   6.2
```

EXAMPLE 2 Multiply: 0.0032×2148.

```
        2 1 4 8   (0 decimal places)
  × 0.0 0 3 2     (4 decimal places)
        4 2 9 6
      6 4 4 4 0
      6.8 7 3 6   (4 decimal places)
```

EXAMPLE 3 Multiply: -0.14×0.867.

First we multiply the absolute values.

```
      0.8 6 7    (3 decimal places)
  ×     0.1 4    (2 decimal places)
      3 4 6 8
      8 6 7 0
  0.1 2 1 3 8    (5 decimal places)
```

Since the product of a negative number and a positive number is negative, the answer is -0.12138.

Do Exercises 2 and 3. ▶

Multiply.

2.
```
      1 2 3 4
  × 0.0 0 4 1
```

GS 3. -42.65×0.804

First we multiply the absolute values.

```
        4 2.6 5
    ×   0.8 0 4
      1 ☐ 0 6 0
  3 4 1 2 0 ☐ 0
  3 4.☐ 9 0 ☐ 0
```

Since the product of a negative number and a positive number is ________ (positive/negative), the answer is ☐.

Multiplying by 0.1, 0.01, 0.001, and So On

Now let's consider some special kinds of products. The first involves multiplying by a tenth, hundredth, thousandth, ten-thousandth, and so on. Let's look at such products.

$$0.1 \times 38 = \frac{1}{10} \times 38 = \frac{38}{10} = 3.8$$

$$0.01 \times 38 = \frac{1}{100} \times 38 = \frac{38}{100} = 0.38$$

$$0.001 \times 38 = \frac{1}{1000} \times 38 = \frac{38}{1000} = 0.038$$

$$0.0001 \times 38 = \frac{1}{10{,}000} \times 38 = \frac{38}{10{,}000} = 0.0038$$

Note in each case that the product is *smaller* than 38. That is, the decimal point in each product is farther to the left than the unwritten decimal point in 38.

Answers

1. 529.48 2. 5.0594 3. −34.2906

Guided Solution:
3. 7, 0, 2, 6; negative, −34.2906

To multiply any number by 0.1, 0.01, 0.001, and so on,

a) count the number of decimal places in the tenth, hundredth, or thousandth, and so on, and

$0.\underline{001} \times 34.45678$ → 3 places

b) move the decimal point in the other number that many places to the left.

$0.001 \times 34.45678 = 0.034.45678$

Move 3 places to the left.

$0.001 \times 34.45678 = 0.03445678$

EXAMPLES Multiply.

4. $0.1 \times 14.605 = 1.4605$ 1.4.605

5. $0.01 \times 14.605 = 0.14605$

6. $0.001 \times 14.605 = 0.014605$ We write an extra zero.

7. $0.0001 \times 14.605 = 0.0014605$ We write two extra zeros.

◀ **Do Exercises 4–7.**

Multiply.

4. 0.1×3.48

5. 0.01×3.48

6. 0.001×3.48

7. 0.0001×3.48

Multiplying by 10, 100, 1000, and So On

Next, let's consider multiplying by 10, 100, 1000, and so on. Let's look at some of these products.

$$10 \times 97.34 = 973.4$$
$$100 \times 97.34 = 9734$$
$$1000 \times 97.34 = 97{,}340$$

Note in each case that the product is *larger* than 97.34. That is, the decimal point in each product is farther to the right than the decimal point in 97.34.

To multiply any number by 10, 100, 1000, and so on,

a) count the number of zeros, and

$1\underline{000} \times 34.45678$ → 3 zeros

b) move the decimal point in the other number that many places to the right.

$1000 \times 34.45678 = 34.456.78$

Move 3 places to the right.

$1000 \times 34.45678 = 34{,}456.78$

EXAMPLES Multiply.

8. $10 \times 14.605 = 146.05$ 14.6.05

9. $100 \times 14.605 = 1460.5$

10. $1000 \times 14.605 = 14{,}605$

11. $10{,}000 \times 14.605 = 146{,}050$ 14.6050. We write an extra zero.

◀ **Do Exercises 8–11.**

Multiply.

8. 10×3.48

9. 100×3.48

10. 1000×3.48

11. $10{,}000 \times 3.48$

Answers

4. 0.348 5. 0.0348 6. 0.00348
7. 0.000348 8. 34.8 9. 348
10. 3480 11. 34,800

b NAMING LARGE NUMBERS; MONEY CONVERSION

Naming Large Numbers

We often see notation like the following in newspapers and magazines and on television and the Internet.

- In 2012, wildfires burned over 9.1 million acres in the United States.
  Source: National Interagency Fire Center
- The number of valid U.S. passports in circulation in 2011 exceeded 113.4 million.
  Source: U.S. State Department
- Each day, about 144.8 billion e-mails are sent worldwide.
  Source: Mashable.com
- At one point in 2013, the U.S. national debt was approximately $16.501 trillion. The national debt has increased, on average, by $3.81 billion per day since 2007.
  Source: www.usdebtclock.org

To understand such notation, consider the information in the following table.

NAMING LARGE NUMBERS

1 hundred $= 100 = 10 \cdot 10 = 10^2$ → 2 zeros

1 thousand $= 1000 = 10 \cdot 10 \cdot 10 = 10^3$ → 3 zeros

1 million $= 1{,}000{,}000 = 10 \cdot 10 \cdot 10 \cdot 10 \cdot 10 \cdot 10 = 10^6$ → 6 zeros

1 billion $= 1{,}000{,}000{,}000 = 10^9$ → 9 zeros

1 trillion $= 1{,}000{,}000{,}000{,}000 = 10^{12}$ → 12 zeros

To convert a large number to standard notation, we proceed as follows.

EXAMPLE 12 *Text Messages.* Worldwide, 8.6 trillion text messages are sent each year. Convert 8.6 trillion to standard notation.

Source: CNNTech, Portio Research

$$8.6 \text{ trillion} = 8.6 \times 1 \text{ trillion}$$
$$= 8.6 \times \underbrace{1{,}000{,}000{,}000{,}000}_{12 \text{ zeros}}$$
$$= 8{,}600{,}000{,}000{,}000$$

Moving the decimal point 12 places to the right

◀ Do Exercises 12 and 13.

Convert the number in each sentence to standard notation.

12. The largest building in the world is the Pentagon, which has 3.7 million square feet of floor space.

13. *Text Messages.* In the United States, 2.2 trillion text messages are sent each year.

Source: CNNTech, Forrester Research

Money Conversion

Converting from dollars to cents is like multiplying by 100. To see why, consider \$19.43.

$\$19.43 = 19.43 \times \$1$	We think of \$19.43 as 19.43×1 dollar, or $19.43 \times \$1$.
$= 19.43 \times 100¢$	Substituting 100¢ for \$1: $\$1 = 100¢$
$= 1943¢$	Multiplying

DOLLARS TO CENTS

To convert from dollars to cents, move the decimal point two places to the right and change the \$ sign in front to a ¢ sign at the end.

Convert from dollars to cents.

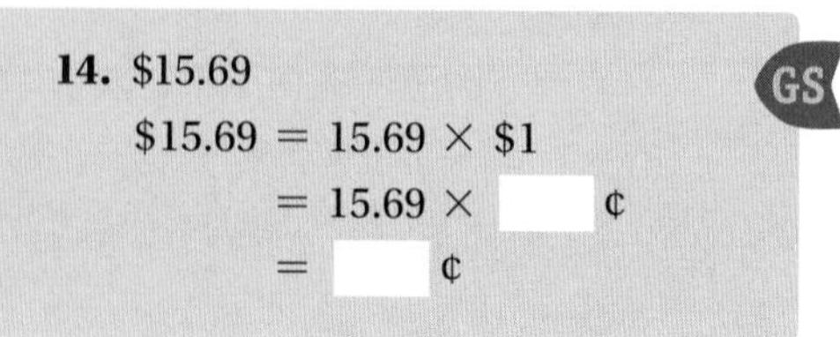
14. \$15.69

$\$15.69 = 15.69 \times \$1$

$= 15.69 \times$ ____ ¢

$=$ ____ ¢

15. \$0.17

EXAMPLES Convert from dollars to cents.

13. $\$189.64 = 18{,}964¢$

14. $\$0.75 = 75¢$

◀ Do Exercises 14 and 15.

Converting from cents to dollars is like multiplying by 0.01. To see why, consider 65¢.

$65¢ = 65 \times 1¢$	We think of 65¢ as 65×1 cent, or $65 \times 1¢$.
$= 65 \times \$0.01$	Substituting \$0.01 for 1¢: $1¢ = \$0.01$
$= \$0.65$	Multiplying

CENTS TO DOLLARS

To convert from cents to dollars, move the decimal point two places to the left and change the ¢ sign at the end to a \$ sign in front.

Convert from cents to dollars.

16. 35¢

17. 577¢

EXAMPLES Convert from cents to dollars.

15. $395¢ = \$3.95$

16. $8503¢ = \$85.03$

◀ Do Exercises 16 and 17.

Answers

12. 3,700,000 **13.** 2,200,000,000,000
14. 1569¢ **15.** 17¢ **16.** \$0.35 **17.** \$5.77

Guided Solution:
14. 100, 1569

5.3 Exercise Set

For Extra Help MyMathLab® MathXL® PRACTICE WATCH READ REVIEW

Reading Check

Match each expression with an equivalent expression from the list at the right.

RC1. 0.001×38 ______ **a)** 3800¢

RC2. 1000×38 ______ **b)** 380

RC3. 38¢ ______ **c)** 0.038

RC4. $38 ______ **d)** $3.80

RC5. 380¢ ______ **e)** 38,000

RC6. 10×38 ______ **f)** $0.38

a Multiply.

1. $\begin{array}{r} 8.6 \\ \times \quad 7 \\ \hline \end{array}$

2. $\begin{array}{r} 5.7 \\ \times 0.8 \\ \hline \end{array}$

3. $\begin{array}{r} 0.84 \\ \times \quad 8 \\ \hline \end{array}$

4. $\begin{array}{r} 9.4 \\ \times 0.6 \\ \hline \end{array}$

5. $\begin{array}{r} 6.3 \\ \times 0.04 \\ \hline \end{array}$

6. $\begin{array}{r} 9.8 \\ \times 0.08 \\ \hline \end{array}$

7. $\begin{array}{r} 87 \\ \times 0.006 \\ \hline \end{array}$

8. $\begin{array}{r} 18.4 \\ \times 0.07 \\ \hline \end{array}$

9. 10×23.76

10. 100×3.8798

11. -1000×583.686852

12. -0.34×1000

13. -7.8×100

14. $0.00238 \times (-10)$

15. 0.1×89.23

16. 0.01×789.235

17. 0.001×97.68

18. 8976.23×0.001

19. -78.2×0.01

20. -0.0235×0.1

21. $\begin{array}{r} 32.6 \\ \times \quad 16 \\ \hline \end{array}$

22. $\begin{array}{r} 9.28 \\ \times \quad 8.6 \\ \hline \end{array}$

23. $\begin{array}{r} 0.984 \\ \times \quad 3.3 \\ \hline \end{array}$

24. $\begin{array}{r} 8.489 \\ \times \quad 7.4 \\ \hline \end{array}$

25. $(374)(-2.4)$

26. $(865)(-1.08)$

27. $-749(-0.43)$

28. $-978(-20.5)$

29. $\begin{array}{r} 0.87 \\ \times\ 64 \\ \hline \end{array}$

30. $\begin{array}{r} 7.25 \\ \times\ 60 \\ \hline \end{array}$

31. $\begin{array}{r} 46.50 \\ \times\ 75 \\ \hline \end{array}$

32. $\begin{array}{r} 8.24 \\ \times\ 703 \\ \hline \end{array}$

33. $\begin{array}{r} 81.7 \\ \times\ 0.612 \\ \hline \end{array}$

34. $\begin{array}{r} 31.82 \\ \times\ 7.15 \\ \hline \end{array}$

35. $\begin{array}{r} 10.105 \\ \times\ 11.324 \\ \hline \end{array}$

36. $\begin{array}{r} 151.2 \\ \times\ 4.555 \\ \hline \end{array}$

37. $\begin{array}{r} 12.3 \\ \times\ 1.08 \\ \hline \end{array}$

38. $\begin{array}{r} 7.82 \\ \times\ 0.024 \\ \hline \end{array}$

39. $\begin{array}{r} 32.4 \\ \times\ 2.8 \\ \hline \end{array}$

40. $\begin{array}{r} 8.09 \\ \times\ 0.0075 \\ \hline \end{array}$

41. $\begin{array}{r} 0.00342 \\ \times\ 0.84 \\ \hline \end{array}$

42. $\begin{array}{r} 2.0056 \\ \times\ 3.8 \\ \hline \end{array}$

43. $\begin{array}{r} 0.347 \\ \times\ 2.09 \\ \hline \end{array}$

44. $\begin{array}{r} 2.532 \\ \times\ 1.067 \\ \hline \end{array}$

45. $3.005 \times (-0.623)$

46. $(-6.4)(-15.6)$

47. 1000×45.678

48. 0.001×45.678

b Convert from dollars to cents.

49. \$28.88

50. \$67.43

51. \$0.66

52. \$1.78

Convert from cents to dollars.

53. 34¢

54. 95¢

55. 3445¢

56. 933¢

57. *Farming Area.* China has 3.48 million hectares of land devoted to farming. This area is approximately 24% of China's total area. Convert 3.48 million to standard notation.

Source: Viking Cruises

58. *Doll Sales.* In the United States, spending on dolls totaled $2.7 billion in 2011. Convert 2.7 billion to standard notation.

Source: The NPD Group

59. *Spending on Pets.* In 2011, Americans spent approximately $50.96 billion on their pets, of which $13.41 billion was for veterinary care. Convert 50.96 billion and 13.41 billion to standard notation.

Source: American Pet Association

60. *Library of Congress.* The Library of Congress is the largest public library in the United States. It holds about 33.5 million books. Convert 33.5 million to standard notation.

Source: American Library Association

61. *Safe Water.* Worldwide, about 2.2 million people die each year because of diseases caused by unsafe water. Convert 2.2 million to standard notation.

Source: United Nations Environment Programme

62. *Overdraft Charges.* Revenue of U.S. banks from checking account overdraft charges rose to $31.5 billion in the fiscal year ending in June 2012. Convert 31.5 billion to standard notation.

Sources: Moebs Services; abcNews, *Consumer Report*, September 27, 2012

Skill Maintenance

Calculate.

63. $2\frac{1}{3} \cdot 4\frac{4}{5}$ [4.6a]

64. $-2\frac{1}{3} \div 4\frac{4}{5}$ [4.6b]

65. $4\frac{4}{5} - 2\frac{1}{3}$ [4.5b]

66. $4\frac{4}{5} + 2\frac{1}{3}$ [4.5a]

Divide. [1.5a], [2.5a]

67. $40\overline{)3480}$

68. $17\overline{)20{,}006}$

69. $49 \div (-7)$

70. $-56 \div 4$

Synthesis

Consider the following names for large numbers in addition to those already discussed in this section:

$$1 \text{ quadrillion} = 1{,}000{,}000{,}000{,}000{,}000 = 10^{15};$$

$$1 \text{ quintillion} = 1{,}000{,}000{,}000{,}000{,}000{,}000 = 10^{18};$$

$$1 \text{ sextillion} = 1{,}000{,}000{,}000{,}000{,}000{,}000{,}000 = 10^{21};$$

$$1 \text{ septillion} = 1{,}000{,}000{,}000{,}000{,}000{,}000{,}000{,}000 = 10^{24}.$$

Find each of the following. Express the answer with a name that is a power of 10.

71. (1 trillion) · (1 billion)

72. (1 million) · (1 billion)

73. (1 trillion) · (1 trillion)

74. Is a billion millions the same as a million billions? Explain.

5.4 Division

OBJECTIVES

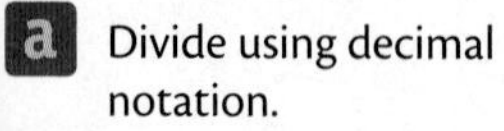

a Divide using decimal notation.

b Solve equations of the type $a \cdot x = b$, where a and b may be in decimal notation.

c Simplify expressions using the rules for order of operations.

SKILL TO REVIEW

Objective 1.5a: Divide whole numbers.

Divide.

1. $5\,\overline{)\,245}$

2. $23\,\overline{)\,1978}$

a DIVISION

Whole-Number Divisors

We use the following method when we divide a decimal quantity by a whole number.

> To divide by a whole number,
>
> **a)** place the decimal point directly above the decimal point in the dividend, and
>
> **b)** divide as though dividing whole numbers.
>
> ```
>             0.8 4 ← Quotient
> Divisor →7 ) 5.8 8 ← Dividend
>              5 6
>                2 8
>                2 8
>                  0 ← Remainder
> ```

EXAMPLE 1 Divide: 379.2 ÷ 8.

```
        ↓ Place the decimal point.
      4 7.4
 8 ) 3 7 9.2
     3 2
       5 9
       5 6          Divide as though dividing whole numbers.
         3 2
         3 2
           0
```

EXAMPLE 2 Divide: 82.08 ÷ (−24).

First we consider 82.08 ÷ 24.

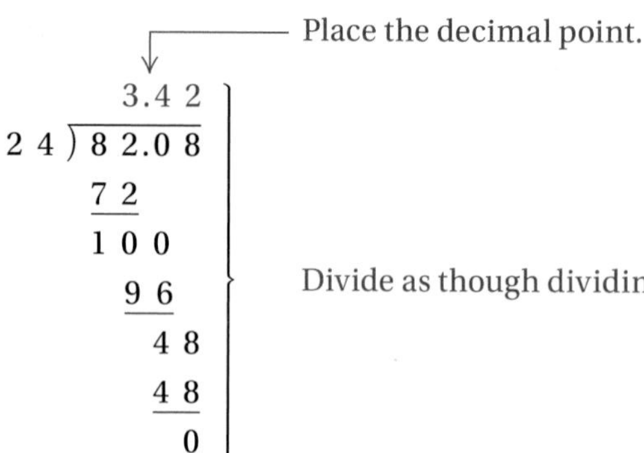

Since a positive number divided by a negative number is negative, the answer is −3.42.

◀ **Do Margin Exercises 1–3.**

Divide.

1. $9\,\overline{)\,5.4}$

2. $15\,\overline{)\,22.5}$

3. 38.54 ÷ (−82)

Answers

Skill to Review:
1. 49 2. 86

Margin Exercises:
1. 0.6 2. 1.5 3. −0.47

We can think of a whole-number dividend as having a decimal point at the end with as many zeros as we wish after the decimal point. For example, $12 = 12. = 12.0 = 12.00 = 12.000$, and so on. We can also add zeros after the last digit in the decimal portion of a number: $3.6 = 3.60 = 3.600$, and so on.

EXAMPLE 3 Divide: $30 \div 8$.

$$8\overline{)30.}$$ with quotient 3., 24, remainder 6 — Place the decimal point and divide to find how many ones.

$$8\overline{)30.0}$$ quotient 3., 24, 60 — Write an extra zero.

$$8\overline{)30.0}$$ quotient 3.7, 24, 60, 56, 4 — Divide to find how many tenths.

$$8\overline{)30.00}$$ quotient 3.7, 24, 60, 56, 40 — Write another zero.

$$8\overline{)30.00}$$ quotient 3.75, 24, 60, 56, 40, 40, 0 — Divide to find how many hundredths.

Check: $3.75 \times 8 = 30.00$ (carries 6 4)

EXAMPLE 4 Divide: $-4 \div 25$.

First we consider $4 \div 25$.

$$25\overline{)4.00}$$ quotient 0.16, 25, 150, 150, 0

Check: 0.16×25: 80, 320, 4.00 (carries 1 3)

Since a negative number divided by a positive number is negative, the answer is -0.16.

Do Exercises 4–6. ▶

CALCULATOR CORNER

Calculations with Decimal Notation To use a calculator to add, subtract, multiply, and divide when any of the numbers are in decimal notation, we use the decimal key, [·], the operation keys, [+], [−], [×], and [÷], the negative number key, [+/−], and the equals key, [=].

EXERCISES Calculate.

1. $1.7 + 14.56 + 0.89$
2. $52.34 - 18.51$
3. $489 - 34.26$
4. 0.04×12.69
5. $49\overline{)125.44}$
6. $1.6 \div 25$

Divide.

4. $25\overline{)8}$
5. $-15 \div 4$

GS 6. $2.15 \div 86$

$$86\overline{)2.150}$$ quotient 0.☐2☐; 172; 43☐; 430; 0

Answers

4. 0.32 5. −3.75 6. 0.025

Guided Solution:

6. 0, 5, 0

Divisors That Are Not Whole Numbers

Consider the division

$0.24\overline{)8.208}$

We write the division as $\frac{8.208}{0.24}$. Then we multiply by 1 to change to a whole-number divisor:

$$\frac{8.208}{0.24} = \frac{8.208}{0.24} \times \frac{100}{100} = \frac{820.8}{24}.$$

The division $0.24\overline{)8.208}$ is the same as $24\overline{)820.8}$.

The divisor is now a whole number.

To divide when the divisor is not a whole number,

a) move the decimal point (multiply by 10, 100, and so on) to make the divisor a whole number;

$0.24\overline{)8.208}$

Move 2 places to the right.

b) move the decimal point in the dividend the same number of places (multiply the same way); and

$0.24\overline{)8.208}$

Move 2 places to the right.

c) place the decimal point directly above the new decimal point in the dividend and divide as though dividing whole numbers.

```
          3 4.2
0.2 4 ) 8.2 0^8  <-
        7 2
        1 0 0
          9 6
            4 8
            4 8
              0
```

(The new decimal point in the dividend is indicated by a caret.)

Divide.

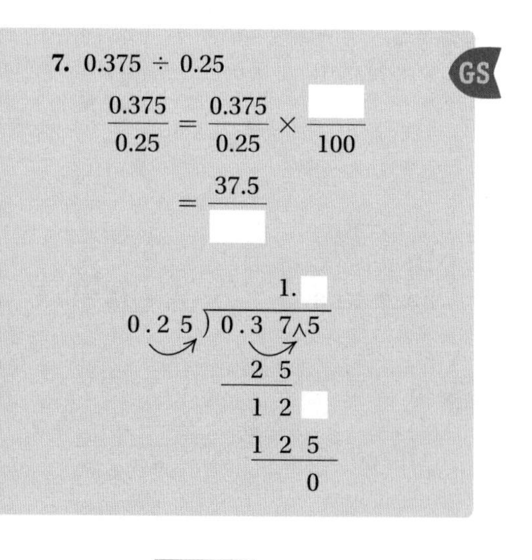

7. $0.375 \div 0.25$

$$\frac{0.375}{0.25} = \frac{0.375}{0.25} \times \frac{\square}{100} = \frac{37.5}{\square}$$

```
           1.□
0.2 5 ) 0.3 7^5
          2 5
          1 2 □
          1 2 5
              0
```

8. $0.83\overline{)4.067}$

9. $3.5\overline{)44.8}$

EXAMPLE 5 Divide: $5.848 \div 8.6$.

$8.6\overline{)5.848}$ Multiply the divisor by 10. (Move the decimal point 1 place.) Multiply the same way in the dividend. (Move 1 place.)

```
       0.6 8
8.6 ) 5.8^4 8
      5 1 6
        6 8 8
        6 8 8
            0
```

Place a decimal point above the new decimal point in the dividend and then divide.

Note: $\frac{5.848}{8.6} = \frac{5.848}{8.6} \cdot \frac{10}{10} = \frac{58.48}{86}$.

Do Exercises 7–9.

Answers

7. 1.5 8. 4.9 9. 12.8

Guided Solution:

7. 100, 25; 5, 5

EXAMPLE 6 Divide: $-12 \div 0.64$.

First we consider $12 \div 0.64$.

```
0.64 ) 12.
```

Place a decimal point at the end of the whole number.

```
0.64 ) 12.00
```

Multiply the divisor by 100 (move the decimal point 2 places). Multiply the same way in the dividend (move 2 places).

```
           1 8.7 5
0.6 4 ) 1 2.0 0^0 0
          6 4
          5 6 0
          5 1 2
            4 8 0
            4 4 8
              3 2 0
              3 2 0
                  0
```

Place a decimal point above the new decimal point in the dividend and then divide.

Since $12 \div 0.64 = 18.75$, we have $-12 \div 0.64 = -18.75$.

Do Exercise 10.

10. Divide: $-25 \div 1.6$.

Dividing by 10, 100, 1000, and So On

It is often helpful to be able to divide quickly by a ten, hundred, or thousand, or by a tenth, hundredth, or thousandth. Each procedure we use is based on multiplying by 1. Consider the following example:

$$\frac{23.789}{1000} = \frac{23.789}{1000} \cdot \frac{1000}{1000} = \frac{23{,}789}{1{,}000{,}000} = 0.023789.$$

We are dividing by a number greater than 1: The result is *smaller* than 23.789.

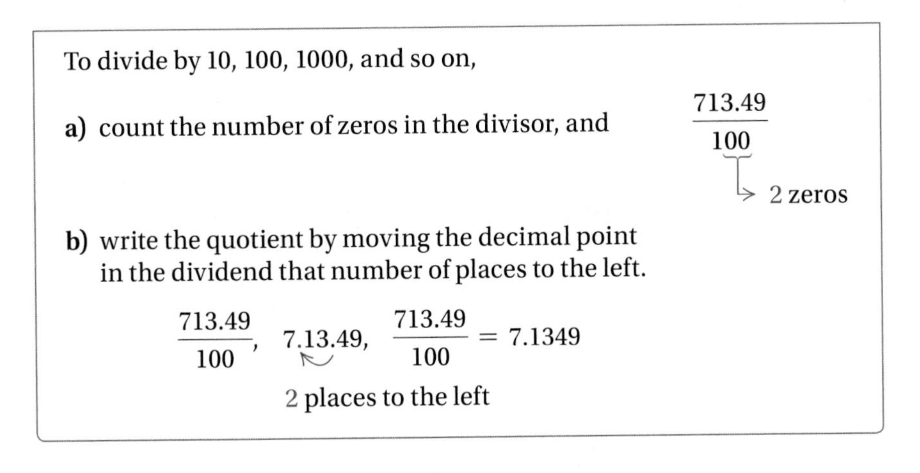

To divide by 10, 100, 1000, and so on,

a) count the number of zeros in the divisor, and

$\frac{713.49}{100}$ → 2 zeros

b) write the quotient by moving the decimal point in the dividend that number of places to the left.

$\frac{713.49}{100}$, 7.13.49, $\frac{713.49}{100} = 7.1349$

2 places to the left

EXAMPLE 7 Divide: $\frac{0.0104}{10}$.

$\frac{0.0104}{10}$, 0.0.0104, $\frac{0.0104}{10} = 0.00104$

1 zero 1 place to the left

Answer

10. 15.625

Dividing by 0.1, 0.01, 0.001, and So On

Now consider the following example:

$$\frac{23.789}{0.01} = \frac{23.789}{0.01} \cdot \frac{100}{100} = \frac{2378.9}{1} = 2378.9.$$

We are dividing by a number less than 1: The result is *larger* than 23.789.
We use the following procedure.

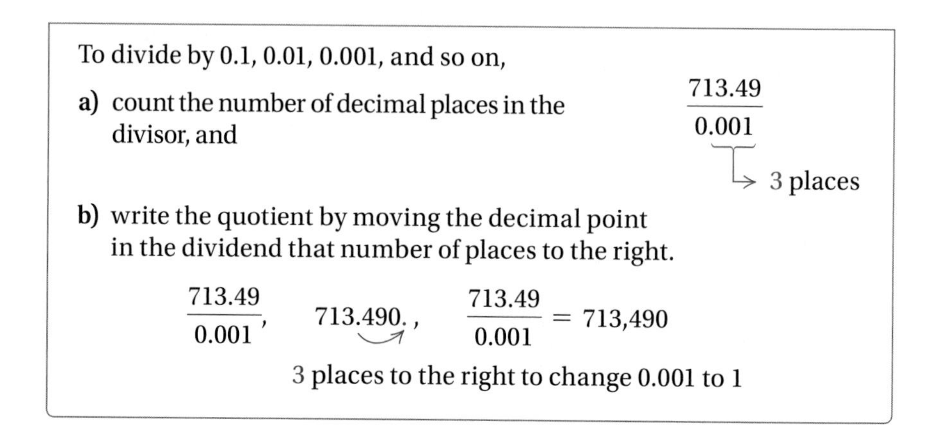

To divide by 0.1, 0.01, 0.001, and so on,

a) count the number of decimal places in the divisor, and

$\frac{713.49}{0.001}$ → 3 places

b) write the quotient by moving the decimal point in the dividend that number of places to the right.

$\frac{713.49}{0.001}$, 713.490., $\frac{713.49}{0.001} = 713{,}490$

3 places to the right to change 0.001 to 1

EXAMPLE 8 Divide: $\frac{23.738}{0.001}$.

$\frac{23.738}{0.001}$, 23.738., $\frac{23.738}{0.001} = 23{,}738$

3 places   3 places to the right

◀ Do Exercises 11–14.

Divide.

11. $\frac{0.1278}{0.01}$ **12.** $\frac{0.1278}{100}$

13. $\frac{98.47}{1000}$ **14.** $\frac{-6.7832}{0.1}$

b SOLVING EQUATIONS

Now let's solve equations of the type $a \cdot x = b$, where a and b may be in decimal notation. Proceeding as before, we divide by a on both sides.

EXAMPLE 9 Solve: $8 \cdot x = 27.2$.

We have

$$\frac{8 \cdot x}{8} = \frac{27.2}{8} \quad \text{Dividing by 8 on both sides}$$

$$x = 3.4.$$

```
     3.4
  8 ) 2 7.2
      2 4
        3 2
        3 2
          0
```

The solution is 3.4. ■

Answers

11. 12.78 **12.** 0.001278 **13.** 0.09847
14. −67.832

EXAMPLE 10 Solve: $2.9 \cdot t = -0.14616$.

We have

$$\frac{2.9 \cdot t}{2.9} = \frac{-0.14616}{2.9} \qquad \text{Dividing by 2.9 on both sides}$$

$$t = -0.0504.$$

```
          0.0 5 0 4
2.9 ) 0.1^4 6 1 6
        1 4 5
            1 1 6
            1 1 6
                0
```

Since $0.14616 \div 2.9 = 0.0504$, then $-0.14616 \div 2.9 = -0.0504$. Thus the solution is -0.0504.

Do Exercises 15 and 16. ▶

Solve.

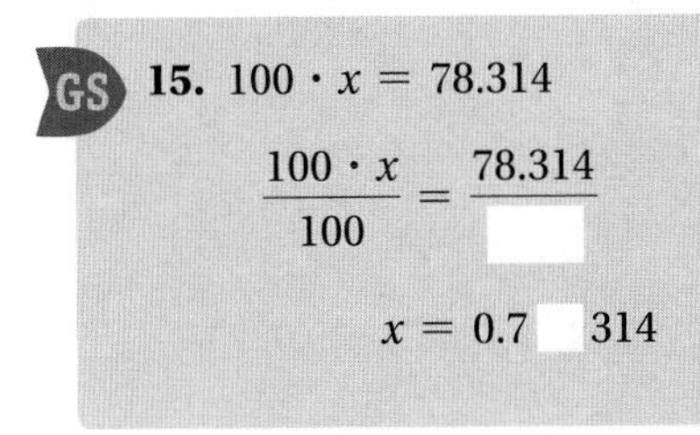

16. $0.25 \cdot y = -276.4$

C ORDER OF OPERATIONS: DECIMAL NOTATION

The same rules for order of operations that we use with integers and fraction notation apply when simplifying expressions with decimal notation.

> **RULES FOR ORDER OF OPERATIONS**
>
> 1. Do all calculations within grouping symbols before operations outside.
> 2. Evaluate all exponential expressions.
> 3. Do all multiplications and divisions in order from left to right.
> 4. Do all additions and subtractions in order from left to right.

EXAMPLE 11 Simplify: $2.56 \times 25.6 \div 25{,}600 \times 256$.

There are no exponents or parentheses, so we multiply and divide from left to right:

$$\begin{aligned} 2.56 \times 25.6 \div 25{,}600 \times 256 &= 65.536 \div 25{,}600 \times 256 \\ &= 0.00256 \times 256 \\ &= 0.65536. \end{aligned}$$

Doing all multiplications and divisions in order from left to right ■

EXAMPLE 12 Simplify: $(5 - 0.06) \div (-2) + 3.42 \times 0.1$.

$(5 - 0.06) \div (-2) + 3.42 \times 0.1 = 4.94 \div (-2) + 3.42 \times 0.1$ Carrying out the operation inside parentheses

$= -2.47 + 0.342$ Doing all multiplications and divisions in order from left to right

$= -2.128$ Adding ■

Answers

15. 0.78314 16. −1105.6

Guided Solution:
15. 100, 8

Simplify.

17. $625 \div 62.5 \times 25 \div 6250$
$= \square \times 25 \div 6250$
$= \square \div 6250$
$= 0.04$

GS

18. $0.25 \cdot (-1 + 0.08) - 0.0274$

19. $20^2 - 3.4^2 + \{2.5[20(9.2 - 5.6)] + 5(10 - 5)\}$

20. *Population Density.* The table below shows the number of residents per square mile in five northwestern states. Find the average number of residents per square mile for this group of states.

Source: 2010 U.S. Census

STATE	RESIDENTS PER SQUARE MILE
Washington	101.2
Oregon	39.9
Idaho	19
Montana	6.8
Wyoming	5.8

Answers

17. 0.04 18. −0.2574 19. 593.44 20. 34.54

Guided Solution:
17. 10, 250

EXAMPLE 13 Simplify: $10^2 \times \{[(3 - 0.24) \div 2.4] - (0.21 - 0.092)\}$.

$10^2 \times \{[(3 - 0.24) \div 2.4] - (0.21 - 0.092)\}$

$= 10^2 \times \{[2.76 \div 2.4] - 0.118\}$ Doing the calculations in the innermost parentheses first

$= 10^2 \times \{1.15 - 0.118\}$ Again, doing the calculations in the innermost grouping symbols

$= 10^2 \times 1.032$ Subtracting inside the grouping symbols

$= 100 \times 1.032$ Evaluating the exponential expression

$= 103.2$

◀ Do Exercises 17–19.

EXAMPLE 14 *Population Density.* The table below shows the number of residents per square mile in the six New England states. Find the average number of residents per square mile for this group of states.

Source: 2010 U.S. Census

STATE	NUMBER OF RESIDENTS PER SQUARE MILE
Maine	43.1
New Hampshire	147.0
Vermont	67.9
Massachusetts	839.4
Rhode Island	1018.1
Connecticut	738.1

The **average** of a set of numbers is the sum of the numbers divided by the number of addends. We find the sum of the population densities per square mile and divide it by the number of addends, 6:

$$\frac{43.1 + 147 + 67.9 + 839.4 + 1018.1 + 738.1}{6} = \frac{2853.6}{6} = 475.6.$$

Thus the average number of residents per square mile for these six states is 475.6.

◀ Do Exercise 20.

5.4 Exercise Set

For Extra Help MyMathLab® MathXL® PRACTICE WATCH READ REVIEW

Reading Check

Name the operation that should be performed first in evaluating each expression. Do not calculate.

RC1. $(2 - 0.04) \div 4 + 8.5$ ____________

RC2. $0.02 + 2.06 \div 0.01$ ____________

RC3. $5 \times 2.1 + 0.1 - 8^3$ ____________

RC4. $18.2 - (4.1 + 6.9)$ ____________

RC5. $16 - 9 \div 3 + 7.3$ ____________

RC6. $4(10 - 5) \times 14.2$ ____________

a Divide.

1. $2\overline{)5.98}$

2. $5\overline{)18}$

3. $4\overline{)95.12}$

4. $8\overline{)25.92}$

5. $12\overline{)89.76}$

6. $23\overline{)25.07}$

7. $33\overline{)237.6}$

8. $54\overline{)448.2}$

9. $-9.144 \div 8$

10. $4.5 \div (-9)$

11. $12.123 \div (-3)$

12. $-12.4 \div 4$

13. $0.06\overline{)3.36}$

14. $0.04\overline{)1.68}$

15. $0.12\overline{)8.4}$

16. $0.36\overline{)2.88}$

17. $3.4\overline{)68}$

18. $0.25\overline{)5}$

19. $-6 \div (-15)$

20. $-1.8 \div (-12)$

21. $36\overline{)14.76}$

22. $52\overline{)119.6}$

23. $3.2\overline{)27.2}$

24. $8.5\overline{)27.2}$

25. $4.2\overline{)\,39.06}$

26. $4.8\overline{)\,0.1104}$

27. $-5 \div 8$

28. $-3 \div 8$

29. $0.47\overline{)\,0.1222}$

30. $1.08\overline{)\,0.54}$

31. $4.8\overline{)\,75}$

32. $0.28\overline{)\,63}$

33. $0.032\overline{)\,0.07488}$

34. $0.017\overline{)\,1.581}$

35. $82\overline{)\,38.54}$

36. $34\overline{)\,0.1462}$

37. $\frac{213.4567}{1000}$

38. $\frac{769.3265}{1000}$

39. $\frac{-23.59}{10}$

40. $\frac{-83.57}{10}$

41. $\frac{426.487}{100}$

42. $\frac{591.348}{100}$

43. $\frac{-16.94}{0.1}$

44. $\frac{-100.7604}{0.1}$

45. $\frac{1.0237}{0.001}$

46. $\frac{3.4029}{0.001}$

47. $\frac{-42.561}{0.01}$

48. $\frac{-98.473}{0.01}$

b Solve.

49. $4.2 \cdot x = 39.06$

50. $36 \cdot y = 14.76$

51. $1000 \cdot y = -9.0678$

52. $-789.23 = 0.25 \cdot q$

53. $1048.8 = 23 \cdot t$

54. $28.2 \cdot x = 423$

 Simplify.

55. $14 \times (82.6 + 67.9)$

56. $(26.2 - 14.8) \times 12$

57. $0.003 - 3.03 \div 0.01$

58. $9.94 + 118.8 \div (-6.2 - 4.6) - 0.9$

59. $42 \times (10.6 + 0.024)$

60. $(18.6 - 4.9) \times 13$

61. $4.2 \times 5.7 + 0.7 \div 3.5$

62. $123.3 - 4.24 \times 1.01$

63. $-9.0072 + 0.04 \div 0.1^2$

64. $-12 \div 0.03 + 12 \times 0.03^2$

65. $(8 - 0.04)^2 \div 4 + 8.7 \times 0.4$

66. $(5 - 2.5)^2 \div 100 + 0.1 \times 6.5$

67. $86.7 + 4.22 \times (9.6 - 0.03)^2$

68. $2.48 \div (1 - 0.504) + 24.3 - 11 \times 2$

69. $4 \div (-0.4) + 0.1 \times 5 - 0.1^2$

70. $6 \times (-0.9) + 0.1 \div 4 - 0.2^3$

71. $5.5^2 \times [(6 - 4.2) \div 0.06 + 0.12]$

72. $12^2 \div (12 + 2.4) - [(2 - 1.6) \div 0.8]$

73. $200 \times \{[(4 - 0.25) \div 2.5] - (4.5 - 4.025)\}$

74. $0.03 \times \{1 \times 50.2 - [(8 - 7.5) \div 0.05]\}$

75. Find the average of \$1276.59, \$1350.49, \$1123.78, and \$1402.58.

76. Find the average weight of two wrestlers who weigh 308 lb and 296.4 lb.

77. *Mountain Peaks in Colorado.* The elevations of four mountain peaks in Colorado are listed in the table below. Find the average elevation of these peaks.

MOUNTAIN PEAK	ELEVATION (in feet)
Mount Elbert	14,440
Mount Evans	14,271
Pikes Peak	14,115
Crested Butte	12,168

78. *Driving Costs.* The table below shows the cost per mile when specific types of vehicles are driven 15,000 miles in a year. Find the average cost per mile for the listed vehicles.

TYPE OF VEHICLE	COST PER MILE (in cents)
Small sedan	44.9
Medium sedan	58.5
Minivan	63.4
Large sedan	75.5
SUV 4WD	75.7

SOURCE: AAA

Skill Maintenance

Simplify. [3.5b]

79. $\frac{38}{146}$

80. $-\frac{92}{124}$

Find the prime factorization. [3.1d]

81. 684

82. 2005

83. Add: $10\frac{1}{2} + 4\frac{5}{8}$. [4.5a]

84. Subtract: $10\frac{1}{2} - 4\frac{5}{8}$. [4.5b]

Evaluate. [1.9b]

85. 7^3

86. 2^6

Solve. [1.7b]

87. $235 = 5 \cdot z$

88. $q + 31 = 72$

Synthesis

Simplify.

89. 🖩 $9.0534 - 2.041^2 \times 0.731 \div 1.043^2$

90. 🖩 $23.042(7 - 4.037 \times 1.46 - 0.932^2)$

In Exercises 91–94, find the missing value.

91. $439.57 \times 0.01 \div 1000 \times \square = 4.3957$

92. $5.2738 \div 0.01 \times 1000 \div \square = 52.738$

93. $0.0329 \div 0.001 \times 10^4 \div \square = 3290$

94. $0.0047 \times 0.01 \div 10^4 \times \square = 4.7$

Mid-Chapter Review

Concept Reinforcement

Determine whether each statement is true or false.

______________ **1.** In the number 308.00567, the digit 6 names the tens place. [5.1a]

______________ **2.** When writing a word name for decimal notation, we write the word "and" for the decimal point. [5.1a]

______________ **3.** To multiply any number by 10, 100, 1000, and so on, count the number of zeros and move the decimal point that many places to the right. [5.3a]

Guided Solutions

Fill in each blank with the number that creates a correct statement or solution.

4. Solve: $y + 12.8 = 23.35$. [5.2d]

$y + 12.8 = 23.35$

$y + 12.8 - \square = 23.35 - \square$ Subtracting 12.8 on both sides

$y + \square = \square$ Carrying out the subtraction

$y = \square$

5. Simplify: $5.6 + 4.3 \times (6.5 - 0.25)^2$. [5.4c]

$5.6 + 4.3 \times (6.5 - 0.25)^2 = 5.6 + 4.3 \times (\square)^2$ Carrying out the operation inside parentheses

$= 5.6 + 4.3 \times \square$ Evaluating the exponential expression

$= 5.6 + \square$ Multiplying

$= \square$ Adding

Mixed Review

6. *Mile Run Record.* The difference between the men's record for the mile run, held by Hicham El Guerrouj of Morocco (with a time of 3 min 43.13 sec) and the women's record for the mile run, held by Svetlana Masterkova of Russia (with a time of 4 min 12.56 sec) is 29.43 sec. Write a word name for 29.43. [5.1a]

Source: International Association of Athletics Federations (iaaf.org)

7. *Skin Allergies.* Skin allergies are the most common allergies among children. In 2010, 9.4 million children in the United States suffered from skin allergies. Convert 9.4 million to standard notation. [5.3b]

Source: CDC National Center for Health Statistics

Write fraction notation. [5.1b]

8. 4.53

9. -0.287

Which number is larger? [5.1c]

10. 0.07, 0.13

11. $-5.2, -5.09$

Write decimal notation. [5.1b]

12. $\frac{7}{10}$

13. $\frac{-639}{100}$

14. $35\frac{67}{100}$

15. $8\frac{2}{1000}$

Round 28.4615 to the nearest: [5.1d]

16. Thousandth.

17. Hundredth.

18. Tenth.

19. One.

Add. [5.2a], [5.2c]

20. $\begin{array}{r} 47.638 \\ +\ 2.457 \\ \hline \end{array}$

21. $\begin{array}{r} 15.6 \\ 234.729 \\ 3.08 \\ +\ 961.453 \\ \hline \end{array}$

22. $-4.5 + 0.728$

23. $16 + 0.34 + 1.9$

Subtract. [5.2b], [5.2c]

24. $\begin{array}{r} 321.57 \\ -\ 49.38 \\ \hline \end{array}$

25. $\begin{array}{r} 5.6 \\ -\ 0.007 \\ \hline \end{array}$

26. $34.3 - 18.75$

27. $-49.07 - 9.7$

Multiply. [5.3a]

28. $\begin{array}{r} 4.6 \\ \times\ 0.9 \\ \hline \end{array}$

29. $\begin{array}{r} 15.3 \\ \times\ 6.07 \\ \hline \end{array}$

30. -100×81.236

31. 0.1×29.37

Divide. [5.4a]

32. $20.24 \div 4$

33. $-21.76 \div 6.8$

34. $76.34 \div 0.1$

35. $914.036 \div 1000$

36. Convert \$20.45 to cents. [5.3b]

37. Convert 147¢ to dollars. [5.3b]

38. Solve: $46.3 + x = -59$. [5.2d]

39. Solve: $42.84 = 5.1 \cdot y$. [5.4b]

Simplify. [5.4c]

40. $6.594 + 0.5318 \div 0.01$

41. $7.3 \times 4.6 - 0.8 \div (-3.2)$

Understanding Through Discussion and Writing

42. A fellow student rounds 236.448 to the nearest one and gets 237. Explain the possible error. [5.1d]

43. Explain the error in the following: [5.2b]
Subtract.

$$73.089 - 5.0061 = 2.3028$$

44. Explain why $10 \div 0.2 = 100 \div 2$. [5.4a]

45. Kayla made these two computational mistakes:

$$0.247 \div 0.1 = 0.0247; \qquad 0.247 \div 10 = 2.47.$$

In each case, how could you convince her that a mistake has been made? [5.4a]

STUDYING FOR SUCCESS *As You Study*

- ☐ Find a quiet place to study.
- ☐ Be disciplined about your use of video games, the Internet, and television. Study first!
- ☐ Pace yourself. It is usually better to study for shorter periods several times a week than to study in one marathon session each week.

Converting from Fraction Notation to Decimal Notation

5.5

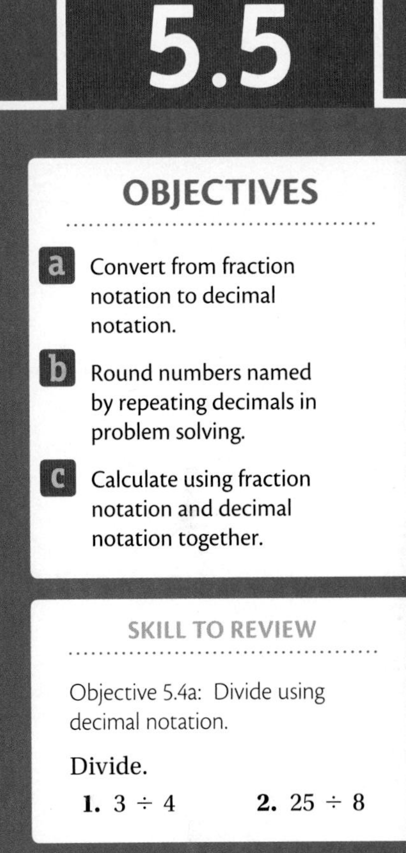

OBJECTIVES

a Convert from fraction notation to decimal notation.

b Round numbers named by repeating decimals in problem solving.

c Calculate using fraction notation and decimal notation together.

SKILL TO REVIEW

Objective 5.4a: Divide using decimal notation.

Divide.

1. $3 \div 4$ **2.** $25 \div 8$

a FRACTION NOTATION TO DECIMAL NOTATION

When a denominator has no prime factors other than 2's and 5's, we can find decimal notation by multiplying by 1. We multiply to get a denominator that is a power of ten, like 10, 100, or 1000.

EXAMPLE 1 Find decimal notation for $\frac{3}{5}$.

$$\frac{3}{5} = \frac{3}{5} \cdot \frac{2}{2} = \frac{6}{10} = 0.6$$

$5 \cdot 2 = 10$, so we use $\frac{2}{2}$ for 1 to get a denominator of 10.

EXAMPLE 2 Find decimal notation for $-\frac{87}{25}$.

$$-\frac{87}{25} = -\frac{87}{25} \cdot \frac{4}{4} = -\frac{348}{100} = -3.48$$

$25 \cdot 4 = 100$, so we use $\frac{4}{4}$ for 1 to get a denominator of 100.

EXAMPLE 3 Find decimal notation for $\frac{9}{40}$.

$$\frac{9}{40} = \frac{9}{40} \cdot \frac{25}{25} = \frac{225}{1000} = 0.225$$

$40 \cdot 25 = 1000$, so we use $\frac{25}{25}$ for 1 to get a denominator of 1000.

Do Margin Exercises 1–3. ▶

We can always divide to find decimal notation.

EXAMPLE 4 Find decimal notation for $\frac{3}{5}$.

$$\frac{3}{5} = 3 \div 5 \qquad \begin{array}{r} 0.6 \\ 5\overline{)3.0} \\ \underline{3\,0} \\ 0 \end{array} \qquad \frac{3}{5} = 0.6$$

EXAMPLE 5 Find decimal notation for $-\frac{7}{8}$.

First we consider $\frac{7}{8}$.

$$\frac{7}{8} = 7 \div 8 \qquad \begin{array}{r} 0.8\,7\,5 \\ 8\overline{)7.0\,0\,0} \\ \underline{6\,4} \\ 6\,0 \\ \underline{5\,6} \\ 4\,0 \\ \underline{4\,0} \\ 0 \end{array} \qquad \frac{7}{8} = 0.875$$

Since $\frac{7}{8} = 0.875$, we have $-\frac{7}{8} = -0.875$.

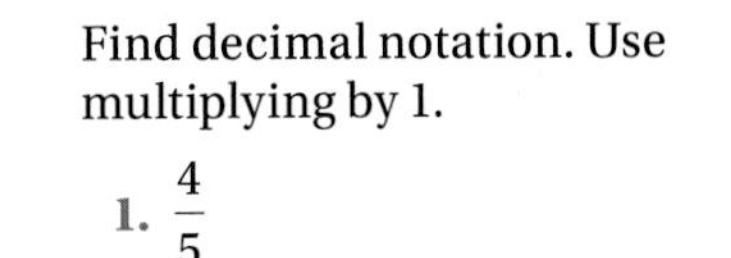

Find decimal notation. Use multiplying by 1.

1. $\frac{4}{5}$

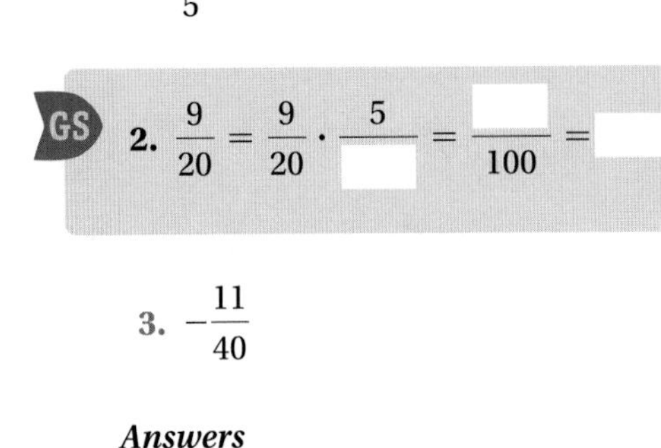

GS **2.** $\frac{9}{20} = \frac{9}{20} \cdot \frac{5}{\square} = \frac{\square}{100} = \square$

3. $-\frac{11}{40}$

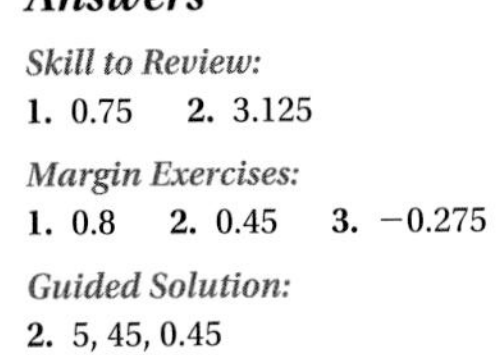

Answers

Skill to Review:
1. 0.75 **2.** 3.125

Margin Exercises:
1. 0.8 **2.** 0.45 **3.** -0.275

Guided Solution:
2. 5, 45, 0.45

Find decimal notation.

4. $-\frac{2}{5}$ **5.** $\frac{3}{8}$

◀ **Do Exercises 4 and 5.**

In Examples 4 and 5, the division *terminated,* meaning that eventually we got a remainder of 0. A **terminating decimal** occurs when the denominator of a fraction has only 2's or 5's, or both, as factors, as in $\frac{17}{25}$, $\frac{5}{8}$, or $\frac{83}{100}$. This assumes that the fraction notation has been simplified.

Consider a different situation: $\frac{5}{6}$, or $5/(2 \cdot 3)$. Since 6 has a 3 as a factor, the division will not terminate. Although we can still use division to get decimal notation, the answer will be a **repeating decimal**, as follows.

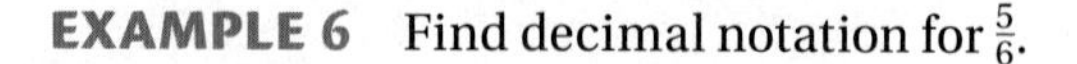

EXAMPLE 6 Find decimal notation for $\frac{5}{6}$.

$\frac{5}{6} = 5 \div 6$

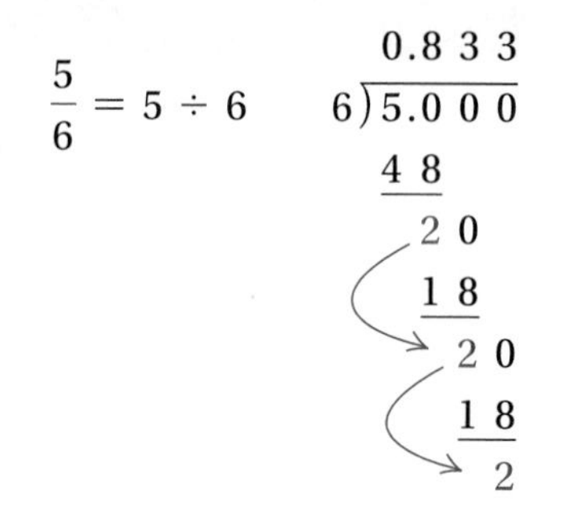

Find decimal notation.

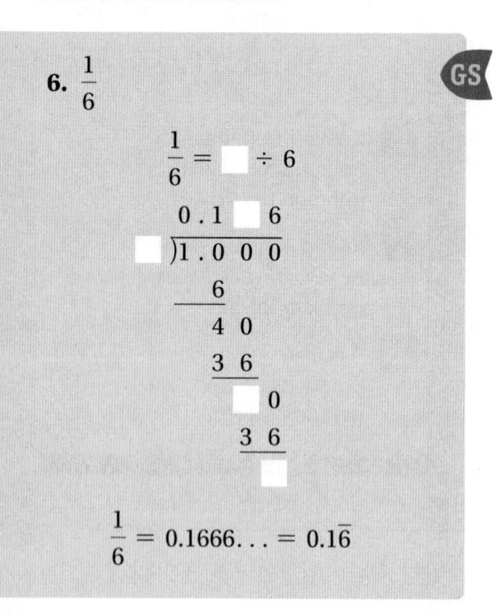

7. $\frac{2}{3}$

Since 2 keeps reappearing as a remainder, the digits repeat and will continue to do so; therefore,

$$\frac{5}{6} = 0.83333\ldots.$$

The red dots indicate an endless sequence of digits in the quotient. When there is a repeating pattern, the dots are often replaced by a bar to indicate the repeating part—in this case, only the 3:

$$\frac{5}{6} = 0.8\overline{3}.$$

◀ **Do Exercises 6 and 7.**

EXAMPLE 7 Find decimal notation for $-\frac{4}{11}$.

First we consider $\frac{4}{11}$.

$$\frac{4}{11} = 4 \div 11$$

```
      0.3 6 3 6
11)4.0 0 0 0
   3 3
     7 0
     6 6
       4 0
       3 3
         7 0
         6 6
           4
```

Since 7 and 4 keep repeating as remainders, the sequence of digits "36" repeats in the quotient, and

$$\frac{4}{11} = 0.363636\ldots, \text{ so we have } -\frac{4}{11} = -0.363636\ldots, \text{ or } -0.\overline{36}.$$

◀ **Do Exercises 8 and 9.**

Find decimal notation.

8. $\frac{5}{11}$ **9.** $-\frac{12}{11}$

Answers

4. -0.4 **5.** 0.375 **6.** $0.1\overline{6}$ **7.** $0.\overline{6}$
8. $0.\overline{45}$ **9.** $-1.\overline{09}$

Guided Solution:
6. 1; 6, 6, 4, 4

EXAMPLE 8 Find decimal notation for $\frac{5}{7}$.

$$
\begin{array}{r}
0.714285 \\
7\overline{)5.000000} \\
\underline{4\,9} \\
1\,0 \\
\underline{7} \\
3\,0 \\
\underline{2\,8} \\
2\,0 \\
\underline{1\,4} \\
6\,0 \\
\underline{5\,6} \\
4\,0 \\
\underline{3\,5} \\
5
\end{array}
$$

Since 5 appears again as a remainder, the sequence of digits "714285" repeats in the quotient, and

$$\frac{5}{7} = 0.714285714285\ldots, \quad \text{or} \quad 0.\overline{714285}.$$

The length of a repeating part can be very long—too long to find on a calculator. An example is $\frac{5}{97}$, which has a repeating part of 96 digits.

Do Exercise 10.

10. Find decimal notation for $\frac{3}{7}$.

b ROUNDING IN PROBLEM SOLVING

In applied problems, repeating decimals are rounded to get approximate answers. To round a repeating decimal, we can extend the decimal notation at least one place past the rounding digit, and then round as usual.

EXAMPLES Round each of the following to the nearest tenth, hundredth, and thousandth.

	Nearest tenth	*Nearest hundredth*	*Nearest thousandth*
9. $0.8\overline{3} = 0.83333\ldots$	0.8	0.83	0.833
10. $0.\overline{09} = 0.090909\ldots$	0.1	0.09	0.091
11. $-0.\overline{714285} = -0.714285714285\ldots$	−0.7	−0.71	−0.714

Do Exercises 11–13.

Round each to the nearest tenth, hundredth, and thousandth.

11. $0.\overline{6}$

12. $0.\overline{80}$

13. $-6.\overline{245}$

Converting Ratios to Decimal Notation

When solving applied problems, we often convert ratios to decimal notation.

EXAMPLE 12 *Gas Mileage.* A car travels 457 mi on 16.4 gal of gasoline. The ratio of number of miles driven to amount of gasoline used is the *gas mileage*. Find the gas mileage and convert the ratio to decimal notation rounded to the nearest tenth.

$$\frac{\text{Miles driven}}{\text{Gasoline used}} = \frac{457}{16.4} \approx 27.86 \qquad \text{Dividing to 2 decimal places}$$

$$\approx 27.9 \qquad \text{Rounding to 1 decimal place}$$

The gas mileage is 27.9 miles to the gallon.

Answers

10. $0.\overline{428571}$ **11.** 0.7; 0.67; 0.667
12. 0.8; 0.81; 0.808 **13.** −6.2; −6.25; −6.245

14. *Coin Tossing.* A coin is tossed 51 times. It lands heads 26 times. Find the ratio of heads to tosses and convert it to decimal notation rounded to the nearest thousandth. (This is also the experimental probability of getting heads.)

15. *Gas Mileage.* A car travels 380 mi on 15.7 gal of gasoline. Find the gasoline mileage and convert the ratio to decimal notation rounded to the nearest tenth.

EXAMPLE 13 *Space Travel.* The space shuttle *Columbia* traveled about 121,700,000 mi in 28 flights. Find the ratio of number of miles traveled to number of flights and convert it to decimal notation. Round to the nearest tenth.

Source: NASA

We have

$$\frac{\text{Miles traveled}}{\text{Number of flights}} = \frac{121{,}700{,}000}{28}$$
$$\approx 4{,}346{,}428.571$$
$$\approx 4{,}346{,}428.6.$$

The ratio is about 4,346,428.6 mi per flight.

Do Exercises 14 and 15.

Averages

EXAMPLE 14 *Price of Gold.* The price of gold more than doubled between 2007 and 2012. The graph below shows the average price of gold per troy ounce for those years. Find the average price of gold over the entire period. Round to the nearest hundredth.

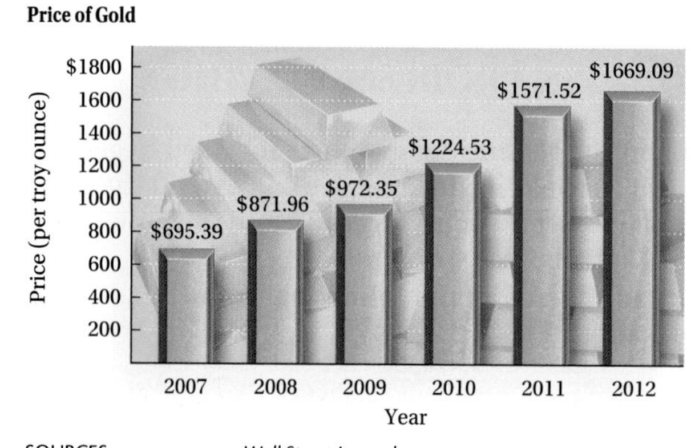

SOURCES: www.nma.org; *Wall Street Journal*

We add the prices of gold and divide by the number of addends, 6. The average is

$$\frac{695.39 + 871.96 + 972.35 + 1224.53 + 1571.52 + 1669.09}{6}$$
$$= \frac{7004.84}{6} = 1167.47\overline{3} \approx 1167.47.$$

From 2007 to 2012, the price of gold averaged about $1167.47 per troy ounce.

Do Exercise 16.

16. *Gold Prices.* Refer to the data on the price of gold in Example 14. Find the average price of gold per troy ounce for the period 2009–2012. Round the answer to the nearest hundredth.

Answers

14. 0.510 **15.** 24.2 mpg **16.** $1359.37

C CALCULATIONS WITH FRACTION NOTATION AND DECIMAL NOTATION TOGETHER

In certain kinds of calculations, fraction notation and decimal notation occur together. In such cases, there are at least three ways in which we might proceed.

EXAMPLE 15 Calculate: $\frac{3}{5} \times 0.56$.

Method 1: One way to do this calculation is to convert the fraction notation to decimal notation so that both numbers are in decimal notation.

$$\frac{3}{5} \times 0.56 = 0.6 \times 0.56 = 0.336$$

Method 2: A second way to do this calculation is to convert the decimal notation to fraction notation so that both numbers are in fraction notation. The answer can be left in fraction notation and simplified, or we can convert to decimal notation and round, if appropriate.

$$\frac{3}{5} \times 0.56 = \frac{3}{5} \cdot \frac{56}{100} = \frac{3 \cdot 2 \cdot 2 \cdot 2 \cdot 7}{5 \cdot 2 \cdot 2 \cdot 5 \cdot 5}$$
$$= \frac{2 \cdot 2}{2 \cdot 2} \cdot \frac{3 \cdot 2 \cdot 7}{5 \cdot 5 \cdot 5} = 1 \cdot \frac{3 \cdot 2 \cdot 7}{5 \cdot 5 \cdot 5}$$
$$= \frac{3 \cdot 2 \cdot 7}{5 \cdot 5 \cdot 5} = \frac{42}{125}, \quad \text{or} \quad 0.336$$

Method 3: A third way to do this calculation is to treat 0.56 as $\frac{0.56}{1}$. Then we multiply 0.56 by 3 and divide the result by 5.

$$\frac{3}{5} \times 0.56 = \frac{3}{5} \times \frac{0.56}{1} = \frac{3 \times 0.56}{5} = \frac{1.68}{5} = 0.336$$

Do Exercise 17. ▶

EXAMPLE 16 Calculate: $\frac{2}{3} \times 0.576 - 3.287 \div \frac{4}{5}$.

We use the rules for order of operations, doing first the multiplication and then the division. Then we add. Since $\frac{2}{3}$ would convert to a repeating decimal that would need to be rounded, method 1 would give an approximation. Here we use method 3.

$$\frac{2}{3} \times 0.576 - 3.287 \div \frac{4}{5} = \frac{2}{3} \times \frac{0.576}{1} - \frac{3.287}{1} \times \frac{5}{4}$$
$$= \frac{2 \times 0.576}{3} - \frac{3.287 \times 5}{4}$$
$$= 0.384 - 4.10875$$
$$= -3.72475$$

Do Exercises 18 and 19. ▶

GS **17.** Calculate: $\frac{3}{4} \times 0.62$.

Method 1:

$$\frac{3}{4} \times 0.62 = 0.\square 5 \times 0.62$$
$$= 0.465$$

Method 2:

$$\frac{3}{4} \times 0.62 = \frac{3}{4} \cdot \frac{62}{\square}$$
$$= \frac{1\square 6}{400}, \quad \text{or} \quad 0.465$$

Method 3:

$$\frac{3}{4} \times 0.62 = \frac{3}{4} \times \frac{0.62}{\square}$$
$$= \frac{1.8\square}{4}, \quad \text{or} \quad 0.465$$

Calculate.

18. $\frac{1}{3} \times 0.384 + \frac{5}{8} \times 0.6784$

19. $\frac{5}{6} \times 0.864 - 14.3 \div \frac{8}{5}$

Answers

17. 0.465 **18.** 0.552 **19.** −8.2175

Guided Solution:
17. 7; 100, 8; 1, 6

5.5 Exercise Set

For Extra Help MyMathLab® MathXL® PRACTICE WATCH READ REVIEW

Reading Check

Determine whether the decimal notation for each fraction is terminating or repeating.

RC1. $\frac{4}{9}$ ____________

RC2. $\frac{3}{32}$ ____________

RC3. $\frac{39}{40}$ ____________

RC4. $\frac{7}{12}$ ____________

RC5. $\frac{2}{11}$ ____________

RC6. $\frac{80}{125}$ ____________

a Find decimal notation.

1. $\frac{23}{100}$

2. $\frac{9}{100}$

3. $\frac{3}{5}$

4. $\frac{19}{20}$

5. $-\frac{13}{40}$

6. $-\frac{3}{16}$

7. $\frac{1}{5}$

8. $\frac{4}{5}$

9. $-\frac{17}{20}$

10. $-\frac{11}{20}$

11. $\frac{3}{8}$

12. $\frac{7}{8}$

13. $-\frac{39}{40}$

14. $-\frac{31}{40}$

15. $\frac{13}{25}$

16. $\frac{61}{125}$

17. $-\frac{2502}{125}$

18. $-\frac{181}{200}$

19. $\frac{1}{4}$

20. $\frac{1}{2}$

21. $-\frac{29}{25}$

22. $-\frac{37}{25}$

23. $\frac{19}{16}$

24. $\frac{5}{8}$

25. $\frac{4}{15}$

26. $\frac{7}{9}$

27. $\frac{1}{3}$

28. $\frac{1}{9}$

29. $-\frac{4}{3}$

30. $-\frac{8}{9}$

31. $-\frac{7}{6}$

32. $-\frac{7}{11}$

33. $\frac{4}{7}$

34. $\frac{14}{11}$

35. $-\frac{11}{12}$

36. $-\frac{5}{12}$

b

37.–47. *Odds.* Round each answer for odd-numbered Exercises 25–35 to the nearest tenth, hundredth, and thousandth.

38.–48. *Evens.* Round each answer for even-numbered Exercises 26–36 to the nearest tenth, hundredth, and thousandth.

Round each to the nearest tenth, hundredth, and thousandth.

49. $0.\overline{18}$

50. $0.\overline{83}$

51. $-0.2\overline{7}$

52. $-3.5\overline{4}$

53. For this set of people, what is the ratio, in decimal notation rounded to the nearest thousandth, where appropriate, of:

a) women to the total number of people?
b) women to men?
c) men to the total number of people?
d) men to women?

54. For this set of pennies and quarters, what is the ratio, in decimal notation rounded to the nearest thousandth, where appropriate, of:

a) pennies to quarters?
b) quarters to pennies?
c) pennies to total number of coins?
d) total number of coins to pennies?

55. *Batting Average.* During the 2012 season, Brandon Phillips, a second baseman for the Cincinnati Reds, had 163 hits in 580 at bats. Find the ratio of number of hits to number of at bats. This is a *batting average*. Give Phillips's batting average in decimal notation rounded to the nearest thousandth.

Source: Major League Baseball

56. *Batting Average.* During the 2012 season, Buster Posey, a catcher for the San Francisco Giants, had 178 hits in 530 at bats. Find the ratio of number of hits to number of at bats. This is a *batting average*. Give Posey's batting average in decimal notation rounded to the nearest thousandth.

Source: Major League Baseball

Gas Mileage. In each of Exercises 57–60, find the gas mileage rounded to the nearest tenth.

57. 285 mi; 18 gal

58. 396 mi; 17 gal

59. 324.8 mi; 18.2 gal

60. 264.8 mi; 12.7 gal

Use the following table for Exercises 61 and 62.

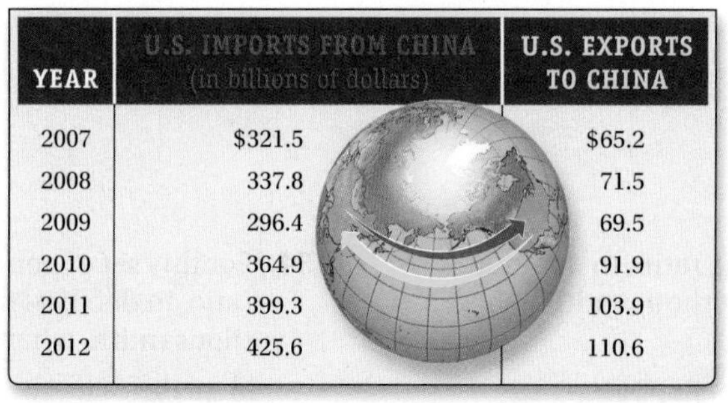

YEAR	U.S. IMPORTS FROM CHINA (in billions of dollars)	U.S. EXPORTS TO CHINA
2007	$321.5	$65.2
2008	337.8	71.5
2009	296.4	69.5
2010	364.9	91.9
2011	399.3	103.9
2012	425.6	110.6

SOURCE: U.S. Census Bureau

61. Find the average amount, in billions of dollars per year, of imports from China to the United States for the period 2007–2012. Round to the nearest tenth.

62. Find the average amount, in billions of dollars per year, of exports from the United States to China for the period 2007–2012. Round to the nearest tenth.

63. *Windy Cities.* Although nicknamed the Windy City, Chicago is not the windiest city in the United States. Listed in the table below are the five windiest cities and their average wind speeds, as well as that on Mt. Washington (the windiest location in the United States). Find the average of these wind speeds and round your answer to the nearest tenth.

Source: *The Handy Geography Answer Book*

CITY	AVERAGE WIND SPEED (in miles per hour)
Mt. Washington, NH	35.3
Boston, MA	12.5
Honolulu, HI	11.3
Dallas, TX	10.7
Kansas City, MO	10.7
Chicago, IL	10.4

64. *Areas of the New England States.* The table below lists the areas of the New England states. Find the average area of this group of states and round your answer to the nearest tenth.

Source: *The New York Times Almanac*

STATE	TOTAL AREA (in square miles)
Maine	33,265
New Hampshire	9,279
Vermont	9,614
Massachusetts	8,284
Connecticut	5,018
Rhode Island	1,211

C Calculate.

65. $\frac{7}{8} \times 12.64$

66. $\frac{4}{5} \times 384.8$

67. $2\frac{3}{4} + 5.65$

68. $4\frac{4}{5} + 3.25$

69. $\frac{47}{9} \times (-79.95)$

70. $\frac{7}{11} \times (-2.7873)$

71. $\frac{1}{2} - 0.5$

72. $3\frac{1}{8} - 2.75$

73. $4.875 - 2\frac{1}{16}$

74. $55\frac{3}{5} - 12.22$

75. $\frac{5}{6} \times 0.0765 - \frac{5}{4} \times 0.1124$

76. $\frac{2}{5} \times 114.8 - \frac{3}{8} \times 156.56$

77. $\frac{4}{5} \times 384.8 + 24.8 \div \frac{8}{3}$

78. $102.4 \div \frac{2}{5} - 12 \times \frac{5}{6}$

79. $\frac{7}{8} \times 0.86 - 0.76 \times \frac{3}{4}$

80. $17.95 \div \frac{5}{8} + \frac{3}{4} \times 16.2$

81. $3.375 \times 5\frac{1}{3}$

82. $2.5 \times 3\frac{5}{8}$

83. $6.84 \div \left(-2\frac{1}{2}\right)$

84. $-8\frac{1}{2} \div 2.125$

Skill Maintenance

Calculate.

85. $9 \cdot 2\frac{1}{3}$ [4.6a]

86. $16\frac{1}{10} - 14\frac{3}{5}$ [4.5b]

87. $-84 \div 8\frac{2}{5}$ [4.6b]

88. $14\frac{3}{5} + 16\frac{1}{10}$ [4.5a]

Solve. [4.5c]

89. A recipe for bread calls for $\frac{2}{3}$ cup of water, $\frac{1}{4}$ cup of milk, and $\frac{1}{8}$ cup of oil. How many cups of liquid ingredients does the recipe call for?

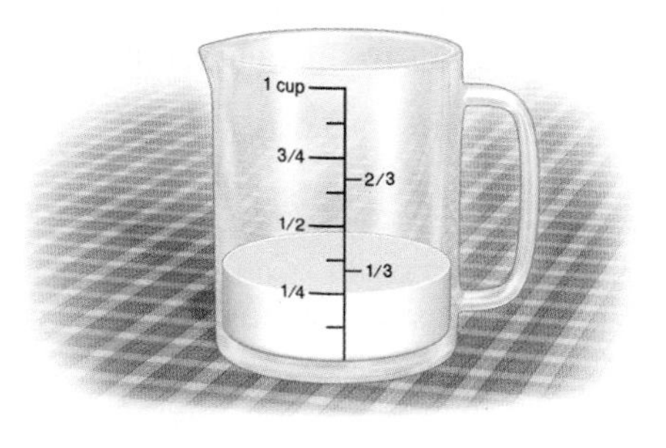

90. A board $\frac{7}{10}$ in. thick is glued to a board $\frac{3}{5}$ in. thick. The glue is $\frac{3}{100}$ in. thick. How thick is the result?

Synthesis

91.  Find decimal notation for $\frac{1}{7}, \frac{2}{7}, \frac{3}{7}, \frac{4}{7}$, and $\frac{5}{7}$. Observe the pattern and guess the decimal notation for $\frac{6}{7}$.

92. Find decimal notation for $\frac{1}{9}, \frac{1}{99}$, and $\frac{1}{999}$. Observe the pattern and guess the decimal notation for $\frac{1}{9999}$.

5.6 Estimating

OBJECTIVE

a Estimate sums, differences, products, and quotients.

SKILL TO REVIEW

Objective 1.6b: Estimate sums, differences, products, and quotients by rounding.

Estimate by first rounding to the nearest ten.

1. $\begin{array}{r} 467 \\ -\ 284 \\ \hline \end{array}$

2. $\begin{array}{r} 54 \\ \times\ 29 \\ \hline \end{array}$

a ESTIMATING SUMS, DIFFERENCES, PRODUCTS, AND QUOTIENTS

Estimating has many uses. It can be done before we even attempt a problem in order to get an idea of the answer. It can be done afterward as a check, even when we are using a calculator. In many situations, an estimate is all we need. We usually estimate by rounding the numbers so that there are one or two nonzero digits, depending on how accurate we want our estimate to be. Consider the following prices for Examples 1–3.

EXAMPLE 1 Estimate by rounding to the nearest ten the total cost of one grill and one TV.

We are estimating the sum

$$\$289.95 + \$449.99 = \text{total cost.}$$

The estimate found by rounding the addends to the nearest ten is

$$\$290 + \$450 = \$740. \qquad \text{(Estimated total cost)}$$

◀ Do Margin Exercise 1.

EXAMPLE 2 About how much more does the TV cost than the camera? Estimate by rounding to the nearest ten.

We are estimating the difference

$$\$449.99 - \$139.97 = \text{price difference.}$$

The estimate, found by rounding the minuend and the subtrahend to the nearest ten, is

$$\$450 - \$140 = \$310. \qquad \text{(Estimated price difference)}$$

◀ Do Exercise 2.

1. Estimate by rounding to the nearest ten the total cost of one grill and one camera. Which of the following is an appropriate estimate?
   a) \$43 b) \$400
   c) \$410 d) \$430

2. About how much more does the TV cost than the grill? Estimate by rounding to the nearest ten. Which of the following is an appropriate estimate?
   a) \$100 b) \$150
   c) \$160 d) \$300

Answers

Skill to Review:
1. 190 2. 1500

Margin Exercises:
1. (d) 2. (c)

EXAMPLE 3 Estimate the total cost of four cameras.

We are estimating the product

$4 \times \$139.97 = \text{total cost}.$

The estimate is found by rounding 139.97 to the nearest ten:

$4 \times \$140 = \$560.$

Do Exercise 3.

EXAMPLE 4 A student government group is planning an event for incoming freshmen. Since the local weather is often rainy, the group decides to purchase umbrellas bearing the university logo for door prizes. About how many umbrellas that cost $39.99 each can be purchased for $356?

We are estimating the quotient

$\$356 \div \$39.99.$

Since we want a whole-number estimate, we need to round appropriately. Rounding $39.99 to the nearest one, we get $40. Since $356 is close to $360, which is a multiple of $40, we estimate

$\$360 \div \$40 = 9.$

The answer is about 9 umbrellas.

Do Exercise 4.

EXAMPLE 5 Estimate: 4.8×52. Do not find the actual product. Which of the following is an appropriate estimate?

a) 25 **b)** 250 **c)** 2500 **d)** 360

We round 4.8 to the nearest one and 52 to the nearest ten:

$5 \times 50 = 250.$ (Estimated product)

Thus an approximate estimate is (b).

Other estimates that we might have used in Example 5 are

$5 \times 52 = 260,$ or $4.8 \times 50 = 240.$

The estimate used in Example 5, $5 \times 50 = 250$, is the easiest to do because the factors have the fewest nonzero digits.

Do Exercises 5–10.

EXAMPLE 6 Estimate: $82.08 \div 24$. Which of the following is an appropriate estimate?

a) 400 **b)** 16 **c)** 40 **d)** 4

This is about $80 \div 20$, so the answer is about 4. Thus an appropriate estimate is (d).

3. Estimate the total cost of six TVs. Which of the following is an appropriate estimate?

a) $450 **b)** $2700
c) $4500 **d)** $27,000

4. Refer to the umbrella price in Example 4. About how many umbrellas can be purchased for $675?

Estimate each product. Do not find the actual product. Which of the choices is an appropriate estimate?

5. 2.4×8

a) 16 **b)** 34
c) 125 **d)** 5

6. 24×0.6

a) 200 **b)** 5
c) 110 **d)** 20

7. 0.86×0.432

a) 0.04 **b)** 0.4
c) 1.1 **d)** 4

8. 0.82×0.1

a) 800 **b)** 8
c) 0.08 **d)** 80

9. 0.12×18.248

a) 180 **b)** 1.8
c) 0.018 **d)** 18

10. 24.234×5.2

a) 200 **b)** 120
c) 12.5 **d)** 234

Answers

3. (b) **4.** 17 umbrellas **5.** (a) **6.** (d)
7. (b) **8.** (c) **9.** (b) **10.** (b)

Estimate each quotient. Which of the choices is an appropriate estimate?

11. $59.78 \div 29.1$

a) 200 **b)** 20
c) 2 **d)** 0.2

12. $82.08 \div 2.4$

a) 40 **b)** 4.0
c) 400 **d)** 0.4

13. $0.1768 \div 0.08$

a) 8 **b)** 10
c) 2 **d)** 20

14. Estimate: $0.0069 \div 0.15$. Which of the following is an appropriate estimate?

a) 0.5 **b)** 50
c) 0.05 **d)** 0.004

Answers

11. (c) 12. (a) 13. (c) 14. (c)

EXAMPLE 7 Estimate: $94.18 \div 3.2$. Which of the following is an appropriate estimate?

a) 30 **b)** 300 **c)** 3 **d)** 60

This is about $90 \div 3$, so the answer is about 30. Thus an appropriate estimate is (a).

EXAMPLE 8 Estimate: $0.0156 \div 1.3$. Which of the following is an appropriate estimate?

a) 0.2 **b)** 0.002 **c)** 0.02 **d)** 20

This is about $0.02 \div 1$, so the answer is about 0.02. Thus an appropriate estimate is (c).

◀ **Do Exercises 11–13.**

In some cases, it is easier to estimate a quotient directly rather than by rounding the divisor and the dividend.

EXAMPLE 9 Estimate: $0.0074 \div 0.23$. Which of the following is an appropriate estimate?

a) 0.3 **b)** 0.03 **c)** 300 **d)** 3

We estimate 3 for a quotient. We check by multiplying.

$$0.23 \times 3 = 0.69$$

We make the estimate smaller. We estimate 0.3 and check by multiplying.

$$0.23 \times 0.3 = 0.069$$

We make the estimate smaller. We estimate 0.03 and check by multiplying.

$$0.23 \times 0.03 = 0.0069$$

This is about 0.0074, so the quotient is about 0.03. Thus an appropriate estimate is (b).

◀ **Do Exercise 14.**

Reading Check

Match each calculation with the most appropriate estimate from the list at the right.

RC1. 0.1003×0.8 ______ **a)** 0.8

RC2. $38.41 + 41.777$ ______ **b)** 0.08

RC3. 0.00152×4025 ______ **c)** 8

RC4. $1632 \div 1.9$ ______ **d)** 80

RC5. $9.054 - 8.3111$ ______ **e)** 800

RC6. $162{,}105 \times 0.0496$ ______ **f)** 8000

a For Exercises 1–6, use the prices shown below to estimate each sum, difference, product, or quotient. Indicate which of the choices is an appropriate estimate.

2 Quart Ice Cream Maker $79.99

Satellite Radio Receiver $149.99

Black & White Multifunction Printer/Copier/Scanner $279.89

1. About how much more does the printer cost than the satellite radio?

a) \$1300 **b)** \$200 **c)** \$130 **d)** \$13

2. About how much more does the satellite radio cost than the ice cream maker?

a) \$7000 **b)** \$70 **c)** \$130 **d)** \$700

3. Estimate the total cost of six ice cream makers.

a) \$480 **b)** \$48 **c)** \$240 **d)** \$4800

4. Estimate the total cost of four printers.

a) \$1200 **b)** \$1120 **c)** \$11,200 **d)** \$600

5. About how many ice cream makers can be purchased for \$830?

a) 120 **b)** 100 **c)** 10 **d)** 1000

6. About how many printers can be purchased for \$5627?

a) 200 **b)** 20 **c)** 1800 **d)** 2000

Estimate by rounding as directed.

7. $0.02 + 1.31 + 0.34$; nearest tenth

8. $0.88 + 2.07 + 1.54$; nearest one

9. $6.03 + 0.007 + 0.214$; nearest one

10. $1.11 + 8.888 + 99.94$; nearest one

11. $52.367 + 1.307 + 7.324$; nearest one

12. $12.9882 + 1.2115$; nearest tenth

13. $2.678 - 0.445$; nearest tenth

14. $12.9882 - 1.0115$; nearest one

15. $198.67432 - 24.5007$; nearest ten

Estimate. Indicate which of the choices is an appropriate estimate.

16. $234.12321 - 200.3223$

a) 600 **b)** 60
c) 300 **d)** 30

17. 49×7.89

a) 400 **b)** 40
c) 4 **d)** 0.4

18. 7.4×8.9

a) 95 **b)** 63
c) 124 **d)** 6

19. 98.4×0.083

a) 80 **b)** 12
c) 8 **d)** 0.8

20. 78×5.3

a) 400 **b)** 800
c) 40 **d)** 8

21. $3.6 \div 4$

a) 10 **b)** 1
c) 0.1 **d)** 0.01

22. $0.0713 \div 1.94$

a) 3.5 **b)** 0.35
c) 0.035 **d)** 35

23. $74.68 \div 24.7$

a) 9 **b)** 3
c) 12 **d)** 120

24. $914 \div 0.921$

a) 10 **b)** 100
c) 1000 **d)** 1

25. *Fence Posts.* A zoo plans to construct a fence around a proposed exhibit featuring animals of the Great Plains. The perimeter of the area to be fenced is 1760 ft. Estimate the number of wooden fence posts needed if the posts are placed 8.625 ft apart.

26. *Ticketmaster.* Recently, Ticketmaster stock sold for $22.25 per share. Estimate how many shares can be purchased for $4400.

27. *Day-Care Supplies.* Helen wants to buy 12 boxes of crayons at $1.89 per box for the day care center that she runs. Estimate the total cost of the crayons.

28. *Batteries.* Oscar buys 6 packages of AAA batteries at $5.29 per package. Estimate the total cost of the purchase.

Skill Maintenance

Simplify. [1.9c]

29. $2^4 \div 4 - 2$

30. $3 \cdot 60 - (12 + 3)$

31. $200 + 40 \div 4$

Solve. [1.7b]

32. $p + 14 = 83$

33. $50 = 5 \cdot t$

34. $270 + y = 800$

Synthesis

The following were done on a calculator. Estimate to determine whether the decimal point was placed correctly.

35. $19.7236 - 1.4738 \times 4.1097 = 1.366672414$

36. $28.46901 \div 4.9187 - 2.5081 = 3.279813473$

37. 🖩 Use one of $+, -, \times,$ and $\div$ in each blank to make a true sentence.

a) $(0.37 \square 18.78) \square 2^{13} = 156{,}876.8$

b) $2.56 \square 6.4 \square 51.2 \square 17.4 = 312.84$

Applications and Problem Solving

5.7

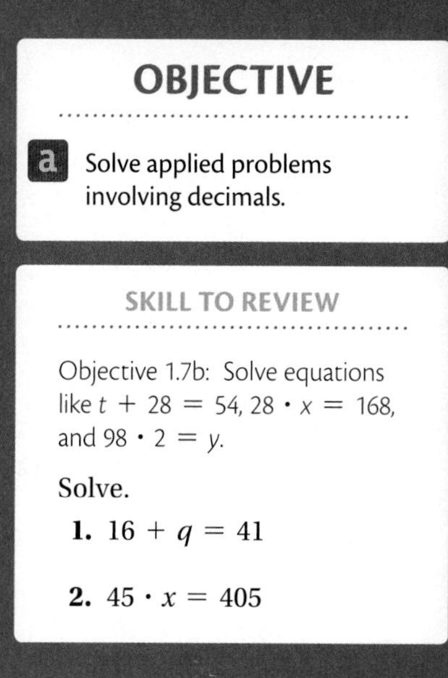

OBJECTIVE

a Solve applied problems involving decimals.

SKILL TO REVIEW

Objective 1.7b: Solve equations like $t + 28 = 54$, $28 \cdot x = 168$, and $98 \cdot 2 = y$.

Solve.

1. $16 + q = 41$

2. $45 \cdot x = 405$

a SOLVING APPLIED PROBLEMS

Solving applied problems with decimals is like solving applied problems with integers. We translate first to an equation that corresponds to the situation. Then we solve the equation.

EXAMPLE 1 *Window Code.* To meet building codes, the area of the opening when the lower part of a double-sash window is raised must be 5 ft^2 (720 in^2) for first-floor windows and 5.7 ft^2 (820.8 in^2) for second-floor windows. A contractor orders windows that have an opening measuring 27.063 in. $\times$ 28.563 in. Find the area of the opening and determine if the windows meet the first-floor code.

Sources: Matt Beecher Builders; Paul's Glass

1. **Familiarize.** We first make a drawing.

2. **Translate.** We use the formula $A = l \cdot w$ and substitute.

$$A = l \cdot w = 27.063 \times 28.563$$

3. **Solve.** We solve by carrying out the multiplication.

$$\begin{array}{r} 2\,8.5\,6\,3 \\ \times\ \ 2\,7.0\,6\,3 \\ \hline 8\,5\,6\,8\,9 \\ 1\,7\,1\,3\,7\,8\,0 \\ 1\,9\,9\,9\,4\,1\,0\,0\,0 \\ 5\,7\,1\,2\,6\,0\,0\,0\,0 \\ \hline 7\,7\,3.0\,0\,0\,4\,6\,9 \end{array}$$

Thus, $A = 773.000469$ in^2.

4. **Check.** We obtain a partial check by estimating the product:

$$A = 27.063 \times 28.563 \approx 25 \times 30 = 750.$$

The estimate is close to 773, so the answer is probably correct.

5. **State.** The area of the opening is about 773 in^2. Since $773 > 720$, the windows meet the first-floor code.

Do Margin Exercise 1.

1. *Window Code.* Refer to Example 1. A contractor orders windows that have an opening measuring 26.535 in. $\times$ 29.265 in. Find the area of the opening and determine if the windows meet code for the second floor. Round to the nearest tenth.

Answers

Skill to Review:
1. 25 2. 9

Margin Exercise:
1. 776.5 in^2; the windows do not meet the second-floor code.

EXAMPLE 2 *Canals.* The Panama Canal in Panama is 50.7 mi long. The Suez Canal in Egypt is 119.9 mi long. How much longer is the Suez Canal?

1. **Familiarize.** We let l = the distance in miles by which the length of the longer canal differs from the length of the shorter canal.
2. **Translate.** We translate as follows, using the given information:

Length of Panama Canal, the shorter canal	plus	Additional length	is	Length of Suez Canal, the longer canal
↓	↓	↓	↓	↓
50.7 mi	$+$	l	$=$	119.9 mi.

3. **Solve.** We solve the equation by subtracting 50.7 mi on both sides:

$$50.7 + l = 119.9$$
$$50.7 + l - 50.7 = 119.9 - 50.7$$
$$l = 69.2.$$

4. **Check.** We can check by adding.

$$\begin{array}{r} 50.7 \\ +\ 69.2 \\ \hline 119.9 \end{array}$$

The answer checks.

5. **State.** The Suez Canal is 69.2 mi longer than the Panama Canal.

Do Exercise 2.

2. *Debit-Card Transactions.* U.S. debit-card transactions in 2009 totaled 38.5 billion. The number of transactions in 2012 was 52.6 billion. How many more debit-card transactions were there in 2012 than in 2009?

Source: *The Nilson Report*

EXAMPLE 3 *iPad Purchase.* A large architectural firm spent $10,399.74 on 26 iPads for its architects. How much did each iPad cost?

1. **Familiarize.** We let c = the cost of each iPad.
2. **Translate.** We translate as follows:

Number of iPads purchased	times	Cost of each iPad	is	Total cost of purchase
↓	↓	↓	↓	↓
26	$\cdot$	c	$=$	$10,399.74.

Answer

2. 14.1 billion transactions

3. **Solve.** We solve the equation by dividing by 26 on both sides, and then we carry out the division.

$$\frac{26 \cdot c}{26} = \frac{10{,}399.74}{26}$$

$$c = \frac{10{,}399.74}{26}$$

$$c = 399.99$$

```
         3 9 9.9 9
26)1 0,3 9 9.7 4
     7 8
     2 5 9
     2 3 4
       2 5 9
       2 3 4
         2 5 7
         2 3 4
           2 3 4
           2 3 4
               0
```

4. **Check.** We check by estimating:

$$10{,}399.74 \div 26 \approx 10{,}000 \div 25 = 400.$$

Since 400 is close to 399.99, the answer is probably correct.

5. **State.** The cost of each iPad was \$399.99.

Do Exercise 3. ▶

3. One pound of lean boneless ham contains 4.5 servings. It costs \$7.99 per pound. What is the cost per serving? Round to the nearest cent.

EXAMPLE 4 *IRS Driving Allowance.* The Internal Revenue Service allowed a tax deduction of 55.5¢ per mile driven for business purposes in 2012. What deduction, in dollars, was allowed for driving 8407 mi during the year?

Source: Internal Revenue Service

1. **Familiarize.** We first make a drawing or at least visualize the situation. Repeated addition fits this situation. We let $d =$ the deduction, in dollars, allowed for driving 8407 mi.

2. **Translate.** We translate as follows, writing 55.5¢ as \$0.555:

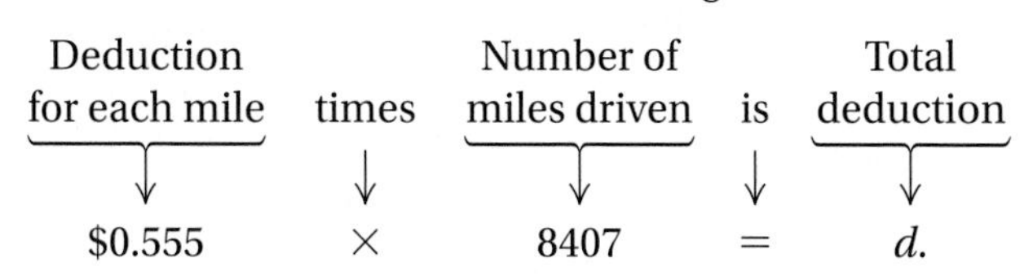

3. **Solve.** To solve the equation, we carry out the multiplication.

```
      8 4 0 7
    × 0.5 5 5
    4 2 0 3 5
  4 2 0 3 5 0
4 2 0 3 5 0 0
4 6 6 5.8 8 5
```

Thus, $d = 4665.885 \approx 4665.89$.

Answer

3. \$1.78 per serving

4. **Check.** We can obtain a partial check by rounding and estimating:

$$8407 \times 0.555 \approx 8000 \times 0.6 = 4800 \approx 4665.89.$$

5. **State.** The total allowable deduction was $4665.89.

◀ **Do Exercise 4.**

4. *Printing Costs.* At Kwik Copy, the cost of copying is 11 cents per page. How much, in dollars, would it cost to make 466 copies?

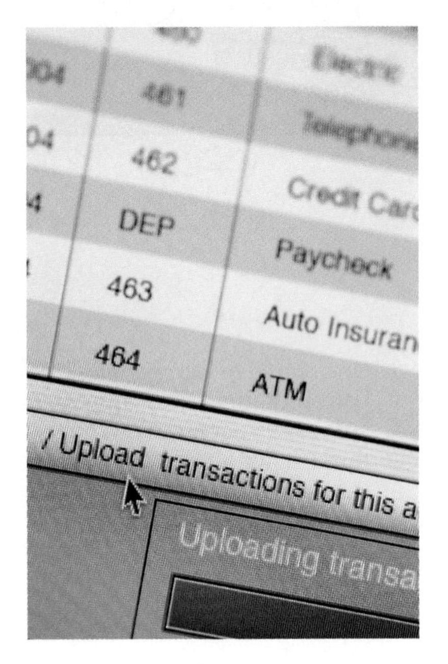

Multistep Problems

EXAMPLE 5 *Tracking a Bank Balance.* Revenue of U.S. banks from checking account overdraft charges was $31.5 billion in a recent fiscal year. (**Source:** Moebs Services) To avoid overdrafts and to track her spending, Maggie keeps a running account of her banking transactions. She checks her balance online, but because of pending amounts that post later, she keeps a separate record. Maggie had $2432.27 in her account. She used her debit card to pay her rent of $835 and make purchases of $14.13, $38.60, and $205.98. She then deposited her weekly pay of $748.35. What was her balance after these transactions?

1. **Familiarize.** We first find the total of the debits. Then we find how much is left in the account after the debits are deducted. Finally, we add to this amount the deposit to find the balance in the account after all the transactions.

2, 3. **Translate and Solve.** We let $d =$ the total amount of the debits. We are combining amounts: $\$835 + \$14.13 + \$38.60 + \$205.98 = d$. To solve the equation, we add.

835.00	First debit
14.13	Second debit
38.60	Third debit
+ 205.98	Fourth debit
1093.71	Total debits

Thus, $d = 1093.71$.

Now let $a =$ the amount in the account after the debits are deducted. We subtract: $\$2432.27 - \$1093.71 = a$.

2432.27	Original amount
− 1093.71	Total debits
1338.56	New amount

Thus, $a = 1338.56$.

Finally, we let $f =$ the amount in the account after the paycheck is deposited.

Amount after debits	plus	Amount of deposit	is	Final amount
↓	↓	↓	↓	↓
1338.56	+	748.35	=	f

To solve the equation, we add.

1338.56	Balance after debits
+ 748.35	Paycheck deposit
2086.91	Final amount

Answer

4. $51.26

4. **Check.** We repeat the computations.
5. **State.** Maggie had \$2086.91 in her account after all the transactions.

Do Exercise 5.

5. *Bank Balance.* Stephen had \$915.22 in his checking account. He used his debit card to pay a charge card minimum payment of \$36 and to make purchases of \$67.50, \$178.23, and \$429.05. He then deposited his weekly pay of \$570.91. How much was in his account after these transactions?

EXAMPLE 6 *Gas Mileage.* Ava filled her gas tank and noted that the odometer read 67,507.8. After the next fill-up, the odometer read 68,006.1. It took 16.5 gal to fill the tank. How many miles per gallon did Ava get?

1. **Familiarize.** We first make a drawing.

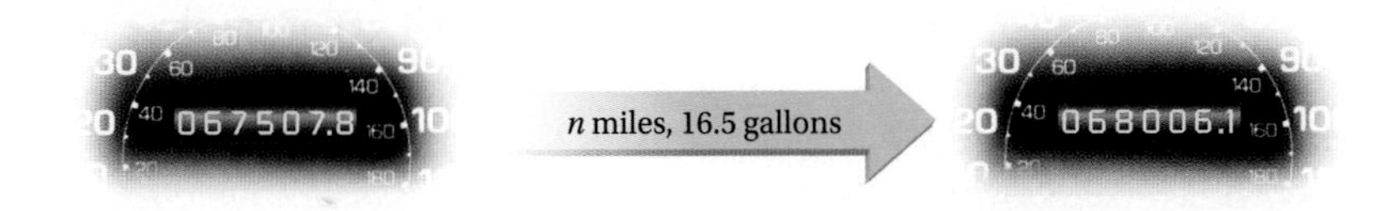

This is a two-step problem. First, we find the number of miles that have been driven between fill-ups. We let n = the number of miles driven.

2, 3. Translate and Solve. We translate and solve as follows:

First odometer reading	plus	Number of miles driven	is	Second odometer reading
↓	↓	↓	↓	↓
67,507.8	+	n	=	68,006.1.

To solve the equation, we subtract 67,507.8 on both sides:

$$n = 68{,}006.1 - 67{,}507.8 = 498.3.$$

$$\begin{array}{r} 68{,}006.1 \\ -\ 67{,}507.8 \\ \hline 498.3 \end{array}$$

Next, we divide the total number of miles driven by the number of gallons. This gives us m = the number of miles per gallon—that is, the gas mileage. The division that corresponds to the situation is $498.3 \div 16.5 = m$. To find the number m, we divide.

$$\begin{array}{r} 30.2 \\ 16.5\overline{)\,498.30} \\ 495 \\ \hline 3\,30 \\ 3\,30 \\ \hline 0 \end{array}$$

Thus, $m = 30.2$.

4. **Check.** To check, we first multiply the number of miles per gallon times the number of gallons to find the number of miles driven:

$$16.5 \times 30.2 = 498.3.$$

Then we add 498.3 to 67,507.8 to find the new odometer reading:

$$67{,}507.8 + 498.3 = 68{,}006.1.$$

The gas mileage of 30.2 checks.

5. **State.** Ava got 30.2 miles per gallon.

Do Exercise 6.

6. *Gas Mileage.* John filled his gas tank and noted that the odometer read 38,320.8. After the next fill-up, the odometer read 38,735.5. It took 14.5 gal to fill the tank. How many miles per gallon did John get?

Answers

5. \$775.35 6. 28.6 mpg

Translating for Success

The goal of these matching questions is to practice step (2), Translate, of the five-step problem-solving process. Translate each word problem to an equation and select a correct translation from equations A–O.

A. $2\frac{1}{2} \cdot n = 1314$

B. $18.4 \times 3.87 = n$

C. $n = 85.2 \times 52.3$

D. $1314.28 - 437 = n$

E. $3 \times 18.40 = n$

F. $2\frac{1}{2} \cdot 1314 = n$

G. $3092 + n = 4638$

H. $18.4 \cdot n = 3.87$

I. $\frac{406.12}{28.4} = n$

J. $52.3 \cdot n = 85.2$

K. $n = 1314.28 + 437$

L. $52.3 + n = 85.2$

M. $3092 + 4638 = n$

N. $3 \cdot n = 18.40$

O. $85.2 + 52.3 = n$

Answers on page A-9

1. ***Gas Mileage.*** Art filled his SUV's gas tank and noted that the odometer read 38,271.8. At the next fill-up, the odometer read 38,677.92. It took 28.4 gal to fill the tank. How many miles per gallon did the SUV get?

2. ***Dimensions of a Parking Lot.*** A store's parking lot is a rectangle that measures 85.2 ft by 52.3 ft. What is the area of the parking lot?

3. ***Game Snacks.*** Three students pay $18.40 for snacks at a football game. What is each student's share of the cost?

4. ***Electrical Wiring.*** An electrician needs 1314 ft of wiring cut into $2\frac{1}{2}$-ft pieces. How many pieces will she have?

5. ***College Tuition.*** Wayne needs $4638 for the fall semester's tuition. On the day of registration, he has only $3092. How much does he need to borrow?

6. ***Cost of Gasoline.*** What is the cost, in dollars, of 18.4 gal of gasoline at $3.87 per gallon?

7. ***Savings Account Balance.*** Margaret has $1314.28 in her savings account. Before using her debit card to buy an office chair, she transferred $437 to her checking account. How much was left in her savings account?

8. ***Acres Planted.*** This season Sam planted 85.2 acres of corn and 52.3 acres of soybeans. Find the total number of acres that he planted.

9. ***Amount Inherited.*** Tara inherited $2\frac{1}{2}$ times as much as her cousin. Her cousin received $1314. How much did Tara receive?

10. ***Travel Funds.*** The athletic department needs travel funds of $4638 for the tennis team and $3092 for the golf team. What is the total amount needed for travel?

5.7 Exercise Set

For Extra Help MyMathLab® MathXL® PRACTICE WATCH READ REVIEW

Reading Check

Complete each step in the five-step problem-solving strategy with the correct word from the following list.

Solve Familiarize State Translate Check

Five-Step Problem Solving Strategy

RC1. ________________ yourself with the problem situation.

RC2. ________________ the problem to an equation.

RC3. ________________ the equation.

RC4. ________________ the solution.

RC5. ________________ the answer using a complete sentence.

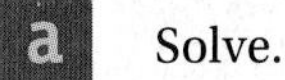

Solve.

International Travel. The chart below shows the numbers of U.S. citizens who traveled internationally in 2007–2011. Use this chart in Exercises 1 and 2.

YEAR	U.S. CITIZENS WHO TRAVELED INTERNATIONALLY (in millions)
2007	64
2008	63.6
2009	61.4
2010	60.3
2011	58.7

SOURCE: Office of Travel and Tourism

1. Find the average number of citizens per year who traveled internationally during 2007–2011.

2. How many more U.S. citizens traveled internationally in 2007 than in 2011?

3. *Lottery Winnings.* A group of 28 employees won \$300,000 in a lottery. After $33\frac{1}{3}\%$ was withheld for taxes, the remaining amount, \$200,000, was split equally among the employees. How much was each winner's share?

4. *Lunch Costs.* A group of 4 students pays \$47.84 for lunch and splits the cost equally. What is each person's share?

5. *Body Temperature.* Normal body temperature is 98.6°F. During an illness, a patient's temperature rose 4.2° higher than normal. What was the new temperature?

6. *Gasoline Cost.* What is the cost, in dollars, of 12.6 gal of gasoline at \$3.79 per gallon? Round the answer to the nearest cent.

7. Find the perimeter.

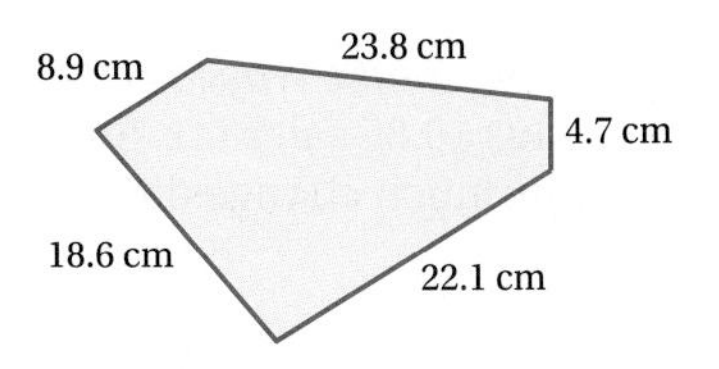

8. Find the length d.

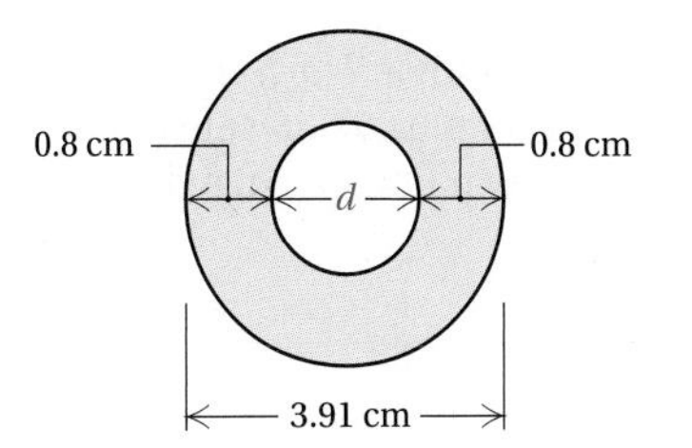

9. *Odometer Reading.* The Lanosga family's odometer reads 22,456.8 at the beginning of a trip. The family's online driving directions tell them that they will be driving 234.7 mi. What will the odometer read at the end of the trip?

10. *Miles Driven.* Petra bought gasoline when the odometer read 14,296.3. At the next gasoline purchase, the odometer read 14,515.8. How many miles had been driven?

11. *100-Meter Record.* The fastest speed clocked for a cheetah running a distance of 100 m is 5.95 sec. The men's world record for the 100-m dash is held by Jamaican Usain Bolt. His time was 3.63 sec more than the cheetah's. What is the men's 100-m record held by Usain Bolt?

Source: "Cheetahs on the Edge," by Roff Smith, *National Geographic,* November 2012.

12. *Record Movie Openings.* The movie *Marvel's The Avengers* took in $207.4 million on its first weekend. This topped the previous record for opening-weekend revenue set by *Harry Potter and the Deathly Hallows: Part II* by $38.2 million. How much did *Harry Potter and the Deathly Hallows: Part II* take in on its opening weekend?

Source: Nash Information Services

13. *Tank Capacity.* The water in a filled tank weighs 748.45 lb. One cubic foot of water weighs 62.5 lb. How many cubic feet of water does the tank hold?

14. Lot A measures 250.1 ft by 302.7 ft. Lot B measures 389.4 ft by 566.2 ft. What is the total area of the two lots?

15. *Cost of Bottled Water.* The cost of a year's supply of a popular brand of bottled water, based on the recommended consumption of 64 oz per day, at the supermarket price of $3.99 for a six-pack of half-liter bottles, is $918.82. This is $918.31 more than the cost of drinking the same amount of tap water for a year. What is the cost of drinking tap water for a year?

Source: American Water Works Association

16. *Highway Routes.* You can drive from home to work using either of two routes:

Route A: Via interstate highway, 7.6 mi, with a speed limit of 65 mph.

Route B: Via a country road, 5.6 mi, with a speed limit of 50 mph.

Assuming you drive at the posted speed limit, which route takes less time? (Use the formula *distance* = *speed* × *time*.)

17. Andrew bought a DVD of the movie *Horton Hears a Who* for his nephew for $23.99 plus $1.68 sales tax. He paid for it with a $50 bill. How much change did he receive?

18. Claire bought a copy of the book *Make Way for Ducklings* for her daughter for $16.95 plus $0.85 sales tax. She paid for it with a $20 bill. How much change did she receive?

19. *Gas Mileage.* Peggy filled her van's gas tank and noted that the odometer read 26,342.8. After the next fill-up, the odometer read 26,736.7. It took 19.5 gal to fill the tank. How many miles per gallon did the van get?

20. *Gas Mileage.* Henry filled his Honda's gas tank and noted that the odometer read 18,943.2. After the next fill-up, the odometer read 19,306.2. It took 13.2 gal to fill the tank. How many miles per gallon did the car get?

21. A rectangular yard is 20 ft by 15 ft. The yard is covered with grass except for an 8.5-ft-square flower garden. How much grass is in the yard?

22. Rita receives a gross pay (before deductions) of $495.72. Her deductions are $59.60 for federal income tax, $29.00 for FICA, and $29.00 for medical insurance. What is her take-home pay?

23. *Pole Vault Pit.* Find the area and the perimeter of the landing area of the pole vault pit shown here.

16.4 ft

16.4 ft

Landing area

24. *Stamp.* Find the area and the perimeter of the stamp shown here.

2.5 cm

3.25 cm

25. *Loose Change.* In 2010, passengers at New York's John F. Kennedy International Airport left $46,918.06 in loose change at airport checkpoints. This was $27,807.23 more than passengers left that year at Los Angeles International Airport. How much change was left at the Los Angeles airport?

Source: Transportation Security Administration

26. *Online Ad Spending.* It is projected that $132.1 billion will be spent for online advertising in 2015. This is $63.7 billion more than in 2010. What was spent on online advertising in 2010?

Source: eMarketer, June 2011

27. *Loan Payment.* In order to make money on loans, financial institutions are paid back more money than they loan. Suppose you borrow $120,000 to buy a house and agree to make monthly payments of $880.52 for 30 years. How much do you pay back altogether? How much more do you pay back than the amount of the loan?

28. *Loan Payment.* In order to make money on loans, financial institutions are paid back more money than they loan. Suppose you borrow $270,000 to buy a house and agree to make monthly payments of $1105.73 for 30 years. How much do you pay back altogether? How much more do you pay back than the amount of the loan?

29. *Calories Burned Mowing.* A person weighing 150 lb burns 7.3 calories per minute while mowing a lawn with a power lawnmower. How many calories would be burned in 2 hr of mowing?

Source: *The Handy Science Answer Book*

30. *Weight Loss.* A person weighing 170 lb burns 8.6 calories per minute while mowing a lawn. One must burn about 3500 calories in order to lose 1 lb. How many pounds would be lost by mowing for 2 hr? Round to the nearest tenth.

31. *Construction Pay.* A construction worker is paid \$18.50 per hour for the first 40 hr of work, and time and a half, or \$27.75 per hour, for any overtime exceeding 40 hr per week. One week she works 46 hr. How much is her pay?

32. *Summer Work.* Zachary worked 53 hr during a week one summer. He earned \$7.50 per hour for the first 40 hr and \$11.25 per hour for overtime (hours exceeding 40). How much did Zachary earn during the week?

33. *Corn Production.* The table below lists 2012 corn production for six of the top ten corn-producing nations. Find the average corn production for the top two countries.

RANK	COUNTRY	CORN PRODUCTION (in millions of metric tons)
1	United States	272.4
2	China	200.0
3	Brazil	70.0
7	Mexico	20.7
9	South Africa	13.5
10	Canada	11.6

SOURCE: www.indexmundi.com

34. *Projected World Population.* Using the information in the bar graph below, determine the average population of the world for the years 1950 through 2050. Round to the nearest thousandth of a billion.

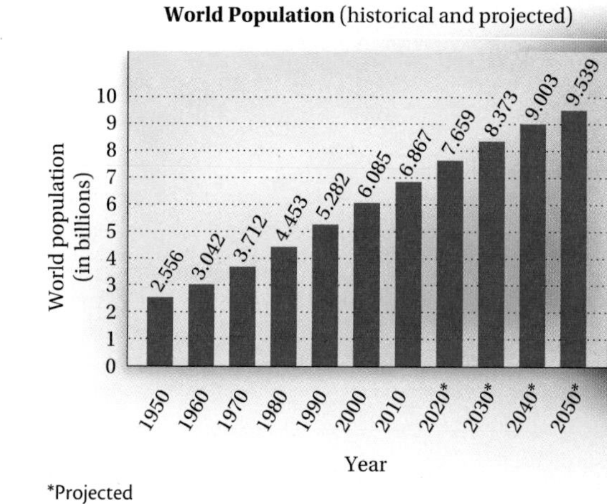

*Projected

SOURCE: U.S. Census Bureau

35. *Checking Account Balance.* Jacob had \$1218.30 in his bank account. After using his debit card for three purchases, writing a check for his rent, and receiving his weekly paycheck by direct deposit, he checked what he had calculated as his balance against his online account balance. They did not match. Find his errors and correct the numbers in the balance column.

DATE	DEBIT CARD	CHECK NUMBER	DESCRIPTION	(−) PAYMENT/ DEBIT	(+) DEPOSIT/ CREDIT	BALANCE
						\$1218.30
1/28	✓		Walgreen's	38 15		1186 15
2/1		2578	Rent	680 00		506 15
2/5			Paycheck		416 27	922 42
2/7	✓		Kroger	40 95		963 37
2/10	✓		Bookstore	110 64		852 73

36. *Checking Account Balance.* Lynn had \$3425.24 in her bank account. After paying her mortgage payment with a check, making two purchases with her debit card, making an online utility payment, and depositing her paycheck, she checked what she had calculated as her balance against her online account balance. They did not match. Find her errors and correct the numbers in the balance column.

DATE	DEBIT CARD	CHECK NUMBER	DESCRIPTION	(−) PAYMENT/ DEBIT	(+) DEPOSIT/ CREDIT	BALANCE \$3425.24
7/20		471	Mortgage payment	905 17		2525 24
7/25			Paycheck		1210 48	3735 72
7/28			PE Energy	89 20		3646 52
8/3	✓		Groceries	32 49		3614 03
8/11	✓		Gas	62 17		3676 20

37. *Body Temperature.* Normal body temperature is 98.6°F. A baby's bath water should be 100°F. How many degrees above normal body temperature is this?

38. *Body Temperature.* Normal body temperature is 98.6°F. The lowest body temperature at which a patient has survived is 69°F. How many degrees below normal is this?

39. *Property Taxes.* The Brunners own a house with an assessed value of \$165,500. For every \$1000 of assessed value, they pay \$8.50 in property taxes each year. How much do they pay in property taxes each year?

40. *Property Taxes.* The Colavitos own a house with an assessed value of \$184,500. For every \$1000 of assessed value, they pay \$7.68 in property taxes each year. How much do they pay in property taxes each year?

Skill Maintenance

Add.

41. $-\frac{5}{6} - \frac{7}{10}$ [4.2a]

42. $4\frac{1}{3} + 2\frac{1}{2}$ [4.5a]

Subtract.

43. $\frac{2}{3} - \frac{5}{8}$ [4.3a]

44. $4569 - 1766$ [1.3a]

Simplify. [3.5b]

45. $\frac{125}{400}$

46. $-\frac{325}{625}$

47. If a bicycle wheel made 480 revolutions at a rate of $66\frac{2}{3}$ revolutions per minute, for how long did it rotate? [4.6c]

Synthesis

48. Suppose you buy a half-dozen packs of basketball cards with a dozen cards in each pack. The cost is twelve dozen cents for each half-dozen cards. How much do you pay for the cards?

CHAPTER

5 Summary and Review

Vocabulary Reinforcement

Complete each statement with the correct word from the list at the right.

1. A ______________ decimal occurs when we convert a fraction to a decimal and the denominator of the fraction has at least one factor other than 2 or 5. [5.5a]
2. A ______________ decimal occurs when we convert a fraction to a decimal and the denominator of the fraction has only 2's or 5's, or both, as factors. [5.5a]
3. One ______________ = 1,000,000,000. [5.3b]
4. One ______________ = 1,000,000. [5.3b]
5. One ______________ = 1,000,000,000,000. [5.3b]
6. The ______________ consist of the integers, . . . , −3, −2, −1, 0, 1, 2, 3, . . . , and fractions like $\frac{1}{2}$, $-\frac{4}{5}$, and $\frac{31}{25}$. [5.1a]

trillion
million
billion
repeating
terminating
whole numbers
rational numbers

Concept Reinforcement

Determine whether each statement is true or false.

______________ 1. One thousand billion is one trillion. [5.3b]

______________ 2. The number of decimal places in the product of two numbers is the product of the numbers of places in the factors. [5.3a]

______________ 3. An estimate found by rounding to the nearest ten is usually more accurate than one found by rounding to the nearest hundred. [5.6a]

______________ 4. For a fraction with a factor other than 2 or 5 in its denominator, decimal notation terminates. [5.5a]

Study Guide

Objective 5.1b Convert between decimal notation and fraction notation.

Example Write fraction notation for 5.347.

5.347 ↑ 3 decimal places; 5.347. Move 3 places to the right; $\frac{5347}{1000}$ ↑ 3 zeros

Practice Exercise

1. Write fraction notation for 50.93.

Example Write decimal notation for $-\frac{29}{1000}$.

$-\frac{29}{1000}$ ↑ 3 zeros; −0.029. Move 3 places to the left; $-\frac{29}{1000} = -0.029$

Practice Exercise

2. Write decimal notation for $-\frac{817}{10}$.

Example Write decimal notation for $4\frac{63}{100}$.

$$4\frac{63}{100} = 4 + \frac{63}{100} = 4 \text{ and } \frac{63}{100} = 4.63$$

Practice Exercise

3. Write decimal notation for $42\frac{159}{1000}$.

Objective 5.1d Round decimal notation to the nearest thousandth, hundredth, tenth, one, ten, hundred, or thousand.

Example Round 19.7625 to the nearest hundredth.

Locate the digit in the hundredths place, 6. Consider the next digit to the right, 2. Since that digit, 2, is 4 or lower, round down.

19.7625
↓
19.76

Practice Exercise

4. Round 153.346 to the nearest hundredth.

Objective 5.2a Add using decimal notation for positive numbers.

Example Add: 14.26 + 63.589.

$$\begin{array}{r} \scriptstyle 1 \\ 14.260 \\ +\ 63.589 \\ \hline 77.849 \end{array}$$

Writing an extra zero

Practice Exercise

5. Add: 5.54 + 33.071.

Objective 5.2b Subtract using decimal notation for positive numbers.

Example Subtract: 67.345 − 24.28.

$$\begin{array}{r} \scriptstyle 2\ 14 \\ 67.\not{3}\not{4}5 \\ -\ 24.280 \\ \hline 43.065 \end{array}$$

Writing an extra zero

Practice Exercise

6. Subtract: 221.04 − 13.192.

Objective 5.3a Multiply using decimal notation.

Example Multiply: 1.8 × 0.04.

$$\begin{array}{rl} 1.8 & (1 \text{ decimal place}) \\ \times\ 0.04 & (2 \text{ decimal places}) \\ \hline 0.072 & (3 \text{ decimal places}) \end{array}$$

Practice Exercise

7. Multiply: 5.46 × 3.5.

Example Multiply: 0.001 × 87.1.

0.001 × 87.1 — 3 decimal places

0.087.1 — Move 3 places to the left. We write an extra zero.

0.001 × 87.1 = 0.0871

Practice Exercise

8. Multiply: −17.6 × 0.01.

Example Multiply: −63.4 × 100.

−63.4 × 100 — 2 zeros

−63.40. — Move 2 places to the right. We write an extra zero.

−63.4 × 100 = −6340

Practice Exercise

9. Multiply: 1000 × 60.437.

Objective 5.4a Divide using decimal notation.

Example Divide: $21.35 \div 6.1$.

$$\begin{array}{r} 3.5 \\ 6.1\overline{)\,21.3{}_\wedge 5} \\ \underline{18\,3} \\ 3\,05 \\ \underline{3\,05} \\ 0 \end{array}$$

Practice Exercise

10. Divide: $26.64 \div 3.6$.

Example Divide: $\frac{16.7}{1000}$.

$\frac{16.7}{1000}$ 0.016.7

3 zeros   Move 3 places to the left.

$$\frac{16.7}{1000} = 0.0167$$

Practice Exercise

11. Divide: $\frac{4.7}{100}$.

Example Divide: $\frac{-42.93}{0.001}$.

$\frac{-42.93}{0.001}$ −42.930.

3 decimal places   Move 3 places to the right.

$$\frac{-42.93}{0.001} = -42{,}930$$

Practice Exercise

12. Divide: $\frac{-156.9}{0.01}$.

Review Exercises

Convert the number in each sentence to standard notation. [5.3b]

1. Russia has the largest total area of any country in the world, at 6.59 million square miles.
2. Americans eat more than 3.1 billion pounds of chocolate each year.

Source: Chocolate Manufacturers' Association

Write a word name. [5.1a]

3. 3.47   **4.** 0.031

5. 27.0001   **6.** −0.9

Write fraction notation. [5.1b]

7. 0.09   **8.** 4.561

9. −0.089   **10.** −3.0227

Write decimal notation. [5.1b]

11. $\frac{34}{1000}$ **12.** $-\frac{42{,}603}{10{,}000}$

13. $27\frac{91}{100}$ **14.** $-867\frac{6}{1000}$

Which number is larger? [5.1c]

15. 0.034, 0.0185   **16.** 0.91, 0.19

17. 0.741, 0.6943   **18.** −1.038, −1.041

Round 17.4287 to the nearest: [5.1d]

19. Tenth.   **20.** Hundredth.

21. Thousandth.   **22.** One.

Add. [5.2a]

23.
$$\begin{array}{r} 2.048 \\ 65.371 \\ +\ 507.1 \\ \hline \end{array}$$

24.
$$\begin{array}{l} 0.6 \\ 0.004 \\ 0.07 \\ +\ 0.0098 \\ \hline \end{array}$$

25. $-219.3 + 2.8 + 7$

26. $0.41 + 4.1 + 41 + 0.041$

Subtract. [5.2b, c]

27.
$$\begin{array}{r} 30.0 \\ -\ 0.7908 \\ \hline \end{array}$$

28.
$$\begin{array}{r} 845.08 \\ -\ 54.79 \\ \hline \end{array}$$

29. $37.645 - (-8.497)$

30. $-70.8 - 0.0109$

Multiply. [5.3a]

31.
$$\begin{array}{r} 48 \\ \times\ 0.27 \\ \hline \end{array}$$

32. $-0.174 \cdot (-0.83)$

33. 100×0.043

34. 0.001×-24.68

Divide. [5.4a]

35. $-60 \div 8$

36. $52\overline{)23.4}$

37. $2.6\overline{)117.52}$

38. $2.14\overline{)2.18708}$

39. $\dfrac{276.3}{1000}$

40. $\dfrac{-13.892}{0.01}$

Solve. [5.2d], [5.4b]

41. $x + 51.748 = 548.0275$

42. $3 \cdot x = -20.85$

43. $10 \cdot y = 425.4$

44. $0.0089 + y = 5$

Solve. [5.7a]

45. Cole earned $620.80 working as a coronary intensive-care nurse for 40 hr one week. What is his hourly wage?

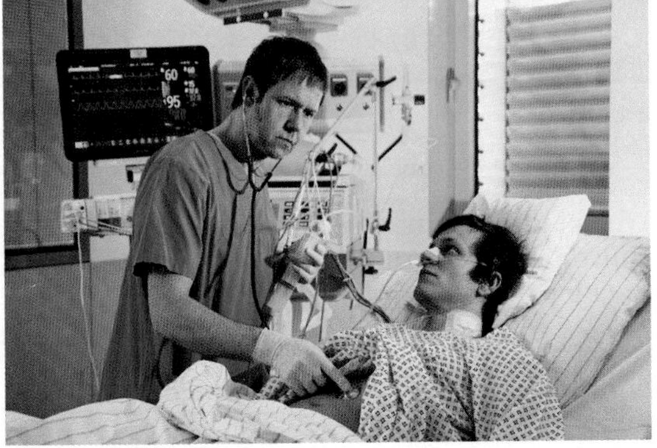

46. *Nutrition.* The average person eats 688.6 lb of fruits and vegetables in a year. What is the average consumption in one day? (Use 1 year = 365 days.) Round to the nearest tenth of a pound.

Source: U.S. Department of Agriculture

47. Derek had $1034.46 in his bank account. He used his debit card to buy a Wii game system for $249.99. How much was left in his account?

48. *Scanning Posters.* A high school club needs to scan posters designed by students and load them onto a flash drive. The copy center charges $12.99 for the flash drive and $1.09 per square foot for scanning. If the club needs to scan 13 posters at 3 sq ft per poster, what will the total cost be?

49. *Gas Mileage.* Ellie wants to estimate the gas mileage of her car. At 36,057.1 mi, she fills the tank with 10.7 gal. At 36,217.6 mi, she fills the tank with 11.1 gal. Find the mileage per gallon. Round to the nearest tenth.

50. *Books in Libraries.* The table below lists the numbers of books, in millions, held in the five largest public libraries in the United States. Find the average number of books per library. Round to the nearest tenth.

LIBRARY	NUMBER OF BOOKS (in millions)
Library of Congress	33.5
Boston Public Library	24.1
New York Public Library	16.6
Harvard University	16.6
University of Illinois–Urbana	12.8

SOURCE: American Library Association

Estimate each of the following. [5.6a]

51. The product 7.82×34.487 by rounding to the nearest one

52. The difference $219.875 - 4.478$ by rounding to the nearest one

53. The sum $\$45.78 + \$78.99$ by rounding to the nearest one

Find decimal notation. Use multiplying by 1. [5.5a]

54. $\frac{13}{25}$ **55.** $-\frac{9}{20}$ **56.** $\frac{11}{4}$

Find decimal notation. Use division. [5.5a]

57. $-\frac{13}{4}$ **58.** $\frac{7}{6}$ **59.** $\frac{17}{11}$

Round the answer to Exercise 59 to the nearest: [5.5b]

60. Tenth. **61.** Hundredth. **62.** Thousandth.

Convert from cents to dollars. [5.3b]

63. 8273¢ **64.** 487¢

Convert from dollars to cents. [5.3b]

65. \$24.93 **66.** \$9.86

Calculate. [5.4c], [5.5c]

67. $(8 - 1.23) \div (-4) + 5.6 \times 0.02$

68. $(1 + 0.07)^2 + 10^3 \div 10^2 + [4(10.1 - 5.6) + 8(11.3 - 7.8)]$

69. $\frac{3}{4} \times (-20.85)$

70. Divide: $\frac{346.295}{0.001}$. [5.4a]

A. 0.346295 **B.** 3.46295
C. 34,629.5 **D.** 346,295

71. Estimate the quotient $82.304 \div 17.287$ by rounding to the nearest ten. [5.6a]

A. 0.4 **B.** 4
C. 40 **D.** 400

Synthesis

72. In each of the following, use one of $+, -, \times,$ and $\div$ in each blank to make a true sentence. [5.4c]

a) 2.56 ☐ 6.4 ☐ 51.2 ☐ 17.4 ☐ 89.7 = 72.62

b) (11.12 ☐ 0.29) ☐ 3^4 = 877.23

73. Use the fact that $\frac{1}{3} = 0.\overline{3}$ to find repeating decimal notation for 1. Explain how you got your answer. [5.5a]

Understanding Through Discussion and Writing

1. Describe in your own words a procedure for converting from decimal notation to fraction notation. [5.1b]

2. A student insists that $346.708 \times 0.1 = 3467.08$. How could you convince him that a mistake had been made without checking on a calculator? [5.3a]

3. When is long division *not* the fastest way to convert from fraction notation to decimal notation? [5.5a]

4. Consider finding decimal notation for $\frac{44}{125}$. Discuss as many ways as you can for finding such notation and give the answer. [5.5a]

For Extra Help For step-by-step test solutions, access the Chapter Test Prep Videos in MyMathLab® or on YouTube (search "BittingerBasicEl" and click on "Channels").

1. In a recent year, 2.6 billion ducks were killed for food worldwide. Convert 2.6 billion to standard notation.

Sources: FAO; *National Geographic*, May 2011, Nigel Holmes

2. Write a word name for 123.0047.

Write fraction notation.

3. 0.91

4. -2.769

Which number is larger?

5. 0.07, 0.162

6. $-0.078, -0.06$

Write decimal notation.

7. $\frac{74}{1000}$

8. $-\frac{37{,}047}{10{,}000}$

9. $756\frac{9}{100}$

Round 5.6783 to the nearest:

10. One.

11. Thousandth.

12. Tenth.

Calculate.

13. $-102.4 + 6.1 + 78$

14. $\begin{array}{r} 52.678 \\ -\ \ 4.321 \\ \hline \end{array}$

15. $\begin{array}{r} 20.0 \\ -\ \ 0.9099 \\ \hline \end{array}$

16. $\begin{array}{r} 0.125 \\ \times\ \ 0.24 \\ \hline \end{array}$

17. $0.001 \times (-213.45)$

18. $-19 \div 4$

19. $3.3\,\overline{)\,100.32}$

20. $82\,\overline{)\,15.58}$

21. $\frac{-346.89}{1000}$

Solve.

22. $-4.8 \cdot y = 404.448$

23. $x + 0.018 = 9$

Calculate.

24. $256 \div 3.2 \div 2 - 1.56 + 78.325 \times 0.02$

25. $-\frac{7}{8} \times (-345.6)$

Find decimal notation.

26. $\frac{7}{20}$

27. $\frac{22}{25}$

28. $\frac{9}{11}$

29. $-\frac{89}{12}$

Round the answer to Exercise 28 to the nearest:

30. Tenth.

31. Hundredth.

32. Thousandth.

Estimate each of the following.

33. The product 8.91×22.457 by rounding to the nearest one

34. The quotient $78.2209 \div 16.09$ by rounding to the nearest ten

35. *Scanning Blueprints.* A building contractor needs to scan and load blueprints onto a flash drive. The copy center charges $10.99 for the flash drive and $1.19 per square foot for scanning. If the contractor needs to scan 5 blueprints at 6 sq ft per print, what will the total cost be?

36. *Gas Mileage.* Tina wants to estimate the gas mileage in her economy car. At 76,843 mi, she fills the tank with 14.3 gal of gasoline. At 77,310 mi, she fills the tank with 16.5 gal of gasoline. Find the mileage per gallon. Round to the nearest tenth.

37. *Life Expectancy.* Life expectancies at birth for seven Asian countries are listed in the table below. Find the average life expectancy for this group of countries. Round to the nearest tenth.

COUNTRY	LIFE EXPECTANCY (in years)
Japan	82.25
South Korea	79.05
People's Republic of China	73.47
India	69.9
Russia	66.03
North Korea	63.81
Afghanistan	44.64

SOURCE: CIA *Factbook* 2011

38. The office manager for the law firm Drake, Smith, and Hartner buys 7 cases of copy paper at $41.99 per case. What is the total cost?

39. Convert from cents to dollars: 949¢.

A. 0.949¢ **B.** $9.49 **C.** $94.90 **D.** $949

Synthesis

40. The Silver's Health Club charges a $79 membership fee and $42.50 a month. Allise has a coupon that will allow her to join the club for $299 for 6 months. How much will Allise save if she uses the coupon?

41. Arrange from smallest to largest.

$$-\frac{2}{3}, -\frac{15}{19}, -\frac{11}{13}, -\frac{5}{7}, -\frac{13}{15}, -\frac{17}{20}$$

CHAPTERS

1–5 Cumulative Review

Convert to fraction notation.

1. $2\frac{2}{9}$

2. -3.051

Find decimal notation.

3. $-\frac{7}{5}$

4. $\frac{6}{11}$

5. Determine whether 43 is prime, composite, or neither.

6. Determine whether 2,053,752 is divisible by 4.

Calculate.

7. $48 + 12 \div 4 - 10 \times 2 + 6892 \div 4$

8. $0.2 - \{0.1[1.2(3.95 - 1.65) + 1.5 \div 2.5]\}$

Round to the nearest hundredth.

9. 584.973

10. $218.\overline{5}$

11. Estimate the product 16.392×9.715 by rounding to the nearest one.

12. Estimate by rounding to the nearest tenth:

$$2.714 + 4.562 - 3.31 - 0.0023.$$

13. Estimate the product 6418×1984 by rounding to the nearest hundred.

14. Estimate the quotient $717.832 \div 124.998$ by rounding to the nearest ten.

Add and simplify.

15. $2\frac{1}{4} + 3\frac{4}{5}$

16. $34{,}921 + 93{,}092 + 11{,}103$

17. $\frac{1}{6} + \frac{2}{3} + \frac{8}{9}$

18. $-143.9 + 2.053$

Subtract and simplify.

19. $723{,}041 - 12{,}904$

20. $19 - 5.903$

21. $5\frac{1}{7} - 4\frac{3}{7}$

22. $\frac{9}{10} - \frac{10}{11}$

Multiply and simplify.

23. $\frac{3}{8} \cdot \left(-\frac{4}{9}\right)$

24. 2532×2100

25. 23.9×0.2

26. 27.9431×0.001

Divide and simplify.

27. $35.013 \div 16.5$ (16.5) 35.013)

28. $47{,}918 \div 26$ (26) 47,918)

29. $13.8621 \div 0.001$

30. $-\frac{4}{9} \div \left(-\frac{8}{15}\right)$

Solve.

31. $8.32 + x = 9.1$

32. $75 \cdot x = 2100$

33. $y \cdot 9.47 = -81.6314$

34. $1062 + y = 368{,}313$

35. $t + \frac{5}{6} = \frac{8}{9}$

36. $-\frac{7}{8} \cdot t = \frac{7}{16}$

37. *Dominant Languages.* The table below lists the numbers of first-language speakers (people who acquired the language as infants) for the top five languages spoken worldwide.

LANGUAGE	FIRST-LANGUAGE SPEAKERS (in millions)
Chinese	1213
Spanish	329
English	328
Arabic	221
Hindi	182

SOURCES: Living Tongues Institute for Endangered Languages; UNESCO; SIL International

How many people in the world are first-language speakers of Spanish, Arabic, or Hindi?

38. Refer to the table in Exercise 37. How many more people in the world are first-language speakers of Chinese than first-language speakers of English?

39. After Lauren made a \$450 down payment on a sofa, $\frac{3}{10}$ of the total cost was paid. How much did the sofa cost?

40. Joshua's tuition was \$3600. He obtained a loan for $\frac{2}{3}$ of the tuition. How much was the loan?

41. The balance in Elliott's bank account is \$314.79. After a debit transaction for \$56.02, what is the balance in the account?

42. A clerk in Leah's Delicatessen sold $1\frac{1}{2}$ lb of ham, $2\frac{3}{4}$ lb of turkey, and $2\frac{1}{4}$ lb of roast beef. How many pounds of meat were sold altogether?

43. A baker used $\frac{1}{2}$ lb of sugar for cookies, $\frac{2}{3}$ lb of sugar for pie, and $\frac{5}{6}$ lb of sugar for cake. How much sugar was used in all?

44. The Currys' rectangular family room measures 19.8 ft by 23.6 ft. What is its area?

Simplify.

45. $\left(\frac{3}{4}\right)^2 - \frac{1}{8} \cdot \left(3 - 1\frac{1}{2}\right)^2$

46. $-1.2 \times 12.2 \div 0.1 \times 3.6$

Synthesis

47. Using a manufacturer's coupon, Lucy bought 2 cartons of orange juice and received a third carton free. The price of each carton was \$3.59. What was the cost per carton with the coupon? Round to the nearest cent.

48. A carton of gelatin mix packages weighs $15\frac{3}{4}$ lb. Each package weighs $1\frac{3}{4}$ oz. How many packages are in the carton? (1 lb = 16 oz)

CHAPTER

6

Ratio and Proportion

STUDYING FOR SUCCESS *Working Exercises*

- ☐ Don't try to solve a homework problem by working backward from the answer given at the back of the text. Remember: quizzes and tests have no answer section!
- ☐ Check answers to odd-numbered exercises at the back of the book.
- ☐ Work some even-numbered exercises, whose answers are not provided, as practice. Check your answers later with a friend or your instructor.

6.1 Introduction to Ratios

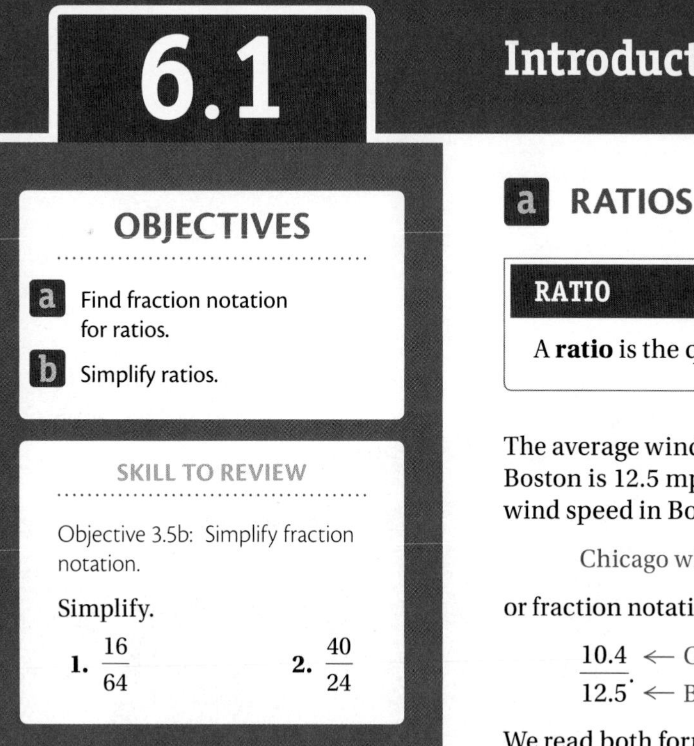

OBJECTIVES

a Find fraction notation for ratios.

b Simplify ratios.

SKILL TO REVIEW

Objective 3.5b: Simplify fraction notation.

Simplify.

1. $\frac{16}{64}$ **2.** $\frac{40}{24}$

a RATIOS

RATIO

A **ratio** is the quotient of two quantities.

The average wind speed in Chicago is 10.4 mph. The average wind speed in Boston is 12.5 mph. The *ratio* of average wind speed in Chicago to average wind speed in Boston is written using colon notation,

Chicago wind speed → 10.4 : 12.5, ← Boston wind speed

or fraction notation,

$\frac{10.4}{12.5}$. ← Chicago wind speed (numerator); ← Boston wind speed (denominator)

We read both forms of notation as "the ratio of 10.4 to 12.5."

RATIO NOTATION

The **ratio** of a to b is expressed by the fraction notation $\frac{a}{b}$, where a is the numerator and b is the denominator, or by the colon notation $a : b$.

EXAMPLE 1 Find the ratio of 7 to 8.

The ratio is $\frac{7}{8}$, or 7 : 8.

EXAMPLE 2 Find the ratio of 31.4 to 100.

The ratio is $\frac{31.4}{100}$, or 31.4 : 100.

EXAMPLE 3 Find the ratio of $4\frac{2}{3}$ to $5\frac{7}{8}$. You need not simplify.

The ratio is $\frac{4\frac{2}{3}}{5\frac{7}{8}}$, or $4\frac{2}{3} : 5\frac{7}{8}$.

◀ Do Margin Exercises 1–3.

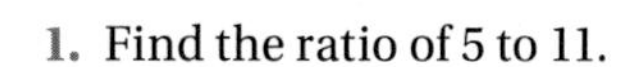

1. Find the ratio of 5 to 11.
2. Find the ratio of 57.3 to 86.1.
3. Find the ratio of $6\frac{3}{4}$ to $7\frac{2}{5}$.

Answers

Skill to Review:

1. $\frac{1}{4}$ 2. $\frac{5}{3}$

Margin Exercises:

1. $\frac{5}{11}$, or 5 : 11 2. $\frac{57.3}{86.1}$, or 57.3 : 86.1
3. $\frac{6\frac{3}{4}}{7\frac{2}{5}}$, or $6\frac{3}{4} : 7\frac{2}{5}$

In most of our work, we will use fraction notation for ratios.

EXAMPLE 4 *Media Usage.* In 2012, Americans spent an average of 6 hr/month on Facebook and 147 hr/month watching TV. Find the ratio of average time spent on Facebook to average time spent watching TV.

Sources: comScore and Nielsen

The ratio is $\frac{6}{147}$. ■

EXAMPLE 5 *Record Rainfall.* The greatest rainfall ever recorded in the United States during a 12-month period was 739 in. in Kukui, Maui, Hawaii, from December 1981 to December 1982. What is the ratio of amount of rainfall, in inches, to time, in months? of time, in months, to amount of rainfall, in inches?

Source: *Time Almanac*

The ratio of amount of rainfall, in inches, to time, in months, is

$\frac{739}{12}$. ← Rainfall / ← Time

The ratio of time, in months, to amount of rainfall, in inches, is

$\frac{12}{739}$. ← Time / ← Rainfall ■

EXAMPLE 6 Refer to the triangle below.

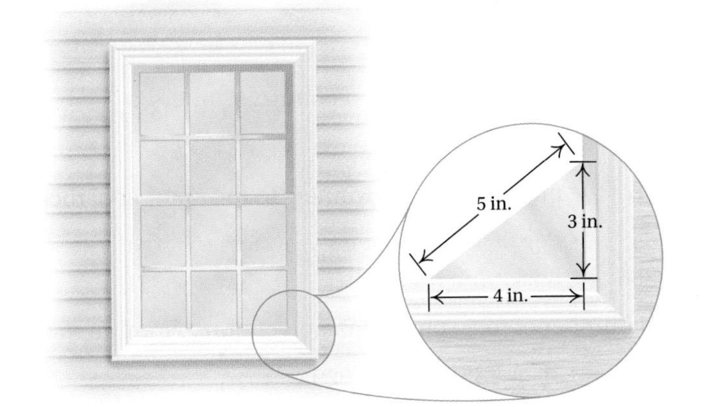

a) What is the ratio of the length of the longest side to the length of the shortest side?

$\frac{5}{3}$ ← Longest side / ← Shortest side

b) What is the ratio of the length of the shortest side to the length of the longest side?

$\frac{3}{5}$ ← Shortest side / ← Longest side

Do Exercises 4–6. ▶

4. *Record Snowfall.* The greatest snowfall recorded in North America during a 24-hr period was 76 in. in Silver Lake, Colorado, on April 14–15, 1921. What is the ratio of amount of snowfall, in inches, to time, in hours?

Source: U.S. Army Corps of Engineers

5. *Frozen Fruit Drinks.* A Berries & Kreme Chiller from Krispy Kreme contains 960 calories, while Smoothie King's MangoFest drink contains 258 calories. What is the ratio of the number of calories in the Krispy Kreme drink to the number of calories in the Smoothie King drink? of the number of calories in the Smoothie King drink to the number of calories in the Krispy Kreme drink?

Source: Physicians Committee for Responsible Medicine

GS **6.** In the triangle below, what is the ratio of the length of the shortest side to the length of the longest side?

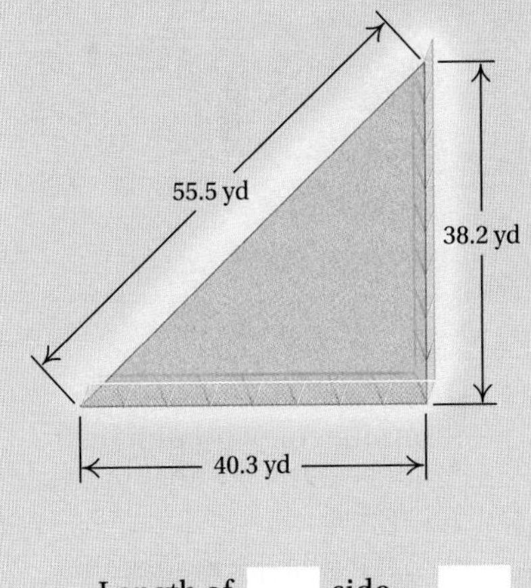

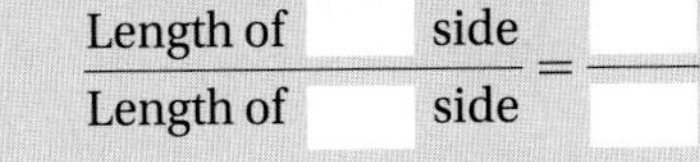

Answers

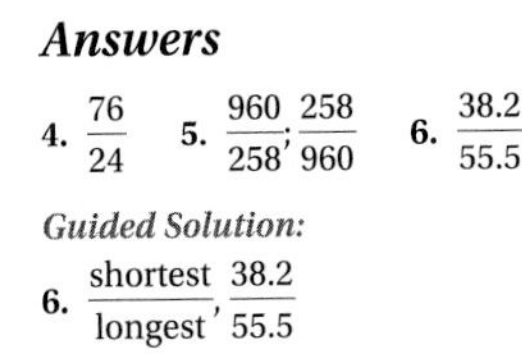

4. $\frac{76}{24}$ **5.** $\frac{960}{258}; \frac{258}{960}$ **6.** $\frac{38.2}{55.5}$

Guided Solution:

6. $\frac{\text{shortest}}{\text{longest}}, \frac{38.2}{55.5}$

EXAMPLE 7 *Grammy Awards.* From 1959 to 2011, a total of 798 musical artists won Grammy awards for Record of the Year, Album of the Year, and Best New Artist. The bar graph below shows the numbers of award winners from four music genres.

a) What is the ratio of the number of country winners to the number of rock winners?

b) What is the ratio of the number of pop, R&B, and country winners to the number of rock winners?

c) What is the ratio of the number of Grammy winners from genres other than rock, pop, R&B, and country to the total number of Grammy winners?

a) The ratio of the number of country winners to the number of rock winners is

$$\frac{57}{286}. \begin{array}{l}\leftarrow \text{Country winners}\\ \leftarrow \text{Rock winners}\end{array}$$

b) The total number of pop, R&B, and country winners is

$$151 + 135 + 57 = 343.$$

The ratio of the number of these winners to the number of rock winners is

$$\frac{343}{286}. \begin{array}{l}\leftarrow \text{Pop, R\&B, and country winners}\\ \leftarrow \text{Rock winners}\end{array}$$

c) The total number of rock, pop, R&B, and country Grammy winners is

$$286 + 151 + 135 + 57 = 629.$$

Thus the number of Grammy winners from other genres is

$$798 - 629 = 169.$$

The ratio of the number of Grammy winners from other genres to the total number of Grammy winners is

$$\frac{169}{798}. \begin{array}{l}\leftarrow \text{Winners from other genres}\\ \leftarrow \text{Total number of winners}\end{array}$$

◀ **Do Exercise 7.**

7. *NBA Playoffs.* In the final game of the 2012 NBA Playoffs, the Miami Heat made a total of 67 baskets. Of these, 27 were free throws, 26 were two-point field goals, and the remainder were three-point field goals.
Source: nba.com

a) What was the ratio of the number of two-point field goals to the number of free throws?

b) What was the ratio of the number of three-point field goals to the total number of baskets?

Answers

7. (a) $\frac{26}{27}$ (b) $\frac{14}{67}$

b SIMPLIFYING NOTATION FOR RATIOS

Sometimes a ratio can be simplified. Simplifying provides a means of finding other numbers with the same ratio.

EXAMPLE 8 Find the ratio of 6 to 8. Then simplify to find two other numbers in the same ratio.

We write the ratio in fraction notation and then simplify:

$$\frac{6}{8} = \frac{2 \cdot 3}{2 \cdot 4} = \frac{2}{2} \cdot \frac{3}{4} = 1 \cdot \frac{3}{4} = \frac{3}{4}.$$

Thus, 3 and 4 have the same ratio as 6 and 8. We can express this by saying "6 is to 8 as 3 is to 4."

Do Exercise 8.

EXAMPLE 9 Find the ratio of 2.4 to 10. Then simplify to find two other numbers in the same ratio.

We first write the ratio in fraction notation. Next, we multiply by 1 to clear the decimal from the numerator. Then we simplify.

$$\frac{2.4}{10} = \frac{2.4}{10} \cdot \frac{10}{10} = \frac{24}{100} = \frac{4 \cdot 6}{4 \cdot 25} = \frac{4}{4} \cdot \frac{6}{25} = \frac{6}{25}$$

Thus, 2.4 is to 10 as 6 is to 25.

Do Exercises 9 and 10.

EXAMPLE 10 An HDTV screen that measures approximately 46 in. diagonally has a width of 40 in. and a height of $22\frac{1}{2}$ in. Find the ratio of width to height and simplify.

The ratio is $\frac{40}{22\frac{1}{2}} = \frac{40}{22.5} = \frac{40}{22.5} \cdot \frac{10}{10} = \frac{400}{225}$

$$= \frac{25 \cdot 16}{25 \cdot 9} = \frac{25}{25} \cdot \frac{16}{9}$$

$$= \frac{16}{9}.$$

Thus we can say that the ratio of width to height is 16 to 9, which can also be expressed as 16 : 9.

Do Exercise 11.

8. Find the ratio of 18 to 27. Then simplify to find two other numbers in the same ratio.

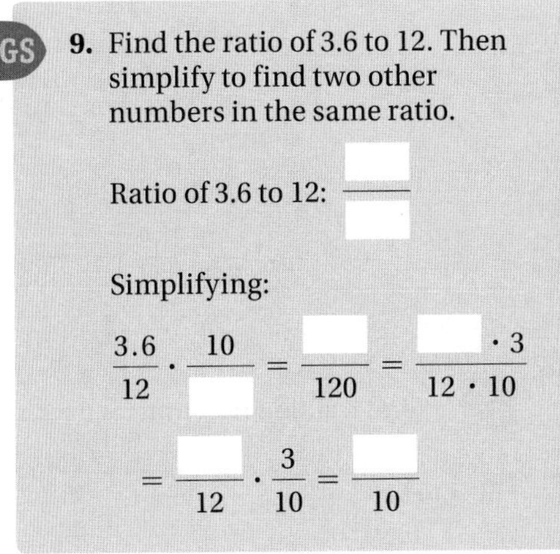

GS **9.** Find the ratio of 3.6 to 12. Then simplify to find two other numbers in the same ratio.

Ratio of 3.6 to 12: $\frac{\square}{\square}$

Simplifying:

$$\frac{3.6}{12} \cdot \frac{10}{\square} = \frac{\square}{120} = \frac{\square \cdot 3}{12 \cdot 10}$$

$$= \frac{\square}{12} \cdot \frac{3}{10} = \frac{\square}{10}$$

10. Find the ratio of 1.2 to 1.5. Then simplify to find two other numbers in the same ratio.

11. An HDTV screen that measures 44 in. diagonally has a width of 38.4 in. and a height of 21.6 in. Find the ratio of height to width and simplify.

Answers

8. 18 is to 27 as 2 is to 3.
9. 3.6 is to 12 as 3 is to 10.
10. 1.2 is to 1.5 as 4 is to 5. **11.** $\frac{9}{16}$

Guided Solution:

9. $\frac{3.6}{12}$; 10, 36, 12, 12, 3

6.1 Exercise Set

For Extra Help
MyMathLab® MathXL® PRACTICE WATCH READ REVIEW

Reading Check

Determine whether each statement is true or false.

_______________ **RC1.** A ratio is a quotient.

_______________ **RC2.** If there are 2 teachers and 27 students in a classroom, the ratio of students to teachers is 27 : 2.

_______________ **RC3.** The ratio 6 : 7 can also be written $\frac{7}{6}$.

_______________ **RC4.** The numbers 2 and 3 are in the same ratio as the numbers 4 and 9.

a Find fraction notation for each ratio. You need not simplify.

1. 4 to 5

2. 3 to 2

3. 178 to 572

4. 329 to 967

5. 0.4 to 12

6. 2.3 to 22

7. 3.8 to 7.4

8. 0.6 to 0.7

9. 56.78 to 98.35

10. 456.2 to 333.1

11. $8\frac{3}{4}$ to $9\frac{5}{6}$

12. $10\frac{1}{2}$ to $43\frac{1}{4}$

Shark Attacks. Of the 75 unprovoked shark attacks recorded worldwide in 2011, 29 occurred in U.S. waters. The bar graph below shows the breakdown by state. Use the graph for Exercises 13–16.

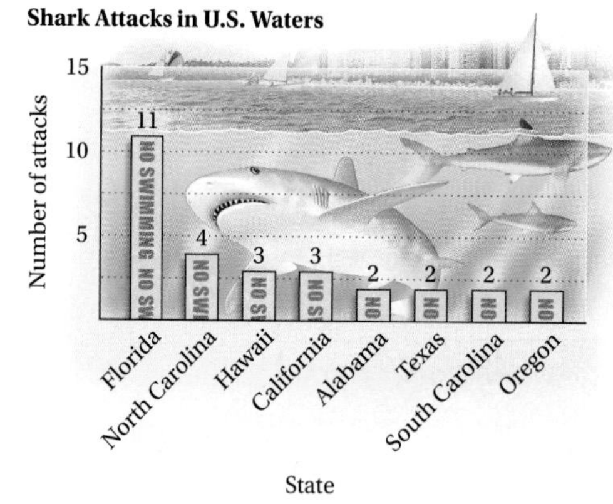

SOURCE: University of Florida

13. What is the ratio of the number of shark attacks in U.S. waters to the number of shark attacks worldwide?

14. What is the ratio of the number of shark attacks in North Carolina to the number of shark attacks in California?

15. What is the ratio of the number of shark attacks in Florida to the total number of shark attacks in the other seven states?

16. What is the ratio of the total number of shark attacks in states bordering the Pacific Ocean (Hawaii, California, and Oregon) to the total number of shark attacks in the other five states?

17. *Tax Freedom Day.* Of the 366 days in 2012 (a leap year), the average American worked 107 days to pay his or her federal, state, and local taxes. Find the ratio of the number of days worked to pay taxes in 2012 to the number of days in the year.

Source: Tax Foundation

18. *Careers in Medicine.* The number of jobs for nurses is expected to increase by 711,900 between 2010 and 2020. During the same decade, the number of jobs for physicians is expected to increase by 168,300. Find the ratio of the increase in jobs for physicians to the increase in jobs for nurses.

Source: Bureau of Labor Statistics

19. *Music Album Sales.* In 2011, there were 104.8 million albums downloaded digitally and 3100.7 million albums sold on CDs. What is the ratio of downloads to CD sales? of CD sales to downloads?

Source: Recording Industry of America

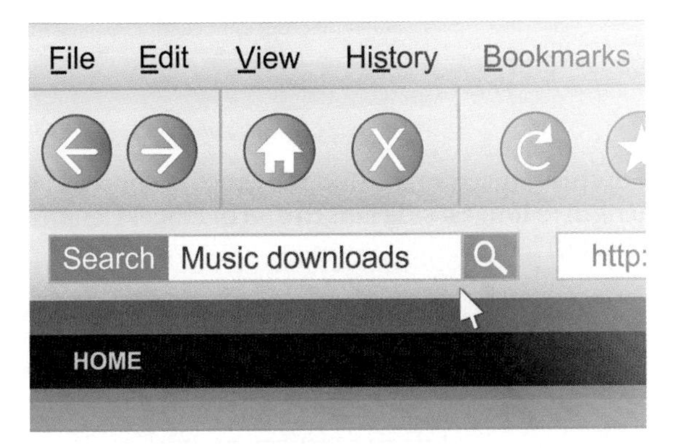

20. *Silicon in the Earth's Crust.* Every 100 tons of the earth's crust contains about 28 tons of silicon. What is the ratio of the weight of silicon to the weight of crust? of the weight of crust to the weight of silicon?

Source: *The Handy Science Answer Book*

21. *Field Hockey.* A diagram of the playing area for field hockey is shown below. What is the ratio of width to length? of length to width?

Source: *Sports: The Complete Visual Reference*

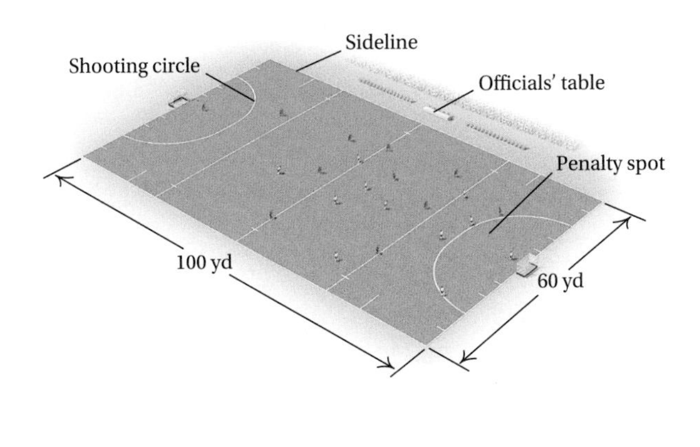

22. *The Leaning Tower of Pisa.* The Leaning Tower of Pisa was reopened to the public in 2001 following a 10-yr stabilization project. The 184.5-ft tower now leans about 13 ft from its base. What is the ratio of the distance that it leans to its height? of its height to the distance that it leans?

Source: CNN

b Find the ratio of the first number to the second and simplify.

23. 4 to 6

24. 6 to 10

25. 18 to 24

26. 28 to 36

27. 4.8 to 10

28. 5.6 to 10

29. 2.8 to 3.6

30. 4.8 to 6.4

31. 20 to 30

32. 40 to 60

33. 56 to 100

34. 42 to 100

35. 128 to 256

36. 232 to 116

37. 0.48 to 0.64

38. 0.32 to 0.96

39. For this rectangle, find the ratios of length to width and of width to length.

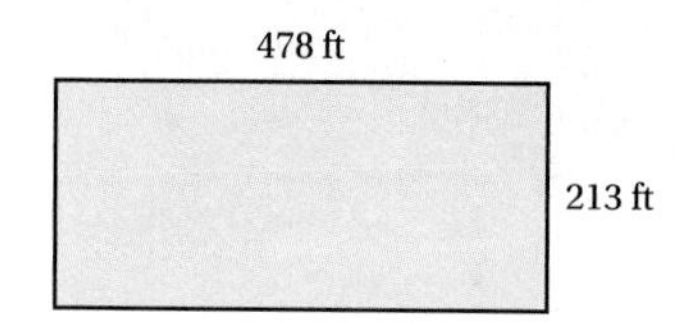

40. For this right triangle, find the ratios of shortest side length to longest side length and of longest length to shortest length.

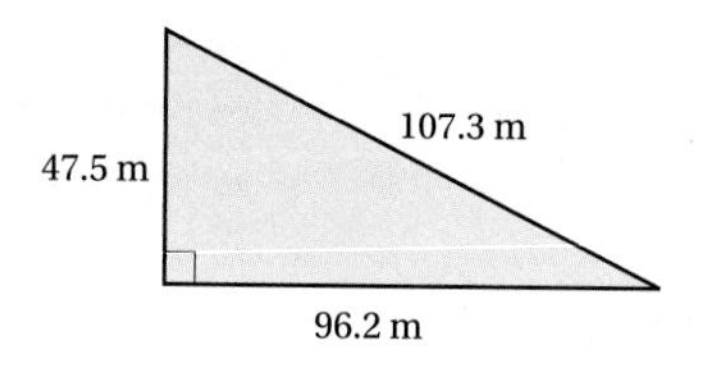

Skill Maintenance

Add. Simplify if possible.

41. 18,468 + 390,082 [1.2a]

42. 24 + 3.006 [5.2a]

43. 4.2 + 28.07 − 365 [5.2c]

44. $\frac{1}{3} + \frac{3}{10}$ [4.2a]

45. $\frac{3}{8} + \left(-\frac{7}{12}\right)$ [4.2a]

46. $4\frac{5}{6} + 9\frac{1}{2}$ [4.5a]

Subtract. Simplify if possible.

47. 982,001 − 39,782 [1.3a]

48. 12.5 − 9.9 [5.2b]

49. −54.1 − 7.29 [5.2c]

50. $\frac{9}{16} - \frac{1}{2}$ [4.3a]

51. $8\frac{2}{3} - 6\frac{4}{5}$ [4.5b]

52. $11 - 6\frac{1}{3}$ [4.5b]

Synthesis

53. Find the ratio of $3\frac{3}{4}$ to $5\frac{7}{8}$ and simplify.

Fertilizer. Exercises 54 and 55 refer to a common lawn fertilizer known as "5, 10, 15." This mixture contains 5 parts of potassium for every 10 parts of phosphorus and 15 parts of nitrogen. (This is often denoted 5 : 10 : 15.)

54. Find and simplify the ratio of potassium to nitrogen and of nitrogen to phosphorus.

55. Simplify the ratio 5 : 10 : 15.

Rates and Unit Prices

6.2

OBJECTIVES

a Give the ratio of two different measures as a rate.

b Find unit prices and use them to compare purchases.

SKILL TO REVIEW

Objective 5.4a: Divide using decimal notation.

Divide.

1. $15\overline{)634.5}$ **2.** $2.4\overline{)6}$

a RATES

A 2013 Honda Odyssey can travel 504 mi on 18 gal of gasoline. Let's consider the ratio of miles to gallons:

Source: Kia Motors America, Inc.

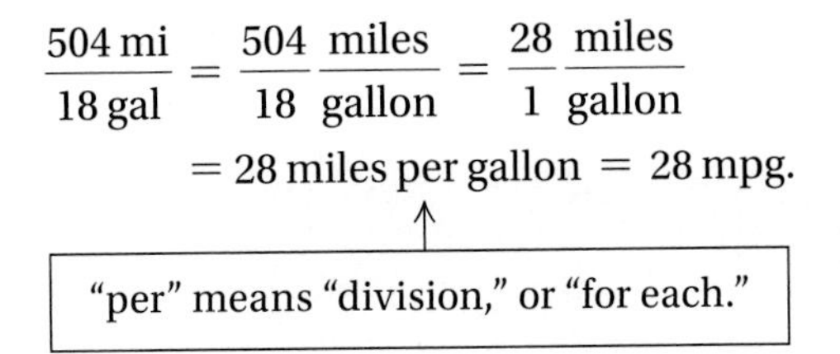

$$\frac{504 \text{ mi}}{18 \text{ gal}} = \frac{504 \text{ miles}}{18 \text{ gallon}} = \frac{28 \text{ miles}}{1 \text{ gallon}}$$

$$= 28 \text{ miles per gallon} = 28 \text{ mpg}.$$

"per" means "division," or "for each."

The ratio

$$\frac{504 \text{ mi}}{18 \text{ gal}}, \quad \text{or} \quad \frac{504 \text{ mi}}{18 \text{ gal}}, \quad \text{or} \quad 28 \text{ mpg},$$

is called a **rate**.

RATE

When a ratio is used to compare two different kinds of measure, we call it a **rate**.

Suppose David's car travels 475.4 mi on 16.8 gal of gasoline. Is the gas mileage (mpg) of his car better than that of the Honda Odyssey above? To determine this, it helps to convert the ratio to decimal notation and perhaps round. Thus we have

$$\frac{475.4 \text{ miles}}{16.8 \text{ gallons}} = \frac{475.4}{16.8} \text{ mpg} \approx 28.298 \text{ mpg}.$$

Since $28.298 > 28$, David's car gets better gas mileage than the Honda Odyssey does.

EXAMPLE 1 It takes 60 oz of grass seed to seed 3000 sq ft of lawn. What is the rate in ounces per square foot?

$$\frac{60 \text{ oz}}{3000 \text{ sq ft}} = \frac{1 \text{ oz}}{50 \text{ sq ft}}, \quad \text{or} \quad 0.02 \frac{\text{oz}}{\text{sq ft}}$$

EXAMPLE 2 Martina bought 5 lb of organic russet potatoes for \$4.99. What was the rate in cents per pound?

$$\frac{\$4.99}{5 \text{ lb}} = \frac{499 \text{ cents}}{5 \text{ lb}} = 99.8¢/\text{lb}$$

Answers

Skill to Review:
1. 42.3 **2.** 2.5

EXAMPLE 3 *Hourly Wage.* In 2011, Walmart employees earned, on average, \$470 for a 40-hr work week. What was the rate of pay per hour?

Source: "Living Wage Policies and Big-Box Retail," UC Berkeley Center for Labor Research and Education, Ken Jacobs, Dave Graham-Squire, and Stephanie Luce, at http://laborcenter.berkeley.edu/retail/bigbox_livingwage_policies11.pdf

The rate of pay is the ratio of money earned to length of time worked, or

$$\frac{\$470}{40\text{ hr}} = \frac{470}{40}\frac{\text{dollars}}{\text{hr}} = 11.75\frac{\text{dollars}}{\text{hr}}, \quad \text{or} \quad \$11.75 \text{ per hr.}$$

EXAMPLE 4 *Ratio of Strikeouts to Home Runs.* In the 2012 baseball season, Miguel Cabrera of the Detroit Tigers had 98 strikeouts and 44 home runs. What was his strikeout to home-run rate?

Source: Major League Baseball

$$\frac{98\text{ strikeouts}}{44\text{ home runs}} = \frac{98}{44}\frac{\text{strikeouts}}{\text{home runs}} = \frac{98}{44}\text{ strikeouts per home run}$$
$$\approx 2.227 \text{ strikeouts per home run}$$

◀ Do Exercises 1–5.

A ratio of distance traveled to time is called a *speed.* What is the rate, or speed, in miles per hour?

1. 45 mi, 9 hr

2. 120 mi, 10 hr

What is the rate, or speed, in feet per second?

3. 2200 ft, 2 sec

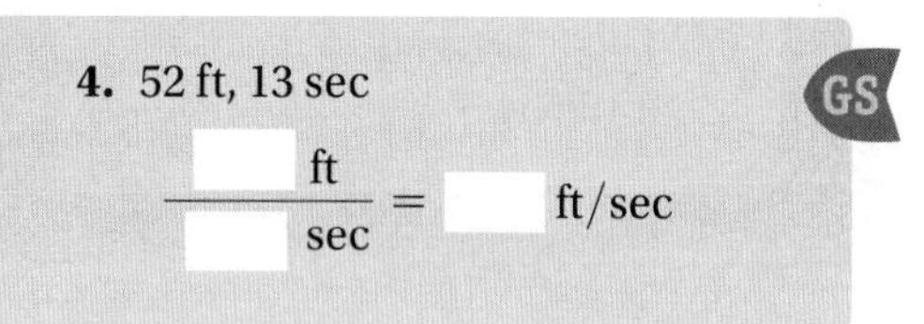

4. 52 ft, 13 sec

GS

$$\frac{\square}{\square}\frac{\text{ft}}{\text{sec}} = \square \text{ ft/sec}$$

5. *Babe Ruth.* In his baseball career, Babe Ruth had 1330 strikeouts and 714 home runs. What was his home-run to strikeout rate?

Source: Major League Baseball

b UNIT PRICING

UNIT PRICE

A **unit price**, or **unit rate**, is the ratio of price to the number of units.

EXAMPLE 5 *Unit Price of Pears.* Ruth bought a 15.2-oz can of pears for \$1.30. What is the unit price in cents per ounce?

$$\text{Unit price} = \frac{\text{Price}}{\text{Number of units}}$$
$$= \frac{\$1.30}{15.2\text{ oz}} = \frac{130\text{ cents}}{15.2\text{ oz}} = \frac{130}{15.2}\frac{\text{cents}}{\text{oz}}$$
$$\approx 8.553 \text{ cents per ounce}$$

◀ Do Exercise 6.

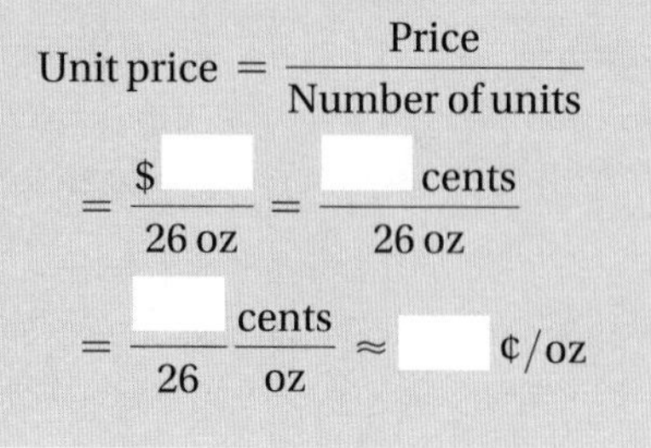

6. Unit Price of Pasta Sauce. Gregory bought a 26-oz jar of pasta sauce for \$2.79. What is the unit price in cents per ounce?

GS

$$\text{Unit price} = \frac{\text{Price}}{\text{Number of units}}$$
$$= \frac{\$\square}{26\text{ oz}} = \frac{\square\text{ cents}}{26\text{ oz}}$$
$$= \frac{\square}{26}\frac{\text{cents}}{\text{oz}} \approx \square \text{ ¢/oz}$$

Answers

1. 5 mi/hr, or 5 mph **2.** 12 mi/hr, or 12 mph **3.** 1100 ft/sec **4.** 4 ft/sec **5.** $\frac{714}{1330}$ home run per strikeout ≈ 0.537 home run per strikeout **6.** 10.731¢/oz

Guided Solutions:

4. $\frac{52}{13}$, 4 **6.** 2.79, 279, 279, 10.731

Unit prices enable us to do comparison shopping and determine the best buy for a product on the basis of price. It is often helpful to change all prices to cents so that we can compare unit prices more easily.

EXAMPLE 6 *Unit Price of Salad Dressing.* At the request of his customers, Angelo started bottling and selling the basil and lemon salad dressing that he serves in his café. The dressing is sold in four sizes of containers as listed in the table below. Compute the unit price of each size of container and determine which size is the best buy on the basis of unit price alone.

Size	Price	Unit Price
10 oz	\$2.49	
16 oz	\$3.59	
20 oz	\$4.09	
32 oz	\$6.79	

We compute the unit price of each size and fill in the chart:

$$10\text{ oz: } \frac{\$2.49}{10\text{ oz}} = \frac{249\text{ cents}}{10\text{ oz}} = \frac{249}{10}\frac{\text{cents}}{\text{oz}} = 24.900¢/\text{oz};$$

$$16\text{ oz: } \frac{\$3.59}{16\text{ oz}} = \frac{359\text{ cents}}{16\text{ oz}} = \frac{359}{16}\frac{\text{cents}}{\text{oz}} \approx 22.438¢/\text{oz};$$

$$20\text{ oz: } \frac{\$4.09}{20\text{ oz}} = \frac{409\text{ cents}}{20\text{ oz}} = \frac{409}{20}\frac{\text{cents}}{\text{oz}} = 20.450¢/\text{oz};$$

$$32\text{ oz: } \frac{\$6.79}{32\text{ oz}} = \frac{679\text{ cents}}{32\text{ oz}} = \frac{679}{32}\frac{\text{cents}}{\text{oz}} \approx 21.219¢/\text{oz}.$$

Size	Price	Unit Price
10 oz	\$2.49	24.900¢/oz
16 oz	\$3.59	22.438¢/oz
20 oz	\$4.09	20.450¢/oz Lowest unit price
32 oz	\$6.79	21.219¢/oz

On the basis of unit price alone, we see that the 20-oz container is the best buy.

Do Exercise 7.

Although we often think that "bigger is cheaper," this is not always the case, as we see in Example 6. In addition, even when a larger package has a lower unit price than a smaller package, it still might not be the best buy for you. For example, some of the food in a large package could spoil before it is used, or you might not have room to store a large package.

7. *Cost of Mayonnaise.* Complete the following table for Hellmann's mayonnaise sold on an online shopping site. Which size has the lowest unit price?

Source: Peapod.com

Size	Price	Unit Price
8 oz	\$2.79	
10 oz	\$3.69	
30 oz	\$5.39	

Answers

7. 34.875¢/oz; 36.900¢/oz; 17.967¢/oz; the 30-oz size has the lowest unit price.

6.2 Exercise Set

For Extra Help
MyMathLab® MathXL® PRACTICE WATCH READ REVIEW

Reading Check

Complete each sentence with the correct quantity from the list at the right, given the following information.

Jason wrote an 8-page essay in 9 hr.

Keri graded 8 essays in 3 hr.

RC1. Jason wrote at a rate of ________ pages/hr.

RC2. Jason wrote at a rate of ________ hr/page.

RC3. Keri graded at a rate of ________ hr/essay.

RC4. Keri graded at a rate of ________ essays/hr.

a) $\frac{8}{3}$

b) $\frac{8}{9}$

c) $\frac{9}{8}$

d) $\frac{3}{8}$

a In Exercises 1–4, find each rate, or speed, as a ratio of distance to time. Round to the nearest hundredth where appropriate.

1. 120 km, 3 hr

2. 18 mi, 9 hr

3. 217 mi, 29 sec

4. 443 m, 48 sec

5. *Chevrolet Malibu LS—City Driving.* A 2013 Chevrolet Malibu LS will travel 312.5 mi on 12.5 gal of gasoline in city driving. What is the rate in miles per gallon?

Source: Chevrolet

6. *Mazda3—City Driving.* A 2013 Mazda3 i SV will travel 348 mi on 14.5 gal of gasoline in city driving. What is the rate in miles per gallon?

Source: mazdausa.com

7. *Mazda3—Highway Driving.* A 2013 Mazda3 i SV will travel 643.5 mi on 19.5 gal of gasoline in highway driving. What is the rate in miles per gallon?

Source: mazdausa.com

8. *Chevrolet Malibu LS—Highway Driving.* A 2013 Chevrolet Malibu LS will travel 499.5 mi on 13.5 gal of gasoline in highway driving. What is the rate in miles per gallon?

Source: Chevrolet

9. *Population Density of Monaco.* Monaco is a tiny country on the Mediterranean coast of France. It has an area of 0.75 sq mi and a population of 35,427 people. What is the rate of number of people per square mile? The rate of people per square mile is called the *population density.* Monaco has the highest population density of any country in the world.

Source: World Bank

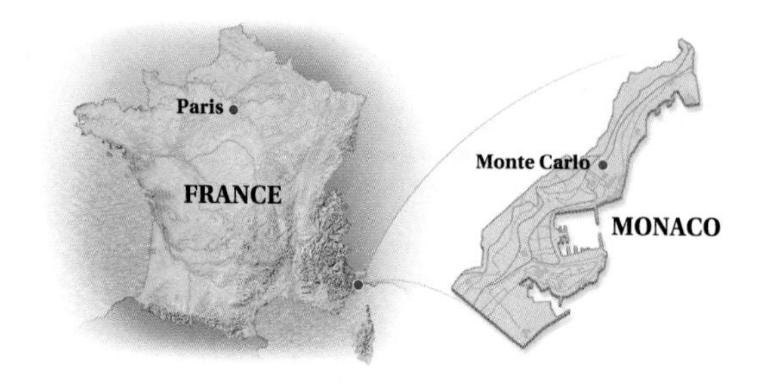

10. *Population Density of Australia.* The continent of Australia, with the island state of Tasmania, has an area of 2,967,893 sq mi and a population of 22,620,600 people. What is the rate of number of people per square mile? The rate of people per square mile is called the *population density.* Australia has one of the lowest population densities in the world.

Source: World Bank

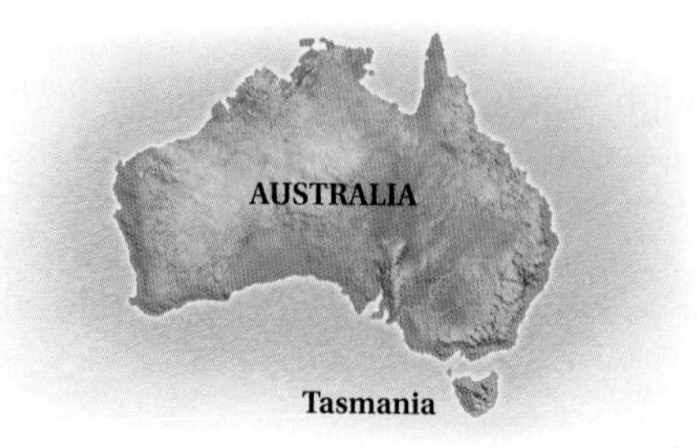

11. A car is driven 500 mi in 20 hr. What is the rate in miles per hour? in hours per mile?

12. A student eats 3 hamburgers in 15 min. What is the rate in hamburgers per minute? in minutes per hamburger?

13. *Broadway Musicals.* In the 17 years from 1987 through 2003, the musical *Les Misérables* was performed on Broadway 6680 times. What was the average rate of performances per year?

Source: broadwaymusicalhome.com

14. *Employment Growth.* In the 10 years from 2010 to 2020, the number of jobs for interpreters and translators is expected to grow by 24,600. What is the expected average rate of growth in jobs per year?

Source: U.S. Bureau of Labor Statistics

15. *Speed of Light.* Light travels 186,000 mi in 1 sec. What is its rate, or speed, in miles per second?

Source: *The Handy Science Answer Book*

16. *Speed of Sound.* Sound travels 1100 ft in 1 sec. What is its rate, or speed, in feet per second?

Source: *The Handy Science Answer Book*

17. Impulses in nerve fibers travel 310 km in 2.5 hr. What is the rate, or speed, in kilometers per hour?

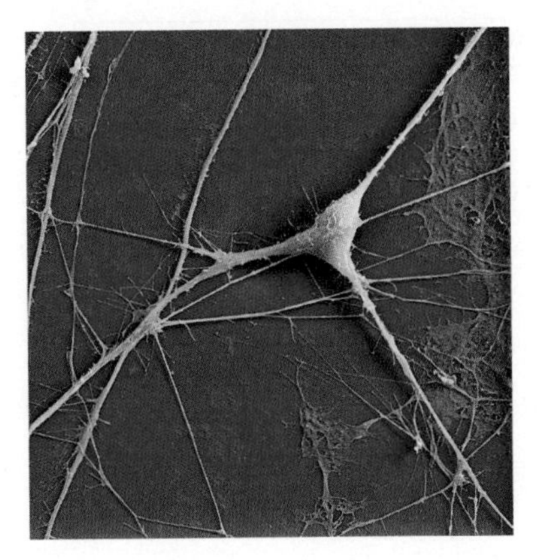

18. A black racer snake can travel 4.6 km in 2 hr. What is its rate, or speed, in kilometers per hour?

19. *Lawn Watering.* Watering a lawn adequately requires 623 gal of water for every 1000 ft^2. What is the rate in gallons per square foot?

20. A car is driven 200 km on 40 L of gasoline. What is the rate in kilometers per liter?

21. *Elephant Heart Rate.* The heart of an elephant, at rest, beats an average of 1500 beats in 60 min. What is the rate in beats per minute?

Source: *The Handy Science Answer Book*

22. *Human Heart Rate.* The heart of a human, at rest, beats an average of 4200 beats in 60 min. What is the rate in beats per minute?

Source: *The Handy Science Answer Book*

b Find each unit price in Exercises 23–30. Then, in each exercise, determine which size is the better buy based on unit price alone.

23. *Hidden Valley Ranch Dressing.*

Size	Price	Unit Price
16 oz	$4.19	
20 oz	$5.29	

24. *Miracle Whip.*

Size	Price	Unit Price
32 oz	$5.29	
48 oz	$8.29	

25. *Cascade Powder Detergent.*

Size	Price	Unit Price
45 oz	$5.90	
75 oz	$8.59	

26. *Bush's Homestyle Baked Beans.*

Size	Price	Unit Price
16 oz	$2.99	
28 oz	$4.51	

27. *Jif Creamy Peanut Butter.*

Size	Price	Unit Price
18 oz	$3.28	
28 oz	$4.89	

28. *Hills Brothers Coffee.*

Size	Price	Unit Price
26 oz	$9.99	
39 oz	$14.99	

29. *Campbell's Condensed Tomato Soup.*

Size	Price	Unit Price
10.7 oz	$1.09	
26 oz	$2.69	

30. *Nabisco Saltines.*

Size	Price	Unit Price
16 oz	$3.59	
32 oz	$5.19	

Find the unit price of each brand in Exercises 31–34. Then, in each exercise, determine which brand is the better buy based on unit price alone.

31. *Vanilla Ice Cream.*

Brand	Size	Price
B	32 oz	\$5.99
E	48 oz	\$6.99

32. *Orange Juice.*

Brand	Size	Price
M	54 oz	\$4.79
T	59 oz	\$5.99

33. *Tomato Ketchup.*

Brand	Size	Price
A	24 oz	\$2.49
B	36 oz	\$3.29
H	46 oz	\$3.69

34. *Yellow Mustard.*

Brand	Size	Price
F	14 oz	\$1.29
G	19 oz	\$1.99
P	20 oz	\$2.49

Skill Maintenance

Multiply. Simplify, if possible.

35. 25×462 [1.4a]

36. 8.4×80.892 [5.3a]

37. 0.01×274.568 [5.3a]

38. $\frac{50}{9} \cdot \left(-\frac{6}{5}\right)$ [3.6a]

39. $3\frac{4}{5} \cdot 2\frac{1}{4}$ [4.6a]

40. $-4\frac{1}{10} \cdot 3\frac{1}{3}$ [4.6a]

Divide. Simplify, if possible.

41. $4000 \div 32$ [1.5a]

42. $95 \div 10$ [5.4a]

43. $-80.892 \div 8.4$ [5.4a]

44. $\frac{50}{9} \div \frac{6}{5}$ [3.7b]

45. $200 \div 1\frac{1}{3}$ [4.6b]

46. $4\frac{6}{7} \div \left(-\frac{1}{4}\right)$ [4.6b]

Synthesis

47. Manufacturers sometimes change the sizes of their containers in such a way that the consumer thinks the price of a product has been lowered when, in reality, a higher unit price is being charged.

Some aluminum juice cans are now concave (curved in) on the bottom. Suppose the volume of the can in the figure has been reduced from 6 fl oz to 5.5 fl oz, and the price of each can has been reduced from 65¢ to 60¢. Find the unit price of each container in cents per ounce.

6.3 Proportions

OBJECTIVES

a Determine whether two pairs of numbers are proportional.

b Solve proportions.

SKILL TO REVIEW

Objective 3.5c: Use the test for equality to determine whether two fractions name the same number.

Use $=$ or $\neq$ for $\square$ to write a true sentence.

1. $\frac{18}{12} \square \frac{9}{6}$ **2.** $\frac{-4}{7} \square \frac{-5}{8}$

Determine whether the two pairs of numbers are proportional.

1. 3, 4 and 6, 8
2. 1, 4 and 10, 39

3. 1, 2 and 20, 39 GS

We compare cross products.

$1 \cdot \square = \square$ $\qquad \frac{1}{2} \stackrel{?}{=} \frac{20}{39}$

$2 \cdot \square = \square$

Since $39 \neq 40$, the numbers ______ proportional.
are/are not

Determine whether the two pairs of numbers are proportional.

4. 6.4, 12.8 and 5.3, 10.6
5. 6.8, 7.4 and 3.4, 4.2

Answers

Skill to Review:
1. $=$ **2.** $\neq$

Margin Exercises:
1. Yes **2.** No **3.** No **4.** Yes **5.** No

Guided Solution:
3. 39, 39, 20, 40; are not

a PROPORTIONS

When two pairs of numbers, such as 3, 2 and 6, 4, have the same ratio, we say that they are **proportional**. The equation

$$\frac{3}{2} = \frac{6}{4}$$

states that the pairs 3, 2 and 6, 4 are proportional. Such an equation is called a **proportion**. We sometimes read $\frac{3}{2} = \frac{6}{4}$ as "3 is to 2 as 6 is to 4."

Since ratios can be written using fraction notation, we can use the test for equality of fractions to determine whether two ratios are the same.

A TEST FOR EQUALITY OF FRACTIONS

Two fractions are equal if their cross products are equal.

EXAMPLE 1 Determine whether 1, 2 and 3, 6 are proportional.

We can use cross products.

$$1 \cdot 6 = 6 \qquad \frac{1}{2} \stackrel{?}{=} \frac{3}{6} \qquad 2 \cdot 3 = 6$$

Since the cross products are the same, $6 = 6$, we know that $\frac{1}{2} = \frac{3}{6}$, so the numbers are proportional. ■

EXAMPLE 2 Determine whether 2, 5 and 4, 7 are proportional.

We can use cross products.

$$2 \cdot 7 = 14 \qquad \frac{2}{5} \stackrel{?}{=} \frac{4}{7} \qquad 5 \cdot 4 = 20$$

Since the cross products are not the same, $14 \neq 20$, we know that $\frac{2}{5} \neq \frac{4}{7}$, so the numbers are not proportional.

◀ **Do Margin Exercises 1–3.**

EXAMPLE 3 Determine whether 3.2, 4.8 and 0.16, 0.24 are proportional.

We can use cross products.

$$3.2 \times 0.24 = 0.768 \qquad \frac{3.2}{4.8} \stackrel{?}{=} \frac{0.16}{0.24} \qquad 4.8 \times 0.16 = 0.768$$

Since the cross products are the same, $0.768 = 0.768$, we know that $\frac{3.2}{4.8} = \frac{0.16}{0.24}$, so the numbers are proportional.

◀ **Do Exercises 4 and 5.**

EXAMPLE 4 Determine whether $4\frac{2}{3}, 5\frac{1}{2}$ and $8\frac{7}{8}, 16\frac{1}{3}$ are proportional.

We can use cross products:

$$\frac{4\frac{2}{3}}{5\frac{1}{2}} \stackrel{?}{=} \frac{8\frac{7}{8}}{16\frac{1}{3}}$$

$$4\frac{2}{3} \cdot 16\frac{1}{3} = \frac{14}{3} \cdot \frac{49}{3} = \frac{686}{9} = 76\frac{2}{9};$$

$$5\frac{1}{2} \cdot 8\frac{7}{8} = \frac{11}{2} \cdot \frac{71}{8} = \frac{781}{16} = 48\frac{13}{16}.$$

Since the cross products are not the same, $76\frac{2}{9} \neq 48\frac{13}{16}$, we know that the numbers are not proportional.

Do Exercise 6. ▶

6. Determine whether $4\frac{2}{3}, 5\frac{1}{2}$ and $14, 16\frac{1}{2}$ are proportional.

b SOLVING PROPORTIONS

One way to solve a proportion is to use cross products. Then we can divide on both sides to get the variable alone.

EXAMPLE 5 Solve the proportion $\frac{x}{3} = \frac{4}{6}$.

$$\frac{x}{3} = \frac{4}{6}$$

$$x \cdot 6 = 3 \cdot 4 \quad \text{Equating cross products (finding cross products and setting them equal)}$$

$$\frac{x \cdot 6}{6} = \frac{3 \cdot 4}{6} \quad \text{Dividing by 6 on both sides}$$

$$x = \frac{3 \cdot 4}{6} = \frac{12}{6} = 2$$

We can check that 2 is the solution by replacing x with 2 and finding cross products:

$$2 \cdot 6 = 12 \qquad \frac{2}{3} \stackrel{?}{=} \frac{4}{6} \qquad 3 \cdot 4 = 12.$$

Since the cross products are the same, it follows that $\frac{2}{3} = \frac{4}{6}$. Thus the pairs of numbers 2, 3 and 4, 6 are proportional, and 2 is the solution of the proportion.

SOLVING PROPORTIONS

To solve $\frac{x}{a} = \frac{c}{d}$ for x, equate *cross products* and divide on both sides to get x alone.

Do Exercise 7. ▶

7. Solve: $\frac{x}{63} = \frac{2}{9}$.

Answers

6. Yes 7. 14

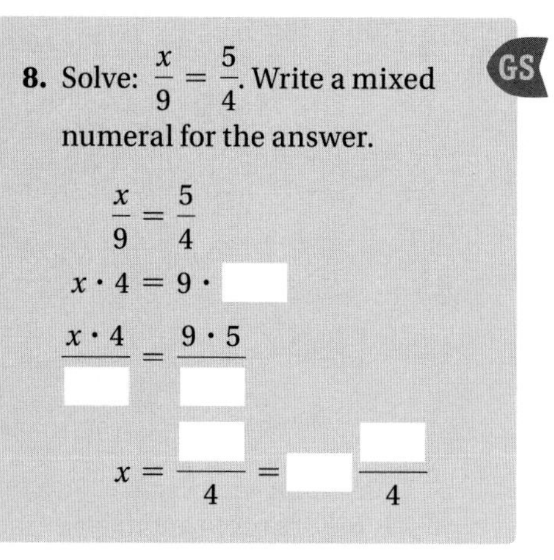

EXAMPLE 6 Solve: $\frac{x}{7} = \frac{5}{3}$. Write a mixed numeral for the answer.

We have

$$\frac{x}{7} = \frac{5}{3}$$

$$x \cdot 3 = 7 \cdot 5 \qquad \text{Equating cross products}$$

$$\frac{x \cdot 3}{3} = \frac{7 \cdot 5}{3} \qquad \text{Dividing by 3}$$

$$x = \frac{7 \cdot 5}{3} = \frac{35}{3}, \text{ or } 11\frac{2}{3}.$$

The solution is $11\frac{2}{3}$.

◀ **Do Exercise 8.**

EXAMPLE 7 Solve: $\frac{7.7}{15.4} = \frac{y}{2.2}$.

We have

$$\frac{7.7}{15.4} = \frac{y}{2.2}$$

$$7.7 \times 2.2 = 15.4 \times y \qquad \text{Equating cross products}$$

$$\frac{7.7 \times 2.2}{15.4} = \frac{15.4 \times y}{15.4} \qquad \text{Dividing by 15.4}$$

$$\frac{7.7 \times 2.2}{15.4} = y$$

$$\frac{16.94}{15.4} = y \qquad \text{Multiplying}$$

$$1.1 = y. \qquad \text{Dividing: } 15.4\overline{)16.94} = 1.1$$

```
         1.1
1 5.4 )1 6.9^4
        1 5 4
          1 5 4
          1 5 4
              0
```

The solution is 1.1.

EXAMPLE 8 Solve: $\frac{8}{x} = \frac{5}{3}$. Write decimal notation for the answer.

We have

$$\frac{8}{x} = \frac{5}{3}$$

$$8 \cdot 3 = x \cdot 5 \qquad \text{Equating cross products}$$

$$\frac{8 \cdot 3}{5} = x \qquad \text{Dividing by 5}$$

$$\frac{24}{5} = x \qquad \text{Multiplying}$$

$$4.8 = x. \qquad \text{Simplifying}$$

The solution is 4.8.

◀ **Do Exercises 9 and 10.**

9. Solve: $\frac{21}{5} = \frac{n}{2.5}$.

10. Solve: $\frac{6}{x} = \frac{25}{11}$. Write decimal notation for the answer.

Answers

8. $11\frac{1}{4}$ **9.** 10.5 **10.** 2.64

Guided Solution:
8. 5, 4, 4, 45, 11, 1

EXAMPLE 9 Solve: $\frac{3.4}{4.93} = \frac{10}{n}$.

We have

$$\frac{3.4}{4.93} = \frac{10}{n}$$

$$3.4 \times n = 4.93 \times 10 \quad \text{Equating cross products}$$

$$\frac{3.4 \times n}{3.4} = \frac{4.93 \times 10}{3.4} \quad \text{Dividing by 3.4}$$

$$n = \frac{4.93 \times 10}{3.4}$$

$$n = \frac{49.3}{3.4} \quad \text{Multiplying}$$

$$n = 14.5. \quad \text{Dividing}$$

The solution is 14.5.

Do Exercise 11. ▶

11. Solve: $\frac{0.4}{0.9} = \frac{4.8}{t}$.

EXAMPLE 10 Solve: $\frac{4\frac{2}{3}}{5\frac{1}{2}} = \frac{14}{x}$.

We have

$$\frac{4\frac{2}{3}}{5\frac{1}{2}} = \frac{14}{x}$$

$$4\frac{2}{3} \cdot x = 5\frac{1}{2} \cdot 14 \quad \text{Equating cross products}$$

$$\frac{14}{3} \cdot x = \frac{11}{2} \cdot 14 \quad \text{Converting to fraction notation}$$

$$\frac{\frac{14}{3} \cdot x}{\frac{14}{3}} = \frac{\frac{11}{2} \cdot 14}{\frac{14}{3}} \quad \text{Dividing by } \frac{14}{3}$$

$$x = \frac{11}{2} \cdot 14 \div \frac{14}{3}$$

$$x = \frac{11}{2} \cdot \cancel{14} \cdot \frac{3}{\cancel{14}} \quad \text{Multiplying by the reciprocal of the divisor}$$

$$x = \frac{11 \cdot 3}{2} \quad \text{Simplifying by removing a factor of 1: } \frac{14}{14} = 1$$

$$x = \frac{33}{2}, \text{ or } 16\frac{1}{2}.$$

The solution is $\frac{33}{2}$, or $16\frac{1}{2}$.

Do Exercise 12. ▶

12. Solve: $\frac{8\frac{1}{3}}{x} = \frac{10\frac{1}{2}}{3\frac{3}{4}}$.

Answers

11. 10.8 **12.** $\frac{125}{42}$, or $2\frac{41}{42}$

6.3 Exercise Set

For Extra Help

MyMathLab® MathXL® PRACTICE WATCH READ REVIEW

Reading Check

Complete each statement with the appropriate word from the following list.

cross products   proportion   proportional   ratio

RC1. The quotient of two quantities is their ________________.

RC2. Two pairs of numbers that have the same ratio are ________________.

RC3. A ________________ states that two pairs of numbers have the same ratio.

RC4. For the equation $\frac{2}{x} = \frac{3}{y}$, the ________________ are $2y$ and $3x$.

a Determine whether the two pairs of numbers are proportional.

1. 5, 6 and 7, 9

2. 7, 5 and 6, 4

3. 1, 2 and 10, 20

4. 7, 3 and 21, 9

5. 2.4, 3.6 and 1.8, 2.7

6. 4.5, 3.8 and 6.7, 5.2

7. $5\frac{1}{3}, 8\frac{1}{4}$ and $2\frac{1}{5}, 9\frac{1}{2}$

8. $2\frac{1}{3}, 3\frac{1}{2}$ and 14, 21

b Solve.

9. $\frac{18}{4} = \frac{x}{10}$

10. $\frac{x}{45} = \frac{20}{25}$

11. $\frac{x}{8} = \frac{9}{6}$

12. $\frac{8}{10} = \frac{n}{5}$

13. $\frac{t}{12} = \frac{5}{6}$

14. $\frac{12}{4} = \frac{x}{3}$

15. $\frac{2}{5} = \frac{8}{n}$

16. $\frac{10}{6} = \frac{5}{x}$

17. $\frac{n}{15} = \frac{10}{30}$

18. $\frac{2}{24} = \frac{x}{36}$

19. $\frac{16}{12} = \frac{24}{x}$

20. $\frac{8}{12} = \frac{20}{x}$

21. $\frac{6}{11} = \frac{12}{x}$

22. $\frac{8}{9} = \frac{32}{n}$

23. $\frac{20}{7} = \frac{80}{x}$

24. $\frac{36}{x} = \frac{9}{5}$

25. $\frac{12}{9} = \frac{x}{7}$

26. $\frac{x}{20} = \frac{16}{15}$

27. $\frac{x}{13} = \frac{2}{9}$

28. $\frac{8}{11} = \frac{x}{5}$

29. $\frac{100}{25} = \frac{20}{n}$

30. $\frac{35}{125} = \frac{7}{m}$

31. $\frac{6}{y} = \frac{18}{15}$

32. $\frac{15}{y} = \frac{3}{4}$

33. $\frac{x}{3} = \frac{0}{9}$

34. $\frac{x}{6} = \frac{1}{6}$

35. $\frac{1}{2} = \frac{7}{x}$

36. $\frac{2}{5} = \frac{12}{x}$

37. $\frac{1.2}{4} = \frac{x}{9}$

38. $\frac{x}{11} = \frac{7.1}{2}$

39. $\frac{8}{2.4} = \frac{6}{y}$

40. $\frac{3}{y} = \frac{5}{4.5}$

41. $\frac{t}{0.16} = \frac{0.15}{0.40}$

42. $\frac{0.12}{0.04} = \frac{t}{0.32}$

43. $\frac{0.5}{n} = \frac{2.5}{3.5}$

44. $\frac{6.3}{0.9} = \frac{0.7}{n}$

45. $\frac{1.28}{3.76} = \frac{4.28}{y}$

46. $\frac{10.4}{12.4} = \frac{6.76}{t}$

47. $\frac{7}{\frac{1}{4}} = \frac{28}{x}$

48. $\frac{5}{\frac{1}{3}} = \frac{3}{x}$

49. $\frac{\frac{1}{5}}{\frac{1}{10}} = \frac{\frac{1}{10}}{x}$

50. $\frac{\frac{1}{4}}{\frac{1}{2}} = \frac{\frac{1}{2}}{x}$

51. $\frac{y}{\frac{3}{5}} = \frac{\frac{7}{12}}{\frac{14}{15}}$

52. $\frac{\frac{5}{8}}{\frac{5}{4}} = \frac{y}{\frac{3}{2}}$

53. $\frac{x}{1\frac{3}{5}} = \frac{2}{15}$

54. $\frac{1}{7} = \frac{x}{4\frac{1}{2}}$

55. $\frac{2\frac{1}{2}}{3\frac{1}{3}} = \frac{x}{4\frac{1}{4}}$

56. $\frac{3\frac{1}{2}}{y} = \frac{6\frac{1}{2}}{4\frac{2}{3}}$

57. $\dfrac{5\frac{1}{5}}{6\frac{1}{6}} = \dfrac{y}{3\frac{1}{2}}$

58. $\dfrac{10\frac{3}{8}}{12\frac{2}{3}} = \dfrac{5\frac{3}{4}}{y}$

Skill Maintenance

59. *Drive-in Movie Theaters.* The number of drive-in movie theaters in the United States has declined steadily since the 1950s. In 2012, there were 365 drive-in theaters. This is 3698 fewer than in 1958. How many drive-in movie theaters were there in 1958? [1.8a]

Source: Drive-Ins.com

60. *Bird Feeders.* After Raena poured a 4-lb bag of birdseed into her new bird feeder, the feeder was $\frac{2}{3}$ full. How much seed does the feeder hold when it is full? [3.7d]

61. Mariah bought $\frac{3}{4}$ lb of cheese at the city market and gave $\frac{1}{2}$ lb of it to Lindsay. How much cheese did Mariah have left? [4.3d]

62. David bought $\frac{3}{4}$ lb of fudge and gave $\frac{1}{2}$ of it to Chris. How much fudge did he give to Chris? [3.6b]

63. Rocky is $187\frac{1}{10}$ cm tall and his daughter is $180\frac{3}{4}$ cm tall. How much taller is Rocky? [4.5c]

64. A serving of fish steak (cross section) is generally $\frac{1}{2}$ lb. How many servings can be prepared from a cleaned $18\frac{3}{4}$-lb tuna? [4.6c]

65. *Expense Needs.* Aaron has $34.97 to spend on a book that costs $49.95, a cap that costs $14.88, and a sweatshirt that costs $29.95. How much more money does Aaron need to make these purchases? [5.7a]

66. *Gas Mileage.* Joanna filled her van's gas tank and noted that the odometer read 42,598.2. After the next fill-up, the odometer read 42,912.1. It took 14.6 gal to fill the tank. How many miles per gallon did the van get? [5.7a]

Synthesis

Solve.

67. $\dfrac{1728}{5643} = \dfrac{836.4}{x}$

68. $\dfrac{328.56}{627.48} = \dfrac{y}{127.66}$

69. *Strikeouts per Home Run.* Baseball Hall-of-Famer Babe Ruth had 1330 strikeouts and 714 home runs in his career. Hall-of-Famer Mike Schmidt had 1883 strikeouts and 548 home runs in his career. Find the rate of strikeouts per home run for each player. (These rates were considered among the highest in the history of the game and yet each player was voted into the Hall of Fame.)

Mid-Chapter Review

Concept Reinforcement

Determine whether each statement is true or false.

______ **1.** A ratio can be written in fraction notation or in colon notation. [6.1a]

______ **2.** A rate is a ratio. [6.2a]

______ **3.** The largest size package of an item always has the lowest unit price. [6.2b]

______ **4.** If $\frac{x}{t} = \frac{y}{s}$, then $xy = ts$. [6.3b]

Guided Solutions

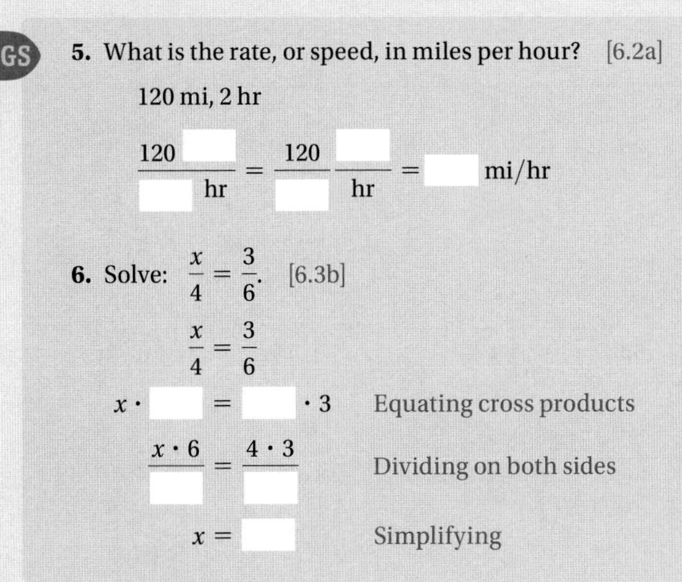

GS **5.** What is the rate, or speed, in miles per hour? [6.2a]

120 mi, 2 hr

$$\frac{120\ \square}{\square\ \text{hr}} = \frac{120}{\square}\ \frac{\square}{\text{hr}} = \square\ \text{mi/hr}$$

6. Solve: $\frac{x}{4} = \frac{3}{6}$. [6.3b]

$$\frac{x}{4} = \frac{3}{6}$$

$x \cdot \square = \square \cdot 3$ Equating cross products

$\frac{x \cdot 6}{\square} = \frac{4 \cdot 3}{\square}$ Dividing on both sides

$x = \square$ Simplifying

Mixed Review

Find fraction notation for each ratio. [6.1a]

7. 4 to 7 **8.** 313 to 199 **9.** 35 to 17 **10.** 59 to 101

Find the ratio of the first number to the second and simplify. [6.1b]

11. 8 to 12 **12.** 25 to 75 **13.** 32 to 28 **14.** 100 to 76

15. 112 to 56 **16.** 15 to 3 **17.** 2.4 to 8.4 **18.** 0.27 to 0.45

Find each rate, or speed, as a ratio of distance to time. Round to the nearest hundredth where appropriate. [6.2a]

19. 243 mi, 4 hr

20. 146 km, 3 hr

21. 65 m, 5 sec

22. 97 ft, 6 sec

23. *Record Snowfall.* The greatest recorded snowfall in a single storm occurred in the Mt. Shasta Ski Bowl in California, when 189 in. fell during a seven-day storm in 1959. What is the rate in inches per day? [6.2a]

Source: U.S. Army Corps of Engineers

24. *Free Throws.* During the 2011–2012 basketball season, Kevin Durant of the Oklahoma City Thunder attempted 501 free throws and made 431 of them. What is the rate in number of free throws made to number of free throws attempted? Round to the nearest thousandth. [6.2a]

Source: National Basketball Association

25. Jerome bought an 18-oz jar of grape jelly for $2.09. What is the unit price in cents per ounce? [6.2b]

26. Martha bought 12 oz of deli honey ham for $5.99. What is the unit price in cents per ounce? [6.2b]

Determine whether the two pairs of numbers are proportional. [6.3a]

27. 3, 7 and 15, 35

28. 9, 7 and 7, 5

29. 2.4, 1.5 and 3.2, 2.1

30. $1\frac{3}{4}, 1\frac{1}{3}$ and $8\frac{3}{4}, 6\frac{2}{3}$

Solve. [6.3b]

31. $\frac{9}{15} = \frac{x}{20}$

32. $\frac{x}{24} = \frac{30}{18}$

33. $\frac{12}{y} = \frac{20}{15}$

34. $\frac{2}{7} = \frac{10}{y}$

35. $\frac{y}{1.2} = \frac{1.1}{0.6}$

36. $\frac{0.24}{0.02} = \frac{y}{0.36}$

37. $\frac{\frac{1}{4}}{x} = \frac{\frac{1}{8}}{\frac{1}{4}}$

38. $\frac{1\frac{1}{2}}{3\frac{1}{4}} = \frac{7\frac{1}{2}}{x}$

Understanding Through Discussion and Writing

39. Can every ratio be written as the ratio of some number to 1? Why or why not? [6.1a]

40. What can be concluded about a rectangle's width if the ratio of length to perimeter is 1 to 3? Make some sketches and explain your reasoning. [6.1a, b]

41. Instead of equating cross products, a student solves $\frac{x}{7} = \frac{5}{3}$ by multiplying on both sides by the least common denominator, 21. Is his approach a good one? Why or why not? [6.3b]

42. An instructor predicts that a student's test grade will be proportional to the amount of time the student spends studying. What is meant by this? Write an example of a proportion that represents the grades of two students and their study times. [6.3b]

STUDYING FOR SUCCESS *Applications*

- ☐ If you find applied problems challenging, don't give up! Your skill will improve with each problem you solve.
- ☐ Make applications real! Look for applications of the math you are studying in newspapers and magazines.

Applications of Proportions

6.4

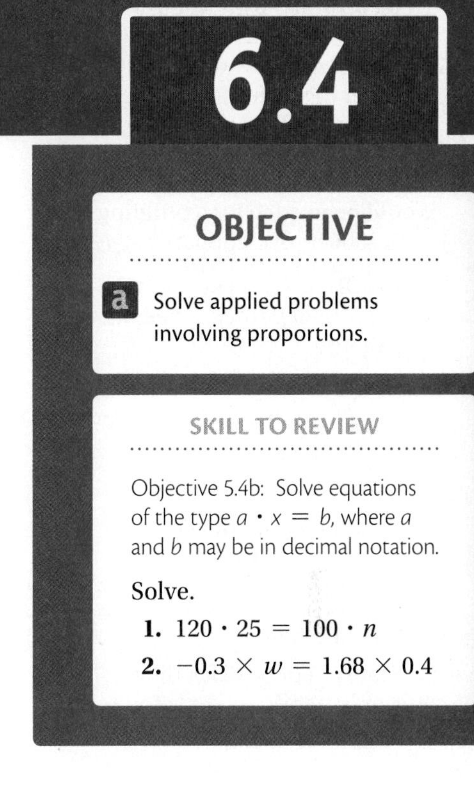

OBJECTIVE

a Solve applied problems involving proportions.

SKILL TO REVIEW

Objective 5.4b: Solve equations of the type $a \cdot x = b$, where a and b may be in decimal notation.

Solve.

1. $120 \cdot 25 = 100 \cdot n$
2. $-0.3 \times w = 1.68 \times 0.4$

a APPLICATIONS AND PROBLEM SOLVING

Proportions have applications in such diverse fields as business, chemistry, health sciences, and home economics, as well as in many areas of daily life. Proportions are useful in making predictions.

EXAMPLE 1 *Predicting Total Distance.* Donna drives her delivery van 800 mi in 3 days. At this rate, how far will she drive in 15 days?

1. **Familiarize.** We let $d =$ the distance traveled in 15 days.
2. **Translate.** We translate to a proportion. We make each side the ratio of distance to time, with distance in the numerator and time in the denominator.

$$\text{Distance in 15 days} \rightarrow \frac{d}{15} = \frac{800}{3} \leftarrow \text{Distance in 3 days}$$
$$\text{Time} \rightarrow 15 \qquad 3 \leftarrow \text{Time}$$

It may help to verbalize the proportion above as "the unknown distance d is to 15 days as the known distance 800 mi is to 3 days."

3. **Solve.** Next, we solve the proportion:

$$d \cdot 3 = 15 \cdot 800 \qquad \text{Equating cross products}$$
$$\frac{d \cdot 3}{3} = \frac{15 \cdot 800}{3} \qquad \text{Dividing by 3 on both sides}$$
$$d = \frac{15 \cdot 800}{3}$$
$$d = 4000. \qquad \text{Multiplying and dividing}$$

4. **Check.** We substitute into the proportion and check cross products:

$$\frac{4000}{15} = \frac{800}{3};$$
$$4000 \cdot 3 = 12{,}000; \qquad 15 \cdot 800 = 12{,}000.$$

The cross products are the same.

5. **State.** Donna will drive 4000 mi in 15 days.

Do Margin Exercise 1. ▶

1. *Burning Calories.* The readout on Mary's treadmill indicates that she burns 108 calories when she walks for 24 min. How many calories will she burn if she walks at the same rate for 30 min?

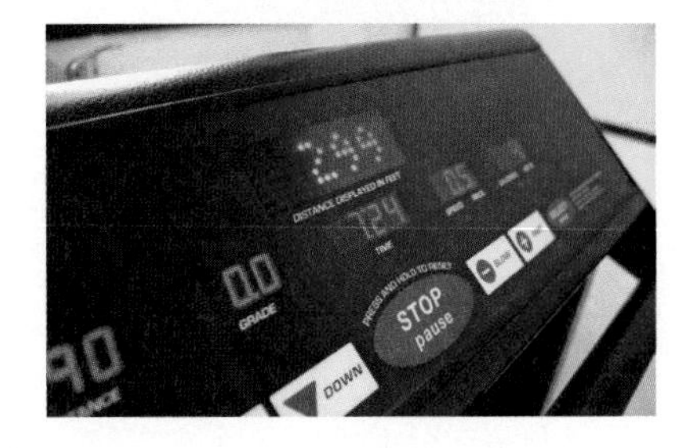

Problems involving proportions can be translated in more than one way. For Example 1, any one of the following is also a correct translation:

$$\frac{15}{d} = \frac{3}{800}, \quad \frac{15}{3} = \frac{d}{800}, \quad \frac{800}{d} = \frac{3}{15}.$$

Equating the cross products in each proportion gives us the equation $d \cdot 3 = 15 \cdot 800$, which is the equation we obtained in Example 1.

Answers

Skill to Review:
1. 30 2. −2.24

Margin Exercise:
1. 135 calories

EXAMPLE 2 *Recommended Dosage.* To control a fever, a doctor suggests that a child who weighs 28 kg be given 320 mg of a liquid pain reliever. If the dosage is proportional to the child's weight, how much of the medication is recommended for a child who weighs 35 kg?

1. **Familiarize.** We let $t =$ the number of milligrams of the liquid pain reliever recommended for a child who weighs 35 kg.
2. **Translate.** We translate to a proportion, keeping the amount of medication in the numerators.

$$\begin{array}{r} \text{Medication suggested} \rightarrow \\ \text{Child's weight} \rightarrow \end{array} \frac{320}{28} = \frac{t}{35} \begin{array}{l} \leftarrow \text{Medication suggested} \\ \leftarrow \text{Child's weight} \end{array}$$

3. **Solve.** Next, we solve the proportion:

$$320 \cdot 35 = 28 \cdot t \qquad \text{Equating cross products}$$

$$\frac{320 \cdot 35}{28} = \frac{28 \cdot t}{28} \qquad \text{Dividing by 28 on both sides}$$

$$\frac{320 \cdot 35}{28} = t$$

$$400 = t. \qquad \text{Multiplying and dividing}$$

4. **Check.** We substitute into the proportion and check cross products:

$$\frac{320}{28} = \frac{400}{35};$$

$$320 \cdot 35 = 11{,}200; \qquad 28 \cdot 400 = 11{,}200.$$

The cross products are the same.

5. **State.** The dosage for a child who weighs 35 kg is 400 mg.

◀ Do Exercise 2.

2. **Determining Paint Needs.** Lowell and Chris run a painting company during the summer to pay for their college expenses. They can paint 1600 ft^2 of clapboard with 4 gal of paint. How much paint would be needed for a building with 6000 ft^2 of clapboard? GS

   1. **Familiarize.** Let $p =$ the amount of paint needed, in gallons.
   2. **Translate.**

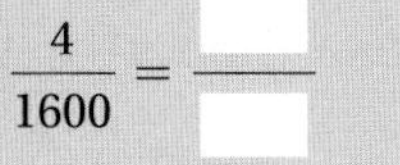

$$\frac{4}{1600} = \frac{\square}{\square}$$

   3. **Solve.**

$$4 \cdot 6000 = 1600 \cdot \square$$
$$\square = p$$

   4. **Check.** The cross products are the same.
   5. **State.** For 6000 ft^2, they would need ____ gal of paint.

Answer

2. 15 gal

Guided Solution:

2. $\frac{p}{6000}$; p, 15; 15

EXAMPLE 3 *Purchasing Tickets.* Carey bought 8 tickets to an international food festival for \$52. How many tickets could she purchase with \$90?

1. **Familiarize.** We let $n =$ the number of tickets that can be purchased with \$90.
2. **Translate.** We translate to a proportion, keeping the number of tickets in the numerators.

$$\begin{array}{r} \text{Tickets} \rightarrow \\ \text{Cost} \rightarrow \end{array} \frac{8}{52} = \frac{n}{90} \begin{array}{l} \leftarrow \text{Tickets} \\ \leftarrow \text{Cost} \end{array}$$

3. **Solve.** Next, we solve the proportion:

$$52 \cdot n = 8 \cdot 90 \quad \text{Equating cross products}$$

$$\frac{52 \cdot n}{52} = \frac{8 \cdot 90}{52} \quad \text{Dividing by 52 on both sides}$$

$$n = \frac{8 \cdot 90}{52}$$

$$n \approx 13.8. \quad \text{Multiplying and dividing}$$

Because it is impossible to buy a fractional part of a ticket, we must round our answer *down* to 13.

4. **Check.** As a check, we use a different approach: We find the cost per ticket and then divide \$90 by that price. Since $52 \div 8 = 6.50$ and $90 \div 6.50 \approx 13.8$, we have a check.
5. **State.** Carey could purchase 13 tickets with \$90.

Do Exercise 3. ▶

3. *Purchasing Shirts.* If 2 shirts can be bought for \$47, how many shirts can be bought with \$200?

EXAMPLE 4 *Waist-to-Hip Ratio.* To reduce the risk of heart disease, it is recommended that a man's waist-to-hip ratio be 0.9 or lower. Mac's hip measurement is 40 in. To meet the recommendation, what should his waist measurement be?

Source: Mayo Clinic

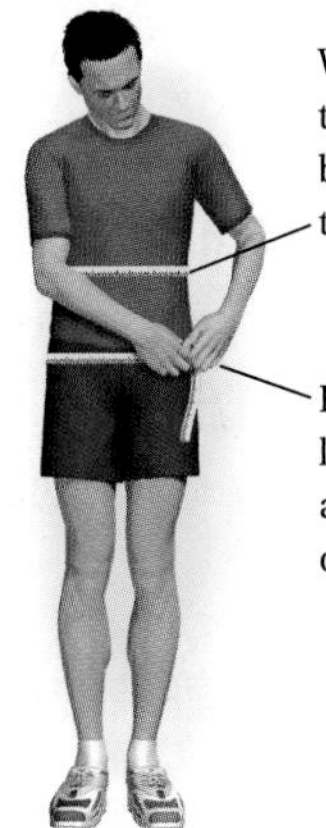

1. **Familiarize.** Note that $0.9 = \frac{9}{10}$. We let w = Mac's waist measurement.
2. **Translate.** We translate to a proportion as follows.

$$\begin{array}{r} \text{Waist measurement} \rightarrow \\ \text{Hip measurement} \rightarrow \end{array} \frac{w}{40} = \frac{9}{10} \begin{array}{l} \text{Recommended} \\ \text{waist-to-hip ratio} \end{array}$$

3. **Solve.** Next, we solve the proportion:

$$w \cdot 10 = 40 \cdot 9 \quad \text{Equating cross products}$$

$$\frac{w \cdot 10}{10} = \frac{40 \cdot 9}{10} \quad \text{Dividing by 10 on both sides}$$

$$w = \frac{40 \cdot 9}{10}$$

$$w = 36. \quad \text{Multiplying and dividing}$$

Answer

3. 8 shirts

4. *Waist-to-Hip Ratio.* It is recommended that a woman's waist-to-hip ratio be 0.85 or lower. Martina's hip measurement is 40 in. To meet the recommendation, what should her waist measurement be?
Source: Mayo Clinic

4. **Check.** As a check, we divide 36 by 40: $36 \div 40 = 0.9$. This is the desired ratio.
5. **State.** Mac's recommended waist measurement is 36 in. or less.

Do Exercise 4.

EXAMPLE 5 *Construction Plans.* Architects make blueprints for construction projects. These are scale drawings in which lengths are in proportion to actual sizes. The Hennesseys are adding a rectangular deck to their house. The architect's blueprint is rendered such that $\frac{3}{4}$ in. on the drawing is actually 2.25 ft on the deck. The width of the deck on the drawing is 4.3 in. How wide is the deck in reality?

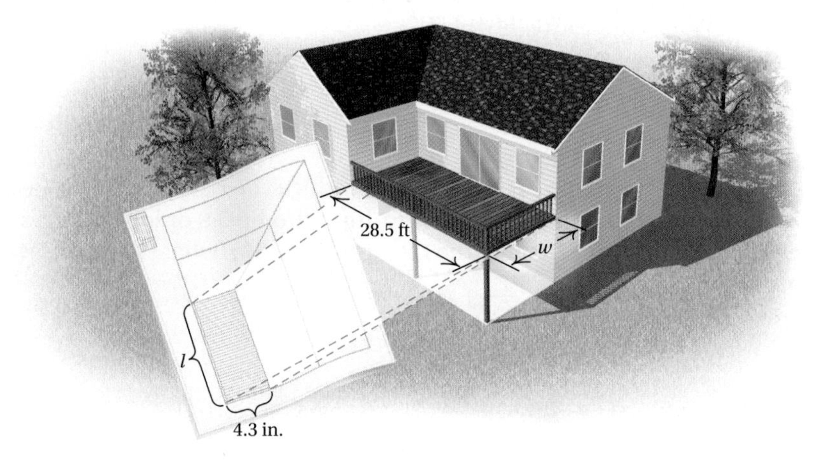

1. **Familiarize.** We let $w =$ the width of the deck.
2. **Translate.** Then we translate to a proportion, using 0.75 for $\frac{3}{4}$ in.

$$\begin{matrix}\text{Measure on drawing} \rightarrow \\ \text{Measure on deck} \rightarrow\end{matrix} \frac{0.75}{2.25} = \frac{4.3}{w} \begin{matrix}\leftarrow \text{Width on drawing} \\ \leftarrow \text{Width on deck}\end{matrix}$$

3. **Solve.** Next, we solve the proportion:

$$0.75 \times w = 2.25 \times 4.3 \qquad \text{Equating cross products}$$

$$\frac{0.75 \times w}{0.75} = \frac{2.25 \times 4.3}{0.75} \qquad \text{Dividing by 0.75 on both sides}$$

$$w = \frac{2.25 \times 4.3}{0.75}$$

$$w = 12.9.$$

4. **Check.** We substitute into the proportion and check cross products:

$$\frac{0.75}{2.25} = \frac{4.3}{12.9};$$

$$0.75 \times 12.9 = 9.675; \qquad 2.25 \times 4.3 = 9.675.$$

The cross products are the same.

5. **State.** The width of the deck is 12.9 ft.

Do Exercise 5.

5. *Construction Plans.* In Example 5, the length of the actual deck is 28.5 ft. What is the length of the deck on the blueprint?

Answers
4. 34 in. or less 5. 9.5 in.

EXAMPLE 6 *Estimating a Wildlife Population.* Scientists often use proportions to estimate the size of a wildlife population. They begin by collecting and marking, or tagging, a portion of the population. This tagged sample is released and mingles with the entire population. At a later date, the scientists collect a second sample from the population. The proportion of tagged individuals in the second sample is estimated to be the same as the proportion of tagged individuals in the entire population.

The marking can be done by using actual tags or by identifying individuals in other ways. For example, marine biologists can identify an individual whale by the patterns on its tail. Recently, scientists have begun using DNA to identify individuals in populations. For example, to identify individual bears in the grizzly bear population of the Northern Continental Divide ecosystem in Montana, geneticists use DNA from fur samples left on branches near the bears' feeding areas.

In one recent large-scale study in this ecosystem, biologists identified 545 individual grizzly bears. If later a sample of 30 bears contained 25 of the previously identified individuals, estimate the total number of bears in the ecosystem.

Source: Based on information from the Northern Divide Grizzly Bear Project

1. **Familiarize.** We let $B =$ the total number of bears in the ecosystem. We assume that the ratio of the number of identified bears to the total number of bears in the ecosystem is the same as the ratio of the number of identified bears in the later sample to the total number of bears in the later sample.
2. **Translate.** We translate to a proportion as follows.

$$\begin{array}{r} \text{Identified bears} \rightarrow \\ \text{Total number of bears} \rightarrow \end{array} \frac{545}{B} = \frac{25}{30} \begin{array}{l} \leftarrow \text{Identified bears in sample} \\ \leftarrow \text{Number of bears in sample} \end{array}$$

3. **Solve.** Next, we solve the proportion:

$$545 \cdot 30 = B \cdot 25 \quad \text{Equating cross products}$$

$$\frac{545 \cdot 30}{25} = \frac{B \cdot 25}{25} \quad \text{Dividing by 25 on both sides}$$

$$\frac{545 \cdot 30}{25} = B$$

$$654 = B. \quad \text{Multiplying and dividing}$$

4. **Check.** We substitute into the proportion and check cross products:

$$\frac{545}{654} = \frac{25}{30};$$

$$545 \cdot 30 = 16{,}350; \qquad 654 \cdot 25 = 16{,}350.$$

The cross products are the same.

5. **State.** We estimate that there are 654 bears in the ecosystem.

Do Exercise 6. ▶

GS **6. Estimating a Deer Population.** To determine the number of deer in a forest, a conservationist catches 153 deer, tags them, and releases them. Later, 62 deer are caught, and it is found that 18 of them are tagged. Estimate how many deer are in the forest.

1. **Familiarize.** Let $D =$ the number of deer in the forest.
2. **Translate.**

$$\frac{153}{D} = \frac{\square}{\square}$$

3. **Solve.**

$$153 \cdot 62 = D \cdot \square$$

$$\square = D$$

4. **Check.** The cross products are the same.
5. **State.** There are about $\square$ deer in the forest.

Answer

6. 527 deer

Guided Solution:

6. $\frac{18}{62}$; 18, 527; 527

Translating for Success

1. *Calories in Cereal.* There are 140 calories in a $1\frac{1}{2}$-cup serving of Brand A cereal. How many calories are there in 6 cups of the cereal?

2. *Calories in Cereal.* There are 140 calories in 6 cups of Brand B cereal. How many calories are there in a $1\frac{1}{2}$-cup serving of the cereal?

3. *Gallons of Gasoline.* Jared's SUV traveled 310 mi on 15.5 gal of gasoline. At this rate, how many gallons would be needed to travel 465 mi?

4. *Gallons of Gasoline.* Elizabeth's fuel-efficient car traveled 465 mi on 15.5 gal of gasoline. At this rate, how many gallons will be needed to travel 310 mi?

5. *Perimeter.* What is the perimeter of a rectangular field that measures 83.7 m by 62.4 m?

The goal of these matching questions is to practice step (2), Translate, of the five-step problem-solving process. Translate each word problem to an equation and select a correct translation from equations A–O.

A. $\dfrac{310}{15.5} = \dfrac{465}{x}$

B. $180 = 1\frac{1}{2} \cdot x$

C. $x = 71\frac{1}{8} - 76\frac{1}{2}$

D. $71\frac{1}{8} \cdot x = 74$

E. $74 \cdot 71\frac{1}{8} = x$

F. $x = 83.7 + 62.4$

G. $71\frac{1}{8} + x = 76\frac{1}{2}$

H. $x = 1\frac{2}{3} \cdot 180$

I. $\dfrac{140}{6} = \dfrac{x}{1\frac{1}{2}}$

J. $x = 2(83.7 + 62.4)$

K. $\dfrac{465}{15.5} = \dfrac{310}{x}$

L. $x = 83.7 \cdot 62.4$

M. $x = 180 \div 1\frac{2}{3}$

N. $\dfrac{140}{1\frac{1}{2}} = \dfrac{x}{6}$

O. $x = 1\frac{2}{3} \div 180$

Answers on page A-10

6. *Electric Bill.* Last month Todd's electric bills for his two rental properties were \$83.70 and \$62.40. What was the total electric bill for the two properties?

7. *Package Tape.* A postal service center uses rolls of package tape that each contain 180 ft of tape. If it takes an average of $1\frac{2}{3}$ ft of tape per package, how many packages can be taped with one roll?

8. *Online Price.* Jane spent \$180 for an area rug in a department store. Later, she saw the same rug for sale online and realized she had paid $1\frac{1}{2}$ times the online price. What was the online price?

9. *Heights of Sons.* Henry's three sons play basketball on three different college teams. Jeff's, Jason's, and Jared's heights are 74 in., $71\frac{1}{8}$ in., and $76\frac{1}{2}$ in., respectively. How much taller is Jared than Jason?

10. *Area of a Lot.* Bradley bought a lot that measured 74 yd by $71\frac{1}{8}$ yd. What was the area of the lot?

6.4 Exercise Set

For Extra Help MyMathLab® MathXL® PRACTICE WATCH READ REVIEW

Reading Check

Complete each proportion to form a correct translation of the following problem.

Christy ran 4 marathons in the first 6 months of the year. At this rate, how many marathons will she run in the first 8 months of the year? Let $m =$ the number of marathons she will run in 8 months.

RC1. $\frac{4}{6} = \frac{\square}{\square}$

RC2. $\frac{6}{4} = \frac{\square}{\square}$

RC3. $\frac{4}{m} = \frac{\square}{\square}$

RC4. $\frac{m}{4} = \frac{\square}{\square}$

a Solve.

1. *Study Time and Test Grades.* An English instructor asserted that students' test grades are directly proportional to the amount of time spent studying. Lisa studies for 9 hr for a particular test and gets a score of 75. At this rate, for how many hours would she have had to study to get a score of 92?

2. *Study Time and Test Grades.* A mathematics instructor asserted that students' test grades are directly proportional to the amount of time spent studying. Brent studies for 15 hr for a final exam and gets a score of 75. At this rate, what score would he have received if he had studied for 18 hr?

3. *Movies.* If *The Hobbit: An Unexpected Journey* is played at the rate preferred by director Peter Jackson, a moviegoer sees 600 frames in $12\frac{1}{2}$ sec. How many frames does a moviegoer see in 160 sec?

4. *Sugaring.* When 20 gal of maple sap are boiled down, the result is $\frac{1}{2}$ gal of maple syrup. How much sap is needed to produce 9 gal of syrup?

Source: University of Maine

5. *Gas Mileage.* A 2013 Ford Mustang GT will travel 403 mi on 15.5 gal of gasoline in highway driving.

a) How many gallons of gasoline will it take to drive 2690 mi from Boston to Phoenix?

b) How far can the car be driven on 140 gal of gasoline?

Source: Ford Motor Company

6. *Gas Mileage.* A 2013 Volkswagen Beetle will travel 495 mi on 16.5 gal of premium gasoline in highway driving.

a) How many gallons of gasoline will it take to drive 1650 mi from Pittsburgh to Albuquerque?

b) How far can the car be driven on 130 gal of gasoline?

Source: Volkswagen of America, Inc.

7. *Overweight Americans.* A recent study determined that of every 100 American adults, 69 are overweight or obese. It is estimated that the U.S. population will be about 337 million in 2020. At the given rate, how many Americans will be considered overweight or obese in 2020?

Source: U.S. Centers for Disease Control and Prevention

8. *Prevalence of Diabetes.* A recent study determined that of every 1000 Americans age 65 or older, 269 have been diagnosed with diabetes. It is estimated that there will be about 55 million Americans in this age group in 2020. At the given rate, how many in this age group will be diagnosed with diabetes in 2020?

Source: American Diabetes Association

9. *Quality Control.* A quality-control inspector examined 100 lightbulbs and found 7 of them to be defective. At this rate, how many defective bulbs will there be in a lot of 2500?

10. *Grading.* A high school math teacher must grade 32 reports on famous mathematicians. She can grade 4 reports in 90 min. At this rate, how long will it take her to grade all 32 reports?

11. *Recommended Dosage.* To control an infection, Dr. Okeke prescribes a dosage of 200 mg of Rocephin every 8 hr for an infant weighing 15.4 lb. At this rate, what would the dosage be for an infant weighing 20.2 lb?

12. *Metallurgy.* In Ethan's white gold ring, the ratio of nickel to gold is 3 to 13. If the ring contains 4.16 oz of gold, how much nickel does it contain?

13. *Painting.* Fred uses 3 gal of paint to cover 1275 ft^2 of siding. How much siding can Fred paint with 7 gal of paint?

14. *Waterproofing.* Bonnie can waterproof 450 ft^2 of decking with 2 gal of sealant. How many gallons of the sealant should Bonnie buy for a 1200-ft^2 deck?

15. *Publishing.* Every 6 pages of an author's manuscript correspond to 5 published pages. How many published pages will a 540-page manuscript become?

16. *Turkey Servings.* An 8-lb turkey breast contains 36 servings of meat. How many pounds of turkey breast would be needed for 54 servings?

17. *Mileage.* Jean bought a new car. In the first 8 months, it was driven 9000 mi. At this rate, how many miles will the car be driven in 1 year?

18. *Coffee Production.* Coffee beans from 18 trees are required to produce enough coffee each year for a person who drinks 2 cups of coffee per day. Jared brews 15 cups of coffee each day for himself and his coworkers. How many coffee trees are required for this?

19. *Cap'n Crunch's Peanut Butter Crunch® Cereal.* The nutritional chart on the side of a box of Quaker Cap'n Crunch's Peanut Butter Crunch® cereal states that there are 110 calories in a $\frac{3}{4}$-cup serving. How many calories are there in 6 cups of the cereal?

Nutrition Facts

Serving Size 3/4 Cup (27g)

Amount Per Serving	Cereal Alone	with 1/2 Cup Vitamin A&D Fortified Skim Milk
Calories	110	150
Calories from Fat	25	25
	% Daily Value	
Total Fat 2.5g	4%	4%
Saturated Fat 1g	5%	6%
Trans Fat 0g		
Polyunsaturated Fat 0.5g		
Monounsaturated Fat 1g		
Cholesterol 0mg	0%	1%
Sodium 200mg	8%	10%
Potassium 65mg	2%	7%
Total Carbohydrate 21g	7%	9%
Dietary Fiber 1g	3%	3%
Sugars 9g		
Other Carbohydrate 11g		
Protein 2g		

20. *Rice Krispies® Cereal.* The nutritional chart on the side of a box of Kellogg's Rice Krispies® cereal states that there are 130 calories in a $1\frac{1}{4}$-cup serving. How many calories are there in 5 cups of the cereal?

Nutrition Facts

Serving Size $1\frac{1}{4}$ Cups (33g/1.2oz)

Amount Per Serving	Cereal	Cereal with 1/2 Cup Vitamins A&D Fat Free Milk
Calories	130	170
Calories from Fat	0	0
	% Daily Value	
Total Fat 0g	0%	0%
Saturated Fat 1g	0%	0%
Trans Fat 0g		
Cholesterol 0mg	0%	0%
Sodium 220mg	9%	12%
Potassium 30mg	1%	7%
Total Carbohydrate 29g	10%	11%
Dietary Fiber 0g	0%	0%
Sugars 4g		
Other Carbohydrate 25g		
Protein 2g		

21. *Lefties.* In a class of 40 students, on average, 6 will be left-handed. If a class includes 9 "lefties," how many students would you estimate are in the class?

22. *Class Size.* A college advertises that its student-to-faculty ratio is 27 to 2. If 81 students register for Introductory Spanish, how many sections of the course would you expect to see offered?

23. *Painting.* Helen can paint 950 ft^2 with 2 gal of paint. How many 1-gal cans does she need in order to paint a 30,000-ft^2 wall?

24. *Snow to Water.* Under typical conditions, $1\frac{1}{2}$ ft of snow will melt to 2 in. of water. How many inches of water will result when $5\frac{1}{2}$ ft of snow melts?

25. *Gasoline Mileage.* Nancy's van traveled 84 mi on 6.5 gal of gasoline. At this rate, how many gallons would be needed to travel 126 mi?

26. *Bicycling.* Roy bicycled 234 mi in 14 days. At this rate, how far would Roy bicycle in 42 days?

27. *Grass-Seed Coverage.* It takes 60 oz of grass seed to seed 3000 ft^2 of lawn. At this rate, how much would be needed for 5000 ft^2 of lawn?

28. *Grass-Seed Coverage.* In Exercise 27, how much seed would be needed for 7000 ft^2 of lawn?

29. *Estimating a Whale Population.* To determine the number of humpback whales in a population, a marine biologist, using tail markings, identifies 27 individual whales. Several weeks later, 40 whales from the population are randomly sighted. Of the 40 sighted, 12 are among the 27 originally identified. Estimate the number of whales in the population.

30. *Estimating a Trout Population.* To determine the number of trout in a lake, a conservationist catches 112 trout, tags them, and throws them back into the lake. Later, 82 trout are caught, and it is found that 32 of them are tagged. Estimate how many trout there are in the lake.

31. *Map Scaling.* On a road atlas map, 1 in. represents 16.6 mi. If two cities are 3.5 in. apart on the map, how far apart are they in reality?

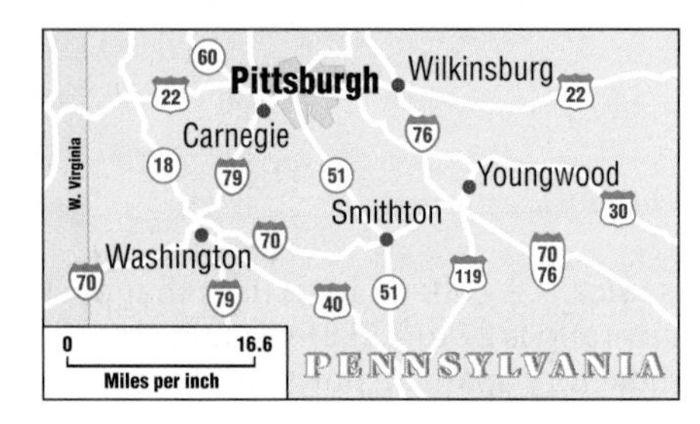

32. *Map Scaling.* On a map, $\frac{1}{4}$ in. represents 50 mi. If two cities are $3\frac{1}{4}$ in. apart on the map, how far apart are they in reality?

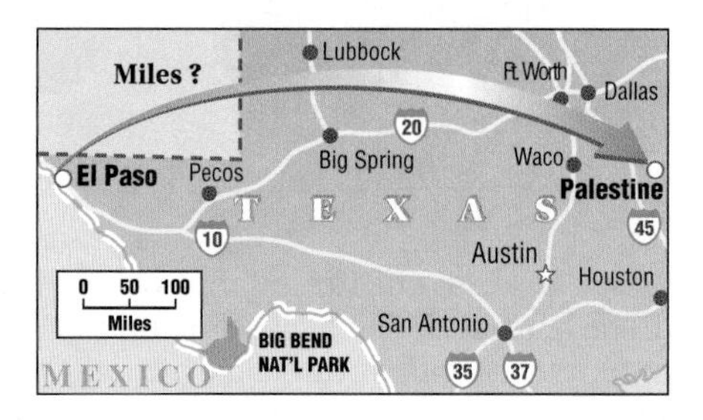

33. *Currency Exchange.* On January 7, 2013, 1 U.S. dollar was worth about 0.76 euro.

a) How much were 50 U.S. dollars worth in euros on that day?
b) How much would a car that costs 8640 euros cost in U.S. dollars?

34. *Currency Exchange.* On January 7, 2013, 1 U.S. dollar was worth about 12.74 Mexican pesos.

a) How much were 150 U.S. dollars worth in Mexican pesos on that day?
b) While traveling in Mexico at that time, Jake bought a watch that cost 3600 Mexican pesos. How much did it cost in U.S. dollars?

35. *Basketball.* After playing 32 games in the 2012–2013 National Basketball Association season, Kobe Bryant of the Los Angeles Lakers had scored 977 points.

a) At this rate, how many games would it take him to score 2000 points?

b) At this rate, how many points would Bryant score in the whole 82-game season?

Source: National Basketball Association

36. *Home Runs.* After playing 118 games in the 2012 Major League Baseball season, Andrew McCutchen of the Pittsburgh Pirates had 24 home runs.

a) At this rate, how many games would it take him to hit 30 home runs?

b) At this rate, how many home runs would McCutchen hit in the entire 162-game season?

Source: Major League Baseball

Skill Maintenance

Solve.

37. $12 \cdot x = 1944$ [1.7b]

38. $6807 = m + 2793$ [1.7b]

39. $t + 4.25 = 8.7$ [5.2d]

40. $-112.5 \cdot p = 45$ [5.4b]

41. $3.7 + y = -18$ [5.2d]

42. $0.078 = 0.3 \cdot t$ [5.4b]

43. $c + \frac{4}{5} = \frac{9}{10}$ [4.3c]

44. $-\frac{5}{6} = \frac{2}{3} \cdot x$ [3.7c]

Synthesis

45. Carney College is expanding from 850 to 1050 students. To avoid any rise in the student-to-faculty ratio, the faculty of 69 professors must also increase. How many new faculty positions should be created?

46. In recognition of her outstanding work, Sheri's salary has been increased from \$26,000 to \$29,380. Tim is earning \$23,000 and is requesting a proportional raise. How much more should he ask for?

47. *Baseball Statistics.* Cy Young, one of the greatest baseball pitchers of all time, gave up an average of 2.63 earned runs every 9 innings. Young pitched 7356 innings, more than anyone else in the history of baseball. How many earned runs did he give up?

48. *Real-Estate Values.* According to Coldwell Banker Real Estate Corporation, a home selling for \$189,000 in Austin, Texas, would sell for \$437,850 in Denver, Colorado. How much would a \$350,000 home in Denver sell for in Austin? Round to the nearest \$1000.

Source: Coldwell Banker Real Estate Corporation

49. The ratio 1 : 3 : 2 is used to estimate the relative costs of a CD player, receiver, and speakers when shopping for a sound system. That is, the receiver should cost three times the amount spent on the CD player and the speakers should cost twice the amount spent on the CD player. If you had \$900 to spend, how would you allocate the money, using this ratio?

6.5 Geometric Applications

OBJECTIVES

a Find lengths of sides of similar triangles using proportions.

b Use proportions to find lengths in pairs of figures that differ only in size.

SKILL TO REVIEW

Objective 6.3b: Solve proportions.

Solve.

1. $\frac{7}{x} = \frac{8}{3}$ 2. $\frac{2}{1\frac{1}{2}} = \frac{p}{\frac{1}{4}}$

a PROPORTIONS AND SIMILAR TRIANGLES

Look at the pair of triangles below. Note that they appear to have the same shape, but their sizes are different. These are examples of **similar triangles**. You can imagine using a magnifying glass to enlarge the smaller triangle so that it looks the same as the larger one. This process works because the corresponding sides of each triangle have the same ratio. That is, the following proportion is true.

$$\frac{a}{d} = \frac{b}{e} = \frac{c}{f}$$

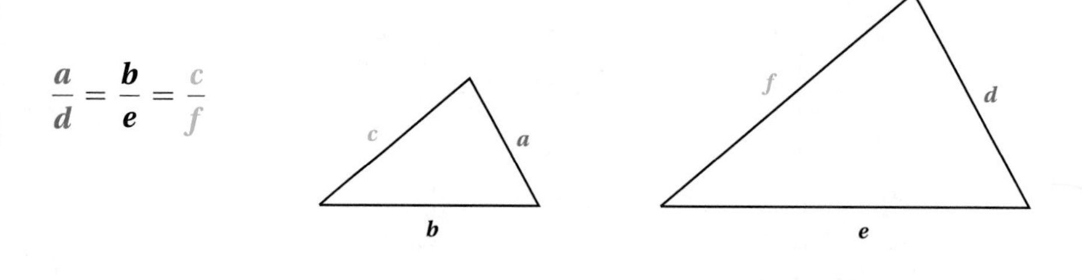

SIMILAR TRIANGLES

Similar triangles have the same shape. The lengths of their corresponding sides have the same ratio—that is, they are proportional.

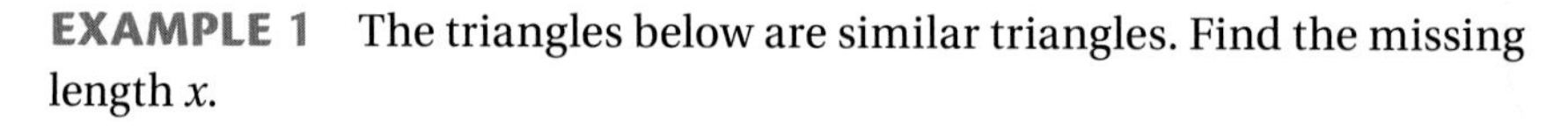

EXAMPLE 1 The triangles below are similar triangles. Find the missing length x.

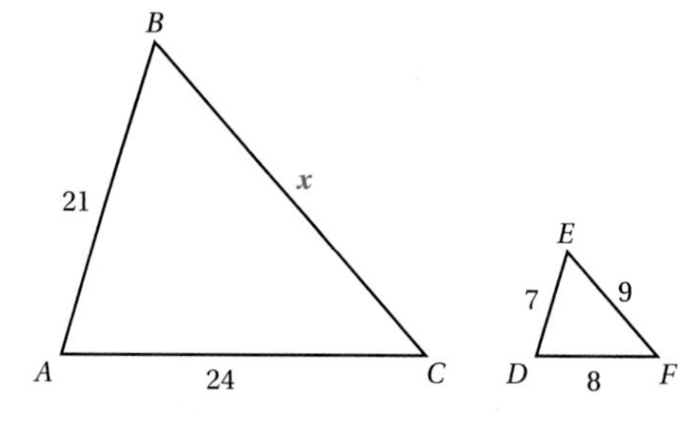

The ratio of x to 9 is the same as the ratio of 24 to 8 or 21 to 7. We get the proportions

$$\frac{x}{9} = \frac{24}{8} \quad \text{and} \quad \frac{x}{9} = \frac{21}{7}.$$

We can solve either one of these proportions. We use the first one:

$$\frac{x}{9} = \frac{24}{8}$$

$x \cdot 8 = 9 \cdot 24$ Equating cross products

$\frac{x \cdot 8}{8} = \frac{9 \cdot 24}{8}$ Dividing by 8 on both sides

$x = 27.$ Simplifying

The missing length x is 27. Other proportions could also be used.

◀ Do Margin Exercise 1.

1. This pair of triangles is similar. Find the missing length x. GS

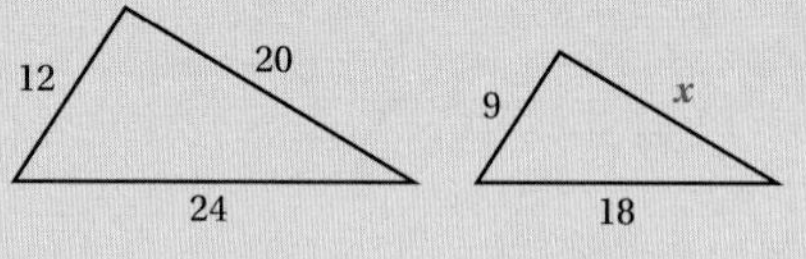

The ratio of x to 20 is the same as the ratio of 9 to ☐.

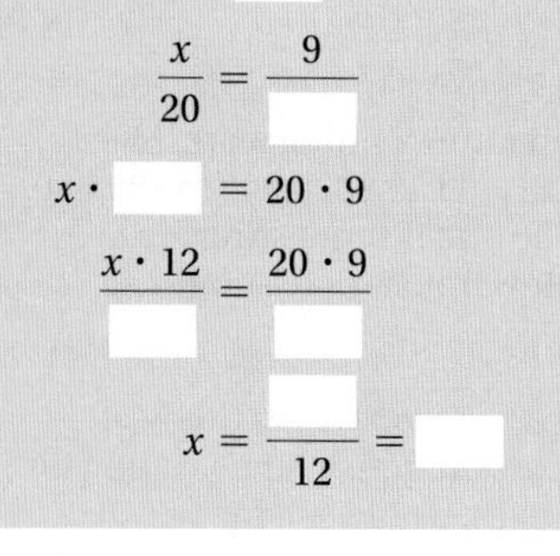

$$\frac{x}{20} = \frac{9}{\square}$$

$$x \cdot \square = 20 \cdot 9$$

$$\frac{x \cdot 12}{\square} = \frac{20 \cdot 9}{\square}$$

$$x = \frac{\square}{12} = \square$$

Answers

Skill to Review:

1. 2.625 2. $\frac{1}{3}$

Margin Exercise:

1. 15

Guided Solution:

1. 12; 12, 12, 12, 12, 180, 15

Similar triangles and proportions can often be used to find lengths that would ordinarily be difficult to measure. For example, we could find the height of a flagpole without climbing it or the distance across a river without crossing it.

EXAMPLE 2 How high is a flagpole that casts a 56-ft shadow at the same time that a 6-ft man casts a 5-ft shadow?

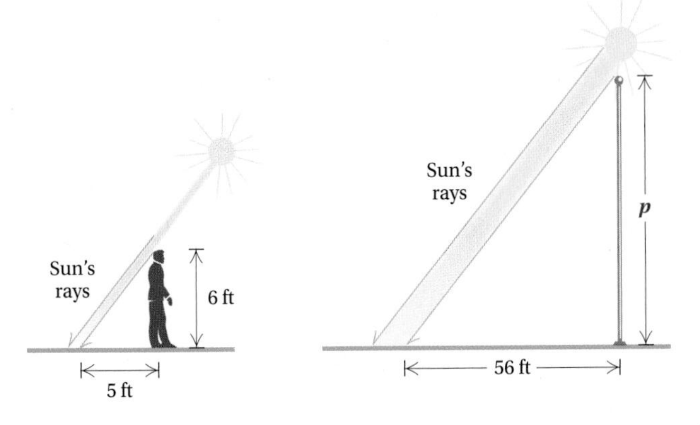

If we use the sun's rays to represent the third side of the triangle in our drawing of the situation, we see that we have similar triangles. Let $p =$ the height of the flagpole. The ratio of 6 to p is the same as the ratio of 5 to 56. Thus we have the proportion

$$\begin{array}{rl} \text{Height of man} \rightarrow & \dfrac{6}{p} = \dfrac{5}{56}. \leftarrow \text{Length of shadow of man} \\ \text{Height of pole} \rightarrow & \qquad\qquad \leftarrow \text{Length of shadow of pole} \end{array}$$

Solve: $6 \cdot 56 = p \cdot 5$ Equating cross products

$$\frac{6 \cdot 56}{5} = \frac{p \cdot 5}{5}$$ Dividing by 5 on both sides

$$\frac{6 \cdot 56}{5} = p$$ Simplifying

$$67.2 = p.$$

The height of the flagpole is 67.2 ft.

Do Exercise 2.

2. How high is a flagpole that casts a 45-ft shadow at the same time that a 5.5-ft woman casts a 10-ft shadow?

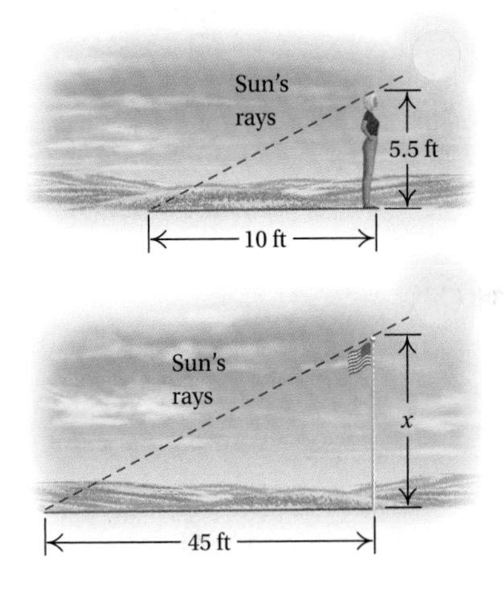

EXAMPLE 3 *Rafters of a House.* Carpenters use similar triangles to determine the length of rafters for a house. They first choose the pitch of the roof, or the ratio of the rise over the run. Then using a triangle with that ratio, they calculate the length of the rafters needed for the house. Loren is making rafters for a roof with a 6/12 pitch on a house that is 30 ft wide. Using a rafter guide, Loren knows that the rafter length corresponding to the 6/12 pitch is 13.4. Find the length x of the rafters for this house to the nearest tenth of a foot.

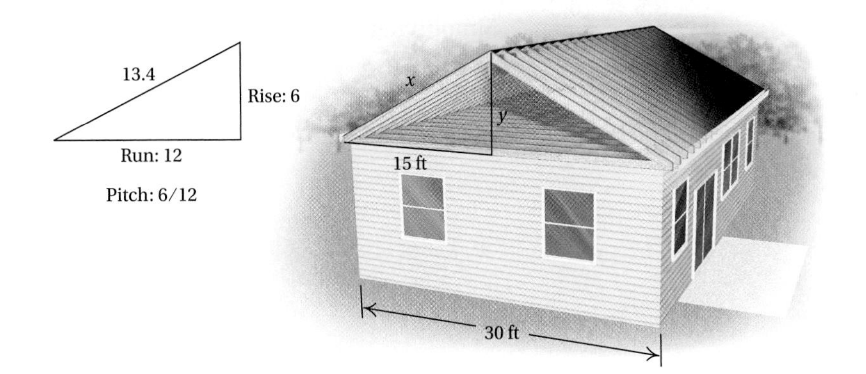

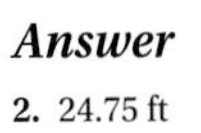

Answer

2. 24.75 ft

We have the proportion

$$\frac{13.4}{x} = \frac{12}{15}.$$

Length of rafter in 6/12 triangle → 13.4; Length of rafter on the house → x; Run in 6/12 triangle → 12; Run in similar triangle on the house → 15

Solve: $13.4 \cdot 15 = x \cdot 12$ Equating cross products

$$\frac{13.4 \cdot 15}{12} = \frac{x \cdot 12}{12}$$ Dividing by 12 on both sides

$$\frac{13.4 \cdot 15}{12} = x$$

$16.8 \text{ ft} \approx x.$ Rounding to the nearest tenth of a foot

The length x of the rafters for the house is about 16.8 ft.

Do Exercise 3.

3. *Rafters of a House.* Referring to Example 3, find the length y of the rise of the rafters of the house to the nearest tenth of a foot.

b PROPORTIONS AND OTHER GEOMETRIC SHAPES

When one geometric figure is a magnification of another, the figures are similar. Thus the corresponding lengths are proportional.

EXAMPLE 4 The sides in the photographs below are proportional. Find the width of the larger photograph.

We let $x =$ the width of the photograph. Then we translate to a proportion.

$$\frac{x}{2.5} = \frac{10.5}{3.5}$$

Larger width → x; Smaller width → 2.5; Larger length ← 10.5; Smaller length ← 3.5

Solve: $x \times 3.5 = 2.5 \times 10.5$ Equating cross products

$$\frac{x \times 3.5}{3.5} = \frac{2.5 \times 10.5}{3.5}$$ Dividing by 3.5 on both sides

$$x = \frac{2.5 \times 10.5}{3.5}$$ Simplifying

$x = 7.5.$

Thus the width of the larger photograph is 7.5 cm.

Do Exercise 4.

4. The sides in the photographs below are proportional. Find the width of the larger photograph.

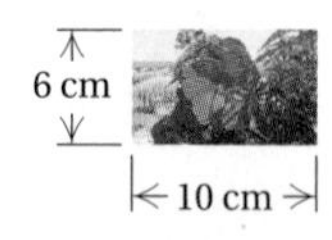

Answers

3. 7.5 ft 4. 21 cm

EXAMPLE 5 A scale model of an addition to an athletic facility is 12 cm wide at the base and rises to a height of 15 cm. If the actual base is to be 52 ft, what will the actual height of the addition be?

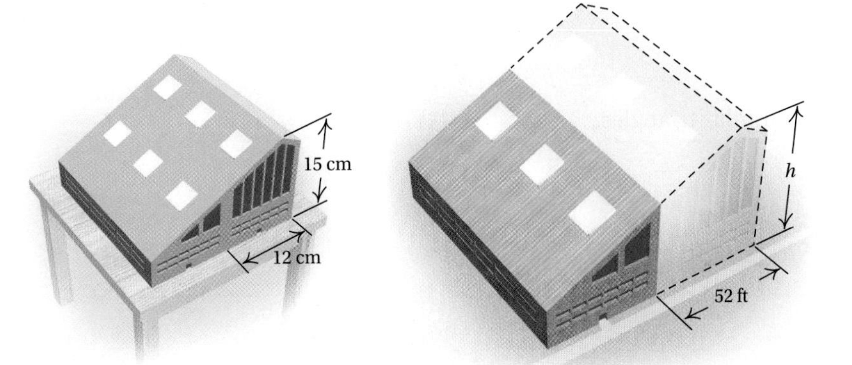

We let h = the height of the addition. Then we translate to a proportion.

$$\begin{matrix}\text{Width in model} \rightarrow \\ \text{Actual width} \rightarrow\end{matrix} \frac{12}{52} = \frac{15}{h} \begin{matrix}\leftarrow \text{Height in model} \\ \leftarrow \text{Actual height}\end{matrix}$$

Solve: $12 \cdot h = 52 \cdot 15$ Equating cross products

$$\frac{12 \cdot h}{12} = \frac{52 \cdot 15}{12}$$ Dividing by 12 on both sides

$$h = \frac{52 \cdot 15}{12} = 65.$$

Thus the height of the addition will be 65 ft.

Do Exercise 5. ▶

GS **5.** Refer to the figure in Example 5. If a skylight on the model is 3 cm wide, how wide will an actual skylight be?

Let w = the width of an actual skylight.

$$\frac{12}{52} = \frac{\square}{w}$$

$$12 \cdot w = 52 \cdot \square$$

$$w = \square$$

The width of an actual skylight will be 13 ____.

EXAMPLE 6 *Bicycle Design.* Two important dimensions to consider when buying a bicycle are *stack* and *reach*, as illustrated in the diagram at the right. The Country Racer bicycle comes in six different frame sizes, each proportional to the others. In the smallest frame size, the stack is 50 cm and the reach is 37.5 cm. In the largest size, the stack is 60 cm. What is the reach in the largest frame size of the Country Racer?

We let r = the reach in the largest frame size. Then we translate to a proportion.

$$\begin{matrix}\text{Stack in smaller frame} \rightarrow \\ \text{Reach in smaller frame} \rightarrow\end{matrix} \frac{50}{37.5} = \frac{60}{r} \begin{matrix}\leftarrow \text{Stack in larger frame} \\ \leftarrow \text{Reach in larger frame}\end{matrix}$$

Solve:

$50 \cdot r = 37.5 \cdot 60$ Equating cross products

$$\frac{50 \cdot r}{50} = \frac{37.5 \cdot 60}{50}$$ Dividing by 50 on both sides

$$r = \frac{37.5 \cdot 60}{50} = 45.$$

Thus the reach in the largest frame size is 45 cm.

Do Exercise 6. ▶

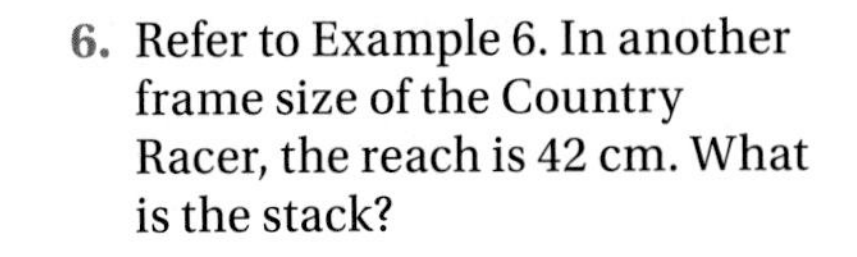

6. Refer to Example 6. In another frame size of the Country Racer, the reach is 42 cm. What is the stack?

Answers

5. 13 ft **6.** 56 cm

Guided Solution:

5. 3, 3, 13; ft

6.5 Exercise Set

For Extra Help

MyMathLab®

Reading Check

Complete each proportion based on the following similar triangles.

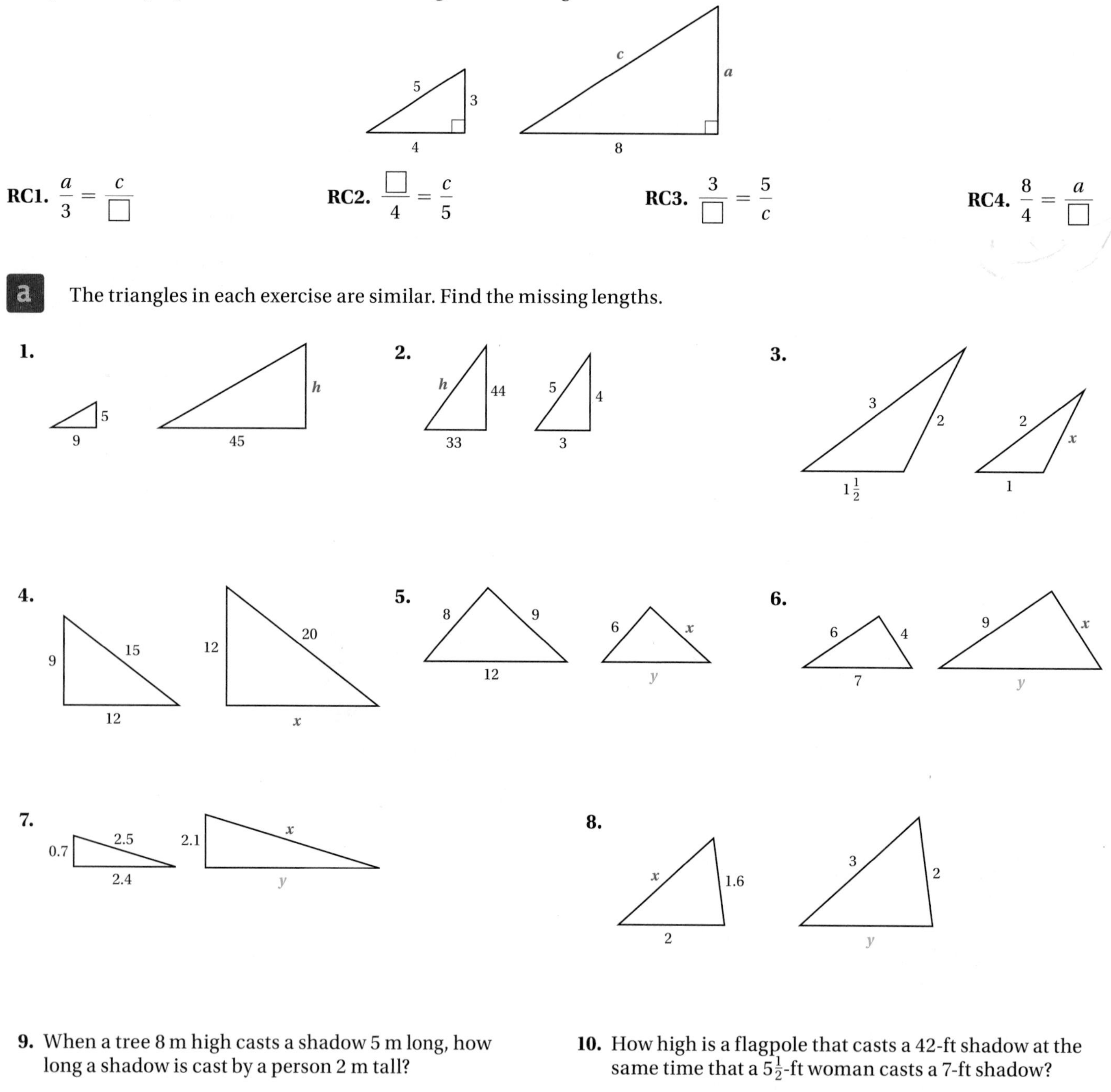

RC1. $\frac{a}{3} = \frac{c}{\square}$ **RC2.** $\frac{\square}{4} = \frac{c}{5}$ **RC3.** $\frac{3}{\square} = \frac{5}{c}$ **RC4.** $\frac{8}{4} = \frac{a}{\square}$

a The triangles in each exercise are similar. Find the missing lengths.

1.

2.

3.

4.

5.

6.

7.

8.

9. When a tree 8 m high casts a shadow 5 m long, how long a shadow is cast by a person 2 m tall?

10. How high is a flagpole that casts a 42-ft shadow at the same time that a $5\frac{1}{2}$-ft woman casts a 7-ft shadow?

11. How high is a tree that casts a 27-ft shadow at the same time that a 4-ft fence post casts a 3-ft shadow?

12. How high is a tree that casts a 32-ft shadow at the same time that an 8-ft light pole casts a 9-ft shadow?

13. Find the height h of the wall.

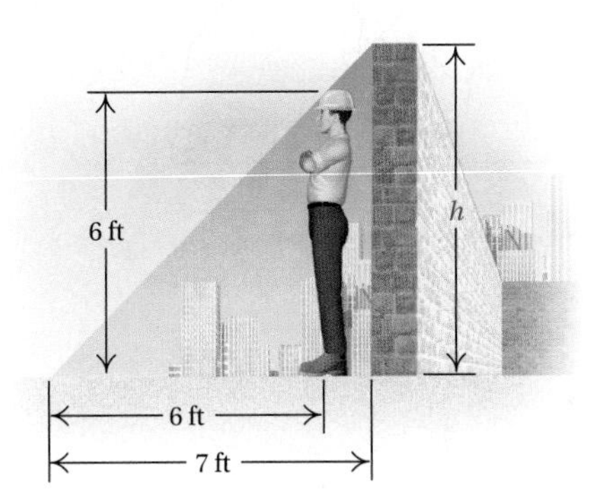

14. Find the length L of the lake. Assume that the ratio of L to 120 yd is the same as the ratio of 720 yd to 30 yd.

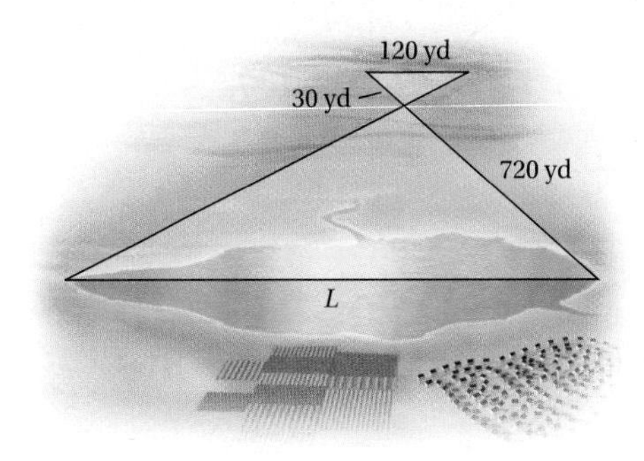

15. Find the distance across the river. Assume that the ratio of d to 25 ft is the same as the ratio of 40 ft to 10 ft.

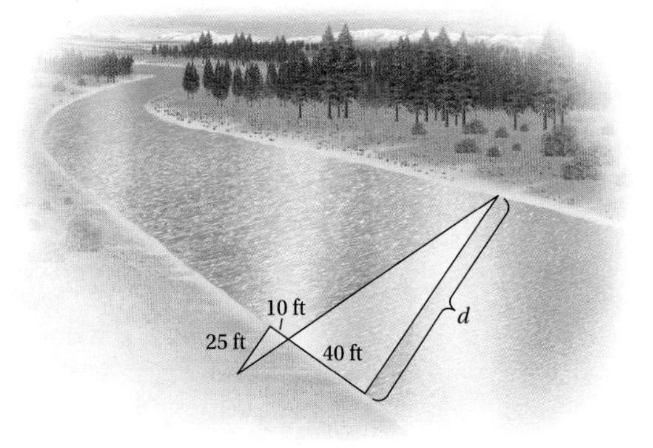

16. To measure the height of a hill, a string is stretched from level ground to the top of the hill. A 3-ft stick is placed under the string, touching it at point P, a distance of 5 ft from point G, where the string touches the ground. The string is then detached and found to be 120 ft long. How high is the hill?

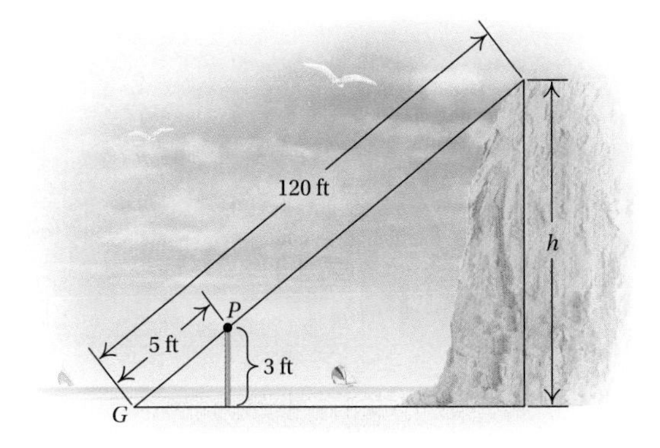

b In each of Exercises 17–26, the sides in each pair of figures are proportional. Find the missing lengths.

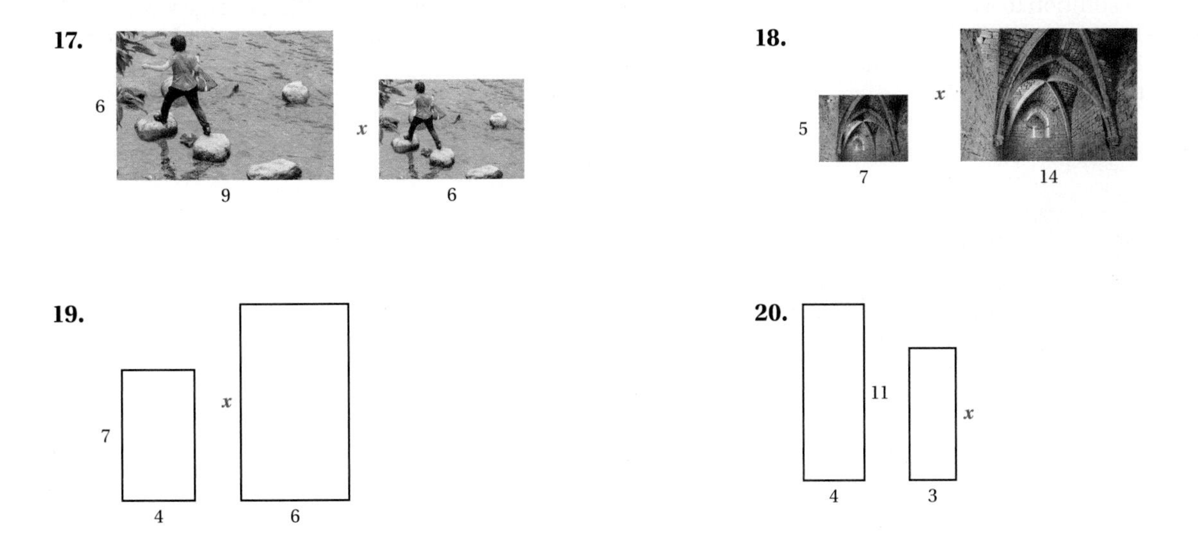

21.

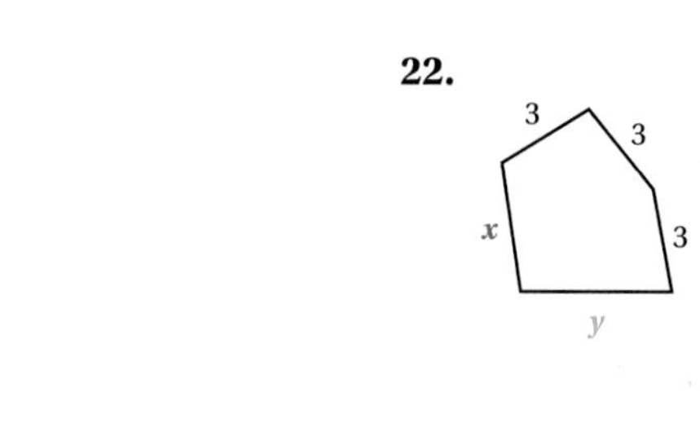

22.

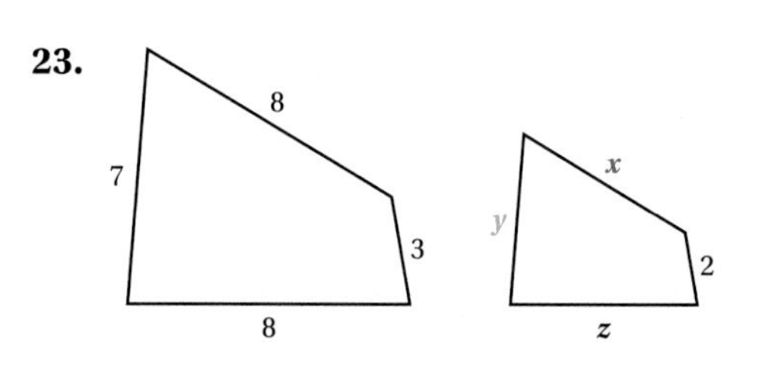

23.

24.

25.

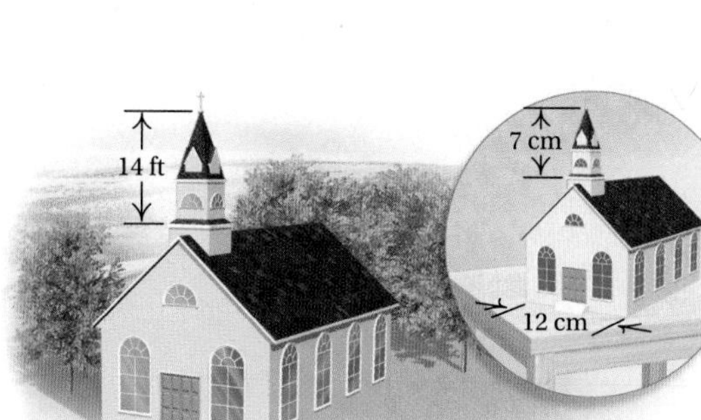

26.

27. A scale model of an addition to a medical clinic is 15 cm wide at the base and rises to a height of 19 cm. If the actual base is to be 120 ft, what will the actual height of the addition be?

28. Refer to the figure in Exercise 27. If a large window on the model is 3 cm wide, how wide will the actual window be?

Skill Maintenance

Determine whether each number is prime, composite, or neither. [3.1c]

29. 83

30. 28

Find the prime factorization of each number. [3.1d]

31. 808

32. 93

Use $=$ or $\neq$ for $\square$ to write a true sentence. [3.5c]

33. $\frac{12}{8} \square \frac{6}{4}$

34. $\frac{-4}{7} \square \frac{-5}{9}$

Use $<$ or $>$ for $\square$ to write a true sentence. [4.3b]

35. $\frac{7}{12} \square \frac{11}{15}$

36. $\frac{-1}{6} \square \frac{-2}{11}$

Simplify.

37. $\left(\frac{1}{2}\right)^2 + \frac{2}{3} \cdot 4\frac{1}{2}$ [4.7a]

38. $18.3 + 2.5 \times (-4.2) - (2.6 - 0.3^2)$ [5.4c]

39. $9 \times 15 - [2^3 \cdot 6 - (2 \cdot 5 + 3 \cdot 10)]$ [1.9d]

40. $2600 \div 13 - 5^3$ [1.9c]

Synthesis

Hockey Goals. An official hockey goal is 6 ft wide. To make scoring more difficult, goalies often position themselves far in front of the goal to "cut down the angle." In Exercises 41 and 42, suppose that a slapshot is attempted from point A and that the goalie is 2.7 ft wide. Determine how far from the goal the goalie should be located if point A is the given distance from the goal. (*Hint*: First find how far the goalie should be from point A.)

41. 25 ft

42. 35 ft

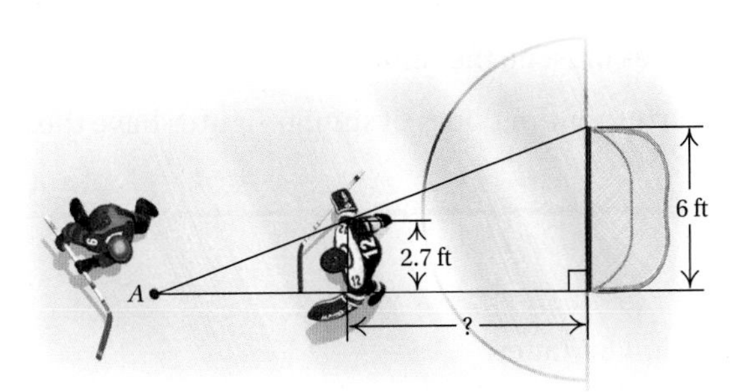

43. A miniature air conditioning unit is built for the model referred to in Exercise 27. An actual unit is 10 ft high. How high should the model unit be?

Solve. Round the answer to the nearest thousandth.

44. $\frac{8664.3}{10{,}344.8} = \frac{x}{9776.2}$

45. $\frac{12.0078}{56.0115} = \frac{789.23}{y}$

The triangles in each exercise are similar triangles. Find the lengths not given.

46. **47.**

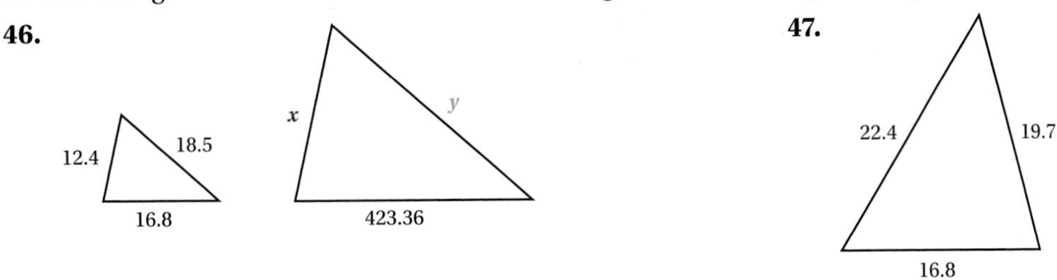

CHAPTER

6 Summary and Review

Vocabulary Reinforcement

Complete each statement with the correct term from the list at the right. Some of the choices may not be used.

1. A ratio is the ______________ of two quantities. [6.1a]
2. Similar triangles have the same ______________. [6.5a]
3. To solve $\frac{x}{a} = \frac{c}{d}$ for x, equate the ______________ and divide on both sides to get x alone on one side. [6.3b]
4. A(n) ______________ is a ratio used to compare two different kinds of measure. [6.2a]
5. A(n) ______________ states that two pairs of numbers have the same ratio. [6.3a]
6. A unit price is the ratio of ______________ to the number of units. [6.2b]

cross products
price
product
proportion
quantity
quotient
rate
shape

Concept Reinforcement

Determine whether each statement is true or false.

______________ 1. When we simplify a ratio like $\frac{8}{12}$, we find two other numbers in the same ratio. [6.1b]

______________ 2. The proportion $\frac{a}{b} = \frac{c}{d}$ can also be written as $\frac{c}{a} = \frac{d}{b}$. [6.3a]

______________ 3. Similar triangles must be the same size. [6.5a]

______________ 4. Lengths of corresponding sides of similar figures have the same ratio. [6.5b]

Study Guide

Objective 6.1a Find fraction notation for ratios.

Example Find the ratio of 7 to 18.

Write a fraction with a numerator of 7 and a denominator of 18: $\frac{7}{18}$.

Practice Exercise

1. Find the ratio of 17 to 3.

Objective 6.1b Simplify ratios.

Example Simplify the ratio 8 to 2.5.

$$\frac{8}{2.5} = \frac{8}{2.5} \cdot \frac{10}{10} = \frac{80}{25} = \frac{5 \cdot 16}{5 \cdot 5}$$

$$= \frac{5}{5} \cdot \frac{16}{5} = \frac{16}{5}$$

Practice Exercise

2. Simplify the ratio 3.2 to 2.8.

Objective 6.2a Give the ratio of two different measures as a rate.

Example A driver travels 156 mi on 6.5 gal of gas. What is the rate in miles per gallon?

$$\frac{156 \text{ mi}}{6.5 \text{ gal}} = \frac{156}{6.5} \frac{\text{mi}}{\text{gal}} = 24 \frac{\text{mi}}{\text{gal}}, \text{ or } 24 \text{ mpg}$$

Practice Exercise

3. A student earned \$120 for working 16 hr. What was the rate of pay per hour?

Objective 6.2b Find unit prices and use them to compare purchases.

Example A 16-oz can of tomatoes costs \$1.00. A 20-oz can costs \$1.23. Which has the lower unit price?

$$16 \text{ oz}: \frac{\$1.00}{16 \text{ oz}} = \frac{100 \text{ cents}}{16 \text{ oz}} = 6.25¢/\text{oz}$$

$$20 \text{ oz}: \frac{\$1.23}{20 \text{ oz}} = \frac{123 \text{ cents}}{20 \text{ oz}} = 6.15¢/\text{oz}$$

Thus the 20-oz can has the lower unit price.

Practice Exercise

4. A 28-oz jar of Brand A spaghetti sauce costs \$2.79. A 32-oz jar of Brand B spaghetti sauce costs \$3.29. Find the unit price of each brand and determine which is a better buy based on unit price alone.

Objective 6.3a Determine whether two pairs of numbers are proportional.

Example Determine whether 3, 4 and 7, 9 are proportional.

$$3 \cdot 9 = 27 \qquad \frac{3}{4} \stackrel{?}{=} \frac{7}{9} \qquad 4 \cdot 7 = 28$$

Since the cross products are not the same $(27 \neq 28)$, $\frac{3}{4} \neq \frac{7}{9}$ and the numbers are not proportional.

Practice Exercise

5. Determine whether 7, 9 and 21, 27 are proportional.

Objective 6.3b Solve proportions.

Example Solve: $\frac{3}{4} = \frac{y}{7}$.

$3 \cdot 7 = 4 \cdot y$ Equating cross products

$\frac{3 \cdot 7}{4} = \frac{4 \cdot y}{4}$ Dividing by 4 on both sides

$\frac{21}{4} = y$

The solution is $\frac{21}{4}$.

Practice Exercise

6. Solve: $\frac{9}{x} = \frac{8}{3}$.

Objective 6.4a Solve applied problems involving proportions.

Example Martina bought 3 tickets to a campus theater production for \$16.50. How much would 8 tickets cost?

We translate to a proportion.

$$\begin{matrix} \text{Tickets} \rightarrow \\ \text{Cost} \rightarrow \end{matrix} \frac{3}{16.50} = \frac{8}{c} \begin{matrix} \leftarrow \text{Tickets} \\ \leftarrow \text{Cost} \end{matrix}$$

$3 \cdot c = 16.50 \cdot 8$ Equating cross products

$c = \frac{16.50 \cdot 8}{3}$

$c = 44$

Eight tickets would cost \$44.

Practice Exercise

7. On a map, $\frac{1}{2}$ in. represents 50 mi. If two cities are $1\frac{3}{4}$ in. apart on the map, how far apart are they in reality?

Objective 6.5a Find lengths of sides of similar triangles using proportions.

Example The triangles below are similar. Find the missing length x.

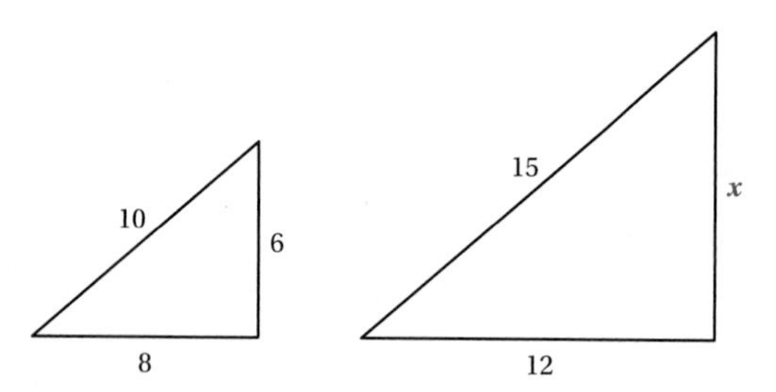

The ratio of 6 to x is the same as the ratio of 8 to 12 (and also as the ratio of 10 to 15). We write and solve a proportion:

$$\frac{6}{x} = \frac{8}{12}$$

$$6 \cdot 12 = x \cdot 8$$

$$\frac{6 \cdot 12}{8} = \frac{x \cdot 8}{8}$$

$$9 = x.$$

The missing length is 9. (We could also have used other proportions, including $\frac{6}{x} = \frac{10}{15}$, to find x.)

Practice Exercise

8. The triangles below are similar. Find the missing length y.

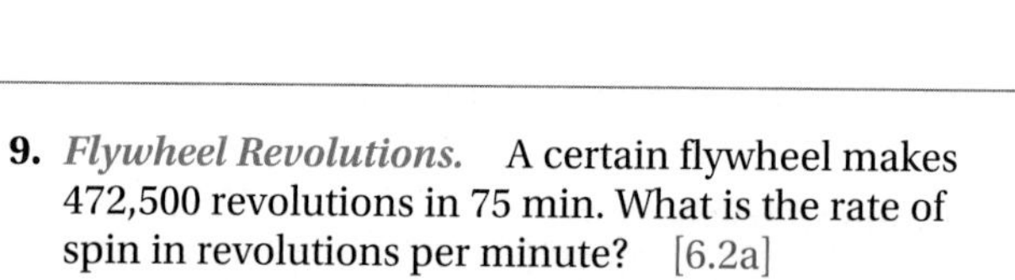

Review Exercises

Write fraction notation for each ratio. Do not simplify. [6.1a]

1. 47 to 84

2. 46 to 1.27

3. 83 to 100

4. 0.72 to 197

5. At Preston Seafood Market, 12,480 lb of tuna and 16,640 lb of salmon were sold one year. [6.1a, b]

a) Write fraction notation for the ratio of tuna sold to salmon sold.

b) Write fraction notation for the ratio of salmon sold to the total number of pounds of both kinds of fish sold.

Find the ratio of the first number to the second number and simplify. [6.1b]

6. 9 to 12

7. 3.6 to 6.4

8. *Gas Mileage.* The Chrysler 200 will travel 406 mi on 14.5 gal of gasoline in highway driving. What is the rate in miles per gallon? [6.2a]

Source: Chrysler Motor Corporation

9. *Flywheel Revolutions.* A certain flywheel makes 472,500 revolutions in 75 min. What is the rate of spin in revolutions per minute? [6.2a]

10. A lawn requires 319 gal of water for every 500 ft^2. What is the rate in gallons per square foot? [6.2a]

11. *Calcium Supplement.* The price for a particular calcium supplement is $18.99 for 300 tablets. Find the unit price in cents per tablet. [6.2b]

12. Raquel bought a 24-oz loaf of 12-grain bread for $4.69. Find the unit price in cents per ounce. [6.2b]

13. *Vegetable Oil.* Find the unit price. Then determine which size has the lowest unit price. [6.2b]

Package	Price	Unit Price
32 oz	$4.79	
48 oz	$5.99	
64 oz	$9.99	

Determine whether the two pairs of numbers are proportional. [6.3a]

14. 9, 15 and 36, 60

15. 24, 37 and 40, 46.25

Solve. [6.3b]

16. $\frac{8}{9} = \frac{x}{36}$

17. $\frac{6}{x} = \frac{48}{56}$

18. $\frac{120}{\frac{3}{7}} = \frac{7}{x}$

19. $\frac{4.5}{120} = \frac{0.9}{x}$

Solve. [6.4a]

20. *Quality Control.* A factory manufacturing computer circuits found 3 defective circuits in a lot of 65 circuits. At this rate, how many defective circuits can be expected in a lot of 585 circuits?

21. *Exchanging Money.* On January 7, 2013, 1 U.S. dollar was worth about 0.99 Canadian dollar.
   a) How much were 250 U.S. dollars worth in Canada on that day?
   b) While traveling in Canada that day, Jamal saw a sweatshirt that cost 50 Canadian dollars. How much would it cost in U.S. dollars?

22. A train travels 448 mi in 7 hr. At this rate, how far will it travel in 13 hr?

23. *Gasoline Consumption.* The United States consumed about 89 billion gallons of gasoline in the first 8 months of 2011. At this rate, how much was consumed in 12 months?

24. *Trash Production.* A study shows that 5 people generate 23 lb of trash each day. The population of Austin, Texas, is 820,611. How many pounds of trash are produced in Austin in one day?

Sources: U.S. Environmental Protection Agency; U.S. Census Bureau

25. *Calories Burned.* Kevin burned 200 calories while playing ultimate frisbee for $\frac{3}{4}$ hr. How many calories would he burn if he played for $1\frac{1}{5}$ hr?

26. *Lawyers in Chicago.* In Illinois, there are about 4.6 lawyers for every 1000 people. The population of Chicago is 2,707,120. How many lawyers would you expect there to be in Chicago?

Sources: American Bar Association; U.S. Census Bureau

Each pair of triangles in Exercises 27 and 28 is similar. Find the missing length(s). [6.5a]

27.

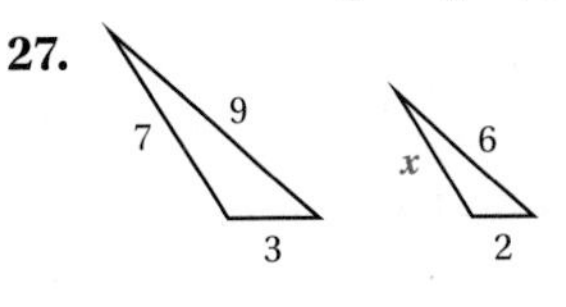

28.

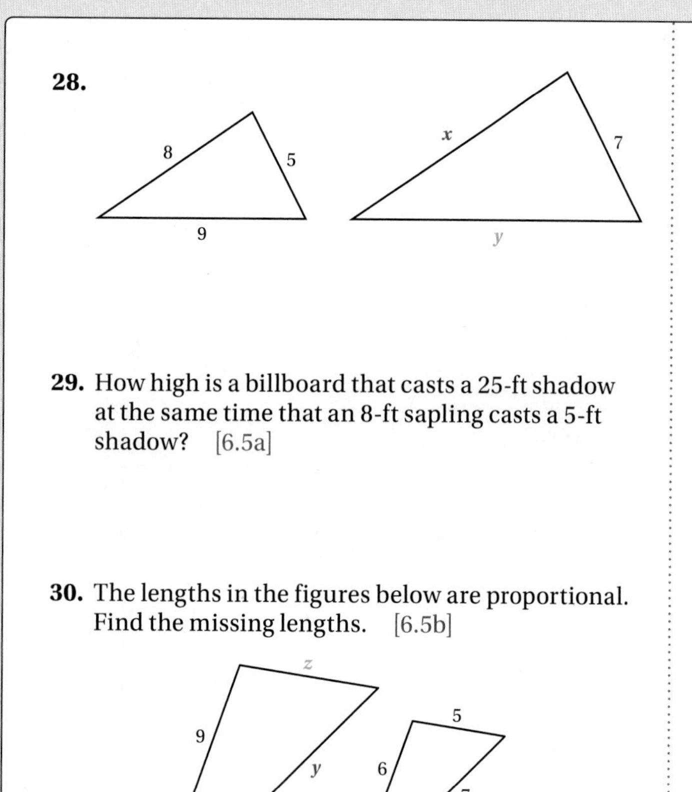

29. How high is a billboard that casts a 25-ft shadow at the same time that an 8-ft sapling casts a 5-ft shadow? [6.5a]

30. The lengths in the figures below are proportional. Find the missing lengths. [6.5b]

31. *Turkey Servings.* A 25-lb turkey serves 18 people. Find the rate in servings per pound. [6.2a]

A. 0.36 serving/lb
B. 0.72 serving/lb
C. 0.98 serving/lb
D. 1.39 servings/lb

32. If 3 dozen eggs cost $5.04, how much will 5 dozen eggs cost? [6.4a]

A. $6.72
B. $6.96
C. $8.40
D. $10.08

Synthesis

33. *Paper Towels.* Find the unit price and determine which item would be the best buy based on unit price alone. [6.2b]

PACKAGE	PRICE	UNIT PRICE PER SHEET
8 rolls, 60 (2 ply) sheets per roll	$8.67	
15 rolls, 60 (2 ply) sheets per roll	$14.92	
6 big rolls, 165 (2 ply) sheets per roll	$13.95	

34. The following triangles are similar. Find the missing lengths. [6.5a]

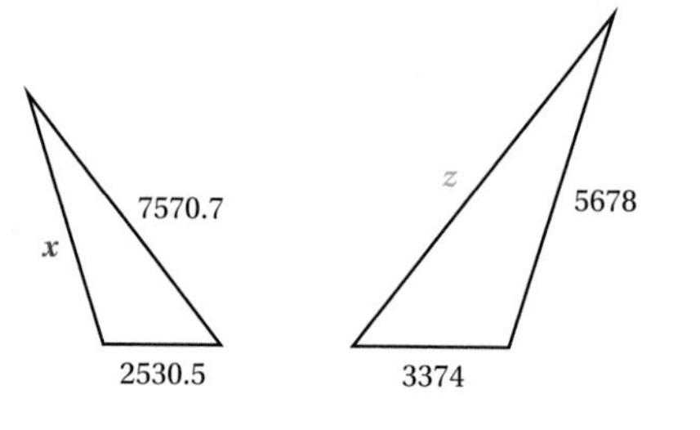

35. Shine-and-Glo Painters uses 2 gal of finishing paint for every 3 gal of primer. Each gallon of finishing paint covers 450 ft^2. If a surface of 4950 ft^2 needs both primer and finishing paint, how many gallons of each should be purchased? [6.4a]

Understanding Through Discussion and Writing

1. If you were a college president, which would you prefer: a low or high faculty-to-student ratio? Why? [6.1a]

2. Can unit prices be used to solve proportions that involve money? Explain why or why not. [6.2b], [6.4a]

3. Write a proportion problem for a classmate to solve. Design the problem so that the solution is "Leslie would need 16 gal of gasoline in order to travel 368 mi." [6.4a]

4. Is it possible for two triangles to have two pairs of sides that are proportional without the triangles being similar? Why or why not? [6.5a]

CHAPTER

6 Test

For Extra Help For step-by-step test solutions, access the Chapter Test Prep Videos in MyMathLab® or on YouTube (search "BittingerBasicEl" and click on "Channels").

Write fraction notation for each ratio. Do not simplify.

1. 85 to 97

2. 0.34 to 124

Find the ratio of the first number to the second number and simplify.

3. 18 to 20

4. 0.75 to 0.96

5. What is the rate in feet per second?

10 feet, 16 seconds

6. *Ham Servings.* A 12-lb shankless ham contains 16 servings. What is the rate in servings per pound?

7. *Gas Mileage.* Jeff's convertible will travel 464 mi on 14.5 gal of gasoline in highway driving. What is the rate in miles per gallon?

8. *Bagged Salad Greens.* Ron bought a 16-oz bag of salad greens for $2.49. Find the unit price in cents per ounce.

9. The table below lists prices for concentrated liquid laundry detergent. Find the unit price of each size in cents per ounce. Then determine which has the lowest unit price.

Size	Price	Unit Price
40 oz	$6.59	
50 oz	$6.99	
100 oz	$11.49	
150 oz	$24.99	

Determine whether the two pairs of numbers are proportional.

10. 7, 8 and 63, 72

11. 1.3, 3.4 and 5.6, 15.2

Solve.

12. $\frac{9}{4} = \frac{27}{x}$

13. $\frac{150}{2.5} = \frac{x}{6}$

14. $\frac{x}{100} = \frac{27}{64}$

15. $\frac{68}{y} = \frac{17}{25}$

Solve.

16. *Distance Traveled.* An ocean liner traveled 432 km in 12 hr. At this rate, how far would the boat travel in 42 hr?

17. *Time Loss.* A watch loses 2 min in 10 hr. At this rate, how much will it lose in 24 hr?

18. *Map Scaling.* On a map, 3 in. represents 225 mi. If two cities are 7 in. apart on the map, how far are they apart in reality?

19. *Tower Height.* A birdhouse on a pole that is 3 m high casts a shadow 5 m long. At the same time, the shadow of a tower is 110 m long. How high is the tower?

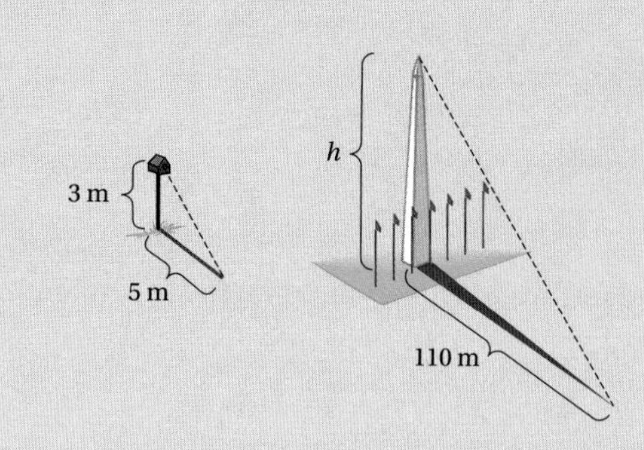

20. *Charity Work.* Kayla is crocheting hats for a charity. She can make 8 hats from 12 packages of yarn.

a) How many hats can she make from 20 packages of yarn?

b) How many packages of yarn does she need to make 20 hats?

21. *Thanksgiving Dinner.* A traditional turkey dinner for 8 people cost about \$33.81 in a recent year. How much would it cost to serve a turkey dinner for 14 people?

Source: American Farm Bureau Federation

The sides in each pair of figures are proportional. Find the missing lengths.

22.

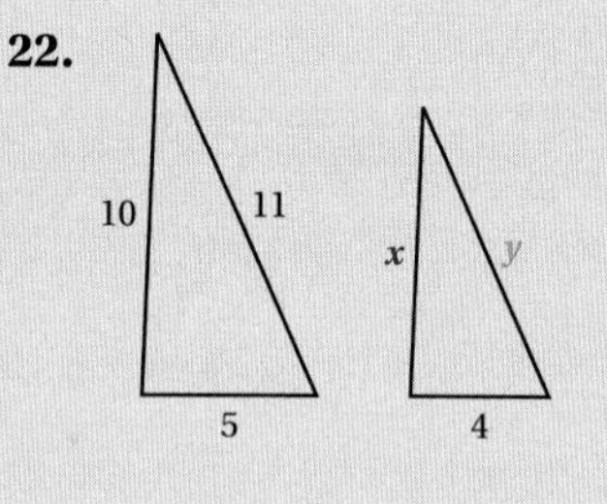

23.

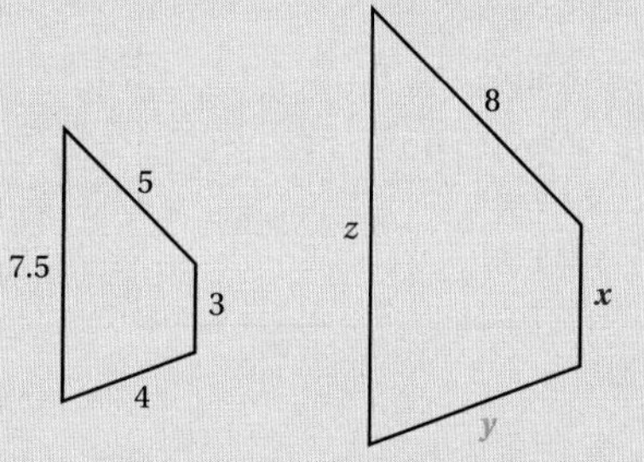

24. Lucita walks $4\frac{1}{2}$ mi in $1\frac{1}{2}$ hr. What is her rate in miles per hour?

A. $\frac{1}{3}$ mph

B. $1\frac{1}{2}$ mph

C. 3 mph

D. $4\frac{1}{2}$ mph

Synthesis

25. Nancy wants to win a gift card from the campus bookstore by guessing the number of marbles in an 8-gal jar. She knows that there are 128 oz in a gallon. She goes home and fills an 8-oz jar with 46 marbles. How many marbles should she guess are in the 8-gal jar?

CHAPTERS

1–6 Cumulative Review

Calculate and simplify.

1. $\begin{array}{r} 27.68 \\ 3.019 \\ +\ 483.297 \\ \hline \end{array}$

2. $\begin{array}{r} 2\frac{1}{3} \\ +\ 4\frac{5}{12} \\ \hline \end{array}$

3. $-\frac{6}{35} + \left(-\frac{5}{28}\right)$

4. $\begin{array}{r} 40.2 \\ -\ 9.709 \\ \hline \end{array}$

5. $73.82 - 0.908$

6. $-\frac{4}{15} - \frac{3}{20}$

7. $\begin{array}{r} 37.64 \\ \times\ 5.9 \\ \hline \end{array}$

8. 5.678×100

9. $2\frac{1}{3} \cdot \left(-1\frac{2}{7}\right)$

10. $2.3\,\overline{)\,98.9}$

11. $54\,\overline{)\,48{,}546}$

12. $-\frac{7}{11} \div \frac{14}{33}$

13. Write expanded notation: 30,074.

14. Write a word name for 120.07.

Which number is larger?

15. 0.7, 0.698

16. −0.8, −0.7999

17. Find the prime factorization of 144.

18. Find the LCM of 18 and 30.

19. What part is shaded?

20. Simplify: $\frac{90}{144}$.

Calculate.

21. $\frac{3}{5} \times 9.53$

22. $\frac{1}{3} \times 0.645 - \frac{3}{4} \times 0.048$

23. Write fraction notation for the ratio 0.3 to 15. Do not simplify.

24. Determine whether the pairs 3, 9 and 25, 75 are proportional.

25. What is the rate in meters per second?

660 meters, 12 seconds

26. *Unit Prices.* An 8-oz can of pineapple chunks costs $0.99. A 24.5-oz jar of pineapple chunks costs $3.29. Which has the lower unit price?

Solve.

27. $\frac{14}{25} = \frac{x}{54}$

28. $423 = 16 \cdot t$

29. $\frac{2}{3} \cdot y = \frac{16}{27}$

30. $\frac{7}{16} = \frac{56}{x}$

31. $34.56 + n = -67.9$

32. $t + \frac{7}{25} = \frac{5}{7}$

33. Ramona's recipe for fettuccini alfredo has 520 calories in 1 cup. How many calories are there in $\frac{3}{4}$ cup?

34. A machine can stamp out 925 washers in 5 min. An order is placed for 1295 washers. How long will it take to stamp them out?

35. A 46-oz juice can contains $5\frac{3}{4}$ cups of juice. A recipe calls for $3\frac{1}{2}$ cups of juice. How many cups are left over?

36. It takes a carpenter $\frac{2}{3}$ hr to hang a door. How many doors can the carpenter hang in 8 hr?

37. *Car Travel.* A car travels 337.62 mi in 8 hr. How far does it travel in 1 hr?

38. *Shuttle Orbits.* A space shuttle made 16 orbits a day during an 8.25-day mission. How many orbits were made during the entire mission?

39. How many even prime numbers are there?

A. 5 **B.** 3 **C.** 2
D. 1 **E.** None

40. The gas mileage of a car is 28.16 mpg. How many gallons per mile is this?

A. $\frac{704}{25}$ **B.** $\frac{25}{704}$ **C.** $\frac{2816}{100}$
D. $\frac{250}{704}$ **E.** None

Synthesis

41. A soccer goalie wishing to block an opponent's shot moves toward the shooter to reduce the shooter's view of the goal. If the goalie can only defend a region 10 ft wide, how far in front of the goal should the goalie be? (See the figure at right.)

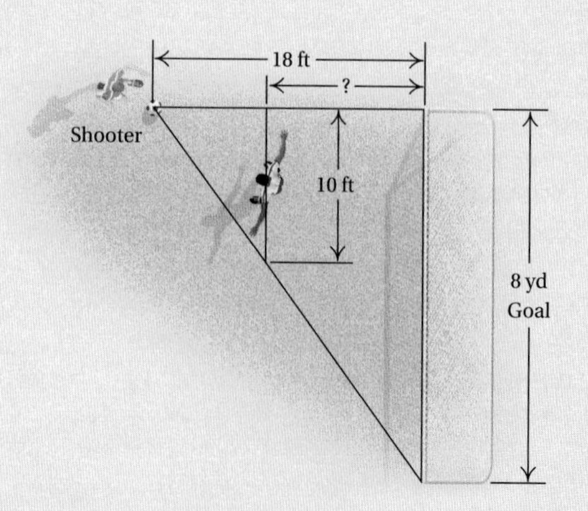

Percent Notation

STUDYING FOR SUCCESS *Quiz and Test Follow-up*

- ☐ Immediately after completing a chapter quiz or test, write out a step-by-step solution for each question that you missed.
- ☐ Meet with your instructor or tutor for help with problems that are still giving you trouble.
- ☐ Keep your tests and quizzes along with their corrections to use as a study guide for the final exam.

Percent Notation

OBJECTIVES

a Write three kinds of notation for a percent.

b Convert between percent notation and decimal notation.

SKILL TO REVIEW

Objective 5.3a: Multiply using decimal notation.

Multiply.

1. 68.3×0.01

2. 3013×2.4

a UNDERSTANDING PERCENT NOTATION

Today almost half of the world's population lives in cities. It is estimated that by 2050, 70% of the population will live in cities. What does 70% mean? It means that of every 100 people on Earth in 2050, 70 will live in cities. Thus, 70% is a ratio of 70 to 100, or $\frac{70}{100}$.

Sources: DuPont, 2011; *National Geographic*, December 2011

70 of 100 squares are shaded.

70% or $\frac{70}{100}$ or 0.70 of the large square is shaded.

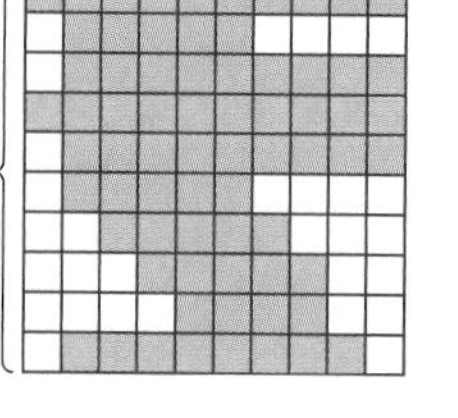

We encounter percent notation frequently. Here are some examples.

- The 1964–2013 Washington quarter is 91.67% copper and 8.33% nickel. The 1946–2013 Jefferson nickel is 75% copper and 25% nickel.
  Source: Coinflation.com
- A blood alcohol level of 0.08% is the standard used by most states as the legal limit for drunk driving.
  Source: The National Highway Safety Administration
- Educational loans of $20,000 or more per year are obtained by 16% of college students.
  Source: TRU Survey
- Of senior citizens aged 76 and older, 31% own a laptop.
  Source: Internet & American Life Project

Percent notation is often represented using a circle graph, or pie chart, to show how the parts of a quantity are related. For example, the circle graph at left illustrates the percentages of coffee drinkers who get their coffee at selected locations.

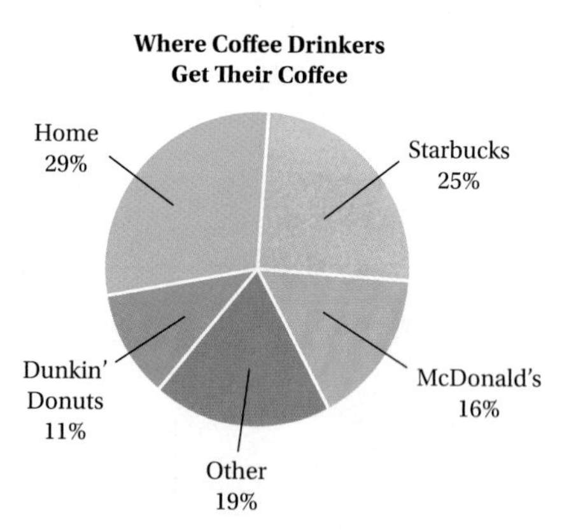

SOURCE: "Starbucks' Big Mug," *TIME*, June 25, 2012

> **PERCENT NOTATION**
>
> The notation ***n*%** means "*n* per hundred."

Answers

Skill to Review:
1. 0.683 **2.** 7231.2

This definition leads us to the following equivalent ways of defining percent notation.

NOTATION FOR $n\%$

Percent notation, $n\%$, can be expressed using:

ratio $\rightarrow n\% =$ the ratio of n to 100 $= \frac{n}{100}$,

fraction notation $\rightarrow n\% = n \times \frac{1}{100}$, or

decimal notation $\rightarrow n\% = n \times 0.01$.

EXAMPLE 1 Write three kinds of notation for 67.8%.

Using ratio: $67.8\% = \frac{67.8}{100}$ — A ratio of 67.8 to 100

Using fraction notation: $67.8\% = 67.8 \times \frac{1}{100}$ — Replacing % with $\times \frac{1}{100}$

Using decimal notation: $67.8\% = 67.8 \times 0.01$ — Replacing % with $\times 0.01$

Do Exercises 1–4.

During 2011, 42.2% of electricity in the United States was generated using coal.

Source: U.S. Energy Information Administration

Write three kinds of notation for each percent, as in Example 1.

1. 70%

2. 23.4%

3. 100%

4. 0.6%

b CONVERTING BETWEEN PERCENT NOTATION AND DECIMAL NOTATION

Consider 78%. To convert to decimal notation, we can think of percent notation as a ratio and write

$$78\% = \frac{78}{100} \quad \text{Using the definition of percent as a ratio}$$
$$= 0.78. \quad \text{Dividing}$$

Similarly,

$$4.9\% = \frac{4.9}{100} = 0.049.$$

We can also convert 78% to decimal notation by replacing "%" with "× 0.01" and writing

$$78\% = 78 \times 0.01 \quad \text{Replacing \% with } \times 0.01$$
$$= 0.78. \quad \text{Multiplying}$$

Similarly,

$$4.9\% = 4.9 \times 0.01 = 0.049.$$

Dividing by 100 amounts to moving the decimal point two places to the left, which is the same as multiplying by 0.01. This leads us to a quick way to convert from percent notation to decimal notation: We drop the percent symbol and move the decimal point two places to the left.

It is thought that the Roman emperor Augustus began percent notation by taxing goods sold at a rate of $\frac{1}{100}$. In time, the symbol "%" evolved by interchanging the parts of the symbol "100" to "0/0" and then to "%."

Answers

1. $\frac{70}{100}$; $70 \times \frac{1}{100}$; 70×0.01
2. $\frac{23.4}{100}$; $23.4 \times \frac{1}{100}$; 23.4×0.01
3. $\frac{100}{100}$; $100 \times \frac{1}{100}$; 100×0.01
4. $\frac{0.6}{100}$; $0.6 \times \frac{1}{100}$; 0.6×0.01

CALCULATOR CORNER

Converting from Percent Notation to Decimal Notation Many calculators have a [%] key that can be used to convert from percent notation to decimal notation. This is often the second operation associated with a particular key and is accessed by first pressing a [2nd] or [SHIFT] key. To convert 57.6% to decimal notation, for example, you might press [5] [7] [·] [6] [2nd] [%] or [5] [7] [·] [6] [SHIFT] [%]. The display would read [0.576], so 57.6% = 0.576.

EXERCISES Use a calculator to find decimal notation.

1. 14% **2.** 0.069%
3. 43.8% **4.** 125%

To convert from percent notation to decimal notation,	36.5%	
a) replace the percent symbol % with × 0.01, and	36.5 × 0.01	
b) multiply by 0.01, which means move the decimal point two places to the left.	0.36.5	Move 2 places to the left.
	36.5% = 0.365	

EXAMPLE 2 Find decimal notation for 99.44%.

a) Replace the percent symbol with × 0.01. 99.44 × 0.01

b) Move the decimal point two places to the left. 0.99.44

Thus, 99.44% = 0.9944.

EXAMPLE 3 The interest rate on a $2\frac{1}{2}$-year certificate of deposit is $6\frac{3}{8}$%. Find decimal notation for $6\frac{3}{8}$%.

a) Convert $6\frac{3}{8}$ to decimal notation and replace the percent symbol with × 0.01. $6\frac{3}{8}$% 6.375 × 0.01

b) Move the decimal point two places to the left. 0.06.375

Thus, $6\frac{3}{8}$% = 0.06375.

Do Exercises 5–8.

Find decimal notation.

5. 34% **6.** 78.9%

Find decimal notation for the percent notation(s) in each sentence.

7. ***Energy Use.*** The United States consumes 19% of the world's energy. Russia consumes only 6% of the world's energy.

Source: U.S. Energy Information Administration

8. ***Blood Alcohol Level.*** A blood alcohol level of 0.08% is the standard used by the most states as the legal limit for drunk driving.

To convert 0.38 to percent notation, we can first write fraction notation, as follows:

$$0.38 = \frac{38}{100} \quad \text{Converting to fraction notation}$$
$$= 38\%. \quad \text{Using the definition of percent as a ratio}$$

Note that 100% = 100 × 0.01 = 1. Thus to convert 0.38 to percent notation, we can multiply by 1, using 100% as a symbol for 1.

$$0.38 = 0.38 \times 1$$
$$= 0.38 \times 100\%$$
$$= 0.38 \times 100 \times 0.01 \quad \text{Replacing 100\% with } 100 \times 0.01$$
$$= (0.38 \times 100) \times 0.01 \quad \text{Using the associative law of multiplication}$$
$$= 38 \times 0.01$$
$$= 38\% \quad \text{Replacing } \times 0.01 \text{ with } \%$$

Even more quickly, since 0.38 = 0.38 × 100%, we can simply multiply 0.38 by 100 and write the % symbol.

To convert from decimal notation to percent notation, we multiply by 100%. That is, we move the decimal point two places to the right and write a percent symbol.

Answers

5. 0.34 **6.** 0.789
7. 0.19; 0.06 **8.** 0.0008

To convert from decimal notation to percent notation, multiply by 100%. That is,	$0.675 = 0.675 \times 100\%$	
a) move the decimal point two places to the right and	0.67.5	Move 2 places to the right.
b) write a % symbol.	67.5% $0.675 = 67.5\%$	

EXAMPLE 4 Of the time off that employees take as sick leave, 0.21 is actually used for family issues. Find percent notation for 0.21.

Source: CCH Inc.

a) Move the decimal point two places to the right. 0.21.

b) Write a % symbol. 21%

Thus, $0.21 = 21\%$.

EXAMPLE 5 Find percent notation for 5.6.

a) Move the decimal point two places to the right, adding an extra zero. 5.60.

b) Write a % symbol. 560%

Thus, $5.6 = 560\%$.

EXAMPLE 6 Of those who play golf, 0.149 play 8–24 rounds per year. Find percent notation for 0.149.

Source: U.S. Golf Association

a) Move the decimal point two places to the right. 0.14.9

b) Write a % symbol. 14.9%

Thus, $0.149 = 14.9\%$.

Do Exercises 9–14. ▶

Find percent notation.

9. 0.24

10. 3.47

11. 1

12. 0.05

Find percent notation for the decimal notation(s) in each sentence.

13. *Women in Congress.* In 2012, 0.19 of the members of the United States Congress were women.

Source: *Wall Street Journal*

14. *Soccer.* Of Americans in the 18–24 age group, 0.311 have played soccer. Of those in the 12–17 age group, 0.396 have played soccer.

Source: ESPN Sports Poll, a service of TNS Sport

Answers

9. 24% **10.** 347% **11.** 100% **12.** 5% **13.** 19% **14.** 31.1%; 39.6%

7.1 Exercise Set

For Extra Help

MyMathLab®

Reading Check

Find percent notation for each shaded area.

RC1.

RC2.

RC3.

RC4.

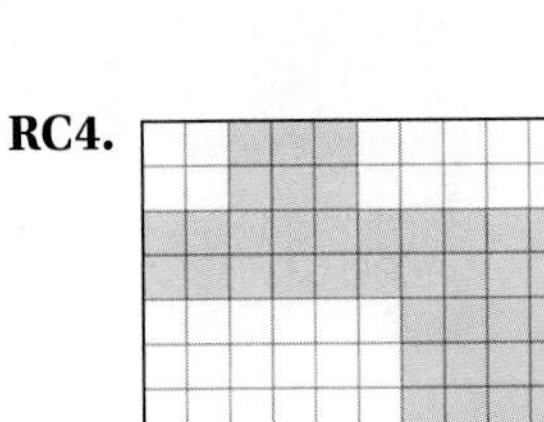

a Write three kinds of notation, as in Example 1 on p. 383.

1. 90% **2.** 58.7% **3.** 12.5% **4.** 130%

b Find decimal notation.

5. 67% **6.** 17% **7.** 45.6% **8.** 76.3%

9. 59.01% **10.** 30.02% **11.** 10% **12.** 80%

13. 1% **14.** 100% **15.** 200% **16.** 300%

17. 0.1% **18.** 0.4% **19.** 0.09% **20.** 0.12%

21. 0.18% **22.** 5.5% **23.** 23.19% **24.** 87.99%

25. $14\frac{7}{8}\%$ **26.** $93\frac{1}{8}\%$ **27.** $56\frac{1}{2}\%$ **28.** $61\frac{3}{4}\%$

Find decimal notation for the percent notation(s) in each sentence.

29. *Daily Calories.* In 2006, American adults got about 13% of their daily calories from fast food. This percentage decreased to 11% in 2010.

Source: *Indianapolis Star*, February 21, 2013, Nanci Hellmich

30. *Bachelor's Degrees.* In 1970, 1% of taxi drivers had a bachelor's degree. By 2010, 15% of taxi drivers had a bachelor's degree.

Source: *USA Today*, January, 28, 2013, Mary Beth Marklein

31. *Fuel Efficiency.* Speeding up by only 5 mph on the highway cuts fuel efficiency by approximately 7% to 8%.

Source: *Wall Street Journal*, "Pain Relief," by A. J. Miranda, September 15, 2008

32. *Foreign-Born Population.* In 2010, the U.S. foreign-born population was 12.9%, the highest since 1920.

Source: U.S. Census Bureau

33. *Credit-Card Debt.* In 2012, approximately 13.9% of American households had credit-card debt that exceeded 40% of their income.

Source: statisticbrain.com

34. *Eating Out.* On a given day, 58% of all Americans eat meals and snacks away from home.

Source: U.S. Department of Agriculture

Find percent notation.

35. 0.47	**36.** 0.87	**37.** 0.03	**38.** 0.01	**39.** 8.7
40. 4	**41.** 0.334	**42.** 0.889	**43.** 0.75	**44.** 0.99
45. 0.4	**46.** 0.5	**47.** 0.006	**48.** 0.008	**49.** 0.017
50. 0.024	**51.** 0.2718	**52.** 0.8911	**53.** 0.0239	**54.** 0.00073

Find percent notation for the decimal notation(s) in each sentence.

55. *Recycling Aluminum Cans.* Over 0.651 of all aluminum cans are recycled.

Source: earth911.com

56. *Wasting Food.* Americans waste an estimated 0.27 of the food available for consumption. The waste occurs in restaurants, supermarkets, cafeterias, and household kitchens.

Source: *New York Times*, "One Country's Table Scraps, Another Country's Meal," by Andrew Martin, May 18, 2008

57. *Dining Together.* In 2012, 0.34 of American families dined together four or five times per week.

Source: unitedfamiliesinternational.wordpress.com

58. *Age 65 and Older.* In Alaska, 0.057 of the residents are age 65 and older. In Florida, 0.176 are age 65 and older.

Source: U.S. Census Bureau

59. *Residents Aged 15 or Younger.* In Haiti, 0.359 of the residents are age 15 or younger. In the United States, 0.2 of the residents are age 15 or younger.

Source: *The World Almanac 2012*

60. *Graduation Rates.* In 2010, the high school graduation rate in the United States was 0.742. The dropout rate was 0.034.

Source: National Center for Educational Statistics

Find decimal notation for each percent notation in the graph.

61. **62.**

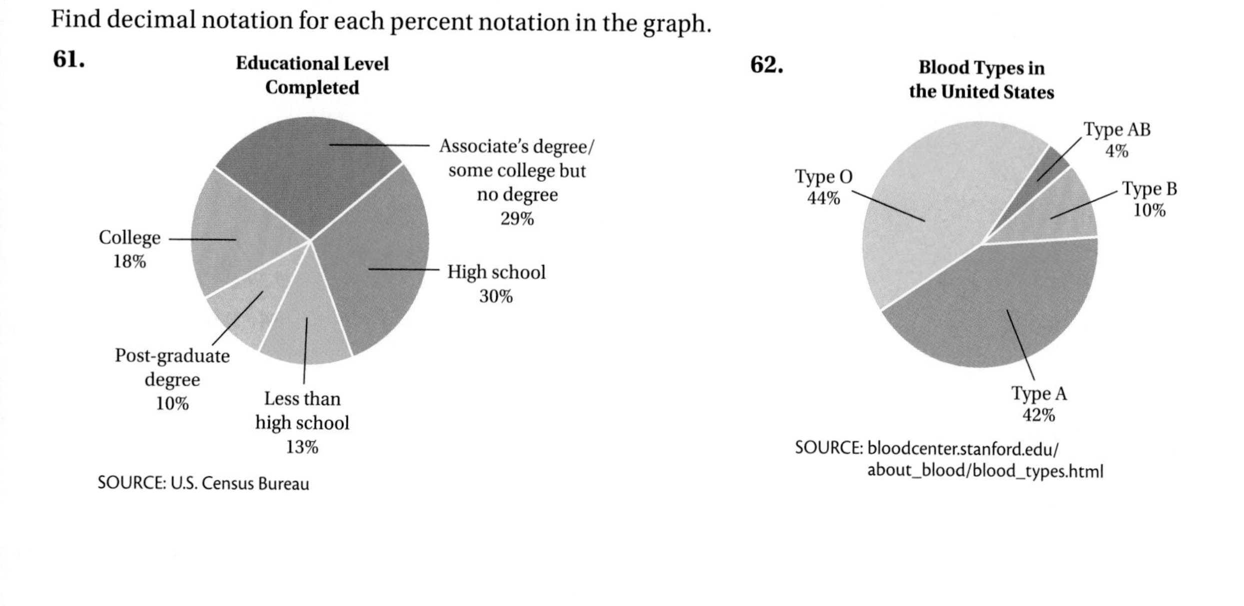

Skill Maintenance

Find all the factors of each number. [3.1a]

63. 84

64. 620

65. Find the LCM of 18 and 60. [4.1a]

66. Find the prime factorization of 90. [3.1d]

67. Solve: $\frac{5}{8} + y = -\frac{13}{16}$. [4.3c]

68. Simplify: $(12 - 3)^2 - 9 + 5^2$. [1.9c]

Synthesis

Find percent notation. (*Hint*: Multiply by a form of 1 and obtain a denominator of 100.)

69. $\frac{1}{2}$

70. $\frac{3}{4}$

71. $\frac{7}{10}$

72. $\frac{2}{5}$

Find percent notation for each shaded area.

73.

74.

Percent Notation and Fraction Notation

7.2

OBJECTIVES

a Convert from fraction notation to percent notation.

b Convert from percent notation to fraction notation.

SKILL TO REVIEW

Objective 5.5a: Convert from fraction notation to decimal notation.

Find decimal notation.

1. $\frac{11}{16}$ **2.** $\frac{5}{9}$

a CONVERTING FROM FRACTION NOTATION TO PERCENT NOTATION

Consider the fraction notation $\frac{7}{8}$. To convert to percent notation, we use two skills that we already have. We first find decimal notation by dividing: $7 \div 8$.

$$\begin{array}{r} 0.875 \\ 8\overline{)7.000} \\ \underline{6\,4} \\ 60 \\ \underline{56} \\ 40 \\ \underline{40} \\ 0 \end{array} \qquad \frac{7}{8} = 0.875$$

Then we convert the decimal notation to percent notation. We move the decimal point two places to the right

0.87.5

and write a % symbol:

$$\frac{7}{8} = 87.5\%, \text{ or } 87\tfrac{1}{2}\%. \qquad 0.5 = \tfrac{1}{2}$$

To convert from fraction notation to percent notation,	$\frac{3}{5}$	Fraction notation
a) find decimal notation by division, and	$\begin{array}{r} 0.6 \\ 5\overline{)3.0} \\ \underline{3\,0} \\ 0 \end{array}$	
b) convert the decimal notation to percent notation.	$0.6 = 0.60 = 60\%$ $\frac{3}{5} = 60\%$	Percent notation

EXAMPLE 1 Find percent notation for $\frac{1}{6}$.

a) We first find decimal notation by division.

$$\begin{array}{r} 0.166 \\ 6\overline{)1.000} \\ \underline{6} \\ 40 \\ \underline{36} \\ 40 \\ \underline{36} \\ 4 \end{array}$$

We get a repeating decimal; $0.16\overline{6}$.

CALCULATOR CORNER

Converting from Fraction Notation to Percent Notation A calculator can be used to convert from fraction notation to percent notation. We simply perform the division on the calculator and then use the percent key. To convert $\frac{17}{40}$ to percent notation, for example, we press [1] [7] [÷] [4] [0] [2nd] [%], or [1] [7] [÷] [4] [0] [SHIFT] [%]. The display reads [42.5], so $\frac{17}{40} = 42.5\%$.

EXERCISES Use a calculator to find percent notation. Round to the nearest hundredth of a percent.

1. $\frac{13}{25}$ **2.** $\frac{5}{13}$

3. $\frac{43}{39}$ **4.** $\frac{12}{7}$

5. $\frac{217}{364}$ **6.** $\frac{2378}{8401}$

Answers

Skill to Review:

1. 0.6875 **2.** $0.\overline{5}$

b) Next, we convert the decimal notation to percent notation. We move the decimal point two places to the right and write a % symbol.

$0.16.\overline{6}$

$\frac{1}{6} = 16.\overline{6}\%$, or $16\frac{2}{3}\%$ $\qquad 0.\overline{6} = \frac{2}{3}$

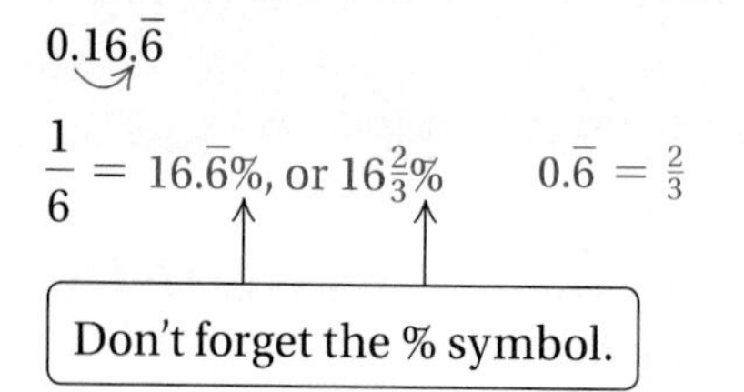

Find percent notation.

1. $\frac{5}{6}$ **2.** $\frac{1}{4}$

◀ **Do Exercises 1 and 2.**

EXAMPLE 2 *First Language.* The first language of approximately $\frac{3}{16}$ of the world's population is Chinese. Find percent notation for $\frac{3}{16}$.

Sources: *National Geographic*, "Languages at Risk," Virginia W. Mason, July 2012; U.S. Census Bureau; *The CIA World Factbook 2012*

a) Find decimal notation by division.

```
     0.1 8 7 5
16 ) 3.0 0 0 0
     1 6
     1 4 0
     1 2 8
       1 2 0
       1 1 2
           8 0
           8 0
             0
```

$\frac{3}{16} = 0.1875$

b) Convert the answer to percent notation.

$0.18.75$

$\frac{3}{16} = 18.75\%$, or $18\frac{3}{4}\%$

3. Water is the single most abundant chemical in the human body. The body is about $\frac{2}{3}$ water. Find percent notation for $\frac{2}{3}$.

4. Find percent notation: $\frac{5}{8}$.

◀ **Do Exercises 3 and 4.**

Answers

1. $83.\overline{3}\%$, or $83\frac{1}{3}\%$ **2.** 25%
3. $66.\overline{6}\%$, or $66\frac{2}{3}\%$ **4.** 62.5%

In some cases, division is not the fastest way to convert a fraction to percent notation. The following are some optional ways in which the conversion might be done.

EXAMPLE 3 Find percent notation for $\frac{69}{100}$.

We use the definition of percent as a ratio.

$$\frac{69}{100} = 69\%$$

EXAMPLE 4 Find percent notation for $\frac{17}{20}$.

We want to multiply by 1 to get 100 in the denominator. We think of what we must multiply 20 by in order to get 100. That number is 5, so we multiply by 1 using $\frac{5}{5}$.

$$\frac{17}{20} \cdot \frac{5}{5} = \frac{85}{100} = 85\%$$

Note that this shortcut works only when the denominator is a factor of 100.

EXAMPLE 5 Find percent notation for $\frac{18}{25}$.

$$\frac{18}{25} = \frac{18}{25} \cdot \frac{4}{4} = \frac{72}{100} = 72\%$$

Do Exercises 5–8.

Find percent notation.

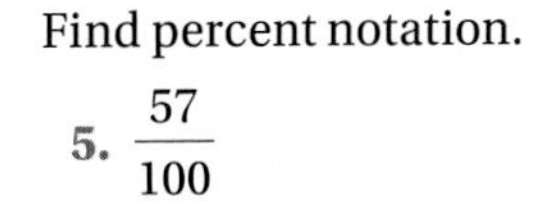

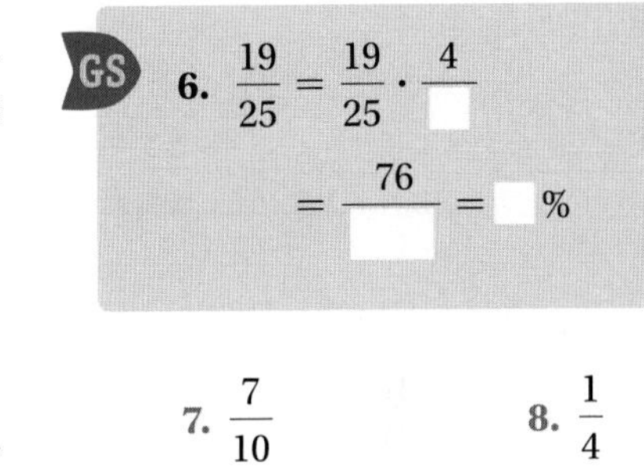

7. $\frac{7}{10}$

8. $\frac{1}{4}$

b CONVERTING FROM PERCENT NOTATION TO FRACTION NOTATION

To convert from percent notation to fraction notation,	30%	Percent notation
a) use the definition of percent as a ratio, and	$\frac{30}{100}$	
b) simplify, if possible.	$\frac{3}{10}$	Fraction notation

EXAMPLE 6 Find fraction notation for 75%.

$$75\% = \frac{75}{100} \quad \text{Using the definition of percent}$$

$$= \frac{3 \cdot 25}{4 \cdot 25} = \frac{3}{4} \cdot \frac{25}{25} = \frac{3}{4} \quad \text{Simplifying}$$

Answers

5. 57% **6.** 76% **7.** 70% **8.** 25%

Guided Solution:

6. 4, 100, 76

Find fraction notation.

9. 60%

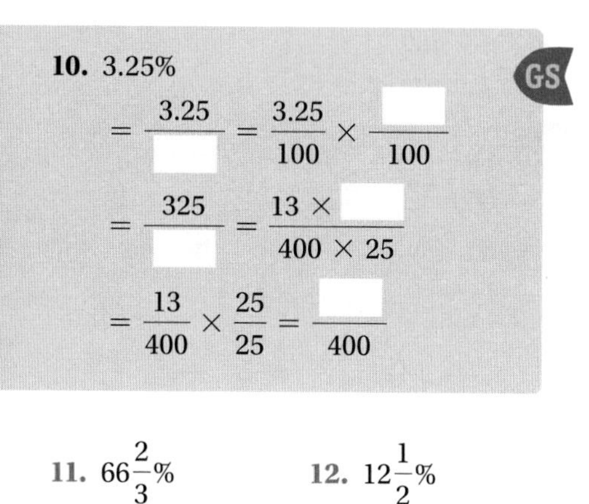

10. 3.25%

$= \frac{3.25}{\square} = \frac{3.25}{100} \times \frac{\square}{100}$

$= \frac{325}{\square} = \frac{13 \times \square}{400 \times 25}$

$= \frac{13}{400} \times \frac{25}{25} = \frac{\square}{400}$

11. $66\frac{2}{3}\%$

12. $12\frac{1}{2}\%$

EXAMPLE 7 Find fraction notation for 62.5%.

$$62.5\% = \frac{62.5}{100} \quad \text{Using the definition of percent}$$

$$= \frac{62.5}{100} \times \frac{10}{10} \quad \text{Multiplying by 1 to eliminate the decimal point in the numerator}$$

$$= \frac{625}{1000}$$

$$\left.\begin{aligned} &= \frac{5 \cdot 125}{8 \cdot 125} = \frac{5}{8} \cdot \frac{125}{125} \\ &= \frac{5}{8} \end{aligned}\right\} \quad \text{Simplifying}$$

EXAMPLE 8 Find fraction notation for $16\frac{2}{3}\%$.

$$16\frac{2}{3}\% = \frac{50}{3}\% \quad \text{Converting from the mixed numeral to fraction notation}$$

$$= \frac{50}{3} \times \frac{1}{100} \quad \text{Using the definition of percent}$$

$$\left.\begin{aligned} &= \frac{50 \cdot 1}{3 \cdot 50 \cdot 2} = \frac{1}{3 \cdot 2} \cdot \frac{50}{50} \\ &= \frac{1}{6} \end{aligned}\right\} \quad \text{Simplifying}$$

◀ **Do Exercises 9–12.**

The table below lists fraction, decimal, and percent equivalents that are used so often it would speed up your work if you memorized them. For example, $\frac{1}{3} = 0.\overline{3}$, so we say that the **decimal equivalent** of $\frac{1}{3}$ is $0.\overline{3}$, or that $0.\overline{3}$ has the **fraction equivalent** $\frac{1}{3}$. This table also appears on the inside back cover of the book.

FRACTION, DECIMAL, AND PERCENT EQUIVALENTS

FRACTION NOTATION	$\frac{1}{10}$	$\frac{1}{8}$	$\frac{1}{6}$	$\frac{1}{5}$	$\frac{1}{4}$	$\frac{3}{10}$	$\frac{1}{3}$	$\frac{3}{8}$	$\frac{2}{5}$	$\frac{1}{2}$	$\frac{3}{5}$	$\frac{5}{8}$	$\frac{2}{3}$	$\frac{7}{10}$	$\frac{3}{4}$	$\frac{4}{5}$	$\frac{5}{6}$	$\frac{7}{8}$	$\frac{9}{10}$	$\frac{1}{1}$
DECIMAL NOTATION	0.1	0.125	$0.16\overline{6}$	0.2	0.25	0.3	$0.\overline{333}$	0.375	0.4	0.5	0.6	0.625	$0.\overline{666}$	0.7	0.75	0.8	$0.83\overline{3}$	0.875	0.9	1
PERCENT NOTATION	10%	12.5%, or $12\frac{1}{2}\%$	$16.\overline{6}\%$, or $16\frac{2}{3}\%$	20%	25%	30%	$33.\overline{3}\%$, or $33\frac{1}{3}\%$	37.5%, or $37\frac{1}{2}\%$	40%	50%	60%	62.5%, or $62\frac{1}{2}\%$	$66.\overline{6}\%$, or $66\frac{2}{3}\%$	70%	75%	80%	$83.\overline{3}\%$, or $83\frac{1}{3}\%$	87.5%, or $87\frac{1}{2}\%$	90%	100%

Find fraction notation.

13. $33.\overline{3}\%$

14. $83.\overline{3}\%$

EXAMPLE 9 Find fraction notation for $16.\overline{6}\%$.

We can use the table above or recall that $16.\overline{6}\% = 16\frac{2}{3}\% = \frac{1}{6}$. We can also recall from our work with repeating decimals in Chapter 5 that $0.\overline{6} = \frac{2}{3}$. Then we have $16.\overline{6}\% = 16\frac{2}{3}\%$ and can proceed as in Example 8.

◀ **Do Exercises 13 and 14.**

Answers

9. $\frac{3}{5}$ 10. $\frac{13}{400}$ 11. $\frac{2}{3}$ 12. $\frac{1}{8}$ 13. $\frac{1}{3}$ 14. $\frac{5}{6}$

Guided Solution:

10. 100, 100, 10,000, 25, 13

7.2 Exercise Set

For Extra Help
MyMathLab® MathXL®
PRACTICE WATCH READ REVIEW

Reading Check

Match each fraction with the equivalent decimal notation from the list at the right. Some choices will not be used.

RC1. $\frac{1}{8}$ ______

RC2. $\frac{3}{8}$ ______

RC3. $\frac{5}{8}$ ______

RC4. $\frac{7}{8}$ ______

RC5. $\frac{1}{5}$ ______

RC6. $\frac{2}{5}$ ______

RC7. $\frac{3}{5}$ ______

RC8. $\frac{4}{5}$ ______

a) 0.875
b) 0.2
c) 0.125
d) 0.4
e) 0.375
f) 0.1
g) 0.625
h) 0.6
i) 0.8
j) 0.675

a Find percent notation.

1. $\frac{41}{100}$
2. $\frac{36}{100}$
3. $\frac{5}{100}$
4. $\frac{1}{100}$
5. $\frac{2}{10}$
6. $\frac{7}{10}$

7. $\frac{3}{10}$
8. $\frac{9}{10}$
9. $\frac{1}{2}$
10. $\frac{3}{4}$
11. $\frac{7}{8}$
12. $\frac{1}{8}$

13. $\frac{4}{5}$
14. $\frac{2}{5}$
15. $\frac{2}{3}$
16. $\frac{1}{3}$
17. $\frac{1}{6}$
18. $\frac{5}{6}$

19. $\frac{3}{16}$
20. $\frac{11}{16}$
21. $\frac{13}{16}$
22. $\frac{7}{16}$
23. $\frac{4}{25}$
24. $\frac{17}{25}$

25. $\frac{1}{20}$
26. $\frac{31}{50}$
27. $\frac{17}{50}$
28. $\frac{3}{20}$

Find percent notation for the fraction notation in each sentence.

29. *Heart Transplants.* In the United States in 2006, $\frac{2}{25}$ of the organ transplants were heart transplants and $\frac{59}{100}$ were kidney transplants.

Source: 2007 OPTN/SRTR Annual Report, Table 1.7

30. *Car Colors.* The four most popular colors for 2006 compact/sports cars were silver, gray, black, and red. Of all cars in this category, $\frac{9}{50}$ were silver, $\frac{3}{20}$ gray, $\frac{3}{20}$ black, and $\frac{3}{20}$ red.

Sources: Ward's Automotive Group; DuPont Automotive Products

In Exercises 31–36, write percent notation for the fractions in the pie chart below.

How Food Dollars Are Spent

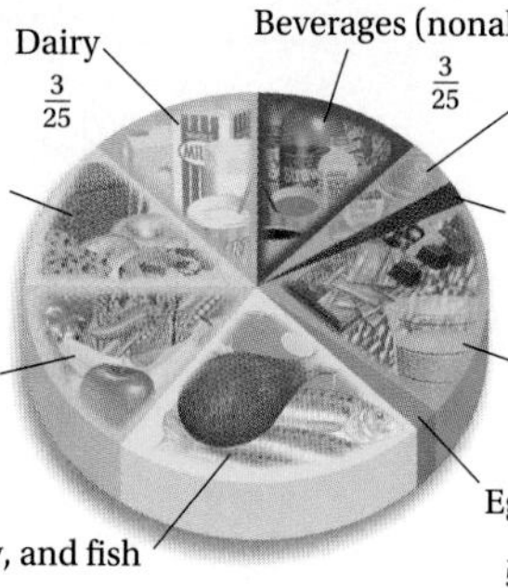

Dairy $\frac{3}{25}$

Beverages (nonalcoholic) $\frac{3}{25}$

Sugar $\frac{1}{25}$

Cereal and baked goods $\frac{13}{100}$

Fats and oils $\frac{3}{100}$

Fruits and vegetables $\frac{3}{20}$

Other $\frac{9}{50}$

Eggs $\frac{1}{50}$

Meat, poultry, and fish $\frac{11}{50}$

SOURCES: U.S. Bureau of Labor Statistics; Consumer Price Index; *The Hoosier Farmer*, Summer 2008

31. $\frac{11}{50}$

32. $\frac{9}{50}$

33. $\frac{3}{25}$

34. $\frac{1}{25}$

35. $\frac{3}{20}$

36. $\frac{13}{100}$

b Find fraction notation. Simplify.

37. 85%

38. 55%

39. 62.5%

40. 12.5%

41. $33\frac{1}{3}\%$

42. $83\frac{1}{3}\%$

43. $16.\overline{6}\%$

44. $66.\overline{6}\%$

45. 7.25%

46. 4.85%

47. 0.8%

48. 0.2%

49. $25\frac{3}{8}\%$

50. $48\frac{7}{8}\%$

51. $78\frac{2}{9}\%$

52. $16\frac{5}{9}\%$

53. $64\frac{7}{11}\%$

54. $73\frac{3}{11}\%$

55. 150%

56. 110%

57. 0.0325%

58. 0.419%

59. $33.\overline{3}\%$

60. $83.\overline{3}\%$

In Exercises 61–66, find fraction notation for the percent notations in the table below.

U.S. POPULATION BY SELECTED AGE CATEGORIES
(Data have been rounded to the nearest percent.)

AGE CATEGORY	PERCENT OF POPULATION
5–17 years	18%
18–24 years	10
15–44 years	41
18 years and older	76
65 years and older	13
75 years and older	6

SOURCES: U.S. Census Bureau; 2010 American Community Survey

61. 6%

62. 18%

63. 13%

64. 41%

65. 76%

66. 10%

Find fraction notation for the percent notation in each sentence.

67. A $\frac{3}{4}$-cup serving of Post Selects Great Grains cereal with $\frac{1}{2}$ cup of fat-free milk satisfies 15% of the minimum daily requirement for calcium.

Source: Kraft Foods Global, Inc.

68. A 1.8-oz serving of Frosted Mini-Wheats®, Blueberry Muffin, with $\frac{1}{2}$ cup of fat-free milk satisfies 35% of the minimum daily requirement for Vitamin B_{12}.

Source: Kellogg, Inc.

69. In 2006, 20.9% of Americans age 18 and older smoked cigarettes.

Sources: *Washington Post*, March 9, 2006; U.S. Centers for Disease Control and Prevention

70. In 2010, 11.9% of Californians age 18 and older smoked cigarettes.

Source: California Department of Public Health

Complete each table.

71.

Fraction Notation	Decimal Notation	Percent Notation
$\frac{1}{8}$		12.5%, or $12\frac{1}{2}$%
$\frac{1}{6}$		
		20%
	0.25	
		$33.\overline{3}$%, or $33\frac{1}{3}$%
		37.5%, or $37\frac{1}{2}$%
		40%
$\frac{1}{2}$		

72.

Fraction Notation	Decimal Notation	Percent Notation
$\frac{3}{5}$		
	0.625	
$\frac{2}{3}$		
	0.75	75%
$\frac{4}{5}$		
$\frac{5}{6}$		$83.\overline{3}$%, or $83\frac{1}{3}$%
$\frac{7}{8}$		87.5%, or $87\frac{1}{2}$%
		100%

73.

Fraction Notation	Decimal Notation	Percent Notation
	0.5	
$\frac{1}{3}$		
		25%
		$16.\overline{6}\%$, or $16\frac{2}{3}\%$
	0.125	
$\frac{3}{4}$		
	$0.8\overline{3}$	
$\frac{3}{8}$		

74.

Fraction Notation	Decimal Notation	Percent Notation
		40%
		62.5%, or $62\frac{1}{2}\%$
	0.875	
$\frac{1}{1}$		
	0.6	
	$0.\overline{6}$	
$\frac{1}{5}$		

Skill Maintenance

Solve.

75. $13 \cdot x = 910$ [1.7b]

76. $15 \cdot y = 75$ [1.7b]

77. $0.05 \times b = -20$ [5.4b]

78. $3 = 0.16 \times b$ [5.4b]

79. $\frac{24}{37} = \frac{15}{x}$ [6.3b]

80. $\frac{17}{18} = \frac{x}{27}$ [6.3b]

81. $\frac{9}{10} = \frac{x}{5}$ [6.3b]

82. $\frac{7}{x} = \frac{4}{5}$ [6.3b]

Convert to a mixed numeral. [4.4a]

83. $\frac{75}{4}$

84. $-\frac{67}{9}$

Convert from a mixed numeral to fraction notation. [4.4a]

85. $-101\frac{1}{2}$

86. $20\frac{9}{10}$

Synthesis

Write percent notation.

87. $2.5\overline{74631}$

88. $\frac{54}{999}$

Write decimal notation.

89. $\frac{729}{7}\%$

90. $\frac{19}{12}\%$

91. Arrange the following numbers from smallest to largest.

$$16\frac{1}{6}\%,\ 1.6,\ \frac{1}{6}\%,\ \frac{1}{2},\ 0.2,\ 1.6\%,\ 1\frac{1}{6}\%,\ 0.5\%,\ \frac{2}{7}\%,\ 0.\overline{54}$$

Solving Percent Problems Using Percent Equations

7.3

OBJECTIVES

a Translate percent problems to percent equations.

b Solve basic percent problems.

SKILL TO REVIEW

Objective 5.4b: Solve equations of the type $a \cdot x = b$, where a and b may be in decimal notation.

Solve.

1. $0.05 \cdot x = 830$
2. $8 \cdot y = 40.648$

a TRANSLATING TO EQUATIONS

To solve a problem involving percents, it is helpful to translate first to an equation. To distinguish the method discussed in this section from that of Section 7.4, we will call these *percent equations.*

KEY WORDS IN PERCENT TRANSLATIONS

"**Of**" translates to "·" or "×".
"**What**" translates to any letter.
"**Is**" translates to "=".
"**%**" translates to "$\times \frac{1}{100}$" or "× 0.01".

EXAMPLES Translate each of the following.

1. 23% of 5 is what?

$$23\% \cdot 5 = a$$

This is a *percent equation.*

2. What is 11% of 49?

$$a = 11\% \cdot 49$$

Any letter can be used.

Do Margin Exercises 1 and 2. ▶

Translate to an equation. Do not solve.

1. 12% of 50 is what?
2. What is 40% of 60?

EXAMPLES Translate each of the following.

3. 3 is 10% of what?

$$3 = 10\% \cdot b$$

4. 45% of what is 23?

$$45\% \times b = 23$$

Do Exercises 3 and 4. ▶

Translate to an equation. Do not solve.

3. 45 is 20% of what?
4. 120% of what is 60?

EXAMPLES Translate each of the following.

5. 10 is what percent of 20?

$$10 = p \times 20$$

6. What percent of 50 is 7?

$$p \cdot 50 = 7$$

Do Exercises 5 and 6. ▶

Translate to an equation. Do not solve.

5. 16 is what percent of 40?
6. What percent of 84 is 10.5?

Answers

Skill to Review:
1. 16,600 **2.** 5.081

Margin Exercises:
1. $12\% \times 50 = a$ **2.** $a = 40\% \times 60$
3. $45 = 20\% \times b$ **4.** $120\% \times b = 60$
5. $16 = p \times 40$ **6.** $p \times 84 = 10.5$

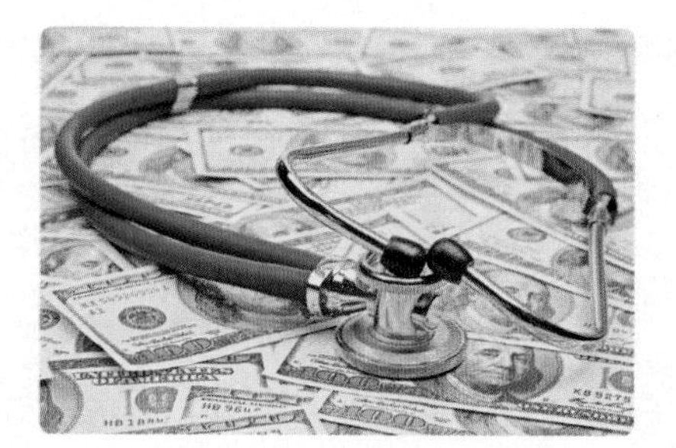

The U.S. gross domestic product for 2010 was approximately $14.66 trillion. Of that amount, Americans spent 17.9% on health care. What was spent on health care? (See Example 7.)

Sources: *EconPost*, April 25, 2011; kaiseredu.org

b SOLVING PERCENT PROBLEMS

In solving percent problems, we use the *Translate* and *Solve* steps in the problem-solving strategy used throughout this text.

Percent problems are actually of three different types. Although the method we present does *not* require that you be able to identify which type you are solving, it is helpful to know them. Each of the three types of percent problems depends on which of the three pieces of information is missing.

1. **Finding the *amount* (the result of taking the percent)**

   Example: What is 25% of 60?

   Translation: $a = 25\% \cdot 60$

2. **Finding the *base* (the number you are taking the percent of)**

   Example: 15 is 25% of what?

   Translation: $15 = 25\% \cdot b$

3. **Finding the *percent number* (the percent itself)**

   Example: 15 is what percent of 60?

   Translation: $15 = p \cdot 60$

Finding the Amount

EXAMPLE 7 What is 17.9% of $14,660,000,000,000?

Translate: $a = 17.9\% \times 14{,}660{,}000{,}000{,}000.$

Solve: The letter is by itself. To solve the equation, we convert 17.9% to decimal notation and multiply:

$$a = 17.9\% \times 14{,}660{,}000{,}000{,}000$$
$$= 0.179 \times 14{,}660{,}000{,}000{,}000 = 2{,}624{,}140{,}000{,}000.$$

Thus, $2,624,140,000,000 is 17.9% of $14,660,000,000,000. The answer is $2,624,140,000,000, or about $2.6 trillion.

7. Solve:

   What is 12% of $50?

◀ **Do Exercise 7.**

EXAMPLE 8 120% of 42 is what?

Translate: $120\% \times 42 = a.$

Solve: The letter is by itself. To solve the equation, we carry out the calculation:

$$a = 120\% \times 42$$
$$a = 1.2 \times 42 \qquad 120\% = 1.2$$
$$a = 50.4.$$

Thus, 120% of 42 is 50.4. The answer is 50.4.

8. Solve:

   64% of 55 is what?

◀ **Do Exercise 8.**

Answers

7. $6 8. 35.2

Finding the Base

EXAMPLE 9 8% of what is 32?

Translate: $8\% \times b = 32.$

Solve: This time the letter is *not* by itself. To solve the equation, we divide by 8% on both sides:

$$\frac{8\% \times b}{8\%} = \frac{32}{8\%} \qquad \text{Dividing by 8\% on both sides}$$

$$b = \frac{32}{0.08} \qquad 8\% = 0.08$$

$$b = 400.$$

Thus, 8% of 400 is 32. The answer is 400.

EXAMPLE 10 \$3 is 16% of what?

Translate: \$3 is 16% of what?

$$3 = 16\% \times b$$

Solve: To solve the equation, we divide by 16% on both sides:

$$\frac{3}{16\%} = \frac{16\% \times b}{16\%} \qquad \text{Dividing by 16\% on both sides}$$

$$\frac{3}{0.16} = b \qquad 16\% = 0.16$$

$$18.75 = b.$$

Thus, \$3 is 16% of \$18.75. The answer is \$18.75.

Do Exercises 9 and 10.

A survey of a group of people found that 8% of the group, or 32 people, chose cookies and cream as their favorite ice cream flavor. How many people were surveyed? (See Example 9.)

Source: Rasmussen Reports Survey

Solve.

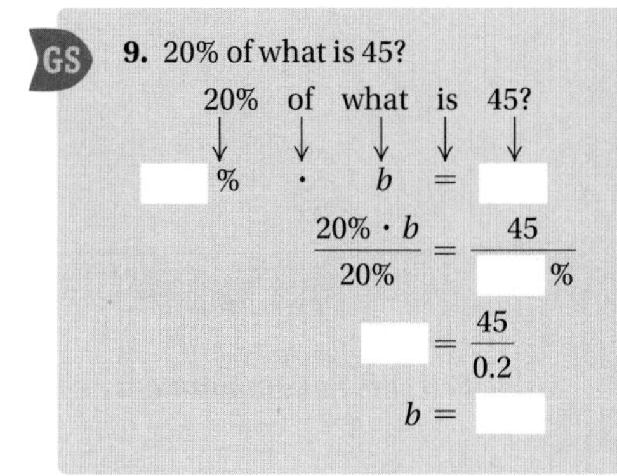

GS **9.** 20% of what is 45?

20% of what is 45?

$$\square\,\% \cdot b = \square$$

$$\frac{20\% \cdot b}{20\%} = \frac{45}{\square\,\%}$$

$$\square = \frac{45}{0.2}$$

$$b = \square$$

10. \$60 is 120% of what?

Finding the Percent Number

In solving these problems, you *must* remember to convert to percent notation after you have solved the equation.

EXAMPLE 11 414,000 is what percent of 621,000?

Translate: 414,000 is what percent of 621,000?

$$414{,}000 = p \times 621{,}000$$

Solve: To solve the equation, we divide by 621,000 on both sides and convert the result to percent notation:

$$p \times 621{,}000 = 414{,}000$$

$$\frac{p \times 621{,}000}{621{,}000} = \frac{414{,}000}{621{,}000} \qquad \text{Dividing by 621,000 on both sides}$$

$$p = 0.666\ldots \qquad \text{Converting to decimal notation}$$

$$p = 66.\overline{6}\%, \text{ or } 66\tfrac{2}{3}\%. \qquad \text{Converting to percent notation}$$

Thus, 414,000 is $66\frac{2}{3}\%$ of 621,000. The answer is $66\frac{2}{3}\%$.

In 2011, 621,000 new housing permits were issued in the United States. Of those new houses, 414,000 have porches. What percent of the houses have porches? (See Example 11.)

Source: U.S. Census Bureau

Answers

9. 225 **10.** \$50

Guided Solution:
9. 20, 45, 20, *b*, 225

Solve.

11. 16 is what percent of 40?

16 is what percent of 40?

$16 \;\square\; p \cdot \square$

$\dfrac{16}{\square} = \dfrac{p \cdot \square}{40}$

$\dfrac{16}{40} = p$

$0.4 = p$

$\square\% = p$

12. What percent of $84 is $10.50?

EXAMPLE 12 What percent of $50 is $16?

Translate: What percent of $50 is $16?

$$p \times 50 = 16$$

Solve: To solve the equation, we divide by 50 on both sides and convert the answer to percent notation:

$$\frac{p \times 50}{50} = \frac{16}{50} \quad \text{Dividing by 50 on both sides}$$

$$p = \frac{16}{50}$$

$$p = 0.32$$

$$p = 32\%. \quad \text{Converting to percent notation}$$

Thus, 32% of $50 is $16. The answer is 32%.

Do Exercises 11 and 12.

Caution!

When a question asks "what percent?", be sure to give the answer in percent notation.

CALCULATOR CORNER

Using Percents in Computations Many calculators have a [%] key that can be used in computations. (See the Calculator Corner on page 384.) For example, to find 11% of 49, we press [1] [1] [2nd] [%] [×] [4] [9] [=], or [4] [9] [×] [1] [1] [SHIFT] [%]. The display reads [5.39], so 11% of 49 is 5.39.

In Example 9, we performed the computation 32/8%. To use the [%] key in this computation, we press [3] [2] [÷] [8] [2nd] [%] [=], or [3] [2] [÷] [8] [SHIFT] [%]. The result is 400.

We can also use the [%] key to find the percent number in a problem. In Example 11, for instance, we answered the question "414,000 is what percent of 621,000?" On a calculator, we press [4] [1] [4] [0] [0] [0] [÷] [6] [2] [1] [0] [0] [0] [2nd] [%] [=], or [4] [1] [4] [0] [0] [0] [÷] [6] [2] [1] [0] [0] [0] [SHIFT] [%]. The result is $66.\overline{6}$, so 414,000 is $66.\overline{6}\%$ of 621,000.

EXERCISES Use a calculator to find each of the following.

1. What is 12.6% of $40?

2. 0.04% of 28 is what?

3. 8% of what is 36?

4. $45 is 4.5% of what?

5. 23 is what percent of 920?

6. What percent of $442 is $53.04?

Answers

11. 40% **12.** 12.5%

Guided Solution:

11. =, 40; 40, 40, 40

7.3 Exercise Set

For Extra Help
MyMathLab®
MathXL® PRACTICE | WATCH | READ | REVIEW

Reading Check

Match each question with the correct translation from the list at the right.

RC1. 18 is 40% of what? __________

RC2. What percent of 45 is 18? __________

RC3. What is 40% of 45? __________

RC4. 0.5% of 1200 is what? __________

RC5. 6 is what percent of 1200? __________

RC6. 6 is 0.5% of what? __________

a) $6 = 0.5\% \cdot b$
b) $6 = p \cdot 1200$
c) $18 = 40\% \cdot b$
d) $0.5\% \cdot 1200 = a$
e) $p \cdot 45 = 18$
f) $a = 40\% \cdot 45$

a Translate to an equation. Do not solve.

1. What is 32% of 78?

2. 98% of 57 is what?

3. 89 is what percent of 99?

4. What percent of 25 is 8?

5. 13 is 25% of what?

6. 21.4% of what is 20?

b Translate to an equation and solve.

7. What is 85% of 276?

8. What is 74% of 53?

9. 150% of 30 is what?

10. 100% of 13 is what?

11. What is 6% of \$300?

12. What is 4% of \$45?

13. 3.8% of 50 is what?

14. $33\frac{1}{3}\%$ of 480 is what?
(*Hint*: $33\frac{1}{3}\% = \frac{1}{3}$.)

15. \$39 is what percent of \$50?

16. \$16 is what percent of \$90?

17. 20 is what percent of 10?

18. 60 is what percent of 20?

19. What percent of \$300 is \$150?

20. What percent of \$50 is \$40?

21. What percent of 80 is 100?

22. What percent of 60 is 15?

23. 20 is 50% of what?

24. 57 is 20% of what?

25. 40% of what is $16?

26. 100% of what is $74?

27. 56.32 is 64% of what?

28. 71.04 is 96% of what?

29. 70% of what is 14?

30. 70% of what is 35?

31. What is $62\frac{1}{2}\%$ of 10?

32. What is $35\frac{1}{4}\%$ of 1200?

33. What is 8.3% of $10,200?

34. What is 9.2% of $5600?

35. 2.5% of what is 30.4?

36. 8.2% of what is 328?

Skill Maintenance

Write fraction notation. [5.1b]

37. -0.9375

38. 0.125

Write decimal notation. [5.1b]

39. $\frac{3}{10}$

40. $-\frac{17}{1000}$

Simplify. [1.9c]

41. $3 + (8 - 6) \cdot 2$

42. $2 \cdot 7 - (5 + 1)$

Synthesis

Solve.

43. $2496 is 24% of what amount?
Estimate ________
Calculate ________

44. What is 38.2% of $52,345.79?
Estimate ________
Calculate ________

45. 40% of $18\frac{3}{4}\%$ of $25,000 is what?

Solving Percent Problems Using Proportions*

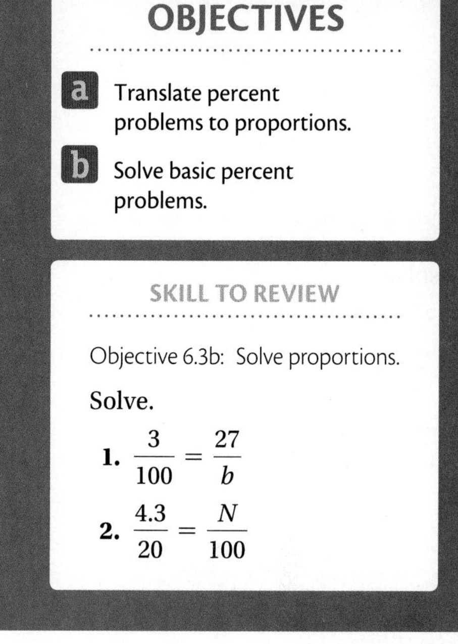

a TRANSLATING TO PROPORTIONS

A percent is a ratio of some number to 100. For example, 5% is the ratio $\frac{5}{100}$. The numbers 7,700,000 and 154,000,000 have the same ratio as 5 and 100.

$$\frac{5}{100} = \frac{7{,}700{,}000}{154{,}000{,}000}$$

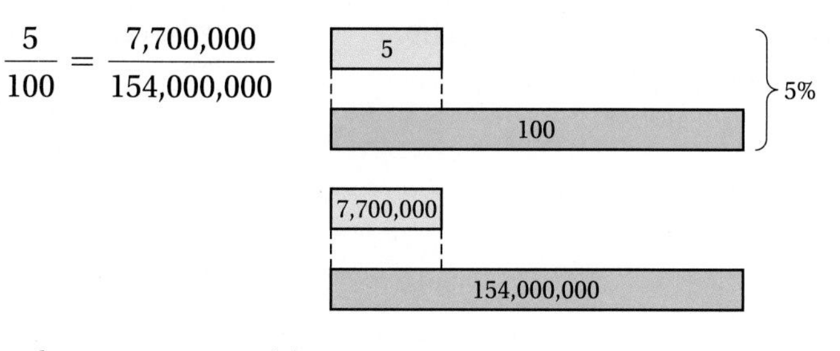

To solve a percent problem using a proportion, we translate as follows:

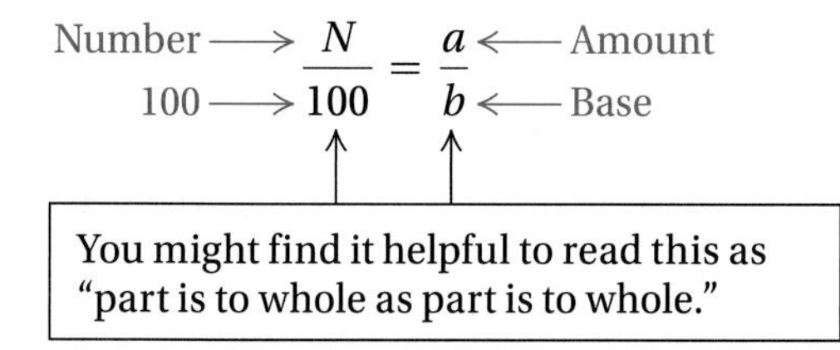

$$\frac{N}{100} = \frac{a}{b}$$

For example, 60% of 25 is 15 translates to

$$\frac{60}{100} = \frac{15}{25} \begin{matrix}\leftarrow \text{Amount} \\ \leftarrow \text{Base}\end{matrix}$$

A clue for translating is that the base, b, corresponds to 100 and usually follows the wording "percent of." Also, $N\%$ always translates to $N/100$. Another aid in translating is to make a comparison drawing. To do this, we start with the percent side and list 0% at the top and 100% near the bottom. Then we estimate where the specified percent—in this case, 60%—is located. The corresponding quantities are then filled in. The base—in this case, 25—always corresponds to 100%, and the amount—in this case, 15—corresponds to the specified percent.

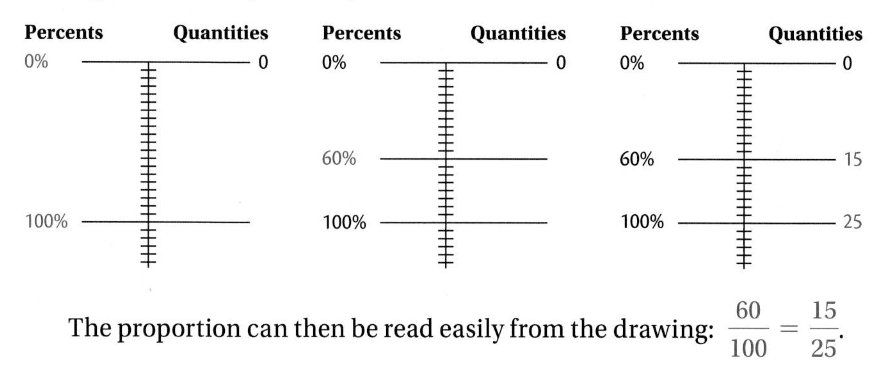

The proportion can then be read easily from the drawing: $\frac{60}{100} = \frac{15}{25}$.

In the United States, 5% of the labor force is age 65 or older. In 2012, there were approximately 154,000,000 people in the labor force. This means that about 7,700,000 workers were age 65 or older.

Sources: U.S. Department of Labor; U.S. Bureau of Labor Statistics

---

**Note*: This section presents an alternative method for solving basic percent problems. You can use either equations or proportions to solve percent problems, but you might prefer one method over the other, or your instructor may direct you to use one method rather than the other.

Answers

Skill to Review:
1. 900 **2.** 21.5

EXAMPLE 1 Translate to a proportion.

23% of 5 is what?

$$\frac{23}{100} = \frac{a}{5}$$

Percents Quantities
0% 0
23% a
100% 5

EXAMPLE 2 Translate to a proportion.

What is 124% of 49?

$$\frac{124}{100} = \frac{a}{49}$$

Percents Quantities
0% 0
100% 49
124% a

Translate to a proportion. Do not solve.

1. 12% of 50 is what?
2. What is 40% of 60?
3. 130% of 72 is what?

◀ **Do Exercises 1–3.**

EXAMPLE 3 Translate to a proportion.

3 is 10% of what?

$$\frac{10}{100} = \frac{3}{b}$$

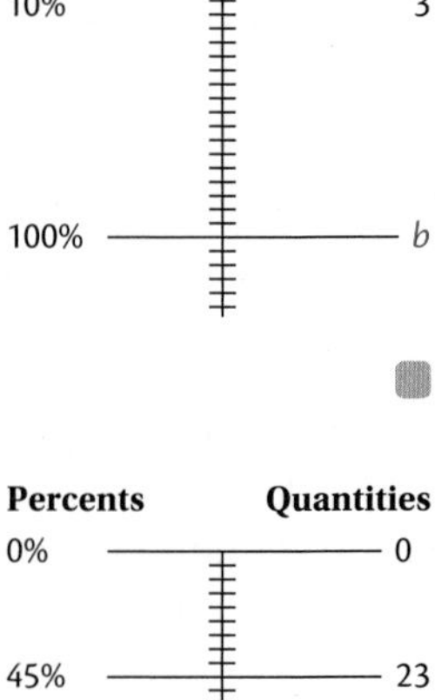

EXAMPLE 4 Translate to a proportion.

45% of what is 23?

$$\frac{45}{100} = \frac{23}{b}$$

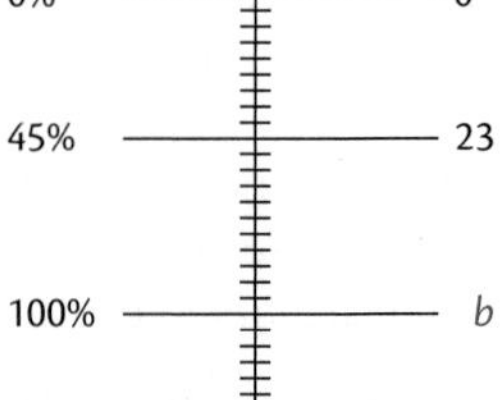

Translate to a proportion. Do not solve.

4. 45 is 20% of what?
5. 120% of what is 60?

◀ **Do Exercises 4 and 5.**

EXAMPLE 5 Translate to a proportion.

10 is what percent of 20?

$$\frac{N}{100} = \frac{10}{20}$$

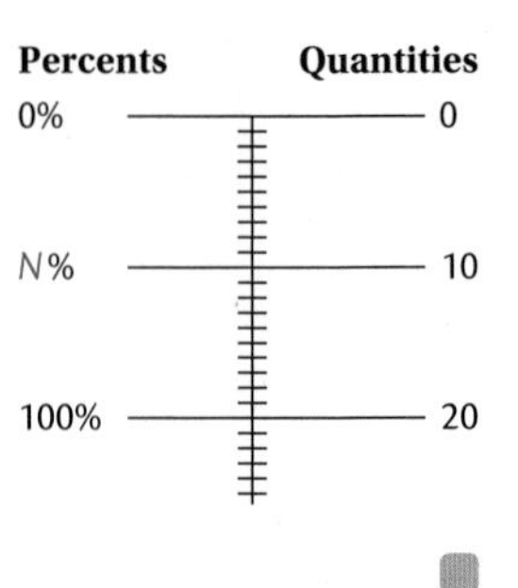

Answers

1. $\frac{12}{100} = \frac{a}{50}$ 2. $\frac{40}{100} = \frac{a}{60}$ 3. $\frac{130}{100} = \frac{a}{72}$
4. $\frac{20}{100} = \frac{45}{b}$ 5. $\frac{120}{100} = \frac{60}{b}$

EXAMPLE 6 Translate to a proportion.

What percent of 50 is 7?

$$\frac{N}{100} = \frac{7}{50}$$

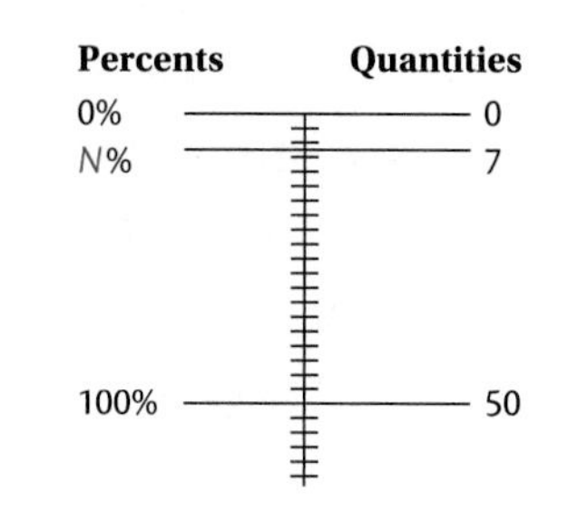

Do Exercises 6 and 7. ▶

Translate to a proportion. Do not solve.

6. 16 is what percent of 40?

7. What percent of 84 is 10.5?

b SOLVING PERCENT PROBLEMS

After a percent problem has been translated to a proportion, we solve as in Section 6.3.

EXAMPLE 7 5% of what is \$20?

Translate: $\frac{5}{100} = \frac{20}{b}$

Solve: $5 \cdot b = 100 \cdot 20$ Equating cross products

$\frac{5 \cdot b}{5} = \frac{100 \cdot 20}{5}$ Dividing by 5

$b = \frac{2000}{5}$

$b = 400$ Simplifying

Thus, 5% of \$400 is \$20. The answer is \$400.

Percents | Quantities: 0% — 0; 5% — 20; 100% — b

Do Exercise 8. ▶

GS **8.** Solve: 20% of what is \$45?

$\frac{20}{\square} = \frac{\square}{b}$

$20 \cdot b = 100 \cdot \square$

$\frac{20 \cdot b}{20} = \frac{100 \cdot 45}{\square}$

$b = \frac{4500}{20}$

$b = \square$

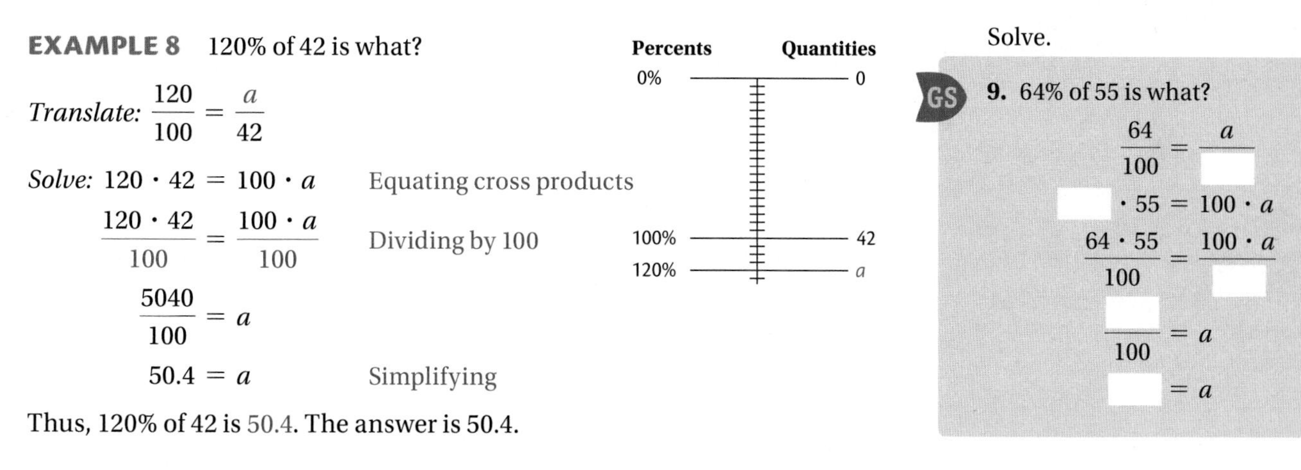

EXAMPLE 8 120% of 42 is what?

Translate: $\frac{120}{100} = \frac{a}{42}$

Solve: $120 \cdot 42 = 100 \cdot a$ Equating cross products

$\frac{120 \cdot 42}{100} = \frac{100 \cdot a}{100}$ Dividing by 100

$\frac{5040}{100} = a$

$50.4 = a$ Simplifying

Thus, 120% of 42 is 50.4. The answer is 50.4.

Percents | Quantities: 0% — 0; 100% — 42; 120% — a

Do Exercises 9 and 10. ▶

Solve.

GS **9.** 64% of 55 is what?

$\frac{64}{100} = \frac{a}{\square}$

$\square \cdot 55 = 100 \cdot a$

$\frac{64 \cdot 55}{100} = \frac{100 \cdot a}{\square}$

$\frac{\square}{100} = a$

$\square = a$

10. What is 12% of 50?

Answers

6. $\frac{N}{100} = \frac{16}{40}$ **7.** $\frac{N}{100} = \frac{10.5}{84}$ **8.** \$225
9. 35.2 **10.** 6

Guided Solutions:

8. 100, 45, 45, 20, 225
9. 55, 64, 100, 3520, 35.2

EXAMPLE 9 210 is $10\frac{1}{2}\%$ of what?

Translate: $\frac{210}{b} = \frac{10.5}{100}$ $\quad 10\frac{1}{2}\% = 10.5\%$

Solve: $210 \cdot 100 = b \cdot 10.5$ Equating cross products

$\frac{210 \cdot 100}{10.5} = \frac{b \cdot 10.5}{10.5}$ Dividing by 10.5

$\frac{21{,}000}{10.5} = b$ Multiplying and simplifying

$2000 = b$ Dividing

Thus, 210 is $10\frac{1}{2}\%$ of 2000. The answer is 2000.

11. Solve:
60 is 120% of what?

◀ **Do Exercise 11.**

EXAMPLE 10 $10 is what percent of $20?

Translate: $\frac{10}{20} = \frac{N}{100}$

Solve: $10 \cdot 100 = 20 \cdot N$ Equating cross products

$\frac{10 \cdot 100}{20} = \frac{20 \cdot N}{20}$ Dividing by 20

$\frac{1000}{20} = N$ Multiplying and simplifying

$50 = N$ Dividing

Note when solving percent problems using proportions that *N* is a percent and need not be converted.

Thus, $10 is 50% of $20. The answer is 50%.

12. Solve: GS
$12 is what percent of $40?

$\frac{12}{40} = \frac{N}{\square}$

$\square \cdot 100 = 40 \cdot N$

$\frac{12 \cdot 100}{\square} = \frac{40 \cdot N}{40}$

$\frac{\square}{40} = N$

$30 = N$

Thus, $12 is 30 $\square$ of $40.

◀ **Do Exercise 12.**

EXAMPLE 11 What percent of 50 is 16?

Translate: $\frac{N}{100} = \frac{16}{50}$

Solve: $50 \cdot N = 100 \cdot 16$ Equating cross products

$\frac{50 \cdot N}{50} = \frac{100 \cdot 16}{50}$ Dividing by 50

$N = \frac{1600}{50}$ Multiplying and simplifying

$N = 32$ Dividing

Thus, 32% of 50 is 16. The answer is 32%.

13. Solve:
What percent of 84 is 10.5?

◀ **Do Exercise 13.**

Answers

11. 50 **12.** 30% **13.** 12.5%

Guided Solution:
12. 100, 12, 40, 1200; %

7.4 Exercise Set

For Extra Help MyMathLab® MathXL® PRACTICE WATCH READ REVIEW

Reading Check

Match each question with the correct translation from the list at the right.

RC1. 70 is 35% of what? ________

RC2. 70 is what percent of 200? ________

RC3. What is 35% of 200? ________

RC4. 74.8 is 110% of what? ________

RC5. What percent of 68 is 74.8? ________

RC6. 110% of 68 is what? ________

a) $\frac{110}{100} = \frac{a}{68}$

b) $\frac{70}{b} = \frac{35}{100}$

c) $\frac{a}{200} = \frac{35}{100}$

d) $\frac{74.8}{68} = \frac{N}{100}$

e) $\frac{70}{200} = \frac{N}{100}$

f) $\frac{74.8}{b} = \frac{110}{100}$

a Translate to a proportion. Do not solve.

1. What is 37% of 74?

2. 66% of 74 is what?

3. 4.3 is what percent of 5.9?

4. What percent of 6.8 is 5.3?

5. 14 is 25% of what?

6. 133% of what is 40?

b Translate to a proportion and solve.

7. What is 76% of 90?

8. What is 32% of 70?

9. 70% of 660 is what?

10. 80% of 920 is what?

11. What is 4% of 1000?

12. What is 6% of 2000?

13. 4.8% of 60 is what?

14. 63.1% of 80 is what?

15. $24 is what percent of $96?

16. $14 is what percent of $70?

17. 102 is what percent of 100?

18. 103 is what percent of 100?

19. What percent of $480 is $120?

20. What percent of $80 is $60?

21. What percent of 160 is 150?

22. What percent of 33 is 11?

23. \$18 is 25% of what?

24. \$75 is 20% of what?

25. 60% of what is 54?

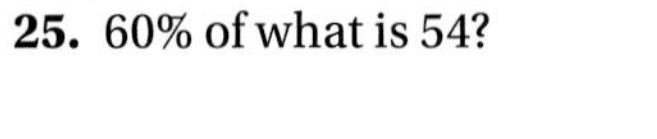

26. 80% of what is 96?

27. 65.12 is 74% of what?

28. 63.7 is 65% of what?

29. 80% of what is 16?

30. 80% of what is 10?

31. What is $62\frac{1}{2}\%$ of 40?

32. What is $43\frac{1}{4}\%$ of 2600?

33. What is 9.4% of \$8300?

34. What is 8.7% of \$76,000?

35. 80.8 is $40\frac{2}{5}\%$ of what?

36. 66.3 is $10\frac{1}{5}\%$ of what?

Skill Maintenance

Solve. [6.3b]

37. $\frac{x}{188} = \frac{2}{47}$

38. $\frac{15}{x} = \frac{3}{800}$

39. $\frac{75}{100} = \frac{n}{20}$

40. $\frac{612}{t} = \frac{72}{244}$

Solve.

41. A recipe for muffins calls for $\frac{1}{2}$ qt of buttermilk, $\frac{1}{3}$ qt of skim milk, and $\frac{1}{16}$ qt of oil. How many quarts of liquid ingredients does the recipe call for? [4.2b]

42. The Ferristown School District purchased $\frac{3}{4}$ ton (T) of clay. If the clay is to be shared equally among the district's 6 art departments, how much will each art department receive? [3.7d]

Synthesis

Solve.

43. 🖩 What is 8.85% of \$12,640?

Estimate ____________

Calculate ____________

44. 🖩 78.8% of what is 9809.024?

Estimate ____________

Calculate ____________

Mid-Chapter Review

Concept Reinforcement

Determine whether each statement is true or false.

__________ **1.** When converting decimal notation to percent notation, move the decimal point two places to the right and write a percent symbol. [7.1b]

__________ **2.** The symbol % is equivalent to $\times$ 0.10. [7.1a]

__________ **3.** Of the numbers $\frac{1}{10}$, 1%, 0.1%, and $\frac{1}{100}$ the smallest number is 0.1%. [7.1b], [7.2a, b]

Guided Solutions

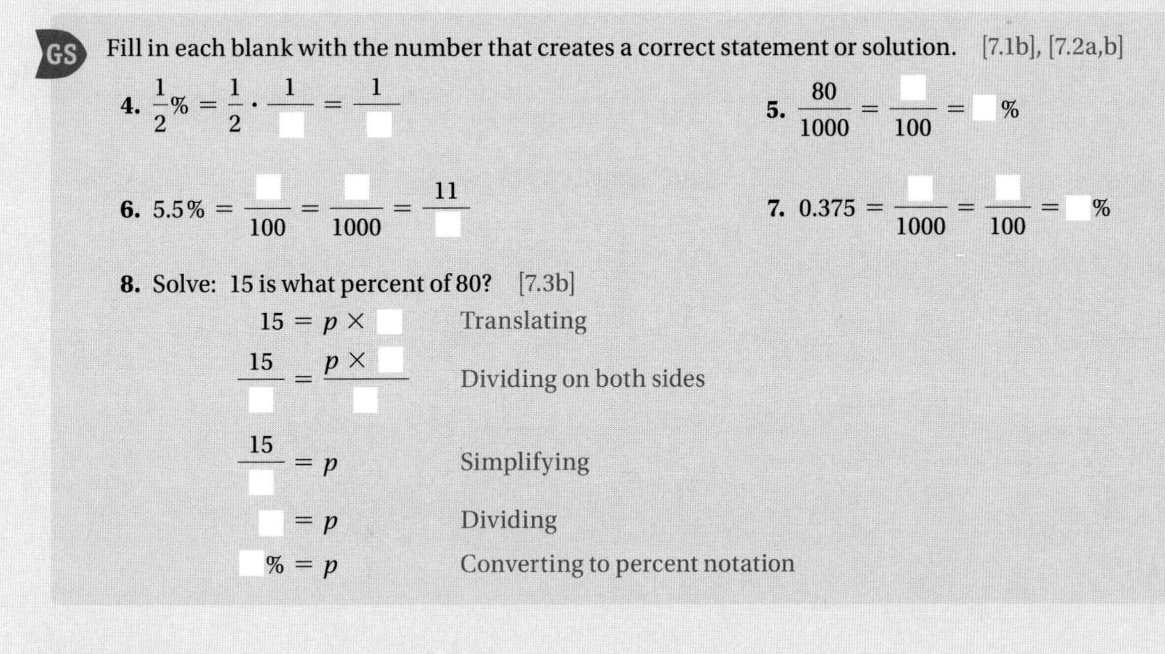

GS Fill in each blank with the number that creates a correct statement or solution. [7.1b], [7.2a,b]

4. $\frac{1}{2}\% = \frac{1}{2} \cdot \frac{1}{\square} = \frac{1}{\square}$

5. $\frac{80}{1000} = \frac{\square}{100} = \square\,\%$

6. $5.5\% = \frac{\square}{100} = \frac{\square}{1000} = \frac{11}{\square}$

7. $0.375 = \frac{\square}{1000} = \frac{\square}{100} = \square\,\%$

8. Solve: 15 is what percent of 80? [7.3b]

$15 = p \times \square$ Translating

$\frac{15}{\square} = \frac{p \times \square}{\square}$ Dividing on both sides

$\frac{15}{\square} = p$ Simplifying

$\square = p$ Dividing

$\square\,\% = p$ Converting to percent notation

Mixed Review

Find decimal notation. [7.1b]

9. 28%

10. 0.15%

11. $5\frac{3}{8}\%$

12. 240%

Find percent notation. [7.1b], [7.2a]

13. 0.71

14. $\frac{9}{100}$

15. 0.3891

16. $\frac{3}{16}$

17. 0.005

18. $\frac{37}{50}$

19. 6

20. $\frac{5}{6}$

Find fraction notation. Simplify. [7.2b]

21. 85%

22. 0.048%

23. $22\frac{3}{4}\%$

24. $16.\overline{6}\%$

Write percent notation for the shaded area. [7.2a]

25.

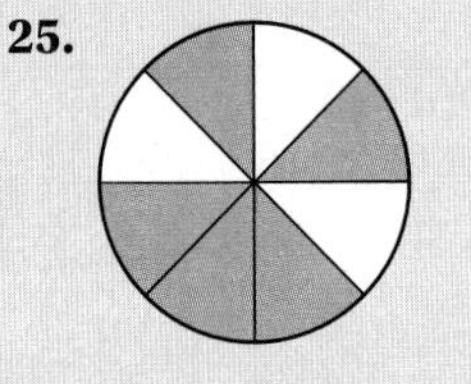

26.

Solve. [7.3b], [7.4b]

27. 25% of what is 14.5?

28. 220 is what percent of 1320?

29. What is 3.2% of 80,000?

30. $17.50 is 35% of what?

31. What percent of $800 is $160?

32. 130% of $350 is what?

33. Arrange the following numbers from smallest to largest. [7.1b], [7.2a, b]

$\frac{1}{2}\%$, 5%, 0.275, $\frac{13}{100}$, 1%, 0.1%, 0.05%, $\frac{3}{10}$, $\frac{7}{20}$, 10%

34. Solve: 8.5 is $2\frac{1}{2}\%$ of what? [7.3b], [7.4b]

A. 3.4 **B.** 21.25

C. 0.2125 **D.** 340

35. Solve: $102,000 is what percent of $3.6 million? [7.3b], [7.4b]

A. $\$2.8\overline{3}$ million **B.** $2\frac{5}{6}\%$

C. $0.028\overline{3}\%$ **D.** $28.\overline{3}\%$

Understanding Through Discussion and Writing

36. Is it always best to convert from fraction notation to percent notation by first finding decimal notation? Why or why not? [7.2a]

37. Suppose we know that 40% of 92 is 36.8. What is a quick way to find 4% of 92? 400% of 92? Explain. [7.3b]

38. In solving Example 10 in Section 7.4 a student simplifies $\frac{10}{20}$ before solving. Is this a good idea? Why or why not? [7.4b]

39. What do the following have in common? Explain. [7.1b], [7.2a, b]

$\frac{23}{16}$, $1\frac{875}{2000}$, 1.4375, $\frac{207}{144}$, $1\frac{7}{16}$, 143.75%, $1\frac{4375}{10,000}$

STUDYING FOR SUCCESS *Make the Most of Your Time in Class*

- ☐ Before each class, try to at least glance at the section in your text that will be covered, so you can concentrate on the instruction in class.
- ☐ Get a great seat! Sitting near the front will help you hear instruction more clearly and avoid distractions.
- ☐ Let your instructor know in advance if you must miss class, and do your best to keep up with any work that you miss.

7.5 Applications of Percent

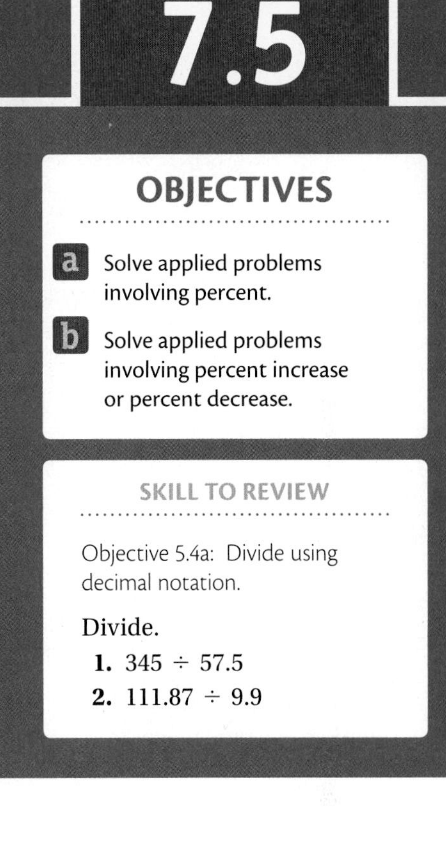

OBJECTIVES

a Solve applied problems involving percent.

b Solve applied problems involving percent increase or percent decrease.

SKILL TO REVIEW

Objective 5.4a: Divide using decimal notation.

Divide.

1. $345 \div 57.5$
2. $111.87 \div 9.9$

a APPLIED PROBLEMS INVOLVING PERCENT

Applied problems involving percent are not always stated in a manner easily translated to an equation. In such cases, it is helpful to rephrase the problem before translating. Sometimes it also helps to make a drawing.

EXAMPLE 1 *Transportation to Work.* In the United States, there are about 154,000,000 workers who are 16 years old or older. Approximately 76.1% of these workers drive to work alone. How many workers drive to work alone?

Transportation to Work in the United States

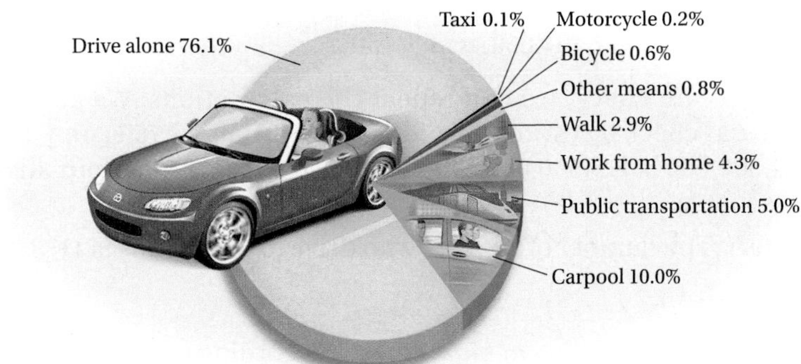

SOURCE: U.S. Bureau of Labor Statistics

1. **Familiarize.** We can simplify the pie chart shown above to help familiarize ourselves with the problem. We let $a =$ the total number of workers who drive to work alone.

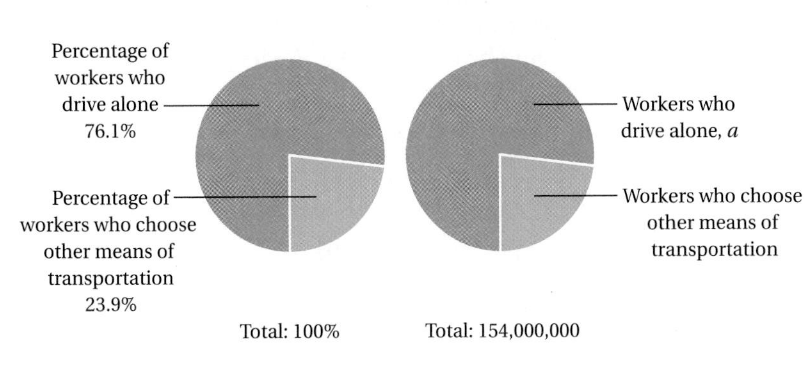

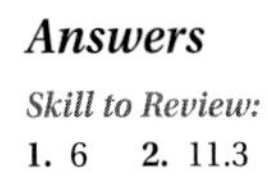

Answers

Skill to Review:
1. 6 2. 11.3

2. **Translate.** There are two ways in which we can translate this problem.

Percent equation (see Section 7.3):

What number is 76.1% of 154,000,000?

$$a = 76.1\% \cdot 154{,}000{,}000$$

Proportion (see Section 7.4):

$$\frac{76.1}{100} = \frac{a}{154{,}000{,}000}$$

3. **Solve.** We now have two ways in which to solve the problem.

Percent equation (see Section 7.3):

$$a = 76.1\% \cdot 154{,}000{,}000$$

We convert 76.1% to decimal notation and multiply:

$$a = 0.761 \times 154{,}000{,}000 = 117{,}194{,}000.$$

Proportion (see Section 7.4):

$$\frac{76.1}{100} = \frac{a}{154{,}000{,}000}$$

$$76.1 \times 154{,}000{,}000 = 100 \cdot a \quad \text{Equating cross products}$$

$$\frac{76.1 \cdot 154{,}000{,}000}{100} = \frac{100 \cdot a}{100} \quad \text{Dividing by 100}$$

$$\frac{11{,}719{,}400{,}000}{100} = a$$

$$117{,}194{,}000 = a \quad \text{Simplifying}$$

4. **Check.** To check, we can repeat the calculations. We also can do a partial check by estimating. Since 76.1% is about 75%, or, $\frac{3}{4}$, and $\frac{3}{4}$ of 154,000,000 is 115,500,000, which is close to 117,194,000, our answer is reasonable.

5. **State.** The number of workers who drive to work alone is 117,194,000.

Do Exercise 1.

1. *Transportation to Work.* There are about 154,000,000 workers 16 years old or older in the United States. Approximately 10.0% of them carpool to work. How many workers carpool to work?

Sources: U.S. Census Bureau; American Community Survey

EXAMPLE 2 *Extinction of Mammals.* According to a study conducted for the International Union for the Conservation of Nature (IUCN), the world's mammals are in danger of an extinction crisis. Of the 5501 species of mammals on Earth, 1139 are on IUCN "vulnerable," "endangered," or "critically endangered" lists. What percent of all mammals are threatened with extinction?

1. **Familiarize.** The question asks for a percent of the world's mammals that are in danger of extinction. We note that 5501 is approximately 5500 and 1139 is approximately 1100. Since 1100 is $\frac{1100}{5500}$, or $\frac{1}{5}$, or 20% of 5500, our answer should be close to 20%. We let p = the percent of mammals that are in danger of extinction.

2. **Translate.** There are two ways in which we can translate this problem.

Percent equation:

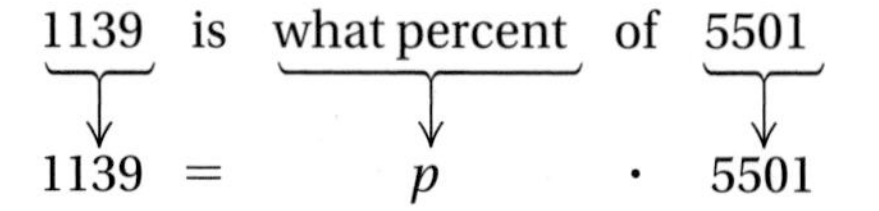

1139 is what percent of 5501

$$1139 = p \cdot 5501$$

Answer

1. 15,400,000

Mammals that are in danger of extinction include those shown here: Barbary macaque, black rhino, Galápagos seal, Malayan tapir, Cuvier's gazelle, Darwin's fox, and indri (a lemur).

Proportion:

$$\frac{N}{100} = \frac{1139}{5501}$$

For proportions, $N\% = p$.

Percents		Quantities
0%		0
N%		1139
100%		5501

3. **Solve.** We now have two ways in which to solve the problem.

Percent equation:

$$1139 = p \cdot 5501$$

$$\frac{1139}{5501} = \frac{p \cdot 5501}{5501} \quad \text{Dividing by 5501 on both sides}$$

$$\frac{1139}{5501} = p$$

$$0.207 \approx p \quad \text{Finding decimal notation and rounding to the nearest thousandth}$$

$$20.7\% = p \quad \text{Remember to find percent notation.}$$

Note here that the solution, p, includes the % symbol.

Proportion:

$$\frac{N}{100} = \frac{1139}{5501}$$

$$N \cdot 5501 = 100 \cdot 1139 \quad \text{Equating cross products}$$

$$\frac{N \cdot 5501}{5501} = \frac{113{,}900}{5501} \quad \text{Dividing by 5501 on both sides}$$

$$N = \frac{113{,}900}{5501}$$

$$N \approx 20.7 \quad \text{Dividing and rounding to the nearest tenth}$$

We use the solution of the proportion to express the answer to the problem as 20.7%. Note that in the proportion method, $N\% = p$.

4. **Check.** To check, we note that the answer 20.7% is close to 20%, as estimated in the *Familiarize* step.
5. **State.** About 20.7% of the world's mammals are threatened with extinction.

Do Exercise 2. ▶

2. ***Presidential Assassinations in Office.*** Of the 43 different U.S. presidents, 4 have been assassinated while in office. These were James A. Garfield, William McKinley, Abraham Lincoln, and John F. Kennedy. What percent have been assassinated in office?

Answer

2. About 9.3%

PERCENT INCREASE OR DECREASE

Percent is often used to state an increase or a decrease. Let's consider an example of each, using the price of a car as the original number.

Percent Increase

One year a car sold for $20,455. The manufacturer decides to raise the price of the following year's model by 6%. The increase is $0.06 \times \$20,455$, or $1227.30. The new price is $\$20,455 + \$1227.30$, or $21,682.30. Note that the *new* price is 106% of the *former* price.

New price: $21,682.30

Increase ↑

Former price: $20,455

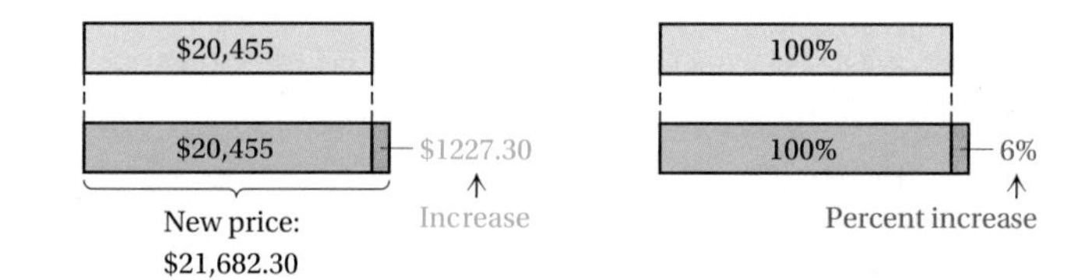

The increase, $1227.30, is 6% of the *former* price, $20,455. The *percent increase* is 6%.

Percent Decrease

Abigail buys the car listed above for $20,455. After one year, the car depreciates in value by 25%. The decrease is $0.25 \times \$20,455$, or $5113.75. This lowers the value of the car to $\$20,455 - \$5113.75$, or $15,341.25. Note that the new value is 75% of the original price. If Abigail decides to sell the car after one year, $15,341.25 might be the most she could expect to get for it.

Original price: $20,455

Decrease ↓

New value: $15,341.25

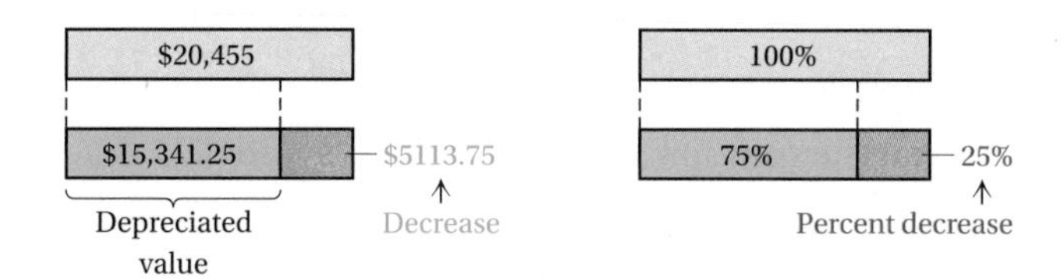

The decrease, $5113.75, is 25% of the *original* price, $20,455. The *percent decrease* is 25%.

◀ **Do Exercises 3 and 4.**

3. *Percent Increase.* The price of a car is $36,875. The price is increased by 4%.
   a) How much is the increase?
   b) What is the new price?

4. *Percent Decrease.* The value of a car is $36,875. The car depreciates in value by 25% after one year.
   a) How much is the decrease?
   b) What is the depreciated value of the car?

When a quantity is decreased by a certain percent, we say that this is a **percent decrease**.

EXAMPLE 3 *Dow Jones Industrial Average.* The Dow Jones Industrial Average (DJIA) plunged from 11,143 to 10,365 on September 29, 2008. This was the largest one-day drop in its history. What was the percent decrease?

Sources: *Nightly Business Reports*, September 29, 2008; DJIA

Answers

3. (a) $1475; (b) $38,350
4. (a) $9218.75; (b) $27,656.25

1. **Familiarize.** We first determine the amount of decrease and then make a drawing.

$$\begin{array}{rl} 11{,}143 & \text{Opening average} \\ -10{,}365 & \text{Closing average} \\ \hline 778 & \text{Decrease} \end{array}$$

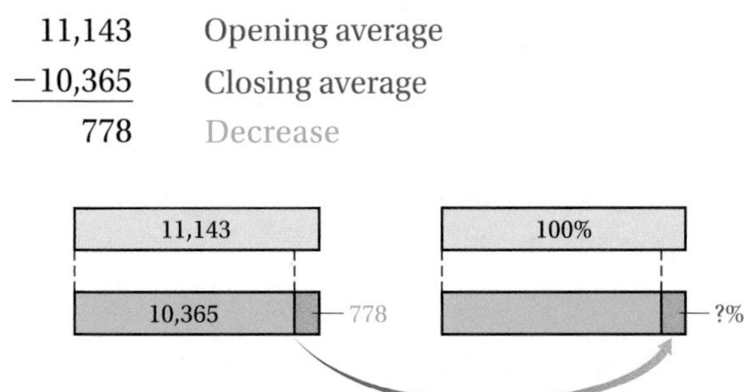

We are asking this question: The decrease is what percent of the opening average? We let $p =$ the percent decrease.

2. **Translate.** There are two ways in which we can translate this problem.

Percent equation:

$$\underbrace{778}\ \text{is}\ \underbrace{\text{what percent}}\ \text{of}\ \underbrace{11{,}143?}$$

$$778 = p \times 11{,}143$$

Proportion:

$$\frac{N}{100} = \frac{778}{11{,}143}$$

Percents	Quantities
0%	0
N%	778
100%	11,143

For proportion, $N\% = p$.

3. **Solve.** We have two ways in which to solve the problem.

Percent equation:

$$778 = p \times 11{,}143$$

$$\frac{778}{11{,}143} = \frac{p \times 11{,}143}{11{,}143} \quad \text{Dividing by 11,143 on both sides}$$

$$\frac{778}{11{,}143} = p$$

$$0.07 \approx p$$

$$7\% = p \quad \text{Converting to percent notation}$$

Proportion:

$$\frac{N}{100} = \frac{778}{11{,}143}$$

$$11{,}143 \times N = 100 \times 778 \quad \text{Equating cross products}$$

$$\frac{11{,}143 \times N}{11{,}143} = \frac{100 \times 778}{11{,}143} \quad \text{Dividing by 11,143 on both sides}$$

$$N = \frac{77{,}800}{11{,}143}$$

$$N \approx 7$$

We use the solution of the proportion to express the answer to the problem as 7%.

5. *Volume of Mail.* The volume of U.S mail decreased from about 102,379 million pieces of mail in 2002 to 68,696 million pieces in 2012. What was the percent decrease?
Source: U.S. Postal Service

4. **Check.** To check, we note that, with a 7% decrease, the closing Dow average should be 93% of the opening average. Since

$$93\% \times 11{,}143 = 0.93 \times 11{,}143 \approx 10{,}363,$$

and 10,363 is close to 10,365, our answer checks. (Remember that we rounded to get 7%.)

5. **State.** The percent decrease in the DJIA was approximately 7%.

Do Exercise 5.

When a quantity is increased by a certain percent, we say that this is a **percent increase**.

EXAMPLE 4 *Costs for Moviegoers.* The average cost of a movie ticket was \$5.80 in 2002. The cost rose to \$8.12 in 2012. What was the percent increase in the cost of a movie ticket?

Sources: National Association of Theatre Owners; theaterseatstore.com

1. **Familiarize.** We first determine the increase in the cost and then make a drawing.

\$8.12	Cost in 2012
− 5.80	Cost in 2002
\$2.32	Increase

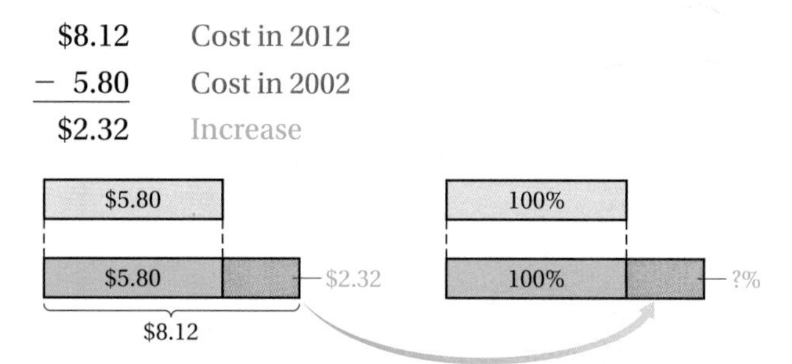

We are asking this question: The increase is what percent of the *original* cost? We let $p =$ the percent increase.

2. **Translate.** There are two ways in which we can translate this problem.

Percent equation:

2.32 is what percent of 5.80?

$$2.32 = p \cdot 5.80$$

Proportion:

$$\frac{N}{100} = \frac{2.32}{5.80}$$

For proportions, $N\% = p$.

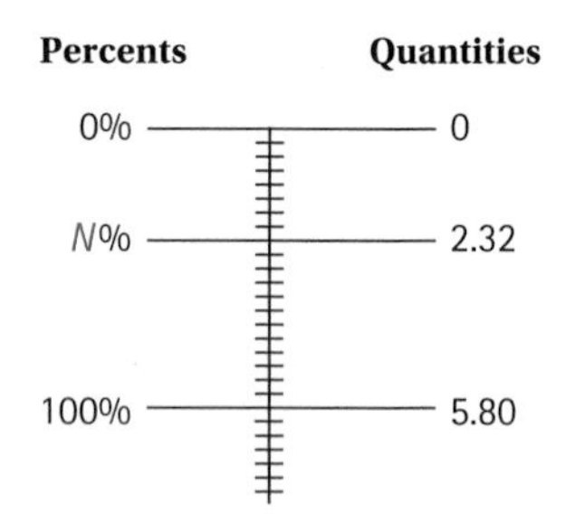

Answer

5. About 32.9%

3. **Solve.** We have two ways in which to solve the problem.

Percent equation:

$$2.32 = p \times 5.80$$

$$\frac{2.32}{5.80} = \frac{p \times 5.80}{5.80} \qquad \text{Dividing by 5.80 on both sides}$$

$$\frac{2.32}{5.80} = p$$

$$0.4 = p$$

$$40\% = p \qquad \text{Converting to percent notation}$$

Proportion:

$$\frac{N}{100} = \frac{2.32}{5.80}$$

$$5.80 \times N = 100 \times 2.32 \qquad \text{Equating cross products}$$

$$\frac{5.80 \times N}{5.80} = \frac{100 \times 2.32}{5.80} \qquad \text{Dividing by 5.80 on both sides}$$

$$N = \frac{232}{5.80}$$

$$N = 40$$

We use the solution of the proportion to express the answer to the problem as 40%.

4. **Check.** To check, we take 40% of 5.80:

$$40\% \times 5.80 = 0.40 \times 5.80 = 2.32.$$

This is the amount of the increase in the price of a movie ticket, so the answer checks.

5. **State.** The percent increase in the cost of a movie ticket was 40%.

Do Exercise 6. ▶

6. *Centenarians.* As of April 1, 2010, the number of centenarians in the United States was 53,364. It is projected that this number will increase to 601,000 by 2050. What is the projected percent increase?

Sources: *National Geographic*, November 2011; Population Reference Bureau, U.S. Census Bureau

Answer

6. About 1026%

Translating for Success

The goal of these matching questions is to practice step (2), Translate, of the five-step problem-solving process. Translate each word problem to an equation and select a correct translation from equations A–O.

1. *Distance Walked.* After a knee replacement, Alex walked $\frac{1}{8}$ mi each morning and $\frac{1}{5}$ mi each afternoon. How much farther did he walk in the afternoon?

2. *Stock Prices.* A stock sold for $5 per share on Monday and only $2.125 per share on Friday. What was the percent decrease from Monday to Friday?

3. *SAT Score.* After attending a class titled *Improving Your SAT Scores,* Jacob raised his total score from 884 to 1040. What was the percent increase?

4. *Change in Population.* The population of a small farming community decreased from 1040 to 884. What was the percent decrease?

5. *Lawn Mowing.* During the summer, brothers Steve and Rob earned money for college by mowing lawns. The largest lawn that they mowed was $2\frac{1}{8}$ acres. Steve can mow $\frac{1}{5}$ acre per hour, and Rob can mow only $\frac{1}{8}$ acre per hour. Working together, how many acres did they mow per hour?

A. $x + \frac{1}{5} = \frac{1}{8}$

B. $250 = x \cdot 1040$

C. $884 = x \cdot 1040$

D. $\frac{250}{16.25} = \frac{1000}{x}$

E. $156 = x \cdot 1040$

F. $16.25 = 250 \cdot x$

G. $\frac{1}{5} + \frac{1}{8} = x$

H. $2\frac{1}{8} = x \cdot 5$

I. $5 = 2.875 \cdot x$

J. $\frac{1}{8} + x = \frac{1}{5}$

K. $1040 = x \cdot 884$

L. $\frac{250}{16.25} = \frac{x}{1000}$

M. $2.875 = x \cdot 5$

N. $x \cdot 884 = 156$

O. $x = 16.25 \cdot 250$

Answers on page A-12

6. *Land Sale.* Cole sold $2\frac{1}{8}$ acres of the 5 acres he inherited from his uncle. What percent of his land did he sell?

7. *Travel Expenses.* A magazine photographer is reimbursed 16.25¢ per mile for business travel, up to 1000 mi per week. In a recent week, he traveled 250 mi. What was the total reimbursement for travel?

8. *Trip Expenses.* The total expenses for Claire's recent business trip were $1040. She put $884 on her credit card and paid the balance in cash. What percent did she place on her credit card?

9. *Cost of Copies.* During the first summer session at a community college, the campus copy center advertised 250 copies for $16.25. At this rate, what is the cost of 1000 copies?

10. *Cost of Insurance.* Following a rise in the cost of health insurance, 250 of a company's 1040 employees canceled their insurance. What percent of the employees canceled their insurance?

7.5 Exercise Set

For Extra Help
MyMathLab® MathXL® PRACTICE WATCH READ REVIEW

Reading Check

Complete the table by filling in the missing numbers.

	Original Price	New Price	Change	Percent Increase or Decrease
RC1.	\$50	\$40	\$ ______	$\frac{\text{Change}}{\text{Original}} = \frac{\$}{\$} =$ ______%
RC2.	\$60	\$75	\$ ______	$\frac{\text{Change}}{\text{Original}} = \frac{\$}{\$} =$ ______%
RC3.	\$360	\$480	\$ ______	$\frac{\text{Change}}{\text{Original}} = \frac{\$}{\$} =$ ______%
RC4.	\$4000	\$2400	\$ ______	$\frac{\text{Change}}{\text{Original}} = \frac{\$}{\$} =$ ______%

a Solve.

1. *Winnings from Gambling.* Pre-tax gambling winnings were \$417 billion worldwide in 2012. Approximately 25.1% of the winnings were in the United States and 5.9% were in Italy. About how much, in dollars, were the gambling winnings in each country?

   Source: H2 Gambling Capital (h2gc.com)

2. *Mississippi River.* The Mississippi River, which extends from its source, at Lake Itasca in Minnesota, to the Gulf of Mexico, is 2348 mi long. Approximately 77% of the river is navigable. How many miles of the river are navigable?

   Source: National Oceanic and Atmospheric Administration

3. A person earns \$43,200 one year and receives an 8% raise in salary. What is the new salary?

4. A person earns \$28,600 one year and receives a 5% raise in salary. What is the new salary?

5. *Test Results.* On a test, Juan got 85%, or 119, of the items correct. How many items were on the test?

6. *Test Results.* On a test, Maj Ling got 86%, or 43, of the items correct. How many items were on the test?

7. *Farmland.* In Kansas, 47,000,000 acres are farmland. About 5% of all the farm acreage in the United States is in Kansas. What is the total number of acres of farmland in the United States?

   Sources: U.S. Department of Agriculture; National Agricultural Statistics Service

8. *World Population.* World population is increasing by 1.2% each year. In 2008, it was 6.68 billion. What will the population be in 2015?

   Sources: U.S. Census Bureau; International Data Base

9. *Car Depreciation.* A car generally depreciates 25% of its original value in the first year. A car is worth $27,300 after the first year. What was its original cost?

10. *Car Depreciation.* Given normal use, an American-made car will depreciate 25% of its original cost the first year and 14% of its remaining value in the second year. What is the value of a car at the end of the second year if its original cost was $36,400? $28,400? $26,800?

11. *Test Results.* On a test of 80 items, Pedro got 95% correct. How many items did he get correct? incorrect?

12. *Test Results.* On a test of 40 items, Christina got 80% correct. How many items did she get correct? incorrect?

13. *Olympic Team.* The 2012 U.S. Summer Olympics team consisted of 529 members. Approximately 49.3% of the athletes were men. How many men were on the team?

Source: United States Olympic Committee

14. *Murder Case Costs.* The cost (for both trial and incarceration) of a life-without-parole case is about 30% of the cost of a death-penalty case. The average cost of a death-penalty case is $505,773. What is the average cost of a life-without-parole case?

Source: Indiana Legislative Services Agency, *2010 Study of Murder Trials*

15. *Spending on Pets.* In 2011, Americans spent approximately $50.8 billion on their pets. Of this amount, $14.1 billion was for veterinarian bills. What percent of the total was spent on veterinary care?

Source: American Pet Products Association

16. *Health Workers' Salaries.* In 2010, the median annual salary of registered nurses was $64,690. The median annual salary of physicians and surgeons was $111,570. What percent of the physician and surgeon's median annual salary is the nurses' median annual salary?

Source: Bureau of Labor Statistics

17. *Tipping.* For a party of 8 or more, some restaurants add an 18% tip to the bill. What is the total amount charged for a party of 10 if the cost of the meal, without tip, is $195?

18. *Tipping.* Diners frequently add a 15% tip when charging a meal to a credit card. What is the total amount charged to a card if the cost of the meal, without tip, is $18? $34? $49?

19. *Wasting Food.* As world population increases and the number of acres of farmland decreases, improvements in food production and packaging must be implemented. Also, wealthy countries need to waste less food. In the United States, consumers waste 39 lb of every 131 lb of fruit purchased. What percent is wasted?

Sources: United States Department of Agriculture; *National Geographic*, July 2011

20. *Credit-Card Debt.* Michael has disposable monthly income of $3400. Each month, he pays $470 toward his credit-card debt. What percent of his disposable income is allotted to paying off credit-card debt?

Monthly Disposable Income: $3400

21. A lab technician has 540 mL of a solution of alcohol and water; 8% is alcohol. How many milliliters are alcohol? water?

22. A lab technician has 680 mL of a solution of water and acid; 3% is acid. How many milliliters are acid? water?

23. *U.S. Armed Forces.* There were 1,384,000 people in the United States in active military service in 2006. The numbers in the four armed services are listed in the table below. What percent of the total does each branch represent? Round the answers to the nearest tenth of a percent.

U.S. ARMED FORCES: 2006

TOTAL	1,384,000*
AIR FORCE	349,000
ARMY	505,000
NAVY	350,000
MARINES	180,000

*Includes National Guard, Reserve, and retired regular personnel on extended or continuous active duty. Excludes Coast Guard.

SOURCES: U.S. Department of Defense; U.S. Census Bureau

24. *Living Veterans.* There were 23,977,000 living veterans in the United States in 2006. Numbers in various age groups are listed in the table below. What percent of the total does each age group represent? Round the answers to the nearest tenth of a percent.

LIVING VETERANS BY AGE: 2006

TOTAL	23,977,000
UNDER 35 YEARS OLD	1,949,000
35–44 YEARS OLD	2,901,000
45–54 YEARS OLD	3,846,000
55–64 YEARS OLD	6,081,000
65 YEARS OLD AND OLDER	9,200,000

SOURCES: U.S. Department of Defense; U.S. Census Bureau

b Solve.

25. *Mortgage Payment Increase.* A monthly mortgage payment increases from $840 to $882. What is the percent increase?

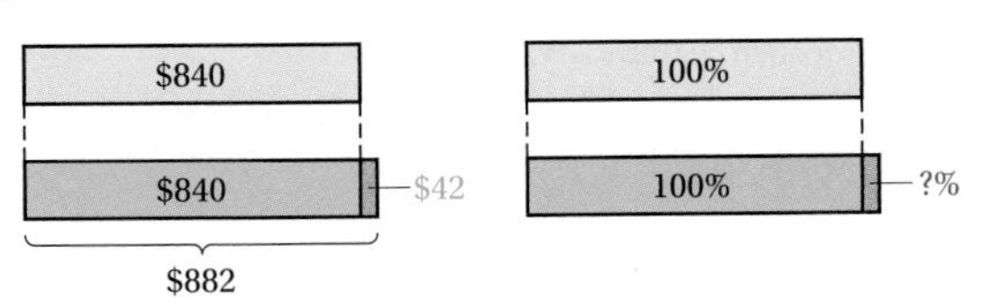

26. *Savings Increase.* The amount in a savings account increased from $200 to $216. What was the percent increase?

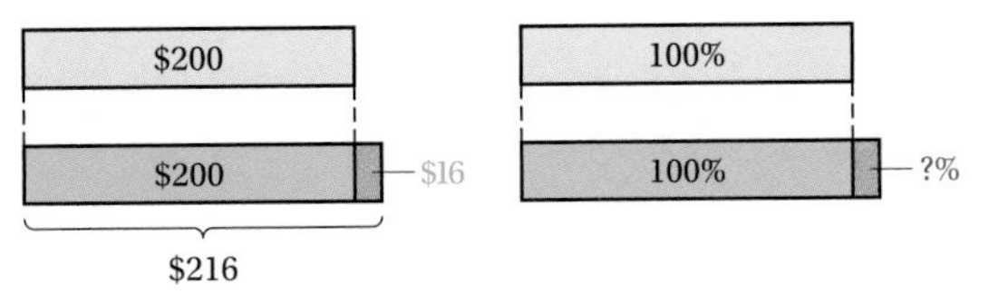

27. A person on a diet goes from a weight of 160 lb to a weight of 136 lb. What is the percent decrease?

28. During a sale, a dress decreased in price from $90 to $72. What was the percent decrease?

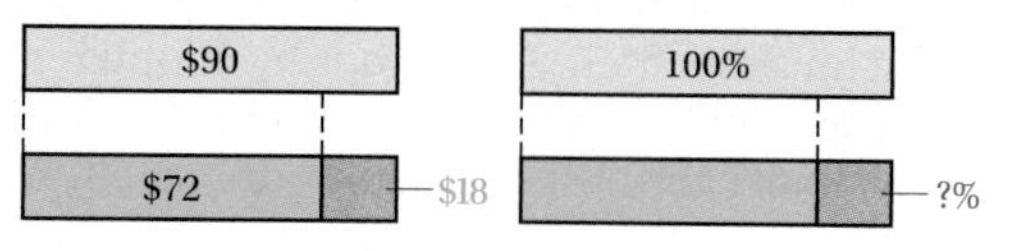

29. *Insulation.* A roll of unfaced fiberglass insulation has a retail price of $23.43. For two weeks, it is on sale for $15.31. What is the percent decrease?

30. *Set of Weights.* A 300-lb weight set retails for $199.95. For its grand opening, a sporting goods store reduced the price to $154.95. What is the percent decrease?

31. *Birds Killed.* In 2009, approximately 440,000 birds were killed by wind turbines in the United States. It is estimated that this number will increase to 1,000,000 per year by 2020. What would the percent increase be?

Sources: National Wind Coordinating Collaborative; American Bird Conservancy; U.S. Fish and Wildlife Service

32. *Miles of Railroad Track.* The greatest combined length of U.S.-owned operating railroad track was 254,037 mi in 1916, when industrial activity increased during World War I. The total length has decreased ever since. By 2006, the number of miles of track had decreased to 140,490 mi. What was the percent decrease from 1916 to 2006?

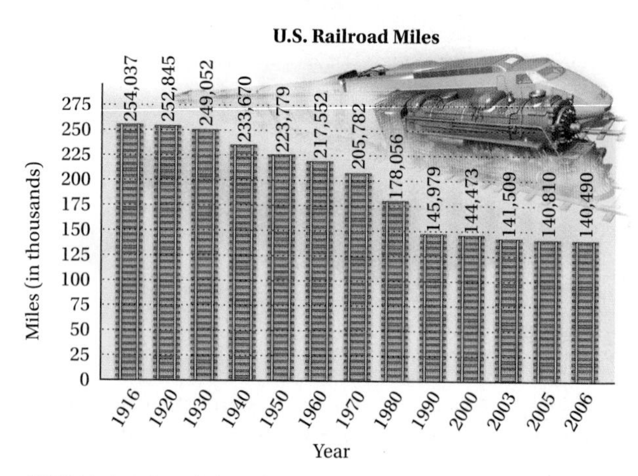

NOTE: The lengths exclude yard tracks, sidings, and parallel tracks.
SOURCE: Association of American Railroads

33. *Overdraft Fees.* Consumers are paying record amounts of fees for overdrawing their bank accounts. In 2007, banks, thrift institutions, and credit unions collected \$45.6 billion in overdraft fees, which is \$15.1 billion more than in 2001. What was the percent increase?

Source: Moebs Services

34. *Credit-Card Debt.* In 2013, the average credit-card debt per household in the United States was \$15,162. In 1990, the average credit-card debt was \$2966. What was the percent increase?

Sources: nerdwallet.com

35. *Population over 1 Million.* In 1950, 74 cities in the world had populations of 1 million or more. In 2010, 442 cities had populations of 1 million or more. What was the percent increase?

Sources: *World Cities*, George Modelski; United Nations

36. *Prescription Drug Sales.* Total spending on prescription drugs in the United States was \$329.2 billion in 2011. This amount dropped to \$325.8 billion in 2012. What was the percent decrease?

Source: IMS Institute for Healthcare Informatics

37. *Patents Issued.* The U.S. Patent and Trademark Office (USPTO) issued a total of 157,284 utility patents in 2007. This number of patents was down from 173,794 in 2006. What was the percent decrease?

Source: IFI Patent Intelligence

38. *Highway Fatalities.* In 2007, there were 41,059 highway fatalities in the United States, which was 1649 fewer than the number in 2006. What was in the percent decrease?

Source: National Highway Traffic Safety Administration

39. *Two-by-Four.* A cross-section of a standard, or nominal, "two-by-four" actually measures $1\frac{1}{2}$ in. by $3\frac{1}{2}$ in. The rough board is 2 in. by 4 in. but is planed and dried to the finished size. What percent of the wood is removed in planing and drying?

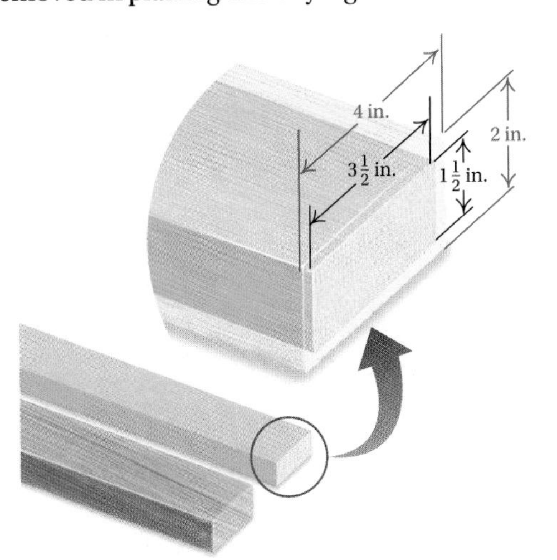

40. *Strike Zone.* In baseball, the *strike zone* is normally a 17-in. by 30-in. rectangle. Some batters give the pitcher an advantage by swinging at pitches thrown out of the strike zone. By what percent is the area of the strike zone increased if a 2-in. border is added to the outside?

Source: Major League Baseball

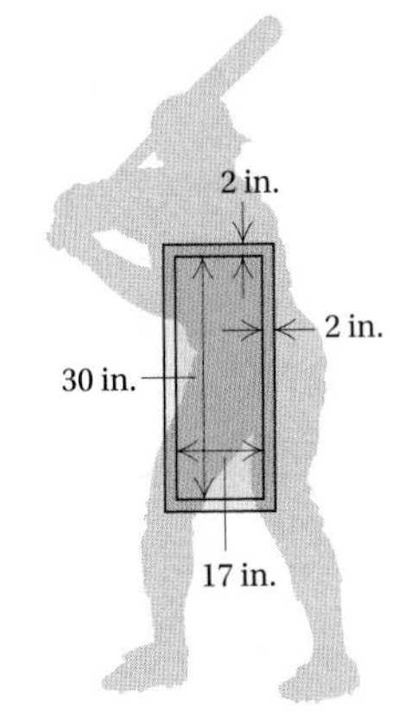

Population Increase. The table below provides data showing how the populations of various states increased from 2000 to 2010. Complete the table by filling in the missing numbers. Round percents to the nearest tenth of a percent.

	State	Population in 2000	Population in 2010	Change	Percent Change
41.	Vermont	608,827	625,741		
42.	Wisconsin	5,363,675		323,311	
43.	Arizona		6,392,017	1,261,385	
44.	Virginia		8,001,024	922,509	
45.	Idaho	1,293,953		273,629	
46.	Georgia	8,186,453	9,535,483		

SOURCE: U.S. Census Bureau

47. *Increase in Population.* Between 2000 and 2010, the population of Utah increased from 2,233,169 to 2,763,885. What was the percent increase?

Sources: U.S. Census Bureau; U.S. Department of Commerce

48. *Decrease in Population.* Between 2000 and 2010, the population of Michigan decreased from 9,938,444 to 9,883,640. What was the percent decrease?

Sources: U.S. Census Bureau; U.S. Department of Commerce

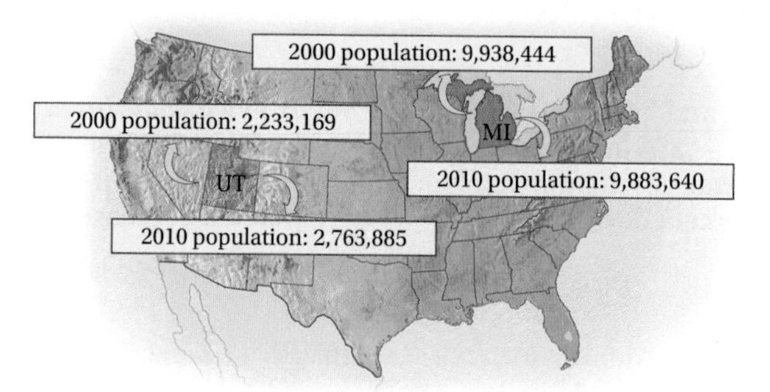

Skill Maintenance

Convert to decimal notation. [5.1b], [5.5a]

49. $\frac{25}{11}$

50. $\frac{11}{25}$

51. $-\frac{27}{8}$

52. $\frac{43}{9}$

Simplify. [3.5b]

53. $\frac{18}{102}$

54. $-\frac{135}{510}$

55. $\frac{192}{1000}$

56. $\frac{70}{406}$

Synthesis

57. A coupon allows a couple to have $10 subtracted from their dinner bill. Before subtracting $10, however, the restaurant adds a tip of 20%. If the couple is presented with a bill for $40.40, how much would the dinner (without tip) have cost without the coupon?

58. If p is 120% of q, then q is what percent of p?

Sales Tax, Commission, and Discount

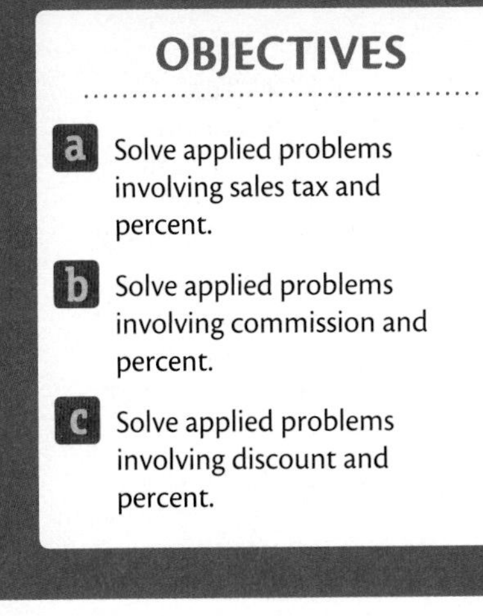

a SALES TAX

Sales tax computations represent a special type of percent increase problem. The sales tax rate in Colorado is 2.9%. This means that the tax is 2.9% of the purchase price. Suppose the purchase price of a canoe is \$749.95. The sales tax is then 2.9% of \$749.95, or $0.029 \times \$749.95$, or \$21.74855, which is about \$21.75.

BILL:

Purchase price	=	\$749.95
Sales tax (2.9% of \$749.95)	=	+ 21.75
Total price		\$771.70

The total that you would pay for the canoe is the purchase price plus the sales tax:

\$749.95 + \$21.75, or \$771.70.

SALES TAX

Sales tax = Sales tax rate × Purchase price
Total price = Purchase price + Sales tax

EXAMPLE 1 *Maine Sales Tax.* The sales tax rate in Maine is 5%. How much tax is charged on the purchase of 3 shrubs at \$42.99 each? What is the total price?

a) We first find the purchase price of the 3 shrubs. It is

$3 \times \$42.99 = \$128.97.$

b) The sales tax on items costing \$128.97 is

$$\underbrace{\text{Sales tax rate}}_{5\%} \times \underbrace{\text{Purchase price}}_{128.97},$$

or 0.05×128.97, or 6.4485. Thus the tax is \$6.45 (rounded to the nearest cent).

c) The total price is given by the purchase price plus the sales tax:

\$128.97 + \$6.45, or \$135.42.

To check, note that the total price is the purchase price plus 5% of the purchase price. Thus the total price is 105% of the purchase price. Since $1.05 \times 128.97 \approx 135.42$, we have a check. The sales tax is \$6.45, and the total price is \$135.42.

1. *Texas Sales Tax.* The sales tax rate in Texas is 6.25%. In Texas, how much tax is charged on the purchase of an ultrasound toothbrush that sells for \$139.95? What is the total price?

2. **Wyoming Sales Tax.** In her hometown, Laramie, Wyoming, Samantha buys 4 copies of *It's All Good* by Gwyneth Paltrow for \$18.95 each. The sales tax rate in Wyoming is 4%. How much sales tax will Samantha be charged? What is the total price?

GS

Sales tax = ☐ % × 4 × \$☐
= 0.04 × \$☐
= \$3.032
≈ \$☐
Total price = \$75.80 + \$☐
= \$☐

◀ Do Exercises 1 and 2.

EXAMPLE 2 The sales tax on the purchase of an eReader that costs \$199 is \$13.93. What is the sales tax rate?

We rephrase and translate as follows:

Rephrase: Sales tax is what percent of purchase price?

Translate: \$13.93 = r × 199.

To solve the equation, we divide by 199 on both sides:

$$\frac{13.93}{199} = \frac{r \times 199}{199}$$

$$\frac{13.93}{199} = r$$

$$0.07 = r$$

$$7\% = r.$$

The sales tax rate is 7%.

3. The sales tax on the purchase of a set of holiday dishes that costs \$449 is \$26.94. What is the sales tax rate?

◀ Do Exercise 3.

EXAMPLE 3 The sales tax on the purchase of a stone-top firepit is \$12.74 and the sales tax rate is 8%. Find the purchase price (the price before the tax is added).

We rephrase and translate as follows:

Rephrase: Sales tax is 8% of what?

Translate: 12.74 = 8% × b,

or 12.74 = 0.08 × b.

Price: ?
\$12.74 tax @ 8%

To solve, we divide by 0.08 on both sides:

$$\frac{12.74}{0.08} = \frac{0.08 \times b}{0.08}$$

$$\frac{12.74}{0.08} = b$$

$$159.25 = b.$$

The purchase price is \$159.25.

4. The sales tax on the purchase of a pair of designer jeans is \$4.84 and the sales tax rate is 5.5%. Find the purchase price (the price before the tax is added).

◀ Do Exercise 4.

Answers

1. \$8.75; \$148.70 2. \$3.03, \$78.83
3. 6% 4. \$88

Guided Solution:
2. 4, 18.95, 75.80, 3.03; 3.03, 78.83

b COMMISSION

When you work for a **salary**, you receive the same amount of money each week or month. When you work for a **commission**, you are paid a percentage of the total sales for which you are responsible.

COMMISSION

Commission = Commission rate × Sales

EXAMPLE 4 *Membership Sales.* A membership salesperson at a fitness club has a commission rate of 3%. What is the commission on the sale of $8300 worth of memberships?

$$\begin{aligned} \text{Commission} &= \text{Commission rate} \times \text{Sales} \\ C &= 3\% \times 8300 \\ C &= 0.03 \times 8300 \\ C &= 249 \end{aligned}$$

The commission is $249.

Do Exercise 5. ▶

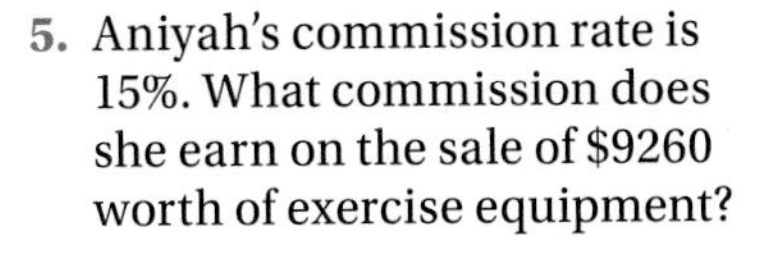

5. Aniyah's commission rate is 15%. What commission does she earn on the sale of $9260 worth of exercise equipment?

EXAMPLE 5 *Earth-Moving Equipment Sales.* Gavin earns a commission of $20,800 for selling $320,000 worth of earth-moving equipment. What is the commission rate?

$$\begin{aligned} \text{Commission} &= \text{Commission rate} \times \text{Sales} \\ 20{,}800 &= r \times 320{,}000 \end{aligned}$$

Answer

5. $1389

To solve this equation, we divide by 320,000 on both sides:

$$\frac{20{,}800}{320{,}000} = \frac{r \times 320{,}000}{320{,}000}$$

$$0.065 = r$$

$$6.5\% = r.$$

The commission rate is 6.5%.

Do Exercise 6.

6. Zion earns a commission of \$2040 for selling \$17,000 worth of concert tickets. What is the commission rate?

EXAMPLE 6 *Cruise Vacations.* Valentina's commission rate on cruise vacation packages is 5.6%. She received a commission of \$2457 on packages that she sold in November. How many dollars worth of packages did she sell?

$$\textit{Commission} = \textit{Commission rate} \times \textit{Sales}$$

$$2457 = 5.6\% \times S, \quad \text{or}$$

$$2457 = 0.056 \times S$$

To solve this equation, we divide by 0.056 on both sides:

$$\frac{2457}{0.056} = \frac{0.056 \times S}{0.056}$$

$$\frac{2457}{0.056} = S$$

$$43{,}875 = S.$$

Valentina sold \$43,875 worth of cruise vacation packages.

Do Exercise 7.

7. Nathan's commission rate on ski passes is 7.5%. He receives a commission of \$2970 from the sale of passes in January. How many dollars worth of ski passes did he sell?

GS

$$\$2970 = \square\,\% \times S$$

$$\$2970 = 0.075 \times S$$

$$\frac{\$2970}{\square} = \frac{0.075 \times S}{0.075}$$

$$\$\square = S$$

C DISCOUNT

Suppose that the regular price of a rug is \$60, and the rug is on sale at 25% off. Since 25% of \$60 is \$15, the sale price is \$60 − \$15, or \$45. We call \$60 the **original**, or **marked, price**, 25% the **rate of discount**, \$15 the **discount**, and \$45 the **sale price**. Note that discount problems are a type of percent decrease problem.

DISCOUNT AND SALE PRICE

Discount = Rate of discount × Original price

Sale price = Original price − Discount

Answers

6. 12% 7. \$39,600

Guided Solution:

7. 7.5, 0.075, 39,600

EXAMPLE 7 A leather sofa marked \$2379 is on sale at $33\frac{1}{3}\%$ off. What is the discount? the sale price?

a) *Discount* = *Rate of discount* × *Original price*

$$D = 33\frac{1}{3}\% \times 2379$$

$$D = \frac{1}{3} \times 2379$$

$$D = \frac{2379}{3} = 793$$

b) *Sale price* = *Original price* − *Discount*

$$S = 2379 - 793$$

$$S = 1586$$

The discount is \$793, and the sale price is \$1586.

Do Exercise 8.

8. A computer marked \$660 is on sale at $16\frac{2}{3}\%$ off. What is the discount? the sale price?

EXAMPLE 8 The price of a snowblower is marked down from \$950 to \$779. What is the rate of discount?

We first find the discount by subtracting the sale price from the original price:

$$950 - 779 = 171.$$

The discount is \$171.

Next, we use the equation for discount:

Discount = *Rate of discount* × *Original price*

$$171 = r \times 950.$$

To solve, we divide by 950 on both sides:

$$\frac{171}{950} = \frac{r \times 950}{950}$$

$$\frac{171}{950} = r$$

$$0.18 = r$$

$$18\% = r.$$

The rate of discount is 18%.

To check, note that an 18% rate of discount means that the buyer pays 82% of the original price:

$$0.82 \times \$950 = \$779.$$

Do Exercise 9.

9. The price of a winter coat is reduced from \$75 to \$60. Find the rate of discount.

Answers

8. \$110; \$550 **9.** 20%

7.6 Exercise Set

For Extra Help
MyMathLab® MathXL® PRACTICE WATCH READ REVIEW

Reading Check

Complete each definition with the word *price*, *rate*, or *tax*.

RC1. Commission = Commission ________ × Sales

RC2. Discount = ________ of discount × Original price

RC3. Sale price = Original ________ − Discount

RC4. Sales tax = Sales ________ rate × Purchase price

RC5. Total price = Purchase price + Sales ________

a Solve.

1. *Wyoming Sales Tax.* The sales tax rate in Wyoming is 4%. How much sales tax would be charged on a fireplace screen with doors that costs $239?

2. *Kansas Sales Tax.* The sales tax rate in Kansas is 6.3%. How much sales tax would be charged on a fireplace screen with doors that costs $239?

3. *New Mexico Sales Tax.* The sales tax rate in New Mexico is 5.125%. How much sales tax is charged on a camp stove that sells for $129.95?

4. *Ohio Sales Tax.* The sales tax rate in Ohio is 5.5%. How much sales tax is charged on a pet carrier that sells for $39.99?

5. *California Sales Tax.* The sales tax rate in California is 7.5%. How much sales tax is charged on a purchase of 4 contour foam travel pillows at $39.95 each? What is the total price?

6. *Illinois Sales Tax.* The sales tax rate in Illinois is 6.25%. How much sales tax is charged on a purchase of 3 wet-dry vacs at $60.99 each? What is the total price?

7. The sales tax is $30 on the purchase of a diamond ring that sells for $750. What is the sales tax rate?

8. The sales tax is $48 on the purchase of a dining room set that sells for $960. What is the sales tax rate?

9. The sales tax is $9.12 on the purchase of a patio set that sells for $456. What is the sales tax rate?

10. The sales tax is $35.80 on the purchase of a refrigerator-freezer that sells for $895. What is the sales tax rate?

11. The sales tax on the purchase of a new fishing boat is \$112 and the sales tax rate is 2%. What is the purchase price (the price before tax is added)?

12. The sales tax on the purchase of a used car is \$100 and the sales tax rate is 5%. What is the purchase price?

13. The sales tax rate in New York City is 4.375% for the city plus 4% for the state. Find the total amount paid for 6 boxes of chocolates at \$17.95 each.

14. The sales tax rate in Nashville, Tennessee, is 2.25% for Davidson County plus 7% for the state. Find the total amount paid for 2 ladders at \$39 each.

15. The sales tax rate in Seattle, Washington, is 2.5% for King County plus 6.5% for the state. Find the total amount paid for 3 ceiling fans at \$84.49 each.

16. The sales tax rate in Miami, Florida, is 1% for Dade County plus 6% for the state. Find the total amount paid for 2 tires at \$49.95 each.

17. The sales tax rate in Atlanta, Georgia, is 1% for the city, 3% for Fulton County, and 4% for the state. Find the total amount paid for 6 basketballs at \$29.95 each.

18. The sales tax rate in Dallas, Texas, is 1% for the city, 1% for Dallas County, and 6.25% for the state. Find the total amount paid for 5 flash drives at \$19.95 each.

b Solve.

19. Benjamin's commission rate is 21%. What commission does he earn on the sale of \$12,500 worth of windows?

20. Olivia's commission rate is 6%. What commission does she earn on the sale of \$45,000 worth of lawn irrigation systems?

21. Alyssa earns \$408 for selling \$3400 worth of shoes. What is the commission rate?

22. Joshua earns \$120 for selling \$2400 worth of television sets. What is the commission rate?

23. *Real Estate Commission.* A real estate agent's commission rate is 7%. She receives a commission of $12,950 from the sale of a home. How much did the home sell for?

24. *Clothing Consignment Commission.* A clothing consignment shop's commission rate is 40%. The shop receives a commission of $552. How many dollars worth of clothing were sold?

25. A real estate commission rate is 8%. What is the commission from the sale of a piece of land for $68,000?

26. A real estate commission rate is 6%. What is the commission from the sale of a $98,000 home?

27. David earns $1147.50 for selling $7650 worth of car parts. What is the commission rate?

28. Jayla earns $280.80 for selling $2340 worth of tee shirts. What is the commission rate?

29. Laila's commission is increased according to how much she sells. She receives a commission of 4% for the first $1000 of sales and 7% for the amount over $1000. What is her total commission on sales of $5500?

30. Malik's commission is increased according to how much he sells. He receives a commission of 5% for the first $2000 of sales and 8% for the amount over $2000. What is his total commission on sales of $6200?

C Complete the table below by filling in the missing numbers.

	Marked Price	Rate of Discount	Discount	Sale Price
31.	$300	10%		
32.	$2000	40%		
33.	$17	15%		
34.	$20	25%		
35.		10%	$12.50	
36.		15%	$65.70	
37.	$600		$240	
38.	$12,800		$1920	

39. Find the marked price and the rate of discount for the surfboard in this ad.

40. Find the marked price and the rate of discount for the steel log rack in this ad.

41. Find the discount and the rate of discount for the pinball machine in this ad.

42. Find the discount and the rate of discount for the amaryllis in this ad.

Skill Maintenance

Solve. [6.3b]

43. $\frac{x}{12} = \frac{24}{16}$

44. $\frac{7}{2} = \frac{11}{x}$

Solve. [5.4b]

45. $0.64 \cdot x = 170$

46. $-29.44 = 25.6 \times y$

Convert to standard notation. [5.3b]

47. 4.03 trillion

48. 5.8 million

49. 42.7 million

50. 6.09 trillion

Synthesis

51. ▦ Sara receives a 10% commission on the first $5000 in sales and 15% on all sales beyond $5000. If Sara receives a commission of $2405, how much did she sell? Use a calculator and trial and error if you wish.

52. Elijah collects baseball memorabilia. He bought two autographed plaques, but then became short of funds and had to sell them quickly for $200 each. On one, he made a 20% profit, and on the other, he lost 20%. Did he make or lose money on the sale?

Simple Interest and Compound Interest; Credit Cards

OBJECTIVES

a Solve applied problems involving simple interest.

b Solve applied problems involving compound interest.

c Solve applied problems involving interest rates on credit cards.

SKILL TO REVIEW

Objective 7.1b: Convert between percent notation and decimal notation.

Find decimal notation.

1. $34\frac{5}{8}\%$ **2.** $5\frac{1}{4}\%$

1. What is the simple interest on $4300 invested at an interest rate of 4% for 1 year?

GS

$I = P \cdot r \cdot t$

$= \$4300 \times$ ____ $\% \times 1$

$= \$$ ____ $\times 0.04 \times 1$

$= \$$ ____

2. What is the simple interest on a principal of $4300 invested at an interest rate of 4% for 9 months?

Answers

Skill to Review:
1. 0.34625 2. 0.0525

Margin Exercises:
1. $172 2. $129

Guided Solution:
1. 4, 4300, 172

a SIMPLE INTEREST

Suppose you put $1000 into an investment for 1 year. The $1000 is called the **principal**. If the **interest rate** is 5%, in addition to the principal, you get back 5% of the principal, which is

$$5\% \text{ of } \$1000, \quad \text{or} \quad 0.05 \times \$1000, \quad \text{or} \quad \$50.00.$$

The $50.00 is called **simple interest**. It is, in effect, the price that a financial institution pays for the use of the money over time.

SIMPLE INTEREST FORMULA

The **simple interest *I*** on principal *P*, invested for *t* years at interest rate *r*, is given by

$$I = P \cdot r \cdot t.$$

EXAMPLE 1 What is the simple interest on $2500 invested at an interest rate of 6% for 1 year?

We use the formula $I = P \cdot r \cdot t$:

$$I = P \cdot r \cdot t = \$2500 \times 6\% \times 1$$
$$= \$2500 \times 0.06$$
$$= \$150.$$

The simple interest for 1 year is $150.

◀ Do Margin Exercise 1.

EXAMPLE 2 What is the simple interest on a principal of $2500 invested at an interest rate of 6% for 3 months?

We use the formula $I = P \cdot r \cdot t$ and express 3 months as a fraction of a year:

$$I = P \cdot r \cdot t = \$2500 \times 6\% \times \frac{3}{12} = \$2500 \times 6\% \times \frac{1}{4}$$
$$= \frac{\$2500 \times 0.06}{4} = \$37.50.$$

The simple interest for 3 months is $37.50.

◀ Do Margin Exercise 2.

When time is given in days, we generally divide it by 365 to express the time as a fractional part of a year.

EXAMPLE 3 To pay for a shipment of lawn furniture, Patio by Design borrows $8000 at $9\frac{3}{4}\%$ for 60 days. Find **(a)** the amount of simple interest that is due and **(b)** the total amount that must be paid after 60 days.

a) We express 60 days as a fractional part of a year:

$$I = P \cdot r \cdot t = \$8000 \times 9\frac{3}{4}\% \times \frac{60}{365}$$

$$= \$8000 \times 0.0975 \times \frac{60}{365}$$

$$\approx \$128.22.$$

The interest due for 60 days is $128.22.

b) The total amount to be paid after 60 days is the principal plus the interest:

$$\$8000 + \$128.22 = \$8128.22.$$

The total amount due is $8128.22.

Do Exercise 3. ▶

GS **3.** The Glass Nook borrows $4800 at $5\frac{1}{2}\%$ for 30 days. Find **(a)** the amount of simple interest due and **(b)** the total amount that must be paid after 30 days.

a) $I = P \cdot r \cdot t$

$= \$4800 \times 5\frac{1}{2}\% \times \frac{\square}{365}$

$= \$4800 \times 0.055 \times \frac{30}{365}$

$\approx \$\square$

b) Total amount

$= \$4800 + \square$

$= \square$

b COMPOUND INTEREST

When interest is paid *on interest*, we call it **compound interest**. This is the type of interest usually paid on investments. Suppose you have $5000 in a savings account at 6%. In 1 year, the account will contain the original $5000 plus 6% of $5000. Thus the total in the account after 1 year will be

106% of $5000, or $1.06 \times \$5000$, or $5300.

Now suppose that the total of $5300 remains in the account for another year. At the end of this second year, the account will contain the $5300 plus 6% of $5300. The total in the account would thus be

106% of $5300, or $1.06 \times \$5300$, or $5618.

Note that in the second year, interest is also earned on the first year's interest. When this happens, we say that interest is **compounded annually**.

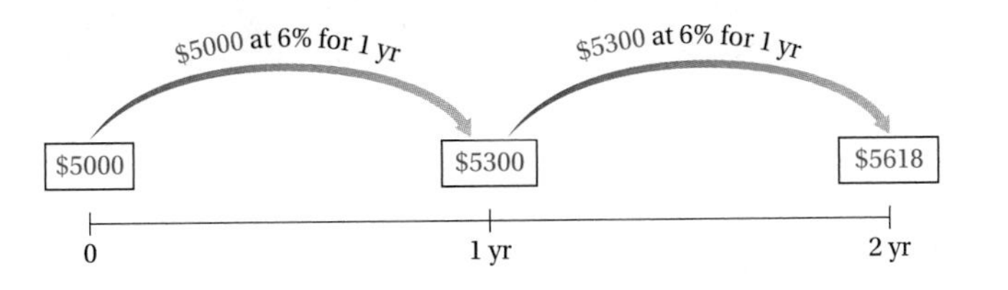

EXAMPLE 4 Find the amount in an account if $2000 is invested at 8%, compounded annually, for 2 years.

a) After 1 year, the account will contain 108% of $2000:

$$1.08 \times \$2000 = \$2160.$$

b) At the end of the second year, the account will contain 108% of $2160:

$$1.08 \times \$2160 = \$2332.80.$$

The amount in the account after 2 years is $2332.80.

Do Exercise 4. ▶

4. Find the amount in an account if $2000 is invested at 2%, compounded annually, for 2 years.

Answers

3. (a) $21.70; **(b)** $4821.70 **4.** $2080.80

Guided Solution:

3. (a) 30, 21.70; **(b)** $21.70, $4821.70

Suppose that the interest in Example 4 were **compounded semi-annually**—that is, every half year. Interest would then be calculated twice a year at a rate of 8% ÷ 2, or 4% each time. The approach used in Example 4 can then be adapted, as follows.

After the first $\frac{1}{2}$ year, the account will contain 104% of $2000:

$$1.04 \times \$2000 = \$2080.$$

After a second $\frac{1}{2}$ year (1 full year), the account will contain 104% of $2080:

$$1.04 \times \$2080 = \$2163.20.$$

After a third $\frac{1}{2}$ year ($1\frac{1}{2}$ full years), the account will contain 104% of $2163.20:

$$1.04 \times \$2163.20 = \$2249.728$$

$$\approx \$2249.73. \qquad \text{Rounding to the nearest cent}$$

Finally, after a fourth $\frac{1}{2}$ year (2 full years), the account will contain 104% of $2249.73:

$$1.04 \times \$2249.73 = \$2339.7192$$

$$\approx \$2339.72. \qquad \text{Rounding to the nearest cent}$$

Let's summarize our results and look at them another way:

End of 1st $\frac{1}{2}$ year $\rightarrow 1.04 \times 2000 = 2000 \times (1.04)^1$;

End of 2nd $\frac{1}{2}$ year $\rightarrow 1.04 \times (1.04 \times 2000) = 2000 \times (1.04)^2$;

End of 3rd $\frac{1}{2}$ year $\rightarrow 1.04 \times (1.04 \times 1.04 \times 2000) = 2000 \times (1.04)^3$;

End of 4th $\frac{1}{2}$ year $\rightarrow 1.04 \times (1.04 \times 1.04 \times 1.04 \times 2000) = 2000 \times (1.04)^4$.

Note that each multiplication was by 1.04 and that

$$\$2000 \times 1.04^4 \approx \$2339.72. \qquad \text{Using a calculator and rounding to the nearest cent}$$

We have illustrated the following result.

COMPOUND INTEREST FORMULA

If a principal P has been invested at interest rate r, compounded n times a year, in t years it will grow to an amount A given by

$$A = P \cdot \left(1 + \frac{r}{n}\right)^{n \cdot t}.$$

Let's apply this formula to confirm our preceding discussion, where the amount invested is $P = \$2000$, the number of years is $t = 2$, and the number of compounding periods each year is $n = 2$. Substituting into the compound interest formula, we have

$$A = P \cdot \left(1 + \frac{r}{n}\right)^{n \cdot t} = 2000 \cdot \left(1 + \frac{8\%}{2}\right)^{2 \cdot 2}$$

$$= \$2000 \cdot \left(1 + \frac{0.08}{2}\right)^4 = \$2000(1.04)^4$$

$$= \$2000 \times 1.16985856 \approx \$2339.72.$$

If you were using a calculator, you could perform this computation in one step.

EXAMPLE 5 The Ibsens invest \$4000 in an account paying $3\frac{5}{8}\%$, compounded quarterly. Find the amount in the account after $2\frac{1}{2}$ years.

The compounding is quarterly, so n is 4. We substitute \$4000 for P, $3\frac{5}{8}\%$, or 0.03625, for r, 4 for n, and $2\frac{1}{2}$, or $\frac{5}{2}$, for t and compute A:

$$A = P \cdot \left(1 + \frac{r}{n}\right)^{n \cdot t} = \$4000 \cdot \left(1 + \frac{3\frac{5}{8}\%}{4}\right)^{4 \cdot 5/2}$$

$$= \$4000 \cdot \left(1 + \frac{0.03625}{4}\right)^{10}$$

$$= \$4000(1.0090625)^{10}$$

$$\approx \$4377.65.$$

The amount in the account after $2\frac{1}{2}$ years is \$4377.65.

Do Exercise 5. ▶

5. A couple invests \$7000 in an account paying $6\frac{3}{8}\%$, compounded semiannually. Find the amount in the account after $1\frac{1}{2}$ years.

CALCULATOR CORNER

Compound Interest A calculator is useful in computing compound interest. Not only does it perform computations quickly but it also eliminates the need to round until the computation is completed. This minimizes round-off errors that occur when rounding is done at each stage of the computation. We must keep order of operations in mind when computing compound interest.

To find the amount due on a \$20,000 loan made for 25 days at 11% interest, compounded daily, we compute $20{,}000\left(1 + \frac{0.11}{365}\right)^{25}$. To do this on a calculator, we press [2] [0] [0] [0] [0] [×] [(] [1] [+] [.] [1] [1] [÷] [3] [6] [5] [)] [y^x] (or [^]) [2] [5] [=]. The result is \$20,151.23, rounded to the nearest cent.

Some calculators have business keys that allow such computations to be done more quickly.

EXERCISES

1. Find the amount due on a \$16,000 loan made for 62 days at 13% interest, compounded daily.
2. An investment of \$12,500 is made for 90 days at 8.5% interest, compounded daily. How much is the investment worth after 90 days?

C CREDIT CARDS

According to nerdwallet.com, the average credit-card debt among U.S. households with such debt was \$15,162 per household in 2012. According to the Aggregate Revolving Consumer Debt Survey in 2012, the average credit-card debt carried by undergraduate college students is \$3137.

The money you obtain through the use of a credit card is not "free" money. There is a price (interest) to be paid for the convenience of using a credit card. A balance carried on a credit card is a type of loan. Comparing interest rates is essential if one is to become financially responsible. A small change in an interest rate can make a large difference in the cost of a loan. When you make a payment on a credit-card balance, do you know how much of that payment is interest and how much is applied to reducing the principal?

Answer

5. \$7690.94

EXAMPLE 6 *Credit Cards.* After the holidays, Addison has a balance of $3216.28 on a credit card with an annual percentage rate (APR) of 19.7%. She decides not to make additional purchases with this card until she has paid off the balance.

a) Many credit cards require a minimum monthly payment of 2% of the balance. At this rate, what is Addison's minimum payment on a balance of $3216.28? Round the answer to the nearest dollar.

b) Find the amount of interest and the amount applied to reduce the principal in the minimum payment found in part (a).

c) If Addison had transferred her balance to a card with an APR of 12.5%, how much of her first payment would be interest and how much would be applied to reduce the principal?

d) Compare the amounts for 12.5% from part (c) with the amounts for 19.7% from part (b).

We solve as follows.

a) We multiply the balance of $3216.28 by 2%:

$0.02 \times \$3216.28 = \$64.3256.$ Addison's minimum payment, rounded to the nearest dollar, is $64.

b) The amount of interest on $3216.28 at 19.7% for one month* is given by

$$I = P \cdot r \cdot t = \$3216.28 \times 0.197 \times \frac{1}{12} \approx \$52.80.$$

We subtract to find the portion of the first payment applied to reduce the principal:

$$\begin{aligned}\text{Amount applied to reduce the principal} &= \text{Minimum payment} - \text{Interest for the month} \\ &= \$64 - \$52.80 \\ &= \$11.20.\end{aligned}$$

Thus the principal of $3216.28 is decreased by only $11.20 with the first payment. (Addison still owes $3205.08.)

c) The amount of interest on $3216.28 at 12.5% for one month is

$$I = P \cdot r \cdot t = \$3216.28 \times 0.125 \times \frac{1}{12} \approx \$33.50.$$

We subtract to find the amount applied to reduce the principal in the first payment:

$$\begin{aligned}\text{Amount applied to reduce the principal} &= \text{Minimum payment} - \text{Interest for the month} \\ &= \$64 - \$33.50 \\ &= \$30.50.\end{aligned}$$

Thus the principal of $3216.28 would have decreased by $30.50 with the first payment. (Addison would still owe $3185.78.)

*Actually, the interest on a credit card is computed daily with a rate called a daily percentage rate (DPR). The DPR for Example 6 would be 19.7%/365 = 0.054%. When no payments or additional purchases are made during a month, the difference in total interest for the month is minimal and we will not deal with it here.

d) Let's organize the information for both rates in the following table.

BALANCE BEFORE FIRST PAYMENT	FIRST MONTH'S PAYMENT	%APR	AMOUNT OF INTEREST	AMOUNT APPLIED TO PRINCIPAL	BALANCE AFTER FIRST PAYMENT
$3216.28	$64	19.7%	$52.80	$11.20	$3205.08
3216.28	64	12.5	33.50	30.50	3185.78

Difference in balance after first payment ⟶ $19.30

At 19.7%, the interest is $52.80 and the principal is decreased by $11.20. At 12.5%, the interest is $33.50 and the principal is decreased by $30.50. Thus the interest at 19.7% is $52.80 − $33.50, or $19.30, greater than the interest at 12.5%. Thus the principal is decreased by $30.50 − $11.20, or $19.30, more with the 12.5% rate than with the 19.7% rate.

Do Exercise 6. ▶

Even though the mathematics of the information in the table below is beyond the scope of this text, it is interesting to compare how long it takes to pay off the balance of Example 6 if Addison continues to pay $64 each month, compared to how long it takes if she pays double that amount, or $128, each month. Financial consultants frequently tell clients that if they want to take control of their debt, they should pay double the minimum payment.

RATE	PAYMENT	NUMBER OF PAYMENTS TO PAY OFF DEBT	TOTAL PAID BACK	ADDITIONAL COST OF PURCHASES
19.7%	$64	107, or 8 yr 11 mo	$6848	$3631.72
19.7	128	33, or 2 yr 9 mo	4224	1007.72
12.5	64	72, or 6 yr	4608	1391.72
12.5	128	29, or 2 yr 5 mo	3712	495.72

As with most loans, if you pay an extra amount toward the principal with each payment, the length of the loan can be greatly reduced. Note that at the rate of 19.7%, it will take Addison almost 9 years to pay off her debt if she pays only $64 per month and does not make additional purchases. If she transfers her balance to a card with a 12.5% rate and pays $128 per month, she can eliminate her debt in approximately $2\frac{1}{2}$ years. You can see how debt can get out of control if you continue to make purchases and pay only the minimum payment each month. The debt will never be eliminated.

6. *Credit Card.* After the holidays, Logan has a balance of $4867.59 on a credit card with an annual percentage rate (APR) of 21.3%. He decides not to make additional purchases with this card until he has paid off the balance.

a) Many credit cards require a minimum monthly payment of 2% of the balance. What is Logan's minimum payment on a balance of $4867.59? Round the answer to the nearest dollar.

b) Find the amount of interest and the amount applied to reduce the principal in the minimum payment found in part (a).

c) If Logan had transferred his balance to a card with an APR of 13.6%, how much of his first payment would be interest and how much would be applied to reduce the principal?

d) Compare the amounts for 13.6% from part (c) with the amounts for 21.3% from part (b).

Answers

6. **(a)** $97; **(b)** interest: $86.40; amount applied to principal: $10.60; **(c)** interest: $55.17; amount applied to principal: $41.83; **(d)** At 13.6%, the principal was reduced by $31.23 more than at the 21.3% rate. The interest at 13.6% is $31.23 less than at 21.3%.

7.7 Exercise Set

For Extra Help MyMathLab® MathXL® PRACTICE WATCH READ REVIEW

Reading Check

In the simple interest formula, $I = P \cdot r \cdot t$, t must be expressed in years. Convert each length of time to years. Use 1 year = 12 months = 365 days.

RC1. 6 months

RC2. 40 days

RC3. 285 days

RC4. 9 months

RC5. 3 months

RC6. 4 months

a Find the simple interest.

	Principal	Rate of Interest	Time	Simple Interest
1.	\$200	4%	1 year	
2.	\$200	7.7%	1 year	
3.	\$4300	10.56%	$\frac{1}{4}$ year	
4.	\$80,000	$6\frac{3}{4}\%$	$\frac{1}{12}$ year	
5.	\$20,000	$4\frac{5}{8}\%$	6 months	
6.	\$8000	9.42%	2 months	
7.	\$50,000	$5\frac{3}{8}\%$	3 months	
8.	\$100,000	$3\frac{1}{4}\%$	9 months	

Solve. Assume that simple interest is being calculated in each case.

9. Mia's Boutique borrows \$10,000 at 9% for 60 days. Find **(a)** the amount of interest due and **(b)** the total amount that must be paid after 60 days.

10. Mason's Drywall borrows \$8000 at 10% for 90 days. Find **(a)** the amount of interest due and **(b)** the total amount that must be paid after 90 days.

11. Animal Instinct, a pet supply shop, borrows \$6500 at $5\frac{1}{4}\%$ for 90 days. Find **(a)** the amount of interest due and **(b)** the total amount that must be paid after 90 days.

12. Andante's Cafe borrows \$4500 at $12\frac{1}{2}\%$ for 60 days. Find **(a)** the amount of interest due and **(b)** the total amount that must be paid after 60 days.

13. Cameron's Garage borrows \$5600 at 10% for 30 days. Find **(a)** the amount of interest due and **(b)** the total amount that must be paid after 30 days.

14. Shear Delights Hair Salon borrows \$3600 at 4% for 30 days. Find **(a)** the amount of interest due and **(b)** the total amount that must be paid after 30 days.

 Interest is compounded annually. Find the amount in the account after the given length of time. Round to the nearest cent.

	Principal	Rate of Interest	Time	Amount in the Account
15.	$400	5%	2 years	
16.	$450	4%	2 years	
17.	$2000	8.8%	4 years	
18.	$4000	7.7%	4 years	
19.	$4300	10.56%	6 years	
20.	$8000	9.42%	6 years	
21.	$20,000	$6\frac{5}{8}\%$	25 years	
22.	$100,000	$5\frac{7}{8}\%$	30 years	

Interest is compounded semiannually. Find the amount in the account after the given length of time. Round to the nearest cent.

	Principal	Rate of Interest	Time	Amount in the Account
23.	$4000	6%	1 year	
24.	$1000	5%	1 year	
25.	$20,000	8.8%	4 years	
26.	$40,000	7.7%	4 years	
27.	$5000	10.56%	6 years	
28.	$8000	9.42%	8 years	
29.	$20,000	$7\frac{5}{8}\%$	25 years	
30.	$100,000	$4\frac{7}{8}\%$	30 years	

Solve.

31. A family invests $4000 in an account paying 6%, compounded monthly. How much is in the account after 5 months?

32. A couple invests $2500 in an account paying 3%, compounded monthly. How much is in the account after 6 months?

33. A couple invests $1200 in an account paying 10%, compounded quarterly. How much is in the account after 1 year?

34. The O'Hares invest $6000 in an account paying 8%, compounded quarterly. How much is in the account after 18 months?

C Solve.

35. *Credit Cards.* Amelia has a balance of \$1278.56 on a credit card with an annual percentage rate (APR) of 19.6%. The minimum payment required in the current statement is \$25.57. Find the amount of interest and the amount applied to reduce the principal in this payment and the balance after this payment.

36. *Credit Cards.* Lawson has a balance of \$1834.90 on a credit card with an annual percentage rate (APR) of 22.4%. The minimum payment required in the current statement is \$36.70. Find the amount of interest and the amount applied to reduce the principal in this payment and the balance after this payment.

37. *Credit Cards.* Antonio has a balance of \$4876.54 on a credit card with an annual percentage rate (APR) of 21.3%.

a) Many credit cards require a minimum monthly payment of 2% of the balance. What is Antonio's minimum payment on a balance of \$4876.54? Round the answer to the nearest dollar.

b) Find the amount of interest and the amount applied to reduce the principal in the minimum payment found in part (a).

c) If Antonio had transferred his balance to a card with an APR of 12.6%, how much of his payment would be interest and how much would be applied to reduce the principal?

d) Compare the amounts for 12.6% from part (c) with the amounts for 21.3% from part (b).

38. *Credit Cards.* Becky has a balance of \$5328.88 on a credit card with an annual percentage rate (APR) of 18.7%.

a) Many credit cards require a minimum monthly payment of 2% of the balance. What is Becky's minimum payment on a balance of \$5328.88? Round the answer to the nearest dollar.

b) Find the amount of interest and the amount applied to reduce the principal in the minimum payment found in part (a).

c) If Becky had transferred her balance to a card with an APR of 13.2%, how much of her payment would be interest and how much would be applied to reduce the principal?

d) Compare the amounts for 13.2% from part (c) with the amounts for 18.7% from part (b).

Skill Maintenance

39. Find the LCM of 32 and 50. [4.1a]

40. Find the prime factorization of 228. [3.1d]

Divide and simplify. [3.7b]

41. $\frac{6}{125} \div \frac{8}{15}$

42. $-\frac{16}{105} \div \frac{5}{14}$

Multiply and simplify. [3.6a]

43. $\frac{4}{15} \times \frac{3}{20}$

44. $\frac{8}{21} \times \frac{49}{800}$

45. Simplify: $4^3 - 6^2 \div 2^2$. [1.9c]

46. Solve: $x + \frac{2}{5} = -\frac{9}{10}$. [4.3c]

Synthesis

Effective Yield. The *effective yield* is the yearly rate of simple interest that corresponds to a rate for which interest is compounded two or more times a year. For example, if P is invested at 12%, compounded quarterly, we multiply P by $(1 + 0.12/4)^4$, or 1.03^4. Since $1.03^4 \approx 1.126$, the 12% compounded quarterly corresponds to an effective yield of approximately 12.6%. In Exercises 47 and 48, find the effective yield for the indicated account.

47. 🖩 The account pays 9% compounded monthly.

48. 🖩 The account pays 10% compounded daily.

CHAPTER

7 Summary and Review

Vocabulary Reinforcement

Complete each statement with the appropriate word or phrase from the list at the right. Some of the choices will not be used.

1. When a quantity is decreased by a certain percent, we say that this is a ______________. [7.5b]
2. The ______________ interest I on principal P, invested for t years at interest rate r, is given by $I = P \cdot r \cdot t$. [7.7a]
3. Sale price = Original price − ______________. [7.6c]
4. Commission = Commission rate × ______________. [7.6b]
5. Discount = ______________ of discount × Original price. [7.6c]
6. When a quantity is increased by a certain percent, we say that this is a ______________. [7.5b]

discount
rate
sales
commission
price
principal
percent increase
percent decrease
simple
compound

Concept Reinforcement

Determine whether each statement is true or false.

______________ 1. A fixed principal invested for 4 years will earn more interest when interest is compounded quarterly than when interest is compounded semiannually. [7.7b]

______________ 2. Of the numbers 0.5%, $\frac{5}{1000}\%$, $\frac{1}{2}\%$, $\frac{1}{5}$, and $0.\overline{1}$, the largest number is $0.\overline{1}$. [7.1b], [7.2a, b]

______________ 3. If principal A equals principal B and principal A is invested for 2 years at 4%, compounded quarterly, while principal B is invested for 4 years at 2%, compounded semiannually, the interest earned from each investment is the same. [7.7b]

Study Guide

Objective 7.1b Convert between percent notation and decimal notation.

Example Find percent notation for 1.3.

We move the decimal point two places to the right and write a percent symbol:

$1.3 = 130\%$.

Practice Exercise

1. Find percent notation for 0.082.

Example Find decimal notation for $12\frac{3}{4}\%$.

We convert $12\frac{3}{4}$ to a decimal. Then we drop the percent symbol and move the decimal point two places to the left:

$12\frac{3}{4}\% = 12.75\% = 0.1275$.

Practice Exercise

2. Find decimal notation for $62\frac{5}{8}\%$.

Objective 7.2a Convert from fraction notation to percent notation.

Example Find percent notation for $\frac{5}{12}$.

$$\begin{array}{r} 0.416 \\ 12\overline{)5.000} \\ \underline{48} \\ 20 \\ \underline{12} \\ 80 \\ \underline{72} \\ 8 \end{array}$$

$\frac{5}{12} = 0.41\overline{6} = 41.\overline{6}\%$, or $41\frac{2}{3}\%$

Practice Exercise

3. Find percent notation for $\frac{7}{11}$.

Objective 7.2b Convert from percent notation to fraction notation.

Example Find fraction notation for 9.5%.

$$\begin{aligned} 9.5\% &= \frac{9.5}{100} = \frac{95}{1000} \\ &= \frac{5 \cdot 19}{5 \cdot 200} \\ &= \frac{5}{5} \cdot \frac{19}{200} \\ &= \frac{19}{200} \end{aligned}$$

Practice Exercise

4. Find fraction notation for 6.8%.

Objective 7.3b Solve basic percent problems.

Example 165 is what percent of 3300?

We have

$165 = p \cdot 3300$ Translating to a percent equation

$$\begin{aligned} \frac{165}{3300} &= \frac{p \cdot 3300}{3300} \\ \frac{165}{3300} &= p \\ 0.05 &= p \\ 5\% &= p. \end{aligned}$$

Thus, 165 is 5% of 3300.

Practice Exercise

5. 12 is what percent of 288?

Objective 7.4b Solve basic percent problems.

Example 18% of what is 1296?

$\frac{18}{100} = \frac{1296}{b}$ Translating to a proportion

$$\begin{aligned} 18 \cdot b &= 100 \cdot 1296 \\ \frac{18 \cdot b}{18} &= \frac{129{,}600}{18} \\ b &= 7200. \end{aligned}$$

Thus, 18% of 7200 is 1296.

Practice Exercise

6. 3% of what is 300?

Objective 7.5b Solve applied problems involving percent increase or percent decrease.

Example The total cost for 16 basic grocery items in the second quarter of 2008 averaged \$46.67 nationally. The total cost of these 16 items in the second quarter of 2007 averaged \$42.95. What was the percent increase?

Source: American Farm Bureau Federation

We first determine the amount of increase:

$\$46.67 - \$42.95 = \$3.72.$

Then we translate to a percent equation or a proportion and solve.

Rewording: \$3.72 is what percent of \$42.95?

Percent Equation:

$$3.72 = p \cdot 42.95$$

$$\frac{3.72}{42.95} = \frac{p \cdot 42.95}{42.95}$$

$$\frac{3.72}{42.95} = p$$

$$0.087 \approx p$$

$$8.7\% = p$$

Proportion:

$$\frac{N}{100} = \frac{3.72}{42.95}$$

$$42.95 \cdot N = 100 \cdot 3.72$$

$$\frac{42.95 \cdot N}{42.95} = \frac{100 \cdot 3.72}{42.95}$$

$$N = \frac{372}{42.95}$$

$$N \approx 8.7$$

The percent increase was about 8.7%.

Practice Exercise

7. In Indiana, the cost for 16 basic grocery items increased from \$40.07 in the second quarter of 2007 to \$46.20 in the second quarter of 2008. What was the percent increase from 2007 to 2008?

Objective 7.6a Solve applied problems involving sales tax and percent.

Example The sales tax is \$34.23 on the purchase of a flat-screen high-definition television that costs \$489. What is the sales tax rate?

Rephrase: Sales tax is what percent of purchase price?

Translate: $34.23 = r \times 489$

Solve: $\frac{34.23}{489} = \frac{r \times 489}{489}$

$$\frac{34.23}{489} = r$$

$$0.07 = r$$

$$7\% = r$$

The sales tax rate is 7%.

Practice Exercise

8. The sales tax is \$1102.20 on the purchase of a new car that costs \$18,370. What is the sales tax rate?

Objective 7.6b Solve applied problems involving commission and percent.

Example A real estate agent's commission rate is $6\frac{1}{2}\%$. She received a commission of $17,160 on the sale of a home. For how much did the home sell?

Rephrase: Commission is $6\frac{1}{2}\%$ of what selling price?

Translate: $17{,}160 = 6\frac{1}{2}\% \times S$

Solve: $17{,}160 = 0.065 \times S$

$$\frac{17{,}160}{0.065} = \frac{0.065 \times S}{0.065}$$

$$264{,}000 = S$$

The home sold for $264,000.

Practice Exercise

9. A real estate agent's commission rate is 7%. He received a commission of $12,950 on the sale of a home. For how much did the home sell?

Objective 7.7a Solve applied problems involving simple interest.

Example To meet its payroll, a business borrows $5200 at $4\frac{1}{4}\%$ for 90 days. Find the amount of simple interest that is due and the total amount that must be paid after 90 days.

$$I = P \cdot r \cdot t = \$5200 \times 4\tfrac{1}{4}\% \times \frac{90}{365}$$

$$= \$5200 \times 0.0425 \times \frac{90}{365}$$

$$\approx \$54.49$$

The interest due for 90 days = $54.49.

The total amount due = $5200 + $54.49 = $5254.49.

Practice Exercise

10. A student borrows $2500 for tuition at $5\frac{1}{2}\%$ for 60 days. Find the amount of simple interest that is due and the total amount that must be paid after 60 days.

Objective 7.7b Solve applied problems involving compound interest.

Example Find the amount in an account if $3200 is invested at 5%, compounded semiannually, for $1\frac{1}{2}$ years.

$$A = P \cdot \left(1 + \frac{r}{n}\right)^{n \cdot t}$$

$$= \$3200\left(1 + \frac{0.05}{2}\right)^{2 \cdot \frac{3}{2}}$$

$$= \$3200(1.025)^3$$

$$= \$3446.05$$

The amount in the account after $1\frac{1}{2}$ years is $3446.05.

Practice Exercise

11. Find the amount in an account if $6000 is invested at $4\frac{3}{4}\%$, compounded quarterly, for 2 years.

Review Exercises

Find percent notation. [7.1b]

1. 1.7

2. 0.065

Find decimal notation for the percent notations in each sentence. [7.1b]

3. In the 2006–2007 school year, about 4% of the 15 million college students in the United States were foreign students. Approximately 14.4% of the foreign students were from India.

Source: Institute of International Education

4. Poland is 62.1% urban; Sweden is 84.2% urban.

Source: *The World Almanac*, 2008

Find percent notation. [7.2a]

5. $\frac{3}{8}$

6. $\frac{1}{3}$

Find fraction notation. [7.2b]

7. 24%

8. 6.3%

Translate to a percent equation. Then solve. [7.3a, b]

9. 30.6 is what percent of 90?

10. 63 is 84% of what?

11. What is $38\frac{1}{2}\%$ of 168?

Translate to a proportion. Then solve. [7.4a, b]

12. 24% of what is 16.8?

13. 42 is what percent of 30?

14. What is 10.5% of 84?

Solve. [7.5a, b]

15. *Favorite Ice Creams.* According to a survey, 8.9% of those interviewed chose chocolate as their favorite ice cream flavor and 4.2% chose butter pecan. At this rate, of the 2000 students in a freshman class, how many would choose chocolate as their favorite ice cream? butter pecan?

Source: International Ice Cream Association

16. *Prescriptions.* Of the 305 million people in the United States in a recent year, 140.3 million took at least one type of prescription drug per day. What percent took at least one type of prescription drug per day?

Source: William N. Kelly, *Pharmacy: What It Is and How It Works*, 2nd ed., CRC Press Pharmaceutical Education, 2006

17. *Water Output.* The average person loses 200 mL of water per day by sweating. This is 8% of the total output of water from the body. How much is the total output of water?

Source: Elaine N. Marieb, *Essentials of Human Anatomy and Physiology*, 6th ed. Boston: Addison Wesley Longman, Inc., 2000

18. *Test Scores.* After Sheila got a 75 on a math test, she was allowed to go to the math lab and take a retest. She increased her score to 84. What was the percent increase?

19. *Test Scores.* James got an 80 on a math test. By taking a retest in the math lab, he increased his score by 15%. What was his new score?

Solve. [7.6a, b, c]

20. A state charges a meals tax of $7\frac{1}{2}\%$. What is the meals tax charged on a dinner party costing $320?

21. In a certain state, a sales tax of $453.60 is collected on the purchase of a used car for $7560. What is the sales tax rate?

22. Kim earns $753.50 for selling $6850 worth of televisions. What is the commission rate?

23. What is the rate of discount of this stepladder?

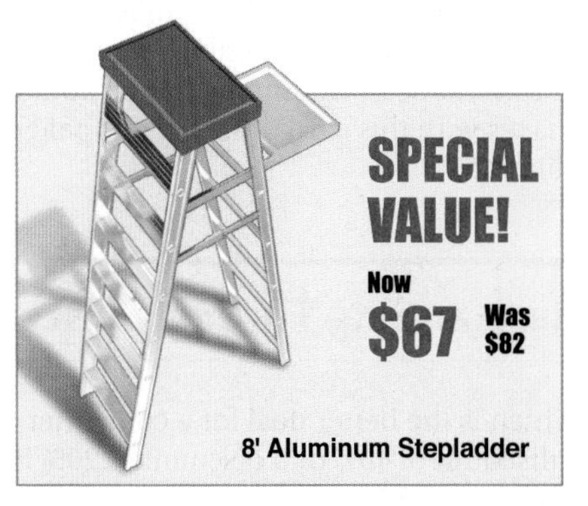

24. An air conditioner has a marked price of $350. It is placed on sale at 12% off. What are the discount and the sale price?

25. The price of a fax machine is marked down from $305 to $262.30. What is the rate of discount?

26. An insurance salesperson receives a 7% commission. If $42,000 worth of life insurance is sold, what is the commission?

Solve. [7.7a, b, c]

27. What is the simple interest on $1800 at 6% for $\frac{1}{3}$ year?

28. The Dress Shack borrows $24,000 at 10% simple interest for 60 days. Find **(a)** the amount of interest due and **(b)** the total amount that must be paid after 60 days.

29. What is the simple interest on a principal of $2200 at an interest rate of 5.5% for 1 year?

30. The Armstrongs invest $7500 in an investment account paying an annual interest rate of 4%, compounded monthly. How much is in the account after 3 months?

31. Find the amount in an investment account if $8000 is invested at 9%, compounded annually, for 2 years.

32. *Credit Cards.* At the end of her junior year of college, Kasha has a balance of $6428.74 on a credit card with an annual percentage rate (APR) of 18.7%. She decides not to make additional purchases with this card until she has paid off the balance.

a) Many credit cards require a minimum payment of 2% of the balance. At this rate, what is Kasha's minimum payment on a balance of $6428.74? Round the answer to the nearest dollar.

b) Find the amount of interest and the amount applied to reduce the principal in the minimum payment found in part (a).

c) If Kasha had transferred her balance to a card with an APR of 13.2%, how much of her payment would be interest and how much would be applied to reduce the principal?

d) Compare the amounts for 13.2% from part (c) with the amounts for 18.7% from part (b).

33. A fishing boat listed at $16,500 is on sale at 15% off. What is the sale price? [7.6c]

A. $14,025 **B.** $2475
C. 85% **D.** $14,225

34. Find the amount in a money market account if $10,500 is invested at 6%, compounded semiannually, for $1\frac{1}{2}$ years. [7.7b]

A. $11,139.45 **B.** $12,505.67
C. $11,473.63 **D.** $10,976.03

Synthesis

35. Mike's Bike Shop reduces the price of a bicycle by 40% during a sale. By what percent must the store increase the sale price, after the sale, to get back to the original price? [7.6c]

36. A worker receives raises of 3%, 6%, and then 9%. By what percent has the original salary increased? [7.5a]

Understanding Through Discussion and Writing

1. Which is the better deal for a consumer and why: a discount of 40% or a discount of 20% followed by another of 22%? [7.6c]

2. Which is better for a wage earner, and why: a 10% raise followed by a 5% raise a year later, or a 5% raise followed by a 10% raise a year later? [7.5a]

3. Ollie buys a microwave oven during a 10%-off sale. The sale price that Ollie paid was $162. To find the original price, Ollie calculates 10% of $162 and adds that to $162. Is this correct? Why or why not? [7.6c]

4. You take 40% of 50% of a number. What percent of the number could you take to obtain the same result making only one multiplication? Explain your answer. [7.5a]

5. A firm must choose between borrowing $5000 at 10% for 30 days and borrowing $10,000 at 8% for 60 days. Give arguments in favor of and against each option. [7.7a]

6. On the basis of the mathematics presented in Section 7.7, discuss what you have learned about interest rates and credit cards. [7.7c]

CHAPTER

7 Test

For Extra Help For step-by-step test solutions, access the Chapter Test Prep Videos in MyMathLab® or on YouTube (search "BittingerBasicEl" and click on "Channels").

1. *Households Owning Pets.* In 2011, about 61.3% of the households in Colorado owned pets. Find decimal notation for 61.3%.

Source: American Veterinary Medical Association

2. *Gravity.* The gravity of Mars is 0.38 as strong as Earth's. Find percent notation for 0.38.

Source: www.marsinstitute.info/epo/mermarsfacts.html

3. Find percent notation for $\frac{11}{8}$.

4. Find fraction notation for 65%.

5. Translate to a percent equation. Then solve.

What is 40% of 55?

6. Translate to a proportion. Then solve. What percent of 80 is 65?

Solve.

7. *Organ Transplants.* In 2011, there were 28,535 organ transplants in the United States. The pie chart below shows the percentages for the main transplants. How many kidney transplants were there in 2011? liver transplants? heart transplants?

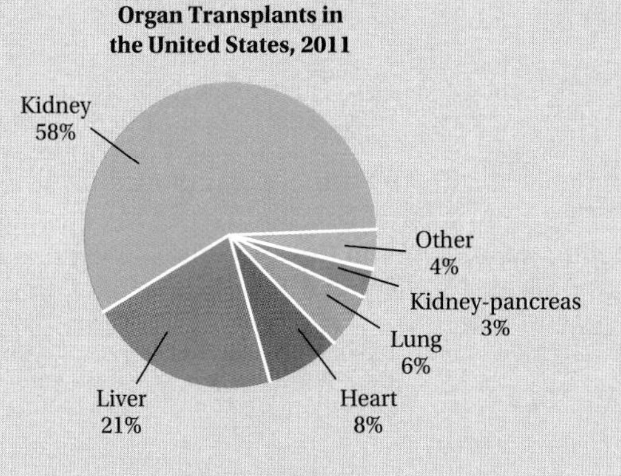

SOURCE: Milliman Research Report

8. *Batting Average.* Garrett Atkins, third baseman for the Colorado Rockies, got 175 hits during the 2008 baseball season. This was about 28.64% of his at-bats. How many at-bats did he have?

Source: Major League Baseball

9. *Foreign Adoptions.* The number of foreign children adopted by Americans declined from 20,679 in 2006 to 19,292 in 2007. Find the percent decrease.

Source: U.S. State Department, *USA TODAY*, August 13, 2008

10. There were about 6,603,000,000 people living in the world in a recent year, and approximately 4,002,000,000 lived in Asia. What percent of people lived in Asia?

Source: Population Division/International Programs Center, U.S. Census Bureau, U.S. Dept. of Commerce

11. *Oklahoma Sales Tax.* The sales tax rate in Oklahoma is 4.5%. How much tax is charged on a purchase of \$560? What is the total price?

12. Noah's commission rate is 15%. What is his commission on the sale of \$4200 worth of merchandise?

13. The marked price of a DVD player is \$200 and the item is on sale at 20% off. What are the discount and the sale price?

14. What is the simple interest on a principal of \$120 at the interest rate of 7.1% for 1 year?

15. A city orchestra invests \$5200 at 6% simple interest. How much is in the account after $\frac{1}{2}$ year?

16. Find the amount in an account if \$1000 is invested at $5\frac{3}{8}\%$, compounded annually, for 2 years.

17. The Suarez family invests $10,000 at an annual interest rate of 4.9%, compounded monthly. How much is in the account after 3 years?

18. *Job Opportunities.* The table below lists job opportunities in 2006 and projected increases for 2016. Complete the table by filling in the missing numbers.

Occupation	Total Employment in 2006	Projected Employment in 2016	Change	Percent of Increase
Dental assistant	280,000	362,000	82,000	29.3%
Plumber	705,000		52,000	
Veterinary assistant	71,000	100,000		
Motorcycle repair technician		24,000	3000	
Fitness professional		298,000		26.8%

SOURCE: EarnMyDegree.com

19. Find the discount and the rate of discount of the television in this ad.

20. *Credit Cards.* Jayden has a balance of $2704.27 on a credit card with an annual percentage rate of 16.3%. The minimum payment required on the current statement is $54. Find the amount of interest and the amount applied to reduce the principal in this payment and the balance after this payment.

21. 0.75% of what number is 300?

A. 2.25 **B.** 40,000 **C.** 400 **D.** 225

Synthesis

22. By selling a home without using a realtor, Juan and Marie can avoid paying a 7.5% commission. They receive an offer of $180,000 from a potential buyer. In order to give a comparable offer, for what price would a realtor need to sell the house? Round to the nearest hundred.

23. Karen's commission rate is 16%. She invests her commission from the sale of $15,000 worth of merchandise at an interest rate of 12%, compounded quarterly. How much is Karen's investment worth after 6 months?

Cumulative Review

1. *Passports.* In 2010, 14,794,604 U.S. passports and passport cards were issued. This number decreased to 12,613,153 in 2011 and then increased to 13,125,829 in 2012. What was the percent decrease from 2010 to 2011? What was the percent increase from 2011 to 2012? Round answers to the nearest tenth of a percent.

Sources: Travel.State.gov; Bureau of Consular Affairs, U.S. Department of State

2. Find percent notation: 0.269.

3. Find percent notation: $\frac{9}{8}$.

4. Find decimal notation: $-\frac{13}{6}$.

5. Write fraction notation for the ratio 5 to 0.5.

6. Find the rate in kilometers per hour: 350 km, 15 hr.

Use $<$, $>$, or $=$ for $\square$ to write a true sentence.

7. $\frac{5}{7} \square \frac{6}{8}$

8. $\frac{-15}{25} \square \frac{-6}{14}$

Estimate the sum or the difference by first rounding to the nearest hundred.

9. 263,961 + 32,090 + 127.89

10. 73,510 − 23,450

Calculate.

11. $46 - [4(6 + 4 \div 2) + 2 \times 3 - 5]$

12. $[-0.8(1.5 - 9.8 \div 49) - (1 + 0.1)^2] \div 1.5$

Compute and simplify.

13. $\frac{6}{5} + 1\frac{5}{6}$

14. $-46.9 + 2.84$

15. $$\begin{array}{r} 487{,}094 \\ 6{,}936 \\ +\ 21{,}120 \\ \hline \end{array}$$

16. $35 - 34.98$

17. $3\frac{1}{3} - 2\frac{2}{3}$

18. $-\frac{8}{9} - \frac{6}{7}$

19. $-\frac{7}{9} \cdot \left(-\frac{3}{14}\right)$

20. $$\begin{array}{r} 236{,}984 \\ \times\ \ 3{,}600 \\ \hline \end{array}$$

21. $$\begin{array}{r} 46.012 \\ \times\ \ 0.03 \\ \hline \end{array}$$

22. $-6\frac{3}{5} \div 4\frac{2}{5}$

23. $431.2 \div 35.2$

24. $15\overline{)1850}$

Solve.

25. $36 \cdot x = 3420$

26. $y + 142.87 = -151$

27. $-\frac{2}{15} \cdot t = \frac{6}{5}$

28. $\frac{3}{4} + x = \frac{5}{6}$

29. $\frac{y}{25} = \frac{24}{15}$

30. $\frac{16}{n} = \frac{21}{11}$

31. *Museum Attendance.* Among leading art museums, the Indianapolis Museum of Art ranks sixth in the nation in the number of visitors as a percentage of the metropolitan population. In 2007, the number of visitors was 30.31% of the metropolitan population, which was 1,525,104. What was the total attendance?

Source: dashboard.imamuseum.org/series/2007+MSA+Attendance

32. *Salaries of Medical Doctors.* In 2007, the starting salary for a doctor in radiology was $172,500 higher than the starting salary of a doctor in neurology. If the starting salary for a radiologist was $350,000, what was the starting salary of a neurologist?

Source: *Journal of the American Medical Association*

33. At one point during the 2008–2009 NBA season, the Cleveland Cavaliers had won 39 out of 49 games. At this rate, how many games would they win in the entire season of 82 games?

Source: National Basketball Association

34. *Shirts.* A total of $424.75 was paid for 5 shirts at an upscale men's store. How much did each shirt cost?

35. *Unit Price.* A 200-oz bottle of liquid laundry detergent costs $14.99. What is the unit price?

36. Patty walked $\frac{7}{10}$ mi to school and then $\frac{8}{10}$ mi to the library. How far did she walk?

37. On a map, 1 in. represents 80 mi. How much does $\frac{3}{4}$ in. represent?

38. *Compound Interest.* The Bakers invest $8500 in an investment account paying 8%, compounded monthly. How much is in the account after 5 years?

39. *Ribbons.* How many pieces of ribbon $1\frac{4}{5}$ yd long can be cut from a length of ribbon 9 yd long?

40. *Auto Repair Technicians.* In 2006, 773,000 people were employed as auto repair technicians. It is projected that by 2016, this number will increase by 110,000. Find the percent increase in the number of auto repair technicians and the number of technicians in 2016.

Source: EarnMyDegree.com

41. Subtract and simplify: $\frac{14}{25} - \frac{3}{20}$.

A. $\frac{11}{500}$ **B.** $\frac{11}{5}$

C. $\frac{41}{100}$ **D.** $\frac{205}{500}$

42. The population of the state of Louisiana decreased from 4,468,976 in 2000 to 4,287,768 in 2006. What was the percent decrease?

Source: U.S. Census Bureau

A. 4.1% **B.** 4.2%

C. 104% **D.** 95.9%

Synthesis

43. If a is 50% of b, then b is what percent of a?

Data, Graphs, and Statistics

STUDYING FOR SUCCESS *Doing Your Homework*

- ☐ Prepare for doing homework by reading explanations of concepts and by following the step-by-step solutions of examples in the text.
- ☐ Include all the steps when solving a problem. This will keep you organized, help you avoid computational errors, and give you a study guide for an exam.
- ☐ Try to do your homework as soon as possible after each class. Avoid waiting until a deadline is near to begin work on an assignment.

8.1 Averages, Medians, and Modes

OBJECTIVES

a Find the average of a set of numbers and solve applied problems involving averages.

b Find the median of a set of numbers and solve applied problems involving medians.

c Find the mode of a set of numbers and solve applied problems involving modes.

SKILL TO REVIEW

Objectives 1.9c and 5.4c: Simplify expressions using the rules for order of operations.

1. Find the average of 282, 137, 5280, and 193.
2. Find the average of \$23.40, \$89.15, and \$148.17 to the nearest cent.

a AVERAGES

A **statistic** is a number describing a set of data. One statistic is a *center point,* or *measure of central tendency,* that characterizes the data. The most common kind of center point is the **arithmetic** (pronounced ăr′ĭth-mĕt′-ĭk) **mean**, or simply the **mean**. This center point is often referred to as the *average.*

AVERAGE

To find the **average** of a set of numbers, add the numbers and then divide by the number of items of data.

EXAMPLE 1 On a 4-day trip, a car was driven the following number of miles: 240, 302, 280, 320. What was the average number of miles per day?

$$\frac{240 + 302 + 280 + 320}{4} = \frac{1142}{4}, \quad \text{or} \quad 285.5$$

The car was driven an average of 285.5 mi per day. Had the car been driven exactly 285.5 mi each day, the same total distance (1142 mi) would have been traveled.

EXAMPLE 2 *Gas Mileage.* The 2013 Volkswagen Jetta TDI is estimated to travel 546 miles on the highway on 13 gal of diesel fuel. What is the expected average number of miles per gallon (mpg)—that is, what is the fuel mileage for highway driving?

Source: vw.com

We divide the total number of miles, 546, by the total number of gallons, 13:

$$\frac{546 \text{ mi}}{13 \text{ gal}} = 42 \text{ mpg}.$$

The Jetta's expected average is 42 mi per gallon for highway driving.

Answers

Skill to Review:
1. 1473 2. \$86.91

Do Exercises 1–4. ▶

Find the average.

1. 14, 175, 36
2. 75, 36.8, 95.7, 12.1
3. In the first five games of the season, a basketball player scored 26, 21, 13, 14, and 23 points. Find the average number of points scored per game.
4. *Home-Run Batting Average.* Babe Ruth hit 714 home runs in 22 seasons in the major leagues. What was his average number of home runs per season? Round to the nearest tenth.

Source: Major League Baseball

In a *weighted average*, more importance, or *weight*, is assigned to some values than to others. For example, a course syllabus may include the following description:

COURSE COMPONENT	WEIGHT FOR GRADE
Quizzes	20
Homework	30
Tests	50

If Allison has scored 70% on quizzes, 100% on homework, and 92% on tests, she cannot calculate her course grade by averaging 70, 100, and 92, because each category is weighted differently. Instead, she must multiply each percentage by its weight, add the results, and divide by the total of the weights:

$$\text{Course grade} = \frac{70 \cdot 20 + 100 \cdot 30 + 92 \cdot 50}{20 + 30 + 50}$$

$$= \frac{9000}{100} = 90.$$

Allison's course grade is 90%.

A grade point average is another example of a weighted average.

EXAMPLE 3 *Grade Point Average.* In many schools, students are assigned grade point values for grades obtained. The **grade point average**, or **GPA**, is the average of the grade point values for each credit hour taken. At Meg's college, grade point values are assigned as follows:

A: 4.0 B: 3.0 C: 2.0 D: 1.0 F: 0.0.

Meg earned the following grades for one semester. What was her grade point average?

COURSE	GRADE	NUMBER OF CREDIT HOURS IN COURSE
Colonial History	B	3
Basic Mathematics	A	4
English Literature	A	3
French	C	4
Time Management	D	1

To find the GPA, we first multiply the grade point value for each grade by the number of credit hours in the course to determine the number of *quality points,* and then add. Here each grade is weighted by the number of credit hours in the course.

Colonial History	$3.0 \cdot 3 =$	9
Basic Mathematics	$4.0 \cdot 4 =$	16
English Literature	$4.0 \cdot 3 =$	12
French	$2.0 \cdot 4 =$	8
Time Management	$1.0 \cdot 1 =$	1
		46 (Total)

Answers

1. 75 2. 54.9 3. 19.4 points per game
4. 32.5 home runs per season

The total number of credit hours taken is $3 + 4 + 3 + 4 + 1$, or 15. We divide the number of quality points, 46, by the number of credit hours, 15, and round to the nearest tenth:

$$\text{GPA} = \frac{46}{15} \approx 3.1.$$

Meg's grade point average was 3.1.

◀ **Do Exercises 5 and 6.**

EXAMPLE 4 *Grading.* To get a B in math, Geraldo must score an average of 80 on five tests. On the first four tests, his scores were 79, 88, 64, and 78. What is the lowest score that Geraldo can get on the last test and still get a B?

We can find the total of the five scores needed as follows:

$$80 + 80 + 80 + 80 + 80 = 5 \cdot 80, \quad \text{or} \quad 400.$$

The total of the scores on the first four tests is

$$79 + 88 + 64 + 78 = 309.$$

Thus Geraldo needs to get at least

$$400 - 309, \quad \text{or} \quad 91$$

in order to get a B. We can check this as follows:

$$\frac{79 + 88 + 64 + 78 + 91}{5} = \frac{400}{5}, \quad \text{or} \quad 80.$$

◀ **Do Exercise 7.**

b MEDIANS

Another type of center-point statistic is the *median*. Medians are useful when we wish to de-emphasize unusually extreme numbers. For example, suppose a small class scored as follows on an exam.

Jae:	78	Pat:	56
Jill:	81	Carmen:	84
Matt:	82		

Let's first list the scores in order from smallest to largest:

56, 78, 81, 82, 84.
↑
Middle score

The middle score—in this case, 81—is called the **median**. Note that because of the extremely low score of 56, the average of the scores is 76.2. In this example, the median may be a more appropriate center-point statistic.

---

5. Soha's sociology professor included the following in the course syllabus:

COURSE COMPONENT	WEIGHT FOR GRADE
Participation	15
Book reports	25
Research paper	40

Soha received 88% on her research paper and 92% on her book reports, and she anticipates a score of 100% for participation. What is her course grade?

GS

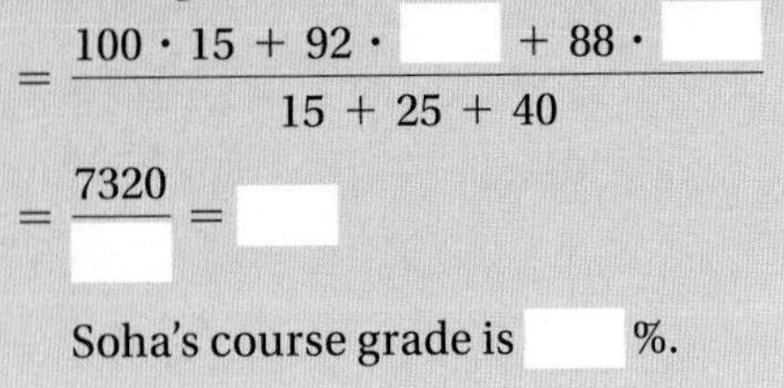

Course grade

$$= \frac{100 \cdot 15 + 92 \cdot \square + 88 \cdot \square}{15 + 25 + 40}$$

$$= \frac{7320}{\square} = \square$$

Soha's course grade is ____ %.

6. *Grade Point Average.* Alex earned the following grades one semester.

GRADE	NUMBER OF CREDIT HOURS IN COURSE
B	3
C	4
C	4
A	2

What was Alex's grade point average? Assume that the grade point values are 4.0 for an A, 3.0 for a B, and so on. Round to the nearest tenth.

7. *Grading.* To get an A in math, Rosa must score an average of 90 on four tests. On the first three tests, her scores were 80, 100, and 86. What is the lowest score that Rosa can get on the last test and still get an A?

Answers

5. 91.5% **6.** 2.5 **7.** 94

Guided Solution:
5. 25, 40, 80, 91.5; 91.5

EXAMPLE 5 What is the median of this set of numbers?

99, 870, 91, 98, 106, 90, 98

We first rearrange the numbers in order from smallest to largest. Then we locate the middle number, 98.

90, 91, 98, 98, 99, 106, 870

↑ Middle number

The median is 98.

Do Exercises 8–10. ▶

Find the median.

8. 17, 13, 18, 14, 19

9. 20, 14, 13, 19, 16, 18, 17

10. 78, 81, 83, 91, 103, 102, 122, 119, 88

MEDIAN

Once a set of data is listed in order, from smallest to largest, the **median** is the middle number if there is an odd number of data items. If there is an even number of items, the median is the number that is the average of the two middle numbers.

EXAMPLE 6 What is the median of this set of numbers?

69, 80, 61, 63, 62, 65

We first rearrange the numbers in order from smallest to largest. There is an even number of numbers. We look for the middle two, which are 63 and 65. The median is halfway between 63 and 65, the number 64.

61, 62, 63, 65, 69, 80

The average of the middle numbers is $\frac{63 + 65}{2} = \frac{128}{2}$, or 64.

The median is 64. ■

EXAMPLE 7 *Salaries.* The following are the salaries of the four highest-paid players in the National Hockey League. What is the median of the salaries?

PLAYER	SALARY
Sidney Crosby	$8,700,000
Alexander Ovechkin	9,500,000
Evgeni Malkin	8,700,000
Eric Staal	8,250,000

SOURCE: Forbes.com

We rearrange the numbers in order from smallest to largest:

$8,250,000, $8,700,000, $8,700,000, $9,500,000.

The two middle numbers are $8,700,000 and $8,700,000. Since they are the same number, their average is $8,700,000. Thus the median salary is $8,700,000.

Do Exercises 11 and 12. ▶

Find the median.

11. *Salaries of Part-Time Typists.* $3300, $4000, $3900, $3600, $3800, $3400

GS **12.** 68, 34, 67, 69, 34, 70

Rearrange the numbers in order from smallest to largest:

34, 34, ____, 68, ____, 70.

The middle numbers are ____ and 68.

The average of 67 and 68 is ____.

The median is ____.

Answers

8. 17 **9.** 17 **10.** 91 **11.** $3700 **12.** 67.5

Guided Solution:

12. 67, 69; 67, 67.5, 67.5

C MODES

The final type of center-point statistic we will consider is the *mode*.

MODE

The **mode** of a set of data is the number or numbers that occur most often. If each number occurs the same number of times, there is *no* mode.

EXAMPLE 8 Find the mode of these data.

17, 13, 18, 17, 14, 19

To find the mode, it is helpful to first rearrange the numbers in order from smallest to largest.

13, 14, 17, 17, 18, 19

The number that occurs most often is 17. Thus the mode is 17.

EXAMPLE 9 Find the mode of these data.

5, 5, 11, 11, 13, 13

The numbers in this set of data are 5, 11, and 13. Each occurs twice, so all the numbers are equally represented. There is *no mode.*

A set of data has just one average (mean) and just one median, but it can have more than one mode.

EXAMPLE 10 Find the modes of these data.

33, 34, 34, 34, 35, 36, 37, 37, 37, 38, 39, 40

There are two numbers that occur most often, 34 and 37. Thus the modes are 34 and 37.

Do Exercises 13–16.

Which statistic is best for a particular situation? If someone is bowling, the *average* from several games is a good indicator of that person's ability. If someone is applying for a job, the *median* salary at that business is often most indicative of what people are earning there because although executives tend to make a lot more money, there are fewer of them. For similar reasons, the selling price of homes is usually reported as a *median* price. Finally, if someone is reordering stock for a clothing store, the *mode* of the sizes sold is probably the most important statistic.

Find the modes of these data.

13. 23, 45, 45, 45, 78

14. 34, 34, 67, 67, 68, 70

15. 24, 89, 13, 28, 67, 27 GS

Rearrange the numbers in order from smallest to largest.

13, 24, ____, 28, 67, ____.

Each number occurs ____ time.

There is no mode.

16. In a lab, Gina determined the mass, in grams, of each of five eggs:

15 g, 19 g, 19 g, 14 g, 18 g.

a) What is the mean?

b) What is the median?

c) What is the mode?

Answers

13. 45 **14.** 34, 67 **15.** No mode exists.
16. **(a)** 17 g; **(b)** 18 g; **(c)** 19 g

Guided Solution:
15. 27, 89; one

8.1 Exercise Set

For Extra Help MyMathLab® MathXL® PRACTICE WATCH READ REVIEW

Reading Check

Complete each sentence with the appropriate word from the list at the right. Not all choices will be used.

RC1. A(n) ________________ is a number describing a set of data.

RC2. To find the ________________ of a set of numbers, add the numbers and then divide by the number of items of data.

RC3. To find the weighted average of a set of numbers, multiply each number by its ________________, add the results, and divide by the total of the weights.

RC4. The ________________ of a set of numbers is the number or numbers that occur most often.

average
median
mode
statistic
weight

a, b, c For each set of numbers, find the average, the median, and any modes that exist.

1. *Great Smoky Mountains National Park.* More people visit Great Smoky Mountains National Park each year than any other U.S. national park. The following bar graph shows the numbers of visitors the park had for 2005 to 2012. What is the average number of visitors for the 8 years? the median? the mode?

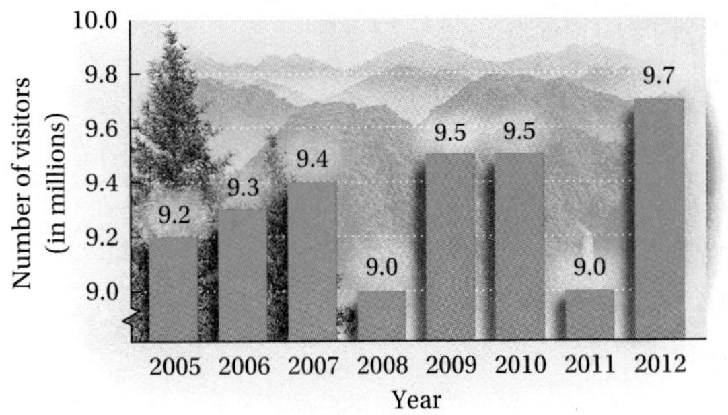

2. *Congestion.* The following bar graph shows the annual number of hours of traffic delay per auto commuter for 8 U.S. cities. What is the average delay time? the median? the mode?

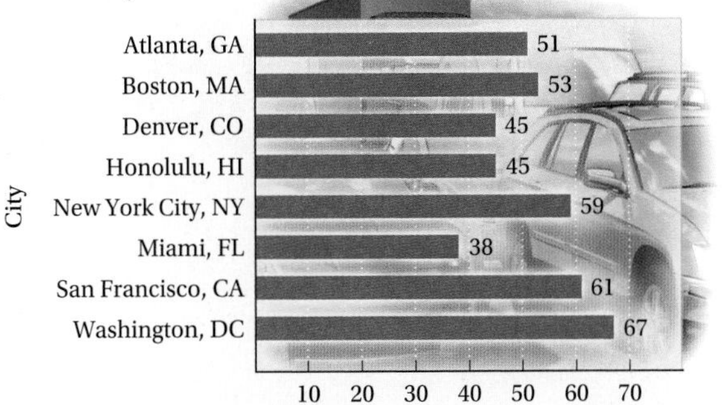

SOURCE: *2012 Annual Urban Mobility Report*, Texas A&M Transportation Institute

3. 17, 19, 29, 18, 14, 29

4. 72, 83, 85, 88, 92

5. 5, 37, 20, 20, 35, 5, 25

6. 13, 32, 25, 27, 13

7. 4.3, 7.4, 1.2, 5.7, 8.3

8. 13.4, 13.4, 12.6, 42.9

9. 234, 228, 234, 229, 234, 278

10. \$29.95, \$28.79, \$30.62, \$28.79, \$29.95

11. *Gas Mileage.* The 2013 Kia Rio does 396 mi of highway driving on 11 gal of gasoline. What is the average number of miles expected per gallon—that is, what is the gas mileage?

Source: fueleconomy.gov

12. *Gas Mileage.* When using gas only, the 2013 Chevrolet Volt does 315 mi of city driving on 9 gal of gasoline. What is the average number of miles expected per gallon—that is, what is the gas mileage?

Source: fueleconomy.gov

Grade Point Average. The tables in Exercises 13 and 14 show the grades of a student for one semester. In each case, find the grade point average. Assume that the grade point values are 4.0 for an A, 3.0 for a B, and so on. Round to the nearest tenth.

13.

GRADE	NUMBER OF CREDIT HOURS IN COURSE
B	4
A	5
D	3
C	4

14.

GRADE	NUMBER OF CREDIT HOURS IN COURSE
A	5
C	4
F	3
B	5

15. *Brussels Sprouts.* The following prices per stalk of Brussels sprouts were found at five farmers' markets:

$3.99, $4.49, $4.99, $3.99, $3.49.

What was the average price per stalk? the median price? the mode?

16. *Mangoes.* The most popular fruit in the world is the mango, which is grown in over 2000 varieties. The following prices per pound of mangoes were found at five supermarkets:

$2.49, $1.59, $2.29, $2.49, $2.29.

What was the average price per pound? the median price? the mode?

17. *Grading.* To get a B in math, Rich must score an average of 80 on five tests. His scores on the first four tests were 80, 74, 81, and 75. What is the lowest score that Rich can get on the last test and still receive a B?

18. *Grading.* To get an A in math, Cybil must score an average of 90 on five tests. Her scores on the first four tests were 90, 91, 81, and 92. What is the lowest score that Cybil can get on the last test and still receive an A?

19. *Length of Pregnancy.* Marta was pregnant 270 days, 259 days, and 272 days for her first three pregnancies. In order for Marta's average pregnancy to equal the worldwide average of 266 days, how long must her fourth pregnancy last?

Source: Vardaan Hospital, Dr. Rekha Khandelwal, M.S.

20. *Male Height.* Jason's brothers are 174 cm, 180 cm, 179 cm, and 172 cm tall. The average male is 176.5 cm tall. How tall is Jason if he and his brothers have an average height of 176.5 cm?

21. *Median Home Prices.* The following table lists the selling prices of homes in two counties during one month.

a) Find the median home price for each county.
b) Which county had the lower median home price?

JEFFERSON COUNTY	HAMILTON COUNTY
$122,587	$387,262
138,291	146,989
121,103	262,105
768,407	253,289
532,194	112,681
129,683	127,092
278,104	131,612
110,329	

22. *Median Salaries.* The following table lists salaries for two small companies.

a) Find the median salary for each company.
b) Which company has the higher median salary?

VALUE SERVICES	DEPENDABLE CARE
$ 48,267	$18,242
32,193	21,607
189,607	98,322
56,189	87,212
28,394	56,812
152,693	42,394
42,681	50,112
	52,987

23. *Movie Ticket Sales.* The following table lists the number of movie tickets sold annually, in billions, from 2002 to 2012.

a) Find the average number of tickets sold for the 8 years from 2002 to 2009.
b) Find the average number of tickets sold for the 8 years from 2005 to 2012.
c) On average, were more tickets sold per year from 2002 to 2009 or from 2005 to 2012?

YEAR	NUMBER OF MOVIE TICKETS SOLD (in billions)	YEAR	NUMBER OF MOVIE TICKETS SOLD (in billions)
2002	1.58	2008	1.39
2003	1.55	2009	1.42
2004	1.49	2010	1.33
2005	1.40	2011	1.30
2006	1.41	2012	1.37
2007	1.40		

SOURCE: the-numbers.com

24. *Movies Released.* The following table lists the number of movies released to U.S. theaters annually from 2005 to 2012.

a) Find the average number of movies released for the 5 years from 2005 to 2009.
b) Find the average number of movies released for the 5 years from 2008 to 2012.
c) On average, were more movies released from 2005 to 2009 or from 2008 to 2012?

YEAR	NUMBER OF MOVIES RELEASED	YEAR	NUMBER OF MOVIES RELEASED
2005	507	2009	557
2006	594	2010	563
2007	611	2011	609
2008	638	2012	677

SOURCE: mpaa.org

Skill Maintenance

Multiply.

25. 12.86×17.5 [5.3a]

26. $-222 \times (-0.5678)$ [5.3a]

27. $\frac{4}{5} \cdot \frac{3}{28}$ [3.6a]

28. $-\frac{28}{45} \cdot \frac{3}{2}$ [3.6a]

Synthesis

29. The ordered set of data 18, 21, 24, a, 36, 37, b has a median of 30 and an average of 32. Find a and b.

30. *Hank Aaron.* Hank Aaron averaged $34\frac{7}{22}$ home runs per year over a 22-year career. After 21 years, Aaron had averaged $35\frac{10}{21}$ home runs per year. How many home runs did Aaron hit in his final year?

31. *Price Negotiations.* Amy offers $6400 for a used Ford Taurus advertised at $8000. The first offer from Jim, the car's owner, is to "split the difference" and sell the car for $(6400 + 8000) \div 2$, or $7200. Amy's second offer is to split the difference between Jim's offer and her first offer. Jim's second offer is to split the difference between Amy's second offer and his first offer. If this pattern continues and Amy accepts Jim's third (and final) offer, how much will she pay for the car?

8.2 Interpreting Data from Tables and Graphs

OBJECTIVES

a Extract and interpret data from tables.

b Extract and interpret data from graphs.

c Extract and interpret data from histograms.

SKILL TO REVIEW

Objective 4.6a: Multiply using mixed numerals.

Multiply.

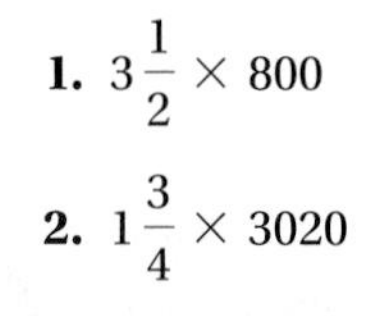

1. $3\frac{1}{2} \times 800$

2. $1\frac{3}{4} \times 3020$

We use tables and graphs to display data and to communicate information about the data. For example, the following table and graphs display data on the resting heart rate for several mammals. Examine each method of presentation. Which method do you like best, and why?

Table

	MOUSE	GIRAFFE	CAT	HUMAN	HORSE	ELEPHANT
Average Resting Heart Rate (in beats per minute)	500	170	130	70	35	28

SOURCES: elephantnaturepark.org; vetmedicine.about.com; giraffeconservation.org; learningabouthorses.com

Pictograph

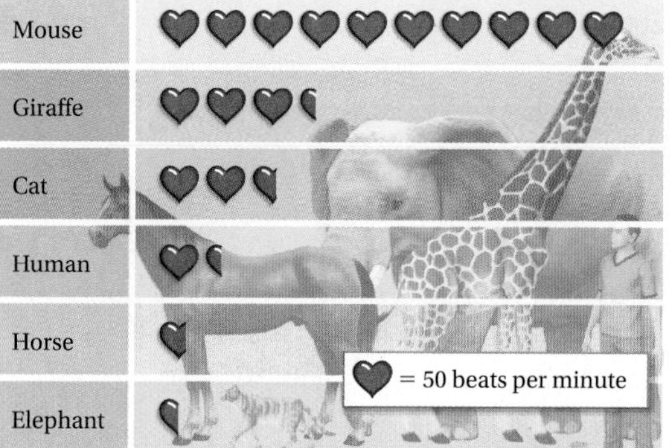

Bar Graph

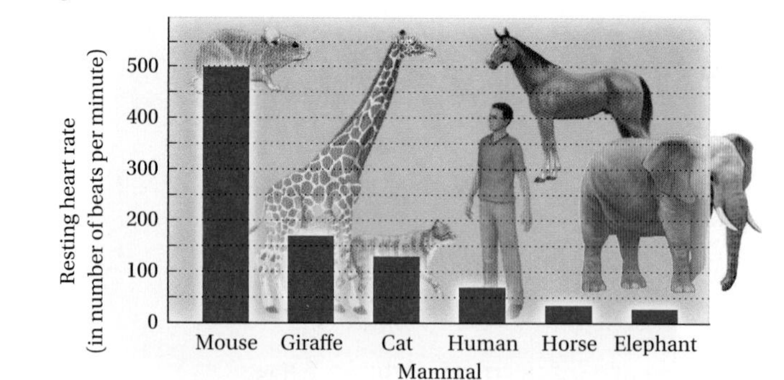

Answers

Skill to Review:
1. 2800 **2.** 5285

Comparing the table and the graphs reveals that the exact data values are most easily read in a table. The fastest and slowest heart rates can be determined easily from the graphs. The graph below communicates additional information about the data. The size of each mammal in this graph indicates the heart rate, not the actual size of the animal. The unexpected relative sizes of the mammals in the graph emphasize the fact that many smaller animals have faster heart rates than larger animals.

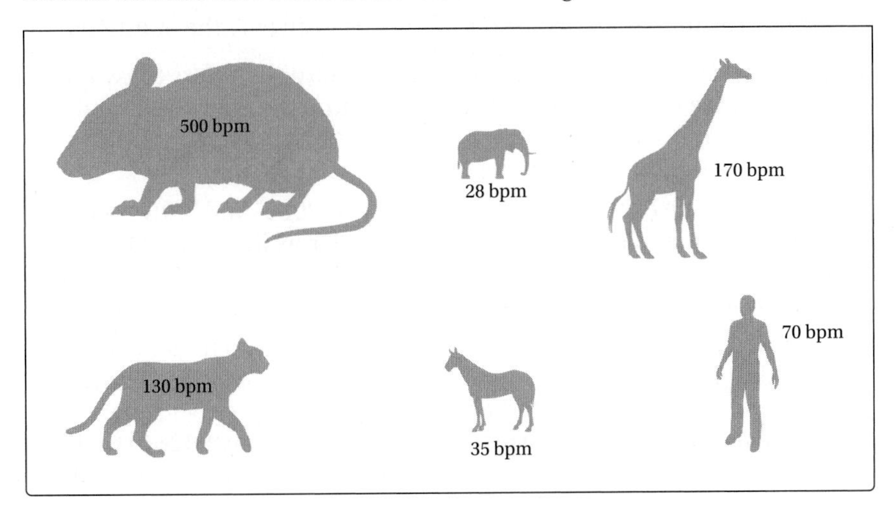

a READING AND INTERPRETING TABLES

A **table** is often used to present data in rows and columns.

EXAMPLE 1 *Population Density.* The following table lists populations and land areas of 10 countries.

COUNTRY	LAND AREA (in square miles)	POPULATION		POPULATION DENSITY (per square mile)	
		2008	2012	2008	2012
Australia	2,941,299	20,434,176	22,015,576	7	7
Brazil	3,265,077	190,010,647	199,321,413	58	61
China	3,600,947	1,321,851,888	1,343,239,923	367	373
Finland	117,558	5,238,460	5,262,930	45	45
Germany	134,836	82,400,996	81,305,856	611	603
India	1,147,955	1,129,866,154	1,205,073,612	984	1050
Japan	144,689	127,433,494	127,368,088	881	880
Kenya	219,789	36,913,721	43,013,341	168	196
Mexico	742,490	108,700,891	114,975,406	146	155
United States	3,537,439	301,139,947	313,847,465	85	89

SOURCES: *World Almanac 2008*; U.S. Census Bureau

a) Which country had the largest population in 2012?

b) In which country or countries did the population decrease from 2008 to 2012?

c) What was the percent increase in population density in India from 2008 to 2012?

d) Find the average land area of the four largest countries in the table.

Careful examination of the table allows us to answer the questions.

a) Note that the column head "Population" actually refers to two table columns. We look down the Population column headed "2012" and find the largest number. That number is 1,343,239,923. Then we look across that row to find the name of the country: China.

b) Comparing the Population columns headed "2008" and "2012," we see that the population decreased in the fifth row (from 82,400,996 to 81,305,856) and in the seventh row (from 127,433,494 to 127,368,088). Looking across these rows, we find the countries: Germany and Japan.

c) We look down the column headed "Country" and find India. Then we look across that row to the columns headed "Population Density." The population density of India in 2008 was 984 people per square mile. This increased to 1050 people per square mile in 2012. To find the percent increase, we find the amount of increase and divide by the population density in 2008.

$$\text{Amount of Increase: } 1050 - 984 = 66$$

$$\text{Percent Increase: } \frac{66}{984} \approx 0.067 = 6.7\%$$

The population density of India increased by 6.7% from 2008 to 2012.

d) By looking down the column headed "Land Area," we determine that the four largest countries in the table are Australia, Brazil, China, and the United States. We find the average land area of these countries:

$$\frac{2{,}941{,}299 + 3{,}265{,}077 + 3{,}600{,}947 + 3{,}537{,}439}{4} = \frac{13{,}344{,}762}{4}$$

$$= 3{,}336{,}190.5 \text{ sq mi.}$$

Use the table in Example 1 to answer Margin Exercises 1–5.

1. Which country has the smallest land area?

2. What was the population of Mexico in 2012?

3. What was the percent decrease in population density in Germany from 2008 to 2012? GS

   The amount of the decrease in population density is

   $611 - 603 = \square$.

   The percent decrease is

   $\frac{8}{\square} \approx 0.013$, or $\square$%.

4. Which country had the greatest increase in population from 2008 to 2012?

5. Find the median population density of these countries in 2012.

Do Exercises 1–5.

b READING AND INTERPRETING GRAPHS

Pictographs (or *picture graphs*) are another way to show information. Instead of actually listing the amounts to be considered, a **pictograph** uses symbols to represent the amounts. A pictograph includes a *key* that tells what each symbol represents.

Answers

1. Finland **2.** 114,975,406 **3.** About 1.3%
4. India, with an increase of 75,207,458
5. 175.5 people per square mile

Guided Solution:
3. 8; 611, 1.3

EXAMPLE 2 *Roller Coasters.* The following pictograph shows the number of roller coasters listed in the Roller Coaster Data Base for six continents. Below the graph is a key that tells you that each symbol represents 100 roller coasters.

Roller Coasters of the World

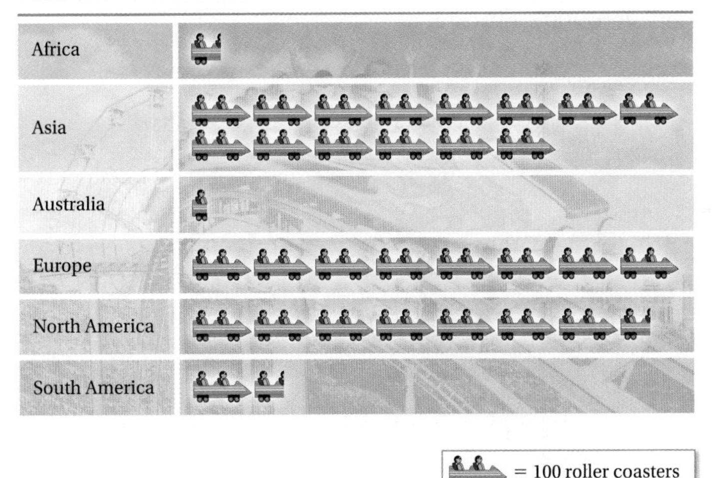

a) Which continent has the greatest number of roller coasters?

b) About how many roller coasters are there in Australia?

c) How many more roller coasters are there in Europe than in North America?

We can determine the answers by reading the pictograph.

a) The continent with the most symbols is Asia, so Asia has the greatest number of roller coasters.

b) The pictograph shows about $\frac{1}{4}$ symbol for Australia. Since each symbol represents 100 roller coasters, there are about $\frac{1}{4} \times 100$, or 25, roller coasters in Australia.

c) From the graph, we see that there are 8×100, or 800, roller coasters in Europe and about $7\frac{1}{2} \times 100$, or 750, roller coasters in North America. Thus there are $800 - 750$, or 50, more roller coasters in Europe than in North America. We could also estimate this difference by noting that Europe has $\frac{1}{2}$ of a symbol more than North America does, and $\frac{1}{2} \times 100 = 50$.

Do Exercises 6–8.

When representing data with graphs, we must be sure that the areas of regions of the graph are proportional to the numbers that the regions represent. For example, in pictographs, each symbol is the same size, and the number of symbols is proportional to the actual data values. Thus the total area of the symbols is proportional to the data values. This area principle is illustrated in the following example.

Use the pictograph in Example 2 to answer Margin Exercises 6–8.

6. Which continent has the smallest number of roller coasters?

7. About how many roller coasters are there in Asia?

8. How many more roller coasters are there in South America than in Africa?

The graph shows $1\frac{1}{2}$ symbols for South America.

This represents ______ roller coasters.

The graph shows $\frac{1}{2}$ symbol for Africa.

This represents ______ roller coasters.

There are ______ more roller coasters in South America than in Africa.

Answers

6. Australia **7.** 1400 roller coasters **8.** About 100 more roller coasters

Guided Solution:
8. 150, 50, 100

EXAMPLE 3 *Electricity Generation.* The following graph illustrates the different methods used in the United States to generate electricity. Each yellow circle represents the amount of electricity generated, in terawatt-hours per year (TWh). Some methods of electricity generation are more efficient than others. The larger circle surrounding each yellow circle represents the amount of electricity that could have been generated if the method were 100% efficient.

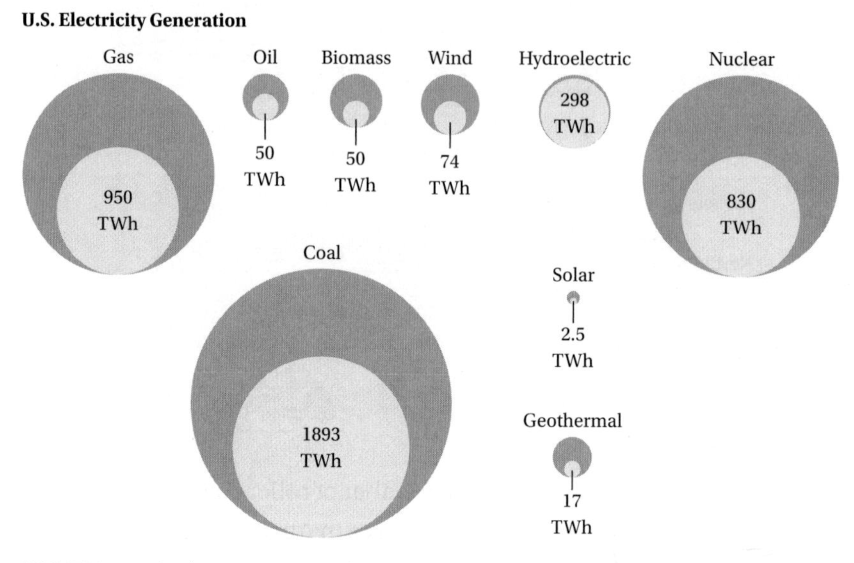

a) How much electricity is generated annually with wind?

b) Which method generates the least electricity?

c) Which method of electricity generation is the most efficient?

d) Is generation of electricity from oil or from biomass more efficient?

We use the information in the graph to answer the questions. Note that the area of the larger circle that is not covered by the yellow circle represents the amount of energy lost or wasted during the generation process.

a) From the yellow circle labeled "Wind," we see that 74 TWh of electricity is generated annually with wind.

b) The smallest yellow circle is labeled "Solar," so the least electricity is generated by solar methods.

c) The yellow circle that most nearly fills its outer circle is labeled "Hydroelectric," so hydroelectric generation is the most efficient.

d) The same amount of electricity, 50 TWh, is generated using oil and biomass. Since the outer circle corresponding to oil is smaller than that corresponding to biomass, generation of electricity from oil is more efficient.

Do Exercises 9–12.

Use the graph in Example 3 to answer Margin Exercises 9–12.

9. How much electricity is generated annually from gas?

10. Which method generates the most electricity?

11. Solar, geothermal, hydroelectric, and wind are all considered renewable energy sources. How much electricity is generated annually from renewable sources?

12. Is nuclear generation of electricity more or less efficient than generation of electricity from gas?

Answers

9. 950 TWh **10.** Coal **11.** 391.5 TWh
12. Nuclear generation of electricity is less efficient than generation from gas.

C HISTOGRAMS

A **histogram** is a special kind of graph that shows how often certain numbers appear in a set of data. Such *frequencies* are usually considered in terms of ranges of values.

EXAMPLE 4 *Fuel Economy.* Listed below are the fuel economy ratings, in miles per gallon, for combined city and highway driving for all 2013 midsize car models sold in the United States.

23, 20, 21, 21, 28, 24, 21, 22, 20, 20, 21, 19, 28, 26, 24, 23, 24, 20, 17, 26,
19, 17, 16, 14, 13, 29, 29, 21, 21, 22, 24, 22, 22, 20, 23, 23, 22, 16, 14, 21,
21, 19, 22, 21, 29, 31, 30, 33, 30, 29, 27, 33, 31, 30, 28, 26, 24, 30, 28, 22,
23, 24, 32, 29, 27, 22, 17, 19, 21, 18, 17, 23, 24, 31, 32, 27, 13, 13, 47, 29,
28, 26, 26, 25, 28, 43, 43, 29, 28, 25, 25, 22, 30, 32, 32, 32, 31, 32, 30, 30,
20, 19, 18, 29, 21, 22, 20, 20, 21, 19, 18, 23, 26, 28, 28, 29, 29, 30, 26, 26,
23, 21, 31, 24, 40, 19, 18, 19, 18, 20, 26, 25, 21, 22, 45, 21, 25, 24, 31, 32,
21, 22, 23, 20, 19, 23, 22, 23, 22, 26, 25, 31, 25, 22, 30, 34, 33, 34, 24, 27,
20, 50, 50, 24, 28, 40, 41, 40, 25, 25, 34, 23, 35, 26, 25, 21, 23

It is difficult to make sense of the 177 numbers in this data set, so the data are displayed below in a histogram.

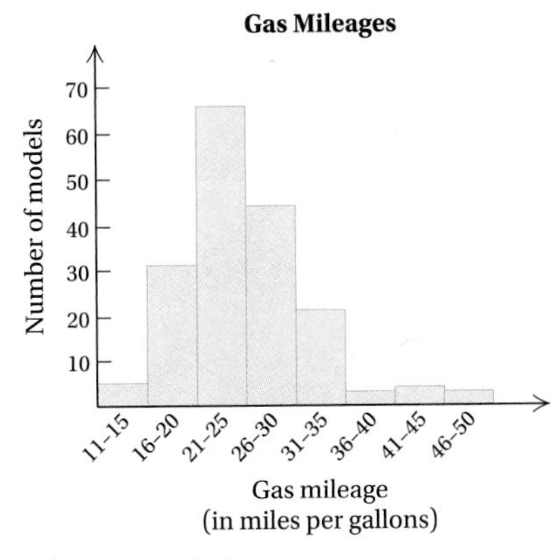

SOURCE: www.fueleconomy.gov

a) In which range of gas mileages did the greatest number of midsize models fall?

b) About how many midsize models had gas mileages that were less than 16 mpg?

c) About how many more midsize models had gas mileages between 26 mpg and 30 mpg than between 31 mpg and 35 mpg?

We use the histogram to answer the questions.

a) The tallest rectangle in the histogram is above the range 21–25, so the range 21 mpg to 25 mpg included the greatest number of midsize models.

b) The rectangle corresponding to 11–15 is 5 units high, so 5 midsize models had gas mileages that were less than 16 mpg.

c) From the histogram, we estimate that about 44 midsize models had gas mileages in the 26–30 range and about 21 models had gas mileages in the 31–35 range. Thus about 44 − 21, or 23, more midsize models had gas mileages between 26 mpg and 30 mpg than between 31 mpg and 35 mpg.

Do Exercises 13–15.

The following histogram illustrates test grades for a class of 100 students. Use the histogram for Margin Exercises 13–15.

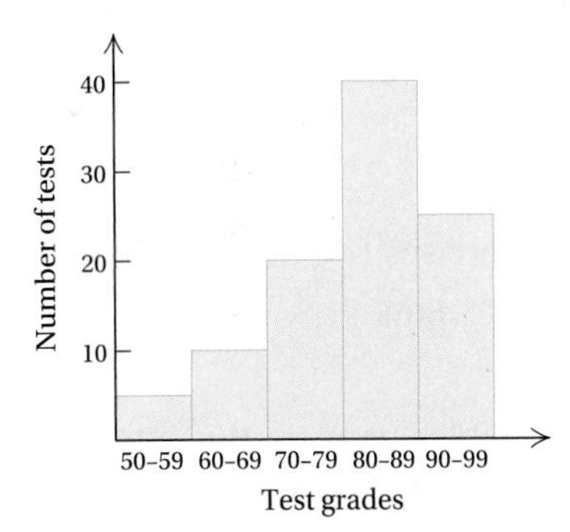

13. Which range of grades included the greatest number of students?

14. About how many students received a test grade between 90 and 99?

15. About how many more students received a grade between 90 and 99 than a grade between 50 and 59?

Answers

13. 80–89 14. 25 students 15. 20 students

8.2 Exercise Set

For Extra Help
MyMathLab® MathXL® PRACTICE WATCH READ REVIEW

Reading Check

Determine whether each statement is true or false.

RC1. ______________ There is only one correct way to represent a set of data.

RC2. ______________ It is usually easy to read exact amounts from a pictograph.

RC3. ______________ Histograms show frequencies.

RC4. ______________ If the same data were displayed in a table and in a pictograph, we would have to use the pictograph to determine a maximum or a minimum.

a

Heat Index. In warm weather, a person can feel hot because of reduced heat loss from the skin caused by higher humidity. The **temperature–humidity index**, or **apparent temperature**, is what the temperature would have to be with no humidity in order to give the same heat effect. The following table lists the apparent temperatures for various actual temperatures and relative humidities. Use this table for Exercises 1–12.

ACTUAL TEMPERATURE (°F)	RELATIVE HUMIDITY									
	10%	20%	30%	40%	50%	60%	70%	80%	90%	100%
	APPARENT TEMPERATURE (°F)									
75°	75	77	79	80	82	84	86	88	90	92
80°	80	82	85	87	90	92	94	97	99	102
85°	85	88	91	94	97	100	103	106	108	111
90°	90	93	97	100	104	107	111	114	118	121
95°	95	99	103	107	111	115	119	123	127	131
100°	100	105	109	114	118	123	127	132	137	141
105°	105	110	115	120	125	131	136	141	146	151

In Exercises 1–4, find the apparent temperature for the given actual temperature and humidity combinations.

1. 80°, 60%

2. 90°, 70%

3. 85°, 90%

4. 95°, 80%

5. Which temperature–humidity combinations give an apparent temperature of 100°?

6. Which temperature–humidity combinations give an apparent temperature of 111°?

7. At a relative humidity of 50%, what actual temperatures give an apparent temperature above 100°?

8. At a relative humidity of 90%, what actual temperatures give an apparent temperature above 100°?

9. At an actual temperature of 95°, what relative humidities give an apparent temperature above 100°?

10. At an actual temperature of 85°, what relative humidities give an apparent temperature above 100°?

11. At an actual temperature of 85°, what is the difference in humidities required to raise the apparent temperature from 94° to 108°?

12. At an actual temperature of 80°, what is the difference in humidities required to raise the apparent temperature from 87° to 102°?

Planets. Use the following table, which lists information about the planets, for Exercises 13–20.

PLANET	AVERAGE DISTANCE FROM SUN (in miles)	DIAMETER (in miles)	LENGTH OF PLANET'S DAY IN EARTH TIME (in days)	TIME OF REVOLUTION IN EARTH TIME (in years)
Mercury	35,983,000	3,031	58.82	0.24
Venus	67,237,700	7,520	224.59	0.62
Earth	92,955,900	7,926	1.00	1.00
Mars	141,634,800	4,221	1.03	1.88
Jupiter	483,612,200	88,846	0.41	11.86
Saturn	888,184,000	74,898	0.43	29.46
Uranus	1,782,000,000	31,763	0.45	84.01
Neptune	2,794,000,000	31,329	0.66	164.78

SOURCE: *The Handy Science Answer Book,* Gale Research, Inc.

13. Find the average distance from the sun to Jupiter.

14. How long is a day on Venus?

15. Which planet has a time of revolution of 164.78 years?

16. Which planet has a diameter of 4221 mi?

17. About how many Earth diameters would it take to equal one Jupiter diameter?

18. How much longer is the longest time of revolution than the shortest?

19. What are the average, the median, and the mode of the diameters of the planets?

20. What are the average, the median, and the mode of the average distances from the sun of the planets?

Nutrition Facts. Most foods are required by law to provide factual information regarding nutrition, like that in the following table of nutrition facts from a box of Frosted Flakes cereal. Use the nutrition data for Exercises 21–26 on the next page.

Nutrition Facts

Serving Size 3/4 Cup (30g/1.1oz)
Servings Per Container About 16

Amount Per Serving	Cereal	Cereal with 1/2 Cup Vitamins A&D Fat Free Milk
Calories	110	150
Calories from Fat	0	0
	% Daily Value**	
Total Fat 0g*	0%	0%
Saturated Fat 0g	0%	0%
Trans Fat 0g		
Cholesterol 0mg	0%	0%
Sodium 140mg	6%	9%
Potassium 20mg	1%	6%
Total Carbohydrate 27g	9%	11%
Dietary Fiber 1g	3%	3%
Sugars 11g		
Other Carbohydrate 15g		
Protein 1g		

Vitamin A	10%	15%
Vitamin C	10%	10%
Calcium	0%	15%
Iron	25%	25%
Vitamin D	10%	25%
Thiamin	25%	30%
Riboflavin	25%	35%
Niacin	25%	25%
Vitamin B_6	25%	25%
Folic Acid	25%	25%
Vitamin B_{12}	25%	35%

*Amount in cereal. One half cup of fat free milk contributes an additional 40 calories, 65mg sodium, 6g total carbohydrates (6g sugars), and 4g protein.

**Percent Daily Values are based on a 2,000 calorie diet. Your daily values may be higher or lower depending on your calorie needs.

SOURCE: © 2013 Kellogg North America Company

21. Suppose your morning bowl of cereal consists of $1\frac{1}{2}$ cups of Frosted Flakes with 1 cup of fat-free milk. How many calories do you consume?

22. Suppose your morning bowl of cereal consists of $1\frac{1}{2}$ cups of Frosted Flakes with 1 cup of fat-free milk. What percent of the daily value of dietary fiber do you consume?

23. A nutritionist recommends that you look for foods that provide 10% or more of the daily value of vitamin C. Do you get that with 1 serving of Frosted Flakes and $\frac{1}{2}$ cup of fat-free milk?

24. Suppose you are trying to limit your daily caloric intake to 2000 calories. How many servings of cereal alone would it take to exceed 2000 calories?

25. Suppose your morning bowl of cereal consists of $1\frac{1}{2}$ cups of Frosted Flakes with 1 cup of fat-free milk. How much sodium do you consume? (*Hint*: Use the data listed in the first footnote below the table of nutrition facts.)

26. Suppose your morning bowl of cereal consists of $1\frac{1}{2}$ cups of Frosted Flakes with 1 cup of fat-free milk. How much protein do you consume? (*Hint*: Use the data listed in the first footnote below the table of nutrition facts.)

b

Rhino Population. The rhinoceros is considered one of the world's most endangered animals. The worldwide total number of rhinoceroses is approximately 20,700. The following pictograph shows the populations of the five remaining rhino species. Located in the graph is a key that tells you that each symbol represents 300 rhinos. Use the pictograph for Exercises 27–32.

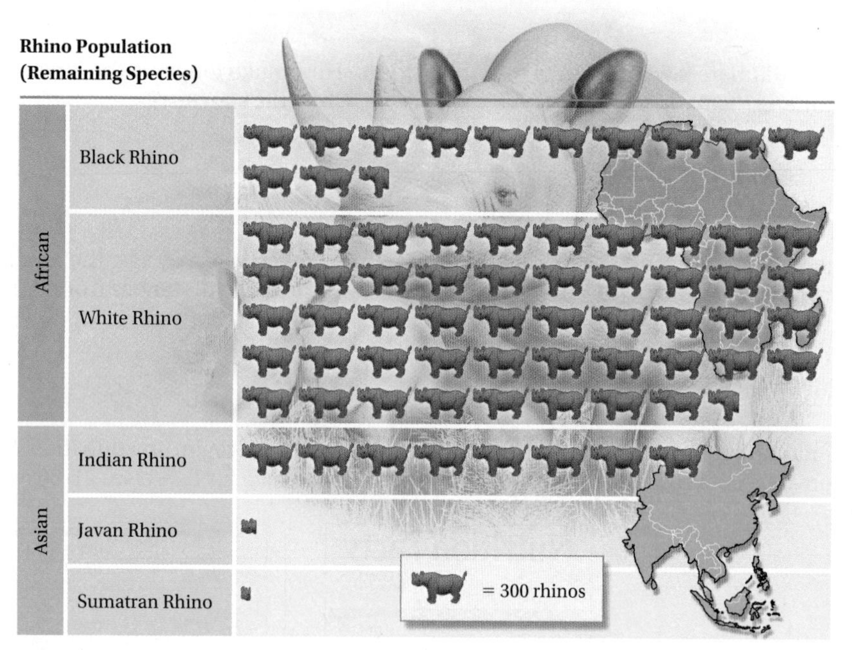

SOURCE: World Wildlife Fund, 2008

27. Which species has the greatest number of rhinos?

28. Which species has the least number of rhinos?

29. How many more black rhinos are there than Indian rhinos?

30. How many more white rhinos are there than black rhinos?

31. What is the average number of rhinos for the five species?

32. How does the white rhino population compare with the Indian rhino population?

Personal Consumption Expenditures. The following graph shows the amounts of personal consumption expenditures, in dollars per person per year, in the United States, for four years. The graph also shows the amounts spent on food and on financial services and insurance for those years, labeled as percents of the personal consumption expenditures. Use the graph for Exercises 33–40.

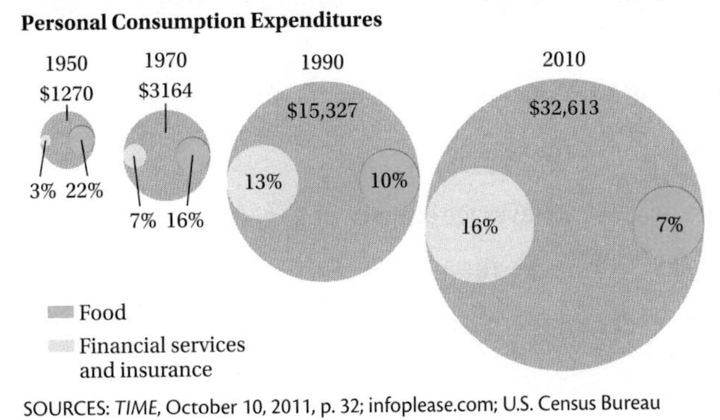

SOURCES: *TIME*, October 10, 2011, p. 32; infoplease.com; U.S. Census Bureau

33. How much were personal consumption expenditures per person in 1950?

34. How much were personal consumption expenditures per person in 2010?

35. For which of the years shown was more spent on food than on financial services and insurance?

36. For which of the years shown was more spent on financial services and insurance than on food?

37. How much per person was spent on food in 1990?

38. How much per person was spent on financial services and insurance in 1970?

39. a) How much less, as a percent of personal consumption expenditures, was spent on food in 2010 than in 1950?

b) How much more, in dollars, was spent on food in 2010 than in 1950?

40. a) How much more, as a percent of personal consumption expenditures, was spent on financial services and insurance in 2010 than in 1950?

b) How much more, in dollars, was spent on financial services and insurance in 2010 than in 1950?

C

Basketball. The following histogram illustrates the number of points scored per game by the Los Angeles Lakers during the 2012–2013 regular basketball season. Use the graph for Exercises 41–44.

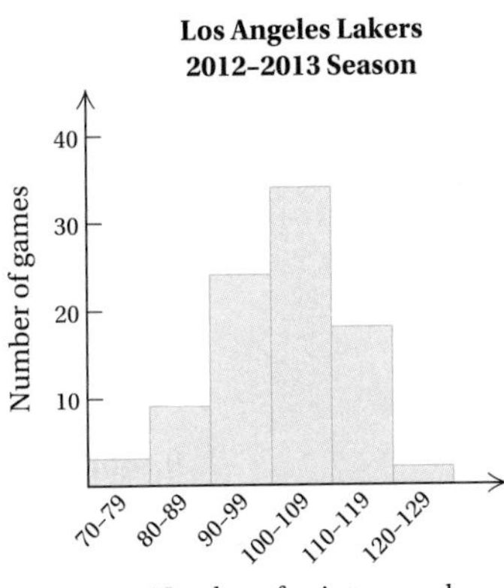

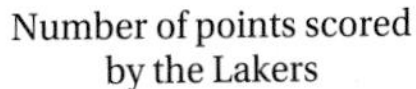

41. In how many games did the Lakers score 90–99 points?

42. In what point range did the highest number of Laker scores lie?

43. In what point range did the lowest number of Laker scores lie?

44. In how many more games did the Lakers score 100–109 points than 90–99 points?

Mid-Chapter Review

Concept Reinforcement

Determine whether each statement is true or false.

________ **1.** A set of data has just one average and just one median, but it can have more than one mode. [8.1a, b, c]

________ **2.** It is possible for the average, the median, and the mode of a set of data to be the same number. [8.1a, b, c]

________ **3.** If there is an even number of items in a set of data, the middle number is the median. [8.1b]

Guided Solutions

GS Fill in each blank with the number that creates a correct solution.

4. The average of 60, 45, 115, 15, and 35 is

$$\frac{60 + 45 + \square + 15 + 35}{\square} = \frac{\square}{5} = \square. \quad [8.1a]$$

5. Find the median of this set of numbers:

2.1, 11.3, 8.7, 6.3, 14.5, 4.8. [8.1b]

We first arrange the numbers from smallest to largest:

$\square$, $\square$, 6.3, $\square$, 11.3, $\square$.

There is an even number of items. The median is the average of

$\square$ and $\square$.

We find that average:

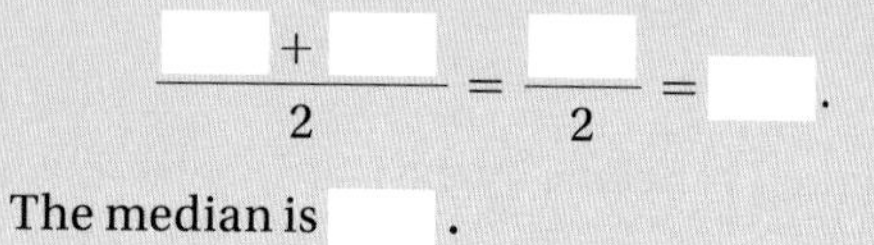

$$\frac{\square + \square}{2} = \frac{\square}{2} = \square.$$

The median is $\square$.

Mixed Review

For each set of numbers, find the average, the median, and any modes that exist. [8.1a, b, c]

6. 56, 29, 45, 240, 175, 7, 29

7. 2.12, 18.42, 9.37, 43.89

8. $\frac{5}{9}, \frac{1}{9}, \frac{8}{9}, \frac{2}{9}, \frac{4}{9}$

9. 160, 102, 102, 116, 160, 116

10. \$4.96, \$5.24, \$4.96, \$10.05, \$5.24

11. $\frac{1}{2}, \frac{3}{4}, \frac{7}{8}, \frac{5}{4}$

12. 2, 5, 7, 7, 8, 5, 5, 7, 8

13. 38.2, 38.2, 38.2, 38.2

Downsizing. Companies sometimes downsize their products. That is, they charge the same price for a package that contains less product. The following table lists products that have been downsized. Use this table for Exercises 14–18. [8.2a]

PRODUCT	SIZE (in ounces)		PERCENT SMALLER
	OLD	NEW	
Breyer's ice cream	56	48	14%
Hellmann's mayonnaise	32	30	6
Hershey's Special Dark chocolate bar	8	6.8	15
Iams cat food	6	5.5	8
Nabisco Chips Ahoy cookies	16	15.25	5
Skippy creamy peanut butter	18	16.3	9
Tropicana orange juice	96	89	7

SOURCE: *Consumer Reports*

14. How much less ice cream is in the new Breyer's ice cream package than in the old package?

15. By what percent has the size of a jar of Hellmann's mayonnaise changed in the downsizing process?

16. Which product in the table showed the greatest percent decrease?

17. How much less orange juice is in the new Tropicana orange juice package than in the old package?

18. Which product in the table showed the smallest percent decrease?

Touchdown Passes. The following pictograph shows the career-high number of touchdown passes in one season for seven quarterbacks in the National Football League. Use the pictograph for Exercises 19–22. [8.2b]

Touchdown Passes (Career high for quarterback)

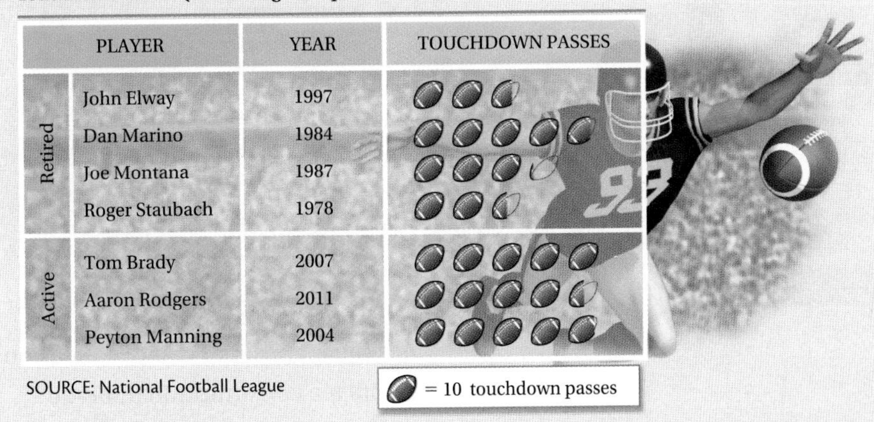

19. Which quarterback threw the greatest number of touchdown passes in one season?

20. About how many touchdown passes did Aaron Rodgers throw in 2011?

21. How many more touchdown passes does Peyton Manning have as his career high than John Elway?

22. What is the average career-high number of touchdown passes in one season for the seven quarterbacks?

Understanding Through Discussion and Writing

23. Is it possible for a driver to average 20 mph on a 30-mi trip and still receive a ticket for driving 75 mph? Why or why not? [8.1a]

24. You are applying for an entry-level job at a large firm. You can be informed of the mean, median, or mode salary. Which of the three figures would you request? Why? [8.1a, b, c]

STUDYING FOR SUCCESS *Making Positive Choices*

- ☐ Choose to improve your attitude and raise your goals.
- ☐ Choose to make a strong commitment to learning.
- ☐ Choose to take the primary responsibility for learning.
- ☐ Choose to allocate the proper amount of time to learn.

8.3 Interpreting and Drawing Bar Graphs and Line Graphs

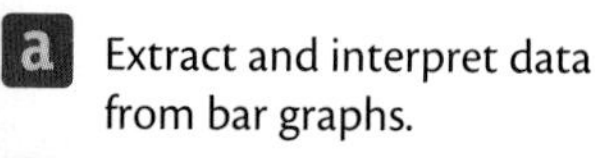

OBJECTIVES

- **a** Extract and interpret data from bar graphs.
- **b** Draw bar graphs.
- **c** Extract and interpret data from line graphs.
- **d** Draw line graphs.

SKILL TO REVIEW

Objective 5.1c: Given a pair of numbers in decimal notation, tell which is larger.

Which number is larger?

1. 0.078, 0.1

2. 36.4, 9.875

a READING AND INTERPRETING BAR GRAPHS

A **bar graph** is convenient for showing comparisons because you can tell at a glance which quantity is the largest or smallest. A *scale* is usually included with a bar graph so that estimates of values can be made with some accuracy. Bar graphs may be drawn horizontally or vertically, and sometimes a double bar graph is used to make comparisons.

EXAMPLE 1 *Coffee and Tea Consumption.* The following horizontal bar graph is a double bar graph, showing per capita consumption, in pounds per person per year, of both coffee and tea for several countries.

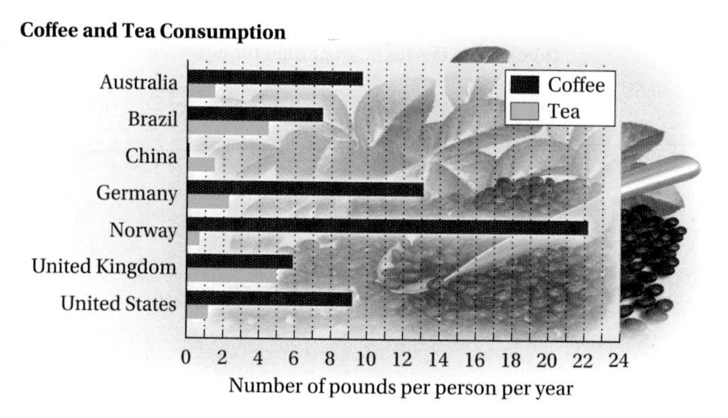

a) Which country has the highest per capita coffee consumption?

b) What is the per capita tea consumption in Brazil?

c) In which country do people consume about the same number of pounds of coffee and pounds of tea per year?

d) In which two countries do people consume about the same amount of tea?

e) In which countries is per capita coffee consumption greater than 10 pounds per year?

We use the graph to answer the questions.

a) The longest brown bar is for Norway. Thus Norway has the highest coffee consumption per capita.

b) We look to the right along the green bar associated with Brazil. Since it ends halfway between 4 and 5, we estimate Brazil's per capita tea consumption to be 4.5 pounds per year.

c) The brown and green bars representing data for the United Kingdom are about the same length, so people in the United Kingdom consume about the same number of pounds of coffee and pounds of tea per year.

Answers

Skill to Review:
1. 0.1 **2.** 36.4

d) The green bars are the same length for Australia and China; both countries have a per capita consumption of 1.5 pounds of tea. Thus people in Australia and China, on average, consume about the same amount of tea per year.

e) We move across the horizontal scale to 10. From there we move up, noting any brown bars that are longer than 10 units. We see that per capita coffee consumption is greater than 10 pounds per year in Germany and Norway.

Do Exercises 1–3. ▶

Use the bar graph in Example 1 to answer Margin Exercises 1–3.

1. What is the per capita coffee consumption in the United Kingdom?

2. In which countries is per capita tea consumption less than 2 pounds per year?

3. How many more pounds of coffee are consumed per person in the United States than pounds of tea?

b DRAWING BAR GRAPHS

EXAMPLE 2 *Population by Age.* Listed below are U.S. population data for selected age groups. Make a vertical bar graph of the data.

AGE GROUP	PERCENT OF POPULATION
5–17 years	17%
18 years and older	76
10–49 years	54
16–64 years	66
55 years and older	25
65 years and older	13
85 years and older	2

SOURCE: U.S. Census Bureau

First, we indicate the age groups in seven equally spaced intervals on the horizontal scale and give the horizontal scale the title "Age category." (See Figure 1 below.)

Next, we scale the vertical axis. To do so, we look over the data and note that it ranges from 2% to 76%. We start the vertical scaling at 0, labeling the marks by 10's from 0 to 80. We give the vertical scale the title "Percent of population" and the graph the overall title "U.S. Population by Age."

Finally, we draw vertical bars to show the various percents, as shown in Figure 2.

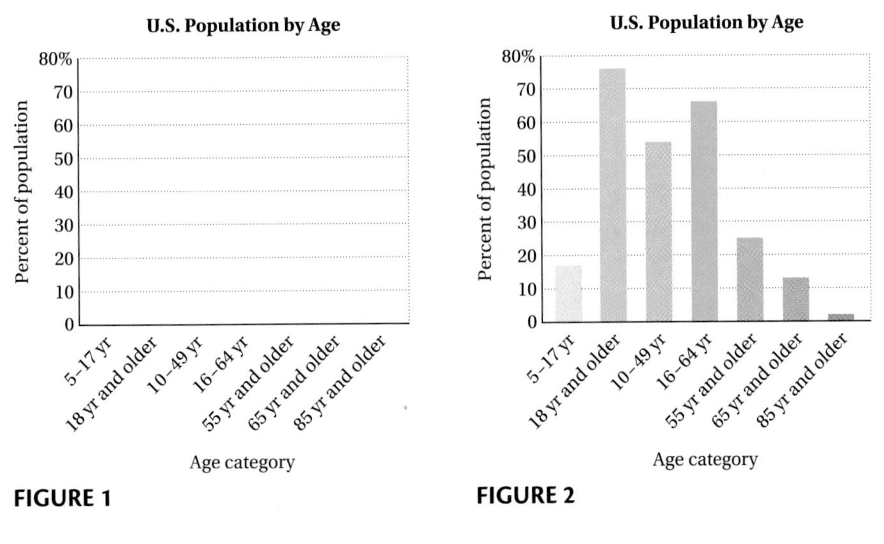

FIGURE 1

FIGURE 2

Do Exercise 4. ▶

4. *Planetary Moons.* Make a horizontal bar graph to show the numbers of moons orbiting the various planets.

PLANET	MOONS
Earth	1
Mars	2
Jupiter	63
Saturn	60
Uranus	27
Neptune	13

SOURCE: National Aeronautics and Space Administration

Answers

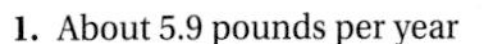

1. About 5.9 pounds per year
2. Australia, China, Norway, and the United States
3. About 8 pounds per person
4.

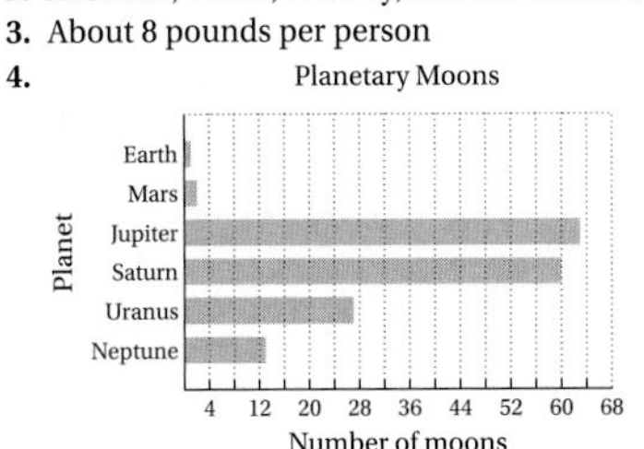

READING AND INTERPRETING LINE GRAPHS

Line graphs are often used to show a change over time as well as to indicate patterns or trends.

EXAMPLE 3 *Gold.* The following line graph shows the average price of gold, in dollars per ounce, for various years from 1970 to 2012.

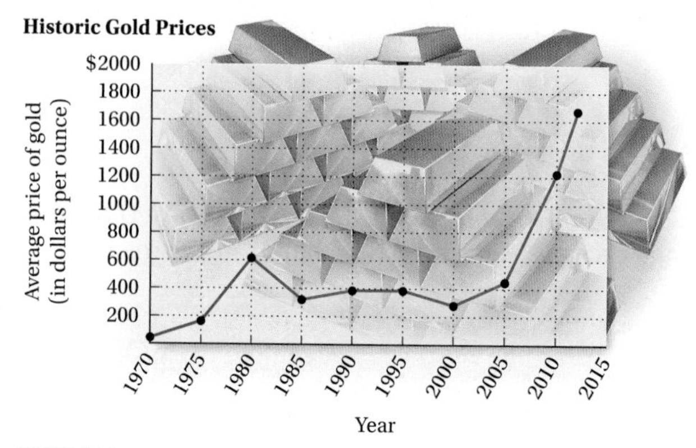

SOURCE: kitco.com

a) For which year before 2000 was the average price of gold the highest?

b) Between which years did the average price of gold decrease?

c) For which year was the average price of gold about $450 per ounce?

d) By how much did the average price of gold increase from 2010 to 2012?

We look at the graph to answer the questions.

a) Before 2000, the highest point on the graph corresponds to 1980. The highest average price of gold was about $610 per ounce in 1980.

b) Reading the graph from left to right, we see that the average price of gold decreased from 1980 to 1985 and from 1995 to 2000.

c) We look from left to right along a line at $450 per ounce. We see that the average price of gold was about $450 per ounce in 2005.

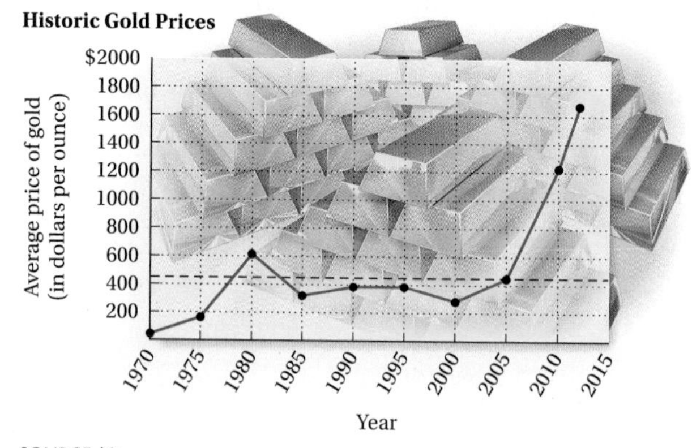

SOURCE: kitco.com

d) The graph shows that the average price of gold was about $1225 per ounce in 2010 and about $1675 per ounce in 2012. Thus the average price of gold increased by $\$1675 - \$1225 = \$450$ per ounce.

Do Exercises 5–7.

Use the line graph in Example 3 to answer Margin Exercises 5–7.

5. For which year after 1980 was the average price of gold the lowest?

6. Between which years did the average price of gold increase by about $800 per ounce?

GS

7. For which years was the average price of gold less than $400 per ounce?

We look from left to right along a line at $ ______ per ounce. The points on the graph that are below this line correspond to the years 1970, 1975, 1985, 1990, 1995, and ______.

Answers

5. 2000 6. Between 2005 and 2010
7. 1970, 1975, 1985, 1990, 1995, 2000

Guided Solution:
7. 400; 2000

d DRAWING LINE GRAPHS

EXAMPLE 4 *Temperature in an Enclosed Vehicle.* The temperature inside an enclosed vehicle increases rapidly with time. Listed in the table below are the inside temperatures of an enclosed vehicle for specified elapsed times when the outside temperature is 80°F. Make a line graph of the data.

ELAPSED TIME	TEMPERATURE IN ENCLOSED VEHICLE WITH OUTSIDE TEMPERATURE 80°F
10 min	99°
20 min	109°
30 min	114°
40 min	118°
50 min	120°
60 min	123°

SOURCES: General Motors; Jan Null, Golden Gate Weather Services

First, we indicate the 10-min elapsed time intervals on the horizontal scale and give the horizontal scale the title "Elapsed time (in minutes)." See the figure on the left below. Next, we scale the vertical axis by 10's beginning with 80 to show the number of degrees and give the vertical scale the title "Temperature (in degrees)." The jagged line at the base of the vertical scale indicates that an unused portion of the scale has been omitted. We also give the graph the overall title "Temperature in Enclosed Vehicle with Outside Temperature 80°F."

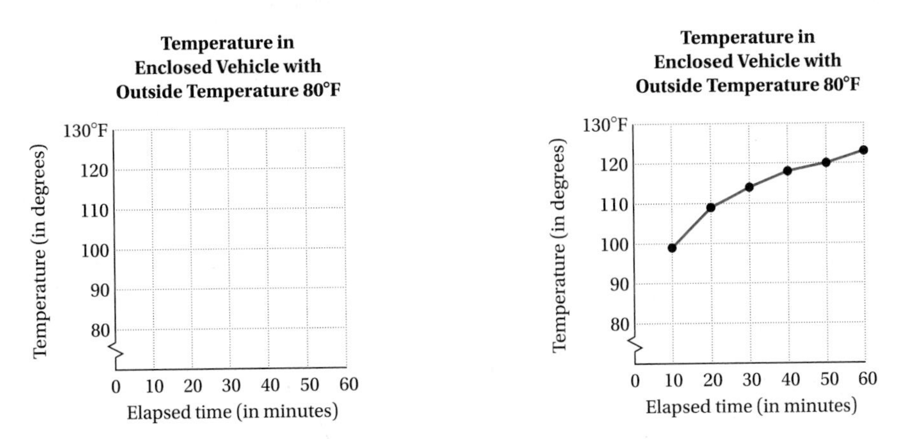

Next, we mark the temperature at the appropriate level above each elapsed time. (See the figure on the right above.) Then we draw line segments connecting the points. The rapid change in temperature can be observed easily from the graph.

Do Exercise 8. ▶

8. *Military Technologies.* Listed below are the numbers of bachelor's degrees in military technologies earned in the United States for the years 2003–2009. Make a line graph of the data.

YEAR	NUMBER OF DEGREES EARNED
2003	6
2004	10
2005	40
2006	33
2007	168
2008	39
2009	55

SOURCE: U.S. Census Bureau

Answer

8. Degrees in Military Technologies

Number of bachelor's degrees earned: 50, 100, 150, 200

Year: 2003 2004 2005 2006 2007 2008 2009

8.3 Exercise Set

For Extra Help MyMathLab® MathXL® PRACTICE WATCH READ REVIEW

Reading Check

Determine whether each statement is true or false.

RC1. ________________ Bar graphs may be drawn horizontally or vertically.

RC2. ________________ A double bar graph indicates two amounts for each category.

RC3. ________________ A line graph is always used to show trends over time.

RC4. ________________ Some data could be illustrated using either a line graph or a bar graph.

a

Bearded Irises. A gardener planted six varieties of bearded iris in a new garden on campus. Students from the horticulture department were assigned to record data on the range of heights for each variety. The vertical bar graph below shows their results. The length of the light green shaded portion of each bar and the blossom illustrates the range of heights for a variety. For example, the range of heights for the miniature dwarf bearded iris is 2 in. to 9 in.

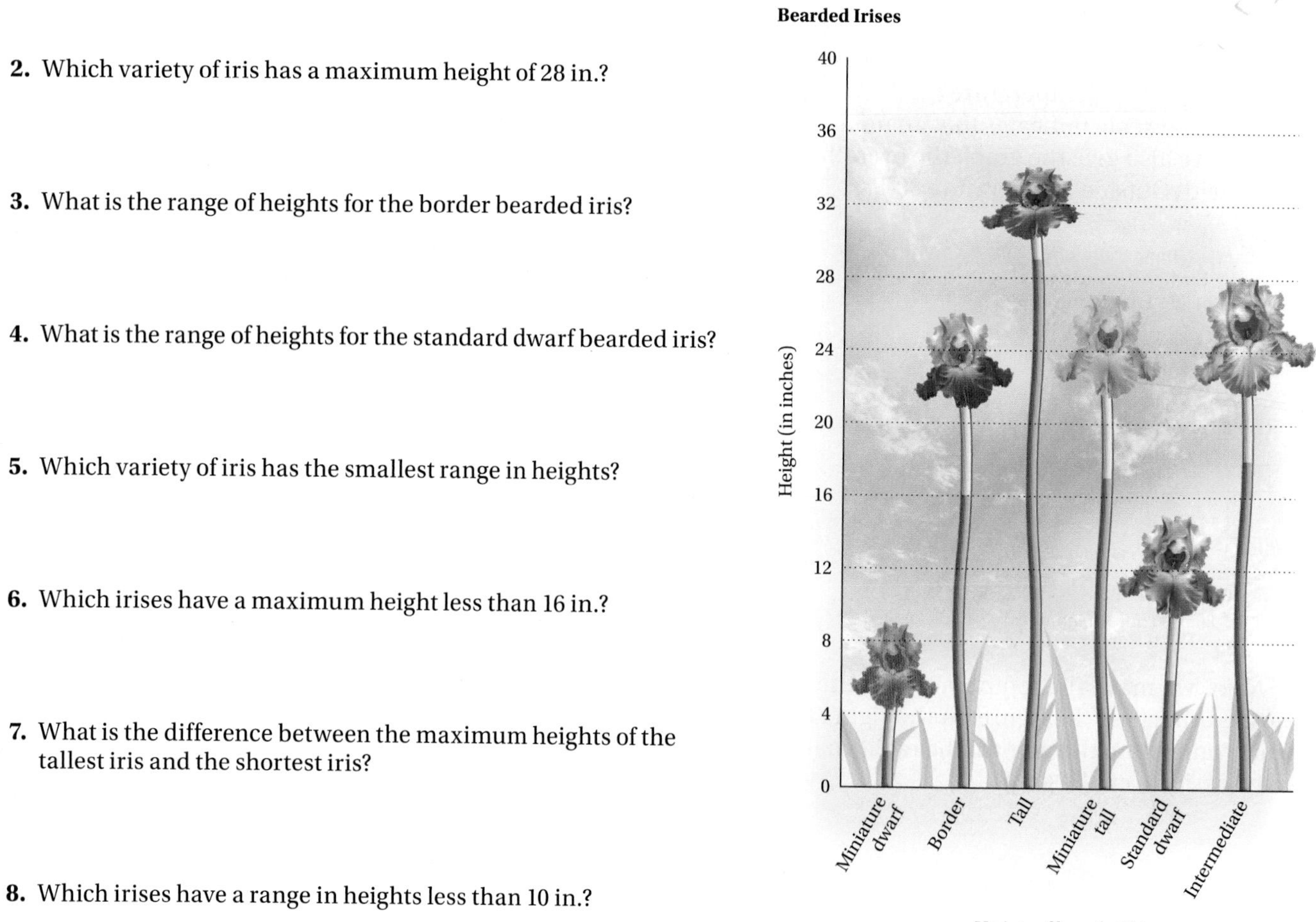

1. Which variety of iris has a minimum height of 17 in.?

2. Which variety of iris has a maximum height of 28 in.?

3. What is the range of heights for the border bearded iris?

4. What is the range of heights for the standard dwarf bearded iris?

5. Which variety of iris has the smallest range in heights?

6. Which irises have a maximum height less than 16 in.?

7. What is the difference between the maximum heights of the tallest iris and the shortest iris?

8. Which irises have a range in heights less than 10 in.?

Chocolate Desserts. The following horizontal bar graph shows the average caloric content of various kinds of chocolate desserts. Use the bar graph for Exercises 9–16.

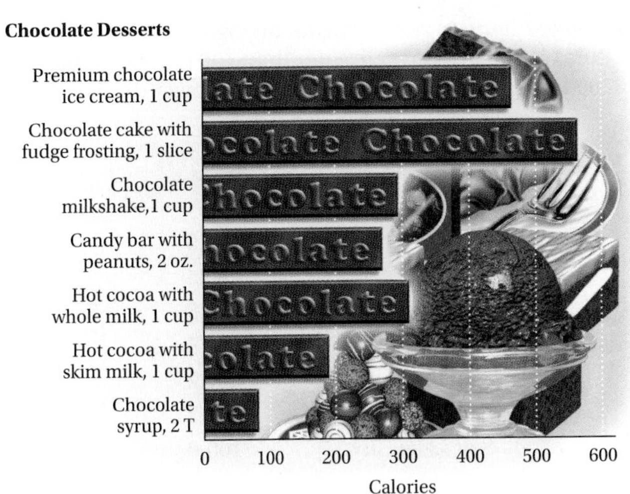

9. Estimate how many calories there are in 1 cup of hot cocoa with skim milk.

10. Estimate how many calories there are in a 2-oz candy bar with peanuts.

11. Which dessert has the highest caloric content?

12. Which dessert has the lowest caloric content?

13. Which dessert contains about 460 calories?

14. Which desserts contain about 300 calories?

15. How many more calories are there in 1 cup of hot cocoa made with whole milk than in 1 cup of hot cocoa made with skim milk?

16. If Emily drinks a 4-cup chocolate milkshake, how many calories does she consume?

Bachelor's Degrees. The graph at right provides data on the numbers of bachelor's degrees conferred on men and on women in selected years. Use the bar graph for Exercises 17–20.

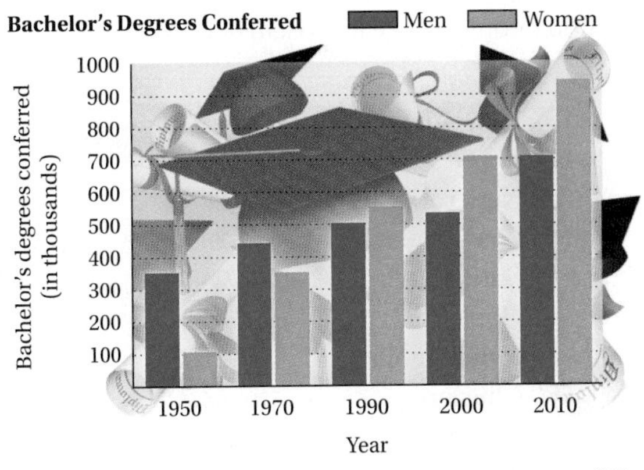

SOURCE: National Center for Education Statistics, U.S. Department of Education

17. In which years were more bachelor's degrees conferred on men than on women?

18. How many more bachelor's degrees were conferred on women in 2010 than in 1970?

19. How many more bachelor's degrees were conferred on women than on men in 2000?

20. In which years were the numbers of bachelor's degrees conferred on men and on women each greater than 500,000?

b

21. *Cost of Living Index.* The following table lists the cost of living index for several cities. The national average of this index is 100. An index greater than 100 indicates that the cost of living is higher than average, and an index less than 100 indicates that the cost of living is lower than average. Make a horizontal bar graph to illustrate the data.

CITY	COST OF LIVING INDEX
Chicago	116.9
Denver	99.4
New York City	185.8
Juneau	136.5
Indianapolis	87.2
San Diego	132.3
Salt Lake City	100.6

SOURCE: U.S. Census Bureau

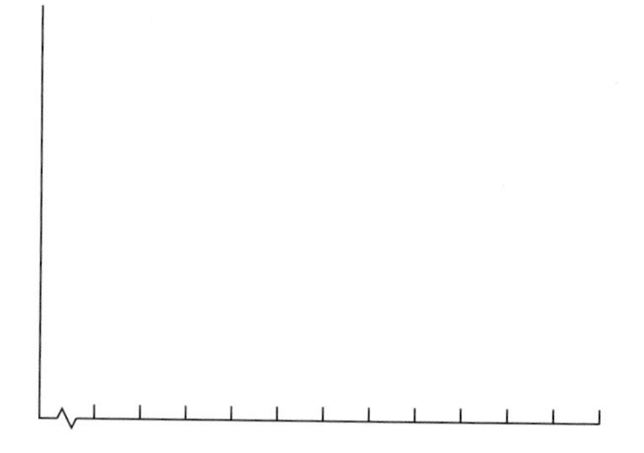

Use the data and the bar graph in Exercise 21 to do Exercises 22–25.

22. Which city has the highest cost of living index?

23. In which cities is the cost of living index less than 100?

24. In which cities is the cost of living approximately the national average?

25. How much higher is the cost of living index in New York City than in Chicago?

26. *Commuting Time.* The following table lists the average commuting time in six metropolitan areas with more than 1 million people. Make a vertical bar graph to illustrate the data.

CITY	COMMUTING TIME (in minutes)
New York City	38.3
Los Angeles	29.0
Phoenix	24.5
Houston	25.8
Indianapolis	21.7
Chicago	33.2

SOURCE: U.S. Census Bureau

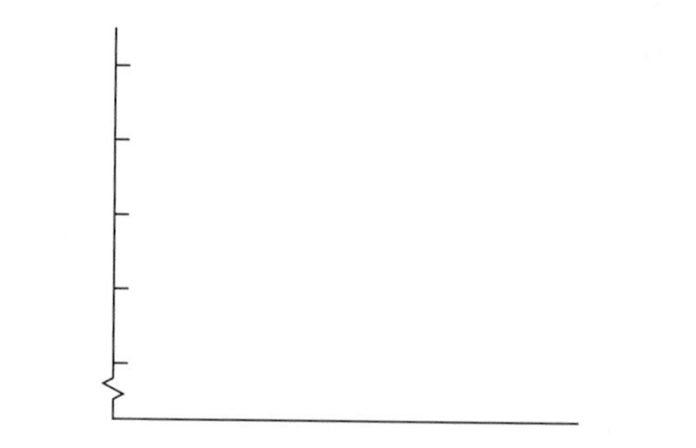

Use the data and the bar graph in Exercise 26 to do Exercises 27–30.

27. Which city has the longest commuting time?

28. Which city has the shortest commuting time?

29. What was the median commuting time for all six cities?

30. What was the average commuting time for the six cities?

C

Facebook Stock. The line graph below shows the price per share of Facebook stock when it was first offered in May 2012 and at the beginning of each month for the remainder of that year. Use the graph for Exercises 31–34.

31. Estimate the opening price per share of Facebook stock in May 2012.

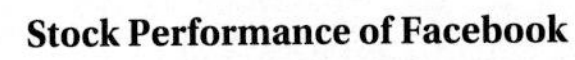
Stock Performance of Facebook

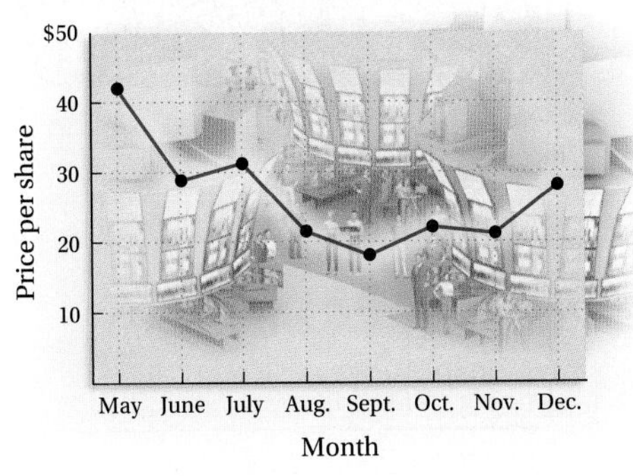

SOURCE: finance.yahoo.com

32. How much higher was the opening price of Facebook stock than its price at the beginning of September?

33. Between which months did the price of Facebook stock increase?

34. Between which months was the decrease in the price of Facebook stock the greatest?

Monthly Loan Payment. Suppose you borrow $110,000 at an interest rate of $5\frac{1}{2}\%$ to buy a condominium. The following graph shows the monthly payment required to pay off the loan, depending on the length of the loan. Use the graph for Exercises 35–42.

35. Estimate the monthly payment for a loan of 15 years.

$110,000 Loan Repayment

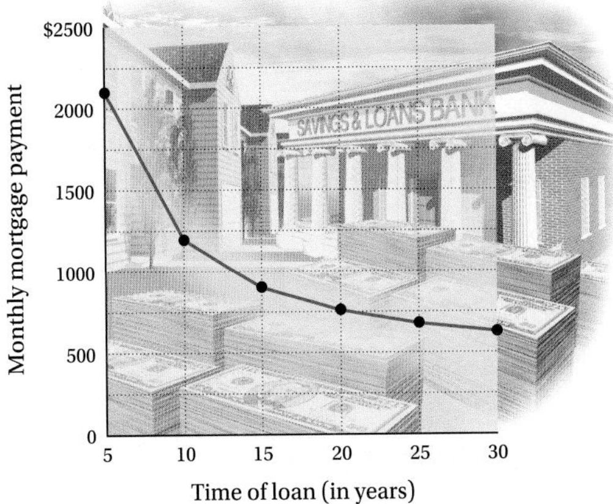

36. Estimate the monthly payment for a loan of 25 years.

37. What time period corresponds to a monthly payment of about $760?

38. What time period corresponds to a monthly payment of about $625?

39. By how much does the monthly payment decrease when the loan period is increased from 10 years to 20 years?

40. By how much does the monthly payment decrease when the loan period is increased from 5 years to 20 years?

41. For a 10-year loan, there are 120 monthly payments. In all, how much will you pay back for a 10-year loan?

42. For a 20-year loan, there are 240 monthly payments. In all, how much will you pay back for a 20-year loan?

d

43. *Longevity Beyond Age 65.* The data in the table below indicate how many years beyond age 65 a male who is 65 in the given year could expect to live. Draw a line graph using the horizontal axis to scale "Year."

YEAR	AVERAGE NUMBER OF YEARS MEN ARE ESTIMATED TO LIVE BEYOND AGE 65
1980	14
1990	15
2000	15.9
2010	16.4
2020	16.9
2030	17.5

SOURCE: 2000 Social Security Report

Use the data and line graph from Exercise 43 to do Exercises 44–47.

44. What was the percent increase in longevity (years beyond 65) between 1980 and 2000?

45. What is the expected percent increase in longevity between 1980 and 2030?

46. What is the expected percent increase in longevity between 2020 and 2030?

47. What is the expected percent increase in longevity between 2000 and 2030?

Skill Maintenance

Solve.

48. $32 + n = 115$ [1.7b]

49. $x \cdot \frac{2}{3} = \frac{8}{9}$ [3.7c]

50. $y + \frac{5}{8} = -\frac{11}{12}$ [4.3c]

51. $-5 \cdot x = 11.3$ [5.4b]

52. $t + 4.752 = 11.1$ [5.2d]

53. $\frac{9}{10} = \frac{x}{8}$ [6.3b]

54. 51.2 is 64% of what? [7.3a, b], [7.4a, b]

55. What is $4\frac{1}{2}$% of 20? [7.3a, b], [7.4a, b]

56. 120 is what percent of 80? [7.3a, b], [7.4a, b]

Calculate.

57. $3 \times [11 + (18 - 10) \div 2^3 - 5]$ [1.9d]

58. $2.56 \div (4 - 3.84) + 6.3 \times 0.2$ [5.4c]

59. $\frac{9}{10} \div \frac{1}{2} \cdot \frac{1}{3} - \left(\frac{1}{4} - \frac{1}{6}\right)$ [4.7a]

60. $-6.25 \times 7\frac{1}{5}$ [5.5c]

Interpreting and Drawing Circle Graphs

8.4

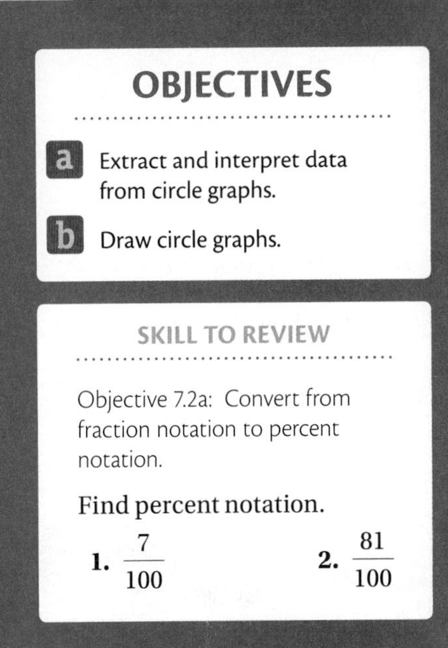

OBJECTIVES

a. Extract and interpret data from circle graphs.

b. Draw circle graphs.

SKILL TO REVIEW

Objective 7.2a: Convert from fraction notation to percent notation.

Find percent notation.

1. $\frac{7}{100}$ 2. $\frac{81}{100}$

We often use **circle graphs**, also called **pie charts**, to show the percent of a quantity in each of several categories. Circle graphs can also be used very effectively to show visually the *ratio* of one category to another. In either case, it is quite often necessary to use mathematics to find the actual amounts represented for each specific category.

a READING AND INTERPRETING CIRCLE GRAPHS

EXAMPLE 1 *Endangered Species.* According to the International Union for Conservation of Nature, seven species of whales are endangered or near-threatened. The following circle graph shows the approximate percentage of the entire population of endangered or near-threatened whales that each species represents.

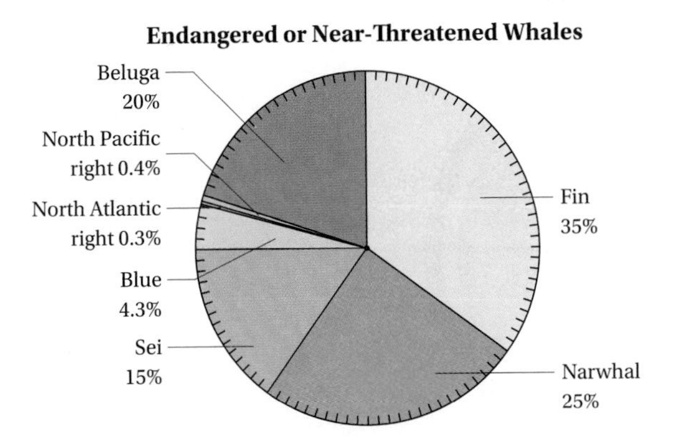

a) Which species has the greatest population?

b) Which species accounts for 25% of the entire population of endangered or near-threatened whales?

c) The total number of whales in these seven species is about 300,000. How many blue whales are there?

d) What percent of the population of endangered or near-threatened whales are right whales?

We look at the sections of the graph to find the answers.

a) The largest section (or *sector*) of the graph represents 35% of the population and corresponds to fin whales.

b) The narwhal accounts for 25% of the endangered or near-threatened whales.

c) The section representing blue whales is 4.3% of the circle. Since 4.3% of 300,000 is 12,900, there are approximately 12,900 blue whales.

d) There are two kinds of right whales represented on the graph: North Pacific right whales and North Atlantic right whales. We add the percents corresponding to these whales:

$$0.4\% + 0.3\% = 0.7\%.$$

Do Margin Exercises 1–3.

Use the circle graph in Example 1 to answer Margin Exercises 1–3.

1. Which species accounts for 20% of the entire population of endangered or near-threatened whales?

2. What percent of the population of endangered or near-threatened whales are fin whales or sei whales?

3. The total number of whales in these seven species is about 300,000. How many fin whales are there?

Answers

Skill to Review:
1. 7% 2. 81%

Margin Exercises:
1. Beluga whales 2. 50% 3. 105,000 whales

DRAWING CIRCLE GRAPHS

EXAMPLE 2 *Education.* The list below shows the percents of students in the United States enrolled in different levels and types of schools. Use this information to draw a circle graph.

Source: *The 2012 Statistical Abstract*, U.S. Census Bureau

Grades K–8, Public:	46%
Grades K–8, Private:	5%
Grades 9–12, Public:	19%
Grades 9–12, Private:	2%
College, Public:	20%
College, Private:	8%

Using a circle with 100 equally spaced tick marks, we start with the 46% of students who are in grades K–8 in public schools. We draw a line from the center of the circle to any tick mark. Then we count off 46 ticks and draw another line. We label the wedge as shown in the figure on the left below. To distinguish the sectors, we can use different colors. We choose blue for this first sector.

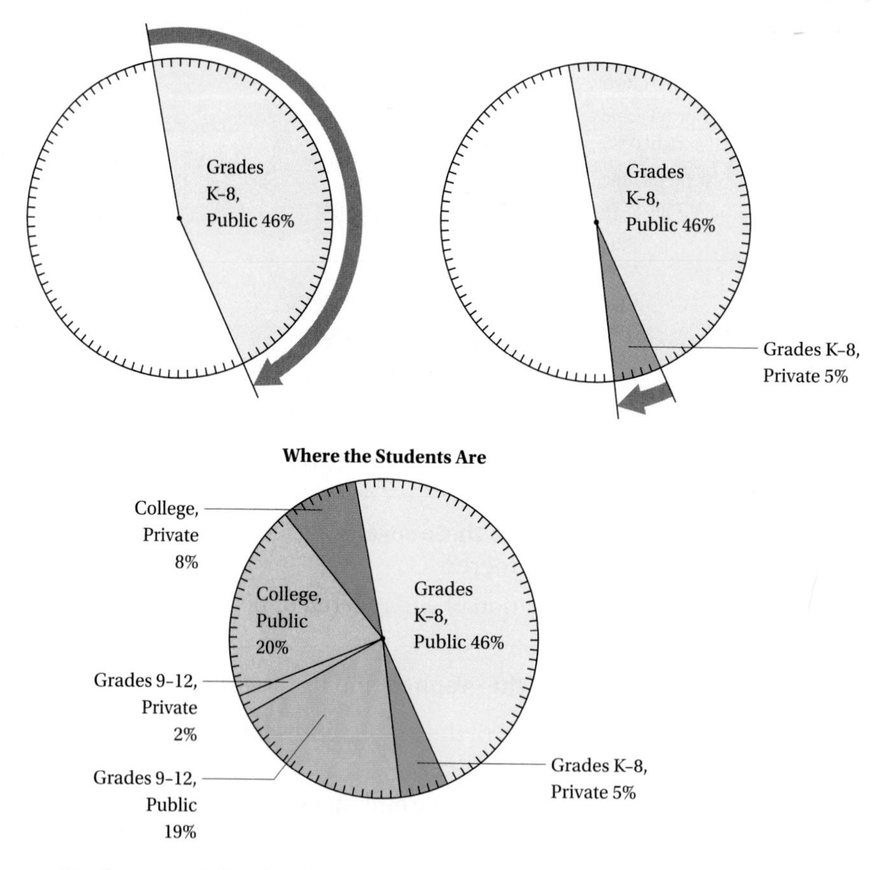

To draw a wedge for the 5% of students who are in grades K–8 in private schools, we start at one side of the wedge for 46%, count off 5 ticks, and draw another line. We label this second wedge as shown in the figure on the right above and shade it using a different color. Continuing in this manner, we obtain the final graph, at the lower middle above, to which we give the overall title "Where the Students Are."

◀ **Do Exercise 4.**

4. *Lengths of Engagement of Married Couples.* The data below show the percents of married couples who were engaged for certain periods of time before marriage. Use this information to draw a circle graph.

ENGAGEMENT PERIOD	PERCENT
Less than 1 year	24%
1–2 years	21
More than 2 years	35
Never engaged	20

SOURCE: Bruskin Goldring Research

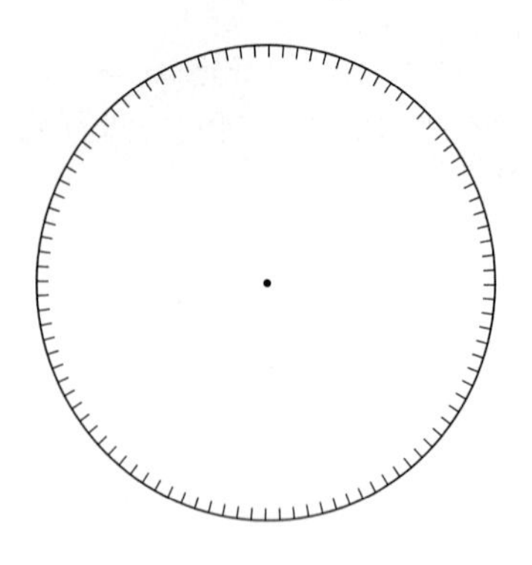

Answer

4. Less than 1 year 24%; More than 2 years 35%; 1–2 years 21%; Never engaged 20%

Translating for Success

1. *Vacation Miles.* The Saenz family drove their new van 13,640.8 mi in the first year. Of this total, 2018.2 mi were driven while on vacation. How many nonvacation miles did they drive?

2. *Rail Miles.* Of the recent $15\frac{1}{2}$ million passenger miles on a rail passenger line, 80% were transportation-to-and-from-work miles. How many rail miles, in millions, were to and from work?

3. *Sales Tax Rate.* The sales tax on the purchase of 10 bath towels that cost $129.50 is $8.42. What is the sales tax rate?

4. *Water Level.* During heavy rains in early spring, the water level in a pond rose 0.5 in. every 35 min. How much did the water rise in 90 min?

5. *Marathon Training.* At one point in his daily training routine for a marathon, Rocco had run $15\frac{1}{2}$ mi. This was 80% of the distance he intended to run that day. How far did Rocco plan to run?

The goal of these matching questions is to practice step (2), Translate, of the five-step problem-solving process. Translate each word problem to an equation and select a correct translation from equations A–O.

A. $8.42 \cdot x = 129.50$

B. $x = 80\% \cdot 15\frac{1}{2}$

C. $x = \dfrac{84 - 68}{84}$

D. $2018.2 + x = 13{,}640.8$

E. $\dfrac{5}{100} = \dfrac{x}{3875}$

F. $2018.2 = x \cdot 13{,}640.8$

G. $4\frac{1}{6} \cdot 73 = x$

H. $\dfrac{x}{5} = \dfrac{100}{3875}$

I. $15\frac{1}{2} = 80\% \cdot x$

J. $8.42 = x \cdot 129.50$

K. $\dfrac{0.5}{35} = \dfrac{x}{90}$

L. $x \cdot 4\frac{1}{6} = 73$

M. $x = \dfrac{84 - 68}{68}$

N. $x = 8.42\% \cdot 129.50$

O. $0.5 \times 35 = 90 \cdot x$

Answers on page A-13

6. *Vacation Miles.* The Ning family drove 2018.2 mi on their summer vacation. If the family put a total of 13,640.8 mi on their new van during that year, what percent were vacation miles?

7. *Sales Tax.* The sales tax rate is 8.42%. Salena purchased 10 pillows at $12.95 each. How much tax was charged on this purchase?

8. *Charity Donations.* Rachel donated $5 to her favorite charity for each $100 she earned. One month, she earned $3875. How much did she donate that month?

9. *Tuxedos.* Emil Tailoring Company purchased 73 yd of fabric for a new line of tuxedos. How many tuxedos can be produced if it takes $4\frac{1}{6}$ yd of fabric for each tuxedo?

10. *Percent Increase.* In a calculus-based physics course, Mime got 68% on the first exam and 84% on the second. What was the percent increase in her score?

Reading Check

The following statements refer to the graph at the right. Determine whether each statement is true or false.

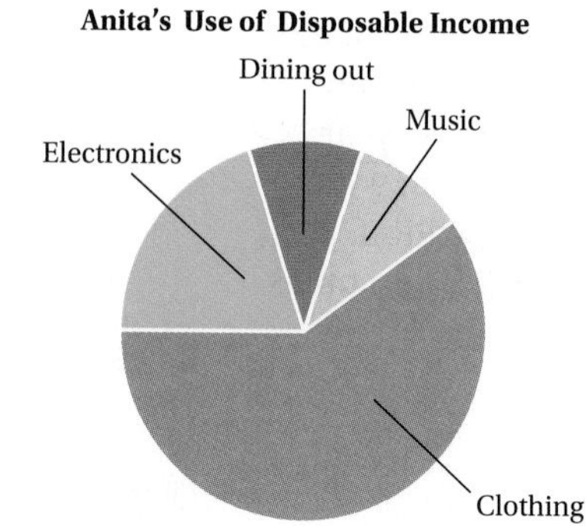

RC1. ________________ The graph is an example of a circle graph, or pie chart.

RC2. ________________ Anita spent 100% of her disposable income on music, clothing, electronics, and dining out.

RC3. ________________ Anita spent more than half of her disposable income on clothing.

RC4. ________________ Anita spent about $\frac{1}{4}$ of her disposable income on music.

RC5. ________________ Anita spent about the same amount on electronics as she spent on dining out and music combined.

RC6. ________________ If Anita has $100 in disposable income, she spends about $50 on electronics.

a *Foreign Students.* The circle graph below shows the foreign countries sending the most students to the United States to attend colleges and universities. Use this graph for Exercises 1–6.

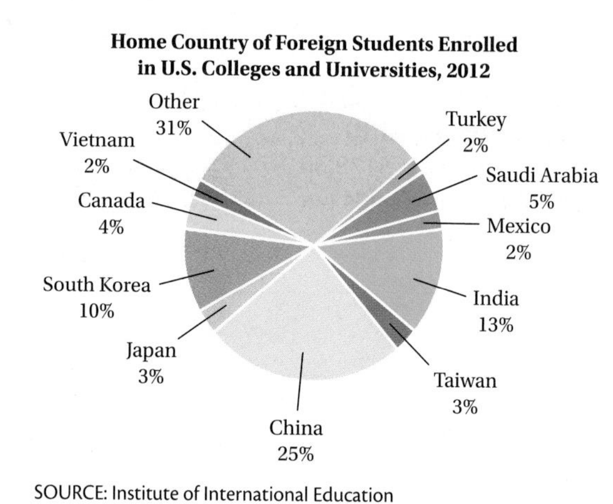

SOURCE: Institute of International Education

1. What percent of foreign students are from South Korea?

2. Together, what percent of foreign students are from China and Taiwan?

3. In 2012, there were approximately 760,000 foreign students studying at colleges and universities in the United States. According to the data in the graph, how many were from India?

4. In 2012, there were approximately 760,000 foreign students studying in the United States. How many were from Saudi Arabia?

5. Which country accounted for 4% of the foreign students?

6. Which country accounted for 13% of the foreign students?

b In Exercises 7–10, use the given information to complete a circle graph. Note that each circle is divided into 100 sections.

7. ***Fruit Juice Sales.*** The table below lists the percentages of various kinds of fruit juice sold.

FRUIT JUICE	PERCENT
Apple	14%
Orange	56
Blends	6
Grape	5
Grapefruit	4
Prune	1
Other	14

SOURCE: Beverage Marketing Corporation

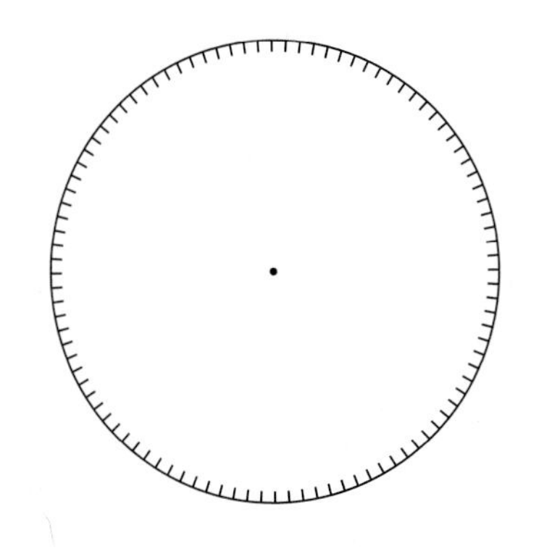

8. ***Population of Continents.*** The table below lists the percentage of the world population on each continent.

CONTINENT	PERCENT
Africa	15%
Asia	60
Europe	10
Oceania, includes Australia	1
North America	9
South America	5

SOURCE: Population Division/International Programs Center, U.S. Census Bureau

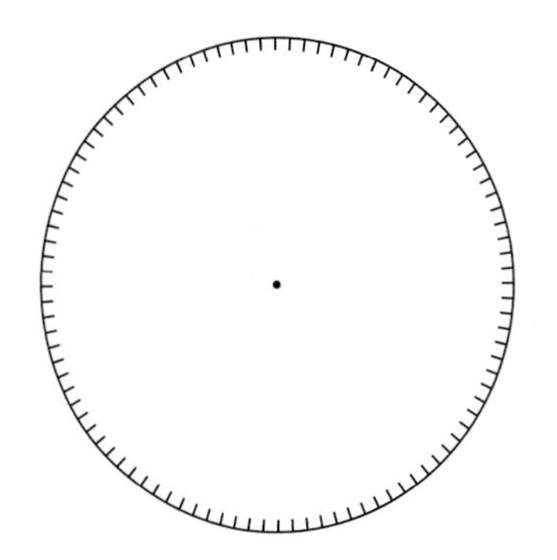

9. ***Substance Abuse.*** The table below lists the types of substances abused by those aged 12 and older who were admitted to substance abuse programs in the United States.

PRIMARY SUBSTANCE(S)	PERCENT
Drugs only	38%
Alcohol only	24
Alcohol with one drug	23
Alcohol with two drugs	14
No primary substance	1

SOURCE: Substance Abuse and Mental Health Services Administration

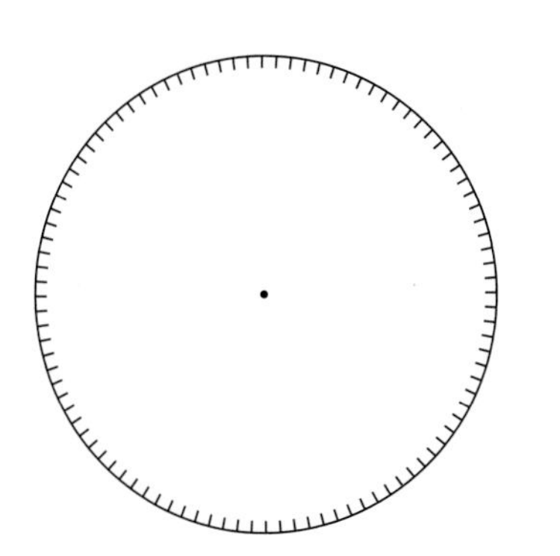

10. ***Causes of Spinal Cord Injuries.*** The table below lists the causes of spinal cord injury.

CAUSES	PERCENT
Motor vehicle accidents	44%
Acts of violence	24
Falls	22
Sports	8
Other	2

SOURCE: National Spinal Cord Injury Association

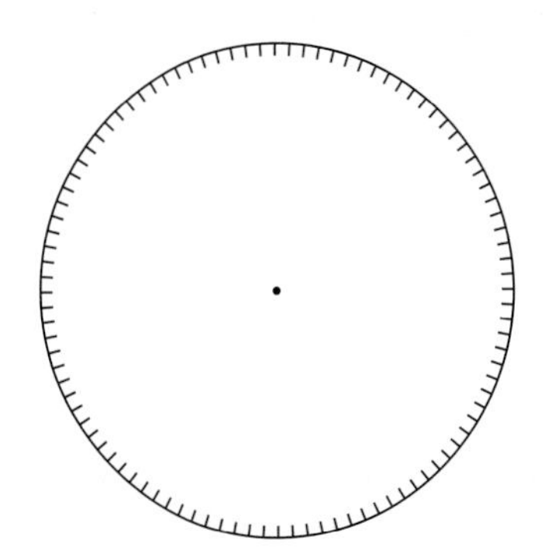

CHAPTER

8 Summary and Review

Vocabulary Reinforcement

Choose the term from the list on the right that best completes each sentence. Not every term will be used.

1. A(n) _______________ presents data in rows and columns. [8.2a]
2. A(n) _______________ illustrates category percentages using different sized sectors or wedges. [8.4a]
3. A(n) _______________ uses symbols to represent amounts. [8.2b]
4. The _______________ of a set of data is the number or numbers that occur most often. [8.1c]
5. The _______________ of a set of data is the sum of the numbers in the set divided by the number of items of data. [8.1a]
6. The _______________ of an ordered set of data is the middle number, or the average of the middle numbers if there is an even number of items of data. [8.1b]

statistic
average
median
mode
table
pictograph
histogram
bar graph
circle graph
line graph

Concept Reinforcement

Determine whether each statement is true or false.

_______________ 1. To find the average of a set of numbers, add the numbers and then multiply by the number of items of data. [8.1a]

_______________ 2. If each number in a set of data occurs the same number of times, there is no mode. [8.1c]

_______________ 3. If there is an odd number of items in a set of data, the middle number is the median. [8.1b]

Study Guide

Objectives 8.1a, b, c Find the average, the median, and the mode of a set of numbers.

Example Find the average, the median, and the mode of this set of numbers:

2.6, 3.5, 61.8, 10.4, 3.5, 21.6, 10.4, 3.5.

Average: We add the numbers and divide by the number of data items:

$$\frac{2.6 + 3.5 + 61.8 + 10.4 + 3.5 + 21.6 + 10.4 + 3.5}{8}$$

$= 14.6625.$

Median: We first rearrange the numbers from smallest to largest:

2.6, 3.5, 3.5, 3.5, 10.4, 10.4, 21.6, 61.8.

The median is halfway between the middle two, which are 3.5 and 10.4. The average of these middle numbers is 6.95.

Mode: The number that occurs most often is 3.5, so it is the mode.

Practice Exercise

1. Find the average, the median, and the mode of this set of numbers:

   8, 13, 1, 4, 8, 7, 15.

Objective 8.2a Extract and interpret data from tables.

Example The table below lists comparative information for oatmeal sold by six companies.

PRODUCT	PER PACKET (instant) OR SERVING (longer-cooking)					
	Cost	Calories	Fat (g)	Fiber (g)	Sugars (g)	Sodium (mg)
Quaker Quick-1 Minute	0.19	150	3.0	4	1	0
Market Pantry Maple & Brown Sugar	0.17	160	2.0	3	13	240
365 Organic Maple Spice	0.42	150	1.5	3	13	200
Kashi Heart to Heart Golden Brown Maple	0.44	160	2.0	5	12	100
McCann's Irish Maple & Brown Sugar	0.45	160	2.0	3	13	240
Quaker Organic Maple & Brown Sugar	0.54	150	2.0	3	12	95
Nature's Path Organic Maple Nut	0.47	200	4.0	4	12	105

SOURCE: *Consumer Reports*, November 2008

a) Which oatmeal has the greatest number of calories?
b) How much sodium is in a serving of Market Pantry Maple & Brown Sugar?

An examination of the table will give the answers.

a) We look down the column headed "Calories" and find the largest number. That number is 200. Then we look left across that row to find the name of the oatmeal: Nature's Path Organic Maple Nut.
b) We look down the column of products and find Market Pantry. Then we move right across that row to the column headed "Sodium" and find the amount of sodium: 240 mg.

Practice Exercises

Use the table in the example shown above for Exercises 2 and 3.

2. Which oatmeal has the greatest cost per serving? What is that cost?

3. How many grams of sugars are in the Kashi oatmeal?

Objective 8.3a Extract and interpret data from bar graphs.

Example The horizontal bar graph below shows the building costs of selected stadiums. When comparing the costs, note the year in which each stadium was built.

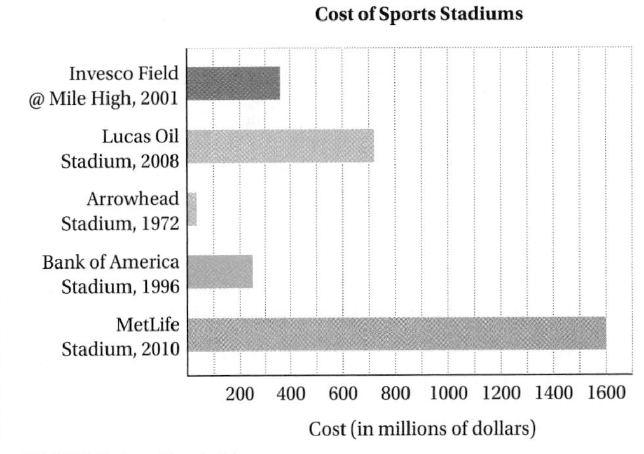

SOURCE: National Football League

a) Estimate how much more Lucas Oil Stadium cost than Invesco Field did.
b) Which stadium cost approximately $250 million to build?

We look at the graph to answer the questions.

a) We move to the right along the bars for Lucas Oil Stadium and Invesco Field and move down to the horizontal scale to estimate the costs: about $720 million for Lucas Oil Stadium and $360 million for Invesco Field. The difference in cost is about $720 million − $360 million, or $360 million. Thus Lucas Oil cost about $360 million more than Invesco Field.
b) We locate the lines representing $200 million and $300 million and go up until we reach a bar that ends close to $250 million. Then we go to the left and read the name of the stadium: Bank of America Stadium.

Practice Exercises

Use the bar graph in the example shown above for Exercises 4 and 5.

4. Which stadium cost less than $100 million?

5. Estimate how much more MetLife Stadium cost than Bank of America Stadium did.

Objective 8.4a Extract and interpret data from circle graphs.

Example The circle graph below shows the percentages of the population of the United States in various age groups.

Population of United States by Age

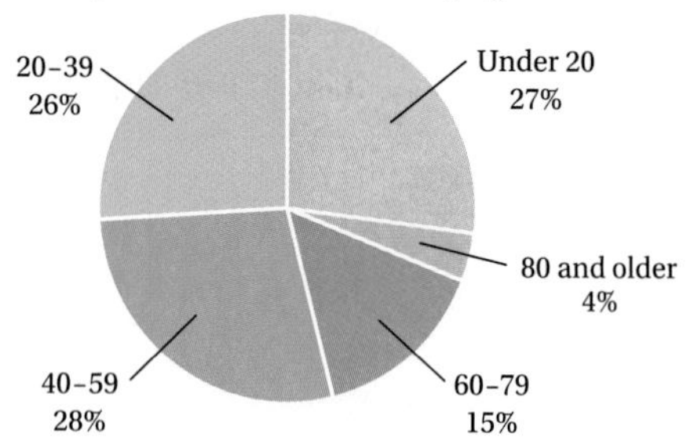

Age group	Percent
Under 20	27%
20–39	26%
40–59	28%
60–79	15%
80 and older	4%

a) What percent of the population is 40–59 years old?
b) How much more of the population is in the 20–39 age group than in the 60–79 age group?

The graph gives us the answers.

a) The graph shows us that the segment of the population that is 40–59 years old is 28% of the total population.
b) From the graph, we see that 26% of the population is 20–39 years old and only 15% is 60–79 years old. We subtract: 26% − 15% = 11%. Thus the percent of the population that is 20–39 years old is 11% greater than the percent that is 60–79 years old.

Practice Exercises

Use the circle graph at left to answer Exercises 6 and 7.

6. Which age group has the fewest people?

7. What percent of the population is under 20 years old?

Review Exercises

Find the average. [8.1a]

1. 26, 34, 43, 51

2. 11, 14, 17, 18, 7

3. 0.2, 1.7, 1.9, 2.4

4. 700, 2700, 3000, 900, 1900

5. \$2, \$14, \$17, \$17, \$21, \$29

6. 20, 190, 280, 470, 470, 500

7. To get an A in math, Naomi must score an average of 90 on four tests. Her scores on the first three tests were 94, 78, and 92. What is the lowest score she can make on the last test and still get an A? [8.1a]

8. *Gas Mileage.* A 2012 Mazda Miata does 336 mi of highway driving on 12 gal of gasoline. What is the gas mileage? [8.1a]

9. *Grade Point Average.* Find the grade point average for one semester given the following grades. Assume the grade point values are 4.0 for A, 3.0 for B, and so on. Round to the nearest tenth. [8.1a]

COURSE	GRADE	NUMBER OF CREDIT HOURS IN COURSE
Math	A	5
English	B	3
Computer Science	C	4
Spanish	B	3
College Skills	B	1

Find the median. [8.1b]

10. 26, 34, 43, 51

11. 7, 11, 14, 17, 18

12. 0.2, 1.7, 1.9, 2.4

13. 700, 900, 1900, 2700, 3000

14. $2, $17, $21, $29, $14, $17

15. 470, 20, 190, 280, 470, 500

16. One summer, a student worked part time as a veterinary assistant. She earned the following weekly amounts over a six-week period: $360, $192, $240, $216, $420, and $132. What was the average amount earned per week? the median? [8.1a, b]

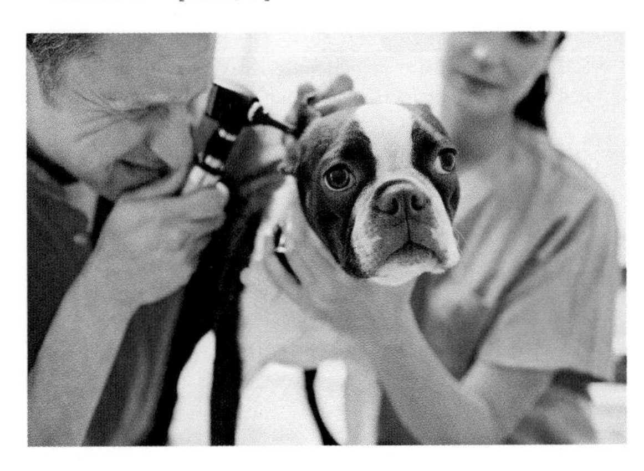

Find the mode. [8.1c]

17. 26, 34, 43, 26, 51

18. 17, 7, 11, 11, 14, 17, 18

19. 0.2, 0.2, 1.7, 1.9, 2.4, 0.2

20. 700, 700, 800, 2700, 800

21. $14, $17, $21, $29, $17, $2

22. 20, 20, 20, 20, 20, 500

Smartphone and Tablet Ownership. The table below lists the percents of the populations of several countries who own a smartphone and who own a tablet. Use this table for Exercises 23–25. [8.2a]

COUNTRY	PERCENT OWNING SMARTPHONE	PERCENT OWNING TABLET
Singapore	56%	18%
China	52	30
United States	35	17
Mexico	15	2
Germany	14	3
Philippines	8	5

SOURCE: Based on information from mastercard.com

23. What percent of China's population owns a smartphone?

24. In what countries does less than 10% of the population own a tablet?

25. In which country do approximately twice as many people own a smartphone as own a tablet?

Major League World Series. Except for four years, the World Series of Major League Baseball has been a best-of-seven series. In 1903, 1919, 1920, and 1921, the championship was a best-of-nine series. The championships have all been decided in 4, 5, 6, 7, or 8 games. The following pictograph shows the number of times the series has extended to each number of games. Use this graph for Exercises 26–28. [8.2b]

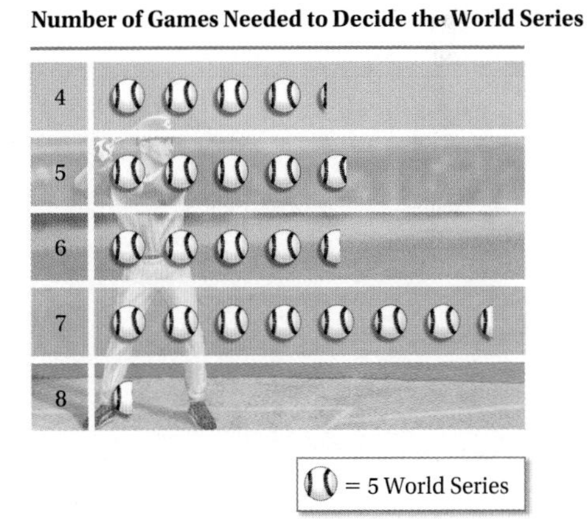

26. How many World Series were decided in 4 games?

27. In what number of games were the most World Series decided?

28. How many more World Series were decided in 7 games than were decided in 4 games?

Governors' Salaries. The histogram below shows the numbers of state governors in the United States who receive annual salaries in the given ranges. Use the graph for Exercises 29–31. [8.2c]

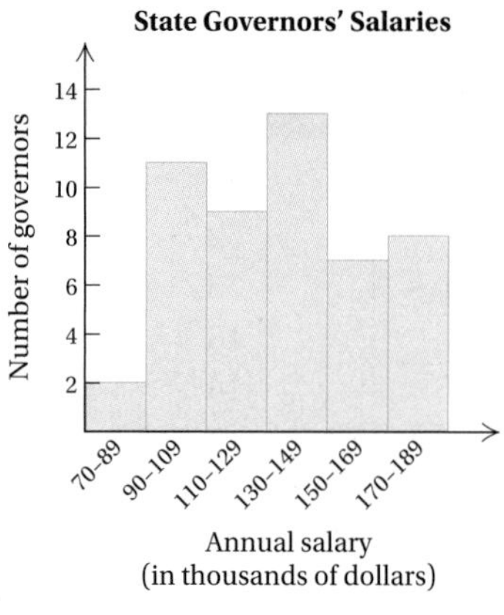

SOURCE: knowledgecenter.csg.org

29. Which salary range has the smallest number of governors?

30. How many more governors make between \$130,000 and \$149,000 than make between \$90,000 and \$109,000?

31. How many governors make less than \$130,000?

Tornadoes. The bar graph below shows the total number of tornadoes that occurred in the United States from 2010 through 2012, by month. Use the graph for Exercises 32–35. [8.3a]

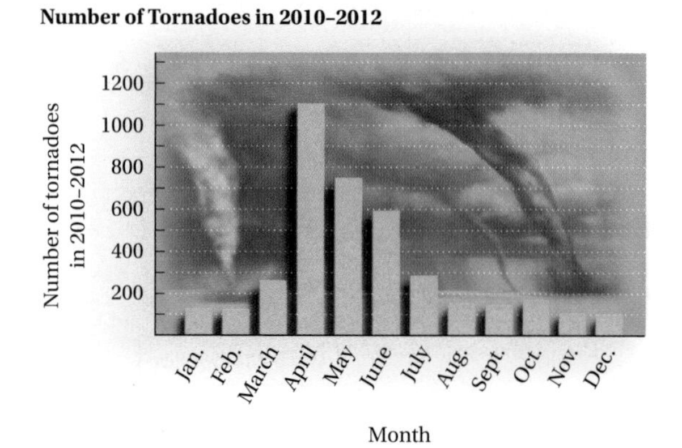

32. Which month had the greatest number of tornadoes?

33. How many tornadoes occurred in August?

34. How many more tornadoes occurred in May than in June?

35. Do more tornadoes occur in the winter or in the spring?

Homelessness. The line graph below shows the average number of homeless children in New York City's shelter system each night for various years. Use the graph for Exercises 36–40. [8.3c]

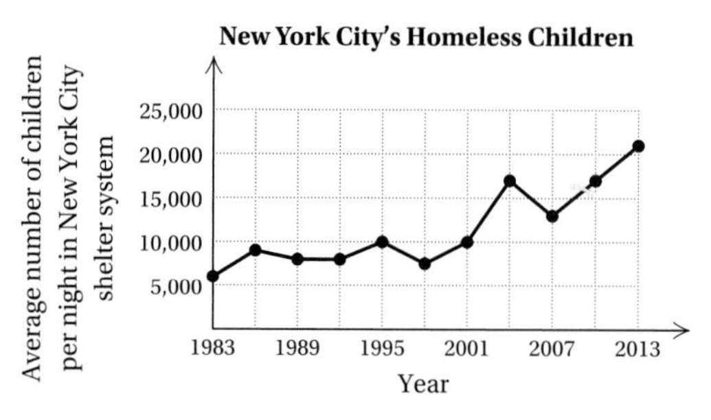

SOURCE: NYC Department of Homeless Services and Human Resources Administration and NYC Stat, shelter census reports

36. During which year after 1990 were there the fewest children in the shelter system?

37. How many children were in the shelter system each night in 2001?

38. In which years were there about 17,000 children each night in the shelter system?

39. Between which years did the number of children in the shelter system decrease?

40. By how much did the number of children in the shelter system each night increase from 2010 to 2013?

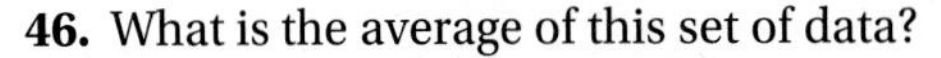

College Costs. The circle graph below shows the various cost categories for a full-time resident student at an Oklahoma regional university and the percentage of the total college cost represented by each category. Use this graph for Exercises 41–44. [8.4a]

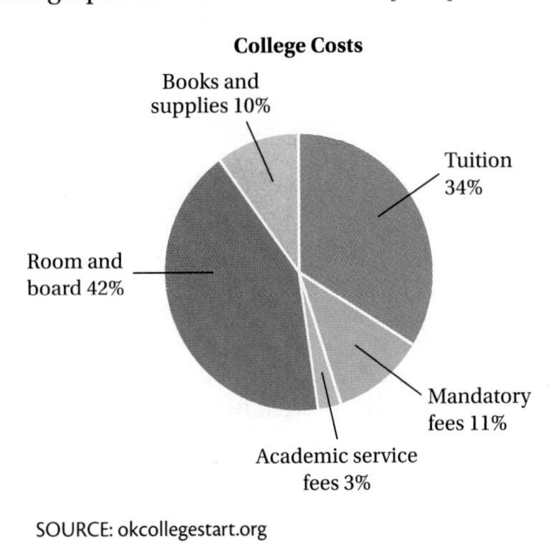

SOURCE: okcollegestart.org

41. What percent of college costs is tuition?

42. Which category accounts for the greatest part of the total college costs?

43. What percent of college costs are fees?

44. In a recent year, the total college cost for a full-time resident student at an Oklahoma regional university was \$11,500. How much did a student pay for room and board?

45. Find the mode(s) of this set of data.

6, 9, 6, 8, 8, 5, 10, 5, 9, 10 [8.1c]

A. 8 **B.** 5, 6, 8, 9, 10
C. 9 **D.** No mode exists.

46. What is the average of this set of data?

$$\frac{1}{2}, \frac{1}{3}, \frac{1}{4}, \frac{1}{5} \quad [8.1a]$$

A. $\frac{77}{240}$ **B.** $\frac{1}{3}$ **C.** $\frac{7}{24}$ **D.** $\frac{1}{4}$

First-Class Postage. The table below lists the cost of first-class postage in various years. Use the table for Exercises 47 and 48.

YEAR	FIRST-CLASS POSTAGE
2001	34¢
2002	37
2006	39
2007	41
2008	42
2009	44
2012	45
2013	46

SOURCE: U.S. Postal Service

47. Make a vertical bar graph of the data. [8.3b]

48. Make a line graph of the data. [8.3d]

49. Construct a circle graph showing the governors' salaries discussed in Exercises 29–31 as percentages of the total:

\$70,000–\$89,000: 4%; \$90,000–\$109,000: 22%;

\$110,000–\$129,000: 18%; \$130,000–\$149,000: 26%;

\$150,000–\$169,000: 14%; \$170,000–189,000: 16%. [8.4b]

Synthesis

50. The ordered set of data 298, 301, 305, a, 323, b, 390 has a median of 316 and an average of 326. Find a and b. [8.1a, b]

Understanding Through Discussion and Writing

1. Find a real-world situation that fits this equation:

$$T = \frac{20{,}500 + 22{,}800 + 23{,}400 + 26{,}000}{4}. \quad [8.1a]$$

2. Can bar graphs always, sometimes, or never be converted to line graphs? Why? [8.3b, d]

3. Discuss the advantages of being able to read a circle graph. [8.4a]

4. Compare bar graphs and line graphs. Discuss why you might use one rather than the other to graph a particular set of data. [8.3b, d]

5. Compare and contrast averages, medians, and modes. Discuss why you might use one over the others to analyze a set of data. [8.1a, b, c]

6. Compare circle graphs to bar graphs. [8.3a], [8.4a]

CHAPTER 8 Test

For Extra Help For step-by-step test solutions, access the Chapter Test Prep Videos in MyMathLab® or on YouTube (search "BittingerBasicEl" and click on "Channels").

Find the average.

1. 45, 49, 52, 52

2. 1, 1, 3, 5, 3

3. 3, 17, 17, 18, 18, 20

Find the median and the mode.

4. 45, 49, 52, 53

5. 1, 1, 3, 5, 3

6. 3, 17, 17, 18, 18, 20

7. *Grades.* To get a C in chemistry, Ted must score an average of 70 on four tests. His scores on the first three tests were 68, 71, and 65. What is the lowest score he can get on the last test and still get a C?

8. *Grade Point Average.* Find the grade point average for one semester given the following grades. Assume the grade point values are 4.0 for A, 3.0 for B, and so on. Round to the nearest tenth.

COURSE	GRADE	NUMBER OF CREDIT HOURS IN COURSE
Introductory Algebra	B	3
English	A	3
Business	C	4
Spanish	B	3
Typing	B	2

Desirable Body Weights. The following tables list the desirable body weights for men and women over age 25. Use the tables for Exercises 9–12.

DESIRABLE WEIGHT OF MEN			
HEIGHT	SMALL FRAME (in pounds)	MEDIUM FRAME (in pounds)	LARGE FRAME (in pounds)
5 ft 7 in.	138	152	166
5 ft 9 in.	146	160	174
5 ft 11 in.	154	169	184
6 ft 1 in.	163	179	194
6 ft 3 in.	172	188	204

DESIRABLE WEIGHT OF WOMEN			
HEIGHT	SMALL FRAME (in pounds)	MEDIUM FRAME (in pounds)	LARGE FRAME (in pounds)
5 ft 1 in.	105	113	122
5 ft 3 in.	111	120	130
5 ft 5 in.	118	128	139
5 ft 7 in.	126	137	147
5 ft 9 in.	134	144	155

SOURCE: U.S. Department of Agriculture

9. What is the desirable weight for a 6 ft 1 in. man with a medium frame?

10. What size woman has a desirable weight of 120 lb?

11. How much more should a 5 ft 3 in. woman with a medium frame weigh than one with a small frame?

12. How much more should a 6 ft 3 in. man with a large frame weigh than one with a small frame?

Waste Generated. The number of pounds of waste generated per person per year varies greatly among countries around the world. In the pictograph at right, each symbol represents approximately 100 lb of waste. Use the pictograph for Exercises 13–16.

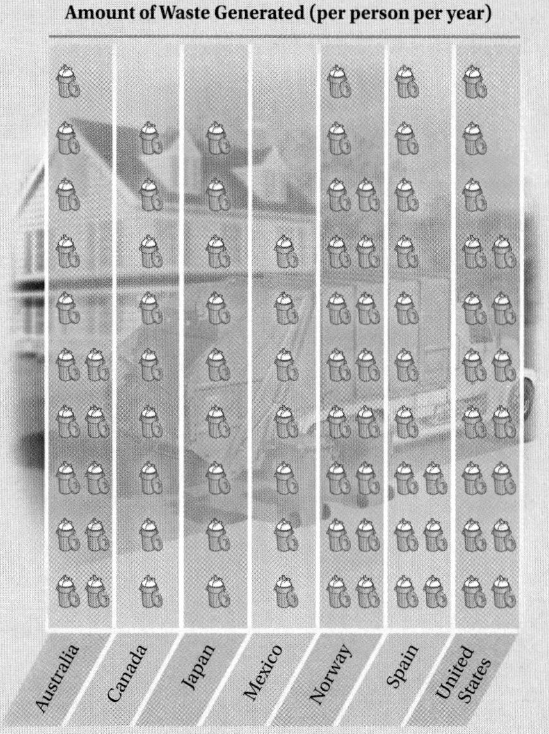

13. In which country does each person generate 1300 lb of waste per year?

14. In which countries does each person generate more than 1500 lb of waste per year?

15. How many pounds of waste per person per year are generated in Canada?

16. How many more pounds of waste per person per year are generated in the United States than in Mexico?

Hurricanes. The following line graph shows the numbers of Atlantic hurricanes for the years 2000–2012. Use the graph for Exercises 17–22.

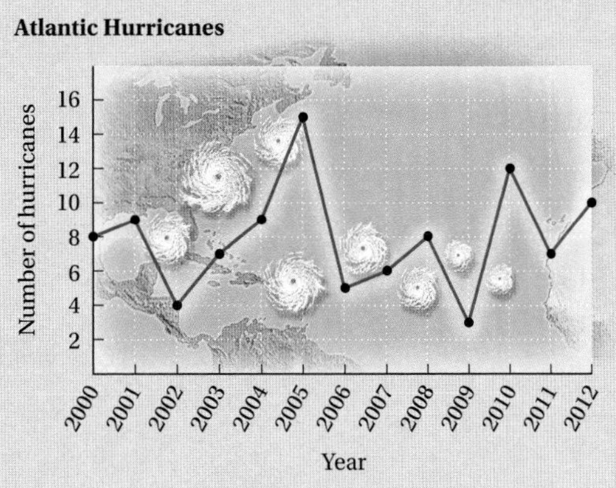

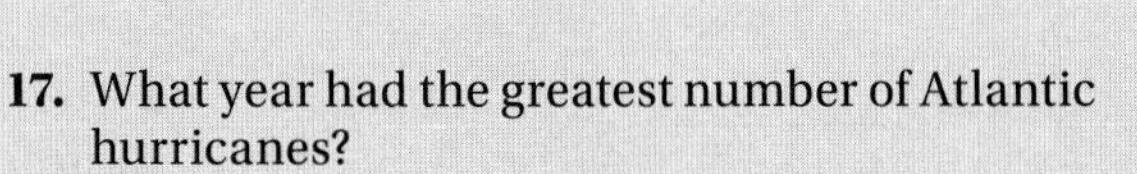

17. What year had the greatest number of Atlantic hurricanes?

18. In what year were there 3 Atlantic hurricanes?

19. How many hurricanes were there in 2012?

20. How many more hurricanes were there in 2005 than in 2006?

21. Find the average number of hurricanes per year for the years 2008–2012.

22. In what years were there 10 or more hurricanes?

Book Circulation. The table below lists the average number of books checked out per day of the week for a branch library. Use this table for Exercises 23 and 24.

DAY	NUMBER OF BOOKS CHECKED OUT
Sunday	210
Monday	160
Tuesday	240
Wednesday	270
Thursday	310
Friday	275
Saturday	420

23. Make a vertical bar graph of the data.

24. Make a line graph of the data.

25. *Food Budget.* The following table lists the percents of a family's food budget spent on selected food categories. Construct a circle graph representing these data.

FOOD CATEGORY	PERCENT OF BUDGET
Meat, poultry, fish, and eggs	23%
Fruits and vegetables	17
Cereals and bakery products	13
Dairy products	11
Other	36

SOURCE: Consumer Expenditure Survey

26. Referring to Exercise 25, consider a family that spends $664 per month on food. Using the percents from the table and the circle graph, find the amount of money spent on cereals and bakery products.

A. $8.63 **B.** $21.58 **C.** $86.32 **D.** $577.68

Synthesis

27. The ordered set of data 69, 71, 73, a, 78, 98, b has a median of 74 and a mean of 82. Find a and b.

CHAPTERS

1–8 Cumulative Review

1. *Net Worth.* In 2013, Warren Buffett of the United States was worth \$53.5 billion. Write standard notation for 53.5 billion.

2. *Gas Mileage.* A 2013 Subaru Outback does 312 mi of city driving on 13 gal of gasoline. What is the gas mileage?

3. In 402,513, what does the digit 5 mean?

4. Evaluate: $3 + 5^3$.

5. Find all the factors of 60.

6. Round 52.045 to the nearest tenth.

7. Convert to fraction notation: $3\frac{3}{10}$.

8. Convert from cents to dollars: 210¢.

9. Find percent notation for $\frac{7}{20}$.

10. Determine whether 11, 30 and 4, 12 are proportional.

Compute and simplify.

11. $2\frac{2}{5} + 4\frac{3}{10}$

12. $41.063 + (-43.5721)$

13. $-\frac{11}{15} - \frac{3}{5}$

14. $350 - 24.57$

15. $3\frac{3}{7} \cdot 4\frac{3}{8}$

16. $12{,}456 \times 220$

17. $-\frac{13}{15} \div \left(-\frac{26}{27}\right)$

18. $104{,}676 \div 24$

Solve.

19. $\frac{5}{8} = \frac{6}{x}$

20. $\frac{2}{5} \cdot y = \frac{3}{10}$

21. $21.5 \cdot y = -146.2$

22. $x = 398{,}112 \div 26$

Solve.

23. Tortilla chips cost \$2.99 for 14.5 oz. Find the unit price in cents per ounce, rounded to the nearest tenth of a cent.

24. A college has a student body of 6000 students. Of these, 55.4% own a car. How many students own a car?

25. A piece of fabric $1\frac{3}{4}$ yd long is cut into 7 equal strips. What is the length of each strip?

26. A recipe calls for $\frac{3}{4}$ cup of sugar. How much sugar should be used for $\frac{1}{2}$ of the recipe?

27. *Peanut Products.* In any given year, the average American eats 2.7 lb of peanut butter, 1.5 lb of salted peanuts, 1.2 lb of peanut candy, 0.7 lb of in-shell peanuts, and 0.1 lb of peanuts in other forms. How many pounds of peanuts and products containing peanuts does the average American eat in one year?

28. *Energy Consumption.* In a recent year, American utility companies generated 1464 billion kilowatt-hours (kWh) of electricity using coal, 455 billion using nuclear power, 273 billion using natural gas, 250 billion using hydroelectric plants, 118 billion using petroleum, and 12 billion using geothermal technology and other methods. How many kilowatt-hours of electricity were produced that year?

29. *Heart Disease.* Of the 301 million people in the United States in a recent year, about 7.5 million had coronary heart disease and about 509,000 died of heart attacks. What percent had coronary heart disease? What percent died of heart attacks? Round your answers to the nearest tenth of a percent.

Source: U.S. Centers for Disease Control

30. *Billionaires.* In 2011, the mean net worth of U.S. billionaires was \$3.72 billion. By 2012, this figure had increased to \$4.2 billion. What was the percent increase?

Source: Forbes

31. A business is owned by four people. One owns $\frac{1}{3}$, the second owns $\frac{1}{4}$, and the third owns $\frac{1}{6}$. How much does the fourth person own?

32. A factory manufacturing valves for engines was discovered to have made 4 defective valves in a lot of 18 valves. At this rate, how many defective valves can be expected in a lot of 5049 valves?

33. A landscaper bought 22 evergreen trees for \$210. What was the cost of each tree? Round to the nearest cent.

34. A salesperson earns \$182 selling \$2600 worth of electronic equipment. What is the commission rate?

FedEx. The following table lists the costs of delivering a package by FedEx Priority Overnight shipping from zip code 46143 to zip code 80403. Use the table for Exercises 35–37.

WEIGHT (in pounds)	COST
1	\$52.55
2	58.23
3	64.47
4	70.48
5	77.12
6	83.08
7	89.32
8	95.00
9	101.13
10	104.19

SOURCE: Federal Express Corporation

35. Find the average and the median of these costs.

36. Make a vertical bar graph of the data.

37. Make a line graph of the data.

Synthesis

38. A photography club meets four times a month. In September, the attendance figures were 28, 23, 26, and 23. In October, the attendance figures were 26, 20, 14, and 28. What was the percent increase or percent decrease in average attendance from September to October?

Measurement

STUDYING FOR SUCCESS *Working with Others*

- ☐ Try being a tutor for a fellow student. You may find that you understand concepts better after explaining them to someone else.
- ☐ Consider forming a study group.
- ☐ Verbalize the math. Often simply talking about a concept with a classmate can help clarify it for you.

9.1 Linear Measures: American Units

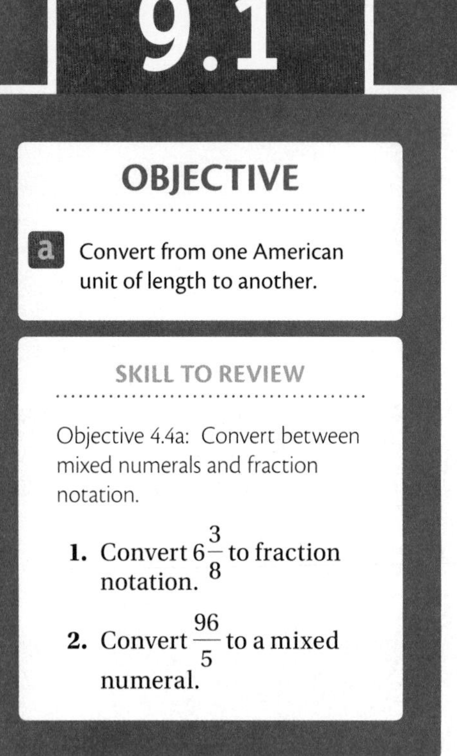

OBJECTIVE

a Convert from one American unit of length to another.

SKILL TO REVIEW

Objective 4.4a: Convert between mixed numerals and fraction notation.

1. Convert $6\frac{3}{8}$ to fraction notation.
2. Convert $\frac{96}{5}$ to a mixed numeral.

Length, or distance, is one kind of measure. To find lengths, we start with some **unit segment** and assign to it a measure of 1. Suppose $\overline{AB}$ below is a unit segment. Let's measure segment $\overline{CD}$, using $\overline{AB}$ as our unit segment.

Unit segment: A ⊢——⊣ B (1)

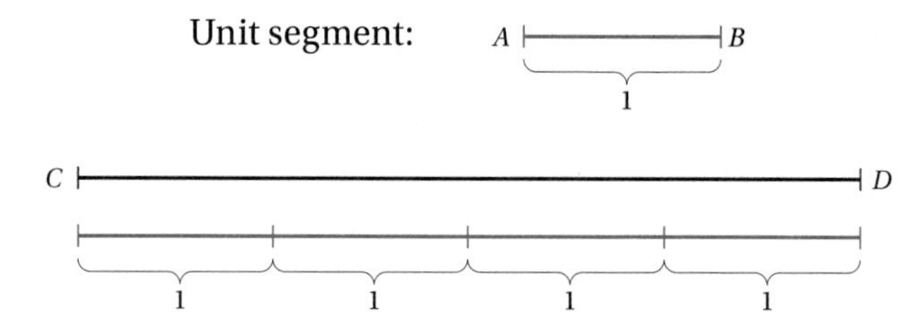

Since we can place 4 unit segments end to end along $\overline{CD}$, the measure of $\overline{CD}$ is 4.

Sometimes we need to use parts of units, called **subunits**. For example, the measure of the segment $\overline{MN}$ below is $1\frac{1}{2}$. We place one unit segment and one half-unit segment end to end.

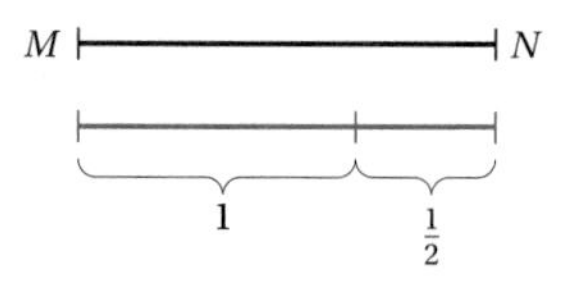

◀ Do Margin Exercises 1–4.

Use the unit below to measure the length of each segment or object.

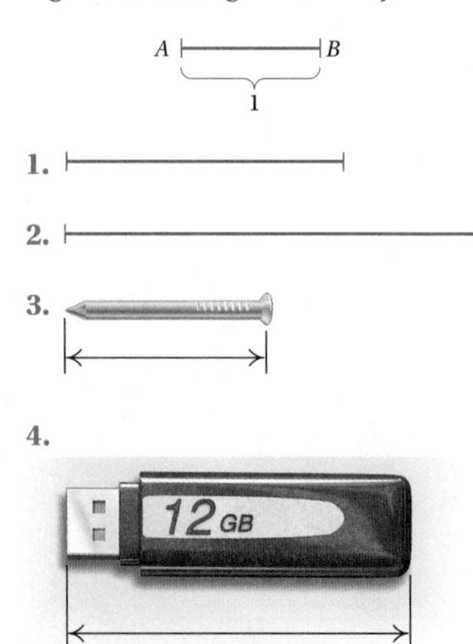

Answers on p. 501

a AMERICAN MEASURES

American units of length are related as follows.

AMERICAN UNITS OF LENGTH	
12 inches (in.) = 1 foot (ft)	3 feet = 1 yard (yd)
36 inches = 1 yard	5280 feet = 1 mile (mi)

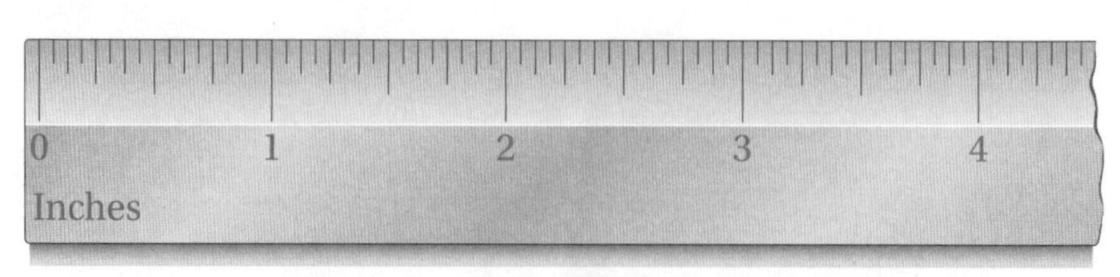

(Actual size, in inches)

We can visualize comparisons of the units as follows:

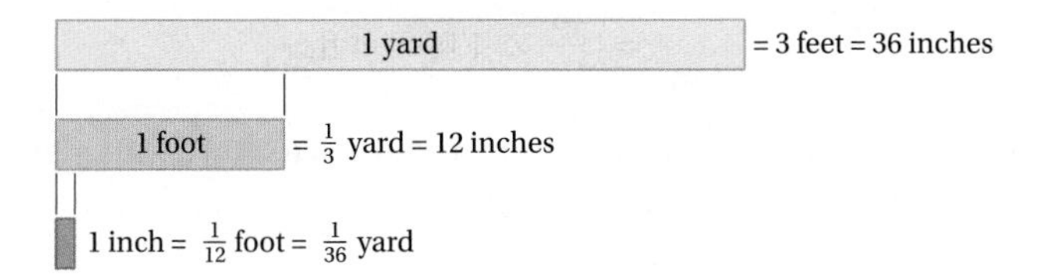

We can also abbreviate the units inches and feet like this: 13 in. = 13″ and 27 ft = 27′. American units have also been called "English," or "British-American," because at one time they were used by both countries. Today, both Canada and England have officially converted to the metric system.

To change from certain American units to others, we make substitutions. Such a substitution is usually helpful when we are converting from a *larger* unit to a *smaller* one.

EXAMPLE 1 Complete: $7\frac{1}{3}$ yd = ________ in.

$$7\frac{1}{3}\text{ yd} = 7\frac{1}{3} \times 1\text{ yd} \qquad \text{We think of } 7\frac{1}{3}\text{ yd as } 7\frac{1}{3} \times \text{yd, or } 7\frac{1}{3} \times 1\text{ yd.}$$

$$= 7\frac{1}{3} \times 36\text{ in.} \qquad \text{Substituting 36 in. for 1 yd}$$

$$= \frac{22}{3} \times 36\text{ in.}$$

$$= 264\text{ in.}$$

Do Exercises 5–7. ▶

Complete.

5. 8 yd = ________ in.

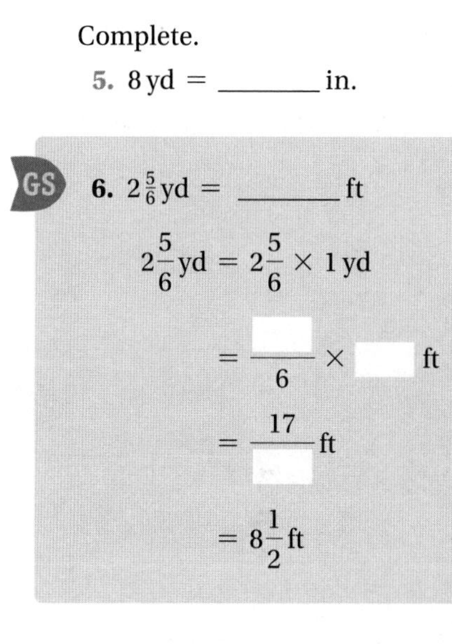

7. 3.8 mi = ________ in.

Sometimes it helps to use multiplying by 1 in making conversions. For example, 12 in. = 1 ft, so

$$\frac{12\text{ in.}}{1\text{ ft}} = 1 \quad \text{and} \quad \frac{1\text{ ft}}{12\text{ in.}} = 1.$$

If we divide 12 in. by 1 ft or 1 ft by 12 in., we get 1 because the lengths are the same. Let's first convert from *smaller* units to *larger* units.

EXAMPLE 2 Complete: 48 in. = ________ ft.

We want to convert from "in." to "ft." We multiply by 1 using a symbol for 1 with "in." in the denominator and "ft" in the numerator to eliminate inches and to convert to feet:

$$48\text{ in.} = \frac{48\text{ in.}}{1} \times \frac{1\text{ ft}}{12\text{ in.}} \qquad \text{Multiplying by 1 using } \frac{1\text{ ft}}{12\text{ in.}} \text{ to eliminate in.}$$

$$= \frac{48\text{ in.}}{12\text{ in.}} \times 1\text{ ft}$$

$$= \frac{48}{12} \times \frac{\text{in.}}{\text{in.}} \times 1\text{ ft}$$

$$= 4 \times 1\text{ ft} \qquad \text{The } \frac{\text{in.}}{\text{in.}} \text{ acts like 1, so we can omit it.}$$

$$= 4\text{ ft.}$$

Answers

Skill to Review:

1. $\frac{51}{8}$ 2. $19\frac{1}{5}$

Margin Exercises:

1. 2 2. 3 3. $1\frac{1}{2}$ 4. $2\frac{1}{2}$
5. 288 6. $8\frac{1}{2}$ 7. 240,768

Guided Solution:

6. 17, 3, 2

Complete.

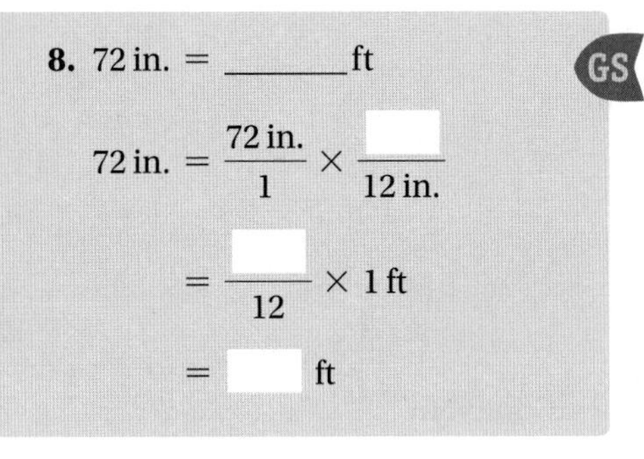

8. 72 in. = _______ ft

$$72\text{ in.} = \frac{72\text{ in.}}{1} \times \frac{\square}{12\text{ in.}}$$

$$= \frac{\square}{12} \times 1\text{ ft}$$

$$= \square\text{ ft}$$

9. 17 in. = _______ ft

Complete.

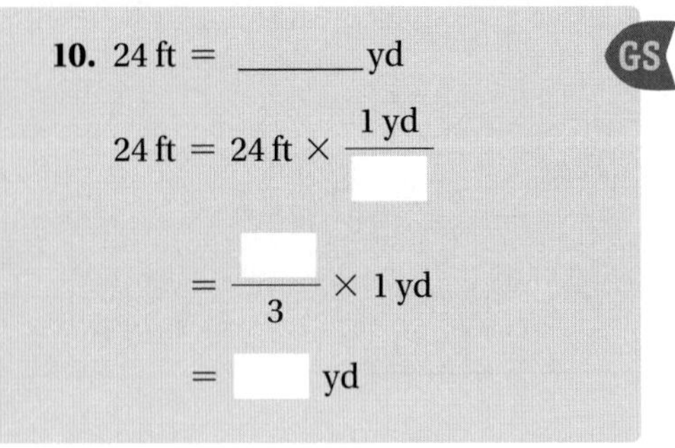

10. 24 ft = _______ yd

$$24\text{ ft} = 24\text{ ft} \times \frac{1\text{ yd}}{\square}$$

$$= \frac{\square}{3} \times 1\text{ yd}$$

$$= \square\text{ yd}$$

11. 35 ft = _______ yd

Complete.

12. 26,400 ft = _______ mi

13. 2640 ft = _______ mi

Answers

8. 6 **9.** $1\frac{5}{12}$ **10.** 8 **11.** $11\frac{2}{3}$, or $11.\overline{6}$
12. 5 **13.** $\frac{1}{2}$, or 0.5

Guided Solutions:

8. 1 ft, 72, 6 **10.** 3 ft, 24, 8

We can also look at this conversion as "canceling" units:

$$48\text{ in.} = \frac{48\text{ in.}}{1} \times \frac{1\text{ ft}}{12\text{ in.}} = \frac{48}{12} \times 1\text{ ft} = 4\text{ ft}.$$

This method is used not only in mathematics, as here, but also in fields such as medicine, chemistry, and physics.

Do Exercises 8 and 9.

EXAMPLE 3 Complete: 25 ft = _______ yd.

Since we are converting from "ft" to "yd," we choose a symbol for 1 with "yd" in the numerator and "ft" in the denominator:

$$25\text{ ft} = 25\text{ ft} \times \frac{1\text{ yd}}{3\text{ ft}}$$

$3\text{ ft} = 1\text{ yd}$, so $\frac{3\text{ ft}}{1\text{ yd}} = 1$ and $\frac{1\text{ yd}}{3\text{ ft}} = 1.$ We use $\frac{1\text{ yd}}{3\text{ ft}}$ to eliminate ft.

$$= \frac{25}{3} \times \frac{\text{ft}}{\text{ft}} \times 1\text{ yd}$$

$$= 8\frac{1}{3} \times 1\text{ yd}$$

The $\frac{\text{ft}}{\text{ft}}$ acts like 1, so we can omit it.

$$= 8\frac{1}{3}\text{ yd, or } 8.\overline{3}\text{ yd}.$$

Again, in this example, we can consider conversion from the point of view of canceling:

$$25\text{ ft} = 25\text{ ft} \times \frac{1\text{ yd}}{3\text{ ft}} = \frac{25}{3} \times 1\text{ yd} = 8\frac{1}{3}\text{ yd, or } 8.\overline{3}\text{ yd}.$$

Do Exercises 10 and 11.

EXAMPLE 4 Complete: 23,760 ft = _______ mi.

We choose a symbol for 1 with "mi" in the numerator and "ft" in the denominator:

$$23{,}760\text{ ft} = 23{,}760\text{ ft} \times \frac{1\text{ mi}}{5280\text{ ft}}$$

$5280\text{ ft} = 1\text{ mi}$, so $\frac{1\text{ mi}}{5280\text{ ft}} = 1.$

$$= \frac{23{,}760}{5280} \times \frac{\text{ft}}{\text{ft}} \times 1\text{ mi}$$

$$= 4.5 \times 1\text{ mi} \qquad \text{Dividing}$$

$$= 4.5\text{ mi}.$$

Let's also consider this example using canceling:

$$23{,}760\text{ ft} = 23{,}760\text{ ft} \times \frac{1\text{ mi}}{5280\text{ ft}}$$

$$= \frac{23{,}760}{5280} \times 1\text{ mi}$$

$$= 4.5 \times 1\text{ mi} = 4.5\text{ mi}.$$

Do Exercises 12 and 13.

We can also use multiplying by 1 to convert from larger units to smaller units. Let's redo Example 1.

EXAMPLE 5 Complete: $7\frac{1}{3}\text{ yd} = \_\_\_\_\_\_ \text{ in.}$

$$7\frac{1}{3}\text{ yd} = \frac{22\text{ yd}}{3} \times \frac{36\text{ in.}}{1\text{ yd}} = \frac{22 \times 36}{3} \times 1\text{ in.} = 264\text{ in.}$$

Do Exercise 14. ▶

14. Complete. Use multiplying by 1.

$2\frac{2}{3}\text{ yd} = \_\_\_\_\_\_ \text{ in.}$

EXAMPLE 6 *Illuminated Bridge.* A computer-controlled light sculpture on the San Francisco–Oakland Bay Bridge will be visible for a two-year period that began March 5, 2013. It consists of 25,000 white LED lights that run for 1.8 mi on the bridge's west span. The artist, Leo Villareal, wrote a program to ensure that the light patterns in the sculpture will never repeat. Convert 1.8 mi to yards.

Source: http://articles.latimes.com/2013/mar/05/news/la-trb-california-san-francisco-bay-bridge-lighting-20130304

We have

$$\begin{aligned} 1.8\text{ mi} &= 1.8\text{ mi} \times \frac{5280\text{ ft}}{1\text{ mi}} \times \frac{1\text{ yd}}{3\text{ ft}} \\ &= \frac{1.8 \times 5280}{1 \times 3} \times 1\text{ yd} \\ &= 3168\text{ yd.} \end{aligned}$$

The illuminated lights run for 3168 yd on the west span of the bridge.

Do Exercise 15. ▶

15. *Pedestrian Paths.* There are 23 mi of pedestrian paths in Central Park in New York City. Convert 23 miles to yards.

Answers

14. 96 **15.** 40,480 yd

9.1 Exercise Set

For Extra Help MyMathLab® MathXL® PRACTICE WATCH READ REVIEW

Reading Check

When converting from smaller units to larger units or from larger units to smaller units, it is convenient to multiply by a symbol for 1. Complete each symbol for 1 by choosing a number from the list at the right. Some choices may not be used; others may be used more than once.

RC1. $1 = \frac{1\text{ ft}}{\square\text{ in.}}$

RC2. $1 = \frac{\square\text{ ft}}{1\text{ mi}}$

RC3. $1 = \frac{1\text{ yd}}{\square\text{ in.}}$

RC4. $1 = \frac{1\text{ mi}}{\square\text{ ft}}$

RC5. $1 = \frac{\square\text{ ft}}{1\text{ yd}}$

RC6. $1 = \frac{\square\text{ in.}}{1\text{ ft}}$

3
12
24
36
1760
5280

a Complete.

1. 1 ft = ______ in.

2. 1 yd = ______ ft

3. 1 in. = ______ ft

4. 1 mi = ______ yd

5. 1 mi = ______ ft

6. 1 ft = ______ yd

7. 3 yd = ______ in.

8. 10 yd = ______ ft

9. 84 in. = ______ ft

10. 48 ft = ______ yd

11. 18 in. = ______ ft

12. 29 ft = ______ yd

13. 5 mi = ______ ft

14. 5 mi = ______ yd

15. 63 in. = ______ ft

16. 11,616 ft = ______ mi

17. 10 ft = ______ yd

18. 9.6 yd = ______ ft

19. 7.1 mi = ______ ft

20. 31,680 ft = ______ mi

21. $4\frac{1}{2}$ ft = ______ yd

22. 48 in. = ______ ft

23. 45 in. = ______ yd

24. $6\frac{1}{3}$ yd = ______ in.

25. 330 ft = ______ yd

26. 5280 yd = ______ mi

27. 3520 yd = ______ mi

28. 25 mi = ______ ft

29. 100 yd = ______ ft

30. 480 in. = ______ ft

31. 360 in. = ______ ft

32. 720 in. = ______ yd

33. 1 in. = ______ yd

34. 25 in. = ______ ft

35. 2 mi = ______ in.

36. 63,360 in. = ______ mi

37. 83 yd = ______ in.

38. 450 in. = ______ yd

Skill Maintenance

Convert to fraction notation. [7.2b]

39. 9.25%

40. $87\frac{1}{2}\%$

Solve. [5.4b]

41. $3.5 \cdot q = 0.2142$

42. $-1.95 \cdot w = 0.078$

43. Divide: $5\frac{1}{6} \div 4\frac{2}{3}$. [4.6b]

44. Find another name for $\frac{4}{15}$ with 90 as the denominator. [3.5a]

Solve. [6.1a, b]

45. *Accessing Facebook.* During the month of February 2013, U.S. users spent an average of 320 minutes accessing Facebook via desktop computers and an average of 785 minutes accessing Facebook via smart phones. What is the ratio of minutes spent accessing Facebook via desktop computers to minutes spent accessing it via smart phones? What is the ratio of minutes spent accessing Facebook via smart phones to minutes spent accessing it via desktop computers?

Sources: www.statista.com; comScore; J. P. Morgan

46. *Radio Stations.* In the United States in 2012, there were 2020 commercial radio stations offering primarily country music programming and 1503 radio stations offering primarily news/talk programming. What is the ratio of the number of stations offering country music to the number of stations offering news/talk programming? What is the ratio of the number of stations offering news/talk programming to the number of stations offering country music?

Source: *The World Almanac and Book of Facts 2013*

Synthesis

47. *Noah's Ark.* It is believed that the biblical measure called a *cubit* was equal to about 18 in.: 1 cubit $\approx$ 18 in. The dimensions of Noah's ark are given as follows: "The length of the ark shall be three hundred cubits, the breadth of it fifty cubits, and the height of it thirty cubits." What were the dimensions of Noah's ark in inches? in feet?

Source: *Holy Bible, King James Version*, Gen. 6:15

48. *Goliath's Height.* The biblical measure called a *span* was considered to be half of a cubit (1 cubit $\approx$ 18 in.; see Exercise 47). The giant Goliath's height "was six cubits and a span." What was the height of Goliath in inches? in feet?

Source: *Holy Bible, King James Version*, 1 Sam. 17:4

9.2 Linear Measures: The Metric System

OBJECTIVE

a Convert from one metric unit of length to another.

SKILL TO REVIEW

Objective 5.3a: Multiply using decimal notation.

Multiply.

1. 0.5603×1000

2. 18.7×100

Although the **metric system** is used in most countries of the world, it is used very little in the United States. The metric system does not use inches, feet, pounds, and so on, but its units for time and electricity are the same as those used now in the United States.

An advantage of the metric system is that it is easier to convert from one unit to another within this system than within the American system. That is because the metric system is based on the number 10.

The basic unit of length is the **meter**. It is just over a yard. In fact, 1 meter $\approx$ 1.1 yd.

(Comparative sizes are shown.)

1 Meter

1 Yard

The other units of length are multiples of the length of a meter:

10 times a meter, 100 times a meter, 1000 times a meter, and so on,

or fractions of a meter:

$\frac{1}{10}$ of a meter, $\frac{1}{100}$ of a meter, $\frac{1}{1000}$ of a meter, and so on.

METRIC UNITS OF LENGTH

1 *kilo*meter (km) = 1000 meters (m)

1 *hecto*meter (hm) = 100 meters (m)

1 *deka*meter (dam) = 10 meters (m)

1 meter (m)

1 *deci*meter (dm) = $\frac{1}{10}$ meter (m)

1 *centi*meter (cm) = $\frac{1}{100}$ meter (m)

1 *milli*meter (mm) = $\frac{1}{1000}$ meter (m)

You should memorize the names and abbreviations for metric units of length. Remember *kilo-* for 1000, *hecto-* for 100, *deka-* for 10, *deci-* for $\frac{1}{10}$, *centi-* for $\frac{1}{100}$, and *milli-* for $\frac{1}{1000}$. (The units dekameter and decimeter are not used often.) We will also use these prefixes when considering units of area, capacity, and mass.

Thinking Metric

To familiarize yourself with metric units, consider the following.

1 kilometer (1000 meters)	is slightly more than $\frac{1}{2}$ mile (0.6 mi).
1 meter	is just over a yard (1.1 yd).
1 centimeter (0.01 meter)	is a little more than the width of a paperclip (about 0.3937 inch).

1 cm

Answers

Skill to Review:
1. 560.3 **2.** 1870

1 inch is 2.54 centimeters.

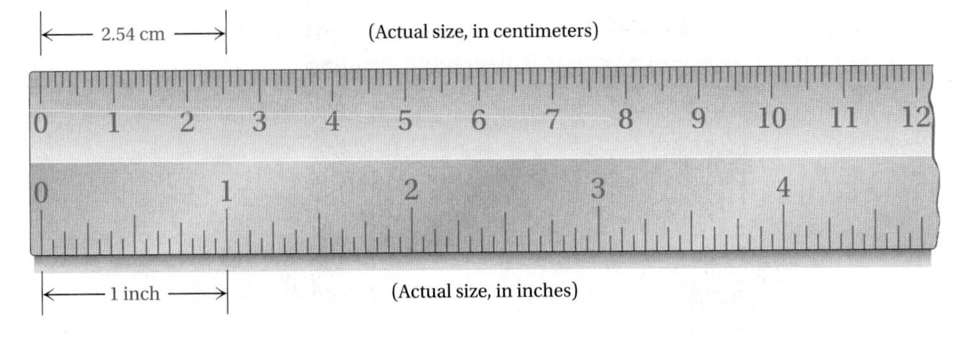

1 millimeter is about the diameter of paperclip wire.

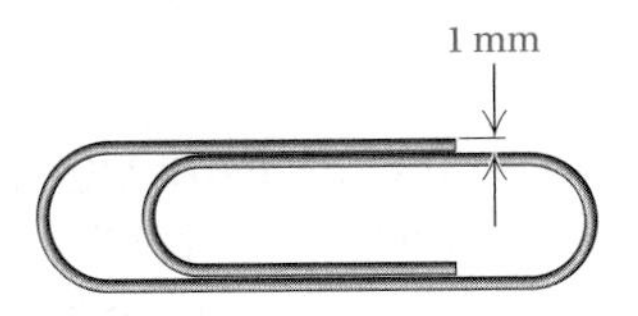

The millimeter (mm) is often used in jewelry making.

In many countries, the centimeter (cm) is used for body dimensions and clothing sizes.

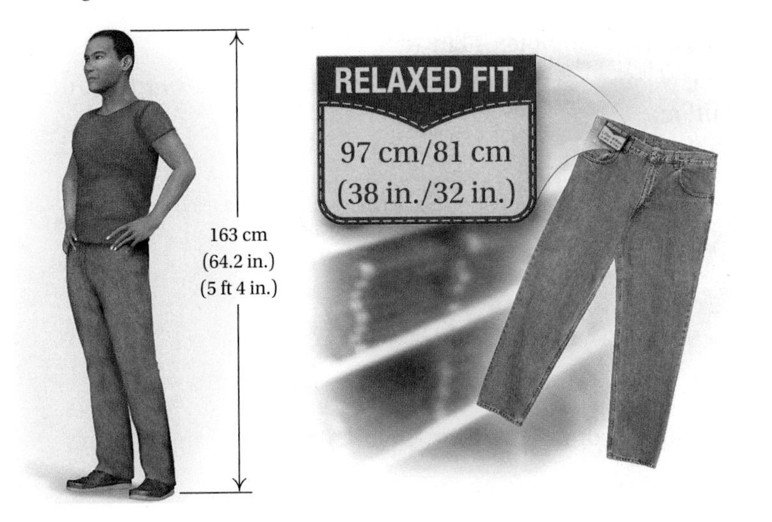

Do Exercises 1–3.

Using a centimeter ruler, measure each object.

1.

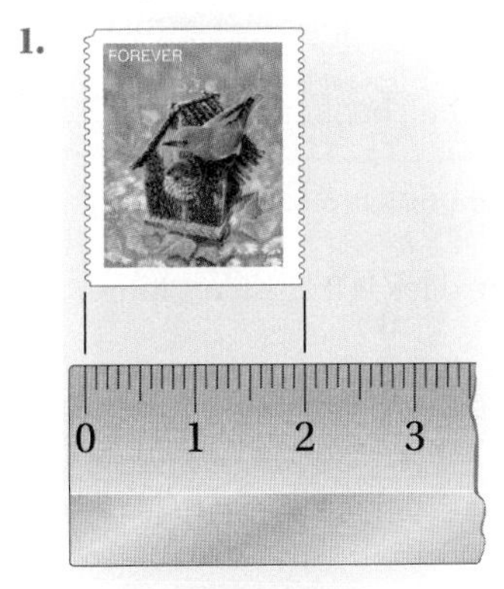

2.

3.

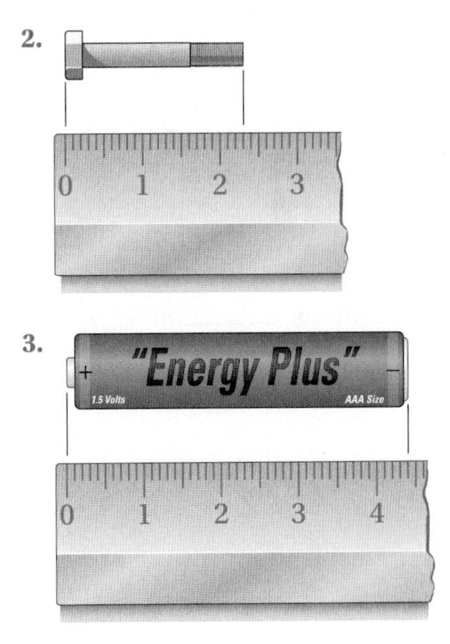

Answers

1. 2 cm, or 20 mm
2. 2.3 cm, or 23 mm
3. 4.4 cm, or 44 mm

The meter (m) is used for expressing dimensions of large objects—say, the height of Hoover dam, 221.4 m, or the distance around a standard athletic track, 400 m—and for expressing somewhat smaller dimensions like the length and width of an organic vegetable garden.

The kilometer (km) is used for longer distances, mostly those that are expressed in miles in American units.

1 mile is about 1.6 km.

1 km

1 mi

Complete with mm, cm, m, or km.

4. A stick of gum is 7 ________ long.

5. Dallas is 1512 ________ from Minneapolis.

6. A penny is 1 ________ thick.

7. The halfback ran 7 ________.

8. The book is 3 ________ thick.

9. The desk is 2 ________ long.

◀ **Do Exercises 4–9.**

a CHANGING METRIC UNITS

As with American units, when changing from a *larger* unit to a *smaller* unit, we usually make substitutions.

EXAMPLE 1 Complete: 4 km = ________ m.

Since we are converting from a *larger* unit to a *smaller* unit, we use substitution.

$$4\text{ km} = 4 \times 1\text{ km}$$
$$= 4 \times 1000\text{ m} \qquad \text{Substituting 1000 m for 1 km}$$
$$= 4000\text{ m}$$

Complete.

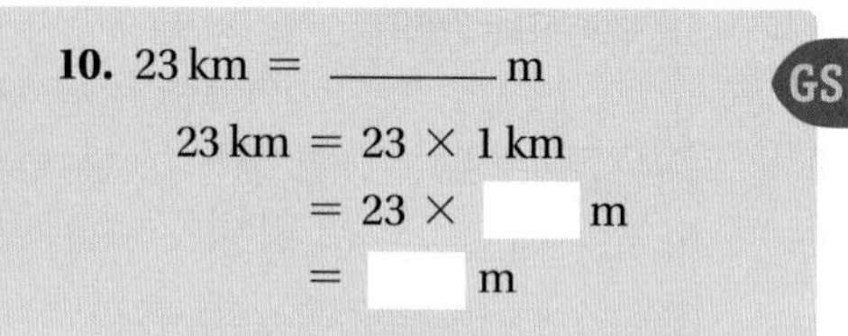

10. 23 km = ________ m

23 km = 23 × 1 km

= 23 × ☐ m

= ☐ m

GS

11. 4 hm = ________ m

◀ **Do Exercises 10 and 11.**

Since

$$\frac{1}{10}\text{ m} = 1\text{ dm}, \quad \frac{1}{100}\text{ m} = 1\text{ cm}, \quad \text{and} \quad \frac{1}{1000}\text{ m} = 1\text{ mm},$$

it follows that

1 m = 10 dm, 1 m = 100 cm, and 1 m = 1000 mm.

EXAMPLE 2 Complete: 93.4 m = ________ cm.

Since we are converting from a *larger* unit to a *smaller* unit, we use substitution. We substitute 100 cm for 1 m:

$$93.4\text{ m} = 93.4 \times 1\text{ m} = 93.4 \times 100\text{ cm} = 9340\text{ cm}.$$

Answers

4. cm 5. km 6. mm 7. m 8. cm
9. m 10. 23,000 11. 400

Guided Solution:
10. 1000, 23,000

EXAMPLE 3 Complete: $0.248\text{ m} = \underline{\qquad}\text{ mm}$.

Since we are converting from a *larger* unit to a *smaller* unit, we use substitution.

$$\begin{aligned} 0.248\text{ m} &= 0.248 \times 1\text{ m} \\ &= 0.248 \times 1000\text{ mm} \qquad \text{Substituting 1000 mm for 1 m} \\ &= 248\text{ mm} \end{aligned}$$

Do Exercises 12 and 13. ▶

Complete.

12. $1.78\text{ m} = \underline{\qquad}\text{ cm}$

13. $9.04\text{ m} = \underline{\qquad}\text{ mm}$

We now convert from "m" to "km." Since we are converting from a *smaller* unit to a *larger* unit, we use multiplying by 1. We choose a symbol for 1 with "km" in the numerator and "m" in the denominator.

EXAMPLE 4 Complete: $2347\text{ m} = \underline{\qquad}\text{ km}$.

$$\begin{aligned} 2347\text{ m} &= 2347\text{ m} \times \frac{1\text{ km}}{1000\text{ m}} \qquad \text{Multiplying by 1 using } \frac{1\text{ km}}{1000\text{ m}} \\ &= \frac{2347}{1000} \times \frac{\text{m}}{\text{m}} \times 1\text{ km} \qquad \text{The } \frac{\text{m}}{\text{m}} \text{ acts like 1, so we omit it.} \\ &= 2.347\text{ km} \qquad \text{Dividing by 1000 moves the decimal point three places to the left.} \end{aligned}$$

Using canceling, we can work this example as follows:

$$2347\text{ m} = 2347\,\cancel{\text{m}} \times \frac{1\text{ km}}{1000\,\cancel{\text{m}}} = \frac{2347}{1000} \times 1\text{ km} = 2.347\text{ km}.$$

Sometimes we multiply by 1 more than once.

Complete.

14. $7814\text{ m} = \underline{\qquad}\text{ km}$

EXAMPLE 5 Complete: $8.42\text{ mm} = \underline{\qquad}\text{ cm}$.

$$\begin{aligned} 8.42\text{ mm} &= 8.42\text{ mm} \times \frac{1\text{ m}}{1000\text{ mm}} \times \frac{100\text{ cm}}{1\text{ m}} \qquad \text{Multiplying by 1 using } \frac{1\text{ m}}{1000\text{ mm}} \text{ and } \frac{100\text{ cm}}{1\text{ m}} \\ &= \frac{8.42 \times 100}{1000} \times \frac{\text{mm}}{\text{mm}} \times \frac{\text{m}}{\text{m}} \times 1\text{ cm} \\ &= \frac{842}{1000}\text{ cm} = 0.842\text{ cm} \end{aligned}$$

Do Exercises 14–17. ▶

GS 15. $7814\text{ m} = \underline{\qquad}\text{ dam}$

$$\begin{aligned} 7814\text{ m} &= 7814\text{ m} \times \frac{1\text{ dam}}{\square} \\ &= \frac{7814}{\square} \times \frac{\text{m}}{\text{m}} \times 1\text{ dam} \\ &= \square\text{ dam} \end{aligned}$$

16. $9.67\text{ mm} = \underline{\qquad}\text{ cm}$

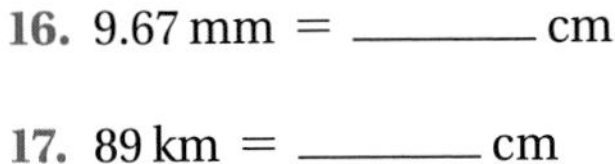

17. $89\text{ km} = \underline{\qquad}\text{ cm}$

Mental Conversion

Changing from one unit of length to another in the metric system amounts to the movement of a decimal point. That is because the metric system is based on 10. Let's find a faster way to convert. Look at the following table.

1000 m	100 m	10 m	1 m	0.1 m	0.01 m	0.001 m
1 km	1 hm	1 dam	1 m	1 dm	1 cm	1 mm

Each place in the table has a value $\frac{1}{10}$ that to the left or 10 times that to the right. Thus moving one place in the table corresponds to moving one decimal place.

Answers

12. 178 **13.** 9040 **14.** 7.814 **15.** 781.4 **16.** 0.967 **17.** 8,900,000

Guided Solution:
15. 10 m, 10, 781.4

Let's convert mentally.

EXAMPLE 6 Complete: 8.42 mm = _______ cm.

Think: To go from mm to cm in the table is a move of one place to the left. Thus we move the decimal point one place to the left.

1000 m	100 m	10 m	1 m	0.1 m	0.01 m	0.001 m
1 km	1 hm	1 dam	1 m	1 dm	1 cm	1 mm

1 place to the left

8.42 0.8.42 8.42 mm = 0.842 cm

EXAMPLE 7 Complete: 1.886 km = _______ cm.

Think: To go from km to cm in the table is a move of five places to the right. Thus we move the decimal point five places to the right.

1000 m	100 m	10 m	1 m	0.1 m	0.01 m	0.001 m
1 km	1 hm	1 dam	1 m	1 dm	1 cm	1 mm

5 places to the right

1.886 1.88600. 1.886 km = 188,600 cm

EXAMPLE 8 Complete: 3 m = _______ cm.

Think: To go from m to cm in the table is a move of two places to the right. Thus we move the decimal point two places to the right.

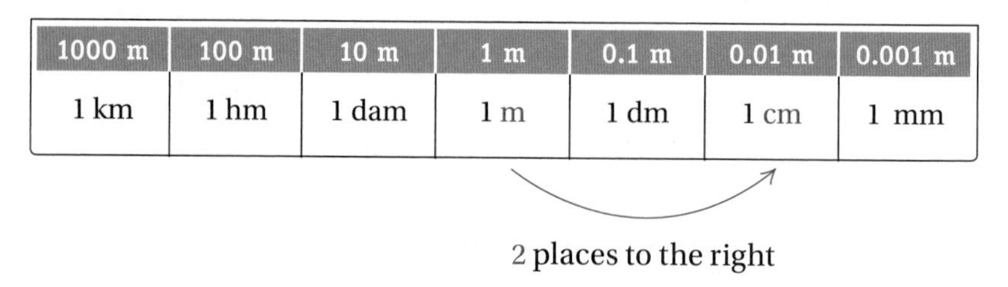

1000 m	100 m	10 m	1 m	0.1 m	0.01 m	0.001 m
1 km	1 hm	1 dam	1 m	1 dm	1 cm	1 mm

2 places to the right

3 3.00. 3 m = 300 cm

You should try to make metric conversions mentally as much as possible.

The fact that conversions can be done so easily is an important advantage of the metric system. The most commonly used metric units of length are km, m, cm, and mm. We have purposely used these more often than the others in the exercises.

Do Exercises 18–21.

Complete. Try to do this mentally using the table.

18. 6780 m = _______ km

19. 9.74 cm = _______ mm

20. 1 mm = _______ cm

21. 845.1 mm = _______ dm

Answers

18. 6.78 **19.** 97.4 **20.** 0.1 **21.** 8.451

9.2 Exercise Set

Reading Check

Complete each sentence using > or < for □.

RC1. 3 dm □ 3 dam

RC2. 3 hm □ 3 cm

RC3. 3 mm □ 3 km

RC4. 3 m □ 3 hm

RC5. 3 dam □ 3 mm

RC6. 3 cm □ 3 m

a Complete. Do as much as possible mentally.

1. a) 1 km = ______ m
b) 1 m = ______ km

2. a) 1 hm = ______ m
b) 1 m = ______ hm

3. a) 1 dam = ______ m
b) 1 m = ______ dam

4. a) 1 dm = ______ m
b) 1 m = ______ dm

5. a) 1 cm = ______ m
b) 1 m = ______ cm

6. a) 1 mm = ______ m
b) 1 m = ______ mm

7. 6.7 km = ______ m

8. 27 km = ______ m

9. 98 cm = ______ m

10. 0.789 cm = ______ m

11. 8921 m = ______ km

12. 8664 m = ______ km

13. 56.66 m = ______ km

14. 4.733 m = ______ km

15. 5666 m = ______ cm

16. 869 m = ______ cm

17. 477 cm = ______ m

18. 6.27 mm = ______ m

19. 6.88 m = ______ cm

20. 6.88 m = ______ dm

21. 1 mm = ______ cm

22. 1 cm = ______ km

23. 1 km = ______ cm

24. 2 km = ______ cm

25. 14.2 cm = ______ mm

26. 25.3 cm = ______ mm

27. 8.2 mm = ______ cm

28. 9.7 mm = ______ cm

29. 4500 mm = ______ cm

30. 8,000,000 m = ______ km

31. 0.024 mm = ______ m

32. 60,000 mm = ______ dam

33. 6.88 m = ______ dam

34. 7.44 m = ______ hm

35. 2.3 dam = ______ dm

36. 9 km = ______ hm

37. 392 dam = ______ km

38. 0.056 mm = ______ dm

Complete the following table.

	Object	Millimeters (mm)	Centimeters (cm)	Meters (m)
39.	Width of a football field		4844	
40.	Length of a football field			109.09
41.	Length of 4 meter sticks			4
42.	Width of a credit card	56		
43.	Thickness of an index card	0.27		
44.	Thickness of a piece of cardboard		0.23	
45.	Height of One World Trade Center, New York, New York			541.3
46.	Height of The Gateway Arch, St. Louis, Missouri	192,000		

Skill Maintenance

Divide. [5.4a]

47. $23.4 \div 100$

48. $23.4 \div 1000$

Multiply.

49. 3.14×4.41 [5.3a]

50. $-4 \times 20\frac{1}{8}$ [4.6a]

Convert to percent notation. [7.2a]

51. $\frac{2}{3}$

52. $\frac{5}{8}$

Calculate.

53. $\frac{7}{15} + \frac{4}{25}$ [4.2a]

54. $-\frac{11}{18} - \frac{5}{24}$ [4.3a]

Find decimal notation for the percent notation in each sentence. [7.1b]

55. Blood is 90% water.

56. Of those accidents with victims requiring medical attention, 10.8% occur on roads.

Synthesis

Each sentence is incorrect. Insert or reposition a decimal point to make the sentence correct.

57. When my right arm is extended, the distance from my left shoulder to the end of my right hand is 10 m.

58. The height of the Shanghai World Financial Center is 49.2 m.

59. A stack of ten quarters is 140 cm high.

60. The width of an adult's hand is 112 cm.

Converting Between American Units and Metric Units

9.3

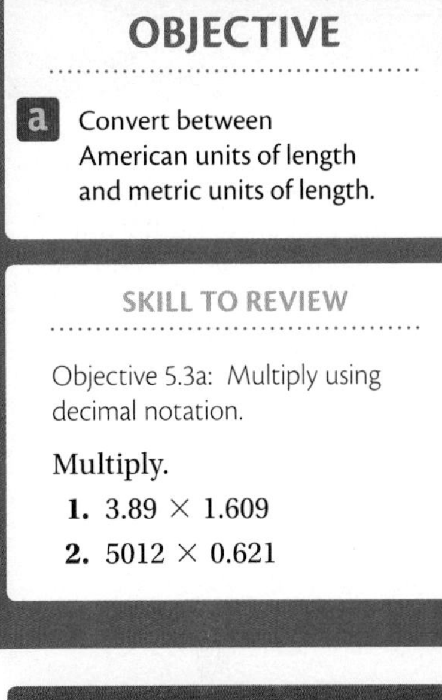

OBJECTIVE

a Convert between American units of length and metric units of length.

SKILL TO REVIEW

Objective 5.3a: Multiply using decimal notation.

Multiply.

1. 3.89×1.609
2. 5012×0.621

a CONVERTING UNITS

We can make conversions between American units and metric units by substituting based on the rounded approximations in the following table.

AMERICAN	METRIC
1 in.	2.54 cm
1 ft	0.305 m
1 yd	0.914 m
1 mi	1.609 km
0.621 mi	1 km
1.094 yd	1 m
3.281 ft	1 m
39.370 in.	1 m

EXAMPLE 1 Complete: 11 in. = ________ cm.
(The wingspan of the world's largest butterfly, the Queen Alexandra)
Source: *Top 10 of Everything 2013*

$$\begin{aligned} 11 \text{ in.} &= 11 \times 1 \text{ in.} \\ &= 11 \times 2.54 \text{ cm} \quad \text{Substituting 2.54 cm for 1 in.} \\ &= 27.94 \text{ cm} \end{aligned}$$

This answer would probably be rounded to the nearest one: 28 cm.

EXAMPLE 2 Complete: 26.2 mi = ________ km.
(The length of the Olympic marathon)

$$\begin{aligned} 26.2 \text{ mi} &= 26.2 \times 1 \text{ mi} \\ &\approx 26.2 \times 1.609 \text{ km} \quad \text{Substituting 1.609 km for 1 mi} \\ &= 42.1558 \text{ km} \end{aligned}$$

EXAMPLE 3 Complete: 100 m = ________ ft.
(The length of the 100-m dash)

$$\begin{aligned} 100 \text{ m} &= 100 \times 1 \text{ m} \\ &\approx 100 \times 3.281 \text{ ft} \quad \text{Substituting 3.281 ft for 1 m} \\ &= 328.1 \text{ ft} \end{aligned}$$

EXAMPLE 4 Complete: 4544 km = ________ mi.
(The distance from New York to Los Angeles)

$$\begin{aligned} 4544 \text{ km} &= 4544 \times 1 \text{ km} \\ &\approx 4544 \times 0.621 \text{ mi} \quad \text{Substituting 0.621 mi for 1 km} \\ &= 2821.824 \text{ mi} \end{aligned}$$

We would probably round this answer to 2822 mi.

Do Margin Exercises 1–3.

Complete.

1. 100 yd = ________ m
(The length of a football field, excluding the end zones)

2. 2.5 mi = ________ km
(The length of the tri-oval track at Daytona International Speedway)

GS 3. 2383 km = ________ mi
(The distance from St. Louis to Phoenix)

$$\begin{aligned} 2383 \text{ km} &= 2383 \times 1 \text{ km} \\ &\approx 2383 \times \square \text{ mi} \\ &= \square \text{ mi} \end{aligned}$$

Answers

Skill to Review:
1. 6.25901 2. 3112.452

Margin Exercises:
1. 91.4 2. 4.0225 3. 1479.843

Guided Solution:
3. 0.621, 1479.843

4. The Pacific Coast Highway, which is part of California State Route 1, is 655.8 mi long. Find this length in kilometers, rounded to the nearest tenth.

5. The height of the Stratosphere Tower in Las Vegas, Nevada, is 1149 ft. Find the height in meters.

6. Complete:
3.175 mm = ________ in.
(The thickness of a quarter)

Answers

4. 1055.2 km 5. 350.445 m 6. 0.125

EXAMPLE 5 *Millau Viaduct.* The Millau viaduct is part of the E11 expressway connecting Paris, France, and Barcelona, Spain. The viaduct has the highest bridge piers ever constructed. The tallest pier is 804 ft high and the overall height including the pylon is 1122 ft, making this the highest bridge in the world. Convert 804 feet and 1122 feet to meters.

Source: www.abelard.org/france/viaduct-de-millau.php

We let $P =$ the height of the pier and $H =$ the overall height of the bridge. To convert feet to meters, we substitute 0.305 m for 1 ft.

$$\begin{aligned} P &= 804 \text{ ft} \\ &= 804 \times 1 \text{ ft} \\ &\approx 804 \times 0.305 \text{ m} \\ &= 245.22 \text{ m} \end{aligned} \qquad \begin{aligned} H &= 1122 \text{ ft} \\ &= 1122 \times 1 \text{ ft} \\ &\approx 1122 \times 0.305 \text{ m} \\ &= 342.21 \text{ m} \end{aligned}$$

◀ **Do Exercises 4 and 5.**

EXAMPLE 6 Complete: 0.10414 mm = ________ in.
(The thickness of a $1 bill)

In this case, we must make two substitutions or multiply by two forms of 1 since the table on the preceding page does not provide a direct way to convert from millimeters to inches. Here we choose to multiply by forms of 1.

$$\begin{aligned} 0.10414 \text{ mm} &= 0.10414 \times 1 \cancel{\text{mm}} \times \frac{1 \text{ cm}}{10 \cancel{\text{mm}}} \\ &= 0.010414 \text{ cm} \\ &= 0.010414 \times 1 \cancel{\text{cm}} \times \frac{1 \text{ in.}}{2.54 \cancel{\text{cm}}} \\ &= 0.0041 \text{ in.} \end{aligned}$$

◀ **Do Exercise 6.**

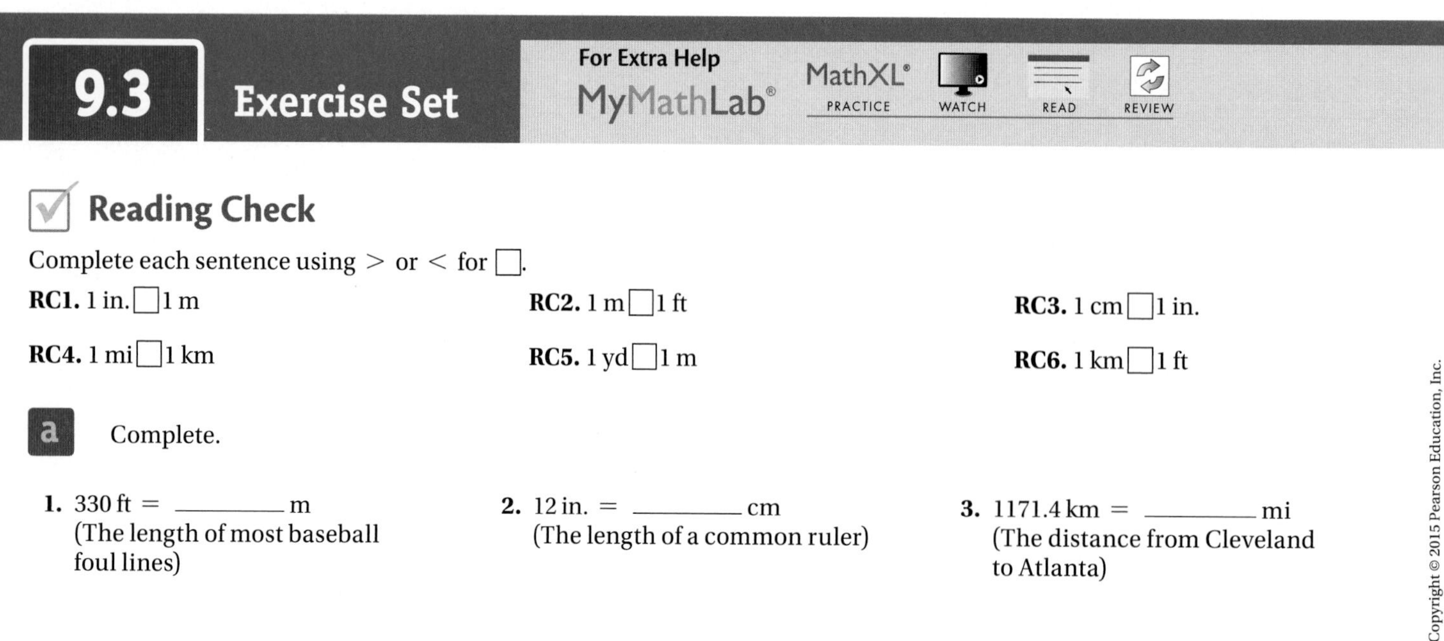

9.3 Exercise Set

For Extra Help
MyMathLab®
MathXL® PRACTICE WATCH READ REVIEW

Reading Check

Complete each sentence using > or < for □.

RC1. 1 in. □ 1 m

RC2. 1 m □ 1 ft

RC3. 1 cm □ 1 in.

RC4. 1 mi □ 1 km

RC5. 1 yd □ 1 m

RC6. 1 km □ 1 ft

a Complete.

1. 330 ft = ________ m
(The length of most baseball foul lines)

2. 12 in. = ________ cm
(The length of a common ruler)

3. 1171.4 km = ________ mi
(The distance from Cleveland to Atlanta)

4. 2 m = ________ ft
(The length of a desk)

5. 65 mph = ________ km/h
(A common speed limit in the United States)

6. 100 km/h = ________ mph
(A common speed limit in Canada)

7. 180 mi = ________ km
(The distance from Indianapolis to Chicago)

8. 141,600,000 mi = ________ km
(The farthest distance of Mars from the sun)

9. 70 mph = ________ km/h
(An interstate speed limit in Arizona)

10. 60 km/h = ________ mph
(A city speed limit in Canada)

11. 10 yd = ________ m
(The length needed for a first down in football)

12. 450 ft = ________ m
(The length of a long home run in baseball)

13. 1.91 m = ________ in.
(The height of Jeremy Lin of the Houston Rockets)

14. 69 in. = ________ m
(The height of Nate Robinson of the Chicago Bulls)

15. 169.41 m = ________ ft
(The height of the Washington Monument)

16. 1671 ft = ________ m
(The height of the Taipei 101 skyscraper)

17. 15.7 cm = ________ in.
(The length of a $1 bill)

18. 7.5 in. = ________ cm
(The length of a pencil)

19. 2216 km = ________ mi
(The distance from Chicago to Miami)

20. 1862 mi = ________ km
(The distance from Seattle to Kansas City)

21. 13 mm = ________ in.
(The thickness of a plastic case for a DVD)

22. 0.25 in. = ________ mm
(The thickness of an eraser on a pencil)

Complete the following table. Answers may vary, depending on the conversion factor used.

	Object	Yards (yd)	Centimeters (cm)	Inches (in.)	Meters (m)	Millimeters (mm)
23.	Width of a piece of typing paper			$8\frac{1}{2}$		
24.	Length of a football field	120				
25.	Width of a football field		4844			
26.	Width of a credit card					56
27.	Length of 4 yardsticks	4				
28.	Length of 3 meter sticks		300			
29.	Thickness of an index card				0.00027	
30.	Thickness of a piece of cardboard		0.23			
31.	The Channel Tunnel connecting France and England				50,500	
32.	Height of Jin Mao Tower, Shanghai	460				

Skill Maintenance

33. *Kangaroos.* Kangaroos are found in Australia, Tasmania, and surrounding islands. There are approximately 1.91 million kangaroos worldwide. Convert 1.91 million to standard notation. [5.3b]

Source: www.kidsplanet.org/factsheets/kangaroo.html

34. *Presidential Libraries.* The National Archives and Records Administration budgeted about $68.7 million for operations and maintenance of the 13 presidential libraries during 2013. Of this amount, approximately 8.6% was designated for the George W. Bush Presidential Center in Dallas, Texas, which opened April 25, 2013. Find the amount spent on operations and maintenance of the George W. Bush Presidental Center. [7.5a]

Source: National Archives and Records Administration

Synthesis

35. Develop a formula to convert from inches to millimeters.

36. Develop a formula to convert from millimeters to inches. How does it relate to the answer for Exercise 35?

37. 🖩 As of August 2012, the men's world record for the 100-m dash was 9.63 sec, set by Usain Bolt of Jamaica in the 2012 Summer Olympics in London. How fast is this in miles per hour? Round to the nearest hundredth of a mile per hour.

Source: en.wikipedia.com

38. 🖩 As of November 2008, the women's world record for the 100-m dash was 10.49 sec, set by Florence Griffith-Joyner in Indianapolis, Indiana, on July 16, 1988. How fast is this in miles per hour? Round to the nearest tenth of a mile per hour.

Source: International Association of Athletics Federations

Mid-Chapter Review

Concept Reinforcement

Determine whether each statement is true or false.

______ **1.** Distances that are measured in miles in the American system would probably be measured in meters in the metric system. [9.2a], [9.3a]

______ **2.** One meter is slightly more than one yard. [9.2a], [9.3a]

______ **3.** One kilometer is longer than one mile. [9.2a], [9.3a]

______ **4.** When converting from meters to centimeters, move the decimal point to the right. [9.2a]

______ **5.** One foot is approximately 30 centimeters. [9.2a], [9.3a]

Guided Solutions

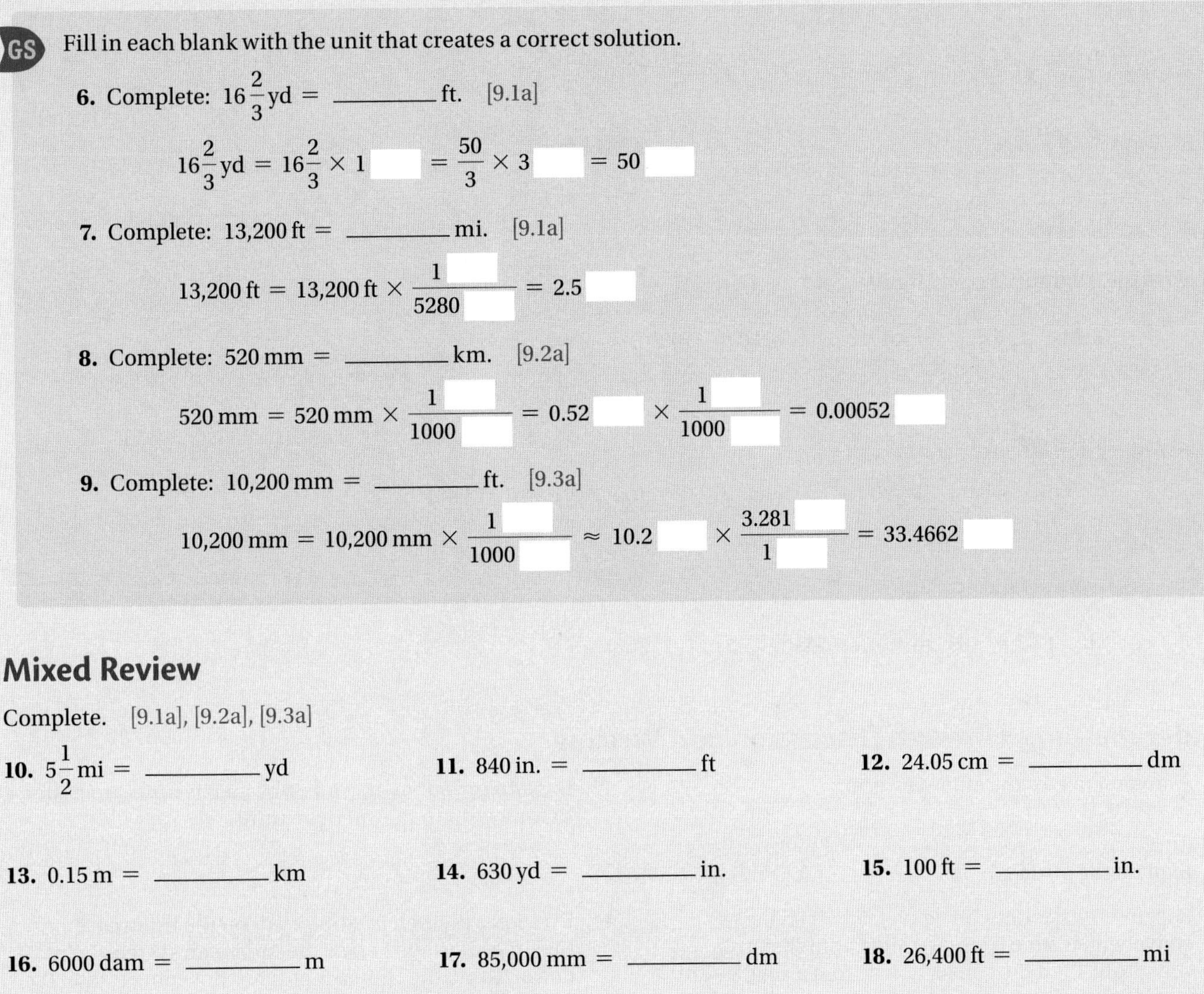

GS Fill in each blank with the unit that creates a correct solution.

6. Complete: $16\frac{2}{3}$ yd = ______ ft. [9.1a]

$$16\frac{2}{3}\text{ yd} = 16\frac{2}{3} \times 1\,\square = \frac{50}{3} \times 3\,\square = 50\,\square$$

7. Complete: 13,200 ft = ______ mi. [9.1a]

$$13{,}200\text{ ft} = 13{,}200\text{ ft} \times \frac{1\,\square}{5280\,\square} = 2.5\,\square$$

8. Complete: 520 mm = ______ km. [9.2a]

$$520\text{ mm} = 520\text{ mm} \times \frac{1\,\square}{1000\,\square} = 0.52\,\square \times \frac{1\,\square}{1000\,\square} = 0.00052\,\square$$

9. Complete: 10,200 mm = ______ ft. [9.3a]

$$10{,}200\text{ mm} = 10{,}200\text{ mm} \times \frac{1\,\square}{1000\,\square} \approx 10.2\,\square \times \frac{3.281\,\square}{1\,\square} = 33.4662\,\square$$

Mixed Review

Complete. [9.1a], [9.2a], [9.3a]

10. $5\frac{1}{2}$ mi = ______ yd

11. 840 in. = ______ ft

12. 24.05 cm = ______ dm

13. 0.15 m = ______ km

14. 630 yd = ______ in.

15. 100 ft = ______ in.

16. 6000 dam = ______ m

17. 85,000 mm = ______ dm

18. 26,400 ft = ______ mi

19. 3753 ft = ______ yd

20. 10 mi = ______ ft

21. 1800 m = ______ cm

22. 8.4 km = ________ dm

23. 0.007 km = ________ cm

24. 40 dm = ________ dam

25. 80.09 cm = ________ m

26. 360 in. = ________ yd

27. 19.2 m = ________ mm

28. 1200 in. = ________ ft

29. 0.0001 mm = ________ hm

30. 4 km = ________ cm

31. 12 mi = ________ in.

32. 36 m = ________ ft

33. 80 dm = ________ dam

34. 2.5 yd = ________ m

35. 6000 mm = ________ dm

36. 0.0635 mm = ________ in.

37. Match each measure in the first column with an equivalent measure in the second column by drawing connecting lines. [9.1a], [9.2a]

$\frac{1}{4}$ yd	24,000 dm
144 in.	1320 yd
2400 m	2400 mm
0.75 mi	9 in.
24 m	0.024 km
240 cm	12 ft

38. Arrange from smallest to largest:

100 in., 430 ft, $\frac{1}{100}$ mi, 3.5 ft, 6000 ft, 1000 in., 2 yd. [9.1a]

39. Arrange from largest to smallest:

3240 cm, 300 m, 250 dm, 150 hm, 33,000 mm, 310 dam, 13 km. [9.2a]

40. Arrange from smallest to largest:

2 yd, 1.5 mi, 65 cm, $\frac{1}{2}$ ft, 3 km, 2.5 m. [9.3a]

Understanding Through Discussion and Writing

41. A student makes the following error:

23 in. = 23 · (12 ft) = 276 ft.

Explain the error. [9.1a]

42. Explain in your own words why metric units are easier to work with than American units. [9.2a]

43. Recall the guidelines for conversion: (1) If the conversion is from a larger unit to a smaller unit, substitute. (2) If the conversion is from a smaller unit to a larger unit, multiply by 1. Explain why each is the easier way to convert in that situation. [9.1a], [9.2a]

44. Do some research in a library or on the Internet about the metric system versus the American system. Why do you think the United States has not converted to the metric system? [9.3a]

STUDYING FOR SUCCESS *Throughout the Semester*

- ☐ Review regularly. A good way to do this is by doing the Skill Maintenance exercises found in each exercise set.
- ☐ Try creating your own glossary. Understanding terminology is essential for success in any math course.
- ☐ Memorizing is a helpful tool in the study of mathematics. Ask your instructor what you are expected to have memorized for tests.

Weight and Mass; Medical Applications 9.4

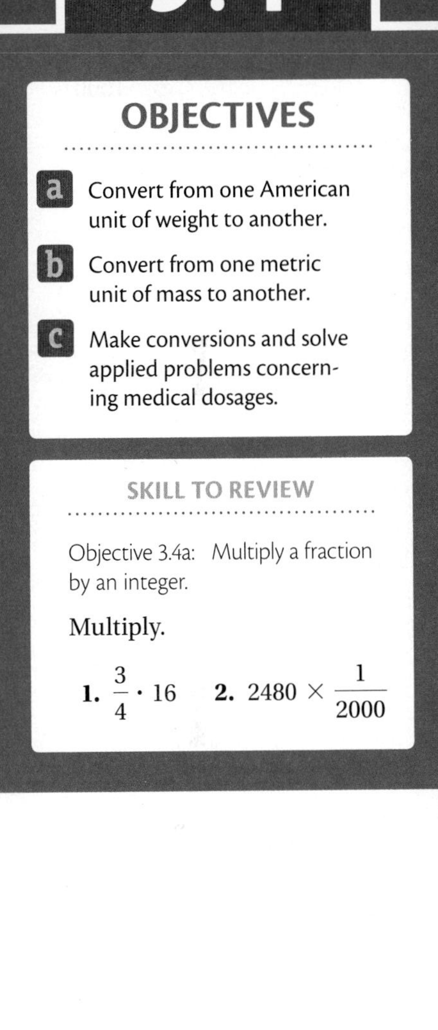

OBJECTIVES

- **a** Convert from one American unit of weight to another.
- **b** Convert from one metric unit of mass to another.
- **c** Make conversions and solve applied problems concerning medical dosages.

SKILL TO REVIEW

Objective 3.4a: Multiply a fraction by an integer.

Multiply.

1. $\frac{3}{4} \cdot 16$ **2.** $2480 \times \frac{1}{2000}$

There is a difference between **mass** and **weight**, but the terms are often used interchangeably. People sometimes use the word "weight" when, technically, they are referring to "mass." Weight is related to the force of gravity. The farther you are from the center of the earth, the less you weigh. Your mass stays the same no matter where you are.

a WEIGHT: THE AMERICAN SYSTEM

AMERICAN UNITS OF WEIGHT	
1 ton (T) = 2000 pounds (lb)	1 lb = 16 ounces (oz)

The term "ounce" used here for weight is different from the "ounce" used for capacity, which we will discuss in Section 9.5. We convert units of weight using the same techniques that we use with linear measures.

EXAMPLE 1 A well-known hamburger is called a "quarter-pounder." Find its name in ounces: a "______ ouncer."

Since we are converting from a larger unit to a smaller unit, we use substitution.

$$\frac{1}{4}\text{ lb} = \frac{1}{4} \cdot 1\text{ lb} = \frac{1}{4} \cdot 16\text{ oz} \qquad \text{Substituting 16 oz for 1 lb}$$
$$= 4\text{ oz}$$

A "quarter-pounder" can also be called a "four-ouncer." ■

EXAMPLE 2 Complete: 15,360 lb = ______ T.

Since we are converting from a smaller unit to a larger unit, we use multiplying by 1.

$$15{,}360\text{ lb} = 15{,}360\text{ lb} \times \frac{1\text{ T}}{2000\text{ lb}} \qquad \text{Multiplying by 1}$$
$$= \frac{15{,}360}{2000}\text{ T} = 7.68\text{ T}$$

Do Margin Exercises 1 and 2. ▶

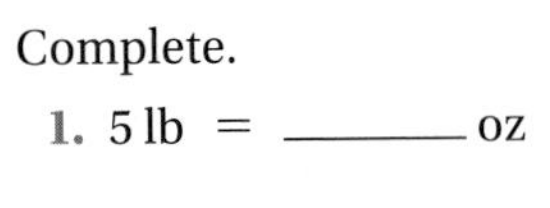

Complete.

1. 5 lb = ______ oz

2. 8640 lb = ______ T

Answers

Skill to Review:
1. 12 **2.** $\frac{31}{25}$, or 1.24

Margin Exercises:
1. 80 **2.** 4.32

METRIC UNITS OF MASS		
1 metric ton (t)	=	1000 kilograms (kg)
1 *kilo*gram (kg)	=	1000 grams (g)
1 *hecto*gram (hg)	=	100 grams (g)
1 *deka*gram (dag)	=	10 grams (g)
1 gram (g)		
1 *deci*gram (dg)	=	$\frac{1}{10}$ gram (g)
1 *centi*gram (cg)	=	$\frac{1}{100}$ gram (g)
1 *milli*gram (mg)	=	$\frac{1}{1000}$ gram (g)

b MASS: THE METRIC SYSTEM

The basic unit of mass is the **gram** (g), which is the mass of 1 cubic centimeter ($1\ \text{cm}^3$) of water. Since a cubic centimeter is small, a gram is a small unit of mass.

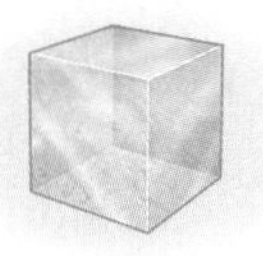
$1\ \text{g} = 1\ \text{cm}^3$ of water

$$1\ \text{g} = 1\ \text{gram} = \text{the mass of } 1\ \text{cm}^3 \text{ of water}$$

The metric units of mass are listed to the left. The prefixes are the same as those for length.

Thinking Metric

One gram is about the mass of 1 raisin or 1 package of artificial sweetener. Since 1 kg is about 2.2 lb, 1000 kg is about 2200 lb, or 1 metric ton (t), which is just a little more than 1 American ton (T), which is 2000 lb.

1 gram

1 kilogram of grapes

1 pound of grapes

Small masses, such as dosages of medicine and vitamins, may be measured in milligrams (mg). The gram (g) is used for objects ordinarily measured in ounces, such as the mass of a letter, a piece of candy, or a coin.

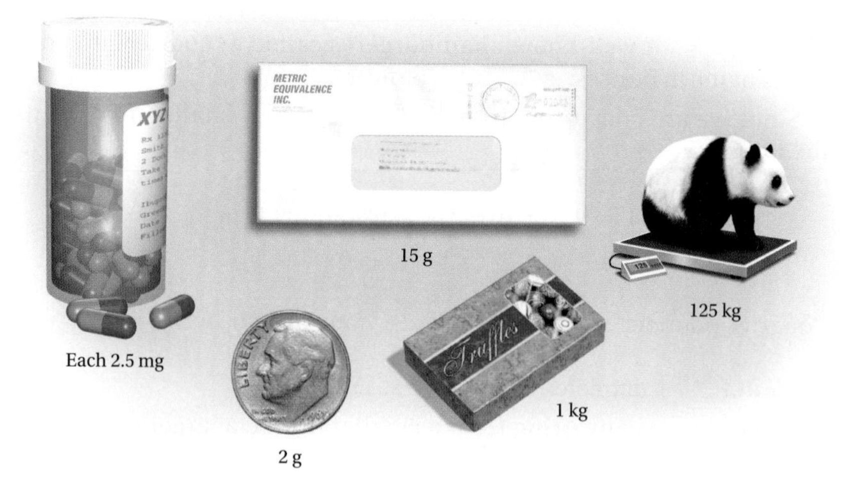

Each 2.5 mg

15 g

2 g

1 kg

125 kg

The kilogram (kg) is used for larger food packages and for body masses. The metric ton (t) is used for very large masses, such as the mass of an automobile, a truckload of gravel, or an airplane.

Do Exercises 3–7.

Complete with mg, g, kg, or t.

3. A laptop computer has a mass of 6 ______.

4. Eric has a body mass of 85.4 ______.

5. This is a 3-______ vitamin.

6. A pen has a mass of 12 ______.

7. A sport utility vehicle has a mass of 3 ______.

Answers

3. kg 4. kg 5. mg 6. g 7. t

Changing Units Mentally

As before, changing from one metric unit of mass to another requires only the movement of a decimal point. We use this table.

1000 g	100 g	10 g	1 g	0.1 g	0.01 g	0.001 g
1 kg	1 hg	1 dag	1 g	1 dg	1 cg	1 mg

EXAMPLE 3 Complete: 8 kg = ______ g.

Think: To go from kg to g in the table is a move of three places to the right. Thus we move the decimal point three places to the right.

1000 g	100 g	10 g	1 g	0.1 g	0.01 g	0.001 g
1 kg	1 hg	1 dag	1 g	1 dg	1 cg	1 mg

3 places to the right

8.0 8.000. 8 kg = 8000 g

EXAMPLE 4 Complete: 4235 g = ______ kg.

Think: To go from g to kg in the table is a move of three places to the left. Thus we move the decimal point three places to the left.

1000 g	100 g	10 g	1 g	0.1 g	0.01 g	0.001 g
1 kg	1 hg	1 dag	1 g	1 dg	1 cg	1 mg

3 places to the left

4235.0 4.235.0 4235 g = 4.235 kg

Do Exercises 8 and 9. ▶

Complete.

8. 6.2 kg = ______ g

9. 304.8 cg = ______ g

EXAMPLE 5 Complete: 6.98 cg = ______ mg.

Think: To go from cg to mg is a move of one place to the right. Thus we move the decimal point one place to the right.

1000 g	100 g	10 g	1 g	0.1 g	0.01 g	0.001 g
1 kg	1 hg	1 dag	1 g	1 dg	1 cg	1 mg

1 place to the right

6.98 6.9.8 6.98 cg = 69.8 mg

Answers

8. 6200 **9.** 3.048

The most commonly used metric units of mass are kg, g, cg, and mg. We have purposely used those more than the others in the exercises.

EXAMPLE 6 Complete: 89.21 mg = ______ g.

Think: To go from mg to g is a move of three places to the left. Thus we move the decimal point three places to the left.

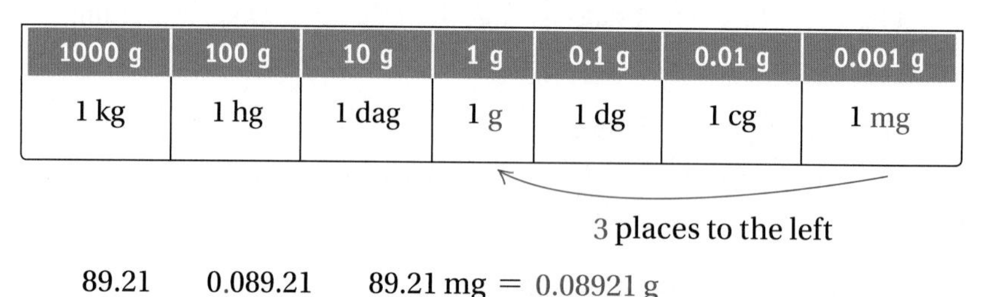

1000 g	100 g	10 g	1 g	0.1 g	0.01 g	0.001 g
1 kg	1 hg	1 dag	1 g	1 dg	1 cg	1 mg

3 places to the left

89.21   0.089.21   89.21 mg = 0.08921 g

◀ Do Exercises 10–12.

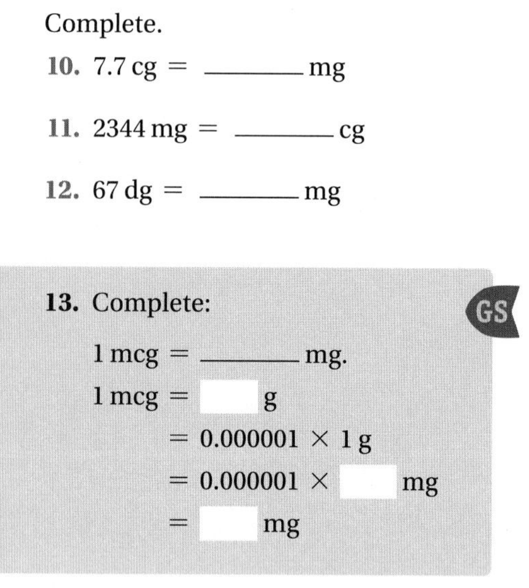

Complete.

10. 7.7 cg = ______ mg

11. 2344 mg = ______ cg

12. 67 dg = ______ mg

13. Complete: GS

1 mcg = ______ mg.

1 mcg = ☐ g

= 0.000001 × 1 g

= 0.000001 × ☐ mg

= ☐ mg

C MEDICAL APPLICATIONS

Another metric unit that is used in medicine is the microgram (mcg). It is defined as follows.

MICROGRAM

$$1 \text{ microgram} = 1 \text{ mcg} = \frac{1}{1{,}000{,}000} \text{ g} = 0.000001 \text{ g}$$

$$1{,}000{,}000 \text{ mcg} = 1 \text{ g}$$

EXAMPLE 7 Complete: 1 mg = ______ mcg.

We convert to grams and then to micrograms:

1 mg = 0.001 g
= 0.001 × 1 g
= 0.001 × 1,000,000 mcg   Substituting 1,000,000 mcg for 1 g
= 1000 mcg.

◀ Do Exercise 13.

14. *Medical Dosage.* A physician prescribes 500 mcg of alprazolam, an antianxiety medication. How many milligrams is this dosage?
Source: Steven R. Smith, M.D.

EXAMPLE 8 *Medical Dosage.* Nitroglycerin sublingual tablets come in 0.4-mg tablets. How many micrograms are in each tablet?
Source: Steven R. Smith, M.D.

We are to complete: 0.4 mg = ______ mcg. Thus,

0.4 mg = 0.4 × 1 mg
= 0.4 × 1000 mcg   From Example 7, substituting 1000 mcg for 1 mg
= 400 mcg.

We can also do this problem in a manner similar to Example 7.

◀ Do Exercise 14.

Answers

10. 77 **11.** 234.4 **12.** 6700 **13.** 0.001
14. 0.5 mg

Guided Solution:
13. 0.000001, 1000, 0.001

9.4 Exercise Set

For Extra Help

MyMathLab® MathXL®  PRACTICE WATCH READ REVIEW

Reading Check

Determine whether each statement is true or false.

RC1. ________ $400 \text{ g} > 40 \text{ dg}$

RC2. ________ $5 \text{ hg} < 400 \text{ g}$

RC3. ________ $0.5 \text{ kg} = 500 \text{ g}$

RC4. ________ $48 \text{ oz} = 4 \text{ lb}$

RC5. ________ $7500 \text{ lb} < 3.5 \text{ T}$

RC6. ________ $800 \text{ cg} > 6 \text{ g}$

a Complete.

1. 1 T = ________ lb

2. 1 lb = ________ oz

3. 6000 lb = ________ T

4. 8 T = ________ lb

5. 4 lb = ________ oz

6. 10 lb = ________ oz

7. 6.32 T = ________ lb

8. 8.07 T = ________ lb

9. 3200 oz = ________ T

10. 6400 oz = ________ T

11. 80 oz = ________ lb

12. 960 oz = ________ lb

13. *Pecans.* In 2011, U.S. farmers produced 269,700,000 pounds of pecans. How many tons of pecans were produced?

Source: National Agricultural Statistics Service, U.S. Department of Agriculture

14. *Peaches.* In 2011, U.S. farmers produced 978,260 tons of peaches. How many pounds of peaches were produced?

Source: National Agricultural Statistics Service, U.S. Department of Agriculture

b Complete.

15. 1 kg = ________ g

16. 1 hg = ________ g

17. 1 dag = ________ g

18. 1 dg = ________ g

19. 1 cg = ________ g

20. 1 mg = ________ g

21. 1 g = _______ mg

22. 1 g = _______ cg

23. 1 g = _______ dg

24. 25 kg = _______ g

25. 234 kg = _______ g

26. 9403 g = _______ kg

27. 5200 g = _______ kg

28. 1.506 kg = _______ g

29. 67 hg = _______ kg

30. 45 cg = _______ g

31. 0.502 dg = _______ g

32. 0.0025 cg = _______ mg

33. 8492 g = _______ kg

34. 9466 g = _______ kg

35. 585 mg = _______ cg

36. 96.1 mg = _______ cg

37. 8 kg = _______ cg

38. 0.06 kg = _______ mg

39. 1 t = _______ kg

40. 2 t = _______ kg

41. 3.4 cg = _______ dag

42. 115 mg = _______ g

43. 60.3 kg = _______ t

44. 15.68 kg = _______ t

C Complete.

45. 1 mg = _______ mcg

46. 1 mcg = _______ mg

47. 325 mcg = _______ mg

48. 0.45 mg = _______ mcg

49. 210.6 mg = _______ mcg

50. 8000 mcg = _______ mg

51. 4.9 mcg = _______ mg

52. 0.075 mg = _______ mcg

Medical Dosage. Solve each of the following. (None of these medications should be taken without consulting your own physician.)

Source: Steven R. Smith, M.D.

53. Digoxin is a medication used to treat heart problems. A physician orders 0.125 mg of digoxin to be taken once daily. How many micrograms of digoxin are there in the daily dosage?

54. Digoxin is a medication used to treat heart problems. A physician orders 0.25 mg of digoxin to be taken once a day. How many micrograms of digoxin are there in the daily dosage?

55. Triazolam is a medication used for the short-term treatment of insomnia. A physician advises her patient to take one of the 0.125-mg tablets each night for 7 nights. How many milligrams of triazolam will the patient have ingested over that 7-day period? How many micrograms?

56. Clonidine is a medication used to treat high blood pressure. The usual starting dose of clonidine is one 0.1-mg tablet twice a day. If a patient is started on this dose by his physician, how many total milligrams of clonidine will the patient have taken before he returns to see his physician 14 days later? How many micrograms?

57. Cephalexin is an antibiotic that frequently is prescribed in a 500-mg tablet form. A physician prescribes 2 g of cephalexin per day for a patient with a skin sore. How many 500-mg tablets would have to be taken in order to achieve this daily dosage?

58. Quinidine gluconate is a liquid mixture, part medicine and part water, that is administered intravenously. There are 80 mg of quinidine gluconate in each cubic centimeter (cc) of the liquid mixture. A physician orders 900 mg of quinidine gluconate to be administered daily to a patient with malaria. How much of the solution would have to be administered in order to achieve the recommended daily dosage?

59. Amoxicillin is an antibiotic commonly prescribed for children as a liquid suspension composed of part amoxicillin and part water. In one formulation of amoxicillin suspension, there are 250 mg of amoxicillin in 5 cc of the liquid suspension. A physician prescribes 400 mg per day for a 2-year-old child with an ear infection. How much of the amoxicillin liquid suspension would the child's parent need to administer in order to achieve the recommended daily dosage of amoxicillin?

60. Albuterol is a medication used for the treatment of asthma. It comes in an inhaler that contains 17 mg of albuterol mixed with a liquid. One actuation (inhalation) from the mouthpiece delivers a 90-mcg dose of albuterol.

a) A physician orders 2 inhalations 4 times per day. How many micrograms of albuterol does the patient inhale per day?

b) How many actuations/inhalations are contained in one inhaler?

c) Danielle is leaving for 4 months of college and wants to take enough albuterol to last for that time. Her physician has prescribed 2 inhalations 4 times per day. How many inhalers will Danielle need to take with her for the 4-month period?

Skill Maintenance

Convert to fraction notation. [7.2b]

61. 35%

62. 99%

63. 85.5%

64. 34.2%

65. $37\frac{1}{2}\%$

66. $66.\overline{6}\%$

67. $83.\overline{3}\%$

68. $16\frac{2}{3}\%$

Solve.

69. A country has a population that is increasing by 4% each year. This year the population is 180,000. What will it be next year? [7.5b]

70. A college has a student body of 1850 students. Of these, 17.5% are seniors. How many students are seniors? [7.5a]

71. A state charges a meals tax of $4\frac{1}{2}\%$. What is the meals tax charged on a dinner party costing $540? [7.6a]

72. The price of a microwave oven was reduced from $350 to $308. Find the percent decrease in price. [7.5b]

Synthesis

73. A case of boxes of cereal weighs $14\frac{7}{8}$ lb. Each individual box of cereal weighs 17 oz. How many boxes of cereal are in the case?

74. At $1.89 a dozen, the cost of eggs is $1.26 per pound. How much does an egg weigh?

75. *Tanzanite.* Tanzanite is a gemstone discovered in 1967 in the East African state of Tanzania, the only place in the world where it has been found. Rarer than diamond, tanzanite ranges in color from ultramarine blue to light violet-blue. The world's biggest piece of tanzanite was unearthed near Tanzania's Mount Kilimanjaro. The gemstone weighed 16,839 carats and was the size of a brick. A **carat** (also spelled **karat**) is a unit of weight for precious stones; 1 carat = 200 mg.

Sources: International Colored Gemstone Association; "World's Biggest Tanzanite Gem Found Near Kilimanjaro," Bloomberg.com, Aug. 3, 2005

a) How many grams is this record tanzanite gemstone?

b) 🖩 Given that 1 lb = 453.6 g, how many ounces does this gemstone weigh?

Capacity; Medical Applications

9.5

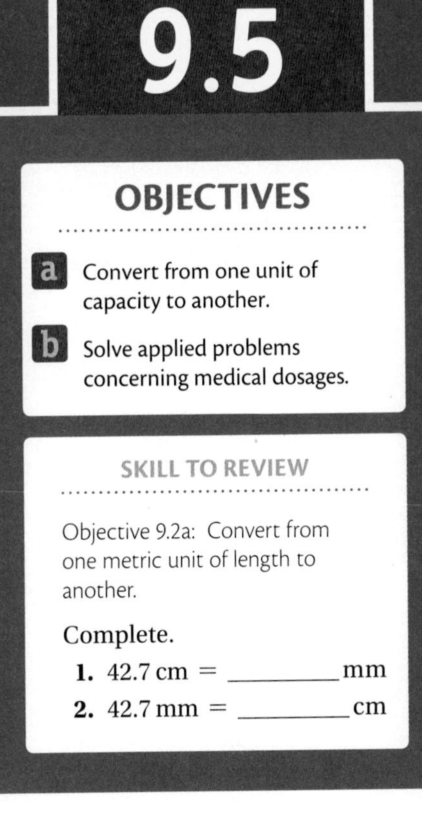

OBJECTIVES

a Convert from one unit of capacity to another.

b Solve applied problems concerning medical dosages.

SKILL TO REVIEW

Objective 9.2a: Convert from one metric unit of length to another.

Complete.

1. 42.7 cm = ________ mm

2. 42.7 mm = ________ cm

a CAPACITY

American Units

To answer a question like "How much soda is in the can?" we need measures of **capacity**. American units of capacity are fluid ounces, cups, pints, quarts, and gallons. These units are related as follows.

AMERICAN UNITS OF CAPACITY

1 gallon (gal) = 4 quarts (qt)	1 pt = 2 cups = 16 fluid ounces (fl oz)
1 qt = 2 pints (pt)	1 cup = 8 fluid oz

Fluid ounces, abbreviated fl oz, are often referred to as ounces, or oz.

EXAMPLE 1 Complete: 9 gal = ______ oz.

Since we are converting from a *larger* unit to a *smaller* unit, we use substitution:

$$\begin{aligned} 9\text{ gal} &= 9 \cdot 1\text{ gal} = 9 \cdot 4\text{ qt} && \text{Substituting 4 qt for 1 gal} \\ &= 9 \cdot 4 \cdot 1\text{ qt} = 9 \cdot 4 \cdot 2\text{ pt} && \text{Substituting 2 pt for 1 qt} \\ &= 9 \cdot 4 \cdot 2 \cdot 1\text{ pt} = 9 \cdot 4 \cdot 2 \cdot 16\text{ oz} && \text{Substituting 16 oz for 1 pt} \\ &= 1152\text{ oz.} \end{aligned}$$

EXAMPLE 2 Complete: 24 qt = ______ gal.

Since we are converting from a *smaller* unit to a *larger* unit, we multiply by 1 using 1 gal in the numerator and 4 qt in the denominator:

$$24\text{ qt} = 24\text{ qt} \cdot \frac{1\text{ gal}}{4\text{ qt}} = \frac{24}{4} \cdot 1\text{ gal} = 6\text{ gal.}$$

Do Margin Exercises 1 and 2. ▶

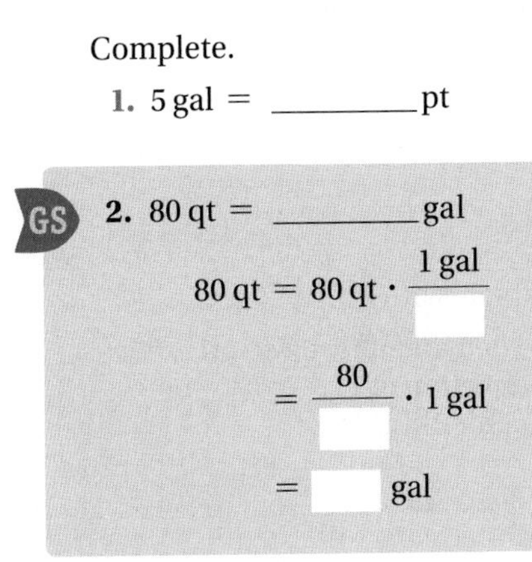

Complete.

1. 5 gal = ________ pt

GS **2.** 80 qt = ________ gal

$$80\text{ qt} = 80\text{ qt} \cdot \frac{1\text{ gal}}{\square}$$

$$= \frac{80}{\square} \cdot 1\text{ gal}$$

$$= \square\text{ gal}$$

Metric Units

One unit of capacity in the metric system is a **liter**. A liter is just a bit more than a quart. It is defined as follows.

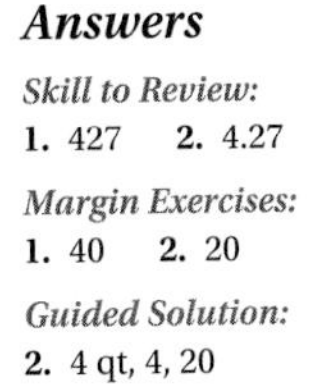

Answers

Skill to Review:
1. 427 **2.** 4.27

Margin Exercises:
1. 40 **2.** 20

Guided Solution:
2. 4 qt, 4, 20

METRIC UNITS OF CAPACITY

1 liter (L) $=$ 1000 cubic centimeters (1000 cm^3)

The script letter ℓ is also used for "liter."

The metric prefixes are also used with liters. The most common is **milli-**. The milliliter (mL) is, then, $\frac{1}{1000}$ liter. Thus,

$$1\text{ L} = 1000\text{ mL} = 1000\text{ cm}^3;$$
$$0.001\text{ L} = 1\text{ mL} = 1\text{ cm}^3.$$

Although the other metric prefixes are rarely used for capacity, we display them in the following table as we did for linear measure.

1000 L	100 L	10 L	1 L	0.1 L	0.01 L	0.001 L
1 kL	1 hL	1 daL	1 L	1 dL	1 cL	1 mL (cc)

A preferred unit for drug dosage is the milliliter (mL) or the cubic centimeter (cm^3). The notation "cc" is also used for cubic centimeter, especially in medicine. The milliliter and the cubic centimeter represent the same measure of capacity. A milliliter is about $\frac{1}{5}$ of a teaspoon.

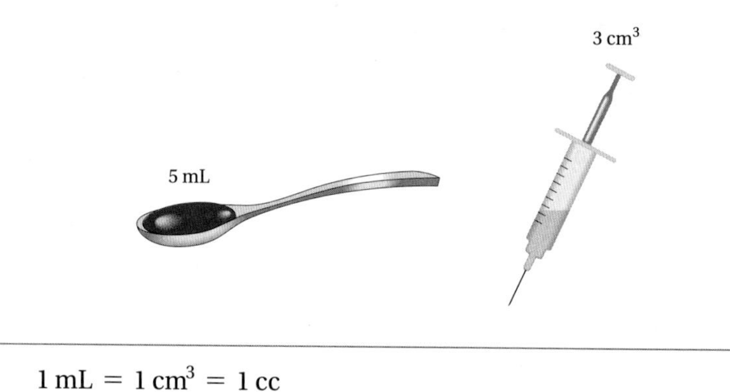

$$1\text{ mL} = 1\text{ cm}^3 = 1\text{ cc}$$

Volumes for which quarts and gallons are used are expressed in liters. Large volumes in business and industry are expressed using measures of cubic meters (m^3).

◀ Do Exercises 3–6.

Complete with mL or L.

3. The patient received an injection of 2 ________ of penicillin.

4. There are 250 ________ in a coffee cup.

5. The gas tank holds 80 ________.

6. Bring home 8 ________ of milk.

EXAMPLE 3 Complete: 4.5 L $=$ ________ mL.

$4.5\text{ L} = 4.5 \times 1\text{ L} = 4.5 \times 1000\text{ mL}$ Substituting 1000 mL for 1 L

$= 4500\text{ mL}$

1000 L	100 L	10 L	1 L	0.1 L	0.01 L	0.001 L
1 kL	1 hL	1 daL	1 L	1 dL	1 cL	1 mL (cc)

3 places to the right

Answers

3. mL 4. mL 5. L 6. L

EXAMPLE 4 Complete: 280 mL = _______ L.

$$280 \text{ mL} = 280 \times 1 \text{ mL}$$
$$= 280 \times 0.001 \text{ L} \quad \text{Substituting 0.001 L for 1 mL}$$
$$= 0.28 \text{ L}$$

1000 L	100 L	10 L	1 L	0.1 L	0.01 L	0.001 L
1 kL	1 hL	1 daL	1 L	1 dL	1 cL	1 mL (cc)

3 places to the left

We do find metric units of capacity in frequent use in the United States—for example, in sizes of soda bottles and automobile engines.

Do Exercises 7 and 8. ▶

Complete.

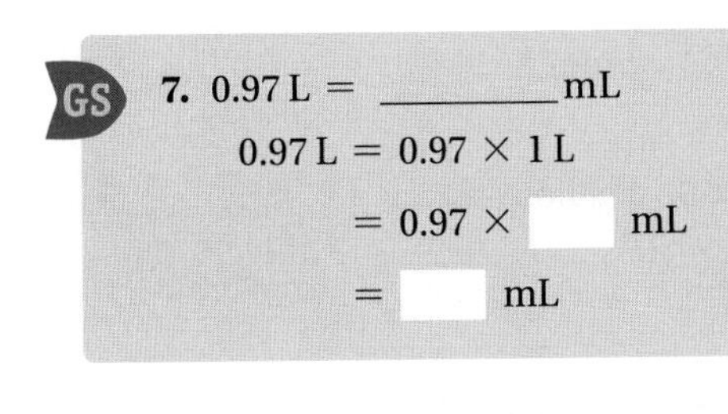

8. 8990 mL = _______ L

b MEDICAL APPLICATIONS

The metric system is used extensively in medicine.

EXAMPLE 5 *Medical Dosage.* A physician orders 3.5 L of 5% dextrose in water (abbreviated as D5W) to be administered over a 24-hr period. How many milliliters were ordered?

We convert 3.5 L to milliliters:

$$3.5 \text{ L} = 3.5 \times 1 \text{ L} = 3.5 \times 1000 \text{ mL} = 3500 \text{ mL}.$$

The physician ordered 3500 mL of D5W.

Do Exercise 9. ▶

9. *Medical Dosage.* A physician orders 2400 mL of 0.9% saline solution to be administered intravenously over a 24-hr period. How many liters were ordered?

EXAMPLE 6 *Medical Dosage.* Liquids at a pharmacy are often labeled in liters or milliliters. This means that if a physician's prescription is given in ounces, it must be converted. For conversion, a pharmacist knows that 1 fluid oz ≈ 29.57 mL.* A prescription calls for 3 fluid oz of theophylline. How many milliliters does the prescription call for?

We convert as follows:

$$3 \text{ oz} = 3 \times 1 \text{ oz} \approx 3 \times 29.57 \text{ mL} = 88.71 \text{ mL}.$$

The prescription calls for 88.71 mL of theophylline.

Do Exercise 10. ▶

10. *Medical Dosage.* A prescription calls for 2 oz of theophylline.

a) How many milliliters does the prescription call for?

b) How many liters does the prescription call for?

*In practice, most pharmacists use 30 mL as an approximation to 1 oz.

Answers

7. 970 8. 8.99 9. 2.4 L
10. (a) About 59.14 mL; (b) about 0.059 L

Guided Solution:
7. 1000, 970

9.5 Exercise Set

For Extra Help

MyMathLab® MathXL® PRACTICE WATCH READ REVIEW

Reading Check

Complete each sentence using one of the two words listed below the blank.

RC1. 1 pint = 2 ________ (cups/quarts)

RC2. 1 quart = 2 ________ (cups/pints)

RC3. 1 gallon = 8 ________ (cups/pints)

RC4. 1 cup = 8 ________ (fluid ounces/pints)

a Complete.

1. 1 L = ________ mL = ________ cm^3

2. ________ L = 1 mL = ________ cm^3

3. 87 L = ________ mL

4. 806 L = ________ mL

5. 49 mL = ________ L

6. 19 mL = ________ L

7. 0.401 mL = ________ L

8. 0.816 mL = ________ L

9. 78.1 L = ________ cm^3

10. 99.6 L = ________ cm^3

11. 10 qt = ________ oz

12. 9.6 oz = ________ pt

13. 20 cups = ________ pt

14. 1 gal = ________ oz

15. 8 gal = ________ qt

16. 1 gal = ________ cups

17. 5 gal = ________ qt

18. 11 gal = ________ qt

19. 56 qt = ________ gal

20. 84 qt = ________ gal

21. 11 gal = ________ pt

22. 5 gal = ________ pt

Complete.

	Object	Gallons (gal)	Quarts (qt)	Pints (pt)	Cups	Ounces (oz)
23.	12-can package of 12-oz sodas					144
24.	Large container of milk			8		
25.	Full tank of gasoline	16				
26.	Dove shampoo					12
27.	Downy fabric softener					51
28.	Williams Lectric Shave					7

Complete.

	Object	Liters (L)	Milliliters (mL)	Cubic Centimeters (cc)	Cubic Centimeters (cm^3)
29.	2-L bottle of soda	2			
30.	Heinz vinegar		3755		
31.	Full tank of gasoline in Europe	64			
32.	Williams Lectric Shave				207
33.	Dove shampoo			355	
34.	Newman's Own salad dressing		473		

b *Medical Dosage.* Solve each of the following.

Source: Steven R. Smith, M.D.

35. An emergency-room physician orders 2.0 L of Ringer's lactate to be administered over 2 hr for a patient in shock. How many milliliters is this?

36. An emergency-room physician orders 2.5 L of 0.9% saline solution over 4 hr for a patient suffering from dehydration. How many milliliters is this?

37. A physician orders 320 mL of 5% dextrose in water (D5W) solution to be administered intravenously over 4 hr. How many liters of D5W is this?

38. A physician orders 40 mL of 5% dextrose in water (D5W) solution to be administered intravenously over 2 hr to an elderly patient. How many liters of D5W is this?

39. A physician orders 0.5 oz of magnesia and alumina oral suspension antacid 4 times per day for a patient with indigestion. How many milliliters of the antacid is the patient to ingest in a day?

40. A physician orders 0.25 oz of magnesia and alumina oral suspension antacid 3 times per day for a child with upper abdominal discomfort. How many milliliters of the antacid is the child to ingest in a day?

41. A physician orders 0.5 L of normal saline solution. How many milliliters are ordered?

42. A physician has ordered that his patient receive 60 mL per hour of normal saline solution intravenously. How many liters of the saline solution is the patient to receive in a 24-hr period?

43. A physician wants her patient to receive 3.0 L of normal saline intravenously over a 24-hr period. How many milliliters per hour must the nurse administer?

44. A physician tells a patient to purchase 0.5 L of hydrogen peroxide. Commercially, hydrogen peroxide is found on the shelf in bottles that hold 4 oz, 8 oz, and 16 oz. Which bottle has a capacity closest to 0.5 L?

Medical Dosage. Because patients do not always have a working knowledge of the metric system, physicians often prescribe dosages in teaspoons (t or tsp) and tablespoons (T or Tbsp). The units are related to the metric system and to each other as follows:

$5 \text{ mL} \approx 1 \text{ tsp}, \quad 3 \text{ tsp} = 1 \text{ T}.$

Complete.

45. 45 mL = _____ tsp

46. 3 T = _____ tsp

47. 1 mL = _____ tsp

48. 18.5 mL = _____ tsp

49. 2 T = _____ tsp

50. 8.5 tsp = _____ T

51. 1 T = _____ mL

52. 18.5 mL = _____ T

Skill Maintenance

Convert to percent notation. [7.1b], [7.2a]

53. 0.452

54. 0.999

55. $\frac{1}{3}$

56. $\frac{2}{3}$

57. $\frac{11}{20}$

58. $\frac{21}{20}$

59. $\frac{22}{25}$

60. $\frac{2}{25}$

61. *Lumber Consumption.* Of the 39.6 billion board feet of lumber consumed in 2010 in the United States, 33.1 billion board feet were softwood and 6.5 billion board feet were hardwood. What is the ratio of hardwood board feet to total amount of board feet? What is the ratio of softwood board feet to hardwood board feet? [6.1a]

Source: U.S. Forest Service

62. *Cigarette Exports.* In 2000, the United States exported 148 billion cigarettes. That number decreased to 19 billion in 2012. What was the percent decrease? [7.5b]

Source: Tobacco Merchants Association

Synthesis

63. *Wasting Water.* Many people leave the water running while they are brushing their teeth. Suppose that one person wastes 32 oz of water in this way each day. How much water, in gallons, is wasted by one person in a week? in a month (30 days)? in a year (365 days)? Assuming each of the 314 million people in the United States wastes water in this way, estimate how much water is wasted in the United States in a day; in a year.

64. *Bees and Honey.* The average bee produces only $\frac{1}{8}$ teaspoon of honey in its lifetime. It takes 60,000 honeybees to produce 100 lb of honey. How much does a teaspoon of honey weigh? Express the answer in ounces.

65. *Cost of Gasoline.* Suppose that premium gasoline is selling for about \$3.79/gal. Using the fact that 1 L = 1.057 qt, determine the price of the gasoline in dollars per liter.

Time and Temperature

9.6

OBJECTIVES

a Convert from one unit of time to another.

b Convert between Celsius and Fahrenheit temperatures using the formulas

$$F = \frac{9}{5} \cdot C + 32$$

and

$$C = \frac{5}{9} \cdot (F - 32).$$

SKILL TO REVIEW

Objective 4.7a: Simplify expressions using the rules for order of operations.

Simplify.

1. $\frac{9}{5} \cdot 10 + 32$
2. $\frac{5}{9} \cdot (100 - 32)$

a TIME

A table of units of time is shown below. The metric system uses "h" for hour and "s" for second, but we will use the more familiar "hr" and "sec."

UNITS OF TIME

1 day = 24 hours (hr)	1 year (yr) = $365\frac{1}{4}$ days
1 hr = 60 minutes (min)	
1 min = 60 seconds (sec)	1 week (wk) = 7 days

The earth revolves completely around the sun in $365\frac{1}{4}$ days. Since we cannot have $\frac{1}{4}$ day on the calendar, we give each year 365 days and every fourth year 366 days (a leap year), unless it is a year at the beginning of a century not divisible by 400.

EXAMPLE 1 Complete: 1 hr = ______ sec.

$$\begin{aligned} 1 \text{ hr} &= 60 \text{ min} \\ &= 60 \cdot 1 \text{ min} \\ &= 60 \cdot 60 \text{ sec} \qquad \text{Substituting 60 sec for 1 min} \\ &= 3600 \text{ sec} \end{aligned}$$

EXAMPLE 2 Complete: 5 years = ______ days.

$$\begin{aligned} 5 \text{ years} &= 5 \cdot 1 \text{ year} \\ &= 5 \cdot 365\tfrac{1}{4} \text{ days} \qquad \text{Substituting } 365\tfrac{1}{4} \text{ days for 1 year} \\ &= 5 \cdot \frac{1461}{4} \text{ days} \\ &= \frac{7305}{4} \text{ days} \\ &= 1826\frac{1}{4} \text{ days} \end{aligned}$$

EXAMPLE 3 Complete: 4320 min = ______ days.

$$4320 \text{ min} = 4320 \cancel{\text{min}} \cdot \frac{1 \cancel{\text{hr}}}{60 \cancel{\text{min}}} \cdot \frac{1 \text{ day}}{24 \cancel{\text{hr}}} = \frac{4320}{60 \cdot 24} \text{ days} = 3 \text{ days}$$

Do Margin Exercises 1–4. ▶

Complete.

1. 2 hr = ______ min
2. 4 years = ______ days
3. 1 day = ______ min

GS 4. 168 hr = ______ wk

$$\begin{aligned} 168 \text{ hr} &= 168 \text{ hr} \times \frac{1 \text{ day}}{\square} \times \frac{1 \text{ wk}}{\square} \\ &= \frac{\square}{24 \cdot 7} \text{ wk} \\ &= \square \text{ wk} \end{aligned}$$

Answers

Skill to Review:
1. 50 **2.** $\frac{340}{9}$, or $37\frac{7}{9}$, or $37.\overline{7}$

Margin Exercises:
1. 120 **2.** 1461 **3.** 1440 **4.** 1

Guided Solution:
4. 24 hr, 7 days, 168, 1

b TEMPERATURE

Below are two temperature scales: **Fahrenheit** for American measure and **Celsius** for metric measure.

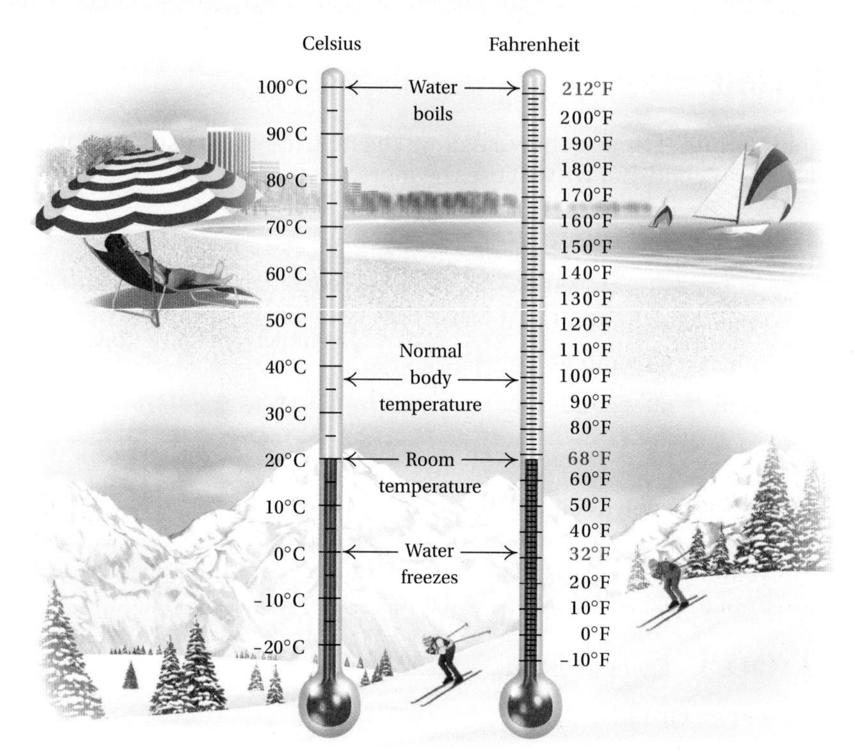

By laying a straight edge horizontally between the scales, we can make an approximate conversion from one measure of temperature to the other and get an idea of how the temperature scales compare.

EXAMPLES Convert to Celsius using the scales shown above. Approximate to the nearest ten degrees.

4. 212°F (Boiling point of water)	100°C	This is exact.
5. 32°F (Freezing point of water)	0°C	This is exact.
6. 105°F	40°C	This is approximate.

◀ **Do Exercises 5–7.**

EXAMPLES Make an approximate conversion to Fahrenheit using the scales shown above.

7. 44°C (Hot bath)	110°F	This is approximate.
8. 20°C (Room temperature)	68°F	This is exact.
9. 83°C	180°F	This is approximate.

◀ **Do Exercises 8–10.**

Convert to Celsius. Approximate to the nearest ten degrees.

5. 180°F (Brewing coffee)

6. 25°F (Cold day)

7. −10°F (Miserably cold day)

Convert to Fahrenheit. Approximate to the nearest ten degrees.

8. 25°C (Warm day at the beach)

9. 40°C (Temperature of a patient with a high fever)

10. 10°C (Cold bath)

Answers

5. 80°C **6.** 0°C **7.** −20°C **8.** 80°F
9. 100°F **10.** 50°F

The following formula allows us to make exact conversions from Celsius to Fahrenheit.

CELSIUS TO FAHRENHEIT

$$F = \frac{9}{5} \cdot C + 32, \text{ or } F = 1.8 \cdot C + 32$$

$\left(\text{Multiply the Celsius temperature by } \frac{9}{5}, \text{ or } 1.8, \text{ and add } 32.\right)$

EXAMPLES Convert to Fahrenheit.

10. 0°C (Freezing point of water)

$$F = \frac{9}{5} \cdot C + 32 = \frac{9}{5} \cdot 0 + 32 = 0 + 32 = 32$$

Thus, 0°C = 32°F.

11. 37°C (Normal body temperature)

$$F = 1.8 \cdot C + 32 = 1.8 \cdot 37 + 32 = 66.6 + 32 = 98.6$$

Thus, 37°C = 98.6°F.

Check the answers to Examples 10 and 11 using the scales on p. 534.

Do Exercises 11 and 12. ▶

The following formula allows us to make exact conversions from Fahrenheit to Celsius.

FAHRENHEIT TO CELSIUS

$$C = \frac{5}{9} \cdot (F - 32), \text{ or } C = \frac{F - 32}{1.8}$$

$\left(\text{Subtract 32 from the Fahrenheit temperature and multiply by } \frac{5}{9} \text{ or divide by } 1.8.\right)$

EXAMPLES Convert to Celsius.

12. 212°F (Boiling point of water)

$$C = \frac{5}{9} \cdot (F - 32)$$
$$= \frac{5}{9} \cdot (212 - 32)$$
$$= \frac{5}{9} \cdot 180 = 100$$

Thus, 212°F = 100°C.

13. 77°F

$$C = \frac{F - 32}{1.8}$$
$$= \frac{77 - 32}{1.8}$$
$$= \frac{45}{1.8} = 25$$

Thus, 77°F = 25°C.

Check the answers to Examples 12 and 13 using the scales on p. 534.

Do Exercises 13 and 14. ▶

Convert to Fahrenheit.

11. 80°C

12. 35°C

CALCULATOR CORNER

Temperature Conversions Temperature conversions can be done quickly using a calculator. To convert 37°C to Fahrenheit, for example, we press [1] [.] [8] [×] [3] [7] [+] [3] [2] [=]. The calculator displays [98.6], so 37°C = 98.6°F. We can convert 212°F to Celsius by pressing [(] [2] [1] [2] [−] [3] [2] [)] [÷] [1] [.] [8] [=]. The display reads [100], so 212°F = 100°C. Note that we must use parentheses when converting from Fahrenheit to Celsius in order to get the correct result.

EXERCISES Use a calculator to convert each temperature to Fahrenheit.

1. 5°C

2. 50°C

Use a calculator to convert each temperature to Celsius.

3. 68°F

4. 113°F

Convert to Celsius.

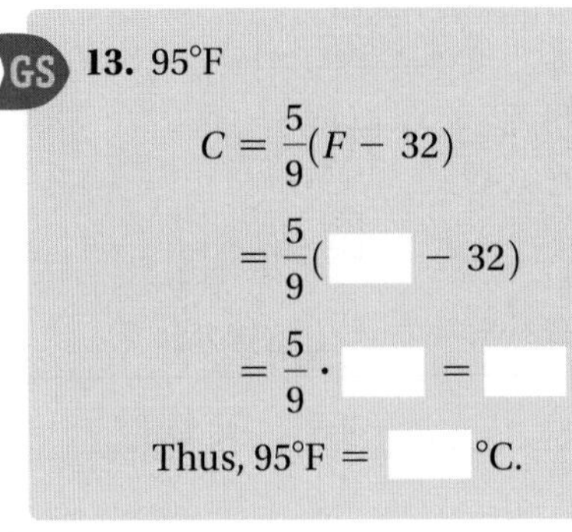

GS **13.** 95°F

$$C = \frac{5}{9}(F - 32)$$
$$= \frac{5}{9}(\square - 32)$$
$$= \frac{5}{9} \cdot \square = \square$$

Thus, 95°F = $\square$ °C.

14. 113°F

Answers

11. 176°F **12.** 95°F **13.** 35°C **14.** 45°C

Guided Solution:
13. 95, 63, 35; 35

9.6 Exercise Set

For Extra Help
MyMathLab®
MathXL® PRACTICE WATCH READ REVIEW

✓ Reading Check

Select the temperature in each pair that is higher.

RC1. $-10°C$, $-10°F$

RC2. 80°C, 200°F

RC3. 100°F, 30°C

RC4. 0°F, 0°C

a Complete.

1. 1 day = _______ hr

2. 1 hr = _______ min

3. 1 min = _______ sec

4. 1 wk = _______ days

5. 1 year = _______ days

6. 2 years = _______ days

7. 180 sec = _______ hr

8. 60 sec = _______ hr

9. 492 sec = _______ min (The amount of time it takes for the rays of the sun to reach the earth)

10. 18,000 sec = _______ hr

11. 156 hr = _______ days

12. 444 hr = _______ days

13. 645 min = _______ hr

14. 375 min = _______ hr

15. 2 wk = _______ hr

16. 4 hr = _______ sec

17. 756 hr = _______ wk

18. 166,320 min = _______ wk

19. 2922 wk = _______ years

20. 623 days = _______ wk

21. *Actual Time in a Day.* Although we round it to 24 hr, the actual length of a day is 23 hr, 56 min, and 4.2 sec. How many seconds are there in an actual day?

Source: *The Handy Geography Answer Book*

22. *Time Length.* What length of time is 86,400 sec? Is it 1 hr, 1 day, 1 week, or 1 month?

b Convert to Fahrenheit. Use the formula $F = \frac{9}{5} \cdot C + 32$ or $F = 1.8 \cdot C + 32$.

23. 25°C

24. 85°C

25. 40°C

26. 90°C

27. 86°C

28. 93°C

29. −20°C

30. −25°C

31. 2°C

32. 78°C

33. −24°C

34. −28°C

35. 3000°C
(The melting point of iron)

36. 1000°C
(The melting point of gold)

Convert to Celsius. Use the formula $C = \frac{5}{9} \cdot (F - 32)$ or $C = \frac{F - 32}{1.8}$.

37. 86°F

38. 59°F

39. −13°F

40. −4°F

41. 178°F

42. 195°F

43. 140°F

44. 107°F

45. 68°F

46. 50°F

47. 10°F

48. 0°F

49. 98.6°F
(Normal body temperature)

50. 104°F
(High-fever body temperature)

51. *Record High Temperature.* The record high temperature in Arizona through 2010 occurred on June 29, 1994. The record was 128°F at Lake Havasu City. Convert 128°F to Celsius.

Source: National Climatic Data Center, NESDIS, NOAA, U.S. Department of Commerce

52. *Record High Temperature.* The record high temperature in Utah through 2010 occurred on July 5, 1985. The record was 117°F at Saint George. Convert 117°F to Celsius.

Source: National Climatic Data Center, NESDIS, NOAA, U.S. Department of Commerce

53. *Highest Temperatures.* The highest temperature ever recorded in the world is 136°F in the desert of Libya in 1922. The highest temperature ever recorded in the United States is $56\frac{2}{3}$°C in California's Death Valley in 1913.

Source: *The Handy Geography Answer Book*

a) Convert each temperature to the other scale.
b) How much higher in degrees Fahrenheit was the world record than the U.S. record?

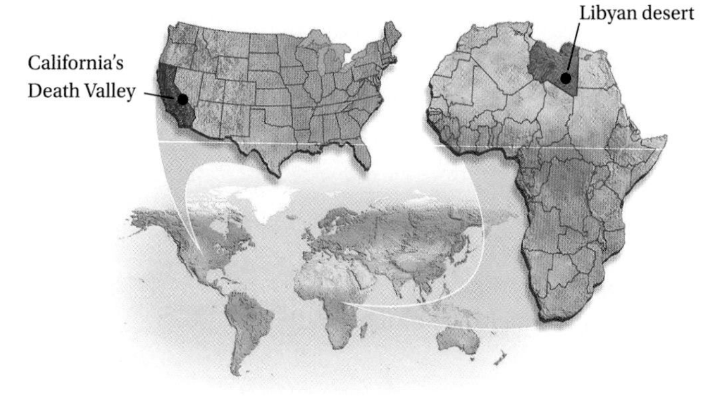

54. *Boiling Point and Altitude.* The boiling point of water actually changes with altitude. The boiling point is 212°F at sea level, but lowers about 1°F for every 500 ft that the altitude increases above sea level.

Sources: *The Handy Geography Answer Book; The New York Times Almanac*

a) What is the boiling point at an elevation of 1500 ft above sea level?
b) The elevation of Tucson is 2564 ft above sea level and that of Phoenix is 1117 ft. What is the boiling point in each city?
c) How much lower is the boiling point in Denver, whose elevation is 5280 ft, than in Tucson?
d) What is the boiling point at the top of Mt. McKinley in Alaska, the highest point in the United States, at 20,320 ft?

Skill Maintenance

55. Divide: $739 \div 13$. [1.5a]

56. Multiply: 8.03×0.001. [5.3a]

Solve.

57. $0.05 + x = 2.525$ [5.2d]

58. $\frac{5}{13}y = 130$ [3.7c]

59. $4 \cdot q = -16.2$ [5.4b]

Simplify. [4.7a]

60. $1 + \frac{1}{2} + \left(\frac{1}{2}\right)^2 + \left(\frac{1}{2}\right)^3$

61. $-\frac{2}{5} \div \frac{1}{5} \cdot \frac{3}{10}$

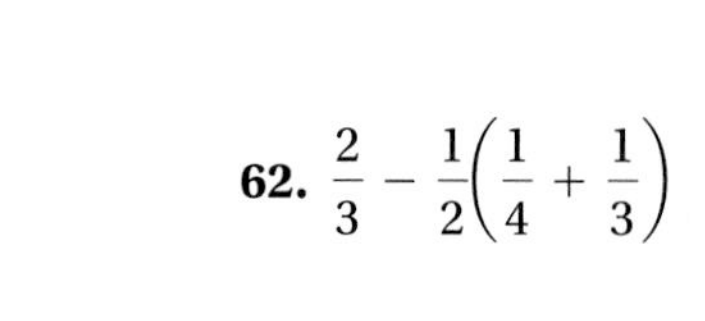

62. $\frac{2}{3} - \frac{1}{2}\left(\frac{1}{4} + \frac{1}{3}\right)$

Synthesis

63. Estimate the number of years in one million seconds.

64. Estimate the number of years in one billion seconds.

65. Estimate the number of years in one trillion seconds.

Complete.

66. $88\frac{\text{ft}}{\text{sec}} = \underline{\qquad\qquad}\frac{\text{mi}}{\text{hr}}$

67. $0.9\frac{\text{L}}{\text{hr}} = \underline{\qquad\qquad}\frac{\text{mL}}{\text{sec}}$

Converting Units of Area

9.7

a AMERICAN UNITS

Let's do some conversions from one American unit of area to another.

EXAMPLE 1 Complete: $1\text{ ft}^2 = \_\_\_\_\_\_\_\_\text{ in}^2$.

$$1\text{ ft}^2 = 1 \cdot (12\text{ in.})^2 \quad \text{Substituting 12 in. for 1 ft}$$
$$= 1 \cdot 12\text{ in.} \cdot 12\text{ in.} = 144\text{ in}^2$$

EXAMPLE 2 Complete: $8\text{ yd}^2 = \_\_\_\_\_\_\_\_\text{ ft}^2$.

$$8\text{ yd}^2 = 8 \cdot (3\text{ ft})^2 \quad \text{Substituting 3 ft for 1 yd}$$
$$= 8 \cdot 3\text{ ft} \cdot 3\text{ ft} = 8 \cdot 3 \cdot 3 \cdot \text{ft} \cdot \text{ft} = 72\text{ ft}^2$$

Do Margin Exercises 1–3. ▶

AMERICAN UNITS OF AREA

1 square yard (yd^2) = 9 square feet (ft^2)
1 square foot (ft^2) = 144 square inches (in^2)
1 square mile (mi^2) = 640 acres
1 acre = $43{,}560\text{ ft}^2$

EXAMPLE 3 Complete: $36\text{ ft}^2 = \_\_\_\_\_\_\_\_\text{ yd}^2$.

We are converting from "ft^2" to "yd^2." Thus we choose a symbol for 1 with yd^2 in the numerator and ft^2 in the denominator.

$$36\text{ ft}^2 = 36\text{ ft}^2 \times \frac{1\text{ yd}^2}{9\text{ ft}^2} \quad \text{Multiplying by 1 using } \frac{1\text{ yd}^2}{9\text{ ft}^2}$$
$$= \frac{36}{9} \times \frac{\text{ft}^2}{\text{ft}^2} \times 1\text{ yd}^2 = 4\text{ yd}^2$$

EXAMPLE 4 Complete: $7\text{ mi}^2 = \_\_\_\_\_\_\_\_$ acres.

$$7\text{ mi}^2 = 7 \cdot 1\text{ mi}^2$$
$$= 7 \cdot 640\text{ acres} \quad \text{Substituting 640 acres for 1 mi}^2$$
$$= 4480\text{ acres}$$

Do Exercises 4 and 5. ▶

b METRIC UNITS

Let's now convert from one metric unit of area to another.

EXAMPLE 5 Complete: $1\text{ km}^2 = \_\_\_\_\_\_\_\_\text{ m}^2$.

$$1\text{ km}^2 = 1 \cdot (1000\text{ m})^2 \quad \text{Substituting 1000 m for 1 km}$$
$$= 1 \cdot 1000\text{ m} \cdot 1000\text{ m} = 1{,}000{,}000\text{ m}^2$$

OBJECTIVES

a Convert from one American unit of area to another.

b Convert from one metric unit of area to another.

SKILL TO REVIEW

Objective 1.9b: Evaluate exponential notation.

Evaluate.

1. 9^2 **2.** 10^3

Complete.

1. $1\text{ yd}^2 = \_\_\_\_\_\_\_\_\text{ ft}^2$

2. $5\text{ yd}^2 = \_\_\_\_\_\_\_\_\text{ ft}^2$

GS **3.** $20\text{ ft}^2 = \_\_\_\_\_\_\_\_\text{ in}^2$

$$20\text{ ft}^2 = 20 \times 1\text{ ft}^2$$
$$= 20 \times \square\text{ in}^2$$
$$= \square\text{ in}^2$$

Complete.

GS **4.** $360\text{ in}^2 = \_\_\_\_\_\_\_\_\text{ ft}^2$

$$360\text{ in}^2 = 360\text{ in}^2 \times \frac{1\text{ ft}^2}{\square}$$
$$= \frac{360}{\square} \times \frac{\text{in}^2}{\text{in}^2} \times 1\text{ ft}^2$$
$$= \square\text{ ft}^2$$

5. $5\text{ mi}^2 = \_\_\_\_\_\_\_\_$ acres

Answers

Skill to Review:
1. 81 **2.** 1000

Margin Exercises:
1. 9 **2.** 45 **3.** 2880 **4.** 2.5 **5.** 3200

Guided Solutions:
3. 144, 2880 **4.** 144 in^2, 144, 2.5

EXAMPLE 6 Complete: $10{,}000\ \text{cm}^2 = \_\_\_\_\_\_\_\_\ \text{m}^2$.

$$10{,}000\ \text{cm}^2 = 10{,}000\ \text{cm}^2 \cdot \frac{1\ \text{m}}{100\ \text{cm}} \cdot \frac{1\ \text{m}}{100\ \text{cm}}$$
$$= 10{,}000\ \text{cm}^2 \cdot \frac{1\ \text{m}^2}{10{,}000\ \text{cm}^2}$$
$$= 1\ \text{m}^2$$

Do Exercises 6 and 7.

Complete.

6. $1\ \text{m}^2 = \_\_\_\_\_\_\_\_\ \text{mm}^2$

7. $100\ \text{mm}^2 = \_\_\_\_\_\_\_\_\ \text{cm}^2$

Mental Conversion

To convert mentally, we first note that $10^2 = 100$, $100^2 = 10{,}000$, and $0.1^2 = 0.01$. We use the table as before and multiply the number of places we move for the simple unit conversion by 2 to determine the number of places to move the decimal point for conversion of squared units.

1000 m	100 m	10 m	1 m	0.1 m	0.01 m	0.001 m
1 km	1 hm	1 dam	1 m	1 dm	1 cm	1 mm

EXAMPLE 7 Complete: $3.48\ \text{km}^2 = \_\_\_\_\_\_\_\_\ \text{m}^2$.

Think: To go from km to m in the table is a move of 3 places to the right.

1000 m	100 m	10 m	1 m	0.1 m	0.01 m	0.001 m
1 km	1 hm	1 dam	1 m	1 dm	1 cm	1 mm

3 moves to the right

So we move the decimal point $2 \cdot 3$, or 6, places to the right.

3.48 3.480000. $3.48\ \text{km}^2 = 3{,}480{,}000\ \text{m}^2$

6 places to the right

EXAMPLE 8 Complete: $586.78\ \text{cm}^2 = \_\_\_\_\_\_\_\_\ \text{m}^2$.

Think: To go from cm to m in the table is a move of 2 places to the left.

1000 m	100 m	10 m	1 m	0.1 m	0.01 m	0.001 m
1 km	1 hm	1 dam	1 m	1 dm	1 cm	1 mm

2 moves to the left

So we move the decimal point $2 \cdot 2$, or 4, places to the left.

586.78 0.0586.78 $586.78\ \text{cm}^2 = 0.058678\ \text{m}^2$

4 places to the left

Do Exercises 8–10.

Complete.

8. $2.88\ \text{m}^2 = \_\_\_\_\_\_\_\_\ \text{cm}^2$

9. $4.3\ \text{mm}^2 = \_\_\_\_\_\_\_\_\ \text{cm}^2$

10. $678{,}000\ \text{m}^2 = \_\_\_\_\_\_\_\_\ \text{km}^2$

Answers

6. 1,000,000 **7.** 1 **8.** 28,800 **9.** 0.043 **10.** 0.678

Translating for Success

1. *Test Items.* On a test of 90 items, Sally got 80% correct. How many items did she get correct?

2. *Suspension Bridge.* The San Francisco–Oakland Bay suspension bridge is 0.4375 mi long. Convert this distance to yards.

3. *Population Growth.* City A's growth rate per year is 0.9%. If the population was 1,500,000 in 2005, what was the population in 2006?

4. *Roller Coaster Drop.* The Manhattan Express Roller Coaster at the New York–New York Hotel and Casino, Las Vegas, Nevada, has a 144-ft drop. The California Screamin' Roller Coaster at Disney's California Adventure, Anaheim, California, has a 32.635-m drop. How much larger, in meters, is the drop of the Manhattan Express than the drop of the California Screamin'?

5. *Driving Distance.* Nate drives the company car 675 mi in 15 days. At this rate, how far will he drive in 20 days?

The goal of these matching questions is to practice step (2), Translate, of the five-step problem-solving process. Translate each word problem to an equation and select a correct translation from equations A–O.

A. $32.635\text{ m} + x = 144\text{ ft} \cdot \frac{0.305\text{ m}}{1\text{ ft}}$

B. $x = 0.4375\text{ mi} \times \frac{5280\text{ ft}}{1\text{ mi}} \times \frac{12\text{ in.}}{1\text{ ft}}$

C. $80\% \cdot x = 90$

D. $x = 0.89\text{ km} \cdot \frac{1000\text{ m}}{1\text{ km}} \cdot \frac{1\text{ m}}{3.281\text{ ft}}$

E. $x = 80\% \cdot 90$

F. $x = 420\text{ m} + 75\text{ ft} \cdot \frac{0.305\text{ m}}{1\text{ ft}}$

G. $\frac{x}{20} = \frac{675}{15}$

H. $x = 0.4375\text{ mi} \times \frac{5280\text{ ft}}{1\text{ mi}} \times \frac{1\text{ yd}}{3\text{ ft}}$

I. $x = 0.89\text{ km} \cdot \frac{0.621\text{ mi}}{1\text{ km}} \cdot \frac{5280\text{ ft}}{1\text{ mi}}$

J. $\frac{x}{15} = \frac{675}{20}$

K. $x = 420\text{ m} \cdot \frac{3.281\text{ ft}}{1\text{ m}} + 75\text{ ft}$

L. $144\text{ ft} + x = 32.635\text{ m} \cdot \frac{1\text{ ft}}{0.305\text{ m}}$

M. $x = 1{,}500{,}000 - 0.9\%(1{,}500{,}000)$

N. $20 \cdot x = 675$

O. $x = 1{,}500{,}000 + 0.9\%(1{,}500{,}000)$

Answers on page A-16

6. *Test Items.* Jason correctly answered 90 items on a recent test. These items represented 80% of the total number of questions. How many items were on the test?

7. *Population Decline.* City B's growth rate per year is −0.9%. If the population was 1,500,000 in 2005, what was the population in 2006?

8. *Bridge Length.* The Tatara Bridge in Onomichi-Imabari, Japan, is 0.89 km long. Convert this distance to feet.

9. *Height of Tower.* The Willis Tower in Chicago is 75 ft taller than the Jin Mao Building in Shanghai. The height of the Jin Mao Building is 420 m. What is the height of the Willis Tower in feet?

10. *Gasoline Usage.* Nate's company car gets 20 mi to the gallon in city driving. How many gallons will it use in 675 mi of city driving?

9.7 Exercise Set

For Extra Help MyMathLab® MathXL® PRACTICE WATCH READ REVIEW

Reading Check

Determine whether each equation is true or false.

RC1. ________ $9\ \text{ft}^2 = 1\ \text{yd}^2$

RC2. ________ $1\ \text{mi}^2 = 640\ \text{ft}^2$

RC3. ________ $1\ \text{acre} = 43{,}560\ \text{ft}^2$

RC4. ________ $36\ \text{in}^2 = 1\ \text{yd}^2$

RC5. ________ $1\ \text{km}^2 = 1{,}000{,}000\ \text{m}^2$

RC6. ________ $1\ \text{ft}^2 = 144\ \text{in}^2$

a Complete.

1. $1\ \text{ft}^2 =$ ________ in^2

2. $1\ \text{yd}^2 =$ ________ ft^2

3. $1\ \text{mi}^2 =$ ________ acres

4. $1\ \text{acre} =$ ________ ft^2

5. $1\ \text{in}^2 =$ ________ ft^2

6. $1\ \text{ft}^2 =$ ________ yd^2

7. $22\ \text{yd}^2 =$ ________ ft^2

8. $40\ \text{ft}^2 =$ ________ in^2

9. $44\ \text{yd}^2 =$ ________ ft^2

10. $144\ \text{ft}^2 =$ ________ yd^2

11. $20\ \text{mi}^2 =$ ________ acres

12. $576\ \text{in}^2 =$ ________ ft^2

13. $1\ \text{mi}^2 =$ ________ ft^2

14. $1\ \text{mi}^2 =$ ________ yd^2

15. $720\ \text{in}^2 =$ ________ ft^2

16. $27\ \text{ft}^2 =$ ________ yd^2

17. $144\ \text{in}^2 =$ ________ ft^2

18. $72\ \text{in}^2 =$ ________ ft^2

19. $1\ \text{acre} =$ ________ mi^2

20. $4\ \text{acres} =$ ________ ft^2

21. $40.3\ \text{mi}^2 =$ ________ acres

22. $1080\ \text{in}^2 =$ ________ ft^2

23. $333\ \text{ft}^2 =$ ________ yd^2

24. $18\ \text{yd}^2 =$ ________ in^2

b Complete.

25. $5.21\ km^2 =$ ________ m^2

26. $65\ km^2 =$ ________ m^2

27. $0.014\ m^2 =$ ________ cm^2

28. $0.028\ m^2 =$ ________ mm^2

29. $2345.6\ mm^2 =$ ________ cm^2

30. $8.38\ cm^2 =$ ________ mm^2

31. $852.14\ cm^2 =$ ________ m^2

32. $125\ mm^2 =$ ________ m^2

33. $250,000\ mm^2 =$ ________ cm^2

34. $2400\ mm^2 =$ ________ cm^2

35. $472,800\ m^2 =$ ________ km^2

36. $1.37\ cm^2 =$ ________ mm^2

Skill Maintenance

In Exercises 37 and 38, find the simple interest. [7.7a]

37. On $2000 at an interest rate of 8% for 1.5 years

38. On $2000 at an interest rate of 5.3% for 2 years

In each of Exercises 39–42, find **(a)** the amount of simple interest due and **(b)** the total amount that must be paid back. [7.7a]

39. A firm borrows $15,500 at 9.5% for 120 days.

40. A firm borrows $8500 at 10% for 90 days.

41. A firm borrows $6400 at 8.4% for 150 days.

42. A firm borrows $4200 at 11% for 30 days.

Synthesis

Complete.

43. $1\ m^2 =$ ________ ft^2

44. $1\ in^2 =$ ________ cm^2

45. $2\ yd^2 =$ ________ m^2

46. $1\ acre =$ ________ m^2

47. *Aalsmeer Flower Auction.* The fourth-largest building in the world in terms of floor space houses the Aalsmeer Flower Auction in Aalsmeer, Netherlands. It covers approximately 990,000 m^2. Each day, over 20 million flowers are sold there. Convert 990,000 m^2 to square feet.

Sources: www.amsterdamlogue.com; www.youTube.com/Aalsmeer, Netherlands, Flower Auction

48. *The Palazzo.* The largest building in the United States in terms of floor space is the Palazzo, a hotel and casino on the Las Vegas Strip in Paradise, Nevada. It contains approximately 6,953,000 ft^2. Convert 6,953,000 ft^2 to square meters. Round the answer to the nearest thousand.

Source: *Top 10 of Everything 2013*

Summary and Review

Units of Measure: Conversions

American Units of Length:	12 in. = 1 ft; 3 ft = 1 yd; 36 in. = 1 yd; 5280 ft = 1 mi
Metric Units of Length:	1 km = 1000 m; 1 hm = 100 m; 1 dam = 10 m; 1 dm = 0.1 m; 1 cm = 0.01 m; 1 mm = 0.001 m
American and Metric:	1 m = 39.370 in.; 1 m = 3.281 ft; 1 ft = 0.305 m; 1 in. = 2.540 cm; 1 km = 0.621 mi; 1 mi = 1.609 km; 1 yd = 0.914 m; 1 m = 1.094 yd
American Units of Weight:	1 T = 2000 lb; 1 lb = 16 oz
Metric Units of Mass:	1 t = 1000 kg; 1 kg = 1000 g; 1 hg = 100 g; 1 dag = 10 g; 1 dg = 0.1 g; 1 cg = 0.01 g; 1 mg = 0.001 g; 1 mcg = 0.000001 g
American Units of Capacity:	1 gal = 4 qt; 1 qt = 2 pt; 1 pt = 16 fluid oz; 1 pt = 2 cups; 1 cup = 8 fluid oz
Metric Units of Capacity:	1 L = 1000 mL = 1000 cm^3 = 1000 cc
American and Metric:	1 oz = 29.57 mL; 1 L = 1.06 qt
Units of Time:	1 min = 60 sec; 1 hr = 60 min; 1 day = 24 hr; 1 wk = 7 days; 1 year = $365\frac{1}{4}$ days
Temperature Conversion:	$F = \frac{9}{5} \cdot C + 32$, or $F = 1.8 \cdot C + 32$; $C = \frac{5}{9} \cdot (F - 32)$, or $C = \frac{F - 32}{1.8}$
American Units of Area:	1 yd^2 = 9 ft^2; 1 ft^2 = 144 in^2; 1 mi^2 = 640 acres; 1 acre = 43,560 ft^2

Concept Reinforcement

Determine whether each statement is true or false.

______ **1.** Distances measured in feet in the American system would probably be measured in meters in the metric system. [9.3a]

______ **2.** When converting from grams to milligrams, we move the decimal point two places to the left. [9.4b]

______ **3.** To convert mm^2 to cm^2, move the decimal point two places to the left. [9.7b]

______ **4.** Since 1 yd = 3 ft, we multiply by 3 to convert square yards to square feet. [9.7a]

______ **5.** You would probably use your furnace when the temperature outside was 40°C. [9.6b]

______ **6.** You could go ice fishing when the temperature outside was 10°C. [9.6b]

Study Guide

Objective 9.1a Convert from one American unit of length to another.

Examples Complete: 126 in. = _________ yd and $5\frac{2}{3}$ yd = _________ ft.

$$126 \text{ in.} = \frac{126 \text{ in.}}{1} \times \frac{1 \text{ yd}}{36 \text{ in.}}$$
$$= \frac{126 \text{ in.}}{36 \text{ in.}} \times 1 \text{ yd}$$
$$= \frac{126}{36} \times \frac{\text{in.}}{\text{in.}} \times 1 \text{ yd}$$
$$= 3.5 \times 1 \text{ yd} = 3.5 \text{ yd};$$
$$5\frac{2}{3} \text{ yd} = 5\frac{2}{3} \times 1 \text{ yd} = \frac{17}{3} \times 3 \text{ ft} = 17 \text{ ft}$$

Practice Exercises

Complete.

1. 7 ft = _________ yd

2. $2\frac{1}{2}$ mi = _________ ft

Objective 9.2a Convert from one metric unit of length to another.

Example Complete: 38 km = _________ cm and 2.9 mm = _________ m.

To go from km to cm, we move the decimal point 5 places to the right.

38 38.00000. 38 km = 3,800,000 cm

To go from mm to m, we move the decimal point 3 places to the left.

2.9 0.002.9 2.9 mm = 0.0029 m

Practice Exercises

Complete.

3. 12 hm = _________ m

4. 4.6 cm = _________ km

Objective 9.3a Convert between American units of length and metric units of length.

Example Complete: 42 ft = _________ m. (Note: 1 ft ≈ 0.305 m.)

$$42 \text{ ft} = 42 \times 1 \text{ ft} \approx 42 \times 0.305 \text{ m}$$
$$= 12.81 \text{ m}$$

Practice Exercise

5. Complete: 10 m = _________ yd. (Note: 1 m ≈ 1.094 yd.)

Objective 9.4a Convert from one American unit of weight to another.

Example Complete: 4020 oz = _________ lb.

$$4020 \text{ oz} = 4020 \cancel{\text{oz}} \times \frac{1 \text{ lb}}{16 \cancel{\text{oz}}} = \frac{4020}{16} \text{ lb} = 251.25 \text{ lb}$$

Practice Exercise

6. Complete: 10,280 lb = _________ T.

Objective 9.4b Convert from one metric unit of mass to another.

Example Complete: 5.62 cg = _________ g.

To go from cg to g, we move the decimal point 2 places to the left.

5.62 0.05.62 5.62 cg = 0.0562 g

Practice Exercise

7. Complete: 9.78 mg = _________ g.

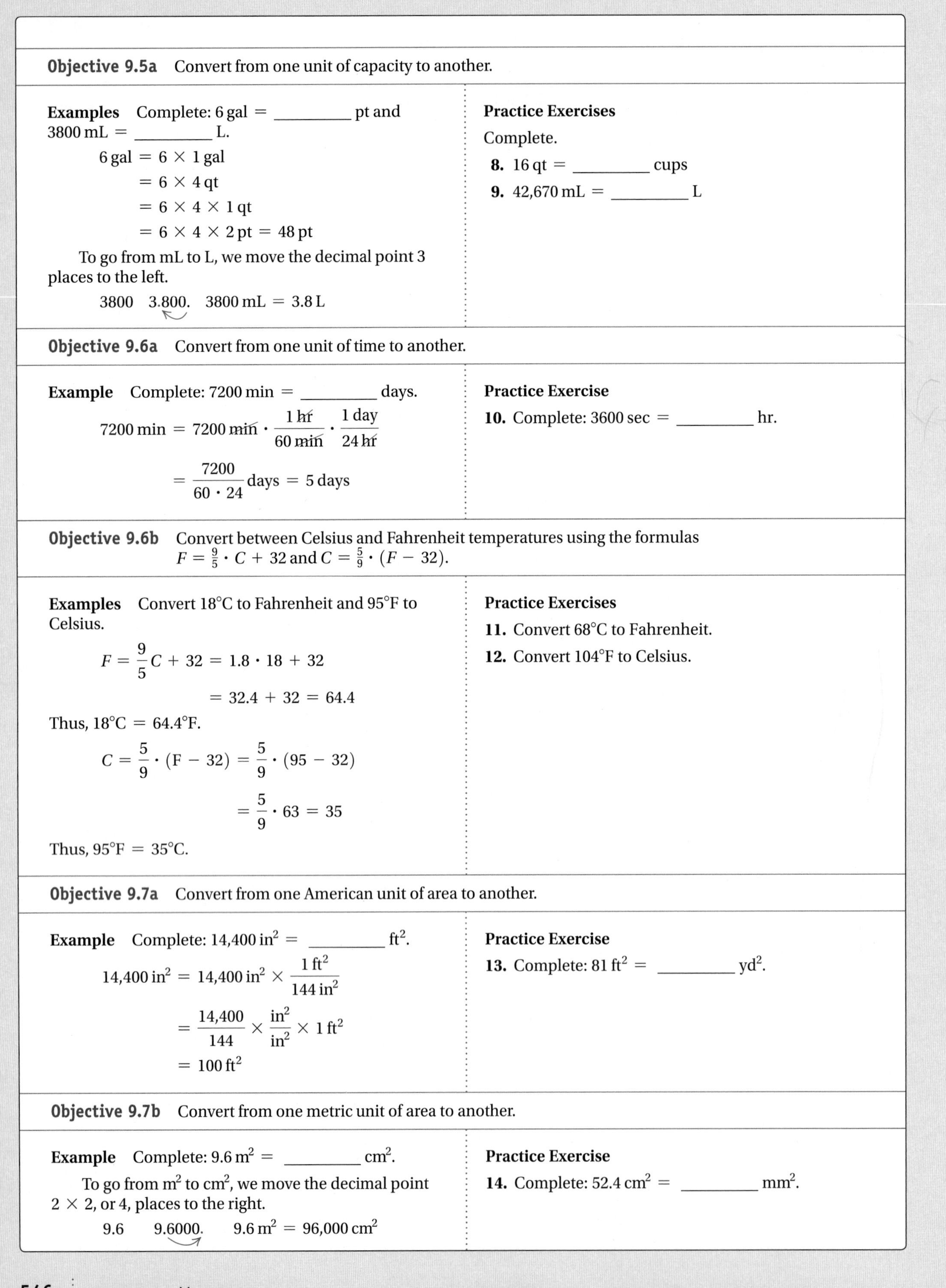

Objective 9.5a Convert from one unit of capacity to another.

Examples Complete: 6 gal = _________ pt and 3800 mL = _________ L.

$$6 \text{ gal} = 6 \times 1 \text{ gal}$$
$$= 6 \times 4 \text{ qt}$$
$$= 6 \times 4 \times 1 \text{ qt}$$
$$= 6 \times 4 \times 2 \text{ pt} = 48 \text{ pt}$$

To go from mL to L, we move the decimal point 3 places to the left.

3800 3.800. 3800 mL = 3.8 L

Practice Exercises

Complete.

8. 16 qt = _________ cups

9. 42,670 mL = _________ L

Objective 9.6a Convert from one unit of time to another.

Example Complete: 7200 min = _________ days.

$$7200 \text{ min} = 7200 \text{ min} \cdot \frac{1 \text{ hr}}{60 \text{ min}} \cdot \frac{1 \text{ day}}{24 \text{ hr}}$$
$$= \frac{7200}{60 \cdot 24} \text{ days} = 5 \text{ days}$$

Practice Exercise

10. Complete: 3600 sec = _________ hr.

Objective 9.6b Convert between Celsius and Fahrenheit temperatures using the formulas $F = \frac{9}{5} \cdot C + 32$ and $C = \frac{5}{9} \cdot (F - 32)$.

Examples Convert 18°C to Fahrenheit and 95°F to Celsius.

$$F = \frac{9}{5}C + 32 = 1.8 \cdot 18 + 32$$
$$= 32.4 + 32 = 64.4$$

Thus, 18°C = 64.4°F.

$$C = \frac{5}{9} \cdot (F - 32) = \frac{5}{9} \cdot (95 - 32)$$
$$= \frac{5}{9} \cdot 63 = 35$$

Thus, 95°F = 35°C.

Practice Exercises

11. Convert 68°C to Fahrenheit.

12. Convert 104°F to Celsius.

Objective 9.7a Convert from one American unit of area to another.

Example Complete: $14{,}400 \text{ in}^2$ = _________ ft^2.

$$14{,}400 \text{ in}^2 = 14{,}400 \text{ in}^2 \times \frac{1 \text{ ft}^2}{144 \text{ in}^2}$$
$$= \frac{14{,}400}{144} \times \frac{\text{in}^2}{\text{in}^2} \times 1 \text{ ft}^2$$
$$= 100 \text{ ft}^2$$

Practice Exercise

13. Complete: 81 ft^2 = _________ yd^2.

Objective 9.7b Convert from one metric unit of area to another.

Example Complete: 9.6 m^2 = _________ cm^2.

To go from m^2 to cm^2, we move the decimal point 2×2, or 4, places to the right.

9.6 9.6000. $9.6 \text{ m}^2 = 96{,}000 \text{ cm}^2$

Practice Exercise

14. Complete: 52.4 cm^2 = _________ mm^2.

Review Exercises

Complete. [9.1a], [9.2a], [9.3a]

1. 8 ft = _______ yd

2. $\frac{5}{6}$ yd = _______ in.

3. 0.3 mm = _______ cm

4. 4 m = _______ km

5. 2 yd = _______ in.

6. 4 km = _______ cm

7. 14 in. = _______ ft

8. 15 cm = _______ m

9. 200 m = _______ yd

10. 20 mi = _______ km

Complete the table below. [9.2a]

	Millimeters (mm)	Centimeters (cm)	Meters (m)
11.		1	
12.			305

Complete. [9.4a,b], [9.5a], [9.6a]

13. 7 lb = _______ oz

14. 4 g = _______ kg

15. 16 min = _______ hr

16. 464 mL = _______ L

17. 3 min = _______ sec

18. 4.7 kg = _______ g

19. 8.07 T = _______ lb

20. 0.83 L = _______ mL

21. 6 hr = _______ days

22. 4 cg = _______ g

23. 0.2 g = _______ mg

24. 0.0003 kg = _______ cg

25. 0.7 mL = _______ L

26. 60 mL = _______ L

27. 0.8 T = _______ lb

28. 0.4 L = _______ mL

29. 20 oz = _______ lb

30. $\frac{5}{6}$ min = _______ sec

31. 20 gal = _______ pt

32. 960 oz = _______ gal

33. 54 qt = _______ gal

34. 2.5 days = _______ hr

35. 3020 cg = _______ kg

36. 10,500 lb = _______ T

Medical Dosage. Solve.

37. Amoxicillin is an antibiotic obtainable in a liquid suspension form, part medication and part water, that is frequently used to treat infections in infants. One formulation of the drug contains 125 mg of amoxicillin per 5 mL of liquid. A pediatrician orders 150 mg per day for a 4-month-old child with an ear infection. How much of the amoxicillin suspension would the parent need to administer to the infant in order to achieve the recommended daily dose? [9.4c], [9.5b]

38. An emergency-room physician orders 3 L of Ringer's lactate to be administered over 4 hr for a patient suffering from shock and severe low blood pressure. How many milliliters is this? [9.5b]

39. A physician prescribes 0.25 mg of alprazolam, an antianxiety medication. How many micrograms are in this dose? [9.4c]

40. Convert $-6°C$ to Fahrenheit. [9.6b]

41. Convert 45°C to Fahrenheit. [9.6b]

42. Convert 68°F to Celsius. [9.6b]

43. Convert $-20°F$ to Celsius. [9.6b]

Complete. [9.7a, b]

44. $4 \text{ yd}^2 =$ ________ ft^2

45. $0.3 \text{ km}^2 =$ ________ m^2

46. $2070 \text{ in}^2 =$ ________ ft^2

47. $600 \text{ cm}^2 =$ ________ m^2

48. Complete: 172.6 cm = ________ hm. [9.2a]

A. 0.1726
B. 1,726,000
C. 0.01726
D. 17.26

49. Complete: 0.16 gal = ________ cups. [9.5a]

A. 1.28
B. 2.56
C. 0.64
D. 160

Synthesis

50. *Running Record.* The men's world record for the 200-m dash is 19.30 sec, set by Usain Bolt of Jamaica in the 2008 Summer Olympics in Beijing. How should the record be changed if the run were a 200-yd dash? [9.3a]

Source: *The New York Times*, August 17, 2008

51. It is known that 1 gal of water weighs 8.3453 lb. Which weighs more: an ounce of pennies or an ounce (as capacity) of water? Explain. [9.4a], [9.5a]

Understanding Through Discussion and Writing

1. Give at least two reasons why someone might prefer the use of grams to the use of ounces. [9.4a, b]

2. Why do you think most containers for liquids list both metric and American units of measure? [9.5a]

3. Explain the difference between the way we move the decimal point for area conversion and the way we do so for length conversion. [9.2a], [9.7b]

4. What advantages does the use of metric units of capacity have over the use of American units? [9.5a]

5. a) The temperature is 23°C. Would you want to play golf? Explain.
   b) Your bathwater has a temperature of 10°C. Would you want to take a bath?
   c) The nearby lake has a temperature of $-10°C$. Would it be safe to go ice skating? [9.6b]

6. Which is larger and why: one square meter or one square yard? [9.3a], [9.7a, b]

CHAPTER 9

Test

For Extra Help For step-by-step test solutions, access the Chapter Test Prep Videos in MyMathLab® or on YouTube (search "BittingerBasicEl" and click on "Channels").

Complete.

1. 4 ft = ________ in.

2. 4 in. = ________ ft

3. 6 km = ________ m

4. 8.7 mm = ________ cm

5. 200 yd = ________ m

6. 2400 km = ________ mi

Complete the table below.

	Object	Millimeters (mm)	Centimeters (cm)	Meters (m)
7.	Width of a key on a calculator		0.5	
8.	Height of one of your authors			1.8542

Complete.

9. 3080 mL = ________ L

10. 0.24 L = ________ mL

11. 4 lb = ________ oz

12. 4.11 T = ________ lb

13. 3.8 kg = ________ g

14. 4.325 mg = ________ cg

15. 2200 mg = ________ g

16. 5 hr = ________ min

17. 15 days = ________ hr

18. 64 pt = ________ qt

19. 10 gal = ________ oz

20. 5 cups = ________ oz

21. 0.37 mg = ________ mcg

22. Convert 95°F to Celsius.

23. Convert 59°C to Fahrenheit.

Complete the table below.

	Object	Yards (yd)	Centimeters (cm)	Inches (in.)	Meters (m)	Millimeters (mm)
24.	Length of a meter stick				1	
25.	Height of Xiamen Posts and Telecommunications Building, Xiamen, China	398				

Medical Dosage. Solve.

26. An emergency-room physician prescribes 2.5 L of normal saline given intravenously over 8 hr for a patient who is severely dehydrated. How many milliliters is this?

27. A physician prescribes 0.5 mg of alprazolam to be taken 3 times a day by a patient suffering from anxiety. How many micrograms of alprazolam is the patient to ingest each day?

28. A prescription calls for 4 oz of dextromethorphan, a cough-suppressant medication. For how many milliliters is the prescription? (Use 1 oz = 29.57 mL.)

Complete.

29. $12\ \text{ft}^2 = $ ________ in^2

30. $3\ \text{cm}^2 = $ ________ m^2

31. Convert 45.5°C to Fahrenheit.

A. 49.9°F

B. 24.3°F

C. 7.5°F

D. 113.9°F

Synthesis

32. *Running Record.* The world's record for the 400-m dash is 43.18 sec, set by Michael Johnson of the United States in Seville, Spain, on August 26, 1999. How should the record be changed if the run were a 400-yd run?

Source: The *World Almanac*, 2008

Cumulative Review

Solve.

1. ***Population 65 and Older.*** The U.S. population 65 and older was about 40.4 million in 2010. It is projected to be 54.8 million in 2020. Find standard notation for 40.4 million and 54.8 million.

   Sources: Decennial Censuses, Annual Population Estimates, U.S. Interim Projections, U.S. Census Bureau; U.S. Department of Commerce

2. ***Gas Mileage.*** A Honda Fit travels 561 mi on the highway on 17 gal of gasoline. What is the gas mileage?

   Source: *Car and Driver*, December 2008

3. ***Corn Production.*** Iowa, the top corn-producing state in the United States, produced 2.4 billion bushels of corn in 2011. Indiana produced 839 million bushels of corn that year. How many more bushels of corn were produced in Iowa than in Indiana?

   Source: U.S. Department of Agriculture

4. ***Glaziers.*** Glaziers install glass in windows, skylights, and storefronts. In 2010, approximately 41,900 people in the United States were employed as glaziers. It is projected that this number will increase to 59,600 by 2020. What is the percent increase in the number of employed glaziers?

   Sources: *Occupational Outlook Handbook 2012–2013*; U.S. Department of Labor

Perform the indicated operation and simplify.

5. $46{,}231 \times 1100$

6. $\frac{1}{10} \cdot \left(-\frac{5}{6}\right)$

7. $-14.5 + \frac{4}{5} - 0.1$

8. $-2\frac{3}{5} \div \left(-3\frac{9}{10}\right)$

9. $0.1\overline{)3.56}$

10. $3\frac{1}{2} - 2\frac{2}{3}$

11. Determine whether 1,298,032 is divisible by 8.

12. Determine whether 5,024,120 is divisible by 3.

13. Find the prime factorization of 99.

14. Find the LCM of 35 and 49.

15. Round $35.\overline{7}$ to the nearest tenth.

16. Write a word name for 103.064.

17. Find the average and the median of this set of numbers:

    29, 21, 9, 13, 17, 18.

Find percent notation.

18. 0.08

19. $\frac{3}{5}$

Complete.

20. 2 yd = ______ ft

21. 6 oz = ______ lb

22. 15°C = ______ °F

23. 0.087 L = ______ mL

24. 9 sec = ______ min

25. 17 cm = ______ m

26. 2200 mi = ______ km

27. 2000 mL = ______ L

28. 0.23 mg = ______ mcg

29. $12\ \text{yd}^2$ = ________ ft^2

Solve.

30. $0.07 \cdot x = -10.535$

31. $x + 12{,}843 = 32{,}091$

32. $\frac{2}{3} \cdot y = 5$

33. $\frac{4}{5} + y = -\frac{6}{7}$

Solve.

34. A mechanic spent $\frac{1}{3}$ hr changing a car's oil, $\frac{1}{2}$ hr rotating the tires, $\frac{1}{10}$ hr changing the air filter, $\frac{1}{4}$ hr adjusting the idle speed, and $\frac{1}{15}$ hr checking the brake and transmission fluids. How many hours did the mechanic spend working on the car?

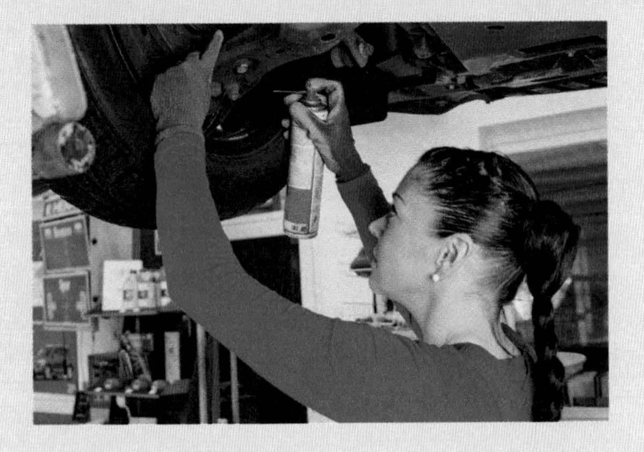

35. ***Milk Production.*** There are 9.15 million dairy cows in the United States, each producing, on average, 19,950 lb of milk per year. How many pounds of milk are produced each year in the United States?

Source: U.S. Department of Agriculture

36. A driver filled the gas tank when the odometer read 86,897.2. At the next gasoline purchase, the odometer read 87,153.0. How many miles had been driven? The tank was filled with 16 gal. What was the gas mileage?

37. ***Real Estate Commission.*** A real estate commission rate is $7\frac{1}{2}\%$. What is the commission on the sale of a property for $215,000?

38. ***Compound Interest.*** A student invests $2000 in an account paying 6%, compounded semiannually. How much is in the account after 3 years?

39. A man on a diet loses $3\frac{1}{2}$ lb in 2 weeks. At this rate, how many pounds will he lose in 5 weeks?

40. A family has an annual income of $52,800. Of this, $\frac{1}{4}$ is spent for food. How much does the family spend each year for food?

41. ***Seed Production.*** The U.S. Department of Agriculture requires that 80% of the seeds that a company produces must sprout. To determine the quality of the seeds it has produced, a company plants 500 seeds. It finds that 417 of the seeds sprout. Do the seeds meet government standards?

Source: U.S. Department of Agriculture

Sundae's Homemade Ice Cream & Coffee Co. This company in Indianapolis, Indiana, makes ice cream, sorbet, and frozen yogurt.

42. Ice cream is packaged in 15-lb tubs. How many ounces are in one tub?

43. Although the process is not perfect, Sundae's attempts to have about 4 oz in each dip of ice cream. How many dips are there in a tub of ice cream?

44. By weighing each tub, the owner can determine how many dips have been sold of that flavor. The weight of a tub changes from 15 lb to $8\frac{5}{8}$ lb over a busy weekend. How many dips of ice cream were served from the tub?

45. A one-dip ice cream cone sells for $2.99. If the entire contents of a tub of ice cream were used to make one-dip cones, how much money would be taken in from the sale of the whole tub?

46. A two-dip ice cream cone sells for $4.05. If the entire contents of a tub of ice cream were used to make two-dip cones, how much money would be taken in from the sale of the whole tub?

Synthesis

47. If r is $\frac{2}{5}$ of q, then q is what fractional part of r?

CHAPTER

10

Geometry

STUDYING FOR SUCCESS *Take Good Care of Yourself*

- ☐ Get plenty of rest, especially the night before a test.
- ☐ Try an exercise break when studying. Often a walk or a bike ride will improve your concentration.
- ☐ Plan leisure time in your schedule. Rest and a change of pace will help you avoid burn-out.

10.1 Perimeter

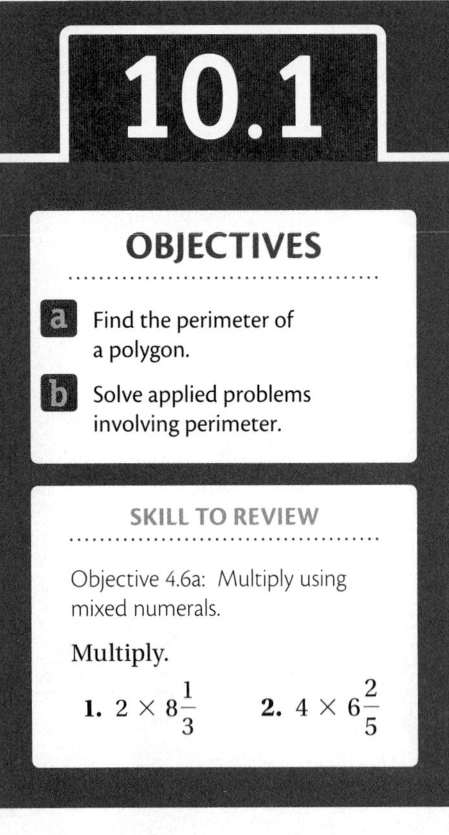

OBJECTIVES

a Find the perimeter of a polygon.

b Solve applied problems involving perimeter.

SKILL TO REVIEW

Objective 4.6a: Multiply using mixed numerals.

Multiply.

1. $2 \times 8\frac{1}{3}$ **2.** $4 \times 6\frac{2}{5}$

Find the perimeter of each polygon.

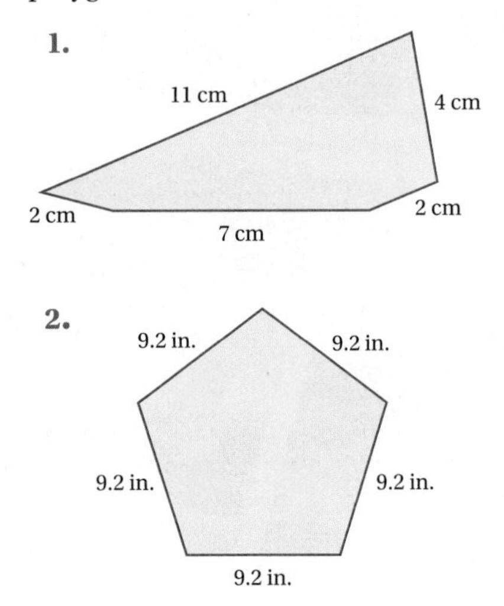

Answers

Skill to Review:

1. $\frac{50}{3}$, or $16\frac{2}{3}$ **2.** $\frac{128}{5}$, or $25\frac{3}{5}$

Margin Exercises:

1. 26 cm **2.** 46 in.

a FINDING PERIMETERS

PERIMETER OF A POLYGON

A **polygon** is a closed geometric figure with three or more sides. The **perimeter of a polygon** is the distance around it, or the sum of the lengths of its sides.

EXAMPLE 1 Find the perimeter of this polygon.

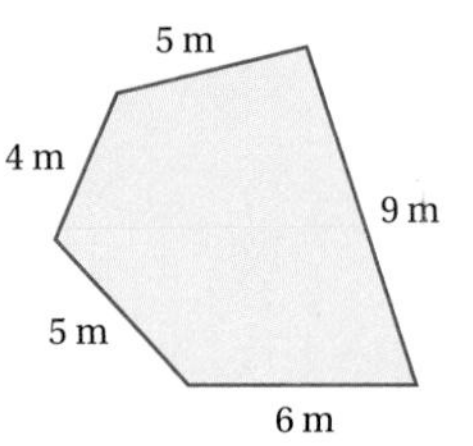

We add the lengths of the sides. Since all units are the same, we add the numbers, keeping meters (m) as the unit.

$$\begin{aligned}\text{Perimeter} &= 6\text{ m} + 5\text{ m} + 4\text{ m} + 5\text{ m} + 9\text{ m}\\ &= (6 + 5 + 4 + 5 + 9)\text{ m}\\ &= 29\text{ m}\end{aligned}$$

◀ **Do Margin Exercises 1 and 2.**

A **rectangle** is a polygon with four sides and four 90° angles.

EXAMPLE 2 Find the perimeter of a rectangle that is 3 cm by 4 cm. The symbol ⌉ in the corner indicates an angle of 90°.

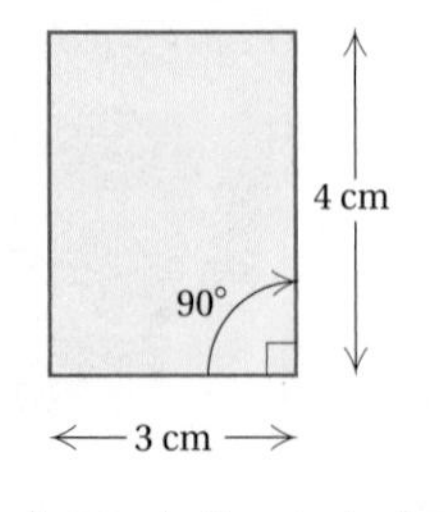

$$\begin{aligned}\text{Perimeter} &= 3\text{ cm} + 4\text{ cm} + 3\text{ cm} + 4\text{ cm}\\ &= (3 + 4 + 3 + 4)\text{ cm}\\ &= 14\text{ cm}\end{aligned}$$

■

PERIMETER OF A RECTANGLE

The **perimeter of a rectangle** is twice the sum of the length and the width, or 2 times the length plus 2 times the width:

$$\mathbf{P = 2 \cdot (l + w)}, \quad \text{or} \quad \mathbf{P = 2 \cdot l + 2 \cdot w}.$$

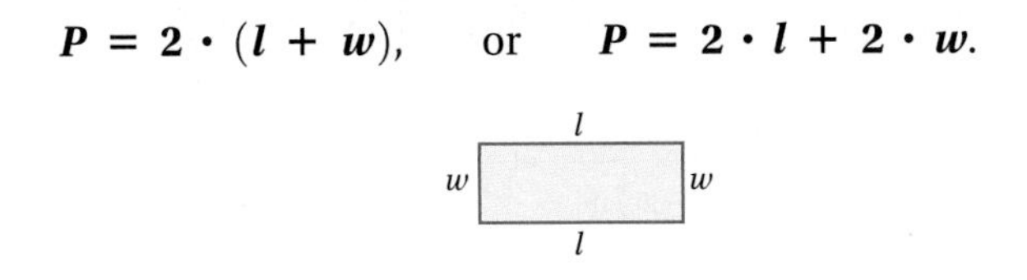

EXAMPLE 3 Find the perimeter of a rectangle that is 7.8 ft by 4.3 ft.

$$\begin{aligned} P &= 2 \cdot (l + w) \\ &= 2 \cdot (7.8 \text{ ft} + 4.3 \text{ ft}) \\ &= 2 \cdot (12.1 \text{ ft}) \\ &= 24.2 \text{ ft} \end{aligned}$$

Do Exercises 3–5. ▶

A **square** is a rectangle with all sides the same length.

EXAMPLE 4 Find the perimeter of a square whose sides are 9 mm long.

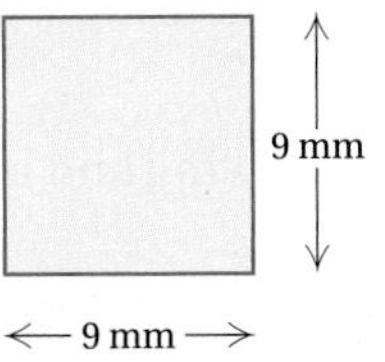

$$\begin{aligned} P &= 9 \text{ mm} + 9 \text{ mm} + 9 \text{ mm} + 9 \text{ mm} \\ &= (9 + 9 + 9 + 9) \text{ mm} \\ &= 36 \text{ mm} \end{aligned}$$

Do Exercise 6. ▶

PERIMETER OF A SQUARE

The **perimeter of a square** is four times the length of a side:

$$\mathbf{P = 4 \cdot s}.$$

s s s s

3. Find the perimeter of a rectangle that is 2 cm by 4 cm.

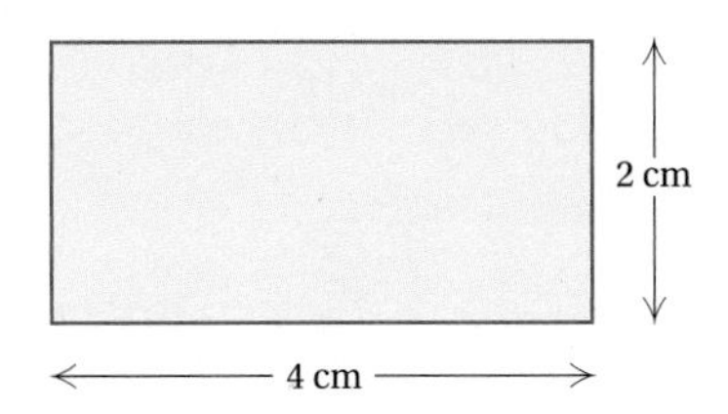

4. Find the perimeter of a rectangle that is 5.25 yd by 3.5 yd.

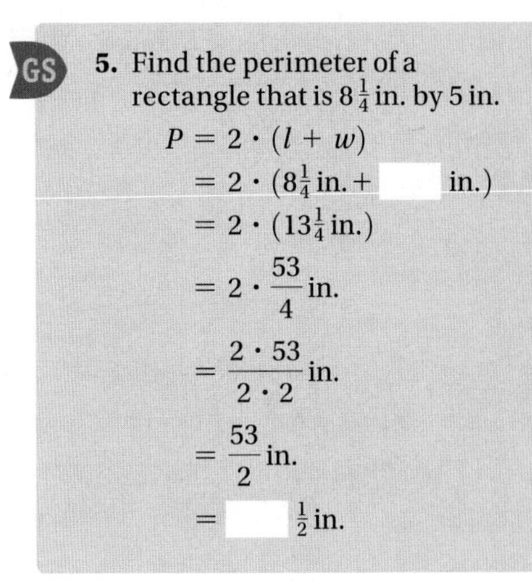

GS **5.** Find the perimeter of a rectangle that is $8\frac{1}{4}$ in. by 5 in.

$$\begin{aligned} P &= 2 \cdot (l + w) \\ &= 2 \cdot (8\tfrac{1}{4} \text{ in.} + \underline{\quad} \text{ in.}) \\ &= 2 \cdot (13\tfrac{1}{4} \text{ in.}) \\ &= 2 \cdot \frac{53}{4} \text{ in.} \\ &= \frac{2 \cdot 53}{2 \cdot 2} \text{ in.} \\ &= \frac{53}{2} \text{ in.} \\ &= \underline{\quad}\tfrac{1}{2} \text{ in.} \end{aligned}$$

6. Find the perimeter of a square with sides of length 10 km.

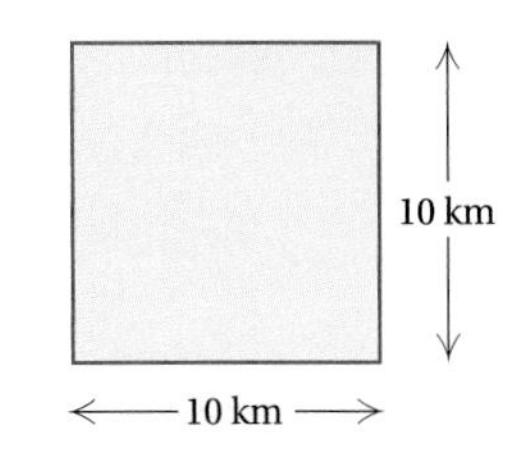

Answers

3. 12 cm **4.** 17.5 yd **5.** $26\frac{1}{2}$ in. **6.** 40 km

Guided Solution:
5. 5, 26

EXAMPLE 5 Find the perimeter of a square whose sides are $20\frac{1}{8}$ in. long.

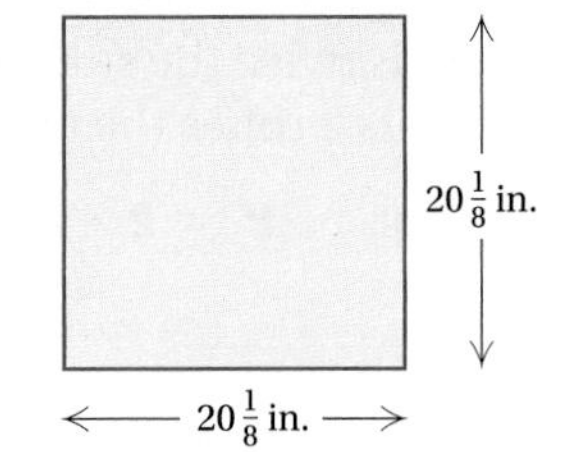

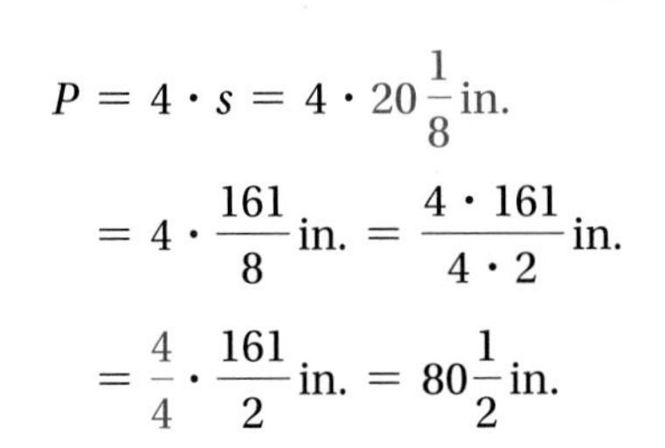

$$P = 4 \cdot s = 4 \cdot 20\frac{1}{8}\text{ in.}$$
$$= 4 \cdot \frac{161}{8}\text{ in.} = \frac{4 \cdot 161}{4 \cdot 2}\text{ in.}$$
$$= \frac{4}{4} \cdot \frac{161}{2}\text{ in.} = 80\frac{1}{2}\text{ in.}$$

Do Exercises 7 and 8.

7. Find the perimeter of a square with sides of length $5\frac{1}{4}$ yd.

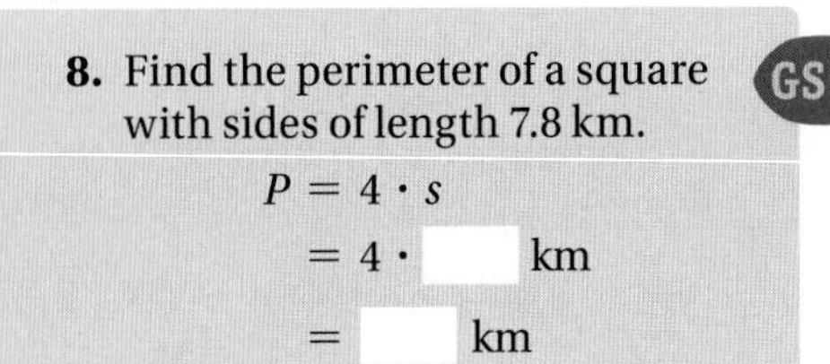

8. Find the perimeter of a square with sides of length 7.8 km. GS

$P = 4 \cdot s$

$= 4 \cdot$ ____ km

$=$ ____ km

b SOLVING APPLIED PROBLEMS

EXAMPLE 6 Jaci is adding crown molding along the top of the walls of her rectangular dining room, which measures 14 ft by 12 ft. How many feet of molding will be needed? If the molding sells for \$3.25 per foot, what will its total cost be?

1. **Familiarize.** We make a drawing and let $P =$ the perimeter.

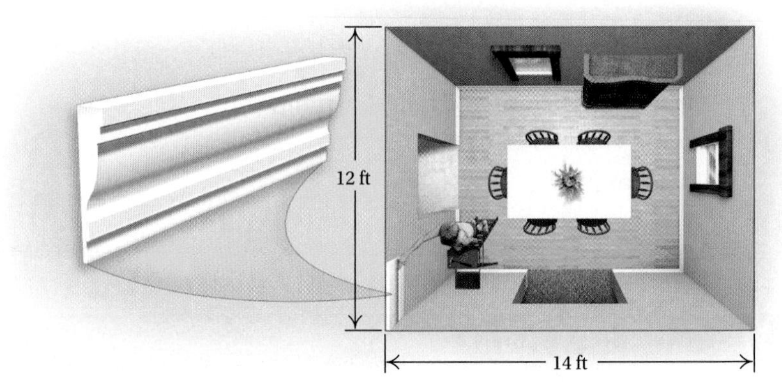

2. **Translate.** The perimeter of the room is given by

$$P = 2 \cdot (l + w) = 2 \cdot (14\text{ ft} + 12\text{ ft}).$$

3. **Solve.** We calculate the perimeter as follows:

$$P = 2 \cdot (14\text{ ft} + 12\text{ ft}) = 2 \cdot (26\text{ ft}) = 52\text{ ft}.$$

Then we multiply by \$3.25 to find the cost of the crown molding:

$$\text{Cost} = \$3.25 \times \text{Perimeter} = \$3.25 \times 52\text{ ft} = \$169.$$

4. **Check.** The check is left to the student.
5. **State.** The 52 ft of crown molding that is needed will cost \$169.

Do Exercise 9.

9. A fence is to be built around a vegetable garden that measures 20 ft by 15 ft. How many feet of fence will be needed? If fencing sells for \$2.95 per foot, what will the fencing cost?

Answers

7. 21 yd 8. 31.2 km 9. 70 ft; \$206.50

Guided Solution:
8. 7.8, 31.2

10.1 Exercise Set

For Extra Help MyMathLab® MathXL® PRACTICE WATCH READ REVIEW

Reading Check

Complete each statement with the correct word from the following list. A word may be used more than once or not at all.

closed	perimeter	rectangle
open	polygon	square

RC1. A polygon is a(n) _______________ figure with three or more sides.

RC2. The distance around a polygon is its _______________.

RC3. The formula $P = 2 \cdot l + 2 \cdot w$ gives the _______________ of a rectangle.

RC4. The perimeter of a(n) _______________ is given by the formula $P = 4 \cdot s$.

a Find the perimeter of each polygon.

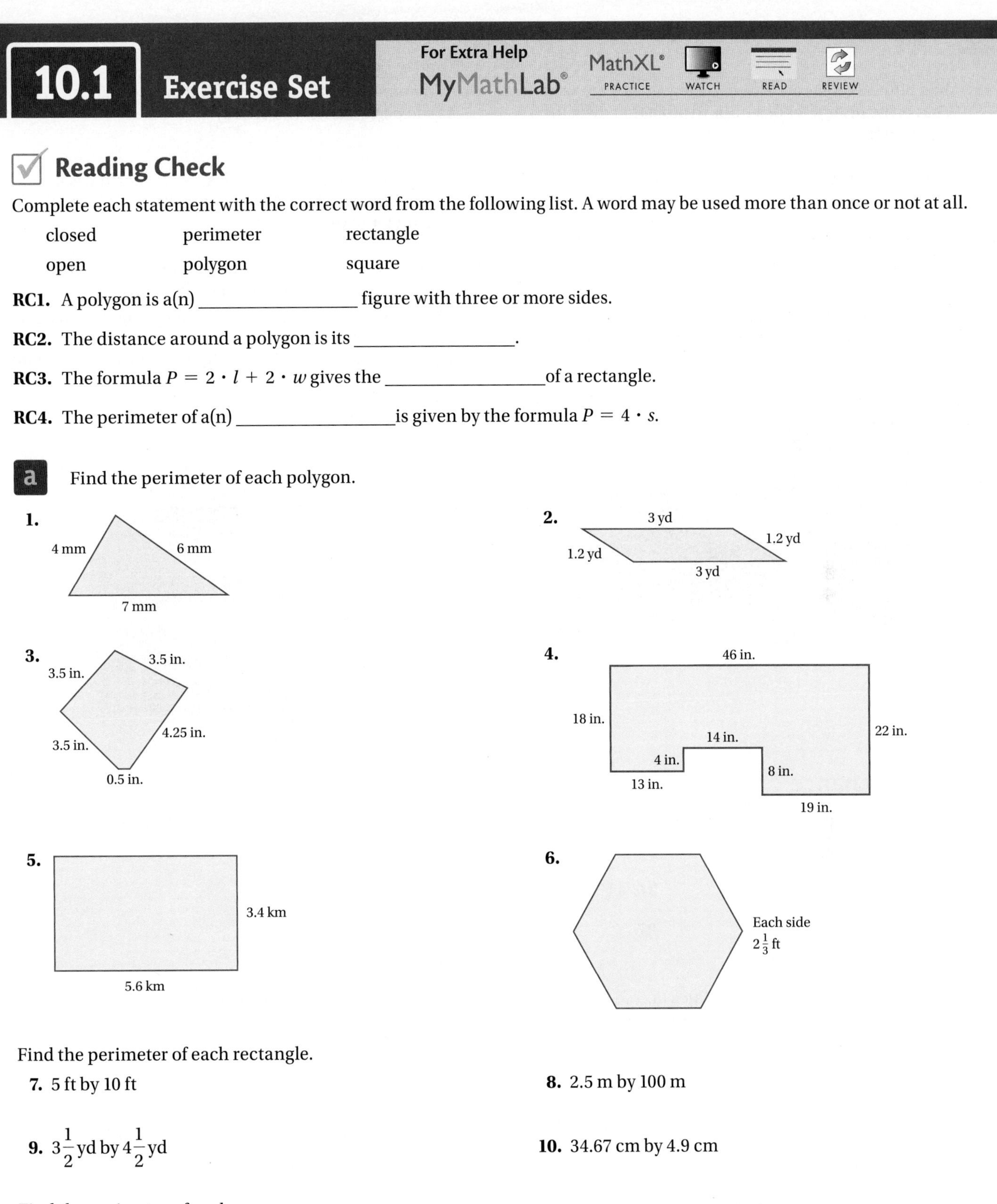

Find the perimeter of each rectangle.

7. 5 ft by 10 ft

8. 2.5 m by 100 m

9. $3\frac{1}{2}$ yd by $4\frac{1}{2}$ yd

10. 34.67 cm by 4.9 cm

Find the perimeter of each square.

11. 22 ft on a side

12. 56.9 km on a side

13. 45.5 mm on a side

14. $3\frac{1}{8}$ yd on a side

b Solve.

15. Most billiard tables are twice as long as they are wide. What is the perimeter of a billiard table that measures 4.5 ft by 9 ft?

16. A rectangular posterboard is 61.8 cm by 87.9 cm. What is the perimeter of the board?

17. A piece of flooring tile is a square with sides of length 30.5 cm. What is the perimeter of the piece of tile?

18. The Plaza de Balcarce in Balcarce, Argentina, is a public square with sides of length 300 m. What is the perimeter of the square?

19. A rain gutter is to be installed around the office building shown in the figure.

a) Find the perimeter of the office building.
b) If the gutter costs $4.59 per foot, what is the total cost of the gutter?

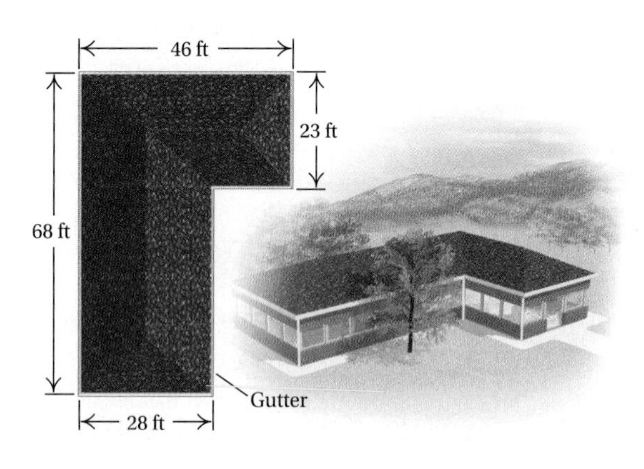

20. Robbin plans to string lights around the lower level of the roof of the gazebo shown in the figure.

a) If all sides of the roof are the same length, find the perimeter of the roof.
b) How many 6-ft strands of lights will Robbin need to buy?

Skill Maintenance

21. Find the simple interest on $600 at 6.4% for $\frac{1}{2}$ year. [7.7a]

22. Find the simple interest on $600 at 8% for 2 years. [7.7a]

Evaluate. [1.9b]

23. 10^3

24. 11^3

25. 15^2

26. 22^2

27. 7^2

28. 4^3

Solve.

29. *Sales Tax.* In a certain state, a sales tax of $878 is collected when a car is purchased for $17,560. What is the sales tax rate? [7.6a]

30. *Commission Rate.* Rich earns $1854.60 selling $16,860 worth of cell phones. What is the commission rate? [7.6b]

Synthesis

31. If it takes 18 in. to make the bow, how much ribbon is needed for the entire package shown here?

32. A carpenter is to build a fence around a 9-m by 12-m garden.

a) The posts are 3 m apart. How many posts will be needed?
b) The posts cost $8.65 each. How much will the posts cost?
c) The fence will surround all but 3 m of the garden, which will be a gate. How long will the fence be?
d) The fencing costs $3.85 per meter. What will the cost of the fencing be?
e) The gate costs $69.95. What is the total cost of the materials?

Area

a RECTANGLES AND SQUARES

A polygon and its interior form a plane region. We can find the area of a *rectangular region*, or *rectangle*, by filling it in with square units. Two such units, a *square inch* and a *square centimeter*, are shown below.

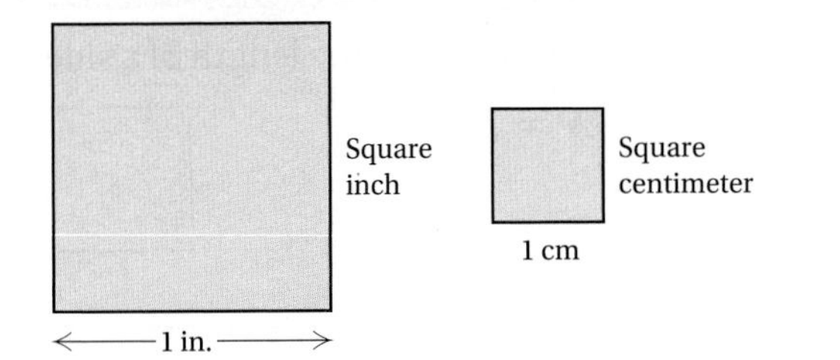

EXAMPLE 1 What is the area of this region?

We have a rectangular array. Since the region is filled with 12 square centimeters, its area is 12 square centimeters (sq cm), or 12 cm^2. The number of units is 3×4, or 12.

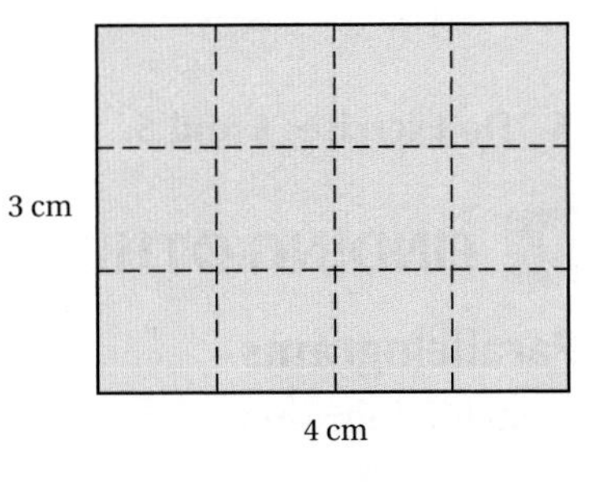

Do Margin Exercise 1.

AREA OF A RECTANGLE

The **area of a rectangle** is the product of the length l and the width w:

$$A = l \cdot w.$$

EXAMPLE 2 Find the area of a rectangle that is 7 yd by 4 yd.

We have

$$A = l \cdot w = 7\text{ yd} \cdot 4\text{ yd}$$
$$= 7 \cdot 4 \cdot \text{yd} \cdot \text{yd} = 28\text{ yd}^2.$$

We think of yd · yd as $(\text{yd})^2$ and denote it yd^2. Thus we read "28 yd^2" as "28 square yards."

Do Exercises 2 and 3.

OBJECTIVES

a Find the area of a rectangle and a square.

b Find the area of a parallelogram, a triangle, and a trapezoid.

c Solve applied problems involving areas of rectangles, squares, parallelograms, triangles, and trapezoids.

SKILL TO REVIEW

Objective 5.5c: Calculate using fraction notation and decimal notation together.

Calculate.

1. $\frac{1}{2} \times 16.243$

2. $0.5 \times \frac{3}{8}$

1. What is the area of this region? Count the number of square centimeters.

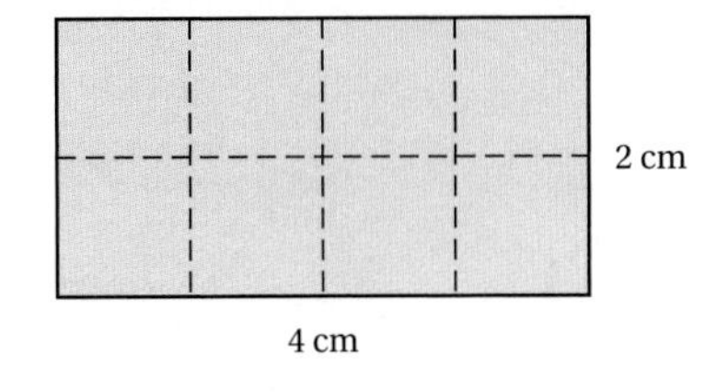

2. Find the area of a rectangle that is 7 km by 8 km.

3. Find the area of a rectangle that is $5\frac{1}{4}$ yd by $3\frac{1}{2}$ yd.

Answers

Skill to Review:

1. 8.1215 **2.** 0.1875, or $\frac{3}{16}$

Margin Exercises:

1. 8 cm^2 **2.** 56 km^2 **3.** $18\frac{3}{8}$ yd^2

4. Find the area of a square with sides of length 12 km.

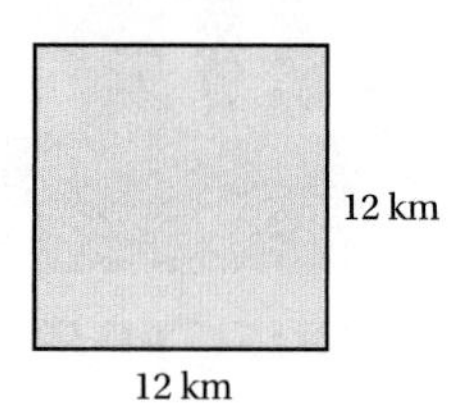

5. Find the area of a square with sides of length 10.9 m.

6. Find the area of a square with sides of length $3\frac{1}{2}$ yd.

GS

$A = s \cdot s$

$= 3\frac{1}{2}\text{ yd} \times \underline{\quad}\text{ yd}$

$= \frac{7}{2}\text{ yd} \times \frac{7}{2}\text{ yd}$

$= \frac{49}{4}\text{ yd}^2$

$= \underline{\quad}\frac{1}{4}\text{ yd}^2$

EXAMPLE 3 Find the area of a square with sides of length 9 mm.

$$A = (9\text{ mm}) \cdot (9\text{ mm})$$
$$= 9 \cdot 9 \cdot \text{mm} \cdot \text{mm}$$
$$= 81\text{ mm}^2$$

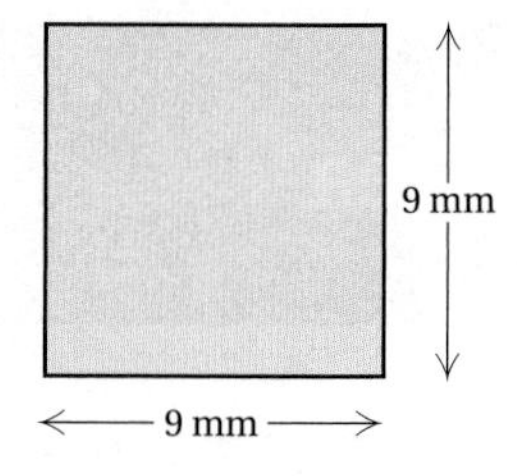

Do Exercise 4.

AREA OF A SQUARE

The **area of a square** is the square of the length of a side:

$$A = s \cdot s, \quad \text{or} \quad A = s^2.$$

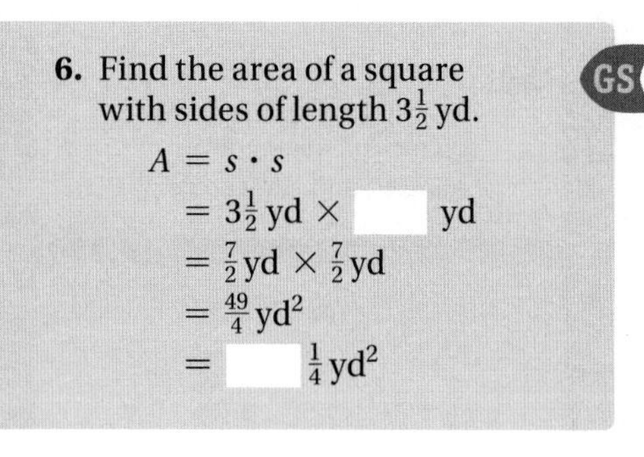

EXAMPLE 4 Find the area of a square with sides of length 20.3 m.

$$A = s \cdot s = 20.3\text{ m} \times 20.3\text{ m} = 20.3 \times 20.3 \times \text{m} \times \text{m} = 412.09\text{ m}^2$$

Do Exercises 5 and 6.

b FINDING OTHER AREAS

Parallelograms

A **parallelogram** is a four-sided figure with two pairs of parallel sides, as shown below.

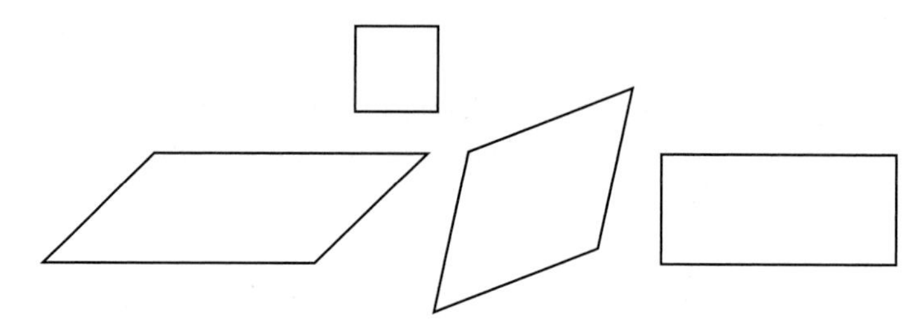

To find the area of a parallelogram, consider the one below.

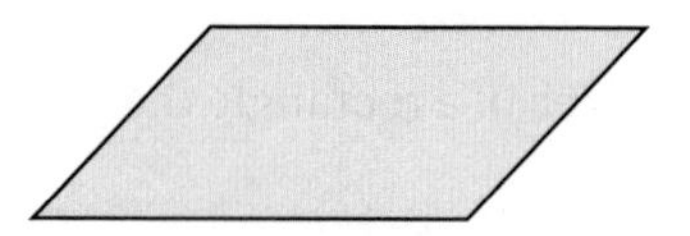

If we cut off a piece and move it to the other end, we get a rectangle.

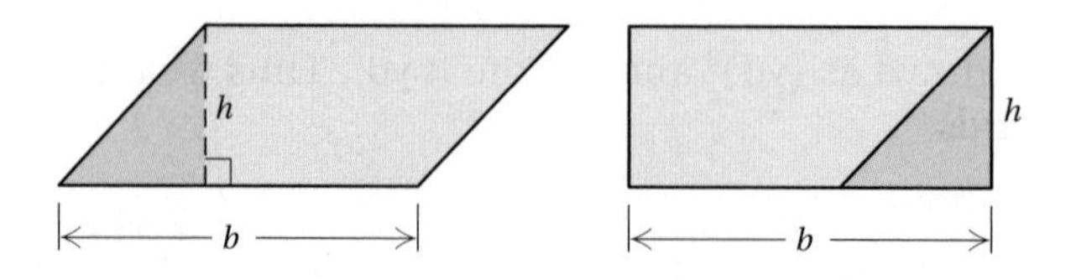

We can find the area by multiplying the length b, called a **base**, by h, called the **height**.

Answers

4. 144 km^2 5. 118.81 m^2 6. $12\frac{1}{4}\text{ yd}^2$

Guided Solution:
6. $3\frac{1}{2}$, 12

AREA OF A PARALLELOGRAM

The **area of a parallelogram** is the product of the length of the base b and the height h:

$$A = b \cdot h.$$

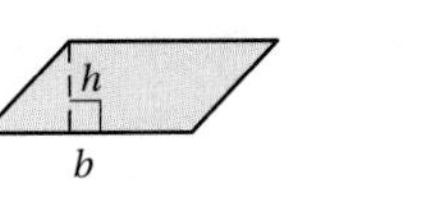

EXAMPLE 5 Find the area of this parallelogram.

$A = b \cdot h$
$= 7 \text{ km} \cdot 5 \text{ km}$
$= 35 \text{ km}^2$

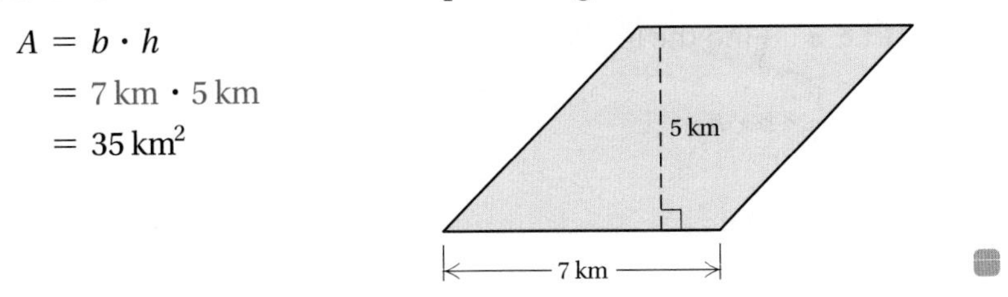

EXAMPLE 6 Find the area of this parallelogram.

$A = b \cdot h$
$= 1.2 \text{ m} \times 6 \text{ m}$
$= 7.2 \text{ m}^2$

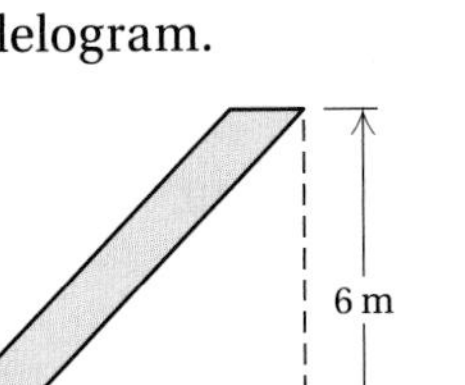

Do Exercises 7 and 8.

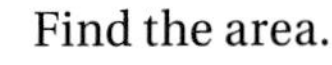

Find the area.

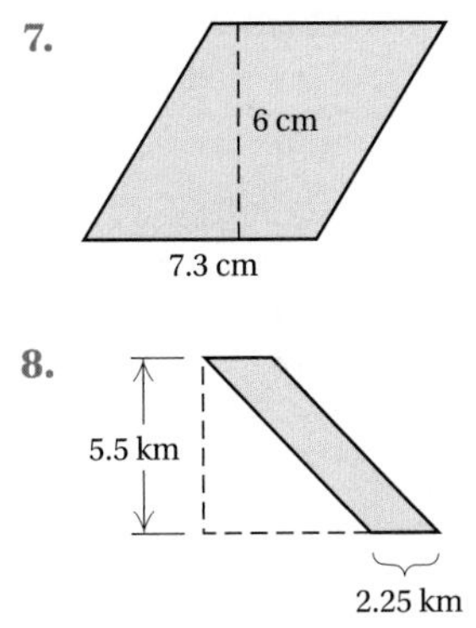

Triangles

A **triangle** is a polygon with three sides. To find the area of a triangle like the one shown on the left below, think of cutting out another just like it and placing it as shown on the right below.

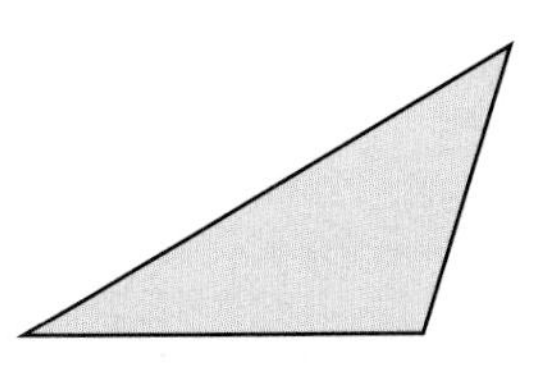

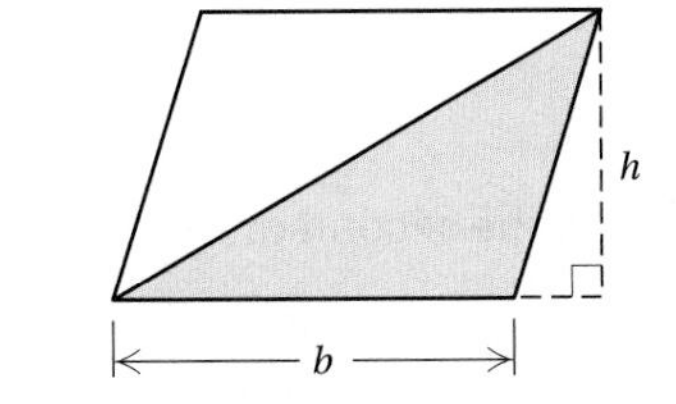

The resulting figure is a parallelogram whose area is

$$b \cdot h.$$

The triangle we began with has half the area of the parallelogram, or

$$\frac{1}{2} \cdot b \cdot h.$$

AREA OF A TRIANGLE

The **area of a triangle** is half the length of the base times the height:

$$A = \frac{1}{2} \cdot b \cdot h.$$

Answers

7. 43.8 cm^2 8. 12.375 km^2

Find the area.

9.

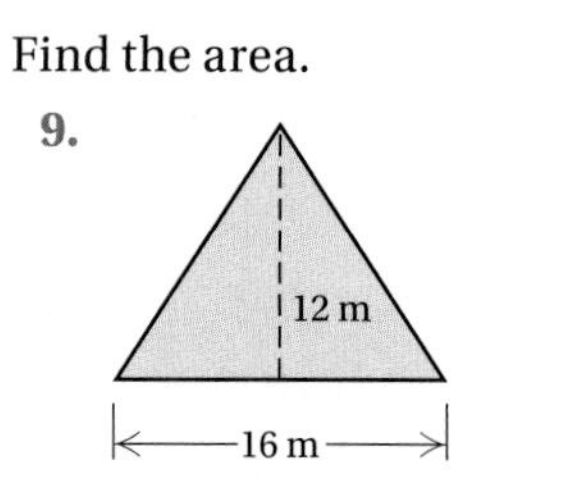

10.

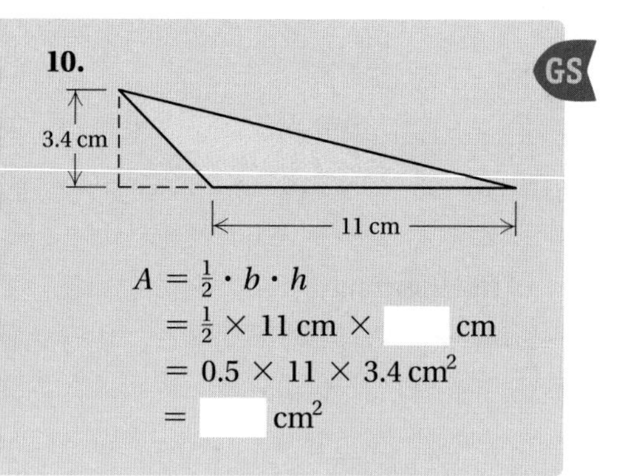

$A = \frac{1}{2} \cdot b \cdot h$

$= \frac{1}{2} \times 11 \text{ cm} \times \underline{\qquad} \text{ cm}$

$= 0.5 \times 11 \times 3.4 \text{ cm}^2$

$= \underline{\qquad} \text{ cm}^2$

EXAMPLE 7 Find the area of this triangle.

$$A = \frac{1}{2} \cdot b \cdot h$$

$$= \frac{1}{2} \cdot 9 \text{ m} \cdot 6 \text{ m}$$

$$= \frac{9 \cdot 6}{2} \text{ m}^2$$

$$= 27 \text{ m}^2$$

6 m

9 m

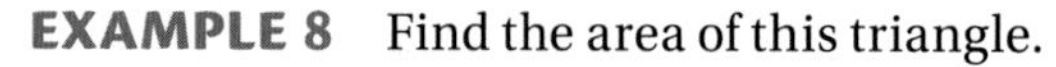

GS **EXAMPLE 8** Find the area of this triangle.

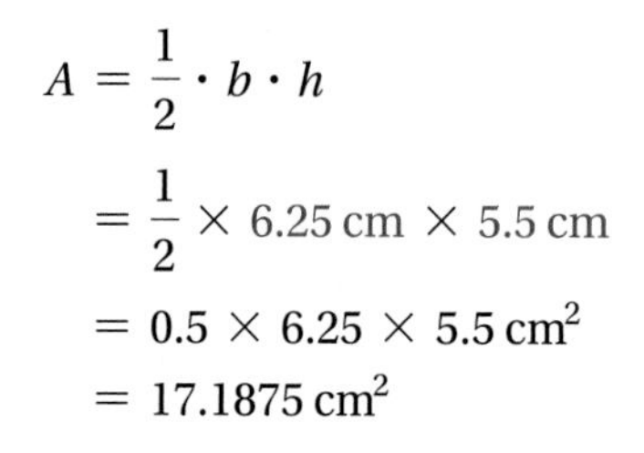

$$A = \frac{1}{2} \cdot b \cdot h$$

$$= \frac{1}{2} \times 6.25 \text{ cm} \times 5.5 \text{ cm}$$

$$= 0.5 \times 6.25 \times 5.5 \text{ cm}^2$$

$$= 17.1875 \text{ cm}^2$$

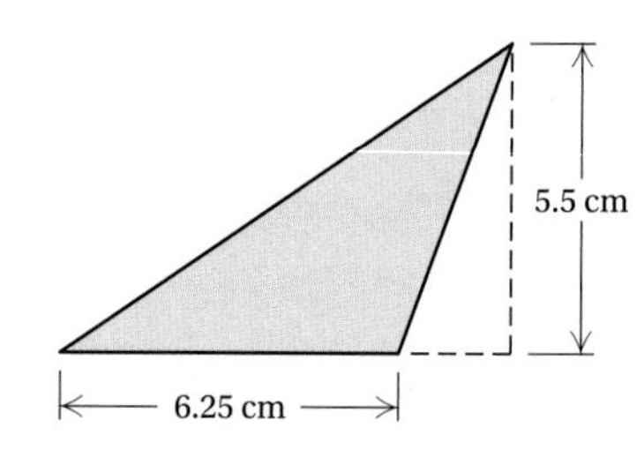

◀ **Do Exercises 9 and 10.**

Trapezoids

A **trapezoid** is a polygon with four sides, two of which, the **bases**, are parallel to each other.

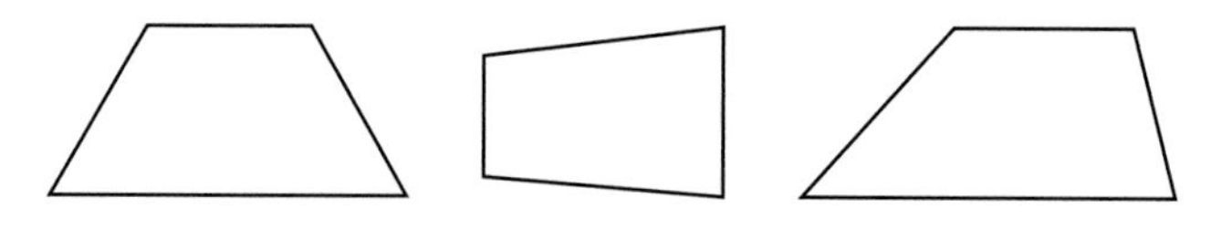

To find the area of a trapezoid, think of cutting out another just like it.

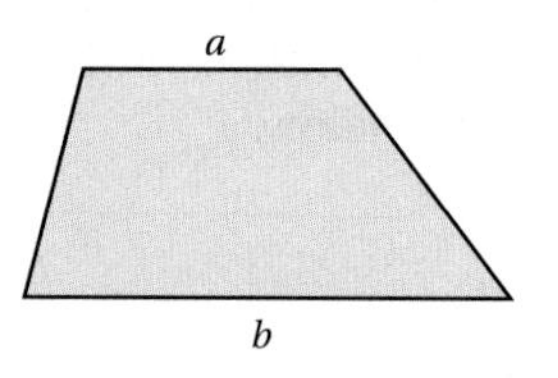

Then place the second one like this.

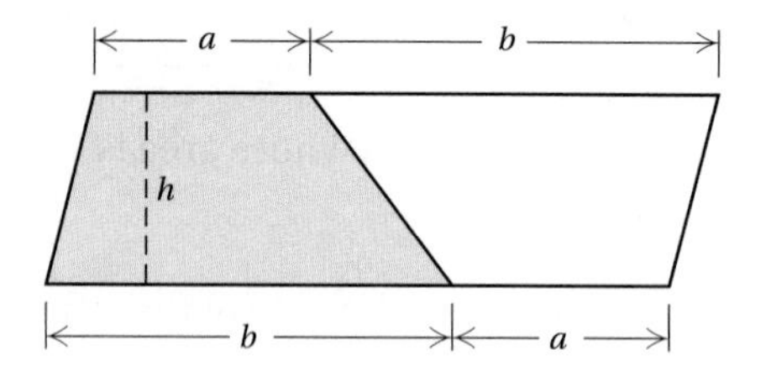

The resulting figure is a parallelogram whose area is

$$h \cdot (a + b). \qquad \text{The base is } a + b.$$

The trapezoid we began with has half the area of the parallelogram, or

$$\frac{1}{2} \cdot h \cdot (a + b).$$

Answers

9. 96 m^2 **10.** 18.7 cm^2

Guided Solution:

10. 3.4, 18.7

AREA OF A TRAPEZOID

The **area of a trapezoid** is half the product of the height and the sum of the lengths of the parallel sides (bases):

$$\mathbf{A = \frac{1}{2} \cdot h \cdot (a + b)}, \quad \text{or} \quad \mathbf{A = \frac{a + b}{2} \cdot h.}$$

a, *h*, *b*

EXAMPLE 9 Find the area of this trapezoid.

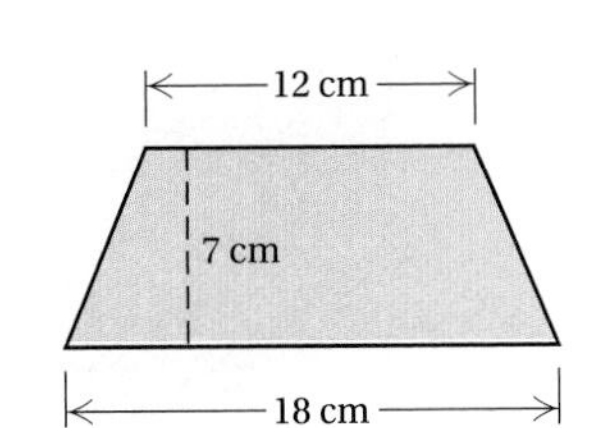

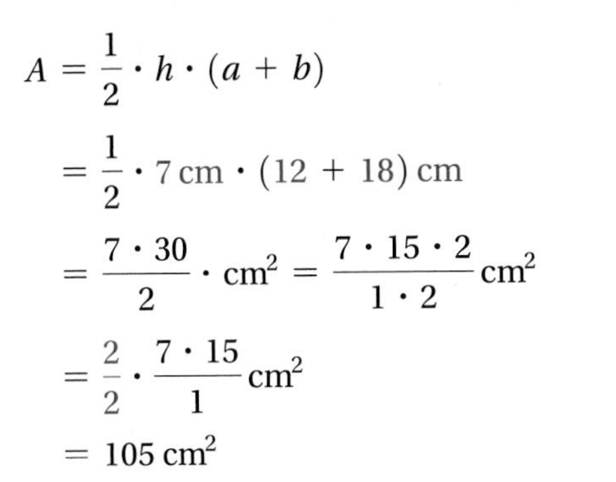

$$\begin{aligned} A &= \frac{1}{2} \cdot h \cdot (a + b) \\ &= \frac{1}{2} \cdot 7\text{ cm} \cdot (12 + 18)\text{ cm} \\ &= \frac{7 \cdot 30}{2} \cdot \text{cm}^2 = \frac{7 \cdot 15 \cdot 2}{1 \cdot 2}\text{cm}^2 \\ &= \frac{2}{2} \cdot \frac{7 \cdot 15}{1}\text{cm}^2 \\ &= 105\text{ cm}^2 \end{aligned}$$

Do Exercises 11 and 12. ▶

Find the area.

11.

12.

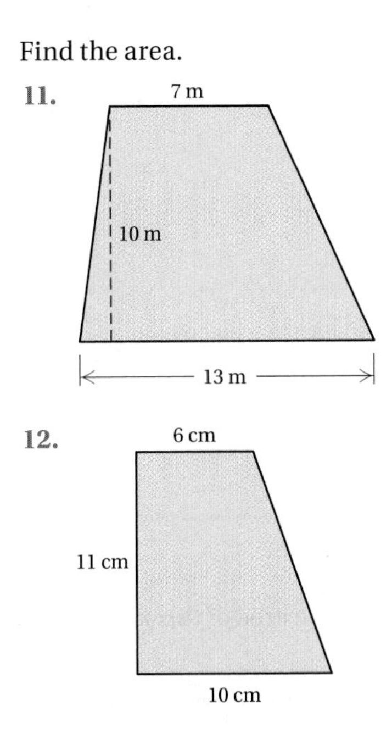

C SOLVING APPLIED PROBLEMS

EXAMPLE 10 *Mosquito Netting.* Malaria is the leading cause of death among children in Africa. Bed nets prevent malaria transmission by creating a protective barrier against mosquitoes at night. In November 2006, the United Nations Foundation, the United Methodist Church, and the National Basketball Association launched the Nothing But Nets campaign to distribute mosquito netting in Africa. In the next six years, nets were sent to more than 25 countries in Africa. A medium-sized net measures approximately 9.843 ft by 8.2025 ft. A large-sized net measures approximately 13.124 ft by 8.2025 ft. Find the area of each net. How much larger is the area of the large net than that of the medium net?

Source: www.nothingbutnets.net

We find the area of each net using the area formula $A = l \cdot w$ and substituting values for l and w. Then we subtract the area of the medium net from the area of the large net.

Area of Medium Net	*Area of Large Net*
$A = l \times w$	$A = l \times w$
$A \approx 9.843\text{ ft} \times 8.2025\text{ ft}$	$A \approx 13.124\text{ ft} \times 8.2025\text{ ft}$
$A \approx 80.74\text{ ft}^2$	$A \approx 107.65\text{ ft}^2$

$$\begin{aligned} \text{Area of Large Net} - \text{Area of Medium Net} &= 107.65\text{ ft}^2 - 80.74\text{ ft}^2 \\ &= 26.91\text{ ft}^2 \end{aligned}$$

The area of the large net is approximately 26.91 ft^2 larger than that of the medium net. ■

Answers

11. 100 m^2 **12.** 88 cm^2

EXAMPLE 11 *Lucas Oil Stadium.* The retractable roof of Lucas Oil Stadium, the home of the Indianapolis Colts football team, divides lengthwise. Each half measures 588 ft by 160 ft. The roof opens and closes in approximately 9–11 min. The opening measures 300 ft across. What is the total area of the retractable roof? What is the area of the opening?

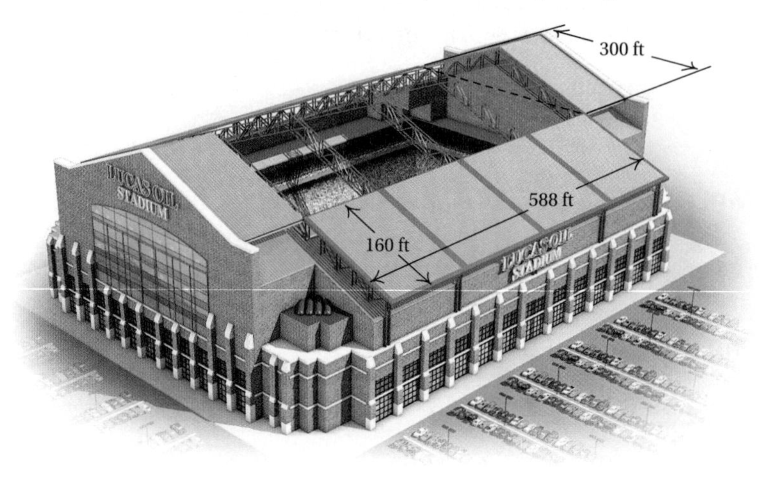

Each half of the retractable roof is a rectangle that measures 588 ft by 160 ft. The area of a rectangle is length times width, so we have

$$
\begin{aligned}
A &= l \cdot w \\
&= 588 \text{ ft} \times 160 \text{ ft} \\
&= 94{,}080 \text{ ft}^2.
\end{aligned}
$$

The total area of the two halves of the retractable roof is

$$
\begin{aligned}
\text{Total area} &= 2 \times 94{,}080 \text{ ft}^2 \\
&= 188{,}160 \text{ ft}^2.
\end{aligned}
$$

When the retractable roof is open, the dimensions of the opening are 588 ft by 300 ft. The area of this rectangle is

$$
\begin{aligned}
A &= l \cdot w \\
&= 588 \text{ ft} \times 300 \text{ ft} \\
&= 176{,}400 \text{ ft}^2.
\end{aligned}
$$

When the roof is open, the area of the opening is $176{,}400 \text{ ft}^2$.

◀ **Do Exercise 13.**

13. Find the area of this kite.

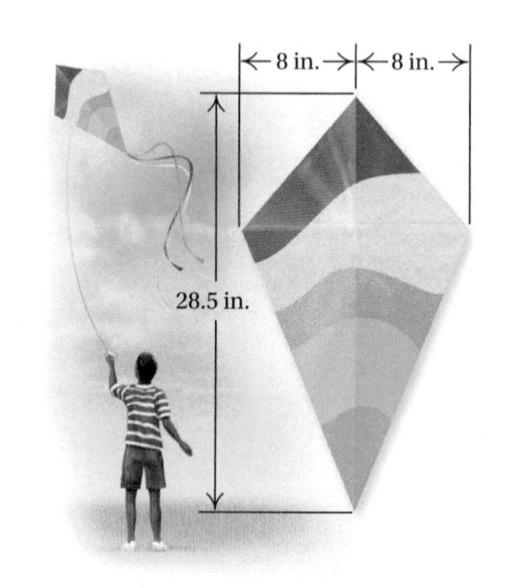

Answer

13. 228 in^2

Reading Check

Complete each statement with the correct phrase from the list at the right.

RC1. The area of a square is ______________.

RC2. The area of a rectangle is ______________.

RC3. The area of a triangle is ______________.

RC4. The area of a trapezoid is ______________.

a) half the length of the base times the height

b) the square of the length of a side

c) the product of the length and the width

d) half the product of the height and the sum of the lengths of the bases

a Find the area.

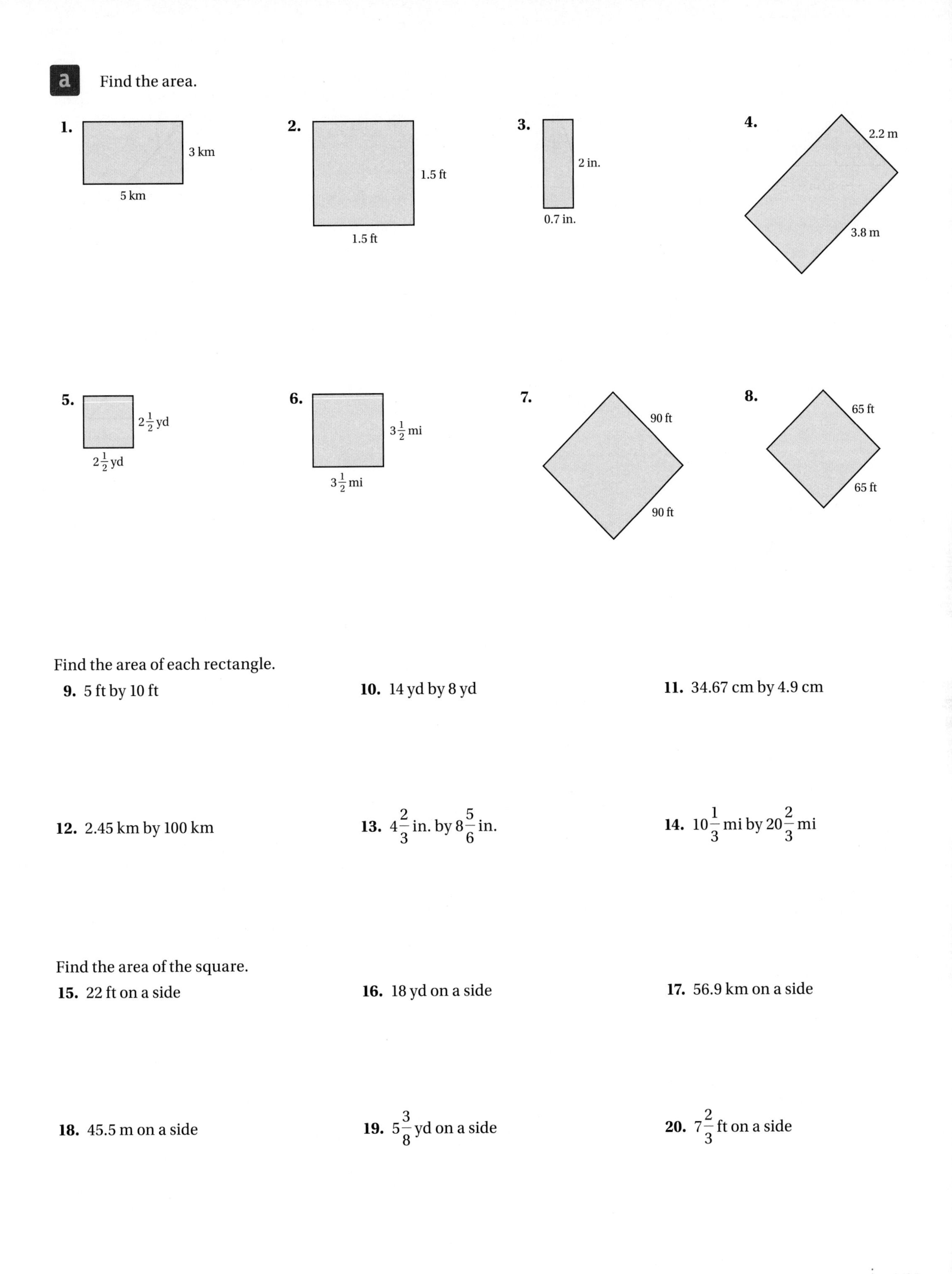

Find the area of each rectangle.

9. 5 ft by 10 ft

10. 14 yd by 8 yd

11. 34.67 cm by 4.9 cm

12. 2.45 km by 100 km

13. $4\frac{2}{3}$ in. by $8\frac{5}{6}$ in.

14. $10\frac{1}{3}$ mi by $20\frac{2}{3}$ mi

Find the area of the square.

15. 22 ft on a side

16. 18 yd on a side

17. 56.9 km on a side

18. 45.5 m on a side

19. $5\frac{3}{8}$ yd on a side

20. $7\frac{2}{3}$ ft on a side

b Find the area.

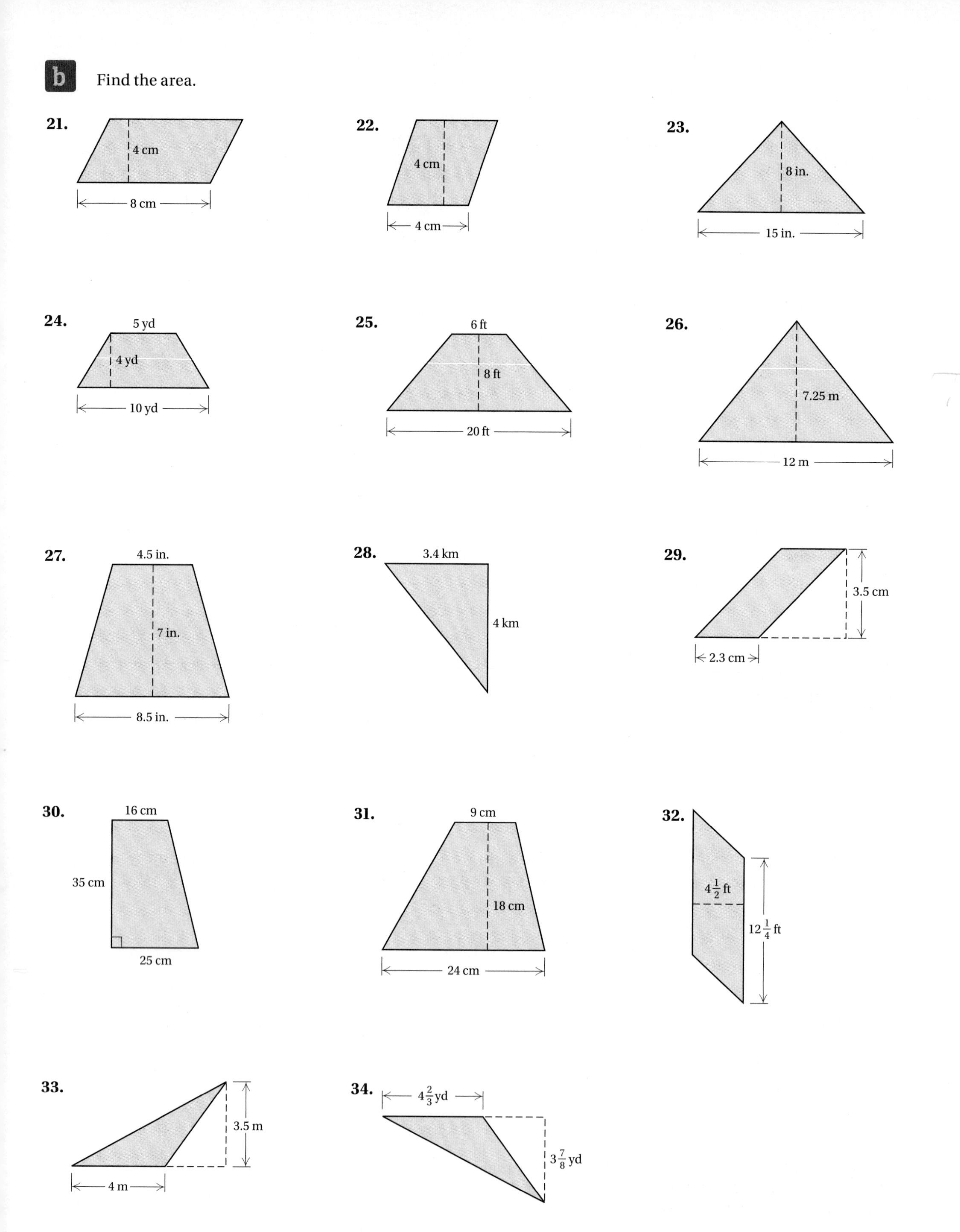

C Solve.

35. *Area of a Lawn.* A lot is 40 m by 36 m. A house 27 m by 9 m is built on the lot. How much area is left over for a lawn?

36. *Area of a Field.* A field is 240.8 m by 450.2 m. A rectangular area that measures 160.4 m by 90.6 m is paved for a parking lot. How much area is unpaved?

37. For a performance at an outdoor event, a folk music group rented a triangular tent. The base of the tent was 20 ft and the height was $17\frac{1}{2}$ ft, and it was placed in a corner of a small park that measured 200 ft by 200 ft. Approximately how much of the park space was left for the audience.

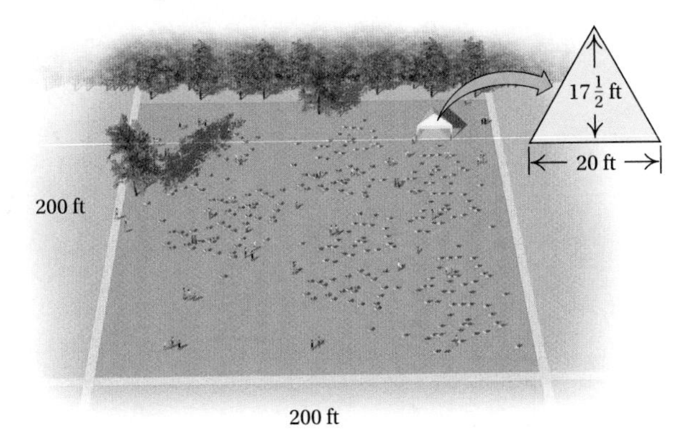

38. Becky's rectangular swimming pool measures 27 ft by 14.6 ft. She likes to relax while floating on an inflatable mattress, which measures 6.5 ft by 2.75 ft. What area of the pool is left around her for her nieces and nephews to play in?

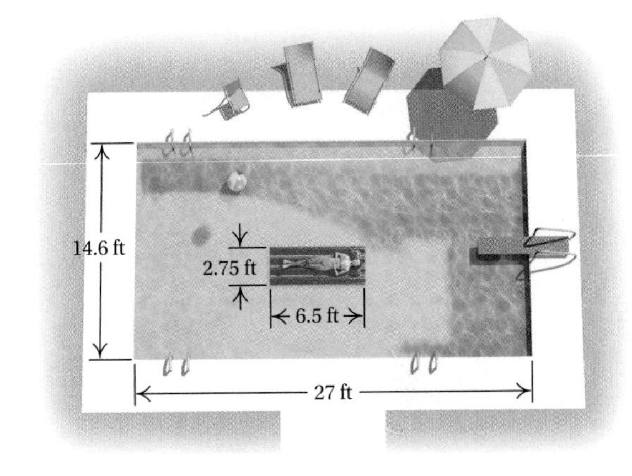

39. *Area of a Sidewalk.* Franklin Construction Company builds a sidewalk around two sides of a new library, as shown in the figure.

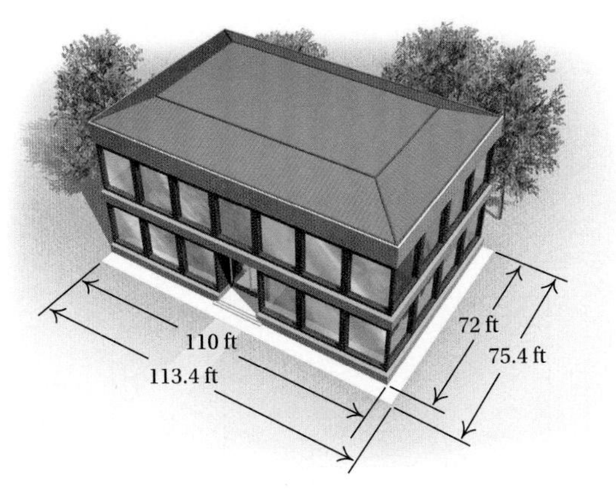

a) What is the area of the sidewalk?
b) The concrete for the sidewalk will cost the library \$12.50 per square foot. How much will the concrete for the project cost?

40. Maravene is planning a wildflower border around three sides of her backyard, as shown in the figure. She will use wildflower mats to seed the border. Each mat covers 7.5 ft^2 and costs \$4.99.

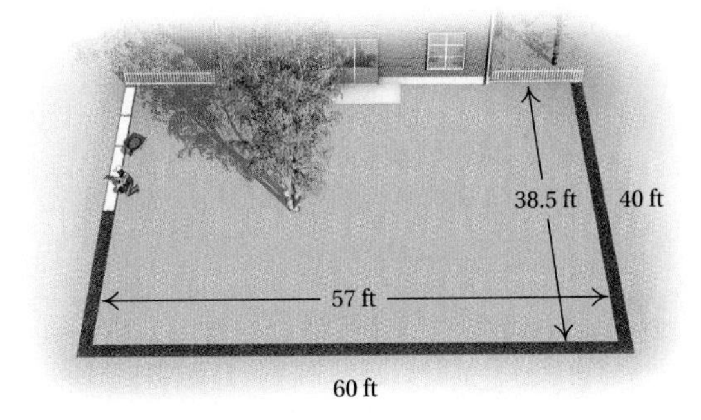

a) What is the area of the border?
b) How many wildflower mats will Maravene need to complete the job?
c) What will be the total cost of the wildflower mats?

41. *Painting Costs.* A room is 15 ft by 20 ft. The ceiling is 8 ft above the floor. There are two windows in the room, each 3 ft by 4 ft. The door is $2\frac{1}{2}$ ft by $6\frac{1}{2}$ ft.

a) What is the total area of the walls and the ceiling?
b) A gallon of paint will cover 360.625 ft^2. How many gallons of paint are needed for the room, including the ceiling?
c) Paint costs \$34.95 a gallon. How much will it cost to paint the room?

42. *Carpeting Costs.* A restaurant owner wants to carpet a 15-yd by 20-yd room.

a) How many square yards of carpeting are needed?
b) The carpeting she wants is \$28.50 per square yard, including installation. How much will it cost to carpet the room?

Find the area of the shaded region in each figure.

43.

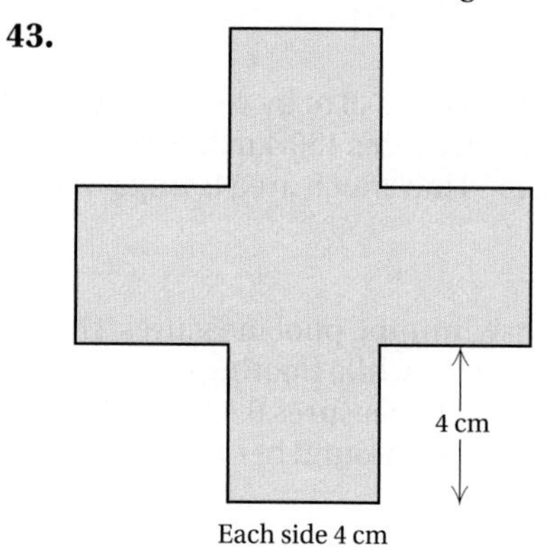

44.

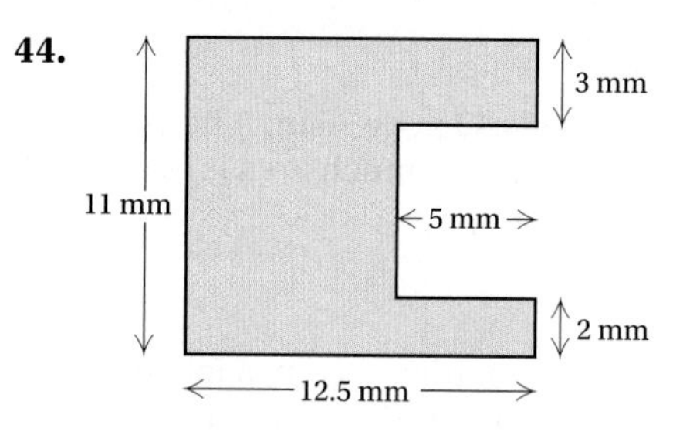

45.

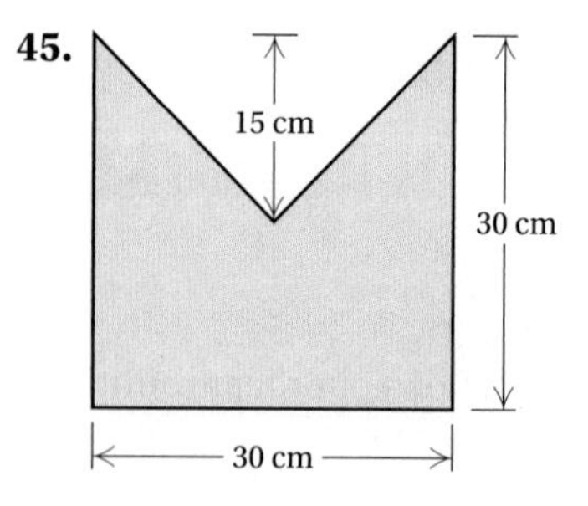

46.

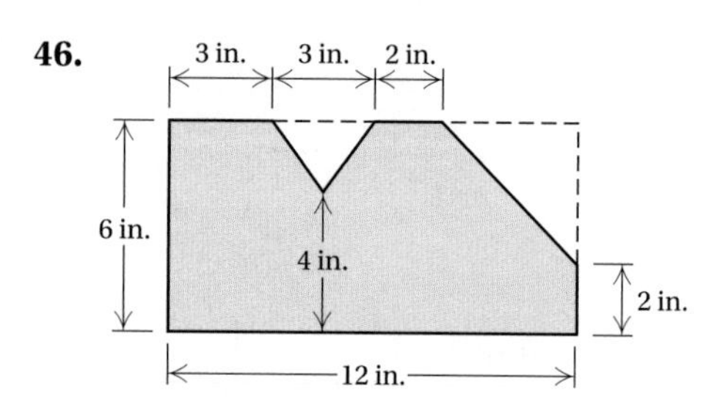

47.

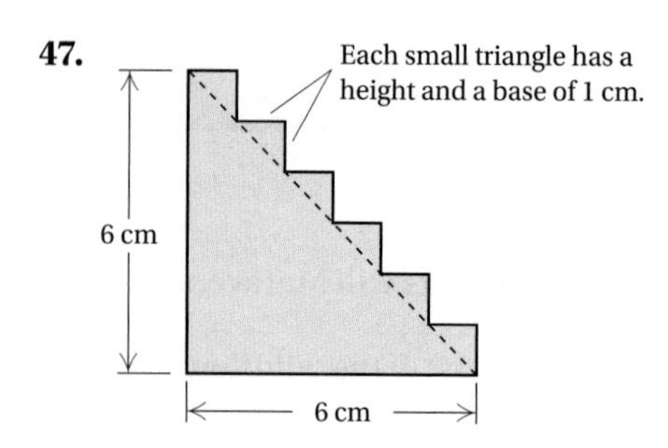

48.

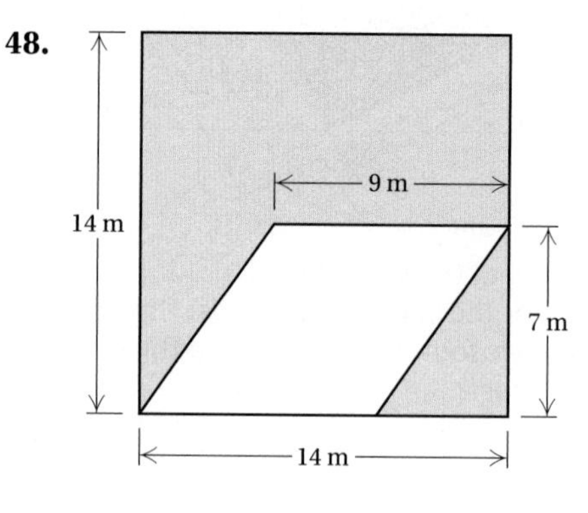

49. *Triangular Sail.* Jane's Custom Sails is making a custom sail for a laser sailboat. From a rectangular piece of dacron sailcloth that measures 18 ft by 12 ft, Jane cuts out a right triangular area plus a rectangular extension on each side for the hems, with the dimensions shown below. How much fabric (area) is left over?

50. *Building Area.* Find the total area of the sides and the ends of the building.

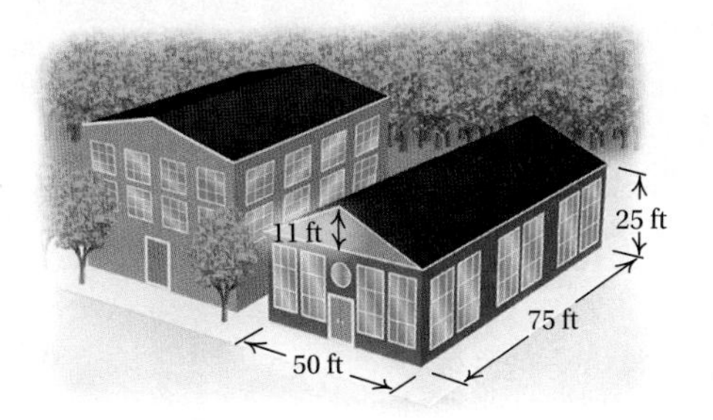

Skill Maintenance

Complete. [9.1a], [9.2a]

51. 23.4 cm = ________ mm

52. 0.23 km = ________ m

53. 28 ft = ________ in.

54. 72 ft = ________ yd

55. 72.4 cm = ________ m

56. 72.4 m = ________ km

57. 70 yd = ________ in.

58. 31,680 ft = ________ mi

59. 84 ft = ________ yd

60. $7\frac{1}{2}$ yd = ________ ft

61. 144 in. = ________ ft

62. 0.73 mi = ________ in.

Interest is compounded semiannually. Find the amount in the account after the given length of time. Round to the nearest cent. [7.7b]

	Principal	Rate of Interest	Time	Amount in the Account
63.	$25,000	4%	5 years	
64.	$150,000	$6\frac{7}{8}$%	15 years	
65.	$150,000	7.4%	20 years	
66.	$160,000	5%	20 years	

Synthesis

67. Find the area, in square inches, of the shaded region.

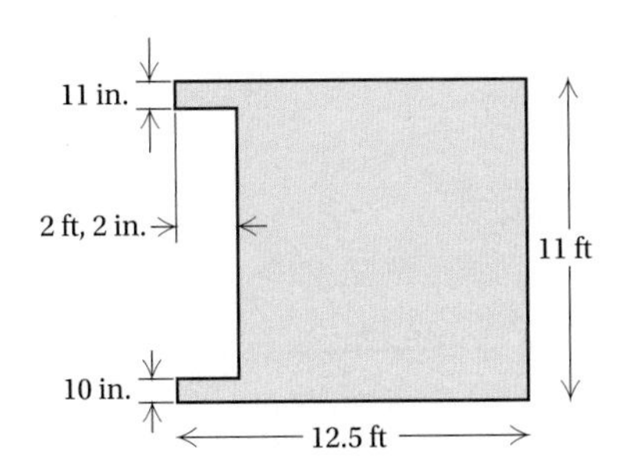

68. Find the area, in square feet, of the shaded region.

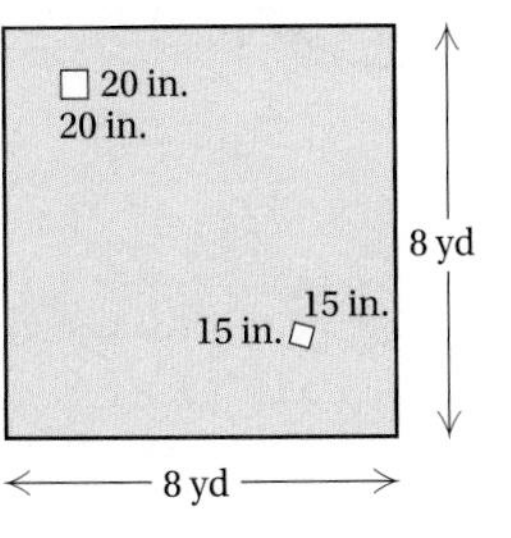

10.3 Circles

OBJECTIVES

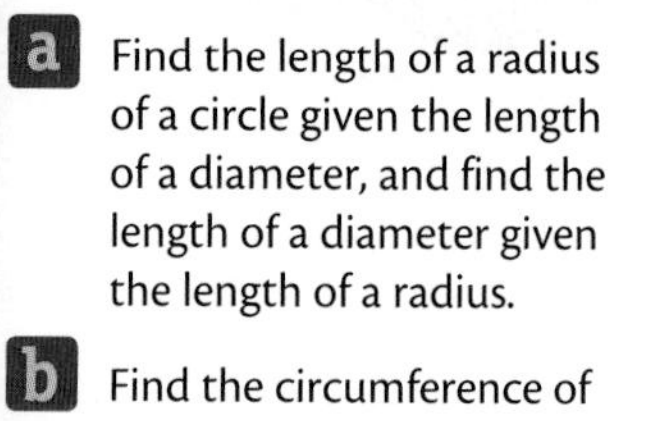

a Find the length of a radius of a circle given the length of a diameter, and find the length of a diameter given the length of a radius.

b Find the circumference of a circle given the length of a diameter or a radius.

c Find the area of a circle given the length of a diameter or a radius.

d Solve applied problems involving circles.

SKILL TO REVIEW

Objective 3.6a: Multiply and simplify using fraction notation.

Multiply and simplify.

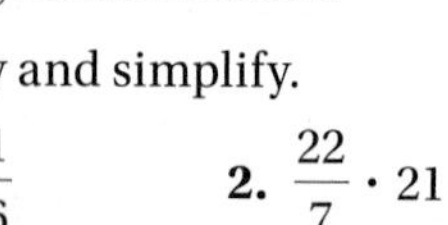

1. $2 \cdot \frac{1}{6}$ **2.** $\frac{22}{7} \cdot 21$

1. Find the length of a radius.

20 m

2. Find the length of a diameter.

$2\frac{1}{2}$ ft

Answers

Skill to Review:

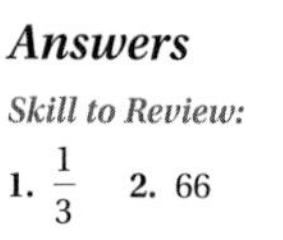

1. $\frac{1}{3}$ **2.** 66

Margin Exercises:

1. 10 m **2.** 5 ft

a RADIUS AND DIAMETER

Shown below is a circle with center O. Segment $\overline{AC}$ is a *diameter*. A **diameter** is a segment that passes through the center of the circle and has endpoints on the circle. Segment $\overline{OB}$ is called a *radius*. A **radius** is a segment with one endpoint on the center and the other endpoint on the circle.

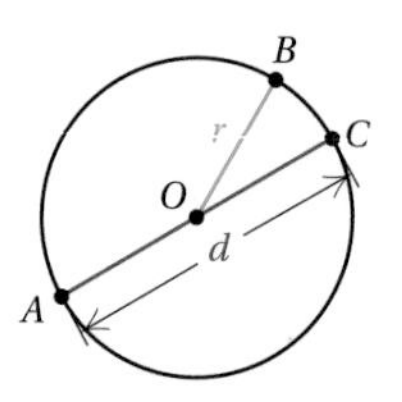

DIAMETER AND RADIUS

Suppose that d is the length of a diameter of a circle and r is the length of a radius. Then

$$\boldsymbol{d = 2 \cdot r} \quad \text{and} \quad \boldsymbol{r = \frac{d}{2}}.$$

EXAMPLE 1 Find the length of a radius of this circle.

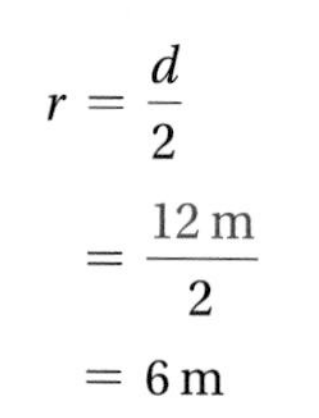

$$r = \frac{d}{2} = \frac{12\text{ m}}{2} = 6\text{ m}$$

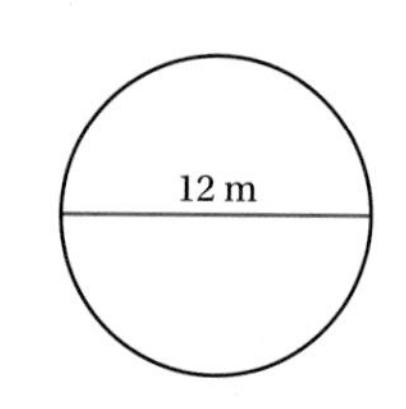

The radius is 6 m.

EXAMPLE 2 Find the length of a diameter of this circle.

$$d = 2 \cdot r = 2 \cdot \frac{1}{4}\text{ft} = \frac{1}{2}\text{ft}$$

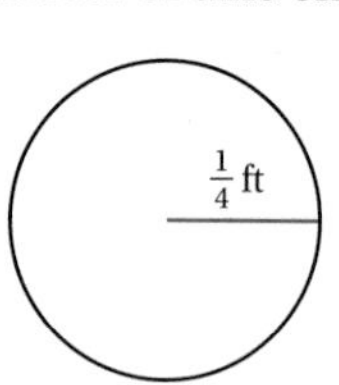

The diameter is $\frac{1}{2}$ ft.

◀ **Do Margin Exercises 1 and 2.**

b CIRCUMFERENCE

The **circumference** of a circle is the distance around it. Calculating the circumference is similar to finding the perimeter of a polygon.

To find a formula for the circumference C of any circle given its diameter d, we consider the ratio C/d. Take a dinner plate and measure the circumference C with a tape measure. Also measure the diameter d. The results for a specific plate are shown in the figure below.

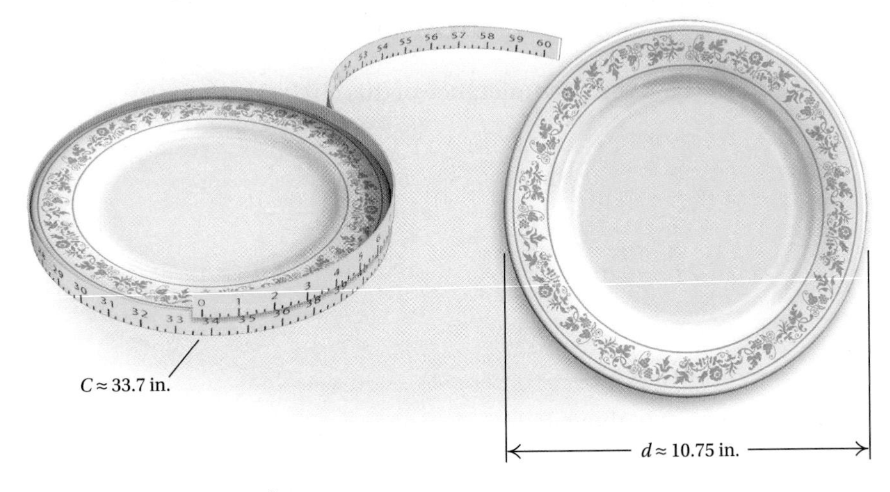

Finding the ratio, we have

$$\frac{C}{d} = \frac{33.7 \text{ in.}}{10.75 \text{ in.}} \approx 3.1.$$

Suppose we do this with plates and circles of several sizes. We get different values for C and d, but always a number close to 3.1 for C/d. For any circle, if we divide the circumference C by the diameter d, we get the same number. We call this number π (pi). The *exact* value of the ratio C/d is π; 3.14 and $22/7$ are approximations of π. If $C/d = \pi$, then $C = \pi \cdot d$.

CIRCUMFERENCE AND DIAMETER

The circumference C of a circle of diameter d is given by

$$C = \pi \cdot d.$$

The number π is about 3.14, or about $\frac{22}{7}$.

EXAMPLE 3 Find the circumference of this circle. Use 3.14 for π.

$$C = \pi \cdot d$$
$$\approx 3.14 \times 6 \text{ cm}$$
$$= 18.84 \text{ cm}$$

6 cm

The circumference is about 18.84 cm.

Do Exercise 3.

GS **3.** Find the circumference of the circle. Use 3.14 for π.

$C = \pi \cdot d$

$\approx 3.14 \times$ ____ in.

$=$ ____ in.

Answer

3. 56.52 in.

Guided Solution:

3. 18, 56.52

CALCULATOR CORNER

The* $\boxed{\pi}$ *Key Many calculators have a $\boxed{\pi}$ key that can be used to enter the value of π in a computation. Results obtained using the $\boxed{\pi}$ key might be slightly different from those obtained when 3.14 or $\frac{22}{7}$ is used for the value of π in a computation.

When we use the $\boxed{\pi}$ key to find the circumference of the circle in Example 3, we get a result of approximately 18.85. Note that this is slightly different from the result found using 3.14 for the value of π.

EXERCISES

1. Use a calculator with a $\boxed{\pi}$ key to perform the computations in Examples 4 and 5.
2. Use a calculator with a $\boxed{\pi}$ key to perform the computations in Margin Exercises 3–5.

---

4. Find the circumference of this circle. Use $\frac{22}{7}$ for π.

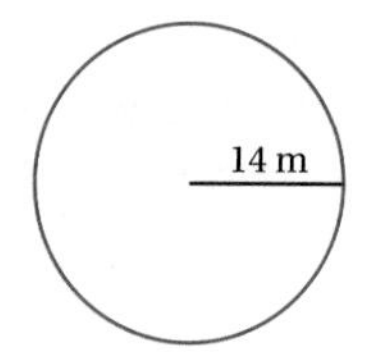

5. Find the perimeter of this figure. Use 3.14 for π.

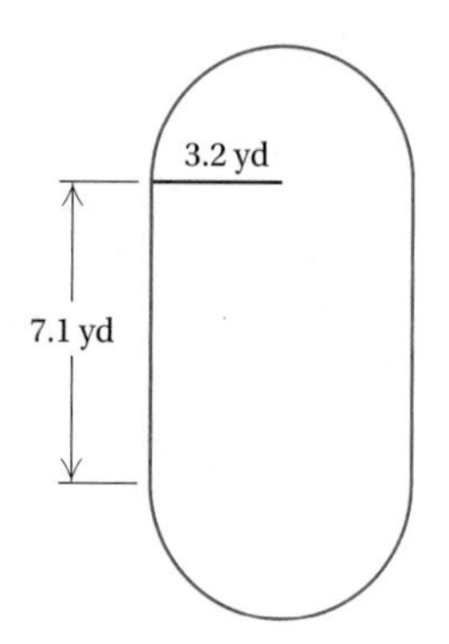

Answers

4. 88 m 5. 34.296 yd

Since $d = 2 \cdot r$, where r is the length of a radius, it follows that

$$C = \pi \cdot d = \pi \cdot (2 \cdot r), \quad \text{or} \quad 2 \cdot \pi \cdot r.$$

CIRCUMFERENCE AND RADIUS

The circumference C of a circle of radius r is given by

$$\boldsymbol{C = 2 \cdot \pi \cdot r}.$$

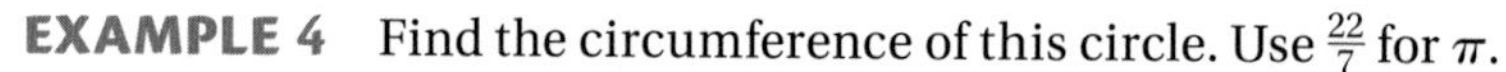

EXAMPLE 4 Find the circumference of this circle. Use $\frac{22}{7}$ for π.

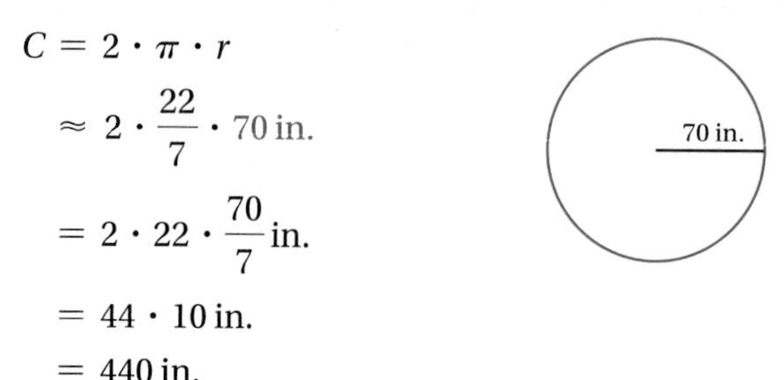

$$\begin{aligned} C &= 2 \cdot \pi \cdot r \\ &\approx 2 \cdot \frac{22}{7} \cdot 70 \text{ in.} \\ &= 2 \cdot 22 \cdot \frac{70}{7} \text{ in.} \\ &= 44 \cdot 10 \text{ in.} \\ &= 440 \text{ in.} \end{aligned}$$

The circumference is about 440 in.

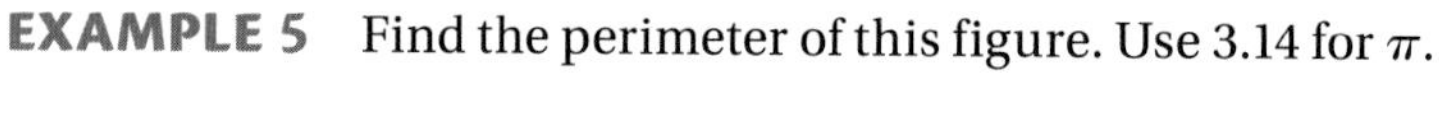

EXAMPLE 5 Find the perimeter of this figure. Use 3.14 for π.

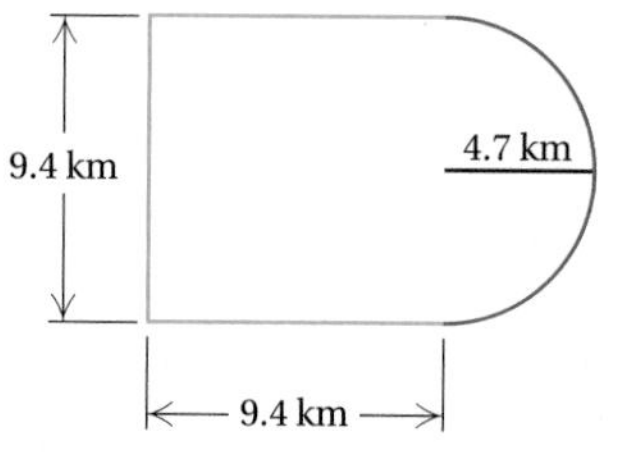

We let $P =$ the perimeter. We see that we have half a circle attached to a square. Thus we add half the circumference of the circle to the lengths of the three sides of the square.

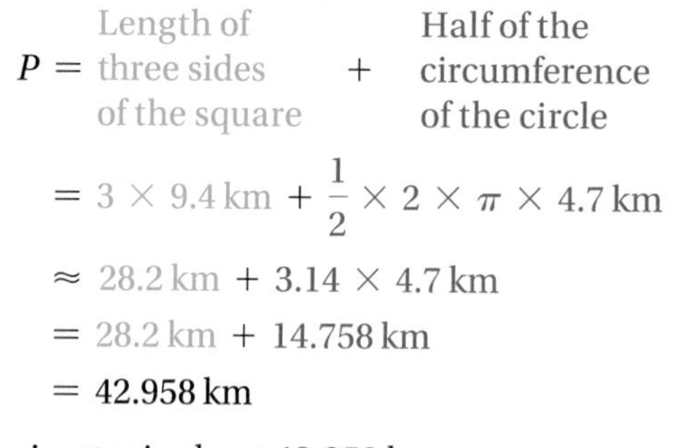

$$P = \begin{matrix}\text{Length of} \\ \text{three sides} \\ \text{of the square}\end{matrix} + \begin{matrix}\text{Half of the} \\ \text{circumference} \\ \text{of the circle}\end{matrix}$$

$$\begin{aligned} &= 3 \times 9.4 \text{ km} + \frac{1}{2} \times 2 \times \pi \times 4.7 \text{ km} \\ &\approx 28.2 \text{ km} + 3.14 \times 4.7 \text{ km} \\ &= 28.2 \text{ km} + 14.758 \text{ km} \\ &= 42.958 \text{ km} \end{aligned}$$

The perimeter is about 42.958 km.

Do Exercises 4 and 5.

C AREA

To find the area of a circle of radius r, think of cutting half the circular region into small pieces and arranging them as shown below.

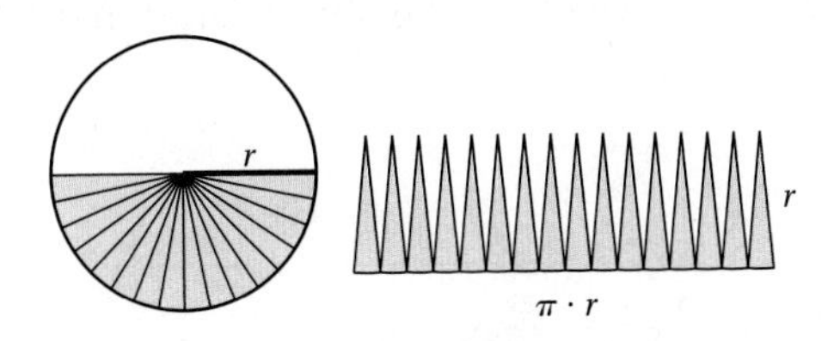

Then imagine cutting the other half of the circular region and arranging the pieces in with the others as shown below.

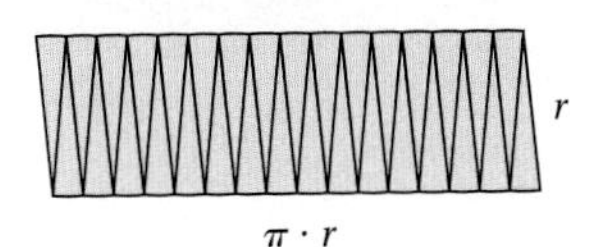

This is almost a parallelogram. The base has length $\frac{1}{2} \cdot 2 \cdot \pi \cdot r$, or $\pi \cdot r$ (half the circumference), and the height is r. Thus the area is

$$(\pi \cdot r) \cdot r.$$

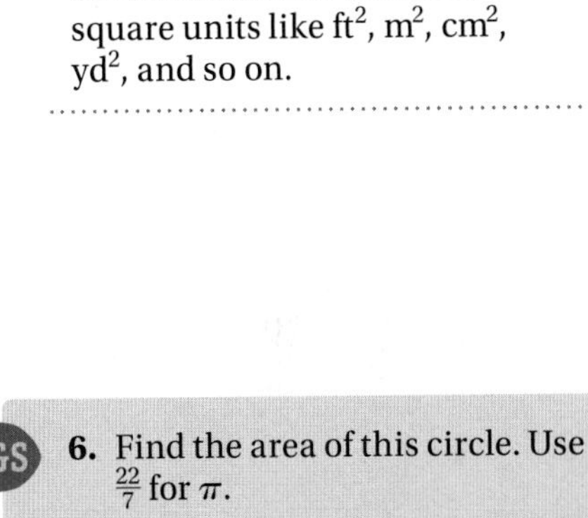

Caution!

Remember that circumference is always measured in linear units like ft, m, cm, yd, and so on. But area is measured in square units like ft^2, m^2, cm^2, yd^2, and so on.

AREA OF A CIRCLE

The **area of a circle** with radius of length r is given by

$$\boldsymbol{A = \pi \cdot r \cdot r}, \quad \text{or} \quad \boldsymbol{A = \pi \cdot r^2}.$$

EXAMPLE 6 Find the area of this circle. Use $\frac{22}{7}$ for π.

$$\begin{aligned} A &= \pi \cdot r \cdot r \\ &\approx \frac{22}{7} \cdot 14\text{ cm} \cdot 14\text{ cm} \\ &= \frac{22}{7} \cdot 196\text{ cm}^2 \\ &= 616\text{ cm}^2 \end{aligned}$$

14 cm

The area is about 616 cm^2.

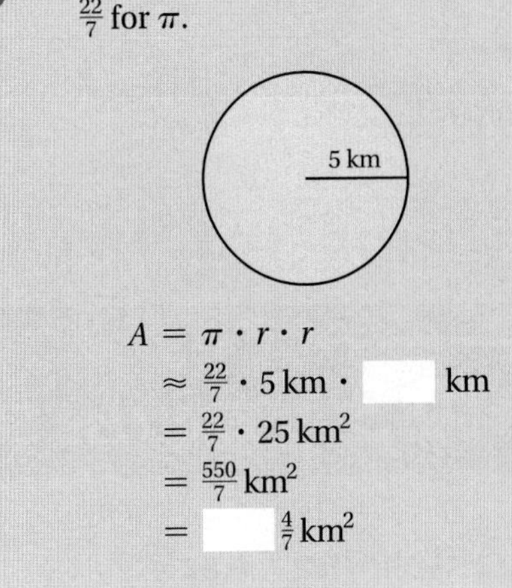

GS **6.** Find the area of this circle. Use $\frac{22}{7}$ for π.

5 km

$$\begin{aligned} A &= \pi \cdot r \cdot r \\ &\approx \tfrac{22}{7} \cdot 5\text{ km} \cdot \underline{\quad}\text{ km} \\ &= \tfrac{22}{7} \cdot 25\text{ km}^2 \\ &= \tfrac{550}{7}\text{ km}^2 \\ &= \underline{\quad}\tfrac{4}{7}\text{ km}^2 \end{aligned}$$

EXAMPLE 7 Find the area of this circle. Use 3.14 for π. Round to the nearest hundredth.

The diameter is 4.2 m; the radius is 4.2 m $\div$ 2, or 2.1 m.

$$\begin{aligned} A &= \pi \cdot r \cdot r \\ &\approx 3.14 \times 2.1\text{ m} \times 2.1\text{ m} \\ &= 3.14 \times 4.41\text{ m}^2 \\ &= 13.8474\text{ m}^2 \\ &\approx 13.85\text{ m}^2 \end{aligned}$$

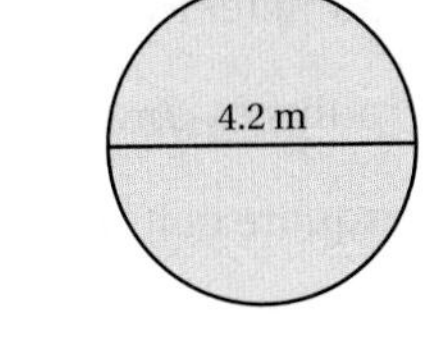

The area is about 13.85 m^2.

Do Exercises 6 and 7.

7. Find the area of this circle. Use 3.14 for π. Round to the nearest hundredth.

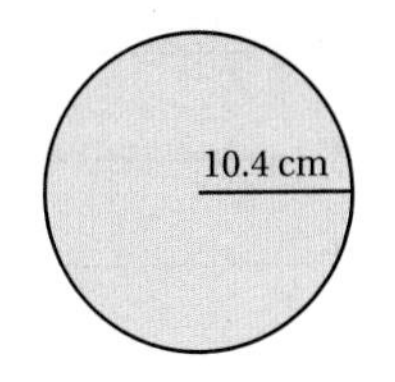

Answers

6. $78\frac{4}{7}\text{ km}^2$ **7.** 339.62 cm^2

Guided Solution:
6. 5, 78

d SOLVING APPLIED PROBLEMS

EXAMPLE 8 *Areas of Cake Pans.* Tyler can make either a 9-in. round cake or a 9-in. square cake for a party. If he makes the square cake, how much more area will he have on the top for decorations?

The area of the square is

$$A = s \cdot s$$
$$= 9 \text{ in.} \times 9 \text{ in.} = 81 \text{ in}^2.$$

The diameter of the circle is 9 in., so the radius is 9 in./2, or 4.5 in. The area of the circle is

$$A = \pi \cdot r \cdot r$$
$$\approx 3.14 \times 4.5 \text{ in.} \times 4.5 \text{ in.} = 63.585 \text{ in}^2.$$

The area of the square cake is larger by about

$$81 \text{ in}^2 - 63.585 \text{ in}^2, \quad \text{or} \quad 17.415 \text{ in}^2.$$

◀ Do Exercise 8.

8. Which is larger and by how much: a 10-ft-square flower bed or a 12-ft-diameter round flower bed?

Answer

8. 12-ft-diameter round flower bed, by about 13.04 ft^2

10.3 Exercise Set

For Extra Help
MyMathLab® MathXL® PRACTICE WATCH READ REVIEW

Reading Check

Complete each statement with the correct word from the following list. A word may be used more than once or not at all.

area   circumference   diameter   radius

RC1. The ______________ of a circle is half the length of its diameter.

RC2. The ______________ of a circle is found by multiplying its diameter by π.

RC3. The ______________ of a circle is found by multiplying its radius by 2π.

RC4. The ______________ of a circle is found by multiplying the square of its radius by π.

 For each circle, find the length of a diameter, the circumference, and the area. Use $\frac{22}{7}$ for π.

1.

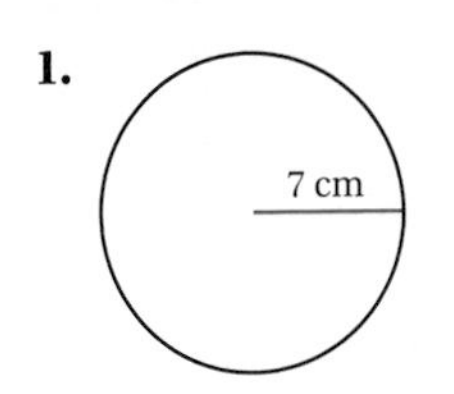

2.

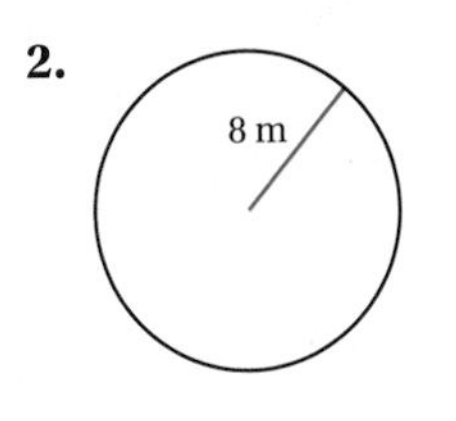

3.

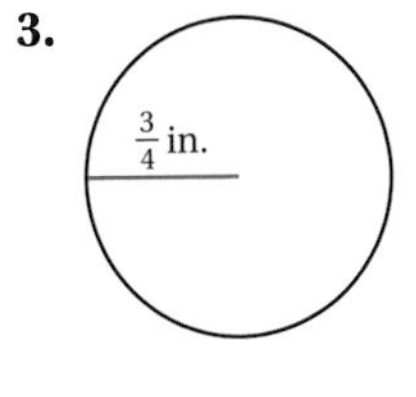

4.

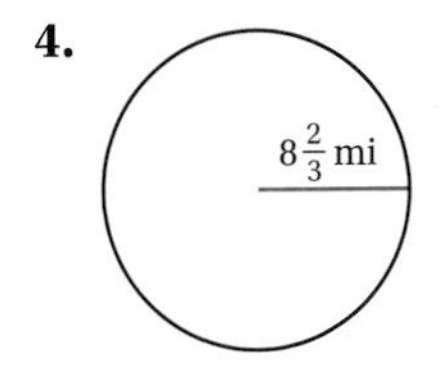

For each circle, find the length of a radius, the circumference, and the area. Use 3.14 for π.

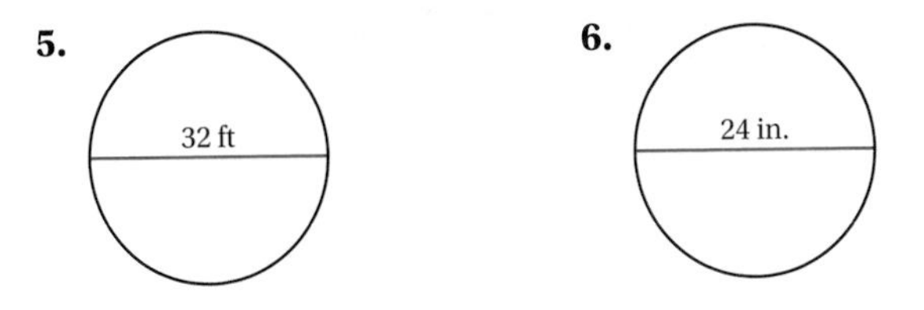

7.

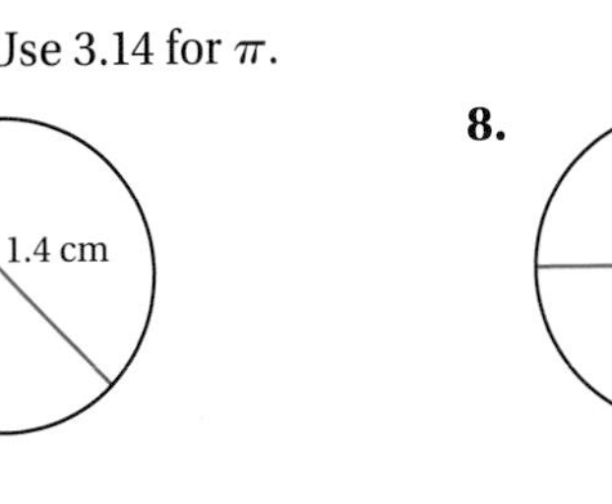

8.

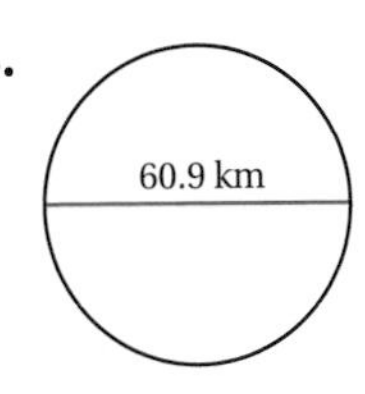

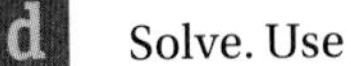 Solve. Use 3.14 for π.

9. *Pond Edging.* Quiet Designs plans to incorporate a circular pond with a diameter of 30 ft in a landscape design. The pond will be edged using stone pavers. How many feet of pavers will be needed?

10. *Gypsy-Moth Tape.* To protect an elm tree in your backyard, you decide to attach gypsy moth caterpillar tape around the trunk. The tree has a 1.1-ft diameter. What length of tape is needed?

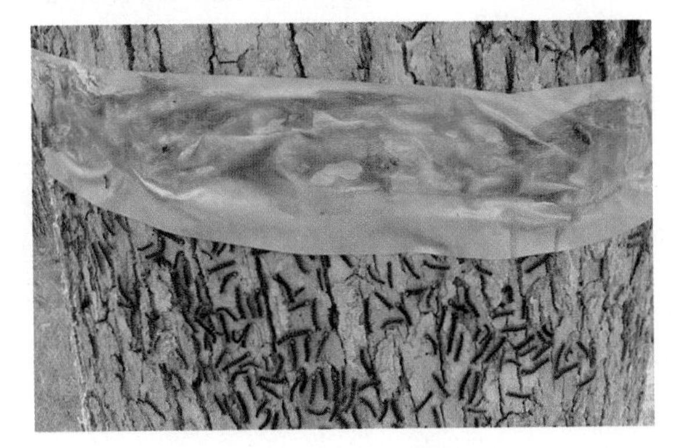

11. *Areas of Pizza Pans.* How much larger is a pizza made in a 16-in.-square pizza pan than a pizza made in a 16-in.-diameter circular pan?

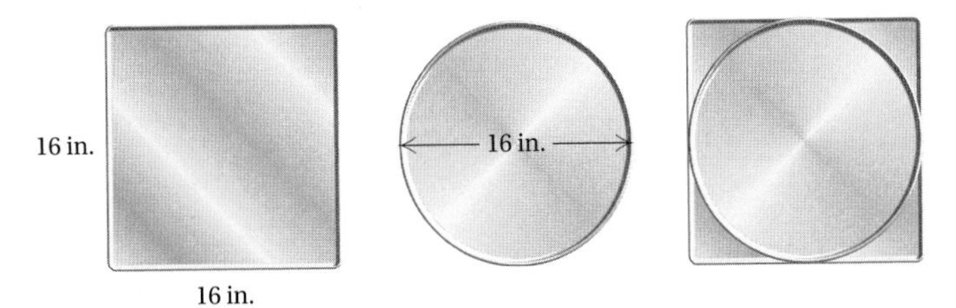

12. *Penny.* A penny has a 1-cm radius. What is its diameter? its circumference? its area?

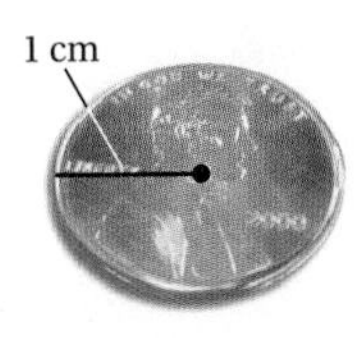

13. *Earth.* The circumference of the earth at the equator is 24,901 mi. What is the diameter of the earth at the equator? the radius?

14. *Dimensions of a Quarter.* The circumference of a quarter is 7.85 cm. What is the diameter? the radius? the area?

15. *Circumference of a Baseball Bat.* In Major League Baseball, the diameter of the barrel of a bat cannot be more than $2\frac{3}{4}$ in., and the diameter of the bat handle cannot be less than $\frac{16}{19}$ in. Find the maximum circumference of the barrel of a bat and the minimum circumference of the bat handle. Use $\frac{22}{7}$ for π.

Source: Major League Baseball

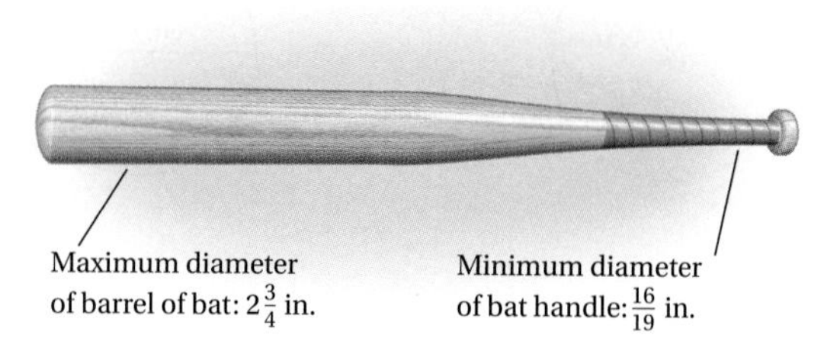

16. *Trampoline.* The standard backyard trampoline has a diameter of 14 ft. What is its area?

Source: International Trampoline Industry Association, Inc.

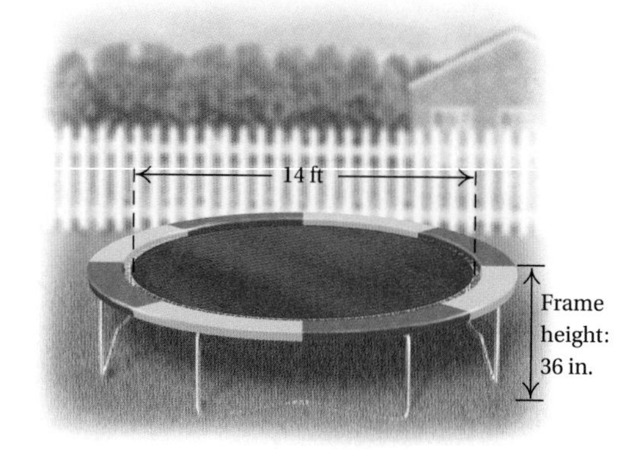

17. *Swimming-Pool Walk.* You want to install a 1-yd-wide walk around a circular swimming pool. The diameter of the pool is 20 yd. What is the area of the walk?

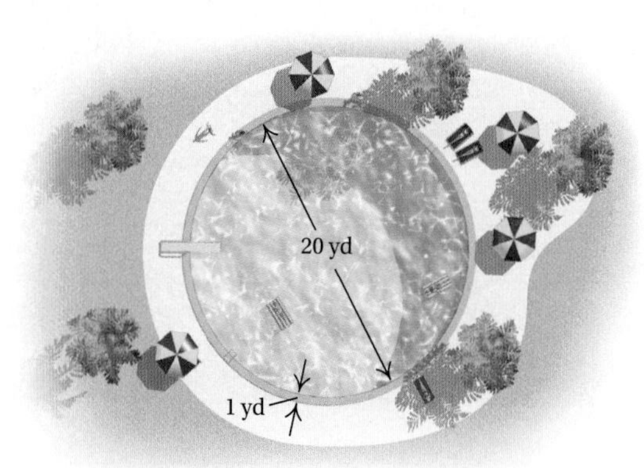

18. *Roller-Rink Floor.* A roller-rink floor is shown below. Each end is a semicircle. What is its area? If hardwood flooring costs $32.50 per square meter, how much will the flooring cost?

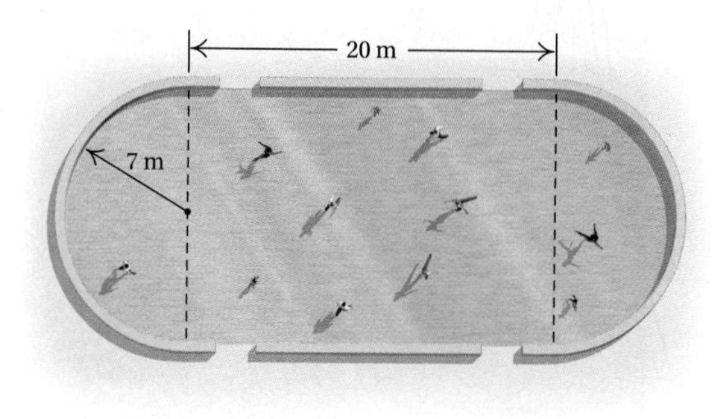

Find the perimeter of each figure. Use 3.14 for π.

19.

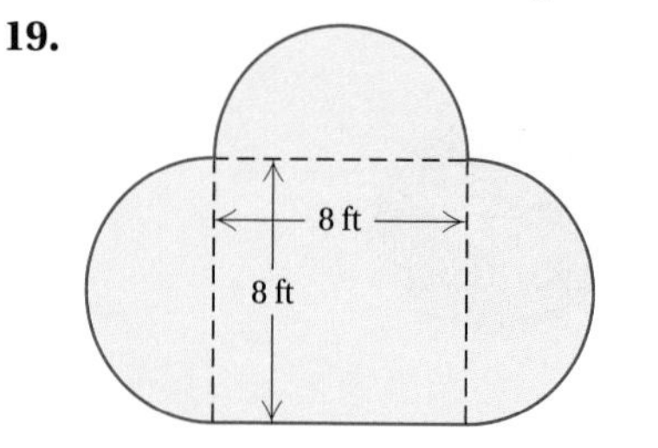

20.

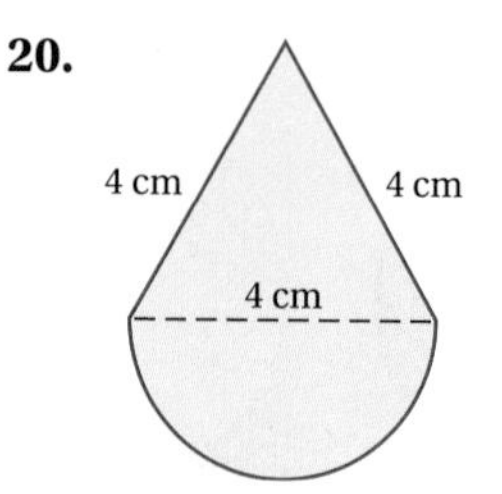

21.

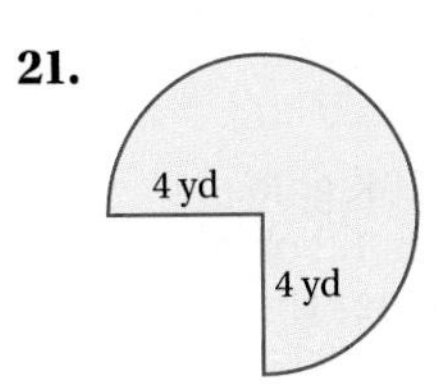

22.

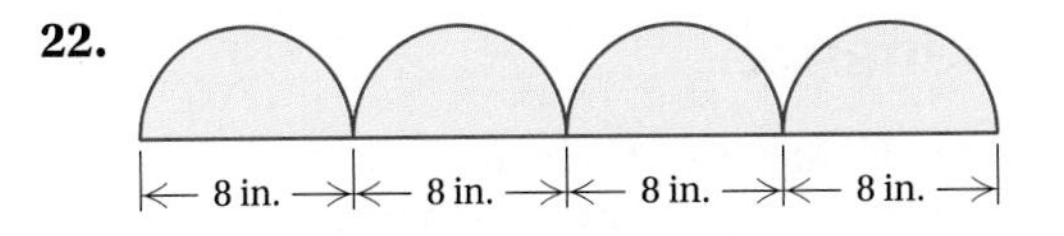

23.

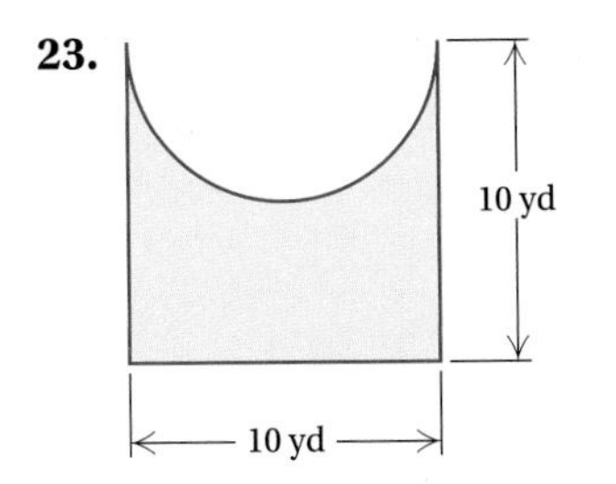

24.

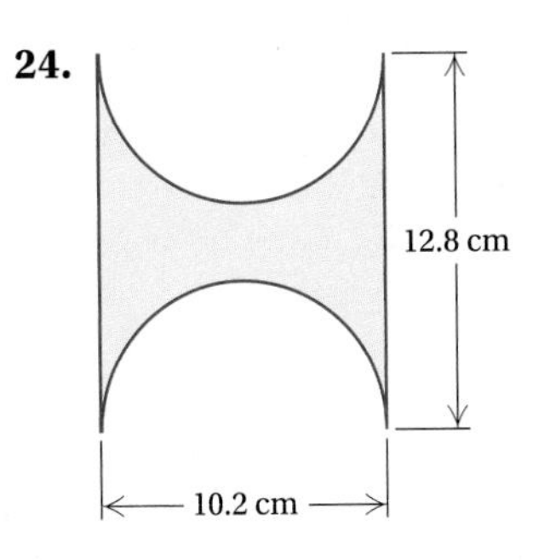

Find the area of the shaded region in each figure. Use 3.14 for π.

25.

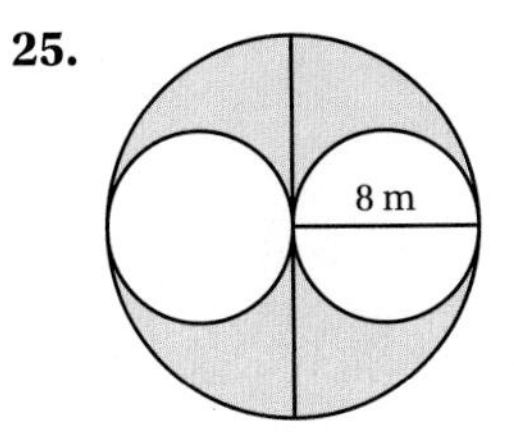

26.

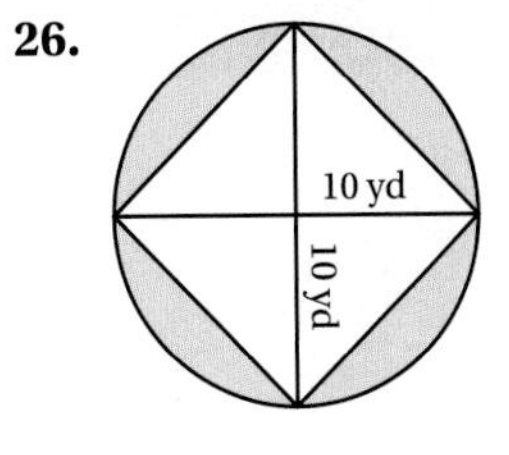

27.

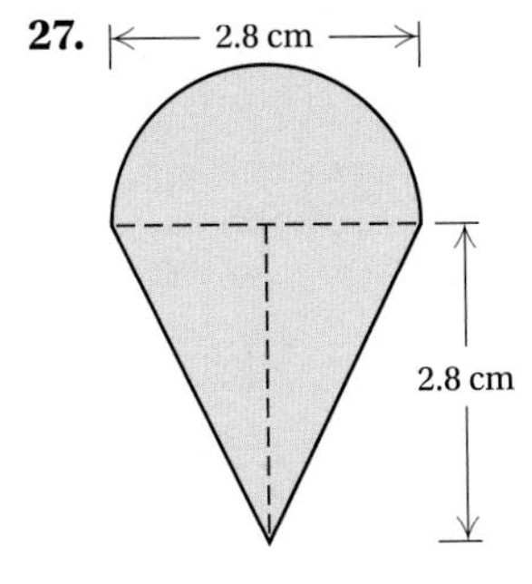

28.

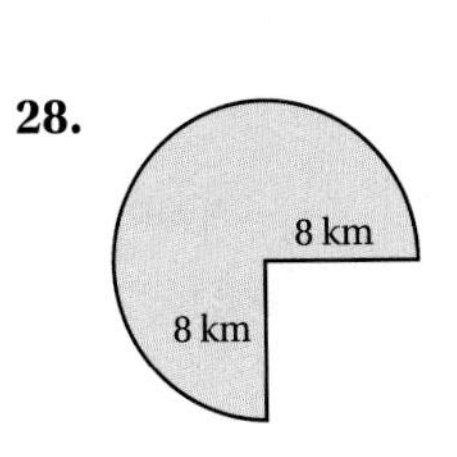

29.

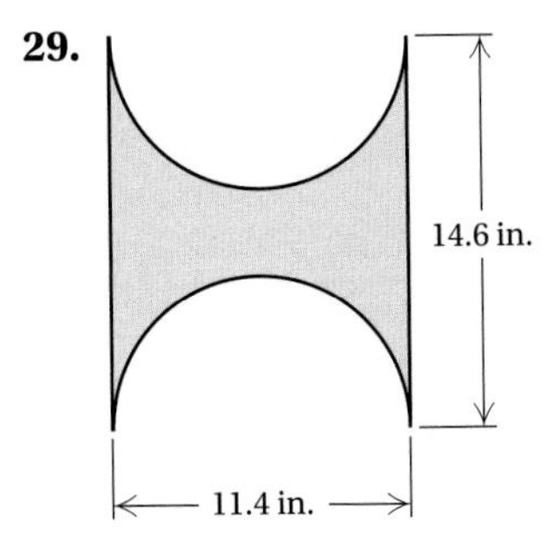

30.

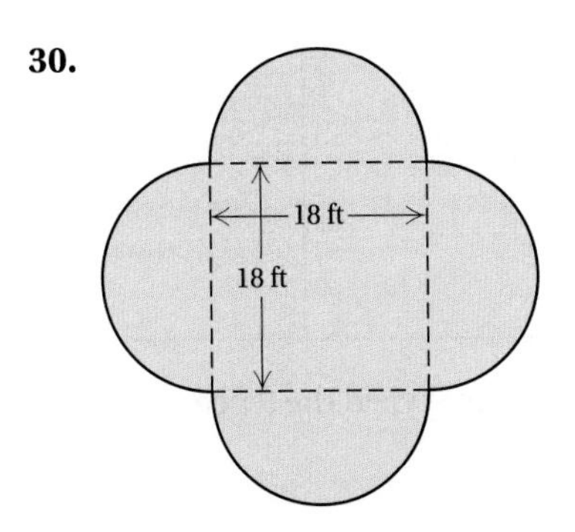

Skill Maintenance

31. A Ford C-Max Hybrid can travel 611 mi on 13 gal of gasoline. What is the rate in miles per gallon? [6.2a]

Source: edmunds.com

32. The ratio of gold to other metals in 18K gold is 18 to 6. If an 18K gold ring contains 1.2 oz of gold, what amount of other metals does the ring contain? [6.4a]

33. The weight of a human brain is 2.5% of total body weight. A person weighs 200 lb. What does the person's brain weigh? [7.5a]

34. Jack's commission is increased according to how much he sells. He receives a commission of 6% for the first $3000 and 10% on the amount over $3000. What is the total commission on sales of $8500? [7.6b]

Synthesis

Comparing Perimeters and Fencing Costs. An **acre** is a unit of area that is defined to be 43,560 ft^2. A farmer needs to fence an acre of land. She is using 32-in. fencing that costs $149.99 for a 330-ft roll. Complete the following table for Exercises 35–39 and then use the data to answer Exercise 40. Use 3.14 for π.

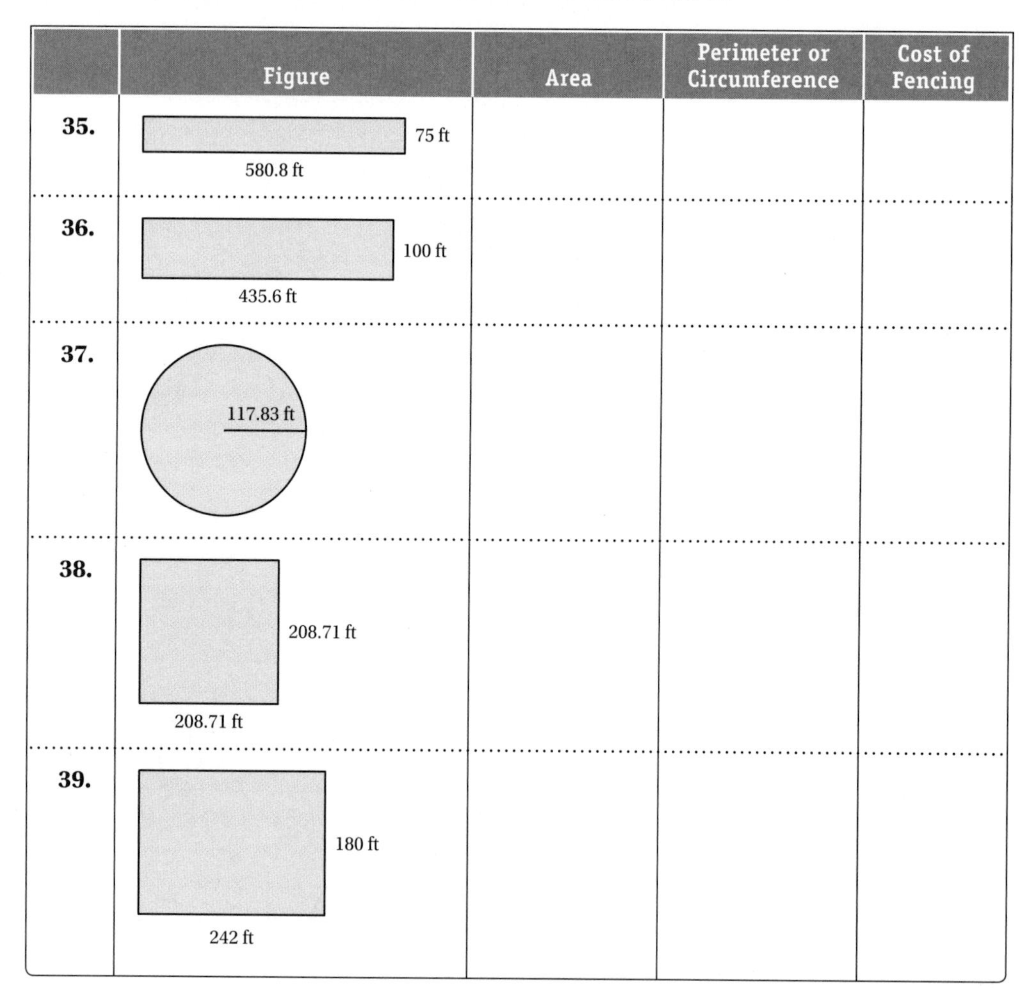

	Figure	Area	Perimeter or Circumference	Cost of Fencing
35.	75 ft; 580.8 ft			
36.	100 ft; 435.6 ft			
37.	117.83 ft			
38.	208.71 ft; 208.71 ft			
39.	180 ft; 242 ft			

40. Which dimensions of the acre yield the fence with **(a)** the shortest perimeter? **(b)** the least area? **(c)** the lowest cost and the largest area?

Mid-Chapter Review

Concept Reinforcement

Determine whether each statement is true or false.

_______ **1.** The area of a square is four times the length of a side. [10.2a]

_______ **2.** The area of a parallelogram with base 8 cm and height 5 cm is the same as the area of a rectangle with length 8 cm and width 5 cm. [10.2a, b]

_______ **3.** The area of a square that is 4 in. on a side is less than the area of a circle whose radius is 4 in. [10.2a], [10.3c]

_______ **4.** The perimeter of a rectangle that is 6 ft by 3 ft is greater than the circumference of a circle whose radius is 3 ft. [10.1a], [10.3b]

_______ **5.** The exact value of the ratio C/d is π. [10.3b]

Guided Solutions

GS Fill in each blank with the number that creates a correct solution.

6. Find the perimeter and the area. [10.1a], [10.2a]

3 ft

10 ft

$P = 2 \cdot (l + w)$

$P = 2 \cdot (\square\text{ ft} + \square\text{ ft})$

$P = 2 \cdot (\square\text{ ft})$

$P = \square\text{ ft}$

$A = l \cdot w$

$A = \square\text{ ft} \cdot \square\text{ ft}$

$A = \square \cdot \square \cdot \text{ft} \cdot \text{ft}$

$A = \square\text{ ft}^{\square}$

7. Find the area. [10.2b]

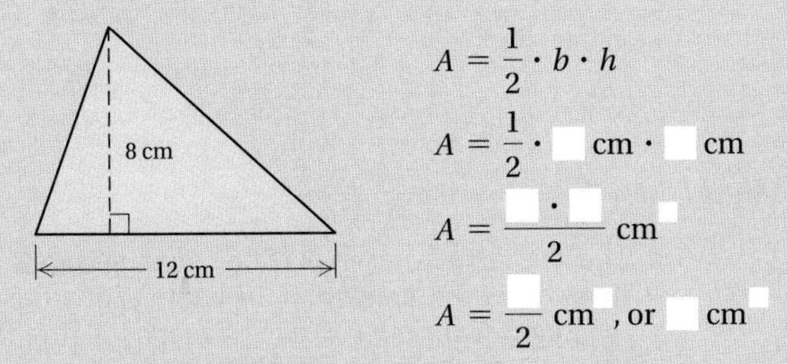

$A = \frac{1}{2} \cdot b \cdot h$

$A = \frac{1}{2} \cdot \square\text{ cm} \cdot \square\text{ cm}$

$A = \frac{\square \cdot \square}{2}\text{ cm}^{\square}$

$A = \frac{\square}{2}\text{ cm}^{\square}$, or $\square\text{ cm}^{\square}$

8. Find the circumference and the area. Use 3.14 for π. [10.3b, c]

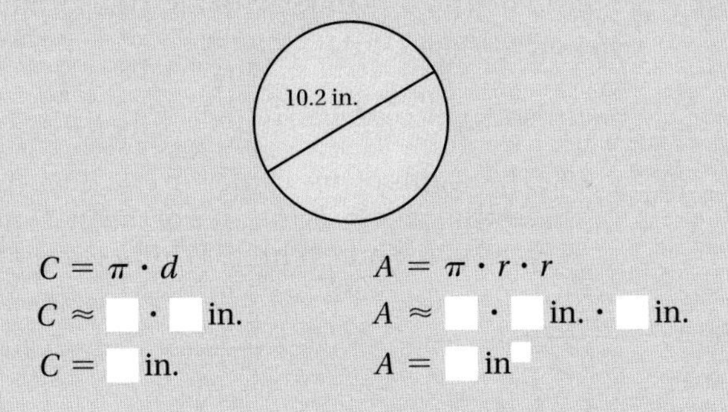

$C = \pi \cdot d$

$C \approx \square \cdot \square\text{ in.}$

$C = \square\text{ in.}$

$A = \pi \cdot r \cdot r$

$A \approx \square \cdot \square\text{ in.} \cdot \square\text{ in.}$

$A = \square\text{ in}^{\square}$

Mixed Review

9. Find the perimeter. [10.1a]

10. Find the perimeter and the area. [10.1a], [10.2a]

11. Find the area. [10.2b]

Find the area. [10.2b]

Find the circumference and the area. Use 3.14 for π. [10.3b, c]

12.

13.

14.

15.

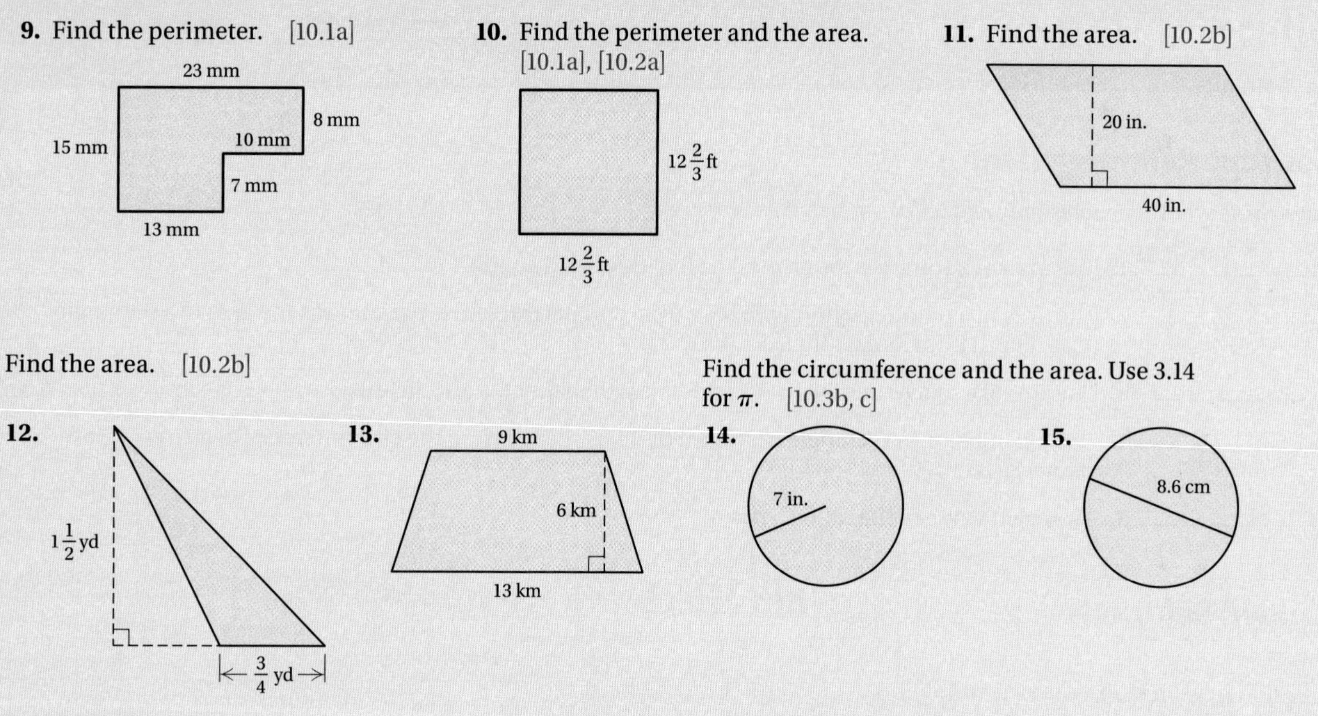

16. *Matching.* Match each item in the first column with the appropriate item in the second column by drawing connecting lines. Some expressions in the second column might be used more than once. Some expressions might not be used. [10.1a], [10.2a, b], [10.3b, c]

Area of a circle with radius 4 ft	24 ft
Area of a square with side 4 ft	16 ft
Circumference of a circle with radius 4 ft	$16 \cdot \pi$ ft^2
Area of a rectangle with length 8 ft and width 4 ft	$8 \cdot \pi$ ft^2
Area of a triangle with base 4 ft and height 8 ft	32 ft^2
Perimeter of a square with side 4 ft	$4 \cdot \pi$ ft
Perimeter of a rectangle with length 8 ft and width 4 ft	$8 \cdot \pi$ ft
	64 ft
	16 ft^2

Understanding Through Discussion and Writing

17. Explain why a 16-in.-diameter pizza that costs $16.25 is a better buy than a 10-in.-diameter pizza that costs $7.85. [10.3d]

18. The length and the width of one rectangle are each three times the length and the width of another rectangle. Is the area of the first rectangle three times the area of the other rectangle? Why or why not? [10.2a]

19. The length of a side of a square is $\frac{1}{2}$ the length of a side of another square. Is the perimeter of the first square $\frac{1}{2}$ the perimeter of the other square? Why or why not? [10.1a]

20. For a fellow student, develop the formula for the perimeter of a rectangle:

$$P = 2 \cdot (l + w) = 2 \cdot l + 2 \cdot w. \quad [10.1a]$$

21. Explain how the area of a triangle can be found by considering the area of a parallelogram. [10.2b]

22. The radius of one circle is twice the length of that of another circle. Is the area of the first circle twice the area of the other circle? Why or why not? [10.3c]

STUDYING FOR SUCCESS *Beginning to Study for the Final Exam*

- ☐ Take a few minutes each week to review highlighted information.
- ☐ Prepare a few pages of notes for the course, then try to condense the notes to just one page.
- ☐ Use the Mid-Chapter Reviews, Summary and Reviews, Chapter Tests, and Cumulative Reviews.

Volume 10.4

OBJECTIVES

- **a** Find the volume of a rectangular solid using the formula $V = l \cdot w \cdot h$.
- **b** Given the radius and the height, find the volume of a circular cylinder.
- **c** Given the radius, find the volume of a sphere.
- **d** Given the radius and the height, find the volume of a circular cone.
- **e** Solve applied problems involving volumes of rectangular solids, circular cylinders, spheres, and cones.

a RECTANGULAR SOLIDS

The **volume** of a **rectangular solid** is the number of unit cubes needed to fill it.

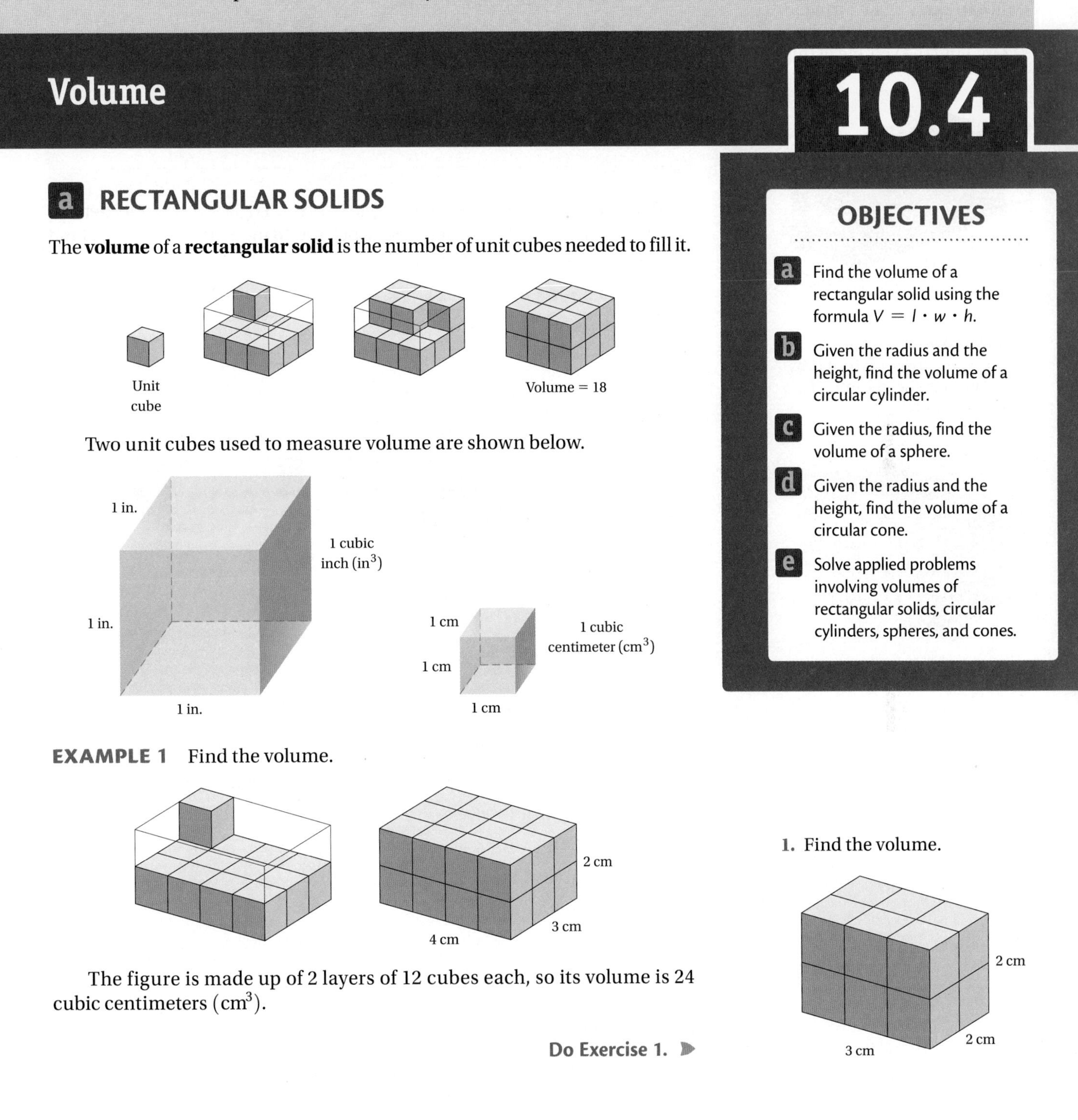

Two unit cubes used to measure volume are shown below.

EXAMPLE 1 Find the volume.

The figure is made up of 2 layers of 12 cubes each, so its volume is 24 cubic centimeters (cm^3).

Do Exercise 1. ▶

1. Find the volume.

2 cm
2 cm
3 cm

Answer

1. $12\ cm^3$

2. *Carry-on Luggage.* The largest piece of luggage that you can carry on an airplane measures 23 in. by 10 in. by 13 in. Find the volume of this solid.

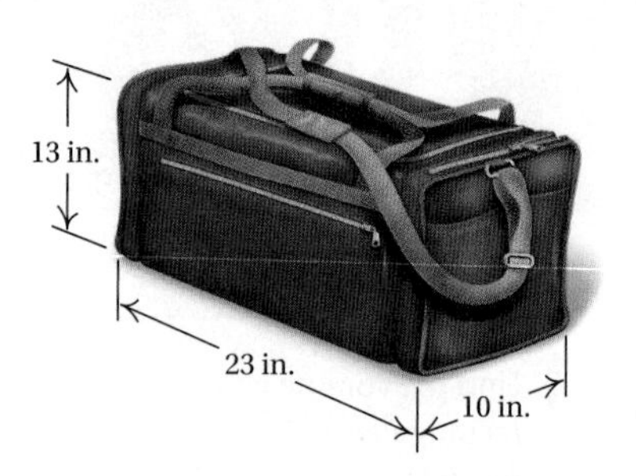

3. *Cord of Wood.* A cord of wood measures 4 ft by 4 ft by 8 ft. What is the volume of a cord of wood?

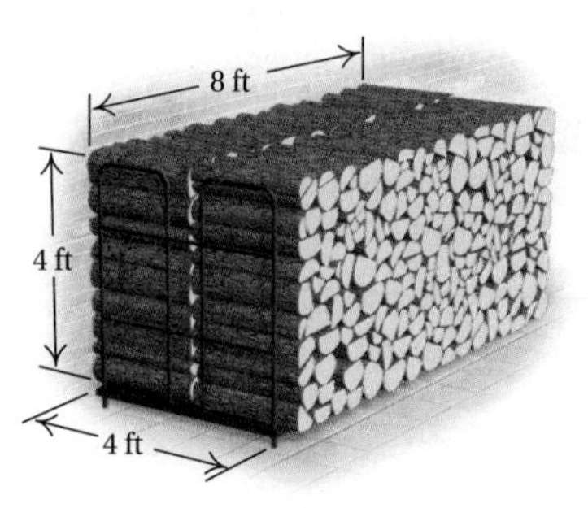

Answers

2. 2990 in^3 3. 128 ft^3

VOLUME OF A RECTANGULAR SOLID

The **volume of a rectangular solid** is found by multiplying length by width by height:

$$V = l \cdot w \cdot h.$$

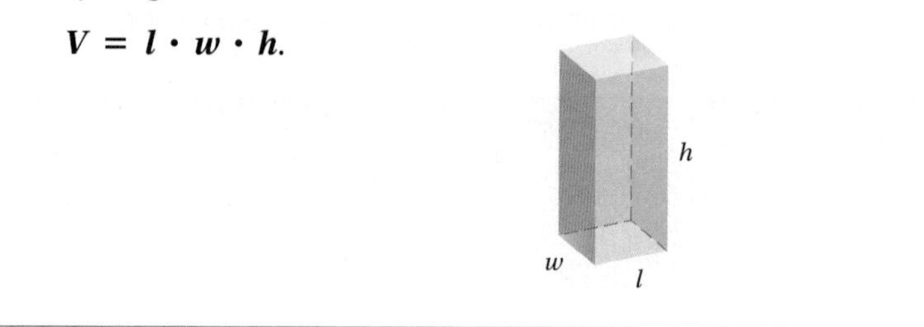

EXAMPLE 2 *Volume of a Safety Deposit Box.* Tricia rents a safety deposit box at her bank. The dimensions of the box are 18 in. × 10.5 in. × 5 in. Find the volume of this rectangular solid.

$$\begin{aligned} V &= l \cdot w \cdot h \\ &= 18\text{ in.} \times 10.5\text{ in.} \times 5\text{ in.} \\ &= 945\text{ in}^3 \end{aligned}$$

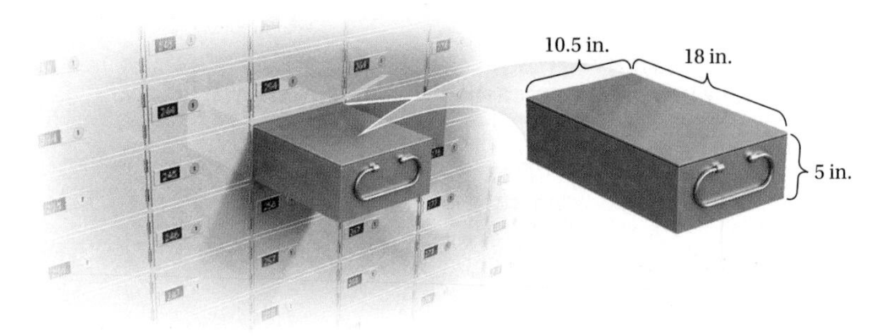

◀ Do Exercises 2 and 3.

b CYLINDERS

A rectangular solid is shown below. Note that we can think of the volume as the product of the area of the base times the height:

$$\begin{aligned} V &= l \cdot w \cdot h \\ &= (l \cdot w) \cdot h \\ &= (\text{Area of the base}) \cdot h \\ &= B \cdot h, \end{aligned}$$

h

w

l

Area of base $= B = l \cdot w$

where B represents the area of the base.

Like rectangular solids, **circular cylinders** have bases of equal area that lie in parallel planes. The bases of circular cylinders are circular regions.

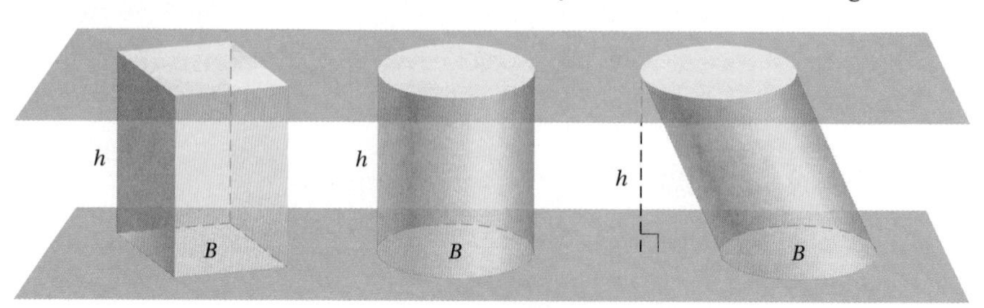

The volume of a circular cylinder is found in a manner similar to the way the volume of a rectangular solid is found. The volume is the product of the area of the base times the height. The height is always measured perpendicular to the base.

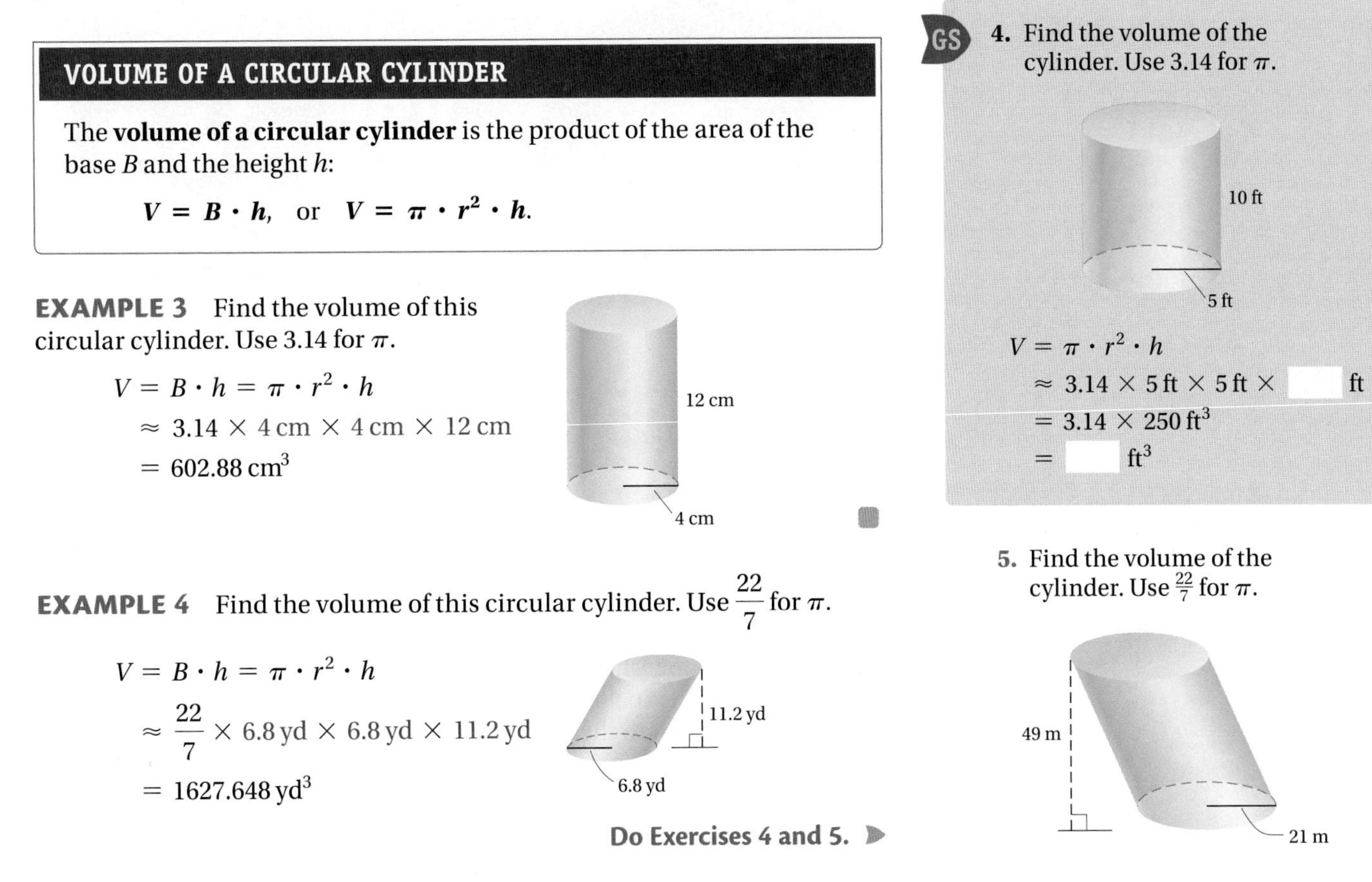

VOLUME OF A CIRCULAR CYLINDER

The **volume of a circular cylinder** is the product of the area of the base B and the height h:

$$\mathbf{V = B \cdot h}, \quad \text{or} \quad \mathbf{V = \pi \cdot r^2 \cdot h}.$$

EXAMPLE 3 Find the volume of this circular cylinder. Use 3.14 for π.

$$\begin{aligned} V &= B \cdot h = \pi \cdot r^2 \cdot h \\ &\approx 3.14 \times 4\text{ cm} \times 4\text{ cm} \times 12\text{ cm} \\ &= 602.88\text{ cm}^3 \end{aligned}$$

EXAMPLE 4 Find the volume of this circular cylinder. Use $\frac{22}{7}$ for π.

$$\begin{aligned} V &= B \cdot h = \pi \cdot r^2 \cdot h \\ &\approx \frac{22}{7} \times 6.8\text{ yd} \times 6.8\text{ yd} \times 11.2\text{ yd} \\ &= 1627.648\text{ yd}^3 \end{aligned}$$

Do Exercises 4 and 5.

GS 4. Find the volume of the cylinder. Use 3.14 for π.

$$\begin{aligned} V &= \pi \cdot r^2 \cdot h \\ &\approx 3.14 \times 5\text{ ft} \times 5\text{ ft} \times \underline{\quad}\text{ ft} \\ &= 3.14 \times 250\text{ ft}^3 \\ &= \underline{\quad}\text{ ft}^3 \end{aligned}$$

5. Find the volume of the cylinder. Use $\frac{22}{7}$ for π.

C SPHERES

A **sphere** is the three-dimensional counterpart of a circle. It is the set of all points in space that are a given distance (the radius) from a given point (the center).

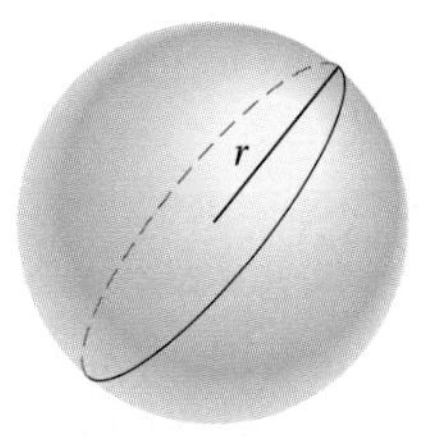

We find the volume of a sphere as follows.

VOLUME OF A SPHERE

The **volume of a sphere** of radius r is given by

$$\mathbf{V = \frac{4}{3} \cdot \pi \cdot r^3}.$$

Answers

4. 785 ft^3 **5.** 67,914 m^3

Guided Solution:

4. 10, 785

6. Find the volume of the sphere. Use $\frac{22}{7}$ for π.

GS

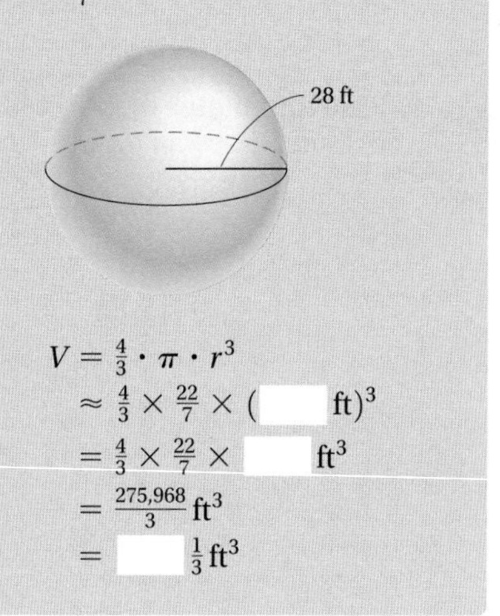

$V = \frac{4}{3} \cdot \pi \cdot r^3$
$\approx \frac{4}{3} \times \frac{22}{7} \times (\quad \text{ft})^3$
$= \frac{4}{3} \times \frac{22}{7} \times \quad \text{ft}^3$
$= \frac{275,968}{3} \text{ft}^3$
$= \quad \frac{1}{3} \text{ft}^3$

7. The radius of a standard-sized golf ball is 2.1 cm. Find its volume. Use 3.14 for π.

8. Find the volume of this cone. Use 3.14 for π.

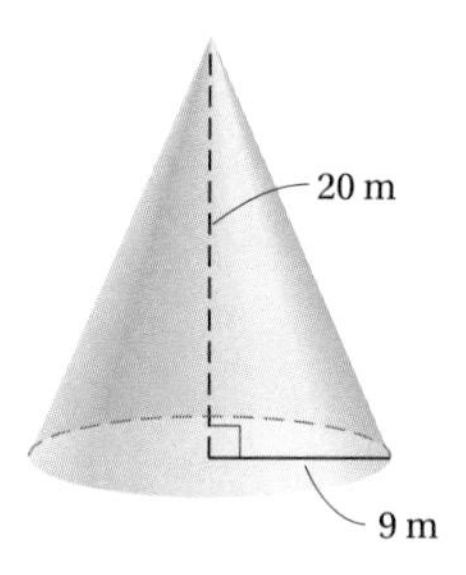

9. Find the volume of this cone. Use $\frac{22}{7}$ for π.

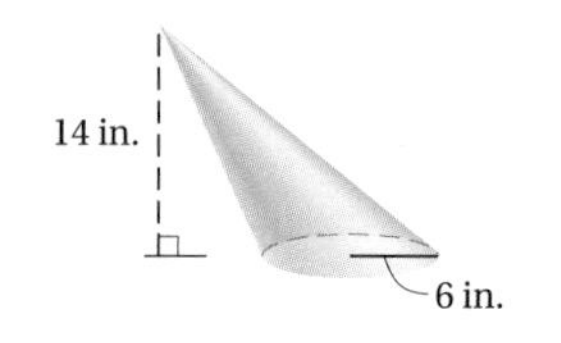

Answers

6. $91,989\frac{1}{3}\text{ft}^3$ 7. 38.77272 cm^3
8. 1695.6 m^3 9. 528 in^3

Guided Solution:
6. 28, 21,952, 91,989

EXAMPLE 5 *Bowling Ball.* The radius of a standard-sized bowling ball is 4.2915 in. Find the volume of a standard-sized bowling ball. Round to the nearest hundredth of a cubic inch. Use 3.14 for π.

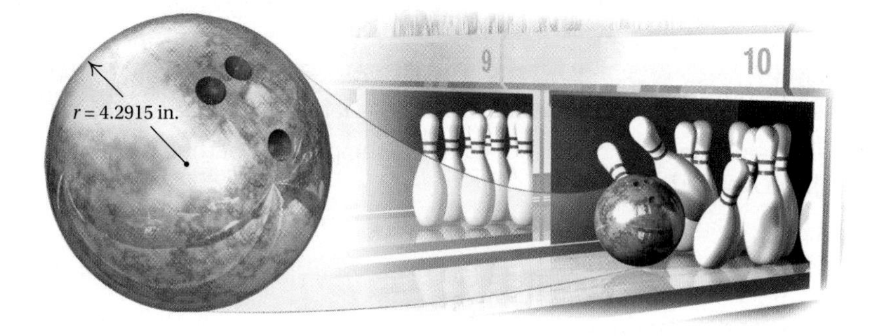

We have

$V = \frac{4}{3} \cdot \pi \cdot r^3 \approx \frac{4}{3} \times 3.14 \times (4.2915 \text{ in.})^3$
$\approx 330.90 \text{ in}^3$. Using a calculator

Do Exercises 6 and 7.

d CONES

Consider a circle in a plane and choose any point P not in the plane. The circular region, together with the set of all segments connecting P to a point on the circle, is called a **circular cone**. The height of the cone is measured perpendicular to the base.

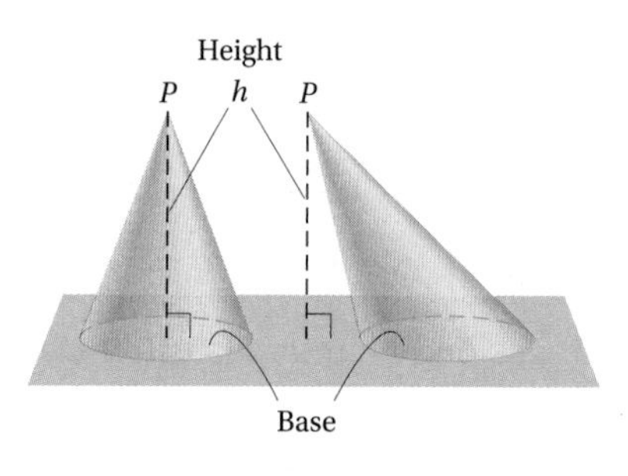

We find the volume of a cone as follows.

VOLUME OF A CIRCULAR CONE

The **volume of a circular cone** with base radius r is one-third the product of the area of the base and the height:

$$V = \frac{1}{3} \cdot B \cdot h = \frac{1}{3} \cdot \pi \cdot r^2 \cdot h.$$

EXAMPLE 6 Find the volume of this circular cone. Use 3.14 for π.

$V = \frac{1}{3} \cdot \pi \cdot r^2 \cdot h$
$\approx \frac{1}{3} \times 3.14 \times 3 \text{ cm} \times 3 \text{ cm} \times 7 \text{ cm}$
$= 65.94 \text{ cm}^3$

7 cm
3 cm

Do Exercises 8 and 9.

e SOLVING APPLIED PROBLEMS

EXAMPLE 7 *Propane Gas Tank.* A propane gas tank is shaped like a circular cylinder with half of a sphere at each end. Find the volume of the tank if the cylindrical section is 5 ft long with a 4-ft diameter. Use 3.14 for π.

1. **Familiarize.** We first make a drawing.

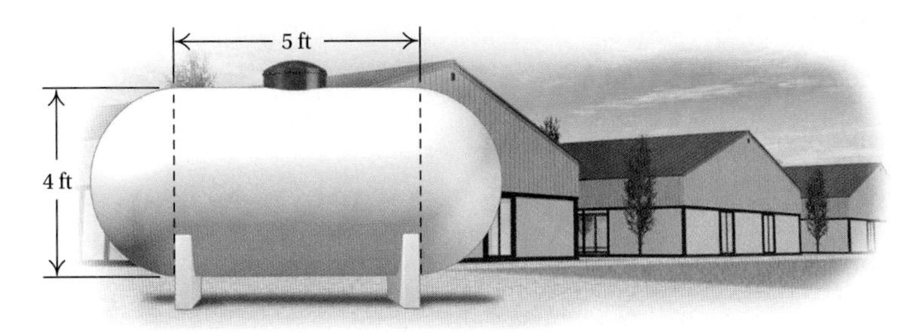

2. **Translate.** This is a two-step problem. We first find the volume of the cylindrical portion. Then we find the volume of the two ends and add. Note that together the two ends make a sphere with a radius of 2 ft. We have

Total volume	is	Volume of the cylinder	plus	Volume of the sphere
V	$=$	$\pi \cdot r^2 \cdot h$	$+$	$\frac{4}{3} \cdot \pi \cdot r^3$,

where V is the total volume. Then

$$V \approx 3.14 \cdot (2\text{ ft})^2 \cdot 5\text{ ft} + \frac{4}{3} \cdot 3.14 \cdot (2\text{ ft})^3.$$

3. **Solve.** The volume of the cylinder is approximately

$$\begin{aligned} 3.14 \cdot (2\text{ ft})^2 \cdot 5\text{ ft} &= 3.14 \cdot 2\text{ ft} \cdot 2\text{ ft} \cdot 5\text{ ft} \\ &= 62.8\text{ ft}^3. \end{aligned}$$

The volume of the two ends is approximately

$$\begin{aligned} \frac{4}{3} \cdot 3.14 \cdot (2\text{ ft})^3 &= \frac{4}{3} \cdot 3.14 \cdot 2\text{ ft} \cdot 2\text{ ft} \cdot 2\text{ ft} \\ &\approx 33.5\text{ ft}^3. \end{aligned}$$

The total volume is about

$$62.8\text{ ft}^3 + 33.5\text{ ft}^3 = 96.3\text{ ft}^3.$$

4. **Check.** We can repeat the calculations. The answer checks.
5. **State.** The volume of the tank is about 96.3 ft^3.

Do Exercise 10. ▶

10. *Medicine Capsule.* A cold capsule is 8 mm long and 4 mm in diameter. Find the volume of the capsule. Use 3.14 for π. (*Hint*: First find the length of the cylindrical section.)

Answer

10. $83.7\overline{3}\text{ mm}^3$

10.4 Exercise Set

For Extra Help

MyMathLab® MathXL® PRACTICE WATCH READ REVIEW

Reading Check

Match each formula with the correct phrase from the list on the right.

RC1. ________ $V = l \cdot w \cdot h$

RC2. ________ $V = \pi \cdot r^2 \cdot h$

RC3. ________ $V = \frac{4}{3} \cdot \pi \cdot r^3$

RC4. ________ $V = \frac{1}{3} \cdot \pi \cdot r^2 \cdot h$

a) the volume of a cylinder

b) the volume of a rectangular solid

c) the volume of a sphere

d) the volume of a cone

a Find the volume of the rectangular solid.

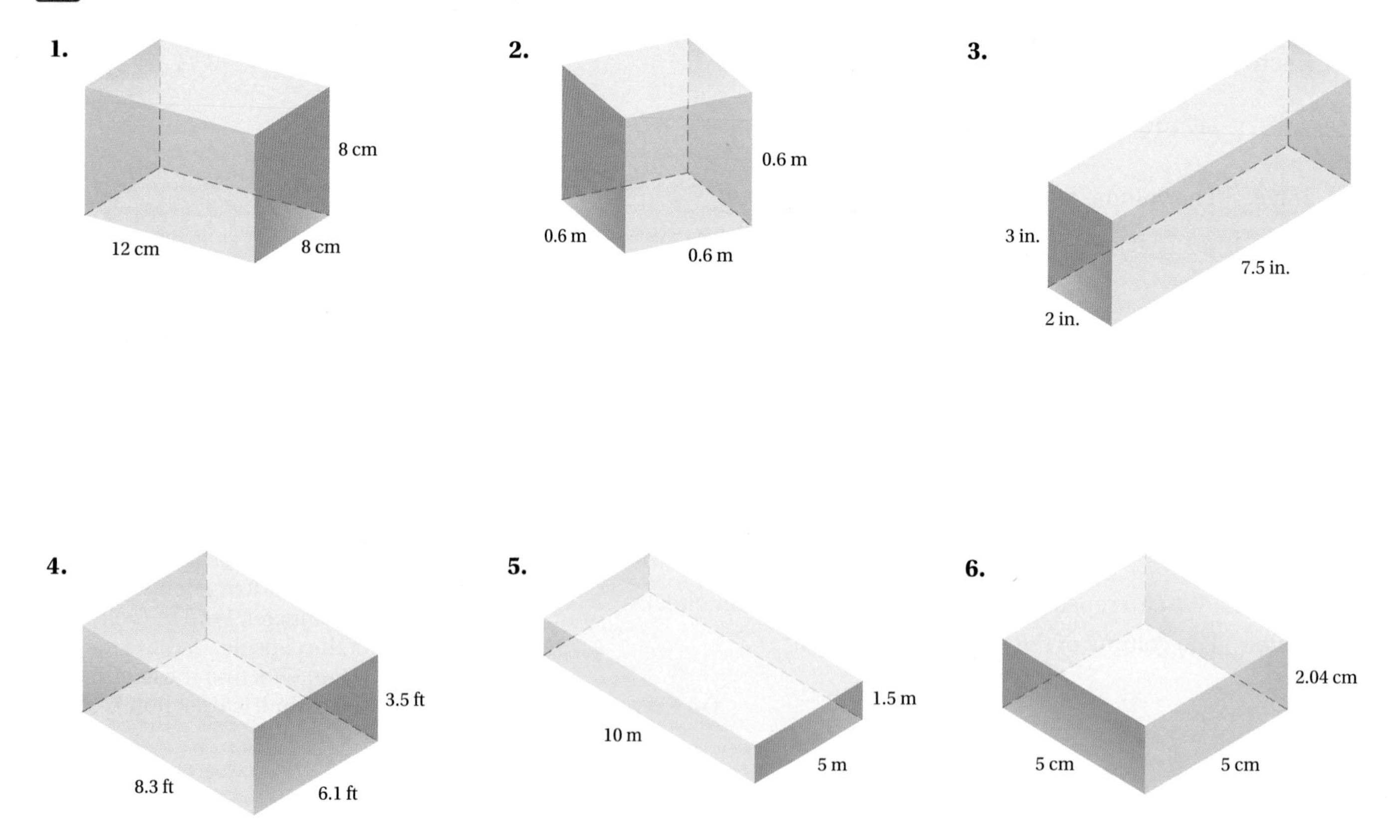

7. **8.**

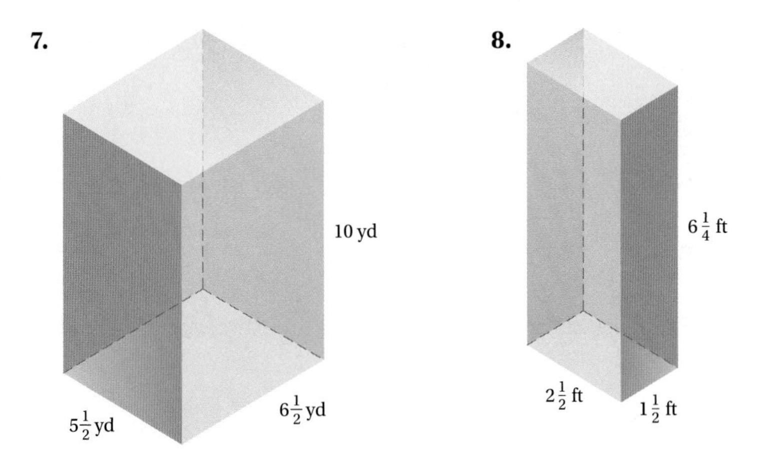

b Find the volume of the circular cylinder. Use 3.14 for π in Exercises 9–12. Use $\frac{22}{7}$ for π in Exercises 13 and 14.

9. **10.** **11.** **12.** **13.** **14.**

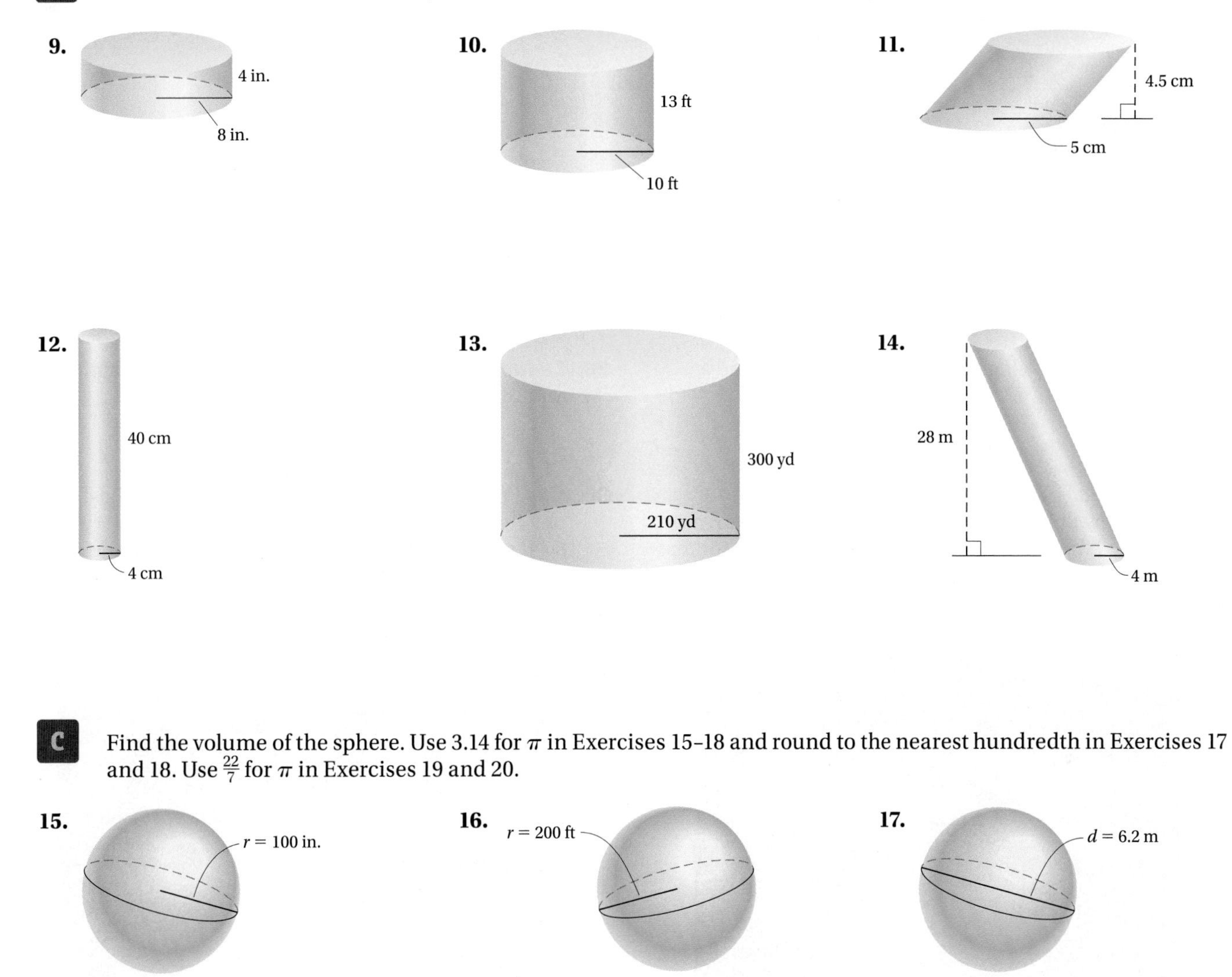

c Find the volume of the sphere. Use 3.14 for π in Exercises 15–18 and round to the nearest hundredth in Exercises 17 and 18. Use $\frac{22}{7}$ for π in Exercises 19 and 20.

15. **16.** **17.**

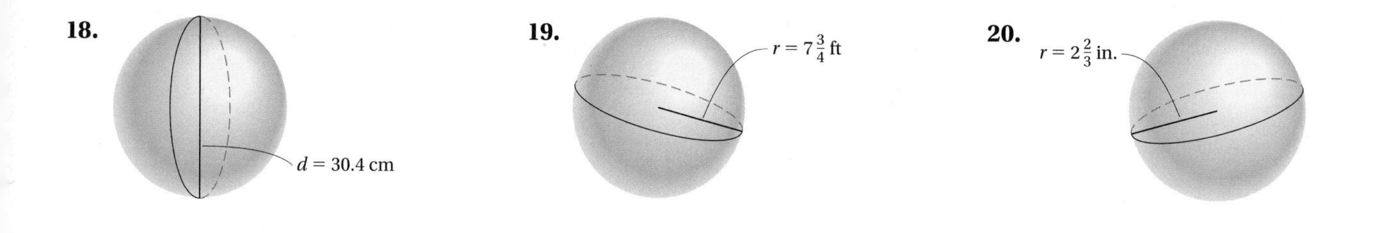

d Find the volume of the circular cone. Use 3.14 for π in Exercises 21, 22, and 26. Use $\frac{22}{7}$ for π in Exercises 23, 24, and 25.

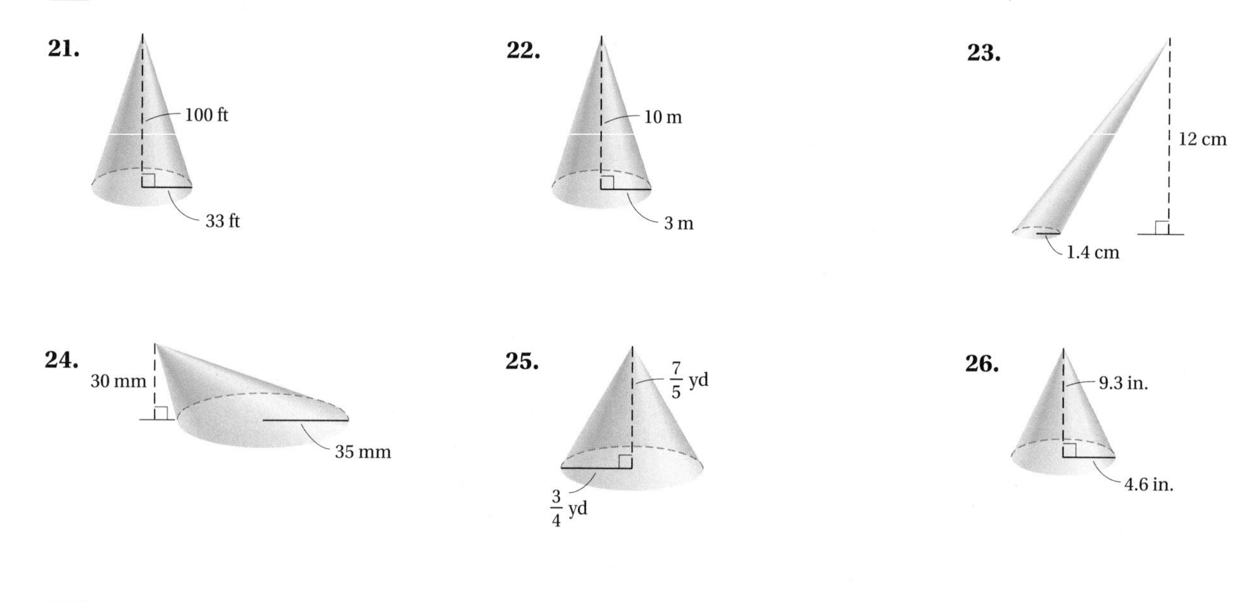

e Solve.

27. ***Oak Log.*** An oak log has a diameter of 12 cm and a length (height) of 42 cm. Find the volume. Use 3.14 for π.

28. ***Ladder Rung.*** A rung of a ladder is 2 in. in diameter and 16 in. long. Find the volume. Use 3.14 for π.

29. ***Architecture.*** The largest sphere in the Oriental Pearl TV tower in Shanghai, China, measures 50 m in diameter. Find the volume of the sphere. Use 3.14 for π, and round to the nearest cubic meter.

Source: www.emporis.com

30. ***Gas Pipeline.*** The 638-mi Rockies Express–East pipeline from Colorado to Ohio is constructed with 80-ft sections of 42-in., or $3\frac{1}{2}$-ft, steel gas pipeline. Find the volume of one section. Use $\frac{22}{7}$ for π.

Source: Rockies Express Pipeline

31. *Volume of a Candle.* Find the approximate volume of a candle that is a circular cone. The diameter of the base of the candle is 4.875 in., and the height is 12.5 in. Use 3.14 for π.

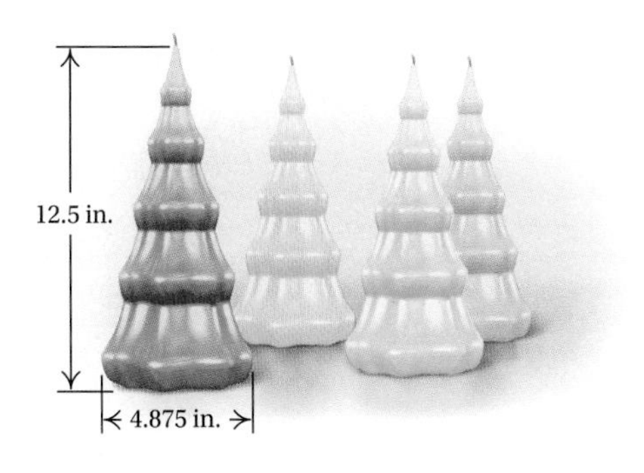

32. *Culinary Arts.* Raena often makes individual soufflés in cylindrical baking dishes called *ramekins*. The diameter of each ramekin is 3.5 in., and the height is 1.75 in. Find the approximate volume of a ramekin. Use 3.14 for π.

33. *Precious Metals.* If all the gold in the world could be gathered together, it would form a cube 18 yd on a side. Find the volume of the world's gold.

34. *Containers.* Medical Hope collects and ships medical supplies to areas of the world that have been damaged by storms or earthquakes. The containers in which the supplies are shipped measure 6 ft by 5 ft by 4.8 ft. Find the volume of a container.

35. *Architecture.* The Westhafen Tower in Frankfort, Germany, is a cylindrical building with a height of 110 m and a radius of 21 m. Find the volume of the building. Use 3.14 for π. Round to the nearest cubic meter.

36. *Roof of a Turret.* The roof of a turret is often in the shape of a circular cone. Find the volume of this circular cone structure if the radius is 2.5 m and the height is 4.6 m. Use 3.14 for π.

37. *Volume of Earth.* The diameter of the earth is about 3980 mi. Find the volume of the earth. Use 3.14 for π. Round to the nearest ten thousand cubic miles.

38. *Astronomy.* The diameter of the largest moon of Uranus is about 1578 km. Find the volume of this satellite. Use $\frac{22}{7}$ for π. Round to the nearest ten thousand cubic kilometers.

39. The volume of a ball is 36π cm^3. Find the dimensions of a rectangular box that is just large enough to hold the ball.

40. *Oceanography.* A research submarine is capsule-shaped. Find the volume of the submarine if it has a length of 10 m and a diameter of 8 m. Use 3.14 for π and round the answer to the nearest hundredth. (*Hint*: First find the length of the cylindrical section.)

41. *Toys.* Toy stores often sell capsules that dissolve in water allowing a toy inside the capsule to expand. One such capsule is 40 mm long with a diameter of 8 mm.

a) What is the volume of the capsule? Use 3.14 for π.

b) The manufacturer claims that the toy in the capsule will expand 600%. What is the volume of the toy after expansion?

42. *Golf-Ball Packaging.* The box shown is just big enough to hold 3 golf balls. If the radius of a golf ball is 2.1 cm, how much air surrounds the three balls? Use 3.14 for π.

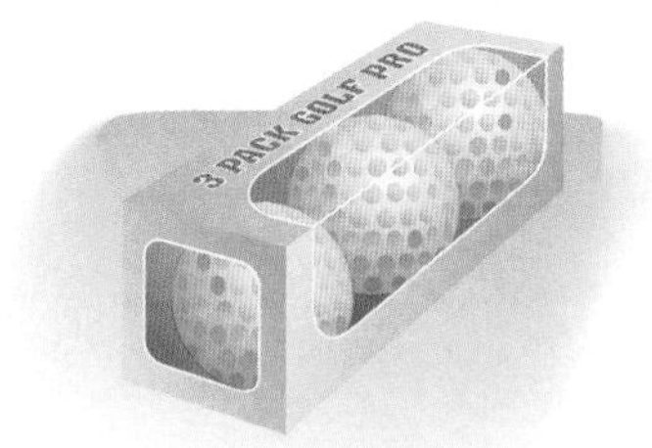

Skill Maintenance

Great Lakes. The Great Lakes contain about 5500 mi^3 (23,000 km^3) of water that covers a total area of about 94,000 mi^2 (244,000 km^2). The Great Lakes are the largest system of fresh surface water on Earth. Use the following table for Exercises 43–48.

		GREAT LAKE				
FEATURE	UNITS	SUPERIOR	MICHIGAN	HURON	ERIE	ONTARIO
Average depth	ft	483	279	195	62	283
Volume	mi^3	2,900	1,180	850	116	393
Water area	mi^2	31,700	22,300	23,000	9,910	7,340

SOURCES: http://www.epa.gov/glnpo/factsheet.html; http://earth1.epa.gov/glnpo/statrefs.html

43. How much greater is the volume of water in Lake Michigan than in Lake Erie? [1.8a], [8.2a]

44. How much less is the water area of Lake Ontario than that of Lake Superior? [1.8a], [8.2a]

45. Find the average of the average depths of the five Great Lakes. [8.1a], [8.2a]

46. Find the average volume of water in the five Great Lakes. [8.1a], [8.2a]

Synthesis

Use the data in the table above for Exercises 47 and 48.

47. Convert the volume of water in Lake Huron from cubic miles to cubic kilometers. Round to the nearest hundredth of a cubic kilometer.

48. Convert the water area of Lake Superior from square miles to square kilometers. Round to the nearest hundredth of a square kilometer.

49. A sphere with diameter 1 m is circumscribed by a cube. How much greater is the volume of the cube than the volume of the sphere? Use 3.14 for π.

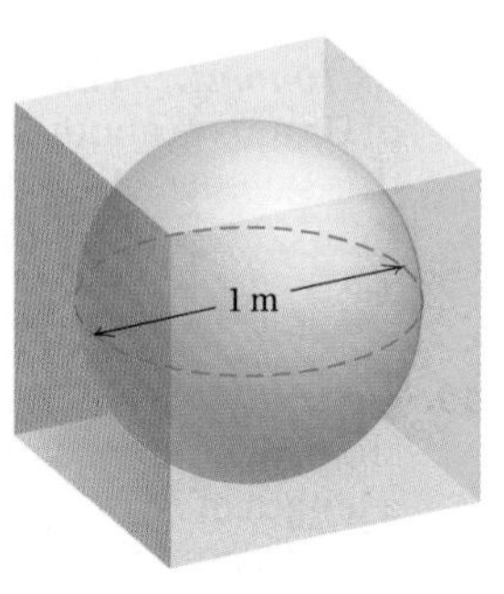

Angles and Triangles

10.5

a MEASURING ANGLES

We see a real-world application of *angles* of various types in the different back postures of the bicycle riders illustrated below.

Style of Biking Determines Cycling Posture

Road	**Mountain**	**Comfort**
About 180° flat	About 45°	About 90°

Riders prefer a more aerodynamic flat-back position.	Riders prefer a semi-upright position to help lift the front wheel over obstacles.	Riders prefer an upright position that lessens stress on the lower back and neck.

SOURCE: USA TODAY research

An **angle** is a set of points consisting of two **rays**, or half-lines, with a common endpoint. The endpoint is called the **vertex** of the angle. The rays are called the **sides** of the angle.

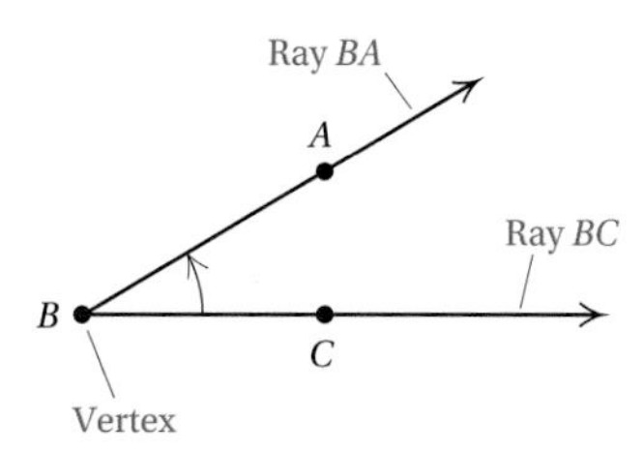

The angle above can be named

angle ABC, angle CBA, $\angle ABC$, $\angle CBA$, or $\angle B$.

Note that the vertex is written in the middle of the name. If there is only one angle with a given vertex in a drawing, the angle may be named using simply its vertex.

Do Margin Exercises 1 and 2.

OBJECTIVES

a Name an angle in five different ways, and measure an angle with a protractor.

b Classify an angle as right, straight, acute, or obtuse.

c Find the measure of a complement or a supplement of a given angle.

d Classify a triangle as equilateral, isosceles, or scalene, and as right, obtuse, or acute.

e Given two of the angle measures of a triangle, find the third.

SKILL TO REVIEW

Objective 1.7b: Solve equations like $t + 28 = 54$, $28 \cdot x = 168$, and $98 \cdot 2 = y$.

Solve.

1. $x + 38 = 180$
2. $x + 37 = 90$

Name the angle in five different ways.

1.

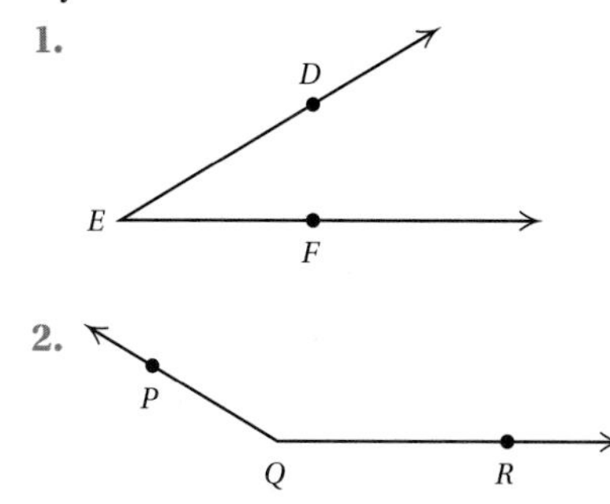

2.

P
Q
R

Answers

Skill to Review:
1. 142 2. 53

Margin Exercises:
1. Angle DEF, angle FED, $\angle DEF$, $\angle FED$, or $\angle E$
2. Angle PQR, angle RQP, $\angle PQR$, $\angle RQP$, or $\angle Q$

To measure angles, we start with some arbitrary angle and assign to it a measure of 1. We call it a *unit angle*. Suppose that $\angle U$, shown below, is a unit angle. Let's measure $\angle DEF$. If we made 3 copies of $\angle U$, they would "fill up" $\angle DEF$. Thus the measure of $\angle DEF$ would be 3.

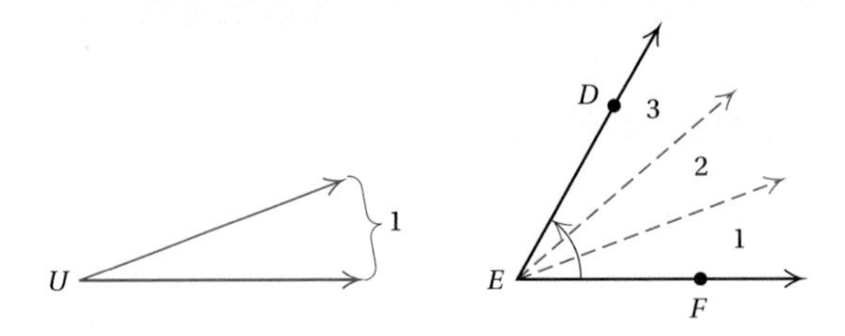

The unit most commonly used for angle measure is the degree. Below is such a unit angle. Its measure is 1 degree, or 1°. There are 360 degrees in a circle.

A 1° angle:

Here are some other angles with their degree measures.

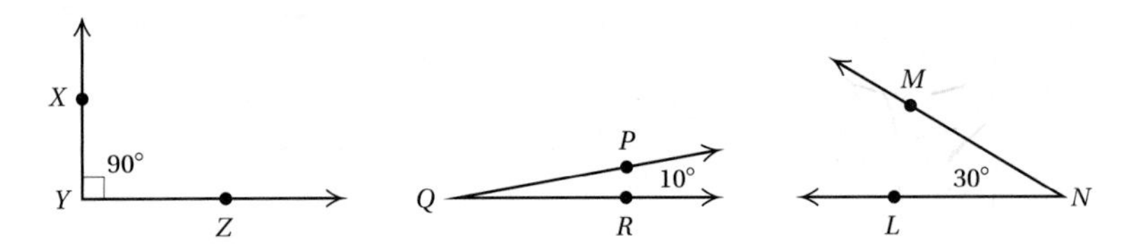

To indicate the *measure* of $\angle XYZ$, we write $m\,\angle XYZ = 90°$. Recall that the symbol ⌉ is sometimes drawn on a figure to indicate a 90° angle.

A device called a **protractor** is used to measure angles. Protractors have two scales (inside and outside). To measure an angle like $\angle Q$ below, we place the protractor's ▲ at the vertex and line up one of the angle's sides at 0°. Then we check where the angle's other side crosses the scale. In the figure below, the side $\overrightarrow{QR}$ lines up with 0° on the *inside* scale, so we check where the angle's other side, $\overrightarrow{QP}$, crosses the *inside* scale. We see that $m\,\angle Q = 145°$.

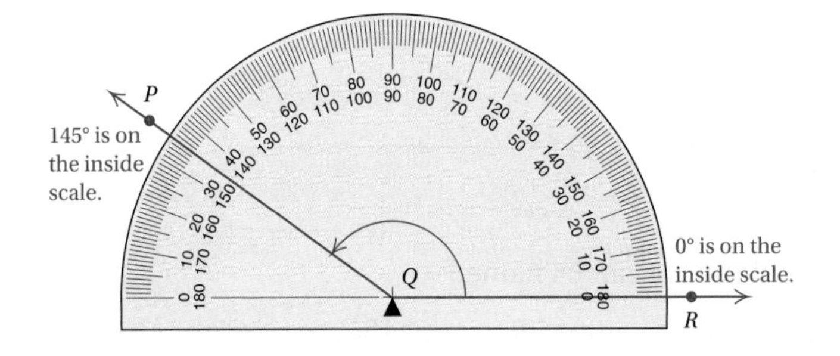

◀ **Do Exercise 3.**

3. Use a protractor to measure this angle.

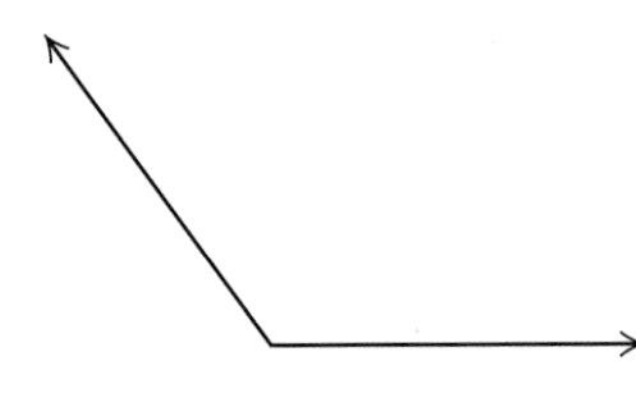

Answer

3. 127°

Let's find the measure of $\angle ABC$. This time we line up one of the angle's sides, $\overrightarrow{BC}$, with 0° on the *outside* scale. Then we check where the angle's other side, $\overrightarrow{BA}$, crosses the *outside* scale. We see that $m\ \angle ABC = 42°$.

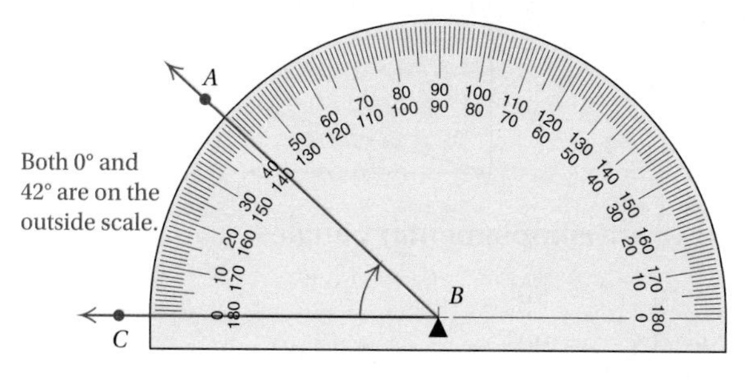

Do Exercise 4.

4. Use a protractor to measure this angle.

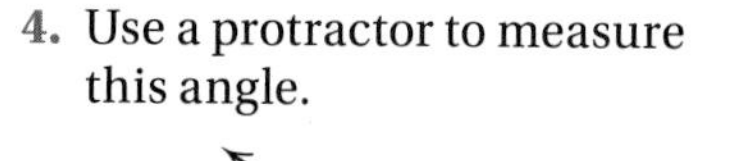

b CLASSIFYING ANGLES

The following are ways in which we classify angles.

TYPES OF ANGLES

Right angle: An angle whose measure is 90°.

Straight angle: An angle whose measure is 180°.

Acute angle: An angle whose measure is greater than 0° and less than 90°.

Obtuse angle: An angle whose measure is greater than 90° and less than 180°.

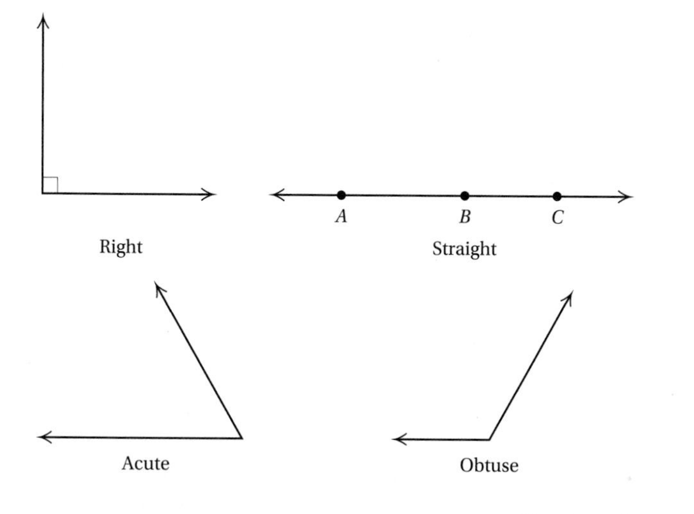

Do Exercises 5–8.

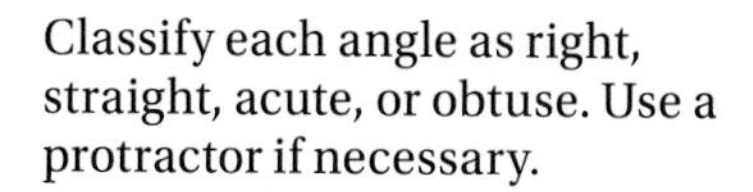
Classify each angle as right, straight, acute, or obtuse. Use a protractor if necessary.

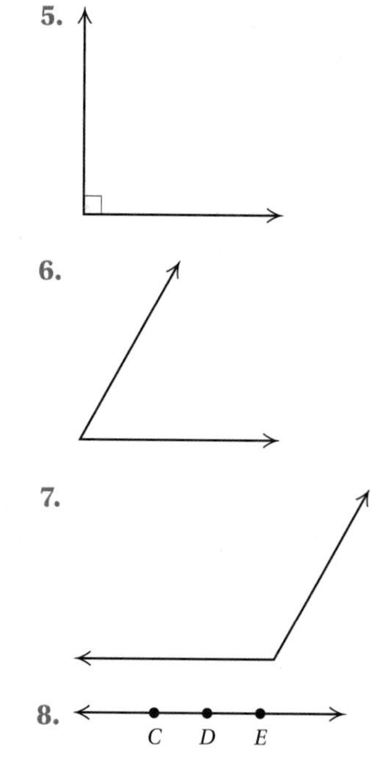

Answers

4. 33° 5. Right 6. Acute 7. Obtuse 8. Straight

COMPLEMENTARY AND SUPPLEMENTARY ANGLES

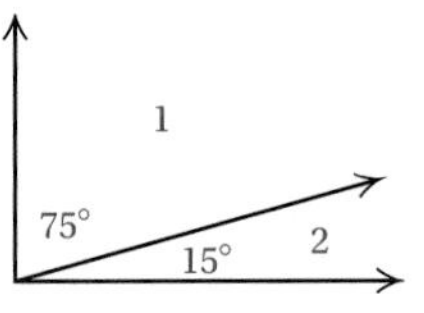

$\angle 1$ and $\angle 2$ above are **complementary** angles.

$$m\angle 1 + m\angle 2 = 90^\circ$$
$$75^\circ + 15^\circ = 90^\circ$$

COMPLEMENTARY ANGLES

Two angles are **complementary** if the sum of their measures is 90°. Each angle is called a **complement** of the other.

If two angles are complementary, each is an acute angle. When complementary angles are adjacent to each other, they form a right angle.

EXAMPLE 1 Identify each pair of complementary angles.

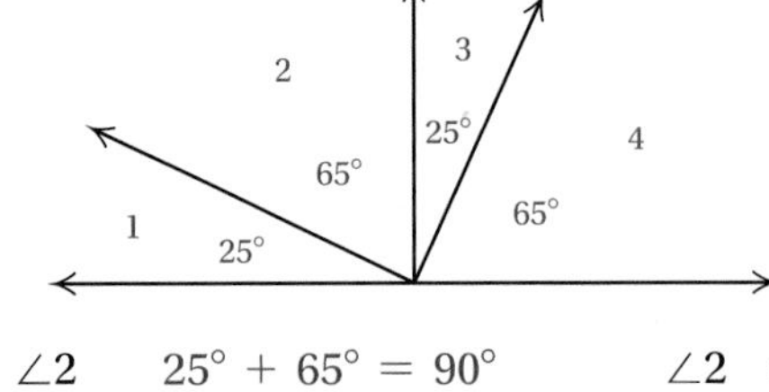

$\angle 1$ and $\angle 2$	$25^\circ + 65^\circ = 90^\circ$	$\angle 2$ and $\angle 3$
$\angle 1$ and $\angle 4$		$\angle 3$ and $\angle 4$

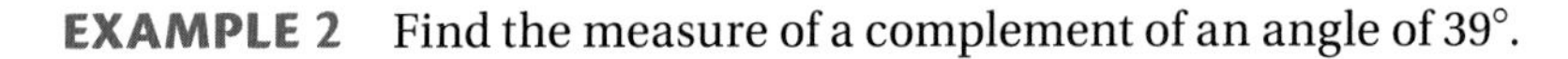

EXAMPLE 2 Find the measure of a complement of an angle of 39°.

$(90-39)^\circ$, 39°

$$90^\circ - 39^\circ = 51^\circ$$

The measure of a complement is 51°.

◀ **Do Exercises 9–13.**

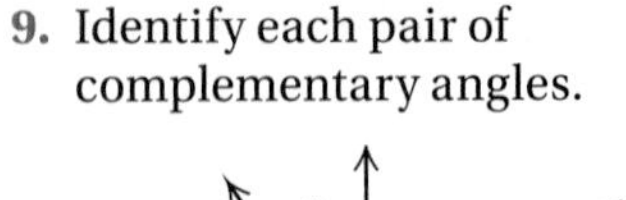

9. Identify each pair of complementary angles.

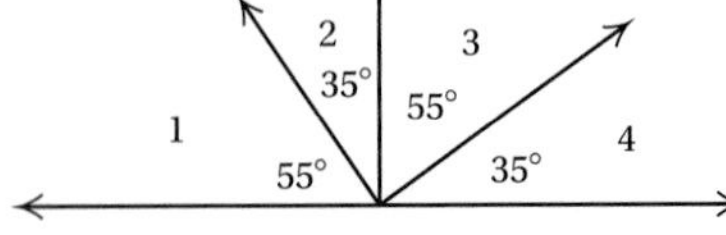

Find the measure of a complement of each angle.

10. 45°

11. 18°

12. 85°

13. 67° GS

$90^\circ - 67^\circ = \square^\circ$

Next, consider $\angle 1$ and $\angle 2$ as shown below. Because the sum of their measures is 180°, $\angle 1$ and $\angle 2$ are said to be **supplementary**. Note that when supplementary angles are adjacent, they form a straight angle.

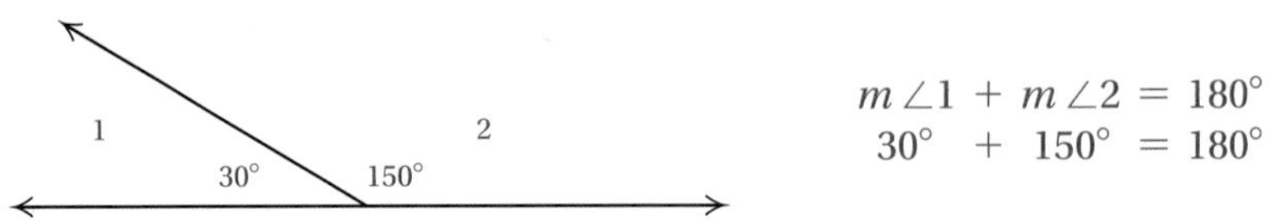

$$m\angle 1 + m\angle 2 = 180^\circ$$
$$30^\circ + 150^\circ = 180^\circ$$

Answers

9. $\angle 1$ and $\angle 2$; $\angle 1$ and $\angle 4$; $\angle 2$ and $\angle 3$; $\angle 3$ and $\angle 4$ **10.** 45° **11.** 72° **12.** 5° **13.** 23°

Guided Solution:
13. 23

SUPPLEMENTARY ANGLES

Two angles are **supplementary** if the sum of their measures is 180°. Each angle is called a **supplement** of the other.

EXAMPLE 3 Identify each pair of supplementary angles.

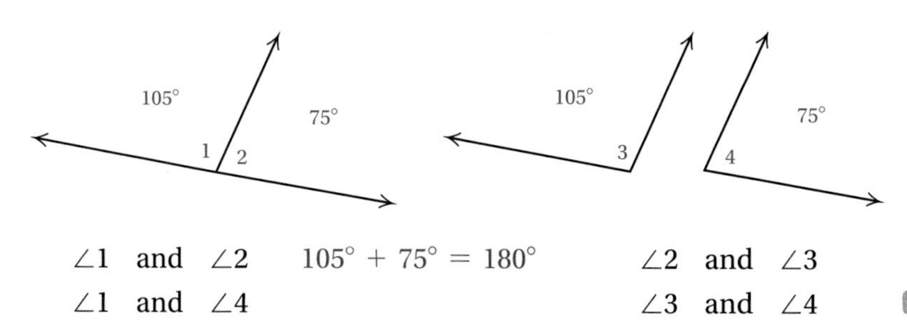

$\angle 1$ and $\angle 2$ $\quad 105° + 75° = 180°$ $\quad \angle 2$ and $\angle 3$

$\angle 1$ and $\angle 4$ $\quad \angle 3$ and $\angle 4$

EXAMPLE 4 Find the measure of a supplement of an angle of 112°.

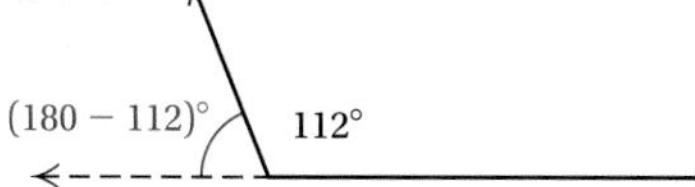

$180° - 112° = 68°$

The measure of a supplement is 68°.

Do Exercises 14–18.

14. Identify each pair of supplementary angles.

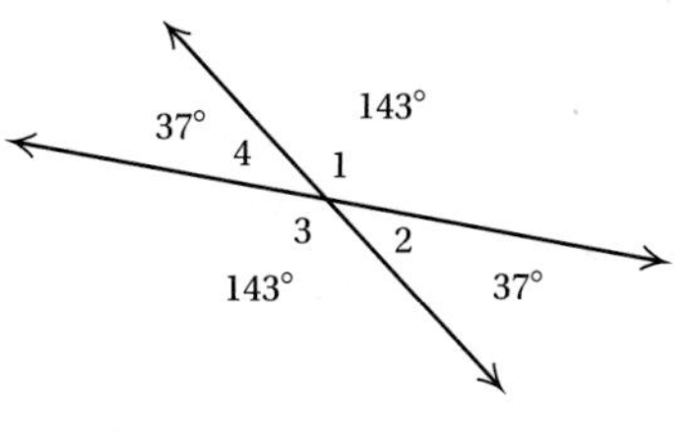

Find the measure of a supplement of each angle.

15. 38° **16.** 157°

17. 90°

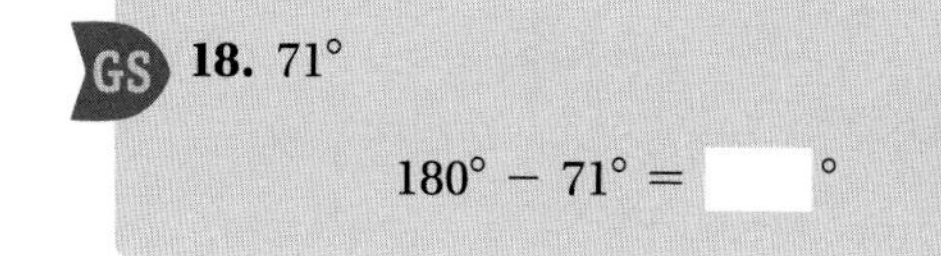

d TRIANGLES

A **triangle** is a polygon made up of three segments, or sides. Consider these triangles. The triangle with vertices *A*, *B*, and *C* can be named $\triangle ABC$.

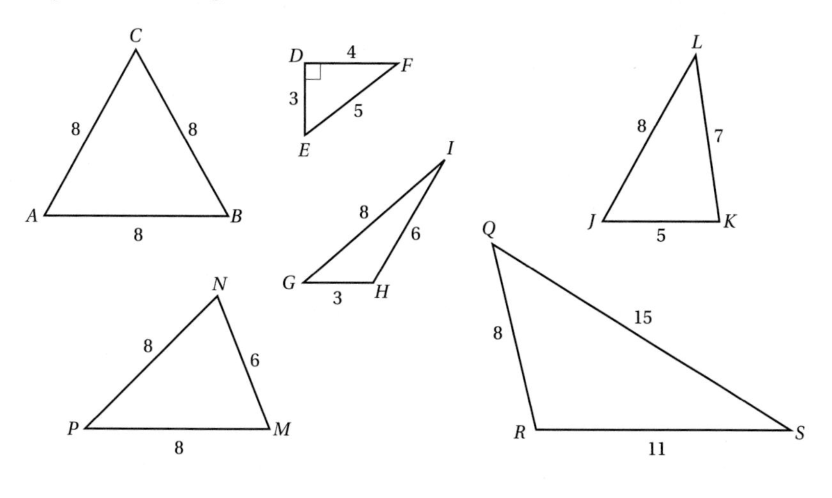

Answers

14. $\angle 1$ and $\angle 2$; $\angle 1$ and $\angle 4$; $\angle 2$ and $\angle 3$; $\angle 3$ and $\angle 4$ **15.** 142° **16.** 23° **17.** 90° **18.** 109°

Guided Solution:
18. 109

19. Which triangles on p. 595 are:
   a) equilateral?
   b) isosceles?
   c) scalene?

20. Are all equilateral triangles isosceles?

21. Are all isosceles triangles equilateral?

22. Which triangles on p. 595 are:
   a) right triangles?
   b) obtuse triangles?
   c) acute triangles?

23. Find $m\angle P + m\angle Q + m\angle R$.

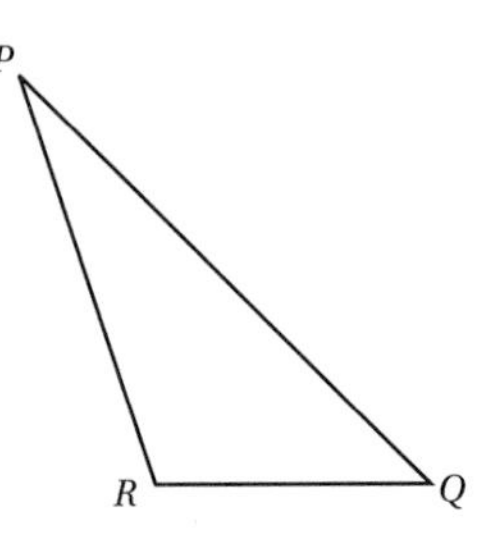

24. Find the missing angle measure.

B
61°
55°
x
A
C

Answers

19. (a) $\triangle ABC$; (b) $\triangle ABC$, $\triangle MPN$; (c) $\triangle DEF$, $\triangle GHI$, $\triangle JKL$, $\triangle QRS$ 20. Yes 21. No 22. (a) $\triangle DEF$; (b) $\triangle GHI$, $\triangle QRS$; (c) $\triangle ABC$, $\triangle MPN$, $\triangle JKL$ 23. 180° 24. 64°

We can classify triangles according to sides and according to angles.

TYPES OF TRIANGLES

Equilateral triangle: All sides are the same length.
Isosceles triangle: Two or more sides are the same length.
Scalene triangle: All sides are of different lengths.
Right triangle: One angle is a right angle.
Obtuse triangle: One angle is an obtuse angle.
Acute triangle: All three angles are acute.

Do Exercises 19–22.

e SUM OF THE ANGLE MEASURES OF A TRIANGLE

The sum of the angle measures of a triangle is 180°. To see this, note that we can think of cutting apart a triangle as shown on the left below. If we reassemble the pieces, we see that a straight angle is formed.

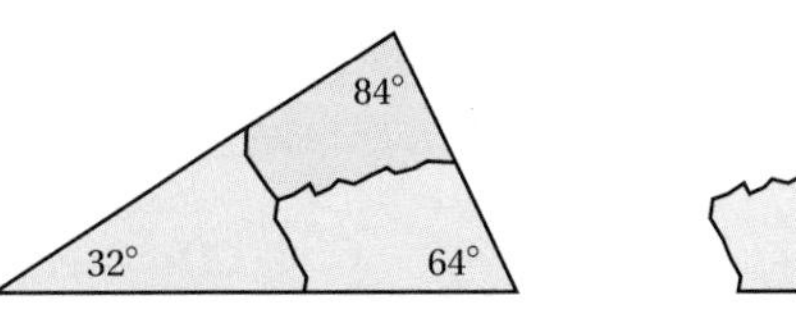

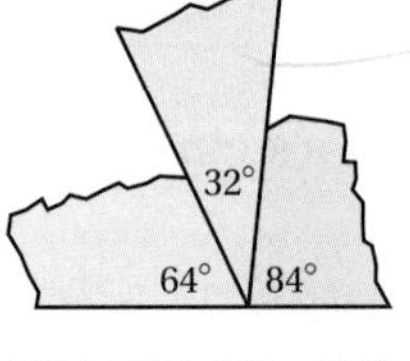

$64° + 32° + 84° = 180°$

SUM OF THE ANGLE MEASURES OF A TRIANGLE

In any $\triangle ABC$, the sum of the measures of the angles is 180°:

$$m\angle A + m\angle B + m\angle C = 180°.$$

Do Exercise 23.

If we know the measures of two angles of a triangle, we can calculate the measure of the third angle.

EXAMPLE 5 Find the missing angle measure.

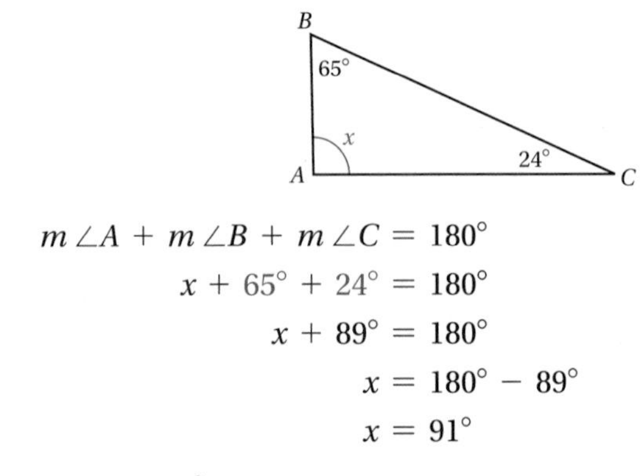

$$\begin{aligned} m\angle A + m\angle B + m\angle C &= 180° \\ x + 65° + 24° &= 180° \\ x + 89° &= 180° \\ x &= 180° - 89° \\ x &= 91° \end{aligned}$$

Thus, $m\angle A = 91°$.

Do Exercise 24.

10.5 Exercise Set

For Extra Help
MyMathLab® MathXL® PRACTICE WATCH READ REVIEW

Reading Check

Match each definition with the correct term from the list on the right.

RC1. ________ An angle whose measure is 90°

RC2. ________ An angle whose measure is 180°

RC3. ________ An angle whose measure is greater than 0° and less than 90°

RC4. ________ An angle whose measure is greater than 90° and less than 180°

RC5. ________ A pair of angles whose measures add to 90°

RC6. ________ A pair of angles whose measures add to 180°

RC7. ________ A triangle with three sides of the same length

RC8. ________ A triangle with all sides of different lengths

RC9. ________ A triangle with two or more sides of the same length

RC10. ________ A triangle containing a 90° angle

a) acute angle
b) complementary angles
c) equilateral triangle
d) isosceles triangle
e) obtuse angle
f) right angle
g) right triangle
h) scalene triangle
i) straight angle
j) supplementary angles

a Name each angle in five different ways.

1.

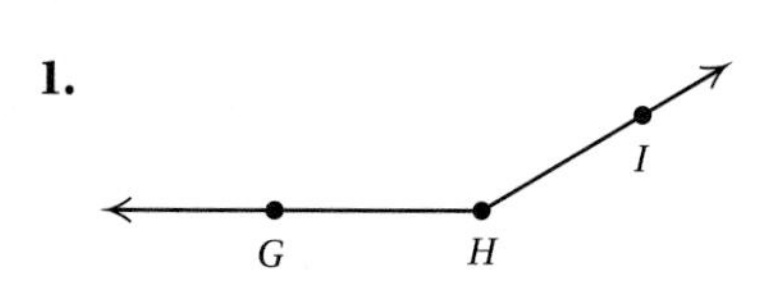

2.

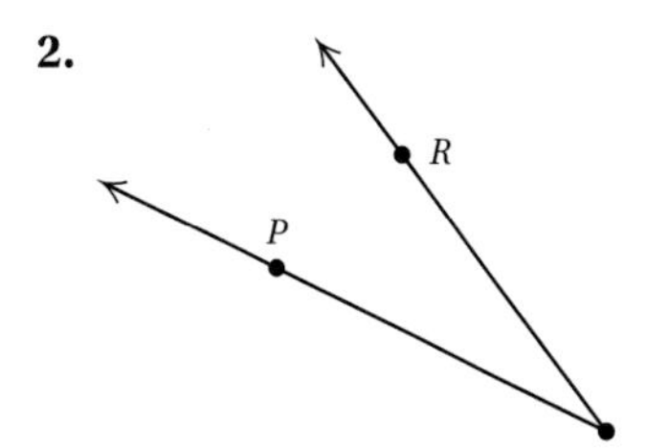

Use a protractor to measure each angle.

3. **4.** **5.** **6.** **7.** **8.**

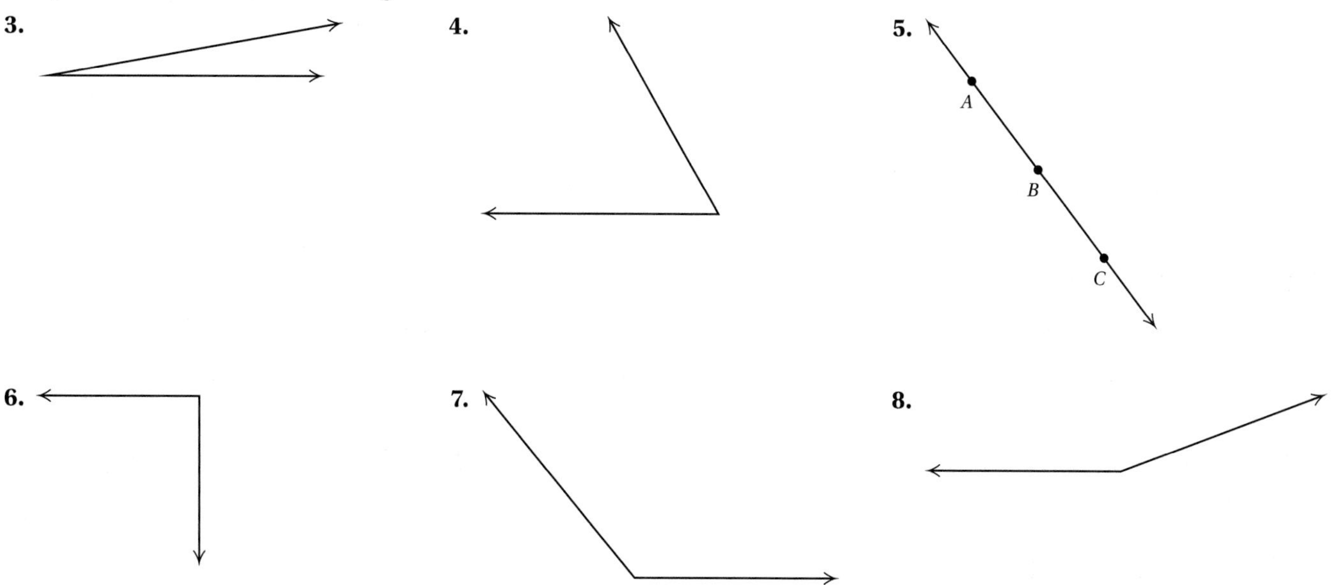

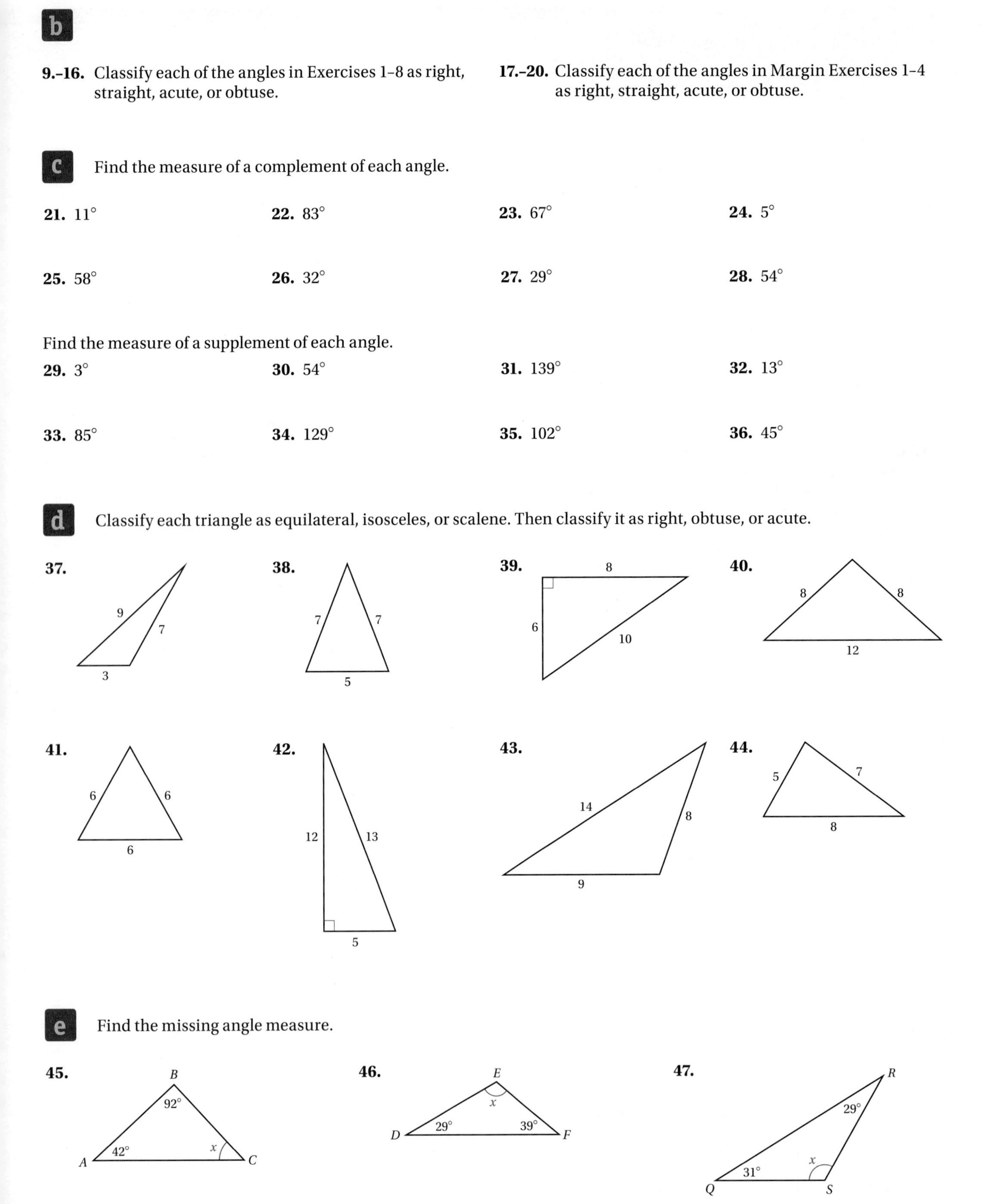

b

9.–16. Classify each of the angles in Exercises 1–8 as right, straight, acute, or obtuse.

17.–20. Classify each of the angles in Margin Exercises 1–4 as right, straight, acute, or obtuse.

c Find the measure of a complement of each angle.

21. 11°

22. 83°

23. 67°

24. 5°

25. 58°

26. 32°

27. 29°

28. 54°

Find the measure of a supplement of each angle.

29. 3°

30. 54°

31. 139°

32. 13°

33. 85°

34. 129°

35. 102°

36. 45°

d Classify each triangle as equilateral, isosceles, or scalene. Then classify it as right, obtuse, or acute.

37.

38.

39.

40.

41.

42.

43.

44.

e Find the missing angle measure.

45.

46.

47.

48. **49.** **50.**

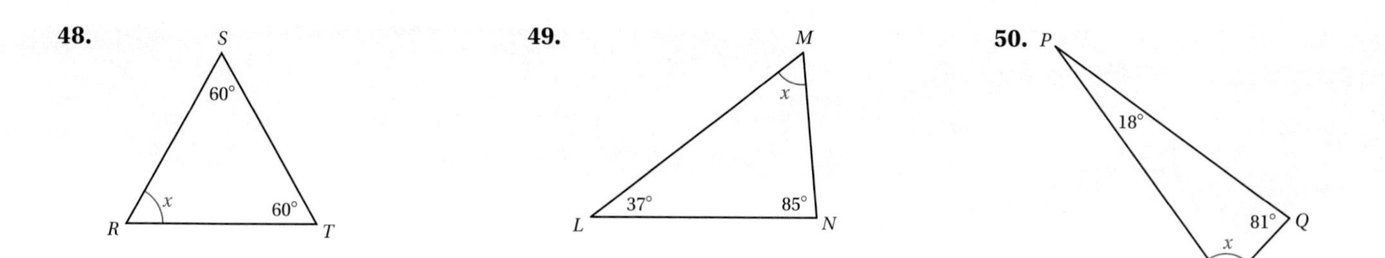

Skill Maintenance

Perform the indicated operation and simplify.

51. Subtract: $3.8 - 1.0875$. [5.2b]

52. Add: $2\frac{1}{3} + 5\frac{3}{4}$. [4.5a]

53. Add: $\frac{3}{10} + \left(-\frac{5}{12}\right)$. [4.2a]

54. Multiply: $\frac{1}{4} \cdot 2\frac{2}{3}$. [4.6a]

55. Divide: $-18 \div \frac{2}{3}$. [3.7b]

56. Divide: $16.8 \div 0.02$. [5.4a]

Solve.

57. $\frac{2}{5} + t = \frac{7}{10}$ [4.3c]

58. $\frac{2}{3} \cdot y = -\frac{1}{8}$ [3.7c]

Synthesis

59. In the figure, $m\angle 1 = 79.8°$ and $m\angle 6 = 33.07°$. Find $m\angle 2$, $m\angle 3$, $m\angle 4$, and $m\angle 5$.

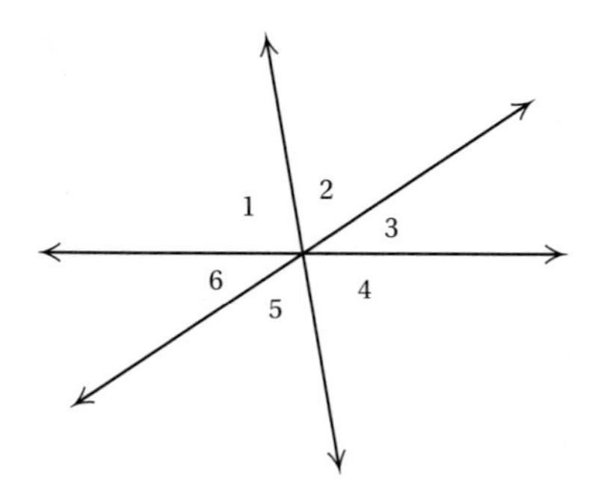

60. In the figure, $m\angle 2 = 42.17°$ and $m\angle 3 = 81.9°$. Find $m\angle 1$, $m\angle 4$, $m\angle 5$, and $m\angle 6$.

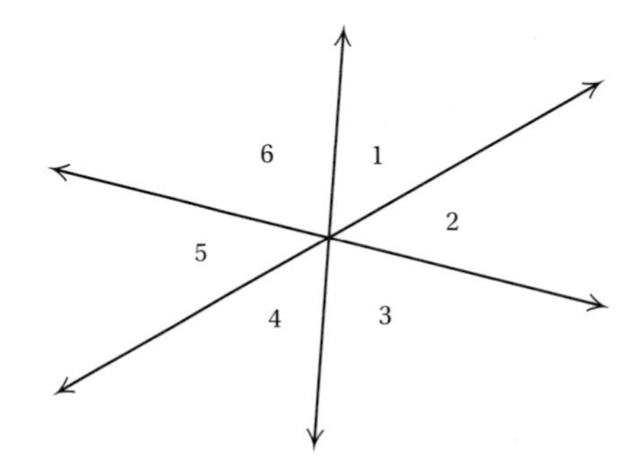

61. Find $m\angle ACB$, $m\angle CAB$, $m\angle EBC$, $m\angle EBA$, $m\angle AEB$, and $m\angle ADB$ in the rectangle shown below.

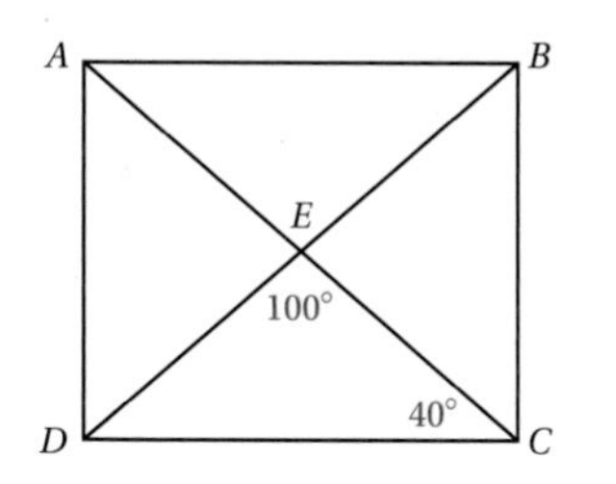

10.6 Square Roots and the Pythagorean Theorem

OBJECTIVES

- **a** Simplify square roots of squares such as $\sqrt{25}$.
- **b** Approximate square roots.
- **c** Given the lengths of any two sides of a right triangle, find the length of the third side.
- **d** Solve applied problems involving right triangles.

SKILL TO REVIEW

Objective 1.9b: Evaluate exponential notation.

Evaluate.

1. 5^2 **2.** 8^2

Find the square.

1. 9^2 **2.** 10^2

3. 11^2 **4.** 12^2

5. 13^2 **6.** 14^2

7. 15^2 **8.** 16^2

9. 17^2 **10.** 18^2

11. 20^2 **12.** 25^2

Simplify. The results of Exercises 1–12 above may be helpful here.

13. $\sqrt{9}$ **14.** $\sqrt{16}$

15. $\sqrt{121}$ **16.** $\sqrt{100}$

17. $\sqrt{81}$ **18.** $\sqrt{64}$

19. $\sqrt{324}$ **20.** $\sqrt{400}$

21. $\sqrt{225}$ **22.** $\sqrt{169}$

23. $\sqrt{1}$ **24.** $\sqrt{0}$

Answers

Answers are on p. 601.

a SQUARE ROOTS

SQUARE ROOT

If a number is a product of two identical factors, then either factor is called a **square root** of the number. (If $a = c^2$, then c is a square root of a.) The symbol $\sqrt{\ }$ (called a **radical sign**) is used in naming square roots.

For example, $\sqrt{36}$ is the square root of 36. It follows that

$\sqrt{36} = \sqrt{6 \cdot 6} = 6$ The square root of 36 is 6.

because $6^2 = 36$.

EXAMPLE 1 Simplify: $\sqrt{25}$.

$\sqrt{25} = \sqrt{5 \cdot 5} = 5$ The square root of 25 is 5 because $5^2 = 25$. ■

EXAMPLE 2 Simplify: $\sqrt{144}$.

$\sqrt{144} = \sqrt{12 \cdot 12} = 12$ The square root of 144 is 12 because $12^2 = 144$. ■

Caution!

It is common to confuse squares and square roots. A number squared is that number multiplied by itself. For example, $16^2 = 16 \cdot 16 = 256$. A square root of a number is a number that when multiplied by itself gives the original number. For example, $\sqrt{16} = 4$, because $4 \cdot 4 = 16$.

EXAMPLES Simplify.

3. $\sqrt{4} = 2$

4. $\sqrt{256} = 16$

5. $\sqrt{49} = 7$

◀ **Do Margin Exercises 1–24.**

b APPROXIMATING SQUARE ROOTS

Many square roots can't be written as whole numbers or fractions. For example, $\sqrt{2}$, $\sqrt{3}$, $\sqrt{39}$, and $\sqrt{70}$ cannot be precisely represented in decimal notation. To see this, consider the following decimal approximations for $\sqrt{2}$. Each gives a closer approximation, but none is exactly $\sqrt{2}$:

$$\sqrt{2} \approx 1.4 \quad \text{because} \quad (1.4)^2 = 1.96;$$
$$\sqrt{2} \approx 1.41 \quad \text{because} \quad (1.41)^2 = 1.9881;$$
$$\sqrt{2} \approx 1.414 \quad \text{because} \quad (1.414)^2 = 1.999396;$$
$$\sqrt{2} \approx 1.4142 \quad \text{because} \quad (1.4142)^2 = 1.99996164.$$

Decimal approximations like these are commonly found by using a calculator.

EXAMPLE 6 Use a calculator to approximate $\sqrt{3}$, $\sqrt{27}$, and $\sqrt{180}$ to three decimal places.

We use a calculator to find each square root. Since the calculator displays more than three decimal places, we round back to three places.

$$\sqrt{3} \approx 1.732, \qquad \sqrt{27} \approx 5.196, \qquad \sqrt{180} \approx 13.416$$

As a check, note that $1 \cdot 1 = 1$ and $2 \cdot 2 = 4$, so we expect $\sqrt{3}$ to be between 1 and 2. Similarly, we expect $\sqrt{27}$ to be between 5 and 6 and $\sqrt{180}$ to be between 13 and 14.

Do Exercises 25–28. ▶

c THE PYTHAGOREAN THEOREM

A **right triangle** is a triangle with a 90° angle, as shown here. In a right triangle, the longest side is called the **hypotenuse**. It is the side opposite the right angle. The other two sides are called **legs**. We generally use the letters a and b for the lengths of the legs and c for the length of the hypotenuse. They are related as follows.

THE PYTHAGOREAN THEOREM

In any right triangle, if a and b are the lengths of the legs and c is the length of the hypotenuse, then

$$a^2 + b^2 = c^2, \quad \text{or}$$
$$(\text{Leg})^2 + (\text{Other leg})^2 = (\text{Hypotenuse})^2.$$

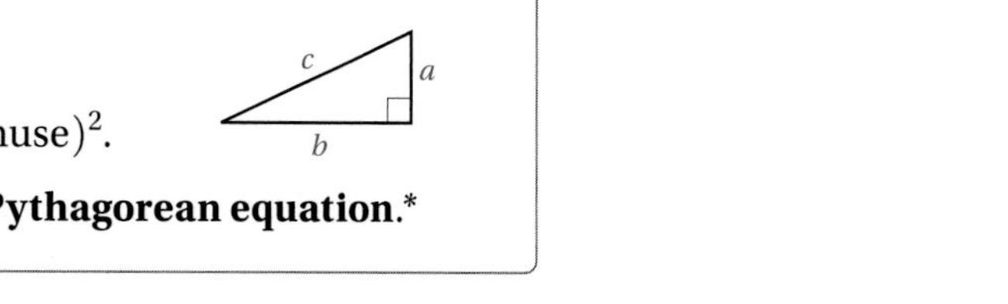

The equation $a^2 + b^2 = c^2$ is called the **Pythagorean equation**.*

*The *converse* of the Pythagorean theorem is also true. That is, if $a^2 + b^2 = c^2$, then the triangle is a right triangle.

CALCULATOR CORNER

Finding Square Roots Many calculators have a square root key, [√]. Often we enter the number whose square root we are finding first and then press the [√] key.

It is always best to wait until calculations are complete before rounding. For example, to find $9 \cdot \sqrt{30}$ rounded to the nearest tenth, we do not first determine that $\sqrt{30} \approx 5.5$ and then multiply by 9 to get 49.5. Rather, we press [9] [×] [3] [0] [2nd] [√] [=] or [9] [×] [3] [0] [SHIFT] [√] [=]. The result is 49.29503018, so $9 \cdot \sqrt{30} \approx 49.3$.

EXERCISES Use a calculator to find each of the following. Round to the nearest tenth.

1. $\sqrt{43}$
2. $7 \cdot \sqrt{8}$
3. $\sqrt{47} - 5$
4. $17 + \sqrt{57}$
5. $13\sqrt{68} + 14$
6. $7 \cdot \sqrt{90} + 3 \cdot \sqrt{40}$

Use a calculator to approximate to three decimal places.

25. $\sqrt{5}$ **26.** $\sqrt{78}$

27. $\sqrt{168}$ **28.** $\sqrt{321}$

Answers

Skill to Review:
1. 25 **2.** 64

Margin Exercises:
1. 81 **2.** 100 **3.** 121 **4.** 144 **5.** 169 **6.** 196 **7.** 225 **8.** 256 **9.** 289 **10.** 324 **11.** 400 **12.** 625 **13.** 3 **14.** 4 **15.** 11 **16.** 10 **17.** 9 **18.** 8 **19.** 18 **20.** 20 **21.** 15 **22.** 13 **23.** 1 **24.** 0 **25.** 2.236 **26.** 8.832 **27.** 12.961 **28.** 17.916

The Pythagorean theorem is named for the Greek mathematician Pythagoras (569?–500? B.C.E.). We can think of this relationship as adding areas.

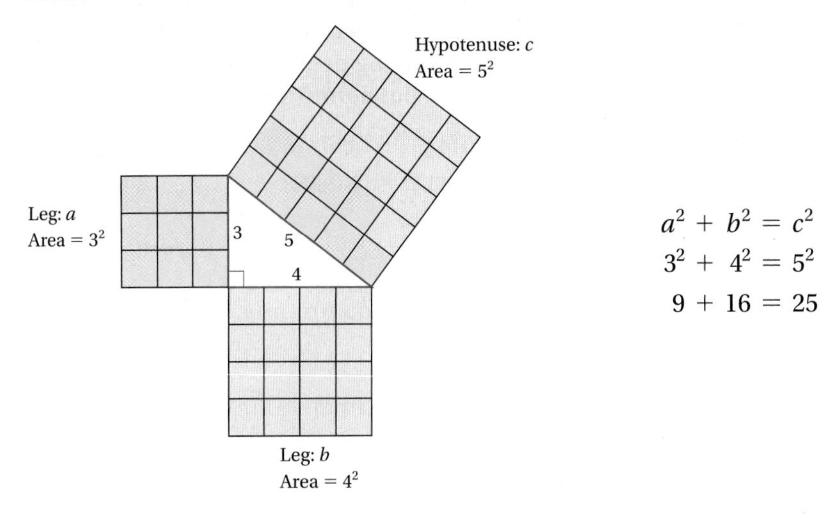

$$a^2 + b^2 = c^2$$
$$3^2 + 4^2 = 5^2$$
$$9 + 16 = 25$$

If we know the lengths of any two sides of a right triangle, we can use the Pythagorean equation to determine the length of the third side.

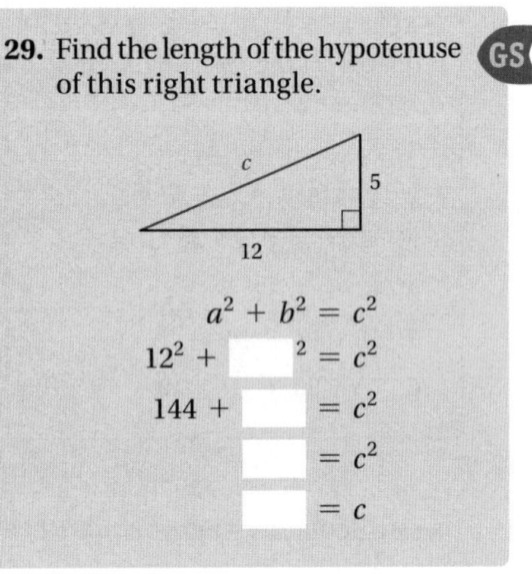

29. Find the length of the hypotenuse of this right triangle. GS

$a^2 + b^2 = c^2$

$12^2 + \square^2 = c^2$

$144 + \square = c^2$

$\square = c^2$

$\square = c$

EXAMPLE 7 Find the length of the hypotenuse of this right triangle.

We substitute in the Pythagorean equation:

$a^2 + b^2 = c^2$

$6^2 + 8^2 = c^2$ Substituting

$36 + 64 = c^2$

$100 = c^2.$

The solution of this equation is the square root of 100, which is 10:

$$c = \sqrt{100} = 10.$$

Do Exercise 29.

EXAMPLE 8 Find the length b for the right triangle shown. Give an exact answer and an approximation to three decimal places.

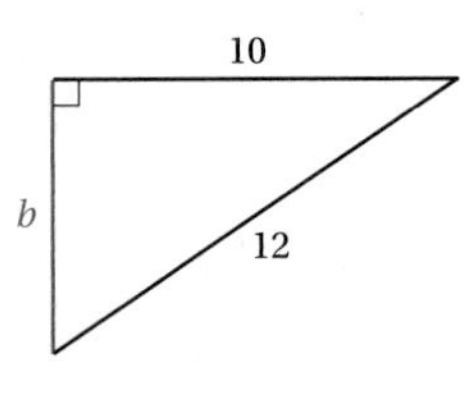

We substitute in the Pythagorean equation. Next, we solve for b^2 and then b, as follows:

$a^2 + b^2 = c^2$

$10^2 + b^2 = 12^2$ Substituting

$100 + b^2 = 144.$

Answer

29. $c = 13$

Guided Solution:

29. 5, 25, 169, 13

Then

$$
\begin{aligned}
100 + b^2 - 100 &= 144 - 100 && \text{Subtracting 100 on both sides} \\
b^2 &= 144 - 100 \\
b^2 &= 44 \\
\textit{Exact answer:} \quad b &= \sqrt{44} \\
\textit{Approximation:} \quad b &\approx 6.633. && \text{Using a calculator}
\end{aligned}
$$

Do Exercises 30–32. ▶

Find the length of the leg of each right triangle. Give an exact answer and an approximation to three decimal places.

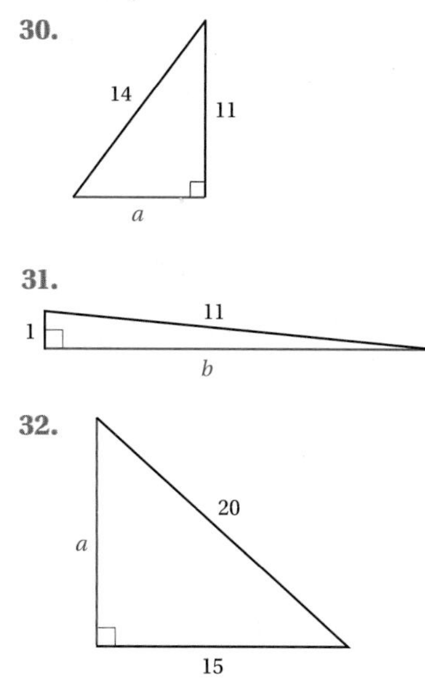

d APPLICATIONS

EXAMPLE 9 *Height of Ladder.* A 12-ft ladder leans against a building. The bottom of the ladder is 7 ft from the building. How high is the top of the ladder? Give an exact answer and an approximation to the nearest tenth of a foot.

1. **Familiarize.** We first make a drawing. In it we see a right triangle. We let $h =$ the unknown height.

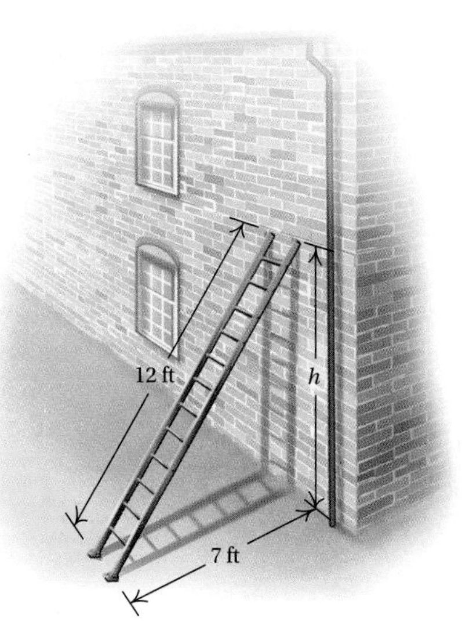

2. **Translate.** We substitute 7 for a, h for b, and 12 for c in the Pythagorean equation:

$$
\begin{aligned}
a^2 + b^2 &= c^2 && \text{Pythagorean equation} \\
7^2 + h^2 &= 12^2.
\end{aligned}
$$

3. **Solve.** We solve for h^2 and then h.

$$
\begin{aligned}
49 + h^2 &= 144 \\
49 + h^2 - 49 &= 144 - 49 \\
h^2 &= 144 - 49 \\
h^2 &= 95 \\
\textit{Exact answer:} \quad h &= \sqrt{95}\text{ ft} \\
\textit{Approximation:} \quad h &\approx 9.7\text{ ft}
\end{aligned}
$$

4. **Check.** $7^2 + (\sqrt{95})^2 = 49 + 95 = 144 = 12^2.$
5. **State.** The top of the ladder is $\sqrt{95}$ ft, or about 9.7 ft, from the ground.

Do Exercise 33. ▶

33. How long is a guy wire reaching from the top of an 18-ft pole to a point on the ground 10 ft from the pole? Give an exact answer and an approximation to the nearest tenth of a foot.

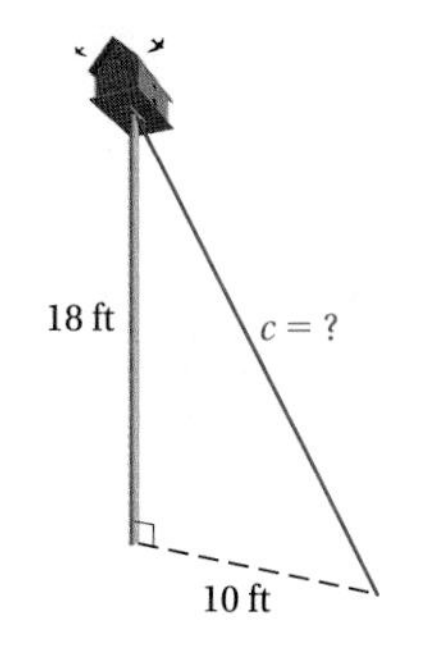

Answers

30. $a = \sqrt{75}$; $a \approx 8.660$ **31.** $b = \sqrt{120}$; $b \approx 10.954$ **32.** $a = \sqrt{175}$; $a \approx 13.229$ **33.** $\sqrt{424}$ ft ≈ 20.6 ft

Translating for Success

1. *Servings of Pork.* An 8-lb pork roast contains 37 servings of meat. How many pounds of pork would be needed for 55 servings?

2. *Height of a Ladder.* A 14.5-ft ladder leans against a house. The bottom of the ladder is 9.4 ft from the house. How high is the top of the ladder?

3. *Cruise Cost.* A group of 6 college students pays $4608 for a spring break cruise. What is each person's share?

4. *Sales Tax Rate.* The sales tax is $14.95 when a new ladder is purchased for $299. What is the sales tax rate?

5. *Volume of a Sphere.* Find the volume of a sphere whose radius is 7.2 cm.

The goal of these matching questions is to practice step (2), Translate, *of the five-step problem-solving process. Translate each word problem to an equation and select a correct translation from equations A–O.*

A. $x = \frac{1}{3}\pi \cdot 6^2 \cdot (7.2)$

B. $6 \cdot x = \$4608$

C. $x = \frac{4}{3} \cdot \pi \cdot 6^2 \cdot (7.2)$

D. $x = \pi \cdot \left(5\frac{1}{2} \div 2\right)^2 \cdot 7$

E. $x = 6\% \times 5 \times \$14.95$

F. $x = \frac{1}{3}\pi\left(5\frac{1}{2}\right)^2$

G. $(9.4)^2 + x^2 = (14.5)^2$

H. $\$14.95 = x \cdot \$299$

I. $x = 2(14.5 + 9.4)$

J. $(9.4 + 14.5)^2 = x$

K. $\frac{8}{37} = \frac{x}{55}$

L. $x = 4(14.5 + 9.4)$

M. $x = 6 \cdot \$4608$

N. $8 \cdot 37 = 55 \cdot x$

O. $x = \frac{4}{3} \cdot \pi \cdot (7.2)^3$

Answers on page A-17

6. *Inheritance.* Six children each inherit $4608 from their mother's estate. What is the total inheritance?

7. *Sales Tax.* Erica buys 5 pairs of earrings at $14.95 each. The sales tax rate is 6%. How much sales tax will be charged?

8. *Volume of a Cone.* Find the volume of a circular cone with a base radius of 6 cm and a height of 7.2 cm.

9. *Volume of a Storage Tank.* The diameter of a cylindrical grain-storage tank is $5\frac{1}{2}$ yd. Its height is 7 yd. Find its volume.

10. *Perimeter of a Photo.* A rectangular photo is 14.5 cm by 9.4 cm. What is the perimeter of the photo?

10.6 Exercise Set

For Extra Help
MyMathLab®   MathXL®   PRACTICE   WATCH   READ   REVIEW

Reading Check

Determine whether each statement is true or false.

__________ **RC1.** $\sqrt{100} = 10$

__________ **RC2.** $\sqrt{3} = 9$

__________ **RC3.** In a right triangle, the side opposite the right angle is the hypotenuse.

__________ **RC4.** In a right triangle, the sum of the lengths of the legs is the length of the hypotenuse.

a Simplify.

1. $\sqrt{100}$ **2.** $\sqrt{25}$ **3.** $\sqrt{441}$ **4.** $\sqrt{225}$

5. $\sqrt{625}$ **6.** $\sqrt{576}$ **7.** $\sqrt{361}$ **8.** $\sqrt{484}$

9. $\sqrt{529}$ **10.** $\sqrt{169}$ **11.** $\sqrt{10{,}000}$ **12.** $\sqrt{4{,}000{,}000}$

b Approximate to three decimal places.

13. $\sqrt{48}$ **14.** $\sqrt{17}$ **15.** $\sqrt{8}$ **16.** $\sqrt{3}$

17. $\sqrt{18}$ **18.** $\sqrt{7}$ **19.** $\sqrt{6}$ **20.** $\sqrt{61}$

21. $\sqrt{10}$ **22.** $\sqrt{21}$ **23.** $\sqrt{75}$ **24.** $\sqrt{220}$

25. $\sqrt{196}$ **26.** $\sqrt{123}$ **27.** $\sqrt{183}$ **28.** $\sqrt{300}$

c Find the length of the third side of each right triangle. Give an exact answer and, where appropriate, an approximation to three decimal places.

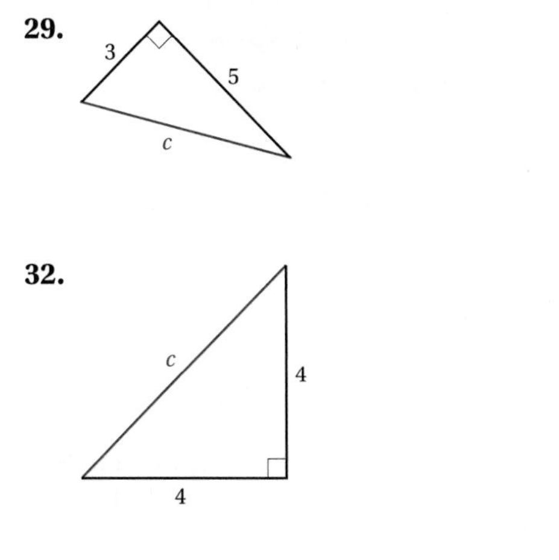

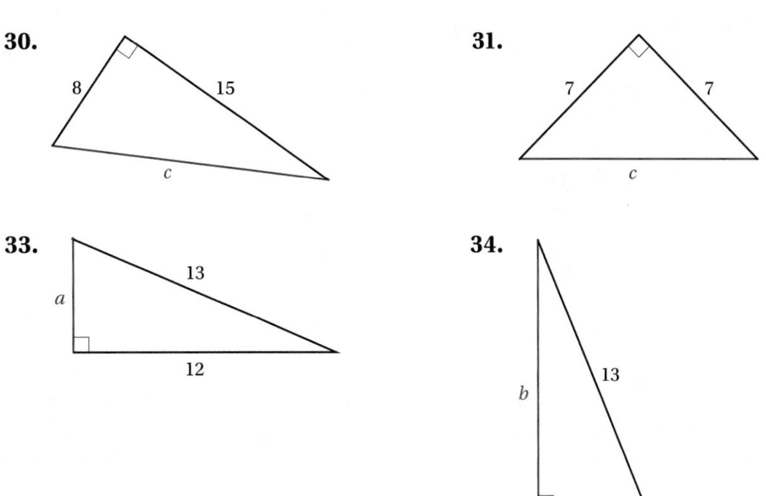

35.

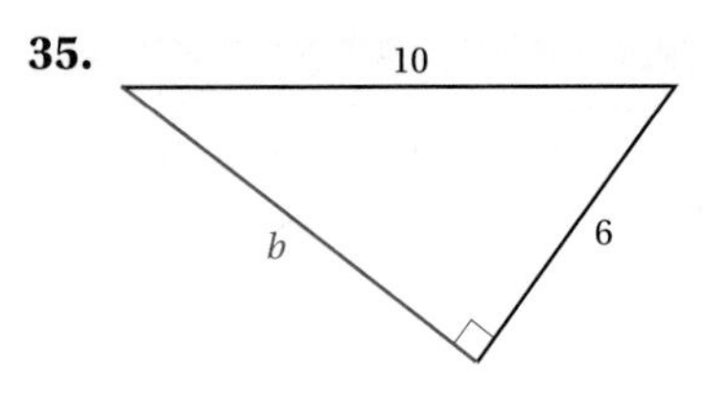

36.

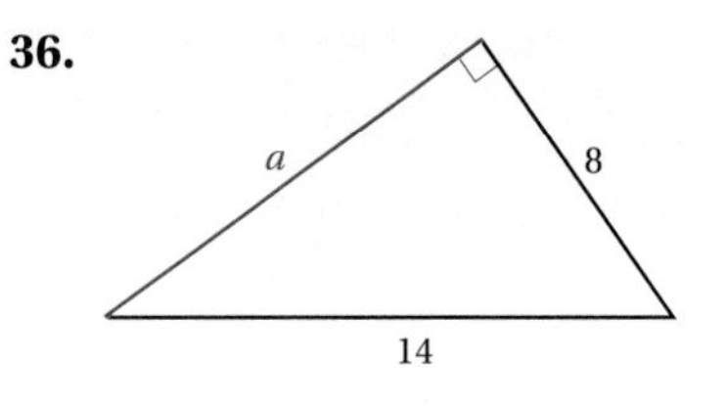

For each right triangle, find the length of the side not given. Assume that c represents the length of the hypotenuse. Give an exact answer and, when appropriate, an approximation to three decimal places.

37. $a = 10, b = 24$

38. $a = 5, b = 12$

39. $a = 9, c = 15$

40. $a = 18, c = 30$

41. $a = 1, c = 32$

42. $b = 1, c = 20$

43. $a = 4, b = 3$

44. $a = 1, c = 15$

d In Exercises 45–52, give an exact answer and an approximation to the nearest tenth.

45. *Softball Diamond.* A slow-pitch softball diamond is actually a square 65 ft on a side. How far is it from home plate to second base?

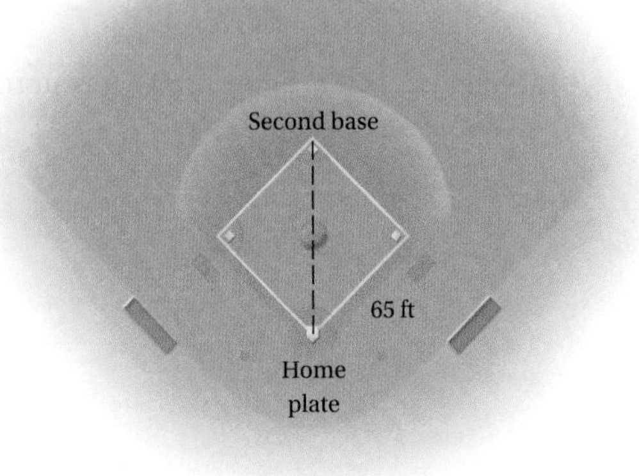

46. *Baseball Diamond.* A baseball diamond is actually a square 90 ft on a side. How far is it from home plate to second base?

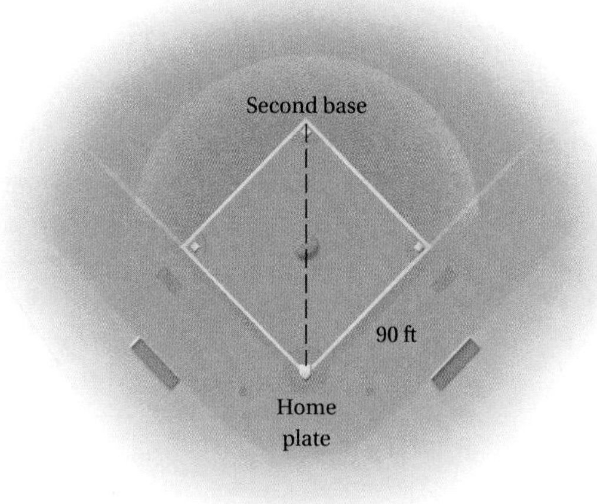

47. A 30-ft string of lights reaches from the top of a pole to a point on the ground 16 ft from the base of the pole. How tall is the pole?

48. A 25-ft wire reaches from the top of a telephone pole to a point on the ground 18 ft from the base of the pole. How tall is the pole?

49. *Great Pyramid.* The Pyramid of Cheops is 146 m high. The distance at ground level from the center of the pyramid to the middle of one of the faces is 115 m, as shown below. What is the slant height of a side of the pyramid?

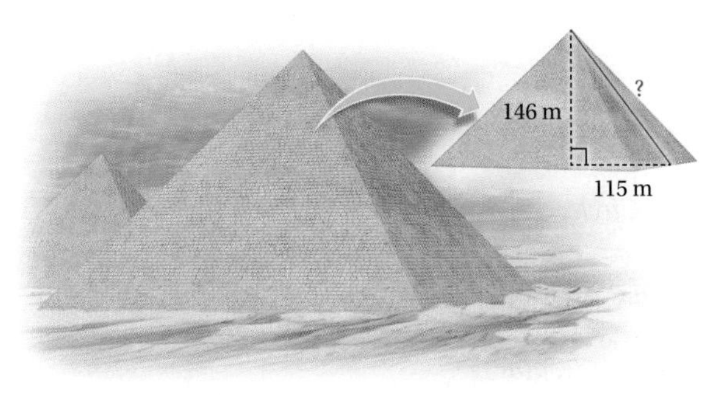

50. *Construction.* In order to support a masonry wall, Matthew erects braces at a height of 12 ft on the wall. The braces are anchored to the ground 15 ft from the base of the wall. How long are the braces?

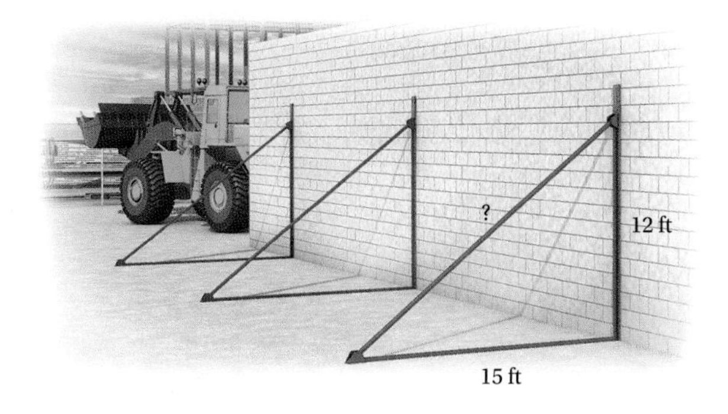

51. An airplane is flying at an altitude of 4100 ft. The slanted distance directly to the point where it will touch down is 15,100 ft. How far is the airplane horizontally from that point?

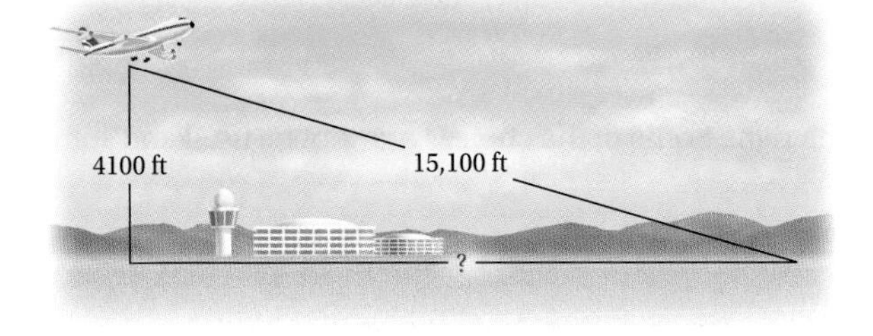

52. A surveyor had poles located at points P, Q, and R around a lake. The distances that the surveyor was able to measure are marked on the drawing. What is the approximate distance from P to R across the lake?

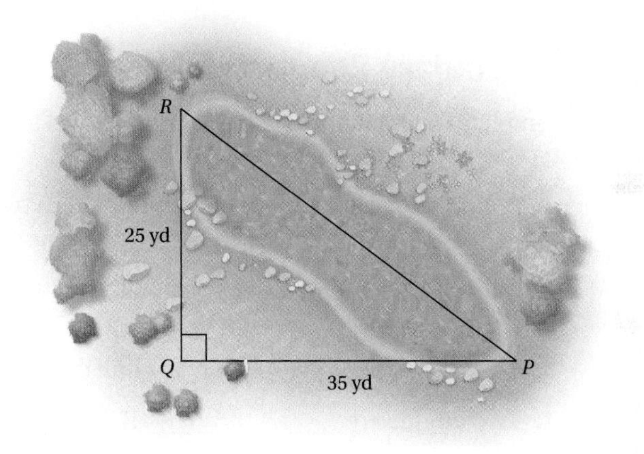

Skill Maintenance

Evaluate. [1.9b]

53. 10^3

54. 10^2

55. 10^5

56. 10^4

Simplify. [1.9c, d]

57. $90 \div 15 \cdot 2 - (1 + 2)^2$

58. $10^3 - \{2 \times [5 \times 3 - (4 + 2)]\}$

Synthesis

59. Which of the triangles below has the larger area?

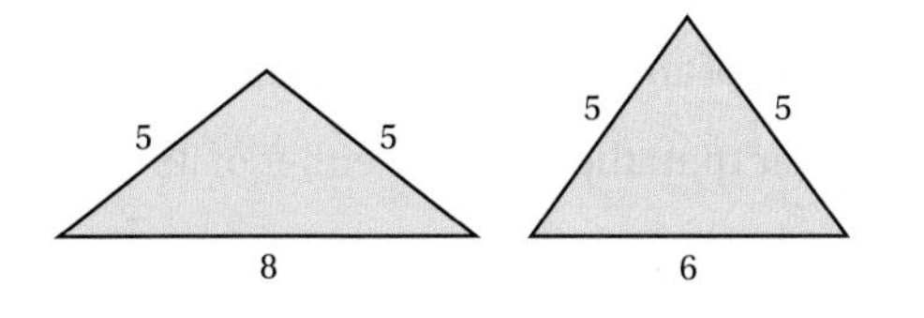

60. Find the area of the trapezoid shown. Round to the nearest hundredth.

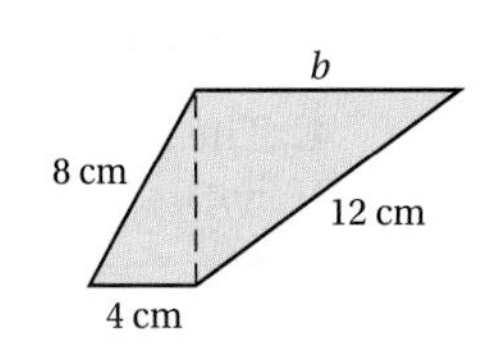

CHAPTER

10 Summary and Review

Formulas

Perimeter of a Rectangle: $P = 2 \cdot (l + w)$, or $P = 2 \cdot l + 2 \cdot w$

Perimeter of a Square: $P = 4 \cdot s$

Area of a Rectangle: $A = l \cdot w$

Area of a Square: $A = s \cdot s$, or $A = s^2$

Area of a Parallelogram: $A = b \cdot h$

Area of a Triangle: $A = \frac{1}{2} \cdot b \cdot h$

Area of a Trapezoid: $A = \frac{1}{2} \cdot h \cdot (a + b)$

Radius and Diameter of a Circle: $d = 2 \cdot r$, or $r = \dfrac{d}{2}$

Circumference of a Circle: $C = \pi \cdot d$, or $C = 2 \cdot \pi \cdot r$

Area of a Circle: $A = \pi \cdot r \cdot r$, or $A = \pi \cdot r^2$

Volume of a Rectangular Solid: $V = l \cdot w \cdot h$

Volume of a Circular Cylinder: $V = \pi \cdot r^2 \cdot h$

Volume of a Sphere: $V = \frac{4}{3} \cdot \pi \cdot r^3$

Volume of a Cone: $V = \frac{1}{3} \cdot \pi \cdot r^2 \cdot h$

Pythagorean Equation: $a^2 + b^2 = c^2$

Vocabulary Reinforcement

Complete each statement with the correct word from the list at the right. Some of the choices may not be used and some may be used more than once.

1. A parallelogram is a four-sided figure with two pairs of _______________ sides. [10.2b]
2. The _______________ of a polygon is the sum of the lengths of its sides. [10.1a]
3. The _______________ of a circle is half the length of its diameter. [10.3a]
4. Two angles are _______________ if the sum of their measures is 180°. [10.5c]
5. A(n) _______________ triangle has all sides of different lengths. [10.5d]
6. The _______________ of a right triangle is the side opposite the right angle. [10.6c]

circumference
radius
perimeter
isosceles
scalene
parallel
perpendicular
hypotenuse
leg
complementary
supplementary

Concept Reinforcement

Determine whether each statement is true or false.

_______________ 1. The acute angles of a right triangle are complementary. [10.5c, d]

_______________ 2. The volume of a sphere with diameter 6 ft is less than the volume of a rectangular solid that measures 6 ft by 6 ft by 6 ft. [10.4a, c]

_______________ 3. The measure of any obtuse angle is larger than the measure of any acute angle. [10.5b]

_______________ 4. The length of the hypotenuse of a right triangle is greater than the length of either of its legs. [10.6c]

Study Guide

Objectives 10.1a and 10.2a Find the perimeter of a polygon; find the area of a rectangle.

Example Find the perimeter and the area of this rectangle.

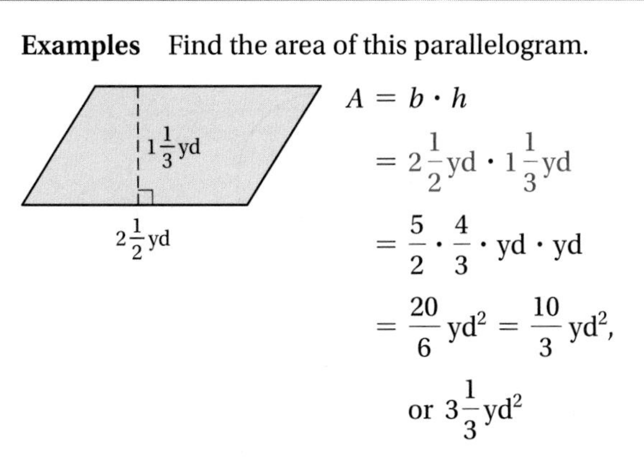

$$P = 2 \cdot (l + w)$$
$$= 2 \cdot (4.3\text{ m} + 2.7\text{ m})$$
$$= 2 \cdot (7\text{ m}) = 14\text{ m}$$
$$A = l \cdot w$$
$$= 4.3\text{ m} \cdot 2.7\text{ m} = 11.61\text{ m}^2$$

Practice Exercise

1. Find the perimeter and the area of this rectangle.

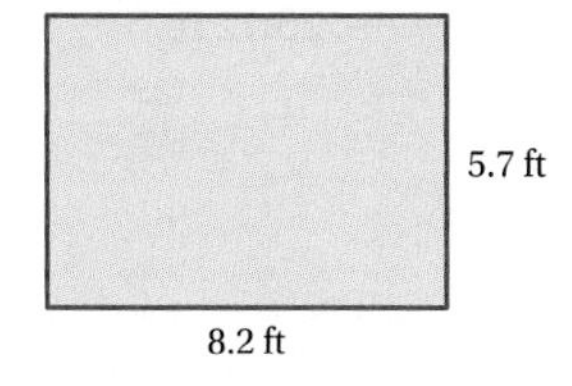

Objective 10.2b Find the area of a parallelogram, a triangle, and a trapezoid.

Examples Find the area of this parallelogram.

$$A = b \cdot h$$
$$= 2\frac{1}{2}\text{ yd} \cdot 1\frac{1}{3}\text{ yd}$$
$$= \frac{5}{2} \cdot \frac{4}{3} \cdot \text{yd} \cdot \text{yd}$$
$$= \frac{20}{6}\text{ yd}^2 = \frac{10}{3}\text{ yd}^2,$$
$$\text{or } 3\frac{1}{3}\text{ yd}^2$$

Find the area of this triangle.

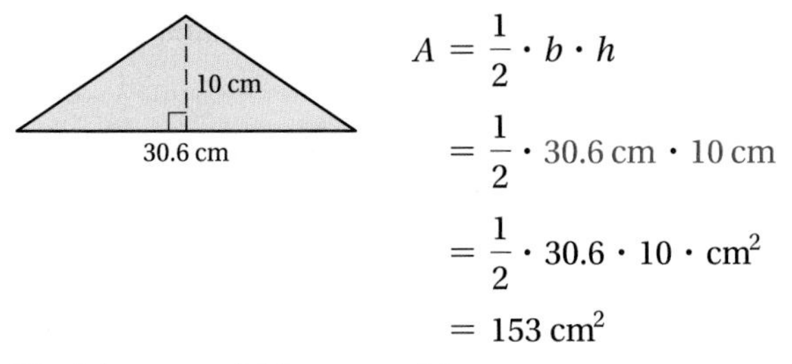

$$A = \frac{1}{2} \cdot b \cdot h$$
$$= \frac{1}{2} \cdot 30.6\text{ cm} \cdot 10\text{ cm}$$
$$= \frac{1}{2} \cdot 30.6 \cdot 10 \cdot \text{cm}^2$$
$$= 153\text{ cm}^2$$

Find the area of this trapezoid.

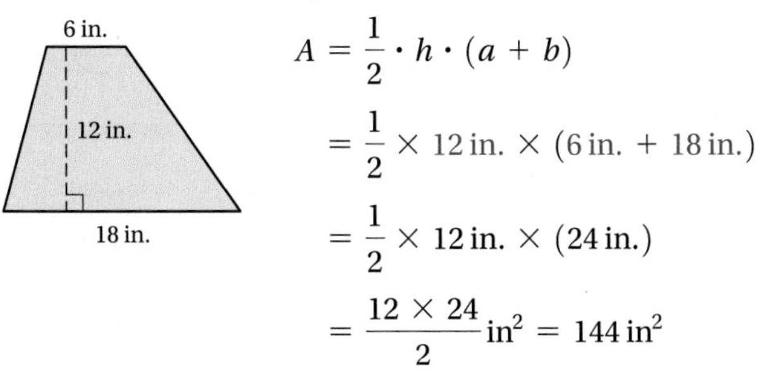

$$A = \frac{1}{2} \cdot h \cdot (a + b)$$
$$= \frac{1}{2} \times 12\text{ in.} \times (6\text{ in.} + 18\text{ in.})$$
$$= \frac{1}{2} \times 12\text{ in.} \times (24\text{ in.})$$
$$= \frac{12 \times 24}{2}\text{ in}^2 = 144\text{ in}^2$$

Practice Exercises

2. Find the area of this parallelogram.

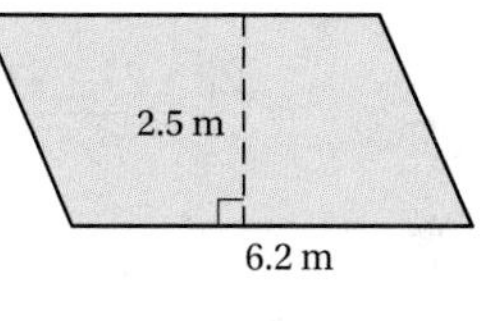

3. Find the area of this triangle.

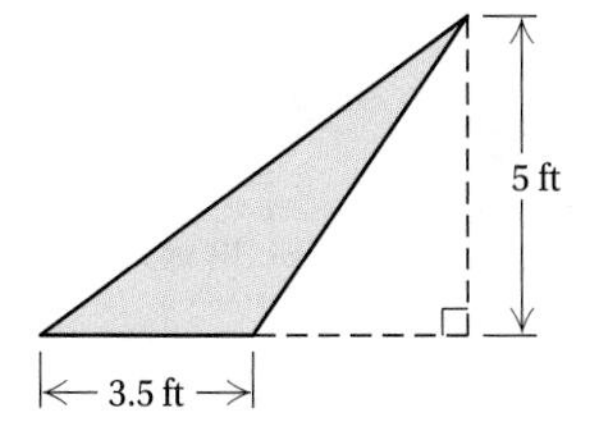

4. Find the area of this trapezoid.

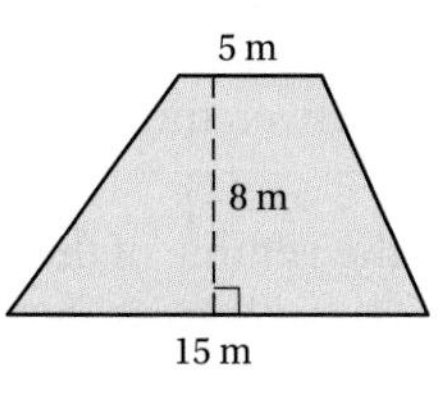

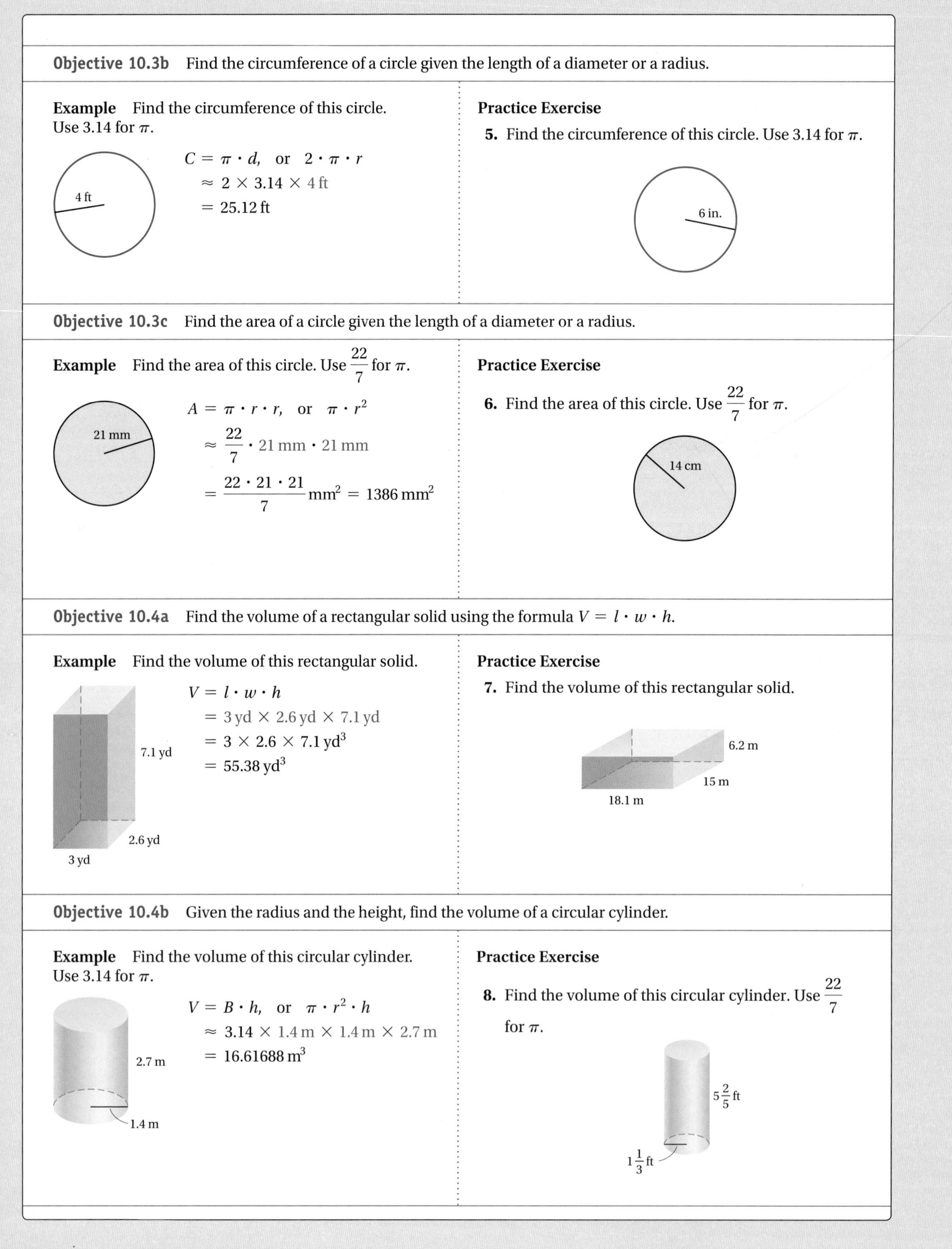

Objective 10.3b Find the circumference of a circle given the length of a diameter or a radius.

Example Find the circumference of this circle. Use 3.14 for π.

$$C = \pi \cdot d, \quad \text{or} \quad 2 \cdot \pi \cdot r$$
$$\approx 2 \times 3.14 \times 4\text{ ft}$$
$$= 25.12\text{ ft}$$

Practice Exercise

5. Find the circumference of this circle. Use 3.14 for π.

Objective 10.3c Find the area of a circle given the length of a diameter or a radius.

Example Find the area of this circle. Use $\frac{22}{7}$ for π.

$$A = \pi \cdot r \cdot r, \quad \text{or} \quad \pi \cdot r^2$$
$$\approx \frac{22}{7} \cdot 21\text{ mm} \cdot 21\text{ mm}$$
$$= \frac{22 \cdot 21 \cdot 21}{7}\text{ mm}^2 = 1386\text{ mm}^2$$

Practice Exercise

6. Find the area of this circle. Use $\frac{22}{7}$ for π.

Objective 10.4a Find the volume of a rectangular solid using the formula $V = l \cdot w \cdot h$.

Example Find the volume of this rectangular solid.

$$V = l \cdot w \cdot h$$
$$= 3\text{ yd} \times 2.6\text{ yd} \times 7.1\text{ yd}$$
$$= 3 \times 2.6 \times 7.1\text{ yd}^3$$
$$= 55.38\text{ yd}^3$$

Practice Exercise

7. Find the volume of this rectangular solid.

Objective 10.4b Given the radius and the height, find the volume of a circular cylinder.

Example Find the volume of this circular cylinder. Use 3.14 for π.

$$V = B \cdot h, \quad \text{or} \quad \pi \cdot r^2 \cdot h$$
$$\approx 3.14 \times 1.4\text{ m} \times 1.4\text{ m} \times 2.7\text{ m}$$
$$= 16.61688\text{ m}^3$$

Practice Exercise

8. Find the volume of this circular cylinder. Use $\frac{22}{7}$ for π.

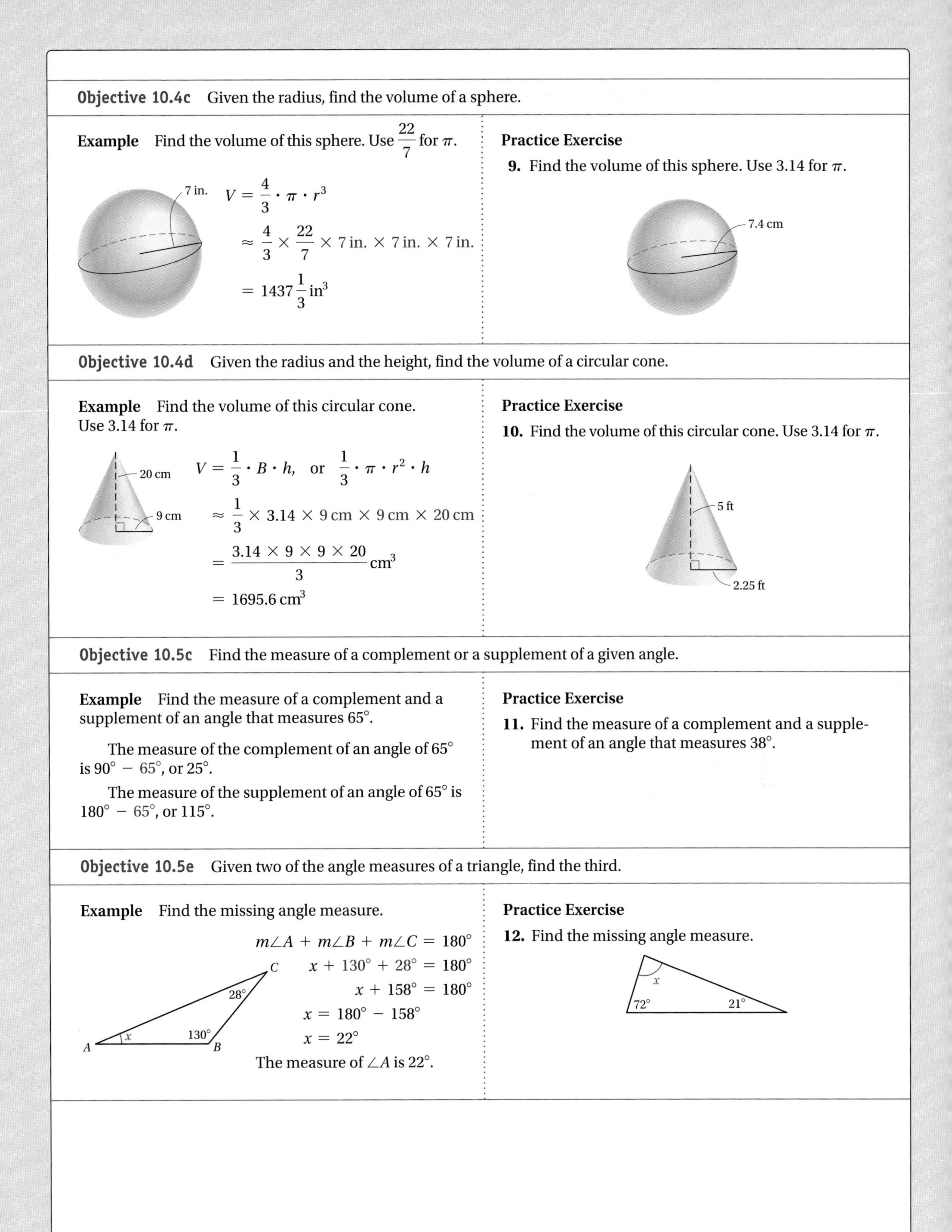

Objective 10.4c Given the radius, find the volume of a sphere.

Example Find the volume of this sphere. Use $\frac{22}{7}$ for π.

$$V = \frac{4}{3} \cdot \pi \cdot r^3$$

$$\approx \frac{4}{3} \times \frac{22}{7} \times 7 \text{ in.} \times 7 \text{ in.} \times 7 \text{ in.}$$

$$= 1437\frac{1}{3} \text{ in}^3$$

Practice Exercise

9. Find the volume of this sphere. Use 3.14 for π.

Objective 10.4d Given the radius and the height, find the volume of a circular cone.

Example Find the volume of this circular cone. Use 3.14 for π.

$$V = \frac{1}{3} \cdot B \cdot h, \quad \text{or} \quad \frac{1}{3} \cdot \pi \cdot r^2 \cdot h$$

$$\approx \frac{1}{3} \times 3.14 \times 9 \text{ cm} \times 9 \text{ cm} \times 20 \text{ cm}$$

$$= \frac{3.14 \times 9 \times 9 \times 20}{3} \text{ cm}^3$$

$$= 1695.6 \text{ cm}^3$$

Practice Exercise

10. Find the volume of this circular cone. Use 3.14 for π.

Objective 10.5c Find the measure of a complement or a supplement of a given angle.

Example Find the measure of a complement and a supplement of an angle that measures 65°.

The measure of the complement of an angle of 65° is 90° − 65°, or 25°.

The measure of the supplement of an angle of 65° is 180° − 65°, or 115°.

Practice Exercise

11. Find the measure of a complement and a supplement of an angle that measures 38°.

Objective 10.5e Given two of the angle measures of a triangle, find the third.

Example Find the missing angle measure.

$$m\angle A + m\angle B + m\angle C = 180°$$

$$x + 130° + 28° = 180°$$

$$x + 158° = 180°$$

$$x = 180° - 158°$$

$$x = 22°$$

The measure of $\angle A$ is 22°.

Practice Exercise

12. Find the missing angle measure.

Objective 10.6c Given the lengths of any two sides of a right triangle, find the length of the third side.

Example Find the length of the third side of this triangle. Give an exact answer and an approximation to three decimal places.

$$a^2 + b^2 = c^2 \quad \text{Pythagorean equation}$$
$$a^2 + 11^2 = 15^2$$
$$a^2 + 121 = 225$$
$$a^2 = 225 - 121$$
$$a^2 = 104$$
$$a = \sqrt{104} \approx 10.198$$

Practice Exercise

13. Find the length of the third side of this right triangle. Give an exact answer and an approximation to three decimal places.

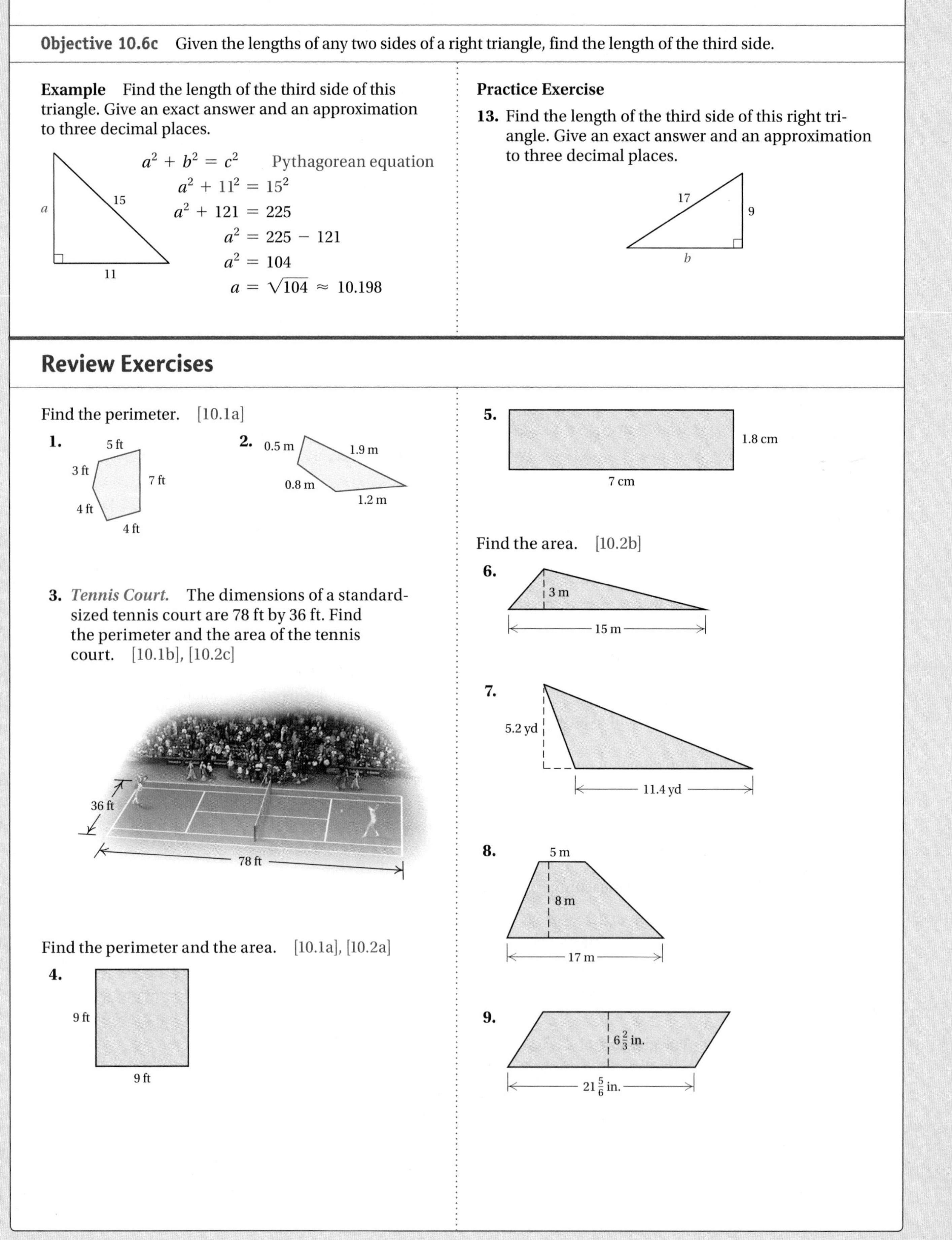

Review Exercises

Find the perimeter. [10.1a]

1.

2.

3. *Tennis Court.* The dimensions of a standard-sized tennis court are 78 ft by 36 ft. Find the perimeter and the area of the tennis court. [10.1b], [10.2c]

Find the perimeter and the area. [10.1a], [10.2a]

4.

5.

Find the area. [10.2b]

6.

7.

8.

9.

10. *Seeded Area.* A grassy area around three sides of a building has equal width on the three sides, as shown below, and is going to be reseeded. What is the total area to be reseeded? [10.2c]

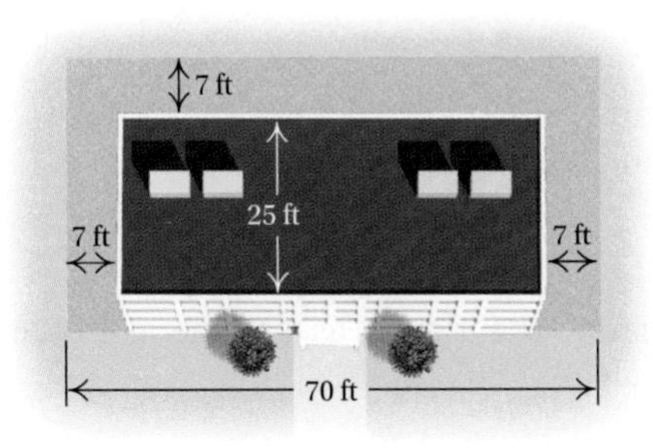

Find the length of a radius of each circle. [10.3a]

11. **12.**

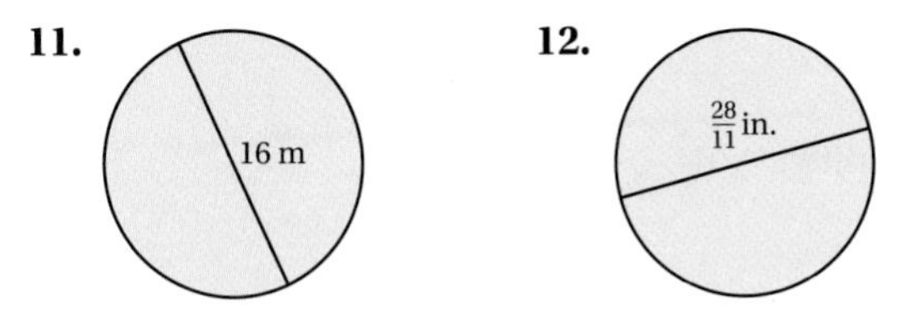

Find the length of a diameter of each circle. [10.3a]

13. **14.**

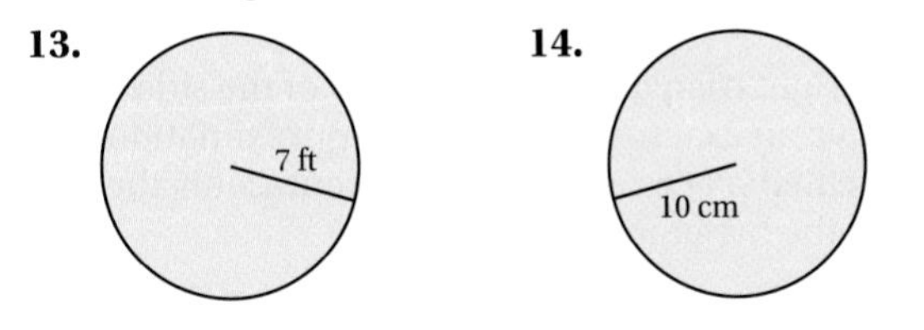

15. Find the circumference of the circle in Exercise 11. Use 3.14 for π. [10.3b]

16. Find the circumference of the circle in Exercise 12. Use $\frac{22}{7}$ for π. [10.3b]

17. Find the area of the circle in Exercise 11. Use 3.14 for π. [10.3c]

18. Find the area of the circle in Exercise 12. Use $\frac{22}{7}$ for π. [10.3c]

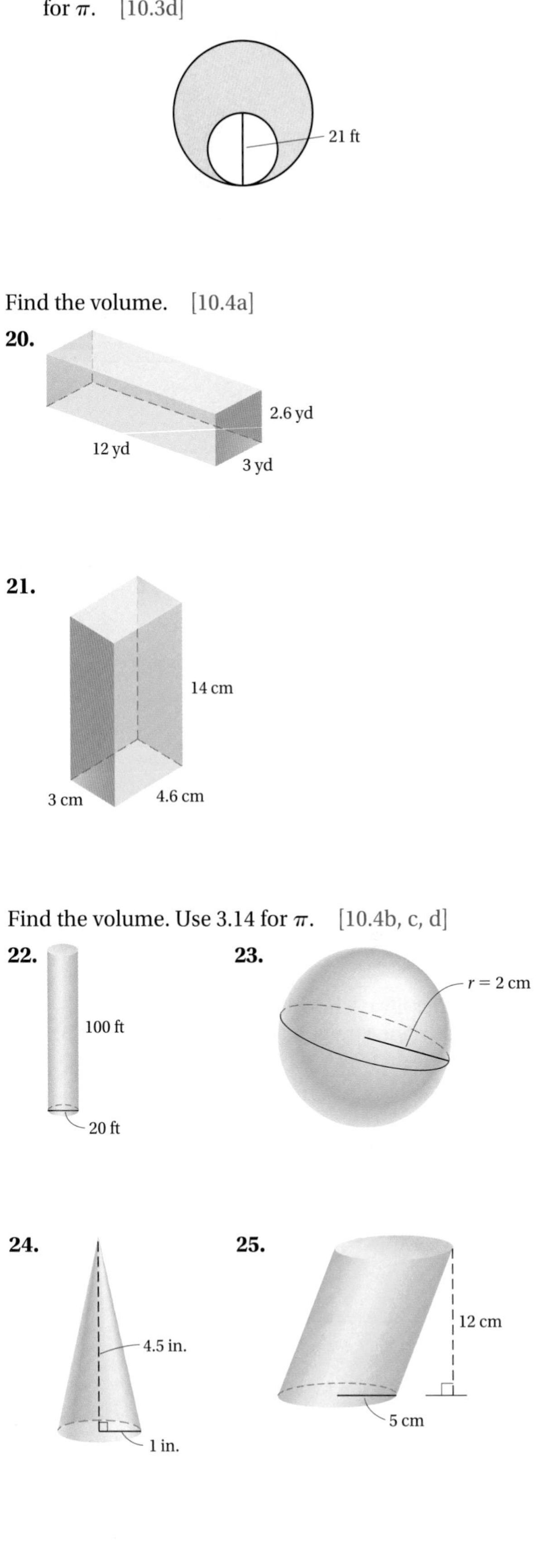

19. Find the area of the shaded region. Use 3.14 for π. [10.3d]

Find the volume. [10.4a]

20.

21.

Find the volume. Use 3.14 for π. [10.4b, c, d]

22. **23.**

24. **25.**

26. A Norman window is designed with dimensions as shown. Find its area and its perimeter. Use 3.14 for π. [10.1b], [10.3d]

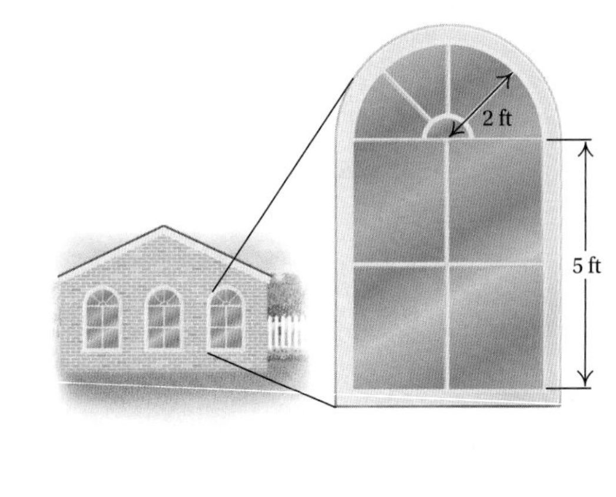

Use a protractor to measure each angle. [10.5a]

27. **28.** **29.** **30.**

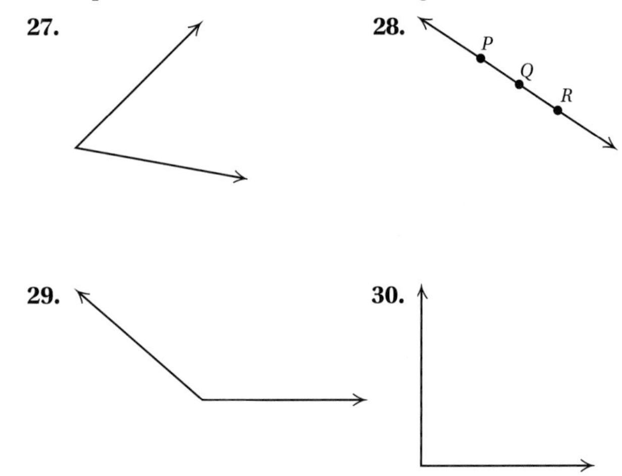

31.–34. Classify each of the angles in Exercises 27–30 as right, straight, acute, or obtuse. [10.5b]

35. Find the measure of a complement of $\angle BAC$. [10.5c]

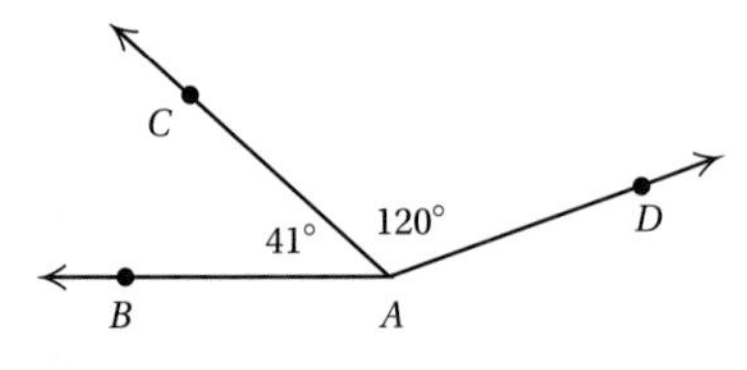

36. Find the measure of a supplement of a 44° angle. [10.5c]

Use the following triangle for Exercises 37–39.

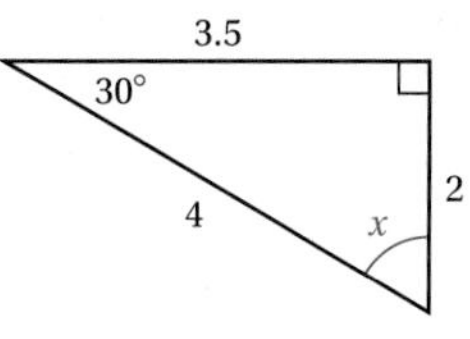

37. Find the missing angle measure. [10.5e]

38. Classify the triangle as equilateral, isosceles, or scalene. [10.5d]

39. Classify the triangle as right, obtuse, or acute. [10.5d]

40. Simplify: $\sqrt{64}$. [10.6a]

41. Use a calculator to approximate $\sqrt{83}$ to three decimal places. [10.6b]

For each right triangle, find the length of the side not given. Give an exact answer and an approximation to three decimal places. Assume that c represents the length of the hypotenuse. [10.6c]

42. $a = 15, b = 25$

43. $a = 7, c = 10$

Find the length of the side not given. Give an exact answer and an approximation to three decimal places. [10.6c]

44. **45.**

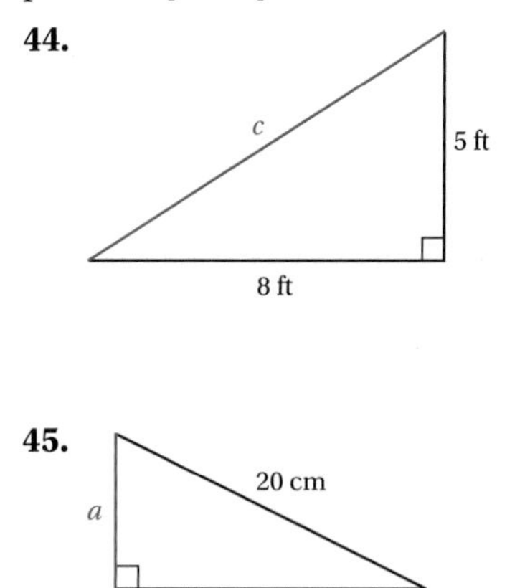

Solve. [10.6d]

46. A wire 24 ft long reaches from the top of a pole to a point on the ground 16 ft from the base of the pole. How tall is the pole? Round to the nearest tenth of a foot.

47. *Construction.* Chloe is designing rafters for a house. The rise of each rafter will be 6 ft and the run 12 ft. What is the rafter length? Round to the nearest hundredth of a foot.

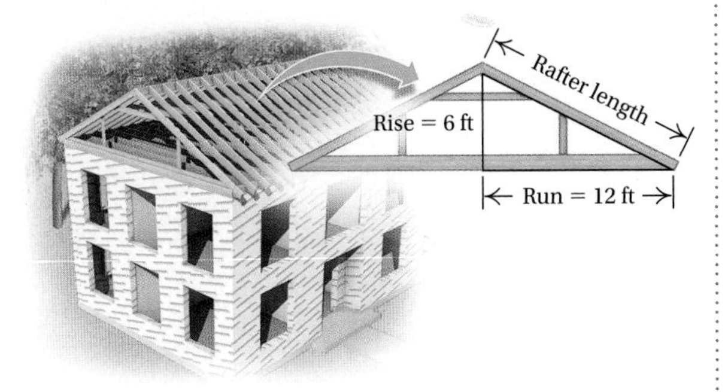

48. Find the length of a diagonal from one corner to another of the tennis court in Exercise 3. Round to the nearest tenth of a foot.

49. Find the measure of a supplement of a $20\frac{3}{4}^\circ$ angle. [10.5c]

A. $339\frac{1}{4}^\circ$ **B.** $159\frac{1}{4}^\circ$ **C.** $69\frac{1}{4}^\circ$ **D.** $70\frac{1}{4}^\circ$

50. Find the area of a circle whose diameter is $\frac{7}{9}$ in. Use $\frac{22}{7}$ for π. [10.3c]

A. $\frac{11}{9}\text{ in}^2$ **B.** $\frac{77}{162}\text{ in}^2$ **C.** $\frac{22}{9}\text{ in}^2$ **D.** $\frac{154}{81}\text{ in}^2$

Synthesis

51. A square is cut in half so that the perimeter of the resulting rectangle is 30 ft. Find the area of the original square. [10.1a], [10.2a]

52. Find the area, in square meters, of the shaded region. [9.2a], [10.2c]

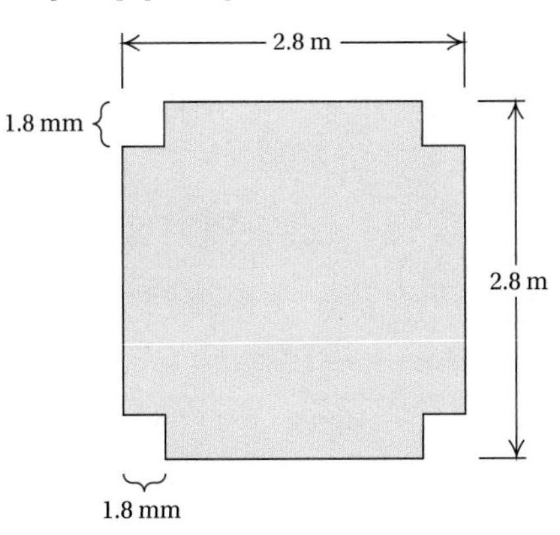

53. Find the area, in square centimeters, of the shaded region. [9.2a], [10.2c]

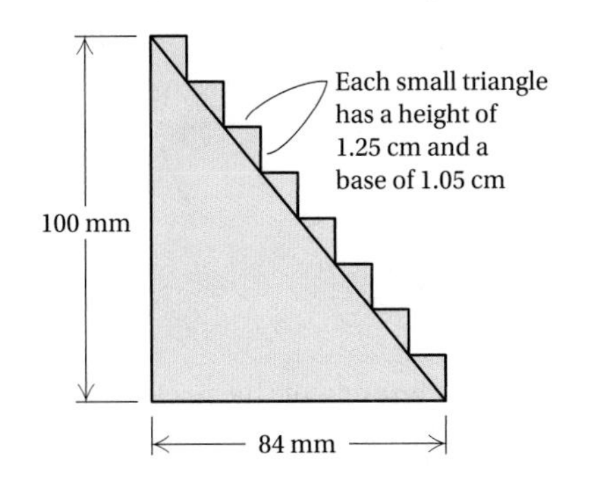

Understanding Through Discussion and Writing

1. Explain a procedure that could be used to determine the measure of an angle's supplement from the measure of the angle's complement. [10.5c]

2. How could you use the volume formulas given in Section 10.4 to help estimate the volume of an egg? [10.4a, b, c, e]

3. Explain how the Pythagorean theorem can be used to prove that a triangle is a *right* triangle. [10.6c]

4. Explain how you might use triangles to find the sum of the angle measures of this figure. [10.5e]

5. The design of a home includes a cylindrical tower that will be capped with either a 10-ft-high dome (half of a sphere) or a 10-ft-high cone. Which type of cap would be more energy-efficient and why? [10.4c, d]

6. Which occupies more volume: two spheres, each with radius r, or one sphere with radius $2r$? Explain why. [10.4c]

CHAPTER 10 Test

For Extra Help For step-by-step test solutions, access the Chapter Test Prep Videos in MyMathLab® or on YouTube (search "BittingerBasicEl" and click on "Channels").

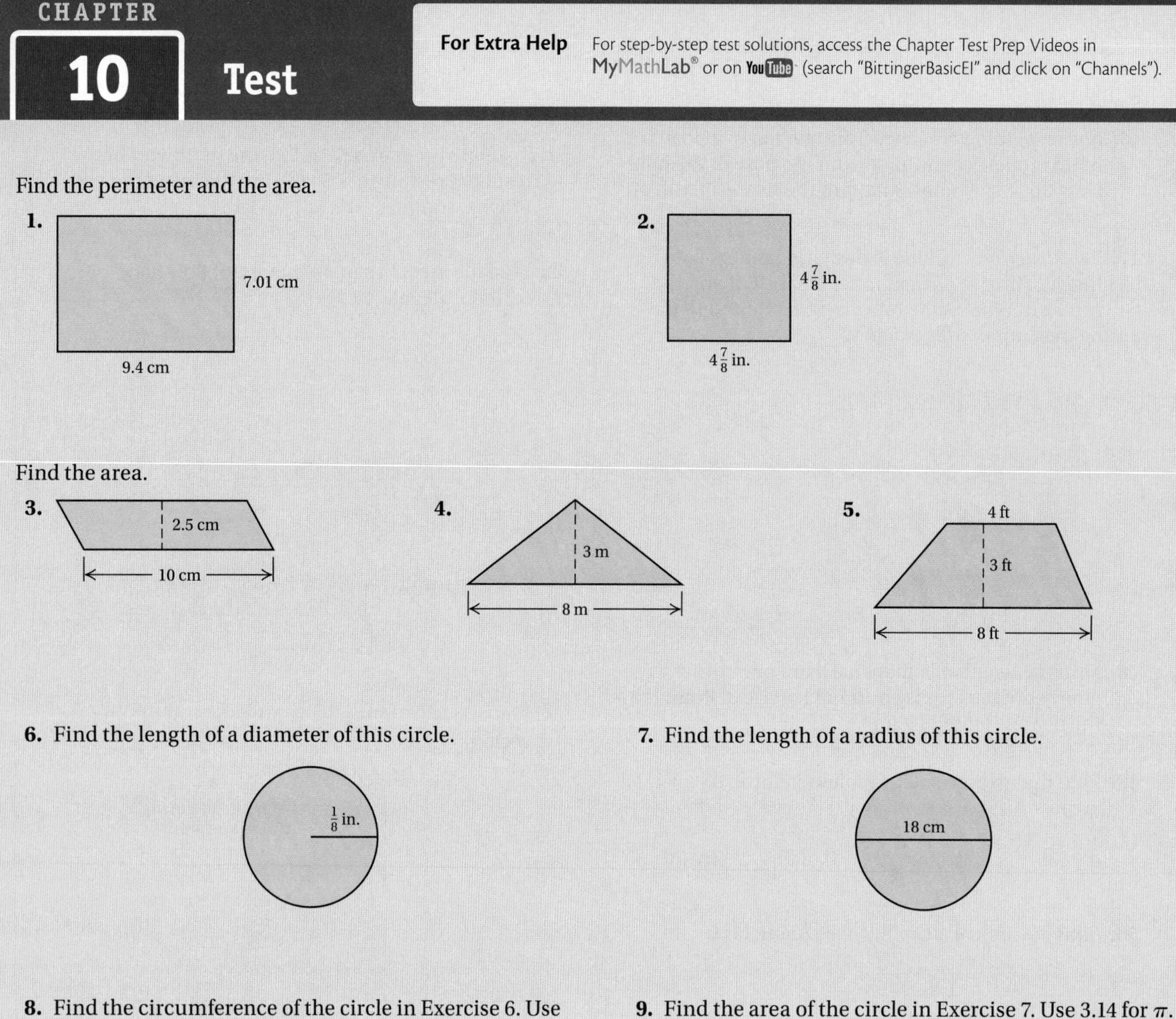

8. Find the circumference of the circle in Exercise 6. Use $\frac{22}{7}$ for π.

9. Find the area of the circle in Exercise 7. Use 3.14 for π.

10. Find the perimeter and the area of the shaded region. Use 3.14 for π.

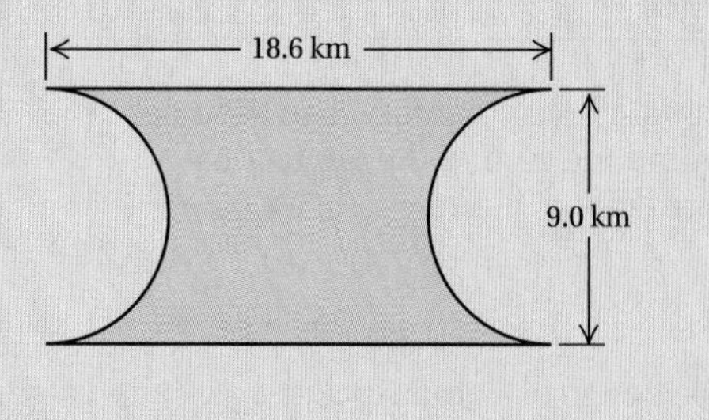

11. Find the volume.

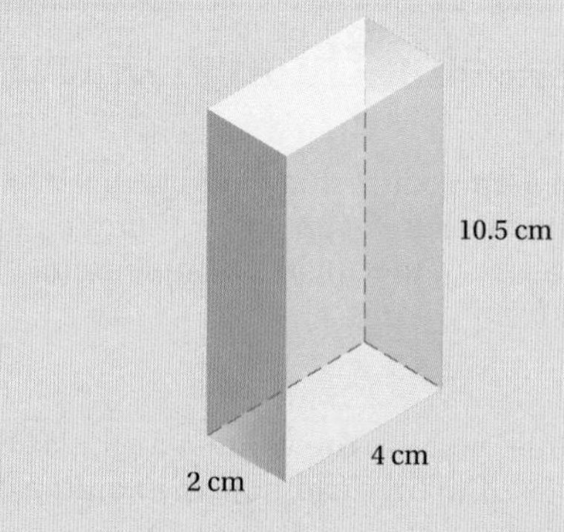

12. A twelve-box rectangular carton of 12-oz juice boxes measures $10\frac{1}{2}$ in. by 8 in. by 5 in. What is the volume of the carton?

Find the volume. Use 3.14 for π.

13. **14.** **15.**

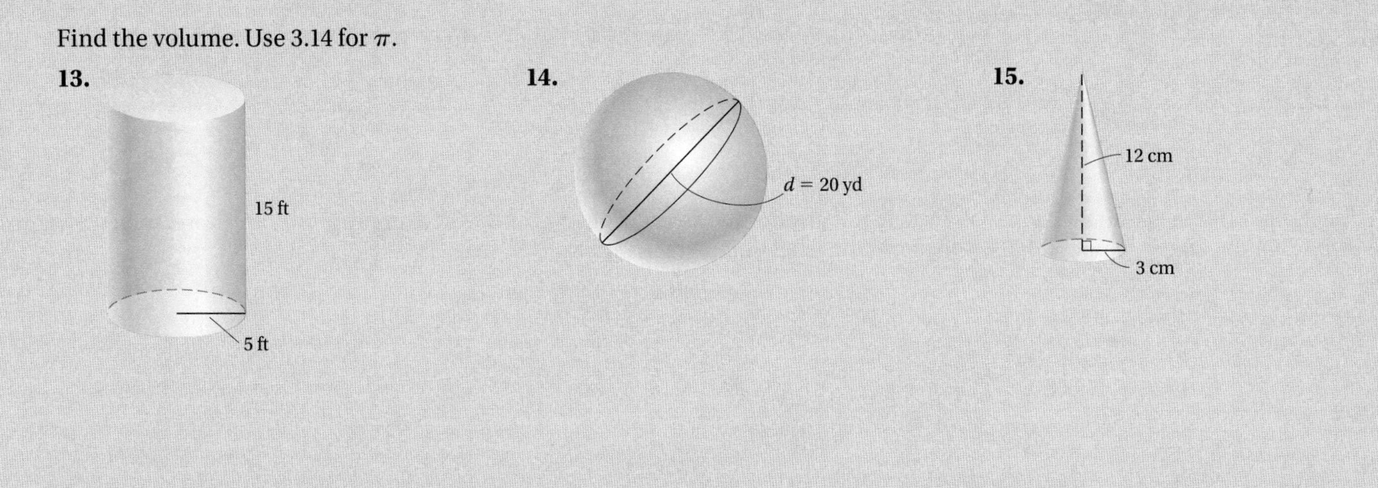

Use a protractor to measure each angle.

16. **17.** **18.** **19.**

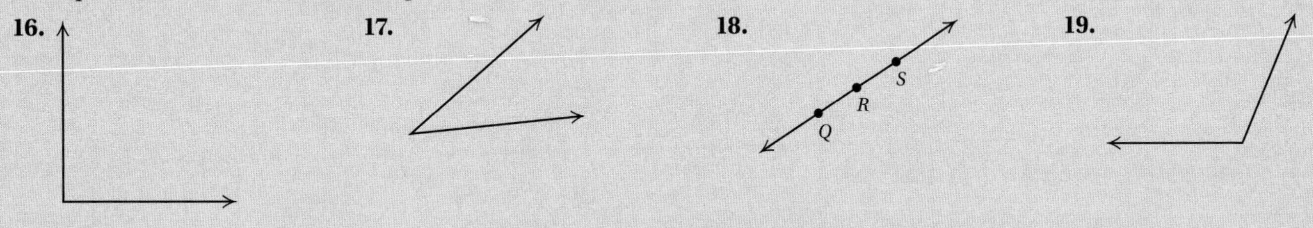

20.–23. Classify each of the angles in Exercises 16–19 as right, straight, acute, or obtuse.

Use the following triangle for Exercises 24–26.

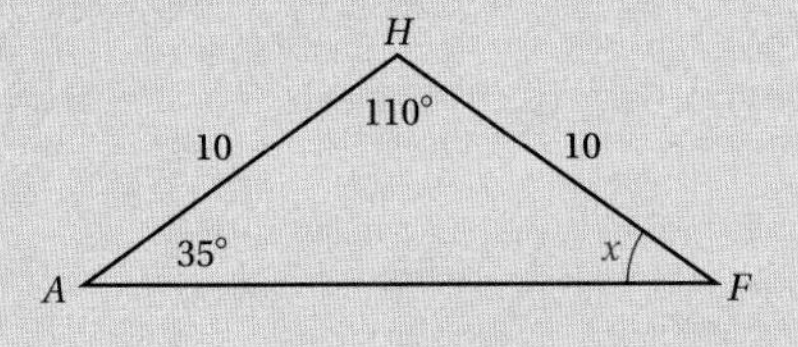

24. Find the missing angle measure.

25. Classify the triangle as equilateral, isosceles, or scalene.

26. Classify the triangle as right, obtuse, or acute.

27. Find the measure of a complement and a supplement of $\angle CAD$.

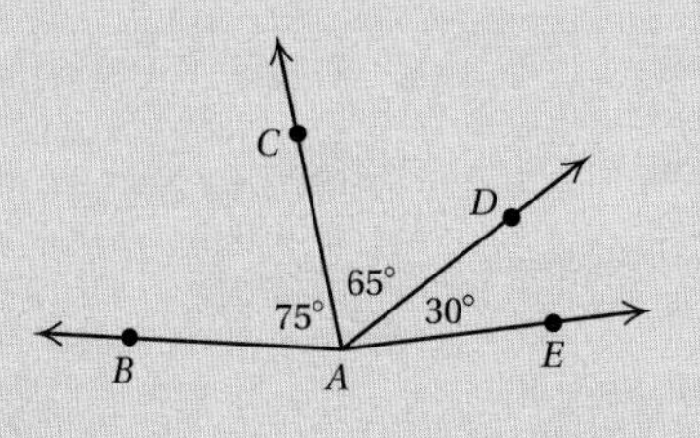

28. Simplify: $\sqrt{225}$.

29. Approximate to three decimal places: $\sqrt{87}$

For each right triangle, find the length of the side not given. Give an exact answer and, where appropriate, an approximation to three decimal places. Assume that c represents the length of the hypotenuse.

30. $a = 24,\ b = 32$

31. $a = 2,\ c = 8$

32.

33.

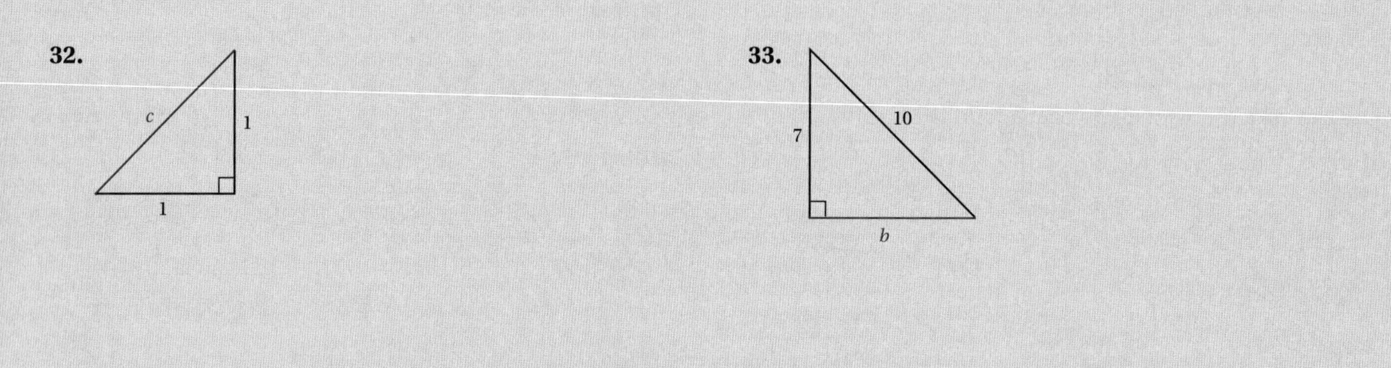

34. How long must a wire be in order to reach from the top of a 13-m antenna to a point on the ground 9 m from the base of the antenna? Round to the nearest tenth of a meter.

35. Find the volume of a sphere whose diameter is 42 cm.

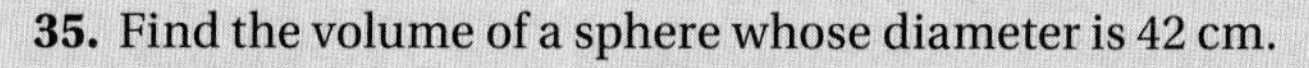

Use $\frac{22}{7}$ for π.

A. 310,464 cm^3
B. 9702 cm^3
C. 1848 cm^3
D. 38,808 cm^3

Synthesis

Find the area of the shaded region. (Note that the figures are not drawn in perfect proportion.) Give the answer in square feet.

36.

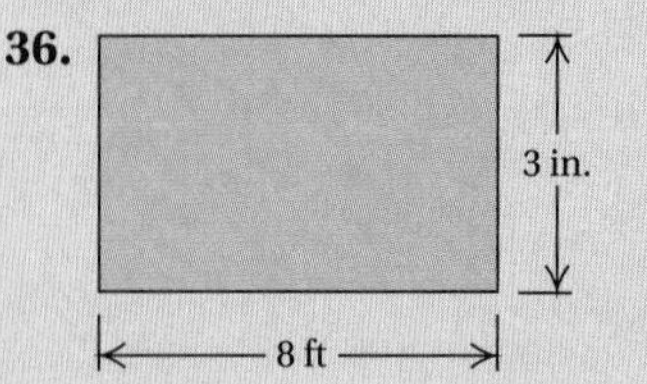

37.

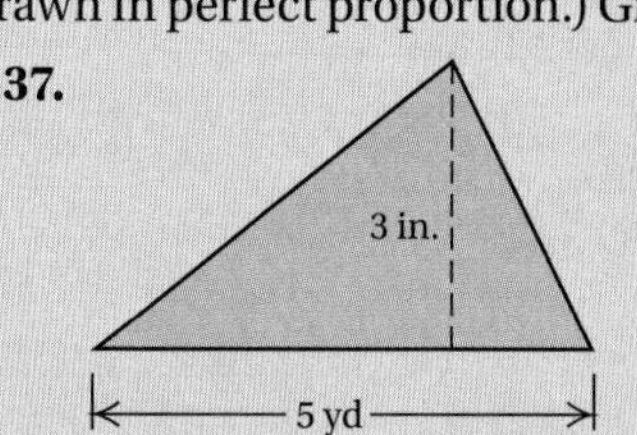

Find the volume of the solid. (Note that the solids are not drawn in perfect proportion.) Give the answer in cubic feet. Use 3.14 for π and round to the nearest thousandth in Exercises 39 and 40.

38.

39.

40.

Cumulative Review

Solve.

1. *Energy.* The Indianapolis Resource Recovery Facility produces steam from trash. This steam is used to generate electricity. The facility produces at least 9.8 million tons of steam a day, enough to power the downtown Indianapolis heating loop. Find standard notation for 9.8 million.

   Source: covantaenergy.com

2. *Firefighting.* During a fire, the firefighters get a 1-ft layer of water on the 25-ft by 60-ft first floor of a 5-floor building. Water weighs $62\frac{1}{2}$ lb per cubic foot. What is the total weight of the water on the floor?

Calculate.

3. $1\frac{1}{2} + 2\frac{2}{3}$

4. $120.5 - 32.98$

5. $22\overline{)27,148}$

6. $8^3 - 45 \cdot 24 - 9^2 \div 3$

7. $\left(\frac{1}{4}\right)^2 \div \left(\frac{1}{2}\right)^3 \times 2^4 - (10.3)(4)$

8. $14 \div [33 \div 11 + 8 \times 2 - (15 - 3)]$

Find fraction notation.

9. 1.209

10. 17%

Use <, >, or = for ☐ to write a true sentence.

11. $\frac{5}{6} \square \frac{7}{8}$

12. $\frac{-15}{18} \square \frac{-10}{12}$

Complete.

13. 12 c = ______ qt

14. 9 sec = ______ min

15. 15°C = ______ °F

16. 0.087 L = ______ mL

17. 3 yd^2 = ______ ft^2

18. 17 cm = ______ m

Solve.

19. $x + \frac{3}{4} = -\frac{7}{8}$

20. $\frac{3}{x} = \frac{7}{10}$

21. $25 \cdot x = 2835$

22. $\frac{12}{15} = \frac{x}{18}$

23. Find the perimeter and the area.

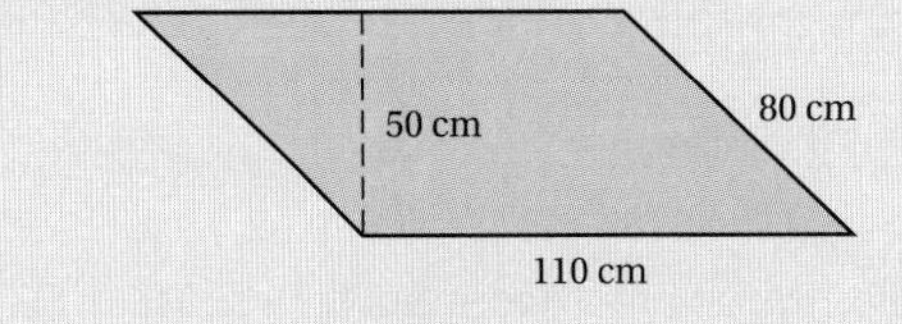

24. Find the diameter, the circumference, and the area of this circle. Use $\frac{22}{7}$ for π.

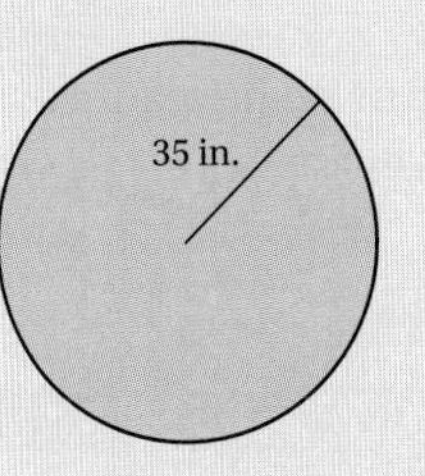

25. Find the volume of this sphere. Use $\frac{22}{7}$ for π.

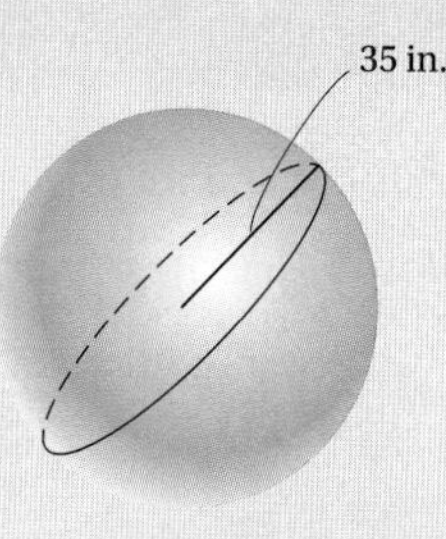

Solve.

26. To get an A in math, a student must score an average of 90 on five tests. On the first four tests, the scores were 85, 92, 79, and 95. What is the lowest score that the student can get on the last test and still get an A?

27. What is the simple interest on $8000 at 4.2% for $\frac{1}{4}$ year?

28. What is the amount in an account after 25 years if $8000 is invested at 4.2%, compounded annually?

29. How long must a rope be in order to reach from the top of an 8-m tree to a point on the ground 15 m from the bottom of the tree?

30. The sales tax on an office supply purchase of $5.50 is $0.33. What is the sales tax rate?

31. A bolt of fabric in a fabric store has $10\frac{3}{4}$ yd on it. A customer purchases $8\frac{5}{8}$ yd. How many yards remain on the bolt?

32. What is the cost, in dollars, of 15.6 gal of gasoline at 239.9¢ per gallon? Round to the nearest cent.

33. A box of powdered milk that makes 20 qt costs $4.99. A box that makes 8 qt costs $1.99. Which size has the lower unit price?

34. It is $\frac{7}{10}$ km from Maria's dormitory to the library. Maria starts to walk from the dorm to the library, changes her mind after going $\frac{1}{4}$ of the distance, and returns to the dorm. How far did she walk?

35. Find the missing angle measure.

130° 20° x

36. Classify the triangle as equilateral, isosceles, or scalene.

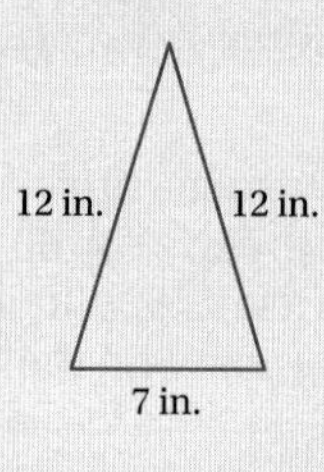

37. Classify the triangle in Exercise 35 as right, obtuse, or acute.

Synthesis

Find the volume in cubic feet. Use 3.14 for π.

38. **39.**

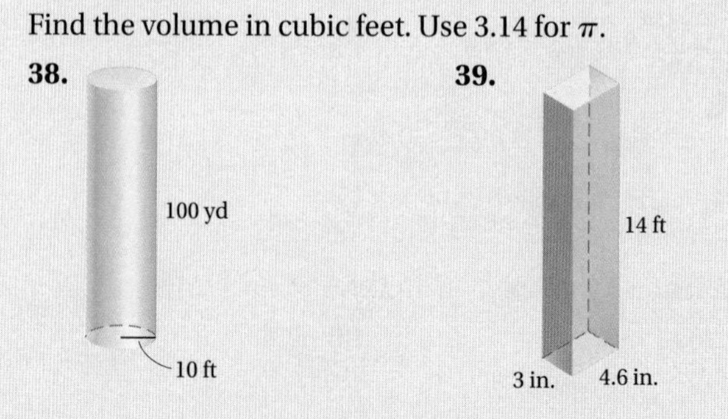

Algebra: Solving Equations and Problems

STUDYING FOR SUCCESS *Preparing for the Final Exam*

- ☐ Browse through each chapter, reviewing highlighted or boxed information and noting important formulas.
- ☐ Attend any exam tutoring sessions offered by your college or university.
- ☐ Retake the chapter tests that you took in class, or take the chapter tests in the text.
- ☐ Work through the Cumulative Review for Chapters 1–11 as a sample final exam.

11.1 Introduction to Algebra

OBJECTIVES

a Evaluate an algebraic expression by substitution.

b Use the distributive laws to multiply expressions like 8 and $x - y$.

c Use the distributive laws to factor expressions like $4x - 12 + 24y$.

d Collect like terms.

SKILL TO REVIEW

Objective 3.1a: Find the factors of a number.

Find all the factors of each number.

1. 124 **2.** 140

The study of algebra involves the use of equations to solve problems. Equations are constructed from algebraic expressions.

a EVALUATING ALGEBRAIC EXPRESSIONS

In arithmetic, you have worked with expressions such as

$$37 + 86, \quad 7 \times 8, \quad 19 - 7, \quad \text{and} \quad \frac{3}{8}.$$

In algebra, we can use letters for numbers and work with *algebraic expressions* such as

$$x + 86, \quad 7 \times t, \quad 19 - y, \quad \text{and} \quad \frac{a}{b}.$$

Expressions like these should be familiar from the equation and problem solving that we have already done.

Sometimes a letter can stand for various numbers. In that case, we call the letter a **variable.** Let $a =$ your age. Then a is a variable since a changes from year to year. Sometimes a letter can stand for just one number. In that case, we call the letter a **constant**. Let $b =$ your birth year. Then b is a constant.

An **algebraic expression** consists of variables, constants, numerals, operation signs, and/or grouping symbols. When we replace a variable with a number, we say that we are **substituting** for the variable. When we replace all the variables in an expression with numbers and carry out the operations in the expression, we are **evaluating the expression.**

EXAMPLE 1 Evaluate $x + y$ for $x = 37$ and $y = 29$.

We substitute 37 for x and 29 for y and carry out the addition:

$$x + y = 37 + 29 = 66.$$

The number 66 is called the **value** of the expression when $x = 37$ and $y = 29$. ■

Answers

Skill to Review:

1. 1, 2, 4, 31, 62, 124 **2.** 1, 2, 4, 5, 7, 10, 14, 20, 28, 35, 70, 140

Algebraic expressions involving multiplication can be written in several ways. For example, "8 times a" can be written as

$$8 \times a, \quad 8 \cdot a, \quad 8(a), \quad \text{or simply} \quad 8a.$$

Two letters written together without an operation sign, such as ab, also indicate a multiplication.

EXAMPLE 2 Evaluate $3y$ for $y = 14$.

$$3y = 3(14) = 42$$

Do Exercises 1–3. ▶

1. Evaluate $a + b$ when $a = 38$ and $b = 26$.

2. Evaluate $x - y$ when $x = 57$ and $y = 29$.

3. Evaluate $4t$ when $t = 15$ and when $t = -6.8$.

EXAMPLE 3 *Area of a Rectangle.* The area A of a rectangle of length l and width w is given by the formula $A = lw$. Find the area when l is 24.5 in. and w is 16 in.

We substitute 24.5 in. for l and 16 in. for w and then carry out the multiplication:

$$\begin{aligned} A = lw &= (24.5\text{ in.})(16\text{ in.}) \\ &= (24.5)(16)(\text{in.})(\text{in.}) \\ &= 392\text{ in}^2\text{, or 392 square inches.} \end{aligned}$$

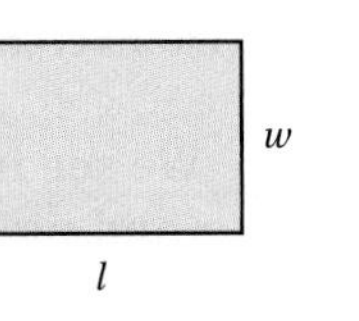

Do Exercise 4. ▶

4. Find the area of a rectangle when l is 24 ft and w is 8 ft.

$$\begin{aligned} A &= lw \\ A &= (24\text{ ft})(\quad\quad) \\ &= (24)(\quad\quad)(\text{ft})(\text{ft}) \\ &= 192 \quad\quad\text{, or} \\ &\quad 192\text{ square feet} \end{aligned}$$

Algebraic expressions involving division can also be written in several ways. For example, "8 divided by t" can be written as

$$8 \div t, \quad \frac{8}{t}, \quad 8/t, \quad \text{or} \quad 8 \cdot \frac{1}{t},$$

where the fraction bar is a division symbol.

EXAMPLE 4 Evaluate $\frac{a}{b}$ for $a = 63$ and $b = 9$.

We substitute 63 for a and 9 for b and carry out the division:

$$\frac{a}{b} = \frac{63}{9} = 7.$$

EXAMPLE 5 Evaluate $\frac{m + n}{12}$ for $m = 8$ and $n = 16$.

$$\frac{m + n}{12} = \frac{8 + 16}{12} = \frac{24}{12} = 2$$

Do Exercises 5 and 6. ▶

5. Evaluate $\frac{a}{b}$ when $a = -200$ and $b = 8$.

6. Evaluate $\frac{p + q}{13}$ when $p = 40$ and $q = 25$.

Answers

1. 64 2. 28 3. 60; −27.2
4. 192 ft^2 5. −25 6. 5

Guided Solution:
4. 8 ft, 8, ft^2

Complete each table by evaluating each expression for the given values.

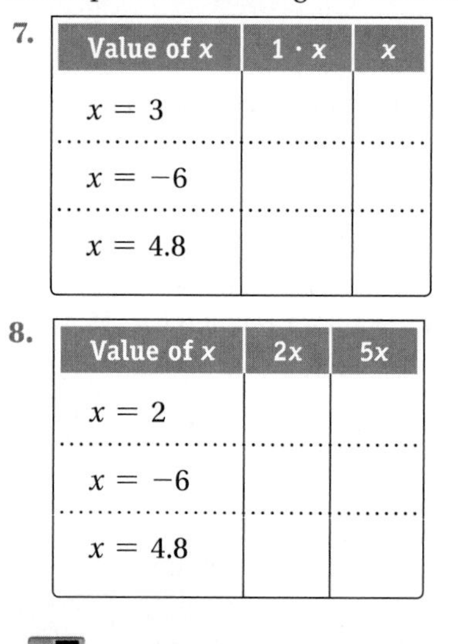

7.

Value of x	1 · x	x
$x = 3$		
$x = -6$		
$x = 4.8$		

8.

Value of x	2x	5x
$x = 2$		
$x = -6$		
$x = 4.8$		

CALCULATOR CORNER

Evaluating Algebraic Expressions We can use a calculator to evaluate algebraic expressions. To evaluate $x - y$ when $x = 48$ and $y = -19$, for example, we press [4] [8] [−] [1] [9] [+/−] [=]. The calculator displays the result, 67.

When we evaluate an expression like $\frac{3x + y}{4}$, we must enclose the numerator in parentheses. To evaluate this expression when $x = 24$ and $y = 16$, we press [(] [3] [×] [2] [4] [+] [1] [6] [)] [÷] [4] [=]. The calculator displays the result, 22.

EXERCISES Evaluate.

1. $\frac{a}{b}$, when $a = 54$ and $b = -9$
2. $\frac{2m}{n}$, when $m = 38$ and $n = -4$
3. $\frac{x - y}{7}$, when $x = 94$ and $y = 31$
4. $\frac{2p + q}{12}$, when $p = 47$ and $q = 50$

Answers

7. 3, 3; −6, −6; 4.8, 4.8
8. 4, 10; −12, −30; 9.6, 24

b EQUIVALENT EXPRESSIONS AND THE DISTRIBUTIVE LAWS

In solving equations and doing other kinds of work in algebra, we manipulate expressions in various ways. To see how to do this, let's consider some examples in which we evaluate expressions.

EXAMPLE 6 Evaluate $1 \cdot x$ when $x = 5$ and when $x = -8$ and compare the results to x.

We substitute 5 for x:

$$1 \cdot x = 1 \cdot 5 = 5.$$

Then we substitute −8 for x:

$$1 \cdot x = 1 \cdot (-8) = -8.$$

We see that $1 \cdot x$ and x represent the same number.

◀ **Do Exercises 7 and 8.**

We see in Example 6 and in Margin Exercise 7 that the expressions $1 \cdot x$ and x represent the same number for any allowable replacement of x. In that sense, the expressions $1 \cdot x$ and x are **equivalent**.

EQUIVALENT EXPRESSIONS

Two expressions that have the same value for all allowable replacements are called **equivalent**.

In the expression $3/x$, the number 0 is not an allowable replacement because $3/0$ is not defined. Even so, the expressions $6/(2x)$ and $3/x$ are *equivalent* because they represent the same number for any allowable (not 0) replacement of x. For example, when $x = 5$,

$$\frac{6}{2x} = \frac{6}{2 \cdot 5} = \frac{6}{10} = \frac{3}{5} \quad \text{and} \quad \frac{3}{x} = \frac{3}{5}.$$

We see in Margin Exercise 8 that the expressions $2x$ and $5x$ are *not* equivalent.

The fact that $1 \cdot x$ and x are equivalent is a law of real numbers called the **identity property of 1**. We often refer to the use of the identity property of 1 as "multiplying by 1." We have used multiplying by 1 many times in this text.

THE IDENTITY PROPERTY OF 1 (MULTIPLICATIVE IDENTITY)

For any real number a,

$$a \cdot 1 = 1 \cdot a = a.$$

We now consider two other laws of real numbers called the **distributive laws**. They are the basis of many procedures in both arithmetic and algebra and are probably the most important laws that we use to manipulate algebraic expressions. The first distributive law involves two operations: addition and multiplication.

Let's begin by considering a multiplication problem from arithmetic:

$$\begin{array}{r l} 4\ 5 & \\ \times\ \ 7 & \\ \hline 3\ 5 & \leftarrow \text{This is } 7 \cdot 5. \\ 2\ 8\ 0 & \leftarrow \text{This is } 7 \cdot 40. \\ \hline 3\ 1\ 5 & \leftarrow \text{This is the sum } 7 \cdot 40 + 7 \cdot 5. \end{array}$$

To carry out the multiplication, we actually added two products. That is,

$$7 \cdot 45 = 7(40 + 5) = 7 \cdot 40 + 7 \cdot 5.$$

Let's examine this further. If we wish to multiply a sum of several numbers by a factor, we can either add and then multiply or multiply and then add.

EXAMPLE 7 Evaluate $5(x + y)$ and $5x + 5y$ when $x = 2$ and $y = 8$, and compare the results.

We substitute 2 for x and 8 for y in each expression. Then we use the rules for order of operations to calculate.

a) $5(x + y) = 5(2 + 8) = 5(10) = 50$

b) $5x + 5y = 5 \cdot 2 + 5 \cdot 8 = 10 + 40 = 50$

The results of (a) and (b) are the same.

Do Exercises 9–11. ▶

9. Complete this table.

Values of x and y	5(x + y)	5x + 5y
$x = 6, y = 7$		
$x = -3, y = 4$		
$x = -10, y = 5$		

10. Evaluate $6x + 6y$ and $6(x + y)$ when $x = 10$ and $y = 5$.

11. Evaluate $4(x + y)$ and $4x + 4y$ when $x = 11$ and $y = 5$.

The expressions $5(x + y)$ and $5x + 5y$, in Example 7 and in Margin Exercise 9, are equivalent. They illustrate the distributive law of multiplication over addition. Margin Exercises 10 and 11 also illustrate this distributive law.

THE DISTRIBUTIVE LAW OF MULTIPLICATION OVER ADDITION

For any numbers a, b, and c,

$$a(b + c) = ab + ac.$$

In the statement of the distributive law, we know that in the expression $ab + ac$, the multiplications are to be done first according to the rules for order of operations. So, instead of writing $(4 \cdot 5) + (4 \cdot 7)$, we can write $4 \cdot 5 + 4 \cdot 7$. However, in $a(b + c)$, we cannot omit the parentheses. If we did, we would have $ab + c$, which means $(ab) + c$. For example, $3(4 + 2) = 18$, but $3 \cdot 4 + 2 = 14$.

The second distributive law relates multiplication and subtraction. This law says that to multiply by a difference, we can either subtract and then multiply or multiply and then subtract.

THE DISTRIBUTIVE LAW OF MULTIPLICATION OVER SUBTRACTION

For any numbers a, b, and c,

$$a(b - c) = ab - ac.$$

Answers

9. 65, 65; 5, 5; −25, −25 **10.** 90; 90
11. 64; 64

12. Evaluate $7(x - y)$ and $7x - 7y$ when $x = 9$ and $y = 7$.

13. Evaluate $6x - 6y$ and $6(x - y)$ when $x = 10$ and $y = 5$.

14. Evaluate $2(x - y)$ and $2x - 2y$ when $x = 11$ and $y = 5$.

What are the terms of each expression?

15. $5x - 4y + 3$

16. $-4y - 2x + 3z$

Multiply.

17. $3(x - 5)$

18. $5(x + 1)$

19. $\frac{5}{4}(x - y + 4)$

20. $-2(x - 3)$
$= -2 \cdot x - (\quad) \cdot 3$
$= -2x - (\quad)$
$= -2x + \quad$

GS

21. $-5(x - 2y + 4z)$

Answers

12. 14; 14 13. 30; 30 14. 12; 12
15. $5x, -4y, 3$ 16. $-4y, -2x, 3z$
17. $3x - 15$ 18. $5x + 5$
19. $\frac{5}{4}x - \frac{5}{4}y + 5$ 20. $-2x + 6$
21. $-5x + 10y - 20z$

Guided Solution:
20. $-2, -6, 6$

We often refer to "*the* distributive law" when we mean *either or both* of these laws.

◀ **Do Exercises 12–14.**

What do we mean by the *terms* of an expression? **Terms** are separated by addition signs. If there are subtraction signs, we can find an equivalent expression that uses addition signs.

EXAMPLE 8 What are the terms of $3x - 4y + 2z$?

$$3x - 4y + 2z = 3x + (-4y) + 2z \qquad \text{Separating parts with + signs}$$

The terms are $3x$, $-4y$, and $2z$.

◀ **Do Exercises 15 and 16.**

The distributive laws are the basis for a procedure in algebra called **multiplying**. In an expression such as $8(a + 2b - 7)$, we multiply each term inside the parentheses by 8:

$$8(a + 2b - 7) = 8 \cdot a + 8 \cdot 2b - 8 \cdot 7 = 8a + 16b - 56.$$

EXAMPLES Multiply.

9. $9(x - 5) = 9x - 9(5)$ Using the distributive law of multiplication over subtraction
$= 9x - 45$

10. $\frac{2}{3}(w + 1) = \frac{2}{3} \cdot w + \frac{2}{3} \cdot 1$ Using the distributive law of multiplication over addition
$= \frac{2}{3}w + \frac{2}{3}$

EXAMPLE 11 Multiply: $-4(x - 2y + 3z)$.

$$-4(x - 2y + 3z) = -4 \cdot x - (-4)(2y) + (-4)(3z) \qquad \text{Using both distributive laws}$$
$$= -4x - (-8y) + (-12z) \qquad \text{Multiplying}$$
$$= -4x + 8y - 12z$$

We can also do this problem by first finding an equivalent expression with all plus signs and then multiplying:

$$-4(x - 2y + 3z) = -4[x + (-2y) + 3z]$$
$$= -4 \cdot x + (-4)(-2y) + (-4)(3z) = -4x + 8y - 12z.$$

◀ **Do Exercises 17–21.**

C FACTORING

Factoring is the reverse of multiplying. To factor, we can use the distributive laws in reverse:

$$ab + ac = a(b + c) \quad \text{and} \quad ab - ac = a(b - c).$$

FACTOR

To **factor** an expression is to find an equivalent expression that is a product.

Look at Example 9. To *factor* $9x - 45$, we find an equivalent expression that is a product, $9(x - 5)$. When all the terms of an expression have a factor in common, we can "factor it out" using the distributive laws. Note the following.

$9x$ has the factors $9, -9, 3, -3, 1, -1, x, -x, 3x, -3x, 9x, -9x$;

-45 has the factors $1, -1, 3, -3, 5, -5, 9, -9, 15, -15, 45, -45$.

We remove the *greatest common factor.* In this case, that factor is 9. Thus,

$$\begin{aligned} 9x - 45 &= 9 \cdot x - 9 \cdot 5 \\ &= 9(x - 5). \end{aligned}$$

Remember that an expression is factored when we find an equivalent expression that is a product.

EXAMPLES Factor.

12. $5x - 10 = 5 \cdot x - 5 \cdot 2$ Try to do this step mentally.

$= 5(x - 2)$ ⟵ You can check by multiplying.

13. $9x + 27y - 9 = 9 \cdot x + 9 \cdot 3y - 9 \cdot 1$

$= 9(x + 3y - 1)$

Caution!

Note in Example 13 that although $3(3x + 9y - 3)$ is also equivalent to $9x + 27y - 9$, it is *not* the desired form. We can find the desired form by factoring out another factor of 3:

$$\begin{aligned} 9x + 27y - 9 &= 3(3x + 9y - 3) \\ &= 3 \cdot 3(x + 3y - 1) \\ &= 9(x + 3y - 1). \end{aligned}$$

Remember to factor out the *greatest common factor.*

EXAMPLES Factor. Try to write just the answer, if you can.

14. $5x - 5y = 5(x - y)$

15. $-3x + 6y - 9z = -3 \cdot x - 3(-2y) - 3(3z) = -3(x - 2y + 3z)$

We usually factor out a negative factor when the first term is negative. The way we factor can depend on the context in which we are working. We might also factor the expression in this example as follows:

$$-3x + 6y - 9z = 3(-x + 2y - 3z).$$

16. $18z - 12x - 24 = 6(3z - 2x - 4)$

Remember that you can always check factoring by multiplying. Keep in mind that an expression is factored when it is written as a product.

Do Exercises 22–25. ▶

Factor.

22. $6z - 12$

23. $3x - 6y + 9$

GS **24.** $16a - 36b + 42$

$= 2 \cdot 8a - \square \cdot 18b + 2 \cdot 21$

$= \square(8a - 18b + 21)$

25. $-12x + 32y - 16z$

Answers

22. $6(z - 2)$ **23.** $3(x - 2y + 3)$

24. $2(8a - 18b + 21)$

25. $-4(3x - 8y + 4z)$, or $4(-3x + 8y - 4z)$

Guided Solution:

24. 2, 2

d COLLECTING LIKE TERMS

Terms such as $5x$ and $-4x$, whose variable factors are exactly the same, are called **like terms.** Similarly, numbers, such as -7 and 13, are like terms. Also, $3y^2$ and $9y^2$ are like terms because the variables are the same and have the same exponent. Terms such as $4y$ and $5y^2$ are not like terms, and $7x$ and $2y$ are not like terms.

The distributive laws are used to **collect like terms.** This procedure can also be called **combining like terms.**

EXAMPLES Collect like terms. Try to write just the answer, if you can.

17. $4x + 2x = (4 + 2)x = 6x$ Factoring out the x using a distributive law

18. $2x + 3y - 5x - 2y = 2x - 5x + 3y - 2y$

$= (2 - 5)x + (3 - 2)y = -3x + y$

19. $3x - x = 3x - 1x = (3 - 1)x = 2x$

20. $x - 0.24x = 1 \cdot x - 0.24x = (1 - 0.24)x = 0.76x$

21. $x - 6x = 1 \cdot x - 6 \cdot x = (1 - 6)x = -5x$

22. $4x - 7y + 9x - 5 + 3y - 8 = 13x - 4y - 13$

23. $\frac{2}{3}a - b + \frac{4}{5}a + \frac{1}{4}b - 10 = \frac{2}{3}a - 1 \cdot b + \frac{4}{5}a + \frac{1}{4}b - 10$

$= \left(\frac{2}{3} + \frac{4}{5}\right)a + \left(-1 + \frac{1}{4}\right)b - 10$

$= \left(\frac{10}{15} + \frac{12}{15}\right)a + \left(-\frac{4}{4} + \frac{1}{4}\right)b - 10$

$= \frac{22}{15}a - \frac{3}{4}b - 10$

◀ Do Exercises 26–32.

Collect like terms.

26. $6x - 3x$

27. $7x - x$

28. $x - 9x$

29. $x - 0.41x$

30. $5x + 4y - 2x - y$

31. $3x - 7x - 11 + 8y + 4 - 13y$ GS
$= (3 - \underline{\quad})x + (8 - 13)y + (\underline{\quad} + 4)$
$= \underline{\quad}x + (\underline{\quad})y + (\underline{\quad})$
$= -4x - 5y - 7$

32. $-\frac{2}{3} - \frac{3}{5}x + y + \frac{7}{10}x - \frac{2}{9}y$

Answers

26. $3x$ **27.** $6x$ **28.** $-8x$ **29.** $0.59x$
30. $3x + 3y$ **31.** $-4x - 5y - 7$
32. $\frac{1}{10}x + \frac{7}{9}y - \frac{2}{3}$

Guided Solution:
31. 7, −11, −4, −5, −7

Reading Check

Classify each algebraic expression as involving either multiplication or division.

RC1. $3/q$ ____________

RC2. $3q$ ____________

RC3. $3 \cdot q$ ____________

RC4. $\frac{3}{q}$ ____________

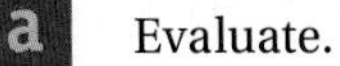

Evaluate.

1. $6x$, when $x = 7$

2. $9t$, when $t = 8$

3. $\frac{x}{y}$, when $x = 9$ and $y = 3$

4. $\frac{m}{n}$, when $m = 18$ and $n = 3$

5. $\frac{3p}{q}$, when $p = -2$ and $q = 6$

6. $\frac{5y}{z}$, when $y = -15$ and $z = -25$

7. $\frac{x + y}{5}$, when $x = 10$ and $y = 20$

8. $\frac{p - q}{2}$, when $p = 17$ and $q = 3$

9. ab, when $a = -5$ and $b = 4$

10. ba, when $a = -5$ and $b = 4$

b Evaluate.

11. $10(x + y)$ and $10x + 10y$, when $x = 20$ and $y = 4$

12. $5(a + b)$ and $5a + 5b$, when $a = 16$ and $b = 6$

13. $10(x - y)$ and $10x - 10y$, when $x = 20$ and $y = 4$

14. $5(a - b)$ and $5a - 5b$, when $a = 16$ and $b = 6$

Multiply.

15. $2(b + 5)$

16. $4(x + 3)$

17. $7(1 - t)$

18. $4(1 - y)$

19. $6(5x + 2)$

20. $9(6m + 7)$

21. $7(x + 4 + 6y)$

22. $4(5x + 8 + 3p)$

23. $-7(y - 2)$

24. $-9(y - 7)$

25. $-9(-5x - 6y + 8)$

26. $-7(-2x - 5y + 9)$

27. $\frac{3}{4}(x - 3y - 2z)$

28. $\frac{2}{5}(2x - 5y - 8z)$

29. $3.1(-1.2x + 3.2y - 1.1)$

30. $-2.1(-4.2x - 4.3y - 2.2)$

c Factor. Check by multiplying.

31. $2x + 4$

32. $5y + 20$

33. $30 + 5y$

34. $7x + 28$

35. $14x + 21y$

36. $18a + 24b$

37. $5x + 10 + 15y$

38. $9a + 27b + 81$

39. $8x - 24$

40. $10x - 50$

41. $32 - 4y$

42. $24 - 6m$

43. $8x + 10y - 22$

44. $9a + 6b - 15$

45. $-18x - 12y + 6$

46. $-14x + 21y + 7$

d Collect like terms.

47. $9a + 10a$

48. $14x + 3x$

49. $10a - a$

50. $-10x + x$

51. $2x + 9z + 6x$

52. $3a - 5b + 4a$

53. $41a + 90 - 60a - 2$

54. $42x - 6 - 4x + 20$

55. $23 + 5t + 7y - t - y - 27$

56. $95 - 90d - 87 - 9d + 3 + 7d$

57. $11x - 3x$

58. $9t - 13t$

59. $6n - n$

60. $10t - t$

61. $y - 17y$

62. $5m - 8m + 4$

63. $-8 + 11a - 5b + 6a - 7b + 7$

64. $8x - 5x + 6 + 3y - 2y - 4$

65. $9x + 2y - 5x$

66. $8y - 3z + 4y$

67. $\frac{11}{4}x + \frac{2}{3}y - \frac{4}{5}x - \frac{1}{6}y + 12$

68. $\frac{13}{2}a + \frac{9}{5}b - \frac{2}{3}a - \frac{3}{10}b - 42$

69. $2.7x + 2.3y - 1.9x - 1.8y$

70. $6.7a + 4.3b - 4.1a - 2.9b$

Skill Maintenance

For a circle with the given radius, find the diameter, the circumference, and the area. Use 3.14 for π. [10.3a, b, c]

71. $r = 15$ yd

72. $r = 8.2$ m

73. $r = 9\frac{1}{2}$ mi

74. $r = 2400$ cm

For a circle with the given diameter, find the radius, the circumference, and the area. Use 3.14 for π. [10.3a, b, c]

75. $d = 20$ mm

76. $d = 264$ km

77. $d = 4.6$ ft

78. $d = 10.3$ m

Synthesis

Collect like terms, if possible, and factor the result, if possible.

79. $q + qr + qrs + qrst$

80. $21x + 44xy + 15y - 16x - 8y - 38xy + 2y + xy$

Solving Equations: The Addition Principle

11.2

a USING THE ADDITION PRINCIPLE

Consider the equation $x = 7$. We can easily see that the solution of this equation is 7. If we replace x with 7, we get $7 = 7$ which is true. Now consider the equation $x + 6 = 13$. The solution of this equation is also 7, but the fact that 7 is the solution is not as obvious. We now begin to consider principles that allow us to start with an equation like $x + 6 = 13$ and end up with an equation like $x = 7$, in which the variable is alone on one side and for which the solution is easier to find. The equations $x + 6 = 13$ and $x = 7$ are **equivalent**.

EQUIVALENT EQUATIONS

Equations with the same solutions are called **equivalent equations**.

One principle that we use to solve equations is the addition principle, which we have used throughout this text.

THE ADDITION PRINCIPLE

For any real numbers a, b, and c,

$$a = b \text{ is equivalent to } a + c = b + c.$$

Let's solve $x + 6 = 13$ using the addition principle. We want to get x alone on one side. To do so, we use the addition principle, choosing to add -6 on both sides because $6 + (-6) = 0$:

$$x + 6 = 13$$

$$x + 6 + (-6) = 13 + (-6) \quad \text{Using the addition principle: adding } -6 \text{ on both sides}$$

$$x + 0 = 7 \quad \text{Simplifying}$$

$$x = 7. \quad \text{Identity property of 0}$$

The solution of $x + 6 = 13$ is 7.

Do Margin Exercise 1.

When we use the addition principle, we sometimes say that we "add the same number on both sides of the equation." This is also true for subtraction, since we can express every subtraction as an addition. That is, since

$$a - c = b - c \quad \text{is equivalent to} \quad a + (-c) = b + (-c),$$

the addition principle tells us that we can "subtract the same number on both sides of an equation."

OBJECTIVE

a Solve equations using the addition principle.

SKILL TO REVIEW

Objective 1.7b: Solve equations like $t + 28 = 54$.

Solve.

1. $x + 5 = 9$
2. $y + 15 = 24$

GS 1. Solve $x + 2 = 11$ using the addition principle.

$$x + 2 = 11$$

$$x + 2 + (-2) = 11 + (\quad)$$

$$x + \quad = 9$$

$$x = \quad$$

Answers

Skill to Review:
1. 4 2. 9

Margin Exercise:
1. 9

Guided Solution:
1. −2, 0, 9

EXAMPLE 1 Solve: $x + 5 = -7$.

$$x + 5 = -7$$
$$x + 5 - 5 = -7 - 5 \quad \text{Using the addition principle: adding } -5 \text{ on both sides or subtracting 5 on both sides}$$
$$x + 0 = -12 \quad \text{Simplifying}$$
$$x = -12 \quad \text{Identity property of 0}$$

To check the answer, we substitute -12 in the original equation.

Check: $x + 5 = -7$

$-12 + 5 \ ? \ -7$

-7 TRUE

The solution of the original equation is -12. The equations $x + 5 = -7$ and $x = -12$ are *equivalent*.

Solve.

2. $x + 7 = 2$

3. $y + 9 = 13$

Do Exercises 2 and 3.

Now we solve an equation that involves a subtraction by using the addition principle.

EXAMPLE 2 Solve: $-6.5 = y - 8.4$.

$$-6.5 = y - 8.4$$
$$-6.5 + 8.4 = y - 8.4 + 8.4 \quad \text{Using the addition principle: adding 8.4 to eliminate } -8.4 \text{ on the right}$$
$$1.9 = y$$

Check: $-6.5 = y - 8.4$

$-6.5 \ ? \ 1.9 - 8.4$

-6.5 TRUE

The solution is 1.9.

Note that equations are reversible. That is, if $a = b$ is true, then $b = a$ is true. Thus to solve $-6.5 = y - 8.4$, we can reverse it and solve $y - 8.4 = -6.5$ if we wish.

Solve.

4. $8.7 = n - 4.5$

5. $x - 6 = -9$

Do Exercises 4 and 5.

Answers

2. -5 3. 4 4. 13.2 5. -3

Reading Check

Match each equation with the correct first step for solving it from the list at the right.

RC1. $9 = x - 4$ ______

RC2. $3 + x = -15$ ______

RC3. $x - 3 = 9$ ______

RC4. $x + 4 = 3$ ______

a) Add -4 on both sides.
b) Add 15 on both sides.
c) Subtract 3 on both sides.
d) Subtract 9 on both sides.
e) Add 3 on both sides.
f) Add 4 on both sides.

a Solve using the addition principle. Don't forget to check!

1. $x + 2 = 6$

Check: $x + 2 = 6$ | ?

2. $y + 4 = 11$

Check: $y + 4 = 11$ | ?

3. $x + 15 = -5$

Check: $x + 5 = -5$ | ?

4. $t + 10 = -4$

Check: $t + 10 = -4$ | ?

5. $x + 6 = 8$

6. $y + 8 = 37$

7. $x + 5 = 12$

8. $x + 3 = 7$

9. $11 = y + 7$

10. $14 = y + 8$

11. $-22 = t + 4$

12. $-14 = t + 8$

13. $x + 16 = -2$

14. $y + 34 = -8$

15. $y + 9 = -9$

16. $x + 13 = -13$

17. $x - 9 = 6$

18. $x - 9 = 2$

19. $t - 3 = 16$

20. $t - 5 = 12$

21. $y - 8 = -9$

22. $y - 6 = -12$

23. $x - 7 = -21$

24. $x - 5 = -16$

25. $5 + t = 7$

26. $6 + y = 22$

27. $-7 + y = 13$

28. $-8 + z = 16$

29. $-3 + t = -9$

30. $-8 + y = -23$

31. $r + \frac{1}{3} = \frac{8}{3}$

32. $t + \frac{3}{8} = \frac{5}{8}$

33. $m + \frac{5}{6} = -\frac{11}{12}$

34. $x + \frac{2}{3} = -\frac{5}{6}$

35. $x - \frac{5}{6} = \frac{7}{8}$

36. $y - \frac{3}{4} = \frac{5}{6}$

37. $-\frac{1}{5} + z = -\frac{1}{4}$

38. $-\frac{1}{8} + y = -\frac{3}{4}$

39. $7.4 = x + 2.3$

40. $9.3 = 4.6 + x$

41. $7.6 = x - 4.8$

42. $9.5 = y - 8.3$

43. $-9.7 = -4.7 + y$

44. $-7.8 = 2.8 + x$

45. $5\frac{1}{6} + x = 7$

46. $5\frac{1}{4} = 4\frac{2}{3} + x$

47. $q + \frac{1}{3} = -\frac{1}{7}$

48. $47\frac{1}{8} = -76 + z$

Skill Maintenance

Add. [2.2a], [4.2a], [5.2c]

49. $-3 + (-8)$

50. $-\frac{2}{3} + \frac{5}{8}$

51. $-14.3 + (-19.8)$

52. $3.2 + (-4.9)$

Subtract. [2.3a], [4.3a], [5.2c]

53. $-3 - (-8)$

54. $-\frac{2}{3} - \frac{5}{8}$

55. $-14.3 - (-19.8)$

56. $3.2 - (-4.9)$

Multiply. [2.4a], [3.6a], [5.3a]

57. $-3(-8)$

58. $-\frac{2}{3} \cdot \frac{5}{8}$

59. $-14.3 \times (-19.8)$

60. $3.2(-4.9)$

Divide. [2.5a] [3.7b], [5.4a]

61. $\frac{-24}{-3}$

62. $-\frac{2}{3} \div \frac{5}{8}$

63. $\frac{283.14}{-19.8}$

64. $\frac{-15.68}{3.2}$

Synthesis

Solve.

65. 🖩 $-356.788 = -699.034 + t$

66. $-\frac{4}{5} + \frac{7}{10} = x - \frac{3}{4}$

67. $x + \frac{4}{5} = -\frac{2}{3} - \frac{4}{15}$

68. $8 - 25 = 8 + x - 21$

69. $16 + x - 22 = -16$

70. $x + x = x$

71. $-\frac{3}{2} + x = -\frac{5}{17} - \frac{3}{2}$

72. $|x| = 5$

Solving Equations: The Multiplication Principle

11.3

 USING THE MULTIPLICATION PRINCIPLE

OBJECTIVE

a Solve equations using the multiplication principle.

SKILL TO REVIEW

Objective 1.7b: Solve equations like $28 \cdot x = 168$.

Solve.

1. $8x = 32$

2. $3x = 48$

Suppose that $a = b$ is true, and we multiply a by some number c. We get the same answer if we multiply b by c, because a and b are the same number.

THE MULTIPLICATION PRINCIPLE

For any real numbers a, b, and c, where $c \neq 0$,

$$a = b \quad \text{is equivalent to} \quad a \cdot c = b \cdot c.$$

When using the multiplication principle, we sometimes say that we "multiply on both sides of the equation by the same number."

EXAMPLE 1 Solve: $5x = 70$.

To get x alone, we multiply by the *multiplicative inverse*, or *reciprocal*, of 5. Then we get the *multiplicative identity* 1 times x, or $1 \cdot x$, which simplifies to x. This allows us to eliminate 5 on the left.

$5x = 70$	The reciprocal of 5 is $\frac{1}{5}$.
$\frac{1}{5} \cdot 5x = \frac{1}{5} \cdot 70$	Multiplying by $\frac{1}{5}$ to get $1 \cdot x$ and eliminate 5 on the left
$1 \cdot x = 14$	Simplifying
$x = 14$	Identity property of 1: $1 \cdot x = x$

Check: $5x = 70$

$5 \cdot 14 \ ? \ 70$

70 | TRUE

The solution is 14.

The multiplication principle also tells us that we can "divide on both sides of the equation by the same nonzero number." This is because dividing is the same as multiplying by a reciprocal. That is,

$$\frac{a}{c} = \frac{b}{c} \quad \text{is equivalent to} \quad a \cdot \frac{1}{c} = b \cdot \frac{1}{c}, \quad \text{when } c \neq 0.$$

In an expression like $5x$ in Example 1, the number 5 is called the **coefficient**. Example 1 could be done as follows, dividing on both sides by 5, the coefficient of x.

EXAMPLE 2 Solve: $5x = 70$.

$5x = 70$	
$\frac{5x}{5} = \frac{70}{5}$	Dividing by 5 on both sides.
$1 \cdot x = 14$	Simplifying
$x = 14$	Identity property of 1

The solution is 14.

Do Margin Exercises 1 and 2. ▶

GS **1.** Solve $6x = 90$ by multiplying on both sides.

$6x = 90$

$\frac{1}{6} \cdot 6x = \square \cdot 90$

$1 \cdot x = 15$

$\square = 15$

Check: $6x = 90$

$6 \cdot \square \ ? \ 90$

90 | TRUE

2. Solve $4x = -7$ by dividing on both sides.

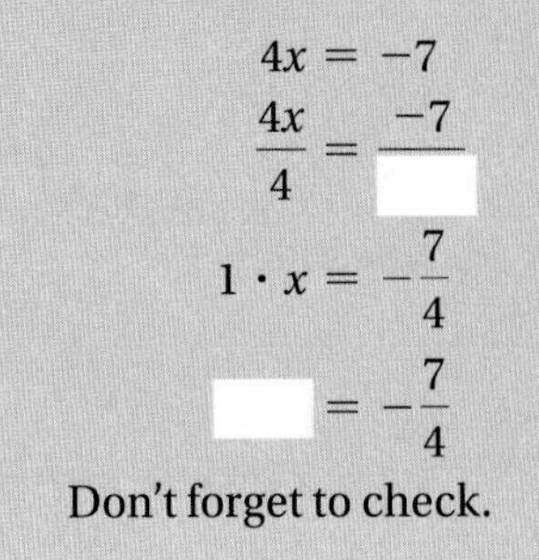

$4x = -7$

$\frac{4x}{4} = \frac{-7}{\square}$

$1 \cdot x = -\frac{7}{4}$

$\square = -\frac{7}{4}$

Don't forget to check.

Answers

Skill to Review:
1. 4 **2.** 16

Margin Exercises:
1. 15 **2.** $-\frac{7}{4}$

Guided Solutions:
1. $\frac{1}{6}$, x, 15 **2.** 4, x

EXAMPLE 3 Solve: $-4x = 92$.

$$-4x = 92$$

$$\frac{-4x}{-4} = \frac{92}{-4}$$ Using the multiplication principle. Dividing by -4 on both sides is the same as multiplying by $-\frac{1}{4}$.

$$1 \cdot x = -23$$ Simplifying

$$x = -23$$ Identity property of 1

Check:

$$\begin{array}{r|l} -4x & = 92 \\ \hline -4(-23) & ?\ 92 \\ 92 & \end{array}$$ TRUE

The solution is -23.

3. Solve: $-6x = 108$.

◀ **Do Exercise 3.**

EXAMPLE 4 Solve: $-x = 9$.

$$-x = 9$$

$$-1 \cdot (-x) = -1 \cdot 9$$ Multiplying by -1 on both sides

$$-1 \cdot (-1) \cdot x = -9$$ $-x = (-1) \cdot x$

$$1 \cdot x = -9$$

$$x = -9$$

Check:

$$\begin{array}{r|l} -x & = 9 \\ \hline -(-9) & ?\ 9 \\ 9 & \end{array}$$ TRUE

The solution is -9.

4. Solve: $-x = -10$.

◀ **Do Exercise 4.**

In practice, it is generally more convenient to divide on both sides of the equation if the coefficient of the variable is in decimal notation or is an integer. If the coefficient is in fraction notation, it is more convenient to multiply by a reciprocal.

EXAMPLE 5 Solve: $\frac{3}{8} = -\frac{5}{4}x$.

$$\frac{3}{8} = -\frac{5}{4}x$$

The reciprocal of $-\frac{5}{4}$ is $-\frac{4}{5}$. There is no sign change.

$$-\frac{4}{5} \cdot \frac{3}{8} = -\frac{4}{5} \cdot \left(-\frac{5}{4}x\right)$$ Multiplying by $-\frac{4}{5}$ to get $1 \cdot x$ and eliminate $-\frac{5}{4}$ on the right

$$-\frac{12}{40} = 1 \cdot x$$

$$-\frac{3}{10} = 1 \cdot x$$ Simplifying

$$-\frac{3}{10} = x$$ Identity property of 1

Answers

3. -18 4. 10

Check: $\dfrac{3}{8} = -\dfrac{5}{4}x$

$$\frac{3}{8} \stackrel{?}{=} -\frac{5}{4}\left(-\frac{3}{10}\right)$$
$$\frac{3}{8} \qquad \text{TRUE}$$

The solution is $-\dfrac{3}{10}$.

As noted in Section 11.2, if $a = b$ is true, then $b = a$ is true. Thus we can reverse the equation $\frac{3}{8} = -\frac{5}{4}x$ and solve $-\frac{5}{4}x = \frac{3}{8}$ if we wish.

Do Exercise 5.

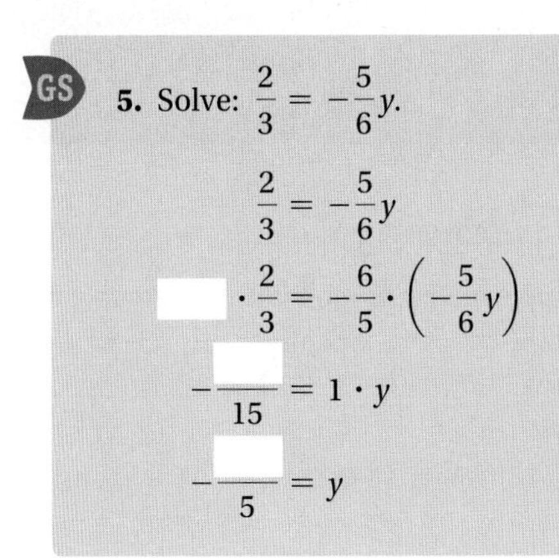

5. Solve: $\dfrac{2}{3} = -\dfrac{5}{6}y$.

$$\frac{2}{3} = -\frac{5}{6}y$$
$$\boxed{} \cdot \frac{2}{3} = -\frac{6}{5} \cdot \left(-\frac{5}{6}y\right)$$
$$-\frac{\boxed{}}{15} = 1 \cdot y$$
$$-\frac{\boxed{}}{5} = y$$

EXAMPLE 6 Solve: $1.16y = 9744$.

$$\frac{1.16y}{1.16} = \frac{9744}{1.16} \qquad \text{Dividing by 1.16 on both sides}$$
$$y = \frac{9744}{1.16}$$
$$y = 8400 \qquad \text{Simplifying}$$

Check: $1.16y = 9744$

$$1.16(8400) \stackrel{?}{=} 9744$$
$$9744 \qquad \text{TRUE}$$

The solution is 8400.

Do Exercises 6 and 7.

Solve.

6. $1.12x = 8736$

7. $6.3 = -2.1y$

Now we use the multiplication principle to solve an equation that involves division.

EXAMPLE 7 Solve: $\dfrac{-y}{9} = 14$.

$$9 \cdot \frac{-y}{9} = 9 \cdot 14 \qquad \text{Multiplying by 9 on both sides}$$
$$-y = 126$$
$$-1 \cdot (-y) = -1 \cdot 126 \qquad \text{Multiplying by } -1 \text{ on both sides}$$
$$y = -126$$

Check: $\dfrac{-y}{9} = 14$

$$\frac{-(-126)}{9} \stackrel{?}{=} 14$$
$$\frac{126}{9}$$
$$14 \qquad \text{TRUE}$$

The solution is -126.

Do Exercise 8.

8. Solve: $-14 = \dfrac{-y}{2}$.

Answers

5. $-\frac{4}{5}$ **6.** 7800 **7.** -3 **8.** 28

Guided Solution:

5. $-\frac{6}{5}$, 12, 4

11.3 Exercise Set

For Extra Help
MyMathLab® MathXL® PRACTICE WATCH READ REVIEW

Reading Check

Match each equation with the correct first step for solving it from the list at the right.

RC1. $3 = -\frac{1}{12}x$ ________

RC2. $-6x = 12$ ________

RC3. $12x = -6$ ________

RC4. $\frac{1}{6}x = 12$ ________

a) Divide by 12 on both sides.
b) Multiply by 6 on both sides.
c) Multiply by 12 on both sides.
d) Divide by -6 on both sides.
e) Divide by 6 on both sides.
f) Multiply by -12 on both sides.

a Solve using the multiplication principle. Don't forget to check!

1. $6x = 36$

2. $4x = 52$

3. $5x = 45$

4. $8x = 56$

5. $84 = 7x$

6. $63 = 7x$

7. $-x = 40$

8. $50 = -x$

9. $6x = -42$

10. $8x = -72$

11. $7x = -49$

12. $9x = -54$

13. $-12x = 72$

14. $-15x = 105$

15. $-9x = 45$

16. $-7x = 56$

17. $-21x = -126$

18. $-13x = -104$

19. $-2x = -10$

20. $-78 = -39p$

21. $\frac{1}{7}t = -9$

22. $-\frac{1}{8}y = 11$

23. $\frac{3}{4}x = 27$

24. $\frac{4}{5}x = 16$

25. $-\frac{1}{3}t = 7$

26. $-\frac{1}{6}x = 9$

27. $-\frac{1}{3}m = \frac{1}{5}$

28. $\frac{1}{5} = -\frac{1}{8}z$

29. $-\frac{3}{5}r = \frac{9}{10}$

30. $\frac{2}{5}y = -\frac{4}{15}$

31. $-\frac{3}{2}r = -\frac{27}{4}$

32. $-\frac{5}{7}x = -\frac{10}{14}$

33. $6.3x = 44.1$

34. $2.7y = -54$

35. $-3.1y = 21.7$

36. $-3.3y = 6.6$

37. $-38.7m = 309.6$

38. $29.4m = 235.2$

39. $-\frac{2}{3}y = -10.6$

40. $-\frac{9}{7}y = 12.06$

41. $\frac{-x}{5} = 10$

42. $\frac{-x}{8} = -16$

43. $\frac{t}{-2} = 7$

44. $\frac{m}{-3} = 10$

Skill Maintenance

45. Find the circumference, the diameter, and the area of a circle whose radius is 10 ft. Use 3.14 for π. [10.3a, b, c]

46. Find the circumference, the radius, and the area of a circle whose diameter is 24 cm. Use 3.14 for π. [10.3a, b, c]

47. Find the volume of a rectangular block of granite of length 25 ft, width 10 ft, and height 32 ft. [10.4a]

48. Find the volume of a rectangular solid of length 1.3 cm, width 10 cm, and height 2.4 cm. [10.4a]

Find the area of each figure. [10.2b]

49.

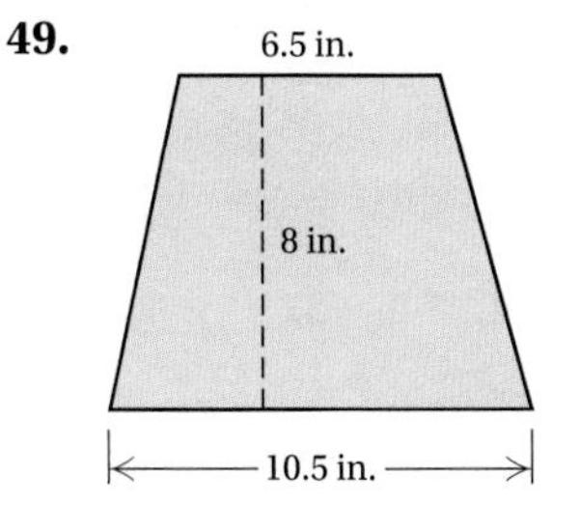

50.

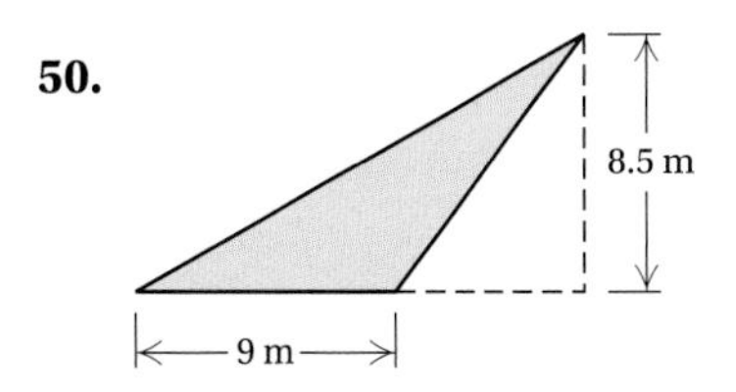

Synthesis

Solve.

51. $-0.2344m = 2028.732$

52. $0 \cdot x = 0$

53. $0 \cdot x = 9$

54. $4|x| = 48$

55. $2|x| = -12$

56. A student makes a calculation and gets an answer of 22.5. On the last step, the student multiplied by 0.3 when she should have divided by 0.3. What should the correct answer be?

Mid-Chapter Review

Concept Reinforcement

Determine whether each statement is true or false.

______ **1.** The expression $2(x + 3)$ is equivalent to the expression $2 \cdot x + 3$. [11.1b]

______ **2.** To factor an expression is to find an equivalent expression that is a product. [11.1c]

______ **3.** Collecting like terms is based on the distributive laws. [11.1d]

______ **4.** $3 - x = 4x$ and $5x = -3$ are equivalent equations. [11.2a]

Guided Solutions

GS Fill in each blank with the number, variable, or expression that creates a correct statement or solution.

5. Factor: $6x - 3y + 18$. [11.1c]

$$6x - 3y + 18 = 3 \cdot \square - 3 \cdot \square + 3 \cdot \square = \square(\square - \square + 6)$$

Solve. [11.2a], [11.3a]

6.

$$\begin{aligned} x + 5 &= -3 \\ x + 5 - 5 &= -3 - \square \\ x + \square &= -8 \\ x &= \square \end{aligned}$$

7.

$$\begin{aligned} -6x &= 42 \\ \frac{-6x}{-6} &= \frac{42}{\square} \\ \square \cdot x &= -7 \\ x &= \square \end{aligned}$$

Mixed Review

Evaluate. [11.1a]

8. $4x$, when $x = -7$

9. $\frac{a}{b}$, when $a = 56$ and $b = 8$

10. $\frac{m - n}{3}$, when $m = 17$ and $n = 2$

Multiply. [11.1b]

11. $3(x + 5)$

12. $4(2y - 7)$

13. $6(3x + 2y - 1)$

14. $-2(-3x - y + 8)$

Factor. [11.1c]

15. $3y + 21$

16. $5z + 45$

17. $9x - 36$

18. $24a - 8$

19. $4x + 6y - 2$

20. $12x - 9y + 3$

21. $4a - 12b + 32$

22. $30a - 18b - 24$

Collect like terms. [11.1d]

23. $7x + 8x$

24. $3y - y$

25. $5x - 2y + 6 - 3x + y - 9$

Solve. [11.2a], [11.3a]

26. $x + 5 = 11$

27. $x + 9 = -3$

28. $8 = t + 1$

29. $-7 = y + 3$

30. $x - 6 = 14$

31. $y - 7 = -2$

32. $3 + t = 10$

33. $-5 + x = 5$

34. $y + \frac{1}{3} = -\frac{1}{2}$

35. $-\frac{3}{2} + z = -\frac{3}{4}$

36. $4.6 = x + 3.9$

37. $-3.3 = -1.9 + t$

38. $7x = 42$

39. $144 = 12y$

40. $17 = -t$

41. $6x = -54$

42. $-5y = -85$

43. $-8x = 48$

44. $\frac{2}{3}x = 12$

45. $-\frac{1}{5}t = 3$

46. $\frac{3}{4}x = -\frac{9}{8}$

47. $-\frac{5}{6}t = -\frac{25}{18}$

48. $1.8y = -5.4$

49. $\frac{-y}{7} = 5$

Understanding Through Discussion and Writing

50. Determine whether $(a + b)^2$ and $a^2 + b^2$ are equivalent for all real numbers. Explain. [11.1a]

51. The distributive law is introduced before the material on collecting like terms. Why do you think this is? [11.1d]

52. Explain the following mistake made by a fellow student. [11.2a]

$$x + \frac{1}{3} = -\frac{5}{3}$$
$$x = -\frac{4}{3}$$

53. Explain the following mistake made by a fellow student. [11.3a]

$$\frac{2}{3}x = -\frac{5}{3}$$
$$x = \frac{10}{9}$$

STUDYING FOR SUCCESS *Looking Ahead*

- ☐ As you register for next semester's courses, evaluate your work and family commitments.
- ☐ If you are registering for another math course, consider keeping your notes, tests, and text from this course as a resource.

11.4 Using the Principles Together

OBJECTIVES

a Solve equations using both the addition principle and the multiplication principle.

b Solve equations in which like terms may need to be collected.

c Solve equations by first removing parentheses and collecting like terms.

SKILL TO REVIEW

Objective 2.5a: Divide integers.

Divide.

1. $40 \div (-5)$
2. $\dfrac{-64}{-16}$

a APPLYING BOTH PRINCIPLES

Consider the equation $3x + 4 = 13$. It is more complicated than those in the two preceding sections. In order to solve such an equation, we first isolate the x-term, $3x$, using the addition principle. Then we apply the multiplication principle to get x by itself.

EXAMPLE 1 Solve: $3x + 4 = 13$.

$$3x + 4 = 13$$

$$3x + 4 - 4 = 13 - 4 \quad \text{Using the addition principle: adding } -4 \text{ or subtracting 4 on both sides}$$

$$3x = 9 \quad \text{Simplifying}$$

$$\frac{3x}{3} = \frac{9}{3} \quad \text{Using the multiplication principle: multiplying by } \tfrac{1}{3} \text{ or dividing by 3 on both sides}$$

$$x = 3 \quad \text{Simplifying}$$

Check:

$3x + 4 = 13$	
$3 \cdot 3 + 4$	? 13
$9 + 4$	
13	TRUE

The solution is 3.

◀ Do Margin Exercise 1.

1. Solve: $9x + 6 = 51$.

EXAMPLE 2 Solve: $-5x - 6 = 16$.

$$-5x - 6 = 16$$

$$-5x - 6 + 6 = 16 + 6 \quad \text{Adding 6 on both sides}$$

$$-5x = 22$$

$$\frac{-5x}{-5} = \frac{22}{-5} \quad \text{Dividing by } -5 \text{ on both sides}$$

$$x = -\frac{22}{5}, \text{ or } -4\frac{2}{5} \quad \text{Simplifying}$$

Answers

Skill to Review:
1. −8 2. 4

Margin Exercise:
1. 5

Check:

$$\begin{array}{c|c} -5x - 6 & 16 \\ \hline -5\left(-\frac{22}{5}\right) - 6 \ ? & 16 \\ 22 - 6 & \\ 16 & \end{array} \quad \text{TRUE}$$

The solution is $-\frac{22}{5}$.

Do Exercises 2 and 3. ▶

Solve.

2. $8x - 4 = 28$

3. $-\frac{1}{2}x + 3 = 1$

EXAMPLE 3 Solve: $45 - x = 13$.

$$\begin{aligned} 45 - x &= 13 \\ -45 + 45 - x &= -45 + 13 && \text{Adding } -45 \text{ on both sides} \\ -x &= -32 \\ -1 \cdot (-x) &= -1 \cdot (-32) && \text{Multiplying by } -1 \text{ on both sides} \\ -1 \cdot (-1) \cdot x &= 32 \\ x &= 32 \end{aligned}$$

Check:

$$\begin{array}{c|c} 45 - x & 13 \\ \hline 45 - 32 \ ? & 13 \\ 13 & \end{array} \quad \text{TRUE}$$

The solution is 32.

Do Exercise 4. ▶

GS 4. Solve: $-18 - m = -57$.

$$\begin{aligned} 18 - 18 - m &= \square - 57 \\ \square &= -39 \\ \square(-m) &= -1(-39) \\ \square &= 39 \end{aligned}$$

EXAMPLE 4 Solve: $16.3 - 7.2y = -8.18$.

$$\begin{aligned} 16.3 - 7.2y &= -8.18 \\ -16.3 + 16.3 - 7.2y &= -16.3 + (-8.18) && \text{Adding } -16.3 \text{ on both sides} \\ -7.2y &= -24.48 \\ \frac{-7.2y}{-7.2} &= \frac{-24.48}{-7.2} && \text{Dividing by } -7.2 \text{ on both sides} \\ y &= 3.4 \end{aligned}$$

Check:

$$\begin{array}{c|c} 16.3 - 7.2y & -8.18 \\ \hline 16.3 - 7.2(3.4) \ ? & -8.18 \\ 16.3 - 24.48 & \\ -8.18 & \end{array} \quad \text{TRUE}$$

The solution is 3.4.

Do Exercises 5 and 6. ▶

Solve.

5. $-4 - 8x = 8$

6. $41.68 = 4.7 - 8.6y$

Answers

2. 4 **3.** 4 **4.** 39 **5.** $-\frac{3}{2}$

6. -4.3

Guided Solution:

4. 18, $-m$, -1, m

b COLLECTING LIKE TERMS

If there are like terms on one side of the equation, we collect them before using the addition principle or the multiplication principle.

EXAMPLE 5 Solve: $3x + 4x = -14$.

$$
\begin{aligned}
3x + 4x &= -14 \\
7x &= -14 && \text{Collecting like terms} \\
\frac{7x}{7} &= \frac{-14}{7} && \text{Dividing by 7 on both sides} \\
x &= -2
\end{aligned}
$$

The number -2 checks, so the solution is -2.

Solve.

7. $4x + 3x = -21$

8. $x - 0.09x = 728$

Do Exercises 7 and 8.

If there are like terms on opposite sides of the equation, we get them on the same side by using the addition principle. Then we collect them. In other words, we get all terms with a variable on one side of the equation and all terms without a variable on the other side.

EXAMPLE 6 Solve: $2x - 2 = -3x + 3$.

$$
\begin{aligned}
2x - 2 &= -3x + 3 \\
2x - 2 + 2 &= -3x + 3 + 2 && \text{Adding 2} \\
2x &= -3x + 5 && \text{Collecting like terms} \\
2x + 3x &= -3x + 5 + 3x && \text{Adding } 3x \\
5x &= 5 && \text{Simplifying} \\
\frac{5x}{5} &= \frac{5}{5} && \text{Dividing by 5} \\
x &= 1 && \text{Simplifying}
\end{aligned}
$$

Check:

$2x - 2$	$= -3x + 3$	
$2 \cdot 1 - 2$	? $-3 \cdot 1 + 3$	Substituting in the original equation
$2 - 2$	$-3 + 3$	
0	0	TRUE

The solution is 1.

Solve.

9. $7y + 5 = 2y + 10$

10. $5 - 2y = 3y - 5$

Do Exercises 9 and 10.

In Example 6, we used the addition principle to get all terms with a variable on one side of the equation and all terms without a variable on the other side. Then we collected like terms and proceeded as before. If there are like terms on one side at the outset, they should be collected first.

Answers

7. -3 8. 800 9. 1 10. 2

EXAMPLE 7 Solve: $6x + 5 - 7x = 10 - 4x + 3$.

$$\begin{aligned} 6x + 5 - 7x &= 10 - 4x + 3 && \\ -x + 5 &= 13 - 4x && \text{Collecting like terms} \\ 4x - x + 5 &= 13 - 4x + 4x && \text{Adding } 4x \\ 3x + 5 &= 13 && \text{Simplifying} \\ 3x + 5 - 5 &= 13 - 5 && \text{Subtracting 5} \\ 3x &= 8 && \text{Simplifying} \\ \frac{3x}{3} &= \frac{8}{3} && \text{Dividing by 3} \\ x &= \frac{8}{3} && \text{Simplifying} \end{aligned}$$

The number $\frac{8}{3}$ checks, so $\frac{8}{3}$ is the solution.

Do Exercises 11 and 12. ▶

Solve.

GS **11.** $7x - 17 + 2x = 2 - 8x + 15$

$\boxed{}\, x - 17 = 17 - 8x$

$8x + 9x - 17 = 17 - 8x + \boxed{}$

$\boxed{}\, x - 17 = 17$

$17x - 17 + 17 = 17 + \boxed{}$

$17x = 34$

$\frac{17x}{17} = \frac{34}{\boxed{}}$

$\boxed{} = 2$

12. $3x - 15 = 5x + 2 - 4x$

Clearing Fractions and Decimals

For the equations considered so far, we generally used the addition principle first. There are, however, some situations in which it is to our advantage to use the multiplication principle first. Consider, for example,

$$\frac{1}{2}x + 5 = \frac{3}{4}.$$

The LCM of the denominators is 4. If we multiply by 4 on both sides, we get

$$\begin{aligned} 4\left(\frac{1}{2}x + 5\right) &= 4 \cdot \frac{3}{4} \\ 4 \cdot \frac{1}{2}x + 4 \cdot 5 &= 4 \cdot \frac{3}{4} \\ 2x + 20 &= 3. \end{aligned}$$

We have "cleared" the fractions. Now consider

$$2.3x + 8 = 15.43.$$

If we multiply by 100 on both sides, we get

$$\begin{aligned} 100(2.3x + 8) &= 100 \cdot 15.43 \\ 100 \cdot 2.3x + 100 \cdot 8 &= 100 \cdot 15.43 \\ 230x + 800 &= 1543, \end{aligned}$$

which has no decimal points. We have "cleared" the decimals. The equations $2x + 20 = 3$ and $230x + 800 = 1543$ are easier to solve than the original equations. It is your choice whether to clear fractions or decimals, but doing so often eases computations.

Answers

11. 2 **12.** $\frac{17}{2}$

Guided Solution:

11. 9, $8x$, 17, 17, 17, x

In what follows, we use the multiplication principle first to "clear," or eliminate, fractions or decimals. For fractions, the number by which we multiply is the **least common multiple of all the denominators.**

EXAMPLE 8 Solve:

$$\frac{2}{3}x - \frac{1}{6} + \frac{1}{2}x = \frac{7}{6} + 2x.$$

The denominators are 3, 6, and 2. The number 6 is the least common multiple of all the denominators. We multiply by 6 on both sides of the equation:

$$6\left(\frac{2}{3}x - \frac{1}{6} + \frac{1}{2}x\right) = 6\left(\frac{7}{6} + 2x\right) \qquad \text{Multiplying by 6 on both sides}$$

$$6 \cdot \frac{2}{3}x - 6 \cdot \frac{1}{6} + 6 \cdot \frac{1}{2}x = 6 \cdot \frac{7}{6} + 6 \cdot 2x \qquad \text{Using the distributive laws. (\textit{Caution}! Be sure to multiply \textit{all} terms by 6.)}$$

$$4x - 1 + 3x = 7 + 12x \qquad \text{Simplifying. Note that the fractions are cleared.}$$

$$7x - 1 = 7 + 12x \qquad \text{Collecting like terms}$$

$$7x - 1 - 12x = 7 + 12x - 12x \qquad \text{Subtracting } 12x$$

$$-5x - 1 = 7 \qquad \text{Simplifying}$$

$$-5x - 1 + 1 = 7 + 1 \qquad \text{Adding 1}$$

$$-5x = 8 \qquad \text{Collecting like terms}$$

$$\frac{-5x}{-5} = \frac{8}{-5} \qquad \text{Dividing by } -5$$

$$x = -\frac{8}{5}.$$

Check:

$$\begin{array}{c|c}
\multicolumn{2}{c}{\frac{2}{3}x - \frac{1}{6} + \frac{1}{2}x = \frac{7}{6} + 2x} \\
\hline
\frac{2}{3}\left(-\frac{8}{5}\right) - \frac{1}{6} + \frac{1}{2}\left(-\frac{8}{5}\right) \; ? & \frac{7}{6} + 2\left(-\frac{8}{5}\right) \\
-\frac{16}{15} - \frac{1}{6} - \frac{8}{10} & \frac{7}{6} - \frac{16}{5} \\
-\frac{32}{30} - \frac{5}{30} - \frac{24}{30} & \frac{35}{30} - \frac{96}{30} \\
\frac{-32 - 5 - 24}{30} & \frac{35 - 96}{30} \\
-\frac{61}{30} & -\frac{61}{30}
\end{array} \qquad \text{TRUE}$$

The solution is $-\frac{8}{5}$.

◀ **Do Exercise 13.**

13. Solve: $\frac{7}{8}x - \frac{1}{4} + \frac{1}{2}x = \frac{3}{4} + x$. GS

$$8 \cdot \left(\frac{7}{8}x - \frac{1}{4} + \frac{1}{2}x\right) = \square \cdot \left(\frac{3}{4} + x\right)$$

$$8 \cdot \frac{7}{8}x - \square \cdot \frac{1}{4} + 8 \cdot \frac{1}{2}x = 8 \cdot \frac{3}{4} + \square \cdot x$$

$$\square x - \square + 4x = 6 + 8x$$

$$\square x - 2 = 6 + 8x$$

$$11x - 2 - 8x = 6 + 8x - \square$$

$$3x - 2 = \square$$

$$3x - 2 + \square = 6 + 2$$

$$3x = \square$$

$$\frac{3x}{3} = \frac{8}{\square}$$

$$x = \frac{8}{3}$$

Answer

13. $\frac{8}{3}$

Guided Solution:

13. 8, 8, 8, 7, 2, 11, 8*x*, 6, 2, 8, 3

To illustrate clearing decimals, we repeat Example 4, but this time we clear the decimals first.

EXAMPLE 9 Solve: $16.3 - 7.2y = -8.18$.

16.3	−	7.2y	=	−8.18
1 decimal place		1 decimal place		2 decimal places

The greatest number of decimal places in any one number is *two*. Multiplying by 100, which has *two* zeros, will clear the decimals.

$$100(16.3 - 7.2y) = 100(-8.18)$$ Multiplying by 100 on both sides

$$100(16.3) - 100(7.2y) = 100(-8.18)$$ Using a distributive law

$$1630 - 720y = -818$$ Simplifying

$$1630 - 720y - 1630 = -818 - 1630$$ Subtracting 1630 on both sides

$$-720y = -2448$$ Collecting like terms

$$\frac{-720y}{-720} = \frac{-2448}{-720}$$ Dividing by −720 on both sides

$$y = \frac{17}{5}, \text{ or } 3.4$$

The number $\frac{17}{5}$, or 3.4, checks, as shown in Example 4, so it is the solution.

Do Exercise 14. ▶

14. Solve: $41.68 = 4.7 - 8.6y$.

C EQUATIONS CONTAINING PARENTHESES

To solve certain kinds of equations that contain parentheses, we first use the distributive laws to remove the parentheses. Then we proceed as before.

EXAMPLE 10 Solve: $8x = 2(12 - 2x)$.

$$8x = 2(12 - 2x)$$

$$8x = 24 - 4x$$ Using the distributive law to multiply and remove parentheses

$$8x + 4x = 24 - 4x + 4x$$ Adding $4x$ to get all x-terms on one side

$$12x = 24$$ Collecting like terms

$$\frac{12x}{12} = \frac{24}{12}$$ Dividing by 12

$$x = 2$$

Check:

$8x = 2(12 - 2x)$	
$8 \cdot 2$?	$2(12 - 2 \cdot 2)$
16	$2(12 - 4)$
	$2 \cdot 8$
	16 TRUE

We use the rules for order of operations to carry out the calculations on each side of the equation.

The solution is 2.

Do Exercises 15 and 16. ▶

Solve.

15. $2(2y + 3) = 14$

16. $5(3x - 2) = 35$

Answers

14. $-\frac{43}{10}$, or -4.3 15. 2 16. 3

Here is a procedure for solving the types of equations discussed in this section.

AN EQUATION-SOLVING PROCEDURE

1. If the equation contains parentheses, multiply using the distributive laws to remove them.
2. Multiply on both sides to clear the equation of fractions or decimals. (This is optional, but it can ease computations.)
3. Collect like terms on each side, if necessary.
4. Get all terms with variables on one side and all constant terms on the other side, using the *addition principle.*
5. Collect like terms again, if necessary.
6. Multiply or divide to solve for the variable, using the *multiplication principle.*
7. Check all possible solutions in the original equation.

EXAMPLE 11 Solve: $2 - 5(x + 5) = 3(x - 2) - 1$.

$$2 - 5(x + 5) = 3(x - 2) - 1$$

$$2 - 5x - 25 = 3x - 6 - 1$$ Using the distributive laws to multiply and remove parentheses

$$-5x - 23 = 3x - 7$$ Collecting like terms

$$-5x - 23 + 5x = 3x - 7 + 5x$$ Adding $5x$

$$-23 = 8x - 7$$ Collecting like terms

$$-23 + 7 = 8x - 7 + 7$$ Adding 7

$$-16 = 8x$$ Collecting like terms

$$\frac{-16}{8} = \frac{8x}{8}$$ Dividing by 8

$$-2 = x$$

Check:

$2 - 5(x + 5) = 3(x - 2) - 1$		
$2 - 5(-2 + 5)$?	$3(-2 - 2) - 1$	
$2 - 5(3)$	$3(-4) - 1$	
$2 - 15$	$-12 - 1$	
-13	-13	TRUE

The solution is -2.

Note that the solution of $-2 = x$ is -2, which is also the solution of $x = -2$.

Do Exercises 17 and 18.

Solve.

17. $3(7 + 2x) = 30 + 7(x - 1)$

18. $4(3 + 5x) - 4 = 3 + 2(x - 2)$

Answers

17. -2 18. $-\frac{1}{2}$

11.4 Exercise Set

Reading Check

Choose from the column on the right the operation that will clear each equation of fractions or decimals.

RC1. $\frac{2}{3}x - 5 + \frac{1}{2}x = \frac{3}{10} + x$ ________

RC2. $0.003y - 0.1 = 0.03 + y$ ________

RC3. $\frac{1}{4} - 8t + \frac{5}{6} = t - \frac{1}{12}$ ________

RC4. $0.5 + 2.15y = 1.5y - 10$ ________

RC5. $\frac{3}{5} - x = \frac{2}{7}x + 4$ ________

a) Multiply by 1000 on both sides.
b) Multiply by 35 on both sides.
c) Multiply by 12 on both sides.
d) Multiply by 30 on both sides.
e) Multiply by 100 on both sides.

a Solve. Don't forget to check!

1. $5x + 6 = 31$

2. $8x + 6 = 30$

3. $8x + 4 = 68$

4. $8z + 7 = 79$

5. $4x - 6 = 34$

6. $4x - 11 = 21$

7. $3x - 9 = 33$

8. $6x - 9 = 57$

9. $7x + 2 = -54$

10. $5x + 4 = -41$

11. $-45 = 3 + 6y$

12. $-91 = 9t + 8$

13. $-4x + 7 = 35$

14. $-5x - 7 = 108$

15. $-7x - 24 = -129$

16. $-6z - 18 = -132$

b Solve.

17. $5x + 7x = 72$

18. $4x + 5x = 45$

19. $8x + 7x = 60$

20. $3x + 9x = 96$

21. $4x + 3x = 42$

22. $6x + 19x = 100$

23. $-6y - 3y = 27$

24. $-4y - 8y = 48$

25. $-7y - 8y = -15$

26. $-10y - 3y = -39$

27. $10.2y - 7.3y = -58$

28. $6.8y - 2.4y = -88$

29. $x + \frac{1}{3}x = 8$

30. $x + \frac{1}{4}x = 10$

31. $8y - 35 = 3y$

32. $4x - 6 = 6x$

33. $8x - 1 = 23 - 4x$

34. $5y - 2 = 28 - y$

35. $2x - 1 = 4 + x$

36. $5x - 2 = 6 + x$

37. $6x + 3 = 2x + 11$

38. $5y + 3 = 2y + 15$

39. $5 - 2x = 3x - 7x + 25$

40. $10 - 3x = 2x - 8x + 40$

41. $4 + 3x - 6 = 3x + 2 - x$

42. $5 + 4x - 7 = 4x - 2 - x$

43. $4y - 4 + y + 24 = 6y + 20 - 4y$

44. $5y - 7 + y = 7y + 21 - 5y$

Solve. Clear fractions or decimals first.

45. $\frac{7}{2}x + \frac{1}{2}x = 3x + \frac{3}{2} + \frac{5}{2}x$

46. $\frac{7}{8}x - \frac{1}{4} + \frac{3}{4}x = \frac{1}{16} + x$

47. $\frac{2}{3} + \frac{1}{4}t = \frac{1}{3}$

48. $-\frac{3}{2} + x = -\frac{5}{6} - \frac{4}{3}$

49. $\frac{2}{3} + 3y = 5y - \frac{2}{15}$

50. $\frac{1}{2} + 4m = 3m - \frac{5}{2}$

51. $\frac{5}{3} + \frac{2}{3}x = \frac{25}{12} + \frac{5}{4}x + \frac{3}{4}$

52. $1 - \frac{2}{3}y = \frac{9}{5} - \frac{y}{5} + \frac{3}{5}$

53. $2.1x + 45.2 = 3.2 - 8.4x$

54. $0.96y - 0.79 = 0.21y + 0.46$

55. $1.03 - 0.62x = 0.71 - 0.22x$

56. $1.7t + 8 - 1.62t = 0.4t - 0.32 + 8$

57. $\frac{2}{7}x - \frac{1}{2}x = \frac{3}{4}x + 1$

58. $\frac{5}{16}y + \frac{3}{8}y = 2 + \frac{1}{4}y$

C Solve.

59. $3(2y - 3) = 27$

60. $4(2y - 3) = 28$

61. $40 = 5(3x + 2)$

62. $9 = 3(5x - 2)$

63. $2(3 + 4m) - 9 = 45$

64. $3(5 + 3m) - 8 = 88$

65. $5r - (2r + 8) = 16$

66. $6b - (3b + 8) = 16$

67. $6 - 2(3x - 1) = 2$

68. $10 - 3(2x - 1) = 1$

69. $5(d + 4) = 7(d - 2)$

70. $3(t - 2) = 9(t + 2)$

71. $8(2t + 1) = 4(7t + 7)$

72. $7(5x - 2) = 6(6x - 1)$

73. $3(r - 6) + 2 = 4(r + 2) - 21$

74. $5(t + 3) + 9 = 3(t - 2) + 6$

75. $19 - (2x + 3) = 2(x + 3) + x$

76. $13 - (2c + 2) = 2(c + 2) + 3c$

77. $0.7(3x + 6) = 1.1 - (x + 2)$

78. $0.9(2x + 8) = 20 - (x + 5)$

79. $a + (a - 3) = (a + 2) - (a + 1)$

80. $0.8 - 4(b - 1) = 0.2 + 3(4 - b)$

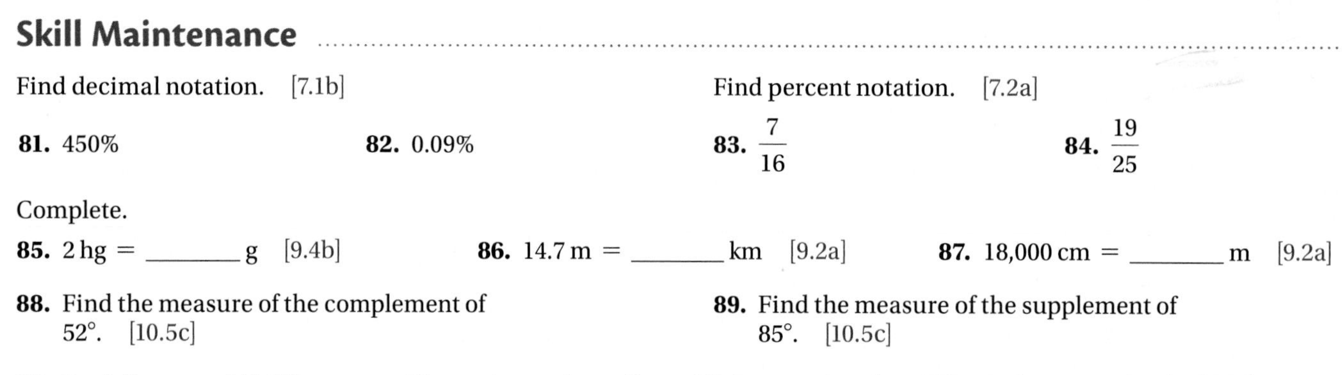

Skill Maintenance

Find decimal notation. [7.1b]

81. 450%

82. 0.09%

Find percent notation. [7.2a]

83. $\frac{7}{16}$

84. $\frac{19}{25}$

Complete.

85. 2 hg = ________ g [9.4b]

86. 14.7 m = ________ km [9.2a]

87. 18,000 cm = ________ m [9.2a]

88. Find the measure of the complement of 52°. [10.5c]

89. Find the measure of the supplement of 85°. [10.5c]

90. Fredrika earns $42,100 one year. The next year, she suffers a 6% decrease in salary. What is the new salary? [7.5a]

Synthesis

Solve.

91. $\frac{y - 2}{3} = \frac{2 - y}{5}$

92. $3x = 4x$

93. $\frac{5 + 2y}{3} = \frac{25}{12} + \frac{5y + 3}{4}$

94. 🖩 $0.05y - 1.82 = 0.708y - 0.504$

95. $\frac{2}{3}(2x - 1) = 10$

96. $\frac{2}{3}\left(\frac{7}{8} - 4x\right) - \frac{5}{8} = \frac{3}{8}$

97. The perimeter of the figure shown is 15 cm. Solve for x.

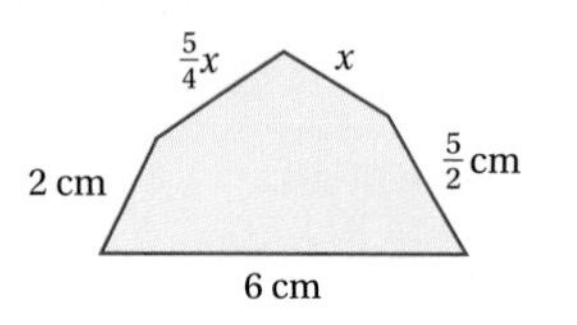

Applications and Problem Solving

11.5

OBJECTIVES

a Translate phrases to algebraic expressions.

b Solve applied problems by translating to equations.

a TRANSLATING TO ALGEBRAIC EXPRESSIONS

In algebra, we translate problems to equations. The different parts of an equation are translations of word phrases to algebraic expressions. To translate, it helps to learn which words translate to certain operation symbols.

Key Words

ADDITION (+)	SUBTRACTION (−)	MULTIPLICATION (·)	DIVISION (÷)
add	subtract	multiply	divide
added to	subtracted from	multiplied by	quotient
sum	difference	product	divided by
total	minus	times	
plus	less than	of	
more than	decreased by		
increased by	take away		

EXAMPLE 1 Translate to an algebraic expression:

Twice (or two times) some number.

Think of some number—say, 8. We can write 2 times 8 as 2×8, or $2 \cdot 8$. We multiplied by 2. To translate to an algebraic expression, we do the same thing using a variable. We can use any variable we wish, such as x, y, m, or n. Let's use y to stand for some number. If we multiply y by 2, we get

$2 \times y$, $2 \cdot y$, or $2y$.

In algebra, $2y$ is the expression used most often.

EXAMPLE 2 Translate to an algebraic expression:

Seven less than some number.

We let $x =$ the number. If the number were 10, then 7 less than 10 would be $10 - 7$, or 3. If we knew the number to be 34, then 7 less than the number would be $34 - 7$. Thus if the number is x, then the translation is

$x - 7.$

Caution!

Note that $7 - x$ is not a correct translation of the expression in Example 2. The expression $7 - x$ is a translation of "seven minus some number."

EXAMPLE 3 Translate to an algebraic expression:

Eighteen more than a number.

We let $t =$ the number. Now if the number were 26, the translation would be $18 + 26$, or $26 + 18$. If we knew the number to be 174, then the translation would be $18 + 174$, or $174 + 18$. The translation we want is

Eighteen more than a number

$18 + t$, or $t + 18$.

Translate to an algebraic expression.

1. Twelve less than some number
2. Twelve more than some number
3. Four less than some number
4. Half of some number
5. Six more than eight times some number
6. The difference of two numbers
7. Fifty-nine percent of some number
8. Two hundred less than the product of two numbers
9. The sum of two numbers

EXAMPLE 4 Translate to an algebraic expression:

A number divided by 5.

We let $m =$ the number. If the number were 8, then the translation would be $8 \div 5$, or $8/5$, or $\frac{8}{5}$. If the number were 213, then the translation would be $213 \div 5$, or $213/5$, or $\frac{213}{5}$. The translation is found as follows:

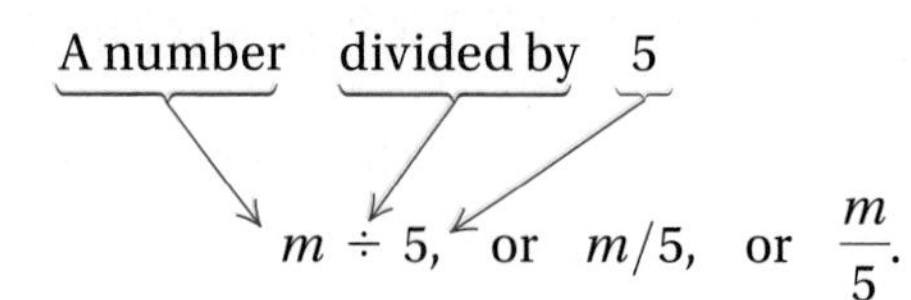

$$m \div 5, \text{ or } m/5, \text{ or } \frac{m}{5}.$$

EXAMPLE 5 Translate to an algebraic expression.

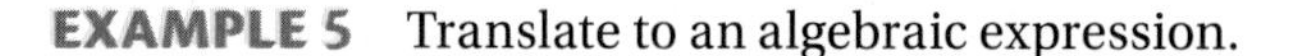

PHRASE	ALGEBRAIC EXPRESSION
Five more than some number	$5 + n$, or $n + 5$
Half of a number	$\frac{1}{2}t$, or $\frac{t}{2}$
Five more than three times some number	$5 + 3p$, or $3p + 5$
The difference of two numbers	$x - y$
Six less than the product of two numbers	$mn - 6$
Seventy-six percent of some number	$76\%z$, or $0.76z$

◀ **Do Exercises 1–9.**

b FIVE STEPS FOR SOLVING PROBLEMS

We have introduced many new equation-solving tools in this chapter. We now apply them to problem solving. We have purposely used the following strategy throughout this text in order to introduce you to algebra.

FIVE STEPS FOR PROBLEM SOLVING IN ALGEBRA

1. *Familiarize* yourself with the problem situation.
2. *Translate* to an equation.
3. *Solve* the equation.
4. *Check* your possible answer in the original problem.
5. *State* the answer clearly.

Of the five steps, the most important is probably the first one: becoming familiar with the problem situation. The following box lists some hints for familiarization.

Answers

1. $x - 12$ 2. $y + 12$, or $12 + y$
3. $m - 4$ 4. $\frac{1}{2} \cdot p$, or $\frac{p}{2}$
5. $6 + 8x$, or $8x + 6$ 6. $a - b$
7. $59\%x$, or $0.59x$ 8. $xy - 200$
9. $p + q$

To familiarize yourself with a problem:

- If a problem is given in words, read it carefully. Reread the problem, perhaps aloud. Try to verbalize the problem as though you were explaining it to someone else.
- Choose a variable (or variables) to represent the unknown and clearly state what the variable represents. Be descriptive! For example, let $L =$ the length in feet, $d =$ the distance in miles, and so on.
- Make a drawing and label it with known information, using specific units if given. Also, indicate unknown information.
- Find further information. Look up formulas or definitions with which you are not familiar. (Geometric formulas appear on the inside back cover of this text.) Consult a reference librarian or the Internet.
- Create a table that lists all the information you have available. Look for patterns that may help in the translation to an equation.
- Guess what the answer might be and check the guess. Note what you do in checking your guess. This will probably help you in translating to an equation.

EXAMPLE 6 *Cycling in Vietnam.* National Highway 1, which runs along the coast of Vietnam, is considered one of the top routes for avid bicyclists. While on sabbatical, a history professor spent six weeks biking 1720 km on National Highway 1 from Hanoi through Ha Tinh to Ho Chi Minh City. At Ha Tinh, he was four times as far from Ho Chi Minh City as he was from Hanoi. How far had he biked, and how far did he still need to bike in order to reach Ho Chi Minh City?

Sources: www.smh.com; *Lonely Planet's Best in 2010*

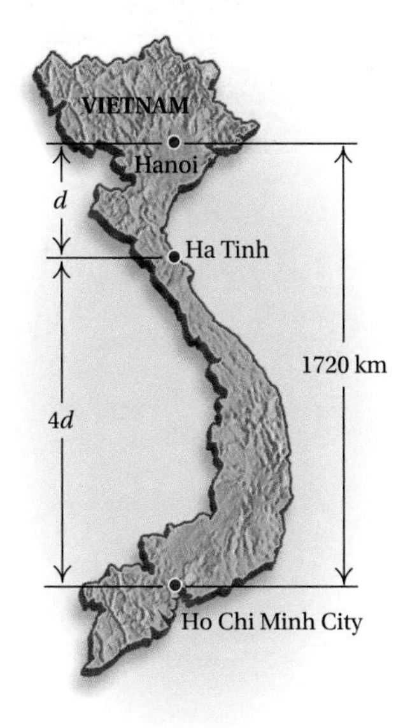

1. **Familiarize.** Let's look at a map, shown at right. To become familiar with the problem, let's guess a possible distance that the professor is from Hanoi—say, 400 km. Four times 400 km is 1600 km. Since 400 km + 1600 km = 2000 km and 2000 km is greater than 1720 km, we see that our guess is too large. Rather than guess again, let's use the equation-solving tools that we learned in this chapter. We let

   $d =$ the distance, in kilometers, from Ha Tinh to Hanoi, and

   $4d =$ the distance, in kilometers, from Ha Tinh to Ho Chi Minh City.

   (We also could let $d =$ the distance from Ha Tinh to Ho Chi Minh City and $\frac{1}{4}d =$ the distance from Ha Tinh to Hanoi.)

2. **Translate.** From the map, we see that the lengths of the two parts of the trip must add up to 1720 km. This leads to our translation.

Distance to Hanoi	plus	Distance to Ho Chi Minh City	is	1720 km.
↓	↓	↓	↓	↓
d	$+$	$4d$	$=$	1720

10. *Running.* Yiannis Kouros of Australia set the record for the greatest distance run in 24 hr by running 188 mi. After 8 hr, he was approximately twice as far from the finish line as he was from the start. How far had he run?

Source: Australian Ultra Runners Association

Answer

10. $62\frac{2}{3}$ mi

3. **Solve.** We solve the equation:

$$
\begin{aligned}
d + 4d &= 1720 \\
5d &= 1720 && \text{Collecting like terms} \\
\frac{5d}{5} &= \frac{1720}{5} && \text{Dividing by 5} \\
d &= 344.
\end{aligned}
$$

4. **Check.** As we expected, d is less than 400 km. If $d = 344$ km, then $4d = 1376$ km. Since 344 km + 1376 km = 1720 km, the answer checks.

5. **State.** At Ha Tinh, the professor had biked 344 km from Hanoi and had 1376 km to go to reach Ho Chi Minh City.

Do Exercise 10.

EXAMPLE 7 *Knitted Scarf.* Lilly knitted a scarf with orange and red yarn, starting with an orange section, then a medium-red section, and finally a dark-red section. The medium-red section is one-half the length of the orange section. The dark-red section is one-fourth the length of the orange section. The scarf is 7 ft long. Find the length of each section of the scarf.

1. **Familiarize.** Because the lengths of the medium-red section and the dark-red section are expressed in terms of the length of the orange section, we let

x = the length of the orange section.

Then $\frac{1}{2}x$ = the length of the medium-red section,

and $\frac{1}{4}x$ = the length of the dark-red section.

We make a drawing and label it.

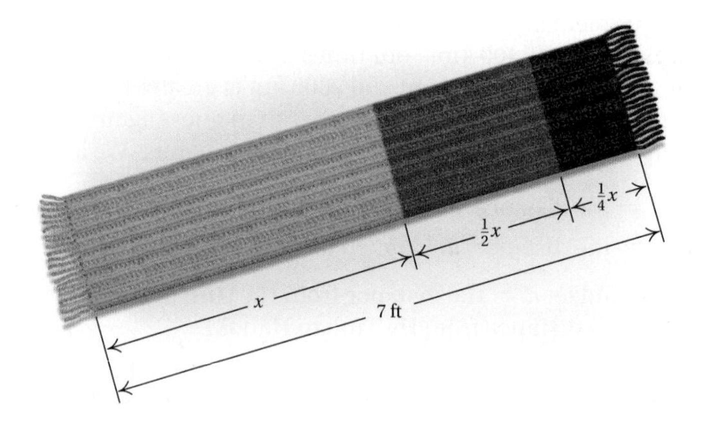

2. **Translate.** From the statement of the problem and the drawing, we know that the lengths add up to 7 ft. This gives us our translation:

Length of orange section	plus	Length of medium-red section	plus	Length of dark-red section	is	Total length
x	$+$	$\frac{1}{2}x$	$+$	$\frac{1}{4}x$	$=$	$7.$

3. **Solve.** First, we clear fractions and then carry out the solution as follows:

$x + \frac{1}{2}x + \frac{1}{4}x = 7$	The LCM of the denominators is 4.
$4\left(x + \frac{1}{2}x + \frac{1}{4}x\right) = 4 \cdot 7$	Multiplying by the LCM, 4
$4 \cdot x + 4 \cdot \frac{1}{2}x + 4 \cdot \frac{1}{4}x = 4 \cdot 7$	Using the distributive law
$4x + 2x + x = 28$	Simplifying
$7x = 28$	Collecting like terms
$\frac{7x}{7} = \frac{28}{7}$	Dividing by 7
$x = 4.$	

4. **Check.** Do we have an answer to the *original problem*? If the length of the orange section is 4 ft, then the length of the medium-red section is $\frac{1}{2} \cdot 4$ ft, or 2 ft, and the length of the dark-red section $\frac{1}{4} \cdot 4$ ft, or 1 ft. The sum of these lengths is 7 ft, so the answer checks.
5. **State.** The length of the orange section is 4 ft, the length of the medium-red section is 2 ft, and the length of the dark-red section is 1 ft. (Note that we must include the unit, feet, in the answer.)

Do Exercise 11. ▶

11. *Gourmet Sandwiches.* A sandwich shop specializes in sandwiches prepared in buns of length 18 in. Suppose Jenny, Emma, and Sarah buy one of these sandwiches and take it back to their apartment. Since they have different appetites, Jenny cuts the sandwich in such a way that Emma gets one-half of what Jenny gets and Sarah gets three-fourths of what Jenny gets. Find the length of each person's sandwich.

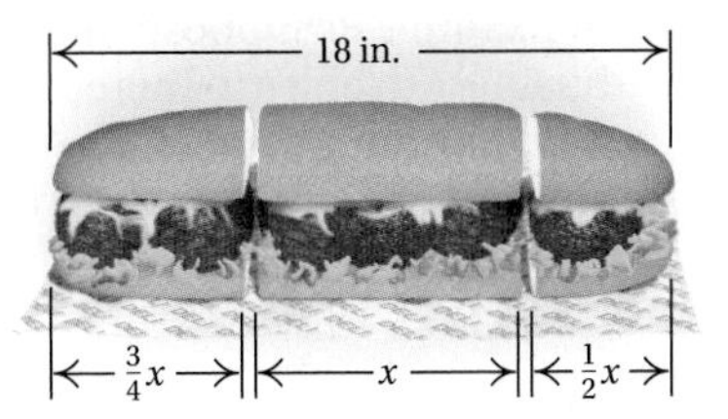

EXAMPLE 8 *Delivery Truck Rental.* An appliance business needs to rent a delivery truck for 6 days while one of its trucks is being repaired. The cost of renting a 16-ft truck is \$29.95 per day plus \$0.29 per mile. If \$550 is budgeted for the rental, how many miles can the truck be driven without exceeding the budget?

1. **Familiarize.** Suppose the truck is driven 1100 mi. Then since the cost is given by the daily charge plus the mileage charge, we have

Daily cost	plus	cost per mile	times	Number of miles
↓	↓	↓	↓	↓
6(\$29.95)	+	\$0.29	·	1100,

which is \$498.70. We see that the truck can be driven more than 1100 mi without exceeding the budget of \$550. This process of guessing familiarizes us with the way the calculation is done.

We let m = the number of miles that can be driven on the budgeted amount of \$550.

Answer

11. Jenny: 8 in.; Emma: 4 in; Sarah: 6 in.

12. *Delivery Truck Rental.* Refer to Example 8. Suppose the business decides to increase its 6-day rental budget to \$625. How many miles can the truck be driven for \$625?

2. **Translate.** We reword the problem and translate as follows:

Daily cost	plus	Cost per mile	times	Number of miles	is	Budgeted amount
6(\$29.95)	$+$	\$0.29	$\cdot$	m	$=$	\$550.

3. **Solve.** We solve the equation:

$$6(29.95) + 0.29m = 550$$
$$179.70 + 0.29m = 550$$
$$0.29m = 370.30 \quad \text{Subtracting 179.70}$$
$$\frac{0.29m}{0.29} = \frac{370.30}{0.29} \quad \text{Dividing by 0.29}$$
$$m = 1277. \quad \text{Rounding to the nearest one}$$

4. **Check.** We check our answer in the original problem. The cost of driving 1277 mi is 1277(\$0.29) = \$370.33. The rental charge for 6 days is 6(\$29.95) = \$179.70. The total cost is then \$370.33 + \$179.70 ≈ \$550, which is the budgeted amount.
5. **State.** The truck can be driven 1277 mi for the budgeted amount of \$550.

◀ **Do Exercise 12.**

EXAMPLE 9 *Perimeter of a Lacrosse Field.* The perimeter of a lacrosse field is 340 yd. The length is 50 yd more than the width. Find the dimensions of the field.

Source: www.sportsknowhow.com

1. **Familiarize.** We first make a drawing.

We let w = the width of the rectangle. Then $w + 50$ = the length. The perimeter P of a rectangle is the distance around the rectangle and is given by the formula $2l + 2w = P$, where

l = the length and w = the width.

2. **Translate.** To translate the problem, we substitute $w + 50$ for l and 340 for P:

$$2l + 2w = P$$
$$2(w + 50) + 2w = 340.$$

Caution! Parentheses are necessary here.

Answer

12. 1536 mi

3. Solve. We solve the equation:

$$2(w + 50) + 2w = 340.$$
$$2 \cdot w + 2 \cdot 50 + 2w = 340 \quad \text{Using the distributive law}$$
$$4w + 100 = 340 \quad \text{Collecting like terms}$$
$$4w + 100 - 100 = 340 - 100 \quad \text{Subtracting 100}$$
$$4w = 240$$
$$\frac{4w}{4} = \frac{240}{4} \quad \text{Dividing by 4}$$
$$w = 60.$$

Thus possible dimensions are

$$w = 60 \text{ yd} \quad \text{and} \quad l = w + 50 = 60 + 50, \text{ or } 110 \text{ yd}.$$

4. Check. If the width is 60 yd and the length is 110 yd, then the perimeter is $2(60\text{ yd}) + 2(110\text{ yd})$, or 340 yd. This checks.

5. State. The width is 60 yd, and the length is 110 yd.

Do Exercise 13. ▶

13. *Perimeter of a High School Basketball Court.* The perimeter of a standard high school basketball court is 268 ft. The length is 34 ft longer than the width. Find the dimensions of the court.

Source: Indiana High School Athletic Association

EXAMPLE 10 *Angles of a Triangle.* The second angle of a triangle is twice as large as the first. The measure of the third angle is 20° greater than that of the first angle. How large are the angles?

1. Familiarize. We first make a drawing. Since the second and third angles are described in terms of the first angle, we let

the measure of the first angle $= x$.

Then the measure of the second angle $= 2x$,

and the measure of the third angle $= x + 20$.

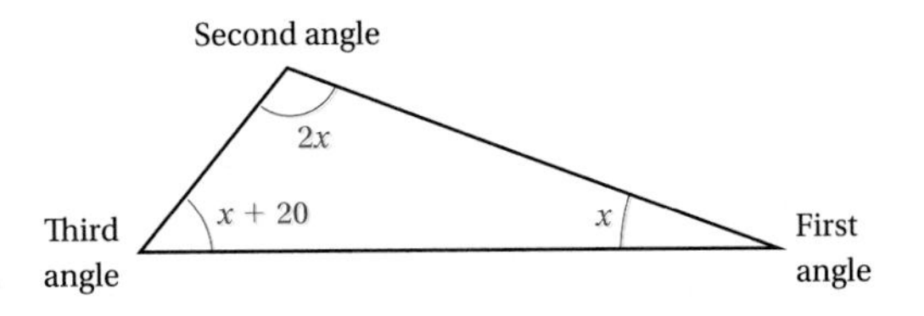

2. Translate. To translate, we recall from Section 10.5 that the sum of the measures of the angles of a triangle is 180°.

Measure of first angle	+	Measure of second angle	+	Measure of third angle	=	180°
x	$+$	$2x$	$+$	$(x + 20)$	$=$	180

3. Solve. We solve the equation:

$$x + 2x + (x + 20) = 180$$
$$4x + 20 = 180$$
$$4x + 20 - 20 = 180 - 20$$
$$4x = 160$$
$$\frac{4x}{4} = \frac{160}{4}$$
$$x = 40.$$

Answer

13. Length: 84 ft; width: 50 ft

Possible measures for the angles are as follows:

First angle: $x = 40°$;
Second angle: $2x = 2(40) = 80°$;
Third angle: $x + 20 = 40 + 20 = 60°$.

4. Check. Consider 40°, 80°, and 60°. The second is twice the first, and the third is 20° greater than the first. The sum is 180°. These numbers check.

5. State. The measures of the angles are 40°, 80°, and 60°.

14. *Angles of a Triangle.* The second angle of a triangle is three times as large as the first. The third angle measures 30° more than the first angle. Find the measures of the angles.

Do Exercise 14.

EXAMPLE 11 *Fastest Roller Coasters.* The average top speed of the three fastest steel roller coasters in the United States is 116 mph. The third-fastest roller coaster, Superman: The Escape (located at Six Flags Magic Mountain, Valencia, California), reaches a top speed that is 28 mph less than that of the fastest roller coaster, Kingda Ka (located at Six Flags Great Adventure, Jackson, New Jersey). The second-fastest roller coaster, Top Thrill Dragster (located at Cedar Point, Sandusky, Ohio), has a top speed of 120 mph. What is the top speed of the fastest steel roller coaster?

Source: Coaster Grotto

1. Familiarize. The average of a set of numbers is the sum of the numbers divided by the number of addends.

We are given that the second-fastest speed is 120 mph. Suppose the three top speeds are 109, 120, and 125. The average is then

$$\frac{109 + 120 + 125}{3} = \frac{354}{3} = 118,$$

which is too high. Instead of continuing to guess, let's use our equation-solving skills. We let $x =$ the top speed of the fastest roller coaster. Then $x - 28 =$ the top speed of the third-fastest roller coaster.

2. Translate. We reword the problem and translate as follows:

$$\frac{\text{Speed of fastest coaster} + \text{Speed of second-fastest coaster} + \text{Speed of third-fastest coaster}}{\text{Number of roller coasters}} = \text{Average speed of three fastest roller coasters}$$

$$\frac{x + 120 + (x - 28)}{3} = 116.$$

3. Solve. We solve as follows:

$$\frac{x + 120 + (x - 28)}{3} = 116$$

$$3 \cdot \frac{x + 120 + (x - 28)}{3} = 3 \cdot 116 \quad \text{Multiplying by 3 on both sides to clear the fraction}$$

$$x + 120 + (x - 28) = 348$$

$$2x + 92 = 348 \quad \text{Collecting like terms}$$

$$2x = 256 \quad \text{Subtracting 92}$$

$$x = 128. \quad \text{Dividing by 2}$$

Answer

14. 30°, 90°, 60°

4. **Check.** If the top speed of the fastest roller coaster is 128 mph, then the top speed of the third-fastest is $128 - 28$, or 100 mph. The average of the top speeds of the three fastest is $(128 + 120 + 100) \div 3 = 348 \div 3$, or 116 mph. The answer checks.

5. **State.** The top speed of the fastest steel roller coaster in the United States is 128 mph.

Do Exercise 15. ▶

15. *Average Test Score.* Sam's average score on his first three math tests is 77. He scored 62 on the first test. On the third test, he scored nine points more than he scored on his second test. What did he score on the second and third test?

EXAMPLE 12 *Simple Interest.* An investment is made at 3% simple interest for 1 year. It grows to \$746.75. How much was originally invested (the principal)?

1. **Familiarize.** Suppose that \$100 was invested. Recalling the formula for simple interest, $I = Prt$, we know that the interest for 1 year on \$100 at 3% simple interest is given by $I = \$100 \cdot 0.03 \cdot 1 = \$3$. Then, at the end of the year, the amount in the account is found by adding the principal and the interest:

$$\begin{array}{ccccc} \text{Principal} & + & \text{Interest} & = & \text{Amount} \\ \downarrow & & \downarrow & & \downarrow \\ \$100 & + & \$3 & = & \$103. \end{array}$$

In this problem, we are working backward. We are trying to find the principal, which is the original investment. We let $x =$ the principal. Then the interest earned is $3\%x$.

2. **Translate.** We reword the problem and then translate:

$$\begin{array}{ccccc} \text{Principal} & + & \text{Interest} & = & \text{Amount} \\ \downarrow & & \downarrow & & \downarrow \\ x & + & 3\%x & = & 746.75. \end{array}$$

Interest is 3% of the principal.

3. **Solve.** We solve the equation:

$$x + 3\%x = 746.75$$

$$x + 0.03x = 746.75 \quad \text{Converting to decimal notation}$$

$$1x + 0.03x = 746.75 \quad \text{Identity property of 1}$$

$$(1 + 0.03)x = 746.75 \quad \text{Collecting like terms}$$

$$1.03x = 746.75$$

$$\frac{1.03x}{1.03} = \frac{746.75}{1.03} \quad \text{Dividing by 1.03}$$

$$x = 725.$$

4. **Check.** We check by taking 3% of \$725 and then adding it to \$725:

$$3\% \times \$725 = 0.03 \times 725 = \$21.75.$$

Then $\$725 + \$21.75 = \$746.75$, so \$725 checks.

5. **State.** The original investment was \$725.

Do Exercise 16. ▶

GS 16. **Simple Interest.** An investment is made at 5% simple interest for 1 year. It grows to \$2520. How much was originally invested (the principal)? Let $x =$ the principal. Then the interest earned is $5\%x$.

Translate and *Solve*:

$$\begin{array}{ccccc} \text{Principal} & + & \text{Interest} & = & \text{Amount} \\ \downarrow & \downarrow & \downarrow & \downarrow & \downarrow \\ x & + & \square & = & 2520 \end{array}$$

$$x + 0.05x = 2520$$

$$(1 + \square)x = 2520$$

$$\square x = 2520$$

$$\frac{1.05x}{1.05} = \frac{2520}{\square}$$

$$x = 2400.$$

The principal is \$ $\square$.

Answers

15. Second: 80; third: 89 **16.** 2400

Guided Solution:

16. $5\%x$, 0.05, 1.05, 1.05; 2400

Translating for Success

1. *Angle Measures.* The measure of the second angle of a triangle is 51° more than that of the first angle. The measure of the third angle is 3° less than twice that of the first angle. Find the measures of the angles.

2. *Sales Tax.* Tina paid $3976 for a used car. This amount included 5% for sales tax. How much did the car cost before tax?

3. *Perimeter.* The perimeter of a rectangle is 2347 ft. The length is 28 ft greater than the width. Find the length and the width.

4. *Fraternity or Sorority Membership.* At Arches Tech University, 3976 students belong to a fraternity or a sorority. This is 35% of the total enrollment. What is the total enrollment at Arches Tech?

5. *Fraternity or Sorority Membership.* At Moab Tech University, 35% of the students belong to a fraternity or a sorority. The total enrollment of the university is 11,360 students. How many students belong to either a fraternity or a sorority?

The goal of these matching questions is to practice step (2), Translate, of the five-step problem-solving process. Translate each word problem to an equation and select a correct translation from equations A–O.

A. $x + (x - 3) + \frac{4}{5}x = 384$

B. $x + (x + 51) + (2x - 3) = 180$

C. $x + (x + 96{,}000) = 180{,}000$

D. $2 \cdot 96 + 2x = 3976$

E. $x + (x + 1) + (x + 2) = 384$

F. $3976 = x \cdot 11{,}360$

G. $2x + 2(x + 28) = 2347$

H. $3976 = x + 5\%x$

I. $x + (x + 28) = 2347$

J. $x = 35\% \cdot 11{,}360$

K. $x + 96 = 3976$

L. $x + (x + 3) + \frac{4}{5}x = 384$

M. $x + (x + 2) + (x + 4) = 384$

N. $35\% \cdot x = 3976$

O. $2x + (x + 28) = 2347$

Answers on page A-19.

6. *Island Population.* There are 180,000 people living on a small Caribbean island. The women outnumber the men by 96,000. How many men live on the island?

7. *Wire Cutting.* A 384-m wire is cut into three pieces. The second piece is 3 m longer than the first. The third is four-fifths as long as the first. How long is each piece?

8. *Locker Numbers.* The numbers on three adjoining lockers are consecutive integers whose sum is 384. Find the integers.

9. *Fraternity or Sorority Membership.* The total enrollment at Canyonlands Tech University is 11,360 students. Of these, 3976 students belong to a fraternity or a sorority. What percent of the students belong to a fraternity or a sorority?

10. *Width of a Rectangle.* The length of a rectangle is 96 ft. The perimeter of the rectangle is 3976 ft. Find the width.

11.5 Exercise Set

For Extra Help MyMathLab® MathXL® PRACTICE WATCH READ REVIEW

Reading Check

Complete each of the five steps for problem solving using one of the words in the list at the right.

RC1. _______________ yourself with the problem situation.

RC2. _______________ to an equation.

RC3. _______________ the equation.

RC4. _______________ your possible answer in the original problem.

RC5. _______________ the answer clearly.

Solve
Familiarize
State
Translate
Check

a Translate to an algebraic expression.

1. Three less than twice a number

2. Three times a number divided by a

3. The product of 97% and some number

4. 43% of some number

5. Four more than five times some number

6. Seventy-five less than eight times a number

b Solve.

7. What number added to 85 is 117?

8. Eight times what number is 2552?

9. *Medals of Honor.* In 1863, the U.S. Secretary of War presented the first Medals of Honor for valor. The two wars for which the most Medals of Honor were given are the Civil War and World War II. There were 464 recipients of the medal for World War II. This number is 1058 fewer than the number of recipients for the Civil War. How many Medals of Honor were awarded in the Civil War?

Sources: U.S. Army Center of Military History; U.S. Department of Defense

10. *Milk Alternatives.* Milk alternatives, such as rice, soy, almond, and flax milk, are becoming more available and increasingly popular. A cup of almond milk contains only 60 calories. This number is 89 calories less than the number of calories in a cup of whole milk. How many calories are in a cup of whole milk?

Source: "Nutrition Udder Chaos," by Janet Kinosian, *AARP Magazine*, August/September, 2012

11. When 17 is subtracted from 4 times a certain number, the result is 211. What is the number?

12. When 36 is subtracted from 5 times a certain number, the result is 374. What is the number?

13. If you double a number and then add 16, you get $\frac{2}{3}$ of the original number. What is the original number?

14. If you double a number and then add 85, you get $\frac{3}{4}$ of the original number. What is the original number?

15. *500 Festival Mini-Marathon.* On May 4, 2013, 35,000 runners participated in the 13.1-mi One America 500 Festival Mini-Marathon. If a runner stopped at a water station that was twice as far from the start as from the finish, how far was the runner from the finish line? Round the answer to the nearest hundredth of a mile.

Source: www.500festival.com

16. *Airport Control Tower.* At a height of 385 ft, the FAA traffic control tower in the Atlanta airport is the tallest such tower in the United States. Its height is 59 ft more than the height of the tower at the Memphis airport. How tall is the control tower at the Memphis airport?

Source: Federal Aviation Administration

17. *Car Rental.* Value Rent-A-Car rents a family-sized car at a daily rate of $69.95 plus 40¢ per mile. Rick is allotted a daily budget of $200. How many miles can he drive per day and stay within his budget?

18. *Van Rental.* Value Rent-A-Car rents a van at a daily rate of $84.95 plus 60¢ per mile. Molly rents a van to deliver electrical parts to her customers. She is allotted a daily budget of $250. How many miles can she drive per day and stay within her budget?

19. *Average Test Score.* Mariana averaged 84 on her first three history exams. The first score was 67. The second score was 7 less than the third score. What did she score on the second and third exams?

20. *Average Price.* David paid an average of $34 per shirt for a recent purchase of three shirts. The price of one shirt was twice as much as another, and the remaining shirt cost $27. What were the prices of the other two shirts?

21. ***Photo Size.*** A hotel purchases a large photo for its newly renovated lobby. The perimeter of the photo is 292 in. The width is 2 in. more than three times the height. Find the dimensions of the photo.

22. ***Two-by-Four.*** The perimeter of a cross section of a "two-by-four" piece of lumber is 10 in. The length is 2 in. more than the width. Find the actual dimensions of the cross section of a two-by-four.

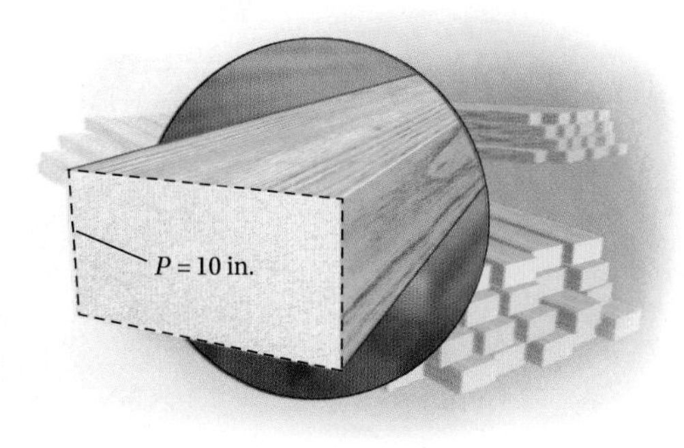

23. ***Public Transit Systems.*** In the first quarter of 2012, the ridership for the public transit system in Boston was 99.2 million. This number is 77.4 million more than the ridership in San Diego over the same period of time. What was the ridership in San Diego during the first quarter of 2012?

Source: American Public Transportation Association

24. ***Home Listing Prices.*** In 2011, the average listing price of a home in Hawaii was $72,000 more than three times the average listing price of a home in Arizona. The average listing price of a home in Hawaii was $876,000. What was the average listing price of a home in Arizona?

Source: Trulia

25. ***Statue of Liberty.*** The height of the Eiffel Tower is 974 ft, which is about 669 ft higher than the Statue of Liberty. What is the height of the Statue of Liberty?

26. ***Area of Lake Ontario.*** The area of Lake Superior is about four times the area of Lake Ontario. The area of Lake Superior is 30,172 mi^2. What is the area of Lake Ontario?

27. ***Pipe Cutting.*** A 480-m pipe is cut into three pieces. The second piece is three times as long as the first. The third piece is four times as long as the second. How long is each piece?

28. ***Rope Cutting.*** A 180-ft rope is cut into three pieces. The second piece is twice as long as the first. The third piece is three times as long as the second. How long is each piece of rope?

29. ***Angles of a Triangle.*** The second angle of a triangular field is three times as large as the first. The third angle is 40° greater than the first. How large are the angles?

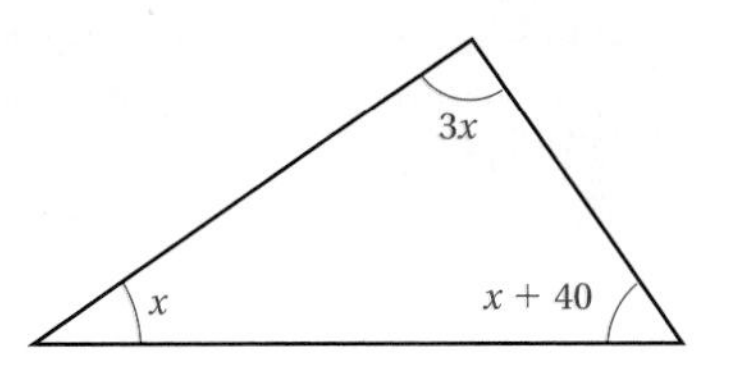

30. ***Angles of a Triangle.*** The second angle of a triangular parking lot is four times as large as the first. The third angle is 45° less than the sum of the other two angles. How large are the angles?

31. *Taxi Fares.* In New Orleans, Louisiana, taxis charge an initial fee of \$3.50 plus \$2.00 per mile. How far can a passenger travel for \$39.50?

Source: www.taxifarefinders.com

32. *Taxi Fares.* In Baltimore, Maryland, taxis charge an initial fee of \$1.80 plus \$2.20 per mile. How far can a passenger travel for \$26?

Source: www.taxifarefinders.com

33. *Stock Prices.* Diego's investment in a technology stock grew 28% to \$448. How much did he invest?

34. *Savings Interest.* Ella invested money in a savings account at a rate of 6% simple interest. After 1 year, she has \$6996 in the account. How much did Ella originally invest?

35. *Credit Cards.* The balance in Will's Mastercard® account grew 2%, to \$870, in one month. What was his balance at the beginning of the month?

36. *Loan Interest.* Alvin borrowed money from a cousin at a rate of 10% simple interest. After 1 year, \$7194 paid off the loan. How much did Alvin borrow?

37. *Price of a Security Wallet.* Carla paid \$26.70, including a 7% sales tax, for a security wallet. How much did the wallet itself cost?

38. *Price of a Car Battery.* Tyler paid \$117.15, including a 6.5% sales tax, for a car battery. How much did the battery itself cost?

39. *Tipping.* Leon left a 15% tip for a meal. The total cost of the meal, including the tip, was $41.40. What was the cost of the meal before the tip was added?

40. *Tipping.* Selena left an 18% tip for a meal. The total cost of the meal, including the tip, was $40.71. What was the cost of the meal before the tip was added?

41. *Hancock Building Dimensions.* The ground floor of the John Hancock Building in Chicago is a rectangle whose length is 100 ft more than the width. The perimeter is 860 ft. Find the length, the width, and the area of the ground floor.

42. *Hancock Building Dimensions.* The top floor of the John Hancock Building in Chicago is in the shape of a rectangle whose length is 60 ft more than the width. The perimeter is 520 ft. Find the length, the width, and the area of the top floor.

Skill Maintenance

Calculate.

43. $-\frac{4}{5} - \frac{3}{8}$ [4.3a]

44. $-\frac{4}{5} + \frac{3}{8}$ [4.2a]

45. $-\frac{4}{5} \cdot \frac{3}{8}$ [3.6a]

46. $-\frac{4}{5} \div \left(\frac{3}{8}\right)$ [3.7b]

47. $-25.6 \div (-16)$ [5.4a]

48. $-25.6(-16)$ [5.3a]

49. $-25.6 - (-16)$ [5.2c]

50. $-25.6 + (-16)$ [5.2c]

Synthesis

51. The width of a rectangle is $\frac{3}{4}$ of the length. The perimeter of the rectangle becomes 50 cm when the length and the width are each increased by 2 cm. Find the original length and the original width.

52. Cookies are set out on a tray for six people to take home. One-third, one-fourth, one-eighth, and one-fifth are given to four people, respectively. The fifth person is given ten cookies, leaving one cookie remaining for the sixth person. Find the original number of cookies on the tray.

53. Susanne went to the bank to get $20 in quarters, dimes, and nickels to use to make change at her yard sale. She got twice as many quarters as dimes and 10 more nickels than dimes. How many of each type of coin did she get?

54. A student has an average score of 82 on three tests. His average score on the first two tests is 85. What was the score on the third test?

11 Summary and Review

Vocabulary Reinforcement

Complete each statement with the appropriate word or phrase from the column on the right. Some of the choices may not be used.

1. When we replace a variable with a number, we say that we are ________________ for the variable. [11.1a]
2. A letter that stands for just one number is called a ________________. [11.1a]
3. The ________________ states that for any real number a, $a \cdot 1 = 1 \cdot a = a$. [11.1b]
4. The ________________ for solving equations states that for any real numbers a, b, and c, where $c \neq 0$, $a = b$ is equivalent to $a \cdot c = b \cdot c$. [11.3a]
5. The ________________ states that for any numbers a, b, and c, $a(b - c) = ab - ac$. [11.1b]
6. The ________________ for solving equations states that for any real numbers a, b, and c, $a = b$ is equivalent to $a + c = b + c$. [11.2a]
7. Equations with the same solutions are called ________________ equations. [11.2a]

addition principle
multiplication principle
identity property of 1
distributive law of multiplication over subtraction
distributive law of multiplication over addition
equivalent
substituting
variable
constant

Concept Reinforcement

Determine whether each statement is true or false.

________ 1. The expression $x - 7$ is not equivalent to the expression $7 - x$. [11.5a]

________ 2. $3y$ and $3y^2$ are like terms. [11.1d]

________ 3. The equations $x + 5 = 2$ and $x = 3$ are equivalent. [11.2a]

________ 4. We can use the multiplication principle to divide on both sides of an equation by the same nonzero number. [11.3a]

Study Guide

Objective 11.1a Evaluate an algebraic expression by substitution.

Example Evaluate $\frac{a - b}{6}$, when $a = 21$ and $b = -15$.

$$\frac{a - b}{6} = \frac{21 - (-15)}{6} = \frac{36}{6} = 6$$

Practice Exercise

1. Evaluate $\frac{ab - 2}{7}$, when $a = -5$ and $b = 8$.

Objective 11.1b Use the distributive laws to multiply expressions like 8 and $x - y$.

Example Multiply: $3(5x - 2y + 4)$.

$$3(5x - 2y + 4) = 3 \cdot 5x - 3 \cdot 2y + 3 \cdot 4$$
$$= 15x - 6y + 12$$

Practice Exercise

2. Multiply: $4(x + 5y - 7)$.

Objective 11.1c Use the distributive laws to factor expressions like $4x - 12 + 24y$.

Example Factor: $12x - 6y + 9$.

$$12x - 6y + 9 = 3 \cdot 4x - 3 \cdot 2y + 3 \cdot 3$$
$$= 3(4x - 2y + 3)$$

Practice Exercise

3. Factor: $24a - 8b + 16$.

Objective 11.1d Collect like terms.

Example Collect like terms: $4a - 2b - 2a + b$.

$$4a - 2b - 2a + b = 4a - 2a - 2b + b$$
$$= 4a - 2a - 2b + 1 \cdot b$$
$$= (4 - 2)a + (-2 + 1)b$$
$$= 2a - b$$

Practice Exercise

4. Collect like terms: $7x + 3y - x - 6y$.

Objective 11.2a Solve equations using the addition principle.

Example Solve: $x + 6 = 8$.

$$x + 6 = 8$$
$$x + 6 - 6 = 8 - 6$$
$$x + 0 = 2$$
$$x = 2$$

The solution is 2.

Practice Exercise

5. Solve: $y - 4 = -2$.

Objective 11.3a Solve equations using the multiplication principle.

Example Solve: $45 = -5y$.

$$45 = -5y$$
$$\frac{45}{-5} = \frac{-5y}{-5}$$
$$-9 = 1 \cdot y$$
$$-9 = y$$

The solution is -9.

Practice Exercise

6. Solve: $9x = -72$.

Objective 11.4a Solve equations using both the addition principle and the multiplication principle.

Example Solve: $3x - 2 = 7$.

$$3x - 2 = 7$$
$$3x - 2 + 2 = 7 + 2$$
$$3x = 9$$
$$\frac{3x}{3} = \frac{9}{3}$$
$$x = 3$$

The solution is 3.

Practice Exercise

7. Solve: $5y + 1 = 6$.

Objective 11.4b Solve equations in which like terms may need to be collected.

Example Solve: $2y - 1 = -3y - 8 + 2$.

$$2y - 1 = -3y - 8 + 2$$
$$2y - 1 = -3y - 6$$
$$2y - 1 + 1 = -3y - 6 + 1$$
$$2y = -3y - 5$$
$$2y + 3y = -3y - 5 + 3y$$
$$5y = -5$$
$$\frac{5y}{5} = \frac{-5}{5}$$
$$y = -1$$

The solution is -1.

Practice Exercise

8. Solve: $6x - 4 - x = 2x - 10$.

Objective 11.4c Solve equations by first removing parentheses and collecting like terms.

Example Solve: $8b - 2(3b + 1) = 10$.

$$8b - 2(3b + 1) = 10$$
$$8b - 6b - 2 = 10$$
$$2b - 2 = 10$$
$$2b - 2 + 2 = 10 + 2$$
$$2b = 12$$
$$\frac{2b}{2} = \frac{12}{2}$$
$$b = 6$$

The solution is 6.

Practice Exercise

9. Solve: $2(y - 1) = 5(y - 4)$.

Objective 11.5a Translate phrases to algebraic expressions.

Example Translate to an algebraic expression: Three less than some number.

We let $n =$ the number. Now if the number were 5, then the translation would be $5 - 3$. Similarly, if the number were 35, then the translation would be $35 - 3$. Thus we see from these numerical examples that if the number were n, the translation would be

$n - 3$.

Practice Exercise

10. Translate to an algebraic expression: Five more than some number.

Review Exercises

1. Evaluate $\frac{x - y}{3}$ when $x = 17$ and $y = 5$. [11.1a]

Multiply. [11.1b]

2. $5(3x - 7)$

3. $-2(4x - 5)$

4. $10(0.4x + 1.5)$

5. $-8(3 - 6x + 2y)$

Factor. [11.1c]

6. $2x - 14$

7. $6x - 6$

8. $5x + 10$

9. $12 - 3x + 6z$

Collect like terms. [11.1d]

10. $11a + 2b - 4a - 5b$

11. $7x - 3y - 9x + 8y$

12. $6x + 3y - x - 4y$

13. $-3a + 9b + 2a - b$

Solve. [11.2a], [11.3a]

14. $x + 5 = -17$

15. $-8x = -56$

16. $-\frac{x}{4} = 48$

17. $n - 7 = -6$

18. $15x = -35$

19. $x - 11 = 14$

20. $-\frac{2}{3} + x = -\frac{1}{6}$

21. $\frac{4}{5}y = -\frac{3}{16}$

22. $y - 0.9 = 9.09$

23. $5 - x = 13$

Solve. [11.4a, b, c]

24. $5t + 9 = 3t - 1$

25. $7x - 6 = 25x$

26. $\frac{1}{4}x - \frac{5}{8} = \frac{3}{8}$

27. $14y = 23y - 17 - 10$

28. $0.22y - 0.6 = 0.12y + 3 - 0.8y$

29. $\frac{1}{4}x - \frac{1}{8}x = 3 - \frac{1}{16}x$

30. $4(x + 3) = 36$

31. $3(5x - 7) = -66$

32. $8(x - 2) - 5(x + 4) = 20x + x$

33. $-5x + 3(x + 8) = 16$

34. Translate to an algebraic expression: [11.5a]

Nineteen percent of some number.

Solve. [11.5b]

35. *Dimensions of Wyoming.* The state of Wyoming is roughly in the shape of a rectangle whose perimeter is 1280 mi. The length is 90 mi more than the width. Find the dimensions.

36. A 21-ft carpet runner is cut into two pieces. One piece is 5 ft longer than the other. Find the lengths of the pieces.

37. The Johnsons bought a lawnmower for \$2449 in June. They paid \$332 more than they would have if they had purchased the mower in February. Find what it would have cost in February.

38. Ty is paid a commission of \$8 for each small appliance he sells. One week, he received \$216 in commissions. How many appliances did he sell?

39. The measure of the second angle of a triangle is 50° more than that of the first angle. The measure of the third angle is 10° less than twice that of the first angle. Find the measures of the angles.

40. After a 30% reduction, a bread maker is on sale for \$154. What was the marked price (the price before the reduction)?

41. A tax-exempt organization received a bill of \$145.90 for office supplies. The bill incorrectly included sales tax of 5%. How much does the organization actually owe?

42. Sam's salary as an executive chef is \$71,400, which represents a 5% increase over his previous year's salary. What was his previous year's salary?

43. ***HDTV Price.*** An HDTV television sold for \$829 in May. This was \$38 less than the cost in January. What was the cost in January?

44. ***Writing Pad.*** The perimeter of a rectangular writing pad is 56 cm. The width is 6 cm less than the length. Find the width and the length.

45. ***Nile and Amazon Rivers.*** The total length of the Nile and Amazon Rivers is 13,108 km. If the Amazon were 234 km longer, it would be as long as the Nile. Find the length of each river.

Source: *The Handy Geography Answer Book*

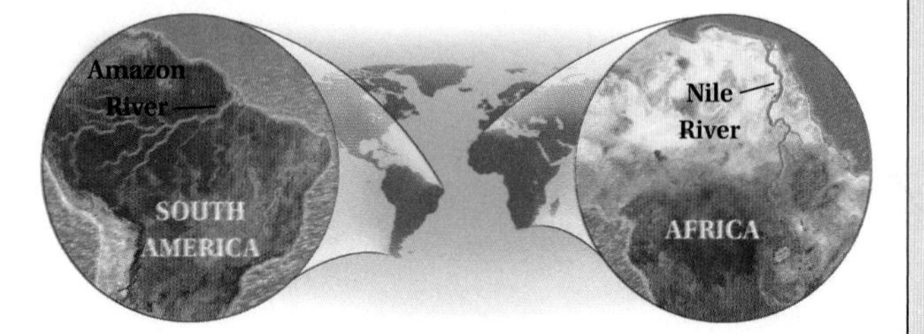

46. Factor: $6a - 30b + 3$. [11.1c]

A. $6(a - 5b)$
B. $3(2a - 10b)$
C. $3(2a - 10b + 1)$
D. $3(2a - b + 1)$

47. Collect like terms: $3x - 2y + x - 5y$. [11.1d]

A. $4x - 7y$
B. $x - 4y$
C. $4x + 3y$
D. $4x + 7y$

Synthesis

Solve. [2.1d], [11.4a]

48. $2|n| + 4 = 50$

49. $|3n| = 60$

Understanding Through Discussion and Writing

1. Explain at least three uses of the distributive laws considered in this chapter. [11.1b, c, d]

2. Explain the role of the opposite of a number when using the addition principle. [11.2a]

3. Explain the role of the reciprocal of a number when using the multiplication principle. [11.3a]

4. Describe a procedure that a classmate could use to solve the equation $ax + b = c$ for x. [11.4a]

CHAPTER 11 Test

For Extra Help For step-by-step test solutions, access the Chapter Test Prep Videos in MyMathLab® or on YouTube (search "BittingerBasicEI" and click on "Channels").

1. Evaluate $\frac{3x}{y}$ when $x = 10$ and $y = 5$.

Multiply.

2. $3(6 - x)$

3. $-5(y - 1)$

Factor.

4. $12 - 22x$

5. $7x + 21 + 14y$

Collect like terms.

6. $9x - 2y - 14x + y$

7. $-a + 6b + 5a - b$

Solve.

8. $x + 7 = 15$

9. $t - 9 = 17$

10. $3x = -18$

11. $-\frac{4}{7}x = -28$

12. $3t + 7 = 2t - 5$

13. $\frac{1}{2}x - \frac{3}{5} = \frac{2}{5}$

14. $8 - y = 16$

15. $-\frac{2}{5} + x = -\frac{3}{4}$

16. $0.4p + 0.2 = 4.2p - 7.8 - 0.6p$

17. $3(x + 2) = 27$

18. $-3x - 6(x - 4) = 9$

19. Translate to an algebraic expression:

Nine less than some number.

Solve.

20. ***Perimeter of a Photograph.*** The perimeter of a rectangular photograph is 36 cm. The length is 4 cm greater than the width. Find the width and the length.

21. ***Amount Spent on Food.*** The Ragers spent $7840 on food in a recent year. This was approximately 17% of their yearly income. What was the Ragers' income that year? Round to the nearest ten dollars.

22. ***Board Cutting.*** An 8-m board is cut into two pieces. One piece is 2 m longer than the other. How long are the pieces?

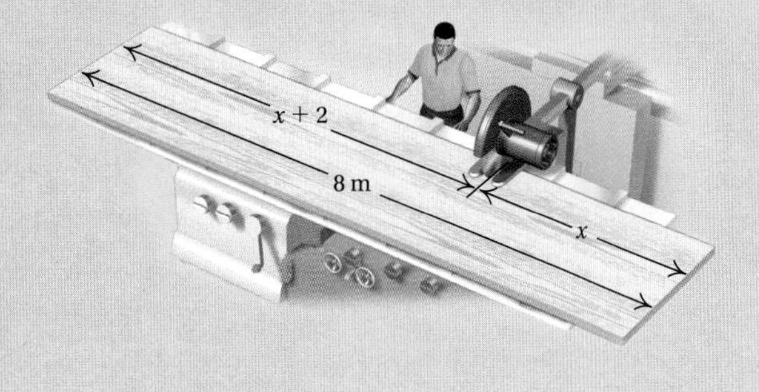

23. ***Tuition Dollars.*** Tuition paid by foreign students attending U.S. universities totaled $14.3 billion in 2010–2011. This number represents a 52% increase over the amount paid in 2005–2006. Find the amount of tuition U.S. universities received from foreign students in 2005–2006.

Source: "A U.S. Degree at Any Cost," by Justin Bergman, *Time*, August 20, 2012

24. If you triple a number and then subtract 14, you get $\frac{2}{3}$ of the original number. What is the original number?

25. The second angle of a triangle is three times as large as the first. The third angle is 25° less than the sum of the other two angles. Find the measure of the first angle.

26. Solve: $5y - 1 = 3y + 7$.

A. -4 **B.** 1
C. 3 **D.** 4

Synthesis

27. Solve: $3|w| - 8 = 37$.

28. A movie theater had a certain number of tickets to give away. Five people got the tickets. The first person got $\frac{1}{3}$ of the tickets, the second got $\frac{1}{4}$ of the tickets, and the third got $\frac{1}{5}$ of the tickets. The fourth person got 8 tickets, and there were 5 tickets left for the fifth person. Find the total number of tickets given away.

1–11 Cumulative Review

This cumulative review also serves as a review for a final examination covering the entire book. A question that may occur at this point is what notation to use for a particular problem or exercise. Although there is no particular rule, especially as you use mathematics outside the classroom, here is the guideline that we follow: Use the notation given in the problem. That is, if the problem is given using mixed numerals, give the answer in mixed numerals. If the problem is given in decimal notation, give the answer in decimal notation.

1. In 47,201, what digit tells the number of thousands?

2. Write expanded notation for 7405.

3. Write a word name for 7.463.

Add and simplify, if appropriate.

4. $\begin{array}{r} 741 \\ +271 \\ \hline \end{array}$

5. $\begin{array}{r} 4903 \\ 5278 \\ 6391 \\ +4513 \\ \hline \end{array}$

6. $-\frac{2}{13} + \frac{1}{26}$

7. $\begin{array}{r} 2\frac{4}{9} \\ +\ 3\frac{1}{3} \\ \hline \end{array}$

8. $\begin{array}{r} 2.048 \\ 63.914 \\ +428.009 \\ \hline \end{array}$

9. $34.56 + 2.783 + 0.433 + 765.1$

Subtract and simplify, if possible.

10. $\begin{array}{r} 674 \\ -522 \\ \hline \end{array}$

11. $\begin{array}{r} 9465 \\ -8791 \\ \hline \end{array}$

12. $\frac{7}{8} - \frac{2}{3}$

13. $\begin{array}{r} 4\frac{1}{3} \\ -\ 1\frac{5}{8} \\ \hline \end{array}$

14. $\begin{array}{r} 20.0 \\ -\ 0.0027 \\ \hline \end{array}$

15. $40.03 - (-5.789)$

Simplify.

16. $\frac{21}{30}$

17. $\frac{275}{5}$

Multiply and simplify, if possible.

18. $\begin{array}{r} 297 \\ \times\ 16 \\ \hline \end{array}$

19. $\begin{array}{r} 349 \\ \times 763 \\ \hline \end{array}$

20. $1\frac{3}{4} \cdot 2\frac{1}{3}$

21. $\frac{9}{7} \cdot \frac{14}{15}$

22. $-12 \cdot \frac{5}{6}$

23. $\begin{array}{r} 34.09 \\ \times\ 7.6 \\ \hline \end{array}$

24. Convert to a mixed numeral: $\frac{18}{5}$.

Divide and simplify. State the answer using a whole-number quotient and a remainder.

25. $6\overline{)3438}$

26. $34\overline{)1914}$

27. Write a mixed numeral for the quotient in Exercise 26.

Divide and simplify, if possible.

28. $\frac{4}{5} \div \frac{8}{15}$

29. $2\frac{1}{3} \div (-30)$

30. $2.7\overline{)105.3}$

31. Round 68,489 to the nearest thousand.

32. Round 0.4275 to the nearest thousandth.

33. Round $21.\overline{83}$ to the nearest hundredth.

34. Determine whether 1368 is divisible by 6.

35. Find all the factors of 15.

36. Find the LCM of 16, 25, and 32.

37. Use = or ≠ for □ to write a true sentence:

$\frac{4}{7} \square \frac{3}{5}$.

38. Use < or > for □ to write a true sentence:

$\frac{4}{7} \square \frac{3}{5}$.

39. Which number is greater, −1.001 or −0.9976?

40. *Pears.* Find the unit price of each brand of canned pears listed in the table below. Then determine which brand has the lowest unit price.

Brand	Size	Price	Unit Price
A	$8\frac{1}{2}$ oz	$0.95	
B	15 oz	$1.66	
C	$15\frac{1}{4}$ oz	$1.86	
D	24 oz	$2.54	
E	29 oz	$3.07	

41. *Pluto.* The dwarf planet Pluto has a diameter of 1400 mi. Use $\frac{22}{7}$ for π.

a) Find the circumference of Pluto.

b) Find the volume of Pluto.

Kitchen Remodeling. The Reisters spent $26,888 to remodel their kitchen. Complete the table below, which relates percents and costs.

	Item	Percent of Cost	Cost
42.	Cabinets	40%	
43.	Countertops		$4033.20
44.	Appliances	13%	
45.	Fixtures		$8066.40
46.	Flooring	2%	

47. Use $<$ or $>$ for $\square$ to write a true sentence:

$987 \square 879$.

48. What part is shaded?

Convert to decimal notation.

49. $\dfrac{37}{1000}$

50. $-\dfrac{13}{25}$

51. $\dfrac{8}{9}$

52. 7%

Convert to fraction notation.

53. 4.63

54. $-7\dfrac{1}{4}$

55. 40%

Convert to percent notation.

56. $\dfrac{17}{20}$

57. 1.5

Solve.

58. $234 + y = 789$

59. $-3.9 \times y = 249.6$

60. $\dfrac{2}{3} \cdot t = \dfrac{5}{6}$

61. $\dfrac{8}{17} = \dfrac{36}{x}$

62. *Late to Work.* The table below shows the results of a survey of 7780 workers, who were asked "How often are you late to work?"

RESPONSE	PERCENT
Never	61%
Once a year	12
Once a month	11
At least once a week	16

SOURCE: *CareerBuilder*

Make a vertical bar graph of the data, showing the percent who gave each response.

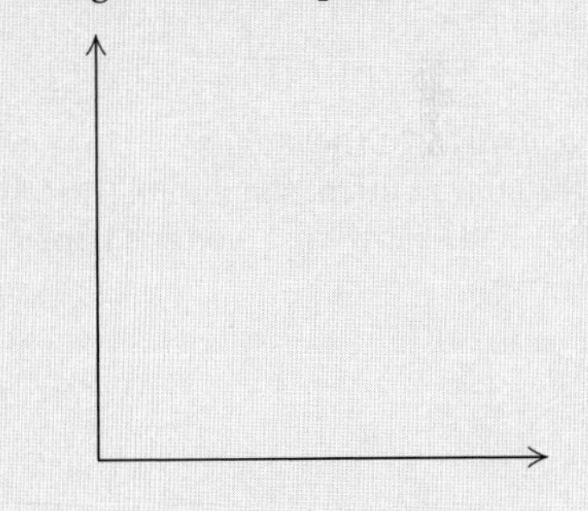

63. Find the missing angle measure.

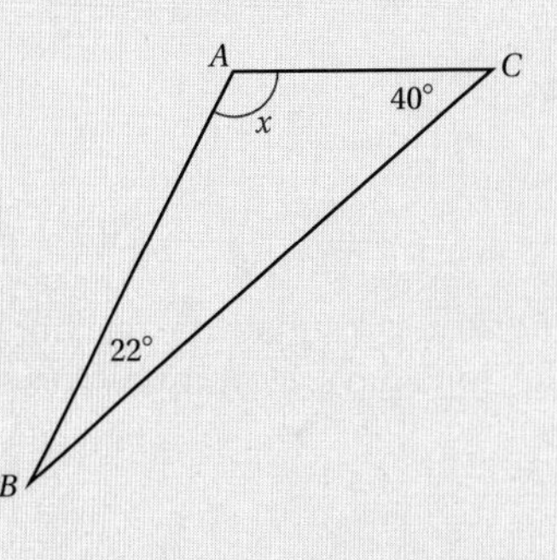

64. Classify the triangle in Exercise 63 as right, obtuse, or acute.

Solve.

65. *Donations.* Lorenzo made donations of \$627 and \$48 to a charity. What was the total donation?

66. *Candy Bars.* A machine wraps 134 candy bars per minute. How long does it take this machine to wrap 8710 bars?

67. *Stock Prices.* A share of stock bought for \$29.63 dropped \$3.88 before it was resold. What was the price when it was resold?

68. *Length of Trip.* At the start of a trip, a car's odometer read 27,428.6 mi, and at the end of the trip, the reading was 27,914.5 mi. How long was the trip?

69. *Taxes.* From an income of \$12,000, amounts of \$2300 and \$1600 are paid for federal and state taxes. How much remains after these taxes have been paid?

70. *Teacher Salary.* A substitute teacher was paid \$87 per day for 9 days. How much was she paid altogether?

71. *Walking Distance.* Celeste walks $\frac{3}{5}$ km per hour. At this rate, how far would she walk in $\frac{1}{2}$ hr?

72. *Sweater Costs.* Eight identical sweaters cost a total of \$679.68. What is the cost of each sweater?

73. *Paint Needs.* Eight gallons of exterior paint covers 400 ft^2. How much paint is needed to cover 650 ft^2?

74. *Simple Interest.* What is the simple interest on \$4000 principal at 5% for $\frac{3}{4}$ year?

75. *Commission Rate.* A real estate agent received \$5880 commission on the sale of an \$84,000 home. What was the rate of commission?

76. *Population Growth.* The population of Waterville is 29,000 this year and is increasing at 4% per year. What will the population be next year?

77. *Student Ages.* The ages of students in a math class at a community college are as follows:

18, 21, 26, 31, 32, 18, 50.

Find the average, the median, and the mode of their ages.

Evaluate.

78. 18^2

79. 7^3

Simplify.

80. $\sqrt{9}$

81. $\sqrt{121}$

82. Approximate to three decimal places: $\sqrt{20}$.

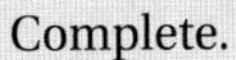

Complete.

83. $\frac{1}{3}$ yd = ______ in.

84. 4280 mm = ______ cm

85. 3 days = ______ hr

86. 20,000 g = ______ kg

87. 5 lb = ______ oz

88. 0.008 cg = ______ mg

89. 8190 mL = ______ L

90. 20 qt = ______ gal

91. Find the length of the third side of this right triangle. Give an exact answer and an approximation to three decimal places.

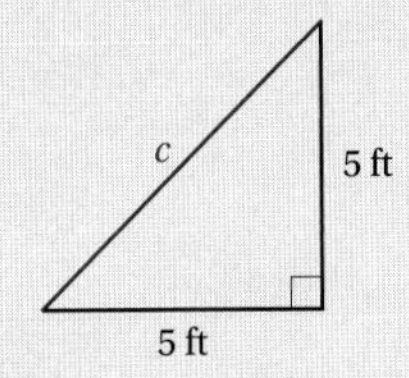

92. Find the diameter, the circumference, and the area of this circle. Use 3.14 for π.

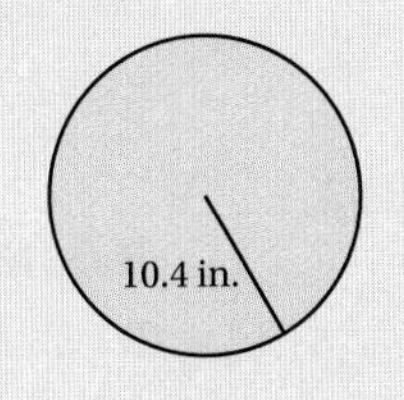

93. Find the perimeter and the area.

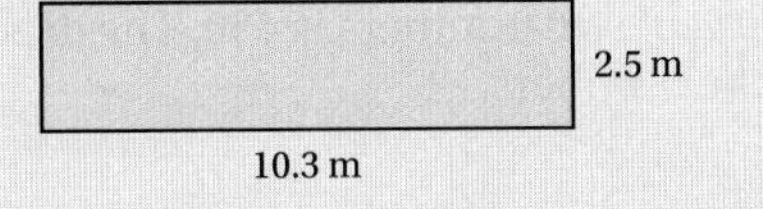

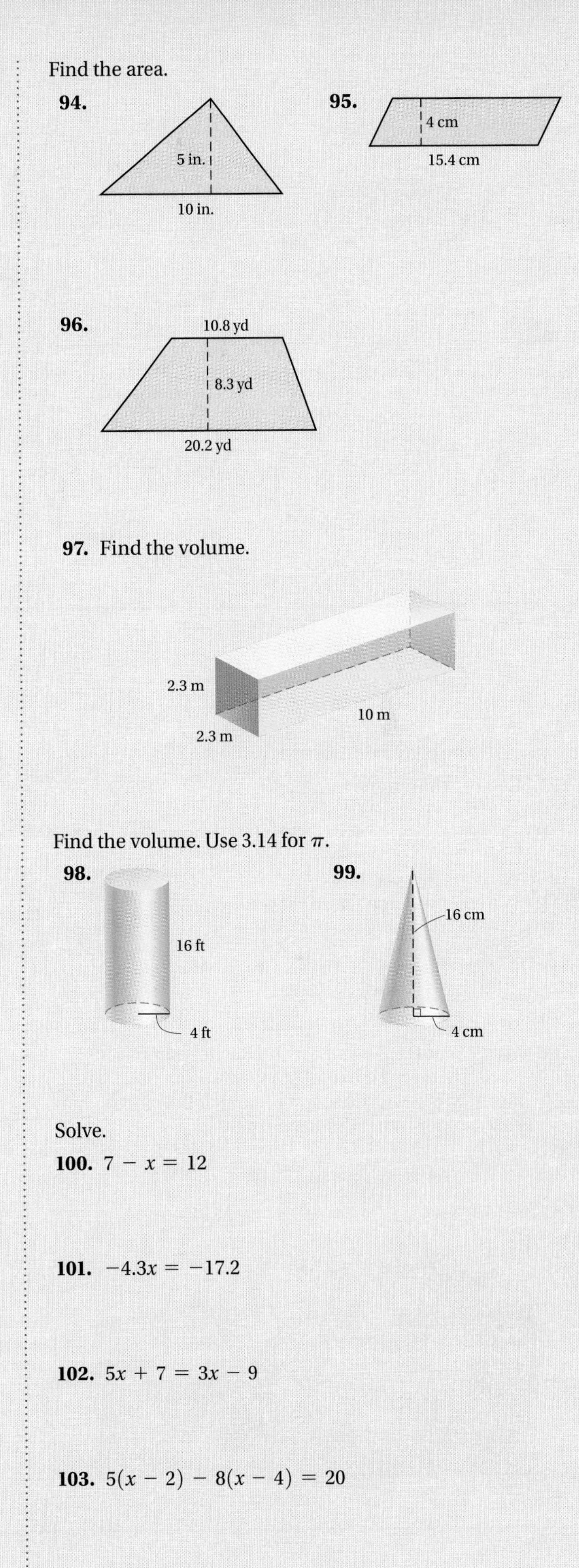

Find the area.

94.

95.

96.

97. Find the volume.

Find the volume. Use 3.14 for π.

98.

99.

Solve.

100. $7 - x = 12$

101. $-4.3x = -17.2$

102. $5x + 7 = 3x - 9$

103. $5(x - 2) - 8(x - 4) = 20$

Compute and simplify.

104. $12 \times 20 - 10 \div 5$

105. $4^3 - 5^2 + (16 \cdot 4 + 23 \cdot 3)$

106. $|(-1) \cdot 3|$

107. $17 + (-3)$

108. $-\frac{1}{3} - \left(-\frac{2}{3}\right)$

109. $(-6) \cdot (-5)$

110. $-\frac{5}{7} \cdot \frac{14}{35}$

111. $\frac{48}{-6}$

Translate to an algebraic expression.

112. 17 more than some number

113. 38 percent of some number

Solve.

114. *Razor Scooters.* Nathan and Rachel purchased Razor scooters for a total of $192. Nathan paid $98 more for his electric scooter than Rachel paid for her kick scooter. What did Rachel pay?

115. *Savings Investment.* Money is invested in a savings account at 4% simple interest. After 1 year, there is $2288 in the account. How much was originally invested?

116. *Wire Cutting.* A 143-m wire is cut into three pieces. The second piece is 3 m longer than the first. The third is four-fifths as long as the first. How long is each piece?

Solve.

117. $\frac{2}{3}x + \frac{1}{6} - \frac{1}{2}x = \frac{1}{6} - 3x$

118. $29.966 - 8.673y = -8.18 + 10.4y$

119. Collect like terms: $\frac{1}{4}x - \frac{3}{4}y + \frac{1}{4}x - \frac{3}{4}y$.

A. 0

B. $-\frac{2}{3}y$

C. $\frac{1}{2}x - \frac{3}{2}y$

D. $-\frac{1}{2}x - \frac{1}{2}y$

120. Factor out the greatest common factor: $8x + 4y - 12z$.

A. $2(4x - 2y - 6z)$

B. $4(2x + y - 3z)$

C. $4(2x - 3z)$

D. $2(4x + 4y - 12z)$

121. Divide: $-\frac{13}{25} \div \left(-\frac{13}{5}\right)$.

A. $\frac{169}{125}$

B. 5

C. $\frac{125}{169}$

D. $\frac{1}{5}$

122. Add: $-27 + (-11)$.

A. -38

B. -16

C. 16

D. 38

Synthesis

123. The sum of two numbers is 430. The difference is 40. Find the numbers.

Answers

CHAPTER 1

Exercise Set 1.1, p. 6

RC1. digit **RC2.** period **RC3.** expanded **RC4.** standard
1. 5 thousands **3.** 5 hundreds **5.** 1 **7.** 2
9. 2 thousands + 0 hundreds + 5 tens + 8 ones, or 2 thousands + 5 tens + 8 ones **11.** 1 thousand + 5 hundreds + 7 tens + 6 ones **13.** 5 thousands + 7 hundreds + 0 tens + 2 ones, or 5 thousands + 7 hundreds + 2 ones **15.** 9 ten thousands + 3 thousands + 9 hundreds + 8 tens + 6 ones
17. 1 billion + 3 hundred millions + 4 ten millions + 3 millions + 2 hundred thousands + 3 ten thousands + 9 thousands + 9 hundreds + 2 tens + 3 ones **19.** 2 hundred millions + 4 ten millions + 8 millions + 6 hundred thousands + 4 ten thousands + 5 thousands + 0 hundreds + 0 tens + 8 ones, or 2 hundred millions + 4 ten millions + 8 millions + 6 hundred thousands + 4 ten thousands + 5 thousands + 8 ones **21.** Eighty-five **23.** Eighty-eight thousand **25.** One hundred twenty-three thousand, seven hundred sixty-five **27.** Seven billion, seven hundred fifty-four million, two hundred eleven thousand, five hundred seventy-seven
29. Seven hundred one thousand, seven hundred ninety-nine
31. Two million, four hundred seventy-four thousand, two hundred eighty **33.** 632,896 **35.** 50,324 **37.** 2,233,812
39. 8,000,000,000 **41.** 40,000,000 **43.** 30,000,103
45. 64,186,000 **47.** 138

Calculator Corner, p. 10

1. 121 **2.** 1602 **3.** 1932 **4.** 864

Exercise Set 1.2, p. 12

RC1. addends **RC2.** sum **RC3.** 0 **RC4.** perimeter
1. 387 **3.** 164 **5.** 5198 **7.** 100 **9.** 8503 **11.** 5266
13. 4466 **15.** 6608 **17.** 34,432 **19.** 101,310 **21.** 230
23. 18,424 **25.** 31,685 **27.** 132 yd **29.** 1661 ft **31.** 570 ft
33. 8 ten thousands **34.** Nine billion, three hundred forty-six million, three hundred ninety-nine thousand, four hundred sixty-eight **35.** $1 + 99 = 100, 2 + 98 = 100, \ldots, 49 + 51 = 100$. Then 49 100's = 4900 and $4900 + 50 + 100 = 5050$.

Calculator Corner, p. 15

1. 28 **2.** 47 **3.** 67 **4.** 119 **5.** 2128 **6.** 2593

Exercise Set 1.3, p. 17

RC1. minuend **RC2.** subtraction symbol **RC3.** subtrahend
RC4. difference
1. 44 **3.** 533 **5.** 39 **7.** 14 **9.** 369 **11.** 26 **13.** 234
15. 417 **17.** 5382 **19.** 2778 **21.** 3069 **23.** 1089
25. 7748 **27.** 4144 **29.** 56 **31.** 454 **33.** 3749
35. 2191 **37.** 43,028 **39.** 95,974 **41.** 4418 **43.** 1305
45. 9989 **47.** 48,017 **49.** 1345 **50.** 924 **51.** 22,692
52. 10,920 **53.** Six million, three hundred seventy-five thousand, six hundred two **54.** 9 thousands + 1 hundred + 0 tens + 3 ones, or 9 thousands + 1 hundred + 3 ones **55.** 3; 4

Calculator Corner, p. 21

1. 448 **2.** 21,970 **3.** 6380 **4.** 39,564 **5.** 180,480
6. 2,363,754

Exercise Set 1.4, p. 23

RC1. factors **RC2.** product **RC3.** 0 **RC4.** 1
1. 520 **3.** 564 **5.** 1527 **7.** 64,603 **9.** 4770 **11.** 3995
13. 870 **15.** 1920 **17.** 46,296 **19.** 14,652 **21.** 258,312
23. 798,408 **25.** 20,723,872 **27.** 362,128 **29.** 302,220
31. 49,101,136 **33.** 25,236,000 **35.** 20,064,048
37. 529,984 sq mi **39.** 8100 sq ft **41.** 12,685 **42.** 10,834
43. 8889 **44.** 254,119 **45.** 4 hundred thousands **46.** 0
47. 1 ten thousand + 2 thousands + 8 hundreds + 4 tens + 7 ones
48. Seven million, four hundred thirty-two thousand
49. 247,464 sq ft

Calculator Corner, p. 30

1. 28 **2.** 123 **3.** 323 **4.** 36

Exercise Set 1.5, p. 32

RC1. quotient **RC2.** dividend **RC3.** remainder
RC4. divisor
1. 12 **3.** 1 **5.** 22 **7.** 0 **9.** Not defined **11.** 6
13. 55 R 2 **15.** 108 **17.** 307 **19.** 753 R 3 **21.** 74 R 1
23. 92 R 2 **25.** 1703 **27.** 987 R 5 **29.** 12,700 **31.** 127
33. 52 R 52 **35.** 29 R 5 **37.** 40 R 12 **39.** 90 R 22 **41.** 29
43. 105 R 3 **45.** 1609 R 2 **47.** 1007 R 1 **49.** 23
51. 107 R 1 **53.** 370 **55.** 609 R 15 **57.** 304
59. 3508 R 219 **61.** 8070 **63.** 1241 **64.** 66,444
65. 19,800 **66.** 9380 **67.** 40 ft **68.** 99 sq ft
69. 54, 122; 33, 2772; 4, 8 **71.** 30 buses

Mid-Chapter Review: Chapter 1, p. 35

1. False **2.** True **3.** True **4.** False **5.** True
6. False **7.** Ninety-five million, four hundred six thousand, two hundred thirty-seven

8. $\begin{array}{r} {\scriptstyle 5\ 9\ 14} \\ \not{6}\,\not{0}\,\not{4} \\ -\ 4\ 9\ 7 \\ \hline 1\ 0\ 7 \end{array}$ **9.** 6 hundreds **10.** 6 ten thousands

11. 6 thousands **12.** 6 ones **13.** 2 **14.** 6 **15.** 5 **16.** 1
17. 5 thousands + 6 hundreds + 0 tens + 2 ones, or 5 thousands + 6 hundreds + 2 ones **18.** 6 ten thousands + 9 thousands + 3 hundreds + 4 tens + 5 ones **19.** One hundred thirty-six
20. Sixty-four thousand, three hundred twenty-five **21.** 308,716
22. 4,567,216 **23.** 798 **24.** 1030 **25.** 7922 **26.** 7534
27. 465 **28.** 339 **29.** 1854 **30.** 4328 **31.** 216
32. 15,876 **33.** 132,275 **34.** 5,679,870 **35.** 253

36. 112 R 5 **37.** 23 R 19 **38.** 144 R 31 **39.** 25 m **40.** 8 sq in. **41.** When numbers are being added, it does not matter how they are grouped. **42.** Subtraction is not commutative. For example, $5 - 2 = 3$, but $2 - 5 \neq 3$. **43.** Answers will vary. Suppose one coat costs \$150. Then the multiplication $4 \cdot \$150$ gives the cost of four coats. Or, suppose one ream of copy paper costs \$4. Then the multiplication $\$4 \cdot 150$ gives the cost of 150 reams. **44.** If we use the definition of division, $0 \div 0 = a$ such that $a \cdot 0 = 0$. We see that a could be *any* number since $a \cdot 0 = 0$ for any number a. Thus we cannot say that $0 \div 0 = 0$. This is why we agree not to allow division by 0.

Exercise Set 1.6, p. 43

RC1. True **RC2.** False **RC3.** False **RC4.** True **1.** 50 **3.** 460 **5.** 730 **7.** 900 **9.** 100 **11.** 1000 **13.** 9100 **15.** 32,800 **17.** 6000 **19.** 8000 **21.** 45,000 **23.** 373,000 **25.** $80 + 90 = 170$ **27.** $8070 - 2350 = 5720$ **29.** 220; incorrect **31.** 890; incorrect **33.** $7300 + 9200 = 16{,}500$ **35.** $6900 - 1700 = 5200$ **37.** 1600; correct **39.** 1500; correct **41.** $10{,}000 + 5000 + 9000 + 7000 = 31{,}000$ **43.** $92{,}000 - 23{,}000 = 69{,}000$ **45.** $50 \cdot 70 = 3500$ **47.** $30 \cdot 30 = 900$ **49.** $900 \cdot 300 = 270{,}000$ **51.** $400 \cdot 200 = 80{,}000$ **53.** $350 \div 70 = 5$ **55.** $8450 \div 50 = 169$ **57.** $1200 \div 200 = 6$ **59.** $8400 \div 300 = 28$ **61.** \$800 **63.** \$1200; no **65.** Answers will vary depending on the options chosen. **67. (a)** \$309,600; **(b)** \$360,000 **69.** 90 people **71.** $<$ **73.** $>$ **75.** $<$ **77.** $>$ **79.** $>$ **81.** $>$ **83.** $1{,}335{,}475 < 4{,}134{,}519$, or $4{,}134{,}519 > 1{,}335{,}475$ **85.** $97{,}382 < 98{,}817$, or $98{,}817 > 97{,}382$ **87.** 86,754 **88.** 13,589 **89.** 48,824 **90.** 4415 **91.** 1702 **92.** 17,748 **93.** 54 R 4 **94.** 208 **95.** Left to the student **97.** Left to the student

Exercise Set 1.7, p. 52

RC1. (c) **RC2.** (a) **RC3.** (d) **RC4.** (b) **1.** 14 **3.** 0 **5.** 90,900 **7.** 450 **9.** 352 **11.** 25 **13.** 29 **15.** 0 **17.** 79 **19.** 45 **21.** 8 **23.** 14 **25.** 32 **27.** 143 **29.** 17,603 **31.** 37 **33.** 1035 **35.** 66 **37.** 324 **39.** 335 **41.** 18,252 **43.** 104 **45.** 45 **47.** 4056 **49.** 2847 **51.** 15 **53.** 205 **55.** 457 **57.** 142 R 5 **58.** 142 **59.** 334 **60.** 334 R 11 **61.** $<$ **62.** $>$ **63.** $>$ **64.** $<$ **65.** 6,376,000 **66.** 6,375,600 **67.** 347

Translating for Success, p. 62

1. E **2.** M **3.** D **4.** G **5.** A **6.** O **7.** F **8.** K **9.** J **10.** H

Exercise Set 1.8, p. 63

RC1. Familiarize. **RC2.** Translate. **RC3.** Solve. **RC4.** Check. **1.** 1962 ft **3.** 1450 ft **5.** 95 milligrams **7.** 18 rows **9.** 43 events **11.** 2054 mi **13.** 2,073,600 pixels **15.** 168 hr **17.** \$273 per month **19.** \$197 **21.** \$7092 **23.** 151,500 **25.** \$78 **27.** \$40 per month **29.** \$24,456 **31.** 35 weeks; 2 episodes **33.** 21 columns **35.** 236 gal **37. (a)** 4200 sq ft; **(b)** 268 ft **39.** 56 cartons **41.** 645 mi; 5 in. **43.** \$247 **45.** 525 min, or 8 hr 45 min **47.** 168,300 jobs **49.** 104 seats **51.** 32 \$10 bills **53.** \$400 **55.** 106 bones **57.** 8273 **58.** 7759 **59.** 806,985 **60.** 147 R 4 **61.** 34 m **62.** 9706 sq ft **63.** $200 \times 600 = 120{,}000$ **64.** 66 **65.** 792,000 mi; 1,386,000 mi

Calculator Corner, p. 71

1. 243 **2.** 15,625 **3.** 20,736 **4.** 2048

Calculator Corner, p. 73

1. 49 **2.** 85 **3.** 36 **4.** 0 **5.** 73 **6.** 49

Exercise Set 1.9, p. 75

RC1. exponent **RC2.** squared **RC3.** multiplication **RC4.** 3 **1.** 3^4 **3.** 5^2 **5.** 7^5 **7.** 10^3 **9.** 49 **11.** 729 **13.** 20,736 **15.** 243 **17.** 22 **19.** 20 **21.** 100 **23.** 1 **25.** 49 **27.** 5 **29.** 434 **31.** 41 **33.** 88 **35.** 4 **37.** 303 **39.** 20 **41.** 70 **43.** 295 **45.** 32 **47.** 906 **49.** 62 **51.** 102 **53.** 32 **55.** \$94 **57.** 401 **59.** 110 **61.** 7 **63.** 544 **65.** 708 **67.** 27 **69.** 452 **70.** 835 **71.** 13 **72.** 37 **73.** 4898 **74.** 100 **75.** 104,286 sq mi **76.** 98 gal **77.** 24; $1 + 5 \cdot (4 + 3) = 36$ **79.** 7; $12 \div (4 + 2) \cdot 3 - 2 = 4$

Summary and Review: Chapter 1, p. 78

Vocabulary Reinforcement

1. perimeter **2.** minuend **3.** digits; periods **4.** dividend **5.** factors; product **6.** additive **7.** associative **8.** divisor; remainder; dividend

Concept Reinforcement

1. True **2.** True **3.** False **4.** False **5.** True **6.** False

Important Concepts

1. 2 thousands **2.** 65,302 **3.** 3237 **4.** 225,036 **5.** 315 R 14 **6.** 36,000 **7.** $<$ **8.** 36 **9.** 216

Review Exercises

1. 8 thousands **2.** 3 **3.** 2 thousands + 7 hundreds + 9 tens + 3 ones **4.** 5 ten thousands + 6 thousands + 0 hundreds + 7 tens + 8 ones, or 5 ten thousands + 6 thousands + 7 tens + 8 ones **5.** 4 millions + 0 hundred thousands + 0 ten thousands + 7 thousands + 1 hundred + 0 tens + 1 one, or 4 millions + 7 thousands + 1 hundred + 1 one **6.** Sixty-seven thousand, eight hundred nineteen **7.** Two million, seven hundred eighty-one thousand, four hundred twenty-seven **8.** 476,588 **9.** 1,640,000,000 **10.** 14,272 **11.** 66,024 **12.** 21,788 **13.** 98,921 **14.** 5148 **15.** 1689 **16.** 2274 **17.** 17,757 **18.** 5,100,000 **19.** 6,276,800 **20.** 506,748 **21.** 27,589 **22.** 5,331,810 **23.** 12 R 3 **24.** 5 **25.** 913 R 3 **26.** 384 R 1 **27.** 4 R 46 **28.** 54 **29.** 452 **30.** 5008 **31.** 4389 **32.** 345,800 **33.** 345,760 **34.** 346,000 **35.** 300,000 **36.** $>$ **37.** $<$ **38.** $41{,}300 + 19{,}700 = 61{,}000$ **39.** $38{,}700 - 24{,}500 = 14{,}200$ **40.** $400 \cdot 700 = 280{,}000$ **41.** 8 **42.** 45 **43.** 58 **44.** 0 **45.** 4^3 **46.** 10,000 **47.** 36 **48.** 65 **49.** 233 **50.** 260 **51.** 165 **52.** \$502 **53.** \$484 **54.** 1982 **55.** 19 cartons **56.** \$13,585 **57.** 14 beehives **58.** 98 sq ft; 42 ft **59.** 137 beakers filled; 13 mL left over **60.** \$27,598 **61.** B **62.** A **63.** D **64.** 8 **65.** $a = 8, b = 4$ **66.** 6 days

Understanding Through Discussion and Writing

1. No; if subtraction were associative, then $a - (b - c) = (a - b) - c$ for any a, b, and c. But, for example,

$$12 - (8 - 4) = 12 - 4 = 8,$$

whereas

$$(12 - 8) - 4 = 4 - 4 = 0.$$

Since $8 \neq 0$, this example shows that subtraction is not associative. **2.** By rounding prices and estimating their sum, a shopper can estimate the total grocery bill while shopping. This is particularly useful if the shopper wants to spend no more than a certain amount.

3. Answers will vary. Anthony is driving from Kansas City to Minneapolis, a distance of 512 mi. He stops for gas after driving 183 mi. How much farther must he drive? **4.** The parentheses are not necessary in the expression $9 - (4 \cdot 2)$. Using the rules for order of operations, the multiplication would be performed before the subtraction even if the parentheses were not present. The parentheses are necessary in the expression $(3 \cdot 4)^2$; $(3 \cdot 4)^2 = 12^2 = 144$, but $3 \cdot 4^2 = 3 \cdot 16 = 48$.

Test: Chapter 1, p. 83

1. [1.1a] 5 **2.** [1.1b] 8 thousands + 8 hundreds + 4 tens + 3 ones **3.** [1.1c] Thirty-eight million, four hundred three thousand, two hundred seventy-seven **4.** [1.2a] 9989 **5.** [1.2a] 63,791 **6.** [1.2a] 3165 **7.** [1.2a] 10,515 **8.** [1.3a] 3630 **9.** [1.3a] 1039 **10.** [1.3a] 6848 **11.** [1.3a] 5175 **12.** [1.4a] 41,112 **13.** [1.4a] 5,325,600 **14.** [1.4a] 2405 **15.** [1.4a] 534,264 **16.** [1.5a] 3 R 3 **17.** [1.5a] 70 **18.** [1.5a] 97 **19.** [1.5a] 805 R 8 **20.** [1.6a] 35,000 **21.** [1.6a] 34,530 **22.** [1.6a] 34,500 **23.** [1.6b] 23,600 + 54,700 = 78,300 **24.** [1.6b] 54,800 − 23,600 = 31,200 **25.** [1.6b] 800 · 500 = 400,000 **26.** [1.6c] > **27.** [1.6c] < **28.** [1.7b] 46 **29.** [1.7b] 13 **30.** [1.7b] 14 **31.** [1.7b] 381 **32.** [1.8a] 83 calories **33.** [1.8a] 20 staplers **34.** [1.8a] 1,256,615 sq mi **35. (a)** [1.2b], [1.4b] 300 in., 5000 sq in.; 264 in., 3872 sq in.; 228 in., 2888 sq in.; **(b)** [1.8a] 2112 sq in. **36.** [1.8a] 1852 12-packs; 7 cakes left over **37.** [1.8a] $95 **38.** [1.9a] 12^4 **39.** [1.9b] 343 **40.** [1.9b] 100,000 **41.** [1.9c] 31 **42.** [1.9c] 98 **43.** [1.9c] 2 **44.** [1.9c] 18 **45.** [1.9d] 216 **46.** [1.9c] A **47.** [1.4b], [1.8a] 336 sq in. **48.** [1.9c] 9 **49.** [1.8a] 80 payments

CHAPTER 2

Exercise Set 2.1, p. 90

RC1. *G* **RC2.** *C* **RC3.** *B* **RC4.** *E* **RC5.** True **RC6.** True **RC7.** False **RC8.** False **1.** 24; −2 **3.** 950,000,000; −460 **5.** 1454; −55 **7.** −6 −5 −4 −3 −2 −1 0 1 2 3 4 5 6 **9.** −6 −5 −4 −3 −2 −1 0 1 2 3 4 5 6 **11.** > **13.** < **15.** > **17.** < **19.** > **21.** < **23.** < **25.** > **27.** 3 **29.** 18 **31.** 325 **33.** 29 **35.** 300 **37.** 53 **39.** 13,549 **40.** 1107 **41.** 50,609 **42.** 219 **43.** 259 **44.** 8778 **45.** > **47.** =

Calculator Corner, p. 95

1. 13 **2.** −8 **3.** −12

Exercise Set 2.2, p. 96

RC1. right, right **RC2.** left, left **RC3.** right, left **RC4.** left, right **1.** −7 **3.** −4 **5.** 0 **7.** −8 **9.** −7 **11.** −27 **13.** 0 **15.** −42 **17.** 0 **19.** 0 **21.** 3 **23.** −9 **25.** 7 **27.** 0 **29.** 45 **31.** −2 **33.** −7 **35.** −19 **37.** −6 **39.** −1 **41.** 39 **43.** 43 **45.** −1093 **47.** −24 **49.** 26 **51.** −9 **53.** 14 **55.** −65 **57.** 5 **59.** 14 **61.** −10 **63.** 3456 **64.** 137,016 **65.** 4,533,324 **66.** 408 R 3 **67.** 127 R 46 **68.** 221 R 331 **69.** 641,540 **70.** 641,500 **71.** 642,000 **73.** All positive **75.** Negative

Exercise Set 2.3, p. 101

RC1. (c) **RC2.** (b) **RC3.** (d) **RC4.** (a) **1.** −4 **3.** −7 **5.** −6 **7.** 0 **9.** −4 **11.** −7 **13.** −6 **15.** 0 **17.** 11 **19.** −14 **21.** 5 **23.** −7 **25.** −5 **27.** −3 **29.** −23 **31.** −68 **33.** −73 **35.** 116 **37.** −2 **39.** −5 **41.** −1 **43.** −5 **45.** 22 **47.** 19 **49.** −20 **51.** −8 **53.** −10 **55.** −3 **57.** −7 **59.** −30 **61.** 0 **63.** −1 **65.** 37 **67.** −62 **69.** 6 **71.** 107 **73.** 219 **75.** Profit of $4300 **77.** 3780 m **79.** −3° **81.** $348 **83.** 13,796 ft above sea level **85.** 64 **86.** 125 **87.** 29 **88.** 41 **89.** 66 **90.** 3 **91.** 8 cans **92.** 288 oz **93.** True **95.** True **97.** True **99.** True

Mid-Chapter Review: Chapter 2, p. 104

1. True **2.** False **3.** True **4.** $-x = -(-4) = 4$; $-(-x) = -(-(-4)) = -(4) = -4$ **5.** $5 - 13 = 5 + (-13) = -8$ **6.** $-6 - (-7) = -6 + 7 = 1$ **7.** 450; −79 **8.** 20; −23 **9.** −6 −5 −4 −3 −2 −1 0 1 2 3 4 5 6 **10.** −6 −5 −4 −3 −2 −1 0 1 2 3 4 5 6 **11.** < **12.** < **13.** > **14.** > **15.** 15 **16.** 18 **17.** 0 **18.** 12 **19.** 5 **20.** −7 **21.** 0 **22.** 49 **23.** 19 **24.** 2 **25.** −2 **26.** −2 **27.** 0 **28.** −17 **29.** −10 **30.** −7 **31.** −9 **32.** −2 **33.** −10 **34.** 16 **35.** 2 **36.** −12 **37.** −4 **38.** −6 **39.** −2 **40.** 13 **41.** 9 **42.** −23 **43.** 75 **44.** 14 **45.** 33°C **46.** $48 **47.** Answers will vary. **48.** The absolute value of a number is its distance from 0, and distance is always nonnegative. **49.** Answers may vary. If we think of the addition on the number line, we start at a negative number and move to the left. This always brings us to a point on the negative portion of the number line. **50.** Yes; consider $m - (-n)$, where both m and n are positive. Then, $m - (-n) = m + n$. Now $m + n$, the sum of two positive numbers, is positive.

Exercise Set 2.4, p. 108

RC1. negative **RC2.** positive **RC3.** positive **RC4.** negative **1.** −16 **3.** −24 **5.** −72 **7.** 16 **9.** 42 **11.** −120 **13.** −238 **15.** 1200 **17.** 84 **19.** −12 **21.** 24 **23.** 21 **25.** −69 **27.** 27 **29.** −18 **31.** −45 **33.** 420 **35.** 36 **37.** −60 **39.** 150 **41.** −90 **43.** 1911 **45.** 56 **47.** 30 **49.** −960 **51.** 1764 **53.** −240 **55.** −30,240 **57.** 237,460 **58.** 237,500 **59.** 237,000 **60.** 13 **61.** 3 **62.** 21 **63. (a)** One must be negative and one must be positive. **(b)** Either or both must be zero. **(c)** Both must be negative or both must be positive.

Translating for Success, p. 114

1. I **2.** C **3.** G **4.** L **5.** B **6.** O **7.** A **8.** F **9.** D **10.** M

Exercise Set 2.5, p. 115

RC1. True **RC2.** True **RC3.** False **RC4.** False **1.** −6 **3.** −13 **5.** −2 **7.** 4 **9.** −8 **11.** 2 **13.** −12 **15.** −8 **17.** Not defined **19.** −9 **21.** 38°F **23.** $26 **25.** 32 m below sea level **27.** −7 **29.** −7 **31.** −334 **33.** 14 **35.** 1880 **37.** 12 **39.** 8 **41.** −86 **43.** 37 **45.** −1 **47.** −10 **49.** −67 **51.** −7988 **53.** −3000 **55.** 60 **57.** 1 **59.** 10 **61.** −2 **63.** −4 **65.** 8 thousands **66.** 8 millions **67.** 8 ones **68.** 8 hundreds **69.** 2203 **70.** 848 **71.** 37,239 **72.** 11,851 **73.** 4992 sq ft; 284 ft **74.** $784 **75.** −159 **77.** Negative **79.** Negative **81.** Positive

Summary and Review: Chapter 2, p. 118

Vocabulary Reinforcement

1. integers **2.** absolute **3.** opposites, additive **4.** difference **5.** positive **6.** negative

Concept Reinforcement

1. False **2.** True **3.** False **4.** True

Study Guide

1. −6 −5 −4 −3 −2 −1 0 1 2 3 4 5 6 **2.** $<$ **3.** **(a)** 17; **(b)** 14 **4.** −3 **5.** −8 **6.** 14 **7.** 72 **8.** −90 **9.** 4 **10.** −4 **11.** −12

Review Exercises

1. 620; −125 **2.** 38 **3.** 7 **4.** 0 **5.** −2 **6.** $<$ **7.** $>$ **8.** $>$ **9.** $<$ **10.** −6 −5 −4 −3 −2 −1 0 1 2 3 4 5 6
11. −6 −5 −4 −3 −2 −1 0 1 2 3 4 5 6 **12.** −8 **13.** 14 **14.** 0 **15.** 23 **16.** 34 **17.** 5 **18.** −3 **19.** −7 **20.** −4 **21.** −5 **22.** 4 **23.** −14 **24.** −8 **25.** 54 **26.** −39 **27.** −56 **28.** −210 **29.** −7 **30.** −3 **31.** 6 **32.** −62 **33.** 26 **34.** −5 **35.** 2 **36.** −180 **37.** −$360 **38.** $62 per share **39.** 8-yd gain **40.** $19 **41.** C **42.** B
43. **(a)** $-7 + (-6) + (-5) + (-4) + (-3) + (-2) + (-1) + 0 + 1 + 2 + 3 + 4 + 5 + 6 + 7 + 8 = 8$; **(b)** 0
44. $9 - (3 - 4) + 5 = 15$ **45.** −5 **46.** −5

Understanding Through Discussion and Writing

1. We know that the product of an even number of negative numbers is positive, and the product of an odd number of negative numbers is negative. Since $(-7)^8$ is equivalent to the product of eight negative numbers, it will be a positive number. Similarly, since $(-7)^{11}$ is equivalent to the product of eleven negative numbers, it will be a negative number. **2.** If the negative integer has the greater absolute value, the sum is negative. **3.** Jake is expecting the multiplication to be performed before the division.
4. Answers will vary. At 4 P.M., the temperature in Circle City was 23°F. By 11 P.M., the temperature had dropped 32°F. What was the temperature at 11 P.M.?

Test: Chapter 2, p. 123

1. [2.1c] $<$ **2.** [2.1c] $>$ **3.** [2.1c] $>$ **4.** [2.1c] $<$ **5.** [2.1d] 7 **6.** [2.1d] 94 **7.** [2.1d] −27 **8.** [2.2b] −23 **9.** [2.2b] 14 **10.** [2.2b] 8 **11.** [2.1b] −6 −5 −4 −3 −2 −1 0 1 2 3 4 5 6
12. [2.3a] 78 **13.** [2.2a] −8 **14.** [2.2a] 2 **15.** [2.3a] 10 **16.** [2.3a] −25 **17.** [2.2a] 15 **18.** [2.4a] −48 **19.** [2.4a] 24 **20.** [2.5a] −9 **21.** [2.5a] 9 **22.** [2.5a] −4 **23.** [2.5c] 109 **24.** [2.3b] 2244 m **25.** [2.3b] Up 15 pts **26.** [2.3b], [2.5b] 16,080 **27.** [2.5b] −2°C each minute **28.** [2.2b] D **29.** [2.1d], [2.3a] 15 **30.** [2.3b] 2385 m
31. **(a)** [2.3a] −4, −9, −15; **(b)** [2.3a] −2, −6, −10; **(c)** [2.3a] −18, −24, −31; **(d)** [2.5a] 4, −2,1

CHAPTER 3

Calculator Corner, p. 129

1. No **2.** Yes **3.** Yes **4.** No

Exercise Set 3.1, p. 131

RC1. True **RC2.** True **RC3.** False **RC4.** True **RC5.** False **RC6.** False
1. No **3.** Yes **5.** 1, 2, 3, 6, 9, 18 **7.** 1, 2, 3, 6, 9, 18, 27, 54 **9.** 1, 2, 4 **11.** 1 **13.** 1, 2, 7, 14, 49, 98 **15.** 1, 3, 5, 15, 17, 51, 85, 255 **17.** 4, 8, 12, 16, 20, 24, 28, 32, 36, 40 **19.** 20, 40, 60, 80, 100, 120, 140, 160, 180, 200 **21.** 3, 6, 9, 12, 15, 18, 21, 24, 27, 30 **23.** 12, 24, 36, 48, 60, 72, 84, 96, 108, 120 **25.** 10, 20, 30, 40, 50, 60, 70, 80, 90, 100 **27.** 9, 18, 27, 36, 45, 54, 63, 72, 81, 90 **29.** No **31.** Yes **33.** Yes **35.** No **37.** Neither **39.** Composite **41.** Prime **43.** Prime **45.** $2 \cdot 2 \cdot 2$ **47.** $2 \cdot 7$ **49.** $2 \cdot 3 \cdot 7$ **51.** $5 \cdot 5$ **53.** $2 \cdot 5 \cdot 5$ **55.** $13 \cdot 13$ **57.** $2 \cdot 2 \cdot 5 \cdot 5$ **59.** $5 \cdot 7$ **61.** $2 \cdot 2 \cdot 2 \cdot 3 \cdot 3$ **63.** $7 \cdot 11$ **65.** $2 \cdot 2 \cdot 7 \cdot 103$ **67.** $3 \cdot 17$ **69.** $2 \cdot 2 \cdot 2 \cdot 2 \cdot 3 \cdot 5 \cdot 5$ **71.** $3 \cdot 7 \cdot 13$ **73.** $2 \cdot 3 \cdot 11 \cdot 17$ **75.** 4 thousands **76.** 4 millions **77.** 4 tens **78.** 4 ten thousands **79.** 34,560 **80.** 34,600 **81.** 2,428,000 **82.** 2,428,500 **83.** Row 1: 48, 90, 432, 63; row 2: 7, 2, 2, 10, 8, 6, 21, 10; row 3: 9, 18, 36, 14, 12, 11, 21; row 4: 29, 19, 42

Exercise Set 3.2, p. 137

RC1. (c) **RC2.** (a) **RC3.** (g) **RC4.** (d) **RC5.** (f) **RC6.** (h) **RC7.** (b) **RC8.** (e)
1. Yes; the sum of the digits is 12, which is divisible by 3. **3.** No; the ones digit is not 0 or 5. **5.** Yes; the ones digit is 0. **7.** Yes; the sum of the digits is 18, which is divisible by 9. **9.** No; the ones digit is not even. **11.** No; the ones digit is not even. **13.** No; 30 is not divisible by 4. **15.** Yes; 840 is divisible by 8. **17.** 6825 is divisible by 3 and 5. **19.** 119,117 is divisible by none of these numbers. **21.** 127,575 is divisible by 3, 5, and 9. **23.** 9360 is divisible by 2, 3, 4, 5, 6, 8, 9, and 10. **25.** 324, 42, 501, 3009, 75, 2001, 402, 111,111, 1005 **27.** 55,555, 200, 75, 2345, 35, 1005 **29.** 56, 784, 200 **31.** 200 **33.** 313,332, 7624, 111,126, 876, 1110, 5128, 64,000, 9990 **35.** 313,332, 111,126, 876, 1110, 9990 **37.** 9990 **39.** 1110, 64,000, 9990 **41.** 138 **42.** 139 **43.** 874 **44.** 56 **45.** 26 **46.** 13 **47.** 45 gal **48.** 4320 min **49.** $2 \cdot 2 \cdot 2 \cdot 3 \cdot 5 \cdot 5 \cdot 13$ **51.** $2 \cdot 2 \cdot 3 \cdot 3 \cdot 7 \cdot 11$ **53.** 95,238

Exercise Set 3.3, p. 144

RC1. (b) **RC2.** (a) **RC3.** (c) **RC4.** (e) **RC5.** (f) **RC6.** (d)
1. Numerator: 3; denominator: 4 **3.** Numerator: 11; denominator: 2 **5.** Numerator: 0; denominator: 7 **7.** $\frac{6}{12}$ **9.** $\frac{1}{8}$ **11.** $\frac{3}{4}$ **13.** $\frac{4}{8}$ **15.** $\frac{12}{12}$ **17.** $\frac{9}{8}$ **19.** $\frac{4}{3}$ **21.** $\frac{4}{3}$ **23.** $\frac{5}{8}$ **25.** $\frac{4}{7}$ **27.** $-\frac{1}{3}$ **29.** $\frac{12}{16}$ **31.** $\frac{38}{16}$ **33.** **(a)** $\frac{2}{8}$; **(b)** $\frac{6}{8}$ **35.** **(a)** $\frac{3}{8}$; **(b)** $\frac{5}{8}$ **37.** **(a)** $\frac{5}{7}$; **(b)** $\frac{5}{2}$; **(c)** $\frac{2}{7}$; **(d)** $\frac{2}{5}$ **39.** **(a)** $\frac{4}{15}$; **(b)** $\frac{4}{11}$; **(c)** $\frac{11}{15}$ **41.** **(a)** $\frac{1060}{100{,}000}$; **(b)** $\frac{743}{100{,}000}$; **(c)** $\frac{865}{100{,}000}$; **(d)** $\frac{1026}{100{,}000}$; **(e)** $\frac{905}{100{,}000}$; **(f)** $\frac{1728}{100{,}000}$ **43.** $\frac{4}{7}$ **45.** 0 **47.** 7 **49.** 1 **51.** 1 **53.** 0 **55.** 1 **57.** 1 **59.** 1 **61.** 1 **63.** −18 **65.** Not defined **67.** Not defined **69.** 90,283 **70.** 29,364 **71.** 4673 **72.** 5338 **73.** 6510 **74.** 14,526 **75.** −7 **76.** −15 **77.** $\frac{1}{6}$ **79.** $\frac{2}{16}$, or $\frac{1}{8}$
81.
83.

Exercise Set 3.4, p. 153

RC1. True **RC2.** True **RC3.** True **RC4.** True
1. $\frac{4}{15}$ **3.** $\frac{70}{9}$ **5.** $\frac{49}{64}$ **7.** $-\frac{2}{15}$ **9.** $\frac{40}{21}$ **11.** $-\frac{6}{5}$ **13.** $\frac{1}{6}$ **15.** $\frac{85}{6}$ **17.** $-\frac{7}{100}$ **19.** $\frac{2}{5}$ **21.** $-\frac{14}{39}$ **23.** $\frac{5}{8}$ **25.** $\frac{1}{40}$ **27.** $\frac{160}{27}$ **29.** $\frac{182}{285}$ **31.** $\frac{9}{16}$ **33.** $-\frac{8}{11}$ **35.** $\frac{40}{3}$ yd **37.** $\frac{1}{2625}$ **39.** $\frac{1}{16}$ **41.** $\frac{9}{20}$ **43.** $946 **44.** 201 min, or 3 hr 21 min **45.** 4^5 **46.** 16 **47.** 50 **48.** 6399 **49.** $\frac{71{,}269}{180{,}433}$ **51.** $-\frac{56}{1125}$

Calculator Corner, p. 158

1. $\frac{14}{15}$ **2.** $\frac{7}{8}$ **3.** $\frac{138}{167}$ **4.** $\frac{7}{25}$

Exercise Set 3.5, p. 160

RC1. Equivalent **RC2.** simplify **RC3.** common **RC4.** cross
1. $\frac{5}{10}$ **3.** $\frac{20}{32}$ **5.** $-\frac{27}{30}$ **7.** $\frac{20}{48}$ **9.** $\frac{-51}{-54}$ **11.** $\frac{42}{132}$ **13.** $\frac{1}{2}$ **15.** $-\frac{3}{4}$ **17.** $\frac{1}{5}$ **19.** −3 **21.** $\frac{3}{4}$ **23.** $\frac{7}{8}$ **25.** $\frac{6}{5}$ **27.** $\frac{1}{3}$ **29.** −6 **31.** $-\frac{1}{3}$ **33.** $\frac{2}{3}$ **35.** $-\frac{7}{90}$ **37.** $=$ **39.** $\neq$ **41.** $=$ **43.** $\neq$ **45.** $\neq$ **47.** $=$ **49.** $\frac{2}{5}$ **51.** $\frac{11}{50}$ **53.** $<$ **54.** $<$ **55.** $>$ **56.** $<$ **57.** 3520 **58.** 89 **59.** 6498 **60.** 85
61. No; $\frac{63}{82} \neq \frac{77}{100}$ because $63 \cdot 100 \neq 82 \cdot 77$.

Mid-Chapter Review: Chapter 3, p. 162

1. True **2.** False **3.** False **4.** True **5.** $\frac{25}{25} = 1$ **6.** $\frac{0}{-9} = 0$ **7.** $\frac{-8}{1} = -8$ **8.** $\frac{6}{13} = \frac{18}{39}$ **9.** $\frac{70}{225} = \frac{2 \cdot 5 \cdot 7}{3 \cdot 3 \cdot 5 \cdot 5} = \frac{5}{5} \cdot \frac{2 \cdot 7}{3 \cdot 3 \cdot 5} = 1 \cdot \frac{14}{45} = \frac{14}{45}$ **10.** 84, 17,576, 224, 132, 594, 504, 1632 **11.** 84, 300, 132, 500, 180 **12.** 17,576, 224, 500 **13.** 84, 300, 132, 120, 1632 **14.** 300, 180, 120 **15.** Prime **16.** Prime **17.** Composite **18.** Neither **19.** 1, 2, 4, 5, 8, 10, 16, 20, 32, 40, 80, 160; $2 \cdot 2 \cdot 2 \cdot 2 \cdot 2 \cdot 5$ **20.** 1, 2, 3, 6, 37, 74, 111, 222; $2 \cdot 3 \cdot 37$ **21.** 1, 2, 7, 14, 49, 98; $2 \cdot 7 \cdot 7$ **22.** 1, 3, 5, 7, 9, 15, 21, 35, 45, 63, 105, 315; $3 \cdot 3 \cdot 5 \cdot 7$ **23.** $\frac{8}{24}$, or $\frac{1}{3}$ **24.** $\frac{8}{6}$, or $\frac{4}{3}$ **25.** $\frac{7}{9}$ **26.** $\frac{8}{45}$ **27.** $-\frac{40}{11}$ **28.** $\frac{21}{32}$ **29.** $\frac{2}{5}$ **30.** $\frac{11}{3}$ **31.** 1 **32.** 0 **33.** $-\frac{9}{31}$ **34.** $\frac{9}{5}$ **35.** $\frac{5}{42}$ **36.** $\frac{21}{29}$ **37.** Not defined **38.** $=$ **39.** $\neq$ **40.** $\frac{25}{200}$, or $\frac{1}{8}$ **41.** $\frac{21}{10{,}000}\ \text{mi}^2$ **42.** Find the product of two prime numbers. **43.** If we use the divisibility tests, it is quickly clear that none of the even-numbered years is a prime number. In addition, the divisibility tests for 5 and 3 show that 2001, 2005, 2007, 2013, 2015, and 2019 are not prime numbers. Then the years 2003, 2009, 2011, and 2017 can be divided by prime numbers to determine whether they are prime. When we do this, we find that 2003, 2011, and 2017 are prime numbers. If the divisibility tests are not used, each of the numbers from 2000 to 2020 can be divided by prime numbers to determine if it is prime. **44.** It is possible to cancel only when identical *factors* appear in the numerator and the denominator of a fraction. Situations in which it is not possible to cancel include the occurrence of identical *addends* or *digits* in the numerator and the denominator. **45.** No; since the only factors of a prime number are the number itself and 1, two different prime numbers cannot contain a common factor (other than 1).

Exercise Set 3.6, p. 166

RC1. products **RC2.** Factor **RC3.** 1 **RC4.** Carry out **1.** $\frac{1}{3}$ **3.** $\frac{1}{8}$ **5.** $-\frac{1}{10}$ **7.** $\frac{1}{6}$ **9.** $\frac{27}{10}$ **11.** $-\frac{14}{9}$ **13.** 1 **15.** 1 **17.** 1 **19.** 1 **21.** 2 **23.** -4 **25.** 9 **27.** -9 **29.** $\frac{15}{2}$ **31.** $\frac{98}{5}$ **33.** 60 **35.** -30 **37.** $\frac{1}{5}$ **39.** $-\frac{9}{25}$ **41.** $\frac{11}{40}$ **43.** $\frac{5}{14}$ **45.** $\frac{5}{8}$ in. **47.** 260 million ounces **49.** 480 addresses **51.** $\frac{1}{3}$ cup **53.** \$115,500 **55.** 160 mi **57.** Food: \$8400; housing: \$10,500; clothing: \$4200; savings: \$3000; taxes: \$8400; other expenses: \$7500 **59.** 8587 **60.** 2707 **61.** -9 **62.** -13 **63.** 2203 **64.** 848 **65.** -4 **66.** 0 **67.** 26 **68.** 256 **69.** -425 **70.** 30 **71.** 0 **72.** 22 **73.** -204 **74.** 8 **75.** $\frac{129}{485}$ **77.** $\frac{1}{12}$ **79.** $\frac{1}{168}$

Translating for Success, p. 174

1. C **2.** H **3.** A **4.** N **5.** O **6.** F **7.** I **8.** L **9.** D **10.** M

Exercise Set 3.7, p. 175

RC1. True **RC2.** False **RC3.** True **RC4.** False **1.** $\frac{6}{5}$ **3.** $\frac{1}{6}$ **5.** 6 **7.** $-\frac{3}{10}$ **9.** $\frac{4}{5}$ **11.** $-\frac{4}{15}$ **13.** 4 **15.** 2 **17.** $\frac{1}{8}$ **19.** $\frac{3}{7}$ **21.** -8 **23.** 35 **25.** -1 **27.** $-\frac{2}{3}$ **29.** $\frac{9}{4}$ **31.** 144 **33.** 75 **35.** -2 **37.** $\frac{3}{5}$ **39.** -315 **41.** 960 extension cords **43.** 32 pairs **45.** 24 bowls **47.** 16 L **49.** 288 km; 108 km **51.** $\frac{1}{16}$ in. **53.** 59,960 members **54.** 71 gal **55.** 209,339 degrees **56.** 1,650,014 degrees **57.** 526,761 sq yd; 3020 yd **58.** \$928 **59.** $\frac{9}{19}$ **61.** 36 **63.** $\frac{3}{8}$

Summary and Review: Chapter 3, p. 178

Vocabulary Reinforcement

1. multiplicative **2.** factors **3.** prime **4.** denominator **5.** equivalent **6.** reciprocals **7.** factorization **8.** multiple

Concept Reinforcement

1. True **2.** False **3.** True **4.** True

Study Guide

1. 1, 2, 4, 8, 13, 26, 52, 104 **2.** $2 \cdot 2 \cdot 2 \cdot 13$ **3.** 0, 1, 18 **4.** $\frac{5}{14}$ **5.** $\neq$ **6.** $-\frac{70}{9}$ **7.** $\frac{7}{10}$ **8.** $\frac{7}{3}$ cups

Review Exercises

1. 1, 2, 3, 4, 5, 6, 10, 12, 15, 20, 30, 60 **2.** 1, 2, 4, 8, 11, 16, 22, 44, 88, 176 **3.** 8, 16, 24, 32, 40, 48, 56, 64, 72, 80 **4.** Yes **5.** No **6.** Prime **7.** Neither **8.** Composite **9.** $2 \cdot 5 \cdot 7$ **10.** $2 \cdot 3 \cdot 5$ **11.** $3 \cdot 3 \cdot 5$ **12.** $2 \cdot 3 \cdot 5 \cdot 5$ **13.** $2 \cdot 2 \cdot 2 \cdot 3 \cdot 3 \cdot 3 \cdot 3$ **14.** $2 \cdot 3 \cdot 5 \cdot 5 \cdot 5 \cdot 7$ **15.** 4344, 600, 93, 330, 255,555, 780, 2802, 711 **16.** 140, 182, 716, 2432, 4344, 600, 330, 780, 2802 **17.** 140, 716, 2432, 4344, 600, 780 **18.** 2432, 4344, 600 **19.** 140, 95, 475, 600, 330, 255,555, 780 **20.** 4344, 600, 330, 780, 2802 **21.** 255,555, 711 **22.** 140, 600, 330, 780 **23.** Numerator: 2; denominator: 7 **24.** $\frac{3}{5}$ **25.** $\frac{7}{6}$ **26.** $\frac{2}{7}$ **27.** (a) $\frac{3}{5}$; (b) $\frac{5}{3}$; (c) $\frac{3}{8}$ **28.** $\frac{2}{5}$ **29.** $\frac{1}{4}$ **30.** 1 **31.** 0 **32.** $\frac{39}{40}$ **33.** 18 **34.** $-\frac{1}{3}$ **35.** $-\frac{11}{23}$ **36.** Not defined **37.** 6 **38.** $\frac{2}{7}$ **39.** $-\frac{32}{225}$ **40.** $\frac{15}{100} = \frac{3}{20}$; $\frac{38}{100} = \frac{19}{50}$; $\frac{23}{100} = \frac{23}{100}$; $\frac{24}{100} = \frac{6}{25}$ **41.** $\neq$ **42.** $=$ **43.** $\neq$ **44.** $=$ **45.** $\frac{3}{2}$ **46.** 56 **47.** $-\frac{5}{2}$ **48.** -24 **49.** $\frac{2}{3}$ **50.** $\frac{1}{14}$ **51.** $-\frac{2}{3}$ **52.** $\frac{1}{22}$ **53.** $\frac{3}{20}$ **54.** $\frac{10}{7}$ **55.** $\frac{5}{4}$ **56.** $-\frac{1}{3}$ **57.** 9 **58.** $-\frac{36}{47}$ **59.** $\frac{9}{2}$ **60.** -2 **61.** $\frac{11}{6}$ **62.** $-\frac{1}{4}$ **63.** $\frac{9}{4}$ **64.** 300 **65.** 1 **66.** $\frac{4}{9}$ **67.** $\frac{3}{10}$ **68.** -240 **69.** 9 days **70.** \$32,085 **71.** 1000 km **72.** $\frac{1}{3}$ cup; 2 cups **73.** \$15 **74.** 60 bags **75.** D **76.** B **77.** $a = 11{,}176$; $b = 9887$ **78.** 13, 11, 101, 37

Understanding Through Discussion and Writing

1. The student is probably multiplying the divisor by the reciprocal of the dividend rather than multiplying the dividend by the reciprocal of the divisor. **2.** $9432 = 9 \cdot 1000 + 4 \cdot 100 + 3 \cdot 10 + 2 \cdot 1 = 9(999 + 1) + 4(99 + 1) + 3(9 + 1) + 2 \cdot 1 = 9 \cdot 999 + 9 \cdot 1 + 4 \cdot 99 + 4 \cdot 1 + 3 \cdot 9 + 3 \cdot 1 + 2 \cdot 1$. Since 999, 99, and 9 are each a multiple of 9, $9 \cdot 999$, $4 \cdot 99$, and $3 \cdot 9$ are multiples of 9. This leaves $9 \cdot 1 + 4 \cdot 1 + 3 \cdot 1 + 2 \cdot 1$, or $9 + 4 + 3 + 2$. If $9 + 4 + 3 + 2$, the sum of the digits, is divisible by 9, then 9432 is divisible by 9. **3.** Taking $\frac{1}{2}$ of a number is equivalent to multiplying the number by $\frac{1}{2}$. Dividing by $\frac{1}{2}$ is equivalent to multiplying by the reciprocal of $\frac{1}{2}$, or 2. Thus taking $\frac{1}{2}$ of a number is not the same as dividing by $\frac{1}{2}$. **4.** We first consider an object, and take $\frac{4}{7}$ of it. We divide the object into 7 parts and take 4 of them, as shown by the shading below.

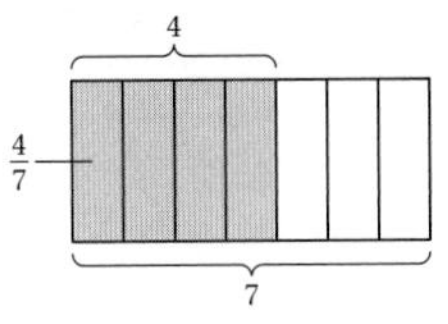

Next, we take $\frac{2}{3}$ of the shaded area above. We divide it into 3 parts and take two of them, as shown below.

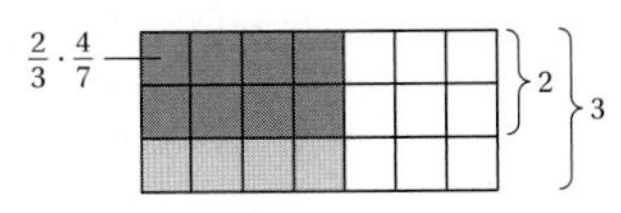

The entire object has been divided into 21 parts, 8 of which have been shaded twice. Thus, $\frac{2}{3} \cdot \frac{4}{7} = \frac{8}{21}$. **5.** Since $\frac{1}{7}$ is a smaller number than $\frac{2}{3}$, there are more $\frac{1}{7}$'s in 5 than $\frac{2}{3}$'s. Thus, $5 \div \frac{1}{7}$ is a greater number than $5 \div \frac{2}{3}$. **6.** No; in order to simplify a fraction, we must be able to remove a factor of the type n/n, $n \neq 0$, where n is a factor that the numerator and the denominator have in common.

Test: Chapter 3, p. 183

1. [3.1a] 1, 2, 3, 4, 5, 6, 10, 12, 15, 20, 25, 30, 50, 60, 75, 100, 150, 300 **2.** [3.1c] Prime **3.** [3.1c] Composite **4.** [3.1d] $2 \cdot 3 \cdot 3$ **5.** [3.1d] $2 \cdot 2 \cdot 3 \cdot 5$ **6.** [3.2a] Yes **7.** [3.2a] No **8.** [3.2a] No **9.** [3.2a] Yes **10.** [3.3a] Numerator: 4; denominator: 5 **11.** [3.3a] $\frac{3}{4}$ **12.** [3.3a] $\frac{3}{7}$ **13.** [3.3a] **(a)** $\frac{259}{365}$; **(b)** $\frac{106}{365}$ **14.** [3.3b] 26 **15.** [3.3b] 1 **16.** [3.3b] 0 **17.** [3.5b] $-\frac{1}{2}$ **18.** [3.5b] 6 **19.** [3.3b] Not defined **20.** [3.3b] Not defined **21.** [3.5b] $-\frac{2}{3}$ **22.** [3.5c] $=$ **23.** [3.5c] $\neq$ **24.** [3.6a] 32 **25.** [3.6a] $-\frac{3}{2}$ **26.** [3.6a] $\frac{5}{2}$ **27.** [3.6a] $\frac{2}{9}$ **28.** [3.7a] $\frac{8}{5}$ **29.** [3.7a] -4 **30.** [3.7a] $\frac{1}{18}$ **31.** [3.7b] $\frac{8}{5}$ **32.** [3.7b] -18 **33.** [3.7b] $\frac{18}{7}$ **34.** [3.7c] -64 **35.** [3.7c] $\frac{7}{4}$ **36.** [3.6b] 4375 students **37.** [3.7d] $\frac{3}{40}$ m **38.** [3.7d] 5 qt **39.** [3.6b] $\frac{3}{4}$ in. **40.** [3.3a] C **41.** [3.6b] $\frac{7}{48}$ acre **42.** [3.6a], [3.7b] $-\frac{7}{960}$

Cumulative Review: Chapters 1–3, p. 185

1. [1.1c] Seven million, four hundred fifty-three thousand, sixty-two **2.** [1.1a] 4 ten thousands **3.** [3.3b] 5 **4.** [3.3b] 1 **5.** [3.5b] $-\frac{4}{3}$ **6.** [3.3b] Not defined **7.** [3.7a] $\frac{1}{8}$ **8.** [3.7a] $-\frac{2}{3}$ **9.** [1.2a] 5387 **10.** [2.2a] -12 **11.** [1.3a] 1384 **12.** [2.3a] -16 **13.** [1.4a] 16,452 **14.** [2.4a] -96 **15.** [3.6a] $-\frac{25}{2}$ **16.** [3.6a] $\frac{2}{3}$ **17.** [1.5a] 451 R 22 **18.** [2.5a] 8 **19.** [3.7b] $-\frac{1}{49}$ **20.** [3.7b] $\frac{4}{3}$ **21.** [1.9c] 0 **22.** [1.9c] 3 **23.** [2.5c] -14 **24.** [2.5c] -4 **25.** [1.9c] 86 **26.** [1.6a] 166,000 **27.** [1.9b] 81 **28.** [1.9b] 125 **29.** [1.9b] 16 **30.** [2.1c] $<$ **31.** [1.6c] $>$ **32.** [2.1d] 33 **33.** [2.1d] 86 **34.** [2.1d] 0 **35.** [2.2b] -29 **36.** [2.2b] 144 **37.** [2.2b] -7 **38.** [2.1b]

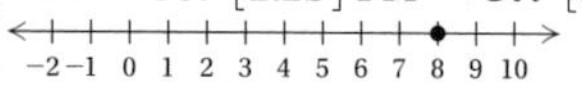

39. [3.1c] Prime **40.** [3.1c] Neither **41.** [3.1c] Composite **42.** [3.1d] $2 \cdot 2 \cdot 3 \cdot 3$ **43.** [3.1d] $3 \cdot 3 \cdot 3 \cdot 5$ **44.** [3.1d] $2 \cdot 3 \cdot 3 \cdot 3 \cdot 5 \cdot 5$ **45.** [3.2a] Divisible by 2, 3, 5, 6, 10; not divisible by 4, 8, 9 **46.** [3.5c] $\neq$ **47.** [3.5c] $=$ **48.** [1.7b] 44 **49.** [3.7c] $-\frac{3}{5}$ **50.** [1.8a] 2062 **51.** [1.8a] \$36 **52.** [2.3b], [2.5b] $-17°$C **53.** [3.6b] 14 students **54.** **(a)** [1.2a], [1.4a] 5, 20; **(b)** [2.2a], [2.4a] -4, -25; **(c)** [1.3a], [1.4a], [2.3a], [2.4a] 10, 10; -10, -10

CHAPTER 4

Exercise Set 4.1, p. 193

RC1. True **RC2.** True **RC3.** True **RC4.** False **1.** 4 **3.** 50 **5.** 40 **7.** 54 **9.** 150 **11.** 120 **13.** 72 **15.** 420 **17.** 144 **19.** 288 **21.** 30 **23.** 90 **25.** 72 **27.** 60 **29.** 36 **31.** 900 **33.** 48 **35.** 50 **37.** 143 **39.** 420 **41.** 378 **43.** 810 **45.** 2160 **47.** 9828 **49.** 6000 **51.** Every 60 years **53.** Every 420 years **55.** -18 **56.** 59 R 77 **57.** $\frac{8}{7}$ **58.** 33,135 **59.** 6,356,118 **60.** 77,699 **61.** 5 in. by 24 in.

Exercise Set 4.2, p. 199

RC1. True **RC2.** False **RC3.** False **RC4.** True **1.** 1 **3.** $\frac{3}{4}$ **5.** $-\frac{1}{6}$ **7.** $\frac{7}{24}$ **9.** $-\frac{1}{10}$ **11.** $\frac{19}{24}$ **13.** $\frac{9}{10}$ **15.** $\frac{1}{18}$ **17.** $\frac{31}{100}$ **19.** $\frac{41}{60}$ **21.** $\frac{9}{100}$ **23.** $\frac{7}{8}$ **25.** $\frac{13}{24}$ **27.** $\frac{17}{24}$ **29.** $\frac{3}{4}$ **31.** $\frac{437}{500}$ **33.** $\frac{13}{40}$ **35.** $\frac{391}{144}$ **37.** $\frac{37}{12}$ mi **39.** $\frac{13}{16}$ lb **41.** $\frac{27}{32}''$ **43.** $\frac{51}{32}$ in. **45.** $\frac{7}{8}$ in. **47.** $\frac{33}{20}$ mi **49.** $\frac{4}{5}$ qt; $\frac{8}{5}$ qt; $\frac{2}{5}$ qt **51.** $\frac{1}{24}$ **52.** 286 cartons; 2 oz left over **53.** 18,940,000 lb **54.** $\frac{3}{64}$ acre **55.** $\frac{4}{15}$; \$320

Translating for Success, p. 206

1. J **2.** E **3.** D **4.** B **5.** I **6.** N **7.** A **8.** C **9.** L **10.** F

Exercise Set 4.3, p. 207

RC1. numerators; denominator **RC2.** denominators **RC3.** denominators **RC4.** numerators **1.** $\frac{2}{3}$ **3.** $\frac{3}{4}$ **5.** $-\frac{5}{8}$ **7.** $\frac{1}{24}$ **9.** $-\frac{1}{2}$ **11.** $\frac{9}{14}$ **13.** $\frac{3}{5}$ **15.** $-\frac{7}{10}$ **17.** $\frac{17}{60}$ **19.** $\frac{53}{100}$ **21.** $\frac{26}{75}$ **23.** $\frac{9}{100}$ **25.** $-\frac{19}{24}$ **27.** $\frac{1}{10}$ **29.** $-\frac{1}{24}$ **31.** $\frac{13}{16}$ **33.** $-\frac{31}{75}$ **35.** $\frac{13}{75}$ **37.** $<$ **39.** $>$ **41.** $<$ **43.** $<$ **45.** $>$ **47.** $>$ **49.** $<$ **51.** $\frac{1}{15}$ **53.** $-\frac{22}{15}$ **55.** $\frac{1}{2}$ **57.** $\frac{5}{12}$ hr **59.** $\frac{1}{32}$ in. **61.** $\frac{11}{20}$ lb **63.** $\frac{1}{4}$ tub **65.** 1 **66.** Not defined **67.** Not defined **68.** 4 **69.** $-\frac{4}{21}$ **70.** $\frac{3}{2}$ **71.** 21 **72.** $\frac{1}{32}$ **73.** 12 **74.** $-\frac{5}{2}$ **75.** $\frac{1}{4}$ **76.** $\frac{2}{5}$ **77.** $\frac{14}{3553}$ **79.** $\frac{21}{40}$ km **81.** $\frac{19}{24}$ **83.** $-\frac{17}{144}$ **85.** $>$ **87.** *Day 1:* Cut off $\frac{1}{7}$ of bar and pay him. *Day 2:* Cut off $\frac{2}{7}$ of bar. Trade him for the $\frac{1}{7}$. *Day 3:* Give him back the $\frac{1}{7}$. *Day 4:* Trade him the $\frac{4}{7}$ for his $\frac{3}{7}$. *Day 5:* Give him the $\frac{1}{7}$ again. *Day 6:* Trade him the $\frac{2}{7}$ for the $\frac{1}{7}$. *Day 7:* Give him the $\frac{1}{7}$ again. This assumes that he does not spend parts of the gold bar immediately.

Exercise Set 4.4, p. 213

RC1. True **RC2.** True **RC3.** False **RC4.** True **1.** $\frac{5}{2}, \frac{15}{8}, \frac{7}{4}$ **3.** $4\frac{1}{4}, 3\frac{1}{3}, 1\frac{1}{8}$ **5.** $\frac{17}{3}$ **7.** $\frac{13}{4}$ **9.** $-\frac{81}{8}$ **11.** $\frac{51}{10}$ **13.** $\frac{103}{5}$ **15.** $-\frac{59}{6}$ **17.** $\frac{73}{10}$ **19.** $\frac{13}{8}$ **21.** $-\frac{51}{4}$ **23.** $-\frac{43}{10}$ **25.** $\frac{203}{100}$ **27.** $\frac{200}{3}$ **29.** $-\frac{279}{50}$ **31.** $\frac{1621}{16}$ **33.** $3\frac{3}{5}$ **35.** $4\frac{2}{3}$ **37.** $-4\frac{1}{2}$ **39.** $5\frac{7}{10}$ **41.** $-7\frac{4}{7}$ **43.** $7\frac{1}{2}$ **45.** $11\frac{1}{2}$ **47.** $-1\frac{1}{2}$ **49.** $7\frac{57}{100}$ **51.** $-43\frac{1}{8}$ **53.** $108\frac{5}{8}$ **55.** $618\frac{1}{5}$ **57.** $40\frac{4}{7}$ **59.** $55\frac{1}{51}$ **61.** $2292\frac{23}{35}$ **63.** 45,800 **64.** 45,770 **65.** $\frac{8}{15}$ **66.** $\frac{21}{25}$ **67.** $-\frac{16}{27}$ **68.** $-\frac{583}{669}$ **69.** $=$ **70.** $\neq$ **71.** $\frac{7}{9}$ **72.** -8 **73.** 35 **74.** 6407 **75.** $237\frac{19}{541}$ **77.** $8\frac{2}{3}$ **79.** $52\frac{2}{7}$

Mid-Chapter Review: Chapter 4, p. 216

1. True **2.** True **3.** False **4.** False

5.
$$\begin{aligned}\frac{11}{42} - \frac{3}{35} &= \frac{11}{2 \cdot 3 \cdot 7} - \frac{3}{5 \cdot 7}\\ &= \frac{11}{2 \cdot 3 \cdot 7} \cdot \left(\frac{5}{5}\right) - \frac{3}{5 \cdot 7} \cdot \left(\frac{2 \cdot 3}{2 \cdot 3}\right)\\ &= \frac{11 \cdot 5}{2 \cdot 3 \cdot 7 \cdot 5} - \frac{3 \cdot 2 \cdot 3}{5 \cdot 7 \cdot 2 \cdot 3}\\ &= \frac{55}{2 \cdot 3 \cdot 5 \cdot 7} - \frac{18}{2 \cdot 3 \cdot 5 \cdot 7}\\ &= \frac{55 - 18}{2 \cdot 3 \cdot 5 \cdot 7} = \frac{37}{210}\end{aligned}$$

6.
$$\begin{aligned}x + \frac{1}{8} &= \frac{2}{3}\\ x + \frac{1}{8} - \frac{1}{8} &= \frac{2}{3} - \frac{1}{8}\\ x + 0 &= \frac{2}{3} \cdot \frac{8}{8} - \frac{1}{8} \cdot \frac{3}{3}\\ x &= \frac{16}{24} - \frac{3}{24}\\ x &= \frac{13}{24}\end{aligned}$$

7.

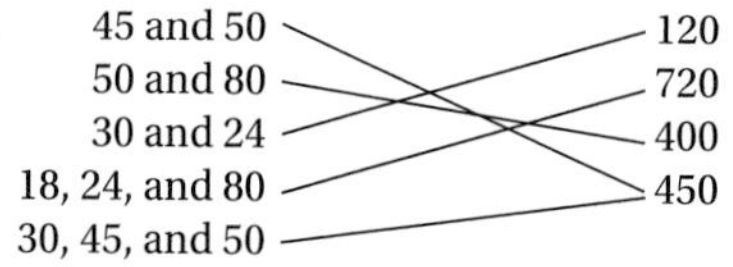

8. $\frac{16}{45}$ **9.** $\frac{25}{12}$ **10.** $-\frac{1}{18}$ **11.** $-\frac{19}{90}$ **12.** $\frac{7}{240}$ **13.** $\frac{156}{119}$ **14.** $\frac{79}{720}$ **15.** $-\frac{6}{91}$ **16.** $\frac{22}{15}$ mi **17.** $\frac{101}{20}$ hr **18.** $\frac{1}{5}, \frac{2}{7}, \frac{3}{10}, \frac{4}{9}$ **19.** $\frac{13}{80}$ **20.** $17\frac{8}{15}$ **21.** C **22.** C **23.** No; if one number is a multiple of the other, for example, the LCM is the larger of the numbers. **24.** We multiply by 1, using the notation n/n, to express each fraction in terms of the least common denominator. **25.** Write $\frac{8}{5}$ as $\frac{16}{10}$ and $\frac{8}{2}$ as $\frac{40}{10}$ and since taking 40 tenths away from 16 tenths would give a result less than 0, it cannot possibly be $\frac{8}{3}$. You could also find the sum $\frac{8}{3} + \frac{8}{2}$ and show that it is not $\frac{8}{5}$. **26.** No; $2\frac{1}{3} = \frac{7}{3}$ but $2 \cdot \frac{1}{3} = \frac{2}{3}$.

Exercise Set 4.5, p. 223

RC1. (d) **RC2.** (a) **RC3.** (b) **RC4.** (c) **1.** $28\frac{3}{4}$ **3.** $185\frac{7}{8}$ **5.** $6\frac{1}{2}$ **7.** $2\frac{11}{12}$ **9.** $14\frac{7}{12}$ **11.** $12\frac{1}{10}$ **13.** $16\frac{5}{24}$ **15.** $21\frac{1}{2}$ **17.** $27\frac{7}{8}$ **19.** $27\frac{13}{24}$ **21.** $1\frac{3}{5}$ **23.** $4\frac{1}{10}$ **25.** $21\frac{17}{24}$ **27.** $12\frac{1}{4}$ **29.** $15\frac{3}{8}$ **31.** $7\frac{5}{12}$ **33.** $13\frac{3}{8}$ **35.** $11\frac{5}{18}$ **37.** $14\frac{13}{24}$ flats **39.** $7\frac{5}{12}$ lb **41.** $\frac{15}{16}$ in. **43.** $20\frac{1}{12}$ yd **45.** $5\frac{3}{8}$ yd **47.** $95\frac{1}{5}$ mi **49.** $134\frac{1}{4}$ in. **51.** $78\frac{1}{12}$ in. **53.** $28\frac{3}{4}$ yd

55. $7\frac{3}{8}$ ft **57.** $5\frac{7}{8}$ in. **59.** $20\frac{1}{8}$ in. **61.** $3\frac{4}{5}$ hr **63.** $66\frac{5}{6}$ ft **65.** Yes **66.** No **67.** No **68.** Yes **69.** No **70.** Yes **71.** Yes **72.** Yes **73.** 3 ten thousands + 8 thousands + 1 hundred + 2 tens + 5 ones **74.** Two million, five thousand, six hundred eighty-nine **75.** 9^4 **76.** 81 **77.** $-8\frac{5}{8}$ **79.** $-11\frac{1}{24}$

Calculator Corner, p. 231

1. $\frac{7}{12}$ **2.** $\frac{11}{10}$ **3.** $\frac{35}{16}$ **4.** $\frac{3}{10}$ **5.** $10\frac{2}{15}$ **6.** $1\frac{1}{28}$ **7.** $10\frac{11}{15}$ **8.** $2\frac{91}{115}$

Translating for Success, p. 232

1. O **2.** K **3.** F **4.** D **5.** H **6.** G **7.** L **8.** E **9.** M **10.** J

Exercise Set 4.6, p. 233

RC1. True **RC2.** True **RC3.** True **RC4.** True **1.** $22\frac{2}{3}$ **3.** $-2\frac{5}{12}$ **5.** $8\frac{1}{6}$ **7.** $9\frac{31}{40}$ **9.** $-24\frac{91}{100}$ **11.** $975\frac{4}{5}$ **13.** $6\frac{1}{4}$ **15.** $1\frac{1}{5}$ **17.** $-3\frac{9}{16}$ **19.** $1\frac{1}{8}$ **21.** $1\frac{8}{43}$ **23.** $\frac{9}{40}$ **25.** About $24 million **27.** About 4,800,000 **29.** $62\frac{1}{2}$ sq ft **31.** $12\frac{4}{5}$ tiles **33.** $343\frac{3}{4}$ lb **35.** 68°F **37.** About $42,000,000,000 **39.** 75 mph **41.** $5\frac{1}{2}$ cups of flour, $2\frac{2}{3}$ cups of sugar **43.** 15 mpg **45.** 400 cu ft **47.** $16\frac{1}{2}$ servings **49.** $35\frac{115}{256}$ sq in.
51. $13\frac{1}{4}$ in. × $13\frac{1}{4}$ in.: perimeter = 53 in.,
area = $175\frac{9}{16}$ sq in.;
$13\frac{1}{4}$ in. × $3\frac{1}{4}$ in.: perimeter = 33 in.,
area = $43\frac{1}{16}$ sq in.
53. $408 **54.** 9 bills **55.** 126 slices **56.** 9 cups **57.** $36,500 a year **58.** 16 packages **59.** 22 m **60.** 4500 sq yd **61.** $360\frac{60}{473}$ **63.** $-2\frac{25}{64}$ **65.** $\frac{4}{9}$ **67.** $1\frac{4}{5}$

Exercise Set 4.7, p. 242

RC1. (c) **RC2.** (d) **RC3.** (a) **RC4.** (b) **1.** $\frac{59}{30}$, or $1\frac{29}{30}$ **3.** $\frac{3}{20}$ **5.** $\frac{211}{8}$, or $26\frac{3}{8}$ **7.** $\frac{7}{16}$ **9.** $\frac{1}{36}$ **11.** $\frac{3}{8}$ **13.** $\frac{17}{6}$, or $2\frac{5}{6}$ **15.** $\frac{8395}{84}$, or $99\frac{79}{84}$ **17.** $\frac{3}{11}$ **19.** $-\frac{14}{3}$, or $-4\frac{2}{3}$ **21.** $\frac{5}{4}$, or $1\frac{1}{4}$ **23.** $\frac{1}{100}$ **25.** $-\frac{1}{6}$ **27.** $\frac{7}{12}$ **29.** $\frac{1}{3}$ **31.** $\frac{37}{48}$ **33.** $\frac{25}{72}$ **35.** $\frac{103}{16}$, or $6\frac{7}{16}$ **37.** $16\frac{7}{96}$ mi **39.** $9\frac{19}{40}$ lb **41.** 0 **43.** 1 **45.** $\frac{1}{2}$ **47.** $\frac{1}{2}$ **49.** 0 **51.** 1 **53.** 3 **55.** 2 **57.** 3 **59.** $271\frac{1}{2}$ **61.** $29\frac{1}{2}$ **63.** 20 **64.** 2 **65.** 84 **66.** 100 **67.** 1, 2, 3, 6, 7, 14, 21, 42 **68.** No **69.** Prime: 5, 7, 23, 43; composite: 9, 14; neither: 1 **70.** $2 \cdot 3 \cdot 5 \cdot 5$ **71.** $a = 2, b = 8$ **73.** The largest is $\frac{4}{3} + \frac{5}{2} = \frac{23}{6}$.

Summary and Review: Chapter 4, p. 245

Vocabulary Reinforcement

1. least common multiple **2.** mixed numeral **3.** fraction **4.** complex fraction **5.** denominators **6.** least common multiple **7.** greatest **8.** numerators

Concept Reinforcement

1. True **2.** True **3.** False **4.** True

Study Guide

1. 156 **2.** $\frac{112}{180}$, or $\frac{28}{45}$ **3.** $\frac{4}{35}$ **4.** $<$ **5.** $-\frac{103}{99}$ **6.** $\frac{26}{3}$ **7.** $-7\frac{5}{6}$ **8.** $7\frac{27}{28}$ **9.** $-14\frac{14}{25}$ **10.** About 4,500,000 **11.** $\frac{9}{2}$, or $4\frac{1}{2}$ **12.** $4\frac{1}{2}$

Review Exercises

1. 36 **2.** 90 **3.** 30 **4.** 1404 **5.** $\frac{63}{40}$ **6.** $-\frac{11}{48}$ **7.** $\frac{25}{12}$ **8.** $\frac{891}{1000}$ **9.** $\frac{1}{3}$ **10.** $\frac{1}{8}$ **11.** $-\frac{5}{27}$ **12.** $-\frac{19}{18}$ **13.** $>$ **14.** $>$ **15.** $\frac{19}{40}$ **16.** $-\frac{7}{5}$ **17.** $\frac{15}{2}$ **18.** $\frac{67}{8}$ **19.** $\frac{13}{3}$ **20.** $-\frac{75}{7}$ **21.** $2\frac{1}{3}$ **22.** $6\frac{3}{4}$ **23.** $-12\frac{3}{5}$ **24.** $3\frac{1}{2}$ **25.** $877\frac{1}{3}$ **26.** $456\frac{5}{23}$ **27.** $10\frac{2}{5}$ **28.** $11\frac{11}{15}$ **29.** $10\frac{2}{3}$ **30.** $8\frac{1}{4}$ **31.** $7\frac{7}{9}$ **32.** $4\frac{11}{15}$ **33.** $4\frac{3}{20}$ **34.** $13\frac{3}{8}$ **35.** 16 **36.** $-3\frac{1}{2}$ **37.** $2\frac{21}{50}$ **38.** -6 **39.** -12 **40.** $1\frac{7}{17}$ **41.** $\frac{1}{8}$ **42.** $\frac{9}{10}$ **43.** $4\frac{1}{4}$ yd **44.** $177\frac{3}{4}\text{ in}^2$ **45.** $50\frac{1}{4}\text{ in}^2$ **46.** $1\frac{73}{100}$ in. **47.** 24 lb **48.** $8\frac{3}{8}$ cups **49.** $850 **50.** $3\frac{3}{4}$ mi **51.** $63\frac{2}{3}$ pies; $19\frac{1}{3}$ pies **52.** 1 **53.** $\frac{7}{40}$ **54.** 3 **55.** $-\frac{4}{5}$ **56.** 30 **57.** $\frac{77}{240}$ **58.** $\frac{1}{2}$ **59.** 0 **60.** 1 **61.** 7 **62.** 10 **63.** $5\frac{1}{2}$ **64.** 0 **65.** $2\frac{1}{2}$ **66.** $28\frac{1}{2}$ **67.** A **68.** D **69.** 12 min **70.** $\frac{6}{3} + \frac{5}{4} = 3\frac{1}{4}$

Understanding Through Discussion and Writing

1. No; if the sum of the fractional parts of the mixed numerals is n/n, then the sum of the mixed numerals is an integer. For example, $1\frac{1}{5} + 6\frac{4}{5} = 7\frac{5}{5} = 8$. **2.** Answers may vary. A wheel makes $33\frac{1}{3}$ revolutions per minute. It rotates for $4\frac{1}{2}$ min. How many revolutions does it make? **3.** The student is multiplying the whole numbers to get the whole-number portion of the answer and multiplying fractions to get the fraction part of the answer. The student should have converted each mixed numeral to fraction notation, multiplied, simplified, and then converted back to a mixed numeral. The correct answer is $4\frac{6}{7}$. **4.** It might be necessary to find the least common denominator before adding or subtracting. The least common denominator is the least common multiple of the denominators. **5.** Answers may vary. Suppose that a room has dimensions $15\frac{3}{4}$ ft by $28\frac{5}{8}$ ft. The equation $2 \cdot 15\frac{3}{4} + 2 \cdot 28\frac{5}{8} = 88\frac{3}{4}$ gives the perimeter of the room, in feet. **6.** Note that $5 \cdot 3\frac{2}{7} = 5\left(3 + \frac{2}{7}\right) = 5 \cdot 3 + 5 \cdot \frac{2}{7}$. The products $5 \cdot 3$ and $5 \cdot \frac{2}{7}$ should be added rather than multiplied together. The student could also have converted $3\frac{2}{7}$ to fraction notation, multiplied, simplified, and converted back to a mixed numeral. The correct answer is $16\frac{3}{7}$.

Test: Chapter 4, p. 251

1. [4.1a] 48 **2.** [4.1a] 600 **3.** [4.2a] 3 **4.** [4.2a] $-\frac{5}{24}$ **5.** [4.2a] $\frac{921}{1000}$ **6.** [4.3a] $\frac{1}{3}$ **7.** [4.3a] $\frac{1}{12}$ **8.** [4.3a] $-\frac{31}{40}$ **9.** [4.3c] $\frac{15}{4}$ **10.** [4.3c] $\frac{1}{4}$ **11.** [4.3b] $>$ **12.** [4.4a] $\frac{7}{2}$ **13.** [4.4a] $-\frac{79}{8}$ **14.** [4.4a] $4\frac{1}{2}$ **15.** [4.4a] $-8\frac{2}{9}$ **16.** [4.4b] $162\frac{7}{11}$ **17.** [4.5a] $14\frac{1}{5}$ **18.** [4.5a] $14\frac{5}{12}$ **19.** [4.5b] $4\frac{7}{24}$ **20.** [4.5b] $6\frac{1}{6}$ **21.** [4.6a] 39 **22.** [4.6a] $-4\frac{1}{2}$ **23.** [4.6b] 2 **24.** [4.6b] $-\frac{1}{36}$ **25.** [4.6c] About 105 kg **26.** [4.6c] 80 books **27.** [4.5c] **(a)** 3 in.; **(b)** $4\frac{1}{2}$ in. **28.** [4.3d] $\frac{1}{16}$ in. **29.** [4.7b] $6\frac{11}{36}$ ft **30.** [4.7a] $3\frac{1}{2}$ **31.** [4.7a] $\frac{3}{4}$ **32.** [4.7b] $\frac{1}{6}$ **33.** [4.7c] 0 **34.** [4.7c] 1 **35.** [4.7c] 16 **36.** [4.1a] D **37.** [4.1a] **(a)** 24, 48, 72; **(b)** 24 **38.** [4.3b], [4.5c] Rebecca walks $\frac{17}{56}$ mi farther.

Cumulative Review: Chapters 1–4, p. 253

1. (a) [4.5c] $14\frac{13}{24}$ mi; **(b)** [4.7b] $4\frac{61}{72}$ mi **2.** [1.8a] 31 people **3.** [3.6b] $\frac{2}{5}$ tsp; 4 tsp **4.** [4.6c] 16 pieces **5.** [1.8a] $108 **6.** [4.2b] $\frac{33}{20}$ mi **7.** [3.3a] $\frac{5}{16}$ **8.** [3.3a] $\frac{4}{3}$ **9.** [1.2a] 8982 **10.** [1.3a] 4518 **11.** [2.2a] -15 **12.** [4.2a] $\frac{5}{12}$ **13.** [2.3a] -14 **14.** [4.3a] $-\frac{13}{12}$ **15.** [4.5a] $8\frac{1}{4}$ **16.** [4.5b] $1\frac{1}{6}$ **17.** [2.4a] -75 **18.** [3.6a] $\frac{3}{2}$ **19.** [3.6a] 15 **20.** [4.6b] $7\frac{1}{3}$ **21.** [1.5a] 715 **22.** [1.5a] 56 R 11 **23.** [4.4b] $56\frac{11}{45}$ **24.** [1.1a] 5 **25. (a)** [4.6c] $142\frac{1}{4}\text{ ft}^2$; **(b)** [4.5c] 54 ft **26.** [1.6a] 38,500 **27.** [4.1a] 72 **28.** [4.7a] $\frac{1377}{100}$, or $13\frac{77}{100}$ **29.** [4.3b] $>$ **30.** [3.5c] $=$ **31.** [4.3b] $<$ **32.** [4.7c] 1 **33.** [4.7c] $\frac{1}{2}$ **34.** [4.7c] 0 **35.** [3.5b] $\frac{4}{5}$ **36.** [3.3b] 0 **37.** [3.5b] 32 **38.** [4.4a] $\frac{37}{8}$ **39.** [4.4a] $5\frac{2}{3}$ **40.** [1.7b] 93 **41.** [4.3c] $\frac{5}{9}$ **42.** [3.7c] $-\frac{12}{7}$ **43.** [1.7b] 905 **44.** [3.1a, c, d], [3.2a] Factors of 68: 1, 2, 4, 17, 34, 68. Factorization of 68: $2 \cdot 2 \cdot 17$, or $2 \cdot 34$. Prime factorization of 68: $2 \cdot 2 \cdot 17$. Numbers divisible by 6: 12, 54, 72, 300. Numbers divisible by 8: 8, 16, 24, 32, 40, 48, 64, 864. Numbers divisible by 5: 70, 95, 215. Prime numbers: 2, 3, 17, 19, 23, 31, 47, 101. **45.** [3.1c] 2003

CHAPTER 5

Exercise Set 5.1, p. 263

RC1. 5 **RC2.** 2 **RC3.** 3 **RC4.** 0 **RC5.** 8 **RC6.** 4 **RC7.** 1 **RC8.** 6
1. One hundred nineteen ten-thousandths
3. One hundred thirty-seven and six tenths
5. Five hundred nineteen and twenty-two hundredths
7. Three and seven hundred eighty-five thousandths **9.** Negative thirty-four and eight hundred ninety-one thousandths
11. $\frac{83}{10}$ **13.** $\frac{356}{100}$ **15.** $\frac{20,003}{1000}$ **17.** $-\frac{10,008}{10,000}$ **19.** $\frac{372}{10}$ **21.** $-\frac{13}{100,000}$ **23.** 0.8 **25.** 3.798 **27.** −8.89 **29.** 0.00019 **31.** 0.0078 **33.** −0.376193 **35.** 2.9 **37.** 3.098 **39.** 99.44 **41.** −2.1739 **43.** 0.58 **45.** 0.91 **47.** −0.0009 **49.** 235.07 **51.** $\frac{4}{100}$ **53.** −0.432 **55.** 0.1 **57.** −0.5 **59.** 2.7 **61.** −123.7 **63.** 0.89 **65.** 0.67 **67.** −1.00 **69.** −0.09 **71.** 0.325 **73.** −17.001 **75.** 10.101 **77.** −9.999 **79.** 800 **81.** 809.573 **83.** 810 **85.** 34.5439 **87.** 34.54 **89.** 35 **91.** 6170 **92.** 6200 **93.** 6000 **94.** $2 \cdot 2 \cdot 2 \cdot 2 \cdot 5 \cdot 5 \cdot 5$, or $2^4 \cdot 5^3$ **95.** $2 \cdot 3 \cdot 3 \cdot 5 \cdot 17$, or $2 \cdot 3^2 \cdot 5 \cdot 17$ **96.** $\frac{5}{21}$ **97.** $\frac{6}{5}$ **99.** −2.109, −2.108, −2.1, −2.0302, −2.018, −2.0119, −2.000001 **101.** 6.78346 **103.** 0.03030

Exercise Set 5.2, p. 270

RC1. 21.824; 23.7 **RC2.** 146.723; 40.9
1. 334.37 **3.** 1576.215 **5.** 132.560 **7.** 50.0248 **9.** 40.007 **11.** 977.955 **13.** 771.967 **15.** 8754.8221 **17.** 49.02 **19.** 85.921 **21.** 2.4975 **23.** 3.397 **25.** 8.85 **27.** 3.37 **29.** 1.045 **31.** 3.703 **33.** 0.9902 **35.** 99.66 **37.** 4.88 **39.** 0.994 **41.** 17.802 **43.** 51.13 **45.** 32.7386 **47.** 4.0622 **49.** −3.29 **51.** −2.5 **53.** −7.2 **55.** 3.379 **57.** −16.6 **59.** 2.5 **61.** −3.519 **63.** 9.601 **65.** 75.5 **67.** 3.8 **69.** −10.292 **71.** −8.8 **73.** 11.65 **75.** 384.68 **77.** −20,033.3 **79.** 582.97 **81.** −533.6 **83.** −6.99 **85.** 35,000 **86.** 34,000 **87.** $\frac{1}{6}$ **88.** $\frac{34}{45}$ **89.** 6166 **90.** 5366 **91.** $16\frac{1}{2}$ servings **92.** $60\frac{1}{5}$ mi **93.** 345.8

Exercise Set 5.3, p. 279

RC1. (c) **RC2.** (e) **RC3.** (f) **RC4.** (a) **RC5.** (d) **RC6.** (b)
1. 60.2 **3.** 6.72 **5.** 0.252 **7.** 0.522 **9.** 237.6 **11.** −583,686.852 **13.** −780 **15.** 8.923 **17.** 0.09768 **19.** −0.782 **21.** 521.6 **23.** 3.2472 **25.** −897.6 **27.** 322.07 **29.** 55.68 **31.** 3487.5 **33.** 50.0004 **35.** 114.42902 **37.** 13.284 **39.** 90.72 **41.** 0.0028728 **43.** 0.72523 **45.** −1.872115 **47.** 45,678 **49.** 2888¢ **51.** 66¢ **53.** \$0.34 **55.** \$34.45 **57.** 3,480,000 **59.** 50,960,000,000; 13,410,000,000 **61.** 2,200,000 **63.** $11\frac{1}{5}$ **64.** $-\frac{35}{72}$ **65.** $2\frac{7}{15}$ **66.** $7\frac{2}{15}$ **67.** 87 **68.** 1176 R 14 **69.** −7 **70.** −14 **71.** 10^{21} = 1 sextillion **73.** 10^{24} = 1 septillion

Calculator Corner, p. 283

1. 17.15 **2.** 33.83 **3.** 454.74 **4.** 0.5076 **5.** 2.56 **6.** 0.064

Exercise Set 5.4, p. 289

RC1. Subtract: 2 − 0.04 **RC2.** Divide: 2.06 ÷ 0.01 **RC3.** Evaluate: 8^3 **RC4.** Add: 4.1 + 6.9 **RC5.** Divide: 9 ÷ 3 **RC6.** Subtract: 10 − 5
1. 2.99 **3.** 23.78 **5.** 7.48 **7.** 7.2 **9.** −1.143 **11.** −4.041 **13.** 56 **15.** 70 **17.** 20 **19.** 0.4 **21.** 0.41 **23.** 8.5 **25.** 9.3 **27.** −0.625 **29.** 0.26 **31.** 15.625 **33.** 2.34 **35.** 0.47 **37.** 0.2134567 **39.** −2.359 **41.** 4.26487 **43.** −169.4 **45.** 1023.7 **47.** −4256.1 **49.** 9.3 **51.** −0.0090678 **53.** 45.6 **55.** 2107 **57.** −302.997 **59.** 446.208 **61.** 24.14 **63.** −5.0072 **65.** 19.3204 **67.** 473.188278 **69.** −9.51 **71.** 911.13 **73.** 205 **75.** \$1288.36 **77.** 13,748.5 ft **79.** $\frac{19}{73}$ **80.** $-\frac{23}{31}$ **81.** $2 \cdot 2 \cdot 3 \cdot 3 \cdot 19$, or $2^2 \cdot 3^2 \cdot 19$ **82.** $5 \cdot 401$ **83.** $15\frac{1}{8}$ **84.** $5\frac{7}{8}$ **85.** 343 **86.** 64 **87.** 47 **88.** 41 **89.** 6.254194585 **91.** 1000 **93.** 100

Mid-Chapter Review: Chapter 5, p. 293

1. False **2.** True **3.** True
4.
$$\begin{aligned} y + 12.8 &= 23.35 \\ y + 12.8 - 12.8 &= 23.35 - 12.8 \\ y + 0 &= 10.55 \\ y &= 10.55 \end{aligned}$$
5.
$$\begin{aligned} 5.6 + 4.3 \times (6.5 - 0.25)^2 &= 5.6 + 4.3 \times (6.25)^2 \\ &= 5.6 + 4.3 \times 39.0625 \\ &= 5.6 + 167.96875 \\ &= 173.56875 \end{aligned}$$
6. Twenty-nine and forty-three hundredths **7.** 9,400,000 **8.** $\frac{453}{100}$ **9.** $-\frac{287}{1000}$ **10.** 0.13 **11.** −5.09 **12.** 0.7 **13.** −6.39 **14.** 35.67 **15.** 8.002 **16.** 28.462 **17.** 28.46 **18.** 28.5 **19.** 28 **20.** 50.095 **21.** 1214.862 **22.** −3.772 **23.** 18.24 **24.** 272.19 **25.** 5.593 **26.** 15.55 **27.** −58.77 **28.** 4.14 **29.** 92.871 **30.** −8123.6 **31.** 2.937 **32.** 5.06 **33.** −3.2 **34.** 763.4 **35.** 0.914036 **36.** 2045¢ **37.** \$1.47 **38.** −105.3 **39.** 8.4 **40.** 59.774 **41.** 33.83 **42.** The student probably rounded over successively from the thousandths place as follows: 236.448 ≈ 236.45 ≈ 236.5 ≈ 237. The student should have considered only the tenths place and rounded down.
43. The decimal points were not lined up before the subtraction was carried out. **44.** $10 \div 0.2 = \frac{10}{0.2} = \frac{10}{0.2} \cdot \frac{10}{10} = \frac{100}{2} = 100 \div 2$
45. For 0.247 ÷ 0.1 = 0.0247, the divisor, 0.1, is smaller than the dividend, 0.247. Thus the answer will be larger than 0.247. The correct answer is 2.47. For 0.247 ÷ 10 = 2.47, the divisor, 10, is larger than the dividend, 0.247. Thus the answer will be smaller than 0.247. The correct answer is 0.0247.

Exercise Set 5.5, p. 300

RC1. Repeating **RC2.** Terminating **RC3.** Terminating **RC4.** Repeating **RC5.** Repeating **RC6.** Terminating
1. 0.23 **3.** 0.6 **5.** −0.325 **7.** 0.2 **9.** −0.85 **11.** 0.375 **13.** −0.975 **15.** 0.52 **17.** −20.016 **19.** 0.25 **21.** −1.16 **23.** 1.1875 **25.** $0.2\overline{6}$ **27.** $0.\overline{3}$ **29.** $-1.\overline{3}$ **31.** $-1.1\overline{6}$ **33.** $0.\overline{571428}$ **35.** $-0.91\overline{6}$ **37.** 0.3; 0.27; 0.267 **39.** 0.3; 0.33; 0.333 **41.** −1.3; −1.33; −1.333 **43.** −1.2; −1.17; −1.167 **45.** 0.6; 0.57; 0.571 **47.** −0.9; −0.92; −0.917 **49.** 0.2; 0.18; 0.182 **51.** −0.3; −0.28; −0.278 **53.** **(a)** 0.571; **(b)** 1.333; **(c)** 0.429; **(d)** 0.75 **55.** 0.281 **57.** 15.8 mpg **59.** 17.8 mpg **61.** \$357.6 billion **63.** 15.2 mph **65.** 11.06 **67.** 8.4 **69.** $-417.51\overline{6}$ **71.** 0 **73.** 2.8125 **75.** −0.07675 **77.** 317.14 **79.** 0.1825 **81.** 18 **83.** −2.736 **85.** 21 **86.** $1\frac{1}{2}$ **87.** −10 **88.** $30\frac{7}{10}$ **89.** $1\frac{1}{24}$ cups **90.** $1\frac{33}{100}$ in. **91.** $0.\overline{142857}, 0.\overline{285714}, 0.\overline{428571}, 0.\overline{571428}, 0.\overline{714285}; 0.\overline{857142}$

Exercise Set 5.6, p. 306

RC1. (b) **RC2.** (d) **RC3.** (c) **RC4.** (e) **RC5.** (a) **RC6.** (f)
1. (c) **3.** (a) **5.** (c) **7.** 1.6 **9.** 6 **11.** 60 **13.** 2.3 **15.** 180 **17.** (a) **19.** (c) **21.** (b) **23.** (b) **25.** 1800 ÷ 9 = 200 posts; answers may vary **27.** \$2 · 12 = \$24; answers may vary **29.** 2 **30.** 165

31. 210 **32.** 69 **33.** 10 **34.** 530 **35.** No
37. **(a)** $+, \times$; **(b)** $+, \times, -$

Translating for Success, p. 314

1. I **2.** C **3.** N **4.** A **5.** G **6.** B **7.** D **8.** O
9. F **10.** M

Exercise Set 5.7, p. 315

RC1. Familiarize **RC2.** Translate **RC3.** Solve
RC4. Check **RC5.** State
1. 61.6 million citizens **3.** About \$7142.86 **5.** 102.8°F
7. 78.1 cm **9.** 22,691.5 mi **11.** 9.58 sec **13.** 11.9752 cu ft
15. \$0.51 **17.** \$24.33 **19.** 20.2 mpg **21.** 227.75 sq ft
23. Area: 268.96 sq ft; perimeter: 65.6 ft **25.** \$19,110.83
27. \$316,987.20; \$196,987.20 **29.** 876 calories **31.** \$906.50
33. 236.2 million metric tons **35.** \$1180.15, 500.15, 916.42, 875.47, 764.83 **37.** 1.4°F **39.** \$1406.75 **41.** $-\frac{23}{15}$ **42.** $6\frac{5}{6}$
43. $\frac{1}{24}$ **44.** 2803 **45.** $\frac{5}{16}$ **46.** $-\frac{13}{25}$ **47.** $7\frac{1}{5}$ min

Summary and Review: Chapter 5, p. 320

Vocabulary Reinforcement

1. repeating **2.** terminating **3.** billion
4. million **5.** trillion **6.** rational numbers

Concept Reinforcement

1. True **2.** False **3.** True **4.** False

Study Guide

1. $\frac{5093}{100}$ **2.** -81.7 **3.** 42.159 **4.** 153.35 **5.** 38.611
6. 207.848 **7.** 19.11 **8.** -0.176 **9.** 60,437 **10.** 7.4
11. 0.047 **12.** $-15{,}690$

Review Exercises

1. 6,590,000 **2.** 3,100,000,000 **3.** Three and forty-seven hundredths **4.** Thirty-one thousandths **5.** Twenty-seven and one ten-thousandth **6.** Negative nine tenths **7.** $\frac{9}{100}$ **8.** $\frac{4561}{1000}$
9. $-\frac{89}{1000}$ **10.** $-\frac{30{,}227}{10{,}000}$ **11.** 0.034 **12.** -4.2603 **13.** 27.91
14. -867.006 **15.** 0.034 **16.** 0.91 **17.** 0.741 **18.** -1.038
19. 17.4 **20.** 17.43 **21.** 17.429 **22.** 17 **23.** 574.519
24. 0.6838 **25.** -209.5 **26.** 45.551 **27.** 29.2092
28. 790.29 **29.** 46.142 **30.** -70.8109 **31.** 12.96
32. 0.14442 **33.** 4.3 **34.** -0.02468 **35.** -7.5 **36.** 0.45
37. 45.2 **38.** 1.022 **39.** 0.2763 **40.** -1389.2
41. 496.2795 **42.** -6.95 **43.** 42.54 **44.** 4.9911
45. \$15.52 **46.** 1.9 lb **47.** \$784.47 **48.** \$55.50 **49.** 14.5 mpg
50. 20.7 million books **51.** 272 **52.** 216 **53.** \$125 **54.** 0.52
55. -0.45 **56.** 2.75 **57.** -3.25 **58.** $1.1\overline{6}$ **59.** $1.\overline{54}$
60. 1.5 **61.** 1.55 **62.** 1.545 **63.** \$82.73 **64.** \$4.87 **65.** 2493¢
66. 986¢ **67.** -1.5805 **68.** 57.1449 **69.** -15.6375 **70.** D
71. B **72.** **(a)** $2.56 \times 6.4 \div 51.2 - 17.4 + 89.7 = 72.62$;
(b) $(11.12 - 0.29) \times 3^4 = 877.23$
73. $1 = 3 \cdot \frac{1}{3} = 3(0.33333333\ldots) = 0.99999999\ldots$, or $0.\overline{9}$

Understanding Through Discussion and Writing

1. Count the number of decimal places. Move the decimal point that many places to the right and write the result over a denominator of 1 followed by that many zeros.
2. $346.708 \times 0.1 = \frac{346{,}708}{1000} \times \frac{1}{10} = \frac{346{,}708}{10{,}000} = 34.6708 \neq 3467.08$
3. When the denominator of a fraction is a multiple of 10, long division is not the fastest way to convert the fraction to decimal notation. Many times when the denominator is a factor of some multiple of 10, this is also the case. The latter situation occurs when the denominator has only 2's or 5's, or both, as factors.
4. Multiply by 1 to get a denominator that is a power of 10:

$$\frac{44}{125} = \frac{44}{125} \cdot \frac{8}{8} = \frac{352}{1000} = 0.352.$$

We can also divide to find that $\frac{44}{125} = 0.352$.

Test: Chapter 5, p. 325

1. [5.3b] 2,600,000,000 **2.** [5.1a] One hundred twenty-three and forty-seven ten-thousandths **3.** [5.1b] $\frac{91}{100}$ **4.** [5.1b] $-\frac{2769}{1000}$
5. [5.1c] 0.162 **6.** [5.1c] -0.06 **7.** [5.1b] 0.074
8. [5.1b] -3.7047 **9.** [5.1b] 756.09 **10.** [5.1d] 6
11. [5.1d] 5.678 **12.** [5.1d] 5.7 **13.** [5.2c] -18.3
14. [5.2b] 48.357 **15.** [5.2b] 19.0901 **16.** [5.3a] 0.03
17. [5.3a] -0.21345 **18.** [5.4a] -4.75 **19.** [5.4a] 30.4
20. [5.4a] 0.19 **21.** [5.4a] -0.34689 **22.** [5.4b] -84.26
23. [5.2d] 8.982 **24.** [5.4c] 40.0065 **25.** [5.5c] 302.4
26. [5.5a] 0.35 **27.** [5.5a] 0.88 **28.** [5.5a] $0.\overline{81}$
29. [5.5a] $-7.41\overline{6}$ **30.** [5.5b] 0.8 **31.** [5.5b] 0.82
32. [5.5b] 0.818 **33.** [5.6a] 198 **34.** [5.6a] 4
35. [5.7a] \$46.69 **36.** [5.7a] 28.3 mpg **37.** [5.7a] 68.5 years
38. [5.7a] \$293.93 **39.** [5.3b] B **40.** [5.7a] \$35
41. [5.1b, c] $-\frac{13}{15}, -\frac{17}{20}, -\frac{11}{13}, -\frac{15}{19}, -\frac{5}{7}, -\frac{2}{3}$

Cumulative Review: Chapters 1–5, p. 327

1. [4.4a] $\frac{20}{9}$ **2.** [5.1b] $-\frac{3051}{1000}$ **3.** [5.5a] -1.4 **4.** [5.5a] $0.\overline{54}$
5. [3.1c] Prime **6.** [3.2a] Yes **7.** [1.9c] 1754 **8.** [5.4c] -0.136
9. [5.1d] 584.97 **10.** [5.5b] 218.56 **11.** [5.6a] 160
12. [5.6a] 4 **13.** [1.6b] 12,800,000 **14.** [5.6a] 6 **15.** [4.5a] $6\frac{1}{20}$
16. [1.2a] 139,116 **17.** [4.2a] $\frac{31}{18}$ **18.** [5.2c] -141.847
19. [1.3a] 710,137 **20.** [5.2b] 13.097 **21.** [4.5b] $\frac{5}{7}$
22. [4.3a] $-\frac{1}{110}$ **23.** [3.6a] $-\frac{1}{6}$ **24.** [1.4a] 5,317,200
25. [5.3a] 4.78 **26.** [5.3a] 0.0279431 **27.** [5.4a] 2.122
28. [1.5a] 1843 **29.** [5.4a] 13,862.1 **30.** [3.7b] $\frac{5}{6}$
31. [5.2d] 0.78 **32.** [1.7b] 28 **33.** [5.4b] -8.62
34. [1.7b] 367,251 **35.** [4.3c] $\frac{1}{18}$ **36.** [3.7c] $-\frac{1}{2}$
37. [1.8a] 732 million **38.** [1.8a] 885 million **39.** [3.7d] \$1500
40. [3.6b] \$2400 **41.** [5.7a] \$258.77 **42.** [4.5c] $6\frac{1}{2}$ lb
43. [4.2b] 2 lb **44.** [5.7a] 467.28 sq ft **45.** [4.7a] $\frac{9}{32}$
46. [5.4c] -527.04 **47.** [5.7a] \$2.39 **48.** [4.6c] 144 packages

CHAPTER 6

Exercise Set 6.1, p. 334

RC1. True **RC2.** True **RC3.** False **RC4.** False
1. $\frac{4}{5}$ **3.** $\frac{178}{572}$ **5.** $\frac{0.4}{12}$ **7.** $\frac{3.8}{7.4}$ **9.** $\frac{56.78}{98.35}$ **11.** $\frac{8\frac{3}{4}}{9\frac{5}{6}}$ **13.** $\frac{29}{75}$
15. $\frac{11}{18}$ **17.** $\frac{107}{366}$ **19.** $\frac{104.8}{3100.7}$; $\frac{3100.7}{104.8}$ **21.** $\frac{60}{100}$; $\frac{100}{60}$ **23.** $\frac{2}{3}$ **25.** $\frac{3}{4}$
27. $\frac{12}{25}$ **29.** $\frac{7}{9}$ **31.** $\frac{2}{3}$ **33.** $\frac{14}{25}$ **35.** $\frac{1}{2}$ **37.** $\frac{3}{4}$ **39.** $\frac{478}{213}$; $\frac{213}{478}$
41. 408,550 **42.** 27.006 **43.** -332.73 **44.** $\frac{19}{30}$ **45.** $-\frac{5}{24}$
46. $14\frac{1}{3}$ **47.** 942,219 **48.** 2.6 **49.** -61.39 **50.** $\frac{1}{16}$
51. $1\frac{13}{15}$ **52.** $4\frac{2}{3}$ **53.** $\frac{30}{47}$ **55.** 1 : 2 : 3

Exercise Set 6.2, p. 340

RC1. (b) **RC2.** (c) **RC3.** (d) **RC4.** (a)
1. 40 km/h **3.** 7.48 mi/sec **5.** 25 mpg **7.** 33 mpg
9. 47,236 people/sq mi **11.** 25 mph; 0.04 hr/mi **13.** About 393 performances/year **15.** 186,000 mi/sec **17.** 124 km/h
19. 0.623 gal/ft^2 **21.** 25 beats/min **23.** 26.188¢/oz; 26.450¢/oz; 16 oz **25.** 13.111¢/oz; 11.453¢/oz; 75 oz
27. 18.222¢/oz; 17.464¢/oz; 28 oz **29.** 10.187¢/oz; 10.346¢/oz; 10.7 oz **31.** B: 18.719¢/oz; E: 14.563¢/oz; Brand E
33. A: 10.375¢/oz; B: 9.139¢/oz; H: 8.022¢/oz; Brand H
35. 11,550 **36.** 679.4928 **37.** 2.74568 **38.** $-\frac{20}{3}$ **39.** $8\frac{11}{20}$
40. $-13\frac{2}{3}$ **41.** 125 **42.** 9.5 **43.** -9.63 **44.** $\frac{125}{27}$ **45.** 150
46. $-19\frac{3}{7}$ **47.** 6-oz: 10.833¢/oz; 5.5 oz: 10.909¢/oz

Exercise Set 6.3, p. 348

RC1. ratio **RC2.** proportional **RC3.** proportion **RC4.** cross products **1.** No **3.** Yes **5.** Yes **7.** No **9.** 45 **11.** 12 **13.** 10 **15.** 20 **17.** 5 **19.** 18 **21.** 22 **23.** 28 **25.** $\frac{28}{3}$, or $9\frac{1}{3}$ **27.** $\frac{26}{9}$, or $2\frac{8}{9}$ **29.** 5 **31.** 5 **33.** 0 **35.** 14 **37.** 2.7 **39.** 1.8 **41.** 0.06 **43.** 0.7 **45.** 12.5725 **47.** 1 **49.** $\frac{1}{20}$ **51.** $\frac{3}{8}$ **53.** $\frac{16}{75}$ **55.** $\frac{51}{16}$, or $3\frac{3}{16}$ **57.** $\frac{546}{185}$, or $2\frac{176}{185}$ **59.** 4063 theaters **60.** 6 lb **61.** $\frac{1}{4}$ lb **62.** $\frac{3}{8}$ lb **63.** $6\frac{7}{20}$ cm **64.** $37\frac{1}{2}$ servings **65.** \$59.81 **66.** 21.5 mpg **67.** Approximately 2731.4 **69.** Ruth: 1.863 strikeouts per home run; Schmidt: 3.436 strikeouts per home run

Mid-Chapter Review: Chapter 6, p. 351

1. True **2.** True **3.** False **4.** False **5.** $\frac{120\text{ mi}}{2\text{ hr}} = \frac{120}{2}\frac{\text{mi}}{\text{hr}} = 60\text{ mi/hr}$

6.
$$\begin{aligned} \frac{x}{4} &= \frac{3}{6} \\ x \cdot 6 &= 4 \cdot 3 \\ \frac{x \cdot 6}{6} &= \frac{4 \cdot 3}{6} \\ x &= 2 \end{aligned}$$

7. $\frac{4}{7}$ **8.** $\frac{313}{199}$ **9.** $\frac{35}{17}$ **10.** $\frac{59}{101}$ **11.** $\frac{2}{3}$ **12.** $\frac{1}{3}$ **13.** $\frac{8}{7}$ **14.** $\frac{25}{19}$ **15.** $\frac{2}{1}$ **16.** $\frac{5}{1}$ **17.** $\frac{2}{7}$ **18.** $\frac{3}{5}$ **19.** 60.75 mi/hr, or 60.75 mph **20.** 48.67 km/h **21.** 13 m/sec **22.** 16.17 ft/sec **23.** 27 in./day **24.** About 0.860 free throw made/attempt **25.** 11.611¢/oz **26.** 49.917¢/oz **27.** Yes **28.** No **29.** No **30.** Yes **31.** 12 **32.** 40 **33.** 9 **34.** 35 **35.** 2.2 **36.** 4.32 **37.** $\frac{1}{2}$ **38.** $\frac{65}{4}$, or $16\frac{1}{4}$ **39.** Yes; every ratio $\frac{a}{b}$ can be written as $\frac{\frac{a}{b}}{1}$. **40.** By making some sketches, we see that the rectangle's length must be twice the width. **41.** The student's approach will work. However, when we use the approach of equating cross products, we eliminate the need to find the least common denominator. **42.** The instructor thinks that the longer a student studies, the higher his or her grade will be. An example is the situation in which one student gets a test grade of 96 after studying for 8 hr while another student gets a score of 78 after studying for $6\frac{1}{2}$ hr. This is represented by the proportion $\frac{96}{8} = \frac{78}{6\frac{1}{2}}$.

Translating for Success, p. 358

1. N **2.** I **3.** A **4.** K **5.** J **6.** F **7.** M **8.** B **9.** G **10.** E

Exercise Set 6.4, p. 359

RC1. $\frac{m}{8}$ **RC2.** $\frac{8}{m}$ **RC3.** $\frac{6}{8}$ **RC4.** $\frac{8}{6}$ **1.** 11.04 hr **3.** 7680 frames **5.** **(a)** About 103 gal; **(b)** 3640 mi **7.** 232.53 million, or 232,530,000 **9.** 175 bulbs **11.** About 262 mg every 8 hr **13.** 2975 ft^2 **15.** 450 pages **17.** 13,500 mi **19.** 880 calories **21.** 60 students **23.** 64 cans **25.** 9.75 gal **27.** 100 oz **29.** 90 whales **31.** 58.1 mi **33.** **(a)** 38 euros; **(b)** \$11,368.42 **35.** **(a)** 66 games; **(b)** About 2504 points **37.** 162 **38.** 4014 **39.** 4.45 **40.** -0.4 **41.** -21.7 **42.** 0.26 **43.** $\frac{1}{10}$ **44.** $-\frac{5}{4}$ **45.** 17 positions **47.** 2150 earned runs **49.** CD player: \$150; receiver: \$450; speakers: \$300

Exercise Set 6.5, p. 368

RC1. 5 **RC2.** 8 **RC3.** a **RC4.** 3 **1.** 25 **3.** $\frac{4}{3}$, or $1\frac{1}{3}$ **5.** $x = \frac{27}{4}$, or $6\frac{3}{4}$; $y = 9$ **7.** $x = 7.5$; $y = 7.2$ **9.** 1.25 m **11.** 36 ft **13.** 7 ft **15.** 100 ft **17.** 4 **19.** $10\frac{1}{2}$ **21.** $x = 6$; $y = 5.25$; $z = 3$ **23.** $x = 5\frac{1}{3}$, or $5.\overline{3}$; $y = 4\frac{2}{3}$, or $4.\overline{6}$; $z = 5\frac{1}{3}$, or $5.\overline{3}$ **25.** 20 ft **27.** 152 ft **29.** Prime **30.** Composite **31.** $2 \cdot 2 \cdot 2 \cdot 101$, or $2^3 \cdot 101$ **32.** $3 \cdot 31$ **33.** $=$ **34.** $\neq$ **35.** $<$ **36.** $>$ **37.** $\frac{13}{4}$, or $3\frac{1}{4}$ **38.** 5.29 **39.** 127 **40.** 75 **41.** 13.75 ft **43.** 1.25 cm **45.** 3681.437 **47.** $x \approx 0.35$; $y = 0.4$

Summary and Review: Chapter 6, p. 372

Vocabulary Reinforcement

1. quotient **2.** shape **3.** cross products **4.** rate **5.** proportion **6.** price

Concept Reinforcement

1. True **2.** True **3.** False **4.** True

Study Guide

1. $\frac{17}{3}$ **2.** $\frac{8}{7}$ **3.** \$7.50/hr **4.** A: 9.964¢/oz; B: 10.281¢/oz; Brand A **5.** Yes **6.** $\frac{27}{8}$ **7.** 175 mi **8.** 21

Review Exercises

1. $\frac{47}{84}$ **2.** $\frac{46}{1.27}$ **3.** $\frac{83}{100}$ **4.** $\frac{0.72}{197}$ **5.** **(a)** $\frac{12{,}480}{16{,}640}$, or $\frac{3}{4}$; **(b)** $\frac{16{,}640}{29{,}120}$, or $\frac{4}{7}$ **6.** $\frac{3}{4}$ **7.** $\frac{9}{16}$ **8.** 28 mpg **9.** 6300 revolutions/min **10.** 0.638 gal/ft^2 **11.** 6.33¢/tablet **12.** 19.542¢/oz **13.** 14.969¢/oz; 12.479¢/oz; 15.609¢/oz; 48 oz **14.** Yes **15.** No **16.** 32 **17.** 7 **18.** $\frac{1}{40}$ **19.** 24 **20.** 27 circuits **21.** **(a)** 247.50 Canadian dollars; **(b)** 50.51 U.S. dollars **22.** 832 mi **23.** 133.5 billion gallons, or 133,500,000,000 gallons **24.** About 3,774,811 lb **25.** 320 calories **26.** About 12,453 lawyers **27.** $x = \frac{14}{3}$, or $4\frac{2}{3}$ **28.** $x = \frac{56}{5}$, or $11\frac{1}{5}$; $y = \frac{63}{5}$, or $12\frac{3}{5}$ **29.** 40 ft **30.** $x = 3$; $y = \frac{21}{2}$, or $10\frac{1}{2}$; $z = \frac{15}{2}$, or $7\frac{1}{2}$ **31.** B **32.** C **33.** 1.806¢/sheet; 1.658¢/sheet; 1.409¢/sheet; 6 big rolls **34.** $x = 4258.5$; $z \approx 10{,}094.3$ **35.** Finishing paint: 11 gal; primer: 16.5 gal

Understanding Through Discussion and Writing

1. In terms of cost, a low faculty-to-student ratio is less expensive than a high faculty-to-student ratio. In terms of quality of education and student satisfaction, a high faculty-to-student ratio is more desirable. A college president must balance the cost and quality issues. **2.** Yes; unit prices can be used to solve proportions involving money. In Example 3 of Section 6.4, for instance, we could have divided \$90 by the unit price, or the price per ticket, to find the number of tickets that could be purchased for \$90. **3.** Leslie used 4 gal of gasoline to drive 92 mi. At the same rate, how many gallons would be needed to travel 368 mi? **4.** Yes; consider the following pair of triangles.

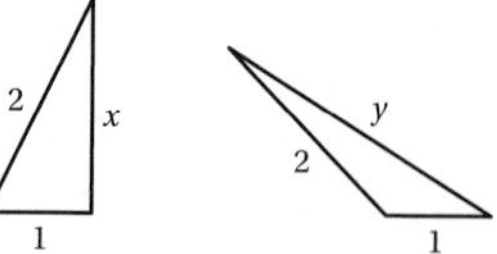

Two pairs of sides are proportional, but we can see that x is shorter than y, so the ratio of x to y is clearly not the same as the ratio of 1 to 1 (or 2 to 2).

Test: Chapter 6, p. 377

1. [6.1a] $\frac{85}{97}$ **2.** [6.1a] $\frac{0.34}{124}$ **3.** [6.1b] $\frac{9}{10}$ **4.** [6.1b] $\frac{25}{32}$ **5.** [6.2a] 0.625 ft/sec **6.** [6.2a] $1\frac{1}{3}$ servings/lb **7.** [6.2a] 32 mpg **8.** [6.2b] About 15.563¢/oz **9.** [6.2b] 16.475¢/oz; 13.980¢/oz; 11.490¢/oz; 16.660¢/oz; 100 oz **10.** [6.3a] Yes **11.** [6.3a] No **12.** [6.3b] 12 **13.** [6.3b] 360 **14.** [6.3b] 42.1875 **15.** [6.3b] 100 **16.** [6.4a] 1512 km **17.** [6.4a] 4.8 min **18.** [6.4a] 525 mi **19.** [6.5a] 66 m **20.** [6.4a] **(a)** 13 hats; **(b)** 30 packages **21.** [6.4a] About \$59.17 **22.** [6.5a] $x = 8$; $y = 8.8$ **23.** [6.5b] $x = \frac{24}{5}$, or 4.8; $y = \frac{32}{5}$, or 6.4; $z = 12$ **24.** [6.2a] C **25.** [6.4a] 5888 marbles

Cumulative Review: Chapters 1–6, p. 379

1. [5.2a] 513.996 **2.** [4.5a] $6\frac{3}{4}$ **3.** [4.2a] $-\frac{7}{20}$ **4.** [5.2b] 30.491 **5.** [5.2b] 72.912 **6.** [4.3a] $-\frac{5}{12}$ **7.** [5.3a] 222.076 **8.** [5.3a] 567.8 **9.** [4.6a] −3 **10.** [5.4a] 43 **11.** [1.5a] 899 **12.** [3.7b] $-\frac{3}{2}$ **13.** [1.1b] 3 ten thousands + 0 thousands + 0 hundreds + 7 tens + 4 ones, or 3 ten thousands + 7 tens + 4 ones **14.** [5.1a] One hundred twenty and seven hundredths **15.** [5.1c] 0.7 **16.** [5.1c] −0.799 **17.** [3.1d] $2 \cdot 2 \cdot 2 \cdot 2 \cdot 3 \cdot 3$, or $2^4 \cdot 3^2$ **18.** [4.1a] 90 **19.** [3.3a] $\frac{5}{8}$ **20.** [3.5b] $\frac{5}{8}$ **21.** [5.5c] 5.718 **22.** [5.5c] 0.179 **23.** [6.1a] $\frac{0.3}{15}$ **24.** [6.3a] Yes **25.** [6.2a] 55 m/sec **26.** [6.2b] 8-oz can **27.** [6.3b] 30.24 **28.** [5.4b] 26.4375 **29.** [3.7c] $\frac{8}{9}$ **30.** [6.3b] 128 **31.** [5.2d] −102.46 **32.** [4.3c] $\frac{76}{175}$ **33.** [3.6b], [6.4a] 390 cal **34.** [6.4a] 7 min **35.** [4.5c] $2\frac{1}{4}$ cups **36.** [3.7d], [6.4a] 12 doors **37.** [5.7a], [6.4a] 42.2025 mi **38.** [5.7a] 132 orbits **39.** [3.1c] D **40.** [6.2a] B **41.** [6.5a] $10\frac{1}{2}$ ft

CHAPTER 7

Calculator Corner, p. 384

1. 0.14 **2.** 0.00069 **3.** 0.438 **4.** 1.25

Exercise Set 7.1, p. 385

RC1. 43% **RC2.** 86% **RC3.** 19% **RC4.** 50%
1. $\frac{90}{100}$; $90 \times \frac{1}{100}$; 90×0.01 **3.** $\frac{12.5}{100}$; $12.5 \times \frac{1}{100}$; 12.5×0.01 **5.** 0.67 **7.** 0.456 **9.** 0.5901 **11.** 0.1 **13.** 0.01 **15.** 2 **17.** 0.001 **19.** 0.0009 **21.** 0.0018 **23.** 0.2319 **25.** 0.14875 **27.** 0.565 **29.** 0.13; 0.11 **31.** 0.07; 0.08 **33.** 0.139; 0.4 **35.** 47% **37.** 3% **39.** 870% **41.** 33.4% **43.** 75% **45.** 40% **47.** 0.6% **49.** 1.7% **51.** 27.18% **53.** 2.39% **55.** 65.1% **57.** 34% **59.** 35.9%; 20% **61.** 0.29; 0.3; 0.13; 0.1; 0.18 **63.** 1, 2, 3, 4, 6, 7, 12, 14, 21, 28, 42, 84 **64.** 1, 2, 4, 5, 10, 20, 31, 62, 124, 155, 310, 620 **65.** 180 **66.** $2 \cdot 3 \cdot 3 \cdot 5$ **67.** $-\frac{23}{16}$ **68.** 97 **69.** 50% **71.** 70% **73.** 20%

Calculator Corner, p. 389

1. 52% **2.** 38.46% **3.** 110.26% **4.** 171.43% **5.** 59.62% **6.** 28.31%

Exercise Set 7.2, p. 393

RC1. (c) **RC2.** (e) **RC3.** (g) **RC4.** (a) **RC5.** (b) **RC6.** (d) **RC7.** (h) **RC8.** (i)
1. 41% **3.** 5% **5.** 20% **7.** 30% **9.** 50% **11.** 87.5%, or $87\frac{1}{2}\%$ **13.** 80% **15.** $66.\overline{6}\%$, or $66\frac{2}{3}\%$ **17.** $16.\overline{6}\%$, or $16\frac{2}{3}\%$ **19.** 18.75%, or $18\frac{3}{4}\%$ **21.** 81.25%, or $81\frac{1}{4}\%$ **23.** 16% **25.** 5% **27.** 34% **29.** 8%; 59% **31.** 22% **33.** 12% **35.** 15% **37.** $\frac{17}{20}$ **39.** $\frac{5}{8}$ **41.** $\frac{1}{3}$ **43.** $\frac{1}{6}$ **45.** $\frac{29}{400}$ **47.** $\frac{1}{125}$ **49.** $\frac{203}{800}$ **51.** $\frac{176}{225}$ **53.** $\frac{711}{1100}$ **55.** $\frac{3}{2}$ **57.** $\frac{13}{40{,}000}$ **59.** $\frac{1}{3}$ **61.** $\frac{3}{50}$ **63.** $\frac{13}{100}$ **65.** $\frac{19}{25}$ **67.** $\frac{3}{20}$ **69.** $\frac{209}{1000}$

71.

Fraction Notation	Decimal Notation	Percent Notation
$\frac{1}{8}$	0.125	12.5%, or $12\frac{1}{2}\%$
$\frac{1}{6}$	$0.1\overline{6}$	$16.\overline{6}\%$, or $16\frac{2}{3}\%$
$\frac{1}{5}$	0.2	20%
$\frac{1}{4}$	0.25	25%
$\frac{1}{3}$	$0.\overline{3}$	$33.\overline{3}\%$, or $33\frac{1}{3}\%$
$\frac{3}{8}$	0.375	37.5%, or $37\frac{1}{2}\%$
$\frac{2}{5}$	0.4	40%
$\frac{1}{2}$	0.5	50%

73.

Fraction Notation	Decimal Notation	Percent Notation
$\frac{1}{2}$	0.5	50%
$\frac{1}{3}$	$0.\overline{3}$	$33.\overline{3}\%$, or $33\frac{1}{3}\%$
$\frac{1}{4}$	0.25	25%
$\frac{1}{6}$	$0.1\overline{6}$	$16.\overline{6}\%$, or $16\frac{2}{3}\%$
$\frac{1}{8}$	0.125	12.5%, or $12\frac{1}{2}\%$
$\frac{3}{4}$	0.75	75%
$\frac{5}{6}$	$0.8\overline{3}$	$83.\overline{3}\%$, or $83\frac{1}{3}\%$
$\frac{3}{8}$	0.375	37.5%, or $37\frac{1}{2}\%$

75. 70 **76.** 5 **77.** −400 **78.** 18.75 **79.** 23.125 **80.** 25.5 **81.** 4.5 **82.** 8.75 **83.** $18\frac{3}{4}$ **84.** $-7\frac{4}{9}$ **85.** $-\frac{203}{2}$ **86.** $\frac{209}{10}$ **87.** $257.\overline{46317}\%$ **89.** $1.04\overline{142857}$ **91.** $\frac{1}{6}\%$, $\frac{2}{7}\%$, 0.5%, $1\frac{1}{6}\%$, 1.6%, $16\frac{1}{6}\%$, 0.2, $\frac{1}{2}$, $0.\overline{54}$, 1.6

Calculator Corner, p. 400

1. $5.04 **2.** 0.0112 **3.** 450 **4.** $1000 **5.** 2.5% **6.** 12%

Exercise Set 7.3, p. 401

RC1. (c) **RC2.** (e) **RC3.** (f) **RC4.** (d) **RC5.** (b) **RC6.** (a)
1. $a = 32\% \times 78$ **3.** $89 = p \times 99$ **5.** $13 = 25\% \times b$ **7.** 234.6 **9.** 45 **11.** $18 **13.** 1.9 **15.** 78% **17.** 200% **19.** 50% **21.** 125% **23.** 40 **25.** $40 **27.** 88 **29.** 20 **31.** 6.25 **33.** $846.60 **35.** 1216 **37.** $-\frac{9375}{10{,}000}$, or $-\frac{15}{16}$ **38.** $\frac{125}{1000}$, or $\frac{1}{8}$ **39.** 0.3 **40.** −0.017 **41.** 7 **42.** 8 **43.** $10,000 (can vary); $10,400 **45.** $1875

Exercise Set 7.4, p. 407

RC1. (b) **RC2.** (e) **RC3.** (c) **RC4.** (f) **RC5.** (d) **RC6.** (a)
1. $\frac{37}{100} = \frac{a}{74}$ **3.** $\frac{N}{100} = \frac{4.3}{5.9}$ **5.** $\frac{25}{100} = \frac{14}{b}$ **7.** 68.4 **9.** 462 **11.** 40 **13.** 2.88 **15.** 25% **17.** 102% **19.** 25% **21.** 93.75%, or $93\frac{3}{4}\%$ **23.** $72 **25.** 90 **27.** 88 **29.** 20 **31.** 25 **33.** $780.20 **35.** 200 **37.** 8 **38.** 4000 **39.** 15 **40.** 2074 **41.** $\frac{43}{48}$ qt **42.** $\frac{1}{8}$ T **43.** $1170 (can vary); $1118.64

Mid-Chapter Review: Chapter 7, p. 409

1. True **2.** False **3.** True **4.** $\frac{1}{2}\% = \frac{1}{2} \cdot \frac{1}{100} = \frac{1}{200}$ **5.** $\frac{80}{1000} = \frac{8}{100} = 8\%$ **6.** $5.5\% = \frac{5.5}{100} = \frac{55}{1000} = \frac{11}{200}$ **7.** $0.375 = \frac{375}{1000} = \frac{37.5}{100} = 37.5\%$

8.
$$\begin{aligned} 15 &= p \times 80 \\ \frac{15}{80} &= \frac{p \times 80}{80} \\ \frac{15}{80} &= p \\ 0.1875 &= p \\ 18.75\% &= p \end{aligned}$$

9. 0.28 **10.** 0.0015 **11.** 0.05375 **12.** 2.4 **13.** 71% **14.** 9% **15.** 38.91% **16.** 18.75%, or $18\frac{3}{4}\%$ **17.** 0.5% **18.** 74% **19.** 600% **20.** $83.\overline{3}\%$, or $83\frac{1}{3}\%$ **21.** $\frac{17}{20}$ **22.** $\frac{3}{6250}$ **23.** $\frac{91}{400}$ **24.** $\frac{1}{6}$ **25.** 62.5%, or $62\frac{1}{2}\%$ **26.** 45% **27.** 58 **28.** $16.\overline{6}\%$, or $16\frac{2}{3}\%$ **29.** 2560 **30.** $50 **31.** 20% **32.** $455 **33.** 0.05%, 0.1%, $\frac{1}{2}\%$, 1%, 5%, 10%, $\frac{13}{100}$, 0.275, $\frac{3}{10}$, $\frac{7}{20}$ **34.** D **35.** B **36.** Some will say that the conversion will be done most accurately by first finding decimal notation. Others will say that it is more efficient to become familiar with some or all of the fraction and percent equivalents that appear inside the back cover and to make the conversion by going directly from fraction notation to percent notation. **37.** Since $40\% \div 10 = 4\%$, we can divide 36.8 by 10, obtaining 3.68. Since 400% $= 40\% \times 10$, we can multiply 36.8 by 10, obtaining 368. **38.** Answers may vary. Some will say this is a good idea since it makes the computations in the solution easier. Others will say it is a poor idea since it adds an extra step to the solution. **39.** They all represent the same number.

Translating for Success, p. 418

1. J **2.** M **3.** N **4.** E **5.** G **6.** H **7.** O **8.** C **9.** D **10.** B

Exercise Set 7.5, p. 419

RC1. \$10, $\frac{\$10}{\$50}$, 20% **RC2.** \$15, $\frac{\$15}{\$60}$, 25% **RC3.** \$120, $\frac{\$120}{\$360}$, $33\frac{1}{3}\%$ **RC4.** \$1600, $\frac{\$1600}{\$4000}$, 40% **1.** United States: about \$104.7 billion; Italy: about \$24.6 billion **3.** \$46,656 **5.** 140 items **7.** About 940,000,000 acres **9.** \$36,400 **11.** 76 items correct; 4 items incorrect **13.** 261 men **15.** About 27.8% **17.** \$230.10 **19.** About 29.8% **21.** Alcohol: 43.2 mL; water: 496.8 mL **23.** Air Force: 25.2%; Army: 36.5%; Navy: 25.3%; Marines: 13.0% **25.** 5% **27.** 15% **29.** About 34.7% **31.** About 127% **33.** About 49.5% **35.** About 497% **37.** About 9.5% **39.** 34.375%, or $34\frac{3}{8}\%$ **41.** 16,914; 2.8% **43.** 5,130,632; 24.6% **45.** 1,567,582; 21.1% **47.** About 23.8% **49.** $2.\overline{27}$ **50.** 0.44 **51.** -3.375 **52.** $4.\overline{7}$ **53.** $\frac{3}{17}$ **54.** $-\frac{9}{34}$ **55.** $\frac{24}{125}$ **56.** $\frac{5}{29}$ **57.** \$42

Exercise Set 7.6, p. 430

RC1. rate **RC2.** rate **RC3.** price **RC4.** tax **RC5.** tax **1.** \$9.56 **3.** \$6.66 **5.** \$11.99; \$171.79 **7.** 4% **9.** 2% **11.** \$5600 **13.** \$116.72 **15.** \$276.28 **17.** \$194.08 **19.** \$2625 **21.** 12% **23.** \$185,000 **25.** \$5440 **27.** 15% **29.** \$355 **31.** \$30; \$270 **33.** \$2.55; \$14.45 **35.** \$125; \$112.50 **37.** 40%; \$360 **39.** \$300; 40% **41.** \$849; 21.2% **43.** 18 **44.** $\frac{22}{7}$ **45.** 265.625 **46.** -1.15 **47.** 4,030,000,000,000 **48.** 5,800,000 **49.** 42,700,000 **50.** 6,090,000,000,000 **51.** \$17,700

Calculator Corner, p. 437

1. \$16,357.18 **2.** \$12,764.72

Exercise Set 7.7, p. 440

RC1. $\frac{6}{12}$ year **RC2.** $\frac{40}{365}$ year **RC3.** $\frac{285}{365}$ year **RC4.** $\frac{9}{12}$ year **RC5.** $\frac{3}{12}$ year **RC6.** $\frac{4}{12}$ year **1.** \$8 **3.** \$113.52 **5.** \$462.50 **7.** \$671.88 **9.** **(a)** \$147.95; **(b)** \$10,147.95 **11.** **(a)** \$84.14; **(b)** \$6584.14 **13.** **(a)** \$46.03; **(b)** \$5646.03 **15.** \$441 **17.** \$2802.50 **19.** \$7853.38 **21.** \$99,427.40 **23.** \$4243.60 **25.** \$28,225.00 **27.** \$9270.87 **29.** \$129,871.09 **31.** \$4101.01 **33.** \$1324.58 **35.** Interest: \$20.88; amount applied to principal: \$4.69; balance after the payment: \$1273.87 **37.** **(a)** \$98; **(b)** interest: \$86.56; amount applied to principal: \$11.44; **(c)** interest: \$51.20; amount applied to principal: \$46.80; **(d)** At 12.6%, the principal is reduced by \$35.36 more than at the 21.3% rate. The interest at 12.6% is \$35.36 less than at 21.3%. **39.** 800 **40.** $2 \cdot 2 \cdot 3 \cdot 19$ **41.** $\frac{9}{100}$ **42.** $-\frac{32}{75}$ **43.** $\frac{1}{25}$ **44.** $\frac{7}{300}$ **45.** 55 **46.** $-\frac{13}{10}$ **47.** 9.38%

Summary and Review: Chapter 7, p. 443

Vocabulary Reinforcement

1. percent decrease **2.** simple **3.** discount **4.** sales **5.** rate **6.** percent increase

Concept Reinforcement

1. True **2.** False **3.** True

Study Guide

1. 8.2% **2.** 0.62625 **3.** $63.\overline{63}\%$, or $63\frac{7}{11}\%$ **4.** $\frac{17}{250}$ **5.** $4.1\overline{6}\%$, or $4\frac{1}{6}\%$ **6.** 10,000 **7.** About 15.3% **8.** 6% **9.** \$185,000 **10.** Simple interest: \$22.60; total amount due: \$2522.60 **11.** \$6594.26

Review Exercises

1. 170% **2.** 6.5% **3.** 0.04; 0.144 **4.** 0.621; 0.842 **5.** 37.5%, or $37\frac{1}{2}\%$ **6.** $33.\overline{3}\%$, or $33\frac{1}{3}\%$ **7.** $\frac{6}{25}$ **8.** $\frac{63}{1000}$ **9.** $30.6 = p \times 90$; 34% **10.** $63 = 84\% \times b$; 75 **11.** $a = 38\frac{1}{2}\% \times 168$; 64.68 **12.** $\frac{24}{100} = \frac{16.8}{b}$; 70 **13.** $\frac{42}{30} = \frac{N}{100}$; 140% **14.** $\frac{10.5}{100} = \frac{a}{84}$; 8.82 **15.** 178 students; 84 students **16.** 46% **17.** 2500 mL **18.** 12% **19.** 92 **20.** \$24 **21.** 6% **22.** 11% **23.** About 18.3% **24.** \$42; \$308 **25.** 14% **26.** \$2940 **27.** \$36 **28.** **(a)** \$394.52; **(b)** \$24,394.52 **29.** \$121 **30.** \$7575.25 **31.** \$9504.80 **32.** **(a)** \$129; **(b)** interest: \$100.18; amount applied to principal: \$28.82; **(c)** interest: \$70.72; amount applied to principal: \$58.28; **(d)** At 13.2%, the principal is decreased by \$29.46 more than at the 18.7% rate. The interest at 13.2% is \$29.46 less than at 18.7%. **33.** A **34.** C **35.** $66.\overline{6}\%$, or $66\frac{2}{3}\%$ **36.** About 19%

Understanding Through Discussion and Writing

1. A 40% discount is better. When successive discounts are taken, each is based on the previous discounted price rather than on the original price. A 20% discount followed by a 22% discount is the same as a 37.6% discount off the original price. **2.** Let S = the original salary. After both raises have been given, the two situations yield the same salary: $1.05 \cdot 1.1S = 1.1 \cdot 1.05S$. However, the first situation is better for the wage earner, because $1.1S$ is earned the first year when a 10% raise is given while in the second situation $1.05S$ is earned that year. **3.** No; the 10% discount was based on the original price rather than on the sale price. **4.** For a number n, 40% of 50% of n is $0.4(0.5n)$, or $0.2n$, or 20% of n. Thus, taking 40% of 50% of a number is the same as taking 20% of the number. **5.** The interest due on the 30-day loan will be \$41.10 while that due on 60-day loan will be \$131.51. This could be an argument in favor of the 30-day loan. On the other hand, the 60-day loan puts twice as much cash at the firm's disposal for twice as long as the 30-day loan does. This could be an argument in favor of the 60-day loan. **6.** Answers will vary.

Test: Chapter 7, p. 449

1. [7.1b] 0.613 **2.** [7.1b] 38% **3.** [7.2a] 137.5% **4.** [7.2b] $\frac{13}{20}$ **5.** [7.3a, b] $a = 40\% \cdot 55$; 22 **6.** [7.4a, b] $\frac{N}{100} = \frac{65}{80}$; 81.25% **7.** [7.5a] 16,550 kidney transplants; 5992 liver transplants; 2283 heart transplants **8.** [7.5a] About 611 at-bats **9.** [7.5b] 6.7% **10.** [7.5a] 60.6% **11.** [7.6a] \$25.20; \$585.20 **12.** [7.6b] \$630 **13.** [7.6c] \$40; \$160 **14.** [7.7a] \$8.52 **15.** [7.7a] \$5356 **16.** [7.7b] \$1110.39 **17.** [7.7b] \$11,580.07 **18.** [7.5b] Plumber: 757,000, 7.4%; veterinary assistant: 29,000, 40.8%; motorcycle repair technician: 21,000, 14.3%; fitness professional: 235,000, 63,000 **19.** [7.6c] \$50; about 14.3% **20.** [7.7c] Interest: \$36.73; amount applied to the principal: \$17.27; balance after payment: \$2687 **21.** [7.3a, b], [7.4a, b] B **22.** [7.6b] \$194,600 **23.** [7.6b], [7.7b] \$2546.16

Cumulative Review: Chapters 1–7, p. 451

1. [7.5b] 14.7%; 4.1% **2.** [7.1b] 26.9% **3.** [7.2a] 112.5% **4.** [5.5a] $-2.1\overline{6}$ **5.** [6.1a] $\frac{5}{0.5}$, or $\frac{10}{1}$ **6.** [6.2a] $\frac{70\text{ km}}{3\text{ hr}}$, or $23.\overline{3}$ km/hr, or $23\frac{1}{3}$ km/hr **7.** [3.5c], [4.3b] < **8.** [3.5c], [4.3b] < **9.** [1.6b], [5.6a] 296,200 **10.** [1.6b] 50,000 **11.** [1.9d] 13 **12.** [5.4c] -1.5 **13.** [4.2a] [4.5a] $\frac{91}{30}$, or $3\frac{1}{30}$ **14.** [5.2c] -44.06 **15.** [1.2a] 515,150 **16.** [5.2b] 0.02 **17.** [4.5b] $\frac{2}{3}$ **18.** [4.3a] $-\frac{110}{63}$ **19.** [3.6a] $\frac{1}{6}$ **20.** [1.4a] 853,142,400 **21.** [5.3a] 1.38036

22. [4.6b] $-\frac{3}{2}$, or $-1\frac{1}{2}$ **23.** [5.4a] 12.25 **24.** [1.5a] 123 R 5 **25.** [1.7b] 95 **26.** [5.2d] −293.87 **27.** [3.7c] −9 **28.** [4.3c] $\frac{1}{12}$ **29.** [6.3b] 40 **30.** [6.3b] $\frac{176}{21}$ **31.** [7.5a] 462,259 visitors **32.** [1.8a] $177,500 **33.** [6.4a] About 65 games **34.** [5.7a] $84.95 **35.** [6.2b] 7.495 cents/oz **36.** [4.2b], [4.4a] $\frac{3}{2}$ mi, or $1\frac{1}{2}$ mi **37.** [6.4a] 60 mi **38.** [7.7b] $12,663.69 **39.** [4.6c] 5 pieces **40.** [7.5b] About 14.2%; 883,000 technicians **41.** [4.3a] C **42.** [7.5b] A **43.** [7.3b], [7.4b] 200%

CHAPTER 8

Exercise Set 8.1, p. 459

RC1. statistic **RC2.** average **RC3.** weight **RC4.** mode **1.** Average: 9.325 million visitors; median: 9.35 million visitors; modes: 9.0 million visitors, 9.5 million visitors **3.** Average: 21; median: 18.5; mode: 29 **5.** Average: 21; median: 20; modes: 5, 20 **7.** Average: 5.38; median: 5.7; no mode exists **9.** Average: 239.5; median: 234; mode: 234 **11.** 36 mpg **13.** 2.7 **15.** Average: $4.19; median: $3.99; mode: $3.99 **17.** 90 **19.** 263 days **21.** **(a)** Jefferson County: $133,987; Hamilton County: $146,989; **(b)** Jefferson County **23.** **(a)** 1.455 billion tickets; **(b)** 1.3775 billion tickets; **(c)** 2002 to 2009 **25.** 225.05 **26.** 126.0516 **27.** $\frac{3}{35}$ **28.** $-\frac{14}{15}$ **29.** $a = 30$; $b = 58$ **31.** $6950

Exercise Set 8.2, p. 468

RC1. False **RC2.** False **RC3.** True **RC4.** False **1.** 92° **3.** 108° **5.** 85°, 60%; 90°, 40%; 100°, 10% **7.** 90° and higher **9.** 30% and higher **11.** 90% − 40% = 50% **13.** 483,612,200 mi **15.** Neptune **17.** 11 Earth diameters **19.** Average: 31,191.75 mi; median: 19,627.5 mi; no mode exists **21.** 300 calories **23.** Yes **25.** 410 mg **27.** White rhino **29.** About 1350 rhinos **31.** About 4100 rhinos **33.** $1270 per person **35.** 1950 and 1970 **37.** $1532.70 per person **39.** **(a)** 15% less; **(b)** $2003.51 more per person **41.** About 24 games **43.** 120–129

Mid-Chapter Review: Chapter 8, p. 472

1. True **2.** True **3.** False **4.** $\frac{60 + 45 + 115 + 15 + 35}{5} = \frac{270}{5} = 54$ **5.** 2.1, 4.8, 6.3, 8.7, 11.3, 14.5; 6.3 and 8.7; $\frac{6.3 + 8.7}{2} = \frac{15}{2} = 7.5$; the median is 7.5. **6.** Average: 83; median: 45; mode: 29 **7.** Average: 18.45; median: 13.895; no mode **8.** Average: $\frac{4}{9}$; median: $\frac{4}{9}$; no mode **9.** Average: 126; median: 116; no mode **10.** Average: $6.09; median: $5.24; modes: $4.96 and $5.24 **11.** Average: $\frac{27}{32}$; median: $\frac{13}{16}$; no mode **12.** Average: 6; median: 7; modes: 5 and 7 **13.** Average: 38.2; median: 38.2; no mode **14.** 8 oz **15.** 6% **16.** Hershey's Special Dark chocolate bar **17.** 7 oz **18.** Nabisco Chips Ahoy cookies **19.** Tom Brady **20.** About 45 touchdown passes **21.** About 22 more touchdown passes **22.** About 39 touchdown passes **23.** Yes. At an average speed of 20 mph, the trip would take $1\frac{1}{2}$ hr ($30 \text{ mi} \div 20 \text{ mph} = 1\frac{1}{2} \text{ hr}$). But the driver could have driven at a speed of 75 mph for a brief period during that time and at lower speeds for the remainder of the trip and still have an average speed of 20 mph. **24.** Answers may vary. Some would ask for the average salary since it is a center point that places equal emphasis on all the salaries in the firm. Some would ask for the median salary since it is a center point that deemphasizes the extremely high and extremely low salaries. Some would ask for the mode of the salaries since it might indicate the salary the applicant is most likely to earn.

Exercise Set 8.3, p. 478

RC1. True **RC2.** True **RC3.** False **RC4.** True **1.** Miniature tall bearded **3.** 16 in. to 26 in. **5.** Tall bearded **7.** 25 in. **9.** 190 calories **11.** 1 slice of chocolate cake with fudge frosting **13.** 1 cup of premium chocolate ice cream **15.** About 125 calories **17.** 1950 and 1970 **19.** About 175,000 bachelor's degrees

21.

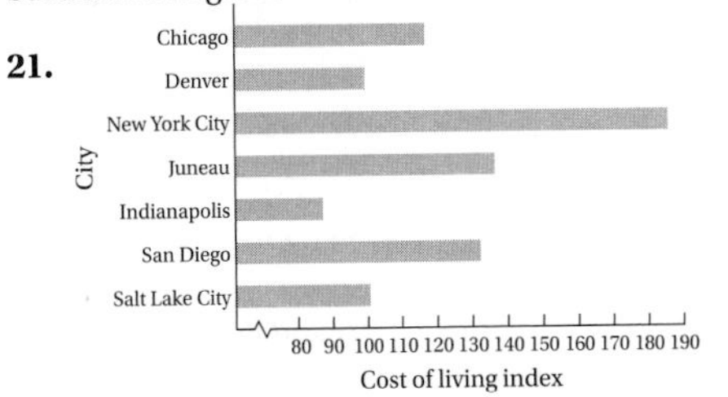

23. Denver and Indianapolis **25.** 68.9 **27.** New York City **29.** 27.4 min **31.** $42 **33.** June to July, September to October, and November to December **35.** About $900 **37.** 20 years **39.** About $440 **41.** About 120($1200) = $144,000

43.

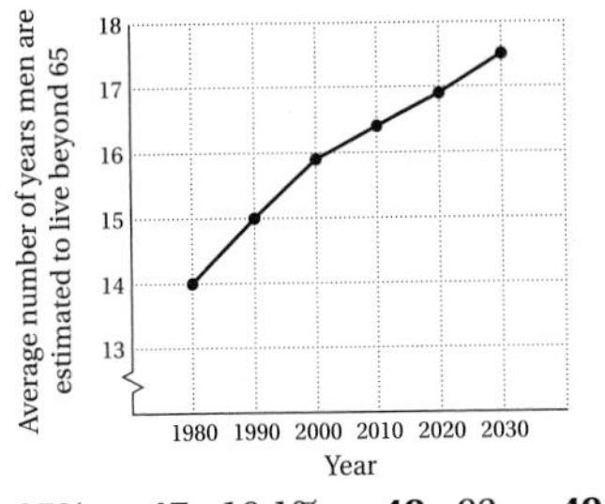

45. 25% **47.** 10.1% **48.** 83 **49.** $\frac{4}{3}$ **50.** $-\frac{37}{24}$ **51.** −2.26 **52.** 6.348 **53.** 7.2 **54.** 80 **55.** 0.9 **56.** 150% **57.** 21 **58.** 17.26 **59.** $\frac{31}{60}$ **60.** −45

Translating for Success, p. 485

1. D **2.** B **3.** J **4.** K **5.** I **6.** F **7.** N **8.** E **9.** L **10.** M

Exercise Set 8.4, p. 486

RC1. True **RC2.** True **RC3.** True **RC4.** False **RC5.** True **RC6.** False **1.** 10% **3.** 98,800 students **5.** Canada **7.**

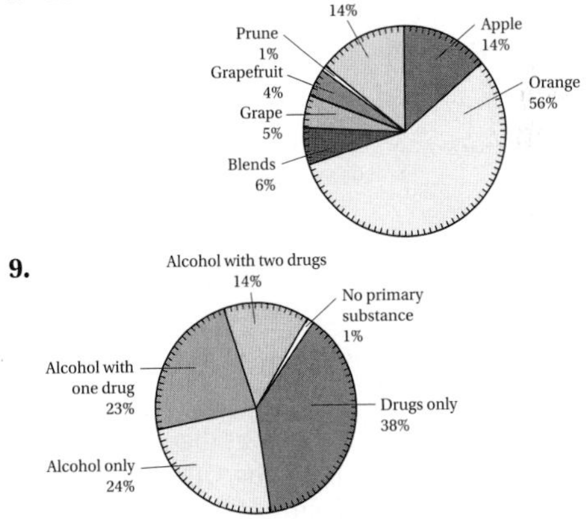

9.

Summary and Review: Chapter 8, p. 488

Vocabulary Reinforcement

1. table **2.** circle graph **3.** pictograph **4.** mode **5.** average **6.** median

Concept Reinforcement

1. False **2.** True **3.** True

Study Guide

1. Average: 8; median: 8; mode: 8 **2.** Quaker Organic Maple & Brown Sugar; \$0.54 per serving **3.** 12 g **4.** Arrowhead Stadium **5.** About \$1350 million more **6.** 80 yr and older **7.** 27%

Review Exercises

1. 38.5 **2.** 13.4 **3.** 1.55 **4.** 1840 **5.** $\$16.\overline{6}$ **6.** $321.\overline{6}$ **7.** 96 **8.** 28 mpg **9.** 3.1 **10.** 38.5 **11.** 14 **12.** 1.8 **13.** 1900 **14.** \$17 **15.** 375 **16.** Average: \$260; median: \$228 **17.** 26 **18.** 11 and 17 **19.** 0.2 **20.** 700 and 800 **21.** \$17 **22.** 20 **23.** 52% **24.** Mexico, Germany, and the Philippines **25.** United States **26.** About 21 World Series **27.** 7 games **28.** About 16 more World Series **29.** \$70,000–\$89,000 **30.** About 2 more governors **31.** About 22 governors **32.** April **33.** About 150 tornadoes **34.** About 150 more tornadoes **35.** In the spring **36.** 1998 **37.** About 10,000 children **38.** 2004 and 2010 **39.** Between 1986 and 1989, between 1995 and 1998, and between 2004 and 2007 **40.** By about 4000 children **41.** 34% **42.** Room and board **43.** 14% **44.** \$4830 **45.** D **46.** A

47.

48.

49.

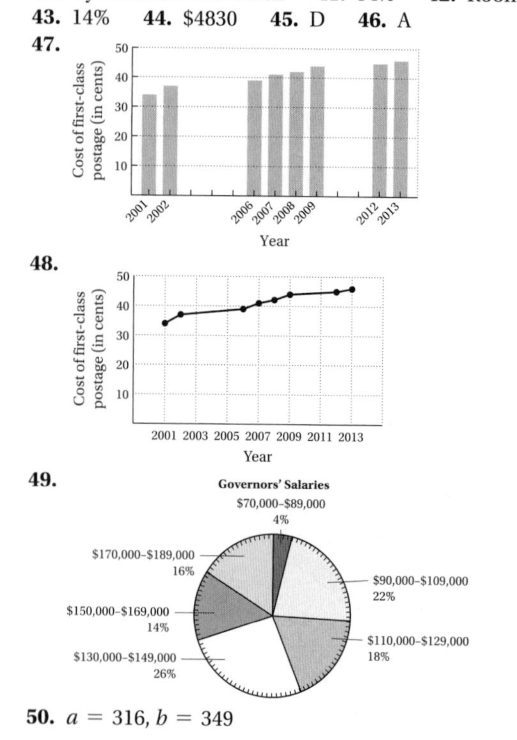

50. $a = 316, b = 349$

Understanding Through Discussion and Writing

1. The equation could represent a person's average income during a 4-yr period. Answers may vary. **2.** Bar graphs that show change over time can be converted to line graphs. Other bar graphs cannot be converted to line graphs. **3.** One advantage is that we can use circle graphs to visualize how the numbers of items in various categories compare in size. **4.** A bar graph is convenient for showing comparisons. A line graph is convenient for showing a change over time as well as to indicate patterns or trends. The choice of which to use to graph a particular set of data would probably depend on the type of data analysis desired. **5.** The average, the median, and the mode are "center points" that characterize a set of data. You might use the average to find a center point that is midway between the extreme values of the data. The median is a center point that is in the middle of all the data. That is, there are as many values less than the median as there are values greater than the median. The mode is a center point that represents the value or values that occur most frequently. **6.** Circle graphs are similar to bar graphs in that both allow us to tell at a glance how items in various categories compare in size. They differ in that circle graphs show percents whereas bar graphs show actual numbers of items in a given category.

Test: Chapter 8, p. 494

1. [8.1a] 49.5 **2.** [8.1a] 2.6 **3.** [8.1a] 15.5 **4.** [8.1b, c] 50.5; no mode exists **5.** [8.1b, c] 3; 1 and 3 **6.** [8.1b, c] 17.5; 17 and 18 **7.** [8.1a] 76 **8.** [8.1a] 2.9 **9.** [8.2a] 179 lb **10.** [8.2a] 5 ft 3 in.; medium frame **11.** [8.2a] 9 lb **12.** [8.2a] 32 lb **13.** [8.2b] Spain **14.** [8.2b] Norway and the United States **15.** [8.2b] 900 lb **16.** [8.2b] 1000 lb **17.** [8.3c] 2005 **18.** [8.3c] 2009 **19.** [8.3c] 10 hurricanes **20.** [8.3c] 10 more hurricanes **21.** [8.3c] About 8 hurricanes/year **22.** [8.3c] 2005, 2010, 2012

23. [8.3b]

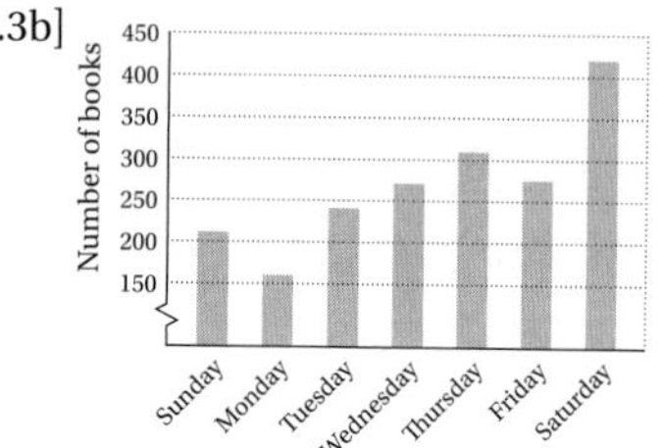

24. [8.3d]

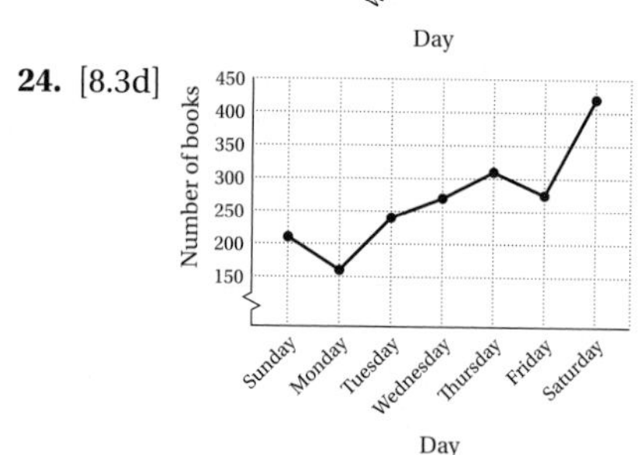

25. [8.4b]

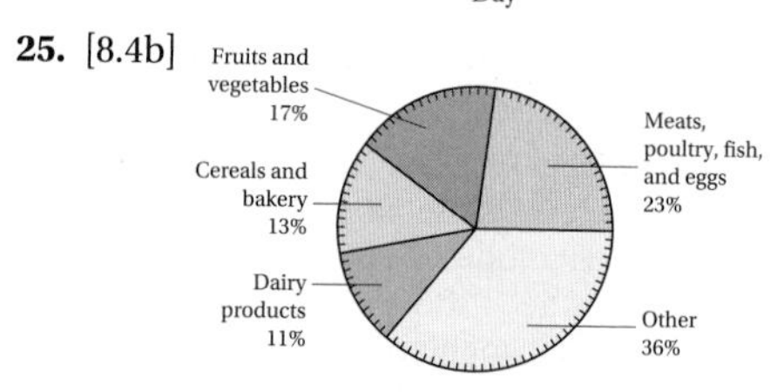

26. [8.4a] C **27.** [8.1a, b] $a = 74, b = 111$

Cumulative Review: Chapters 1–8, p. 497

1. [5.3b] 53,500,000,000 **2.** [8.1a] 24 mpg **3.** [1.1a] 5 hundreds **4.** [1.9c] 128 **5.** [3.1a] 1, 2, 3, 4, 5, 6, 10, 12, 15, 20, 30, 60 **6.** [5.1d] 52.0 **7.** [4.4a] $\frac{33}{10}$ **8.** [5.3b] \$2.10 **9.** [7.2a] 35% **10.** [6.3a] No **11.** [4.5a] $6\frac{7}{10}$ **12.** [5.2c] -2.5091 **13.** [4.3a] $-\frac{4}{3}$ **14.** [5.2b] 325.43 **15.** [4.6a] 15 **16.** [1.4a] 2,740,320 **17.** [3.7b] $\frac{9}{10}$ **18.** [4.4b] $4361\frac{1}{2}$ **19.** [6.3b] $9\frac{3}{5}$ **20.** [3.7c] $\frac{3}{4}$ **21.** [5.4b] -6.8 **22.** [1.7b] 15,312 **23.** [6.2b] 20.6¢/oz **24.** [7.5a] 3324 students **25.** [4.6c] $\frac{1}{4}$ yd **26.** [3.6b] $\frac{3}{8}$ cup **27.** [5.7a] 6.2 lb **28.** [1.8a] 2572 billion kWh **29.** [7.5a] 2.5%; 0.2% **30.** [7.5b] About 12.9% **31.** [4.2b], [4.3d] $\frac{1}{4}$ **32.** [6.4a] 1122 defective valves **33.** [5.7a] \$9.55 **34.** [7.6b] 7% **35.** [8.1a, b] \$79.557; \$80.10

36. [8.3b]

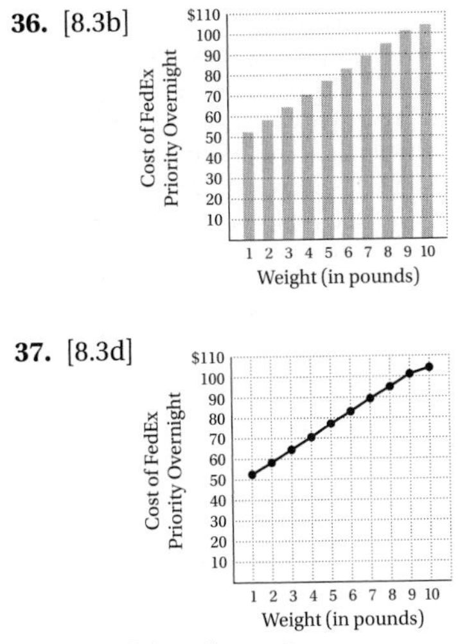

37. [8.3d]

38. [7.5b], [8.1a] 12% decrease

CHAPTER 9

Exercise Set 9.1, p. 503

RC1. 12 **RC2.** 5280 **RC3.** 36 **RC4.** 5280 **RC5.** 3 **RC6.** 12 **1.** 12 **3.** $\frac{1}{12}$ **5.** 5280 **7.** 108 **9.** 7 **11.** $1\frac{1}{2}$, or 1.5 **13.** 26,400 **15.** $5\frac{1}{4}$, or 5.25 **17.** $3\frac{1}{3}$ **19.** 37,488 **21.** $1\frac{1}{2}$, or 1.5 **23.** $1\frac{1}{4}$, or 1.25 **25.** 110 **27.** 2 **29.** 300 **31.** 30 **33.** $\frac{1}{36}$ **35.** 126,720 **37.** 2988 **39.** $\frac{37}{400}$ **40.** $\frac{7}{8}$ **41.** 0.0612 **42.** -0.04 **43.** $1\frac{3}{28}$ **44.** $\frac{24}{90}$ **45.** $\frac{320}{785}$, or $\frac{64}{157}$; $\frac{785}{320}$, or $\frac{157}{64}$ **46.** $\frac{2020}{1503}$; $\frac{1503}{2020}$ **47.** Length: 5400 in., or 450 ft; breadth: 900 in., or 75 ft; height: 540 in., or 45 ft

Exercise Set 9.2, p. 511

RC1. $<$ **RC2.** $>$ **RC3.** $<$ **RC4.** $<$ **RC5.** $>$ **RC6.** $<$ **1. (a)** 1000; **(b)** 0.001 **3. (a)** 10; **(b)** 0.1 **5. (a)** 0.01; **(b)** 100 **7.** 6700 **9.** 0.98 **11.** 8.921 **13.** 0.05666 **15.** 566,600 **17.** 4.77 **19.** 688 **21.** 0.1 **23.** 100,000 **25.** 142 **27.** 0.82 **29.** 450 **31.** 0.000024 **33.** 0.688 **35.** 230 **37.** 3.92 **39.** 48,440; 48.44 **41.** 4000; 400 **43.** 0.027; 0.00027 **45.** 541,300; 54,130 **47.** 0.234 **48.** 0.0234 **49.** 13.8474 **50.** $-80\frac{1}{2}$ **51.** $66.\overline{6}\%$, or $66\frac{2}{3}\%$ **52.** 62.5% **53.** $\frac{47}{75}$ **54.** $-\frac{59}{72}$ **55.** 0.9 **56.** 0.108 **57.** 1.0 m **59.** 1.4 cm

Exercise Set 9.3, p. 514

RC1. $<$ **RC2.** $>$ **RC3.** $<$ **RC4.** $>$ **RC5.** $<$ **RC6.** $>$ **1.** 100.65 **3.** 727.4394 **5.** 104.585 **7.** 289.62 **9.** 112.63 **11.** 9.14 **13.** 75.1967 **15.** 555.83421 **17.** 6.18109 **19.** 1376.136 **21.** 0.51181

	yd	cm	in.	m	mm
23.	0.2361	21.59	$8\frac{1}{2}$	0.2159	215.9
25.	52.9934	4844	1907.0828	48.44	48,440
27.	4	365.6	144	3.656	3656
29.	0.000295	0.027	0.0106299	0.00027	0.27
31.	55,247	5,050,000	1,988,892	50,500	50,500,000

33. 1,910,000 **34.** About $5.91 million **35.** 1 in. = 25.4 mm **37.** 23.21 mph

Mid-Chapter Review: Chapter 9, p. 517

1. False **2.** True **3.** False **4.** True **5.** True **6.** $16\frac{2}{3}$ yd $= 16\frac{2}{3} \times 1$ yd $= \frac{50}{3} \times 3$ ft $= 50$ ft
7. 13,200 ft $=$ 13,200 ft $\times \frac{1\text{ mi}}{5280\text{ ft}} = 2.5$ mi
8. 520 mm $=$ 520 mm $\times \frac{1\text{ m}}{1000\text{ mm}} = 0.52\text{ m} \times \frac{1\text{ km}}{1000\text{ m}} = 0.00052$ km
9. 10,200 mm $=$ 10,200 mm $\times \frac{1\text{ m}}{1000\text{ mm}} \approx 10.2\text{ m} \times \frac{3.281\text{ ft}}{1\text{ m}} =$ 33.4662 ft **10.** 9680 **11.** 70 **12.** 2.405 **13.** 0.00015 **14.** 22,680 **15.** 1200 **16.** 60,000 **17.** 850 **18.** 5 **19.** 1251 **20.** 52,800 **21.** 180,000 **22.** 84,000 **23.** 700 **24.** 0.4 **25.** 0.8009 **26.** 10 **27.** 19,200 **28.** 100 **29.** 0.000000001 **30.** 400,000 **31.** 760,320 **32.** 118.116 **33.** 0.8 **34.** 2.285 **35.** 60 **36.** 0.0025

37.

$\frac{1}{4}$ yd	24,000 dm
144 in.	1320 yd
2400 m	2400 mm
0.75 mi	9 in.
24 m	0.024 km
240 cm	12 ft

38. 3.5 ft, 2 yd, 100 in., $\frac{1}{100}$ mi, 1000 in., 430 ft, 6000 ft
39. 150 hm, 13 km, 310 dam, 300 m, 33,000 mm, 3240 cm, 250 dm
40. $\frac{1}{2}$ ft, 65 cm, 2 yd, 2.5 m, 1.5 mi, 3 km **41.** The student should have multiplied by $\frac{1}{12}$ (or divided by 12) to convert inches to feet. The correct procedure is as follows:

$$23\text{ in.} = 23\text{ in.} \times \frac{1\text{ ft}}{12\text{ in.}} = \frac{23\text{ in.}}{12\text{ in.}} \times 1\text{ ft}$$
$$= \frac{23}{12} \times \frac{\text{in.}}{\text{in.}} \times 1\text{ ft} = \frac{23}{12} \times 1\text{ ft} = \frac{23}{12}\text{ ft}.$$

42. Metric units are based on tens, so computations and conversions with metric units can be done by moving a decimal point. American units, which are not based on tens, require more complicated arithmetic in computations and conversions. **43.** A larger unit can be expressed as an equivalent number of smaller units. Thus when converting from a larger unit to a smaller unit, we can express the quantity as a number times one of the larger units and then substitute the equivalent number of smaller units for the larger unit. When converting from a smaller unit to a larger unit, we multiply by 1 using one larger unit in the numerator and the equivalent number of smaller units in the denominator. This allows us to "cancel" the smaller units, leaving the larger units. **44.** Answers may vary.

Exercise Set 9.4, p. 523

RC1. True **RC2.** False **RC3.** True **RC4.** False **RC5.** False **RC6.** True **1.** 2000 **3.** 3 **5.** 64 **7.** 12,640 **9.** 0.1 **11.** 5 **13.** 134,850 tons **15.** 1000 **17.** 10 **19.** $\frac{1}{100}$, or 0.01 **21.** 1000 **23.** 10 **25.** 234,000 **27.** 5.2 **29.** 6.7 **31.** 0.0502 **33.** 8.492 **35.** 58.5 **37.** 800,000 **39.** 1000 **41.** 0.0034 **43.** 0.0603 **45.** 1000 **47.** 0.325 **49.** 210,600 **51.** 0.0049 **53.** 125 mcg **55.** 0.875 mg; 875 mcg **57.** 4 tablets **59.** 8 cc **61.** $\frac{7}{20}$ **62.** $\frac{99}{100}$ **63.** $\frac{171}{200}$ **64.** $\frac{171}{500}$ **65.** $\frac{3}{8}$ **66.** $\frac{2}{3}$ **67.** $\frac{5}{6}$ **68.** $\frac{1}{6}$ **69.** 187,200 **70.** About 324 students **71.** $24.30 **72.** 12% **73.** 14 boxes **75. (a)** 3367.8 g; **(b)** 118.8 oz

Exercise Set 9.5, p. 530

RC1. cups **RC2.** pints **RC3.** pints **RC4.** fluid ounces **1.** 1000; 1000 **3.** 87,000 **5.** 0.049 **7.** 0.000401 **9.** 78,100 **11.** 320 **13.** 10 **15.** 32 **17.** 20 **19.** 14 **21.** 88

	gal	qt	pt	cups	oz
23.	1.125	4.5	9	18	144
25.	16	64	128	256	2048
27.	0.3984375	1.59375	3.1875	6.375	51

	L	mL	cc	cm^3
29.	2	2000	2000	2000
31.	64	64,000	64,000	64,000
33.	0.355	355	355	355

35. 2000 mL **37.** 0.32 L **39.** 59.14 mL **41.** 500 mL **43.** 125 mL/hr **45.** 9 **47.** $\frac{1}{5}$ **49.** 6 **51.** 15 **53.** 45.2% **54.** 99.9% **55.** 33.$\overline{3}$%, or $33\frac{1}{3}$% **56.** 66.$\overline{6}$%, or $66\frac{2}{3}$% **57.** 55% **58.** 105% **59.** 88% **60.** 8% **61.** $\frac{6.5}{39.6}$, or $\frac{65}{396}$; $\frac{33.1}{6.5}$, or $\frac{331}{65}$ **62.** About 87.2% **63.** 1.75 gal/wk; 7.5 gal/month; 91.25 gal/year; 78.5 million gal/day; 28.6525 billion gal/year **65.** $1.002/L

Calculator Corner, p. 535

1. 41°F **2.** 122°F **3.** 20°C **4.** 45°C

Exercise Set 9.6, p. 536

RC1. −10°C **RC2.** 200°F **RC3.** 100°F **RC4.** 0°C **1.** 24 **3.** 60 **5.** $365\frac{1}{4}$ **7.** 0.05 **9.** 8.2 **11.** 6.5 **13.** 10.75 **15.** 336 **17.** 4.5 **19.** 56 **21.** 86,164.2 sec **23.** 77°F **25.** 104°F **27.** 186.8°F **29.** −4°F **31.** 35.6°F **33.** −11.2°F **35.** 5432°F **37.** 30°C **39.** −25°C **41.** 81.$\overline{1}$°C **43.** 60°C **45.** 20°C **47.** −12.$\overline{2}$°C **49.** 37°C **51.** 53.$\overline{3}$°C **53.** **(a)** 136°F = 57.$\overline{7}$°C, $56\frac{2}{3}$°C = 134°F; **(b)** 2°F **55.** 56 R 11 **56.** 0.00803 **57.** 2.475 **58.** 338 **59.** −4.05 **60.** $\frac{15}{8}$, or $1\frac{7}{8}$ **61.** $-\frac{3}{5}$ **62.** $\frac{3}{8}$ **63.** About 0.03 year **65.** About 31,688 years **67.** 0.25

Translating for Success, p. 541

1. E **2.** H **3.** O **4.** A **5.** G **6.** C **7.** M **8.** I **9.** K **10.** N

Exercise Set 9.7, p. 542

RC1. True **RC2.** False **RC3.** True **RC4.** False **RC5.** True **RC6.** True **1.** 144 **3.** 640 **5.** $\frac{1}{144}$ **7.** 198 **9.** 396 **11.** 12,800 **13.** 27,878,400 **15.** 5 **17.** 1 **19.** $\frac{1}{640}$, or 0.0015625 **21.** 25,792 **23.** 37 **25.** 5,210,000 **27.** 140 **29.** 23.456 **31.** 0.085214 **33.** 2500 **35.** 0.4728 **37.** $240 **38.** $212 **39.** **(a)** $484.11; **(b)** $15,984.11 **40.** **(a)** $209.59; **(b)** $8709.59 **41.** **(a)** $220.93; **(b)** $6620.93 **42.** **(a)** $37.97; **(b)** $4237.97 **43.** 10.76 **45.** 1.67 **47.** 10,657,311 ft^2

Summary and Review: Chapter 9, p. 544

Concept Reinforcement

1. True **2.** False **3.** True **4.** False **5.** False **6.** False

Study Guide

1. $\frac{7}{3}$, or $2\frac{1}{3}$, or 2.$\overline{3}$ **2.** 13,200 **3.** 1200 **4.** 0.000046 **5.** 10.94 **6.** 5.14 **7.** 0.00978 **8.** 64 **9.** 42.67 **10.** 1 **11.** 154.4°F **12.** 40°C **13.** 9 **14.** 5240

Review Exercises

1. $2\frac{2}{3}$ **2.** 30 **3.** 0.03 **4.** 0.004 **5.** 72 **6.** 400,000 **7.** $1\frac{1}{6}$ **8.** 0.15 **9.** 218.8 **10.** 32.18 **11.** 10; 0.01 **12.** 305,000; 30,500 **13.** 112 **14.** 0.004 **15.** $\frac{4}{15}$, or 0.2$\overline{6}$ **16.** 0.464 **17.** 180 **18.** 4700 **19.** 16,140 **20.** 830 **21.** $\frac{1}{4}$, or 0.25 **22.** 0.04 **23.** 200 **24.** 30 **25.** 0.0007 **26.** 0.06 **27.** 1600 **28.** 400 **29.** 1.25 **30.** 50 **31.** 160 **32.** 7.5 **33.** 13.5 **34.** 60 **35.** 0.0302 **36.** 5.25 **37.** 6 mL **38.** 3000 mL **39.** 250 mcg **40.** 21.2°F **41.** 113°F **42.** 20°C **43.** −28.$\overline{8}$°C **44.** 36 **45.** 300,000 **46.** 14.375 **47.** 0.06 **48.** C **49.** B **50.** 17.64 sec **51.** 1 gal = 128 oz, so 1 oz of water (as capacity) weighs $\frac{8.3453}{128}$ lb, or about 0.0652 lb. An ounce of pennies weighs $\frac{1}{16}$ lb, or 0.0625 lb. Thus an ounce of water (as capacity) weighs more than an ounce of pennies.

Understanding Through Discussion and Writing

1. Grams are more easily converted to other units of mass than ounces. Since 1 gram is much smaller than 1 ounce, masses that might be expressed using fractional or decimal parts of ounces can often be expressed by whole numbers when grams are used. **2.** A single container is all that is required for both types of measure. **3.** Consider the table on p. 509 in the text. Moving one place in the table corresponds to moving the decimal point one place. To convert units of length, we determine the corresponding number of moves in the table and then move the decimal point that number of places. Since area involves square units, to convert units of area, we multiply the number of moves in the table on p. 509 by 2 and move the decimal point that number of places. **4.** Since metric units are based on 10, they are more easily converted than American units. **5.** **(a)** 23°C = 73.4°F, so you would want to play golf. **(b)** 10°C = 50°F, so you would not want to take a bath. **(c)** Since 0°C = 32°F, the freezing point of water, the lake would certainly be frozen at the lower temperature of −10°C, and it would be safe to go ice skating. **6.** 1 m ≈ 3.281 ft, so 1 square meter ≈ 3.281 ft × 3.281 ft = 10.764961 square feet. 1 yd = 3 ft, so 1 square yard = 3 ft × 3 ft = 9 square feet. Thus one square meter is larger than 1 square yard.

Test: Chapter 9, p. 549

1. [9.1a] 48 **2.** [9.1a] $\frac{1}{3}$ **3.** [9.2a] 6000 **4.** [9.2a] 0.87 **5.** [9.3a] 182.8 **6.** [9.3a] 1490.4 **7.** [9.2a] 5; 0.005 **8.** [9.2a] 1854.2; 185.42 **9.** [9.5a] 3.08 **10.** [9.5a] 240 **11.** [9.4a] 64 **12.** [9.4a] 8220 **13.** [9.4b] 3800 **14.** [9.4b] 0.4325 **15.** [9.4b] 2.2 **16.** [9.6a] 300 **17.** [9.6a] 360 **18.** [9.5a] 32 **19.** [9.5a] 1280 **20.** [9.5a] 40 **21.** [9.4c] 370 **22.** [9.6b] 35°C **23.** [9.6b] 138.2°F **24.** [9.3a] 1.094; 100; 39.370; 1000 **25.** [9.3a] 36,377.2; 14,328; 363.772; 363,772 **26.** [9.5b] 2500 mL **27.** [9.4c] 1500 mcg **28.** [9.5b] About 118.28 mL **29.** [9.7a] 1728 **30.** [9.7b] 0.0003 **31.** [9.6b] D **32.** [9.3a] About 39.47 sec

Cumulative Review: Chapters 1–9, p. 551

1. [5.3b] 40,400,000; 54,800,000 **2.** [6.2a] 33 mpg **3.** [5.2b], [5.3b] 1.561 billion bushels **4.** [7.5b] About 42.2% **5.** [1.4a] 50,854,100 **6.** [3.6a] $-\frac{1}{12}$ **7.** [5.5c] −13.8 **8.** [4.6b] $\frac{2}{3}$ **9.** [5.4a] 35.6 **10.** [4.5b] $\frac{5}{6}$ **11.** [3.2a] Yes **12.** [3.2a] No **13.** [3.1d] 3 · 3 · 11 **14.** [4.1a] 245 **15.** [5.5b] 35.8 **16.** [5.1a] One hundred three and sixty-four thousandths **17.** [8.1a, b] 17.8$\overline{3}$; 17.5 **18.** [7.1b] 8% **19.** [7.2a] 60% **20.** [9.1a] 6 **21.** [9.4a] 0.375, or $\frac{3}{8}$ **22.** [9.6b] 59 **23.** [9.5a] 87 **24.** [9.6a] 0.15, or $\frac{3}{20}$ **25.** [9.2a] 0.17 **26.** [9.3a] 3539.8 **27.** [9.5a] 2 **28.** [9.4c] 230 **29.** [9.7a] 108 **30.** [5.4b] −150.5 **31.** [1.7b] 19,248 **32.** [3.7c] $\frac{15}{2}$ **33.** [4.3c] $-\frac{58}{35}$ **34.** [4.2b], [4.4a] $1\frac{1}{4}$ hr **35.** [5.3b], [5.7a] 182,542,500,000 lb **36.** [5.7a], [6.2a] 255.8 mi; 15.9875 mpg **37.** [7.6b] $16,125 **38.** [7.7b] $2388.10 **39.** [6.4a] $8\frac{3}{4}$ lb **40.** [3.6b] $13,200 **41.** [7.5a] Yes **42.** [9.4a] 240 oz **43.** [1.8a] 60 dips **44.** [4.5c], [4.6c] About 25 or 26 dips **45.** [5.7a] $179.40 **46.** [5.7a] $121.50 **47.** [3.7c] $\frac{5}{2}$

CHAPTER 10

Exercise Set 10.1, p. 557

RC1. closed **RC2.** perimeter **RC3.** perimeter **RC4.** square **1.** 17 mm **3.** 15.25 in. **5.** 18 km **7.** 30 ft **9.** 16 yd **11.** 88 ft **13.** 182 mm **15.** 27 ft **17.** 122 cm **19.** **(a)** 228 ft; **(b)** $1046.52 **21.** $19.20 **22.** $96 **23.** 1000 **24.** 1331 **25.** 225 **26.** 484 **27.** 49 **28.** 64 **29.** 5% **30.** 11% **31.** 64 in.

Exercise Set 10.2, p. 564

RC1. (b) **RC2.** (c) **RC3.** (a) **RC4.** (d) **1.** 15 km^2 **3.** 1.4 in^2 **5.** $6\frac{1}{4}$ yd^2 **7.** 8100 ft^2 **9.** 50 ft^2 **11.** 169.883 cm^2 **13.** $41\frac{2}{9}$ in^2 **15.** 484 ft^2 **17.** 3237.61 km^2

19. $28\frac{57}{64}$ yd^2 **21.** 32 cm^2 **23.** 60 in^2 **25.** 104 ft^2 **27.** 45.5 in^2 **29.** 8.05 cm^2 **31.** 297 cm^2 **33.** 7 m^2 **35.** 1197 m^2 **37.** 39,825 ft^2 **39.** (a) 630.36 ft^2; (b) $7879.50 **41.** (a) 819.75 ft^2; (b) 3 gal; (c) $104.85 **43.** 80 cm^2 **45.** 675 cm^2 **47.** 21 cm^2 **49.** 144 ft^2 **51.** 234 **52.** 230 **53.** 336 **54.** 24 **55.** 0.724 **56.** 0.0724 **57.** 2520 **58.** 6 **59.** 28 **60.** $22\frac{1}{2}$ **61.** 12 **62.** 46,252.8 **63.** $30,474.86 **64.** $413,458.31 **65.** $641,566.26 **66.** $429,610.21 **67.** 16,914 in^2

Calculator Corner, p. 572

1. Left to the student **2.** Left to the student

Exercise Set 10.3, p. 574

RC1. radius **RC2.** circumference **RC3.** circumference **RC4.** area **1.** 14 cm; 44 cm; 154 cm^2 **3.** $1\frac{1}{2}$ in.; $4\frac{5}{7}$ in.; $1\frac{43}{56}$ in^2 **5.** 16 ft; 100.48 ft; 803.84 ft^2 **7.** 0.7 cm; 4.396 cm; 1.5386 cm^2 **9.** 94.2 ft **11.** About 55.04 in^2 larger **13.** About 7930.25 mi; about 3965.13 mi **15.** Maximum circumference of barrel: $8\frac{9}{14}$ in.; minimum circumference of handle: $2\frac{86}{133}$ in. **17.** 65.94 yd^2 **19.** 45.68 ft **21.** 26.84 yd **23.** 45.7 yd **25.** 100.48 m^2 **27.** 6.9972 cm^2 **29.** 64.4214 in^2 **31.** 47 mpg **32.** 0.4 oz **33.** 5 lb **34.** $730 **35.** 43,560 ft^2; 1311.6 ft; $599.96 **37.** 43,595.47395 ft^2; 739.9724 ft; $449.97 **39.** 43,560 ft^2; 844 ft; $449.97

Mid-Chapter Review: Chapter 10, p. 579

1. False **2.** True **3.** True **4.** False **5.** True

6. $P = 2 \cdot (10\text{ ft} + 3\text{ ft})$
$P = 2 \cdot (13\text{ ft})$
$P = 26$ ft;

$A = 10\text{ ft} \cdot 3\text{ ft}$
$A = 10 \cdot 3 \cdot \text{ft} \cdot \text{ft}$
$A = 30\text{ ft}^2$

7. $A = \frac{1}{2} \cdot 12\text{ cm} \cdot 8\text{ cm}$
$A = \frac{12 \cdot 8}{2}\text{ cm}^2$
$A = \frac{96}{2}\text{ cm}^2$, or 48 cm^2

8. $C \approx 3.14 \cdot 10.2$ in.
$C = 32.028$ in.;

$A \approx 3.14 \cdot 5.1\text{ in.} \cdot 5.1\text{ in.}$
$A = 81.6714\text{ in}^2$

9. 76 mm

10. $P = 50\frac{2}{3}$ ft; $A = 160\frac{4}{9}$ ft^2 **11.** 800 in^2 **12.** $\frac{9}{16}$ yd^2 **13.** 66 km^2 **14.** $C = 43.96$ in.; $A = 153.86$ in^2 **15.** $C = 27.004$ cm; $A = 58.0586$ cm^2 **16.** Area of a circle with radius 4 ft: $16 \cdot \pi$ ft^2; Area of a square with side 4 ft: 16 ft^2; Circumference of a circle with radius 4 ft: $8 \cdot \pi$ ft; Area of a rectangle with length 8 ft and width 4 ft: 32 ft^2; Area of a triangle with base 4 ft and height 8 ft: 16 ft^2; Perimeter of a square with side 4 ft: 16 ft; Perimeter of a rectangle with length 8 ft and width 4 ft: 24 ft **17.** The area of a 16-in.-diameter pizza is approximately $3.14 \cdot 8\text{ in.} \cdot 8\text{ in.}$, or 200.96 in^2. At $16.25, its unit price is $\frac{\$16.25}{200.96\text{ in}^2}$, or about $0.08/in^2. The area of 10-in.-diameter pizza is approximately $3.14 \cdot 5\text{ in.} \cdot 5\text{ in.}$, or 78.5 in^2. At $7.85, its unit price is $\frac{\$7.85}{78.5\text{ in}^2}$, or $0.10/in^2. Since the 16-in.-diameter pizza has the lower unit price, it is a better buy. **18.** No; let l and w represent the length and the width of the smaller rectangle. Then $3 \cdot l$ and $3 \cdot w$ represent the length and the width of the larger rectangle. The area of the first rectangle is $l \cdot w$, but the area of the second is $3 \cdot l \cdot 3 \cdot w = 3 \cdot 3 \cdot l \cdot w = 9 \cdot l \cdot w$, or 9 times the area of the smaller rectangle. **19.** Yes; let s represent the length of a side of the larger square. Then $\frac{1}{2}s$ represents the length of a side of the smaller square. The perimeter of the larger square is $4 \cdot s$, and the perimeter of the smaller square is $4 \cdot \frac{1}{2}s = 2s$, or $\frac{1}{2}$ the perimeter of the larger square.

20. For a rectangle with length l and width w,
$P = l + w + l + w$
$= (l + w) + (l + w)$
$= 2 \cdot (l + w).$
We also have
$P = l + w + l + w$
$= (l + l) + (w + w)$
$= 2 \cdot l + 2 \cdot w.$

21. See p. 561 of the text. **22.** No; let r = radius of the smaller circle. Then its area is $\pi \cdot r \cdot r$, or πr^2. The radius of the larger circle is $2r$, and its area is $\pi \cdot 2r \cdot 2r$, or $4\pi r^2$, or $4 \cdot \pi r^2$. Thus the area of the larger circle is 4 times the area of the smaller circle.

Exercise Set 10.4, p. 586

RC1. (b) **RC2.** (a) **RC3.** (c) **RC4.** (d) **1.** 768 cm^3 **3.** 45 in^3 **5.** 75 m^3 **7.** $357\frac{1}{2}$ yd^3 **9.** 803.84 in^3 **11.** 353.25 cm^3 **13.** 41,580,000 yd^3 **15.** $4,186,666\frac{2}{3}$ in^3 **17.** 124.72 m^3 **19.** $1950\frac{101}{168}$ ft^3 **21.** 113,982 ft^3 **23.** 24.64 cm^3 **25.** $\frac{33}{40}$ yd^3 **27.** 4747.68 cm^3 **29.** 65,417 m^3 **31.** About 77.7 in^3 **33.** 5832 yd^3 **35.** 152,321 m^3 **37.** 32,993,440,000 mi^3 **39.** 6 cm by 6 cm by 6 cm **41.** (a) About 1875.63 mm^3; (b) about 11,253.78 mm^3 **43.** 1064 mi^3 **44.** 24,360 mi^2 **45.** 260.4 ft **46.** 1087.8 mi^3 **47.** 3540.68 km^3 **49.** 0.477 m^3

Exercise Set 10.5, p. 597

RC1. (f) **RC2.** (i) **RC3.** (a) **RC4.** (e) **RC5.** (b) **RC6.** (j) **RC7.** (c) **RC8.** (h) **RC9.** (d) **RC10.** (g) **1.** Angle GHI, angle IHG, $\angle GHI$, $\angle IHG$, or $\angle H$ **3.** 10° **5.** 180° **7.** 130° **9.** Obtuse **11.** Acute **13.** Straight **15.** Obtuse **17.** Acute **19.** Obtuse **21.** 79° **23.** 23° **25.** 32° **27.** 61° **29.** 177° **31.** 41° **33.** 95° **35.** 78° **37.** Scalene; obtuse **39.** Scalene; right **41.** Equilateral; acute **43.** Scalene; obtuse **45.** 46° **47.** 120° **49.** 58° **51.** 2.7125 **52.** $8\frac{1}{12}$ **53.** $-\frac{7}{60}$ **54.** $\frac{2}{3}$ **55.** −27 **56.** 840 **57.** $\frac{3}{10}$ **58.** $-\frac{3}{16}$ **59.** $m\angle 2 = 67.13°$; $m\angle 3 = 33.07°$; $m\angle 4 = 79.8°$; $m\angle 5 = 67.13°$ **61.** $m\angle ACB = 50°$; $m\angle CAB = 40°$; $m\angle EBC = 50°$; $m\angle EBA = 40°$; $m\angle AEB = 100°$; $m\angle ADB = 50°$

Calculator Corner, p. 601

1. 6.6 **2.** 19.8 **3.** 1.9 **4.** 24.5 **5.** 121.2 **6.** 85.4

Translating for Success, p. 604

1. K **2.** G **3.** B **4.** H **5.** O **6.** M **7.** E **8.** A **9.** D **10.** I

Exercise Set 10.6, p. 605

RC1. True **RC2.** False **RC3.** True **RC4.** False **1.** 10 **3.** 21 **5.** 25 **7.** 19 **9.** 23 **11.** 100 **13.** 6.928 **15.** 2.828 **17.** 4.243 **19.** 2.449 **21.** 3.162 **23.** 8.660 **25.** 14 **27.** 13.528 **29.** $c = \sqrt{34}$; $c \approx 5.831$ **31.** $c = \sqrt{98}$; $c \approx 9.899$ **33.** $a = 5$ **35.** $b = 8$ **37.** $c = 26$ **39.** $b = 12$ **41.** $b = \sqrt{1023}$; $b \approx 31.984$ **43.** $c = 5$ **45.** $\sqrt{8450}$ ft ≈ 91.9 ft **47.** $\sqrt{644}$ ft ≈ 25.4 ft **49.** $\sqrt{34,541}$ m ≈ 185.9 m **51.** $\sqrt{211,200,000}$ ft $\approx 14,532.7$ ft **53.** 1000 **54.** 100 **55.** 100,000 **56.** 10,000 **57.** 3 **58.** 982 **59.** The areas are the same.

Summary and Review: Chapter 10, p. 608

Vocabulary Reinforcement

1. parallel **2.** perimeter **3.** radius **4.** supplementary **5.** scalene **6.** hypotenuse

Concept Reinforcement

1. True **2.** True **3.** True **4.** True

Study Guide

1. 27.8 ft; 46.74 ft^2 **2.** 15.5 m^2 **3.** 8.75 ft^2 **4.** 80 m^2 **5.** 37.68 in. **6.** 616 cm^2 **7.** 1683.3 m^3 **8.** $30\frac{6}{35}$ ft^3 **9.** 1696.537813 cm^3 **10.** 26.49375 ft^3 **11.** Complement: 52°; supplement: 142° **12.** 87° **13.** $\sqrt{208} \approx 14.422$

Review Exercises

1. 23 ft **2.** 4.4 m **3.** 228 ft; 2808 ft^2 **4.** 36 ft; 81 ft^2 **5.** 17.6 cm; 12.6 cm^2 **6.** 22.5 m^2 **7.** 29.64 yd^2 **8.** 88 m^2 **9.** $145\frac{5}{9}$ in^2 **10.** 840 ft^2 **11.** 8 m **12.** $\frac{14}{11}$ in., or $1\frac{3}{11}$ in. **13.** 14 ft **14.** 20 cm **15.** 50.24 m **16.** 8 in. **17.** 200.96 m^2 **18.** $5\frac{1}{11}$ in^2 **19.** 1038.555 ft^2 **20.** 93.6 yd^3 **21.** 193.2 cm^3 **22.** 31,400 ft^3 **23.** $33.49\overline{3}$ cm^3 **24.** 4.71 in^3 **25.** 942 cm^3 **26.** 26.28 ft^2; 20.28 ft **27.** 54° **28.** 180° **29.** 140° **30.** 90° **31.** Acute **32.** Straight **33.** Obtuse **34.** Right **35.** 49° **36.** 136° **37.** 60° **38.** Scalene **39.** Right **40.** 8 **41.** 9.110 **42.** $c = \sqrt{850}$; $c \approx 29.155$ **43.** $b = \sqrt{51}$; $b \approx 7.141$ **44.** $c = \sqrt{89}$ ft; $c \approx 9.434$ ft **45.** $a = \sqrt{76}$ cm; $a \approx 8.718$ cm **46.** About 17.9 ft **47.** About 13.42 ft **48.** About 85.9 ft **49.** B **50.** B **51.** 100 ft^2 **52.** 7.83998704 m^2 **53.** 47.25 cm^2

Understanding Through Discussion and Writing

1. Add 90° to the measure of the angle's complement. **2.** This could be done using the technique in Example 7 of Section 10.4. We could also approximate the volume with the volume of a rectangular solid of similar size. Another method is to break the egg and measure the capacity of its contents. **3.** Show that the sum of the squares of the lengths of the legs is the same as the square of the length of the hypotenuse. **4.** Divide the figure into 3 triangles.

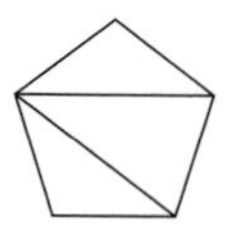

The sum of the measures of the angles of each triangle is 180°, so the sum of the measures of the angles of the figure is $3 \cdot 180°$, or 540°. **5.** The volume of the cone is half the volume of the dome. It can be argued that a cone-cap is more energy-efficient since there is less air under it to be heated and cooled. **6.** Volume of two spheres, each with radius r: $2(\frac{4}{3}\pi r^3) = \frac{8}{3}\pi r^3$; volume of one sphere with radius $2r$: $\frac{4}{3}\pi(2r)^3 = \frac{32}{3}\pi r^3$. The volume of the sphere with radius $2r$ is four times the volume of the two spheres, each with radius r: $\frac{32}{3}\pi r^3 = 4 \cdot \frac{8}{3}\pi r^3$.

Test: Chapter 10, p. 616

1. [10.1a], [10.2a] 32.82 cm; 65.894 cm^2 **2.** [10.1a], [10.2a] $19\frac{1}{2}$ in.; $23\frac{49}{64}$ in^2 **3.** [10.2b] 25 cm^2 **4.** [10.2b] 12 m^2 **5.** [10.2b] 18 ft^2 **6.** [10.3a] $\frac{1}{4}$ in. **7.** [10.3a] 9 cm **8.** [10.3b] $\frac{11}{14}$ in. **9.** [10.3c] 254.34 cm^2 **10.** [10.3d] 65.46 km; 103.815 km^2 **11.** [10.4a] 84 cm^3 **12.** [10.4e] 420 in^3 **13.** [10.4b] 1177.5 ft^3 **14.** [10.4c] $4186.\overline{6}$ yd^3 **15.** [10.4d] 113.04 cm^3 **16.** [10.5a] 90° **17.** [10.5a] 35° **18.** [10.5a] 180° **19.** [10.5a] 113° **20.** [10.5b] Right **21.** [10.5b] Acute **22.** [10.5b] Straight **23.** [10.5b] Obtuse **24.** [10.5e] 35° **25.** [10.5d] Isosceles **26.** [10.5d] Obtuse **27.** [10.5c] Complement: 25°; supplement: 115° **28.** [10.6a] 15 **29.** [10.6b] 9.327 **30.** [10.6c] $c = 40$ **31.** [10.6c] $b = \sqrt{60}$; $b \approx 7.746$ **32.** [10.6c] $c = \sqrt{2}$; $c \approx 1.414$ **33.** [10.6c] $b = \sqrt{51}$; $b \approx 7.141$ **34.** [10.6d] About 15.8 m **35.** [10.4c] D **36.** [9.1a], [10.2a] 2 ft^2 **37.** [9.1a], [10.2b] 1.875 ft^2 **38.** [9.1a], [10.4a] 0.65 ft^3 **39.** [9.1a], [10.4d] 0.033 ft^3 **40.** [9.1a], [10.4b] 0.055 ft^3

Cumulative Review: Chapters 1–10, p. 619

1. [5.3b] 9,800,000 **2.** [10.4e] 93,750 lb **3.** [4.5a] $4\frac{1}{6}$ **4.** [5.2b] 87.52 **5.** [1.5a] 1234 **6.** [2.5c] −595 **7.** [5.5c] −33.2 **8.** [1.9d] 2 **9.** [5.1b] $\frac{1209}{1000}$ **10.** [7.2b] $\frac{17}{100}$ **11.** [4.3b] $<$ **12.** [3.5c] = **13.** [9.5a] 3 **14.** [9.6a] $\frac{3}{20}$ **15.** [9.6b] 59 **16.** [9.5a] 87 **17.** [9.7a] 27 **18.** [9.2a] 0.17 **19.** [4.3c] $-\frac{13}{8}$ **20.** [6.3b] $4\frac{2}{7}$ **21.** [5.4b] 113.4 **22.** [6.3b] $14\frac{2}{5}$ **23.** [10.1a], [10.2b] 380 cm; 5500 cm^2 **24.** [10.3a, b, c] 70 in.; 220 in; 3850 in^2 **25.** [10.4c] $179{,}666\frac{2}{3}$ in^3 **26.** [8.1a] 99 **27.** [7.7a] \$84 **28.** [7.7b] \$22,376.03 **29.** [10.6d] 17 m **30.** [7.6a] 6% **31.** [4.5c] $2\frac{1}{8}$ yd **32.** [5.7a] \$37.42 **33.** [6.2b] The 8-qt box **34.** [3.6b] $\frac{7}{20}$ km **35.** [10.5e] 30° **36.** [10.5d] Isosceles **37.** [10.5d] Obtuse **38.** [9.1a], [10.4b] 94,200 ft^3 **39.** [9.1a], [10.4a] 1.342 ft^3

CHAPTER 11

Calculator Corner, p. 624

1. −6 **2.** −19 **3.** 9 **4.** 12

Exercise Set 11.1, p. 628

RC1. Division **RC2.** Multiplication **RC3.** Multiplication **RC4.** Division

1. 42 **3.** 3 **5.** −1 **7.** 6 **9.** −20 **11.** 240; 240 **13.** 160; 160 **15.** $2b + 10$ **17.** $7 - 7t$ **19.** $30x + 12$ **21.** $7x + 28 + 42y$ **23.** $-7y + 14$ **25.** $45x + 54y - 72$ **27.** $\frac{3}{4}x - \frac{9}{4}y - \frac{3}{2}z$ **29.** $-3.72x + 9.92y - 3.41$ **31.** $2(x + 2)$ **33.** $5(6 + y)$ **35.** $7(2x + 3y)$ **37.** $5(x + 2 + 3y)$ **39.** $8(x - 3)$ **41.** $4(8 - y)$ **43.** $2(4x + 5y - 11)$ **45.** $-6(3x + 2y - 1)$, or $6(-3x - 2y + 1)$ **47.** $19a$ **49.** $9a$ **51.** $8x + 9z$ **53.** $-19a + 88$ **55.** $4t + 6y - 4$ **57.** $8x$ **59.** $5n$ **61.** $-16y$ **63.** $17a - 12b - 1$ **65.** $4x + 2y$ **67.** $\frac{39}{20}x + \frac{1}{2}y + 12$ **69.** $0.8x + 0.5y$ **71.** 30 yd; 94.2 yd; 706.5 yd^2 **72.** 16.4 m; 51.496 m; 211.1336 m^2 **73.** 19 mi; 59.66 mi; 283.385 mi^2 **74.** 4800 cm; 15,072 cm; 18,086,400 cm^2 **75.** 10 mm; 62.8 mm; 314 mm^2 **76.** 132 km; 828.96 km; 54,711.36 km^2 **77.** 2.3 ft; 14.444 ft; 16.6106 ft^2 **78.** 5.15 m; 32.342 m; 83.28065 m^2 **79.** $q(1 + r + rs + rst)$

Exercise Set 11.2, p. 632

RC1. (f) **RC2.** (c) **RC3.** (e) **RC4.** (a)

1. 4 **3.** −20 **5.** 2 **7.** 7 **9.** 4 **11.** −26 **13.** −18 **15.** −18 **17.** 15 **19.** 19 **21.** −1 **23.** −14 **25.** 2 **27.** 20 **29.** −6 **31.** $\frac{7}{3}$ **33.** $-\frac{7}{4}$ **35.** $\frac{41}{24}$ **37.** $-\frac{1}{20}$ **39.** 5.1 **41.** 12.4 **43.** −5 **45.** $1\frac{5}{6}$ **47.** $-\frac{10}{21}$ **49.** −11 **50.** $-\frac{1}{24}$ **51.** −34.1 **52.** −1.7 **53.** 5 **54.** $-\frac{31}{24}$ **55.** 5.5 **56.** 8.1 **57.** 24 **58.** $-\frac{5}{12}$ **59.** 283.14 **60.** −15.68 **61.** 8 **62.** $-\frac{16}{15}$ **63.** −14.3 **64.** −4.9 **65.** 342.246 **67.** $-\frac{26}{15}$ **69.** −10 **71.** $-\frac{5}{17}$

Exercise Set 11.3, p. 638

RC1. (f) **RC2.** (d) **RC3.** (a) **RC4.** (b)

1. 6 **3.** 9 **5.** 12 **7.** −40 **9.** −7 **11.** −7 **13.** −6 **15.** −5 **17.** 6 **19.** 5 **21.** −63 **23.** 36 **25.** −21 **27.** $-\frac{3}{5}$ **29.** $-\frac{3}{2}$ **31.** $\frac{9}{2}$ **33.** 7 **35.** −7 **37.** −8 **39.** 15.9 **41.** −50 **43.** −14 **45.** 62.8 ft; 20 ft; 314 ft^2 **46.** 75.36 cm; 12 cm; 452.16 cm^2 **47.** 8000 ft^3 **48.** 31.2 cm^3 **49.** 68 in^2 **50.** 38.25 m^2 **51.** −8655 **53.** No solution **55.** No solution

Mid-Chapter Review: Chapter 11, p. 640

1. False **2.** True **3.** True **4.** False

5. $6x - 3y + 18 = 3 \cdot 2x - 3 \cdot y + 3 \cdot 6 = 3(2x - y + 6)$

6.
$$\begin{aligned} x + 5 &= -3 \\ x + 5 - 5 &= -3 - 5 \\ x + 0 &= -8 \\ x &= -8 \end{aligned}$$

7.
$$\begin{aligned} -6x &= 42 \\ \frac{-6x}{-6} &= \frac{42}{-6} \\ 1 \cdot x &= -7 \\ x &= -7 \end{aligned}$$

8. -28 **9.** 7 **10.** 5 **11.** $3x + 15$ **12.** $8y - 28$ **13.** $18x + 12y - 6$ **14.** $6x + 2y - 16$ **15.** $3(y + 7)$ **16.** $5(z + 9)$ **17.** $9(x - 4)$ **18.** $8(3a - 1)$ **19.** $2(2x + 3y - 1)$ **20.** $3(4x - 3y + 1)$ **21.** $4(a - 3b + 8)$ **22.** $6(5a - 3b - 4)$ **23.** $15x$ **24.** $2y$ **25.** $2x - y - 3$ **26.** 6 **27.** -12 **28.** 7 **29.** -10 **30.** 20 **31.** 5 **32.** 7 **33.** 10 **34.** $-\frac{5}{6}$ **35.** $\frac{3}{4}$ **36.** 0.7 **37.** -1.4 **38.** 6 **39.** 12 **40.** -17 **41.** -9 **42.** 17 **43.** -6 **44.** 18 **45.** -15 **46.** $-\frac{3}{2}$ **47.** $\frac{5}{3}$ **48.** -3 **49.** -35 **50.** They are not equivalent. For example, let $a = 2$ and $b = 3$. Then $(a + b)^2 = (2 + 3)^2 = 5^2 = 25$, but $a^2 + b^2 = 2^2 + 3^2 = 4 + 9 = 13$. **51.** We use the distributive law when we collect like terms even though we might not always write this step. **52.** The student probably added $\frac{1}{3}$ on both sides of the equation rather than adding $-\frac{1}{3}$ on (or subtracting $\frac{1}{3}$ on) both sides. The correct solution is -2. **53.** The student apparently multiplied by $-\frac{2}{3}$ on both sides rather than dividing by $\frac{2}{3}$ on both sides. The correct solution is $-\frac{5}{2}$.

Exercise Set 11.4, p. 649

RC1. (d) **RC2.** (a) **RC3.** (c) **RC4.** (e) **RC5.** (b)
1. 5 **3.** 8 **5.** 10 **7.** 14 **9.** -8 **11.** -8 **13.** -7 **15.** 15 **17.** 6 **19.** 4 **21.** 6 **23.** -3 **25.** 1 **27.** -20 **29.** 6 **31.** 7 **33.** 2 **35.** 5 **37.** 2 **39.** 10 **41.** 4 **43.** 0 **45.** -1 **47.** $-\frac{4}{3}$ **49.** $\frac{2}{5}$ **51.** -2 **53.** -4 **55.** $\frac{4}{5}$ **57.** $-\frac{28}{27}$ **59.** 6 **61.** 2 **63.** 6 **65.** 8 **67.** 1 **69.** 17 **71.** $-\frac{5}{3}$ **73.** -3 **75.** 2 **77.** $-\frac{51}{31}$ **79.** 2 **81.** 4.5 **82.** 0.0009 **83.** 43.75% **84.** 76% **85.** 200 **86.** 0.0147 **87.** 180 **88.** 38° **89.** 95° **90.** \$39,574 **91.** 2 **93.** -2 **95.** 8 **97.** 2 cm

Translating for Success, p. 662

1. B **2.** H **3.** G **4.** N **5.** J **6.** C **7.** L **8.** E **9.** F **10.** D

Exercise Set 11.5, p. 663

RC1. Familiarize **RC2.** Translate **RC3.** Solve **RC4.** Check **RC5.** State
1. $2x - 3$ **3.** $97\%y$, or $0.97y$ **5.** $5x + 4$, or $4 + 5x$ **7.** 32 **9.** 1522 Medals of Honor **11.** 57 **13.** -12 **15.** 4.37 mi **17.** 325 mi **19.** 89 and 96 **21.** 36 in. × 110 in. **23.** 21.8 million **25.** 305 ft **27.** First: 30 m; second: 90 m; third: 360 m **29.** First: 28°; second: 84°; third: 68° **31.** 18 mi **33.** \$350 **35.** \$852.94 **37.** \$24.95 **39.** \$36 **41.** Length: 265 ft; width: 165 ft; area: 43,725 ft^2 **43.** $-\frac{47}{40}$ **44.** $-\frac{17}{40}$ **45.** $-\frac{3}{10}$ **46.** $-\frac{32}{15}$ **47.** 1.6 **48.** 409.6 **49.** -9.6 **50.** -41.6 **51.** Length: 12 cm; width: 9 cm **53.** Quarters: 60; dimes: 30; nickels: 40

Summary and Review: Chapter 11, p. 668

Vocabulary Reinforcement

1. substituting **2.** constant **3.** identity property of 1 **4.** multiplication principle **5.** distributive law of multiplication over subtraction **6.** addition principle **7.** equivalent

Concept Reinforcement

1. True **2.** False **3.** False **4.** True

Study Guide

1. -6 **2.** $4x + 20y - 28$ **3.** $8(3a - b + 2)$ **4.** $6x - 3y$ **5.** 2 **6.** -8 **7.** 1 **8.** -2 **9.** 6 **10.** $n + 5$, or $5 + n$

Review Exercises

1. 4 **2.** $15x - 35$ **3.** $-8x + 10$ **4.** $4x + 15$ **5.** $-24 + 48x - 16y$ **6.** $2(x - 7)$ **7.** $6(x - 1)$ **8.** $5(x + 2)$ **9.** $3(4 - x + 2z)$ **10.** $7a - 3b$ **11.** $-2x + 5y$ **12.** $5x - y$ **13.** $-a + 8b$ **14.** -22 **15.** 7 **16.** -192 **17.** 1 **18.** $-\frac{7}{3}$ **19.** 25 **20.** $\frac{1}{2}$ **21.** $-\frac{15}{64}$ **22.** 9.99 **23.** -8 **24.** -5 **25.** $-\frac{1}{3}$ **26.** 4 **27.** 3 **28.** 4 **29.** 16 **30.** 6 **31.** -3 **32.** -2 **33.** 4 **34.** $19\%x$, or $0.19x$ **35.** Length: 365 mi; width: 275 mi **36.** 13 ft and 8 ft **37.** \$2117 **38.** 27 appliances **39.** 35°, 85°, 60° **40.** \$220 **41.** \$138.95 **42.** \$68,000 **43.** \$867 **44.** Width: 11 cm; length: 17 cm **45.** Amazon: 6437 km; Nile: 6671 km **46.** C **47.** A **48.** 23, -23 **49.** 20, -20

Understanding Through Discussion and Writing

1. The distributive laws are used to multiply, factor, and collect like terms in this chapter. **2.** For an equation $x + a = b$, we add the opposite of a on both sides of the equation to get x alone. **3.** For an equation $ax = b$, we multiply by the reciprocal of a on both sides of the equation to get x alone. **4.** We add $-b$ (or subtract b) on both sides and simplify. Then we multiply by the reciprocal of a (or divide by a) on both sides and simplify.

Test: Chapter 11, p. 673

1. [11.1a] 6 **2.** [11.1b] $18 - 3x$ **3.** [11.1b] $-5y + 5$ **4.** [11.1c] $2(6 - 11x)$ **5.** [11.1c] $7(x + 3 + 2y)$ **6.** [11.1d] $-5x - y$ **7.** [11.1d] $4a + 5b$ **8.** [11.2a] 8 **9.** [11.2a] 26 **10.** [11.3a] -6 **11.** [11.3a] 49 **12.** [11.4b] -12 **13.** [11.4a] 2 **14.** [11.4a] -8 **15.** [11.2a] $-\frac{7}{20}$ **16.** [11.4b] 2.5 **17.** [11.4c] 7 **18.** [11.4c] $\frac{5}{3}$ **19.** [11.5a] $x - 9$ **20.** [11.5b] Width: 7 cm; length: 11 cm **21.** [11.5b] \$46,120 **22.** [11.5b] 3 m, 5 m **23.** [11.5b] About \$9.4 billion **24.** [11.5b] 6 **25.** [11.5b] 25.625° **26.** [11.4b] D **27.** [2.1d], [11.4a] 15, -15 **28.** [11.5b] 60 tickets

Cumulative Review: Chapters 1–11, p. 675

1. [1.1a] 7 **2.** [1.1b] 7 thousands + 4 hundreds + 0 tens + 5 ones, or 7 thousands + 4 hundreds + 5 ones **3.** [5.1a] Seven and four hundred sixty-three thousandths **4.** [1.2a] 1012 **5.** [1.2a] 21,085 **6.** [4.2a] $-\frac{3}{26}$ **7.** [4.5a] $5\frac{7}{9}$ **8.** [5.2a] 493.971 **9.** [5.2a] 802.876 **10.** [1.3a] 152 **11.** [1.3a] 674 **12.** [4.3a] $\frac{5}{24}$ **13.** [4.5b] $2\frac{17}{24}$ **14.** [5.2b] 19.9973 **15.** [5.2c] 45.819 **16.** [3.5b] $\frac{7}{10}$ **17.** [3.5b] 55 **18.** [1.4a] 4752 **19.** [1.4a] 266,287 **20.** [4.6a] $4\frac{1}{12}$ **21.** [3.6a] $\frac{6}{5}$ **22.** [3.4a] -10 **23.** [5.3a] 259.084 **24.** [4.4a] $3\frac{3}{5}$ **25.** [1.5a] 573 **26.** [1.5a] 56 R 10 **27.** [4.4b] $56\frac{5}{17}$ **28.** [3.7b] $\frac{3}{2}$ **29.** [4.6b] $-\frac{7}{90}$ **30.** [5.4a] 39 **31.** [1.6a] 68,000 **32.** [5.1d] 0.428 **33.** [5.5b] 21.84 **34.** [3.2a] Yes **35.** [3.1a] 1, 3, 5, 15 **36.** [4.1a] 800 **37.** [3.5c] $\neq$ **38.** [4.3b] $<$ **39.** [5.1c] -0.9976 **40.** [6.2b] 11.176¢/oz, 11.067¢/oz, 12.197¢/oz, 10.583 ¢/oz, 10.586 ¢/oz; brand D has the lowest unit price. **41.** [10.3b], [10.4c] **(a)** 4400 mi; **(b)** about 1,437,333,333 mi^3 **42.** [7.5a] \$10,755.20 **43.** [7.5a] 15% **44.** [7.5a] \$3495.44 **45.** [7.5a] 30% **46.** [7.5a] \$537.76 **47.** [1.6c] $>$ **48.** [3.3a] $\frac{3}{5}$ **49.** [5.1b] 0.037 **50.** [5.5a] -0.52 **51.** [5.5a] $0.\overline{8}$ **52.** [7.1b] 0.07 **53.** [5.1b] $\frac{463}{100}$ **54.** [4.4a] $-\frac{29}{4}$ **55.** [7.2b] $\frac{2}{5}$ **56.** [7.2a] 85% **57.** [7.1b] 150% **58.** [1.7b] 555 **59.** [5.4b] -64 **60.** [3.7c] $\frac{5}{4}$ **61.** [6.3b] $\frac{153}{2}$, or $76\frac{1}{2}$, or 76.5 **62.** [8.3b]

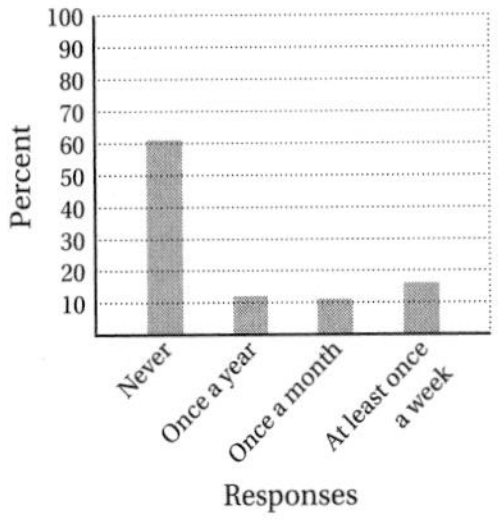

Percent: 100, 90, 80, 70, 60, 50, 40, 30, 20, 10
Responses: Never, Once a year, Once a month, At least once a week

63. [10.5e] 118°

64. [10.5d] Obtuse **65.** [1.8a] \$675 **66.** [1.8a] 65 min
67. [5.7a] \$25.75 **68.** [5.7a] 485.9 mi **69.** [1.8a] \$8100
70. [1.8a] \$783 **71.** [3.4b] $\frac{3}{10}$ km **72.** [5.7a] \$84.96
73. [6.4a] 13 gal **74.** [7.7a] \$150 **75.** [7.6b] 7%
76. [7.5b] 30,160 **77.** [8.la, b, c] 28; 26; 18 **78.** [1.9b] 324
79. [1.9b] 343 **80.** [10.6a] 3 **81.** [10.6a] 11 **82.** [10.6b] 4.472
83. [9.1a] 12 **84.** [9.2a] 428 **85.** [9.6a] 72 **86.** [9.4b] 20
87. [9.4a] 80 **88.** [9.4b] 0.08 **89.** [9.5a] 8.19
90. [9.5a] 5 **91.** [10.6c] $c = \sqrt{50}$ ft; $c \approx 7.071$ ft
92. [10.3a, b, c] 20.8 in.; 65.312 in.; 339.6224 in^2
93. [10.1a], [10.2a] 25.6 m; 25.75 m^2 **94.** [10.2b] 25 in^2
95. [10.2b] 61.6 cm^2 **96.** [10.2b] 128.65 yd^2 **97.** [10.4a] 52.9 m^3
98. [10.4b] 803.84 ft^3 **99.** [10.4d] $267.94\overline{6}$ cm^3 **100.** [11.4a] −5
101. [11.3a] 4 **102.** [11.4b] −8 **103.** [11.4c] $\frac{2}{3}$
104. [1.9c] 238 **105.** [1.9c] 172 **106.** [2.1d], [2.4a] 3
107. [2.2a] 14 **108.** [4.3a] $\frac{1}{3}$ **109.** [2.4a] 30
110. [3.6a] $-\frac{2}{7}$ **111.** [2.5a] −8
112. [11.5a] $y + 17$, or $17 + y$ **113.** [11.5a] 38%x, or 0.38x
114. [11.5b] \$47 **115.** [11.5b] \$2200
116. [11.5b] 50 m; 53 m; 40 m **117.** [11.4b] 0 **118.** [11.4b] 2
119. [11.1d] C **120.** [11.1c] B **121.** [3.7b] D **122.** [2.2a] A
123. [11.5b] 235 and 195

Guided Solutions

CHAPTER 1

Section 1.1

8. $2718 = 2$ thousands $+ 7$ hundreds $+ 1$ ten $+ 8$ ones

17. One million, eight hundred seventy-nine thousand, two hundred four

Section 1.2

2.
$$\begin{array}{r} {\scriptstyle 1\ 1\ 1} \\ 7\,9\,6\,8 \\ +\ 5\,4\,9\,7 \\ \hline 1\,3{,}4\,6\,5 \end{array}$$

5. Perimeter $= 4$ in. $+ 5$ in. $+ 9$ in. $+ 6$ in. $+ 5$ in. $= 29$ in.

Section 1.3

1.
$$\begin{array}{r} 7\,8\,9\,3 \\ -\ 4\,0\,9\,2 \\ \hline 3\,8\,0\,1 \end{array}$$

Check:
$$\begin{array}{r} 3\,8\,0\,1 \\ +\ 4\,0\,9\,2 \\ \hline 7\,8\,9\,3 \end{array}$$

5.
$$\begin{array}{r} {\scriptstyle 4\ \ 9\ 13} \\ \not{5}\,\not{0}\,\not{3} \\ -\ 2\,9\,8 \\ \hline 2\,0\,5 \end{array}$$

Section 1.4

4.
$$\begin{array}{r} {\scriptstyle 1\ 2\ 4} \\ 1\,3\,4\,8 \\ \times\ \ \ \ \ 5 \\ \hline 6\,7\,4\,0 \end{array}$$

20.
$$\begin{aligned} A &= l \cdot w \\ &= 12 \text{ ft} \cdot 8 \text{ ft} \\ &= 96 \text{ sq ft} \end{aligned}$$

Section 1.5

8. $0 \div 2$ means 0 divided by 2.
Since zero divided by any nonzero number is 0, $0 \div 2 = 0$.

9. $7 \div 0$ means 7 divided by 0.
Since division by 0 is not defined, $7 \div 0$ is not defined.

Section 1.6

26. Nearest ten:
$$\begin{array}{r} 840 \\ \times\ 250 \\ \hline 42\ 000 \\ 168\ 000 \\ \hline 210{,}000 \end{array}$$

Nearest hundred:
$$\begin{array}{r} 800 \\ \times\ 200 \\ \hline 160{,}000 \end{array}$$

31. Since 8 is to the left of 12 on the number line, $8 < 12$.

Section 1.7

13.
$$\begin{aligned} x + 9 &= 17 \\ x + 9 - 9 &= 17 - 9 \\ x &= 8 \end{aligned}$$

Check:
$$\begin{array}{c|c} x + 9 & = 17 \\ \hline 8 + 9 & ?\ 17 \\ 17 & \end{array}$$

Since $17 = 17$ is true, the answer checks.
The solution is 8.

19.
$$\begin{aligned} \frac{144}{9} &= \frac{9 \cdot n}{9} \\ 16 &= n \end{aligned}$$

Check:
$$\begin{array}{c|c} 144 & = 9 \cdot n \\ \hline 144 & ?\ 9 \cdot 16 \\ & 144 \end{array}$$

Since $144 = 144$ is true, the answer checks.
The solution is 16.

Section 1.8

4. **1. Familiarize.** Let $p =$ the number of pages William still has to read.

2. Translate.

Pages already read	plus	Number of pages to read	is	Total number of pages
↓	↓	↓	↓	↓
86	+	p	=	234

3. Solve.
$$\begin{aligned} 86 + p &= 234 \\ 86 + p - 86 &= 234 - 86 \\ p &= 148 \end{aligned}$$

4. Check. If William reads 148 more pages, he will have read a total of $86 + 148$ pages, or 234 pages.

5. State. William has 148 more pages to read.

9. **1. Familiarize.** Let $x =$ the number of hundreds in 3500.
Let $t =$ the time it takes to lose one pound.

2. Translate.
$$\begin{aligned} 100 \cdot x &= 3500 \\ x \cdot 2 &= t \end{aligned}$$

3. Solve. From Example 7, we know that $x = 35$.
$$\begin{aligned} x \cdot 2 &= t \\ 35 \cdot 2 &= t \\ 70 &= t \end{aligned}$$

4. Check. Since $70 \div 2 = 35$, there are 35 groups of 2 min in 70 min. Thus you will burn $35 \times 100 = 3500$ calories.

5. State. You must swim for 70 min, or 1 hr 10 min, in order to lose one pound.

Section 1.9

5. $10^4 = 10 \cdot 10 \cdot 10 \cdot 10 = 10{,}000$

15.
$$\begin{aligned} &9 \times 4 - (20 + 4) \div 8 - (6 - 2) \\ &= 9 \times 4 - 24 \div 8 - 4 \\ &= 36 - 24 \div 8 - 4 \\ &= 36 - 3 - 4 \\ &= 33 - 4 \\ &= 29 \end{aligned}$$

25.
$$\begin{aligned} &[18 - (2 + 7) \div 3] - (31 - 10 \times 2) \\ &= [18 - 9 \div 3] - (31 - 10 \times 2) \\ &= [18 - 3] - (31 - 20) \\ &= 15 - 11 \\ &= 4 \end{aligned}$$

CHAPTER 2

Section 2.2

32.
$$\begin{aligned} -x &= -(-5) = 5 \\ -(-x) &= -(-(-5)) \\ &= -(5) = -5 \end{aligned}$$

Section 2.3

11. $2 - 8 = 2 + (-8) = -6$

19. $-12 - (-9) = -12 + 9 = -3$

Section 2.4

20.
$$\begin{aligned} &-2 \cdot (-5) \cdot (-4) \cdot (-3) \\ &= 10 \cdot (-4) \cdot (-3) \\ &= -40 \cdot (-3) \\ &= 120 \end{aligned}$$

Section 2.5

12.
$$\begin{aligned} &32 \div 8 \cdot 2 \div 4 \\ &= 4 \cdot 2 \div 4 \\ &= 8 \div 4 \\ &= 2 \end{aligned}$$

CHAPTER 3

Section 3.1

4. 45

$1 \cdot 45$

2 is not a factor of 45.

$3 \cdot 15$

4 is not a factor of 45.

$5 \cdot 9$

6, 7, and 8 are not factors of 45.

9 is already listed.

The factors of 45 are 1, 3, 5, 9, 15, and 45.

10.
$$\begin{array}{r} 8 \\ 2\overline{)16} \\ \underline{16} \\ 0 \end{array}$$

Since the remainder is 0, 16 is divisible by 2.

Section 3.2

8. Add the digits: $1 + 7 + 2 + 1 + 6 = 17$. Since 17 is not divisible by 3, the number 17,216 is not divisible by 3.

28. The number named by the last two digits is 24. Since 24 is divisible by 4, the number 23,524 is divisible by 4.

Section 3.3

11. Each gallon is divided into 4 equal parts. The unit is $\frac{1}{4}$.

There are 7 equal units shaded. The part that is shaded is $\frac{7}{4}$.

23. $\dfrac{4-4}{567} = \dfrac{0}{567} = 0$

Section 3.4

2.
$$\begin{aligned} \frac{3}{8} \cdot \frac{5}{7} &= \frac{3 \cdot 5}{8 \cdot 7} \\ &= \frac{15}{56} \end{aligned}$$

6.
$$\begin{aligned} 5 \times \frac{2}{3} &= \frac{5}{1} \times \frac{2}{3} \\ &= \frac{5 \times 2}{1 \times 3} \\ &= \frac{10}{3} \end{aligned}$$

Section 3.5

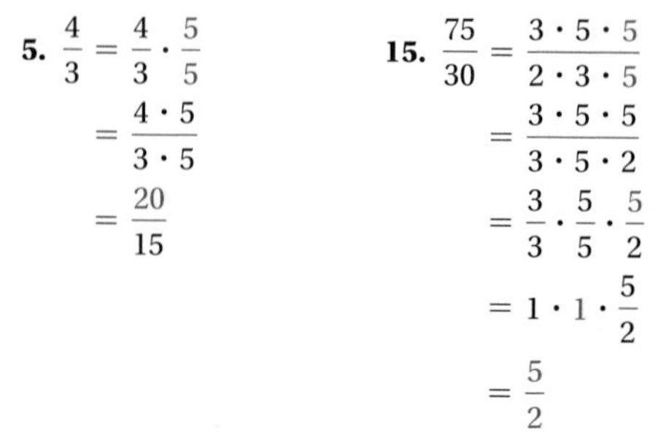

5.
$$\begin{aligned} \frac{4}{3} &= \frac{4}{3} \cdot \frac{5}{5} \\ &= \frac{4 \cdot 5}{3 \cdot 5} \\ &= \frac{20}{15} \end{aligned}$$

15.
$$\begin{aligned} \frac{75}{30} &= \frac{3 \cdot 5 \cdot 5}{2 \cdot 3 \cdot 5} \\ &= \frac{3 \cdot 5 \cdot 5}{3 \cdot 5 \cdot 2} \\ &= \frac{3}{3} \cdot \frac{5}{5} \cdot \frac{5}{2} \\ &= 1 \cdot 1 \cdot \frac{5}{2} \\ &= \frac{5}{2} \end{aligned}$$

20. $2 \cdot 20 = 40 \qquad 3 \cdot 14 = 42$

$$\frac{2}{3} \;\square\; \frac{14}{20}$$

Since $40 \neq 42$, $\frac{2}{3} \neq \frac{14}{20}$.

Section 3.6

1.
$$\begin{aligned} \frac{2}{3} \cdot \frac{7}{8} &= \frac{2 \cdot 7}{3 \cdot 8} \\ &= \frac{2 \cdot 7}{3 \cdot 2 \cdot 2 \cdot 2} \\ &= \frac{2}{2} \cdot \frac{7}{3 \cdot 2 \cdot 2} \\ &= 1 \cdot \frac{7}{3 \cdot 2 \cdot 2} \\ &= \frac{7}{12} \end{aligned}$$

Section 3.7

5.
$$\begin{aligned} \frac{6}{7} \div \frac{3}{4} &= \frac{6}{7} \cdot \frac{4}{3} \\ &= \frac{6 \cdot 4}{7 \cdot 3} \\ &= \frac{2 \cdot 3 \cdot 2 \cdot 2}{7 \cdot 3} \\ &= \frac{3}{3} \cdot \frac{2 \cdot 2 \cdot 2}{7} \\ &= \frac{2 \cdot 2 \cdot 2}{7} \\ &= \frac{8}{7} \end{aligned}$$

10.
$$\begin{aligned} \frac{5}{6} \cdot y &= \frac{2}{3} \\ \frac{\frac{5}{6} \cdot y}{\frac{5}{6}} &= \frac{\frac{2}{3}}{\frac{5}{6}} \\ y &= \frac{2}{3} \cdot \frac{6}{5} \\ &= \frac{2 \cdot 2 \cdot 3}{3 \cdot 5} \\ &= \frac{3}{3} \cdot \frac{2 \cdot 2}{5} \\ &= \frac{4}{5} \end{aligned}$$

CHAPTER 4

Section 4.1

7. a) $18 = 2 \cdot 3 \cdot 3$

$40 = 2 \cdot 2 \cdot 2 \cdot 5$

b) Consider the factor 2. The greatest number of times that 2 occurs in any one factorization is three times. Write 2 as a factor three times.

$2 \cdot 2 \cdot 2 \cdot ?$

Consider the factor 3. The greatest number of times that 3 occurs in any one factorization is two times. Write 3 as a factor two times.

$2 \cdot 2 \cdot 2 \cdot 3 \cdot 3 \cdot ?$

Consider the factor 5. The greatest number of times that 5 occurs in any one factorization is one time. Write 5 as a factor one time.

$2 \cdot 2 \cdot 2 \cdot 3 \cdot 3 \cdot 5$

LCM = 360

16.
$$\begin{array}{r|rrr} 2 & 81 & 90 & 84 \\ 3 & 81 & 45 & 42 \\ 3 & 27 & 15 & 14 \\ & 9 & 5 & 14 \end{array}$$

The LCM is $2 \cdot 3 \cdot 3 \cdot 9 \cdot 5 \cdot 14$, or 11,340.

Section 4.2

7. The LCD is 24.

$$\begin{aligned}\frac{3}{8}+\frac{5}{6} &= \frac{3}{8}\cdot 1+\frac{5}{6}\cdot 1\\ &= \frac{3}{8}\cdot\frac{3}{3}+\frac{5}{6}\cdot\frac{4}{4}\\ &= \frac{9}{24}+\frac{20}{24}\\ &= \frac{29}{24}\end{aligned}$$

12. **1. Familiarize.** Let $T =$ the total amount of berries in the salad.

2. Translate. To find the total amount, we add.

$$\frac{7}{8}+\frac{3}{4}+\frac{5}{16}=T$$

3. Solve. The LCD is 16.

$$\frac{7}{8}\cdot\frac{2}{2}+\frac{3}{4}\cdot\frac{4}{4}+\frac{5}{16}=T$$

$$\frac{14}{16}+\frac{12}{16}+\frac{5}{16}=T$$

$$\frac{31}{16}=T$$

4. Check. The answer is reasonable because it is larger than any of the individual amounts.

5. State. The salad contains a total of $\frac{31}{16}$ qt of berries.

Section 4.3

5. The LCD is 18.

$$\begin{aligned}\frac{5}{6}-\frac{1}{9} &= \frac{5}{6}\cdot\frac{3}{3}-\frac{1}{9}\cdot\frac{2}{2}\\ &= \frac{15}{18}-\frac{2}{18}\\ &= \frac{13}{18}\end{aligned}$$

12. The LCD is 24.

$$\frac{5}{6}=\frac{20}{24}$$

$$\frac{7}{8}=\frac{21}{24}$$

Since $20 < 21$, it follows that

$$\frac{20}{24}<\frac{21}{24}.$$

Thus, $\frac{5}{6}<\frac{7}{8}$.

Section 4.4

6.

$$4\cdot 6 = 24$$

$$24+5=29$$

$$4\frac{5}{6}=\frac{29}{6}$$

10.

$$\begin{array}{r} 2\\ 3\overline{)7}\\ \underline{6}\\ 1\end{array}$$

$$\frac{7}{3}=2\frac{1}{3}$$

Section 4.5

5.

$$\begin{array}{rcr} 8\frac{2}{3} &=& 8\frac{4}{6}\\ -5\frac{1}{2} &=& -5\frac{3}{6}\\ \hline & & 3\frac{1}{6}\end{array}$$

7.

$$\begin{array}{rcr} 5 &=& 4\frac{3}{3}\\ -1\frac{1}{3} &=& -1\frac{1}{3}\\ \hline & & 3\frac{2}{3}\end{array}$$

Section 4.6

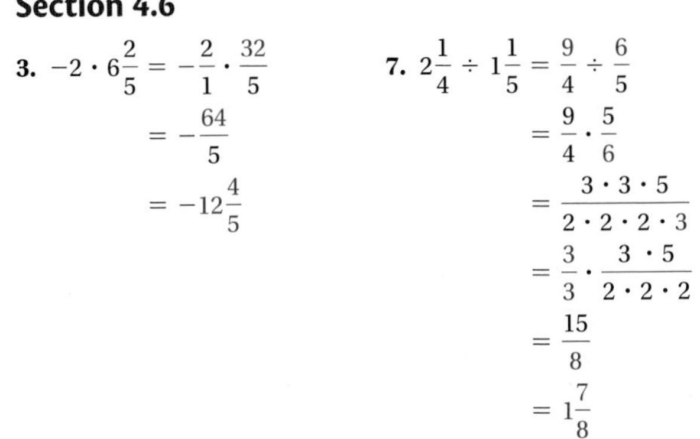

3.

$$\begin{aligned}-2\cdot 6\frac{2}{5} &= -\frac{2}{1}\cdot\frac{32}{5}\\ &= -\frac{64}{5}\\ &= -12\frac{4}{5}\end{aligned}$$

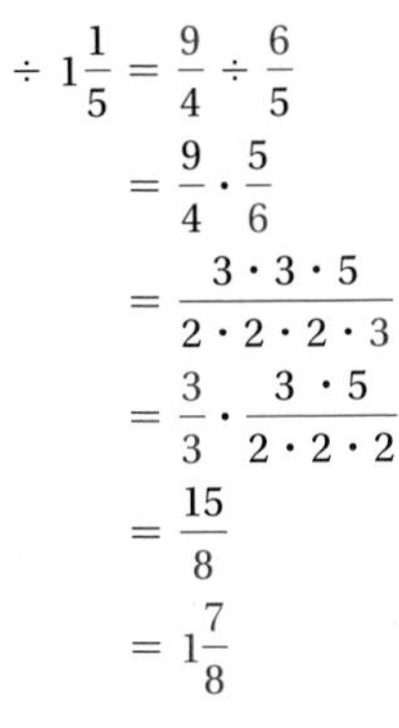

7.

$$\begin{aligned}2\frac{1}{4}\div 1\frac{1}{5} &= \frac{9}{4}\div\frac{6}{5}\\ &= \frac{9}{4}\cdot\frac{5}{6}\\ &= \frac{3\cdot 3\cdot 5}{2\cdot 2\cdot 2\cdot 3}\\ &= \frac{3}{3}\cdot\frac{3\cdot 5}{2\cdot 2\cdot 2}\\ &= \frac{15}{8}\\ &= 1\frac{7}{8}\end{aligned}$$

Section 4.7

2.

$$\begin{aligned}\frac{1}{3}\cdot\frac{3}{4}\div\frac{5}{8}-\frac{1}{10} &= \frac{3}{12}\div\frac{5}{8}-\frac{1}{10}\\ &= \frac{3}{12}\cdot\frac{8}{5}-\frac{1}{10}\\ &= \frac{3\cdot 2\cdot 2\cdot 2}{3\cdot 2\cdot 2\cdot 5}-\frac{1}{10}\\ &= \frac{2}{5}-\frac{1}{10}\\ &= \frac{4}{10}-\frac{1}{10}=\frac{3}{10}\end{aligned}$$

8.

$$\begin{aligned}\frac{\frac{3}{5}}{\frac{7}{10}-\frac{2}{3}} &= \frac{\frac{3}{5}}{\frac{7}{10}\cdot\frac{3}{3}-\frac{2}{3}\cdot\frac{10}{10}}\\ &= \frac{\frac{3}{5}}{\frac{21}{30}-\frac{20}{30}}\\ &= \frac{\frac{3}{5}}{\frac{1}{30}}=\frac{3}{5}\div\frac{1}{30}\\ &= \frac{3}{5}\cdot\frac{30}{1}=\frac{3\cdot 6\cdot 5}{5\cdot 1}=18\end{aligned}$$

CHAPTER 5

Section 5.1

7. 0.896. 3 places

$$0.896=\frac{896}{1000}$$

11. $\frac{743}{100}$ 7.43.

2 zeros 2 places

$$\frac{743}{100}=7.43$$

Section 5.2

7.

$$\begin{array}{r} {\scriptstyle 1\,1\,1\,1}\\ 45.780\\ 2467.000\\ +\quad 1.993\\ \hline 2514.773\end{array}$$

8.

$$\begin{array}{r} {\scriptstyle 6\;\;\;\; 13\; 12}\\ 37.428\\ -\,26.674\\ \hline 10.754\end{array}$$

14.

$$\begin{array}{r} {\scriptstyle 4\;\; 9\;\; 9\;\; 9\; 10}\\ 5.0000\\ -\,0.0089\\ \hline 4.9911\end{array}$$

Section 5.3

3. First we multiply the absolute values.

$$\begin{array}{r} 42.65 \\ \times\ 0.804 \\ \hline 17060 \\ 3412000 \\ \hline 34.29060 \end{array}$$

Since the product of a negative number and a positive number is negative, the answer is −34.2906.

14. $\$15.69 = 15.69 \times \$1$
$= 15.69 \times 100¢$
$= 1569¢$

Section 5.4

6. $$\begin{array}{r} 0.025 \\ 86\overline{)2.150} \\ \underline{172} \\ 430 \\ \underline{430} \\ 0 \end{array}$$

7. $\frac{0.375}{0.25} = \frac{0.375}{0.25} \times \frac{100}{100} = \frac{37.5}{25}$

$$\begin{array}{r} 1.5 \\ 0.25\overline{)0.375} \\ \underline{25} \\ 125 \\ \underline{125} \\ 0 \end{array}$$

15. $\frac{100 \cdot x}{100} = \frac{78.314}{100}$
$x = 0.78314$

17. $625 \div 62.5 \times 25 \div 6250$
$= 10 \times 25 \div 6250$
$= 250 \div 6250$
$= 0.04$

Section 5.5

2. $\frac{9}{20} = \frac{9}{20} \cdot \frac{5}{5} = \frac{45}{100} = 0.45$

6. $\frac{1}{6} = 1 \div 6$

$$\begin{array}{r} 0.166 \\ 6\overline{)1.000} \\ \underline{6} \\ 40 \\ \underline{36} \\ 40 \\ \underline{36} \\ 4 \end{array}$$

$\frac{1}{6} = 0.1666\ldots = 0.1\overline{6}$

17. Method 1:
$\frac{3}{4} \times 0.62 = 0.75 \times 0.62$
$= 0.465$

Method 2:
$\frac{3}{4} \times 0.62 = \frac{3}{4} \times \frac{62}{100}$
$= \frac{186}{400}$, or 0.465

Method 3:
$\frac{3}{4} \times 0.62 = \frac{3}{4} \times \frac{0.62}{1}$
$= \frac{1.86}{4}$, or 0.465

CHAPTER 6

Section 6.1

6. $\frac{\text{Length of shortest side}}{\text{Length of longest side}} = \frac{38.2}{55.5}$

9. Ratio of 3.6 to 12: $\frac{3.6}{12}$

Simplifying:
$\frac{3.6}{12} \cdot \frac{10}{10} = \frac{36}{120} = \frac{12 \cdot 3}{12 \cdot 10} = \frac{12}{12} \cdot \frac{3}{10} = \frac{3}{10}$

Section 6.2

4. $\frac{52 \text{ ft}}{13 \text{ sec}} = 4 \text{ ft/sec}$

6. Unit price $= \frac{\text{Price}}{\text{Number of units}}$
$= \frac{\$2.79}{26 \text{ oz}} = \frac{279 \text{ cents}}{26 \text{ oz}}$
$= \frac{279}{26} \frac{\text{cents}}{\text{oz}} \approx 10.731¢/\text{oz}$

Section 6.3

3. We compare cross products.

$1 \cdot 39 = 39 \quad \frac{1}{2} \overset{?}{=} \frac{20}{39} \quad 2 \cdot 20 = 40$

Since $39 \neq 40$, the numbers are not proportional.

8. $\frac{x}{9} = \frac{5}{4}$
$x \cdot 4 = 9 \cdot 5$
$\frac{x \cdot 4}{4} = \frac{9 \cdot 5}{4}$
$x = \frac{45}{4} = 11\frac{1}{4}$

Section 6.4

2. 1. **Familiarize.** Let p = the amount of paint needed, in gallons.

2. **Translate.** $\frac{4}{1600} = \frac{p}{6000}$

3. **Solve.**
$4 \cdot 6000 = 1600 \cdot p$
$15 = p$

4. **Check.** The cross products are the same.

5. **State.** For 6000 ft^2, they would need 15 gal of paint.

6. 1. **Familiarize.** Let D = the number of deer in the forest.

2. **Translate.** $\frac{153}{D} = \frac{18}{62}$

3. **Solve.**
$153 \cdot 62 = D \cdot 18$
$527 = D$

4. **Check.** The cross products are the same.

5. **State.** There are about 527 deer in the forest.

Section 6.5

1. The ratio of x to 20 is the same as the ratio of 9 to 12.

$\frac{x}{20} = \frac{9}{12}$
$x \cdot 12 = 20 \cdot 9$
$\frac{x \cdot 12}{12} = \frac{20 \cdot 9}{12}$
$x = \frac{180}{12} = 15$

5. Let w = the width of an actual skylight.

$\frac{12}{52} = \frac{3}{w}$
$12 \cdot w = 52 \cdot 3$
$w = 13$

The width of an actual skylight will be 13 ft.

CHAPTER 7

Section 7.2

6. $\frac{19}{25} = \frac{19}{25} \cdot \frac{4}{4}$
$= \frac{76}{100} = 76\%$

10. $3.25\% = \frac{3.25}{100} = \frac{3.25}{100} \times \frac{100}{100}$
$= \frac{325}{10{,}000} = \frac{13 \times 25}{400 \times 25}$
$= \frac{13}{400} \times \frac{25}{25} = \frac{13}{400}$

Section 7.3

9. 20% of what is 45?

$$20\% \cdot b = 45$$

$$\frac{20\% \cdot b}{20\%} = \frac{45}{20\%}$$

$$b = \frac{45}{0.2}$$

$$b = 225$$

11. 16 is what percent of 40?

$$16 = p \cdot 40$$

$$\frac{16}{40} = \frac{p \cdot 40}{40}$$

$$\frac{16}{40} = p$$

$$0.4 = p$$

$$40\% = p$$

Section 7.4

8.
$$\frac{20}{100} = \frac{45}{b}$$
$$20 \cdot b = 100 \cdot 45$$
$$\frac{20 \cdot b}{20} = \frac{100 \cdot 45}{20}$$
$$b = \frac{4500}{20}$$
$$b = 225$$

9.
$$\frac{64}{100} = \frac{a}{55}$$
$$64 \cdot 55 = 100 \cdot a$$
$$\frac{64 \cdot 55}{100} = \frac{100 \cdot a}{100}$$
$$\frac{3520}{100} = a$$
$$35.2 = a$$

12.
$$\frac{12}{40} = \frac{N}{100}$$
$$12 \cdot 100 = 40 \cdot N$$
$$\frac{12 \cdot 100}{40} = \frac{40 \cdot N}{40}$$
$$\frac{1200}{40} = N$$
$$30 = N$$

Thus, \$12 is 30% of \$40.

Section 7.6

2.
$$\begin{aligned} \text{Sales tax} &= 4\% \times 4 \times \$18.95 \\ &= 0.04 \times \$75.80 \\ &= \$3.032 \\ &\approx \$3.03 \\ \text{Total price} &= \$75.80 + \$3.03 \\ &= \$78.83 \end{aligned}$$

7.
$$\$2970 = 7.5\% \times S$$
$$\$2970 = 0.075 \times S$$
$$\frac{\$2970}{0.075} = \frac{0.075 \times S}{0.075}$$
$$\$39{,}600 = S$$

Section 7.7

1.
$$\begin{aligned} I &= P \cdot r \cdot t \\ &= \$4300 \times 4\% \times 1 \\ &= \$4300 \times 0.04 \times 1 \\ &= \$172 \end{aligned}$$

3. a)
$$\begin{aligned} I &= P \cdot r \cdot t \\ &= \$4800 \times 5\tfrac{1}{2}\% \times \frac{30}{365} \\ &= \$4800 \times 0.055 \times \frac{30}{365} \\ &\approx \$21.70 \end{aligned}$$

b) Total amount
$$= \$4800 + \$21.70$$
$$= \$4821.70$$

CHAPTER 8

Section 8.1

5.
$$\text{Course grade} = \frac{100 \cdot 15 + 92 \cdot 25 + 88 \cdot 40}{15 + 25 + 40} = \frac{7320}{80} = 91.5$$

Soha's course grade is 91.5%.

12. Rearrange the numbers in order from smallest to largest:

34, 34, 67, 68, 69, 70.

The middle numbers are 67 and 68.
The average of 67 and 68 is 67.5.
The median is 67.5.

15. Rearrange the numbers in order from smallest to largest.

13, 24, 27, 28, 67, 89.

Each number occurs one time.
There is no mode.

Section 8.2

3. The amount of the decrease in population density is $611 - 603 = 8$.

The percent decrease is $\frac{8}{611} \approx 0.013$, or 1.3%.

8. The graph shows $1\frac{1}{2}$ symbols for South America.

This represents 150 roller coasters.

The graph shows $\frac{1}{2}$ symbol for Africa.

This represents 50 roller coasters.
There are 100 more roller coasters in South America than in Africa.

Section 8.3

7. We look from left to right along a line at \$400 per ounce. The points on the graph that are below this line correspond to the years 1970, 1975, 1985, 1990, 1995, and 2000.

CHAPTER 9

Section 9.1

6.
$$\begin{aligned} 2\tfrac{5}{6}\text{ yd} &= 2\tfrac{5}{6} \times 1\text{ yd} \\ &= \frac{17}{6} \times 3\text{ ft} \\ &= \frac{17}{2}\text{ ft} \\ &= 8\tfrac{1}{2}\text{ ft} \end{aligned}$$

8.
$$\begin{aligned} 72\text{ in.} &= \frac{72\text{ in.}}{1} \times \frac{1\text{ ft}}{12\text{ in.}} \\ &= \frac{72}{12} \times 1\text{ ft} \\ &= 6\text{ ft} \end{aligned}$$

10.
$$\begin{aligned} 24\text{ ft} &= 24\text{ ft} \times \frac{1\text{ yd}}{3\text{ ft}} \\ &= \frac{24}{3} \times 1\text{ yd} \\ &= 8\text{ yd} \end{aligned}$$

Section 9.2

10.
$$\begin{aligned} 23\text{ km} &= 23 \times 1\text{ km} \\ &= 23 \times 1000\text{ m} \\ &= 23{,}000\text{ m} \end{aligned}$$

15.
$$\begin{aligned} 7814\text{ m} &= 7814\text{ m} \times \frac{1\text{ dam}}{10\text{ m}} \\ &= \frac{7814}{10} \times \frac{\text{m}}{\text{m}} \times 1\text{ dam} \\ &= 781.4\text{ dam} \end{aligned}$$

Section 9.3

3.
$$\begin{aligned} 2383\text{ km} &= 2383 \times 1\text{ km} \\ &\approx 2383 \times 0.621\text{ mi} \\ &= 1479.843\text{ mi} \end{aligned}$$

Section 9.4

13. $$\begin{aligned} 1\text{ mcg} &= 0.000001\text{ g} \\ &= 0.000001 \times 1\text{ g} \\ &= 0.000001 \times 1000\text{ mg} \\ &= 0.001\text{ mg} \end{aligned}$$

Section 9.5

2. $$\begin{aligned} 80\text{ qt} &= 80\text{ qt} \cdot \frac{1\text{ gal}}{4\text{ qt}} \\ &= \frac{80}{4} \cdot 1\text{ gal} \\ &= 20\text{ gal} \end{aligned}$$

7. $$\begin{aligned} 0.97\text{ L} &= 0.97 \times 1\text{ L} \\ &= 0.97 \times 1000\text{ mL} \\ &= 970\text{ mL} \end{aligned}$$

Section 9.6

4. $$\begin{aligned} 168\text{ hr} &= 168\text{ hr} \times \frac{1\text{ day}}{24\text{ hr}} \times \frac{1\text{ wk}}{7\text{ days}} \\ &= \frac{168}{24 \cdot 7}\text{wk} \\ &= 1\text{ wk} \end{aligned}$$

13. $$\begin{aligned} C &= \tfrac{5}{9}(F - 32) \\ &= \tfrac{5}{9}(95 - 32) \\ &= \tfrac{5}{9} \cdot 63 = 35 \end{aligned}$$
Thus, 95°F = 35°C.

Section 9.7

3. $$\begin{aligned} 20\text{ ft}^2 &= 20 \times 1\text{ ft}^2 \\ &= 20 \times 144\text{ in}^2 \\ &= 2880\text{ in}^2 \end{aligned}$$

4. $$\begin{aligned} 360\text{ in}^2 &= 360\text{ in}^2 \times \frac{1\text{ ft}^2}{144\text{ in}^2} \\ &= \frac{360}{144} \times \frac{\text{in}^2}{\text{in}^2} \times 1\text{ ft}^2 \\ &= 2.5\text{ ft}^2 \end{aligned}$$

CHAPTER 10

Section 10.1

5. $$\begin{aligned} P &= 2 \cdot (l + w) \\ &= 2 \cdot (8\tfrac{1}{4}\text{ in.} + 5\text{ in.}) \\ &= 2 \cdot (13\tfrac{1}{4}\text{ in.}) \\ &= 2 \cdot \frac{53}{4}\text{ in.} \\ &= \frac{2 \cdot 53}{2 \cdot 2}\text{ in.} \\ &= \frac{53}{2}\text{ in.} \\ &= 26\tfrac{1}{2}\text{ in.} \end{aligned}$$

8. $$\begin{aligned} P &= 4 \cdot s \\ &= 4 \cdot 7.8\text{ km} \\ &= 31.2\text{ km} \end{aligned}$$

Section 10.2

6. $$\begin{aligned} A &= s \cdot s \\ &= 3\tfrac{1}{2}\text{ yd} \times 3\tfrac{1}{2}\text{ yd} \\ &= \tfrac{7}{2}\text{ yd} \times \tfrac{7}{2}\text{ yd} \\ &= \tfrac{49}{4}\text{ yd}^2 \\ &= 12\tfrac{1}{4}\text{ yd}^2 \end{aligned}$$

10. $$\begin{aligned} A &= \tfrac{1}{2} \cdot b \cdot h \\ &= \tfrac{1}{2} \times 11\text{ cm} \times 3.4\text{ cm} \\ &= 0.5 \times 11 \times 3.4\text{ cm}^2 \\ &= 18.7\text{ cm}^2 \end{aligned}$$

Section 10.3

3. $$\begin{aligned} C &= \pi \cdot d \\ &\approx 3.14 \times 18\text{ in.} \\ &= 56.52\text{ in.} \end{aligned}$$

6. $$\begin{aligned} A &= \pi \cdot r \cdot r \\ &\approx \tfrac{22}{7} \cdot 5\text{ km} \cdot 5\text{ km} \\ &= \tfrac{22}{7} \cdot 25\text{ km}^2 \\ &= \tfrac{550}{7}\text{ km}^2 \\ &= 78\tfrac{4}{7}\text{ km}^2 \end{aligned}$$

Section 10.4

4. $$\begin{aligned} V &= \pi \cdot r^2 \cdot h \\ &\approx 3.14 \times 5\text{ ft} \times 5\text{ ft} \times 10\text{ ft} \\ &= 3.14 \times 250\text{ ft}^3 \\ &= 785\text{ ft}^3 \end{aligned}$$

6. $$\begin{aligned} V &= \tfrac{4}{3} \cdot \pi \cdot r^3 \\ &\approx \tfrac{4}{3} \times \tfrac{22}{7} \times (28\text{ ft})^3 \\ &= \tfrac{4}{3} \times \tfrac{22}{7} \times 21{,}952\text{ ft}^3 \\ &= \tfrac{275{,}968}{3}\text{ ft}^3 \\ &= 91{,}989\tfrac{1}{3}\text{ ft}^3 \end{aligned}$$

Section 10.5

13. $90° - 67° = 23°$

18. $180° - 71° = 109°$

Section 10.6

29. $$\begin{aligned} a^2 + b^2 &= c^2 \\ 12^2 + 5^2 &= c^2 \\ 144 + 25 &= c^2 \\ 169 &= c^2 \\ 13 &= c \end{aligned}$$

CHAPTER 11

Section 11.1

4. $$\begin{aligned} A &= lw \\ A &= (24\text{ ft})(8\text{ ft}) \\ &= (24)(8)(\text{ft})(\text{ft}) \\ &= 192\text{ ft}^2\text{, or 192 square feet} \end{aligned}$$

20. $$\begin{aligned} &-2(x - 3) \\ &= -2 \cdot x - (-2) \cdot 3 \\ &= -2x - (-6) \\ &= -2x + 6 \end{aligned}$$

24. $$\begin{aligned} &16a - 36b + 42 \\ &= 2 \cdot 8a - 2 \cdot 18b + 2 \cdot 21 \\ &= 2(8a - 18b + 21) \end{aligned}$$

31. $$\begin{aligned} &3x - 7x - 11 + 8y + 4 - 13y \\ &= (3 - 7)x + (8 - 13)y + (-11 + 4) \\ &= -4x + (-5)y + (-7) \\ &= -4x - 5y - 7 \end{aligned}$$

Section 11.2

1. $$\begin{aligned} x + 2 &= 11 \\ x + 2 + (-2) &= 11 + (-2) \\ x + 0 &= 9 \\ x &= 9 \end{aligned}$$

Section 11.3

1. $$\begin{aligned} 6x &= 90 \\ \frac{1}{6} \cdot 6x &= \frac{1}{6} \cdot 90 \\ 1 \cdot x &= 15 \\ x &= 15 \end{aligned}$$

Check:

$6x = 90$	
$6 \cdot 15 \overset{?}{=} 90$	
90	TRUE

2. $$\begin{aligned} 4x &= -7 \\ \frac{4x}{4} &= \frac{-7}{4} \\ 1 \cdot x &= -\frac{7}{4} \\ x &= -\frac{7}{4} \end{aligned}$$

5. $$\begin{aligned} \frac{2}{3} &= -\frac{5}{6}y \\ -\frac{6}{5} \cdot \frac{2}{3} &= -\frac{6}{5} \cdot \left(-\frac{5}{6}y\right) \\ -\frac{12}{15} &= 1 \cdot y \\ -\frac{4}{5} &= y \end{aligned}$$

Section 11.4

4.
$$\begin{aligned} -18 - m &= -57 \\ 18 - 18 - m &= 18 - 57 \\ -m &= -39 \\ -1(-m) &= -1(-39) \\ m &= 39 \end{aligned}$$

11.
$$\begin{aligned} 7x - 17 + 2x &= 2 - 8x + 15 \\ 9x - 17 &= 17 - 8x \\ 8x + 9x - 17 &= 17 - 8x + 8x \\ 17x - 17 &= 17 \\ 17x - 17 + 17 &= 17 + 17 \\ 17x &= 34 \\ \frac{17x}{17} &= \frac{34}{17} \\ x &= 2 \end{aligned}$$

13.
$$\begin{aligned} \frac{7}{8}x - \frac{1}{4} + \frac{1}{2}x &= \frac{3}{4} + x \\ 8 \cdot \left(\frac{7}{8}x - \frac{1}{4} + \frac{1}{2}x\right) &= 8 \cdot \left(\frac{3}{4} + x\right) \\ 8 \cdot \frac{7}{8}x - 8 \cdot \frac{1}{4} + 8 \cdot \frac{1}{2}x &= 8 \cdot \frac{3}{4} + 8 \cdot x \\ 7x - 2 + 4x &= 6 + 8x \\ 11x - 2 &= 6 + 8x \\ 11x - 2 - 8x &= 6 + 8x - 8x \\ 3x - 2 &= 6 \\ 3x - 2 + 2 &= 6 + 2 \\ 3x &= 8 \\ \frac{3x}{3} &= \frac{8}{3} \\ x &= \frac{8}{3} \end{aligned}$$

Section 11.5

16. Principal + Interest = Amount

$$\begin{array}{ccccc} \text{Principal} & + & \text{Interest} & = & \text{Amount} \\ \downarrow & \downarrow & \downarrow & \downarrow & \downarrow \\ x & + & 5\%x & = & 2520 \end{array}$$

$$\begin{aligned} x + 0.05x &= 2520 \\ (1 + 0.05)x &= 2520 \\ 1.05x &= 2520 \\ \frac{1.05x}{1.05} &= \frac{2520}{1.05} \\ x &= 2400 \end{aligned}$$

The principal is $2400.

Glossary

A

Absolute value The distance that a number is from zero on the number line

Acute angle An angle whose measure is greater than 0° and less than 90°

Acute triangle A triangle in which all three angles are acute

Addend In addition, a number being added

Additive identity The number 0

Additive inverse A number's opposite. Two numbers are additive inverses of each other if their sum is 0.

Algebraic expression An expression consisting of variables, constants, numerals, operation signs, and/or grouping symbols

Angle A set of points consisting of two rays (half-lines) with a common endpoint (vertex)

Area The number of square units that fill a plane region

Arithmetic mean A center point of a set of numbers found by adding the numbers and dividing by the number of items of data; also called mean or average

Arithmetic numbers The whole numbers and the positive fractions; also called the nonnegative rational numbers

Associative law of addition The statement that when three numbers are added, regrouping the addends does not affect the sum

Associative law of multiplication The statement that when three numbers are multiplied, regrouping the factors does not affect the product

Average A center point of a set of numbers found by adding the numbers and dividing by the number of items of data; also called the arithmetic mean or mean

B

Bar graph A graphic display of data using bars proportional in length to the numbers represented

Base In exponential notation, the number being raised to a power

C

Celsius A temperature scale for metric measure

Circle graph A graphic means of displaying data using sectors of a circle; often used to show the percent of a quantity in different categories or to show visually the ratio of one category to another; also called a pie chart

Circumference The distance around a circle

Coefficient The numerical multiplier of a variable

Commission A percent of total sales paid to a salesperson

Commutative law of addition The statement that when two numbers are added, changing the order in which the numbers are added does not affect the sum

Commutative law of multiplication The statement that when two numbers are multiplied, changing the order in which the numbers are multiplied does not affect the product

Complementary angles Two angles for which the sum of their measures is 90°

Composite number A natural number, other than 1, that is not prime

Compound interest Interest paid on interest

Constant A known number

Cross products Given an equation with a single fraction on each side, the products formed by multiplying the left numerator and the right denominator, and the left denominator and the right numerator

D

Decimal notation A representation of a number containing a decimal point

Denominator The bottom number in a fraction

Diameter A segment that passes through the center of a circle and has its endpoints on the circle

Difference The result of subtracting one number from another

Digit A number 0, 1, 2, 3, 4, 5, 6, 7, 8, or 9 that names a place-value location

Discount The amount subtracted from the original price of an item to find the sale price

Distributive law of multiplication over addition The statement that multiplying a factor by the sum of two numbers gives the same result as multiplying the factor by each of the two numbers and then adding

Distributive law of multiplication over subtraction The statement that multiplying a factor by the difference of two numbers gives the same result as multiplying the factor by each of the two numbers and then subtracting

Dividend In division, the number being divided

Divisible The number a is divisible by another number b if there exists a number c such that $a = b \cdot c$

Divisor In division, the number dividing another number

E

Equation A number sentence that says that the expressions on either side of the equals sign, =, represent the same number

Equilateral triangle A triangle in which all sides are the same length

Equivalent equations Equations with the same solutions

Equivalent expressions Expressions that have the same value for all allowable replacements

Equivalent fractions Fractions that represent the same number
Even number A number that is divisible by 2; that is, it has an even ones digit
Exponent In expressions of the form a^n, the number n; for n a natural number, a^n represents n factors of a
Exponential notation A representation of a number using a base raised to a power

F

Factor *Verb:* to write an equivalent expression that is a product. *Noun:* a multiplier
Factorization A number expressed as a product of natural numbers
Fahrenheit A temperature scale for American measure
Fraction notation A number written using a numerator and a denominator

H

Histogram A special kind of graph that shows how often certain numbers appear in a set of data
Hypotenuse In a right triangle, the side opposite the right angle

I

Identity property of 1 The statement that the product of a number and 1 is always the original number
Identity property of 0 The statement that the sum of a number and 0 is always the original number
Inequality A mathematical sentence using $<, >, \leq, \geq$, or $\neq$
Integers The whole numbers and their opposites
Interest A percentage of an amount invested or borrowed
Irrational number A real number that cannot be named as a ratio of two integers
Isosceles triangle A triangle in which two or more sides are the same length

L

Least common denominator (LCD) The least common multiple of the denominators of two or more fractions
Least common multiple (LCM) The smallest number that is a multiple of two or more numbers
Legs In a right triangle, the two sides that form the right angle
Like terms Terms that have exactly the same variable factors
Line graph A graphic means of displaying data by connecting adjacent data points with line segments

M

Marked price The original price of an item
Mean A center point of a set of numbers found by adding the numbers and dividing by the number of items of data; also called the arithmetic mean or average
Median In a set of data listed in order from smallest to largest, the middle number if there are an odd number of data items, or the average of the two middle numbers if there are an even number of data items
Minuend The number from which another number is being subtracted
Mixed numeral A number represented by a whole number and a fraction less than 1
Mode The number or numbers that occur most often in a set of data
Multiple A product of a number and some natural number
Multiplicative identity The number 1
Multiplicative inverses Reciprocals; two numbers whose product is 1

N

Natural numbers The numbers 1, 2, 3, 4, 5, ...
Negative integers The integers to the left of 0 on the number line
Nonnegative rational numbers The whole numbers and the positive fractions; also called the arithmetic numbers
Numerator The top number in a fraction

O

Obtuse angle An angle whose measure is greater than 90° and less than 180°
Obtuse triangle A triangle in which one angle is an obtuse angle
Opposite The opposite, or additive inverse, of a number a can be named $-a$. Opposites are the same distance from 0 on the number line but on different sides of 0.
Original price The price of an item before a discount is deducted

P

Palindrome prime A prime number that remains a prime number when its digits are reversed
Parallelogram A four-sided polygon with two pairs of parallel sides
Percent notation A representation of a number as parts per 100
Perimeter The distance around a polygon, or the sum of the lengths of its sides
Pi (π) The number that results when the circumference of a circle is divided by its diameter; $\pi \approx 3.14$, or $22/7$
Pictograph A graphic means of displaying data using pictorial symbols
Pie chart A graphic means of displaying data using sectors of a circle; often used to show the percent of a quantity used in different categories or to show visually the ratio of one category to another; also called a circle graph
Polygon A closed geometric figure with three or more sides
Positive integers The natural numbers, or the integers to the right of 0 on the number line
Prime factorization A factorization of a composite number as a product of prime numbers
Prime number A natural number that has exactly two *different* factors: only itself and 1
Principal The amount invested
Product The result in multiplication
Proportion An equation that states that two ratios are equal
Protractor A device used to measure angles

Purchase price The price of an item before sales tax is added

Pythagorean equation The equation $a^2 + b^2 = c^2$, where a and b are the lengths of the legs of a right triangle and c is the length of the hypotenuse

Q

Quotient The result when one number is divided by another

R

Radical sign The symbol $\sqrt{}$

Radius A segment with one endpoint on the center of a circle and the other endpoint on the circle

Rate A ratio used to compare two different kinds of measure

Ratio The quotient of two quantities

Rational numbers All numbers that can be named in the form a/b, where a and b are integers and b is not 0

Ray A half-line

Real numbers All rational and irrational numbers

Reciprocal A multiplicative inverse; two numbers are reciprocals if their product is 1

Rectangle A four-sided polygon with four 90° angles

Repeating decimal A decimal that cannot be written using a finite number of decimal places

Right angle An angle whose measure is 90°

Right triangle A triangle that includes a right angle

Rounding Approximating the value of a number; used when estimating

S

Sale price The price of an item after a discount has been deducted

Sales tax A tax added to the purchase price of an item

Scalene triangle A triangle in which each side is a different length

Similar triangles Triangles that have the same shape because the lengths of their corresponding sides have the same ratio—that is, they are proportional

Simple interest A percentage of an amount P invested or borrowed for t years, computed by calculating principal $\times$ interest rate $\times$ time

Solution of an equation A replacement for the variable that makes the equation true

Sphere The set of all points in space that are a given distance from a given point

Square A four-sided polygon with four right angles and all sides of equal length

Square of a number A number multiplied by itself

Square root of a number A number that when multiplied by itself yields the given original number

Statistic A number describing a set of data

Straight angle An angle whose measure is 180°

Substitute To replace a variable with a number

Subtrahend In subtraction, the number being subtracted

Sum The result in addition

Supplementary angles Two angles for which the sum of their measures is 180°

T

Table A means of displaying data in rows and columns

Term A number, a variable, or a product or a quotient of numbers and/or variables

Terminating decimal A decimal that can be written using a finite number of decimal places

Total price The sum of the purchase price of an item and the sales tax on the item

Trapezoid A four-sided polygon with two parallel sides

Triangle A three-sided polygon

U

Unit price The ratio of price to the number of units; also called unit rate

Unit rate The ratio of price to the number of units; also called unit price

V

Variable A letter that represents an unknown number

Vertex The common endpoint of the two rays that form an angle

Volume The number of cubic units that fill a solid region

W

Whole numbers The natural numbers and 0: 0, 1, 2, 3, ...

Index